MOLECULAR AND BIOCHEMICAL TOXICOLOGY

MOLECULAR AND BIOCHEMICAL TOXICOLOGY

Fourth Edition

Robert C. Smart
Professor of Molecular and Cellular Toxicology

Ernest Hodgson
William Neal Reynolds Professor of Toxicology

A JOHN WILEY & SONS, INC., PUBLICATION

Published by John Wiley & Sons, Inc., Hoboken, New Jersey
Published simultaneously in Canada

Library of Congress Cataloging-in-Publication Data:

Molecular and biochemical toxicology / [edited by] Robert C. Smart, Ernest Hodgson. – 4th ed.
   p. ; cm.
  Rev. ed of: Introduction to biochemical toxicology. 2001.
  Includes bibliographical references and index.
  ISBN 978-0-470-10211-4 (cloth)
  1. Biochemical toxicology. I. Smart, Robert C., 1954– II. Hodgson, Ernest, 1932– III. Introduction to biochemical toxicology.
  [DNLM: 1. Poisoning–metabolism. 2. Poisons–metabolism. 3. Molecular Biology–methods. 4. Toxicology–methods. QV 600 M7178 2008]
   RA1219.5.I58 2008
   615.9–dc22

                                2007044577

10 9 8 7 6 5 4

CONTENTS

Loan Receipt
Liverpool John Moores University
Library Services

Borrower Name: Hayward,Sophie
Borrower ID: ********

Molecular and biochemical toxicology /
31111015021270
Due Date: 04/05/2017 23:59

Total Items: 1
26/04/2017 21:05

Please keep your receipt in case of
dispute.

32. Immunotoxicology — page 765

*MaryJane K. Selgrade, Dori R. Germolec, Robert W. Luebke,
Ralph J. Smialowicz, Marsha D. Ward, and Christal C. Bowman*

This is the 4th edition of a series that was initiated in 1980 as *Introduction to Biochemical Toxicology*, edited by Ernest Hodgson and Frank E. Guthrie and based on a course at North Carolina State University that has now been offered for almost 30 years. The 2nd and 3rd editions, with the same title, were edited by Hodgson and Levi and by Hodgson and Smart. The changes in both scope and specific content are greater between this edition and the previous one and, in fact, are so extensive as to require several important changes. First, to recognize the changing roles of the co-editors, they are now listed as Smart and Hodgson. Second, the title has been changed to *Molecular and Biochemical Toxicology*.

The incorporation of the principles and methods of molecular biology into mechanistic toxicology has continued apace and is reflected not only in the change in title but also in the addition of new chapters and the integration of molecular material into most, if not all, of the chapters represented in the 3rd edition and extensively revised for the 4th. The section on *Methodology* (Chapters 2–8) now contains additional chapters on Toxicogenomics, Proteomics, Metabolomics, and Bioinformatics, and the section on *Toxicant Processing* (Chapters 9–15) has been reorganized to better emphasize the role of polymorphisms in xenobiotic-metabolizing enzymes, transporters, and so on. The section on *Mechanisms* (Chapters 16–26) now contains chapters on Cell Death and separate chapters on DNA Damage and DNA Repair. The final section on *Target Organs* (Chapters 27–35), although containing essentially the same chapters as the 3rd edition, has several new authors and all chapters are extensively revised.

We believe that both instructors and advanced students will continue to find this series a usable, integral component of their graduate curriculum, with the broad scope providing a background that will enable all instructors to select material to suit their particular needs. Many thanks to all of the contributing authors and to all at John Wiley & Sons who made this edition possible.

Raleigh, North Carolina   ROBERT C. SMART
ERNEST HODGSON

CONTRIBUTORS

Bonita L. Blake, Departments of Pharmacology and Psychiatry, University of North Carolina at Chapel Hill, Chapel Hill, North Carolina 27599

James C. Bonner, Department of Environmental and Molecular Toxicology, North Carolina State University, Raleigh, North Carolina 27695

Thomas W. Bouldin, Department of Pathology and Laboratory Medicine, University of North Carolina at Chapel Hill, Chapel Hill, North Carolina 27599

Christal C. Bowman, Immunotoxicology Branch, U.S. Environmental Protection Agency, Research Triangle Park, North Carolina 27711

David B. Buchwalter, Department of Environmental and Molecular Toxicology, North Carolina State University, Raleigh, North Carolina 27695

Taehyeon M. Cho, Department of Environmental and Molecular Toxicology, North Carolina State University, Raleigh, North Carolina 27695

John F. Couse, Taconic Inc., Albany Operations, Rensselaer, New York 12144

Edward L. Croom, Department of Environmental and Molecular Toxicology, North Carolina State University, Raleigh, North Carolina 27695

Helen C. Cunny, National Institute of Environmental Health Sciences, Research Triangle Park, North Carolina 27709

Parikshit C. Das, Department of Environmental and Molecular Toxicology, North Carolina State University, Raleigh, North Carolina 27695

Nigel Deighton, Department of Environmental and Molecular Toxicology, North Carolina State University, Raleigh, North Carolina 27695

Susan Elmore, National Institute of Environmental Health Sciences, Research Triangle Park, North Carolina 27709

Sarah J. Ewing, Penn State Erie, The Behrend College, Erie, PA 16563

Dori R. Germolec, National Institute of Environmental Health Sciences, Research Triangle Park, North Carolina 27709

Jeffry F. Goodrum, (retired) Department of Pathology and Laboratory Medicine, University of North Carolina at Chapel Hill, Chapel Hill, North Carolina 27599

Ernest Hodgson, Department of Environmental and Molecular Toxicology, North Carolina State University, Raleigh, North Carolina 27695

Mac Law, Department of Population Health and Pathobiology, College of Veterinary Medicine, North Carolina State University, Raleigh, North Carolina 27695

Gerald A. LeBlanc, Department of Environmental and Molecular Toxicology, North Carolina State University, Raleigh, North Carolina 27695

Kari D. Loomis, Functional Genomics Program, North Carolina State University, Raleigh, North Carolina 27695

Robert W. Luebke, Immunotoxicology Branch, U.S. Environmental Protection Agency, Research Triangle Park, North Carolina 27711

Ruth M. Lunn, National Institute of Environmental Health Sciences, Research Triangle Park, North Carolina 27709

Richard B. Mailman, Departments of Psychiatry, Pharmacology, Neurology and Medicinal Chemistry, University of North Carolina School of Medicine, Chapel Hill, North Carolina 27599

Elizabeth L. MacKenzie, Education and Training Systems International, Chapel Hill, North Carolina 27514

Isabel Mellon, Graduate Center for Toxicology, University of Kentucky, Lexington, Kentucky 40536

B. Alex Merrick, National Institute of Environmental Health Sciences, Research Triangle Park, North Carolina 27709

Sharon A. Meyer, College of Pharmacy, University of Louisiana at Monroe, Monroe, Louisiana 71209

David S. Miller, National Institute of Environmental Health Sciences, Research Triangle Park, North Carolina 27709

Nancy A. Monteiro-Riviere, Center for Chemical Toxicology Research and Pharmacokinetics, North Carolina State University, Raleigh, North Carolina 27606

Dahlia M. Nielsen, Department of Genetics, North Carolina State University, Raleigh, North Carolina 27695

Ninomiya-Tsuji, Jun, Department of Environmental and Molecular Toxicology, North Carolina State University, Raleigh, North Carolina 27695

Marjorie F. Oleksiak, Rosenstiel School of Marine and Atmospheric Sciences, University of Miami, Miami, Florida, 33149

R. Julian Preston, National Health and Environmental Effects Research Laboratory, U.S. Environmental Protection Agency, Research Triangle Park, North Carolina 27711

Donald J. Reed, Department of Biochemistry and Biophysics, Oregon State University, Corvallis, Oregon 97331

Martin J.J. Ronis, Department of Pharmacology and Toxicology, Arkansas Children's Nutrition Center, University of Arkansas for Medical Sciences, Little Rock, Arkansas, 72202

Randy L. Rose, (deceased) Department of Environmental and Molecular Toxicology, North Carolina State University, Raleigh, North Carolina 27695

MaryJane K. Selgrade, Immunotoxicology Branch, U. S. Environmental Protection Agency, Research Triangle Park, North Carolina 27711

John M. Seubert, Faculty of Pharmacy and Pharmaceutical Sciences, University of Alberta, Edmonton AB, Canada

Robert C. Smart, Department of Environmental and Molecular Toxicology, North Carolina State University, Raleigh, North Carolina 27695-7633.

Ralph J. Smialowicz, (retired) Immunotoxicology Branch, U. S. Environmental Protection Agency, Research Triangle Park, North Carolina 27711

Mariana C. Stern, Department of Preventive Medicine, Keck School of Medicine, University of Southern California, Los Angeles, California 90089

Eric A. Stone, Department of Statistics, North Carolina State University, Raleigh, North Carolina 27695

Joan B. Tarloff, Department of Pharmaceutical Sciences, Philadelphia College of Pharmacy, University of the Sciences in Philadelphia, Philadelphia, Pennsylvania 10104

Arrel D. Toews, Department of Biochemistry and Biophysics, University of North Carolina at Chapel Hill, Chapel Hill, North Carolina 27599

Yoshiaki Tsuji, Department of Environmental and Molecular Toxicology, North Carolina State University, Raleigh, North Carolina 27695

Andrew D. Wallace, Department of Environmental and Molecular Toxicology, North Carolina State University, Raleigh, North Carolina 27695

Zhigang Wang, Graduate Center for Toxicology, University of Kentucky, Lexington, Kentucky 40536

Marsha D. Ward, Immunotoxicology Branch, U.S. Environmental Protection Agency, Research Triangle Park, North Carolina 27711

Darryl C. Zeldin, National Institute of Environmental Health Sciences, Research Triangle Park, North Carolina 27709

CHAPTER 1

Molecular and Biochemical Toxicology: Definition and Scope

ERNEST HODGSON

Department of Environmental and Molecular Toxicology, North Carolina State University, Raleigh, North Carolina 27695

ROBERT C. SMART

Department of Environmental and Molecular Toxicology, North Carolina State University, Raleigh, North Carolina 27695-7633

1.1 INTRODUCTION

After the publication of the previous edition, toxicology saw a dramatic increase in the application of the principles and methods of molecular biology. Biochemical and molecular toxicology are concerned with the definition, at the molecular and cellular levels, of the cascade of events that is initiated by exposure to a toxicant and culminates in the expression of a toxic endpoint. Molecular techniques have provided a wealth of mechanistic information about the role of gene function in the interaction of xenobiotics and living organisms. The development of "knockout" mice with genes of interest deleted, along with the development of "humanized mice" with human genes inserted into their genome, has proven extremely valuable in investigations of toxicant metabolism and modes of toxic action. This edition, the fourth (retitled Molecular and Biochemical Toxicology), reflects this by the inclusion of chapters on molecular methods and the inclusion, in essentially every chapter, of molecular studies and approaches currently used in understanding the metabolic processing and mode of toxic action of xenobiotics.

Toxicology can be defined as the branch of science dealing with poisons. Having said that, attempts to define all of the various parameters lead to difficulties. The first difficulty is seen in the definition of a poison. Broadly speaking, a poison is any substance causing harmful effects in an organism to which it is administered, either deliberately or by accident. Clearly, this effect is dose-related inasmuch as any substance, at a low enough dose, is without effect, while many, if not most, substances have deleterious effects at some higher dose. Much of toxicology deals with compounds exogenous to the normal metabolism of the organism, with such compounds being referred to as xenobiotics. However, many endogenous compounds,

Molecular and Biochemical Toxicology, Fourth Edition, edited by Robert C. Smart and Ernest Hodgson
Copyright © 2008 John Wiley & Sons, Inc.

including metabolic intermediates such as glutamate, or hormones such as thyroxine, are toxic when administered in unnaturally high doses. Similarly, trace nutrients such as selenium, which are essential in the diet at low concentrations, are frequently toxic at higher levels. Such effects are properly included in toxicology, while the endogenous generation of high levels of metabolic intermediates due to disease or metabolic defect is not, although the effects on the organism may be similar.

The expression of toxicity, hence the assessment of toxic effects, is another parameter of considerable complexity. *Acute toxicity*, usually measured as mortality and expressed as the lethal dose or concentration required to kill 50% of an exposed population under defined conditions (LD50 or LC50), is probably the simplest measure of toxicity. Nevertheless, it varies with age, gender, diet, the physiological condition of the animals, environmental conditions, and the method of administration. *Chronic toxicity* may be manifested in a variety of ways, including cancer, cataracts, peptic ulcers, and reproductive effects, to name only a few. Furthermore, chemicals may have different effects at different doses. For example, vinyl chloride is a potent hepatotoxicant at high doses and is a carcinogen with a very long latent period at low doses. Considerable variation also exists in the toxic effects of the same chemical administered to different animal species, or even to the same animal when administered via different routes. Malathion, for example, has relatively low toxicity to mammals, but is toxic enough to insects to be a widely used commercial insecticide.

1.2   TOXICOLOGY

Toxicology is clearly related to two of the applied biologies; medicine and agriculture. In medicine, clinical diagnosis and treatment of poisoning as well as the management of toxic side effects of clinical drugs are areas of significance. In agriculture, the development of selective biocides such as insecticides, herbicides, and fungicides is important, and their nontarget effects are of considerable public health significance. Toxicology may also be considered an area of fundamental science because the adaptation of organisms to toxic environments has important implications for ecology and evolution.

The tools of chemistry, biochemistry, and molecular biology are the primary tools of toxicology, and progress in toxicology is closely linked to the development of new methodology in these sciences. Those of chemistry provide analytical methods for toxicants and their metabolites, particularly for forensic toxicology, residue analysis, and toxicant metabolism; those of biochemistry provide methods for the investigation of metabolism and modes of toxic action; and those of molecular biology provide methods for investigations of the roles of genes and gene expression in toxicity.

1.3   BIOCHEMICAL TOXICOLOGY

Biochemical toxicology deals with processes that occur at the cellular and molecular levels when toxic chemicals interact with living organisms. Defining these interactions is fundamental to our understanding of toxic effects, both acute and chronic, and is essential for the development of new therapies, for the determination of toxic hazards, and for the development of new clinical drugs for medicine and biocides for agriculture.

The poisoning process may be thought of as a cascade of more or less distinct events. While biochemical and molecular toxicology are involved in all of these, their involvement in exposure analysis is restricted to the discovery and use of bio-markers of exposure (see Chapter 26). Following exposure, uptake involves the biochemistry of cell membranes and distribution, or transport processes within the body (Chapters 15 and 35). Metabolism, which may take place at portals of entry or, following distribution, in other organs, particularly the liver, may either detoxify toxicants or activate them to reactive metabolites more toxic than the parent chemical (see Chapters 9 through 14). Chemicals with intrinsic toxicity or reactive metabolites are involved in various modes of toxic action, usually initiated by interactions with macromolecules such as proteins and DNA. The study of modes of toxic action is a critically important area of toxicology (see Chapters 16 through 26). The final phase of detoxication, namely excretion (see Chapters 15 and 29), is studied at the cellular, organ, and intact organism levels.

Many of these aspects are studied at the organ level (discussed in Chapters 27 through 35), including portals of entry, respiratory toxicology, hepatotoxicology, nephrotoxicology, toxicology of the peripheral and central nervous systems, immunotoxicity, reproductive and developmental toxicity, and dermatotoxicity.

1.4   CELLULAR TOXICOLOGY

The culture of cells isolated from living organisms has been known since the early years of the twentieth century. By the 1950s the development of standardized culture media and the development of immortalized cell lines increased the utility of cultured cells in many areas of experimental biology, including toxicology. The use of cell culture in toxicological research is an established and useful approach for a number of reasons, including its use in investigating toxic effects on intact cellular systems in a situation less complex than that in the intact organism and its potential utility for routine toxicity testing systems for regulatory evaluations.

Some cells, such as hepatocytes, must be used in primary culture since they will not divide in culture and are relatively short-lived, while other cell lines are capable of division and can, in suitable media, be maintained indefinitely. In other cases, cells have been "immortalized" by fusion with tumor cells and thereafter retain the ability to divide in culture while, at the same time, maintaining many of the properties of the original nontumor cells. All of the various approaches to the use of cultured cells in biochemical and molecular toxicology are summarized in Chapter 8. The relatively recent union of the techniques of cell and molecular biology has been enormously productive for experimental toxicology since cells can be used for the expression of genetic constructs, reproduction of recombinant enzymes, and so on.

1.5   MOLECULAR TOXICOLOGY

The field of molecular biology is usually held to have begun with the description of the double helical structure of DNA by Watson and Crick in 1953, followed by the elucidation of the genetic code in the 1960s. In the subsequent half-century the techniques of molecular biology have expanded exponentially as has its importance in many, if not most, fields of biology. The success of the human genome project has given rise to an entire field devoted to the description of the complete genomes of organisms at all levels in the evolutionary tree. An overview of molecular techniques is presented in Chapter 2, and a review of toxicogenomics is presented in Chapter 3.

The techniques that have proven most valuable in toxicology include those of molecular cloning, the polymerase chain reaction, and the production of genetically modified mice. Microarrays, used to evaluate gene expression under various conditions, including exposure to toxicants, are becoming more important and, in concert with other molecular techniques, are being considered as potentially useful in such applied areas as hazard assessment and risk analysis.

Bioinformatics, which deals with the maintenance, analysis, and integration of genomic data, is discussed in Chapter 6.

1.6   PROTEOMICS AND METABOLOMICS

Since molecular biology is often held to be restricted to events involving nucleic acids, mention must be made of (a) proteomics, the analysis of all proteins in a sample of biological material, and (b) metabolomics, the analysis of all metabolites in a sample of biological material. These fields are discussed in Chapters 4 and 5.

1.7   CONCLUSIONS

The preceding brief description of the nature and scope of biochemical and molecular toxicology should make clear that the study of toxic action is a many-faceted subject, covering all aspects from the initial environmental contact with a toxicant to its toxic endpoints and to its ultimate excretion back into the environment. A considerable amount of material is summarized in the chapters following, but many essentials still remain to be discovered.

SUGGESTED READING

Alberts, B., Roberts, K., Lewis, J., Raff, M., Walter, P., and Johnson, A. *Molecular Biology of the Cell*, 4th ed., Garland Publishing, New York, 2002.

Ausubel, F. M., Brent, R., Kingston, R. E., Moore, D. D., Seidman, J. G., Smith, J. A., and Struhl, K. *Current Protocols in Molecular Biology*, John Wiley & Sons, Hoboken NJ, 1987–2007.

Bus, J. S., Costa, L. G., Hodgson, E., Lawrence, D. A., and Reed, D. J. *Current Protocols in Toxicology*, John Wiley & Sons, Hoboken, NJ, 1999–2007.

Hodgson, E. (Ed.). *A Textbook of Modern Toxicology*, 3rd ed., John Wiley & Sons, Hoboken NJ, 2004.

Klaassen, C. D. (Ed.). *Casarett and Doull's Toxicology, The Basic Science of Poisons*, 6th ed., McGraw-Hill, New York, 2001.

Watson, J. D., Baker, Tania A., Bell, S. P., Gann, A., Levine, M., and Losick, R. *Molecular Biology of the Gene*, 5th ed., Benjamin Cummings, Menlo Park, CA, 2003.

CHAPTER 2

Overview of Molecular Techniques in Toxicology: Genes and Transgenes

ROBERT C. SMART

Department of Environmental and Molecular Toxicology, North Carolina State University, Raleigh, North Carolina 27695-7633

2.1 APPLICABILITY OF MOLECULAR TECHNIQUES TO TOXICOLOGY

Molecular cloning and techniques involving the manipulation of DNA and RNA have revolutionized the biological sciences. These techniques have allowed for the elucidation of gene function in complex biological processes such as cell growth, differentiation, development, and cancer as well as chemical toxicity. In a much broader sense, these techniques have allowed for the successful completion of the Human, Mouse, and Rat Genome Projects. Genomic information and bioinformatic technologies derived from these projects will have a major impact on our understanding of chemical-induced toxicity as it relates to mechanisms, species differences, individual susceptibility, and the use of appropriate animal models for hazard characterization.

Molecular techniques have wide applicability in understanding the mechanisms and the responses of an organism to xenobiotic insult. For example, the response of a host organism to a xenobiotic often results in adaptive responses involving alterations in gene expression. Using various molecular approaches, such genes can be identified and their roles elucidated. Expression of these genes could be used as biomarkers of exposure. Chemical-induced alterations in gene expression and pathways can be elucidated using DNA microarrays where the expression of hundreds to thousands of genes can be studied in a single experiment. Additional examples where molecular biology can be applied to toxicology include the identification of chemical carcinogen-induced mutations in oncogenes and tumor suppressor genes. In certain cases, the mutation spectra in onogene/tumor suppressor gene in human tumors can be used as a molecular fingerprint to aid in the identification of the responsible carcinogen. This approach is currently utilized in the field of Molecular Epidemiology. Genetic susceptibility to xenobiotic insult can be the result of certain genetic polymorphisms. Molecular techniques can be used to identify these genetic

Molecular and Biochemical Toxicology, Fourth Edition, edited by Robert C. Smart and Ernest Hodgson
Copyright © 2008 John Wiley & Sons, Inc.

polymorphisms and to characterize their toxicological significance. Molecular approaches can also be employed to understand species and sex differences to toxicant responses as they relate to differences in gene structure, function, and expression.

The introduction of a gene of interest into a cell or animal is a powerful approach to characterize the function of a gene in a toxicological response. For example, this approach can be used to express human xenobiotic metabolizing genes in recipient cells in culture. Such ectopically expressed genes can facilitate the determination of substrate specificity and the role of individual enzymes in the production and detoxification of toxic metabolites as well as species differences. The ability to generate transgenic animals and gene knockout and knockin animals allows for study of gene function in vivo. The creation of transgenic, knockout, and knockin mice provide a powerful approach in mechanistic toxicological studies and in the development of in vivo models designed to be more predictive of human toxicity. This chapter is intended to provide a conceptual overview/primer of molecular techniques/ approaches. For more comprehensive and detailed information the reader is referred to the textbook *Genes IX*. For technical information and protocols the reader is referred to the following laboratory manuals: *Molecular Cloning; A Laboratory Manual* and *Current Protocols in Molecular Biology*.

2.2   OVERVIEW OF THE GENETIC CODE AND FLOW OF GENETIC INFORMATION

Cells are composed primarily of proteins, nucleic acids, lipids, and carbohydrates. The genetic material of the cell, deoxyribonucleic acid (DNA), is composed of individual deoxyribonucleotide building blocks. Each deoxyribonucleotide or "nucleotide" is composed of a nitrogenous base such as cytosine, adenine, thymine, or guanine linked to a deoxyribose sugar moiety that has one or more phosphate groups on the 5′ carbon of the sugar. These nucleotides are covalently linked together by phosphodiester bonds to form a linear strand that contains millions of nucleotides (Figure 2.1A). A typical DNA molecule is composed of two antiparallel strands forming the Watson–Crick double helix with the bases of one strand forming hydrogen bonds with the bases of the other strand (Figure 2.1B). Adenosine always pairs, or hydrogen bonds, to thymine to form A–T pairs, while cytosine always pairs with guanine to form C–G pairs. Within the coding sequence of genes, the order of the nucleotides in DNA dictates the order of amino acids in proteins. Three adjacent nucleotides, or codon, code for a specific amino acid. Thus, the linear sequence of DNA ultimately specifies the order of amino acids for all proteins.

The flow of information from DNA to protein is transmitted through messenger ribonucleic acid (mRNA). Ribonucleic acid is composed of individual ribonucleotide building blocks, each made up of nitrogenous base, such as cytosine, adenine, uracil, or guanine linked to a ribose sugar moiety that has a phosphate group on the 5′ carbon of the sugar. Transcription of DNA results in a copy of the DNA sequence of a given gene in the form of mRNA (Figure 2.2). The mRNA contains the sequence of bases that are complementary to the bases of the transcribed (antisense) strand of DNA. Eukaryotic genes contain introns (interruptions) in the coding sequence (exons) of the gene. Therefore, eukaryotic DNA is first copied into a primary transcript called heterogeneous nuclear RNA (hnRNA). hnRNA under-

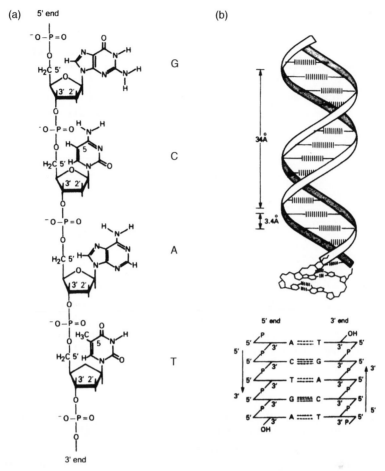

Figure 2.1. A polydeoxynucleotide chain and the structure of DNA. (A) A polydeoxynucleotide chain composed of four different deoxynucleotides covalently linked to one another by a phosphodiester bonds (note orientation of the 5′ and 3′ end of the molecule) (B) Watson–Crick double helix (top) and simple representation of the antiparallel strands (bottom). [Adapted from Volgelstein, B., and Kinzler, K. W. (Eds.), *The Genetic Basis of Human Cancer*, McGraw-Hill, New York, 1998.]

goes extensive processing within the nucleus involving removal of introns and the splicing together of exons, the capping of the 5′ end of mRNA with 7-methylguanosine, and polyadenylation of the 3′end with 100–200 adenosines. These last two modifications tend to stabilize the mRNA and prevent its degradation. The mature mRNA is transported to the cytosol, where it is translated into protein by ribosomal RNA (rRNA) using transfer RNA (tRNA) to supply amino acids to the peptide chain. Translation of the mRNA involves the ordered and sequential covalent incorporation of amino acids into a polypeptide chain to form a functional protein (Figure 2.2). The order of amino acids is determined by the order of the codons present in the mRNA and the subsequent interactions with specific tRNA. Each tRNA is covalently linked to a specific amino acid and has a sequence complementary (anticodon) to the codons in the mRNA. Through the aid of complex ribosomal

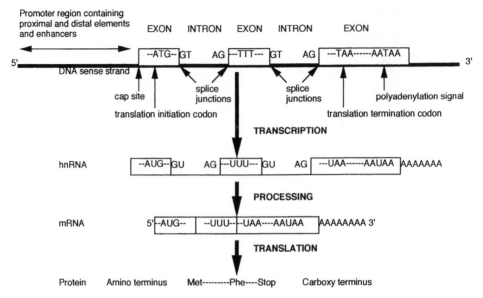

Figure 2.2. Transcription, mRNA processing, and translation. DNA-sense strand is designated by a bold line, while hnRNA and mRNA are designated by thinner lines. Exons are shown as rectangles, while introns are shown as the intervening spaces between exons. [Adapted from Volgelstein, B., and Kinzler, K. W. (Eds.), *The Genetic Basis of Human Cancer*, McGraw-Hill, New York, 1998.]

structures and the specific interaction of the mRNA codon and the tRNA anticodon, the correct amino acid is aligned and released from the tRNA and subsequently covalently linked to the growing polypeptide chain.

Regulation of gene expression can occur at the transcriptional and posttranscriptional levels and may involve mRNA initiation, mRNA stability, mRNA splicing, polyadenylation, and transport of the mRNA from the nucleus to cytoplasm as well as translation of the mRNA. The regulation of transcription is accomplished by the binding of sequence-specific DNA-binding proteins, or transcription factors to specific nucleotides (cis element) within the 5′ noncoding sequence or promoter region of a gene. Transcription factors bound to promoters interact with the RNA polymerase complex to regulate gene expression.

2.3 MOLECULAR CLONING

The human haploid genome contains approximately 3 billion base pairs encoding 20,000 to 25,000 individual genes. In order to study the structure, function, and regulation of genes and their protein products, one must have the capability isolate the gene of interest. The general idea of molecular cloning is to insert a DNA segment (a gene of interest or some part thereof) into a vector. A vector is a DNA molecule originating from a plasmid or virus that has the capacity to self-replicate when placed into a host cell. The vector containing the DNA segment of interest is then placed into a host, such as a bacterium or yeast. Inside the host cell, the vector containing the DNA insert is replicated many times, producing hundreds of exact copies of itself per cell, effectively cloning the inserted DNA. The cloned gene can be used for a large variety of purposes. For example, the cloned gene can be sequenced to determine the nucleotide and corresponding amino acid sequence. In addition, the cloned gene can be placed into a mammalian expression vector and expressed in mammalian cells to characterize its function.

2.3.1   Vectors

There are many kinds of cloning vectors, including plasmids, bacteriophages, viruses, bacteria artificial chromosomes (BAC), and yeast artificial chromosomes (YAC), and there are also many variations on the cloning theme. Generally, plasmids, bacteriophages, cosmids, BACs, and YACs are employed to clone DNA fragments of up to 10 kbp (kilobase pair = 1×10^3 base pairs), 20 kbp, 40 kbp, 150 kbp, and 1 Mbp (megabase pairs = 1×10^6 base pairs) in length, respectively. Bacteriophages and cosmids are frequently used to make genomic libraries, while BACs and YACs are frequently used in chromosomal mapping, positional cloning studies, and genomic sequencing studies. Plasmids such as pBR322 and pUC (Figure 2.3) are very useful for cloning smaller DNA fragments and making cDNA libraries. Bacterial plasmids are double-stranded closed circular DNA molecules that range in size from 1 kb to more than 200 kb and have an origin of DNA replication that allows vector replication independent of the bacterial chromosome (autonomous replication). Plasmids have also been engineered to contain a multiple cloning site that contains numerous unique restriction endonuclease sites. When the insert DNA of interest and the vector have been cut with the same restriction endonuclease, both molecules have "sticky" or complementary ends that enable the molecules to anneal. The annealed molecules can then be covalently joined with DNA ligase. In this manner, the insert DNA is placed into the plasmid to form a recombinant DNA molecule (Figure 2.4). Plasmids also contain genes that give a bacterium a selective advantage, such as antibiotic resistance to ampicillin. Plasmids are introduced into the bacterium through a process referred to as transformation. This can be accomplished by creating bacteria that are "competent" to take up DNA by incubating with a mixture of divalent cations such as $CaCl_2$, which makes the cell permeable to small DNA molecules such as plasmids. The transformed bacteria are then incubated on plates (Petri dishes) containing media and the appropriate antibiotic. As mentioned above, if the recombinant plasmid contains a gene that provides resistance to ampicillin, then the media would contain ampicillin to select for bacteria that have taken up plasmid. Inside the bacterium, the vector will replicate to produce many copies and the bacterium will grow to form a colony. Colonies can subsequently be isolated and examined for the presence of the DNA insert. The positive colony can then be grown in a larger-scale culture to produce milligram quantities of recombinant DNA.

2.3.2   Identification of Bacterial Colonies that Contain Recombinant DNA

There are several methods to identify bacterial colonies harboring recombinant plasmid. One of the more commonly used methods is referred to as α-complementation. The pUC series of plasmids contains the coding sequence for the first 146 amino acids of the amino terminus of the β-galactosidase gene (*lacZ*). Within this

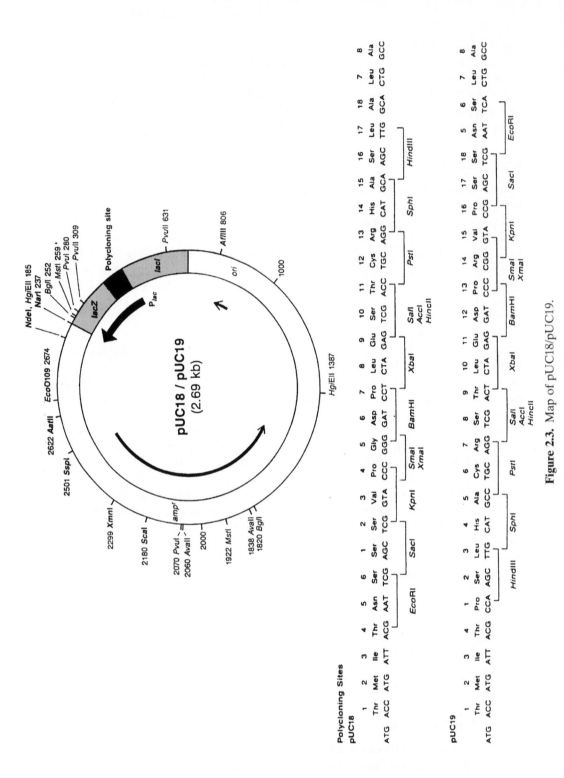

Figure 2.3. Map of pUC18/pUC19.

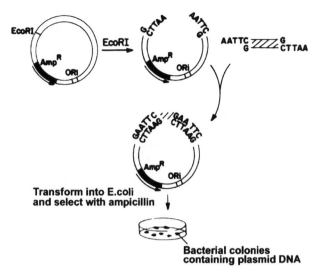

Figure 2.4. Molecular cloning using a plasmid vector.

region is the polycloning site, however the site is not large enough to disrupt the reading frame and results in the harmless incorporation of a small number of amino acids into the amino terminal fragment of β-galactosidase (Figure 2.3). These plasmids are used in the *E. coli* strain, JM103, a strain that has the genetic information to produce the carboxy terminal region of β-galactosidase but not the amino terminal region. Neither the host alone nor plasmid alone can form an active β-galactosidase enzyme; however, together they form an active β-galactosidase enzyme. Thus, α-complementation results in an active β-galactosidase enzyme or Lac⁺ bacteria. These Lac⁺ colonies are easily identified in the presence of the chromogenic substrate 5-bromo-4-chloro-3-indolyl-β-D-galactoside (X-gal) because β-galactosidase metabolizes X-gal to form a blue product and Lac⁺ colonies are blue. However, insertion of foreign DNA into the multiple cloning site results in the disruption of the coding sequence of β-galactosidase amino terminus and an amino terminal fragment that is not capable of α-complementation. These colonies appear white and represent the bacterial colonies that contain recombinant DNA.

2.3.3 Construction of cDNA Library

For the construction of a cDNA library, total RNA is isolated from the cells or tissue of interest. Total RNA contains only 1–2% mRNA and approximately 80% rRNA. mRNA can be isolated from rRNA and tRNA using an oligo(dT)-cellulose column or oligo(dT)-coated beads. The poly (A) 3′ tail of mRNA binds to the immobilized oligo dT, and tRNA and rRNA pass through the column or can be separated from the coated bead. The poly(A) mRNA is then eluted from the column or beads and is utilized to make complementary DNA (cDNA) with the aid of reverse transcriptase, an RNA-dependent DNA polymerase (Figure 2.5). The cDNA is inserted into a vector and the recombinant vector is then transformed into bacteria and plated as described above. For the construction of a genomic library, the genomic DNA is digested with a restriction enzyme to produce an overlapping set of DNA fragments 12–20 kbp in length. These fragments are cloned into a bacteriophage vector and

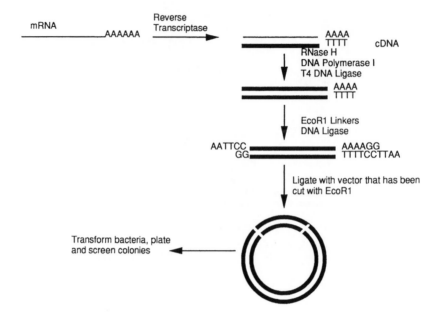

Figure 2.5. Generation of cDNA library using a plasmid vector. RNA is designated with a thin line, and DNA is indicated by a bold line.

encapsidated with viral coat protein, a process often referred to as in vitro packaging. The recombinant infectious phages are then used to infect *E. coli*.

Different methods are available to identify the recombinant clones containing the specific cDNA or genomic DNA of interest. One of the most common methods involves screening by nucleic acid hybridization. Plasmid or phage-containing bacterial colonies are replica plated from a master plate onto a nitrocellulose or nylon membrane (Figure 2.6). The cells are lysed on the membrane and the DNA is immobilized on it. The membrane is then incubated with a radiolabeled DNA fragment or "probe" that is complementary to the desired DNA sequence. Specific annealing conditions allow for the hybridization of probe to the desired DNA sequence. The membrane is then washed and subjected to autoradiography. The radioactive probe will bind to the DNA of the colonies that contain the complementary DNA; and when X-ray film is placed on the membrane, the radioactive annealed probe will expose the film only in this specific area. The corresponding clone from the master plate can then be identified, retrieved, and grown to obtain large quantities of the inserted DNA. The probe used can be the entire cDNA or a portion of the cDNA or genomic sequence. Reducing the stringency of hybridization so that the probe will bind to closely related sequences can also identify related cDNA sequences. In this way, members of a gene family can be identified. If the cDNA or genomic sequences are not available, but a portion of the protein sequence is known, then synthetic antisense degenerate oligonucleotides encoding possible codon combinations can be synthesized and used to screen the library. If no genetic sequence information or protein structural information is known, then special methods are available to express the protein in bacterial or mammalian cells and the library can be screened with an antibody to the protein of interest.

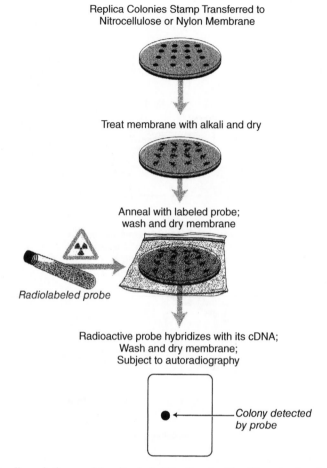

Replica Colonies Stamp Transferred to
Nitrocellulose or Nylon Membrane

Treat membrane with alkali and dry

Anneal with labeled probe;
wash and dry membrane

Radiolabeled probe

Radioactive probe hybridizes with its cDNA;
Wash and dry membrane;
Subject to autoradiography

Colony detected
by probe

Figure 2.6. Replica plating and in situ hybridization to identify colonies that contain the DNA of interest.

2.4 SOUTHERN AND NORTHERN BLOT ANALYSES

Southern analysis of DNA is frequently used to (i) determine if a gene of interest is present, (ii) determine the copy number of a particular gene, (iii) examine restriction fragment length polymorphisms (RFLP) (for example, changes in restriction sites are an indication of genetic differences between individuals or samples), (iv) determine loss of heterozygosity or reduction to homozygosity, and (v) (determine) the genotype of genetically modified animals. Northern analysis is most frequently used to examine mRNA size, splice variants, and levels.

Northern and Southern blot analyses are carried out as shown in Figure 2.7 and as described below. Restriction enzyme digested DNA (Southern) or total RNA or poly(A) mRNA (Northern) is subjected to agarose gel electrophoresis, which separates DNA and RNA molecules based on their size. The separated molecules are then transferred to a nylon or nitrocellulose membrane by capillary action transfer

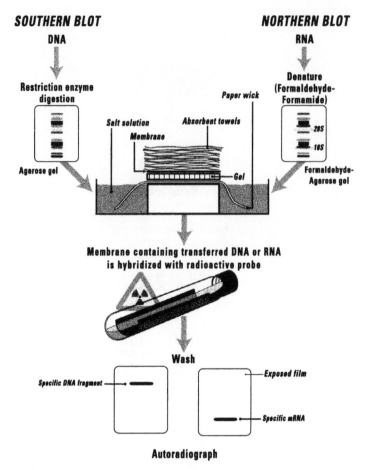

Figure 2.7. Generalized procedure for Southern and Northern analysis. [Adapted from Vol-gelstein, B., and Kinzler, K. W. (Eds.), *The Genetic Basis of Human Cancer*, McGraw-Hill, New York, 1998.]

or electroblotting. The immobilized DNA or RNA on the membrane is incubated with a radiolabeled probe to allow the probe to hybridize to its complementary target sequence. The membrane is then gently washed to remove unbound probe and subjected to autoradiography in which an autoradiographic film is placed over the membrane. Following exposure, the film is developed and the blackened or exposed bands correspond to the binding of the radioactive probe to its comple-mentary sequence. Many methods are currently available that do not use radioactive probes but instead use fluorescent or chemiluminscent probes.

2.5 POLYMERASE CHAIN REACTION (PCR)

Polymerase chain reaction is a powerful technique that can amplify specific DNA sequences. Amplification of DNA by PCR can be used for mutational analysis,

RFLP analysis, allele-specific hybridization, genotyping, or to simply clone a gene of interest. To use PCR, the sequence of the gene of interest must be known, or at the very least the sequence in the area where the primers are to bind must be known. Using the known sequence, primers can be constructed that will flank the area to be amplified and these primers will be starting points for DNA synthesis ($5' \rightarrow 3'$) on each strand. The target DNA of interest is incubated with thermostable DNA polymerase, all four dNTPs and a set of flanking primers. PCR is conducted in a thermal cycler in which the temperature is raised to separate the double-stranded DNA (denaturation step), then the temperature is lowered to allow the primers to anneal to the complementary sequences in the target DNA (annealing step), and finally the temperature is raised to allow the polymerase to synthesize DNA (synthesis step). This cycle of denaturation and annealing followed by synthesis is repeated 20–40 times. Using this PCR approach, the area of the gene of interest can be amplified $10^5–10^9$ times, providing a large amount of the DNA of interest. The procedure is so powerful that DNA from a single cell can be amplified, for example, cells from histological sections can be scraped from a microscope slide, and DNA can be amplified for mutation analysis. Recently, the use of laser capture microdissection (LCM) has allowed for the analysis of DNA collected from just a few cells on a histological section. This is especially useful in mutational analysis of tumor cells and in studies of chemical carcinogenesis and tumor progression.

2.6 SOME METHODS TO EVALUATE GENE EXPRESSION AND REGULATION

2.6.1 Northern Analysis

As described above, Northern blot analysis is commonly used to determine levels of mRNAs of interest. mRNA levels can be compared between experimental and control samples and should be normalized to the expression of a gene whose expression is invariant. For example, housekeeping genes such as actin or GAPDH (3-glyceraldehyde-3-phosphate dehydrogenase) are commonly used.

2.6.2 Nuclear Run-On

While Northern analysis can reveal an increase in the level of mRNA for a specific gene, this increase could be due to mRNA stabilization or could be due an increased rate of transcription or both. To determine if the rate of transcription is increased, a nuclear run-on assay can be performed (Figure 2.8). Nuclei are isolated from the control and treated cells of interest at various times and incubated with [α-^{32}P]UTP. The ^{32}P-labeled uridine will be incorporated into elongating hnRNA. At various time points the hnRNA is isolated and spotted onto a nylon or nitrocellulose membrane containing cDNA of the gene of interest immobilized on it. The newly synthesized radiolabeled hnRNA hybridizes to the immobilized cDNA, the membrane is washed to remove unincorporated [α-^{32}P]UTP, and nonspecific binding of other radiolabeled hnRNAs and autoradiography is conducted. The [α-^{32}P]UTP incorporation can be compared between experimental and control cells. If the experimental group demonstrates increased exposure of the film, it indicates that the rate of mRNA synthesis of the gene of interest is increased.

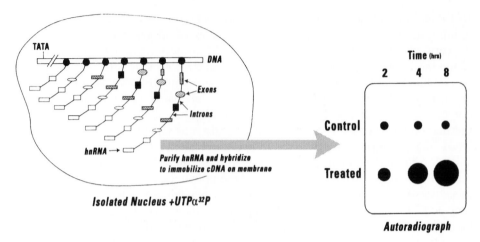

Figure 2.8. Nuclear run-on assay (see text for details).

2.6.3 Promoter Deletion Analysis/Reporter Gene Assays

If after conducting a nuclear run-on assay the rate of transcription is found to be increased, sites within a promoter that are required for transcription and corresponding transcription factors that contribute to the increased rate of transcription can be identified. In order to do this, one must have the promoter region of the gene of interest in hand. The promoter region is inserted into a plasmid such that the promoter regulates the transcription of a reporter gene such as luciferase or chloramphenicol acetyl transferase (CAT). The reporter construct is transfected into a mammalian cells of choice, and the cells are treated with an agent or conditions that alter the expression of the gene. Under these conditions the reporter will be transcribed, mRNA will be translated, and the resultant luciferase activity or CAT activity can be easily quantified. Generally, cells are harvested 24–48 hr after transfection, cell homogenates are prepared, and luciferase or CAT activity is determined. The amount of luciferace activity is directly proportional to the transcription of the reporter. Promoter deletion analysis is conducted to identify the specific cis elements or specific transcription factor binding sites within the promoter that are responsible for the increase in the expression of the reporter gene. Small areas of the promoter are deleted, and these constructs are then transfected and the activity of the promoter is determined as described above. Through this type of analysis, the specific enhancer elements and transcription factors that are important in the regulation of the gene under study can be identified (Figure 2.9).

2.6.4 Microarrays

Microarrays are utilized to study the expression of hundreds or even thousands of genes in a single experiment. A microarray is a solid support to which hundreds to thousands of individual cDNAs or DNA oligomers (each specific to a unique gene) have been immobilized. The cDNAs or oligomers are applied to the support in a grid pattern so that each individual cDNA has a unique position within the grid. mRNA is isolated from the experimental and control tissue or cells and cDNA is

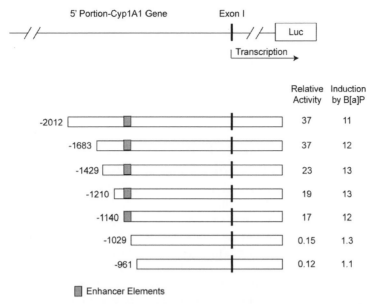

Figure 2.9. Luciferase reporter gene assay/promoter deletion analysis. Cells are transfected with the CYP1A1 promoter reporter (luciferase) constructs in which the promoter region is increasing deleted. The cells are then treated with benzo[a]pyrene (an inducer of CYP1A1), and cells are harvested 24–48 hr later and luciferase activity is measured. Promoter deletion analysis revealed enhancer element between −1140 and −1029. Contained within this region is xenobiotic response element (XRE) to which the aryl hydrocarbon receptor (AHR) binds.

made using reverse transcriptase in the presence of fluorescent or radiolabeled nucleotides. The labeled cDNAs are hybridized to the complementary DNA sequences that are immobilized on the microarray. The degree of hybridization of the cDNA is quantitiated using a specialized scanner or a phosphoimager, and the expression of each gene is normalized to housekeeping genes or other genes whose expression is not altered between the experimental and control group. Through this approach, one can examine the expression of thousands of genes in a single experiment. See Chapter 3 for more details.

2.6.5   Reverse Transcriptase-PCR (RT-PCR) and Real-Time PCR

PCR can also be used to amplify mRNA through a technique called reverse transcriptase PCR (RT-PCR). mRNA is first converted into DNA using reverse transcriptase, and then the corresponding cDNAs are amplified using gene specific primers by PCR. This technique is suitable for studies to determine level of gene expression. To compare levels of mRNA between samples, a semiquantitative RT-PCR technique has been developed whereby PCR products produced are directly related to the amount of starting target sequence. The amount of the PCR product is normalized to the PCR product levels of a housekeeping gene that is amplified in the same PCR reaction using its gene-specific primers. By using specific primers to multiple genes, changes in the expression of numerous genes can be determined and normalized to the expression of a housekeeping gene within a single sample, provided that the PCR product formation is in the linear range. Real-time PCR is modification of the standard RT-PCR described above. Real-time PCR utilizes an instrument that measures PCR product formation in real time so that the investigator can monitor the linear range of PCR product formation. In this linear range, an investigator can compare different samples or assess changes in gene expression with a high degree of accuracy.

2.6.6   RNase Protection Assay (RPA)

Ribonuclease protection assay (RPA) is a sensitive method for the detection and quantitation of mRNA species as well as the detection of alternatively-spliced mRNAs. The assay uses high-specific-activity ^{32}P-, ^{33}P-, or ^{35}S-labeled antisense RNA probes expressed from a T7 or T3 polymerase promoter containing vector using DNA-dependent RNA polymerase. The antisense RNA probe is hybridized to the target mRNA in solution, after which the free probe and other single-stranded RNAs are digested with RNases. The double-stranded RNA is protected from digestion, and the RNase protected probes are purified and resolved on a denaturing polyacrylamide gel and quantified by autoradiography or phosphorimage analysis. The quantity of the target mRNA is based on the level of intensity of the signal.

2.6.7   Electrophoretic Mobility Shift Assay (EMSA)

Electrophoretic mobility shift assay is used to determine the level of binding of a transcription factor to its specific DNA consensus sequence. It is based on the observation that the movement of a DNA molecule through a non-denaturing polyacrylamide gel is slowed when a protein is bound to it. Nuclear extracts are incubated with the radiolabeled DNA consensus sequence of the transcription factor under study and then subjected to electrophoresis on a non-denaturing polyacrylamide gel. If the nuclear extract contains the transcription factor that binds to the specific consensus sequence, the mobility of the consensus sequence will be retarded or slowed on the gel. Autoradiography of the gel will reveal that the free DNA consensus sequence (not bound to its transcription factor) has migrated to the bottom of the gel. However, if the transcription factor specific to the DNA consensus sequence is present in the nuclear extract and it is capable of binding, then a radiolabeled protein–DNA complex with retarded mobility will appear in the top portion of the gel (a mobility shift) (Figure 2.10). The identity of the binding proteins can be confirmed by incubating the nuclear extract, radiolabeled DNA consensus sequence with an antibody to the suspected transcription factor thought to be responsible for the binding. The samples are then subjected to the procedure described above with the end result demonstrating a "supershift" upward of the transcription factor/DNA complex due to the binding of the antibody to the transcription factor.

2.7   METHODS TO EVALUATE GENE FUNCTION

2.7.1   Eukaryotic Expression Systems

Numerous eukaryotic expression systems are employed to study gene function in an intact cell, and these most often use a plasmid or viral expression system. In

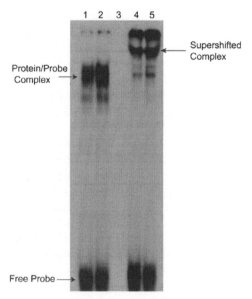

Figure 2.10. Electorphoretic mobility shift assay (EMSA). C/EBPβ is a basic leucine transcription factor. An DNA oligomer consisting of the C/EBP binding site was labeled with ^{32}P and incubated with a nuclear extract (lanes 1 and 2). The same oligomer and nuclear extract were incubated in the presence of a C/EBPβ antibody (lanes 4 and 5). In lanes 1 and 2, the unbound free probe can be seen at the bottom of the autoradiogram, while probe protein complexes can be visualized in the upper section. In lanes 4 and 5 the presence of the antibody resulted in the supershift of the protein–probe complex; the ternary complex consisting of the probe, C/EBPβ protein, and C/EBPβ antibody are now visualized at the top of the autoradiogram. The bands that were supershifted with the antibody represent C/EBPβ.

choosing an expression system, one has to choose between (1) viral transduction versus plasmid transfection systems, (2) a constitutively active promoter versus an inducible promoter, and (3) transient expression versus stable integration into the hosts' genome. The advantage of viral vectors including retroviruses, adenoviruses, and lentiviruses is that they all transduce cells with high efficiency. However, viral vectors are more labor-intensive to use than plasmid vectors; for example, retroviruses and lentiviruses require packaging cell lines, and adenovirus are complicated to engineer. Plasmid-based vectors require transfection; and depending upon the cell type employed, efficiencies of transfection can range from 5% to 80%. Low transfection efficiency may preclude plasmid vector use in some cells. Plasmid eukaryotic expression systems often produce very high levels of expression of the gene of interest due to their use of powerful viral promoters and their ability to autonomously replicate within the host cell.

In general, a plasmid-based expression system can be divided into two types: constitutive expression vectors and inducible expression vectors. In both cases, the gene of interest is cloned into the expression plasmid such that the promoter provided by the vector regulates it. The construct is then transfected into mammalian cells using a variety of methods including those that use cationic lipids, $CaCl_2$, or electroporation. A constitutively active promoter is fully active, driving the transcription of the gene under its regulation independent of extracellular stimuli. The

CMV (cytomegalovirus) promoter is a commonly used promoter in this type of expression vector because it is a potent promoter in many cell types. Inducible expression vectors utilize promoters that can be turned off and on. For example, the metallothionein promoter is inactive in the absence of metals such as zinc. The addition of zinc to tissue culture media triggers the promoter to become active and transcription of the inserted gene ensues. The transfection experiments described above are referred to as transient transfection or transient expression experiments in that the vector is not incorporated into the host genome and cells are generally studied within 48 hr of transfection. However, stable or genetically modified cell lines that have incorporated the expression vector into the host genome can be generated. Selection for cells that have stably incorporated expression constructs into their genome can be accomplished by the presence of a drug resistance marker gene in the vector. The neomycin resistance gene, which is expressed from a constitutive promoter, is commonly used and facilitates the selection of stable transformants in the presence of the antibiotic G418 (an aminoglycoside related to gentamicin which is inactivated by the neomycin resistance gene product). G418 blocks translation and kills cells that do not contain the neomysin resistance gene whose product inactivates G418. Cells that have stably integrated the vector DNA into their genome survive G418 selection forming colonies of drug-resistant cells that may be isolated as clones or grown as populations. Many excellent mammalian expression systems, both constitutive and inducible, are commercially available and are constantly being refined.

2.7.2 Transgenic Mice

A transgenic organism is one that has exogenous DNA in its genome. In order to achieve inheritance of exogenous DNA or transgene, integration of the DNA must occur in the cells that can give rise to germ cells. This is accomplished by introducing the DNA into a fertilized egg cell or into embryonic stem cells. Embryonic stem cells are commonly used to make knockout mice that are deficient in a single gene, while injection of the fertilized egg with DNA is most commonly used to make transgenic mice that overexpress a specific gene. In order to understand how transgenic and knockout mice are made, it is necessary to briefly review early postfertilization and preimplantation development (Figure 2.11).

2.7.2a Procedure for Making Transgenic Mice Using Zygote Injection. Within a few hours after mating, zygotes can be removed from a female mouse before the female pronucleus and male pronucleus fuse. The pronucleus from the male is injected (male pronucleus is larger and provides bigger target for injection) with a construct containing the gene of interest (Figure 2.12). These constructs can be targeted to specific tissues in the adult animal by the use of specific promoters. For example, an albumin promoter will target the expression to the liver, certain keratin promoters will target the expression to the epidermis, and an actin promoter (housekeeping gene) will direct the expression of the transgene in all cells. After injection of the DNA into the pronucleus of the male, the zygotes are placed in the oviducts of a 0.5-day psuedopregnant female (a female that has mated with a sterile male and is physiologically capable of allowing for the development of the injected zygotes). About 15% of the offspring will contain the transgene and will be

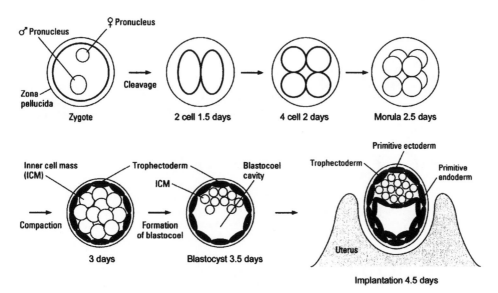

Figure 2.11. Fertilization and preimplantion development. (Adapted from Sedivy, J. M., and Joyner, A. L. *Gene Targeting*, Oxford University Press, New York, 1994.)

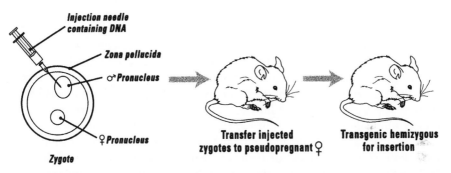

Figure 2.12. Procedure of making transgenic mice using zygote injection. (Adapted from Sedivy, J. M., and Joyner, A. L. *Gene Targeting*, Oxford University Press, New York, 1994.)

transgenic hemizygous animals. Transgenic mice can be identified by Southern analysis of a DNA sample purified from a clipping of a small piece of their tail. Transgene positive mice are referred to as founders and can be mated with wild-type mice to produce offspring that also contain the transgene, which also confirms that the transgene is transmitted by the germ cells. Generally, the approach outlined above to make transgenic mice is used to study the overexpression of a gene (gain of function); however, antisense as well as dominant negative approaches can be utilized and can result in a decrease in the level of the endogenous protein.

2.7.2b *Procedure for Making Knockout Mice Using Embryonic Stem Cells.*
Embryonic stem cells (ES cells) and homologous recombination are utilized to inactivate an endogenous gene from a host's genome. ES cell lines are derived from a 3-day embryo (ICM cells) and are undifferentiated but remain totipotent. Mouse

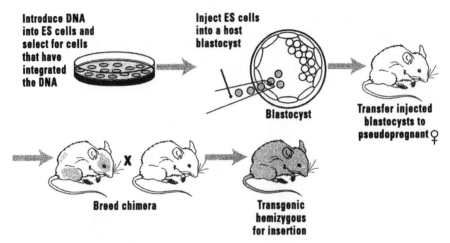

Figure 2.13. Procedure for making nullizygous mice using embryonic stem cells. (Adapted from Sedivy, J. M., and Joyner, A. L. *Gene Targeting*, Oxford University Press, New York, 1994.)

ES cells are now commercially available. These cells are transfected in cell culture with a vector that generally contains a homologous portion of the gene (not cDNA) of interest with a neoresistance gene inserted into an exon to disrupt its function. This piece of DNA can undergo homologous recombination with the ES cell gene and result in the replacement of the ES gene with the defective gene (this is an uncommon event, but it does happen at low frequencies). Through positive and negative selection, homologous recombinants can be identified and then character- ized by PCR analysis to confirm the presence of the defective gene. These ES cells of interest can then be injected into a blastocyst and then transferred into a pseu- dopregnant female (Figure 2.13). If the ES cells were derived from mice with black fur and the blastocyst is from an albino mouse, then offspring will be chimeras con- taining both white and black fur (cells of such chimera mice are from ES cells as well as cells derived from the blastocyst into which they were injected). These chimera can be bred with albino (recessive hair phenotype) animals, and black (dominant hair phenotype) offspring will have inherited one inactive allele of the particular gene of interest. By mating heterozygous mice (one allele inactivated), knockout mice can be produced.

2.7.2c   Conditional Knockout. It is possible that the deletion of a single gene (knockout) can be embryonic lethal as is the case with many genes. To circumvent this problem, several strategies have been developed to produce a conditional knockout. Depending upon the methods used, a conditional knockout can be one in which the gene of interest is specifically deleted in the tissue of interest or one in which the gene is deleted at a specific time—for example, during development or in the adult mouse.

2.7.2d   Cre-loxP Recombination System.   Cre is a bacteriophage-derived recom- binase that recognizes a specific DNA sequence (34 bp) referred to as a *loxP* site. If a gene of interest or the area to be deleted is engineered to be flanked on each end by a *loxP* site (referred to as a floxed gene), Cre will catalyze a site-specific recombination resulting in the deletion of the floxed gene. If Cre is under the control of a tissue-specific promoter, then the floxed gene will only be deleted in that tissue. If Cre is regulated by a promoter that can be controlled, for example, by tetracycline (tet off system), then, when tetracycline is removed from the diet, Cre will be expressed, resulting in deletion of the floxed gene.

In order to use this system, two strains of mice are required; one strain must have the gene of interest floxed. To create this strain of mouse, embryonic stem cells are transfected with the floxed gene of interest, and cells that have undergone a site-specific homologous recombination resulting in replacement of the endogenous gene with the floxed gene are identified through positive and negative selection and PCR analysis. These ES cells of interest can then be injected into a blastocyst and then transferred into a pseudopregnant female, and the genetically modified floxed mice are created in a method similar to making a knockout mouse. The other strain of mouse that is required is a transgenic mouse in which the Cre recombinase is under the control of a tissue-specific promoter—for example, the keratin 5 promoter that can be used to direct the expression of the Cre enzyme to the epidermis. In this particular case when the two strains of mice are mated, the expression of Cre in the epidermis results in the deletion of the floxed gene in the epidermis but not in other tissues where keratin 5 is not expressed. Thus, tissue-specific promoters result in tissue-specific deletion of the floxed gene.

As mentioned above, the Cre-loxP system can also be modified to place Cre under the regulation of an inducible promoter such as one regulated by tetracycline as in the tet-off system. Thus, one can temporally control the deletion of the gene of interest, depending upon when tetracycline is removed from the diet. By combining the tet off system with a tissue-specific promoter, one can control the timing of the gene deletion as well as the tissue the gene will be deleted from. For example, placing the Cre gene under the regulation of the tet-off promoter, which is placed under the regulation of the K5 promoter, will result in both temporal and tissue-specific control. Another variation on this theme is that fusion proteins have been developed which allow for inducible expression. For example, a transgenic mouse carrying a gene encoding a fusion protein that consists of the Cre recombinase coupled to the tamoxifen responsive hormone-binding domain of the estrogen receptor produces a Cre protein, which is only catalytically active in the presence of tamoxifen. Thus, when these mice are mated with mice carrying a floxed gene, the floxed gene would only be deleted if the mice are treated with tamoxifen. This system can also be coupled to a tissue-specific promoter.

SUGGESTED READING

Sambrook, J., and Russel, D. W. *Molecular Cloning: A Laboratory Manual*, 3rd ed., Cold Spring Harbor Press, Cold Spring Harbor, NY, 2001.

Ausubel, F., et al., *Current Protocols in Molecular Biology*, John Wiley & Sons, Hoboken, NJ, 2007.

Joyner, A. L., (Ed.), *Gene Targeting: A Practical Approach*, Oxford University Press, New York, 2000.

Sedivy, J. M., and Joyner, A. L. *Gene Targeting*, Freeman and Co., San Francisco, 1994.

Nagy, A., et al. *Manipulating the Mouse Embryo: A Laboratory Manual*, 3rd ed., Cold Spring Harbor Press, Cold Spring Harbor, NY, 2003.

Lewin, B. *Genes IX*, Oxford Univeristy Press, New York, 2007.

Watson, J. D., et al. *Molecular Biology of the Genes*, 5th ed., Benjamin Cummings, Menlo Park, CA, 2003.

CHAPTER 3

Toxicogenomics

MARJORIE F. OLEKSIAK

Rosenstiel School of Marine and Atmospheric Sciences, University of Miami, Miami, Florida, 33149

3.1 INTRODUCTION

Toxicogenomics is the study of the complex interaction between an organism's genome (the sum total of all genes within the nucleus, mitochondria, and chloroplast), chemicals in the environment, and disease. The goals of toxicogenomic studies are to understand (a) how toxic insults affect an organism's genome and (b) the genetic responses to these insults. The genomic response to toxic exposure can be adaptive (minimizing the toxicant's effects on organismal function) or maladaptive (resulting in mutations that affect gene functions and cause cancers and diseases). This chapter will discuss basic concepts and approaches in genomics as well as how these can be applied in toxicology studies.

Using genomic technologies, scientists are in a powerful position to identify and better understand most, if not all, of the genes affected by chemical exposures and diseases and the biological variation in response. Genomics is the field of research (or set of tools) that examines most, if not all, of the DNA, genes, proteins, or other products derived from the heritable material of the cell. This definition includes "transcriptomics" [studies of all or most mRNAs], "proteomics" [studies of all or most proteins (Chapter 4)], and "metabolomics" [studies of all or most metabolites (Chapter 5)]. The overarching goal is to integrate data derived from these different genomic studies. Herein we address genomics as it relates to DNA structure and transcription.

Unlike earlier studies that investigated one or a few genes encoding specific proteins, channels, or enzymes, genomics examines hundreds to thousands of genes to discover which genes affect physiological performance. By investigating many genes, genomics brings a broader perspective about physiology and toxicology: it addresses how many genes respond to environmental or disease stress and provides insights into the interactions among genes. By defining the variation within and among species and asking whether is it functionally important, toxicogenomics can begin to address fundamental questions about how organisms deal with diseases and chemicals and stresses in their environment. For the first time in toxicological

Molecular and Biochemical Toxicology, Fourth Edition, edited by Robert C. Smart and Ernest Hodgson
Copyright © 2008 John Wiley & Sons, Inc.

studies, we can use genomic approaches to begin to understand the number of changes that occur in a cancer cell versus a normal cell or that occur in an organ upon exposure to a complex suite of chemicals. This global investigation, facilitating the discovery of the unknown and the formation of new testable hypotheses, is one of the greatest promises of toxicogenomics.

3.2   A PRIMER OF GENOMICS

Genomics is steeped in a specialized vocabulary. Table 3.1 summarizes much of this vocabulary.

The genome is an organism's heritable, genetic material. It is made up of DNA in all eukaryotes but can be made up of RNA for many viruses. Vertebrate genome sizes range from the smallest vertebrate genomes [pufferfish *Fugu* and *Tetraodon*: 0.4 billion base pairs (bp)] to the largest (marble lungfish: 130 billion bp). The human genome is 3 billion bp ($0.02\times$–$10\times$ the size of these fish genomes) but has few differences in the numbers of genes. The human mitochondrial genome is 16,569 bp and encodes 13 proteins, 22 transfer RNAs, and 2 ribosomal RNAs.

Genomic DNA includes (Figure 3.1) transcribed DNA, regulatory DNA, and nonfunctional DNA. Transcribed DNA, or genes, are transcribed into one of three major RNAs: messenger RNA (mRNA) is synthesized by RNA polymerase II and later translated into a protein; ribosomal RNA (rRNA) is synthesized by RNA polymerase I and participates in protein assembly; and transfer RNA (tRNA) is synthesized by RNA polymerase III and binds and transfers amino acids during the polymerization of proteins. There are other small RNAs that downregulate gene expression, form a product, or catalyze a reaction (e.g., microRNAs, short interfering RNAs, small nucleolar RNAs, and small nuclear RNAs).

TABLE 3.1. Genomic Vocabulary

bp: Base pairs of DNA (e.g., A–T, G–C).

cDNA: Copy of RNA reverse-transcribed into DNA. cDNAs represent copies of mRNA that can be cloned into vectors and replicated in bacteria.

***cis*-acting:** Acting only on the gene located on the same chromosome, genetically linked or in close proximity. For example, TFIID is a *trans*-acting factor that binds the *cis*-acting sequence immediately 5′ to the transcriped portion of a gene.

Enhancer: DNA sequence that binds a transcription factor. Enhancers typically are *not* part of the proximal promoter and often are effective from a long distance, in either orientation, 5′ or 3′ to a gene.

EST: Expressed sequence tag, a short (100s bp) partial nucleotide sequence from a full-length or nearly full-length cDNA. These sequences are usually from the 5′ or 3′ ends of directionally cloned inserts as part of high-throughput sequencing projects. The partial cDNA sequence provides annotation by similarity searches and the putative identification of function.

FDR: False discovery rate, the expected proportion of rejected null hypotheses that are false positives.

Genome: The DNA that makes up the heritable material for eukaryotes. These genomes include the nuclear genome (0.2–145 billion bp for fishes), where the majority of genes and gene products are produced, and mitochondrial genomes, which among vertebrates is approximately 18,000 bps.

TABLE 3.1. *Continued*

Genome assembly: Once a genome has been sequenced, it has to be assembled or put into the correct order so that the DNA sequences reflect both chromosomes and the gene order on these chromosomes.

Library normalization: A majority of mRNAs are copies of a few hundred genes. To reduce this frequency bias, the most abundant genes in cDNA library are reduced by normalization. Normalization produces a library where most genes are within 10-fold concentration of each other.

miRNAs: MicroRNAs are 22–24 nucleotides in length and downregulate gene expression by attaching themselves to mRNAs and preventing them from being translated into proteins.

mRNA: Messenger RNA typically has a long poly-adenylated 3′ end (as added on after transcription of DNA), a 5′ untranslated region of 10s to 100s of bp, a coding region which consists of 3 bp codons that are translated into proteins and a 100- to 1000s-bp 3′ untranslated tail.

Proximal promoter: The first 200- to 300-bp 5′ to the start of transcription. Typically, the proximal promoter binds "general" transcription factors (not tissue, inducible, or gene-specific). These include binding sites for TFIID (TATAA), Sp1 and related proteins (GGGCGG), and other factors.

RNA Polymerase I, II III. RNA polymerase reads DNA and transcribes this into RNA. The three forms synthesize the formation of rRNA, mRNA, or tRNA, respectively.

rRNA: Ribosomal RNA and include 28S (4000–5000 bp) and 18S (1800–1900 bp) subunits of ribosomal RNA. rRNAs are synthesized by RNA I polymerase.

siRNAs: Short interfering RNAs are approximately 22 nucleotides in length and downregulate gene expression by binding to specific mRNAs and labeling them for destruction by enzymes called endonucleases.

snoRNAs: Small nucleolar RNAs modify ribosomal RNAs (rRNAs) by orchestrating the cleavage of the long pre-rRNA into its functional subunits (18S, 5.8S, and 28S molecules). snoRNAs also add finishing modifications to the rRNA subunits.

SNP: Single nucleotide polymorphism.

snRNAs: Small nuclear RNAs are part of the spliceosome, the cellular machinery that helps to produce mRNA by removing the noncoding regions (introns) of genes and piecing together the coding regions (exons) to be translated into proteins.

TBP: TATA binding protein, the protein subunit of TFIID that binds TATAA sequence at −30 bp from the start of transcription.

TFIID: A basic transcription factor required for all mRNA expression. TFIID binds TATAA sequence or other sequence 30 bp upstream from the start of transcription.

***trans*-acting:** On different chromosomes. A factor or gene that acts on another unlinked gene, a gene on a separate chromosome or genetically unlinked. For example, TFIID is a *trans*-acting factor that binds the *cis*-acting sequence immediately 5′ to the transcriped portion of a gene.

Transcription factor: A protein or complex set of proteins that affects the rate of transcription. Many transcription factors affect transcription by binding DNA and increasing either the probability of transcription or the stability of transcriptional machinery through protein–protein interactions. Other transcription factors act as linkers between DNA-binding transcription factors or RNA polymerase.

tRNA: Transfer RNA that binds amino acids and transfers these amino acids in the polymerization of proteins. tRNAs are recognized by the ribosomal protein–RNA complex by their secondary structure. tRNAs are synthesized by RNA polymerase III.

UTR: Untranslated regions of mRNA. mRNAs have both 5′ leader sequences (10–100s bps) and 3′ sequences (100–1000s bp) that are not translated to proteins.

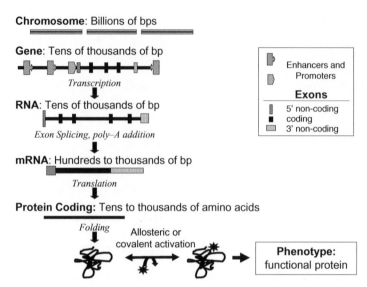

Figure 3.1. Genome to phenotype. Approximately 1.5 billion base pairs of DNA in a vertebrate's chromosomes encode approximately 30,000 genes. These genes have transcribed regions that are regulated by enhancer and promoters. The transcribed genes include large introns and relatively smaller exons. Transcription, processing, and splicing of exons together produces mRNAs that are translated into proteins. The concentration and activity of these proteins produce the phenotypic differences within and among species.

Regulatory sequences are regions of DNA that are involved in transcriptional regulation of coding DNA (Figure 3.1). For example, regulatory sequences define the beginnings and ends of genes, provide sites for the binding of regulatory proteins (transcription factors), and are involved in the accessibility of the DNA. Nonfunctional DNA is not involved in transcription, regulation, or production of any product. Nonfunctional DNA may reflect our ignorance of its function, represent remnants of DNA parasites and duplicated and mutated DNA, or result from neutral mutational processes that allow the accumulation of "junk DNA." One can begin to understand the genome by examining its different levels of organization and function.

Coding regions of DNA make up about 1% of the total DNA in the human genome. In eukaryotes, genes are made up of both exons and introns, interweaved at a location (*locus*) in the genome (Figure 3.1). For most genomes, introns are bigger and make up much more of the nuclear genome than do the coding exons. Both exons and introns are transcribed into messenger RNA by RNA polymerase II. RNA polymerase II recognizes the transcription initiation start site, and its binding is affected by proteins binding to the promoter region of the gene. Transcription stops when the polymerase reaches a termination or stop signal. These transcribed mRNAs are modified. Immediately after transcription, a poly(A) sequence is added to most mRNAs. In addition, the 5' end of the mRNA is capped: a 7-methylguanylate is covalently linked to the first nucleotide of the mRNA. Further editing removes the intron sequences and joins together the exons. This editing process can result in alternatively spliced mRNAs, mRNAs that are transcribed from the same DNA but with different exons spliced together. In summary, a gene is transcribed into RNA and is modified and processed so that the exons are spliced together forming (a) a relatively short 5′ untranslated region (5′UTR), (b) the protein coding region or translated region, which forms the majority of a typical transcript, and (c) the 3′UTR, which tends to be hundreds to thousands of base pairs long. Of the approximately 1% of the human genome that is expressed, at least half of the genes are alternatively spliced to produce multiple transcripts and subsequently multiple proteins from a single gene. For example, a large number of apoptotic factors are regulated via alternative splicing.

Processed mRNAs are the basis for Expressed Sequence Tag (EST) projects. ESTs represent a partial sequence (a few hundred base pairs) of the much longer mRNA expressed in a cell. Due to the final mRNA editing, ESTs encode structural genes without intervening intron sequences and thus can be more informative about the ultimate function of the gene. Alternative splicing is partially responsible for the hundreds of thousands of ESTs in many EST databases; for example, while humans have between 20,000 and 30,000 genes, The Institute for Genomic Research (TIGR) human database lists 835,426 total unique sequences for humans. The large number of "unique" ESTs reflects both differential splicing and weaknesses in gene annotations.

The amount of a gene's mRNA can be regulated prior to initiation, during transcription, and posttranscriptionally. Prior to initiation, modification of the genomic DNA can regulate access to the DNA surrounding the transcribed region: if the gene is not accessible, then it will not be transcribed. Gene silencing also can occur through other modifications like methylation. This is the basis for parental imprinting. Often, when RNA polymerase binds to the promoter region of a gene, transcription initiation as well as transcription involves other factors, transcription factors, that regulate both the process of initiation and transcription. Typically, variation in mRNA expression is thought to be regulated by the rate of transcription, which in turn is regulated by the number, type, and modification of transcription factors.

3.3 TOOLS AND APPROACHES

There are two general approaches to investigate nuclear genome sequences. (1) Production of large DNA fragments in artificial chromosomes (bacterial and yeast artificial chromosomes: BAC and YAC, respectively). Because sequences through these constructs provide both sequence information and gene order, sequencing projects relying on these BAC and YAC are more readily assembled. (2) Shotgun sequencing, which involves sequencing short fragments from DNA libraries containing up to tens of kilobases. Shotgun sequencing is molecularly simpler and faster but requires considerably more computation. Both procedures rely on genetic mapping for complete assembly. The NIH/NIHGR sequence of the human genome relied on the first approach, whereas the human genome produced by Celera Genomics (C. Venter) pioneered the second, shotgun sequence approach. Using these different techniques, the first two, competing, published drafts of the human genome produced similar results.

A complementary, high-throughput approach to sequencing the genome is to identify mRNAs. Expressed mRNAs are much less complex than genomic DNA because they lack introns, promoters, and other nontranscribed DNA and only a subset of genes are expressed in a single tissue. This simplicity allows one to more readily define the proteins expressed by a gene and provides some hints about their functions. For these reasons, isolation and sequencing of mRNAs often are used to investigate genomes. The most common high-throughput method to investigate expressed mRNAs is to make a cDNA library and sequence only a small portion of each cDNA. A cDNA library is made by using reverse transcriptase to convert RNA to DNA (copy DNA or cDNA), and these cDNAs are cloned into circular DNA (vectors or plasmids) that contain antibiotic-resistant genes and are self-replicating in bacterial cells. When transformed into bacteria, only cells with the antibiotic resistant vector survive. Each bacterium has a separate cloned cDNA, and thus one needs 100,000 to millions of transformed bacteria for a complete library of all expressed genes. One simply isolates the clones from the bacteria and sequences the cDNA. These partial sequences of mRNAs are ESTs.

In a sample of mRNA, most genes are relatively rare. A few genes are highly expressed (there are many copies of the transcript). Thus, mRNA from a few (5–10) genes represent 20% of all mRNAs, and 1000 genes represent 60–80%, while the remaining 5000–10,000 different genes represent <20–40%. This bias (a few genes have much greater expression frequency than most genes) in the frequency of individual genes is also found in cDNA libraries, and thus random isolation of cDNAs would produce an abundance of only a few genes. To reduce this problem, cDNA libraries are normalized so that the most prevalent genes represented no more than 10 times more than the most rare genes. Normalization takes advantage of Cot curve hybridizations where the most abundant nucleic acids hybridize fastest. Thus, when both separated strands of cDNA are hybridized in solution, the most abundant cDNAs find their complements first. These double-stranded DNAs are selectively removed, leaving a more even frequency of most cDNAs. One can increase the frequency of finding rare genes by subtracting previously isolated ESTs. For subtraction, one or hundreds of genes are selectively removed from a library by hybridizing all cDNAs in the library to an overabundance of select genes (the subtractor). Typically, the "subtractors" have a tag (e.g., biotin) that can be used to remove them once they hybridize to their target cDNA. Subtraction of the most prevalent cDNAs further reduces or eliminates the discovery of previously isolated genes.

3.4   GENOMES

In 2001, two drafts of the human genome were published. The *Drosophila* genome was published a year earlier; and two years before *Drosophila*, the nematode genome of *Caenorhabditis elegans* was published. These genomes have provided insights into the complexity, architecture, and evolution of genomes. What might be less appreciated is that these genomes are still being analyzed. For example, in 2004 approximately 41,000 human genes were reduced to 21,000 by both computational and manual analyses. Another major effort utilizing the human genome is the identification of single nucleotide polymorphisms (SNPs) among populations and individuals (see Chapters 6, 11, and 13). One goal is to identify individual SNPs or sets of SNPs (haplotypes) that are diagnostic of disease or susceptibility.

Comparative genomics, the use of other genomes from diverse species, is one of the most effective methods to address the difficulties in understanding genomes. Evolutionary comparisons among genomes are used to better annotate genes (providing information on function), define genomic architecture and roles of genes, and develop insights into the creation of novel structures. For instance, fish genomic sequences were used to identify many human genes, and these were used as one of the primary pieces of evidence that the vertebrate genome had 30,000–40,000 genes and not 70,000–140,000. The NCBI UniGene databases are an attempt to distill the large number of ESTs encoding the same protein to a single entry. Each UniGene entry is a set of transcript sequences that appear to come from the same transcription locus (gene or expressed pseudogene). Unigene databases do reduce the number of "genes" but fail to put all appropriate sequences together. For example, there are nearly 400,000 human Unigenes for the approximately 25,000 genes.

3.5 FUNCTIONAL GENOMICS

Figure 3.1 illustrates the processes associated with forming the functional molecule in the cell: an active protein. RNA has to be transcribed, poly-adenylated, spliced to remove introns, and translated into protein. The protein must be properly folded and in many cases activated by allosteric regulators or covalent interactions. Functional genomic studies attempt to relate the genome to functional changes in an organism. Much of Functional Genomics is just defining the genes present in the genome. This process of understanding genomes, what genes they code for, what regulates their expression, and how this expression varies among tissues, environments/disease, or individuals is far from finished. Much of the difficulties arise from relatively small exons and exons that code for only a few, if any, amino acids (e.g., 5′ UTRs). With so much junk DNA and so few coding triplet codons, it is easy to see the difficulty in defining what constitutes a gene. What has helped is Comparative Genomics: genomic structure (chromosomal arrangement of genes), gene identification, transcriptional regulatory elements, and patterns of expression shared among mice and men, flies and worms, fish and mammals. Comparative genomics requires resources from many taxonomic groups and bioinformatic approaches (see Chapter 6).

3.5.1 Microarrays

Microarrays are thousands of 150- to 250-micron spots of DNA bound to microscope slides in a precise and known pattern. Each DNA spot quantitatively hybridizes to a specific mRNA so that expression of thousands of individual genes can be measured simultaneously. Importantly, microarray techniques are sensitive: typically, twofold differences in mRNA concentration are determined; and each gene/DNA spot has a sensitivity of 15 attomoles, or approximately one out of 300,000 transcripts can be measured. Application of statistical analyses with appropriate replication has improved these analyses such that less than 1.5-fold differences are readily discerned. The most common use of microarrays is to profile transcript levels on a genome-wide basis. The idea is that knowledge of where (what tissue) and when and how much a gene is expressed will inform its function. Comparisons of gene expression between different states (normal and diseased) also are used to understand the underlying molecular mechanisms affecting the diseased state.

Microarrays quantitatively measure the expression of hundreds to thousands of genes simultaneously. Microarrays only measure the amount of mRNA, and thus there is an implicit assumption that altering the amount of mRNA changes the amount of active protein. This ignores all other steps that affect protein activity (translation rates, folding, protein activation, Figure 3.1). However, patterns of mRNA expression for some genes do relate to different physiological or disease states; and although not all the activities of proteins are influenced by their mRNA, it appears that the variation in expression of many mRNAs affects protein activities and the phenotypes they create.

Since microarrays are simply spots of DNA arrayed in a known pattern, they also can be used for DNA–DNA hybridization studies. This is the basis for using microarrays for comparative DNA hybridizations, single nucleotide polymorphism (SNP) discovery, and chromatin immunoprecipiation followed by microarray "chip" hybridization (ChIP on chip analyses) to identify transcription factor binding sites (discussed below).

There are two common microarray approaches: two-dye and Affymetrix. Both methods rely on "printing" or synthesizing DNA onto a solid substrate (typically a glass slide) and quantitatively hybridizing and thus capturing labeled RNA (or copies of the RNA as cDNA). For the two-dye approach, long cDNAs, partial cDNAs, or long oligos (50–75 bp) are printed as "spots" onto a glass slide. Each spot ($\approx$150 microns) represents a separate gene, and the identity of these genes often relies on EST sequence information. These DNAs are bound to the glass slide by a poly-amine (e.g., poly-lysine) or an aldehyde derivative coating on the slide. With a poly-amine-coated slide, thymines in the DNA are cross-linked to the amine groups, and thus a portion of the DNA lies flat on the slide. On slides coated with aldehyde derivatives, the 5′ end of a single strand of DNA is labeled with an amine group, and this 5′ end covalently binds the aldehyde group; thus the DNA is attached only at the 5′ end, and little of the DNA is in contact with the slide. Two samples of RNA are hybridized to the slide: typically, one sample is labeled with the fluorescent dye Cy3, and the other sample is labeled with the fluorescent dye Cy5. Other fluorescent dyes and other labels (e.g., biotin) also can be used to label the samples. The fluorescent dyes have different excitation and emission wavelengths and thus can be independently quantified. Thus, two samples of labeled RNA are hybridized to one microarray in a small volume, where each spot is a specific gene that quantitatively captures the corresponding mRNA. The fluorescence from each DNA spot is determined with a scanning confocal laser. Pixels from the resulting tiff image are quantified, and it is these pixel counts that ultimately are analyzed.

When two dyes are used, there are two experimental approaches (Figure 3.2). In the most common approach ("reference design," Figure 3.2), the experimental sample is labeled with one dye while the common reference sample is labeled with the second dye, and both are hybridized to the same microarray. Alternatively, for a more statistically balanced and efficient experiment, samples can be hybridized with other samples in a loop design (Figure 3.2). In the loop design, the two experimental samples (no common reference) are labeled with the two different dyes and hybridized to the microarray.

The second approach for microarray analyses is based on Affymetrix slides and consists of hundreds to thousands of sets of 12–25 pairs of short oligos synthesized

Reference Design   **Loop Design**

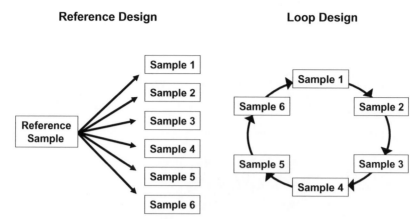

Figure 3.2. Reference versus loop design for two-dye microarrays. Two samples labeled with different fluorescent dyes are hybridized to a microarray with hundreds to thousands of different genes printed as spots. Arrows represent the microarrays and the hybridization scheme: the sample at the base of the arrow is labeled with one dye, and the sample at the head of the arrow is labeled with a different dye. In a reference design, each sample is hybridized with the same "reference" sample. This reference might be a pool of different RNA samples or a control RNA sample. Each sample is measured once while the reference is measured n times, where n is the number of samples. In a loop design, samples are hybridized with other samples. Thus, each sample is measured twice in a balanced design (once with each dye).

directly on the slide. The 12–25 oligo pairs in each set correspond to different sites of a single mRNA. For each pair, one member of the pair is the correct sequence and the other has a single base-pair mismatch. On Affymetrix slides, a single sample of RNA is hybridized. Thus one measures a single experimental sample at a time and compares the signal between the "perfect" and mismatch oligo for 12–25 pairs of oligos.

3.5.1a   Wet and Dry Lab Procedures. Microarray analyses involve both wet and dry lab procedures. The wet lab portion of microarray analyses includes performing the treatment (e.g., dosing with a drug or isolating a tumor), isolating the RNA from the samples, labeling the RNA, hybridizing the labeled RNA to the microarray, and scanning the microarray (Figure 3.3). Just as in any experiment, one needs both technical and biological replicates in order to distinguish these sources of variation from the variation due to the treatment (Figure 3.3A). In other words, without biological replication, one cannot know whether a measured effect is due to the treatment or due to biological variation of an animal (e.g., one animal might constitutively express more cytochrome P4503A than another); without technical replication, one cannot know whether differences between two individuals are due to individual differences or to measurement error.

The dry lab portion of microarray analyses involves identifying the hybridized spots on the scanned tiff images (gridding) and defining the area and portion of the spot that should be quantified. Most often, mean or median pixel intensities for each spot are extracted from the tiff images. The result is a separate numeric value for each spot and each sample that represents the relative amount of sample hybridized to that spot. Thus, the raw results from an array experiment can be thousands

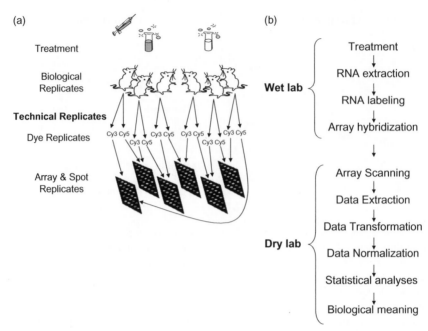

(a)   (b)

Figure 3.3. Schematic of a microarray experiment. (A) Experimental design incorporating both biological and technical replication. There are three treated mice and three control mice, providing biological replication. RNA from each mouse is labeled with each dye and hybridized more than once, providing technical replication. The hybridization design uses a loop design as shown in Figure 3.2. (B) Outline of the wet and dry lab steps involved in a microarray experiment. See color insert.

of values for every sample. With two technical replicates, this number is doubled. The most straightforward way to look at this amount of data is graphically, and the simplest graph is one sample versus another sample. Since most often the null hypothesis is that there is no difference in gene expression levels between the two samples, the expectation is that these data will form a 45-degree line.

One major goal of the dry lab analysis is to reduce nonbiological sources of variation from the wet lab portion of microarrays. These include variation due to differential RNA extraction and labeling efficiencies, spatial heterogeneity in the hybridizations (e.g., a scratch on the array), differences in the dyes' emissions, and differences in the laser power used when scanning the arrays. Data adjustments to account for and minimize the variability within and across different experiments are known as data normalizations. Data transformations also are commonly performed before analyzing microarray experiments. A data transformation is the application of a mathematical function (e.g., a $\log_2$ transformation) often so that the data satisfies assumptions of statistical tests. The most common data transformation used in microarray analyses is a $\log_2$ transformation so that the data are more normally distributed.

The second major goal of the dry lab analyses is to make statistical inferences and discern biological meaning from the expression data. Statistical analyses provide a straightforward approach to analyze microarray data. Genes that are statistically differentially expressed are identified, often using a *t*-test or analysis of variance (ANOVA) to compare treatment means. The application of statistical analyses has not always been applied for microarray analyses; instead, fold-changes between reference and treatment have been used. There are two significant problems with using a fold-change. One is that fold-changes ignore the variation within a sample or treatment and thus large fold-changes between samples could include the normal variation for a gene. Second, many biologically important changes can be relatively small, and fold-changes would ignore these changes. Statistical analyses are becoming the norm and often are required for publication; they identify changes unlikely to occur by chance given the variation in expression for each gene.

A common way to visualize statistically significant data is to plot the change in gene expression on the *x*-axis and the significance as $-\log_{10}$ on the *y*-axis. The resulting plot is known as a volcano plot (Figure 3.4), and it allows one to easily identify both the genes with significantly altered expression and the amount and direction (greater than or less than) of that change.

One way to discern biological patterns among differences in gene expression is to group microarray data based on similarities in gene expression patterns and then to visualize these patterns of expression. This data often are shown as "heat maps" (Figure 3.5) in which either the individual samples or the genes are grouped or clustered by similarities in expression patterns and then visualized using one color to represent relatively greater expression (often red) and a different color to represent relatively lesser expression (often green). This heat map facilitates identification of genes or samples with similar expression patterns. For the samples, this can be used to identify the effects of treatment or disease. For the genes, this can be used to identify affected pathways or genes that may be similarly regulated.

The vast amount of data that can result from one microarray experiment has led to analyses incorporating data reduction. Two mechanisms to reduce the data include

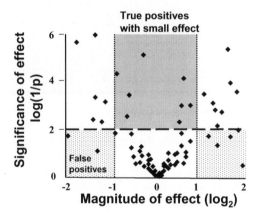

Figure 3.4. Simple volcano plot. For each gene, difference in expression is plotted on the *x*-axis as a $\log_2$ value [a value of 1 is a twofold difference (dotted line)] versus the statistical significance plotted on the *y*-axis as $\log(1/p)$ [a value of 2 represents a *p*-value of 0.01 or a 1-in-100 probability of occurring by chance (dashed line)]. Note that just because there is a large difference in expression, a gene may not be statistically significantly differently expressed (dotted areas) while small-fold differences can be significant (light gray area).

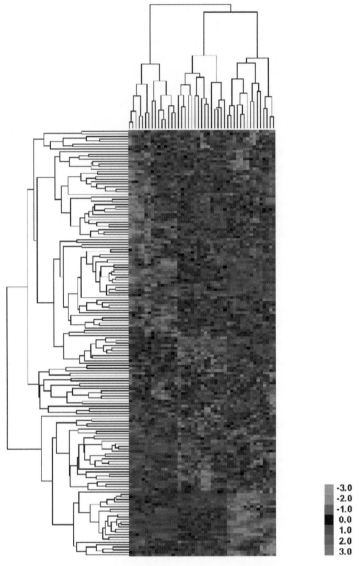

Figure 3.5. Representative heat map. Hierarchical clustering of 45 individuals clustered across the top and 169 metabolic genes clustered along the side. Notice that the 45 individuals form three groups of 15 individuals each. Each group of 15 individuals shares similar patterns of gene expression that are different among the three groups. It is the shared patterns that can be used to diagnose exposure and disease or classify cancers. The relative expression levels range from −3 to +3 as shown by the scale bar. See color insert.

principal component analysis (PCA) and functional grouping using gene ontogeny or GO terms. A principal component (PC) is a linear equation that weights the different variables (for example, gene expression levels in a specific pathway) to maximize the explained variation among individuals or samples without reference to the treatment or physiology of interest—for example, drug exposure or disease. Thus, rather than 100 gene expression levels representing the 100 genes in the TCA cycle, the data are reduced to one or more principal components (PC1, PC2, PC3) that explain most of the variation among individuals or samples in that set of genes.

Functional grouping using GO terms also is used to reduce the dimensions of the data (not look at so many gene expression levels). GO terms describe gene products based on associated biological processes, cellular components, and molecular functions and are species-independent. Thus, rather than analyzing individual genes, profiles of all genes involved in a particular biological process or molecular function might be analyzed together. For example, one can inquire if more genes involved in oxidative reactions are affected, or if a signaling pathway has more changes in gene expression than due to random expectation.

One advantage of these data reduction methods is that fewer tests are done. Rather than 20,000 tests of significance for 20,000 genes, perhaps only 2000 tests will be done on sets of ~100 genes. This is relevant when considering the false discovery rate (FDR). The FDR is the expected proportion of rejected null hypotheses that are false positives: more tests will result in more false positives. There are different ways to adjust for false positives. One approach is to permute (shuffle) the data to create multiple pseudo-data sets and statistically test these new data sets to evaluate whether the departure from the null hypothesis in the observed data is greater than in the pseudo-data sets. Bootstrapping analyses are similar where the pseudo-data sets are created by sampling the observed data with replacement (after a pseudo-data set is sampled from the observed data, the sampled data are returned to the original data set prior to the formation of another pseudo-data set).

3.5.1b Expression Profiling. In toxicogenomics, microarrays often are used as a diagnostic tool: a set of altered genes or a pattern of genes' expressions is used to identify an individual who has been exposed to a specific chemical, disease, or infection. The enhanced or decreased expression of a few or few hundred genes indicates the severity or type of disease. For example, expression profiles have been used to classify different cancer types or whether a specific cancer type is more or less virulent. This has bearing on which chemotherapies should be pursued. Patterns of expression also might be used to establish chemical specific profiles that in turn can be used to classify exposure to unknown chemicals. This is particularly useful in environmental toxicology where exposure usually is not limited to a discrete chemical, but more often is the result of a mixture of chemicals. For instance, field workers most often are exposed to more than one pesticide. An important concept for these diagnostic uses is that of a training set (e.g., particular expression profiles from known tumor types or known chemical exposures) that can be used to identify altered gene expression patterns that are predictive of an unknown disease or exposure. In other words, a gene expression signature or fingerprint is developed that can be used for diagnostic and predictive purposes.

3.5.2 DNA Arrays

As mentioned earlier, all of the DNA (e.g., genomic DNA), not just the expressed genes, can be arrayed. These arrays often are called tiling arrays because they can be made up of millions of DNA probes evenly spaced, or "tiled," across the genome. This tiled DNA includes both coding and noncoding regions, and these tiling arrays can be used for comparative DNA hybridizations, single nucleotide polymorphism (SNP) analyses, and ChIP on chip analyses.

3.5.2a Comparative DNA Hybridizations. Comparative DNA or genomic hybridizations (CGH) can be used to analyze DNA copy-number variations. Altered DNA copy number can change gene expression and function, and there is some variation in DNA copy number among normal individuals. However, altered DNA copy number also can be involved in different disease states. For example, amplification of oncogenes and deletion of tumor suppressors are common in cancer development, and many defects in development are due to gains and losses of chromosomes and chromosomal segments. Similar to gene expression analyses, in comparative genomic hybridizations, separately labeled reference and test DNA (for example, DNA from breast tumor and normal tissue) are hybridized against a large panel of DNA probes to look for regions of DNA gain or loss. This facilitates the identification of crucial genes and pathways involved in specific biological processes and disease.

3.5.2b SNP Discovery. Single nucleotide polymorphisms (SNPs) are base-pair differences between any two individuals. SNP chips contain oligonucleotide probes for particular SNPs and can be used to genotype individuals (identify which polymorphisms and sets of polymorphisms an individual has). Like gene expression microarrays, the power of a SNP chip is that thousands to ten thousands of SNPs can be scored simultaneously for each individual. Shared polymorphisms between individuals with a disease can be used to identify causative disease factors (e.g., identify a common mutation). Again, this is a genomic approach and one does not have to target particular genes one at a time. However, for this approach to work and identify a causative mutation, the SNP chip has to contain the SNP or set of SNPs that are linked to the disease locus. Yet, remember, linkage is not necessarily identification.

3.5.2c ChIP on Chip. A more specialized use of DNA arrays is to identify active transcription factor binding sites in the genomic DNA. This might be used to identify the functional (active) transcription factor binding sites in diseased versus normal or treated versus untreated and is similar to in vivo foot-printing analyses. The basic idea is to isolate genomic DNA with bound proteins, crosslink the proteins to the DNA, get rid of unprotected (protein-free) DNA, immunoprecipitate the protein–DNA complexes of interest using specific antibodies (chromatin immunoprecipitation, or ChIP), release the DNA from the bound proteins, and label and hybridize this protected DNA to the DNA array (ChIP on chip). Hybridized DNA will identify specific protein-binding sites in the genome. For toxicogenomics, one might use this approach to identify binding sites that are occupied only during a specific disease state or upon chemical exposure.

3.6   CONCLUSIONS

The underlying premise of microarray analyses is oligonucleotide hybridizations that are used to gain genomic profiles of individual samples. These profiles can be gene expression profiles, transcription factor binding site profiles, and genomic content at the nucleotide level (SNPs) or at the level of copy number (CGH). The power in these analyses is that they are global. They give the entire picture as opposed to changes in one or a few genes or *loci*. In toxicogenomics, this genomic profile is used to identify biomarkers of disease and chemical exposures, better understand the effects of environmental stresses and individual susceptibility, and clarify the molecular mechanisms of toxicity. The challenges of toxicogenomics are to integrate genomic changes with changes at the organismal level. These genomic changes will include changes in the transcriptome and address how genomic changes affect and interact with changes in the proteome and subsequently with changes in the metabolome. Ultimately, how do these changes affect a tissue, an organism, and a population?

SUGGESTED READING

Churchill, G. A. Fundamentals of experimental design for cDNA microarrays. *Nat. Geneti.* **32**(Suppl.), 490–495, 2002.

DeRisi, J., Penland, L., Brown, P. O., Bittner, M. L., Meltzer, P. S., Ray, M., Chen, Y., Su, Y. A., and Trent, J. M., Use of a cDNA microarray to analyse gene expression patterns in human cancer. *Nat. Genet.* **14**(4), 457–60, 1996.

Jaillon, O., Aury, J. M., Brunet, F., Petit, J. L., Stange-Thomann, N., Mauceli, E., Bouneau, L., Fischer, C., Ozouf-Costaz, C., Bernot, A., Nicaud, S., Jaffe, D., Fisher, S., Lutfalla, G., Dossat, C., Segurens, B., Dasilva, C., Salanoubat, M., Levy, M., Boudet, N., Castellano, S., Anthouard, V., Jubin, C., Castelli, V., Katinka, M., Vacherie, B., Biemont, C., Skalli, Z., Cattolico, L., Poulain, J., De Berardinis, V., Cruaud, C., Duprat, S., Brottier, P., Coutanceau, J. P., Gouzy, J., Parra, G., Lardier, G., Chapple, C., McKernan, K. J., McEwan, P., Bosak, S., Kellis, M., Volff, J. N., Guigo, R., Zody, M. C., Mesirov, J., Lindblad-Toh, K., Birren, B., Nusbaum, C., Kahn, D., Robinson-Rechavi, M., Laudet, V., Schachter, V., Quetier, F., Saurin, W., Scarpelli, C., Wincker, P., Lander, E. S., Weissenbach, J., and Roest Crollius, H. Genome duplication in the teleost fish *Tetraodon nigroviridis* reveals the early vertebrate proto-karyotype. *Nature* **431**(7011), 946–957, 2004.

Perou, C. M., Sorlie, T., Eisen, M. B., van, d. R. M., Jeffrey, S. S., Rees, C. A., Pollack, J. R., Ross, D. T., Johnsen, H., Akslen, L. A., Fluge, O., Pergamenschikov, A., Williams, C., Zhu, S. X., Lonning, P. E., Borresen-Dale, A. L., Brown, P. O., and Botstein, D. Molecular portraits of human breast tumours. *Nature* **406**(6797), 747–752, 2000.

Pritchard, C. C., Hsu, L., Delrow, J., and Nelson, P. S. Project normal: Defining normal variance in mouse gene expression, *Proc. Natl. Acad. Sci. USA* **98**(23), 13266–13271, 2001.

Sorlie, T., Perou, C. M., Tibshirani, R., Aas, T., Geisler, S., Johnsen, H., Hastie, T., Eisen, M. B., van de Rijn, M., Jeffrey, S. S., Thorsen, T., Quist, H., Matese, J. C., Brown, P. O., Botstein, D., Lonning, P. E., and Borresen-Dale, A. L. Gene expression patterns of breast carcinomas distinguish tumor subclasses with clinical implications. *Proc. Natl. Acad. Sci. USA* **98**(19), 10869–10874, 2001.

Waters, M. D., and Fostel, J. M. Toxicogenomics and systems toxicology: Aims and prospects, *Nat. Rev. Genet.* **5**, 936–948, 2004.

CHAPTER 4

Proteomics

B. ALEX MERRICK

National Institute of Environmental Health Sciences, Research Triangle Park, North Carolina 27709

4.1 INTRODUCTION TO PROTEOMICS

In cells and tissues of complex organisms, only a portion of the genome is expressed at any one time. At the mRNA level, the transcriptome represents all genes transcribed or proteins represented by the proteome. Gene expression constantly changes during health, adaptation, disease, aging, and toxicity. The field for describing protein expression on a global scale is termed "proteomics," which aims to detail the structure and functions of all proteins in an organism over time. The wide application of proteomics has generated great interest in many established disciplines of basic biology and medicine, including the field of toxicology creating a relatively new area, "toxicoproteomics." Toxicoproteomics is the application of global protein measurement technologies to toxicology research. Chemical or toxicant exposure can bind to or modify proteins directly, produce changes in protein expression, and dysregulate critical biological pathways and processes that lead to toxicity. Primary aims in toxicoproteomics are the discovery of key modified proteins, the determination of affected pathways, and the development of biomarkers for eventual prediction of toxicity. Key definitions in proteomics are described in Table 4.1.

4.1.1 Attributes of Proteins

Knowledge of protein primary sequence, quantities, posttranslational modifications (PTMs), structures, protein–protein (P–P) interactions, cellular spatial relationships, and functions are seven important attributes (see Table 4.2) needed for comprehensive protein expression analysis. It is this multifold and complex nature of protein attributes that has spawned the development of so different many proteomic technologies. Some of these challenges in proteomic analysis include defining the identities and quantities of an entire proteome in a particular spatial location (i.e., serum, liver mitochondria, brain), the existence of multiple protein forms and complexes, the evolving structural and functional annotations of the human and rodent

Molecular and Biochemical Toxicology, Fourth Edition, edited by Robert C. Smart and Ernest Hodgson
Copyright © 2008 John Wiley & Sons, Inc.

TABLE 4.1. Definitions in Proteomics

Proteome: That portion of the genome expressed as proteins in cell or organism over time.
Proteomics: Global protein analysis to study the structure and function of the proteome.
   All proteomic platforms generally involve separation and identification of proteins.
Toxicoproteomics: Proteomics applied to toxicology with the aims of identifying critical
   proteins and pathways affected by chemical and environmental exposures.

TABLE 4.2. Seven Protein Attributes

Identity: Based on amino acid sequence of protein
Quantity: Absolute (molar) or relative amount (proportion) of protein.
PTMs: Posttranslational modifications (PTMs) are the addition of chemical structures to
   –R groups of proteins. PTMs also include other processing of proteins such as
   proteolytic removal of chemical groups like phosphates by phosphatases or sugars by
   glycosidases or also of primary sequence by proteases.
Structure: Three-dimensional structure in relation to other proteins.
Protein–protein interactions: Physical contact and structural relationships with other
   proteins.
Cellular spatial relationships: Occupation of proteins with intracellular structures,
   organelles or intracellular relationships with other cells in tissue.
Function: Enzymatic, structural, regulatory, transcriptional, translational functions and
   many other roles.

proteomes, and integration of proteomics data with transcriptomics or other expression data. Primary aims of proteomic analysis are intended to achieve maximal proteome depth, high throughput, quantitative protein measurement, and timely analysis through use of discovery-oriented platforms.

4.1.2 History of Proteomics

A brief history of the development of proteomics is presented in Figure 4.1. In 1975, a method was reported for resolution of proteins to near homogeneity from cell and tissue lysates. This new protein separation method was called two-dimensional polyacrylamide gel electrophoresis (2D PAGE) and is now often referred to as 2D gel electrophoresis. Proteins are separated first by charge using isoelectric focusing (IEF) as the first dimension of separation and then by mass using sodium dodecyl-sulfate polyacrylamide gel electrophoresis (SDS PAGE) as the second dimension. An improvement in isoelectric focusing was reported in 1982 and took several years to refine and commercialize as the standard for IEF. Continued evolution of 2D gel separation continues as evidenced by the fluorescent tag labeling of proteins from various treatment groups in a method called differential fluorescence gel electrophoresis (DIGE). Despite continued improvements in the technique, proteins separated by 2D gels could not be readily identified for almost two decades. It was not until the mid-1990s that the resolution power of 2D gels began to be satisfactorily linked with various types of mass spectrometry (MS) for protein identification.

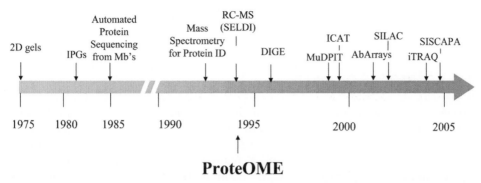

ProteOME

Figure 4.1. History of proteomics. Important developments during the history of proteomics are depicted by the timeline. The development of two-dimensional (2D) gels in 1975 was linked to protein identification by mass spectrometry in the mid-1990s. The introduction of the term proteome occurred in 1994. More recent developments from 1999 to the present involve development liquid chromatography-based separations linked to mass spectrometry. Abbreviations used are IPG, immobilized pH gradients; Mbs, membranes; ID, identification; RC-MS, retentate chromatography–mass spectrometry; SELDI, surface-enhanced laser–desorption–ionization; DIGE, differential gel electrophoresis; MuDPIT, multidimensional protein identification technology; ICAT, isotope-coded affinity tag; Ab arrays, antibody arrays; SILAC, stable isotopic labeling with amino acids in cell culture; iTRAQ, isobaric tags for relative and absolute quantitation; SISCAPA, stable isotope standards, capture by anti-peptide antibodies. ProteOME is a hybrid term from protein and genome.

During this dynamic period, the concept of "proteome" emerged in 1994 at an international conference on 2D gel separations. A proteome was defined as the total set of proteins expressed by a genome in a cell, tissue, or organism at a given time. The term, a hybrid from *prote*in and gen*ome*, encapsulates the complex and dynamic nature of protein expression from signaling to transcription that takes place within a specific spatial reference point spanning from a single cell up to a complete organism. Although 2D gel electrophoresis has biophysical limits that constrain its resolution of proteins at the extremes of pH, hydrophobicity, and mass which will be reviewed shortly, it remains an extremely versatile separation technique that continues to be widely used in proteomics research.

Recent technological developments since 1999 have produced proteomics platforms to supplant the two-dimensional gel-mass spectrometry (2D-MS) platform. Instead of gel-based protein separations, newer proteomic technologies involve liquid chromatography (LC) instruments for separation of proteins or peptides that are connected to mass spectrometers. Such platforms would include ICAT (isotope-coded affinity tag), SILAC (stable isotope labeling with amino acids in cell culture), MuDPIT (multidimensional protein identification technology), and iTRAQ (isobaric tags for relative and absolute quantitation) (see Figure 4.1 for explanation). An immunoaffinity chromatography system linked to mass spectrometry has also been recently developed called SISCAPA (stable isotope standards and capture by anti-protein antibodies) (see Figure 4.1). Other systems like antibody arrays or antibody-bound fluorescent microbeads rely completely upon immunoaffinity for separation and identification. Still in development are nanotechnologies that use less than one microliter of sample for protein analysis. The rapid pace of technical innovation is evident from the new proteomic platforms introduced since 2000 and is likely to continue.

4.2   PROPERTIES OF PROTEINS

Proteins are made up of 20 different amino acids ($R-CHNH_2-COOH$) linked by peptide bonds ($O=C-NH$) between carboxy and amino groups of adjacent amino acids. Generally, amino acids have been greatly conserved in evolution. What makes each amino acid unique is the chemical nature of the R group such as aliphatic, hydroxyaliphatic, sulfhydryl, acidic, basic, aromatic, and imino. The amino acid sequence of each protein is derived from the base pair sequence of a gene through "codons" or sets of three nucleotide sequences that represent amino acids. The four nucleotides (A, T, C, G) of DNA allow for 64 possible combinations of three base sets of codons which permit redundancy in the coding for some amino acids. For example, glycine can be genomically coded by the codons GGA, GGC, GGG, or GGT. The sequence of amino acids in a protein or "polypeptide" is known as primary structure. Secondary structures are regular and repeating local configurations held together by intramolecular hydrogen bonding most commonly as an α-helix or β-sheet. Tertiary structure or folds represent a protein's overall shape as a composite of secondary structures and motifs (recognizable short amino acid sequence perfoming a common function such as helx–loop–helix) that are often maintained by disulfide bridges between cysteine residues. Those complex proteins assembled from multiple polypeptides or subunits (i.e., hemoglobin) contain quaternary structure.

4.2.1   Posttranslational Modifications

Protein structure and function can be greatly influenced by posttranslational modifications (PTMs) of individual amino acids. Some of the more common posttranslational modifications of proteins of amino acid residues and their average mass addition to the polypeptide are shown in Table 4.2. Posttranslational modifications are those changes made to the polypeptide chain by enzymatic processing after translation of the polypeptide from messenger RNA and after peptide bond formation has occurred. Many PTMs involve enzymatic addition of a chemical group like a phosphate, carbohydrate, or lipid group to specific amino acids. Common PTM's like phosphorylation, glycosylation, or lipolyation can occur at a single amino acid or multiple amino acids of the same protein. Alternately, enzymatic modification of proteins can also remove functional groups (i.e., dephosphorylation by phosphatases) or even portions of a protein by proteolytic cleavage to remove one or more amino acids from amino or carboxy termini.

4.2.2   Mass and Charge

These considerations about protein composition and structure are helpful for understanding key properties and attributes of proteins in proteomic analysis. The isoelectric point (pI) and the molecular mass are two important properties used in

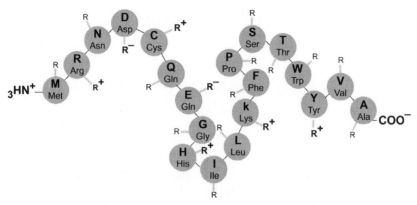

Figure 4.2. Model polypeptide of 20 amino acids. The three-letter and one-letter abbreviations for amino acids are shown. Ionizable amino acids (–R) are shown as positively or negatively charged in bold. Note that the amino-terminal and carboxy-terminal ends are charged. All charges contribute to an overall charge state of the protein at a specific pH. The pI (isoelectric point) is the pH at which a proteins carries no net charge. The pI is helpful in classifying proteins as acidic, neutral, or basic. The addition of posttranslational modifications can also add charges to proteins and greatly affect the pI value.

proteomics. First, the isoelectric point represents the net charge that is useful in classifying proteins as acidic, basic, or neutral in an electric field. Proteins typically contain multiple charges based on their amino acid composition from acidic or basic side chains (–R groups), from the amino and carboxy termini, and from posttranslational modifications (see model polypeptide, Figure 4.2). The charge of a protein can also vary, depending upon the particular pH environment it inhabits. For example, under acidic conditions, amino groups may be positive (C–NH_3^+) while carboxyl (–COO^-) and phosphate ($H_2PO_4^-$) are negative. Isoelectric focusing (IEF) is an electrophoretic method that separates proteins according to their pI. The IEF method typically uses an isoelectric gradient made up of ampholytes (from the word "amphoteric"). Ampholytes are complex commercial mixtures of molecules that carry both positive and negative charges formulated to make acid to basic pH gradients (i.e., pH 3–10). The pH at which a protein migrates in an ampholyte gradient where positive and negative charges are balanced to neutrality is the protein's pI. Second, the molecular mass of a protein can be estimated by its migration in an electrophoresis gel relative to known standards. The usual method for separation of proteins by mass under reducing conditions to remove disulfide bridges is SDS PAGE. Another method to measure protein mass which is much more accurate than gel electrophoresis is mass spectrometry, as will be discussed later in more detail.

Aside from molecular and biophysical properties of proteins such as mass and charge, all proteins share common attributes that are directly related to the ability or inability to perform global protein analysis. In particular, the spatial context of a particular "proteome" represents a large part of the complexity for proteomic analysis (Figure 4.3) since total protein expression may relate to an organism, a specific organ, a tissue, a cell, a subcellular organelle, or even biofluid.

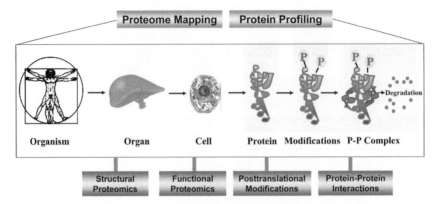

Figure 4.3. Fields of proteomic research. Proteomic research can be classified into six general research fields. Proteomic mapping and proteomic profiling constitute the first tier of proteomic analysis based upon identification and quantitation of proteins within a defined space of interest that can range from the entire organism to the protein level. The second tier of proteomic analyses is shown below involving global characterization of structure, function, posttranslational modifications, and association with other proteins (or other biochemical components).

4.3   FIELDS OF PROTEOMIC RESEARCH

Proteomic research can be classified into six general categories as shown in Figure 4.3. The spatial context of proteins is an implied attribute in the top tier of proteomic analysis: proteomic mapping and proteomic profiling. Proteomic mapping ideally identifies all proteins in a proteome within a specific spatial area. Protein profiling not only identifies but also measures all proteins in a specific proteome. The second tier of proteomic analysis involves more specialized areas for global protein analysis. Structural proteomics determines the molecular structures of proteins. Protein–protein interactions proteomics identifies and measures the mutual association of proteins. Proteomic platforms represent combinations of technologies intended to comprehensively describe protein attributes of all proteins in a biological sample underpinned by their separation, quantitation, and identification. The following proteomic platforms represent some of the primary technologies being used for separating and identifying proteins during toxicoproteomic studies.

4.4   MASS SPECTROMETERS AND PROTEIN IDENTIFICATION

There are several methods that have been developed to identify proteins. Common methods include immunoreactivity typified by Western blot, enzymatic activity based on substrate specificity, receptor-ligand affinity, pharmacologic response, and many others. However, mass spectrometry can more precisely identify proteins using biophysical approaches. The first approach is the *de novo* sequencing of peptides to generate "sequence tags" from enzymatic digests (i.e., tryptic peptides). The primary amino acid sequence is deduced from the assembly of nested sets of N-terminal peptides (b′ ions) or C-terminal peptides (y ions) that differ in length by one amino acid. The second approach is by highly accurate mass analysis of proteo-lytic digest fragments (i.e., tryptic peptides) of proteins that collectively form a diagnostic "peptide fingerprint" for protein identity. Primary protein sequence is the primary venue into searching the more extensive nucleic acid databases like NCBI which have been more highly characterized across many species and phyla com-pared to protein databases like Swiss-Prot. However, peptide finger printing is more rapid and sensitive than *de novo* sequencing. The details of how each method is performed are important to understand modern proteomic platforms which are primarily based upon mass spectrometrometry.

Mass spectrometers measure the mass-to-charge ratio (m/z) of ions from analytes that range in mass from drugs to large polypeptides. Any molecule that can be ionized can theoretically be analyzed by a mass spectrometer, but charged molecules must be able to enter the gas phase to be detected. For small molecules, like many drugs, chemicals, and small peptides, this presents little problem when they are energized to do so. However, there are some natural barriers to be overcome for protein analysis by mass spectrometry. First, it is difficult to sublimate or volatilize peptides or proteins greater than 5–10 kD. Proteins are often digested into smaller fragments at <3 kD to help them ionize and volatilize more easily. Second, ionic moieties on peptides or proteins are extremely sensitive to the pH of the surround-ing environment. Ionic character of proteins are also subject to interference by salts and detergents. As a result, proteins are often gel isolated or the salts and detergents are removed prior to liquid chromatography. Third, the amino acids in peptides or proteins may destroyed by energizing forces that are too strong, for example, by high collision energies. This final factor had proved particularly troublesome until major innovations took place in the 1990s.

4.4.1 Ionization Sources

A major challenge in adapting mass spectrometry as the pivotal technology for protein identification in proteomics was met by development of so-called "soft ion-ization" methods that do not destroy peptides or proteins in the process of ionizing and volatilizing them. The two primary ionization sources are "matrix-assisted laser desorption ionization" (MALDI), as shown in Figure 4.4 and "electrospray ioniza-tion" (ESI) depicted in Figure 4.5. In MALDI, a protein can be isolated by 2D gel or by LC. The protein is digested into small fragments, and this peptide mixture is dried along with a solid matrix (i.e., CHCA or α-cyano-4-hydroxycinnamic acid) that absorbs ultraviolet (UV) laser light. When irradiated by a UV laser, the matrix volatilizes, carrying the peptides with it. Singly charged peptide ions are mainly formed by acquiring an additional proton at the basic amino and the N-terminal alpha-amino groups. MALDI is the so-called "dry method" for peptide ionization.

4.4.2 MALDI-TOF and Peptide Fingerprints

A laser provides the energy, and a matrix [such as α-cyanohydroxycinnamic acid (CHCA)] provides the environment to effectively ionize and volatilize peptides that are dried on an inert target surface (i.e., gold target surface). Once in gas phase, the positively charged peptides are attracted to "fly" through a vacuum tube toward a negatively charged detector source, so that the TOF can be calibrated in proportion

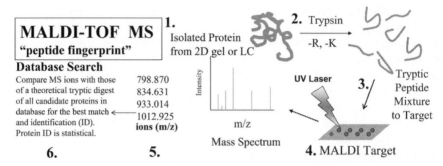

Figure 4.4. Protein identification by MALDI-TOF. Steps in protein identification by MALDI-TOF are shown. Prior separation by gel or liquid chromatography to a single protein is needed prior to enzymatic digestion with trypsin. Digested peptides are spotted onto a MALDI target with an appropriate matrix that assists desorption of peptides upon activaton by laser. Peaks from the resulting mass spectrum represent peptide ions that can be searched in a database to match a theoretical tryptic digest from known proteins. The more proteins that are matched, the greater the statistical confidence in assignment of a protein identification.

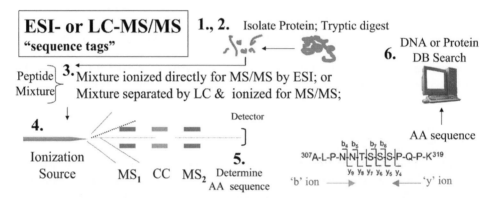

Figure 4.5. Protein identification by ESI or LC-MS/MS. Steps in protein identification by electrospray ionization (ESI) or liquid chromatography (LC)-based electrospray ionization are present. Prior separation to a single protein by gel electrophoresis or liquid chromatography is an advantage prior to enzymatic digestion with trypsin. Digested peptides can be electrosprayed under pressure through a needle or also by liquid chromatography to create microdroplets. Ionized peptide ions form under vacuum. The first mass analyzer (MS_1) measures ionized peptides (parent ions) from the sample. The collision chamber (CC) contains an inert gas to fragment a single, selected ion from which the fragments ("b" and "y" daughter ions) are determined by the second mass analyzers (MS_2) before contacting the detector. The amino acid sequence of the peptide is constructed from the masses of the fragmented ions. The amino acid sequence can be searched from protein or gene databases for identification.

to peptide mass. A TOF analyzer accelerates the ions in a vacuum tube and measures the time it takes each ion to meet the detector and register a signal. Signal is displayed as intensity on *y*-axis and as *m/z* on the *x*-axis. Particle velocity is directly related to mass for ions of the same charge. Note that some peptides may be multiply charged (+1, +2, +3) and that these ions will travel at different speeds.

TABLE 4.3. Common Posttranslational Modifications of Proteins

Modification	Amino Acid	Average Mass Change
Methylation	L,S,T,N	13
Sulfoxidation	M	16
Formylation		28
Carbamylation		41
Acetylation	L,S	42
Phosphorylation	S,T,Y,D,K	80
N-Glycosylation	L	162
O-Glycosylation	S,T	162
N-Glucuronyl		176
Farnesylation		204
Myristolation		210
Palmitoylation		238
Geranylgeranylation		272
Sialylation	K	291
Glutathionation	C	305
ADP-ribosylation	C,R	541
Ubiquitination	K	8565 kD
Sumolation; SUMO-1;SUMO-2;SUMO-3	K	11,557; 10,871; 11,637

Common posttranslational modifications of proteins. Frequently modified amino acids for each moiety are described. Most modifications involve low-mass substituents (<500 amu), but some modifications are small polypeptides of considerable mass such as ubiquitin or highly related peptide isomers like SUMO-1, SUMO-2, and SUMO 3. SUMO is the abbreviation for "similar to ubiquitin methyl organizer", SUMO has different family members (SUMO 1, 2, and 3 that can combine with proteins).

Protein identification is indirect using MALDI-TOF MS because the "peptide fingerprint" may be incomplete or the measured peptide ions may be almost the same value as peptides generated by other different proteins (Figure 4.4). The confidence in the match of measured ions obtained from the mass spectrum becomes greater with an increasing percentage of matches with the theortical masses obtained by a simulated digestion of a protein contained in a database. Of the two MS instruments, MALDI-TOF MS is the more sensitive for peptide fingerprinting, but tandem mass spectrometry (see next section) can provide amino acid sequence for more definitive information in identification. Under different analytical settings, tandem mass spectrometry can also provide find structural information such as specfic posttranslational modifications (Table 4.3) at specific amino acid residues.

4.4.3 ESI-MS/MS, LC-MS/MS, and Sequence Tags

Alternately, for isolated proteins purified to homogeneity that are tryptic digested and remain in aqueous solution, ions can be formed by ESI (Figure 4.5). In ESI, the solubilized tryptic peptides are forced through a small charged capillary that disperses them into an aerosol forming microscopic droplets that contain the charged peptides. The liquid evaporates in a vacuum tube, leaving only the peptide ions. In both MALDI and ESI, the resulting ions are guided into the mass detector for measurement. In a mass spectrum, each peak represents an ionized peptide, originating from a protein in the sample, with the height (also called intensity) of the peak being proportional to the abundance of the peptide. It should be clearly noted that specific strategies for quantitation by mass spectrometry have been developed because the peak height of a peptide ion can vary greatly due to influence of other ions in the mixture. ESI is the so-called "wet method" for peptide ionization.

In electrospray ionization tandem mass spectrometry (ESI–MS), a fine spray of a few microliters of solubilized peptides are vaporized in a high-voltage field from a glass capillary needle under pressure. It is important to note that the protein separation, purification, and digestion can be performed beforehand (offline) and is not being done in real time during online liquid chromatography where the effluent is directly connected to the mass spectrometer. Therefore, peptides separated by liquid chromatography in LC–MS/MS will elute from the column in a fine stream of effluent that is heated to vaporize and ionize separated peptides from a digest mixture in a field of high voltage. Once ionized by either ESI or LC, the first mass analyzer isolates a particular ion for dissociation in a gas phase collision cell whereupon the collision product ions are then analyzed in a second or "tandem" mass analyzer prior to striking the detector.

In tandem mass spectrometry, peptides from protein enzymatic digests are ionized into a mass spectrum, and ions from this first spectrum are selected (manually or by computer algorithm) and further broken apart in a collision cell for analysis by a second (tandem) mass spectrometer. In the collision chamber, fragmentation of the selected peptide occurs at or around the peptide bond to create a ladder of fragments that can be reconstructed into an amino acid sequence or "sequence tag" of the peptide. These peptide sequences derived from the digested protein can be searched in both protein and nucleic acid databases to assign a more definitive protein identification than peptide fingerprinting. Although more sample is required for tandem MS analysis, it has the major advantage of generating the amino acid sequence for a more definitive identification after a search in nucleic acid, protein, or expressed sequence tag (EST) nucleotide databases.

It is important to note that each of these ionization sources, either a laser in MALDI-TOF or the high voltage ionization of droplets in ESI–MS/MS or LC–MS/MS, will each produce a different spectrum of detectable ions and intensities because the effectiveness and nature of peptide ionization is quite different for each source. In addition, the presence of multiple peptides that influence each other's ionization potential notably through "ion suppression" makes most peptide ion measurements only semiquantitative.

4.4.4 Mass Spectrometers: Three Components

Mass spectrometers consist of three components: an ionization source, a mass analyzer, and a detector. First, the ionization source adds charge to the proteins or peptides in the sample, typically in the form of a proton to produce positively charged particles, and injects them into a vacuum chamber. Second, a mass analyzer uses an electromagnetic field to separate and sort the ionized peptides. Third, a detector registers the number of ions at each mass-to-charge value. The two most technically demanding components are the ionization sources, which we have discussed in the previous section, and the mass analyzers.

4.4.5 Mass Analyzers

The most common mass analyzers for protein analysis are time-of-flight (TOF), triple-quadrupole, quadrupole-TOF (hybrid), ion trap devices, and Fourier transform ion cyclotron resonance (FT–ICR) mass analyzers. Mass analyzers resolve the ions formed in the ionization source by mass-to-charge (m/z) ratios. Each type of mass analyzer has different features, most importantly the m/z range that can be covered, the mass accuracy, and the achievable resolution. The compatibility of different analyzers with different ionization methods also varies such that all analyzers can be coupled with electrospray ionization but MALDI is not usually joined to a quadrupole analyzer. Single analyzer instruments produce a mass spectrum based primarily on parent ions from the ionizing source. However, tandem or "MS/MS" mass spectrometers are instruments that generally have more than one mass analyzer which provide greater capabilities for structural information and amino acid sequencing analysis. With the presence of a collision chamber (quadrupole 2) to further fragment parent ions, the second mass analyzer in quadrupole 3 can resolve the resulting daughter ions to reveal more information about the parent ions resolved in quadrupole 1 originating from the sample. The following section will briefly describe common and newer mass analyzers.

A *time-of-flight (TOF)* analyzer allows positively charged peptides to "fly" through a vacuum tube toward a negatively charged detector source, so that the time of flight can be calibrated in proportion to peptide mass. Typically, the longer the vacuum tube, the more accurate the mass will be because of increased resolution.

A *quadrupole* consists of four parallel metal rods connected together by electric current and a radio-frequency voltage. Only ions of a certain m/q traveling down the quadrupole will actually reach the detector for a given ratio setting of voltages, while other ions will collide with the rods and will not be detected. This allows selection of a particular ion, or scanning by varying the voltages. A *triple quadrupole* is a linear series of three quadrupoles in which the first (Q1) and third (Q3) quadrupoles act as mass analyzers, and the middle (q2) quadrupole is employed as a collision cell. This collision cell is filled with an inert gas such as nitrogen or helium to induce collisional dissociation of selected parent ion(s) from Q1. Subsequent daughter fragments are passed through to Q3 where they may be filtered or scanned fully as the second mass spectrum (see Figure 4.5).

A *quadrupole ion trap* analyzer traps and sequentially ejects ions. The ion trap can hold one ion species by ejecting all others; and then by collisional activation, daughter fragments can be detected and analyzed for amino acid sequence. A principal advantage of quadrupole ion traps is that multiple collision-induced dissociation experiments can be performed without having multiple analyzers. The linear ion trap confines ions along the axis of a quadrupole mass analyzer using a two-dimensional radio-frequency (RF) field with potentials applied to end electrodes. The primary advantage to the linear trap is that the larger analyzer volume permits greater dynamic ranges and an improved range of quantitative analysis. The Orbitrap is a newly introduced mass analyzer in which ions are electrostatically trapped in an orbit around a central, spindle-shaped electrode and also oscillate back and forth. It is the oscillation that generates an image current in the detector plates at frequencies relating to the mass-to-charge ratio of the ions in the trap. Mass spectra are obtained by Fourier transformation of the recorded image currents. Fourier transform ion cyclotron resonance (FT-ICR) is a type of mass analyzer for determining the m/z ratio of ions based on the cyclotron frequency of the ions in a fixed magnetic field. Unlike other analyzers that rely on detectors, FT–ICR resolves mass only in frequency. The advantage of FT–ICR analyzers is that masses can be determined with very high accuracy permitting high resolution or narrow peak width. The high resolution can be exploited by finding similar ions in complex peptide mixtures or in study of large macromolecules such as proteins with multiple charges producing a series of isotopic peaks that can more easily be resolved.

Recent innovations in mass spectrometry have provided incorporation of two, three, and four analyzers into commercially available tandem instruments. In addition, different mass analyzers may be combined to form a "hybrid mass" spectrometer such as the quadrupole-TOF (Q-TOF). Various types of tandem mass spectrometers include the quadruopole-TOF, time-of-flight–time-of-flight (TOF–TOF), triple-quadrupole, and Orbitrap-FTICR configurations.

4.5 PROTEOMIC PLATFORMS

Proteomic platforms combine separation and identification technologies that can be completed as an experiment in a timely fashion. Proteomic platforms strive to accomplish five critical objectives: (1) a high level of information about one or more protein attributes; (2) a large number of samples per analysis session (sample throughput); (3) quantitative and comparative protein expression among samples; (4) timely analysis of raw and processed data; and (5) use of a "discovery-oriented" or "open" platform system.

The many attributes of proteins mentioned in Table 4.2 make complete global analysis of proteins from cells, tissues, organs, or organisms a formidable analytical challenge. A primary aim of proteomic analysis is to provide researchers with as much data as possible about one or more protein attributes. However, to many researchers, proteomic analysis usually involves the three attributes of identification and quantitation of all proteins in a defined sample space. Only the highlights of each platform will be summarized here. A critical appraisal of capabilities, advantages, and drawbacks are summarized in Figure 4.13 (Section 4.5.6).

4.5.1 Two-Dimensional Gel Separation or DIGE and Mass Spectrometry

Two-dimensional polyacrylamide gel electrophoresis (2D PAGE) is an established and rapid proteomic platform to separate thousands of proteins from complex lysates derived from almost any source of organelle, cell, tissue, organ, or phyla. The equipment and materials required are relatively inexpensive and widely accessible to researchers. As described in Figure 4.6, proteins are first resolved by charge using isoelectric focusing and then by mass with SDS PAGE. The size and resolution power of 2D gels increases with increasing gel sizes from 7 cm to 24 cm. Larger gels of 18 cm or more (i.e., 18-cm isoelectric focusing gel; 20-cm × 20-cm SDS PAGE gel) can regularly separate 1–2000 proteins, depending upon analytical conditions. Several 2D gels can be run in parallel at 12–20 gels per session. Four important advances in the 2D gel field are (1) development of immobilized pH gradient (IPG) strips for

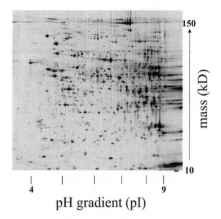

2D Gel-MS
Proteomic Platform

- **Isoelectric Focusing** (IEF) separation by charge to isoelectric point (pI).
- **Mass separation** using SDS PAGE (kD)
- **Detection** by fluorescent dye staining; image analysis and quantitation; 1–2000 proteins per 18- or 24-cm IPG strip
- **Excision** of proteins for tryptic digestion, MS analysis

mass (kD)

150

10

4   9

pH gradient (pI)

Figure 4.6. Two-dimensional (2D) gel electrophoresis and mass spectrometry. Proteins can be separated by charge using isoelectric focusing as a first dimension, and then by mass using SDS polyacrylamide gel electrophoresis (PAGE) as a second dimension to resolve complex protein lysates. Detection by staining proteins with visible or fluorescent dyes shows presence of resolved proteins. Multiple 2D gels can be compared among control and treatment groups for differences aided by use of specialized imaging software. Desired proteins can be excised for digestion and mass spectrometry identification as described in Figures 4.4 or 4.5.

improved isoelectric focusing and larger sample loads for 100- to 1000-µg protein; (2) use of narrow pH range IPG gels such as pI 5–6 or pI 8–11; (3) the introduction of MS-compatible, reversible, sensitive, fluorescent dyes like Sypro Ruby; and (4) a new protein display method termed "differential gel electrophoresis" (DIGE) to more easily find protein expression differences in 2D gels (see below).

The work flow for 2D gel–MS analysis is shown in Figure 4.6. Fluorescent staining of 2D gel-separated proteins with Sypro ruby permits sensitive, quantitative detection that is followed by comparative image analysis of many gels to determine which subset of proteins are differentially expressed among control and treated samples. DIGE 2D gel–MS (Figure 4.7) can improve the determination of up- or downregulated proteins by separately prelabeling each protein preparation from control and treatment groups with Cy2-, Cy3-, or C5-tagged crosslinking agents. An internal standard gel comprised of a mixture of all sample groups is labeled separately for a representation of all proteins in the sample and for quantitative comparision. The control and treatment groups are each labeled with a separate Cy dye. The labeled protein samples are mixed together, separated on a single 2D gel, scanned at different wavelengths for separate dye detection, and the images are electronically merged to determine common and differentially expressed proteins.

Whichever staining method is used (either a single stain or a multiplex DIGE Cy dye staining), proteins found to be differentially expressed can be excised from each gel, enzymatically digested into smaller fragments (i.e., trypsin), and then analyzed by mass spectrometry. Mass spectrometry generally involves spotting onto MALDI targets, or protein digests are made ready for liquid chromatography (LC) separation and MS/MS analysis and protein identification.

Reasons for the popularity of the 2D gel proteomic approach are its adaptability to almost any protein sample from any organism, simultaneous separation of large

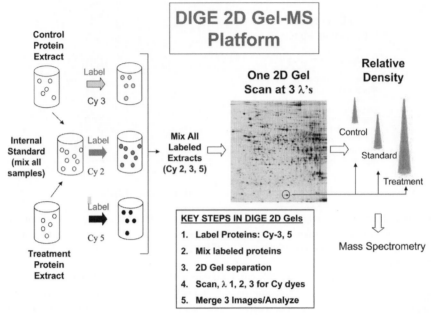

Figure 4.7. DIGE two-dimensional (2D) gel electrophoresis and mass spectrometry. DIGE or differential gel electrophoresis uses fluorescent dyes to provide easier detection of protein expression differences in 2D gels. Proteins are labeled with fluorescent dyes prior to 2D gel analysis. Up to three dyes can be used, typically using Cy2 labeling of a mixture of proteins from all sample to act as an internal standard, Cy 3 labeling of control sample, and Cy 5 labeling of treatment sample. Prelabeled samples are mixed and subjected to 2D gel analysis, thereby eliminating gel-to-gel variation encountered in single staining detection as described in the traditional approach in Figure 4.6. Dyes are scanned at three different wavelengths, and the electronic images are combined. Assignment of color to each electronic image more easily shows differences or proportional changes in protein expression attributed to control or treatment.

numbers of intact proteins by pI and MW, differential display capabilities, and potential for indicating posttranslational modifications due to charge or mass differences. There are some limitations to 2D gel-based methods. The primary drawback of 2D gel separations is that not all proteins will focus or solubilize equally well since hydrophobic and basic proteins focus with difficulty. Also, most low-copy regulatory proteins such as transcription factor or secreted proteins like cytokines cannot be detected over the abundant structural and maintenance proteins in complex samples. While most steps downstream of protein separation can be automated, complete robotic handling of 2D gels has not been achieved and they are labor-intensive.

4.5.2 MudPIT

Chromatographic devices connected online to mass spectrometers have become powerful tools for identifying large number of proteins from proteolytically digested

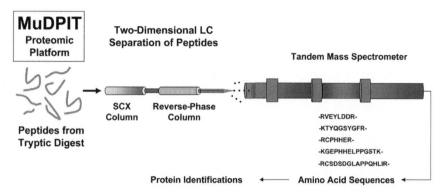

Figure 4.8. MuDPIT platform. The MuDPIT or multidimensional gel protein identification system is an multidimensional liquid chromatography (LC)-based system of separation prior to tandem mass spectrometry (MS/MS). It is not necessary to purify proteins prior MuDPIT, although a reduction in protein complexity by some prior purification is helpful in obtaining interpretable spectra. Protein lysates are digested into peptides that are loaded onto a strong cation exchange (SCX) support. Peptides are sequentially eluted onto a reverse-phase column for a second separation. Eluted peptides from the reverse-phase column are electrosprayed into a tandem mass spectrometer for amino acid sequencing and identification of proteins in the sample.

protein samples. Reverse-phase liquid chromatography (LC) separation by itself cannot resolve complex protein digests since many peptides coelute under one peak. By analogy, incomplete gel separation occurs by SDS PAGE separation of proteins by mass alone since many proteins are of similar molecular weight. However, the use of more than one chromatographic method in sequence, termed "multidimensional LC," can achieve high resolution of peptides from fractionated digests. The best-known system that has been commercialized into a proteomic platform is called multidimensional protein identification technique (MudPIT; see Figure 4.8). In the MudPIT approach, a reduced and denatured protein isolate from any source can be proteolytically digested into peptides that are loaded onto a microcapillary column containing a strong cation exchange (SCX) resin. Small fractions of peptides are eluted in a stepwise salt gradient from the SCX column onto a second reversed-phase LC column whereupon salts are removed and the peptides are eluted by a slow organic solvent gradient directly connected to an MS/MS instrument. The sequence tags generated for each peptide are searched in protein, nucleic acid, and EST databases to identify the proteins from which they were originally derived. This cycle is repeated 10–20 times to separate and identify small groups of peptides in each fraction rather than separate the complex digest all at once. For example, nearly 1500 unique proteins can be identified from a total yeast cell lysate; and in another experiment, nearly all the proteins comprising the yeast 80S ribosomal complex were similarly identified using the MuDPIT platform. Despite the powerful descriptive power of multidimensional LC–MS/MS for identifying proteins, protein levels are difficult to quantify and compare between samples because of the variable nature of peak heights and retention times during liquid chromatography of complex protein digests.

4.5.3 ICAT, SILAC, iTRAQ

In addition to analyzing as many proteins as possible in a sample, the prospect of quantifying protein levels in protein samples has been a difficult aspect of proteomic analysis. Protein quantitation can be thought of in relative terms as a proportion of treatment/test sample compared to control sample or in absolute terms as the number of molecules (moles) or concentration (molarity). Internal standards are useful but not realistic for complex protein samples of unknown composition in most proteomic studies. Quantitative measurement of intensity peak height of signals generated from peptide ions is difficult because of variable ionization efficiency, particularly with MALDI, and due to ion suppression in electrospray ionization. Ion suppression can result from the presence of less volatile compounds in a sample that vary the efficiency of droplet formation or evaporation, ultimately affecting the amount of charged ions available to reach the detector. However, a unique feature of high-resolution mass spectrometers is the ability to finely distinguish between small differences in mass even to the point of resolving the relative abundance of stable isotopes, in otherwise identical samples (Table 4.4). A number of proteomic platforms for protein quantitation and identification have been built around the use of isotopic tagging of proteins (ICAT) or peptides (iTRAQ) or metabolic incorporation of isotopically tagged amino acids (SILAC).

The ICAT proteomic platform or "isotope-coded affinity tagging" permits comparative quantitation between samples which is sufficiently sensitive to identify lower expression proteins (Figure 4.9). The original method used heavy (deuterium, ^{2}H) and light (hydrogen, ^{1}H) bifunctional linking reagents to label native proteins from control and treatment samples. ICAT labels now in use include carbon isotopes such as $[^{12}$C$]_8$-linker and $[^{13}$H$]_8$-linker, and they also have acid-cleavable biotin moieties with cysteine-reactive groups to label native proteins. ICAT-tagged proteins are mixed and trypic-digested, and the resulting peptides are collected with streptavidin beads by affinity chromatography of biotin-labeled ICAT-reagent-tagged peptides. LC–MS analysis is performed for relative quantitation of the isotopes on identical peptides. The ratio of the peaks areas for specific ICAT-labeled pairs defines the relative abundance of its parent proteins between the two cell states (control and treatment). Amino acid sequencing of individual peptides by LC–MS/MS is used to

TABLE 4.4. Proteomic Platforms Using Stable Isotopes

Stable isotope: Stable isotopes of an element differ in mass due to the number of neutrons but have the same elemental and chemical characterisitics as the element. Stable isotopes are not radioactive. Common stable elements and their stable isotopes are ^{1}H and ^{2}H; ^{12}C and ^{13}C; ^{14}N and ^{15}N; ^{16}O and ^{18}O; ^{32}S and ^{34}S.

Element: Number of protons within the nucleus of an atom.

Atomic mass: Number of protons and neutrons in the nucleus.

AMU: Atomic mass units. Hydrogen = 1; Deuterium (^{2}H) = 2.

Stable isotope proteomic platforms: Peptides carrying different stable isotopes can be distinguished by high-resolution mass spectrometry. 1 amu can be resolved. ICAT, iTRAQ, and SILAC are examples.

ICAT: Isotope-coded affinity tag.

iTRAQ: isotope tag for relative and absolute quantitation.

SILAC: Stable isotope labeling of amino acids in culture.

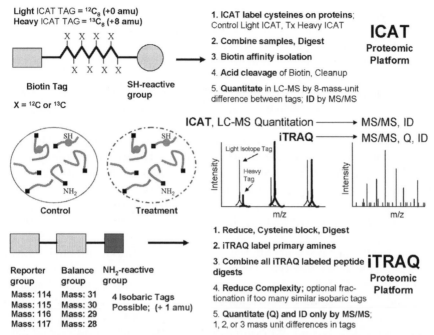

Figure 4.9. Proteomic analysis by ICAT and iTRAQ. ICAT (isotope-coded affinity tags) and iTRAQ (isobaric tags for real and absolute quantitation) are two approaches to protein identification and quantitation using LC-MS/MS. In ICAT, proteins are labeled with either light or heavy linker reagents that also have a biotin tag and sulfhydryl reactive group to react with reduced cysteine groups in proteins prior to digestion. After labeling of control and treatment groups in separate reactions, the protein lysates are mixed and enzymatically digested. Strepavidin beads bind to the biotin end of all tags (light and heavy) for removal of untagged peptides. After acid cleavage to remove biotin, samples can be separated by LC-MS/MS. The first mass spectrum is used for quantitation by comparison of peak heights separated by 8 amu (atomic mass units). These peaks are fragmented and identified by amino acid sequencing. iTRAQ analysis uses a slightly different approach. Four iTRAQ tags are available that can be used in pairs or groups of 3 or 4. A separate tag is used to label proteins from each sample group, so up to four groups can be analyzed (unlike ICAT which is two groups). A chemically reactive end of the iTRAQ tag can bind to all free primary amine groups occurring at amino-termini of proteins or basic amino acid residues (lysine, arginine). iTRAQ-tagged proteins from all groups are mixed and digested. The balance group and reporter group on each tag is balanced for mass so that the overall mass of each iTRAQ tag, and therefore each tagged peptide, is isobaric (same mass). When peptides are separated and analyzed by MS/MS, the first mass spectrum shows the ionized peptides. However, only upon fragmentation of each ionized peptide does the reporter group become evident for quantitative comparison of peak heights as well determination of amino acid sequence. The process of quantitation and identification are essentially performed in the same step. The difference between each iTRAQ tag could be 1 amu (although use of different reporter group tags could also show differences of 2 amu or 3 amu), which is distinguishable by high-resolution mass spectrometers.

obtain sequence tags for protein identification. One disadvantage of ICAT is that not all proteins contain cysteines, which may limit protein coverage in a sample. In some cases, shared cysteine-containing peptides in sequences among highly homologous proteins may not distinguish among different protein (or gene) family members.

Isotope tag for relative and absolute quantitation (iTRAQ) is a related proteomic platform to ICAT but differs in its use of stable isotopes and binding strategies. The iTRAQ approach uses isobaric-tags containing three features: a reporter, a mass balancer and primary amine reactive groups as shown in Figure 4.9. The peptide reactive group, an *N*-hydroxysuccinimide ester, reacts with primary amine groups so that N-terminal peptides should always be able to be tagged, as well as basic amino acid residues like lysine and arginine. Instead of tagging native proteins like ICAT, the control or treatment samples are first enzyme digested (i.e., trypsin) and then labeled with iTRAQ reagents. In this way, more proteins are tagged with iTRAQ reagents.

Commercially, at least four different iTRAQ are available although more tags are technically feasible. The reporter of the iTRAQ reagent is a tag with a mass of 114, 115, 116, or 117 daltons, depending on the differential isotopic combinations of $^{12}C/^{13}C$ and $^{16}O/^{18}O$ in each individual reagent. The iTRAQ reagent contains a balance (carbonyl) group and a pendant reporter group (*N*-methylpiperazine). The distribution of ^{13}C, ^{15}N, and ^{18}O isotopes between the reporter group and balance group differs in the four versions of each iTRAQ tag such that the reporter group masses differ successively by 1 dalton. When an iTRAQ-labeled peptide is subjected to collision-induced dissociation, the iTRAQ tag fragments between the reporter group, balance group, and the peptide. The balance group is uncharged and, therefore, is not present in product ion spectra. The reporter group, however, retains a proton, and the resulting reporter ions derived from the four versions of the iTRAQ reagent appear at 114.1 *m/z*, 115.1 *m/z*, 116.1 *m/z*, or 117.1 *m/z* in each product ion spectrum. A minimum of two iTRAQ-tagged samples (control and treatment) can be performed, but up to four samples could be tagged and analyzed at the same time. Peptide quantitation is achieved by comparing the peak areas of these reporter ions and identification by MS/MS analysis of individual ions. While iTRAQ may be particularly useful because of the possibility of labeling and distinguishing up to four samples, the quantitative differences in peptides levels can only be determined after MS/MS. As a consequence in iTRAQ analysis of some complex samples, there may easily be an overabundance of isobaric tagged peptides in the LC–MS phase (first dimension) that makes quantitation during MS/MS difficult. Therefore, a strategy to reduce complexity by affinity, chromatographic, or electrophoretic separations prior to MS analysis in iTRAQ may be very helpful in limiting the number of first MS dimension, isobaric-tagged peptides. Another plus is that iTRAQ can be used for detection of posttranslational modifications such as phosphorylation.

Stable-isotope labeling with amino acids in cell culture (SILAC), is a proteomic approach for metabolic incorporation of stable-isotope-labeled tags into proteins during normal cellular processing (Figure 4.10). It is potentially more inclusive for tagging all proteins than ex vivo labeling strategies. However, steady-state labeling must be accomplished prior to experimentation that typically involves five or six cell doublings. For this reason, SILAC is not useful for acute labeling circumstances like ICAT or iTRAQ. SILAC relies on metabolic incorporation of a given "light"

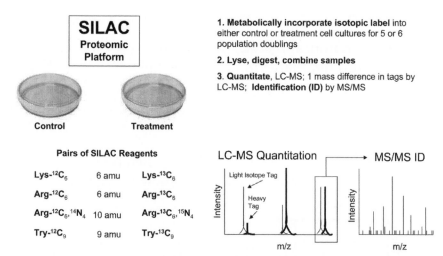

Figure 4.10. Proteomic analysis by SILAC. Proteomic analysis by SILAC or stable isotope labeling of amino acids in cell culture utilize *de novo* metabolic incorporation of stable-isotope-labeled amino acids during protein synthesis. Cells can be cultured with various combinations of stable-isotope-labeled amino acids such as lysine or arginine. Tyrosine has been used in phosphoprotein studies of tyrosine residues. About five or six cell divisions are needed for complete labeling of proteins in cell cultures prior to experimentation. Labeled cells from control and treatment(s) lysates are combined and digested. Quantitation and identification are performed by LC-MS/MS.

or "heavy" form of the amino acid into proteins. Any number of stable isotopic amino acids are possible. The method relies on the incorporation of amino acids with substituted stable isotopes (e.g., ^2H, ^{13}C, ^{15}N). Thus in an experiment, two cell populations are grown in culture media that are identical except that one of them contains a "light" and the other a "heavy" form of a particular amino acid (e.g., ^{12}C- and ^{13}C-labeled L-lysine, respectively). When the labeled analog of an amino acid is supplied to cells in culture instead of the natural amino acid, it is incorporated into all newly synthesized proteins. After a number of cell divisions, each instance of this particular amino acid will be replaced by its isotope-labeled analog. Since there is hardly any chemical difference between the labeled amino acid and the natural amino acid isotopes, the cells behave exactly like the control cell population grown in the presence of normal amino acid. It is efficient and reproducible because the incorporation of the isotope label is nearly 100%.

4.5.4   Retentate Chromatography Mass Spectrometry (RC–MS)

RC-MS stems from the observation that mass spectra of peptides and proteins can be generated by laser activation of chromatographic surfaces placed on MALDI targets. The product application of RC–MS, called surface-enhanced laser desorption ionization–time of flight (SELDI-TOF), refers to a mass spectrometry method for measuring native proteins adsorbed (retained) on various chemical or biochemical surfaces (Figure 4.11). Although the concept of laser desorption of intact proteins is well established, it has been commercially exploited only recently as a means

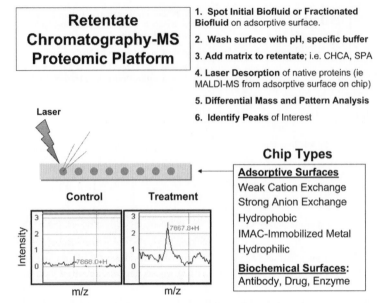

Figure 4.11. Retentate chromatography mass spectrometry. Retentate chromatography (RC) mass spectrometry (MS) is a type of MALDI-based ionization of peptides after adsorption and washing of proteins from a adsorptive active surface. The most popular commercial form is SELDI-TOF or surface-enhanced laser desorption ionization time-of-flight mass spectrometry. The SELDI approach to proteomics has found the greatest use in comparing mass spectral profiles of biofluids such as serum among patients with disease, particularly cancer. The RC-MS approach does not attempt to globally analyze all proteins or even to identify them during the initial phase of study. A common objective is find a set of protein peaks that distinguish one group of samples from another for classification. For example, serum samples (either native or prefractionated) are spotted onto an adsorptive surface that retains a subset of proteins based on their chemical or biochemical attraction and the remaining nonadsorbed proteins are washed away. A mass spectrum is produced using a MALDI-TOF instrument adapted to accept chips. Several hundred samples using various combinations of chip types and washing conditions can be processed in a few days. Specialized software examines the data for distinguishing peaks and peak heights to classify groups of samples. Some tandem mass spectrometers can be adapted to perform MS/MS from a chip for protein identification, or scale-up measures can be used to isolate enough selected protein for conventional MS identification.

of rapid protein profiling by combining mass spectrometry and protein chromatography. A key feature of the SELDI-TOF instrument is its ability to provide a rapid protein expression profile from a variety of biological and clinical samples. In particular, protein profiling can be performed using only a few microliters of accessible biological fluids such as serum, plasma, and urine. It has been used for biomarker identification in projects ranging from prostate, bladder, breast, and ovarian cancers.

The SELDI instrument is a laser desorption ionization mass spectrometer capable of ionizing proteins from an adsorptive surface, or "biochip." The instrument has a customized carriage to feed biochips into what is essentially a MALDI instrument that are read rapidly. The instrument is equipped with pulsed UV nitrogen laser with a low-to-medium mass accuracy range and specialized software to normalize and compare spectra from hundreds of samples and many treatment groups. In a manner analogous to gel electrophoresis separation of denatured proteins according to size, a spectrum of protein masses is measured after laser energy ionizes and volatilizes proteins from the target surface in the presence of an overlaying organic matrix such as CHCA (α-cyano-4-hydroxycinnamic acid) and SPA (sinnapinic acid). The system is most effective in profiling relatively low-molecular-weight proteins (<20 kD), although higher-molecular-weight molecules can be detected by increasing laser intensity and the organic matrix (CHCA for low-molecular-weight range; 3,5-dimethoxy-4-hydroxycinnamic acid or SPA for high-molecular-weight protein range). The analysis or "reading" of each sample is relatively rapid and can be automated so that 8-surface-biochip or 16-surface-biochip surfaces can be read on a single biochip in about 1 hr. Each 2-mm surface on a biochip (10 mm × 80 mm) adsorbs proteins from a complex protein mixture by chemical affinity (anionic, cationic, hydrophobic, hydrophilic, immobilized metal affinity chromatography (IMAC)] or by more selective biochemical attraction (antibody, receptor, oligonucleotide, peptide). Selectivity on the binding surface can be varied by pH, ionic strength, detergent, organic solvent, and other modifiers. Nonspecific bound proteins are removed by repeated washing steps.

The capabilities of the SELDI technology are selectivity of proteins for specific chemical or biochemical surfaces; sensitivity for displaying subsets of proteins from complex mixtures; rapid analysis by reading hundreds of samples per day; automation by use of liquid handling workstations for accurate microsample processing; measurement of native proteins without enzymatic digestion; and production of biomarker patterns to define the presence of disease without exact knowledge of protein identities. Limitations of this technology are the nonglobal assessment of protein mixtures; the limited mass range that SELDI-TOF effectively measures; and the inability to directly identify proteins. A recent exciting development for protein identification from biochips is the construction of a biochip interface on tandem quadrupole-TOF MS instrument that permits sequence tag generation directly off the chip.

Innovative approaches to sample preparation and data analysis have led to major some major advances in the field biomarker identification of disease using SELDI analysis despite some limitations in definitive protein identification like that achieved by protein fingerprinting and sequence tags by traditional MS applications. In terms of sample preparation, complex samples such as serum can be analyzed by direct application of 1–3 µl of serum upon a chemical surface, or subsets of serum proteins can be analyzed by multidimensional chromatography. Adsorption of proteins to strong anion exchange (SAX) resin and sequential pH elution of proteins results in fractionated subsets of proteins that can be further subdivided by adsorption onto the different SELDI surfaces. The approach produces a composite protein profile that can isolate abundant serum proteins like serum albumin to one or two fractions and improve detection of other low-abundance proteins that may be more informative of the disease state or chemical exposure. Analysis of SELDI data examines each spectrum as a series of hundreds of data points. The use of serum samples from well-characterized patients as training sets can result in development of a critical set of protein ions diagnostic of toxicity or disease. However, the difficulty in obtaining structural information such as amino acid sequence to identify peaks of interest have proved a major limitation of this platform.

4.5.5 Antibody Arrays

Antibody arrays are the most common form of the general category of protein arrays that can include arrays of peptides, antibodies, recombinant proteins, and even tissue sections. Protein arrays represent miniaturized, highly parallel assays for multiplexed protein identification, protein–protein interactions, enzyme activities, and affinity studies for a great number of ligands and substrates. The term "protein microarray" has often been used to collectively refer to immobilized peptides, enzymatic substrates, polypeptides, and antibodies. Protein arrays hold great promise for revolutionizing protein profiling, functional proteomics, and proteoinformatics because of the massive parallel processing capabilities, quantitative data, and ability to integrate into established hardware and software systems already developed for DNA microarrays.

Development of high-throughput detection of proteins or specific protein features has been facilitated by formatting 9- or 16-plex antibodies into each well of 96-well plates in a format that easily links to ELISA technology. Specialized functional assays such as kinase, gel shift, electrophoresis mobility shift (EMSA) in 96-well formats have also been useful in validating phosphorylation or transcription factor activation. Proteins can be arrayed on glass or filter surfaces (i.e., nitrocellulose) or can also be organized in capillary systems to form microfluidic arrays. The primary challenge is to preserve proteins in lysates in their biologically active form to sufficiently recognize and bind to their immobilized protein or ligand.

Many assays that have been bead-captured and gel-based have been converted to 96-well polystyrene microtiter plates to standardize the format for automation and higher throughput. In many ways, plate formats are a bridge to the higher-density microarray formats and remain quite useful until full conversion to microarray or microfluidic formats can be made. Microtiter plate assays or activated microbeads (i.e., Luminex) for panels of cytokines are widely available because of their use in clinical medicine.

Antibody arrays immobilized on glass surfaces mimic DNA microarrays in format and spot size. The biggest challenge in protein profiling using antibody microarrays is selection of validated antibodies that are useful in the desired sample environment. Many of the initial reports used antibody arrays assayed for cytokines because serum presents a relatively simple sample assay environment compared to tissue and also because there are numerous validated antibodies available for this clinically important set of proteins. Tissue and cell lysates present more complex assay environments with more opportunities for antibody cross-reactivity and other interferences which erode the biological meaningfulness of the data.

Figure 4.12 describes two popular antibody microarrays formats that are constructed for antigen capture in small sample volumes with detection by either sandwich immunoassay or antigen labeling. In sandwich immunoassay, capture antibodies are arrayed and immobilized to select specific proteins which are then found by a second labeled detection antibody. In protein target labeling, all proteins in the sample are prelabeled (i.e., fluorescent dyes) prior to capture by immobilized antibody arrays. In direct assay systems, sample proteins are directly immobilized onto

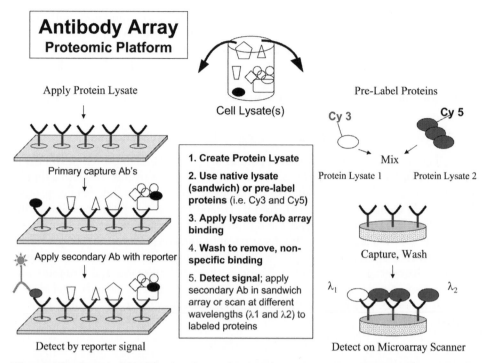

Figure 4.12. Protein identification by antibody microarrays. Two schemes are shown for capture of proteins by specific antibodies immobilized on solid surfaces in microarray format. On the left, a protein lysate is applied to an array of immobilized antibodies that have been validated for selective capture of specific proteins. For detection, a second, signal-generating antibody must selectively bind to the captured protein, creating a "sandwich assay." The scheme on the right of the figure uses a different approach. Proteins from each lysate (protein lysates 1 and 2) are prelabeled with different fluorescent dyes such as Cy3 and Cy5, then mixed and applied to a collection of immobilized capture antibodies. After incubation and washing to remove unbound protein, the array is scanned at wavelengths (λ1 and λ2) to detect fluorescent signal and a ratio is created from the intensities. If more protein is present in one sample, a greater signal will result. In the protein prelabeling scheme on the right, it is sometimes difficult to label all proteins stoichiometrically in each sample with fluorescently tagged crosslinking reagents. Note that in the sandwich assay approach, the labeled secondary antibody can detect its free target protein directly or sometimes as part of a complex (far right Ab-complex). Sometimes, target proteins in an antibody are hidden which can lead to underrepresentation of protein signal. Overrepresentation of signal may occur with either scheme from nonspecific binding or cross-reactivity of antibodies with common epitopes, which confounds interpretation of results.

surfaces in much the same way a dot-blot is conducted, and selected proteins are then detected by prelabeled specific antibodies.

The capabilities of antibody microarray technology are similar to those for DNA array methods: selectivity of immunoreagents in complex protein lysates; rapid, massively parallel analysis of proteins; small sample volume requirement; and automation and compatibility with DNA microarray technologies (in hardware, software, and bioinformatics); also, native proteins are analyzed, which affords information on specific structure and protein–protein interactions. Limitations of this technology are (a) the requirement for a multitude of validated antibodies; (b) potential cross-reactivity with related proteins, which complicates biological interpretation, nonglobal assessment of protein expression defined by selection of antibody arrays; and (c) protein–protein interactions that can either mask desired epitopes and reduce signal or, alternately, inflate target protein signal by detection of attached nontarget proteins bound to the target.

4.5.6   Advantages and Disadvantages of Proteomic Platforms

In their capabilities to both separate and identify proteins, each proteomic platform brings an emphasis upon complex attributes of proteins. Platforms strive for a timely, sensitive, and comprehensive analysis of the proteome. The life cycle of each protein is extremely complex, ranging (a) from its assembly as a polypeptide chain during its translation to (b) folding, (c) combination with other proteins, and (d) trafficking in the cellular milieu or extracellular secretion to finds its place within the organism. If each structural change of a protein constitutes an analyte, then each cell easily contains over one million analytes. Faced with such challenges, the strengths and weakness of each platform as summarized in Figure 4.13 are put forth in relation to each other as much as to their more challenging aim of true global protein analysis.

2D Gel-MS	DIGE 2D Gel-MS	RC-MS	Ab Arrays
Medium resolution	Medium resolution	Low resolution	Immunoaffinity separation
Semi-quantitative	Semi-quantitative	Rapid profiling biofluids	Rapid multiplex analysis
Differential expression	Improved differential	Signatures	Immunoaffinity protein ID
Detects PTMs	Detects PTMs	Small sample size	Small sample size

2D Gel-MS	DIGE 2D Gel-MS	RC-MS	Ab Arrays
Low sensitivity	Improved sensitivity	Protein ID difficult	Limited validated Ab's
Low-global separation	Low-global separation	Low-global separation	Cross-reactivity epitopes
Solubilization problems	Solubilization problems	Limited mass range	Semi-quantitative
Basic, hydrophobic proteins poorly soluble	Basic, hydrophobic proteins poorly soluble	Prefractionation needed for better resolution	Protein-protein interactions may mask epitope sites

MuDPIT	ICAT	iTRAQ	SILAC
Sensitive, semi-quantitative	Sensitive, quantitative	Sensitive, quantitative	Sensitive, quantitative
High resolution	In vivo or in vitro labeling	In vivo or in vitro labeling	Cell culture labeling
No labeling necessary	Bidifferential expression	4 stable isotope labels	Bidifferential expression
Small sample size	Detects PTMs	Detects PTMs	Detects PTMs

MuDPIT	ICAT	iTRAQ	SILAC
Non-native, protein digests	Non-native, protein digests	Non-native, protein digests	Non-native, protein digests
No differential expression between samples	Not all proteins cross-link; alternate linkers available	Protein isoforms interfere with isobaric tags	Long, labeling time needed for steady state
PTMs not readily viewed	Ionization efficiency varies	Complex spectra	De novo labeling only
Low throughput	Sample recovery varies	Low throughput	Low throughput

Figure 4.13. Summary of advantages and limitations of representative proteomic platforms for protein expression analysis. The capabilities and advantages (solid line) are compared to the drawbacks (dotted lines) for each proteomic platform. Explanation of advantages and limitations are briefly highlighted here and more thoroughly discussed in the text. Abbreviations for each platform are provided in the Figure 4.1 legend and in the text. Other abbreviations used are: PTMs, posttranslational modifications; ID, identification.

4.6 PROTEOMES AND SUBPROTEOMES: EXPECTATIONS AND REALITY

In DNA microarray analysis, substantially up- or downregulated transcripts are of primary interest in toxicogenomics. Similarly, the primary object in toxicoproteomic studies is to determine the group of differentially expressed proteins as the basis of protein selection for excision and identification. Given the large dynamic range of protein expression and preponderance of structural and metabolic proteins, finding differentially expressed proteins in whole-cell or whole-organ homogenates is presently out of reach.

The dynamic range of protein expression in a cell is about 10^7. In fluids such as blood serum or plasma, it is even greater at 10^{12} to 10^{14}. The two main detection methods used, mass spectrometry and fluorescence, can cover up to 10^4 orders of magnitude. Antibody or other ligand affinity can extend that range somewhat. However, in general there is no tool, chip, gel, or LC method that can cover the range of protein expression found in nature in cells, tissue, and biofluids of any organism in nature. Therefore, any given proteomic platform can only separate, detect, and identify a subset of the proteins, a subproteome, that represent the proteome of a cell, organ, tissue, or biofluid.

A "reduction in complexity" strategy of the proteome is often required and involves some form of protein enrichment, purification, to discriminate differentially expressed proteins from those with "no change" in expression. The isolation of "subproteomes" enriched by subcellular fractionation or by immunoaffinity capture, ligand binding, or adsorption chromatography are common and effective strategies for drilling down to lower copy proteins at a particular cellular spatial location or for those proteins containing special functional attributes.

The top tier of proteomic analysis categories—proteomic mapping and proteomic profiling—will continue to dominate the field of proteomics because investigators are interested in what proteins are present in their sample and how much. "How much" is typically a proportion of experimental treatment to control, but proteomic technologies are improving on their abilities to provide amounts and concentrations. Therefore, an important consideration in proteomic analysis is a realistic sense of how much of the proteome can actually be measured with the proteomic platform available in order to best answer the scientific problem at hand.

4.7 SUMMARY

A description of the complex nature of protein structures, actions, and integration of organizational hierarchies in health and disease has often been described as "systems biology." Multiple high-throughput technologies in gene expression at the protein and transcript levels will be needed to meet the challenge of defining clinical and experimental proteomes and their response to toxic reagents. Protein separation and enrichment strategies conducted beforehand or connected directly online to mass spectrometers remain a central technology for high-throughput proteomic systems, but chip-based systems and various protein capture microarrays will soon make high-throughput proteomics more accessible to the scientific and clinical communities. High-throughput strategies in protein expression will produce large volumes but little scientific advancement unless these technologies are accompanied by validation of high-throughput results, informed hypothesis development, and rigorous experimentation. Progress in miniaturized, mass parallel and more comprehensive protein measurements will greatly enrich proteomic analysis.

SUGGESTED READING

Baldwin, M. A. Mass spectrometers for the analysis of biomolecules. *Methods Enzymol.* **402**, 3–48, 2005.

Carr, S. A., Annan, R. S., and Huddleston, M. J. Mapping posttranslational modifications of proteins by MS-based selective detection: Application to phosphoproteomics. *Methods Enzymol.* **405**, 82–115, 2005.

Delahunty, C., and Yates, J. R., 3rd. Protein identification using 2D-LC-MS/MS. *Methods* **35**, 248–255, 2005.

Engwegen, J. Y., Gast, M. C., Schellens, J. H., and Beijnen, J. H. Clinical proteomics: searching for better tumour markers with SELDI-TOF mass spectrometry. *Trends Pharmacol. Sci.* **27**, 251–259, 2006.

Merrick, B. A., and Bruno, M. E. Genomic and proteomic profiling for biomarkers and signature profiles of toxicity. *Curr. Opin. Mol. Ther.* **6**, 600–607, 2004.

Morandell, S., Stasyk, T., Grosstessner-Hain, K., Roitinger, E., Mechtler, K., Bonn, G. K., and Huber, L. A. Phosphoproteomics strategies for the functional analysis of signal transduction. *Proteomics* **6**, 4047–4056, 2006.

Morelle, W., Canis, K., Chirat, F., Faid, V., and Michalski, J. C. The use of mass spectrometry for the proteomic analysis of glycosylation. *Proteomics* **6**, 3993–4015, 2006.

Ong, S. E., and Mann, M. Mass spectrometry-based proteomics turns quantitative. *Nat. Chem. Biol.* **1**, 252–262, 2005.

Simpson, R. J. *Proteins and Proteomics: A Laboratory Manual*, Cold Spring Harbor Laboratory Press, Cold Spring Harbor, NY, 2003.

Turner, S. M. Stable isotopes, mass spectrometry, and molecular fluxes: Applications to toxicology. *J. Pharmacol. Toxicol. Methods* **53**, 75–85, 2006.

Twyman, R. M. *Principles of Proteomics* BIOS Scientific Publishers, Taylor and Francis Group, Andover, Hampshire, UK, 2004.

Uhlen, M., and Ponten, F. Antibody-based proteomics for human tissue profiling. *Mol. Cell. Proteomics* **4**, 384–393, 2005.

Wetmore, B. A., and Merrick, B. A. Toxicoproteomics: proteomics applied to toxicology and pathology. *Toxicol. Pathol.* **32**, 619–642, 2004.

Yates, J. R., 3rd, Gilchrist, A., Howell, K. E., and Bergeron, J. J. Proteomics of organelles and large cellular structures. *Nat. Rev. Mol. Cell. Biol.* **6**, 702–714, 2005.

CHAPTER 5

Metabolomics

NIGEL DEIGHTON

Department of Environmental and Molecular Toxicology, North Carolina State University, Raleigh, North Carolina 27695

5.1 INTRODUCTION

5.1.1 Definitions

The metabolome is considered an experimentally accessible feature of the cell that manifests important phenotypic information, and it is also [arguably] the least characterized of the "omes" in all biological systems. The study or observation of the metabolome—namely metabolomics—represents a logical progression from large-scale RNA and proteomic analysis in systems biology, and its application to biological questions will likely become as extensive as either of the aforementioned approaches in the coming years. The development and application of metabolomics offers for the first time the opportunity to gain an holistic view of the interaction between genes, RNA, proteins, metabolism, and biological effect (phenotype). Two long-standing questions in the life sciences are: Do increases in mRNA levels indicate an increase in the level of the gene product (protein?) and, once translated, is that gene product active? The development of metabolomics potentially provides an answer to the next logical question in this series: How does the expression of a gene directly impact metabolism, both in the manner that would be expected, and also to what extent is seemingly unrelated metabolism affected?

Fiehn has suggested definitions for *metabolomics* and the related, yet distinct, approaches of *metabolite profiling* and *metabolite fingerprinting*. Metabolomics should be considered the identification and quantitation of all of the metabolites within a biological system. As such, metabolomics approaches must aim to avoid the exclusion of any metabolite through the application of well-conceived sample preparation procedures and analytical techniques. Consequently, due to the chemical diversity of the constituents of the metabolome, along with the fact that their concentrations will differ by many orders of magnitude, it is inconceivable that a single-sample preparation regimen, or indeed a single analytical tool, will suffice in metabolomics studies. *Metabolite profiling* may be used to characterize (or eluci-

Molecular and Biochemical Toxicology, Fourth Edition, edited by Robert C. Smart and Ernest Hodgson
Copyright © 2008 John Wiley & Sons, Inc.

date) a whole biochemical pathway, or even intersecting pathways. For metabolite profiling, it is not necessary to observe the entire metabolome, and it is thus possible to use sample preparation and analytical approaches that are best suited to the predefined metabolites. *Metabolite fingerprinting* is often used to screen large sample numbers with the aim of classifying those samples by biological origin or some other relevant metric. This approach will highlight similar samples along with dissimilar samples and does not require extensive data treatment, or indeed identification of the observed metabolites.

The aim of the present chapter is not an exhaustive treatise on metabolomics and these other approaches (often misrepresented as metabolomics), but more to provide an appetizer. The following sections will provide an entry-level description of some of the approaches used in metabolomics, then highlight some of the early successes of metabolomics. Some of the pitfalls/practicalities of such a global approach will be considered, and finally I'll attempt to tie metabolomics into present-day toxicology.

5.1.2 Nonbiased Approaches in Metabolomics

In what many consider to be a landmark publication on metabolomics, Fiehn et al. (2000) state "it is crucial to perform unbiased (metabolite) analyses in order to define precisely the biochemical function of plant metabolism." The authors argue that for metabolomics/metabolite profiling to become a robust and sensitive method suited to automation, a mature technology such as gas chromatography–mass spectrometry (GC–MS) is required as an analytical technique. The authors go on to describe a simple sample preparation and analysis regime that allowed for the detection and quantification of more than 300 compounds from a single-leaf sample extract.

Of course, in addition to the requirement for an unbiased extraction methodology, it is imperative that all enzymic activity is quenched throughout the extraction and sample preparation. Typically, this is assumed to be the case when extracting in organic solvents, although numerous enzymes may retain activity in such environments. The physical nature of the samples themselves has an impact upon the efficiency of extraction of metabolites. For example, eukaryotic and prokaryotic samples behave very differently during the several steps of classic sample preparation methods. Even within the eukaryotes alone, there is a vast diversity of cellular structures that would seem to make it imprudent to blindly adopt protocols that were devised for other organisms or tissues.

Additional sample preparation is often necessary. In the case of GC–MS, this requires derivatization such that the analytes are volatile; ideally this must allow for maximum applicability across all metabolites, regardless of size, polarity, and functional groups within the metabolite. The most popular derivatization approach is based upon silylation of organic compounds, which are converted to their alkyl–silyl derivatives. Figure 5.1 depicts a simple schematic for the derivatization of glucose. Some disadvantages of this approach are the fact that derivatization must be done under anhydrous conditions, and also that this derivatization requires heat. Numerous alternatives have been proposed (e.g., methyl chloroformate derivatives); but again, in all likelihood, each alternative will suffer its own drawbacks. In short, all researchers acknowledge the shortfalls in these approaches, and perhaps quite rightly place great emphasis on the success of the approach, warts and all.

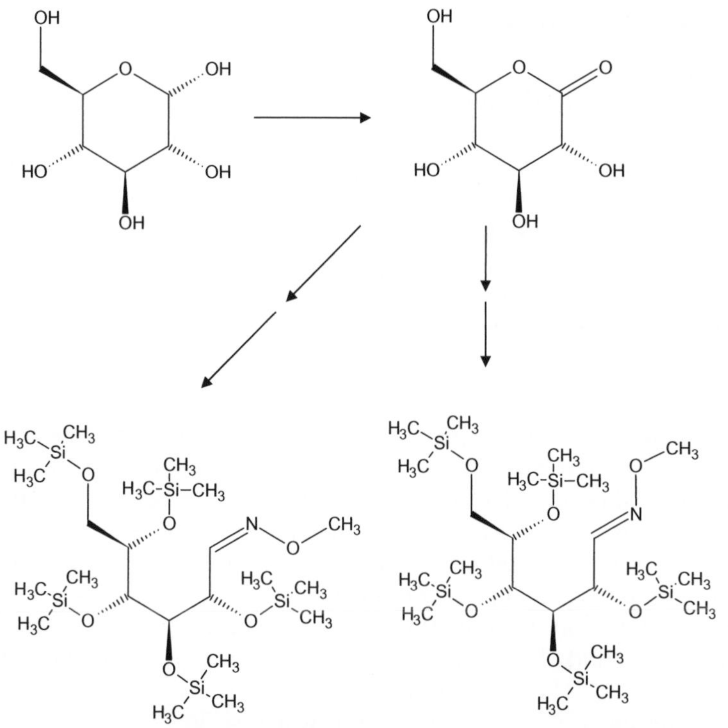

Figure 5.1. Derivatization of glucose for GC–MS analysis. Aldehyde and keto groups are converted into oximes using methoxyamine, followed by conversion of hydroxyl groups into trimethylsilyl (TMS) groups. This example depicts the derivatization of two tautomeric forms, *syn* and *anti*, formed during rotation along the C=N bond.

5.1.3 Pipelined Approaches to Metabolomics

An obvious alternative to the use of single, unbiased methodologies is the construction of a pipelined approach consisting of multiple analyses upon the same sample. When completed, the pipeline could be considered a series of targeted analyses for which both extraction and analytical aspects have been optimized for the analytes in question. Such a pipelined approach might consist of multiple analytical tools, such as gas chromatography–mass spectrometry (GC–MS), liquid chromatography–mass spectrometry (LC–MS), nuclear magnetic resonance spectroscopy (NMR), and a host of other spectroscopic techniques. The development of pipelined approaches for metabolomics remains in its infancy, yet it is appealing because such approaches can also fully utilize advanced features of the analytical tools applied to metabolomics—for example, advanced scan functions of mass spectrometers to lower limits of detection.

5.2 METHODS

As stated above, any number of analytical tools lend themselves to the acquisition of metabolomics data. A brief description of two major tools, mass spectrometry and nuclear magnetic resonance, follow.

5.2.1 Mass Spectrometry

A mass spectrometer creates charged particles (ions) from molecules. It then analyzes those ions to provide information about the molecular weight of the compound and its chemical structure. The information that all spectrometers generate is the ratio of mass to charge (e.g., m/z). In most cases, $z = 1$; and, as such, m/z output equals the true mass of the molecule/fragment ion under investigation. There are many types of mass spectrometers and sample introduction techniques which allow a wide range of analyses. This discussion will focus on mass spectrometry because it is used in the powerful and widely used methods of interfacing gas chromatography with mass spectrometry and also liquid chromatography with mass spectrometry.

5.2.1a   Ionization. All mass spectrometers operate under high vacuum to allow for the free flight of ions from the source of their generation, through a mass selective filter and then on to a detector. As such, the interfacing of a chromatographic front end operating in either a gaseous or liquid phase introduces a problem: namely, the molecules require ionization within the gas/liquid state and then need to be introduced into a high vacuum environment.

In the case of gas chromatography, this is achieved through immediate entry of the eluting compounds into a small chamber (source) that causes ionization by electron impact (EI). The gas molecules exiting the GC are bombarded by a high-energy electron beam (70 eV). An electron which strikes a molecule may impart enough energy to remove another electron from that molecule, generating an ion. Electron impact generates radical ions (i.e., bearing unpaired electrons); these ions themselves are prone to further fragmentation, and, as such, additional structural information is obtained that, in favorable cases, allows for the identification of the ionized substance. An example mass spectrum (of eicosane, an alkane $C_{20}H_{42}$) is depicted in Figure 5.2. The molecular ion (i.e., the whole molecule less an electron)

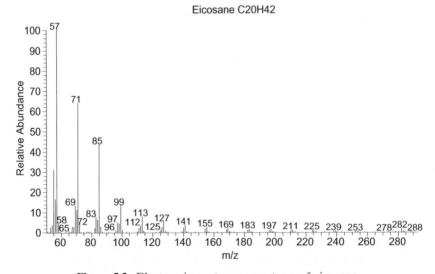

Figure 5.2. Electron impact mass spectrum of eicosane.

is visible at 282; below this mass is a series of fragment ions with Δm of 14, characteristic of —(CH_2)— units.

Interfacing mass spectrometers with liquid chromatography introduces additional constraints, both in instrumental and in spectral terms. The introduction of a liquid into a high-vacuum environment requires very efficient (expensive) pumping systems to maintain vacuum. In addition, ionization is carried out in the presence of solvent, as opposed to in a partial vacuum as is the case in electron impact ionization (above). Ionization in the presence of solvents (and other reagents) does not occur by the simple removal of an electron, but is typically through the gain or loss of protons, or other cations/anions to form adducts—for example, MH^+, MNH_4^+, $[M-H]-$. Such ionization is referred to as chemical ionization (CI), and the most widespread methods for chemical ionizations are electrospray ionization (ESI), atmospheric pressure chemical ionization (APCI), and atmospheric pressure photoionization (APPI). A detailed treatise of each of these ionization methods falls beyond the scope of this short chapter, yet information on all of these techniques is widely available. Perhaps the greatest difference between ions that are generated by EI and CI is that the latter do not contain unpaired electrons and do not undergo fragmentation to the same extent as those generated by electron impact. Additional structural information can then be gained by further fragmentation of the "stable" molecular ion in the high-vacuum region of the mass spectrometer.

5.2.1b *Mass Selection and Ion Detection.* Having generated ions, the mass spectrometer now needs to sort these according to their m/z values and to detect them. Many mass selective devices exist; three of the most common are quadrupoles (Q) (http://www.chm.bris.ac.uk/ms/theory/quad-massspec.html), quadrupole ion traps (qIT) (http://www.abrf.org/ABRFNews/1996/September1996/sep96iontrap.html), and time of flight (TOF) (http://www.rmjordan.com/tt1.html) devices.

5.2.2 Nuclear Magnetic Resonance Spectroscopy

5.2.2a *Theory.* Nuclear magnetic resonance (NMR) spectroscopy represents a powerful and theoretically complex tool. In this short section, we shall only consider the basic theory behind the technique; a more detailed treatise, along with its application to metabolomics, is available.

Subatomic particles (protons, neutrons, and electrons) can all be considered as spinning on their axes. In many atoms (such as ^{12}C) these spins are paired against one another, such that the nucleus of the atom has no overall spin. However, in other nuclei (such as ^1H and ^{13}C) the nucleus does possess an overall spin. The rules for determining the nuclear spin are as follows:

1. If the number of protons and the number of neutrons are both even, then the nucleus has no overall spin.
2. If the number of protons plus the number of neutrons is odd, then the nucleus has a half-integer spin (i.e., $\frac{1}{2}$, $\frac{3}{2}$, $\frac{5}{2}$).
3. If the number of protons and the number of neutrons are both odd, then the nucleus has integer spin (i.e., 1, 2, 3).

Where an overall (or net) spin occurs on a nucleus, I (where $I \neq 0$), the nucleus is amenable to study by NMR. Quantum mechanics tells us that the nucleus of spin I will have $2I + 1$ possible orientations; thus a nucleus with $I = \frac{1}{2}$ will have two possible orientations. In the absence of an electromagnetic field, these two orientations are equivalent, yet when a magnetic field is applied to this system, the two orientations or energy levels are distinct, or separate. Each of these distinct energy levels is given a magnetic quantum number, m (Figure 5.3).

When the nucleus is in a magnetic field, the populations of the energy levels are determined by thermodynamics, according to the Boltzmann distribution. Practically, this means that the lower energy level will contain slightly more nuclei than the higher energy level. By the application of additional electromagnetic radiation, it is possible to excite these nuclei into the higher energy level, and the frequency of the electromagnetic radiation required to excite the nuclei is proportional to the difference in energy between the two levels.

The nucleus (because of protons) has a positive charge and is spinning; this will generate a small magnetic field. The nucleus as such possesses a magnetic moment, μ, which in turn is proportional to the nuclear spin, I:

$$\mu = \frac{\gamma I h}{2\pi}$$

The constant, γ, is called the gyromagnetic ratio (magnetogyric) and is a fundamental constant for each nucleus, h is Planck's constant. The energy of any particular energy level, E, is given by

$$E = \frac{\gamma h}{2\pi} m B$$

Where B is the magnetic field at the nucleus. The difference between energy levels (transition energy) can be derived from

$$\Delta E = \frac{\gamma h B}{2\pi}$$

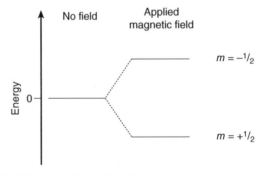

Figure 5.3. The separation of nuclear spin states by a magnetic field.

Consequently, as the magnetic field strength, B, is increased, then so is ΔE. In addition, if a nucleus has a relatively large gyromagnetic ratio, then ΔE is correspondingly large.

5.2.2b   *The NMR Spectrum.*   Simplistically, the NMR spectrum of a molecule is the sum of all of the contributions made by the individual nuclei that possess nonzero spin (^1H, ^{13}C). The spectrum itself comprises a series of peaks, the position and patterns of which are indicative of the molecule under investigation. The position (or chemical shift) of the peak arises due to nuclear shielding. The magnetic field at the nucleus is not equal to that of an applied external field because electrons around the nucleus shield it from the applied field. This difference between the applied and the actual field at the nucleus is termed nuclear shielding.

Consider electrons within an s-orbital. They have a spherical symmetry and circulate in the applied field, producing a magnetic field that opposes the applied field (Figure 5.4). This means that the applied field strength must be increased for the nucleus to absorb at its transition frequency. This upfield shift is also called a diamagnetic shift. As a result, electrons in p-orbitals have no spherical symmetry and produce relatively large magnetic effects at the nucleus, which give rise to a low-field shift. This "deshielding" effect is termed a paramagnetic shift.

The final location of a peak within an NMR spectrum is then a combination of these high and low field shifts and is a function of the nucleus and its environment.

The phenomenon of spin–spin coupling gives rise to multiple peaks in the NMR spectrum from a single nucleus within the molecule under investigation. In the simplistic example depicted in Figure 5.5, the resonant energy of nucleus A is dependent upon the alignment of the nucleus B, and the signal from nucleus A is split into two parts because the spin states of B are approximately equally populated. The size of the splitting (coupling constant, J) is independent of the applied magnetic field and is measured as an absolute frequency in hertz. The number of splittings indicates the number of chemically bonded nuclei in the vicinity of the observed nucleus. There are many good online tutorials and other resources that cover NMR spectral interpretation (http://www.chem.ucla.edu/~webspectra/ and http://www.wfu.edu/~ylwong/chem/nmr/h1/index.html), which will provide the reader with a better grounding in both NMR theory and applications.

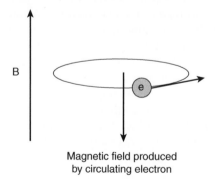

B

Magnetic field produced
by circulating electron

Figure 5.4. An s-orbital electron in an applied magnetic field produces an opposing magnetic field.

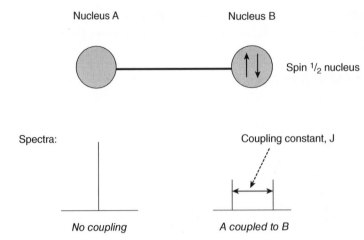

Figure 5.5. Spin–spin coupling arises from the relative alignment of nucleus B to nucleus A.

5.2.3 NMR and Metabolomics

On the face of it, NMR would appear to be an ideal tool for metabolomics; it has great application in the solution of structures, it is amenable to most biological molecules (due to the presence of ^{13}C and ^{1}H nuclei, albeit with only 1% natural abundance in the case of ^{13}C), it is well established, and it is a commonly used technique within the biological sciences. NMR alone, however, might not be the ideal tool, ironically because of its own strengths. An NMR spectrum of a single compound will typically contain multiple peaks (Section 5.2.2); it follows then that an NMR spectrum of a mixture of, say, 100 compounds will contain many hundreds of peaks, and in all likelihood many of these will overlap. In addition, the intensities of the peaks themselves are dependent upon (a) the concentrations of the molecules from which they arise and (b) the extent to which the peak is split as a result of spin–spin coupling. The use of higher-frequency instruments and two-dimensional NMR experiments has facilitated the continued application to metabolomics, and the construction of NMR metabolomics databases (e.g., http://www.liu.se/hu/mdl/main) will undoubtedly aid further enhancements in metabolomics studies.

It also seems reasonable to assume that the hyphenation of a chromatographic step (e.g., HPLC) with NMR would provide the great utility of mass spectrometry when coupled with GC and LC. LC–NMR would be expected to be one of the most powerful and time-saving methods for the separation and structural elucidation of unknown compounds and mixtures. As was the case for coupling liquid chromatography with mass spectrometry, the coupling of liquid chromatography with NMR requires adjustment of both analytical systems. Capillary LC-NMR™ is a product that resulted from a collaboration between Bruker, Waters, and Protasis, who claim that with on-flow detection, solute detection limits of a few tens of nanograms will be achieved.

It seems reasonable that additional hyphenation of mass spectrometry (e.g., LC–NMR–MS) would be an even more powerful approach when considering complex mixtures. Such an LC–NMR–MS setup was first introduced by Bruker

BioSpin in 1999 with an ion trap mass spectrometer, and since 2004 a time-of-flight option has been available (http://www.bruker-biospin.de/NMR/hyphenation/lcnmrms.html).

5.3 DIAGNOSTICS AND FUNCTIONAL GENOMICS

5.3.1 Metabolomics in Diagnostics

For decades, doctors have been measuring a few metabolites to aid in the diagnosis of patients' ailments; glucose in the case of diabetes is an example. Metabolomics researchers, however, search through hundreds of molecules to tease out single compounds or groups of compounds that serve as a signature of a particular disease state.

One company, Phenomenome Discoveries of Saskatoon, Canada, is developing small molecule biomarkers, based upon metabolomics analysis for six types of cancer and four neurological disorders including Alzheimer's disease and autism (http://www.phenomenome.com/biomarker_diagnostic.htm). They claim that their small molecule-based biomarkers have advantages over protein and genetic biomarkers, because the small molecule components are stable and transportable and actually reflect the functional condition of the patient. Their pipelined approach utilizes Fourier transform ion cyclotron resonance (FT–ICR) mass spectrometry as a high-resolution tool to obtain fingerprints of normal and disease state samples. In this step, the sample is infused directly into the mass spectrometer interface. Principal component analysis (PCA) is then employed to "cluster" data according to their origins (normal versus disease). From the PCA data it is then possible to determine which of the individual components of the fingerprint are responsible for the clustering of the control and disease states. Validation then involves the targeted analysis of a subset of the normal and disease state samples by LC-MS/MS. By defining the precursor ions (MH^+) (determined in the initial infusion step), it is possible to define precursor–product ion transitions (Section 5.2.1b) that are specific to the biomarkers determined in the infusion experiment. The inclusion of internal standards for which precursor–product ion transitions are defined allows for relative quantitation of the biomarkers under study.

After this step, they adapt to a true high-throughput screening method that measures each of the internal standards and biomarker precursor–product transitions by direct injection. By eliminating the chromatographic step, it is possible to reduce sample analysis time to one minute per sample, and the biomarker–internal standard ratio is used to determine to which population the analyzed sample belongs, disease state or control.

Metabolon of North Carolina (http://www.metabolon.com/) has initiated a study of Lou Gehrig's Disease (amyotrophic lateral sclerosis, ALS), in collaboration with the Massachusetts General Hospital. Although the analytical tools used were somewhat different from the direct infusion–mass spectrometrye utilized by Phenomenome (above), they identified key biomarkers that were considered generic to motor neuron disorders as well as biomarkers that appeared specific to ALS. The company aims to expand upon this study to include hundreds of patients in a longitudinal study aimed at following disease progression and to then contrast these findings with similar studies into Alzheimer's and Parkinson's disease.

5.3.2   Metabolomics in Functional Genomics

Now that the entire human genome has been sequenced, it is clearly desirable that the function of each gene be determined (annotated), in an effort to understand the control mechanisms built into biological systems. It is also required that the roles that both genotype and environment play in determining phenotype be elucidated. To understand gene function, researchers will need to apply high-throughput technologies to study functional pathways and networks. With enough data, along with the appropriate chemometric tools, it may well prove possible to do this. Metabolomics clearly represents an important tool in this effort.

Metabolites can be viewed as the end products of gene expression and enzyme activity. Thus, they are indicative of both of these processes. Therefore, it follows that metabolomics has the potential to shed light upon the upstream determinant of each of these two processes, namely gene function. One [quite reasonable] assumption in functional biology is that a change in the transcriptome (the entire collection of transcribed elements of the genome) affects the total catalytic activities of enzymes, causing a change in the metabolome.

The actual size of the metabolome is still subject to conjecture. Yeast cells have been estimated to contain around 550 low-molecular-weight components, single plant cells have been estimated to comprise 4000–20,000 distinct components, and to date approximately 200,000 compounds have been identified in the plant kingdom. The Human Metabolome Project seeks to identify, quantify, catalog, and store all metabolites present in human tissues and biofluids. There are currently around 2500 compounds within this database. Does this mean that the human metabolome consists of just 2500 compounds? Probably not. The full aims of this study are limited to those metabolites that are present at concentrations greater than one micromolar, presumably because these are easier to visualize. This cutoff alone will have direct impact on the size of the "finished metabolome." There are also several entries that on first sight don't seem to belong within the human metabolome (ascorbic acid and ibuprofen are not synthesized in humans), yet they fulfill the criterion of being low-molecular-weight substances found in human fluids and tissues. For metabolomics as it relates to functional genomics, such nonendogenous compounds are confounding constituents of the sample; for other applications such as diagnostics and toxicology, they are likely to be of extreme importance and, in the opinion of the author, warrant their place in such databases.

An important difference of metabolites, compared to mRNAs and proteins, is that it is difficult or impossible to establish a direct association between individual metabolites and genes. Functions (or biological context) have been established for many metabolites, or can be inferred from knowledge of the roles in other organisms; however, because a single metabolite can be a member of numerous biochemical pathways and can have modulatory activities over multiple biological processes, metabolites do not lend themselves to unambiguous association with single genes.

5.4   METABOLOMICS AND TOXICOLOGY

Metabolomics as defined in Section 5.1.1 is a recent addition to accepted technology platforms. As such, when researching this chapter, it became obvious that there is very little in the way of information regarding metabolomics that was specific to toxicology. Metabolic profiling, on the other hand (especially of urine or blood plasma), is a well-established procedure to detect physiological changes caused by toxic insult of a chemical (or mixture of chemicals). Metabolic profiling is also routinely used in the screening of metabolic disorders in neonates. In both cases, the observed changes can be related to specific syndromes (e.g., a specific lesion in an organ). This is of particular relevance to pharmaceutical companies wishing to test the toxicity of potential drug candidates: If a compound can be eliminated before it reaches clinical trials on the grounds of adverse toxicity, it saves the enormous expense of the trials.

5.4.1 Attrition Rates of Drugs

Whilst pharmaceutical ventures remain productive, ever-rising expenditures fuel the continuing mergers and business acquisitions of the major pharmaceutical companies. Intrinsic to this is the escalating costs associated with target validation, screening, and clinical trials; yet in spite of these considerably higher costs, there is no resultant decrease in candidate drug attrition. There is thus a critical need to develop and utilize innovative technologies and strategies to identify drug candidates (or their known metabolites) that will likely succumb to attrition later in clinical trials. Outcome and mechanistic biomarkers will also be needed to fully support this effort; and while metabolomics would appear a promising candidate, it will likely be a combination of metabolomics along with hypothesis-driven analysis that will yield the most valuable information.

Strategies to reduce the attrition rate of candidate drugs might be most cost-effective when implemented early in the process of drug discovery. Consider the case of a protein whose enzymic activity is the target for a drug under development. Since substrate–product ratios are a better measure of target enzyme function than an absolute value of either, a single hypothesis-driven (targeted) analysis might lead to a direct measure of the effect of the substance upon the enzyme. This approach alone, however, will not provide any information on the "surrounding metabolic environment" or on the more distant, unexpected changes in metabolism. The immediate metabolic neighbors of both substrate and product lend themselves to *metabolite profiling* (Section 5.1.1), whereby a whole biochemical pathway is characterized. Alternatively, *metabolite fingerprinting* might provide insight into those changes that were unexpected, or more distant from the targeted enzyme.

One laboratory that many credit with sparking interest in NMR-based metabolomics (metabonomics), particularly with respect to toxicology, is at Imperial College in London (http://www.metabometrix.com). The college recently completed a three-year academic–industry consortium project entitled COMET (the Consortium on Metabonomic Toxicology), which resulted in a database of 150 compounds and biofluid fingerprints that can be used to check for toxicity. The second phase of this project, which aims to perform more mechanistic toxicology studies and to supplement the existing database, is already underway. Metabometrix is a spin-out company from the Imperial College group. Their proprietary technology platform generates, classifies, and interprets metabolic information from biological fluids and tissues.

An industry leader in the application of metabolomics in the discovery of biomarkers is Metabolon, based in the Research Triangle Park of North Carolina. Their

proprietary technology platform is amenable to target identification, target validation, lead prioritization, lead optimization, and mechanism of action studies, with
emphasis throughout on toxicity. Their pipelined approach, utilizing several mass
spectrometer-based analyses, allows integration of the different data types acquired,
which is then analyzed with their own proprietary software. Their website (http://
www.metabolon.com/Disease-biomarkers.htm) gives several worked examples of
their technology.

5.4.2   Mode of Action of Toxicants

In much the same way that metabolomics can be used to infer function of unknown
genes, it is likely to be increasingly used in the determination of mode of action of
biologically active substances, be they naturally occurring or synthetic. Early efforts
in this regard have heavily focused upon the discovery of new herbicides with novel
modes of action, this is considered a priority assignment in plant protection research,
see Grossmann (2005) for an example.

Thousands of novel compounds are developed each year. Invariably, their use
and disposal will result in their accumulation in soils, water, air, and biota within
which they may express toxicity. While the mode of action of toxicity of those compounds with structures related to well-characterized toxic compounds is readily
inferred, those with novel structures present a challenge in determining their mode
of action. Creating advanced, robust, and accessible risk assessment tools for these
emerging pollutants is absolutely necessary to convince Federal Agencies and the
public that potential health and environmental risks have been identified. Omics
tools will be developed and optimized in order to understand and predict the
impacts of existing and novel compounds on organisms. These efforts will likely
concentrate on existing sentinel (or indicator) species and will likely encompass all
of the omics technologies. Metabolomics has an obvious and large role ahead in
studies of this type.

SUGGESTED READING

Blain, J. A., Patterson, J. D., Shaw, C. E., and Akhtar, M. W. Study of bound phospholipase
activities of fungal mycelia using an organic solvent system. *Lipids* **11**, 553–560,
1976.

Fan, T. Metabolite profiling by one- and two-dimensional NMR analysis of complex mixtures.
Progr. NMR Spec. **26**, 161–219, 1996.

Fiehn, O., Kopka, J., Altmann, T., Trethewey, R. N., and Willmitzer, L. Metabolite profiling for
plant functional genomics. *Nat. Biotechnol.* **18**, 1157–1161, 2000.

Grossmann, K. What it takes to get a herbicide's mode of action. Physionomics, a classical
approach in a new complexion. *Pest Manage. Sci.* **5**, 423–431, 2005.

Gygi, S. P., Rochon, Y., Franza, B. R., and Aebersold, R. Gene expression: Correlation between
protein and mRNA abundance in yeast. *Mol. Cell. Biol.* **19**, 1720–1730, 1999.

Harrigan, G. G. Metabolomics: A systems contribution to pharmaceutical discovery and drug
development. *Drug Discovery World* **Spring** 39–46, 2006.

Ke, T., and Klibanov, A. M. On enzymatic activity in organic solvents as a function of enzyme
history. *Biotechnol. Bioeng.* **20**, 746–750, 1998.

Lindon, J. C., Keun, H. C., Ebbels, T .M. D., Pearce, J. M. T., Holmes, E., and Nicholson, J. K. The Consortium for Metabonomic Toxicology (COMET): aims, activities and achievements, *Pharmacogenomics* **6**, 691–699, 2005.

Nikiforova, V., Freitag, J., Kempa, S., Adamik, M., Hesse, H., and Hoefgen, R. Transcriptome analysis of sulfur depletion in *Arabidopsis thaliana*: Interlacing of biosynthetic pathways provides response specificity. *Plant J.* **33**, 633–650, 2003.

Schmid, E. F., and Smith, D. A. Is declining innovation in the pharmaceutical industry a myth? *Drug Discovery Today* **1**, 1031–1039, 2005.

Schwab, W. Metabolome diversity: Too few genes, too many metabolites? *Phytochemistry* **62**, 837–849, 2003.

Sweeley, C. C., Bentley, R., Makita, M., and Wells, W. W. Gas–liquid chromatography of trimethylsilyl derivatives Of sugars and related substances. *J. Am. Chem. Soc.* **85**, 2497–2507, 1963.

Viant, M. R. Improved methods for the acquisition and interpretation of NMR metabolomic data. *Biochem. Biophys. Res. Commun.* **310**, 943–948, 2003.

Villas-Bôas, S. G., Højer-Pedersen, J., Åkesson, M., Smedsgaard, J., and Nielsen, J.. Global metabolite analysis of yeast: Evaluation of sample preparation methods. *Yeast* **22**, 1155–1169, 2005.

Villas-Bôas, S. G., Gutierrez Delicado, D., Åkesson, M., and Nielsen, J. Simultaneous analysis of amino and non-amino organic acids using methyl chloroformate derivatives using gas chromatography–mass spectrometry. *Anal. Biochem.* **322**, 134–138, 2003.

Weljie, A. M., Newton, J., Mercier, P., Carlson, E., and Slupsky, C. M. Targeted profiling: Quantitative analysis of ^{1}H NMR metabolomics data. *Anal. Chem.* **78**, 4430–4442, 2006.

CHAPTER 6

Bioinformatics

ERIC A. STONE

Department of Statistics, North Carolina State University, Raleigh, North Carolina 27695

DAHLIA M. NIELSEN

Department of Genetics, North Carolina State University, Raleigh North Carolina 27695

6.1 INTRODUCTION

The field of bioinformatics has been defined in many ways and encompasses a large variety of areas of study. Within this scope, the common bond is genomic data, including its maintenance, analysis, and integration. In this chapter we retain this focus, providing an overview of various concepts of interest to researchers in the field of Toxicology.

This chapter is informally divided into two parts. The first part is devoted to sequence comparisons and culminates in the database search tool BLAST. The second part considers the analysis of population-level data, providing sufficient background to understand how genes influencing traits such as response to toxicity may be located within the genome.

6.1.1 Introduction to Biological Sequence Analysis

We begin by considering a biological sequence of interest, with the goal of obtaining information about its origin, its function, and its interactions. In some instances, this information has already been collected and stored in one of the many biological databases available to the public. When this is the case, as in Section 6.2, the challenge is simply to identify the appropriate record in the relevant database. More commonly, as with novel sequences, there is no direct information available, and in these cases we rely on annotation transfer. This strategy assumes that the properties of an uncharacterized biological sequence can be gleaned from a similar sequence that has already been characterized. The idea that similarity in sequence suggests similarity in function is pervasive in bioinformatics, and sequence similarity plays a central role in most analyses. It is for this reason that we also focus on sequence similarity, dedicating Section 6.3 to its source and quantification. That background leads to an application in Section 6.4, namely the database search tool BLAST. As

Molecular and Biochemical Toxicology, Fourth Edition, edited by Robert C. Smart and Ernest Hodgson
Copyright © 2008 John Wiley & Sons, Inc.

we will demonstrate, the integration of BLAST and cross-database retrieval at NCBI greatly facilitates the characterization of biological sequences.

6.2   OBTAINING THE GENBANK RECORD OF A KNOWN GENE

The National Center for Biotechnology Information (NCBI) web portal (http:// www.ncbi.nlm.nih.gov) is often the first stop in the search for information about a biological sequence of interest. Through the search engine Entrez, the NCBI supports a cross-database search of a variety of information potentially relevant to a query. Principal among the databases included is GenBank, which contains the vast majority of public DNA and protein sequences complete with biological annotation and references to associated publications. As an illustration of a cross-database search incorporating GenBank, suppose we are interested in retrieving the DNA sequence of the human gene encoding the enzyme CYP2E1 (cytochrome P450, family 2, subfamily E, polypeptide 1). Entering "*Homo sapiens* CYP2E1" in the search field (Figure 6.1a) yields, among other things, 63 nucleotide sequences in the GenBank database (Figure 6.1b). Clicking on the heading "Nucleotide," we obtain a list whose nineteenth entry, NM_000773, is the *accession number* of the human gene we seek (Figure 6.1c). NM_000773 links to a record that contains extensive information about *CYP2E1*, including a 1667-bp cDNA sequence spanning the entire coding region of the gene (Figure 6.1d). The GenBank record also serves as a gateway to external databases using the /db_xref qualifier; for instance, /db_ xref="MIM:124040" links to the entry for CYP2E1 in Online Mendelian Inheritance in Man. For further details on the contents and nomenclature of a GenBank record, view the annotated example provided by the NCBI at http://www.ncbi.nlm. nih.gov/Sitemap/samplerecord.html.

   While the name of a gene or its protein product is usually sufficient to locate its GenBank record, this may first require combing through an extensive list of spurious entries that also match the search criteria. Above, the query "*Homo sapiens* CYP2E1" yielded 63 nucleotide sequences, but our target was only nineteenth on the list. When an accession number is a priori available (as when obtained from a publication), these potential complications can be avoided; searching for NM_000773 returns exactly one GenBank record, that of the human gene encoding the enzyme CYP2E1.

   Thus far we have considered using a gene's name or accession number to retrieve its GenBank record and associated nucleotide sequence. Suppose now we wish to do the reverse and obtain the name and accession number of the gene associated with a particular nucleotide sequence. To put this in terms of our example, can we use the nucleotide sequence of *CYP2E1* as a query and obtain the GenBank record with the accession number NM_000773? While conceptually similar to our previous search strategies, here new challenges arise because in practice the query sequence is unlikely to exactly match the gene sequence stored in the database. Inexact matching may occur for a number of reasons: The query sequence may differ from the recorded sequence due to polymorphism or sequencing errors, or the two sequences may differ in length because one is simply a fragment of the other (e.g., because gene boundaries are nebulously defined). The problem of comparing two sequences that are similar but not identical is fundamental to bioinformatics and

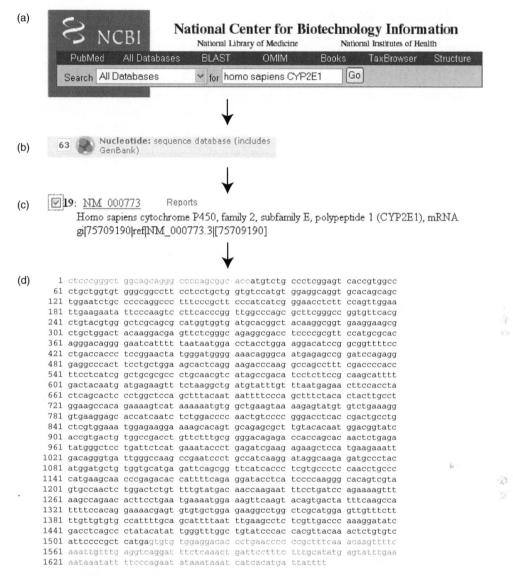

Figure 6.1. Using NCBI. Using the National Center for Biotechnology Information (NCBI) web portal. (a) On the NCBI homepage, the query "*Homo sapiens* CYP2E1" is entered into the search field. (b) The search yields 63 nucleotide sequences. (c) Each of the 63 sequences is referenced by an accession number and a brief description. The nineteenth entry, NM_000773, is the accession number of the human gene we seek. (d) A portion of the GenBank record of NM_000773 containing the cDNA sequence of the gene. The coding sequence (bps 34 to 1515) is shown in black, while the flanking region is shown in gray.

dominates the first half of this chapter. This background enables a discussion of the database search tool BLAST, whose application will solve our "reverse lookup" problem by obtaining the GenBank record(s) most relevant to a nucleotide sequence query.

6.3   SEQUENCE COMPARISON

The previous section established one rationale for comparing two similar DNA (or protein) sequences. In fact, however, the utility of sequence comparison is extensive and rests on the observation that sufficiently similar sequences often share a consistent biological function. A relevant example is the DNA sequence from rhesus monkey with accession number XR_011473: computational methods predict XR_011473 to be a protein-coding gene, while similarity to NM_000773 suggests that the encoded protein may be the rhesus monkey equivalent of the enzyme CYP2E1. This section aims to elucidate the details of a comparison between two sequences such as XR_011473 and NM_000773. Our discussion begins at the source of sequence similarity.

6.3.1   Homology

Why are there genes in both humans and rhesus monkeys which encode the enzyme CYP2E1? One compelling possibility is that a gene encoding the enzyme CYP2E1 was passed on from an ancestral primate species to both humans and rhesus monkeys. In this case, the human and rhesus monkey sequences are said to be *orthologous* to one another (or simply "are orthologs") because they are related by speciation from a common ancestor. More broadly, common ancestry implies that the two sequences are *homologous* to one another; however, in general a pair of homologs need not be orthologs. *Paralogs* are homologs that are related to each other by ancestral gene duplication rather than by speciation. Thus, loosely speaking, paralogs are related sequences in the same species, while orthologs are the same sequence in related species (see Figure 6.2). See Sonnehammer and Koonin (2002) in Suggested Reading for a more precise account of homology and its subtypes.

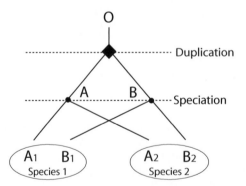

Figure 6.2. Homology, orthology, and paralogy. Two extant species, labeled 1 and 2, are related by speciation from a common ancestor whose genome encodes gene A and gene B. Each extant species retains a copy of gene A and a copy of gene B, with A_1 and B_1 private to species 1 and A_2 and B_2 private to species 2. The genes A and B are themselves related by an ancestral duplication of the gene O. Because each of the genes A_1, B_1, A_2, and B_2 is derived from the ancestral gene O, they are said to be homologous to one another. Additionally, A_1 and A_2 (also B_1 and B_2) are orthologous to one another because they are derived from gene A by ancestral speciation. By contrast, A_1 and B_1 (also A_2 and B_2) are paralogous to one another because they are derived from gene O by ancestral duplication.

For our introduction to sequence comparison, it is enough to recognize that as a result of common ancestry, homologous sequences will tend to be similar. Moreover, as we will elaborate upon in the next section, the extent of similarity between homologs tends to diminish over time. Because other sources of similarity (e.g., convergent evolution) are rare, similar sequences often turn out to be homologous, and when the similarity between two sequences is striking (as with XR_011473 and NM_ 000773), homology is the probable explanation. Though an abuse of language, it is for this reason that "similarity search" and "homology search" are often used interchangeably. More correctly, however, similarity is the symptom while homology is its underlying cause. Though we observe similarity, it is homology which can be directly modeled from our knowledge of how biological sequences change over time. Thus, it is ultimately the evolution of DNA and protein sequences which supplies the framework for sequence comparison.

6.3.2 Pairwise Sequence Alignment

If we suppose that a gene encoding the enzyme CYP2E1 was passed on from an ancestral primate species to both humans and rhesus monkeys, why is it that the orthologous sequences XR_011473 and NM_ 000773 are not identical? The answer is evolution: From the ancestral speciation event forward, the *CYP2E1* gene sequences in both lineages have been independently accumulating changes through the processes of mutation, genetic drift, and natural selection. The majority of these changes, called *substitutions*, involve one nucleotide or amino acid (collectively, *residues*) being replaced by another. The changes generating length variation are *insertions* and *deletions*, which respectively modify the ancestral sequence by the addition or removal of residues. Though unobservable, some past series of evolutionary changes was responsible for the differences we now observe in the human and rhesus monkey orthologs. In particular, the relationship between these two sequences can be explained by some historical pattern of substitutions, insertions, and deletions.*

A *pairwise sequence alignment* (PSA) is an arrangement of two sequences that emphasizes the pairwise relationships between their respective residues. Two residues, one from each sequence, are *aligned* in a PSA to highlight their similarity to one another. More precisely, from an evolutionary perspective, two residues are aligned to indicate that both have evolved from the same residue in an ancestral sequence (that is, to indicate that the residues are homologous). To clarify this, consider two sequences X and Y derived from an ancestral sequence A (Figure 6.3a). In the absence of insertions and deletions along their evolutionary paths, X and Y will have the same length as A and differ only by the replacement of ancestral residues via substitution. Therefore, the first residues of sequences X and Y (which we will call X_1 and Y_1, respectively) must have derived from first residue of sequence A, and in a PSA the residues X_1 and Y_1 should be aligned. Similarly, X_2 and Y_2 should be aligned, and so on, across the lengths of the entire sequences, so that in fact the PSA can be obtained by simply writing the sequence X directly above the sequence Y (Figure 6.3b).

*This discussion discounts the possibility of more complicated evolutionary changes such as inversions. Fortunately, such occurrences appear to be infrequent relative to the intensity of substitutions, insertions, and deletions.

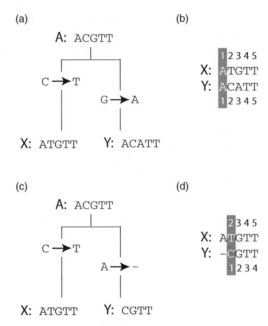

Figure 6.3. Alignments and ancestry. From an evolutionary perspective, alignments are a statement about ancestry. (a) The ancestral sequence A evolves on two separate lineages into X (left) and Y (right). On the lineage leading to X, the second nucleotide has changed from C to T; while on the lineage leading to Y, the third nucleotide has changed from G to A. (b) An alignment consistent with the evolutionary history in (a). Pairs are aligned to indicate that they have evolved from a common ancestral nucleotide. Above each residue in sequence X is its position in the sequence; the equivalent information for Y is shown below. (c) As in (a), however, this time the first nucleotide of A has been deleted on the lineage leading to Y. (d) As in (b), except now the first nucleotide of X must be paired to a gap.

Consider now a minor change to the above scenario. Suppose instead that, subsequent to ancestral speciation, the first residue of sequence A (which we call A_1) was deleted along the lineage leading to sequence Y (Figure 6.3c). In this case, X_1 and Y_1 no longer trace their history to the same ancestral residue, because X_1 is derived from A_1 while Y_1 is derived from A_2. Because X_2 is also derived from A_2, it is clear that X_2 and Y_1 should be aligned; however, X_1 has no complementary residue in sequence Y. This problem is solved by the gap character "-": By pairing X_1 to a gap, the entirety of both sequences can be correctly aligned, with X_2 aligned to Y_1, with X_3 aligned to Y_2, and so on. Put another way, by appending a gap to the beginning of sequence Y, the PSA can once again be obtained by writing the sequence X directly above the sequence Y (Figure 6.3d). This strategy turns out to be quite general: Any pattern of substitutions, insertions, and deletions can be represented in a PSA by the appropriate addition of gap characters. In fact, some texts define the construction of a pairwise sequence alignment as adding gap characters to two sequences so that they have the same length.

We began this section by asking why the human and rhesus monkey genes encoding the enzyme CYP2E1 are not identical. In answering that question, we have also acquired a means to highlight their similarity. Unfortunately, our discussion of align-

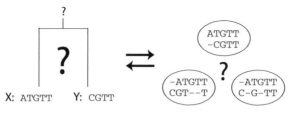

Figure 6.4. Alignments are hypotheses about the unobservable. Each alignment represents a competing hypothesis. When the evolutionary history is unknown (left), so too is the correct alignment (right).

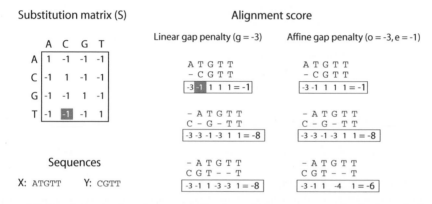

Figure 6.5. Scoring pairwise alignments. Scoring schemes are comprised of a substitution matrix (S) and gap penalty. Here we consider a pairwise comparison of nucleotide sequences, and the matrix S scores +1 for a match and −1 for a mismatch (left). Three alignments of X and Y are shown, and each is scored using both a linear and an affine gap penalty. The score of each residue pair is shown beneath it, and these are summed to produce the alignment score. Note that the affine gap penalty scores neighboring gaps as −3 and −1; the ordering is not determined, but the end result is their sum −4.

ment relied heavily on the unobservable: the series of evolutionary changes which led to the differences in the human and rhesus monkey orthologs. In practice, without that knowledge, how do we construct a PSA of the two CYP2E1 genes? As the alternate definition of PSA suggests, there are as many possible alignments as there are ways to place gaps into to the two sequences so that they have equal length. Each potential alignment can be viewed as a hypothesis of which pairs of residues across the two sequences are related; choosing the best alignment among these requires a method of evaluating these competing hypotheses (Figure 6.4).

6.3.3 Evaluating Sequence Alignments

Pairwise sequence alignments are typically evaluated through their constituent residue pairs. In this approach, each individual pair of residues is assigned a *score*, and the PSA is evaluated by summing the scores of all pairs (Figure 6.5). Because each position in the PSA is considered separately, every occurrence of the residue pair (a, b) receives the same score. Thus, for each possible (a, b) only one score

$s(a, b)$ is required, and it is common practice to tabulate the $s(a, b)$ in a *substitution matrix* S. The meaning of "substitution" here is consistent with its use in the previous section and again refers to the replacement of one residue by another; however, as we have already seen, insertions and deletions may occur and result in the alignment of a residue to a gap. The most basic approach to scoring the alignment of a residue to a gap is to assign the pair a *gap penalty g* (or sometimes $-g$) that reduces the overall score of the PSA by g. In other words, just as (a, b) was assigned the score $s(a, b)$, the pair $(a, -)$ is given the score $-g$. This scheme implicitly assumes a *linear gap penalty* that treats multiple-base insertions and deletions harshly, because gaps of length n will reduce the score of the PSA by the product $g \times n$. Thus, while appealing in its simplicity, proposing a linear gap penalty makes it difficult to align sequences with indels of even moderate length. In practice, it is more common to use an *affine gap penalty* that differentiates between the opening and the extension of a gap: If o and e are the gap opening and gap extension penalties, respectively, then a gap of length n will reduce the score of the PSA by $o + e \times (n - 1)$. The affine gap penalty blends computational tractability with a semblance of biological reality, and this has made it the method of choice for database search applications such as NCBI-BLAST.

A scoring scheme is defined by the combination of a gap treatment and a substitution matrix. Thus, the problem of pairwise sequence alignment can be phrased as follows: "Among all possible pairwise sequence alignments, find the one with the highest score relative to the choice of scoring scheme." This makes it clear that pairwise sequence alignment is fundamentally an optimization problem and that the optimal PSA is (a) not necessarily unique and (b) dependent on the scoring scheme. When the optimal alignment between two sequences is unique, it is this PSA which gets reported as their sequence alignment. Therefore, solving the alignment problem requires identification among all possibilities of the PSA which scores highest.

6.3.4 Finding the Optimal Alignment

Having outlined a method for evaluating alignments, finding the optimal alignment between two sequences is conceptually just a matter of scoring all possible PSAs between them. In practice, however, this exhaustive search strategy is prohibitive because the number of possible alignments between two sequences is exponential in their lengths. Fortunately, an alternate search strategy based on the principle of dynamic programming (DP) can ascertain the optimal alignment much more efficiently. The DP algorithm known as Needleman–Wunsch exploits a substructure property that guarantees that the optimal PSA between two sequences contains a succession of optimal alignments of their substrings. To identify the optimal PSA, Needleman–Wunsch applies this logic in reverse by first aligning substrings to build the optimal PSA in stages. This reduces the computational complexity of the optimization from exponential to quadratic, making pairwise alignment feasible for sequences of most any length. In sum, given a pair of biological sequences, it is tractable to identify their optimal PSA with respect to any scoring scheme of the type we have considered. Our discussion now moves to a second version of the pairwise alignment problem, called local alignment, to contrast the global procedure we have already described.

6.3.5 Local and Global Alignment

Thus far, our discussion of PSA has been implicitly restricted to the problem of *global alignment*. Section 6.3.2 defined a PSA to be an arrangement of two sequences that emphasizes the pairwise relationships between their respective residues, and a global alignment is one that compares the two sequences across their entire lengths. Global alignment is thus applicable when two sequences are expected to share similarity across their entire lengths, but even closely related homologs need not rise to meet that expectation. To illustrate the limitations of global PSA, consider the case of two multidomain proteins that share one domain but not the remainder; in attempting to force the entire lengths of their amino acid sequences to align, global alignment may underestimate or miss entirely the similarity presented by the shared domain. Such examples of regional similarity are endemic among related biological sequences and motivate the need for a regional alignment tool. The solution, *local alignment*, attempts to (globally) align only the most similar regions of a sequence pair.

Perhaps counterintuitively, local alignment is a generalization of global alignment. Whereas the goal of global alignment is to identify the highest-scoring PSA between two sequences X and Y, local alignment seeks to identify the highest-scoring PSA between any *substrings* of the sequences X and Y. Thus, while all global alignments are local, only a fraction of local alignments are global. Figure 6.6 shows both a local alignment and a global alignment for the two sequences X and Y first introduced in Figure 6.3c. The local alignment pairs only the "GTT" substrings common to both sequences, thereby obtaining a score of 3 under the scheme of Figure 6.5. The global alignment also pairs the "GTT" common to both sequences; but in contrast to the local PSA, the global PSA must also account for X_1, X_2, and Y_1 (A, T, and C in the figure, respectively). Because X_1 and X_2 cannot be paired to Y_1 here in any meaningful way, the score of any global alignment suffers; in this case, the local alignment is arguably more successful at emphasizing the pairwise relationships between the residues of X and Y.

Above, we saw that number of possible global alignments between two sequences made an exhaustive search for their optimal PSA prohibitive. Since global alignments make up only a subset of local alignments, it is clear that for local alignments the same holds true. Fortunately, just as the Needleman–Wunsch algorithm uses DP to produce a quadratic-time solution to the global alignment problem, the Smith–

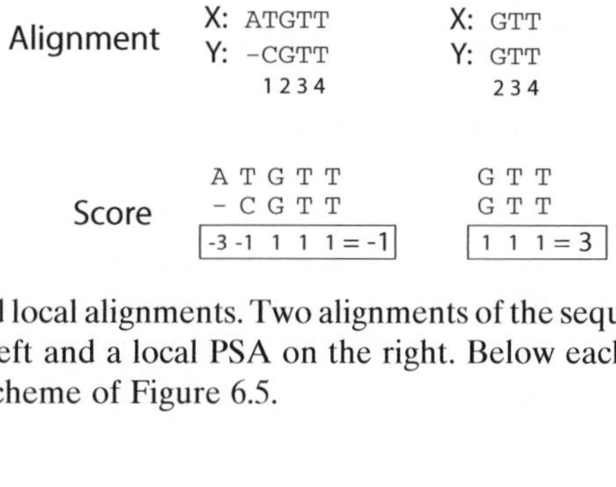

Figure 6.6. Global and local alignments. Two alignments of the sequences X and Y are shown: a global PSA on the left and a local PSA on the right. Below each alignment is its score as calculated using the scheme of Figure 6.5.

Waterman algorithm uses DP to produce a quadratic-time solution to the local alignment problem. Thus, in most cases, both approaches are feasible, and the choice of alignment strategy is driven by the purpose of the pairwise comparison.

6.3.6 Alignments for Sequence Comparison

Though differently motivated, both the local and global approaches to alignment produce PSAs whose scores can be used to quantify sequence similarity and infer pairwise homology. Thus, choosing between local and global alignment depends largely upon which philosophy is better suited to the problem at hand. For example, if the goal is to reconstruct evolutionary history, global alignment should be used; however, local alignment is more sensitive in determining whether or not two sequences share common ancestry at all. To further illustrate the uses of local and global alignment, we return to the human gene encoding the cytochrome P450 enzyme CYP2E1. As a means of obtaining additional cytochromes, suppose that we seek related genes among a database of many sequences from a variety of organisms. A straightforward approach is to search the database for sequences which have high-scoring alignments to CYP2E1, but there are a number of practical issues to consider. Even before deciding which alignment strategy is best, we must first choose whether similarity should be assessed at the nucleotide or amino acid level. Almost invariably, it is better to use the protein than the gene which encodes it, and so we will seek similarity to the amino acid translation of *CYP2E1*. Next, we must decide which philosophy to pursue, and in this case the sensitivity of local alignment is desirable; moreover, in contrast to global alignment, local alignment to CYP2E1 may reveal potential cytochrome sequences despite substantially different length. Having chosen to scan the database for high-scoring local alignments to the protein, we now turn to the results of our search: in particular, how can we recognize other members of cytochrome P450, family 2? Toward this end, global alignment may be preferable, because we expect family members to share similarity across their lengths. In fact, cytochrome P450 families are defined in these terms and group sequences which share at least 40% amino acid identity across their lengths.

Lost in the above discussion is an unfortunate truth: Even with DP, it takes too long to rigorously align a query sequence to each of the member sequences of a database. Because of this, database search tools seek approximations to the optimal local alinment, negotiating a balance between quality and speed with each comparison. This tradeoff and its implications are central to forthcoming discussion of the database search tool BLAST hosted at NCBI.

6.4 DATABASE SEARCHING WITH BLAST

In Section 6.2, we asked whether the nucleotide sequence of CYP2E1 could be used as a query to obtain the GenBank record with the accession number NM_000773. Accomplishing this required a means of comparing two sequences, and Section 6.3 discussed sequence comparison at length. In this section, we will use the concepts of Section 6.3 to search for NM_000773 and its homologs by aligning the nucleotide sequence of CYP2E1 to each sequence in the GenBank database. Not surprisingly, the size of the database necessitates the use of fast, heuristic methods that may produce suboptimal alignments. The most popular such methods, FASTA and BLAST, attempt to find high-scoring local alignments between a query sequence (such as the nucleotide sequence of CYP2E1) and the sequences in a database (such as GenBank). In this chapter we will focus on BLAST, the database search tool supported by the NCBI. We begin by describing the BLAST algorithm in detail.

6.4.1 The BLAST Algorithm

BLAST, short for "Basic Local Alignment Search Tool," seeks high-scoring local alignments between a query sequence and members of a sequence database. The BLAST algorithm begins by reducing the query sequence to a list of *words*, which in this context are sequence substrings of an arbitrary fixed length W (Figure 6.7a). Using a prespecified substitution matrix, for each word of the query, a list of potential high-scoring alignments is compiled; this is accomplished by enumerating the set of all length-W sequences which would receive a high score when aligned without gaps to the query word (Figure 6.7b). A high score is one that exceeds the *neighborhood word score threshold* T; and as with the word length W, T is arbitrary and can be adjusted. Thus, the performance of BLAST is minimally dependent upon the initial choices of W, T, and scoring scheme.

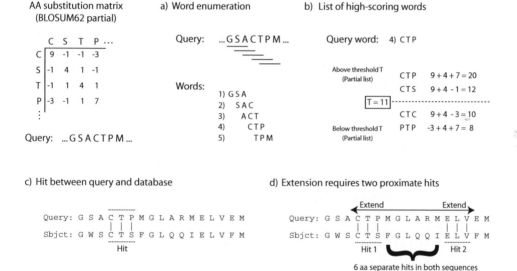

Figure 6.7. Schematic of protein–protein BLAST. BLAST algorithm for a protein query using the substitution matrix BLOSUM62. Only a fragment of the query and a portion of the matrix are shown. (a) The query is parsed into overlapping words of length $W = 3$. (b) A list of high-scoring words is generated for each word in the query. Here, the fourth word of the query fragment, CTP, is used to generate a list of words (including CTP and CTS) which score more than the threshold value of $T = 11$ when aligned to CTP without gaps. (c) A sequence in the database (the subject, abbreviated "Sbjct") contains the high-scoring word CTS, whose alignment to the word CTP in the query ("Query") forms a hit. (d) Two proximate, in-phase hits are found between the subject and the query. The hits are close enough (by default, less than $A = 40$ residues apart) that they combine to trigger an extension.

The next phase of BLAST begins with the identification of all high-scoring alignments between words of the query and words of the database. This is done, for each word of the query, by scanning the database for exact matches to its compiled list of high-scoring words. Once identified, these aligned word pairs, called *hits*, can then be used as seeds from which longer local alignments are grown (Figure 6.7c); however, only a fraction of all hits will proceed to initiate local alignments. The current implementation of NCBI-BLAST requires that two nonoverlapping hits be in close proximity to one another (within an arbitrary distance A), with the further condition that the component words in both sequences be separated by an equal number of residues (Figure 6.7d). This latter requirement limits spurious matching by ensuring that the two hits are "in phase." Upon their discovery, two proximate, in-phase hits are joined to create one ungapped local alignment. This alignment is then extended, one residue at a time, in both directions until the resulting score is sufficiently below the maximum score thus far obtained. The local alignment resulting from extension is called a *high-scoring segment pair* (HSP). BLAST retains for output only those HSPs scoring above a specified cutoff value S.

The cutoff value S is related to the E-value, which quantifies the expected number of high-scoring local alignments between the query and unrelated sequences in the database. In other words, the E-value represents the number of HSPs expected to score S or above by chance. Clearly, the larger the database, the greater the potential for spurious high-scoring local alignments; therefore, it is intuitive that the size of the database factors into the relationship between S and E. Longer sequences also increase the potential for random alignment; accordingly, both the length of the query and the average length of sequences in the database also play a role. To quantify the effect of these and other influences on the relationship between S and E requires a specific null model of the scores of HSPs between unrelated sequences. We explore this model in the ensuing section.

6.4.2 BLAST Statistics

Above we learned that BLAST outputs a list of high-scoring local alignments between a query sequence and the member sequences of a database. This list is ordered by decreasing alignment score, or equivalently by increasing E-value, to prioritize the database sequences according to their similarity to the query. The relationship between an observed alignment score and the number of scores of random alignments expected to exceed it is modeled by the Gumbel extreme value distribution (EVD). Just as the normal distribution is characterized by its mean and standard deviation, the EVD is characterized by two parameters that determine its location and scale. Choosing the parameter values appropriate to a BLAST search depends on the features of the search itself.

The primary features of a BLAST search are the query sequence and the database to which it is compared. The number of chance high-scoring local alignments depends on the number of local alignments considered, and this in turn depends on the product of the length of the query (m) and the length of the database (N, the sum of member sequence lengths). The distribution of chance alignment scores also depends on the scoring scheme and the residue composition of the sequences in the database, both of which can be accounted for in the parameters K and λ of the EVD. The E-value corresponding to an observed alignment score x is a function of m, N,

K, and λ. The functional form is provided by the EVD: $E = 1 - \exp(-KmNe^{-\lambda x})$, which is in practice corrected by BLAST to account for gaps and edge effects.

Because of the many factors on which BLAST scores and statistics depend, the results of disparate searches may not be directly compared. To standardize across scoring schemes and variable database composition, the formula $B = (\lambda x - \ln K)/(\ln 2)$ is used to convert an alignment score x into a *bit score* B. This conversion effectively removes K and λ from subsequent calculations, standardizing all BLAST searches to reference the same EVD. It is the bit score which NCBI-BLAST includes along with the E-value of each reported high-scoring local alignment. We demonstrate in the next section with an example.

6.4.3 A BLAST Example

In Section 6.3.6, we speculated on the use of CYP2E1 as a query to search for potential cytochrome P450 homologs. Now, equipped with BLAST, we can proceed to search the nonredundant (nr) database of GenBank proteins for homologs of the query protein NP_000764. Because here we are searching a protein database with a protein query, the appropriate tool is protein–protein BLAST (blastp); however, the NCBI supports tools for any nucleotide/protein combination of database and query (see http://www.ncbi.nlm.nih.gov/blast/producttable.shtml). Each tool supported by NCBI, including blastp, has default settings that the user can change: For blastp, the default scoring scheme uses the substitution matrix BLOSUM62, with gap opening and gap extension penalties of 11 and 1, respectively. The default word size for blastp is $W = 3$, and the neighborhood word score threshold is set to $T = 11$ (see Figure 6.7). These settings were chosen empirically to balance speed and sensitivity.

Using the cutoff S which corresponds to an E-value of 10, BLAST identifies 634 sequences, the first 100 of which are shown by default. Not surprisingly, the first sequence reported (and thus the highest-scoring) is the query protein NP_000764 itself, and many of the top results are annotated as CYP2E1 primate orthologs. Among these is the protein sequence NP_001035303 from the rhesus monkey, provisionally annotated as "cytochrome P450, family 2, subfamily E, polypeptide 1 [*Macaca mulatta*]". Though there does not appear to have been experimental validation, the BLAST evidence supporting this annotation is compelling: The human and rhesus monkey sequences show similarity across their entire lengths, and their amino acid sequences are 94% identical. The alignment scores 2382 using BLOSUM62, which standardizes in this case to 922 bits; the corresponding E-value is too small to calculate, and it is reported by BLAST as Expect = 0.0 (Figure 6.8).

Given the size of the superfamily of cytochrome P450 enzymes in humans, it may be surprising to find that the BLAST results are dominated by proteins from other species. In fact, however, this is to be expected, and part of the explanation lies in Figure 6.2. The diversity seen among human cytochrome P450 enzymes originates from a remarkable number of ancestral gene duplications, most (if not all) of which occurred prior to the mammalian radiation. Thus, in many cases, the common ancestor of two human cytochrome P450s will be a gene whose duplication occurred prior to a vast number of relevant speciations. As a result, we expect that many CYP2E1 orthologs will be more similar to the human sequence than will be any human

```
>gi|94158996|ref|NP_001035303.1|  cytochrome P450, family 2, subfamily E,
polypeptide 1 [Macaca mulatta]
 gi|55976407|sp|Q6GUQ4|CP2E1_MACMU  Cytochrome P450 2E1 (CYPIIE1)
 gi|49066335|gb|AAT49269.1|  cytochrome P450 CYP2E1 [Macaca mulatta]
Length=493

 Score =  922 bits (2382),  Expect = 0.0, Method: Composition-based stats.
 Identities = 467/493 (94%), Positives = 485/493 (98%), Gaps = 0/493 (0%)

Query  1     MSALGVTvallvwaaflllvSMWRQVHSSWnlppgpfplpiignLFQLELKNIPKSFTRL   60
             MSALGV+VALLVW A LLLVS+WRQVHSSWNLPPGPFPLPIIGNLFQLELKNIPKSFTRL
Sbjct  1     MSALGVSVALLVWVAVLLLVSIWRQVHSSWNLPPGPFPLPIIGNLFQLELKNIPKSFTRL   60

Query  61    AQRFGPVFTLYVGSQRMVVMHGYKAVKEALLDYKDEFSGRGDLPAFHAHRDRGIIFNNGP   120
             AQRFGPVFTLYVGS+R+VV+HGYKAV+E LLD+KDEFSGRGD+PAFHAHRDRGIIFNNGP
Sbjct  61    AQRFGPVFTLYVGSRRVVVVHGYKAVREVLLDHKDEFSGRGDIPAFHAHRDRGIIFNNGP   120

Query  121   TWKDIRRFSLTTLRNYGMGKQGNESRIQREAHFLLEALRKTQGQPFDPTFLIGCAPCNVI   180
             TWKDIRRFSLTTLRNYGMGKQGNESRIQREAHFLLEALRKTQGQPFDPTFLIGCAPCNVI
Sbjct  121   TWKDIRRFSLTTLRNYGMGKQGNESRIQREAHFLLEALRKTQGQPFDPTFLIGCAPCNVI   180

Query  181   ADILFRKHFDYNDEKFLRLMYLFNENFHLLSTPWLQLYNNFPSFLHYLPGSHRKVIKNVA   240
             ADILFRKHFDYNDEKFLRLMYLFNENF LLSTPWLQLYNNFPS LHYLPGSHRKV+KNVA
Sbjct  181   ADILFRKHFDYNDEKFLRLMYLFNENFQLLSTPWLQLYNNFPSLLHYLPGSHRKVMKNVA   240

Query  241   EVKEYVSERVKEHHQSLDPNCPRDLTDCLLVEMEKEKHSAERLYTMDGITVTVADLFFAG   300
             E+KEYVSERVKEH QSLDPNCPRDLTDCLLVEMEKEKHSAERLYTMDGITVTVADLFFAG
Sbjct  241   EIKEYVSERVKEHLQSLDPNCPRDLTDCLLVEMEKEKHSAERLYTMDGITVTVADLFFAG   300

Query  301   TETTSTTLRYGLLILMKYPEIEEKLHEEIDRVIGPSRIPAIKDRQEMPYMDAVVHEIQRF   360
             TETTSTTLRYGLLILMKYPEIEEKLHEEIDRVIGPSRIPAIKDRQEMPYMDAVVHEIQRF
Sbjct  301   TETTSTTLRYGLLILMKYPEIEEKLHEEIDRVIGPSRIPAIKDRQEMPYMDAVVHEIQRF   360

Query  361   ITLVPSNLPHEATRDTIFRGYLIPKGTVVVPTLDSVLYDNQEFPDPEKFKPEHFLNENGK   420
             ITLVPSNLPHEATRDTIFRGY+IPKGTV+VPTLDSVLYDNQEFPDPEKFKPEHFL+E+GK
Sbjct  361   ITLVPSNLPHEATRDTIFRGYIIPKGTVIVPTLDSVLYDNQEFPDPEKFKPEHFLDESGK   420

Query  421   FKYSDYFKPFSTGKRVCAGEGLARMELFLLLCAILQHFNLKPLVDPKDIDLSPIHIGFGC   480
             FKYSDYFKPFS GKRVCAGEGLARMELFLLL AILQHFNLKPLVDPKDID+SP++IGFGC
Sbjct  421   FKYSDYFKPFSAGKRVCAGEGLARMELFLLLSAILQHFNLKPLVDPKDIDISPVNIGFGC   480

Query  481   IPPRYKLCVIPRS   493
             IPPR+KLCVIPRS
Sbjct  481   IPPRFKLCVIPRS   493
```

Figure 6.8. A high-scoring alignment identified by BLAST. Searching with the query NP_ 000764, BLAST finds a high-scoring alignment between the human ("Query") and rhesus monkey ("Sbjct") orthologs of CYP2E1. The alignment spans amino acid positions 1 through 493 and scores 922 bits. The corresponding *E*-value is too small to be calculated and is reported as "Expect = 0.0."

CYP2E1 paralogs (see Figure 6.2); both time and functional divergence have worked to drive the human sequences apart.

Suppose now, in our database search, we wish to recover only the human sequences similar to CYP2E1. The obvious problem is that this requires parsing through the higher-scoring sequences from other species; a more subtle issue is that the comparisons made between the human query and the sequences from other species count against the significance (*E*-value) of the human–human comparisons through the database size. Fortunately, the NCBI makes it straightforward to limit the search to human sequences, and much of the redundancy (e.g., sequences with multiple accession numbers) can be removed by changing the database from "nr" to "refseq". This search reveals over 60 distinct human proteins with high-scoring alignments to the query sequence NP_000764, and nearly all of these are annotated as members of the cytochrome P450 superfamily. The highest-scoring sequences belong to cytochrome P450, family 2, with members of subfamily C showing the most similarity to our subfamily E query. Following family 2 comes family 1, and so on, as the human cytochrome sequences are ranked by their similarity to CYP2E1, and for many of the lower-scoring sequences it is clear that this similarity is only regional. This exemplifies why local alignment is superior to global alignment for database searching: Even when distantly related, P450s recognize other P450s by means of their cytochrome domains.

6.5 EXTENSIONS TO MULTIPLE SEQUENCES

The discussion thus far has concentrated on how the properties of a known biological sequence can be used to elucidate the properties of a similar sequence not yet characterized. A natural extension of this discussion is to consider how knowledge can be transferred by establishing similarity to a family of characterized sequences. Not surprisingly, the inclusion of multiple sequences leads to more powerful inference about the biological role of the query. The essential bioinformatics challenge is also predictable, and concerns how similarity to a family of sequences can be quantified. Just as pairwise similarity is quantified through pairwise local and global alignments, comparisons to sequence families proceed analogously by means of multiple sequence alignment (MSA). What is commonly called "multiple sequence alignment" is in fact the global version of the problem; and just as in the pairwise case, optimality can be quantified. In this case, however, the optimal alignment can rarely be obtained, because even the DP implementation is too slow to be practical. Thus, like BLAST, the competing methods of global MSA available are heuristics that seek a compromise between speed and alignment quality. A good review of these methods is given in Edgar and Batzoglou (2006) in Suggested Reading.

The local MSA problem is most often encountered in the attempt to characterize a sequence *motif*. A motif is a short nucleotide or amino acid sequence pattern that is known or thought to have a biological significance; and importantly, the pattern that characterizes a motif can be sought within one or more candidate sequences. Thus, motifs provide a means for recognizing similarity to a family of sequences, and some characterizations allow that similarity to be quantified. Common approaches to characterize the pattern of a sequence motif include consensus methods, regular expressions, position-specific scoring matrices (PSSMs), and hidden Markov models. Each of these approaches can be used to summarize a local pattern present among a sequence family, and the resulting summary can be used to search for additional sequences which share a similar pattern. We conclude the first half of this chapter by returning to our cytochrome P450 example. The remaining sections demonstrate how known members of the superfamily can be used to recognize future candidates.

6.5.1 Regular Expressions

Regular expressions specify a set of character strings and are useful for pattern matching. In the context of an MSA, regular expressions can be used to define, though not quantify, the variation in a local sequence pattern. Cytochrome P450s, for example, have a signature that can be summarized by a regular expression: The pattern given by PROSITE (http://www.expasy.org/prosite/) is [FW]–[SGNH]–x–[GD]–{F}–[RKHPT]–{P}–C–[LIVMFAP]–[GAD]. The expression above is a pattern of 10 consecutive amino acids found within the sequences of cytochrome P450 proteins. The rules for each position in the pattern are separated by dashes (–), and the letters correspond to amino acids; thus [FW] indicates that the first amino acid must be either F or W (phenylalanine or tryptophan), while {F} indicates that the fifth amino acid cannot be phenylalanine. It should be apparent that such a pattern can be used to search a database of sequences. In addition to the search tool available at PROSITE, there is a variant of BLAST, called pattern-hit initiated BLAST (PHI-BLAST), which is supported by NCBI for this purpose. PHI-BLAST uses the BLAST engine to search for database proteins similar to a query, with the added restriction that each protein must match a specific regular expression pattern.

6.5.2 Position-Specific Scoring Matrices

Position-specific scoring matrices are similar in spirit to regular expressions; both seek to summarize and identify local sequence patterns, but their means to that end differ. Whereas regular expressions consider the presence and absence of residues in a pattern, PSSMs focus on the residue frequencies. These frequencies, computed within each column of an MSA, are then converted to scores that reflect how representative each residue is of each alignment column. Because PSSMs can be constructed from any MSA, and because any sequence can be scored using a PSSM, PSSMs provide a general means to quantifying similarity between a sequence and an MSA. It should be apparent that, like regular expressions, PSSMs can be used to search a database of sequences.

PSSMs are scoring matrices akin to the substitution matrices first described in Section 6.3.3. The difference is that, while substitution matrices assign scores to (a, b) residue pairs, PSSMs assign position-specific scores to each residue a. In a database search, these are effectively the same, because the residue b in each position of the query sequence supplies half of the (a, b) residue pair; in other words, the query and the substitution matrix combine to assign position-specific scores to each residue a, just as accomplished by a PSSM. In light of this, there is no fundamental difference between using a sequence and using a PSSM as a query to search a database, and in fact NCBI-BLAST allows PSSMs to be used as a query. In addition, NCBI supports an iterative procedure called position-specific iterated BLAST (PSI-BLAST), which uses PSSMs to focus a search around an initial sequence query. PSI-BLAST takes a query sequence as usual and performs a BLAST search, but it subsequently forms a PSSM from an MSA of the high-scoring BLAST results. This PSSM is then used as a query for BLAST, and a new list of results is obtained. The procedure can be iterated as desired.

The effect of PSI-BLAST is to attune the scoring scheme to the query, and in practice this may or may not be desirable. On one hand, PSI-BLAST can be very sensitive to remote homology; on the other hand, spurious sequences may be included and their scores may become inflated. Previously we used NP_000764 as a BLAST query to identify CYP2E1 homologs; now we use NP_000764 to initiate a PSI-BLAST search. The first round of PSI-BLAST yields the same BLAST results as before, and by default PSI-BLAST uses those with *E*-values less than 0.005 to construct the PSSM. Using this PSSM as a query, we initiate PSI-BLAST iteration 2, and the results of this search can be used to construct a PSSM for the next iteration of PSI-BLAST. Here, the results of PSI-BLAST iteration 2 are enough to demonstrate that the procedure is a mixed blessing: The list of results now contains many more annotated cytochrome P450s, but the sequence scoring the highest of all is the hypothetical cow protein with accession number NP_00106951.

6.5.3 Contrasting Summaries

Regular expressions and PSSMs have complementary strengths and weaknesses. In contrast to PSSMs, regular expressions do not quantify similarity, nor do they take frequency information into account. Ironically, however, these shortcomings provide flexibility: While PSSMs require an MSA, regular expressions can easily accommodate unaligned sequences of varying length. The strengths of regular expressions and PSSMs are combined in hidden Markov models (HMMs), the details of which are beyond the scope of this chapter. Like regular expressions and PSSMs, HMMs can be used to search a database; perhaps the most famous application is Pfam (http://pfam.wustl.edu/), which has both a searchable protein database and HMMs describing protein families. Pfam also supports methods for nucleotide sequences, and this should not be surprising: The multiple sequence approaches that we have described are equally suited to either class of biological sequence. For example, the identification of regulatory elements, such as transcription factor binding sites, can be accomplished by a summarize-and-search procedure. The fruits of these search procedures can be used to refine the patterns targeted in subsequent searches, with the hope of refining these patterns even further. It is upon this recursive foundation that much of our knowledge of biological sequences rests.

6.6 GENETIC MAPPING

The goal of genetic mapping is to identify, among the tens of thousands of genes in the genome, specific genes that appear to influence a particular trait. For example, the gene Cytochrome P450, Subfamily IID, Polypeptide 6 (CYP2D6) has been implicated in the metabolism and elimination of ingested chemicals and other molecules. By examining the Online Mendelian Inheritance in Man database (OMIM), we see that gene mapping techniques were employed to help identify this gene and its role in metabolism.

6.6.1 Background Principles

To understand the principles behind genetic mapping, it is important first to understand some basic concepts of genetics. One of the most important of these in the context of genetic mapping is that for the usual types of organisms that toxicologists are interested in, including humans, individuals carry two copies of each of their chromosomes. One of these chromosomal copies comes from each of the two parents of that individual. Sex chromosomes are the only ones that deviate somewhat from this rule; individuals still receive one sex chromosome from each parent, but these might not be two copies of the same chromosome (mammalian males receive one copy of a Y chromosome and one copy of an X). In this chapter, we are primarily interested in the autosomal chromosomes; the discussion of sex-linked traits is outside the scope of this chapter.

6.6.1a   Recombination. When a parent transmits a chromosome to an offspring, various things can happen. One possibility is that the copy that the parent received from its own mother is transmitted, and another is that the parent's paternal copy is transmitted. The final possibility is to transmit a chromosome that is a combination of the parent's maternal and paternal copies (Figure 6.9). The act of creating a new combined chromosomal copy is called recombination, and is of fundamental importance in genetic mapping. Biologically, this process occurs when gametes are created.

6.6.1b   Genetic Markers. The goal of genetic mapping is to locate genes of interest. It might be that we start off with no information at all about where these genes might be, or we might already have general target locations in mind. Either way, the goal is to identify, with as much precision as possible, specific regions of the genome that contain the putative genes. In order to do this, we rely on the availability of genetic markers.

Genetic markers are sections of DNA that, to be useful for mapping, satisfy three general requirements. The first of these is that we know where they are in the

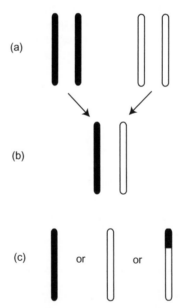

Figure 6.9. Recombination among chromosomes. Chromosomes being transmitted from grandparents (A) to a parent (B), along with examples of gametes that could be passed along to an offspring (C). The third possibility in part C represents a recombined chromosome, which is composed of one piece originally derived from the grandfather and one piece derived from the grandmother.

genome. This allows us to be able to use them as landmarks, as we try to determine if genes we are looking for appear to be nearby any of the known markers. The second general requirement is that the sections of DNA that form the markers vary to some extent from one chromosome to another in the population. We call each of these variant forms an allele. The final requirement is that we can assay individuals for each of these markers and thus determine which alleles any given individual carries for each marker. Since individuals receive two copies of each autosomal chromosome, they receive two copies of the markers that lie on these chromosomes.

There are various different types of genetic markers, some with DNA variations that involve a single base change, while others involve changes over long stretches of DNA. Markers whose alleles are defined by a single base change are called single nucleotide polymorphisms (SNPs).

It is worth noting that the distinction between what we call a gene and what we call a marker is often blurred. Essentially any variable sequence of DNA that can be identified and assayed can be used as a marker. Genes generally cover large stretches of DNA and can contain multiple variable sites. Any of these variable sites can be used as markers, so that markers may exist within genes as well as between them. In general, any site on the genome, including markers, genes, and markers within genes, can be referred to as a locus (plural loci).

6.6.1c *Recombination as a Measure of Distance.* The process of recombination provides us with a way to measure distances between positions on a chromosome. The principle behind this is that the further away two sites are, the higher the chance that a recombination event will occur between them when a gamete is formed. Recombination events can be identified by examining the inheritance of alleles at these sites from one generation to the next. Estimating distance between loci allows us, for example, to gauge the proximity between markers and a gene of interest.

6.6.2 Linkage Analysis

One mapping methodology strives to estimate these recombination distances directly by examining the inheritance of marker alleles and trait values from generation to generation within a collection of families. Various different family structures can be used for this type of analysis, using both human data and data from model organisms.

6.6.2a *Human Pedigrees.* Perhaps we are interested in studying the genetics of drug metabolism rates. One possibility for carrying out this type of study may be to collect families and ask participating family members to ingest a calibrated dose of a test drug, such as dextromethorphan (found in over-the-counter cough and cold medicines). Urine samples can then be collected several hours later, and the amount of dextromethorphan and its metabolic byproduct, destrorphan, can be quantified.

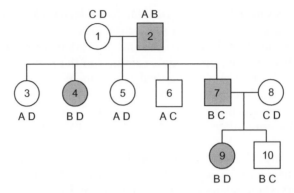

Figure 6.10. A pedigree. Three generations of individuals are represented, males as squares and females as circles. The grandparents have five offspring, the youngest of which has two offspring of his own. Shading represents individuals who express a trait of interest. Characters next to the symbols represent alleles at a genetic marker. We are interested in determining whether marker alleles are passed along from generation to generation in a pattern that appears to be similar to that of the trait.

Based on these measurements, individuals who are poor metabolizers can be identified. This is a trait that might be of concern for drug metabolism in general.

Figure 6.10 shows a diagrammatic illustration of an example family for this type of study. Following convention, males are denoted by squares and females by circles. The letters next to the individuals represent different alleles at a single marker. For instance, in Figure 6.10, individual 3, the first daughter in the second generation, has alleles "A" and "D" at the marker. She appears to have received the "A" allele from her father and the "D" allele from her mother. Shaded boxes represent individuals that display the trait of interest; in Figure 6.10, for example, perhaps these individuals have been determined to be rapid metabolizers (RM) of the drug. In this family, we can see that the grandfather is an RM, as are several of his children and a grandchild. If we examine the inheritance of marker alleles from the grandparents' generation to their children, we see that each of the offspring who receives a "B" allele from the grandfather is also an RM. Those who receive the other allele ("A") are not RMs. If we look down one more generation, we see that individual 7 also has two children. One of these is an RM, the other is not. Both these grandchildren, however, have received the "B" allele from their father. One explanation for finding a pedigree that looks like this could be that this marker is located nearby a gene that influences drug metabolism rates. One of the grandfather's copies of this chromosome carries the "B" allele at the marker and an RM allele at the gene. This section of the chromosome tends to be transmitted intact to his children, because these loci are close enough together that a recombination event is not likely. However, in the inheritance from individual 7 to his son (individual 10), a recombination event does occur, disconnecting marker allele "B" and the rapid metabolism allele. Allele "B" is passed along at the marker, but it is the grandmother's allele that is inherited at the gene.

6.6.2b Inbred Line Crosses. It is often impractical or unethical to collect the type of toxicological information on humans that would be necessary for some important projects. Instead, animal models, such as mouse or rat lines, can be used. For example, we might be interested in understanding drug toxicity for a specific AIDS cocktail. Reliable treatment for HIV/AIDS is of global importance, but it is also necessary that the treatment does not itself prove toxic for the individuals undergoing it. It is known that some genes, such as CYP2D6, influence drug metabolism in general. Perhaps we are interested in identifying other genes that appear to influence toxicity levels of this cocktail. If there is evidence that individuals respond differently to the standard doses, we can hypothesize that there is a genetic component to this trait.

The ultimate goal of this type of experiment might be that, once alleles have been identified, a general genetic assay can be developed so that each individual receiving treatment can be tested and then prescribed an appropriate cocktail dosage, or perhaps a different cocktail.

In the past few decades, many different lines of mice have been bred and are maintained at various facilities. These lines have been selected based on numerous traits, including drug sensitivity. One experiment that might be interesting for us to perform to understand the genetics of sensitivity to our AIDS cocktail would be to subject mice from several high- and low-sensitivity lines to the cocktail. Once the specific lines that display high and low tolerance have been identified, we can begin to examine the genetic differences between these two lines that contribute to this sensitivity.

One of the purposes of developing inbred lines is that individuals within a line are genetically highly similar. This means that the chromosome that a mouse gets from its mother is close to being identical to the copy received from the father. The first step in a mapping study would be to cross the two lines, selecting both males and females from each line to breed to the respective gender of the other line. The resulting offspring will all have one chromosomal copy from each line. It is the step after this that becomes interesting. We want to examine recombination events between loci in order to gauge distances. If we cross two individuals from the second generation to form the third, these recombination events begin to be apparent. Each generation we do this provides more opportunities to examine recombination events. In an actual experiment, the time and cost involved in waiting for each generation must be balanced against the information that may be gained. Once the final generation of mice has been bred, we can expose members of it to the cocktail. Individuals displaying signs of toxicity, along with ones who do not, are identified and assayed for the genetic markers. Those markers whose alleles appear to be inherited together with the trait provide evidence for the location of genes that influence sensitivity to this cocktail.

6.6.3 Association Mapping

While linkage analysis has been very effective in identifying regions containing genes for many different traits, it typically provides quite low resolution, identifying large regions that may contain tens or even hundreds of genes. This is because we rely on detecting recombination events in order to estimate distances, and (by definition) at small genetic distances these events are rare. This means that the resolution of linkage analysis depends on the number and size of the families collected, factors that are often limited in practice.

Association mapping methods attempt to get past the difficulty of low resolution by relying on information on a population level rather than on just the individual family level. The idea is that population history can provide information about historical recombination events occurring far prior to the current generation of individuals. To understand association mapping methods, it is necessary to understand a little about the quantity linkage disequilibrium.

6.6.3a   Linkage Disequilibrium.   Linkage disequilibrium is essentially a measure of correlation between alleles at two (or more) loci in a population. For example, consider two markers, one with alleles designated as "A" and "a" and the second with alleles "B" and "b" (Figure 6.11). Linkage disequilibrium is a measure of whether allele "A" occurs on the same chromosome as allele "B" more often than one would expect by chance (any of the four combinations of alleles at the two loci can be used). In Figure 6.11, the sample of chromosomes shown either carry "A" and "B" alleles together or "a" and "b" alleles together. The fact that we only see two of the four possibilities for alleles on a chromosome indicates that linkage disequilibrium between these two markers is not zero.

There are a number of ways linkage disequilibrium can arise between loci in a population. One possibility is selection: If alleles at two (or more) loci confer an advantage to an individual when the alleles occur together, then chromosomes containing these alleles will start to become more frequent in the population than they would otherwise. This increase in this combination of alleles creates linkage disequilibrium. If two populations in different environments experience different levels of selection, the amount of linkage disequilibrium will also probably differ between the populations. This illustrates an important concept: While the distance between two loci is constant across populations, the amount of linkage

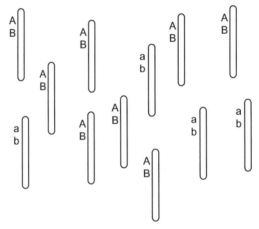

Figure 6.11. Distribution of marker alleles on a collection of chromosomes. A sample of 12 chromosomes are shown, each one having been assayed for two markers. The marker toward the top of the chromosome has alleles "A" and "a", and the marker further down has alleles "B" and "b". In this sample, a chromosome carries an "A" allele and a "B" allele or an "a" allele and a "b" allele. This indicates that there is linkage disequilibrium between alleles at these two markers: alleles do not appear to be independently distributed across chromosomes.

disequilibrium between them is not. There are a number of other ways that linkage disequilibrium can be created in a population, including (among other phenomena) mixing of populations, mutation events, and rapid changes in population size.

While the creation of linkage disequilibrium in a population is important, for genetic mapping purposes, it is the decay of linkage disequilibrium that is of primary interest. This is because there is a connection between how this decay occurs and the distance between loci. For mapping purposes, we are interested in distances between loci in order to map our unknown genes to nearby markers.

The more frequently that specific alleles occur together on a chromosome, the stronger the linkage disequilibrium between those loci. At the same time, the further apart two loci are, the more likely that alleles will be separated by recombination when gametes are created. Because of this, once linkage disequilibrium is created in a population, recombination between loci will act to reduce it. The amount it will be reduced depends on the amount of recombination between the loci, which in turn depends on how close the loci are.

6.6.3b *Using Linkage Disequilibrium to Map Genes.* The relationship between recombination and linkage disequilibrium is the basis of association mapping. The idea is to identify marker alleles that appear to be in linkage disequilibrium with alleles at a gene that influences the trait. Since the alleles at the gene are not themselves being examined (as they are unknown), we must instead evaluate this relationship indirectly. This is done by looking for a connection between alleles at the marker and the trait itself. The presumption is that markers that appear to be correlated with the trait are so because of linkage disequilibrium, the presence of which indicates proximity to a gene of interest. There are various different strategies that have been proposed for detecting this population-level connection between marker alleles and a trait.

It is worth noting that most association mapping techniques are not directed at estimating recombination rates directly, as linkage analysis methods are. Instead, they attempt to examine whether there is an indication that past recombination events have acted to decay linkage disequilibrium from previous generations, many times removed. The general idea is that in most natural populations, such as human ones, there should be little or no linkage disequilibrium over reasonably large regions of the genome. It will instead only be found between loci that are very close together. This forms the basis of how association mapping methods provide high-resolution results.

6.6.3c *Case–Control Tests.* The case–control design is one of the simplest association mapping strategies. In its most basic form, unrelated individuals who present the trait of interest are sampled (these are the cases), together with a group of unrelated, matched controls. The control group contains individuals who do not present the trait of interest, but are as similar as possible in all other respects to the cases. A possible scenario for this type of experiment (keeping along the lines of the drug toxicity examples) might be that we are interested in collecting data on individuals who have had an adverse reaction to a standard treatment for high blood pressure. Again, our ultimate goal may be to gain enough information that we can ultimately prescreen patients prior to treatment in order to determine an appropriate protocol.

We can identify the case individuals through various clinics that treat this condition, and then we must find an appropriate set of individuals for the control group. Once all the individuals have been collected, each is assayed at the genetic markers of interest. We are interested in identifying markers where some of the alleles are more prevalent among the individuals with the adverse response. This indicates that these marker alleles are associated with the trait, possible evidence that there is linkage disequilibrium between the marker alleles and alleles at a gene that influences whether individuals will have an adverse response to the drug. Evidence of linkage disequilibrium, in turn, may indicate that these markers are nearby a gene of interest.

When performing case–control-type tests, it is of crucial importance that the control individuals are well matched with those in the case group. If this does not occur, it is possible that markers that appear to be associated with the trait will be identified, even though these markers may be biologically unrelated to the trait. One often-discussed type of mismatching is discussed in the following section.

6.6.3d Transmission–Disequilibrium-Type Tests. One of the criticisms of the case–control-type design is the possibility of obtaining spurious results, ones for which markers appear to be significantly associated with the trait, even though there is no genetic basis for this association. This generally arises through mismatching the case samples with the controls. One source of mismatching that is of concern for genetic studies derives from cryptic population sub-structure (where groups of individuals live together, but preferentially produce offspring within groups rather than randomly mix across groups). If the two groups have differences in allele frequencies at any of the markers (which is very likely for any populations with different histories), and also have different incidence rates for the trait, spurious results are likely. This is because it is more likely that individuals from the group with the higher incidence rate will be selected for the case sample, and individuals from the other group for the control samples. Any random difference in allele frequencies at the markers between these groups will appear to be associated with the trait, when instead, they are merely associated with the subpopulations.

A number of different strategies have been proposed to counter this problem. One of these is the transmission–disequilibrium-type approach. This methodology combines the principles of association mapping together with linkage analysis. For this type of design, small unrelated families are collected that contain individuals displaying the trait of interest. Evidence of association between marker alleles and the trait is obtained by looking across families, while inheritance patterns within families provide evidence regarding recombination. This strategy provides the high resolution potential of association mapping while controlling for spurious results.

Figure 6.12 provides an example of a collection of families displaying population association, but where the within-family inheritance patterns indicate that the marker is probably not nearby a gene of interest.

6.6.4 Environmental Factors in Gene Mapping

The descriptions of the methods in the previous sections have focused on identifying genetic components of traits such as toxicity response. However, we may also be interested in environmental factors, such as pollutant levels or exposure to cigarette

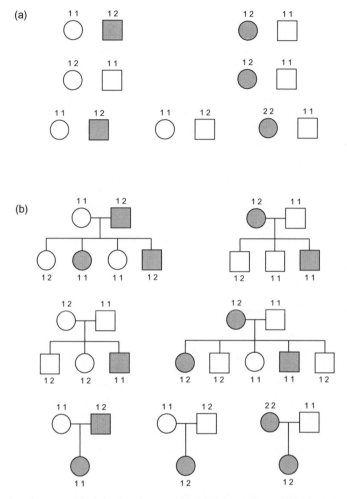

Figure 6.12. Spurious association. A sample of individuals collected for an association study is displayed together with these individuals' respective marker alleles (see Section 6.6.2a and Figure 6.10 for nomenclature). (A) In this diagram, only unrelated individuals are shown. Among these individuals, it appears that the "2" allele at the marker is much more common among individuals who display the trait (shaded) than it is among those that do not (unshaded). This is an indication of population association. (B) In this diagram, offspring of the individuals from part A are also shown. If the marker was nearby a gene influencing the trait, recombination events would be unlikely, and we would expect to find a common inheritance pattern between the marker alleles and the trait status. Because no such pattern is apparent, we could conclude that the association seen in the parents is a spurious result; the marker does not appear to have a genetic connection with the trait.

smoke. Most of the techniques that have been developed for gene mapping studies also allow enviromental factors to be considered. This includes not only the general question of whether these factors influence the trait, but also whether there is evidence that they act as modifiers of genetic components of response. The second point addresses the question of gene by environment interaction effects. An example of this type of effect is the following. Consider a hypothetical gene coding for an enzyme that helps control the rate of cell division in the body. There happen to be two alleles of this gene found in the population, coding for slightly different forms of the enzyme. Both of these forms are capable of functioning adequately in their role in cell cycle regulation under normal conditions. One form of the enzyme, however, is susceptible to being bound with a byproduct of a manufactoring process. How individuals are affected by exposure to this compound will depend on which alleles they carry at this gene. Those who produce only the unbound enzyme are not affected, while those whose enzymes become bound by the compound will probably experience problems with cell cycle regulation. This, in turn, puts these individuals at increased risk for developing tumors. In this example, if there is no exposure to the compound, variations of the gene do not affect risk. Additionally, if there is no genetic variation in the population, all individuals have equal risk of tumor development in the presence (or absence) of the compound. The differences in risk between individuals arise through the interaction of the genetic and environmental factors, where different combinations produce different outcomes.

6.7   CONCLUSION

This chapter has covered various techniques that are of interest in the analysis and integration of genomic data. We began with an overview of methods that allow queries of gene function to be addressed. We ended with techniques designed to locate genes that may influence traits of interest in the field of Toxicology. The information flow back and forth between these topics is continuously evolving, based on the growth of information available in many repositories and databases. It is possible to begin a study with no knowledge regarding the genomics of a trait, perform an appropriate gene mapping experiment, and walk away with a few specific target regions that appear to contain genes of interest. The genes that are located within these regions can then be queried using sequence comparison tools to determine whether any function can be ascribed to them. Information gained from this can then be used to develop new hypotheses regarding their role in the particular study. Similarly, genes may be identified initially by comparative means, and these can then be tested in the context of an association mapping procedure. In this case, the mapping project is initiated with target genes in mind: markers are identified that are within or very nearby these genes, and their roles in a particular trait are investigated. Genes identified in model organisms (through gene mapping studies or by other means) provide a source of information relevant to human studies as well. Sequence similarity procedures provide the basis of identifying human genes based on the findings from these model organisms.

SUGGESTED READING

Durbin, R., Eddy, S. R., Krogh, A., and Mitchison, G. *Biological Sequence Analysis: Probabilistic Models of Proteins and Nucleic Acids*, Cambridge University Press, Cambridge, 1999.

Edgar, R. C., and Batzoglou, S. Multiple sequence alignment. *Curr. Opin. Struct. Biol.* **16**(3), 368–373, 2006.

Ott, J. *Analysis of Human Genetic Linkage*, Johns Hopkins University Press, Baltimore, 1999.

Sonnhammer, E. L., and Koonin, E. Orthology, paralogy and proposed classification for paralog subtypes. *Trends Genet.* **18**(12), 619–620, 2002.

Thomas, D. C. *Statistical Methods in Genetic Epidemiology*, Oxford University Press, New York, 2004.

CHAPTER 7

Immunochemical Techniques In Toxicology

GERALD A. LeBLANC

Department of Environmental and Molecular Toxicology, North Carolina State University, Raleigh, North Carolina 27695

7.1 INTRODUCTION

The advent of antibody technology has revolutionized every branch of biology that involves protein analyses, including toxicology. Proteins serve as both basic structural and functional components of organisms and, as such, are major target sites for the action of many toxic agents. Furthermore, proteins can modulate chemical toxicity through various processes including toxicant activation, deactivation, sequestration, and elimination. Antibody technology provides powerful tools for a variety of toxicological analyses including: (a) assessing the effects of a toxicant on intracellular protein localization, (b) precisely assessing the effects of a toxicant on specific protein levels (i.e., protein induction or suppression), (c) measuring interactions between a specific protein in a mixture of proteins and a toxicant (e.g., contribution of a specific enzyme in a complex cellular mixture to the metabolism of a toxicant), and (d) isolating individual proteins for subsequent characterization. Immunochemical techniques also have provided means for the rapid quantification of toxicants, drugs, hormones, and other small molecules in biological and environmental samples.

7.2 DEFINITIONS

It is not the intent of this chapter to provide a synopsis of the science of immunology. Rather, this chapter is intended to provide a general overview of some basic immunochemical techniques that have proven invaluable to toxicological research. The following immunological terms are integral to the discussion of immunochemical analyses and are routinely used in this chapter.

> *Immunogen*: Any foreign molecule (usually a protein) that elicits an immune response.

Molecular and Biochemical Toxicology, Fourth Edition, edited by Robert C. Smart and Ernest Hodgson
Copyright © 2008 John Wiley & Sons, Inc.

Immune Response: The production of antibodies by an organism in response to exposure to a foreign molecule (immunogen). The antibodies produced will recognize the foreign molecule.

Antibody: Proteins secreted by B lymphocytes that recognize and bind to a specific antigen.

Antigen: Any foreign molecule that is recognized by an antibody. Note that a single protein is typically first an immunogen (elicits the immune response) and is then an antigen (is bound by the resulting antibodies).

Epitope: The specific site on an antigen that is recognized by an antibody.

Immunogenicity: The ability of a foreign molecule to elicit an immune response.

Antigenicity: The ability of a foreign molecule to be recognized and bound by an antibody.

Hapten: A small molecule (e.g., a peptide, hormone, drug, or toxicant) that is conjugated to a larger carrier molecule (a protein) so that an immune response is elicited and antibodies are produced that recognize epitopes on the small molecule.

Adjuvants: Nonspecific stimulators of the immune response that are administered during immunization.

Primary Antibody: The antibody used in an immunoassay that recognizes and binds to the antigen.

Secondary Antibody: An antibody used in an immunoassay that recognizes and binds to the primary antibody. The secondary antibody is equipped with some means of detection that allows for the analyses of the immune complex.

7.3   IMMUNOGENS AND ANTIGENS

Immunogens and antigens are often the same material but with two functionally different definitions. The first step in an immunoassay is often producing the antibodies that are required for the assay. This is accomplished by administering the material that will be measured in the immunoassay to an animal and allowing the animal to produce antibodies against this material. In this capacity, the material is the immunogen—the material that is responsible for the immune response. Antibodies produced as part of the immune response have the ability to bind the material of interest with high affinity and specificity. These binding properties of antibodies are the basis for all immunoassays. In this binding capacity, the material is the antigen—the material to which the antibodies bind with high affinity and specificity. Most often, the immunogen and antigen are the same material. However, exceptions exist.

- An immunogen can be a modified form of the antigen. As discussed below, immunogens can be chemically modified to maximize the yield of antibodies.
- An immunogen can be a specific portion of the antigen. A specific peptide, linked to a larger molecule forming a hapten, may serve as the immunogen for which a full protein, which contains the peptide, is the antigen.

• An immunogen may be a distinct material that shares a common epitope with the antigen. For example, a specific protein may serve as immunogen, while a subpopulation of the antibodies produced recognize a related protein (antigen) in addition to the protein used as the immunogen.

7.4 POLYCLONAL ANTIBODIES

When a mammal is injected with a foreign molecule (immunogen), the immune response elicited includes the generation of antibodies from a variety of B lymphocytes. The antibodies produced from each lymphocyte will recognize a single epitope on the foreign molecule (antigen). However, different subpopulations of antibodies will recognize and bind to multiple epitopes on the antigen because each subpopulation originates from a different B-lymphocyte. The entire population of antibodies, collected from the serum of the immunized animal, consists of polyclonal antibodies (Figure 7.1).

Advantages of polyclonal antibodies over monoclonal antibodies are as follows: (a) They are highly reactive due to the binding of multiple epitopes on the same antigen and (b) they are easy to produce. The disadvantage of polyclonal antibodies as compared to monoclonal antibodies is that they often have a lower degree of specificity (i.e., will bind antigens that are not of interest).

7.4.1 Polyclonal Antibody Production

The generation of polyclonal antibodies involves the following steps and considerations.

7.4.1a Immunogen Preparation. The first step in generating an immune response with the intent of collecting antibodies is to prepare the immunogen to which the response will be generated. Thus, the generation of antibodies for use in protein

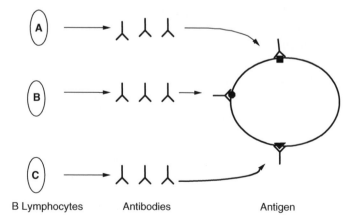

B Lymphocytes   Antibodies   Antigen

Figure 7.1. The generation of antibodies of several clonal origins (polyclonal antibodies) with antibodies from each clonal origin (monclonal antibodies, clones A, B, and C) recognizing a distinct epitope on the antigen.

analyses typically requires that the protein or peptide, which the antibodies will recognize, be available in purified or semipurified form. The following characteristics of the immunogen must be considered to maximize success in generating useful antibodies.

(i) Immunogen Purity. The antigenic specificity of the antibodies produced will only be as good as the purity of the immunogen used to immunize the animal. Partially purified immunogen may be adequate for certain immunoassays; however, fully purified immunogen will yield antibodies containing the greatest specificity for the immunogen. Partially purified immunogen can often be completely purified for immunization by electrophoretically separating the immunogen from the contaminating proteins, cutting the immunogen from the solid matrix with which it is associated (e.g., acrylamide gel, nitrocellulose membrane) and using it for immunization.

(ii) Immunogen Size. Immunogens typically must be greater than approximately 5000 daltons to produce an immune response. If the immunogen is too small, it can be used as a hapten.

(iii) Immunogenicity. Some foreign compounds have high immunogenicity, while others have low immunogenicity. The following should be considered for materials having poor immunogenicity.

- Change the species or strain of animal being immunized. Immunization of a species with an immunogen derived from the same species may not elicit a sufficient immune response since the immunogen will not be recognized as foreign.
- Increase the dose of the immunogen. Within limits, the degree of immune response increases with increasing dose of immunogen. Weakly immunogenic materials may require a higher dose to stimulate a sufficient immune response.
- Alter the physical makeup of the immunogen (e.g., denature the protein, conjugate to dinitrophenol). This may reveal immunogenic sites that were not available on the native protein.

7.4.1b Immunization. Once the immunogen has been adequately prepared, it is then administered to an appropriate species with which an immune response will be generated. The following points should be considered during animal immunization.

(i) Selection of Animal Species to Be Immunized. Both scientific and practical considerations contribute to the selection of the appropriate animal species to be immunized.

- *Self-Recognition.* Administration of an immunogen to the same species from which the immunogen was derived poses the risk that the immunogen will not be recognized as foreign and will accordingly not generate a sufficient immune response. The species selected for immunization should be different from that used to generate the immunogen.

- *Amount of Antibodies Required.* Polyclonal antibodies are generally obtained from the immunized animal by collecting blood and preparing serum from the blood. The antibody-rich serum is used as the immunochemical reagent and contains the polyclonal antibodies to the antigen. The amount of immunochemical reagent derived from an immunization is therefore dependent upon the amount of blood that can be collected from the immunized animal. For example, a maximum of 2 ml of serum can generally be collected from mice, whereas up to 500 ml of serum can be derived from rabbits. Polyclonal antibodies are most frequently produced from rabbits.

- *Amount of Immunogen Available.* A limited amount of immunogen may dictate that a small species be used for immunization. For example, mice typically require 10 times less immunogen than that required by rabbits to elicit an equivalent immune response.

- *Animal Maintenance Capabilities.* Larger species require more extensive animal holding facilities. While goats or horses may be used to generate large volumes of immunochemical reagent, most laboratories are not equipped with adequate facilities to accommodate such species.

(ii) Adjuvant Selection. An adjuvant should be selected that will adequately stimulate the immune system of the immunized animals without posing undue stress to the animals. Historically, Freund's adjuvant was most commonly used in antibody preparation. Freund's adjuvant consists of a water–oil emulsion containing heat-killed bacteria. While very effective in stimulating a prolonged immune response, Freund's adjuvant has fallen into disfavor due to its propensity to induce persistent and aggressive lesions (granulomas) resulting in pain and distress to the immunized animals.

Adjuvant consisting of monophosphoryl lipid A (MPL; a detoxified bacterial endotoxin) plus trehalose dicorynomycolate (TDM) in a water–oil emulsion has proven to be a viable alternative to Freund's adjuvant. MPL + TDM stimulates the immune system without eliciting the degree of toxicity associated with Freund's adjuvant. Aluminum salts also have been used to stimulate the immune response without severe toxic side effects to the immunized animal. Aluminum salts appear to function by trapping the immunogen and providing for a slow, prolonged release of the immunogen following administration.

(iii) Immunogen Dose. If the immunogen is readily available, then one strives to administer a dose that will elicit the maximum immune response (generally 0.5–1.0 mg in rabbits, 10× lower in mice). If the immunogen is in short supply, then one strives to administer the minimum dose that will elicit a sufficient immune response (generally 10–100 µg for rabbits, 10× lower for mice). An animal is generally injected several times with the immunogen to elicit a satisfactory immune response. Thus, total dose must be considered when determining the amount of immunogen required.

7.4.1c *Serum Collection and Screening for Antibody Titer.* Blood samples are generally collected and serum prepared 8–10 weeks after the first injection.

Serum samples are analyzed for antibody titer. When the titer is sufficiently high, the animal is bled. Serum samples are generally screened for the presence of the antibody by enzyme-linked immunosorbent assay (ELISA). ELISA methods are discussed later in this chapter.

7.5   MONOCLONAL ANTIBODIES

After an immune response has been elicited in an immunized animal, the isolation and propagation of individual B lymphocytes in cell culture will provide antibodies that are all of the same clonal origin and thus recognize a single epitope on the antigen. These antibodies, collected from the cell culture medium, are monoclonal antibodies (Figure 7.1). Two important characteristics of monoclonal antibodies are (a) their single-epitope specificity of binding and (b) their ability to be produced in unlimited quantities. The disadvantages of monoclonal antibodies are that (a) their single-epitope specificity often makes them unsuitable for certain immunoassays and (b) their production is time-intensive and requires cell culture facilities and expertise.

7.5.1   Monoclonal Antibody Production

All of the processes and considerations described for polyclonal antibody production are also relevant to monoclonal antibody production. The only variation between the two, with respect to the the immunization process, is species selection. Monoclonal antibodies are most commonly produced in mice because of the commercial availability of mouse myeloma cells that are particularly suited as partner cells in hybridoma preparation (discussed below).

Monoclonal antibody production deviates from polyclonal antibody production once the immune response has been generated. During polyclonal antibody production, high-titer antibodies in the serum denotes the end of the process and the antibody-rich blood is sampled from the animal. During monoclonal antibody production, a high titer of antibodies in the serum denotes the time to begin the generation of monoclonal antibodies. This encompasses the preparation of hybridomas, the production of hybridomas of monoclonal origin, and the screening of the hybridoma clones for the production of the antibodies that recognize the antigen of interest. This process is diagrammed in Figure 7.2.

The B lymphocytes of the immunized animal are the source of the antibodies that recognize the administered immunogen. Removal of the spleen and culture of individual B lymphocytes derived from this organ may seem to be a viable approach to the generation of monoclonal antibodies in culture. However, primary B lymphocytes will not survive in culture. These antibody-producing cells are therefore immortalized by fusion to myeloma cells. These cells are derived from tumors of B lymphocyte origin and will grow indefinitely in culture. B lymphocytes from the immunized mice and the myeloma cells are most commonly fused by stirring and centrifugation in polyethylene glycol.

Following fusion, hybridomas must be separated from fusion products of the same cell type (e.g., fusion of two B lymphocytes or two myeloma cells) and cells that did not undergo fusion. Unfused B lymphocytes or the fusion product of two

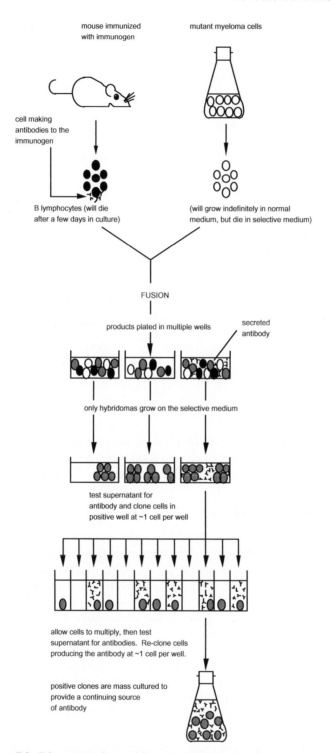

Figure 7.2. Diagrammatic overview of monoclonal antibody production.

B lymphocytes will be eliminated during continuous culture because these cells have a short life span in culture. Commercially available myeloma cells for hybridoma production have mutations in one of the enzymes of the salvage pathway of purine nucleotide biosynthesis. Hybridoma cells are cultured in medium that forces the cells to utilize the salvage pathway for nucleotide synthesis. The mutated myeloma cells or hybridization products of two myeloma cells will die in this selection medium since they are incapable of nucleotide synthesis under these propagation conditions. However, myeloma cells that have fused to the B lymphocytes derived from the spleen of the immunized animal will have an intact salvage pathway and will survive in the selection medium. Thus, only the B lymphocytes–myeloma hybridomas will survive prolonged culture in the selection medium.

Hybridomas are then cultured, and the culture media are assayed for the presence of antibodies to the antigen of interest. Cultures producing desirable antibodies are then aliquoted at dilutions calculated to yield 0.5 to 1.0 cells per aliquot. These aliquoted samples are then cultured. Many of these cultures will be initiated with a single cell; thus the resulting cell population will be of monoclonal origin. Media from these cultures are again assayed for the presence of the antibodies. Positive cultures are again diluted and aliquoted to ≤1 cell per aliquot. This ensures that the resulting cell populations are of monoclonal origin. Media from these cultures are again assayed for the presence of the antibody, and positive clones are used in mass culture for monoclonal antibody production. The ability to produce the antibodies in culture provides for an inexhaustible supply of immunochemical reagent.

7.6   IMMUNOASSAYS

The polyclonal or monoclonal antibodies produced can be used in a variety of immunoassays for the detection or characterization of an antigen. Some of the more common immunoassays used in toxicology are described below.

7.6.1   Immunohistochemistry

This technique is a semiquantitative means of determining the presence of a protein within a cell or tissue section, its abundance, and its subcellular localization. This approach could be used in toxicology, for example, to determine if, upon treating cells with a specific toxicant, a receptor to which the toxicant binds localizes from the cell cytoplasm to the nucleus.

The general approach used to assess the intracellular localization of a protein within a cell is as follows.

- Cells or tissue sections under study are immobilized onto a solid support to facilitate later manipulation. The biological materials are generally secured to a glass slide on which they subsequently can be viewed with a microscope. This can be accomplished by growing adherent cells on a slide or chemically attaching the materials to the slide.
- The cells or tissue sections are then fixed to preserve cellular structure and immobilize proteins in place. Fixation is often accomplished by incubating the cells/tissue in organic solvent, paraformaldehyde, or glutaraldehyde.

- Fixed preparations are incubated with the primary antibody. The primary antibody will recognize and bind to the protein of interest.
- Preparations are incubated with secondary antibody. The secondary antibody will recognize and bind to the primary antibodies. This tertiary (protein–primary antibody–secondary antibody) complex can now be subjected to detection methods.
- Preparations are incubated with appropriate reagents to allow visualization based upon the detection system associated with the secondary antibody. The secondary antibody may be conjugated to a enzyme (e.g., alkaline phosphatase, horseradish peroxidase). Incubation with the appropriate substrate to the enzyme will result in the production of an insoluble colored product that can be detected upon microscopic analyses of the cells. Secondary antibodies can also be conjugated to fluorochromes (e.g., fluorescein, rhodamine) that can be detected using a microscope equipped to detect fluorescence. Immunohistochemistry has proven to be a powerful tool in biochemical toxicology allowing for in situ assessments of protein responses to toxicant exposure.

7.6.2   Immunoaffinity Purification

Immunochemical approaches allow for the purification of a protein from a complex mixture in a single chromatographic step. Since antibodies to the protein targeted for purification are required, immunoaffinity purification techniques are often used for the routine purification of a protein subsequent to its initial purification by traditional chromatographic approaches. Immunoaffinity purification of a protein involves the following general steps:

- Antibodies are bound to a solid-phase matrix. Protein A or G beads are commonly used. Proteins A and G are polypeptides located on the cell wall of some bacteria. These proteins bind the Fc (constant) region of antibodies without affecting the ability of the antibodies to bind antigen. Proteins A and G are commercially available conjugated to sepharose, agarose, or acrylic beads. A chromatographic column is prepared with the tertiary complex (antibody–Protein A/G–solid matrix).
- The mixture containing the protein targeted for purification is added to the column, and sufficient time is allowed for the protein and antibodies to bind.
- Unbound proteins are washed from the column.
- Antibody-bound proteins are dissociated from the antibodies and eluted from the column. This is often the most difficult step since the antibodies and antigen bind with high affinity. Nondestructive approaches should be attempted to dissociate the antibodies and antigen such as by manipulating the salt concentration or pH of the elution buffer. Sometimes, more drastic approaches are required such as adding a denaturent (i.e., detergent) to the elution buffer.

7.6.3   Immunoprecipitation

Immunoprecipitation provides a means by which a protein can be selectively removed from a complex mixture. While conceptually similar to immunopurification, immunoprecipitation is typically used in the direct functional characterization of a protein, not as a means of obtaining protein. For example, a toxicologist may determine that two enzymes (designated A and B) are capable of inactivating a specific chemical carcinogen. Immunoprecipitation could then be used to determine the relative contribution of each enzyme in a tissue preparation to the metabolism of the chemical as follows:

- Incubate increasing concentrations of antibody, that binds to enzyme A, with the tissue homogenate. Increasing amounts of enzyme A, within the homogenate, will be bound with increasing antibody concentration in the assay.
- Precipitate the immune complexes (antibody–antigen) by adding Protein A or G beads to the mixture, incubate, and then centrifuge the mixture. The protein A or G will bind to the Fc region of the antibodies. The beads (e.g., sepharose) associated with the Protein A/G will provide sufficient mass to the complex to allow its precipitation during centrifugation.
- Use the supernatant from the precipitations to measure biotransformation of the carcinogen. With increasing concentration of antibody, greater amounts of enzyme A will be removed from the supernatant. If enzyme A is responsible for the biotransformation of the carcinogen, then progressively less biotransformation will be measured with increasing concentrations of antibody added to the mixture (Figure 7.3).
- Repeat the above procedure, but use antibodies that bind to enzyme B instead of enzyme A. This portion of the experiment will reveal the relative contribution of enzyme B to the biotransformation of the carcinogen in the tissue preparation. Taken together, results will establish the relative contribution of each enzyme to the biotransformation of the chemical carcinogen in the tissue sample (Figure 7.3).

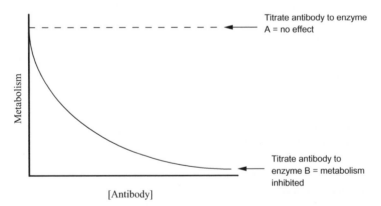

Figure 7.3. The use of immunoprecipitation to assess the relative contribution of enzymes A and B in a cellular preparation to the metabolism of a chemical carcinogen. Metabolism was decreased following immunoprecipitation with increasing concentration of antibodies to enzyme B. These results indicate that enzyme B, and not enzyme A, was responsible for the metabolism of the carcinogen.

Immunoprecipitation also is a powerful and effective means of identifying interactions among proteins. For example, many proteins normally exist in multimeric complexes with other proteins within the cell. Immunoprecipitation can be used to precipitate a protein that is targeted by the antibodies used. Proteins in the precipitate then can be analyzed to determine what other proteins were associated with the targeted protein and co-precipitated by the antibodies.

Co-immunoprecipitation methodologies have similarly been used to evaluate protein–DNA binding interactions. For example, chromatin immunoprecipitation (ChIP) makes use of immunoprecipitation to identify specific locations in a genome at which targeted protein interact. ChIP is commonly used to study histone and other DNA-binding proteins (e.g., transcription factors) in the context of the DNA with which they associate. Typically, protein–nucleic acid binding interactions are fixed within the cells by crosslinking with formaldehyde. After crosslinking, chromatin is released from the nuclei and sheared into fragments by sonication. Chromatin fragments (typically 200–1000 base pairs) are incubated with antibodies to the targeted protein. Antibody–protein–DNA complexes are then immunoprecipitated with protein A- or G-coated beads as described above. Following immunoprecipitation, DNA fragments are released from the complex, and column purified. The identity of the DNA fragments can then be determined using PCR amplification with primer sets that target specific regions in the genome under study (e.g., promoter region of a specific gene).

7.6.4 Immunoblotting (Western Blotting)

This procedure makes use of electrophoretic and immunochemical techniques to identify a specific protein in a complex mixture. Immunoblotting can be used to (a) determine the presence of a specific protein in a biological preparation (b) determine relative amounts of a specific protein in a biological preparation, and (c) estimate the molecular weight of a specific protein.

Consider the following experiment. A toxicologist wants to determine whether administration of a polychlorinated biphenyl (PCB) to rats induces a specific hepatic P450 enzyme (Figure 7.4).

- Liver samples from both PCB treated (T) and untreated (C) rats are subjected to SDS-polyacrylamide gel electrophoresis to separate individual proteins based upon their molecular weights. While separated proteins are depicted in Figure 7.4, these proteins would not be visible without the use of some means of protein detection (i.e., protein staining).
- The separated proteins are transferred from the fragile acrylamide gel used in electrophoresis to a membrane (e.g., nitrocellulose) to which the proteins will bind.
- The remaining protein-binding sites on the membrane are blocked to prevent nonspecific binding of the antibodies to the membrane. This is generally accomplished by incubating the membrane in a solution containing a high concentration of proteins (solution of milk proteins, solution of serum proteins, etc.).
- Membrane is incubated with primary antibody. A binary complex forms that consists of the antigen on the membrane and the primary antibody.

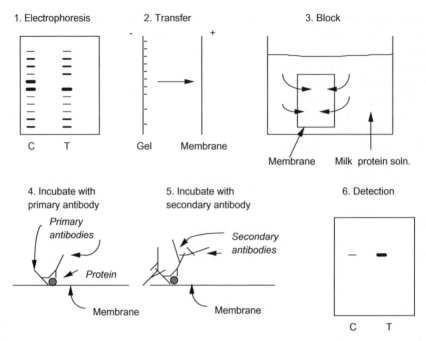

Figure 7.4. Diagrammatic representation of the use of immunoblotting to assess relative levels of a P450 protein following treatment of rats with a PCB. C, hepatic microsomal proteins from a control, untreated rat; T, hepatic microsomal proteins from a rat treated with PCBs.

- Membrane is incubated with secondary antibody. A tertiary complex forms that consists of the antigen–primary antibody–secondary antibody.
- The immune complex on the membrane is visualized using appropriate procedures. Color development or chemiluminescence with alkaline phosphatase or horseradish peroxidase-conjugated secondary antibodies are most commonly used.

7.6.5 Radioimmunassay

Radioimmunassay (RIA) is used to detect and quantify minute quantities of an antigen in biological or environmental samples. RIA is commonly used to measure nonprotein antigens such as drugs and toxicants. This immunoassay is highly quantitative. An example of the use of RIA in toxicology would be the analyses of dioxin levels in the blood of employees from a chlorophenol production plant.

To perform this assay, antibodies to dioxin would be required. Since the molecular weight of dioxin is less than 5000 daltons, it would be necessary to use dioxin as a hapten to generate the antibodies. Various strategies can be used in RIA though the antigen capture strategy is most commonly used (Figure 7.5).

- A known amount of antibody is bound to a solid phase (e.g., PVC test tubes or wells of a microtiter plate).

1. Known amount of dioxin antibodies bound to wells

2. Fixed concentration of radiolabeled dioxin (●) and increasing
   concentrations of unlabelled dioxin (o) added to wells

3. Wash out unbound dioxin and measure radioactivity associated with wells

4. Prepare a standard curve comparing the amount of unlabeled dioxin
   added to the wells and radioactivity (DPMs) measured in the wells

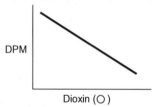

5. Incubate blood samples in wells (step 2) in place of unlabeled dioxin. Process
   through step 3. Determine concentration of dioxin in blood samples by using the
   standard curve to convert DPM to the concentration of dioxin in the sample.

Figure 7.5. Diagrammatic representation of the use of RIA for the quantification of dioxin
levels in the blood of factory workers.

- A known amount of $[^{125}I]$dioxin is added to a series of antibody-containing
  tubes. An increasing amount of unlabeled dioxin is added to each tube.
  The solutions are incubated, during which time the iodinated dioxin and
  the noniodinated dioxin will compete for binding with the immobilized
  antibodies.

- The solutions are removed, the tubes are rinsed, and the radioactivity remaining in each tube is measured. The radioactivity remaining in each tube will be
  equal to the amount of $[^{125}I]$dioxin bound by the antibodies in the tubes. The
  amount of bound $[^{125}I]$dioxin will decrease with increasing concentrations of
  unlabeled dioxin (due to increased competition for the antibody). The amount

of bound radioactivity associated with each concentration of unlabeled dioxin can then be used to construct a standard curve.

- Incubate the blood sample in an antibody-containing tube along with the same amount of [^{125}I]dioxin used to develop the standard curve. Remove the solution and measure radioactivity associated with the tube. The level of radioactivity measured can then be applied to the standard curve to determine the amount of dioxin in the blood sample.

7.6.6 Enzyme-Linked Immunosorbant Assay

The enzyme-linked immunosorbant assay (ELISA) encompasses a variety of assay designs for the analyses of either an antigen or antibody in a mixture using some means of immune complex immobilization and enzyme-mediated detection. The basic assay has been used as a rapid screening method for assessing antibody titer in plasma samples during antibody production, the detection of antigens in biological samples, and the quantification of chemicals (e.g., pesticides) in environmental samples. While historically used as a qualitative measure of antigen or antibody presence, the assay has been modified to provide quantitative measures that rival RIA. The following are three examples of the use of ELISA.

Example 1. A rabbit has been injected with a protein with the intent of generating polyclonal antibodies to the immunogen. After adequate time for the generation of the immune response, a blood sample is drawn from the rabbit and serum is analyzed for the presence of the antibodies by ELISA (Figure 7.6).

Method:

- The antigen is bound to a solid phase (e.g., well of a PVC microtiter plate). This bound antigen will serve as the means of capturing the antibodies of interest from the serum. Remaining unoccupied protein binding sites are blocked with a protein-containing reagent (e.g., milk protein solution).
- Serum from the rabbit is added to the antigen-containing well and incubated. Antibody–antigen complexes will from during this time.
- Serum is removed and the well is thoroughly washed. After washing, the only serum proteins that will remain in the well are those bound to the immobilized antigen (i.e., the antibodies of interest).
- Enzyme-linked secondary antibody is added to the well. The solution is incubated for a sufficient time to allow for the tertiary complexes (antigen–primary antibody–secondary antibody) to form.
- The solution is removed, the well is washed thoroughly (to remove unbound secondary antibody), and the immune complex is assayed for using the appropriate colorimetric or luminescent substrate.
- The presence of the antibodies in the serum samples will be indicated by the generation of color or light.

Example 2. Male fish have been collected from a reservoir suspected of being contaminated with an estrogenic chemical. Blood samples are drawn from the fish and

1. Wells coated with antigen (●) remaining protein
   binding sites are blocked

2. Serum from immunized rabbit added to well.
   ⅄=desired antibodies; ◌ =other serum proteins

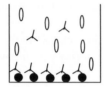

3. Wash out unbound proteins

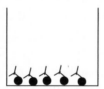

4. Add secondary antibody (Ⴟ)

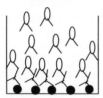

5. Wash out unbound secondary
antibodies and assay for complexes

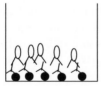

Figure 7.6. Diagrammatic representation of the use of ELISA to assess antibody titer in the serum of an immunized rabbit.

are to be analyzed by ELISA for the presence of a protein that is induced by estrogenic compounds.

Method:

- Sera collected from the fish is added to the wells of a PVC microtiter plate. Serum protein, including the antigen of interest, will bind to the PVC. Remaining unoccupied protein binding sites are blocked with a protein-containing reagent (e.g., milk protein solution).
- Antibodies to the antigen of interest are added to the wells and incubated. Antibody–antigen complexes will form during this time.
- Solutions are removed and the wells are washed thoroughly. After washing, the only antibodies that will remain in the wells are those bound to the immobilized antigen.
- Enzyme-linked secondary antibodies are added to the wells and incubated for a sufficient time to allow for the tertiary complexes (antigen–primary antibody–secondary antibody) to form.
- The antibody solutions are removed, the wells are washed thoroughly (to remove unbound secondary antibody), and the immune complexes are detected using the appropriate colorimetric or luminescent substrate.
- The presence of the estrogen-inducible protein in the serum samples will be indicated by the generation of color or light.

Example 3. Water samples have been collected from a farm pond that was found to contain deformed frogs. The water samples are to be analyzed by ELISA to determine the concentration of an agricultural fungicide that may be responsible for the deformities. A *competitive ELISA* in conjunction with a standard curve would be necessary due to the quantitative nature of the analyses (Figure 7.7).

Standard Curve Preparation:

- Increasing concentrations of the antigen (fungicide) are incubated with a fixed concentration of antibody. With increasing antigen concentration, more of the antibodies in the incubation solution will be complexed and inactivated.
- A fixed amount of antigen is bound to the wells of a microtiter plate. The remaining protein-binding sites in the wells are blocked by incubating with a protein-containing reagent (e.g., milk protein solution).
- Solutions containing the preincubated antibodies and antigen, from step 1, are transferred to the antigen-coated wells. Free (uncomplexed) antibody will bind to the antigen fixed to the well. The amount of free antibody available to bind to the fixed antigen in the well will decrease with increasing concentration of the antigen added to the preincubation solution.
- Wells are washed and secondary antibody is incubated in each well. The tertiary complex forms with the amount of complex determined by the amount of free primary antibody added to the well.
- The solutions are removed from the wells, the wells are washed thoroughly (to remove unbound secondary antibody), and the immune complexes are assayed

1. Standard curve preparation. Incubate increasing concentrations of antigen
   (●) with a fixed concentration of primary antibodies (人).

2. Wells coated with antigen (●)

3. Solutions from step 1 added to wells prepared in step 2

4. Wash out unbound proteins and add secondary antibodies (人).

5. Wash out unbound secondary antibodies and assay complexes.

6. Prepare a standard curve consisting of amount of antigen added to tubes in
   step 1 to signal intensity in step

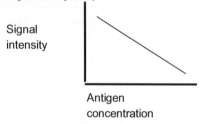

7. Incubate environmental samples in step 1 in place of antigen. Process
   through step 5. Determine concentration of antigen in samples by using
   the standard curve to convert signal intensity (step 5) to antigen
   concentration.

Figure 7.7. Diagrammatic representation of the use of a competitive ELISA for quantifying
the amount of a fungicide in reservoir water samples.

using the appropriate colorimetric or luminescent substrate. The intensity of the detection signal will be inversely proportional to the original concentration of antigen used as part of the standard curve.

Sample Analyses:

- Water samples containing an unknown amount of fungicide (antigen) are analyzed in the same manner as the standard curve with the water samples used in place of the known amount of antigen in the preincubations.
- Resulting intensity of the detection signal is compared to the standard curve and used to calculate the concentration of fungicide in the water sample.

7.6.7   Immunochemical Flow Cytometry

Flow cytometry allows for the rapid analyses of large numbers of individual cells for the presence of fluorochromes associated with the cells. Immunochemical approaches can be exploited in flow cytometry by introducing fluorochromes to the cells attached to antibodies that specifically recognize targeted biomolecules or chemical adducts associated with the cells. The cells suspended in fluid media are passed at high speed through a laser beam. Fluorescence emitted by individual cells are detected and collated to provide a highly quantitative measure of the proportion of the cells that carry the antibody and thus possess the antigen. Applications of immunochemical flow cytometry in toxicology include assessing the proportion of cells in which a protein is induced, adducts are formed, or cells are otherwise modified in response to chemical exposure.

7.6.8   Detection of Toxicant–Biomolecule Adducts

The covalent binding of reactive chemicals with DNA or protein is often a cause of toxicity. Such reactivity can be conferred to a chemical following metabolic activation. The ability to detect such toxicant–biomolecule adducts can be used as a dosimeter of chemical exposure, can be diagnostic of the toxicity of a chemical, and can be informative in establishing the mechanism of toxicity of a chemical. The immunochemical detection of adducts provides a highly sensitive means of detecting chemical–biomolecule interactions in small tissue samples. Accordingly, this approach has been used extensively in the biomonitoring of human populations that are at risk of exposure to reactive chemicals. Adduct formation can be evaluated immunochemically with the availability of antibodies either to the chemical or to the chemical–biomolecule adduct. The biomolecule population (e.g., total protein or DNA) is isolated from the tissue of interest (e.g., liver of dosed rodent, blood samples from a human population, in vitro cultured cells). The preparation is washed extensively to remove unbound chemical. Proteins are typically denatured to dissociate any noncovalent binding between chemical and protein. Chemical–biomolecule adducts can be measured from these samples using many of the assay techniques described in this chapter (immunohistochemistry, electrophoretic separation, ELISA). Immunochemical approaches have been used to detect and quantify adducts formed by drugs (i.e., acetaminophen, cisplatin), carcinogens (e.g., 4-aminobiphenyl, 2-acetylaminofluorene), and environmental chemicals (e.g., acrolein, 1,2-naphthoquinone).

7.7 CONCLUSIONS

A variety of immunoassays are available to toxicologists for the detection, characterization, and quantification of antigens. Antigens are most typically proteins, but through the use of haptens as immunogens, antibodies can be generated to virtually any molecule. Such advances have allowed for the development of methods for the immunoquantification of toxicants, drugs, and other chemicals in biological (e.g., blood) and environmental (e.g., water) samples.

SUGGESTED READING

Harlow, E., and Lane, D. Using Antibodies: A Laboratory Manual, Cold Spring Harbor Laboratory, Cold Spring Harbor, NY, 1999.

Price, C. P., and Newman, D. J. (Eds.). *Principles and Practice of Immunoassay*, Stockton Press, New York, 1997.

Shepherd, P., and Dean, C. (Eds.). *Monoclonal Antibodies: A Practical Approach*, Oxford University Press, New York, 2000.

Wild, D. (Ed.). *The Immunoassay Handbook*, Elsevier, Amsterdam, 2005.

CHAPTER 8

Cellular Techniques

SHARON A. MEYER

College of Pharmacy, University of Louisiana at Monroe, Monroe, Louisiana 71209

8.1 INTRODUCTION

Since Harrison, in 1907, first microscopically observed neurite outgrowth from cultured frog neurons, many have experienced the powerful persuasion of direct visualization of cellular responses in culture. Standardization of media composition, introduction of antibiotics, and development of cryopreservation techniques and immortalized cell lines in the 1950s enabled routine application of cell culture in the biological research laboratory. Utilization of cell culture in toxicology soon followed with development of systems for monitoring chemically induced mammalian genotoxicity.

A variety of cellular techniques are now used as tools for biochemical toxicology; in particular, they relate to (1) molecular changes associated with toxicant exposure, (2) models for study of toxicodynamic mechanisms that are simplified "living" systems relative to intact animals, (3) replacements for whole-animal toxicity testing for some well-validated endpoints, and (4) human cells for comparison to target tissue effects of toxicants identified in animal testing. Techniques enabling evaluation of cellular toxic effects of molecular changes are an indispensable complement to nucleic-acid-based screening techniques. Molecular changes ranging from increased expression of normal or mutated protein to ablation of protein function can be engineered into isolated cells or studied in cells isolated from engineered animals (see Chapter 2). Simplified cellular systems are amenable to relating cause and toxic effect and are especially valuable for characterizing toxicant interactions within multicomponent pathways. Cellular changes that have been specifically associated with organismal toxic effects and exhibit high sensitivity may be useful surrogates allowing replacement of certain animal tests. Determination of similar toxic effects and mechanisms in human cellular systems and those of experimental animals argues relevance to human health and provides information important for toxicant risk assessment.

Limitations of cellular techniques are largely due to two factors: (1) changes that occur in cells upon isolation from the organism and propagation and (2) difficulty in duplicating kinetic aspects of toxicant exposure as occurs in the intact animal.

Molecular and Biochemical Toxicology, Fourth Edition, edited by Robert C. Smart and Ernest Hodgson
Copyright © 2008 John Wiley & Sons, Inc.

The most common alteration that cells exhibit upon cultivation in culture is the loss of differentiated function. This is a consequence of the conflicting need to enhance cell proliferation to provide adequate experimental material. Differentiated function of a particular cell type is partly determined by interaction with other cell types and noncellular tissue constituents; these are lost upon isolation and traditional cell monoculture. Also, undefined biological materials such as fetal calf serum, pituitary extract, and so on, are routinely added to cell culture because they have been shown empirically to have mitogenic effects, but they also have often unappreciated effects on other cellular functions. With regard to toxicokinetics, one of the most challenging problems is relating effective toxicant concentrations in cell culture to that in vivo (Section 8.7).

Although there are many useful attributes of cell culture in toxicological research, it must be borne in mind that the cells are isolated from a multicellular organism (i.e., ex vivo). It is important that, insofar as possible, the culture conditions should closely mimic conditions within the intact organism. A modified toxic effect due to simplified, ex vivo culture conditions can be informative, provided that the basis for this change is understood and allows extrapolation to the in vivo response. Also, considerable progress in addressing many of the above limitations has been guided by discoveries of the fundamental properties of cell growth, differentiation, and their interaction. Many improvements have come from efforts to closely simulate the in vivo environment with the intent of maintenance of the differentiated state of cells in culture. Another approach is to propagate growth competent cells and induce their differentiation in culture. Some tumor-derived cells and nontumorous cells transformed with inducible transforming proteins from DNA viruses, especially SV40 large T antigen, exhibit such conditional proliferation and differentiation within a specific lineage. Pluripotential embryonic stem (ES) cells offer this potential across multiple lineages (Section 8.4.4).

The focus of this chapter is on technical aspects of isolation of cells from intact tissue, their propagation and maintenance in culture, and use in biochemical toxicology. General cellular mechanisms of toxic effect are the subject of Chapters 16–25. The methodology included is treated generally and is only a small collection from applied research on ex vivo cellular biology. For a more thorough description, the reader is referred to an excellent group of books by I. Fresney. An informative website with practical cell culture information is maintained by the American Type Culture Collection (http://www.atcc.org). There are also journals and serials devoted to in vitro toxicology, some of which are listed at the end of this chapter.

8.2 CELLULAR STUDIES IN INTACT TISSUE

8.2.1 Whole-Animal Studies

Although the emphasis of this chapter is on isolated cells in culture, responses at the cellular level can be assessed in intact tissue after exposure of the whole animal. Observations of cellular responses are most often made with in situ detection techniques and microscopic observation, such as immunohistochemisty (Chapter 7) and nucleic acid hybridization (Chapter 2). Preparations used for these in situ techniques are generally tissue that has been fixed after toxicant treatment, then embedded and sliced thinly enough ($\sim$5 μm) to enable observation by microscopy, usually with transmission or fluorescent modes. Immunological and nucleic acid probes are available that allow observation of many proteins as modulated at transcriptional and posttranscriptional steps and of complex cellular processes such as cell proliferation and apoptosis. Probes for protein modification are available for such changes as signal transduction-induced phosphorylation, oxidant-induced reactions, and alkylation by reactive metabolites.

8.2.2 Tissue Slices

Tissue slices are an ex vivo system that offers simplicity relative to the organism, but retains some of the tissue-level complexity relative to isolated, cultured cells. Although earlier preparations suffered from extremely short viability, tissue slices have again become a popular source of cellular material since the introduction of vibrating, precision slicing instruments. Tissue slices have found greatest applicability in assessment of toxicant metabolic profiles, although loss of metabolic enzyme activity with time has limited this application to short-term studies (<24 hr). This limitation also restricts studies on toxic effects to immediate responses to acute exposures. Although the thickness of tissue slices (~200 μm) precludes observation with standard microscopic techniques at the cellular level, the newer development of confocal microscopy allows real-time observation of fluorescently tagged responses and should enable broader application of tissue slices in the study of toxic effects.

8.2.3 Reconstructed/Bioengineered Tissue

Various tissue constructs have been reassembled from isolated constituents, including resident cell types whose numbers have been amplified or modified in culture. A three-dimensional co-culture system for human skin keratinocytes layered upon a synthetic mesh infiltrated with dermal fibroblasts, when floated to allow contact of the uppermost keratinocytes with air, exhibits stratification and cornification remarkably similar to in vivo squamous epithelia. This reconstructed epithelial model has been recommended as an in vitro replacement for dermal corrosivity testing. It has been anticipated that this and a similar noncornified model will have application in dermal and ocular irritation testing, but thus far validation studies have yielded mixed results. Reconstructed tissues can also provide context for basic toxicological research on aberrant cellular interactions with cellular and acellular constituents, as illustrated by invasion of cancerous epithelial cells into underlying "dermis" of a skin equivalent model.

Further potential for bioengineered tissue in toxicology has resulted from adaptation of microfabrication techniques to cell culture. Construction of complex arrays of cells encapsulated in hydrogels separated by minute fluidic channels is now possible. Efficient, high-throughput systems for monitoring toxic effects can then be engineered by incorporation of biosensors in such designs.

8.3 STUDIES WITH DISPERSED, ISOLATED CELLS

Suspensions of freshly isolated, dispersed cells commonly are used in short-term incubations for biochemical studies on intermediary metabolism. Similar systems are useful for biochemical toxicology, although the limited viability of these preparations restrict their application to acute exposures and measurement of relatively rapid responses. Alternately, toxicant exposure in suspension followed by monolayer culture appears to be a suitable method for study of activation-dependent toxic effects that require longer times for expression. Rodent hepatocytes are frequently used in this type of study because of the ease of isolation of large numbers of viable cells and because hepatocytes from numerous species, including humans, are readily available for comparative studies.

8.3.1   Tissue Digestion and Cell Separation

Except for circulating blood cells and cells easily flushed by lavage, such as peritoneal and alveolar macrophages, preparation of isolated cells generally requires some means of release of individual cells from a solid tissue environment and separation from other dislodged tissue constituents. Cell–cell and cell–substratum interactions are the basic forces that maintain tissue architecture. Cell–cell interactions are several types of varying strength, but in general are mediated through protein binding, some of which is Ca^{2+}-dependent. Relatively strong intercellular connections, tight junctions, and desmosomes join cells of tight endothelia and epithelia, like duodenum. Several of the membrane proteins of these junctional complexes have been identified at the molecular level, and their associations have been shown to be Ca^{2+}-dependent. In addition, adherens junctions that are dispersed along the lateral walls of tight epithelia and through the plasma membranes of nonepithelial cells mediate weaker cell–cell associations. The proteins of these structures, the cadherins and associated catenins, also require Ca^{2+} for binding.

Cell-substratum interactions occur at hemidesmosomal complexes. These structures contain the transmembrane α- and β-subunits of cell-type specific integrins. The extracellular ligands for the integrins are anchorage proteins, such as fibronectin, laminin, and osteopontin, that are embedded within a collagenous matrix. There are several α- and β-integrin, anchorage protein, and collagen gene products, and each combination is cell-type-specific (Figure 8.1). This association is Ca^{2+}-dependent and, for fibroblasts, relies on recognition by integrin of the arginine–glycine–aspartatic acid (RGD) tripeptide motif of fibronectin. Extracellular matrix also includes considerable amounts of carbohydrate polymers in glycosaminoglycan chains linked to the polypeptides of proteoglycans and as hyaluronic acid. Cellular integral membrane proteins that bind hyaluronic acid have been identified. Also, the transmembrane proteoglycan syndecans participate in cell–matrix adhesion by bridging cytoskeletal actin to various matrix components bound to their sticky glycosaminoglycans.

Since protein complex formation and Ca^{2+} are critical to cell fixation within a tissue, dissociation media usually contain some type of proteolytic enzyme and the Ca^{2+} chelator, EDTA. The proteolytic enzyme can be of general specificity, such as trypsin, or can be a more targeted enzyme, such as a collagenase selective for the collagen-type characteristic of the tissue of interest. Hyaluronidase has been also used with matrix rich in hyaluronic acid, such as for isolation of duodenal enterocytes. In all cases, the appropriate incubation times and concentrations to achieve cell dispersal, but retain high viability, need to be determined empirically. One factor

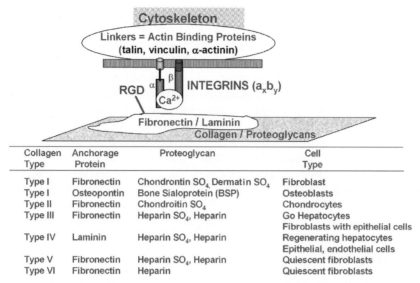

Collagen Type	Anchorage Protein	Proteoglycan	Cell Type
Type I	Fibronectin	Chondrontin SO$_4$ Dermatin SO$_4$	Fibroblast
Type I	Osteopontin	Bone Sialoprotein (BSP)	Osteoblasts
Type II	Fibronectin	Chondroitin SO$_4$	Chondrocytes
Type III	Fibronectin	Heparin SO$_4$, Heparin	Go Hepatocytes
			Fibroblasts with epithelial cells
Type IV	Laminin	Heparin SO$_4$, Heparin	Regenerating hepatocytes
			Epithelial, endothelial cells
Type V	Fibronectin	Heparin SO$_4$, Heparin	Quiescent fibroblasts
Type VI	Fibronectin	Heparin	Quiescent fibroblasts

Figure 8.1. Generalized scheme of components involved in a cell's interaction with extracellular matrix within an intact tissue. Collagen–anchorage protein–proteoglygan combinations specific for various cell types are listed below. Ca^{2+} dependence is also indicated. Preparation of isolated cell suspension requires that these interactions be disrupted, as generally accomplished with proteolysis and Ca^{2+} chelation.

that compromises viability is the variable proteolysis of cells at different depths in the tissue, even when minced into small blocks (~1 mm^3). For some tissues, the inherent structure enables access of dissociation medium to deeper cells. The ease of preparation of rat hepatocyte suspensions is enabled by perfusion through the existing sinusoidal vasculature such that each hepatocyte is bathed on two sides with dissociation medium.

All solid tissues are composed of a variety of cell types. If the cells of interest are fibroblasts, traditional monolayer culture on plastic substratum in medium supplemented with fetal calf serum will select for this type of cell. If differential sensitivity of a tissue's cell types to any of the dispersal conditions exist, then these can be exploited to generate a suspension enriched in a given cell type. An example is the isolation of nonparenchymal cells from liver digested with pronase, which destroys parenchymal hepatocytes. However, isolation of a specific cell type more frequently requires application of separation techniques that exploit differences in physical or biological characteristics of the different cell types after dispersal. The most common method is velocity sedimentation, (i.e., centrifugation at a fixed g-force and time), and separation is based upon size. Hepatocytes are approximately twice as large as nonparenchymal cells and thus can be cleanly separated by low-speed centrifugation. A better technique for cells of similar size entails centrifugation through a density gradient such that a bouyant force counters the centrifugational force. When equilibrium is achieved, cells will band at a density equivalent to the cellular density. Ficoll, Percoll, and Nicodenz are polymers that contribute density, but not excessive osmotic strength, and are commonly used with cell suspensions.

In some cases, the biology of the cell type can be used for selection. For example, hepatocytes are the only liver cells that will survive in arginine-free, ornithine-supplemented medium since arginase, a urea cycle enzyme exclusive to hepatocytes, is needed to synthesize the essential amino acid arginine. Several methods can be used to achieve separation based upon cell-type-specific surface markers. Surface markers can be tagged with fluorescent antibodies and cells separated with a fluorescence activated cell sorter (FACS). Antibodies to cell surface markers bound to magnetic beads allow target cells to be pulled from suspension with a magnet. Ligands of surface receptors can be coated on tissue culture plasticware and adherent cells collected by panning. If only small numbers of cells are required, microarray technology with assorted spotted capture ligands can be used to simultaneously isolate multiple cell types from a mixed population. Numerous high-throughput platforms can then be used to coordinately perform cytotoxicity assays of the various living cell types in response to toxicant.

8.3.2 Limited Maintenance in Defined Media

Once the cell type of interest is isolated, short-term culture in suspension will require choice of an appropriate incubation medium. In general, media requirements are relatively simple compared to formulations for longer-term monolayer culture. Minimally, suspended cells need salts to maintain osmolarity, a pH buffer, and an energy source. Usually NaCl, bicarbonate and glucose serve these needs in the standard balanced salt solutions (e.g., Earle's, Hanks'). It is also necessary to provide a gas phase of oxygen and CO_2. The latter buffers the solution upon equilibration with bicarbonate anion and is the system of choice since cells independently require some bicarbonate for metabolic processes. Oxygen (95%)/5% CO_2 gas is usually provided as the head space in a gas-impermeable container, but this arrangement introduces pH instability if interim samples are to be taken from a common vessel. To compensate, media may contain other, solution-based buffers such as HEPES. Also, oxygen delivery from the gaseous phase can be diffusion-limited, and thus shallow solutions and gentle mixing are required.

8.3.3 Long-Term Suspension Culture

Cell types that are normally nonadherent, such as lymphocytes, and some tumor cells are able to grow in suspension when the medium is supplemented with adequate nutrients. For small-scale cultures, cells can simply be seeded onto standard tissue culture plasticware and maintained in a shallow layer of medium in a CO_2 incubator. If necessary, cell attachment can be prevented by coating the plastic surface and reducing the medium Ca^{2+}. Suspension cultures can also be propagated in roller bottles, and larger vessels (spinner flasks) have been designed for hundreds of milliliters of stirred culture medium. Growth in suspension culture is advantageous for ease of propagation of large quantities of material. This feature is the impetus for adaptation of adherent cells to suspension culture by seeding them onto microcarriers. In this way, their growth requirement for attachment is met, yet large quantities can be efficiently cultured using suspension techniques, such as in industrial bioreactors.

8.4 MONOLAYER CELL CULTURE

Proliferation of most cells in culture requires attachment to a substratum. The most common means to achieve this is to seed cells onto the plastic base of tissue cultureware. Tissue culture plastic is usually polystyrene modified to carry a charged surface to which cells can initially attach and then subsequently elaborate an underlying matrix and spread to form a monolayer. Growth and maintenance of cultured cells for longer than a few hours also requires additional nutrients, which are added to bicarbonate- and glucose-supplemented basal salt solutions in various formulations as standard tissue culture media. Necessary nutrients are essential amino acids, vitamins, additional salts, and trace minerals. Many cells also require serum for optimal growth, although this contributes considerable variability to cell culture because serum is largely undefined and its constituents vary from sample to sample. Serum replacements have been formulated from the identified serum components known to support growth; and for those cells that will adapt to serum-free media, their use can remove some of the empiricism of cell culture. Some of the more important serum constituents are insulin, growth factors, the nutrients ethanolamine and pyruvate, selenium for glutathione peroxidase synthesis, and transferrin for provision of bioavailable iron. All standard culture media, except L15, use bicarbonate buffering; thus, a CO_2 gas phase also is needed for monolayer culture. In this application, CO_2 is mixed at ~5–10% volume with ambient air (~20% O_2) in a tissue culture incubator, a specialized cabinet that maintains humidity and temperature, as well as CO_2. Phenol red is also included in standard media to enable convenient visual assessment of pH during culture.

8.4.1 Propagation of Primary and Passaged Cultures

The monolayer culture that results after seeding a cell suspension from dissociated tissue is called a primary culture. Those cells that are growth competent initially will exhibit "log-phase" growth during which they proliferate exponentially providing that their medium is replenished every few days (Figure 8.2). When the surface area of the vessel bottom is covered (i.e., at confluence), the closely apposed cells will then arrest their growth, a phenomenon named contact inhibition. To resume proliferating, the cells must be replated onto a larger surface area per cell. This procedure, called subculturing, involves detachment of adherent cells from the substratum and dissociation to a single-cell suspension, followed by reseeding into a larger vessel or reseeding a portion of the cells in a vessel of the same size. The detached, dissociated cell suspension is generated using a limited protease digestion, usually with trypsin, and EDTA. This procedure is repeated, with each repetition referred to as a "passage" and with the ratio of the surface area per cell after subculture to that before subculture as the "split ratio."

After numerous passages, cells will no longer proliferate and cultures will begin to deteriorate. This property, "replicative senescence," is thought to be a manifestation of a predetermined limit on the number of proliferative cycles available to a cell. During senescence of the culture, rare cells may arise that exhibit resistance, continue to proliferate, and overgrow the culture.

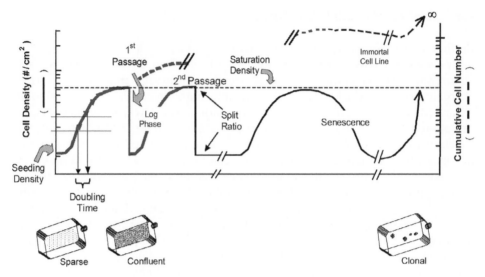

Figure 8.2. Propagation of cells in monolayer culture. Cell proliferation as an increase in number of cells/cm² within a passage (——) and as cumulative cells from all passages (- - -) is shown. Tissue culture flasks at various stages of monolayer growth are diagramed below the graph to indicate the corresponding stage of confluency. Note the logarithmic scaling of the ordinate.

8.4.2 Immortalized Cells

These "immortalized" cells then become the founders of continuous cell lines. Intense research has been conducted to define the molecular basis of senescence since its occurrence has profound implications for aging and carcinogenesis. One of the best supported concepts is that chromosomal ends, telomeres, shorten with each replicative cycle until chromosome loss limits viability. Cells upon immortalization then derepress expression of the enzyme telomerase that restores chromosome length. Rodent cells have a low rate of spontaneous immortalization that can be substantially increased by transfection with the oncogenes of certain DNA tumor viruses. DNA tumor viral oncogenes are not sufficient to immortalize human cells, but they do condition human cell cultures to facilitate additional complementary changes in rare cells that enable them to escape senescence. The most commonly used viral oncogene for immortalization is the large T antigen of SV40. Alternately, human cells can be immortalized directly by engineering the expression of a recombinant telomerase.

Once an immortalized cell line has been established, it is possible to obtain derivatives or "clones" that originate from a single cell. These can be generated by plating a cell suspension diluted to theoretically yield one cell per small vessel or by sparsely plating on a larger plate and subsequently collecting isolated cell clusters with cloning cylinders. Random clones can be isolated from normal medium with the objective of generating a homogenous cell line or can be isolated from conditions that impose a selective pressure, such as resistance to a given toxicant. In the latter case, there will be genotypic changes characteristic of the clonal cell line that enabled its survival in the selective medium. The number of cells that survive in selective medium is increased by mutagenic and clastogenic events, and quantitation of clonal survival is a frequently used parameter of genotoxicity (see Chapter 25).

Immortalized cells can be induced to undergo additional changes in phenotype. Transformation is a change that reflects a further loss of growth control from immortalization and is highly correlated with tumorigenicity of cells when they are transplanted into animal hosts. Transformation has been an invaluable tool for identification of carcinogenic chemicals and oncogenes and for mechanistic studies on carcinogenesis (Chapter 24). Properties of transformed cells in culture are (1) a spindle-shaped morphology, (2) reduced serum requirement, (3) loss of contact inhibition, and (4) anchorage-independent growth. Loss of contact inhibition is apparent as small (~2-mm diameter) piles of cells or foci that overlie the monolayer. Anchorage-independent growth is operationally defined as the ability to grow in soft agar, which precludes contact with a substratum. Another phenotypic change that can be induced in several immortalized cell lines is expression of differentiated properties. Examples include cell lines that express differentiated properties of neurons, hepatocytes, adipocytes, skeletal and adrenocortical cells, and renal tubule and intestinal epithelial cells. Differentiation can be induced by a variety of hormonal and chemical agents such as cyclic AMP, glucocorticoids, butyrate, retinoic acid, and DMSO. Selected examples of application of differentiated cell lines in biochemical toxicology are presented in Table 8.1.

8.4.3   Modifications to Monolayer Cell Culture

The tendency to lose differentiated function in culture has been one driving force for innovative improvements in traditional monolayer cell culture. Provision of substrata to mimic that in vivo has improved maintenance of differentiated function in some cases. For example, primary rodent hepatocytes retain sensitivity to phenobarbital induction of the cytochrome P450 2B1 when cultured on matrix that contains mostly laminin with some fibronectin and proteoglycan. A synthetic polymer with attached repeats of the RGD integrin-binding motif has also been developed. Polarity of transporting epithelial cells, such as MDCK kidney cells, is sustained when they are cultured on permeable membranes inserted into the medium of traditional tissue culture plates. Culture in three-dimensions, such as with multicellular aggregates or on alginate microcarriers, has improved differentiation of some cells. Culture in the presence of other cell types (i.e., co-culture) can also improve viability and differentiation, as can culture of one cell type in medium conditioned by exposure to another cell type.

Co-culture also provides a system in which questions about cell–cell interactions can be addressed. If a toxic effect in a tissue requires cooperation between cell types, then considerable information about the mechanism of this interaction can be obtained through use of co-culture constructs. For example, an effect may be observed when different cell types are admixed, but not when cell types are physically separated and only communicate through a common medium, as can be achieved by plating one type on the vessel bottom and another on a membrane insert. This result would argue against involvement of a soluble mediator and suggest a mechanism involving direct contact of the different cell types.

TABLE 8.1. Examples of Application of Cell Lines Retaining Differentiated Properties in the Study of Toxic Effects

Cell Line	Source	Differentiated Cell type	Toxicant	Measured Endpoint
N1E-115	Mouse neuroblastoma	Cholinergic neuron	Lead	Blockage of voltage-dependent Ca^{2+} channels
			Pyrethroid insecticide	Prolonged open time for voltage-dependent Na^+ channels
PC12	Rat pheochromocytoma (adrenal medullary tumor)	Adrenergic neuron	Tricresyl phosphate (organophosphate)	Inhibition of neurofilament assembly and axonal growth
SK-N-SH	Human neuroblastoma	Neuron	Anesthetic N_2O	Depressed cholinergic Ca^{2+} signaling
Hepa-1	Mouse hepatoma	Hepatocyte	2,3,7,8-Tetrachlorodibenzodioxane (TCDD)	Induction of CYP 1A1 and 1B1
HII4E	Rat hepatoma	Hepatocyte	Polychlorinated biphenyls (PCBs)	Induction of CYP 1A1
			Cyclophosphamide (antineoplastic)	Cytochrome P450-dependent genotoxicity
HepG2	Human hepatoblastoma	Hepatocyte	Rifampicin (PXR ligand)	Inhibition of bile acid synthesis
3T3-L1	Mouse embryo fibroblasts	Adipocyte	TCDD	Inhibition of glucose transport and lipoprotein lipase
Y1	Mouse adrenocortical tumor	Adrenocortical cell	Methyl sulfone metabolites of PCBs	Inhibit corticosterone synthesis by competitive inhibition of CYP
LLC-PK1	Pig kidney	Renal tubule epithelial cell	Cadmium	Cytotoxicity, apoptosis
MDCK	Dog kidney	Renal tubule epithelial cell	Organic mercury compounds	Cytotoxicity, transepithelial leakiness
Caco-2	Human colon adenocarcinoma	Intestinal epithelial cell	Arsenic	Transepithelial leakiness
C2	Mouse skeletal muscle	Skeletal myotubes	Oxidants	Calcium dysregulation

8.4.4   Stem Cell Cultures

Interest in stem cells has heightened because of recent improvements in methods for propagation of human cells in culture and their promise for regenerative medicine. Cell culture techniques are pivotal for expansion of the limited numbers of cells present in vivo and in inducing lineage commitment of progeny cells.

The most versatile stem cell cultures are derived from the inner cell mass of embryonic blastocysts. Cultured, mouse embryonic stem cells have been long available and used primarily for generation of "knockout" mice. The intrinsic capacity of these pluripotential cells to differentiate along several independent lineages is so robust in monolayer culture that intervention, such as propagation on feeder layers of mitotically arrested fibroblasts, is necessary to maintain an undifferentiated state. Cell cultures with a more limited repertoire of induced lineages are those derived from mesencymal cells enriched by monolayer culture of adult bone marrow cell isolates and from human placenta. Minor populations of multipotent, self-renewing stem cells are assumed to exist in regeneration-competent adult tissue as occupants of specialized niches adjacent to localized proliferative, putative daughter cells. Tangible evidence for existence of adult stem cells comes from the well-known clonogenic cell culture assays for colony forming units of hematopoietic lineages from adult bone marrow cells. Direct demonstration in culture of multipotentially of clonally derived glial precursors from rat brain has convincingly supported the presence of stem cells in this tissue.

The value of cultured stem cells to toxicology is in their promise as a continuous source of cells that can be induced to express a chosen cell-type-specific mature phenotype. This achievement would address two current limitations in safety testing: scarcity of metabolically competent human systems and use of experimental animals. Numerous culture additives and conditions have been identified that cause commitment and partial differentiation along specific lineages, such as retinoic-acid-induced neuronal commitment of mouse embryonic stem cells. However, definition of conditions that produce fully differentiated cells with cell-type-specific function quantitatively similar to that of intact tissue remains problematic and is a high-priority research area in toxicology.

8.5   OBSERVATION OF CULTURED CELLS

Real-time observation of living cells is invaluable for monitoring the progress of cell cultures. However, living cells are translucent and thus require techniques other than standard bright-field microscopy for visualization. Phase contrast microscopy is the preferred technique for routine observation of monolayer cultures and relies upon the change in phase that occurs when white light, made coherent by a phase ring in the microscope condenser, passes through living cells due to differences in the refractive index. These phase changes can be observed as differences in brightness when the transmitted light rays constructively or destructively interfere when passed through a second-phase ring in the objective lens of a phase contrast microscope. Also, since cell monolayers are viewed while attached to the bottom of a culture vessel, conventional microscopic design does not provide a large enough working distance. Thus, the inverted phase contrast microscope is used, in which the relative position of objective lenses and condenser is reversed. With this design, light enters from the top and the image is collected underneath the specimen.

The recent development of cell permeable fluorescent tags has provided a means to observe alterations in cell function in addition to morphological changes. Inverted fluorescent microscopes are used for these studies, and various stage attachments can be added to allow control of temperature and other variables. Fluorescent tags are available for monitoring oxidant status, sulfhydryl content, intracellular Ca^{2+}, H^+, Na^+, and K^+, mitochondrial function, and membrane potential. Digital electronic imaging and computerized data analysis can enhance the sensitivity of this technique and provide information on temporal relationships between multiple responses within a single cell. A disadvantage of fluorescence imaging is poor resolution since light emitted above and below the plane of focus is also collected in the objective lens. However, this problem can be circumvented with the use of laser-scanning confocal light microscopy, which can optically limit the image to thin slices within the depth of the monolayer. Confocal microscopy has also enabled visualization of three-dimensional structures and fluorescent imaging of systems other than cell monolayers, such as tissue slices and multicellular aggregates.

8.6   INDICATORS OF TOXICITY

8.6.1   Intact Tissue

Overt cell death in intact tissue is recognized microscopically by loss of definition of cell margins and organelles with the vacated space filled with amorphous material (coagulative necrosis) and/or lymphocyte infiltrate. Nuclear disintegration is easily recognized as karyolysis. Apoptosis is a less extensive form of cell death and can be recognized microscopically as isolated cellular residues, sometime phagocytosed within neighboring cells, or after labeling for associated internucleosomal DNA strand breaks. Biochemical parameters of cytotoxicity are used to assess viability of tissue slices. In general, leakage into the incubation medium of normally intracellular components is used. The most commonly monitored parameters are K^+, ATP, and the cytosolic enzyme lactate dehydrogenase (LDH).

8.6.2   Cell Culture

Parameters used to measure toxicity of cultured cells can be general properties related to loss of vital functions or impairment of cell-type-specific functions. These indicators can be further classified as those occurring shortly after toxicant exposure and those requiring prolonged culture for expression.

Assessment of general vital functions is done within minutes to hours of toxicant exposure and is based upon monitoring performance of critical organelles. These endpoints are schematically represented in Figure 8.3. The most definitive indicator is cell death that occurs upon rupture of the plasma membrane. However, several parameters of dysfunction can be observed prior to this terminal event. Microscopically, various membrane extrusions are observed. The formation of filopodia, pseudopodia, and "blebs" (i.e., ballooned regions devoid of intracellular organelles) are indicative of disruption of plasma membrane interaction with the underlying corti-

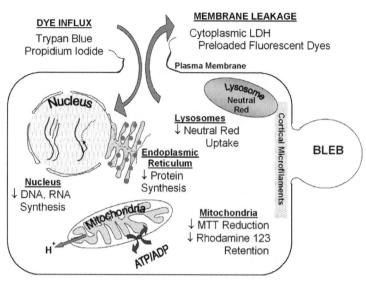

Figure 8.3. Idealized diagram of a cell to illustrate parameters often used to measure cytotoxicity and the corresponding affected subcellular organelle.

cal cytoskeleton. Damaged plasma membranes will also exhibit leakiness. This can be monitored as the egress of normally impermeant molecules, such as endogenous LDH or fluorescent carboxylate dyes, which are preloaded as permeable acetoxymethylesters that become entrapped upon carboxylase hydrolysis in healthy cells. Alternately, influx of the exogenous markers trypan blue or propidium iodide, which concentrate in the nucleus after breaching the plasma membrane barrier, is used.

Function of mitochondria is also commonly monitored as an indicator of cellular toxicity. Mitochondrial uptake and retention of the fluorescent dye, rhodamine 123, can be visualized microscopically. Biochemical measurements of mitochondrial function include the ATP–ADP ratio and dehydrogenase activity with MTT (3-(4,5-dimethylthiazol-2-yl)-2,5-diphenyltetrazolium bromide), which yields a colored formazan product upon reduction. The dye, neutral red (3-amino-7-dimethyl-amino-2-methylphenazine hydrochloride), targets lysosomes, and its retention is inversely related to cytotoxicity. Commercially available versions of the MTT and neutral red assays have been adapted to microtiter plate formats to provide highly efficient screening assays. Examples of how cell-type-specific functions can be followed as indicators of cell toxicity are included in Table 8.1.

Longer-term assays of cytotoxicity include assessment of general cell function. Toxicant effects on cellular protein, RNA, and DNA synthesis can be measured as incorporation of radioactivity in macromolecules of cells prelabeled with cell permeant, radioactive precursors, for example, [35S]methionine, [3H]uracil, and [3H]thymidine. Cell proliferation can simply be monitored by counting cell number with a hemocytometer or Coulter counter. Assessment of cell proliferation data is usually determined as rate of log-phase growth, as expressed as time for cell number to double (doubling time). Number of cells per area at stationary phase (saturation density) may also be affected. These parameters are indicated in Figure 8.2. A more stringent test of growth competency is afforded by assessment of clonal growth in which isolated colonies that survive and grow out from a known number of sparsely plated cells are scored. Apoptosis is measurable in cultured cells using DNA end-labeling or morphology. Cell culture techniques are also used for a subset of genotoxicity tests. Clonal growth in selective media is used for mutagenicity studies, and observations on cells arrested in mitosis can be used to assess larger chromosomal abnormalities. Excision repair of DNA damage is also observable in cultured cells as nonreplicative nuclear incorporation of radiolabeled deoxynucleotide precursor ("unscheduled DNA synthesis," UDS). The theoretical basis and interpretation of these types of genotoxicity assays are detailed in Chapter 25.

8.7   ARTIFACTS AND CONFOUNDERS

Cytotoxicity studies in cultured tumor cells have served as an essential tool for chemotherapeutic drug development for decades. However, cell culture results are sometimes dismissed outright in toxicology because of beliefs that the inability to experimentally replicate a cell's complex in vivo exposure to toxicant precludes expression of a meaningful response. A common critique is that of lack of toxicological relevance because toxic concentrations in culture exceed those thought effective in vivo. One source of decreased sensitivity, that imposed by lesser metabolic activation in cultured cells, can be addressed by demonstration of toxicity upon direct addition of metabolic intermediates at lower concentrations. However, very little toxicokinetic data are currently available for direct comparison of plasma concentrations of toxicant or metabolites and cytotoxic concentrations in cell culture. Such a comparison is further complicated by the need to adjust concentrations by factors affecting activity, such as due to protein binding, which are different in vivo and in culture. Only when quantitative values for these parameters, both in culture and in vivo, become routinely available will it be possible to model activity of chemical at its cellular target in an intoxicated animal for comparison to that in culture. Compilation of such a database can be justified by the recent determination that empirically determined EC_{50}s for cytotoxicity in cultured cells are reasonably correlated ($r^2 \sim$ 0.5) with oral LD_{50}s for acute rat lethality (http://iccvam.niehs.nih.gov/).

Another issue that influences the concentration deemed "toxic" in culture is choice of endpoint. As in experimental animals, lethal and pre-lethal events can be monitored as metrics of adverse effect with the latter occurring at lower doses. Pre-lethal outcomes are predictive of lethality if they are biomarkers of effect, usually as a result of their being more distal events in a sequence of failures leading to organismal death. Modeling of dose responsiveness on the basis of other coincidental events, such as unappreciated adaptive responses, can lead to overestimation of risk of chemically induced adverse effect. Similar relationships exist for cytotoxicity endpoints in culture and likewise affect derivation of "cytotoxic" concentrations. One can thus increase sensitivity through choice of a pre-terminal cellular response, but care must be exercised such that specificity is not compromised. Use of early-onset, pre-terminal metrics of cytotoxicity, combined with metabolite profiling, offers one means for assessing activation-dependent cytotoxicity in cell cultures with limited life spans. This approach may prove especially useful for utilization of suspension cultures of cryopreserved human cells, which retain metabolic competency but are only viable for a few hours.

8.8   REPLACEMENT OF ANIMAL TESTING WITH CELL CULTURE MODELS

The above discussion has largely been focused upon techniques enabling provision of cell culture models for mechanistic study of toxic effects. The greatest priority in model development to achieve this objective is maintenance of target tissue differentiated function. However, another major thrust driving the use of cell culture models in toxicology is the promise that they hold for replacement of animals in toxicity testing. Much of the momentum for in vitro toxicology arose because of the ethical questions inherent in imposition of potential pain, distress, and lethality on nonhuman, higher species for the purpose of predicting human risks from chemical exposure. However, another equally compelling reason to incorporate cell culture models is that they can be used to address the limitations in extrapolation of toxic effects in nonhuman species to humans through the so-called "parallelogram" approach (Figure 8.4). Its application leads to increased confidence in the predictivity of animal test results when shared metabolite profiles and toxic endpoints are demonstrated in culture for both animal and human cells from the target tissue identified in the experimental animal.

Several academic, government, and industrial organizations have been formed to address the potential of "alternative" methods in toxicity testing. Alternative test methods have been defined as procedures that accomplish the "3Rs"—that is, replacement, reduction or refinement of use of higher animals. Several cell culture systems are being considered as alternative test methods and comprehensive, inter-laboratory studies designed to validate their results against historical animal results are in progress. In general, these efforts have revealed that qualitatively there is often good concordance between the in vitro and in vivo methodologies for

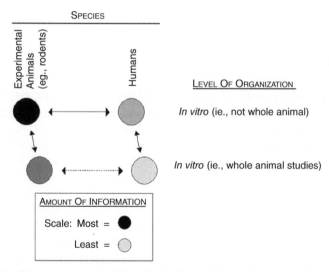

Figure 8.4. Parallelogram approach to relating knowledge obtained with in vitro systems from experimental animals and humans to in vivo observations with experimental animals and extrapolation to predict human in vivo responses. (Modified from Sobels, F. H. *Arch. Toxicol* . **46**, 21–30, 1980.)

chemicals at the extremes of toxicity and nontoxicity. However, ranking of chemicals with more similar toxicities is less well predicted, and quantitative relationships are difficult to relate to in vivo exposures. Thus, it appears that the most beneficial application of cell culture models as an alternative will be as early screens in tiered protocols for product safety testing. These in vitro systems appear adequate to identify highly hazardous compounds for which no further animal testing may be needed to justify curtailment of new product development. A similar application that fulfills the 3R objectives is use of results from chemically induced cytotoxicity in cultured cells to predict starting dose for acute lethality determinations, a protocol currently under review by the Interagency Coordinating Committee on the Validation of Alternative Methods (ICCVAM) of the National Toxicology Program (http://iccvam.niehs.nih.gov/).

Another application of cell culture expected to impact on toxicity testing is the use of engineered cell lines for hazard identification in high-throughput screening protocols. Knowledge gained from basic research identifying mechanisms and molecular mediators of certain toxic effects has been utilized in constructing reporter systems for certain classes of chemicals. A noteworthy example is the use of cell lines designed to detect estrogenic or androgenic activities of chemicals. It has recently been mandated that chemicals be tested for "endocrine disrupting activity." The magnitude of this task, which is predicted to require assessment of over 60,000 chemicals in the near future, has necessitated prioritization based upon prescreening results. For this purpose, cell lines have been engineered to contain a vector with a reporter gene whose expression is responsive to activation of a cotransfected steroid hormone receptor. Activation or inhibition of the steroid receptor activity by test compound can then be efficiently determined by monitoring signal from the reporter gene product. Similar engineered cell lines are available for detecting dioxin-like compounds through their interaction with the Ah receptor and others that are sensitive to heavy metals, oxidative stress, and DNA damage.

8.9 CONCLUSIONS

Acceptance of cell culture techniques in biochemical toxicology has been slow relative to other basic science fields. The most important limitation contributing to this hesitancy has been the loss of differentiated function in culture, primarily loss of components necessary for metabolic activation. However, these problems are being resolved as discoveries about the basic biology of cell differentiation are translating into modifications to cell culture methodology that support maintenance of metabolic function. Consequently, several cell culture systems are now available for study of metabolism-dependent toxic effects. Also, the increased availability of cell culture systems from humans has substantially improved the predictability of chemical toxicity through metabolic profiling. Perhaps the last frontier will be application of cell culture systems in toxicology testing. At present, some limited applications have gained acceptance, and practical and ethical issues concerning animal usage will most certainly stimulate further attempts to develop and validate in vitro toxicology models.

SUGGESTED READING

Ehrich, M., and Sharova, L. Unit 2.6: In vitro methods for detecting cytotoxicity. In: Costa, L.G., et al. (Eds.). *Current Protocols in Toxicology*, John Wiley & Sons, New York, 2004, pp. 2.6.1–2.6.27.

Freshney, R. I. *Culture of Animal Cells. A Manual of Basic Technique*, 5th ed, John Wiley & Sons, New York, 2005.

Freshney, R. I., and Freshney, M. G. *Culture of Epithelial Cells*, 2nd ed. John Wiley & Sons, New York, 2002.

Freshney, R. I., and Freshney, M. G. *Culture of Immortalized Cells*. John Wiley & Sons, New York, 1996.

Freshney, R. I., Pragnell, I. B., and Freshney, M. G. *Culture of Hematopoietic Cells*, John Wiley & Sons, New York, 1994.

Picot, J. *Human Cell Culture Protocols*, 2nd ed., Humana Press, Totowa, NJ, 2004.

Kedderis, G. L. Extrapolation of in vitro enzyme induction data to humans in vivo. *Chem. Biol. Interact.* **107**, 109–121, 1997.

Lemasters, J. J., Gores, G. J., Nieminen, A.-L., Dawson, T. L., Wray, B. E., and Herman, B. Multiparameter digitized video microscopy of toxic and hypoxic injury in single cells. *Environ. Health Perspect.* **84**, 83–94, 1990.

O'Hare, S., and Atterwill, C. K. *In Vitro Toxicity Testing Protocols*. Humana Press, Totowa, NJ, 1995.

Tyson, C. A., and Frazier, J. M. (Ed.). *In Vitro Biological Systems*, Vol. 1A, *Methods in Toxicology*, Academic Press, San Diego, 1993.

Tyson, C. A., and Frazier, J. M. (Ed.). *In Vitro Toxicity Indicators*, Vol. 1B, *Methods in Toxicology*, Academic Press, San Diego, 1994.

Walum, E., Stenberg, K., and Jenssen, D. *Understanding Cell Toxicology, Principles and Practice*. Ellis Horwood, New York, 1990.

Journals

In Vitro Cellular & Developmental Biology—Animal, Journal of the Society for *In Vitro* Biology. Publisher: SIVB, Largo, MD.

In Vitro Toxicology, A Journal of Molecular and Cellular Toxicology; journal of the Industrial In Vitro Toxicology Group. Publisher: Mary Ann Liebert, Larchmont, NY.

Toxicology in Vitro; in association with BIBRA. Publisher, Elsevier.

Cell Biology and Toxicology. Publisher: Kluwer Academic Publishers.

CHAPTER 9

Structure, Mechanism, and Regulation of Cytochromes P450

DARRYL C. ZELDIN

National Institute of Environmental Health Sciences, Research Triangle Park, North Carolina 27709

JOHN M. SEUBERT

Faculty of Pharmacy and Pharmaceutical Sciences, University of Alberta, Edmonton AB, Canada

9.1   INTRODUCTION

The biotransformation of xenobiotics has been classified into phase I (functionalization) and phase II (conjugation) reactions. The enzymes catalyzing these reactions are generally localized in the membranes of the endoplasmic reticulum (microsomes) or mitochondria, as well as in the cytosol. Phase I reactions include (a) oxidation and reduction of xenobiotics that alter and introduce functional groups and (b) hydrolysis, which cleaves esters and amides to release masked functional groups. These changes increase the polarity of the xenobiotic and provide molecular sites that enable phase II enzymes to conjugate moieties such as glucuronic acid, acetic acid, or sulfuric acid. Conjugation markedly increases the water-solubility of phase I metabolites.

The most important enzyme system involved in phase I metabolism is the cytochrome P450 monooxygenase system (CYP or P450). CYPs form a multigene superfamily of heme-thiolate proteins comprised of more than 6000 individual enzymes. These enzymes are ubiquitous and occur in many different organisms including prokaryotes and eukaryotes. The term "cytochrome P450" originates from the appearance of the spectra of these heme-thiolate proteins, specifically the presence of a prominent absorption band near 450 nm when the reduced (ferrous, Fe^{2+}) form of the hemoprotein is ligated with carbon monoxide. There are multiple CYP enzymes that have overlapping substrate selectivity; however, individual isoforms can demonstrate unique regio- or stereoselectivity toward particular substrates—for example, testosterone or arachidonic acid. P450s play a major role in the metabolism of foreign lipophilic compounds (xenobiotics), including drugs and chemical carcinogens, as well as endogenous compounds such as steroids, fat-soluble vitamins, fatty acids, and biogenic amines. CYP expression and activity is under the control of hormones, growth factors, and transcription factors. Indeed, different

Molecular and Biochemical Toxicology, Fourth Edition, edited by Robert C. Smart and Ernest Hodgson
Copyright © 2008 John Wiley & Sons, Inc.

CYP subfamilies can display complex sex-, tissue-, and development-specific expression patterns. In addition, CYP expression and activity can be influenced by genetic polymorphism.

CYP enzymes are predominately localized in the microsomal subcellular fraction of mammalian cells; some P450s are also present in mitochondria. Microsomes are obtained by homogenization and differential centrifugation. They are found in the pellet following ultracentrifugation at $100,000\,g$ for 60 min of the post-mitochondrial supernatant (~$10,000\,g$ for 10 min) fraction and consist of fragments of the endoplasmic reticulum. CYP enzymes are tethered to the membrane of the endoplasmic reticulum through a hydrophobic transmembrane helix at the N-terminus of the protein, with the bulk of the protein exposed on the cytosolic side.

Monooxygenation reactions involve the reduction of one atom of molecular oxygen to water and the incorporation of the other oxygen atom into the substrate [Eq. (9.1)]. The electrons involved in the reduction of CYP are transferred from NADPH by another enzyme called NADPH–cytochrome P450 oxidoreductase (CYPOR).

$$NADPH + H^+ \qquad\qquad NADP^+$$

$$RH + O_2 \longrightarrow ROH + H_2O \qquad (9.1)$$

CYP enzymes play a role in both the detoxification of compounds and the bioactivation of drugs, pesticides and many environmental pollutants into carcinogens, mutagens and cytotoxins. Thus, the P450 superfamily of enzymes is important from medical, pharmaceutical, and agrochemical perspectives.

9.2 COMPLEXITY OF THE CYTOCHROME P450 GENE SUPERFAMILY

9.2.1 CYP Gene Families and Subfamilies

By convention, the *CYP* superfamily of genes are subdivided into families and subfamilies based on amino acid homology. Details of the current classification and nomenclature system are found at http://www.imm.ki.se/CYPalleles/ and http://drnelson.utmem.edu/CytochromeP450.html. *CYP* genes having >40% homology in their deduced amino acid sequence are classified into different families, which are designated by an Arabic number and written in all uppercase letters for all species except mouse (e.g., *CYP1*, *CYP2*, and *CYP4* in human; *Cyp1*, *Cyp2*, and *Cyp4* in mouse). There are 265 *CYP* families, 18 of which exist in mammals, encompassing greater than 2000 distinct mammalian *CYP* genes. *CYP* genes with >55% deduced amino acid sequence homology are further grouped into subfamilies, which are designated by uppercase letters in all species except mouse (e.g., *CYP1A*, *CYP2C*, and *CYP4A* in human; *Cyp1a*, *Cyp2c*, and *Cyp4a* in mouse). Within the 18 mammalian *CYP* families, 26 mammalian subfamilies have been categorized. Each *CYP* subfamily is comprised of one or more individual *CYP* genes. An additional Arabic number designates the individual *CYP* genes (e.g., *CYP1A1*, *CYP2C8*, and *CYP4A11* in human; *Cyp1a1*, *Cyp2c29*, and *Cyp4a12* in mouse) (Figure 9.1). There are generally 50–110 individual genes in a given species; for example, there are 57 human genes and 102 mouse genes.

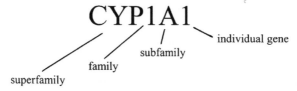

Figure 9.1. Nomenclature of cytochromes P450. Members of the same family share >40% amino acid homology, and members of the same subfamily share >55% amino acid homology.

Italics are used to designate genes, whereas mRNAs and proteins are nonitalicized by convention.

The *CYP* superfamily is very old, with the ancestral gene having existed before the time of prokaryote–eukaryote divergence. The basic characteristics of vertebrate CYPs have not changed substantially in 420 million years. It is thought that the fundamental importance of a particular *CYP* gene increases the farther back it can be traced. Genes within a given subfamily are nonsegregating (i.e., they lie within the same "cluster" on a chromosome); clusters most likely arose via gene duplication events during the last 400 million years. *CYP* genes within a given family have the same number of exons and similar intron–exon boundaries. Spontaneous mutations, selection, and species diversity are contributing factors to the evolution of mammalian *CYPs*. It has been estimated that the mouse–human divergence occurred about 75 million years ago (MYA), the mouse–rat divergence occurred about 20 MYA, and the human–chimpanzee divergence occurred about 6 MYA. This provided enough time for major changes in gene clusters and organization, but not enough time to form new *CYP* families. Approximately 80% of the *CYP* genes in human and mouse have a 1:1 orthologous relationship, but the mouse has undergone significant expansion in seven clusters (Table 9.1).

9.2.2 Pseudogenes and Alternate Splice Variants

CYP enzymes are involved in a vast array of important reactions that play a role in key processes required for life. Mutations in *CYP* genes can lead to inborn errors of metabolism and contribute to clinically relevant diseases. Examples of these associations include *CYP1B1* and primary congenital glaucoma, *CYP7A1* and hypercholesterolemia, and *CYP17A1* and mineralocorticoid excess syndromes, glucocorticoid deficiencies, and sex hormone deficiencies. The human genome, and mammalian genomes in general, have been estimated to contain between 26,000 and 40,000 genes. Exons that contain the RNA coding sequences only cover about 1.4% of the genome length. Pseudogenes are typically found within intergenic genomic areas. These are defunct relatives of known genes that have lost their protein-coding ability or are no longer expressed in the cell. The role of pseudogenes has not been fully elucidated; however, they appear to be involved in conversion and recombination events of nearby functional genes, provide points of reference in the genome, and provide information on biological and evolutionary histories within their sequences due to a shared ancestry with a functional gene. It is thought that pseudogenes contain more single-nucleotide polymorphisms (SNPs), insertions, and deletions than functional genes. Therefore, knowledge of pseudogenes will lead to a better understanding of evolution and will facilitate comparisons among functional genes.

TABLE 9.1. Human and Mouse Putatively Functional Full-Length *CYP* Genes

Human	Mouse	Human	Mouse	Human	Mouse
CYP1A1	Cyp1a1	CYP2E1	Cyp2e1	CYP4F2	Cyp4f13
CYP1A2	Cyp1a2	CYP2F1	Cyp2f2	CYP4F3	Cyp4f14
CYP1B1	Cyp1b1		Cyp2g1[b]	CYP4F8	Cyp4f15
CYP2A6	Cyp2a4		Cyp2j5	CYP4F11	Cyp4f16
CYP2A7	Cyp2a5		Cyp2j6	CYP4F12	Cyp4f17
CYP2A13	Cyp2a12		Cyp2j7	CYP4F22	Cyp4f18
	Cyp2a22		Cyp2j8		Cyp4f37
CYP2B6	Cyp2b9		Cyp2j9		Cyp4f39
	Cyp2b10		Cyp2j11		Cyp4f40
	Cyp2b13		Cyp2j12	CYP4V2	Cyp4v3
	Cyp2b19		Cyp2j13	CYP4X1	Cyp4x1
	Cyp2b23	CYP2R1	Cyp2r1	CYP4Z1	a
CYP2C8	Cyp2c29	CYP2S1	Cyp2s1	CYP5A1	Cyp5a1
CYP2C9	Cyp2c37		Cyp2t4[b]	CYP7A1	Cyp7a1
CYP2C18	Cyp2c38	CYP2U1	Cyp2u1	CYP7B1	Cyp7b1
CYP2C19	Cyp2c39	CYP2W1	Cyp2w1	CYP8A1	Cyp8a1
	Cyp2c40		Cyp2ab1[b]	CYP8B1	Cyp8b1
	Cyp2c44		Cyp2ab1-ps[b]	CYP11A1	Cyp11a1
	Cyp2c50	CYP3A4	Cyp3a11	CYP11B1	Cyp11b1
	Cyp2c54	CYP3A5	Cyp3a13	CYP11B2	Cyp11b2
	Cyp2c55	CYP3A7	Cyp3a16	CYP17A1	Cyp17a1
	Cyp2c65	CYP3A43	Cyp3a25	CYP19A1	Cyp19a1
	Cyp2c66		Cyp3a41	CYP20A1	Cyp20a1
	Cyp2c67		Cyp3a44	CYP21A2	Cyp21a1
	Cyp2c68		Cyp3a57	CYP24A1	Cyp24a1
	Cyp2c69		Cyp3a59	CYP26A1	Cyp26a1
	Cyp2c70	CYP4A11	Cyp4a10	CYP26B1	Cyp26b1
CYP2D6	Cyp2d9	CYP4A22	Cyp4a12a	CYP26C1	Cyp26c1
	Cyp2d10		Cyp4a12b	CYP27A1	Cyp27a1
	Cyp2d11		Cyp4a14	CYP27B1	Cyp27b1
	Cyp2d12		Cyp4a29	CYP27C1	a
	Cyp2d13		Cyp4a30b	CYP39A1	Cyp39a1
	Cyp2d22		Cyp4a31	CYP46A1	Cyp46a1
	Cyp2d26		Cyp4a32	CYP51A1	Cyp51a1
	Cyp2d34	CYP4B1	Cyp4b1		
	Cyp2d40				

[a]Subfamilies absent in mouse.
[b]Subfamilies in mouse with only pseudogene orthologs in human.

CYP2AC1P and *Cyp2ac1-ps* are recently discovered pseudogenes in human and mouse, respectively, but are functional genes in rat. The gene names are listed numerically and alphabetically by subfamilies, and pairing does not necessarily denote orthologous genes; for example, it is not known whether human *CYP2A6* is the ortholog of mouse *Cyp2a4*. Genes having the identical combination of numbers and letters are othologs between the two species (see text).

Source: Nelson, D. R., et al. Comparison of cytochrome P450 genes from the mouse and human genomes including nomenclature recommendations for genes, pseudogenes, and alternative-splice variants. *Pharmacogenetics* **14**, 1–18, 2004.

When the mouse and human *CYP* gene families are combined, there are 146 pseudogenes and 159 putative functional genes; hence, the number of pseudogenes almost equals the number of functional genes. The relative abundance of pseudogenes suggests the importance of their identification and annotation. Recently, a nomenclature system for each of four categories of *CYP* pseudogenes has been proposed; details can be found at the website http://drnelson.utmem.edu/cytochromeP450.html/. The first category of pseudogenes is the full-length, or nearly full-length, pseudogenes, which may be chromosomally secluded from other members, freestanding or found within their gene clusters. The designation "*P*" ("-*ps*" in mouse) is appended to the gene name, as in human *CYP4F9P* or mouse *Cyp4f38-ps*. For example, the human pseudogene *CYP2G2P* is a nearly intact gene that contains two stop-codons in the coding sequence. The human pseudogene *CYP2D7AP* has only one frameshift in exon 1 and one aberrant GT donor splice site at the exon 2–intron 2 boundary. It can be predicted that a polymorphism, gene conversion, or alternative splicing at these sites could result in a functional gene. The second category is solo exon pseudogenes. These are single or groups of exons that are located away from their respective gene cluster. They will have the extension -*sen[x*1:*x*2*xi]*, in which -*se* denotes solo exon, *n* represents unique number, and the values within the *[x*1:*x*2*xi]* refer to which exon(s) has been duplicated, as in human *CYP4F-se1[6:8]* and mouse *Cyp4f-se[6]*. The third category, called detritus exons, are pseudogenes representing fragments caused by gene disintegration or partial gene duplication. Detritus pseudogenes will have the extension -*dex*1*jx*2*j . . . xij*, where *de* denotes detritus exon, *x* represents the number of exons that have been duplicated, and the lowercase Arabic letter (starting with *b, c, d,* etc., because the letter *a* is reserved for the parent-gene normal exons) indicates the unique pseudogene. For example, the mouse *Cyp2j* gene cluster contains three exon 9 pseudogenes that are downstream of the functional gene *Cyp2j7-Cyp2j7-de9b, Cyp2j7-de9c,* and *Cyp2j7-de9d*. The fourth category of pseudogenes is referred to as (a) the internal exons which are duplicated intact or (b) partial exons found within the gene. These pseudogenes will have the extension -*iexij*, in which *ie* denotes internal exons, *xi* refers to the specific exon number that has been duplicated, and the lowercase Arabic letter (starting with *b, c, d,* etc., as the letter *a* is reserved for the parent-gene normal exons) represents the unique pseudogene. Although rare, these types of pseudogenes do exist. For example, the partial duplication of exon 1 of the *CYP3A4* gene is designated *CYP3A4-ie1b*. Because many exons found within pseudogenes are highly similar to functional genes and can interfere with PCR-based genotyping assays, a potentially hazardous outcome can result from incorrect identifications.

Alternative splicing is the process in which the splicing of a pre-mRNA transcribed from one gene can lead to different mature mRNA molecules and therefore to different proteins. The Guidelines for Human Gene Nomenclature suggest the symbol _*v* to denote alternative transcripts, as in *Cyp2c39_v6* and *CYP3A5_v1*. Splice variants may or may not be functionally relevant.

9.3 CYTOCHROME P450 STRUCTURE

Despite less than 20% amino acid sequence homology, along with differences in cellular localization and substrate specificity, P450 enzymes share a remarkably

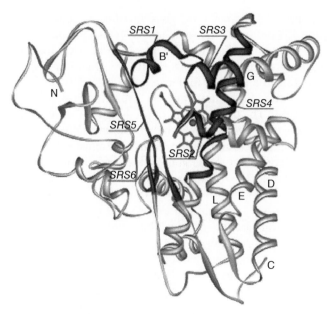

Figure 9.2. Structure of a cytochrome P450. The folding pattern of cytochromes P450 is highly conserved and shown in a ribbon representation (distal face). Substrate recognition sequence (SRS) regions are shown in black and labeled. The α-helixes are labeled with capital letters. (Reprinted with permission from Denisov, I. G., Makris, T. M., Sligar, S. G., and Schlichting, J. Structure and chemistry of cytochrome P450. *Chem. Rev.* **105**, 2253–2278, 2005 American Chemical Society.)

common overall structure. In addition, they contain a number of highly conserved secondary structural elements (Figure 9.2). In general, valuable insight regarding protein structure and function can be obtained from three-dimentional structures; however, structural information is limited as the vast majority of proteins are too large, insufficiently soluble, or too difficult to crystallize. Several X-ray crystal structures are now available for bacterial P450s that are soluble rather than membrane bound. For example, structures are available for CYP101A1 (P450 cam), CYP102A1 (P450 BM-3), CYP108A1 (P450 terp), and CYP 107A1 (P450 eryF). Recent determinations of crystal structures of mammalian P450s have also become available. For example, structures are now available for rabbit CYP2C5 and human CYP2A, CYP2C, CYP2D, and CYP3A. Table 9.2 provides a complete listing of available P450 crystal structures. In general, the information gleaned from structural analysis of mammalian P450s complement that obtained from analysis of bacterial and fungal P450 structures.

The structures demonstrate highly dynamic and flexible regions that enable the active site to adapt to binding of substrates of different size, shape, and polarity. The highest structural conservation is found within the central core around the heme group, reflecting the common mechanism of electron and proton transfer and oxygen activation. The conserved P450 structural core consists of a four-helix bundle (D, E, I, and L), helices J and K, two sets of β sheets, and a coil called the "meander." The prosthetic heme-binding loop, found within this region, contains the signature P450 amino acid sequence (Phe–X–X–Gly–X–Arg–X–Cys–X–Gly–X–X–X–X), located on the proximal face of the heme just before the L helix.

TABLE 9.2. P450 (CYP) Crystal Structures

BACTERIA	FUNGI
CYP51B1 (14α-sterol demethylase)	CYP55A1 (nor)
CYP101A1 (cam)	CYP119
CYP102A1 (BM-3)	
CYP105A3 (sca-2)	**MAMMALS**
CYP107A1 (eryF)	CYP2A6 (human)
CYP108A1 (terp)	CYP2B4 (human)
CYP121	CYP2C5 (rabbit)
CYP152A1 (BSβ)	CYP2C8 (human)
CYP154A1	CYP2C9 (human)
CYP154C1	CYP2D6 (human)
CYP158A2	CYP3A4 (human)
CYP165B3 (OxyB)	
CYP165C4 (OxyC)	
CYP167A1 (epoK)	
CYP175A1	

Adapted from de Groot, M. J. *Drug Discovery Today* **11**, 601–606, 2006.

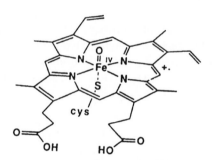

Figure 9.3. The cytochrome P450 active site containing the heme iron with six ligands. Four ligands are occupied by nitrogens from the protoporphyrin IX moiety. The "fifth ligand" is occupied by a sulfur from the invariant cysteine. The "sixth ligand" is occupied by oxygen, carbon monoxide, or water.

Amino acid sequence alignments reveal several highly conserved regions in all mammalian P450s. Conserved in all P450s is a cysteine that is considered the proximal or "fifth" ligand to the heme iron (Figure 9.3). This sulfur-containing amino acid is located within the heme-binding loop in the C-terminal portion of the protein. The other heme ligands include four nitrogens from the protoporphyrin IX moiety and the "sixth ligand," which can be oxygen, carbon monoxide, or water. A conserved motif (Glu-X-X-Arg) on the proximal side of the heme in helix K is probably needed to stabilize the core structure. Another consensus sequence (Ala/Gly-Gly-X-Asp/Glu-Thr-Thr/Ser) in the long I helix is positioned in the active site and is believed to be involved in catalysis.

Although the characteristic fold is highly conserved across the P450 superfamily, enough structural diversity can be found to allow for substrate selectivity. Most of this variability exists within the membrane targeting or anchoring regions and within the six substrate-recognition sites (SRS1-6). The SRSs are located near the substrate-access channel and catalytic sites of P450 enzymes: the B' helix region (SRS1), F and G regions (SRS2 and SRS3), the I helix (SRS4), the K helix β2 connecting region (SRS6), and the β4 hairpin (SRS5). These SRSs provide the P450 enzymes with their substrate specificity by altering the topography of the substrate access channel. For example, large bulky amino acids may limit substrate access. Moreover, mutations within the SRSs may affect substrate binding. SRSs are considered flexible protein regions as they move upon substrate binding to favor the catalytic reaction in an induced-fit mechanism.

Another important cluster for microsomal anchoring frequently consists of a group of prolines (Pro-Pro-X-Pro) that form a hinge between the N-terminal hydrophobic membrane anchoring segment (highly variable) and the globular part of the protein.

9.4 MECHANISMS OF P450 CATALYSIS

9.4.1 Catalytic Cycle

The general catalytic mechanism for the oxygenation of a substrate is common to all CYP enzymes and involves several intermediate steps as depicted in Figure 9.4. Briefly, the process begins when the substrate (RH) binds to the ferric (Fe^{3+}) form of the enzyme (step 1). An electron is then transferred from NADPH via the accessory flavoprotein CYPOR, reducing the iron atom to its ferrous state (Fe^{2+}) (step 2). Ferrous CYP binds molecular oxygen (O_2) producing an unstable complex, which

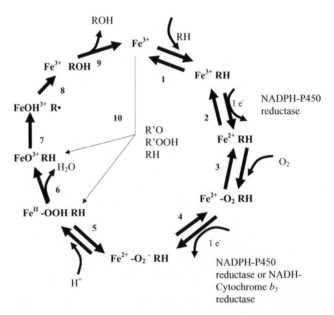

Figure 9.4. Generalized cytochrome P450 catalytic cycle for the oxidation of substrates (Adapted from Guengerich, F. P., Common and uncommon cytochrome P450 reactions related to metabolism and chemical toxicity. *Chem. Res. Toxicol.* **14**, 611–650, 2001.)

can generate ferric iron and superoxide anion radical (step 3). A second electron and a proton (H^+) enter the cycle, via either CYPOR or NADH-cytochrome b_5 reductase (step 4). The Fe O–O bond is then broken releasing H_2O and a high-valent FeO^{3+} complex (step 5). It is the FeO^{3+} complex that directly interacts with the substrate, abstracting either a hydrogen atom or an electron, thereby generating a product that dissociates from the enzyme and thus completing the cycle (steps 6–9). Some CYP enzymes may oxidize substrates in the absence of CYPOR, NADPH, or O_2 by utilization of peroxy compounds such as alkyl hydroperoxides as a source of both atomic oxygen and reducing equivalents (step 10). P450 reactions can result in the formation of reactive electrophilic intermediates or, due to uncoupling between activated oxygenating species of CYP and substrate oxidations, release reactive oxygen species (ROS) including free radicals (e.g., $^{.}O_2^-$, $^{.}OH^-$, H_2O_2) which are capable of causing a toxic response.

9.4.2 Catalytic Requirements

Although the substrates and products of microsomal P450 reactions differ greatly, the utilization of molecular oxygen and electrons from NADPH is a common component. The monooxygenase reaction requires a redox partner to transfer electrons to the P450 so that substrate binding and activation of the iron-bound molecular oxygen can occur. During the oxidation of a substrate, two electrons, normally from NADPH, are transferred to the iron atom in the heme (protoporphyrin IX) group. The thiolate group of the conserved cysteine residue is ligated to iron (Fe^{3+})-protoporphyrin IX, the active center for catalysis. This transfer of electrons to the CYP is almost always mediated by a membrane bound flavoprotein, CYPOR, which contains two subcomponents, flavin adenine dinucleotide (FAD) and flavin mononucleotide (FMN). However, alternative reduction systems can be utilized to transfer electrons to the catalytic site in P450 reactions. As such, P450s can be divided into classes. Class I P450s require both flavin adenine dinucleotide (FAD)-containing reductase and an iron sulfur redoxin to transfer electrons. These proteins are associated with the inner mitochondrial membrane and some bacterial systems. Class II P450s reside in the membrane of eukaryotic endoplasmic reticulum and receive reducing equivalents from NADPH via CYPOR or cytochrome $b5$. Class III P450s are isomerases instead of monooxygenases, and therefore they are self-sufficient and require no electron donors. Class IV P450s can receive electrons directly from NADPH without the intervention of an electron carrier.

9.4.3 Substrate Specificity and Overlap

Various structural elements that are found in all P450s have been identified that undergo large conformational changes to allow substrate access and product release. In addition, flexible regions may enable the active site to adapt to the binding of substrates of different size, shape, and polarity. The catalytic center of P450 enzymes surrounds a porphyrin–heme complex confined within different amino acid sequences. Molecular and chemical interactions will alter the structure of the active site to accommodate a diverse array of substrates. Regio- and stereospecificity of the oxygenation reaction will depend largely on how tightly the substrate is bound in the active site and how the substrate is positioned with respect to the activated oxygen. There are several factors that influence the binding affinity of substrates to P450 enzymes. These include ionic interactions, hydrogen bond interactions, π–π interactions, desolvation processes of the heme hydrophobic "pocket" during binding (the release of the water electrostatically bound to a molecule), and loss in bond rotational freedom on binding. The major factors governing P450 substrate-binding affinity are desolvation and hydrogen-bond formation with key amino acid residues lining the heme pocket. In turn, crucial factors for substrate selectivity include molecular size and shape, number and disposition of hydrogen bond donor/acceptor atoms, aromatic rings in the molecule, and the presence of basic or acidic groups.

Fundamental to substrate selectivity, binding affinity, and clearance is the compounds lipophilicity. This is represented as a log P value, which is the lipophilicity factor, where P is the octanol–water partition coefficient. Mammalian P450 substrates tend to be very lipophilic. The log P values are generally greater than zero, and the greater the value of log P for a given chemical, the more avidly it will bind. This increased affinity reaches a point at which the compound becomes too hydrophobic and affinity begins to decline (Table 9.3). Together, lipophilicity, molecular mass, and hydrogen-bonding potential are important factors that determine substrate selectivity of individual CYP enzymes. Understanding of key physicochemical, structural, and electronic characteristics will help to predict substrate selectivity and binding affinity.

TABLE 9.3. Characteristics of Human P450 Substrates of Families CYP1, CYP2, and CYP3

% Drugs	CYP	Range of log P Values	Average log P Value (n)	Other Characteristics	Typical Substrate and log P
3	1A1	1.39–6.35	3.41 [16]	Planar PAHs and their diols	DMBA-3,4-diol 3.42
10	1A2	0.08–3.61	2.01 [18]	Planar amines and amides	MeIQ 1.98
1	1B1	1.40–6.35	3.73 [12]	Planar PAHs and their diols	BP 7,8 diol 3.87
3	2A6	0.07–2.79	1.44 [18]	Fairly small molecules	Lesigamone 1.46
4	2B6	0.23–4.89	2.54 [16]	Basic (un-ionized)	Buproprion 2.54
	2C8	0.06–6.98	3.38 [12]	Acidic (ionized)	Resigfitazone 3.20
25	2C9	0.89–5.18	3.20 [18]	Acidic (un-ionized)	Naproxon 3.18
	2C1	1.49–4.42	2.56 [16]	Amides and amines	Proguanil 2.53
15	2D6	0.75–5.04	3.08 [16]	Basic (ionized)	Propranolol 3.09
3	2E1	−1.35–3.63	2.07 [20]	Small molecules	4-Nitrophenol 2.04
NA	−2F1	0.37–5.14	2.63 [12]	Fairly small molecules	3-Methylindole 2.72
36	3A4	0.97–7.54	3.10 [50]	Large molecules	Nifedipine 3.17

n, number of compounds investigated; C, calculated value (Pallas Software); NA, data not available. DMBA, dimethylbenzanthracene; MeIQ, 2-amino-3,4-dimethylimidazo[4,5-f]quinoline; BP, benzo(a)pyrene.

Potentially important discriminants of P450 selectivity include: size (diameter), molecular mass or length of the molecule, planarity (a/d^2) or rectangularity (l/w) of the molecule, lipophilicity (log P or log D 7.4), basicity/acidity of the compound (pKa), polarity (dipole moment) of the molecule, and compound lipophilicity for substrate binding to human P450s in drug metabolism.

Source: Lewis, D. F. V., Jacobsa, M. N., and Dickins, M. *Drug Discovery Today* **9**, 530–537, 2004.

9.4.4 Endogenous Substrates

There are many endogenous substrates, of widely different chemical structure, that are metabolized through oxidative, peroxidative, and reductive changes introduced by P450 enzymes. These include saturated and unsaturated fatty acids, eicosanoids, sterols and steroids, bile acids, vitamin D derivatives, retinoids, and uroporphyrinogens (Tables 9.4 and 9.5).

Arachidonic acid, a polyunsaturated fatty acid normally found esterified to cell membrane glycerophospholipids, can be released by phospholipases in response to several stimuli such as hormones or stress. Free arachidonic acid is then available for metabolism by cyclooxygenases, lipoxygenases, and cytochrome P450 monooxygenases to generate numerous metabolites, collectively termed eicosanoids (Figure 9.5). Many cytochrome P450 enzymes in the *CYP1, CYP2, CYP3*, and *CYP4* families are involved in the metabolism of arachidonic acid to a large array of metabolites with numerous cellular effects. Arachidonic acid can be metabolized by CYP epoxygenases (CYP2C, CYP2J) and CYP ω-hydroxylases (CYP4A, CYP4F) to products that have vastly different physiologic effects. The CYP epoxygenases biosynthesize four regioisomeric *cis*-epoxyeicosatrienoic acids (5,6-, 8,9-, 11,12-, and 14,15-EET), all of which are biologically active. EETs are important components of many intracellular signaling pathways in a wide range of tissues. For example, EETs activate Ca^{2+}-sensitive K^+ channels (BKca) in vascular smooth muscle cells resulting in hyperpolarization of the resting membrane potential and vasodilation of the coronary circulation. EETs display anti-inflammatory, thrombolytic, and angiogenic properties within the vasculature. In endothelial cells, EETs activate

TABLE 9.4. Substrates and Functions of Human CYP Gene Families

Family	Number of Subfamilies	Number of Genes	Substrates and Functions
CYP1	2	3	Foreign chemicals, arachidonic acid, eicosanoids
CYP2	13	16	Foreign chemicals, arachidonic acid, eicosanoids
CYP3	1	4	Foreign chemicals, arachidonic acid, eicosanoids
CYP4	5	12	Fatty acids, arachidonic acid, eicosanoids
CYP5	1	1	Thromboxane A_3 synthase
CYP7	2	2	Cholesterol, bile acid synthesis
CYP8	2	2	Prostacyclin synthase, bile-acid synthesis
CYP11	2	3	Steroidogenesis
CYP17	1	1	Steroid 17α-hydroxylase, 17/20-lyase
CYP19	1	1	Aromatase to form estrogen
CYP20	1	1	Unknown
CYP21	1	1	Steroid 21-hydroxylase
CYP24	1	1	Vitamin D_3 24-hydroxylase
CYP26	3	3	Retinoic acid hydroxylation
CYP27	3	3	Bile acid biosynthesis, vitamin D_3 hydroxylations
CYP39	1	1	24-Hydroxycholesterol7α-Hydroxylase
CYP46	1	1	Cholesterol 24-hydroxylase
CYP51	1	1	Lanosterol 14α-desmethylase

Source: Nebert, D. W., and Russell, D. W. Clinical importance of the cytochromes P450. *Lancet* **360**, 1155–1162, 2002.

TABLE 9.5. Classification of Human Cytochrome P450s Based on Major Substrate Class

Sterols	Xenobiotics	Fatty Acids	Eicosanoids	Vitamins	Unknown
1B1	1A1	2J2	4F2	2R1	2A7
7A1	1A2	4A11	4F3	24A1	2S1
7B1	2A6	4B1	4F8	26A1	2U1
8B1	2A13	4F12	5A1	26B1	2W1
11A1	2B6		8A1	26C1	3A43
11B1	2C8			27B1	4A22
11B2	2C9				4F11
17A1	2C18				4F22
19A1	2C19				4V2
21A2	2D6				4X1
27A1	2E1				4Z1
39A1	2F1				20A1
46A1	3A4				27C1
51A1	3A5				
	3A7				

Source: Guengerich, F. P. Cytochrome P450s and other enzymes in drug metabolism and toxicity. *AAPS J.* **8**, E101–E111, 2006.

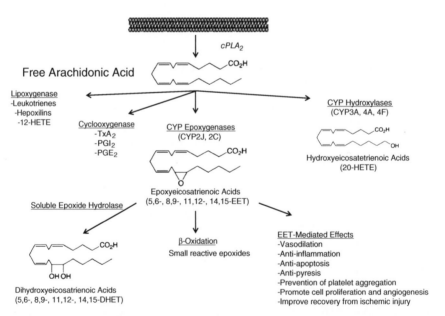

Figure 9.5. CYP epoxygenase- and hydroxylase-mediated metabolism of arachidonic acid. Products include biologically active EETs and HETEs. (Adapted from Seubert *et al. Prostaglandins and Other Lipid Mediators* **82**, 50–59, 2007.)

mitogen-activated protein kinase (MAPK) and phosphatidylinositol-3 kinase (PI3K)–Akt signaling pathways, increase intracellular cAMP levels, upregulate expression of nitric oxide synthase, and protect against hypoxia–reoxygenation injury. The actions of EETs are terminated by conversion to the corresponding and less biologically active dihydroxyeicosatrienoic acids (DHETs) by epoxide hydrolases. CYP ω-hydroxylases biosynthesize hydroxyeicosatetraenoic acids such as 17-, 18-, 19-, and 20-HETE. 20-HETE is a potent vasoconstrictor that acts by inhibiting large conductance, Ca^{2+}-activated K^+ channels, thereby depolarizing the vascular smooth muscle membrane. Renal ω-hydroxylation occurs in the proximal tubule and the thick ascending limb of Henle. In the proximal tubule, 20-HETE inhibits Na^+/K^+-ATPase, whereas in the thick ascending limb, 20-HETE blocks a K^+ channel, which limits K^+ availability for transport by a Na^+-K^+-$2Cl^{2-}$ cotransporter. In the lung microvasculature, 20-HETE has an effect opposite to that seen in other vascular beds, vasodilating pulmonary vessels in an endothelium- and cyclooxygenase-dependent manner. 20-HETE can also act as a vascular oxygen sensor and mediate the mitogenic actions of vasoactive agents and growth factors. Therefore, changes in the expression and/or activity of specific CYP epoxygenase and hydroxylase enzymes can alter the delicate balance between EETs and HETEs.

A number of cytochrome P450 enzymes are involved in the conversion of acetate to sterols and bile acids (Figure 9.6). The participation of P450 enzymes in pathways of cholesterol biosynthesis and elimination demonstrate their important role in cholesterol homeostasis. Lanosterol 14α-desmethylase, encoded by *CYP51A1*, is a pivotal P450 involved in cholesterol biosynthesis. The synthesis of bile acids from

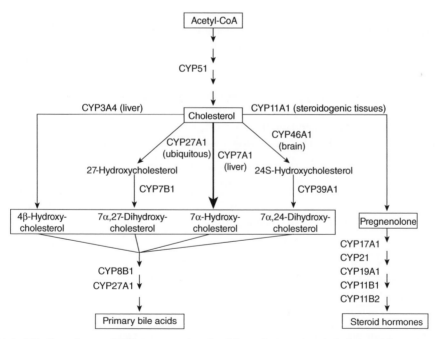

Figure 9.6. Cytochrome P450 enzymes involved in maintenance of cholesterol homeostasis. (From Pikuleva, I. A. Cytochrome P450s and cholesterol homeostasis, *Pharmacol. Ther.* **112**, 761–773, 2006.)

cholesterol by P450s such as *CYP7A1, CYP27A1, CYP46A1, CYP7B1, CYP39A1,* and *CYP8B1* represents the major route of cholesterol elimination in mammals.

P450 enzymes from the families *CYP11, CYP17, CYP19,* and *CYP21* are involved in steroidogenesis (Figure 9.6). Steroid hormone biosynthesis is controlled by the activity of several highly substrate-selective P450 enzymes and a number of steroid dehydrogenases and reductases. All steroid hormone synthesis begins with the conversion of cholesterol to pregnenolone by CYP11A. CYP11A is found in the inner mitochondrial membranes of all steroidogenic tissues such as adrenal, testis, and ovary. Pregnenolone is converted to progesterone by 3β-hydroxysteroid dehydrogenase. Together, pregnenolone and progesterone are the precursors for all other steroid hormones. Two P450s located in the endoplasmic reticulum, CYP17A1 and CYP19A1, are required for biosynthesis of cortisol, testosterone and estrogen, and conversion of androgenic precursors into estrogen. The biosynthesis of glucocorticoids and mineralcorticoids is catalyzed by the hydroxylation of carbon-21 in precursors by CYP21A2.

Other important endogenous pathways involving P450 enzymes include (a) the synthesis and metabolism of vitamin D by members of *CYP2D, CYP24A, CYP27A,* and *CYP27B* families and (b) hydroxylation of retinoic acid by members of the *CYP26* gene family.

9.5 CYTOCHROME P450 REGULATION

9.5.1 Tissue Distribution

In general, CYP enzymes are most abundant in the liver, where they play an essential role in drug metabolism. However, various isoforms are also found in extrahepatic tissues, where their expression is regulated by multiple different mechanisms. Tissues that contain CYP enzymes include the gastrointestinal tract, lung, kidney, olfactory mucosa, heart, and central nervous system (Table 9.6). Tissue-specific differences in CYP-mediated metabolism of certain chemicals are the result of different expression and/or regulation of the relevant P450 enzymes. A compound that may be readily metabolized in the liver may be toxic in some other tissues due to differences in CYP expression which are influenced by both endogenous and exogenous factors.

TABLE 9.6. CYP expression in Hepatic and Extrahepatic Tissues

Liver	Kidney	Heart	Lung
1A2, 2A6, 2B1, 2C8, 2C9, 2D6, 2E1, 3A4, 7A1, 7B1, 27A1	1A1, 2A, 2C, 2J2, 4A1, 7B1, 27A1	1A1, 1B1, 2J2, 2C8, 2C9, 2C11, 4F	1A1, 1B1, 2J2, 2B, 3A

Brain	Olfactory	Intestinal Tract	Steroidogenic Tissues
2C8, 2C18, 2D18, 7B1, 11A1, 27B1, 46A1	1A, 2A, 2B, 2C, 2E, 2F, 2G, 2J, 2S, 3A, 4A, 4B	2J2, 3A4, 2C18, 2C19, 7B1	1A1, 2C8, 3A4, 4A11, 7B1, 8A, 11A1, 19A, 27A1, 51

Interest in the regulation of CYP enzymes in extrahepatic tissues is important from a detoxication as well as a bioactivation perspective. For example, evidence exists that increased expression of CYP1A1 is a risk factor for polycyclic aromatic hydrocarbon (PAH)-mediated carcinogenesis, especially in lung. The lung exhibits several characteristics that increase its susceptibility to xenobiotic toxicity, accounting for the fact this organ is a major site of chemically mediated disease. First, exposure to xenobiotics occurs from contact with both air and the circulation. Second, the lung receives the total cardiac output, thereby increasing exposure to xenobiotics. Third, heterogeneity of cell types in lung, with decreased levels of detoxication enzymes as compared to liver, favor bioactivation. Fourth, a high O_2 tension favors formation of ROS. A classical example of bioactivation is the participation of CYP in the conversion of the pro-carcinogen benzo[a]pyrene (B[a]P) to its electrophilic 7R,8S-dihydrodiol-B[a]P-9S,10R-epoxide metabolite forming a DNA adduct responsible for B[a]P carcinogenicity.

9.5.2 Constitutive and Inducible P450 Enzymes

The ability of a given CYP enzyme to catalyze the oxidation of a particular substrate may be affected by such complex mechanisms as modification of the hemoprotein during synthesis, interference with heme incorporation into the apoenzyme to form holoenzyme, or posttranslational alterations. In addition, the catalytic activities of CYP enzymes depend largely on their constitutive and inducible expression levels in various tissues. Some CYP enzymes are primarily constitutive (e.g., *CYP2C8, CYP2J2*), while other CYP enzymes are also inducible (e.g., *CYP1A1, CYP2B1, CYP2E1, CYP3A4, CYP4A1*). There are several different mechanisms by which CYP enzymes are induced and expressed. Both physiological and pathophysiological factors such as inflammation, infection, and stress can affect CYP enzyme expression.

Most cases of CYP regulation involve increases in gene transcription, which occur via receptor-mediated mechanisms. For example, activation of *CYP2* and *CYP3* family genes is through ligand-activated nuclear receptors (CAR and/or PXR). In contrast, *CYP1A* subfamily genes are regulated via the aryl hydrocarbon receptor (AhR) signaling pathway, which requires heterodimerization of the AhR nuclear translocator (Arnt) with the Per-Arnt-Sim (PAS) transcription factor AhR. In addition to gene transcriptional regulation, there are nontranscriptional mechanisms involved in CYP regulation. For example, increased CYP3A expression in rat following treatment with the antibiotic troleandomycin can result from decreased rate of protein degradation, without changes in protein synthesis. Another example is the regulation of CYP2E1, which can occur by posttranslational stabilization of protein following exposure to alcohol, acetone, or isoniazid.

Unlike inhibition, which is an immediate regulatory response, induction is a slow process that requires new protein synthesis. Thus, it takes time to reach a new steady-state level of enzyme activity following induction. Induction of CYP enzymes can have major implications for multidrug therapy and detoxification processes. Notably, induction of a *CYP* gene by a drug may reduce the therapeutic efficacy of co-medications. For example, induction of CYP3A4 by rifampcin is thought to cause acute transplant rejection in patients co-treated with cyclosporine due to increased metabolism. Another undesirable consequence of induction involves the imbalance between detoxification and activation, resulting in increased formation of reactive metabolites. A notable example of increased toxicity is when induction of CYP1A by polycyclic aromatic hydrocarbons (PAHs) leads to elevated levels of reactive metabolites which can lead to tissue injury.

Mechanisms of P450 suppression are poorly understood but represent important responses to xenobiotics and/or perturbations in homeostatic processes. Various mechanisms of negative transcriptional control include the following: (a) silencing of transcriptional initiation by direct inhibition; (b) inability of transcription factors to bind to DNA due to steric hindrance of adjacent or overlapping factors; (c) sequestration of transcription factors in an inactive form which may block an activating signal required by RNA polymerase (squelching); and (d) recruitment of corepressors and modulation of chromatin structure.

9.5.3 Aromatic Hydrocarbon Receptor Induction of P450s

The transcriptional regulation of the CYP1A and CYP1B subfamilies is mediated by the AhR, a ligand-activated transcription factor belonging to the basic helix–loop–helix (bHLH)/Per, Arnt, and Sim (PAS) family (Figure 9.7). The unliganded AhR resides in the cytoplasm complexed with two heat shock proteins 90 (Hsp90) and Ah receptor-interacting protein (AIP), an immunophilin-type chaperone protein. The binding of a ligand to the AhR initiates a transformation, triggering the release of Hsp90 and AIP, thereby allowing the ligand–AhR complex to translocate into the nucleus, where it forms a heterodimer with the AhR nuclear

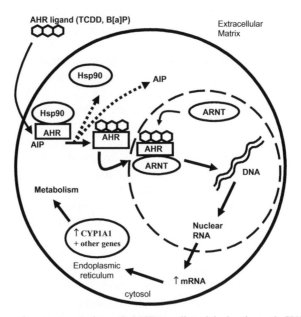

Figure 9.7. Schematic representation of AHR-mediated induction of *CYP1A1*. (Adapted from Whitlock, J. P. Induction of cytochrome P4501A1. *Annu. Rev. Toxicol.* **39**, 103–125, 1999; Ma, Q. Induction of CYP1A1. The Ahr/DRE paradigm: transcription, receptor regulation, and expanding biological roles. *Curr. Drug Metab.* **2**, 149–164, 2001.)

translocator (Arnt) protein. The actual site of dissociation of the AhR:Hsp90:AIP can occur either in the cytosol or in the nucleus. In either case, the AhR:Arnt heterodimer has a high affinity for specific DNA recognition sequences, known as dioxin-responsive elements (DREs). These enhancer sequences are located upstream of the CYP1A or CYP1B transcription start sites and promote gene transcription.

The interaction of hydrophobic, planar, or coplanar PAH ligands such as benzo[a]pyrene and dioxin (TCDD), which fit into a 6.8-Å × 13.7-Å rectangle, have been extensively studied. Several observable differences in the induction of *CYP1A1* following exposure to TCDD or B[a]P demonstrate how different ligands for the AhR result in an altered inductive response. TCDD, as opposed to B[a]P, has a very high binding affinity for the AhR and is not readily metabolized by CYP1A1. Consequently, the long half-life of dioxin permits accumulation within cells and prolongs the inductive response compared to the transient response observed with B[a]P. Molecular biological studies reveal that the AhR is an ancient protein, with both vertebrate and invertebrate homologs. In addition to its importance in response to PAH-type xenobiotics, cell culture and animal studies indicate a role for the AhR in the development of the liver, lung, and the immune system, in reproductive function, and in cell growth and differentiation. Together, these facts suggest that the AhR has a physiological function in addition to its role in responding to environmental chemicals and that the AhR may have an endogenous ligand.

Further evidence suggesting the existence of an endogenous AhR ligand include reports where *CYP1A1* induction occurs in the absence of any known exogenous ligand. These include stress conditions such as hyperoxia and hydrodynamic shearing in keratinocytes and murine hepatoma Hepa 1c1c7 cells, as well as in rat epithelial cells cultured in suspension. A role for the AhR in the induction of *CYP1A1* in the absence of a known exogenous ligand in these studies was inferred by inhibition with AhR antagonists. However, the role and identification of an endogenous ligand for the AhR has yet to be determined. Bilirubin, an end product of heme metabolism, was demonstrated to directly upregulate *Cyp1a1* gene expression and enzymatic activity in hepatoma Hepa 1c1c7 cells in an AhR-dependent manner. Subsequently, it was reported that bilirubin and biliverdin are AhR ligands and can activate *CYP1A1* expression in intact cells from several species.

Several nonpolycyclic and nonplanar compounds, structurally unrelated to the classical CYP1A1 inducers, activate CYP1A1 in various cell culture systems. Inductive responses were observed following exposure to omeprazole, the fungicide thiabendazole, carotenoids, the insecticide carbaryl, and the malaria drug primaquine. Three hypotheses to explain this phenomenon include: (1) the above compounds may act as weak ligands for the AhR, thereby making it extremely difficult to conduct competitive binding experiments; (2) the metabolites of these xenobiotics might be AhR ligands; and (3) these compounds might activate CYP1A1 through another signaling pathway.

9.5.4 Phenobarbital Induction of P450s

Several P450s are induced by phenobarbital (PB) including members of the *CYP2A, CYP2C, CYP2H,* and *CYP3A* subfamilies in mammals and bacterial *CYP102* (P450

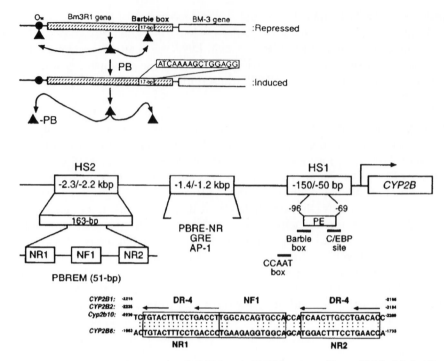

Figure 9.8. PB response elements in *CYP102* and *CYP2B* genes. (From Kakizaki, S., Yamamoto, Y., Ueda, A., Moore, R., Sueyoshi, T., and Negishi, M. Phenobarbital induction of drug/steroid-metabolizing enzymes and nuclear receptor CAR. *Biochim. Biophys. Acta* **1619**, 239–242, 2003.)

BM-3). PB increases the rate of transcription of *CYP102* genes by removal of a repressor (derepression) from a 17-bp regulatory sequence called the Barbie Box (Figure 9.8). Barbie Box-like sequences are also present in the proximal promoter region of mammalian PB-inducible genes, but their role in mediating PB induction is less clear. DNA sequences in the distal regions of the CYP2B promoter contribute to PB induction. These sites include (a) a PB-responsive element (PBRE) that is at –1.4 kb and (b) a PB-responsive enhancer module (PBREM) that is at –2.3 kb (Figure 9.8). The nuclear orphan receptor CAR (constitutively active receptor) and RXR (retinoid X receptor) form a heterodimer that binds to and activates the NR1 site of PBREM in response to PB induction. The CAR-mediated transactivation of PBREM in vivo becomes PB responsive through a nuclear translocation process that is regulated by phosphorylation-dephosphorylation. CAR null mice completely lack PB induction of the *Cyp2b10* gene, demonstrating unequivocally that CAR is an essential transcription factor in this process. However, the precise mechanism whereby PB activates CAR remains unknown. In *CYP2B* genes, PBREM is a composite element consisting of two nuclear receptor-binding sites (NR1 and NR2) flanking an NF1-binding site. Both NR1 and NR2 are direct repeat motifs (also called DR-4). Similar PB response elements are found in other PB responsive genes including *CYP3A4, CYP2H, CYP2C9*, and *UGT1A1*.

9.5.5   Induction of P450s by Peroxisome Proliferators

A variety of structurally dissimilar compounds cause peroxisome proliferation in mammals and lead to induction of enzymes that are important in the metabolism of long-chain fatty acids. For example, clofibrate, a hypolipidemic drug, induces transcription of the *CYP4A1* gene via a process mediated by the peroxisome proliferators activated receptors (PPARs). PPARs have homology with other members of the steroid/thyroid/retinoid nuclear hormone receptor superfamily. Three distinct PPARs have been described: PPARα, PPARγ, and PPARβ/δ. PPARα is widely expressed in numerous tissues including the kidney, liver, heart, adrenal, and adipose tissue. It is activated by Wy-14,643. PPARγ is abundantly expressed in adipose tissue, adrenal, and spleen. It is activated by BRL-49563 and PGJ₂. PPARβ/δ is expressed in heart, adrenal, spleen, and intestine. It is activated by a variety of fatty acids. Like other steroid hormone receptors, PPARs have the following: (a) a highly conserved zinc-finger domain that enables them to bind to specific DNA sequences; and (b) a ligand-binding domain that recognizes the receptor's specific ligand (Figure 9.9). The PPARs form heterodimers with a second protein called the retinoid X receptor (RXR) and bind to specific DNA regulatory sequences called peroxisome proliferator response elements (PPREs) in the 5' flanking region of target genes such as *CYP4A1* (Figure 9.10). Thus, fatty acids can activate PPARs that can regulate genes involved in metabolism of fatty acids; this constitutes a feedback loop in fatty acid homeostasis.

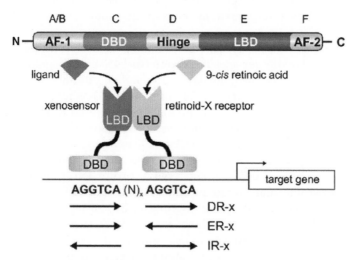

Figure 9.9. PPAR domain structure. Domains of the receptor that are required for certain receptor functions are indicated: a highly variable N-terminal region that in some receptors harbors an activation function (AF-1); a DNA-binding domain (DBD) consisting of two zinc-finger motifs; a flexible hinge domain, and a ligand-binding domain (LBD) that also contains an activation function (AF-2). DBD, DNA-binding domain; LBD, ligand-binding domain. The nuclear receptors bind as heterodimers with the RXR to repeats of the nucleotide hexamer AGG/TTCA with variable spacing. The hexamers can be arranged either as direct repeats (DR), everted repeats (ER), or inverted repeats (IR). (From Handschin, C., and Meyer, U. A. Induction of drug metabolism: The role of nuclear receptors. *Pharmacol. Rev.* **55**, 649–673, 2003.)

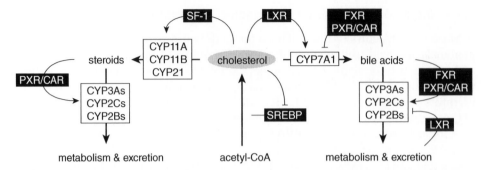

Figure 9.10. Regulation of target genes by PPAR:RXR heterodimers. Fatty acids serve as transcriptional inducers and substrates of enzymes involved in lipid homeostasis. (From Handschin, C., and Meyer, U. A. Induction of drug metabolism: The role of nuclear receptors. *Pharmacol. Rev.* **55**: 649–673, 2003.)

9.5.6 Hormonal Regulation of P450s

Several P450s are expressed in a sex-specific manner and are subject to complex developmental regulation and endocrine control. Regulation of P450 gene expression during development is isoform-dependent and appears to occur in a tissue-specific fashion. Some P450 genes can be directly regulated by androgens and estrogens. One example is regulation of mouse *Cyp2j5* in kidney. Androgen regulation of *Cyp2j5* may be mediated by the androgen receptor via ligand-dependent binding to an androgen responsive element in the promoter region of the *Cyp2j5* gene.

For other hepatic P450 genes, gonadal hormones (testosterone, estrogen) do not act directly to regulate the sex-specific patterns of P450 expression, but rather their effects are mediated via the gonadal–hypothalamic–pituitary axis and its sex-dependent regulation of pituitary growth hormone (GH) secretory patterns (Figure 9.11). For example, *CYP2C11* is not expressed in immature rats but is induced dramatically at puberty in male rats and suppressed in female rats. In contrast, *CYP2C12* is expressed at similar levels in prepubertal male and female rats but is induced at puberty in female rats only. The cellular and molecular mechanisms whereby pituitary GH secretory profiles differentially regulate expression of the sex-specific liver P450s are only partially understood. Pulsatile secretion of GH activates liver STAT (signal transducer and activator of transcription) by causing induction of tyrosine phosphorylation by members of the Janus kinase (JAK) superfamily. Phosphorylation of STAT results in its ability to translocate into the nucleus where it can bind specific DNA *cis*-acting elements and stimulate gene transcription. In contrast, continuous secretion of GH leads to phosphotyrosine dephosphorylation resulting in downregulation of STAT and reduced gene transcription.

For some P450s, developmental regulation involves changes in the accessibility of transcription factors to chromatin. For example, liver *CYP2E1* transcription begins within a few hours after birth and reaches maximal levels by 7 days of age. Evidence suggests that the *CYP2E1* chromatin structure changes after birth coincident with cytosine demethylation in the 5′ flanking region of the *CYP2E1* gene; demethylation leads to increased accessibility of the *CYP2E1* promoter to transcription factors (e.g., HNF-1α), which leads to increased transcription.

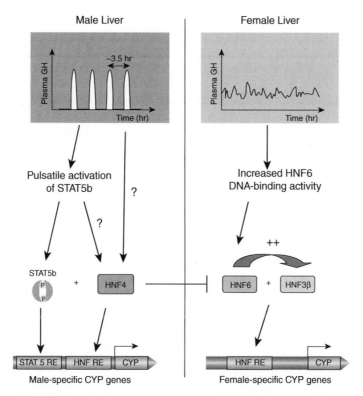

Figure 9.11. Model for sex-dependent regulation of P450 genes by hepatic nuclear factors. Activation of male-specific P450s (e.g. CYP2C11) occurs via the concerted action of growth hormone (GH) pulse-induced STAT5b and the nuclear receptor HNF4α. Regulation of female-specific P450s (e.g. CYP2C12) occurs by the synergistic action of HNF6 and HNF3β which are both elevated in female liver in response to the continuous GH secretion pattern. (From Wiwi, C. A., and Waxman, D. J. Role of hepatocyte nuclear factors in growth hormone-regulated, sexually dimorphic expression of liver cytochromes P450. *Growth Factors* **22**, 79–88, 2004.)

9.5.7 Nutritional Regulation of P450s

The influence of dietary constituents on P450 enzyme expression can have clinically important effects on the pharmacokinetics of numerous drugs. Indirectly, food can alter gastric emptying, increase the solubilization of drugs, form complexes with drugs, and alter hepatic blood flow, all of which can change drug pharmacokinetics. However, dietary components can directly regulate P450 gene expression and function, not only impacting drug elimination but also significantly affecting disease pathogenesis. As the vast majority of xenobiotics enter the body in the diet, such interactions can have dramatic clinical consequences.

Inhibition of human CYP3A4 by grapefruit juice is one the best-characterized food–drug interactions. Grapefruit juice inhibits pre-systemic drug metabolism mediated by CYP3A4 in the small intestine, thereby increasing plasma concentrations of a number of oral medications, such as cyclosporin, terfenadine, and felodipine. Coumarin derivatives found in grapefruit juice, like dihydroxybergamottin, are thought to be the active components, although bioflavonoids such as naringin and quercetin may also be involved. Constituents of cruciferous vegetables, such as brussel sprouts, cabbage, and broccoli, increase CYP1A-related activities in rat intestine and liver. Certain dietary indoles and flavonoids—as well as smoking or charcoal-broiled foods, which generate multiple heterocyclic aromatic amines (HA) and polycyclic aromatic hydrocarbons—can activate *CYP1A* expression. This induction can occur either by direct ligand interaction with the AhR or by augmenting the interaction of the AhR with xenobiotic response elements in *CYP1A1*.

In general, dietary interactions can complicate drug therapy and can contribute to disease pathogenesis; however, inhibition of certain P450 reactions may also protect against toxic metabolites and/or free radicals generated by CYPs. These differences can be observed in experimental models where animals were fed high-fat diets. Increased availability of lipid substrates for CYPs can enhance free radical production and exacerbate tissue injury. Downregulation of *CYP2E1, CYP2C23*, and *CYP4A* occurs after feeding of experimental diets that are high in fat and carbohydrates; these diets promote hepatic lipid infiltration characteristic of liver steatosis and obesity. On the other hand, hepatic downregulation of CYP2E1 from high-fat diet elicited hepatic steatosis is thought to play a role in attenuating acetaminophen-induced hepatotoxicity. The renal regulation of CYP2C23 and CYP4A enzymes by dietary salt intake can lead to altered formation of 20-HETE and EETs and can influence renal function and blood pressure regulation. Thus, a variety of food constituents modulate CYP expression and function, with the potential for both deleterious and beneficial outcomes.

9.5.8 Regulation of P450s by Cytokines

Alterations in the cellular, tissue, or organism homeostasis as a result of exposure to a pathobiological stimuli, such as a injury or infection, may have profound effects on drug metabolism by altering the expression and/or catalytic activity of CYPs. The regulation of CYPs by inflammatory cytokines has been studied; however, the mechanisms of action involved in the modulation of specific enzymes remains incompletely understood. Most investigations have focused on the roles of the major proinflammatory cytokines interleukin 6 (IL-6), tumor necrosis factor alpha (TNFα), and interleukin 1β (IL-1β) on the regulation of CYPs. For example, decreases in CYP2C11- and CYP3A2-dependent catalytic activities, but not those of CYP2A1 or CYP2C6, were observed after TNFα treatment of rats. In rat hepatocytes, CYP2C11 mRNA is rapidly suppressed by IL-1 and TNFα treatment, an effect that is blocked by the competitive IL-1 receptor antagonist, IL-1-ra. Lipopolysaccharide (LPS) treatment of rats also decreased hepatic microsomal CYP1A, CYP2B, CYP3A, and CYP4A proteins and activities in both wild-type and TNFα receptor-deficient animals.

9.5.9 Regulation of P450s by Nitric Oxide

Exposure to pathobiological stressors causes the formation or release of mediators other than cytokines and ROS. These include stress-induced hormones and nitric oxide (NO). Stress responses stimulate the hypothalamic–pituitary axis causing the release of adrenocorticotropin, thereby elevating glucocorticoid levels. Glucocorticoids have been shown to have a bimodal effect on CYP2C11 mRNA expression in rat hepatocytes, causing induction at low (normal) concentrations and inhibition at high (stress) concentrations. The induction of nitric oxide synthase (iNOS) and NO release occurs in many pathobiological states including inflammation. The role of NO in the regulation of CYP catalytic activities, protein levels, and mRNA expression is controversial. NO transiently forms nitrosyl–heme iron complexes in liver microsomes resulting in inhibition of CYPs. Further evidence for NO modulation of CYP expression is found in experiments with iNOS inhibitors, where administration of N^G-[1-iminoethyl]-L-ornithine following LPS treatment resulted in prevention of CYP2C11 and CYP3A2 downregulation in rat liver.

9.6 TRANSGENIC ANIMAL MODELS

There is a tremendous degree of complexity within a number of CYP subfamilies and a high degree of variation between species, especially between humans and rodents. This creates a problem when trying to study the functional role of individual CYPs or CYP subfamilies within an intact organism. To overcome this problem, researchers have developed mouse models that either have targeted disruption of specific *CYP* genes (e.g., *Cyp1a1, Cyp1a2, Cyp1b1, Cyp2e1*) or specific CYP regulators (e.g., *AhR, CAR, PPARα*). Together, these genetic tools have resulted in great advances in understanding CYP regulation and function. However, these models still do not circumvent the problem of species differences. An alternative approach to overcome this dilemma has been to generate "humanized transgenic mice" by introducing human genes into the mouse genome, thus providing a better system to predict the human response to foreign chemicals and to understand the underlying mechanisms. Examples of humanized mouse models include transgenic mice that overexpress human *CYP2D6, CYP2J2, CYP2E1, CYP3A4, AhR, CAR*, and *PPARα*. The most common approach to generate a humanized transgenic mouse has been to fuse the human cDNA to a promoter that drives expression of the cDNA in specific mouse tissues. The development of these mouse models has been invaluable in studying the function and regulation of CYPs in animal systems in vivo.

9.7 REACTIVE OXYGEN SPECIES

One of the consequences of exposure to pathobiological stressors is the overproduction of ROS which alters the redox status within the cell and results in oxidative stress. ROS are formed following incomplete reduction of molecular oxygen (O_2) to water during normal aerobic metabolism. Generation of $^{\cdot}O_2^-$ (superoxide anion radical), H_2O_2 (hydrogen peroxide), or $OH^{\cdot}$ (hydroxyl radical) can result. ROS are able to oxidize biological macromolecules such as DNA, protein, and lipids, causing cytotoxicity and damage. Cellular metabolism continuously produces ROS, a process that occurs predominantly in the mitochondria. Nonmitochondrial sources of ROS, including CYP enzymes, can generate superoxide anion radicals and hydrogen peroxide through uncoupling during catalytic oxidation. The relative amount of ROS generated by CYPs is enzyme-dependent, with CYP2E1 being particularly active. ROS are removed by various antioxidant systems, such as thiol-containing moieties, radical scavenging molecules (i.e., ascorbic acid or α-tocopherol), and enzymes like catalase, NAD(P)H:quinone oxidoreductase (NQOR), or glutathione S-transferase Ya subunit (GST Ya).

Self-inactivation of CYP enzymes may occur during the catalytic cycle when ROS are formed at the active center, as a result of uncoupling of monooxygenase reactions. This may be associated with oxidation of the apoenzyme at several amino acid residues, such as cysteine, histidine, or tryptophan, or oxidation of the heme moiety resulting in inactivation and/or heme loss. In addition to direct inactivation of CYP isozymes, ROS (notably H_2O_2) can suppress transcription of *CYP1A1* and CYP*1A2* in isolated rat hepatocytes. The suppression of *CYP1A1* transcription by H_2O_2 was suggested to be either through the XRE or by repression of the *CYP1A1* gene promoter, by targeting the nuclear factor I/CCAAT box transcription factor (NF1/CTF). Nuclear factor 1 (NF1) protein is a member of a family of ubiquitous transcription factors, which form homo- and heterodimers that regulate promoters of genes including collagen, albumin and *CYP1A1*. Either H_2O_2 at millimolar concentrations, or glutathione depletion in HepG2 cells, represses the transactivating function of NFI/CTF and represses the promoter activity of *CYP1A1*. Recent evidence suggests that Cys427 located within the transactivation domain (TAD) of the C-terminal end of NFI/CTF factor plays a critical role in redox regulation. Thus, evidence indicates that CYP expression may be modulated by endogenously or exogenously generated ROS at the transcriptional level. CYP enzyme activity may also be influenced by ROS through oxidation of critical amino acids in the active site of the enzyme or by heme modifications.

9.8 POSTTRANSLATION MODIFICATION OF P450s

Compared to transcriptional regulation studies, much less is known about post-translational modification of CYP enzymes. Although there are many reports of CYP enzymes being phosphorylated, the physiological significance of these modifications remains elusive. Individual CYPs are phosphorylated by different kinases leading to enzyme-specific alterations in catalytic activity. Currently the best evidence for modulation and regulation of CYP function comes from studies examining cholesterol and steroid homeostasis. Reports suggest that the dephosphorylation of CYP51 (sterol 14α-demethylase) may decrease its catalytic activity, and this could be an important regulatory mechanism for cholesterol biosynthesis and potentially a target for cholesterol lowering drugs. Other studies suggest that the regulation of CYP7A1 (cholesterol 7α-hydroxylase), the enzyme that catalyzes the rate-limiting step in the synthesis of bile acids from cholesterol, may occur by the reversible phosphorylation–dephosphorylation of the enzyme. In addition, evidence exists supporting a role for phosphorylation in increasing the catalytic activity of CYP19A1 (aromatase), the enzyme involved in the biosynthesis of estrogen from androgen. Conversely, phosphorylation of CYP2B by protein kinase A (PKA) has been shown to decrease hydroxylation of testosterone metabolism in the absence of protein degradation.

Substitutions or mutations in phosphorylatable sites, such as serine residues, can have opposite affects. For example, substitution of Ser[129] with Gly[129] in CYP2E1 has been reported to increase the catalytic activity following phosphorylation by PKA.

Some P450 enzymes can be targeted to both the endoplasmic reticulum and mito-chondria (e.g., CYP1A1, CYP2B2, CYP2E1, and CYP2D6). While the mechanism(s) is unclear, evidence suggests that it might be related to cleavage of the N-terminal segment or phosphorylation of key amino acid residues such as a serine. The phos-phorylation–dephosphorylation of an individual CYP can act like a switch, rapidly activating or inactivating the protein, as opposed to the time-dependent inductive process. Thus polymorphisms in primary phosphorylation sites can potential impact drug metabolism and toxicity.

Glycosylation of P450 enzymes (e.g., CYP2B2, CYP2B4) have been known to occur for over 25 years; however, the significance of this modification remains to be clarified. As protein glycosylation is generally involved in cell adhesion, protein targeting, and prevention of proteolysis, it is thought to modulate catalytic activity, such as for CYP19A1. Ubiquitination of several mammalian CYP enzymes (CYP2B1, CYP2E1, CYP3A1, CYP3A2, and CYP3A4) has also been described. This process targets the proteins for subsequent degradation by the 26S proteasomal pathway. It is thought that this represents a regulatory pathway for normal protein turnover or degradation of chemically modified or structurally damaged CYP enzymes. Although there are numerous reports of posttranslational modifications of CYP enzymes, much work remains to further characterize and identify their physiological role.

9.9 SUMMARY

The cytochrome P450 superfamily is complex. Indeed, more than 2000 distinct *CYP* genes have been described and are subdivided into families and subfamilies based on amino acid sequence homology. Cytochromes P450 are involved in the metabo-lism of a large number of xenochemicals and endogenous compounds such as fatty acids, steroids, cholesterol, and bile acids. Information gleaned from the crystal structures of a number of bacterial and mammalian P450s provides a useful frame-work to understand P450 structure–function relationships. Some P450s are consti-tutively expressed, whereas others are inducible by agents such as polycyclic aromatic hydrocarbons (mediated by AhR), phenobarbital (mediated by CAR and RXR), and peroxisome proliferators (mediated by PPARs). P450s are also regulated during development, in a sex-specific manner (mediated by growth hormone secretory pattern), by proinflammatory cytokines and by nutritional factors.

SUGGESTED READING

Johnson, E. F., and Waterman, M. R. (Eds.). *Cytochrome P450*, Academic Press, San Diego, 2002.

Denisov, I. G., Makris, T. M., Sligar, S. G., and Schlichting, I. Structure and chemistry of cyto-chrome P450. *Chem. Rev.* **105**, 2253–2577, 2005.

Domanski, T. L., and Halpert, J. R. Analysis of mammalian cytochrome P450 structure and function by site-directed mutagenesis. *Curr. Drug Metab.* **2**, 117–137, 2001.

Gonzalez, F. J., and Yu, A. M. Cytochrome P450 and xenobiotic receptor humanized mice. *Annu. Rev. Pharmacol. Toxicol.* **46**, 41–64, 2006.

Guengerich, F. P., Hosea, N. A., Parikh, A., Bell-Parikh, L. C., Johnson, W. W., Gillam, E. M., and Shimada, T. Twenty years of biochemistry of human P450s: Purification, expression, mechanism, and relevance to drugs. *Drug Metab. Dispos.* **26**, 1175–1178, 1998.

Lewis, D. F. V. *Guide to cytochrome P450 Structure and Function*, Taylor & Francis, New York, 2001.

Morgan, E. T. Regulation of cytochrome p450 by inflammatory mediators: Why and how? *Drug Metab. Dispos.* **29**, 207–212, 2001.

Nebert, D. W., and Russell, D. W. Clinical importance of the cytochromes P450. *Lancet* **360**, 1155–1162, 2002.

Nelson, D. R., Zeldin, D. C., Hoffman, S. M., Maltais, L. J., Wain, H. M., and Nebert, D. W. Comparison of cytochrome P450 (CYP) genes from the mouse and human genomes, including nomenclature recommendations for genes, pseudogenes and alternative-splice variants. *Pharmacogenetics* **14**, 1–18, 2004.

Raucy, J. L., and Allen, S. W. Recent advances in P450 research. *Pharmacogenomics J.* **1**, 178–186, 2001.

Roman, R. J. P-450 metabolites of arachidonic acid in the control of cardiovascular function. *Physiol Rev.* **82**, 131–185, 2002.

Schuetz, E. G. Induction of cytochromes P450. *Curr. Drug Metab.* **2**, 139–147, 2001.

Sueyoshi, T., and Negishi, M. Phenobarbital response elements of cytochrome P450 genes and nuclear receptors. *Annu. Rev. Pharmacol. Toxicol.* **41**, 123–143, 2001.

Zeldin, D. C. Epoxygenase pathways of arachidonic acid metabolism. *J. Biol. Chem.* **276**, 36059–36062, 2001.

CHAPTER 10

Phase 1 Metabolism of Toxicants and Metabolic Interactions

ERNEST HODGSON

Department of Environmental and Molecular Toxicology, North Carolina State University, Raleigh, North Carolina 27695

PARIKSHIT C. DAS

Department of Environmental and Molecular Toxicology, North Carolina State University, Raleigh, North Carolina 27695

TAEHYEON M. CHO

Department of Environmental and Molecular Toxicology, North Carolina State University, Raleigh, North Carolina 27695

RANDY L. ROSE

(deceased) Department of Environmental and Molecular Toxicology, North Carolina State University, Raleigh, North Carolina 27695

10.1 INTRODUCTION

The majority of xenobiotics that enter the body tissues are lipophilic, a property that enables them to penetrate lipid membranes and to be transported by lipoproteins in body fluids. The metabolism of xenobiotics, carried out by a number of relatively nonspecific enzymes, usually consists of two phases. During phase I, a polar group is introduced into the molecule; and although this increases the molecule's water solubility, the most important effect is to render the xenobiotic a suitable substrate for phase II reactions. In phase II reactions, the altered compounds combine with an endogenous substrate to produce a water-soluble conjugation product that is readily excreted. Although this sequence of events is generally a detoxication mechanism, in some cases the intermediates or final products are more toxic than the parent compound, and the sequence is termed an activation or intoxication mechanism. See Chapter 20 for discussion of activation and toxicity.

Although most phase I reactions are oxidations, reductions, or hydrolyses, the hydration of epoxides and dehydrohalogenations also occur. A summary of the most important reactions is shown in Table 10.1.

Molecular and Biochemical Toxicology, Fourth Edition, edited by Robert C. Smart and Ernest Hodgson
Copyright © 2008 John Wiley & Sons, Inc.

TABLE 10.1. Phase I Xenobiotic-Metabolizing Enzymes with Examples of Substrates

Enzymes	Examples
Cytochrome P450 (CYP)	
Epoxidation and hydroxylation	Aflatoxin, aldrin, benzo[*a*]pyrene, bromobenzene, naphthalene
N-Dealkylation	Ethylmorphine, atrazine, dimethylnitrocarbamate, dimethylaniline
O-Dealkylation	*p*-Nitroanisole, chlorfenvinphos, codeine
S-Dealkylation	Methylmercaptan
S-Oxidation	Thiobenzamide, phorate, endosulfan, methiocarb, chlorpromazine
N-Oxidation	2-Acetylaminofluorene
P-Oxidation	Diethylphenylphosphine
Desulfuration	Parathion, fonofos, carbon disulfide
Dehalogenation	CCl_4, $CHCl_3$
Nitro reduction	Nitrobenzene
Azo reduction	*O*-Aminoazotoluene
Flavin-Containing Monooxygenase (FMO)	
N-Oxidation	Nicotine, dimethylaniline, imipramine
S-Oxidation	Thiobenzamide, phorate, thiourea
P-Oxidation	Diethylphenylphosphine
Desulfuration	Fonofos
Alcohol Dehydrogenase	
Oxidations	Ethanol, methanol, isopropanol, glycols, glycol ethers (2-butoxyethanol)
Reductions	Aldehydes and ketones
Aldehyde Dehydrogenase	
Oxidations	Aldehydes from alcohol and glycol oxidations
Molybdenum Hydroxylases (Aldehyde Oxidase, Xanthine Oxidase)	
Oxidations	Purines, pteridine, methrotrexate, quinolones, 6-deoxycyclovir
Reductions	Aromatic nitro compounds, azo dyes, nitrosamines, *N*-oxides, sulfoxides
Prostaglandin Synthetase (PGS) Cooxidation	
Dehydrogenation	Acetaminophen, benzidine, DES, epinephrine
N-Demethylation	Dimethylaniline, benzphetamine, aminocarb
Hydroxylation	Benzo[*a*]pyrene, 2-aminofluorene, phenylbutazone
Epoxidation	7,8-Dihydrobenzo[*a*]pyrene
Sulfoxidation	Methylphenylsulfide
Oxidations	FANFT, ANFT, bilirubin
Esterases and Amidases	
	Paraoxon, dimethoate, phenyl acetate
Epoxide Hydrolase	
	Benzo(*a*)pyrene epoxide, styrene oxide
DDT-Dehydrochlorinase	
	p,p-DDT
Glutathione Reductase	
	Disulfiram

10.2 MICROSOMAL MONOOXYGENATIONS: GENERAL BACKGROUND

Monooxygenations are those oxidations in which one atom of molecular oxygen is reduced to water while the other is incorporated into the substrate. Microsomal monooxygenation reactions are catalyzed by nonspecific enzymes such as the flavin-containing monooxygenases (FMOs) or the multienzyme system that has cytochrome P450s (CYPs) as the terminal oxidases.

10.2.1 CYP-Dependent Monooxygenase Reactions

The location, structure, classification, regulation, and mechanism of action of CYPs are discussed in Chapter 9. This chapter summarizes the reactions involved in the oxidation of xenobiotics by CYP and other enzymes. Although microsomal monooxygenase reactions are basically similar with respect to the role played by molecular oxygen and in the supply of electrons, the enzymes are markedly nonspecific, with both substrates and products falling into many different chemical classes. It is convenient, therefore, to classify these activities on the basis of chemical reactions, bearing in mind that not only do the classes often overlap, but the same substrate may undergo more than one oxidative reaction.

10.2.1a Epoxidation and Aromatic Hydroxylation. Epoxidation is an extremely important microsomal reaction because not only can stable epoxides be formed but arene oxides, the epoxides of aromatic rings, are intermediates in aromatic hydroxylations. In the case of polycyclic hydrocarbons, the reactive arene oxides are known to be involved in carcinogenesis.

The epoxidation of aldrin to dieldrin is an example of the metabolic formation of a stable epoxide (Figure 10.1A), while the oxidation of naphthalene was one of the earliest understood examples of an epoxide (arene oxide) as an intermediate in aromatic hydroxylation (Figure 10.1B). The arene oxide can rearrange nonenzymatically to yield predominantly 1-naphthol, can interact with the enzyme epoxide hydrolase to yield the dihydrodiol, or can interact with glutathione S-transferase to yield the glutathione conjugate that ultimately is metabolized to a mercapturic acid. This reaction is also of importance in the metabolism of the insecticide carbaryl, which contains the naphthalene nucleus.

The proximate carcinogens arising from the metabolic activation of benzo(a)pyrene are isomers of benzo(a)pyrene 7,8-diol-9,10-epoxide (Figure 10.1C). These metabolites arise by prior formation of the 7,8-epoxide, which gives rise to the 7,8-dihydrodiol through the action of epoxide hydrolase. The diol is further metabolized by the microsomal monooxygenase system to the 7,8-diol-9,10-epoxides, which are both potent mutagens and unsuitable substrates for the further action of epoxide hydrolase.

10.2.1b Aliphatic Hydroxylations. Alkyl side chains of aromatic compounds are readily oxidized, often at more than one position, and provide good examples of this type of oxidation. In the rabbit the *n*-propyl side chain of *n*-propylbenzene can be oxidized at any of the three carbon atoms to yield 3-phenylpropan-l-ol ($C_6H_5CH_2CH_2CH_2OH$) by ω-oxidation, benzylmethylcarbinol ($C_6H_5CH_2CHOHCH_3$) by ω-1-oxidation, and ethylphenylcarbinol ($C_6H_5CHOHCH_2CH_3$) by ω-2-oxidation. Further oxidation of these alcohols is also possible.

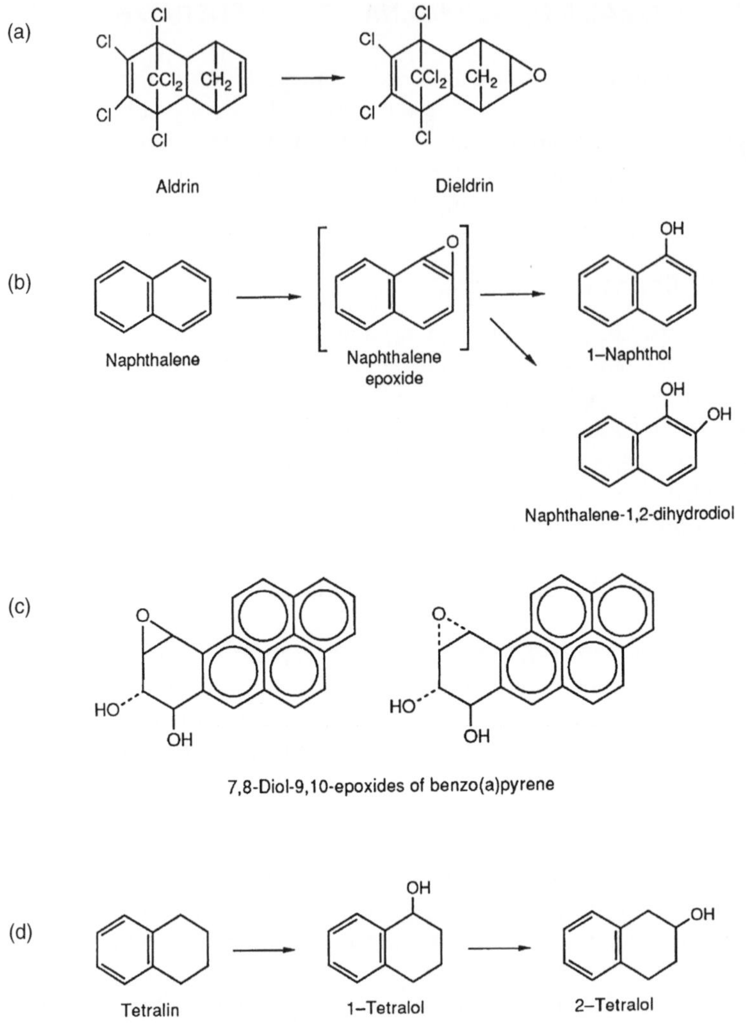

Figure 10.1. Examples of epoxidation and aromatic hydroxylation reactions.

Alicyclic compounds, such as cyclohexane, are also susceptible to oxidation, in this case first to cyclohexanol and then to *trans*-cyclohexane-1,2-diol, cyclohexaone, and adipic acid.

In compounds with both saturated and aromatic rings, the former appears to be the most readily hydroxylated. For example, the major oxidation products of tetralin (5,6,7,8-tetrahydronaphthalene) in the rabbit are 1- and 2-tetralol, whereas only a trace of the phenol, 5,6,7,8-tetrahydro-2-naphthol, is formed (Figure 10.1D).

10.2.1c   Dealkylation: O-, N-, and S-Dealkylation. A well-known example of *O*-dealkylation is the demethylation of *p*-nitroanisole. Due to the ease with which the colored product *p*-nitrophenol can be measured, *p*-nitroanisole has been used to demonstrate monooxygenase activity. The reaction is thought to proceed via an

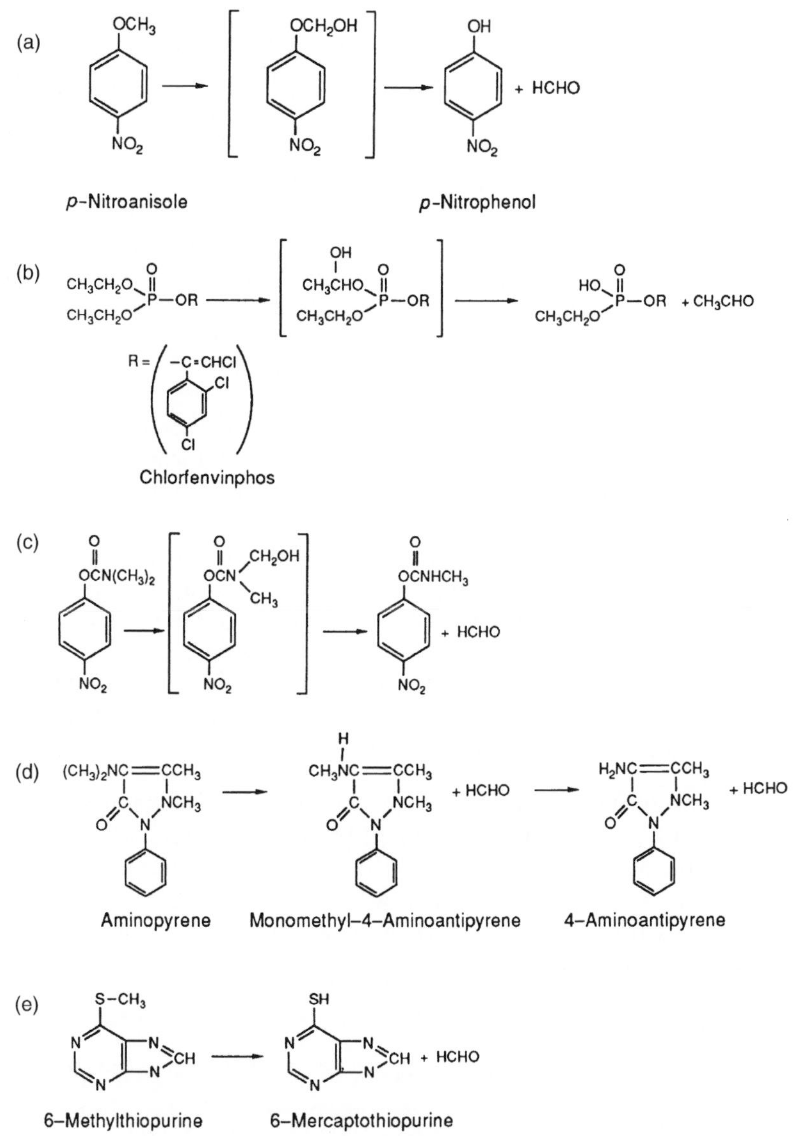

Figure 10.2. Examples of dealkylation reactions.

unstable methylol derivative (Figure 10.2A). Other substrates that undergo *O*-dealkylation include the drugs codeine and phenacetin and the insecticide methoxychlor.

The *O*-dealkylation of organophosphorus triesters differs from the above reactions in that it involves the dealkylation of an ester rather than an ether. The reaction was first described for the insecticide chlorfenvinphos (Figure 10.2B), but is now known to occur with a wide variety of vinyl, phenyl, phenylvinyl, and naphthyl phosphates and the thionophosphate triesters. At least one phosphonate, *O*-ethyl *O*-*p*-nitrophenyl phenylphosphonate (EPNO), is also metabolized by this mechanism.

N-Dealkylation is a common reaction in the metabolism of drugs, insecticides, and other *N*-alkyl xenobiotics. Both the *N*- and *N, N*-dialkyl carbamate insecticides are readily dealkylated, and in some cases the methylol intermediates are stable enough to be isolated. *N, N*-Dimethyl-*p*-nitrophenyl carbamate is a useful model compound for this reaction (Figure 10.2C). Another important example is the insecticide carbaryl, which undergoes several different microsomal oxidation reactions, including an attack on the *N*-methyl group. In this case, the methylol compound is stable enough to be isolated or to be conjugated in vivo. The drug aminopyrene undergoes two *N*-demethylations to form first monomethyl-4-amino-antipyrene and then 4-aminoantipyrene (Figure 10.2D).

S-Dealkylation is known to occur with a number of thioethers such as methyl-mercaptan, 6-methylthiopurine, and so on (Figure 10.2E).

10.2.1d *N*-Oxidation.

N-Oxidation can occur in a number of ways, including hydroxylamine formation, oxime formation, and *N*-oxide formation. The latter is primarily dependent on the FMO also found in the microsomes and is discussed in Section 10.2.2.

Hydroxylamine formation occurs with a number of amines, such as aniline and many of its substituted derivatives (Figure 10.3A). In the case of 2-acetylaminofluorene, the product is a potent carcinogen and thus the reaction, catalyzed by CYP1A2, is an activation reaction (Figure 10.3B).

Oximes can be formed by the *N*-hydroxylation of imines and primary amines. Imines have, furthermore, been suggested as intermediates in the formation of oximes from primary amines (Figure 10.3C).

10.2.1e *Oxidative Deamination.*

Oxidative deamination of amphetamine occurs in the rabbit liver but not to any extent in either the dog or rat, both of which tend to hydroxylate the aromatic ring (Figure 10.3D). A close examination of this reaction indicates that it is probably not an attack on the nitrogen but rather an attack on the adjacent carbon atom, giving rise to a carbinol amine, which produces the ketone by elimination of ammonia:

$$R_2CHNH_2 \xrightarrow{\text{O}} R_2C(OH)NH_2 \xrightarrow{-NH_3} R_2C=O$$

The carbinol, by another reaction sequence, can also give rise to an oxime:

$$R_2C(OH)N_2 \xrightarrow{-H_2O} R_2C=NH \xrightarrow{\text{O}} R_2CNOH$$

The oxime can now be hydrolyzed to yield the ketone, which is thus formed by two different routes:

$$R_2CNOH \xrightarrow{-H_2O} R_2C=O$$

10.2.1f *S*-Oxidation.

Thioethers in general are oxidized to sulfoxides by microsomal monooxygenases, including the FMO (Section 10.2.2) and CYP. Some of the sulfoxides are further metabolized to sulfones. This reaction is very common among insecticides of several different chemical classes including carbamates, organophosphates, and chlorinated hydrocarbons (Figure 10.4A).

Organophosphates include phorate, dimeton, and others, whereas among the chlorinated hydrocarbons, endosulfan is oxidized to endosulfan sulfate and methiochlor to a series of sulfoxides and sulfones to yield eventually the bissulfone.

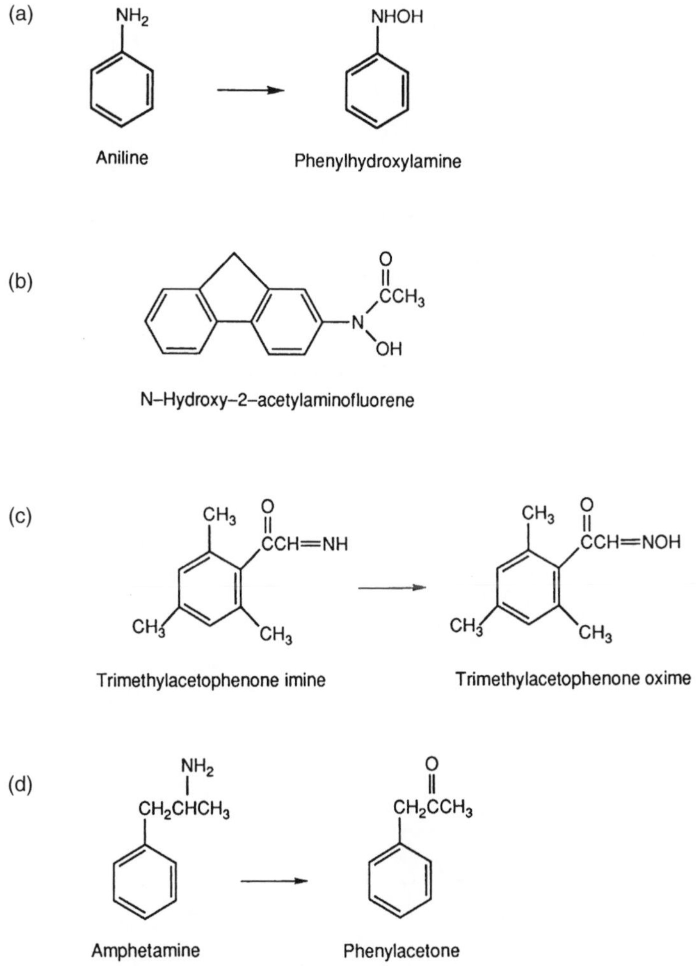

Figure 10.3. Examples of *N*-oxidation reactions.

S-Oxidation is also known among drugs (e.g., chlorpromazine), whereas the solvent dimethylsulfoxide is further oxidized to the sulfone:

$$(CH3)_2 SO \xrightarrow{\text{O}} (CH3)_2 SO_2$$

10.2.1g *P-Oxidation.* This little-known reaction involves the conversion of tri-substituted phosphines to phosphine oxides. It is catalyzed not only by CYP but also by the FMO. Known substrates are diphenylmethylphosphine and 3-dimethyl-aminopropyl-diphenylphosphine (Figure 10.4B).

10.2.1h *Desulfuration and Ester Cleavage.* Organophosphorus insecticides containing the P=S moiety owe their insecticidal activity and their mammalian toxic-ity to an oxidative reaction in which the P=S group is converted to P=O, thereby converting compounds relatively inactive toward cholinesterases into potent cho-

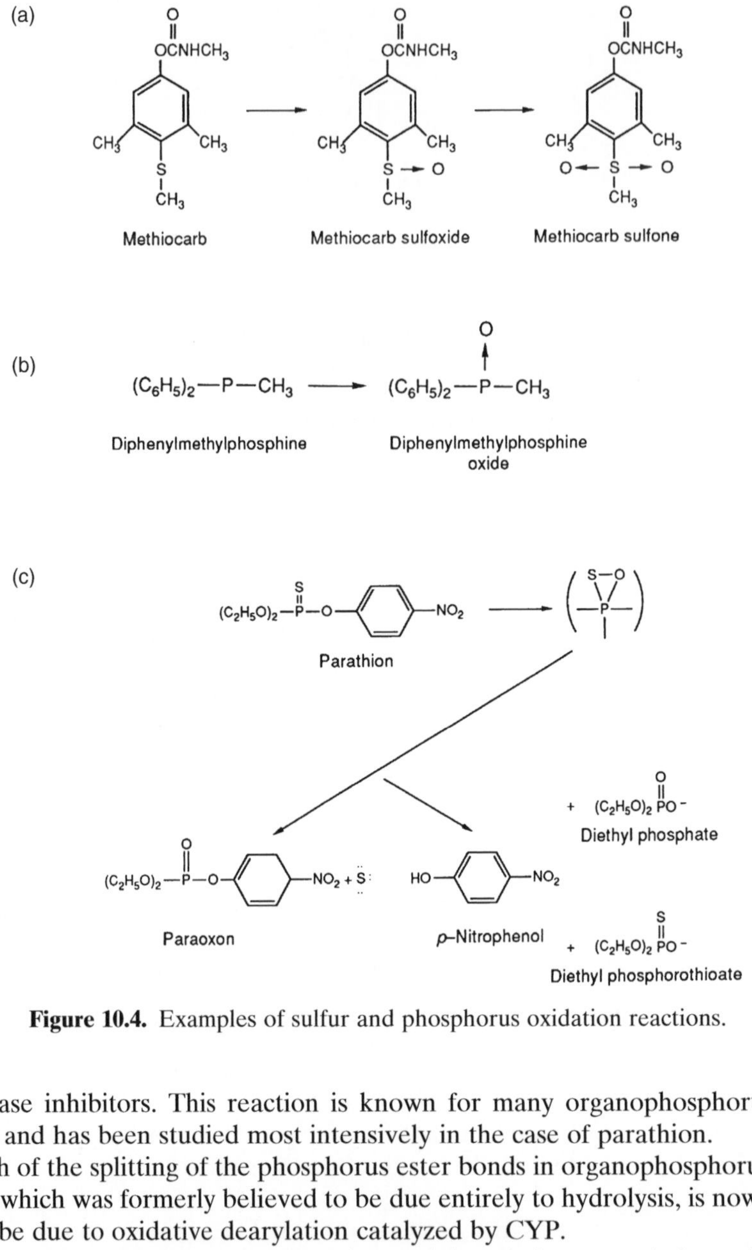

Figure 10.4. Examples of sulfur and phosphorus oxidation reactions.

linesterase inhibitors. This reaction is known for many organophosphorus compounds and has been studied most intensively in the case of parathion.

Much of the splitting of the phosphorus ester bonds in organophosphorus insecticides, which was formerly believed to be due entirely to hydrolysis, is now known also to be due to oxidative dearylation catalyzed by CYP.

The question of whether desulfuration and dearylation occur independently of each other, possibly catalyzed by different P450s, or whether they involve common intermediates is not yet resolved with certainty. However, despite the finding that different individual CYP isoforms may produce widely different ratios of desulfuration and dearylation products, the hypothesis that both reactions involve a common intermediate of the "phosphooxythiiran" type (Figure 10.4C) is still widely held.

10.2.1i   Cyanide Release. Considerable evidence suggests that the toxicity of organonitriles may occur as a result of the release of cyanide from the parent compound. Organonitriles are used in numerous manufacturing processes including production of synthetic fiber, plastic, pharmaceuticals, and dye stuffs, thus presenting the potential for extensive human exposure. Because many of these organonitriles are volatile and water-soluble, they would be absorbed to a large extent in the nasal mucosa. The nasal cavity contains a high concentration of some CYP isoforms, and in fact microsomal preparations of nasal mucosa are more efficient than liver microsomes in catalyzing cyanide release from a number of organonitriles. For primary organonitriles, CYP catalyzes the oxidation of the carbon alpha to the cyano group to produce cyanohydrins that decompose to hydrogen cyanide and an aldehyde.

$$RCH_2C\equiv N \xrightarrow{O} RCHO + HCN$$

In general, substrate affinity and cyanide release increase with increasing size of the R group with benzylcyanide having a K_m as low as 2.3 pM for rat nasal microsomes. Metabolism of acrylonitrile may proceed via an epoxide, formed by CYP, which is then converted to the diol by epoxide hydrolase. The diol would then release cyanide by the same mechanism as other cyanohydrins.

10.2.2 The Flavin-Containing Monooxygenase (FMO)

FMOs, like CYP, are located in the endoplasmic reticulum and are involved in the oxidation of numerous organic xenobiotics containing nitrogen, sulfur, or phosphorus heteroatoms as well as some inorganic ions.

The types of compounds oxidized by FMOs are shown in Table 10.2 with some examples of reactions catalyzed shown in Figure 10.5. Substrates are soft nucleophiles and compounds containing additional charged groups (anionic or cationic) are excluded as substrates. With the exception of cysteamine, there are no known endogenous substrates.

Several unique features of the catalytic cycle of the FMOs are important for understanding the mechanism by which they oxidize xenobiotics. The catalytic mechanism for the FMO has been shown to involve the formation of an enzyme bound 4a-hydroperoxyl-flavin (Figure 10.6) in an NADPH and O_2 dependent reaction. Reduction of the flavin by NADPH occurs before binding of oxygen can occur, and activation of oxygen by the enzyme occurs in the absence of substrate by oxidizing NADPH to form NADP and peroxide. Finally, addition of the substrate to the peroxyflavin complex is the last step prior to oxygenation. This is in contrast to the CYP catalytic cycle in which the substrate binds to the oxidized enzyme which is subsequently reduced.

The flavin hydroperoxide intermediate in the FMO enzyme forms a relatively stable, potent oxygenating species. Thus, any nucleophile that can be oxidized by an organic peroxide and can gain access to the active site is a potential substrate for the FMO. This capability accounts for the wide substrate specificity of the FMO. Although the flavin hydroperoxide of the FMO is a strong electrophile, it exhibits a high degree of selectivity toward certain types of soft nucleophiles, primarily organic compounds with sulfur, nitrogen, and phosphorus heteroatoms. Compounds containing ionized carboxyl groups, which include most physiological sulfur compounds, are not substrates for the FMO, with the exception of cysteamine, which is an excellent substrate.

TABLE 10.2. Substrates Oxidized by Flavin-Containing Monooxygenases

Inorganic

HS-, S_8, I⁻, IO⁻, I_2, CNS-

Organic Nitrogen Compounds

Sec- and *tert-*acyclic and cyclic amines
N-Alkyl and *N,N*-dialkylarylamines
Hydrazines
Primary amines

Organic Sulfur Compounds

Thiols and disulfides
Cyclic and acyclic sulfides
Mercapto-purines, -pyrimidines, -imidazoles
Dithio acids and dithiocarbamides
Thiocarbamides and thioamides

Organic Phosphorus Compounds

Phosphines
Phosphonates

Others

Boronic acids
Selenides, selenocarbamides

Reactivation of the enzyme is considered to be the rate-limiting step in the reaction. Because oxidation of the substrate occurs more rapidly than the regeneration of the active enzyme, the V_{max} values are relatively similar for a variety of substrates even though the K_m values differ.

Often the same substrate is metabolized by both CYP and FMO; this situation is especially prevalent with many N- and S-containing pesticides and drugs (e.g., phorate and nicotine). To study the relative contributions of these two enzymes with common substrates, methods have been developed to measure each separately in microsomal preparations. The most useful of these techniques is the inhibition of CYP activity by using an antibody to CYP reductase, thus permitting measurement of FMO activity alone. A second procedure is heat treatment of microsomal preparations (50 °C for 1 min) which inactivates the FMO, thus allowing determination of CYP activity, which is unchanged by heat treatment. Thermal inactivation, however, is ineffective with lung microsomes, because the lung FMO is more heat stable than the liver FMO.

Although the levels of FMO are not readily altered by classic chemical inducing and inhibiting agents, unlike the CYPs, the balance of enzyme activity between CYP and FMO is easily disturbed, especially in the liver, by compounds that alter the concentration of CYP isoforms. Of special interest is a change in the balance of activity after in vivo exposure of animals to either inducers or inhibitors of CYP activity. For example, microsomes prepared from the livers of mice pretreated with

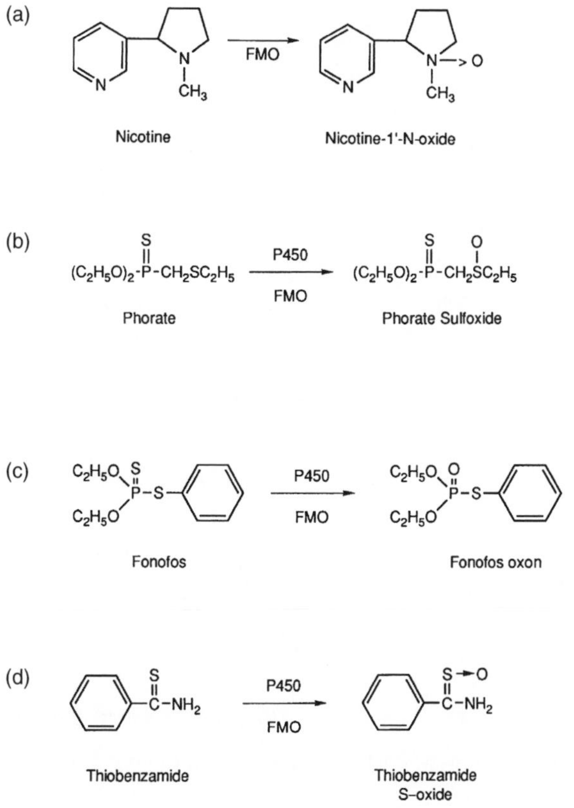

(a) Nicotine → Nicotine-1'-N-oxide

(b) Phorate → Phorate Sulfoxide

(c) Fonofos → Fonofos oxon

(d) Thiobenzamide → Thiobenzamide S-oxide

Figure 10.5. Examples of flavin-containing monooxygenase catalyzed reactions.

phenobarbital showed not only an increase in the total rate of oxidation of the insecticide phorate, but also in the proportion metabolized by CYP. As a result, the percentage of products due to FMO is decreased. The effects of xenobiotics on the relative contributions of FMO and P450 appear to be mediated primarily by P450 because the FMO does not appear to be inducible by xenobiotics.

FMO levels may, however, vary with nutrition, diurnal rhythms, gender, pregnancy, and corticosteroids, although the effects appear to be both species- and tissue-dependent. Such alterations in the relative contributions of the two enzyme systems may assume toxicological importance when the products from the two enzymes differ, particularly when one metabolite is more toxic than the others. Thus prior exposure of animals to environmental agents can have a significant effect on activation/detoxication pathways and the toxicity of other xenobiotics.

Although FMOs, unlike P450s, are not induced by xenobiotics, their level of expression can be dramatically affected by gender and during development. In the mouse, overall FMO activity is higher in female liver than in male liver. FMO1 is two to three times higher in females than in males, whereas FMO3 is not expressed in the liver of male mice, being completely suppressed at puberty by testosterone. FMO5, on the other hand, is expressed to the same extent in both genders. In the rat, FMO1 activity is higher in males than in females, whereas FMOs 3 and 5 show no difference between genders. Development can also affect the level of expression of FMO isoforms. In humans, FMO1 is the predominant form in fetal liver, with FMO3 not being

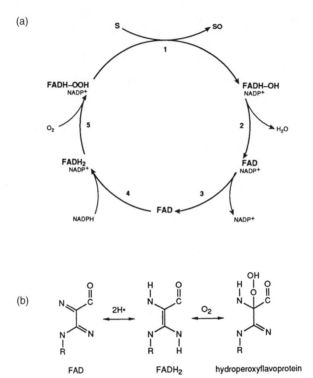

Figure 10.6. Catalytic cycle of the flavin-containing monooxygenase.

expressed, and FMO3 is the predominant form in the adult, with FMO1 not being expressed. This situation appears to be gender-independent. In the mouse, FMO1 is expressed in the liver by gestation day 13, with the gender effect becoming apparent by 4 weeks of age. Hepatic FMO3 is not expressed until about 2 weeks of age, is equivalent in males and females until puberty, and is then completely suppressed in males, this suppression being a function of increased testosterone levels. Thus, the expression of FMOs in the female mouse, with FMO1 expressed in the fetal liver and FMO3 in the adult liver, resembles FMO expression in human.

FMO was first purified to homogeneity from pig liver microsomes and subsequently from the livers of several other mammalian species. It is a highly lipophilic protein containing FAD as the only flavin, with a monomeric molecular mass of 56,000 per mole of FAD. Investigations to date indicate that there are at least five, or possibly six, forms, named FMO1 through FMO6.

The six FMO isoforms are the products of distinct genes, but all fall into a single gene family. The isoforms are 50–58% identical compared with each other and are highly conserved in mammals, with each isoform being over 80% identical across species. The importance of FMO polymorphisms is discussed in Chapter 11. No sequences are known for FMOs from nonmammalian vertebrates or from invertebrates; thus the true extent of the gene family is not readily apparent. FMOs are expressed in several tissues and organs but have been most studied in the liver and lung and, to a lesser extent, in the kidney. Isoforms 1, 3, and 5 are generally expressed in the liver. FMO2 is expressed predominantly in the lung and the kidney.

The expression of FMO2 as the sole isoform in the lung of rabbit is unusual. In other species, although FMO2 is almost always expressed in the lung, other isoforms are also present. The mRNA for FMO4, the least understood of the FMO isoforms, has been demonstrated in the rabbit nervous system, where it appears to be the only FMO message present.

Cellular localization studies of FMO carried out using an immunohistochemical method utilizing peroxidase-labeled antibodies and diaminobenzidine revealed that in the rabbit lung the FMO is highly localized in the nonciliated bronchiolar epithelial (Clara) cells. Similar immunohistochemical studies of FMO distribution in the skin of mice and pigs revealed significant staining in epidermis, sebaceous gland cells, and hair follicles. Since the lung and skin are often a major route of entry for environmental chemicals, such as pesticides, the presence of the FMO in these tissues is of considerable interest.

FMO1, FMO2, and FMO3 are all broadly nonspecific, although substrate specificity does vary from one to another, indicating some differences in the substrate binding regions. Because FMO4 has proven difficult to express in heterologous expression sytems, it has been identified primarily from the cDNA and inferred amino acid sequences and no substrates are yet known. FMO5 has a restricted substrate specificity, oxidizing primary amines such as *n*-octylamine.

10.3 NONMICROSOMAL OXIDATIONS

In addition to the monooxygenases, there are a number of other enzymes that are involved in the oxidation of foreign compounds. These oxidoreductases (Table 10.1) are located in either the mitochondrial fraction or the $100,000\,g$ supernatant of tissue homogenate.

10.3.1 Alcohol Dehydrogenases

As a class of enzymes, the alcohol dehydrogenases catalyze the conversion of alcohols to aldehydes or ketones, for example,

$$RCH_2OH + NAD^+ \leftrightarrow RCHO + NADH + H^+$$

This should not be confused with the CYP-catalyzed monooxygenation of ethanol observed in liver microsomes. The alcohol dehydrogenase reaction is reversible and carbonyl compounds can be reduced to alcohols. Alcohol dehydrogenase is probably the most important dehydrogenase involved in the metabolism of foreign alcohols and carbonyl compounds. The enzyme is found in the soluble fraction of the liver, kidney, and lung and requires NAD or NADP as a coenzyme. The reaction proceeds at a slower rate with NADP. In the intact organism, the reaction proceeds to the right, because aldehydes are further oxidized to acids. The further oxidation of aldehydes to acids is a vital detoxication reaction because aldehydes are usually toxic and, because of their lipid solubility, are not readily excreted.

All of the human alcohol dehydrogenases are dimeric zinc metalloenzymes existing in multiple forms that may be homo- or heterodimers with subunits of aproximately 40,000 daltons. Five classes exist in humans and their distribution varies between organs. Details of the isoforms and their polymorphic forms are presented in Chapter 11.

Alcohol dehydrogenases metabolize primary alcohols to aldehydes, with *n*-butanol being the substrate oxidized at the highest rate. Secondary alcohols are oxidized to ketones but at a reduced rate; for example, butanol-2 is oxidized at one-third the rate of *n*-butanol. Tertiary alcohols are not readily oxidized.

Poisoning by ethylene glycol (CH_2OHCH_2OH) is due to aldehydes, glycolate, oxalate, and lactate, resulting from an initial attack by alcohol dehydrogenase. This is similar to the activation of methanol to formaldehyde and subsequent oxidation by aldehyde dehydrogenase to formic acid.

These two enzymes also activate the solvent 2-butoxyethanol ($CH_3CH_2CH_2CH_2OCH_2CH_2OH$) to 2-butoxyacetic acid, responsible for its hematotoxicity.

10.3.2 Aldehyde Dehydrogenases

The oxidation of aliphatic and aromatic aldehydes to their corresponding acids allows the acids to be excreted or conjugated in phase II reactions:

$$RCHO + NAD^+ + H_2O \leftrightarrow RCOOH + NADH + H^+$$

Aldehyde dehydrogenases have been isolated from a variety of sources, the most important being the liver. With a series of linear aliphatic aldehydes, the rate of oxidation increases as the carbonyl carbon becomes more positive.

In addition to liver aldehyde dehydrogenase, a number of other enzymes present in the soluble fraction of liver homogenates will oxidize aldehydes and certain N-heterocyclic compounds. Among these are aldehyde oxidase and xanthine oxidase (see below), both flavoprotein enzymes containing molybdenum. These enzymes catalyze the oxidation of aldehydes formed by the deamination of endogenous amines by amine oxidases.

10.3.3 Amine Oxidases

The biological function of amine oxidases involves the oxidation of biogenic amines formed during normal biological processes. In mammals, the monoamine oxidases are involved in the control of the serotonin: catecholamine ratios in the brain, which in turn influence sleep and EEG patterns, body temperature, and mental depression. Two groups of amine oxidases are involved in the oxidative deamination of naturally occurring amines as well as foreign compounds.

10.3.3a Monoamine Oxidases. Monoamine oxidases (MAO) are flavoprotein enzymes located in the mitochondrial fraction of liver, kidney, and brain. The enzymes have been found also in blood platelets and intestinal mucosa. The MAO exist as a large group of similar enzymes with overlapping substrate specificities and inhibition patterns. The number is difficult to assess because of the great variety and variation within each tissue as well as in different animal species.

The general reaction catalyzed is shown in Figure 10.7. MAO deaminates primary, secondary, and tertiary aliphatic amines, the reaction rate is faster with the primary amines and slower with the secondary and tertiary amines. Electron-withdrawing

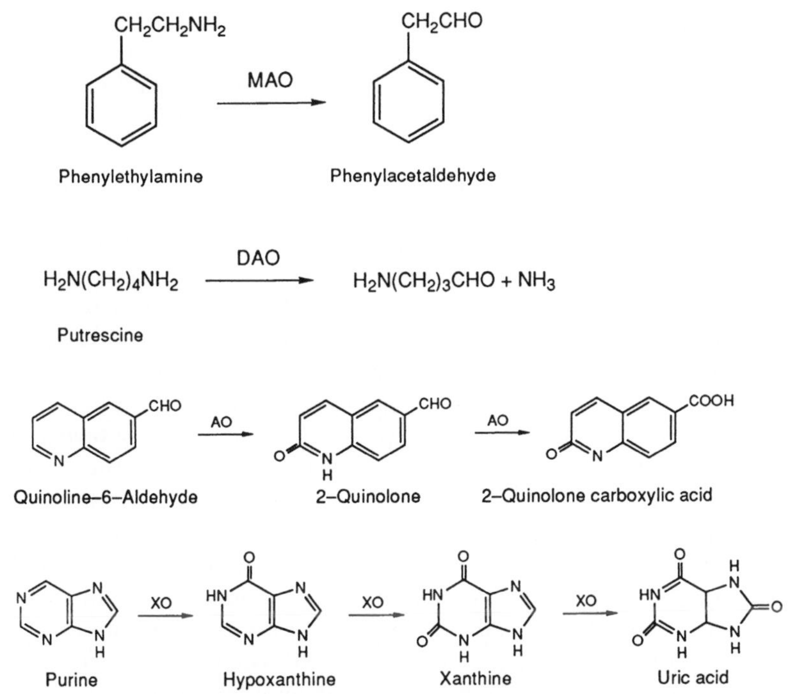

Figure 10.7. Examples of nonmicrosomal oxidations.

substituents on an aromatic ring increase the rate of deamination. Compounds that have a substituted methyl group on the α-carbon atoms are not metabolized by the MAO system (e.g., amphetamine and ephedrine).

10.3.3b Diamine Oxidases. Diamine oxidases (DAO) also oxidize diamines to the corresponding aldehydes in the presence of oxygen. The DAO are pyridoxal phosphate proteins containing copper that are found in the soluble fraction of liver, intestine, kidney, and placenta. A typical DAO reaction is the oxidation of putrescine (Figure 10.7).

The rate of deamination is determined by the chain length; and with polymethylene diamines, $NH_2\text{-}(CH_2)_n\text{-}NH_2$, the maximum rate occurs when $n = 4$ (putrescine) and $n = 5$ (cadaverine) and decreases to zero when $n = 9$ or more. At this chain length, MAO becomes active and can oxidatively deaminate the diamines. It should be noted that although MAO can deaminate both substituted and primary amines, DAO can deaminate only primary amines.

10.3.4 Molybdenum Hydroxylases

There are a number of molybdenum-containing enzymes, but those that are important in carbon oxidation of xenobiotics are aldehyde oxidase (AO) and xanthine oxidase (XO), also referred to as molybdenum hydroxylases (Figure 10.7). Both enzymes catalyze the oxidation of a wide range of aldehydes and N-heterocycles. The name aldehyde oxidase is somewhat misleading, however, because oxidation of heteroaromatics is more significant. The differences in substrate specificities between monooxygenases and molybdenum hydroxylases is based on the reaction mechanism since with monooxygenases, the mechanism involves an electrophilic attack on the carbon, whereas the hydroxylases catalyze nucleophilic addition at an unsaturated carbon, which can be represented as an attack by the hydroxyl ion.

$$RH + OH^- \rightarrow ROH + 2e^- + H^+$$

Although molecular oxygen may be involved in the overall oxidation process, the oxygen atom incorporated into the product is derived from water, whereas with monooxygenations, the oxygen comes from molecular oxygen. In N-heteroaromatic ring compounds, the most electropositive carbon is usually adjacent to a ring N atom, and this is the normal position of nucleophilic molybdenum hydroxylase attack. In contrast, a CYP-catalyzed attack would tend to occur distant from the N, probably in an adjacent carbocyclic ring. Uncharged carbocyclic compounds, such as benzene and naphthalene, which are CYP substrates, are not substrates for the molybdenum hydroxylases. However, as the number of N atoms increases in a ring, the affinity of the hydroxylases for the substrate increases. Thus, the bases purine and pteridine are oxidized exclusively by AO and XO, although the oxidations do not necessarily occur at the same positions. These enzymes are found predominantly in the cytosol and occur in most tissues. However, the highest levels are found in those tissues most often exposed to ingested foreign compounds; AO is most prevalent in the liver, whereas the highest levels of XO occur in the small intestine, milk, and mammary gland. Both enzymes react with endogenous substrates; XO is involved in the final stages of purine catabolism, and AO catalyzes the oxidation of vitamins such as pyridoxal and N-methylnicotinamide. Aldehyde oxidase and xanthine oxidase also catalyze the oxidation of a wide range of xenobiotics including aldehydes, uncharged bases, quaternary N-heterocycles, and carbocyclics. Some examples are shown in Figure 10.7. In addition to their oxidative role, these enzymes may be involved in reductive pathways in vivo, and they are known to catalyze the in vitro reduction of aromatic nitro compounds, azo dyes, nitrosamines, N-oxides, and sulfoxides. Table 10.1 lists some of their known substrates. Because these enzymes metabolize such a wide range of xenobiotics, including many that are also substrates for the monooxygenases, the role of the molybdenum hydroxylases in detoxication cannot be discounted. In addition, they may be important also in the production as well as deactivation of toxic metabolites through reductive pathways.

10.4 COOXIDATION BY PROSTAGLANDIN SYNTHETASE

The most important peroxidase reaction involved in the metabolic oxidation of xenobiotics is probably that catalyzed by prostaglandin synthetase (PGS). PGS catalyzes the oxygenation of polyunsaturated fatty acids to hydroxy endoperoxides, with the preferential substrate in vivo being arachidonic acid (AA). PGS catalyzes two activities, the fatty acid cyclooxygenase activity that brings about the oxygenation of polyunsaturated fatty acids, such as arachidonic acid, to form a cyclic endoperoxide, prostaglandin G (PGG), which is then reduced to a hydroperoxy endoperoxide termed prostaglandin H (PGH). A number of xenobiotics may be cooxidized during this second hydroperoxidase reaction (Figure 10.8).

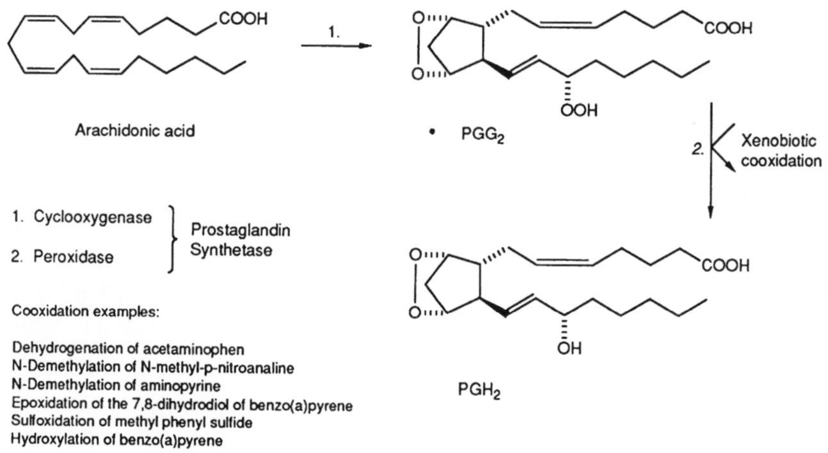

Figure 10.8. Examples of xenobiotic cooxidations during prostaglandin biosynthesis.

The enzyme, which is membrane bound, is present in virtually all mammalian tissues, and is especially concentrated in seminal vesicles, a rich source that has been used extensively in experimental studies. PGS is high also in platelets, lungs, skin, kidney medulla, endothelial cells, and embryonic tissue. It is a glycoprotein with a subunit molecular mass of about 70,000 daltons, containing one heme per subunit. A wide variety of compounds can undergo oxidation during PGH biosynthesis; the types of cooxidations catalyzed include dehydrogenation, demethylation, epoxidation, sulfoxidation, N-oxidation, C-hydroxylation, and dioxygenation. Some of the more important compounds that can be cooxidized are listed in Table 10.1.

Many xenobiotics can be metabolized to reactive metabolites by PGS. For a discussion of activations catalyzed by PGS, the reader is referred to Chapter 20.

It is evident that PGS can serve as an alternate enzyme for xenobiotic metabolism, particularly in tissues with low monooxygenase activity. Consequently, the cooxidation of drugs, chemical carcinogens, and other xenobiotics by PGS is an important area of study. In addition, this enzyme system, like the monooxygenases, may be important experimentally in screening chemicals for mutagenic and teratogenic potential.

10.5 REDUCTION REACTIONS

A number of functional groups, such as nitro, diazo, carbonyls, disulfides, sulfoxides, and alkenes, are susceptible to reduction. In many cases it is difficult to determine whether these reactions proceed nonenzymatically by the action of biological reducing agents such as NADPH, NADH, and FAD or through the mediation of functional enzyme systems. As noted above, the molybdenum hydroxylases can carry out, in vitro, a number of reduction reactions, including nitro, azo, N-oxide, and sulfoxide reduction. Although the in vivo consequences of this are not yet clear, much of the distribution of "reductases" described below may be, in whole or in part, the distribution of molybdenum hydroxylases.

10.5.1 Nitro Reduction

Aromatic nitro compounds such as nitrobenzene are susceptible to reduction by both bacterial and mammalian nitroreductase systems (Figure 10.9A).

Nitroreductase activity has been demonstrated in liver homogenates as well as in the soluble fraction, whereas other studies have reported that nitroreductase activity has been found in all liver fractions evaluated. The reductase appears to be distributed in liver, kidney, lung, heart, and brain. The reaction utilizes both NADPH and NADH and requires anaerobic conditions. The reaction can be inhibited by the addition of oxygen. The reaction is stimulated by FMN and FAD, and at high flavin concentrations they can act simply as nonenzymatic electron donors. The reduction

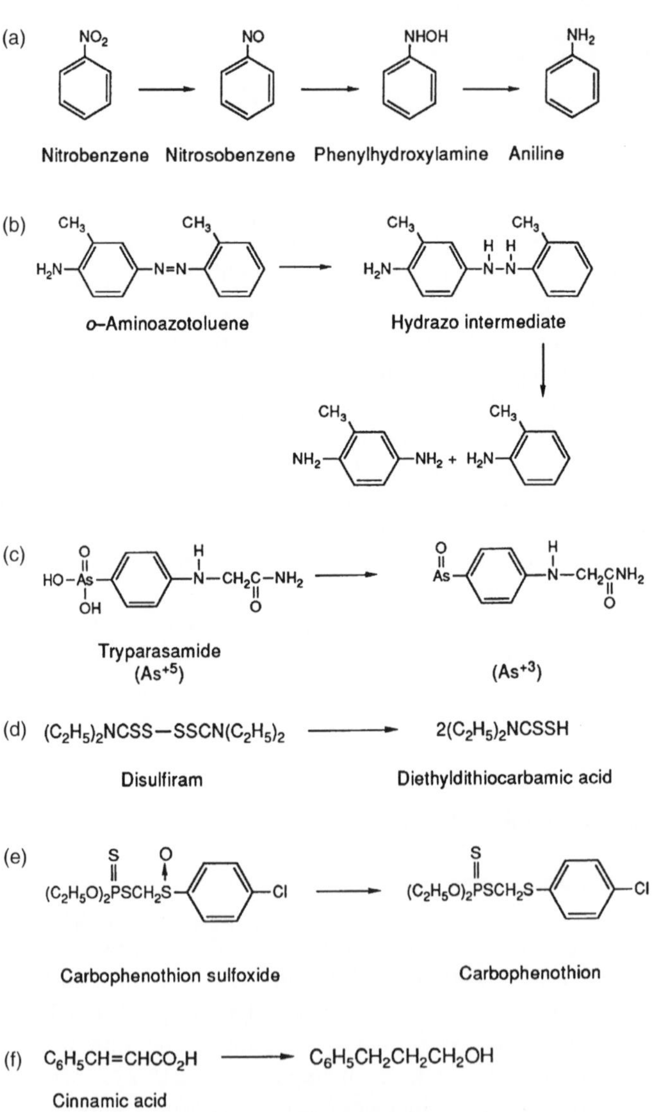

Figure 10.9. Examples of reduction reactions.

process probably proceeds via the nitroso and hydroxylamine intermediates as illustrated above.

10.5.2 Azo Reduction

Requirements for azoreductases are similar to those for nitroreductases, that is, they require anaerobic conditions and NADPH and are stimulated by reduced flavins (Figure 10.9B).

The ability of mammalian tissue to reduce azo bonds is rather poor. With *p*-[2,4-(diaminophenyl)azo]benzenesulfonamide (Prontosil), in vivo reduction forms sulfanilamide. However, pretreatment with antibiotics destroys the intestinal bacteria, which results in a decrease in the formation of the amino compound. Generally, it would appear that both nitro and azo reduction as a function of specific tissues is of minor importance, and intracellular bacteria as well as intestinal bacteria are actually responsible for these reductions.

10.5.3 Reduction of Pentavalent Arsenic to Trivalent Arsenic

Pentavalent arsenic compounds such as tryparsamide appear to require reduction to the trivalent state for antiparasitic activity (Figure 10.9C). Various arsenicals containing pentavalent arsenic have only slight in vitro antiprotozoal activity, whereas compounds containing trivalent arsenic are highly active.

10.5.4 Reduction of Disulfides

A number of disulfides are reduced in mammals to their sulfhydryl compounds. An example is disulfiram (Antabuse), a drug used for the treatment of alcoholism (Figure 10.9D).

10.5.5 Ketone and Aldehyde Reduction

The reduction of ketones and aldehydes occurs through the reverse reaction of alcohol dehydrogenases (Section 10.3.1):

$$R_1R_2CO \rightarrow R_1R_2CHOH$$

10.5.6 Sulfoxide and *N*-Oxide Reduction

The reduction of sulfoxides and *N*-oxides has been reported to occur in a number of mammalian systems. It appears that liver enzymes reduce sulfoxide or sulfonyl compounds to thioether or sulfides under anaerobic conditions (Figure 10.9E). *N*-Oxides have also been reported to be reduced by a bacterial reductase:

$$R_3CNO \rightarrow R_3CN$$

10.5.7 Reduction of Double Bonds

Certain aromatic compounds (e.g., cinnamic acid) have been reported to be reduced by intestinal flora (Figure 10.9F).

10.6   HYDROLYSIS

A large number of xenobiotics, such as esters, amides, or substituted phosphates that include ester-type bonds, are susceptible to hydrolysis. Hydrolytic reactions are the only phase I reactions that do not utilize energy. Numerous hydrolases are found in blood plasma, liver, intestinal muscosa, kidney, muscle, and nervous tissue. Hydrolases are present in both soluble and microsomal fractions. The general reactions are shown in Figure 10.10A.

The acid and alcohol or amine formed on hydrolysis can be eliminated directly or conjugated by phase II reactions. In general, amide analogs are hydrolyzed at slower rates than the corresponding esters.

Tissue esterases have been divided into two classes: the A-type esterases, which are insensitive, and the B-type esterases, which are sensitive to inhibition by organophosphorus esters. The A esterases include the arylesterases, whereas the B esterases include cholinesterases of plasma, acetylcholinesterases of erythrocytes and nervous tissue, carboxylesterases, lipases, and so on. The nonspecific arylesterases that hydrolyze short-chain aromatic esters are activated by Ca^{2+} ions and are responsible for the hydrolysis of certain organophosphate triesters such as paraoxon (Figure 10.10B).

Certain hydrolases also are present in mammalian plasma and tissue, and these enzymes hydrolyze and subsequently detoxify chemical warfare agents such as the nerve gases tabun, sarin, and DFP (Figure 10.10C). A variety of foreign compounds,

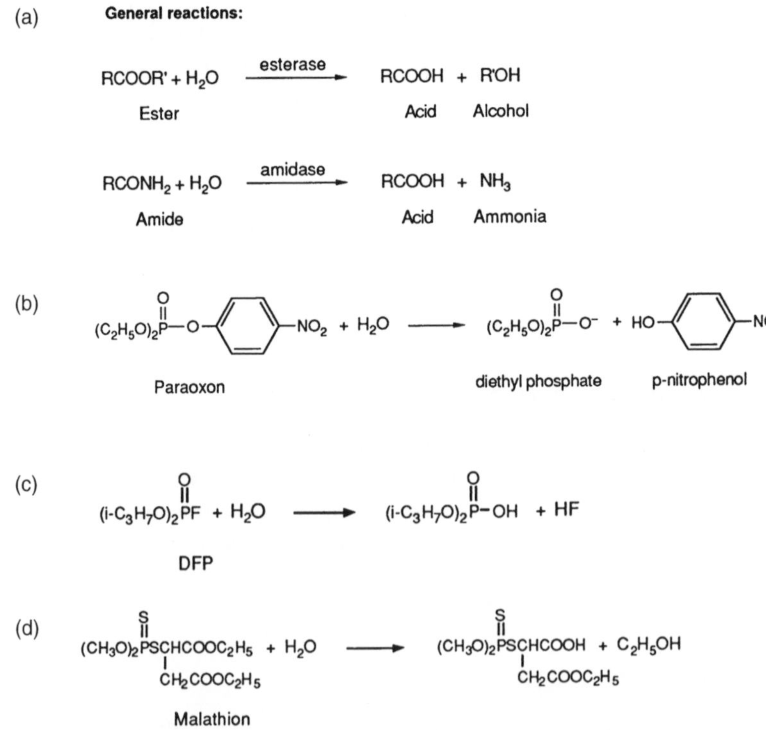

Figure 10.10. Examples of hydrolysis reactions.

such as phthalic acid esters (plasticizers), phenoxyacetic and picolinic acid esters (herbicides), and pyrethroids and their derivatives (insecticides), as well as a variety of ester and amide derivatives of drugs, are detoxified by hydrolases found in plants, animals, and bacteria—that is, the entire plant and animal kingdom.

Several human carboxylesterases have been cloned, sequenced and expressed. These human carboxylesterases are important in the hydrolysis of certain pesticides such as the pyrethroids. In certain strains of insects that are resistant to malathion, the resistance mechanism is associated with a higher level of a carboxylesterase, which detoxifies malathion (Figure 10.10D).

10.7 EPOXIDE HYDRATION

Epoxide rings of certain alkene and arene compounds are hydrated enzymatically by epoxide hydrolases to form the corresponding *trans*-dihydrodiols (Figure 10.11). The epoxide hydrolases are a family of enzymes known to exist both in the endoplasmic reticulum and in the cytosol. In earlier studies they were named epoxide hydratase, epoxide hydrase, or epoxide hydrolase. Epoxide hydrolase, however, has been recommended by the International Union of Biochemists Nomenclature Committee and is now in general use.

10.7.1 Microsomal Epoxide Hydrolase

The microsomal epoxide hydrolase converts arene and alkene oxides to vicinal dihydrodiols by hydrolytic cleavage of the oxirane ring. This is a detoxication reaction in that it converts generally highly reactive electrophilic oxirane species to less reactive nonelectrophilic dihydrodiols. It should be borne in mind, however, that some epoxides are unreactive, both toward macromolecules and toward epoxide hydrolase (e.g., dieldrin), and that dihydrodiols may be further epoxidized to produce metabolites that are even more reactive, as is the case with

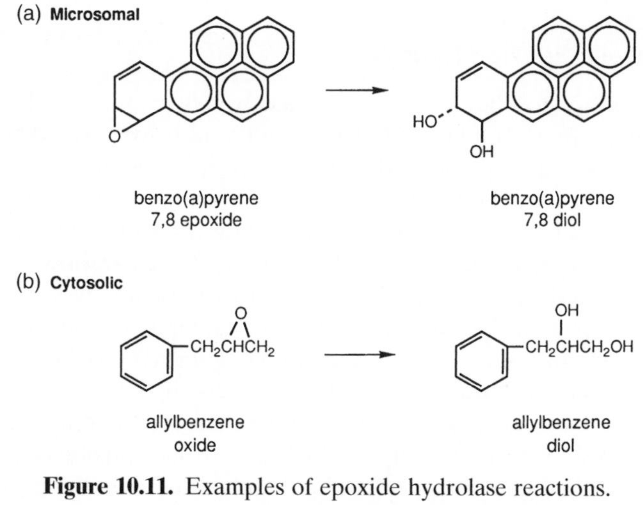

Figure 10.11. Examples of epoxide hydrolase reactions.

benzo(*a*)pyrene. An example of a substrate, benzo(*a*)pyrene 7,8-epoxide, is shown in Figure 10.11A.

Microsomal epoxide hydrolase is widely distributed, having been described from plants, invertabrates, and vertebrates. In vertebrates it has wide organ distribution; for example, in the rat, the most studied species, the enzyme has been found in essentially every organ and tissue. Although predominantly located in the endoplasmic reticulum (microsomes), epoxide hydrolase is also found in the plasma and nuclear membranes and, to some extent, in the cytosolic fraction.

Microsomal epoxide hydrolase has been purified from several sources. It is a hydrophobic protein with a molecular mass of approximately 50,000 daltons. The complete amino acid sequence has been determined directly and has been deduced from the cDNA sequence. It appears to be a single polypeptide of 455 amino acid residues with a high degree of homology between species. Although existence of multiple forms within a single tissue has been suggested, this may be an artifact of purification because only a single gene coding for microsomal epoxide hydrolase has been isolated from rat liver.

The reaction is highly regio- and stereospecific, always proceeding with inversion of configuration. With arene oxides, this means that only *trans*-dihydrodiols are formed. All of the experimental evidence points to a general base-catalyzed nucleophilic (S_N2) addition of water to the oxirane ring.

A number of inhibitors of this enzyme are known. They include epoxides that are hydrolyzed by the enzymes, such as trichloropropene oxide, metal ions such as Hg^{2+}, Zn^{2+}, and Cd^{2+} and 2-bromo-4'-acetophenone, a potent inhibitor that binds to imidazole nitrogen atoms. Microsomal epoxide hydrolase can be induced by compounds such as phenobarbital, Arochlor 1254, 2(3)-*t*-butyl-4-hydroxyanisole (BHA), and 3,5-di-*t*-butyl-hydroxytoluene (BHT). Many microsomal epoxide hydrolase inducers are inducers also of CYP and produce a general proliferation of the endoplasmic reticulum. Induction does, however, involve an increase in the mRNA specific for the hydrolase.

A second unique microsomal epoxide hydrolases has been described. This enzyme appears to have a narrow substrate specificity, being specific for cholesterol 5,6-oxides and related steroid 5,6-oxides.

10.7.2 Cytosolic Epoxide Hydrolase

The existence of a cytosolic epoxide hydrolase was first indicated by its ability to hydrolyze analogs of insect juvenile hormone not readily hydrolyzed by microsomal epoxide hydrolase. Subsequent studies demonstrated a unique cytosolic enzyme catalytically and structurally distinct from the microsomal enzyme. It appears probable that the cytosolic enzyme is peroxisomal in origin. Both enzymes are broadly nonspecific and have many substrates in common. It is clear, however, that many substrates hydrolyzed well by cytosolic epoxide hydrolase are hydrolyzed poorly by microsomal epoxide hydrolase and vice versa. For example, 1-(4'-ethylphenoxy)-3,7-dimethyl-6,7-epoxy-*trans*-2-octene, a substituted geranyl epoxide insect juvenile hormone mimic, is hydrolyzed 10 times more rapidly by the cytosolic enzyme than by the microsomal one. In any series, such as the substituted styrene oxides, the trans configuration is hydrolyzed more rapidly by the cytosolic epoxide hydrolase than is the cis isomer. At the same time, it should remembered that in this and other series, cis as well as trans isomers are hydrolyzed, although generally at lower rates. An example of a substrate, allylbenzene oxide, is shown in Figure 10.11B.

As with substrates, the microsomal epoxide hydrolase and the cytosolic epoxide hydrolase have inhibitors in common; and, again, those that are most effective toward one are less effective toward the other. Inhibitors of cytosolic epoxide hydrolase include poorly metabolized substrates, such as chalcone oxide and its substituted analogs. A group of cyclopropyl oxiranes were specifically designed as suicide inhibitors of cytosolic epoxide hydrolase, binding covalently to the active site through a transient intermediate formed during the enzymatically induced ring opening. All of the known inducers of cytosolic epoxide hydrolase are also peroxisome proliferating agents. They include the drug clofibrate and the plasticizer di-(2-ethylhexyl)phthalate and many of their analogs and related compounds.

Cytosolic epoxide hydrolase has been purified from several sources. It is a dimer consisting of two identical monomers of approximately 60,000 daltons. Its amino acid sequence is not known, but antibodies to the microsomal and cytosolic enzymes do not cross-react. It occurs broadly across species; activity in the liver of those species examined is highest in the mouse and lowest in the rat, with activity in rabbit, guinea pig, and human liver being intermediate. It is also broadly distributed between organs, occurring in all major organs of the species most investigated, the mouse and rat.

Leukotriene A4 hydrolase is a unique cytosolic epoxide hydrolase, structurally dissimilar to the cytosolic enzyme described above. Its substrate specificity is narrow, being restricted to leukotriene A4, (5(S)-trans-5,6-oxido-7,9-cis-11,14-trans-eicosatetraenoic acid), and related fatty acids.

10.8 DDT-DEHYDROCHLORINASE

In the early 1950s, it was demonstrated that DDT-resistant houseflies detoxified DDT mainly to its noninsecticidal metabolite DDE. The rate of dehydrohalogenation of DDT to DDE was found to vary between various insect strains as well as between individuals. The enzyme involved, DDT-dehydrochlorinase, also occurs in mammals but has been studied more intensively in insects.

DDT-dehydrochlorinase, a reduced glutathione (GSH)-dependent enzyme, has been isolated from the 100,000 g supernatant of resistant houseflies. Although the enzyme-mediated reaction requires glutathione, the glutathione levels are not altered at the end of the reaction (Figure 10.12).

The lipoprotein enzyme has a molecular mass of 36,000 daltons as a monomer and 120,000 daltons as the tetramer. The K_m for DDT is 5×10^{-7} M with optimum activity at pH 7.4. This enzyme catalyzes the degradation of p,p-DDT to p,p-DDE or the degradation of p,p-DDD (2,2,-bis(p-chlorophenyl)-1,1-dichloroethane) to the

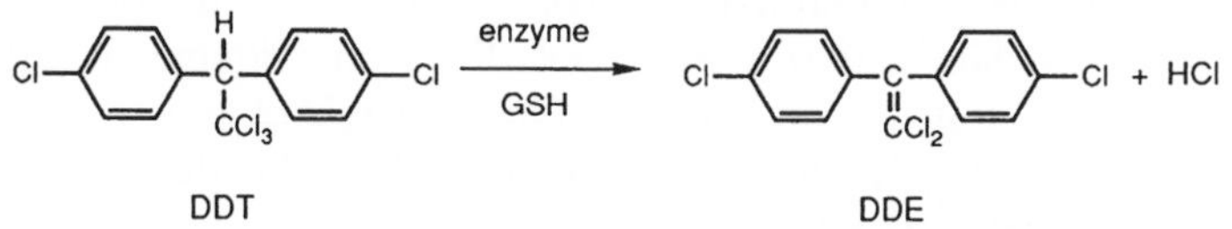

Figure 10.12. Conversion of DDT to DDE.

corresponding DDT ethylene TDEE (2,2-bis(*p*-chlorophenyl)-l-chloroethylene). *o,p*-DDT is not degraded by DDT-dehydrochlorinase, suggesting a *p,p*-orientation requirement for dehalogenation. In general, the DDT resistance of house fly strains is correlated with the activity of DDT-dehydrochlorinase, although other resistance mechanisms are known in certain strains.

10.9 INTERACTIONS INVOLVING XENOBIOTIC METABOLIZING ENZYMES

Xenobiotics, in addition to serving as substrates for a number of enzymes, may serve also as inhibitors or inducers of these or other enzymes, and examples of compounds that first inhibit and subsequently induce enzymes such as CYP isoforms are known. This situation is further complicated by the fact that although some substances have an inherent toxicity and are detoxified in vivo, others without inherent toxicity can be metabolically activated to potent toxicants. Situations that may arise, even when only two chemicals are involved, include the following:

> Compound A, without inherent toxicity, is metabolized to a potent toxicant. In the presence of an inhibitor of its metabolism, its toxicity will be reduced, whereas in the presence of an inducer of the activating enzyme, compound A will show greater toxicity.
>
> Compound B, a toxicant, is metabolically detoxified. In the presence of an inhibitor of the detoxifying enzymes, there will be an increase in toxicity, whereas in the presence of an inducer of these enzymes, toxicity would decrease.

10.9.1 Induction

Induction of XMEs is an increase in the amount of the enzyme and, as a consequence, an increase in the ability to metabolize one or more xenobiotics. This phenomenon is both widespread and relatively nonspecific. It is critically important in the metabolism of drugs and other xenobiotics and is considered in detail in Chapter 9.

10.9.2 Inhibition

As indicated above, inhibition of xenobiotic-metabolizing enzymes can cause either an increase or a decrease in toxicity. Several well-known inhibitors of such enzymes are shown in Figure 10.13. The effects of inhibition can be demonstrated in a number of ways at different organizational levels, from the intact animal to purified enzymes.

Experimental demonstration of inhibition can be carried out by observing in vivo symptoms. The measurement of the effect of an inhibitor on the duration of drug action in vivo is the most common method of demonstrating inhibitory action. The most useful and reliable in vivo tests involve the measurement of effects on the hexobarbital sleeping time or the zoxazolamine paralysis time. Both of these drugs are fairly rapidly deactivated by hepatic microsomal CYP isoforms, and thus CYP inhibitors prolong their action.

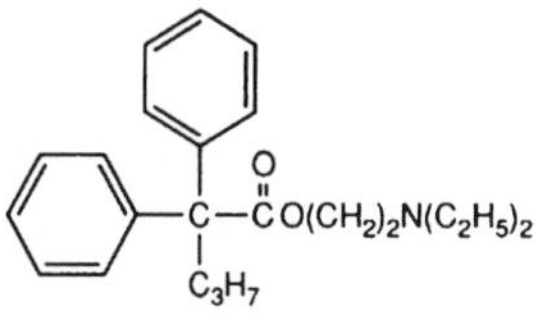

2-(Diethylamino) ethyl-
2,2-diphenylpentanoate (SKF-525A)
[P450]

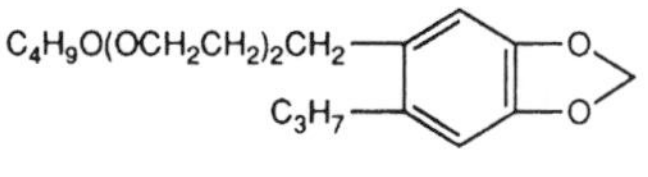

3,4-Methylenedioxy-6-propylbenzyl
n-butyl diethyleneglycol ether
(Piperonyl Butoxide) [P450]

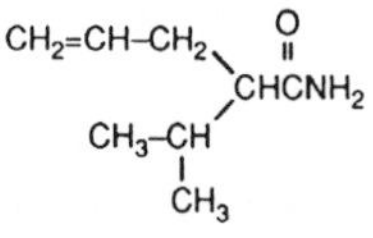

Allylisopropylacetamide
[P450]

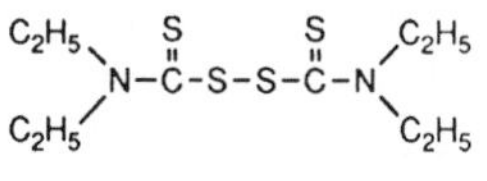

Disulfiram (Antabuse)
[aldehyde dehydrogenase]

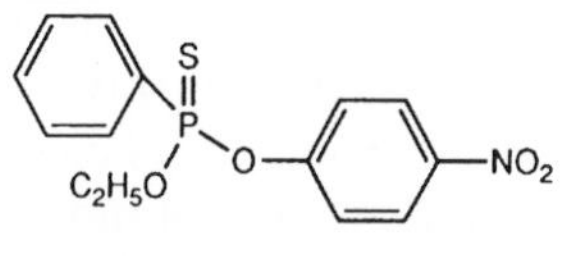

O-Ethyl-O-*p*-nitrophenyl
phenylphosphorothioate (EPN)
[esterases]

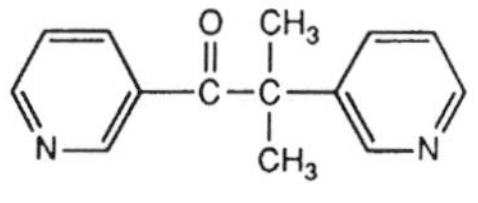

Metyrapone
[P450]

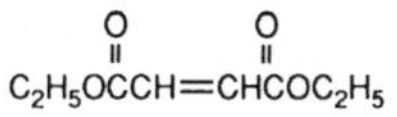

Diethyl maleate
[glutathione s-transferase]

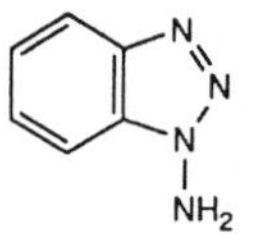

1-Aminobenzotriazole (1-ABT)
[P450]

Figure 10.13. Structures of some XME inhibitors.

Distribution and blood levels may also be affected by inhibition. Treatment of an animal with an inhibitor of xenobiotic metabolism may cause changes in the blood levels of an unmetabolized toxicant and/or its metabolites. This procedure may be utilized in the investigation of the inhibition of detoxication pathways; it has the advantage over in vitro methods in that it yields results of direct physiological or toxicological interest since it deals with the intact animal.

Effects on in vitro metabolism following in vivo treatment as a method of demonstrating inhibition is of variable utility. The preparation of enzymes from animal tissues usually involves considerable dilution with the preparative medium during homogenization, centrifugation, and resuspension; as a result, inhibitors not tightly bound to the enzyme in question are lost, either in whole or in part, during the preparative proceedures. Thus, while negative results have little utility, positive results not only indicate that the compound administered is an inhibitor but also provide a clear indication of excellent binding to the enzyme. The inhibition of esterases following treatment of the animal with organophosphates such as para-oxon is a good example because the phosphorylated enzyme is stable and is still inhibited after the preparative procedures. Inhibition of the same enzymes by carbamates is greatly reduced by the same procedures, however, because the carba-mylated enzyme is unstable and, in addition, any residual carbamate is diluted.

Microsomal monooxygenase inhibitors that form stable inhibitory complexes with CYP, such as SKF-525A, piperonyl butoxide and other methylenedioxyphenyl compounds, and amphetamine and its derivatives, can be readily investigated in this way because the microsomes isolated from pretreated animals have a reduced capacity to oxidize many xenobiotics.

Another chemical interaction resulting from inhibition in vivo that can then be demonstrated in vitro involves those xenobiotics that function by causing destruction of the enzyme in question. Exposure of rats to vinyl chloride results in a loss of CYP and a corresponding reduction in the capacity of microsomes subsequently isolated to metabolize foreign compounds. Allyl isopropylacetamide and other allyl compounds have long been known to have a similar effect.

In vitro measurement of the effect of one xenobiotic on the metabolism of another is the most common type of investigation of interactions involving inhibition. Although it is the most useful method for the study of an inhibitory mechanism, particularly when purified or recombinant enzymes are used, it is of more limited utility in assessing the toxicological implications for the intact animal. The principal reason for this is that it does not assess the effects of factors that affect absorption, distribution, and prior metabolism, all of which occur before the inhibitory event under consideration. Although the kinetics of inhibition of xenobiotic-metabolizing enzymes can be investigated in the same ways as any other enzyme mechanisms, a number of problems arise that may decrease the value of this type of investigation. They include the following:

1. Many investigations have been carried out on a particulate enzyme system, the microsomal monooxygenase system, using methods developed for single soluble enzymes. As a result, Lineweaver–Burke or other reciprocal plots are frequently curvilinear, and the same reaction may appear to have quite different characteristics from laboratory to laboratory, species to species, and organ to organ.

2. The nonspecific binding of substrate and/or inhibitor to membrane components is a further complicating factor affecting inhibition kinetics.

3. Both substrates and inhibitors are frequently lipophilic with low solubility in aqueous media.

4. Xenobiotic-metabolizing enzymes commonly exist in multiple forms (e.g., glutathione S-transferases and CYP) that are all relatively nonspecific but differ from one another in relative affinities for different substrates.

The primary considerations in studies of inhibition mechanisms are reversibility and selectivity. The inhibition kinetics of reversible inhibition give considerable insight into the reaction mechanisms of enzymes, and for that reason they have been well-studied. Generally speaking, reversible inhibition involves no covalent binding, occurs rapidly, and can be reversed by dialysis or, more rapidly, by dilution. Reversible inhibition is usually divided into competitive inhibition, uncompetitive inhibition, and noncompetitive inhibition, although these are not rigidly separated types, and many intermediate classes have been described.

Competitive inhibition is usually due to two substrates competing for the same active site. Following classical enzyme kinetics, there should be a change in the apparent K_m but not in V_{max}. In microsomal monooxygenase reactions, type I ligands, which often appear to bind as substrates but do not bind to the heme iron, might be expected to be competitive inhibitors, and this frequently appears to be the case. Examples include the inhibition of O-demethylation of p-nitroanisole by aminopyrene, aldrin epoxidation by dihydroaldrin, and N-demethylation of aminopyrene by nicotinamide. Some of the polychlorinated biphenyls, notably dichlorobiphenyl, have been shown to have high-affinity type I ligands for rabbit liver CYP and to be competitive inhibitors of the O-demethylation of p-nitroanisole. Pilocarpine, which potentiates nicotine-induced convulsions and hexobarbital hypnosis, has been shown to be a potent competitive inhibitor of the microsomal metabolism of these two compounds. Competitive inhibition may also result from an inhibitor binding to a site on the enzyme other than the active site, which, nevertheless, blocks the active site to the substrate by bringing about a conformational change.

Uncompetitive inhibition has seldom been reported in studies of xenobiotic metabolism. It is seen when an inhibitor interacts with an enzyme-substrate complex but cannot interact with free enzyme. Both K_m and V_{max} change by the same ratio, giving rise to a family of parallel lines in a Lineweaver–Burke plot.

Simple noncompetitive inhibitors can bind to both the enzyme and enzyme–substrate complex to form either an enzyme–inhibitor complex or an enzyme–inhibitor–substrate complex. The net result is a decrease in V_{max} but no change in K_m. Metyrapone, a well-known inhibitor of monooxygenase reactions, can also, under some circumstances, stimulate metabolism in vitro. In either case, the effect is noncompetitive in that the K_m does not change while V_{max} does, decreasing in the case of inhibition and increasing in the case of stimulation.

Irreversible inhibition can arise from a variety of causes, some of which are extremely important toxicologically. In the vast majority of cases, either covalent binding or disruption of the enzyme structure is involved. In neither case can the effect be reversed in vitro by either dialysis or dilution.

Covalent binding may involve the prior formation of a metabolic intermediate that then interacts with the enzyme. An example of this type of inhibition is the effect of the insecticide synergist piperonyl butoxide on hepatic microsomal monooxygenase activity. This compound can form a stable inhibitory complex that blocks carbon monoxide binding to CYP and also prevents substrate oxidation (Figure 10.14). This complex is the result of metabolite formation, probably a carbene, which is shown by the fact that the type of inhibition changes from competitive to irreversible as metabolism, in the presence of NADPH and oxygen, proceeds. Piperonyl butoxide inhibits the in vitro metabolism of many substrates of the monooxygenase system, including aldrin, ethylmorphine, aniline, aminopyrene, carbaryl, biphenyl, hexobarbital, and p-nitroanisole.

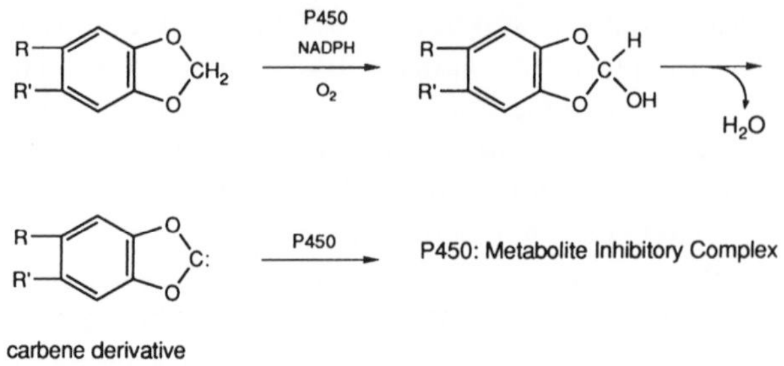

carbene derivative

Figure 10.14. Formation of a piperonyl butoxide metabolite-inhibitory complex.

A number of classes of monooxygenase inhibitors, in addition to methylene-dioxyphenyl compounds, are now known to form "metabolic-intermediate complexes" including amphetamine and its derivatives and SKF-525A and its derivatives.

Disulfiram (Antabuse) inhibits aldehyde dehydrogenase irreversibly, causing an increase in the level of acetaldehyde, which has been formed from ethanol by the enzyme alcohol dehydrogenase. This results in nausea, vomiting, and other symptoms in the human—hence its use as a deterrent of alcohol consumption in the treatment of alcoholism. The inhibition by disulfiram appears to be irreversible, with the level returning to normal only as a result of protein synthesis.

The inhibition by other organophosphate compounds of the carboxylesterase which hydrolyzes malathion is a further example of xenobiotic interaction resulting from irreversible inhibition because, in this case, the enzyme is phosphorylated by the inhibitor. A second type of inhibition involving organophosphorus insecticides involves those containing the P=S moiety. During CYP activation to the esterase-inhibiting oxon, reactive sulfur is released that inhibits CYP isoforms by an irreversible interaction with the heme iron. As a result, these chemicals are inhibitors of the metabolism of other xenobiotics, such as carbaryl and fipronil, and are potent inhibitors of the metabolism of steroid hormones such as testosterone and estradiol.

Another class of irreversible inhibitors of toxicological significance consists of those compounds that bring about the destruction of the xenobiotic-metabolizing enzyme. The drug allylisopropylacetamide, as well as a number of other allyl compounds, has long been known to cause the breakdown of CYP and resultant release of heme. The hepatocarcinogen vinyl chloride has been shown to have a similar effect, probably also mediated through the generation of a highly reactive metabolic intermediate.

10.9.3 Activation in Vitro

There are several examples of activation of xenobiotic-metabolizing enzymes by compounds other than the substrate. This differs from induction (described in Chapter 9) in that it is an immediate effect on a preexisting enzyme, occurring in an enzyme preparation in vitro, that does not involve *de novo* protein synthesis. The occurrence and significance of such stimulation in vivo is not apparent.

CYP-mediated microsomal oxidations may be stimulated by the addition of any of a rather heterogeneous group of compounds, including ethyl isocyanide, acetone,

2,2′-bipyridyl, and, under certain conditions, the inhibitor metyrapone. The effect is not uniform, however, because acetone, for example, stimulates aniline hydroxylation but has no effect on the N-demethylation of N-methylaniline, N,N-dimethylaniline, or ethylmorphine, or the O-demethylation of p-nitroanisole. 2,2′-Bipyridyl, on the other hand, stimulates both aniline hydroxylation and N-demethylation of N-methyl- and N,N-dimethylaniline but, at the same time, inhibits the N-dealkylation of ethylmorphine and aminopyrine. Presumably, these differences are related to isoform specificity for the activator and/or the substrate.

The activation of another membrane-bound enzyme of interest in biochemical toxicology, UDP glucuronyltransferase, is probably due to effects on the membrane. In this case, significant stimulation is brought about by "aging" the enzyme preparation, by sonication, and by such agents as dilute detergents and proteolytic enzymes.

10.9.4 Synergism and Potentiation

The terms "synergism" and "potentiation" have been variously used and defined but, in any case, involve a toxicity that is greater when two compounds are given simultaneously or sequentially than would be expected from a consideration of the individual toxicities of the compounds.

Generally, toxicologists use the term "synergism" for cases that fit the above definition, but only when one compound is toxic alone while the other has little or no intrinsic toxicity. This is the case with the toxicity of insecticides to insects and mammals and the effects on this toxicity of methylenedioxphenyl synergists such as piperonyl butoxide, sesamex, and tropital. The term "potentiation" is then reserved for those cases in which both compounds have appreciable intrinsic toxicity, such as in the case of malathion and EPN. Pharmacologists use the terms in the opposite sense—that is, synergism when both compounds are toxic and potentiation when one is nontoxic. Even more unfortunate is the tendency to use synergism and potentiation as synonyms, leaving no distinction based on relative toxicity.

Although examples are known in which synergistic interactions take place at the receptor site, the majority of such interactions appear to involve the inhibition of xenobiotic-metabolizing enzymes. Two examples involve the insecticide synergists, particularly the methylenedioxyphenyl synergists, and the potentiation of the insecticide malathion by a large number of other organophosphate compounds.

The first example has already been mentioned. Piperonyl butoxide, sesamex, and related compounds increase the toxicity of certain insecticides by inhibiting the insect monooxygenase system. They are of commercial importance in household aerosol formulations containing pyrethrum. This inhibition, which appears to be the same in mammals and insects, involves the formation of a metabolite-inhibitory complex with CYP. The complex probably results from the formation of a carbene (Figure 10.14), which then reacts with the heme iron in a reaction involving n-bonding, as well as the dative σ-bond formed by the free pair of electrons, to form a complex that blocks CO (and presumably O_2) binding and inhibits the metabolism of xenobiotics.

The best-known example of potentiation involving insecticides and an enzyme other than the monooxygenase system is the increase in the toxicity of malathion to mammals brought about by certain other organophosphates. Malathion has a low mammalian toxicity due primarily to its rapid hydrolysis by a carboxylesterase. EPN, a phosphonate insecticide, causes a dramatic increase in malathion toxicity to mammals at dose levels that, given alone, cause essentially no inhibition of cholinesterase. In vitro studies further established the fact that the oxygen analog of EPN, as well as many other organophosphate compounds, increases the toxicity of malathion by inhibiting the carboxylesterase responsible for malathion's degradation.

Synergistic action is often seen with drugs. Almost all cases of increased hexobarbital sleeping time or zoxazolamine paralysis time by other chemicals could be described as synergism or potentiation. Synergism may also result from competition for binding sites on plasma proteins or for the active secretion mechanism in the renal tubule.

SUGGESTED READING

Arand, M., Cronin, A., Adamska, M., and Oesch, F. Epoxide hydrolyses: Structure, function, metabolism and assay. *Methods Enzymol.* **400**, 569–588, 2005.

Arinc, E., Schenkman, J. B., and Hodgson, E. (Eds.). *Molecular and Applied Aspects of Oxidative Drug Metabolizing Enzymes*, Kluwer Academic/Plenum Publishers, New York, 1999.

Engel, P. C., *Enzyme Kinetics*, John Wiley & Sons, New York, 1977.

Gong, B., and Boor, P. J. The role of amine oxidases in xenobiotic metabolism. *Expert Opin. Drug Metab. Toxicol.* **2**, 559–571, 2006.

Gonzalez, F. J. The molecular biology of cytochrome P450s. *Pharmacol. Rev.* **40**, 243–288, 1990.

Guengerich, F. P. Enzymatic oxidation of xenobiotic chemicals. *Crit. Rev. Biochem. Mol. Biol.* **25**, 97–153, 1990.

Guengerich, F. P. Molecular advances for the cytochrome P-450 superfamily. *TIPS* **12**, 281, 1991.

Guengerich, F. P. Oxidation of toxic and carcinogenic chemicals by human cytochrome P450 enzymes. *Chem. Res. Toxicol.* **4**, 391–407, 1991.

Hayes, J. D., Pickett, C. B., and Mantle, T. J. (Eds.). *Glutathione S-Transferase and Drug Resistance*, Taylor and Francis, London, 1990.

Hodgson, E., and Philpot, R. M. Interaction of methylenedioxyphenyl (1,3-benzodioxole) compounds with enzymes and their effects on mammals. *Drug Metab. Rev.* **3**, 231, 1974.

Hodgson, E. Induction and inhibition of pesticide-metabolizing enzymes: Roles in synergism of pesticides and pesticide action. *Toxicol. Ind. Health* **15**, 6, 1999.

Honkakoski, P., and Negishi, M. Regulatory DNA elements of phenobarbital-responsive cytochrome P450 CYP2B genes. *J. Biochem. Molec. Toxicol.* **12**, 3, 1998.

Isin, E. M., and Guengerich, F. P. Complex reactions catalyzed by cytochrome P450 enzymes. *Biochim. Biophys. Acta* **1770**, 314–329, 2007.

Kulkarni, A. P., and Hodgson, E. The metabolism of insecticides: The role of monooxygenase enzymes. *Annu. Rev. Pharmacol. Toxicol.* **24**, 19–42, 1984.

Murray, M., and Reidy, G. F. Selectivity in the inhibition of mammalian cytochromes P-450 by chemical agents. *Pharmacol. Rev.* **42**, 85, 1990.

Parkinson, A. Biotransformation of xenobiotics. In Klaassen, C. D. (Ed.). *Casarett and Doull's Toxicology*, 6th ed., McGraw-Hill, New York, 2001, pp. 133–224.

Satoh, T., and Hosokawa, M. Structure, function and regulation of carboxylesterases. *Chem. Biol Interact.* **162**, 195–211, 2006.

Strolin-Benedetti, M., Whomsley, R., and Baltes, E. Involvement of enzymes other than CYPs in the oxidative metabolism of xenobiotics. *Expert Opin. Drug Metab. Toxicol.* **2**, 895–921, 2006.

Testa, B., and Kramer, S. D. The biochemistry of drug metabolism—An introduction: Part 2. redox reactions and their enzymes. *Chem. Biodiversity* **4**, 257–405, 2007.

Ziegler, D. M. Flavin-containing monooxygenases: enzymes adapted for multisubstrate specificity. *Trends Pharmacol. Sci.* **11**, 321–324, 1990.

CHAPTER 11

Phase I—Toxicogenetics

ERNEST HODGSON

Department of Environmental and Molecular Toxicology, North Carolina State University, Raleigh, North Carolina 27695

EDWARD L. CROOM

Department of Environmental and Molecular Toxicology, North Carolina State University, Raleigh, North Carolina 27695

11.1 INTRODUCTION

Toxicogenetics is the study of the hereditary basis of differences in response to toxicants. Pharmacogenetics is the branch of toxicogenetics that deals with clinical drugs. Both are important because response to toxicants, including drugs, shows considerable interindividual variation, a large part of which is due to variations in xenobiotic metabolism. Some of this variation has been shown to be due to genetic mutations (polymorphisms) at certain gene loci. A *polymorphism* is an inherited monogenetic trait that exists in the population in at least two genotypes (two or more variant alleles) and is stably inherited. Several modified alleles may occur at the same gene locus, and population differences in the incidence of polymorphisms are known to occur.

Pharmacogenetics was first studied intensively during World War II when red cell hemolysis was seen in some soldiers after administration of the antimalarial drug primaquine. The incidence of hemolysis was most frequent in African-American soldiers. Subsequently, this drug-induced hemolysis was shown to be due to a genetic deficiency in glucose-6-phosphate (G6PDH) dehydrogenase. The population differences were explained by an association between G6PDH deficiency and the ability to survive malaria, which conveyed a biological advantage to individuals with G6PDH deficiency in malaria-infested countries. G6PDH comes from an X-linked gene, so that G6PDH deficiency is far more prevalent among males.

The first polymorphism in a drug metabolizing enzyme was documented in the 1950s, when it was found that there was a high incidence of peripheral neuropathy in response to the antituberculosis drug isoniazid due to slow clearance of the parent drug in some individuals. Subsequently, individuals were phenotyped and found to be either rapid or slow acetylators. In family studies the slow metabolizer phenotype was shown to be inherited recessively as a monogenetic trait. In population studies,

Molecular and Biochemical Toxicology, Fourth Edition, edited by Robert C. Smart and Ernest Hodgson
Copyright © 2008 John Wiley & Sons, Inc.

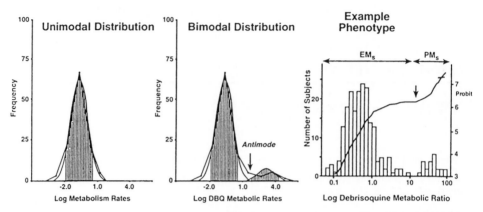

Figure 11.1. Demonstrates a unimodal and bimodal distribution of drug metabolism. The theoretical unimodal (single gaussian distribution) on the left is typical of a nonpolymorphic metabolism, and the bimodal distribution (mixture of two Gaussian distributions) on the right represents the results of phenotyping for a polymorphic substrate, debrisoquine. The intersection of the two bimodal distributions is the *antimode*, or the estimated cutoff point between the two phenotypes used to predict the PM phenotype. Adapted, in part, from Henthorn T. K., Benitez, J., Avram, M. J., Martinez, C., Llerena, A., Cobaleda, J., Krejcie, T., and Gibbons, R. D. *Clin. Pharmacol. Ther.* **45**, 328–333, 1989; and Küpfer, A., and Presage, R. *Eur. J. Clin. Pharmacol.* **26** 753–759, 1984.

Caucasians could be divided into almost equal numbers of rapid and slow acetylators, while only 15% of Japanese and Chinese were slow acetylators. This polymorphism is important because it also affects the metabolism of aromatic amines to which humans are exposed industrially, through the diet, and environmentally; and, as a result, it affects susceptibility to certain cancers. The first known polymorphisms in the phase I, xenobiotic metabolizing enzymes (XMEs) cytochrome P450 (CYP), were reported in the 1970s–1980s. One was a polymorphism in *CYP2D6*, which affects the metabolism of many clinically used drugs including the antihypertensive drug debrisoquine. The second was an effect on the metabolism of an anticonvulsant drug, mephenytoin, which was later shown to be due to polymorphisms in CYP2C19.

Polymorphisms in drug metabolizing enzymes are usually inherited as autosomal recessive traits. In many cases, polymorphisms are indicated by a bimodal distribution of metabolism in population studies (Figure 11.1). Individuals can be divided into two populations, extensive metabolizers (EMs) and poor metabolizers (PMs) of the drug. *Phenotype* refers to the biological traits of the individual (e.g., EMs or PMs), and *genotype* refers to the genetic makeup of the individual. Because of the autosomal recessive nature of most of these traits, PMs usually carry two mutant alleles representing various deleterious mutants of the same enzyme, whereas EMs can be either homozygous or heterozygous wild type.

Metabolic polymorphisms affect the rate of clearance of certain drugs. A good example can be seen in the pharmacogenetics haloperidol in poor and extensive metabolizers of CYP2D6 (Figure 11.2). If a drug has a narrow therapeutic index (ratio of toxic dose/therapeutic dose), the polymorphism may produce toxic levels in the PM (Figure 11.3).

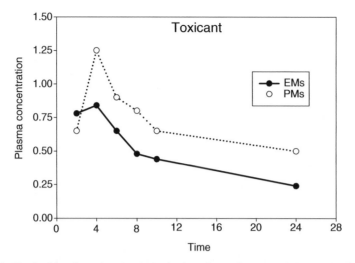

Figure 11.2. Toxicokinetics of a hypothetical toxicant in extensive metabolizers (EMs) and poor metabolizers (PMs) that are either with or without a polymorphism in the xenobiotic metabolizing enzyme primarily responsible for the metabolism of the toxicant in question.

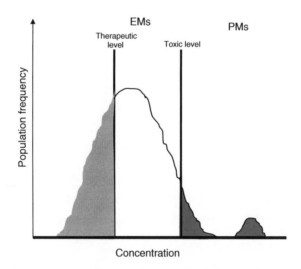

Figure 11.3. Steady-state concentrations of a hypothetical clinical drug and toxicant in a population that includes extensive metabolizers (EMs) and poor metabolizers (PMs).

11.2   POLYMORPHISMS IN CYP ISOFORMS

Table 11.1 summarizes a number of CYP polymorphisms, prototype drugs metabolized by these enzymes, clinical consequences, and examples of genetic defects causing these polymorphisms. There is now a unified nomenclature for polymorphic CYP alleles (http://www.imm.ki.se/CYPalleles/).

TABLE 11.1. Examples of Polymorphisms in Human CYP Enzymes and Their Consequences

Enzyme	Change	Result	Substrate	Clinical Effect
CYP1A2	Silent base	More inducible	Caffeine, activates food mutagens	Uncertain
	Splice variant	Decreased activity	Caffeine, clozapine	Uncertain
CYP1B1	Amino acid	Loss of function	Unknown	Mutations occur in primary congenital glaucoma
	Frameshifts			
CYP2A6	Amino acid	Decreased activity	Nicotine	Reduces risk for tobacco addiction
	Gene deletion	Loss of function		
	Gene conversions			
CYP2B6	Amino acid	Less protein	Efavirenz	Increased CNS effects, risk for resistant HIV from monotherapy
CYP2C9	Amino acid	Altered K_m	Warfarin	Excessive bleeding
			Tolbutamide	Dangerous lowering of blood sugar
			Phenytoin	Phenytoin intoxication
CYP2C19	Splice variants	Loss of function	Mephenytoin	Exacerbated sedation
	Stop codons		Proguanil	Decreased effectiveness
CYP2D6	Splice defects	Loss of function	Debrisoquine	Exaggerated cardiovascular effects
			Sparteine	Fetal deaths
			Codeine	Lack of efficacy
CYP2E1	Gene duplication	Increased activity	Codeine, dolasteron	Increased sedation, vomiting
	Amino acid	Decreased activity	Acetone, nitrosamines	Uncertain
	Repeats	Increased activity after ethanol exposure		Uncertain

11.2.1 CYP2B6

CYP2B6 metabolizes several pesticides, chemotherapeutic agents, and antidepressants. CYP2B6 is highly polymorphic and also inducible. One polymorphism, CYP2B6 516G > T, is linked to higher plasma levels of the non-nucleoside reverse-transcriptase inhibitor, efavirenz. The only known case of efavirenz related psychosis was in an individual homozygous for this mutation. Along with increased risk of CNS effects, individuals with the CYP2B6 516TT genotype are more likely to have efavirenz persist in the body for weeks after treatment ends. This can produce resistant HIV if treatment is not modified. This polymorphism is found in frequencies ranging from 14% in Koreans to 62% in Papua New Guineans.

11.2.2 CYP2C9

CYP2C9 metabolizes many clinically important drugs including the anticoagulant warfarin, the diabetic agent tolbutamide, the antihypertensive losartan, the anticonvulsant phenytoin, as well as numerous antiinflammatory drugs such as ibuprofen. CYP2C9*2 and *CYP2C9*3* are the most important of at least 29 variant alleles. CYP2C9*2 appears to have intermediate activity toward some substrates. CYP2C9*3 has lower activity for drugs such as tolbutamide, and people who are homozygous for this allele are slow metabolizers of warfarin, tolbutamide, and phenytoin. While homozygotes with these variants are exceptionally rare, detecting these alleles can help reduce life-threatening drug effects. The presence of even one variant allele can require a reduction in the warfarin dose.

11.2.3 CYP2C19

The anticonvulsant drug mephenytoin exists as two enantiomers (*R*- and *S*-). Küfer reported the stereoselective metabolism of *S*-mephenytoin in the dog. Later, while studying the metabolism of mephenytoin in humans, Küfer found that the formation and urinary excretion of the 4′-hydroxymetabolite of mephenytoin is rapid during the first 24 hr after administration of the drug and this metabolite is derived almost entirely from the *S*-enantiomer. Unexpectedly, one of the human subjects complained of unacceptable sedation on a dose of mephenytoin. From a family study it was determined that metabolism of mephenytoin in the family of the individual who metabolized the drug poorly was inherited as an autosomal recessive trait. Subsequent population studies showed that while ~3–5% of Caucasians were PMs, as many as 13–23% of Asians were PMs of mephenytoin.

Subsequently it was shown that CYP2C19 was the enzyme responsible for the 4′-hydroxylation and of *S*-mephenytoin. At least seven defective alleles of *CYP2C19* were discovered, accounting for the majority of the PM phenotype. The two most frequent polymorphisms are an aberrant splice site in exon 5 and a premature stop codon in exon 4. CYP2C19 also metabolizes the anti-ulcer drug omeprazole, the HIV protease inhibitor nelfinavir, certain barbiturates, and a number of antidepressants, is partially responsible for metabolism of the anxiolytic drug diazepam, and activates certain antimalarial drugs, such as proguanil.

11.2.4 CYP2D6

CYP2D6 was the first CYP enzyme shown to be polymorphic. During the course of a clinical trial of debrisoquine, which included a study of the metabolism and pharmacokinetics in volunteers, one of the investigators, Robert Smith, himself a volunteer, developed severe orthostatic hypotension in response to debrisoquine, with blood pressure dropping to 70/50 and symptoms persisting for two days. Other colleagues taking a similar dose showed few, if any, significant cardiovascular effects. Urine analysis showed that Smith eliminated the drug essentially unchanged, whereas other participants of the study excreted primarily the major metabolite, 4-hydroxy debrisoquine. This study prompted a search for a polymorphism of debrisoquine metabolism in a study of 94 volunteers and three families of PMs, leading, in 1977, to the first description of a genetic polymorphism in CYP-mediated drug oxidation in humans. The CYP2D6 polymorphism is prevalent in Caucasians and African-Americans, in which ~7% are PMs. In Asian populations, the frequency of PMs is only 1%.

CYP2D6 metabolizes a large number of clinically used drugs. These include antihypertensive drugs such as debrisoquine; antiarrhythmic drugs such as fiecainide; beta blockers such as metoprolol, propranolol, and bufuralol; antidepressants such as nortriptyline, desipramine, and clomipramine; a number of neuroleptics such as halperidol and thioridazine; and opiates such as codeine. Many of these drugs have a narrow therapeutic index, resulting in greater toxicity in PMs (Figure 11.3). The oxytocic drug sparteine was withdrawn from the market because of fetal deaths in women treated with this drug. At least 59 alleles of CYP2D6 are now recognized.

11.2.5 Other CYP Polymorphisms

CYP1B1 metabolizes many premutagenic polycyclic hydrocarbons. Several mutations have been found to be associated with congenital glaucoma, suggesting that this enzyme has a function in maintenance of intraocular pressure. It has been suggested that an upstream polymorphism in the *CYP2E1* gene changes expression of the gene, at least in promotor constructs in vitro, and one of several amino acid changes decreases expression of the protein. A change in the downstream region of CYP1A1 and an amino acid change have been studied for alterations in risk for lung cancer in epidemiology studies. However, it is questionable whether these changes affect catalytic activity. An amino acid change in *CYP1A2* has been recently identified, but its effect on catalytic activity is not yet known. Two rare inactivating alleles of *CYP2A6* have been reported. This enzyme metabolizes coumarin. It metabolizes certain procarcinogens including (methylnitrosamino)-1-(3-pyridyl)-1-butanone (NNK), a nitrosamine found in tobacco smoke, and also metabolizes nicotine to conitine. It has been suggested that individuals with defective CYP2A6 who have an impairment of nicotine metabolism decrease the risk for nicotine addition and are less likely to become tobacco-dependent smokers. There have been many polymorphisms found in the CYP3A4 gene and its promoter region. However, they do not explain most of the variability in CYP3A4 activity levels. Work has begun to identify polymorphisms in the gene for SXR, a receptor that regulates CYP3A4 expression. A polymorphism resulting from an SNP (single nucleotide polymorphism) in the upstream region of the gene (CYP3A4*2) has been characterized, but it does not appear to affect promotor activity. A second polymorphism involves Ser > Pro amino acid change that appears to change the intrinsic clearance of CYP3A4 for nefedifine but not testosterone.

11.3 POLYMORPHISMS IN ALCOHOL DEHYDROGENASE

Alcohol dehydrogenase exists as at least seven genetic loci. However, three genes produce the low-K_m class I ADHs. They produce the subunits a, b, and g. The ADH molecule is a dimer, consisting of homodimers a/a, b/b, and g/g and heterodimers a/b, a/g, and b/g. If an individual is heterozygous for b1 and b2, for example, ADH consists of 10 dimers formed by random association with either of the B alleles substituting randomly to form a/a, b1/b1, b2/b2, b1/b2, g/g, and heterodimers a/b1, a/b2 a, g, b1/g, and b1/g. Several of these genes, including ADH2 (determines subunit b chains) and ADH3 (determines g chains), are polymorphic (Table 11.2). The three b subunits are encoded by the genes ADH2*1, ADH2*2, and ADH2*3, and two g subunits are encoded by ADH3*1 and ADH3*2. ADH2*1 (b1) contains arginine, whereas the ADH2*2 (b2), which contains histidine, has a 40-fold higher V_{max} for ethanol than ADH2*1. The ADH2*2 allele (encoding the b2 subunit, which shows greater efficiency in metabolizing ethanol) may reduce the risk for alcoholism, whereas the ADH2*1 allele with low capacity for metabolizing ethanol is thought to be a risk factor. ADH2*3 encodes b3, which has a high K_m but a high V_{max}, which makes it less efficient at low ethanol concentrations but gives it greater capacity at high concentrations, which can occur during binge drinking. It has been suggested that this allele may be somewhat protective against alcohol-related birth defects. In contrast, g1 and g2 encoded by the ADH3*1 and ADH3*2 alleles have similar catalytic activity toward ethanol. The ADH4 gene (encoding ADH p) has also been found to have several polymorphisms in the upstream region, one of which affected the activity of promotor constructs in vitro approximately twofold and could potentially affect expression of this protein. Both the alcohol and aldehyde dehydrogenase polymorphisms are being widely studied with respect to their infiuence on alcoholism and alcohol-related diseases.

TABLE 11.2. Clinically Important Alcohol Dehydrogenase and Aldehyde Dehydrogenase Polymorphisms

Alcohol Dehydrogenase				
Gene	Allele	Change	Substrate	Clinical Effect
ADH2 (ADH1A2)	ADH1A2*1		Ethanol	↑ Risk of alcoholism
	ADH1A2*2	↑v_{max}	Ethanol	↓ Risk of alcoholism
	ADH1A2*3	↑k_m ↑v_{max}	Ethanol	↓ Risk of alcohol-related birth defects

Aldehyde Dehydrogenase				
Gene	Allele	Change	Substrate	Clinical Effect
ALDH2	ALDH2*1		Ethanol	↑ Risk of alcoholism
	ALDH2*2	Loss of activity	Ethanol	↓ Risk of alcoholism
			Vinyl chloride	Increased genotoxity

11.4   POLYMORPHISMS IN ALDEHYDE DEHYDROGENASE

There are at least 20 human aldehyde dehydrogenase genes. ALDH2 is the major form that metabolizes acetaldehyde. A polymorphism in this enzyme (Table 11.2), which consists of an amino acid substitution Glu504Lys, results in loss of activity of the enzyme. This mutation is particularly prevalent (50%) in Asians and causes high levels of acetaldehyde after ingestion of ethanol, causing flushing and nausea in affected individuals. This phenotype generally causes an aversion to alcohol, and individuals homozygous for this allele are at greatly reduced risk for alcoholism.

11.5   POLYMORPHISMS IN FLAVIN-CONTAINING MONOOXYGENASES

11.5.1   FMO2

FMO2 is one of six forms of FMO2 identified in mammals. It is expressed primarily in the lung and can catalyze N-oxidation of certain primary alkylamines. Surprisingly, analysis of the gene for FMO2 in humans indicates that it contains a $C > T$ nonsense mutation at codon 472, resulting in a gene that produces a peptide that is lacking 64 amino acids at its 3′-terminus compared with the FMO2 gene of closely related primates, including the gorilla and chimpanzee. Heterologous expression of the recombinant protein revealed that it was catalytically inactive. While this is not a polymorphism, it is an example of a gene that is inactivated by a nonsense mutation. As a result, data on lung metabolism by FMO2 in laboratory animals is not predictive of human pulmonary metabolism.

11.5.2   FMO3

Individuals with a defect in FMO3 have a condition known as fish odor syndrome, or trimethylaminurea, which is characterized by increased levels of free trimethylamine. Individuals with this syndrome exhibit an objectionable body odor reminiscent of rotting fish due to increased levels of free trimethylamine in sweat and urine. This syndrome can lead to social isolation, clinical depression, and even suicide. This disease is linked to deficiency in the N-oxidation of trimethyl amine obtained from foods including meat, eggs, and soybeans. A single amino acid mutation has been discovered which leads to a Pro153 > Leu substitution that abolishes the catalytic activity of the enzyme. Although FMOs oxidize a number of drugs, pesticides, and other xenobiotics, the toxicological significance of this polymorphism is still not known. Affected individuals do have a deficiency in the ability to N-oxidize nicotine. There have also been reports that this polymorphism improves the ability of sulindac to treat familial adenomatous polyposis.

11.6   POLYMORPHISMS IN EPOXIDE HYDROLASE

Microsomal epoxide hydrolase matabolizes xenobiotic expoxides including epoxides of promutagenic and procarcinogenic polycyclic hydrocarbons. Mutations in this enzyme could be of potential importance in carcinogenesis. A number of SNPs have been reported that cause amino acid substitutions. One mutation in exon three replaces the tyrosine at position 113 with a histidine. This change causes decreased activity in vitro and has been linked to changes in the risk of developing certain cancers among smokers and people who eat red meat. However, more studies are needed to confirm these results.

11.7 POLYMORPHISMS IN SERUM CHOLINESTERASE

A polymorphism in serum cholinesterase is one of the oldest polymorphisms known. It leads to prolonged muscle relation or prolonged paralysis after administration of the muscle relaxant succinylcholine. Several mutants occur, the most common is a point mutation causing the substitution of glycine for aspartic acid at position 70. This variant shows defective binding of choline esters to the anionic binding site but has normal activity with neutral or positively charged esters. There also numerous other variants, many with partial or complete loss of activity.

11.8 POLYMORPHISMS IN PARAOXONASE (PON1)

Human serum paraoxonase (PON1) detoxifies the toxic oxons of organophosphorus cholinesterases and is also involved in the metabolism of oxidized lipids. Paraoxonase hydrolyzes organophosphate esters, carbamates, and aromatic carboxylic esters. There are two allozymes (the Q and R forms) containing a single amino acid difference: either a glutamine (Q allele) or an arginine (R allele) at position 192. It has been suggested that the R192 polymorphisms may be a risk factor for coronary artery disease. However, since there is a wide interindividual variation in serum paraoxonase levels independent of this polymorphism, correlations with disease based solely on genotyping may be inadequate and phenotyping as well as genotyping is necessary.

The Q and R alloenzymes have similar activity toward diazoxon but differ toward other substrates. Taking blood samples and comparing activity toward diazoxon and paraoxon determines PON1 status, which identifies both genotype and paraoxonase levels. QQ individuals have higher activities toward sarin and soman, while RR individuals have greater activity toward chlorpyrifos-oxon and paraoxon. PON1 status also varies with age and diet. PON1 levels are lower in children than in adults and may be at greater risk from organophosphorus pesticides and nerve agents.

11.9 POLYMORPHISMS: MECHANISTIC CLASSIFICATION

The different types of modified alleles include the following: splicing defects or changes at intron–exon junctions that cause frame shifts, alter the coding sequence, and generally produce premature stop codons and truncated inactive proteins; insertions or deletions of nucleotides that cause frame shifts; gene deletions or rearrangments; single nucleotide changes that can either result in amino acid changes that may affect catalytic activity or produce premature stop codons or silent (noncoding) changes. Alterations in promotor upstream regions have been reported for some genes such as *CYP3A* and *CYP2E1*.

Splicing defects occur because most eukaryotic genes contain exons that code for protein interrupted by noncoding regions or introns. The DNA is transcribed into pre-mRNA, and then the introns are spliced out from the RNA to form mRNA, which is translated into the protein (Figure 11.4). Correct splicing depends partially on conserved splice sites at intron–exon junctions. There is normally a gt at the 5′ end of the donor splice site of the intron and an ag at the 3′ end of the intron (Figure 11.5A). Changes in splice site junctions often are responsible for incorrect splicing. For example, the most common defective *CYP2D6* allele contains a mutation in the acceptor site of a splice junction (*CYP2D6*4*) (Figure 11.5B). There is also a

Normal splicing of eucaryotic genes

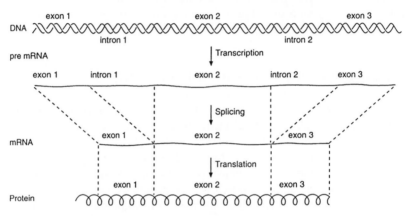

Figure 11.4. Normal splicing mechanism of eukaryotic genes. (Used with permission from (Lewis, B. Ed.). *Genes*, 2nd ed., John Wiley & Sons, New York, 1983.).

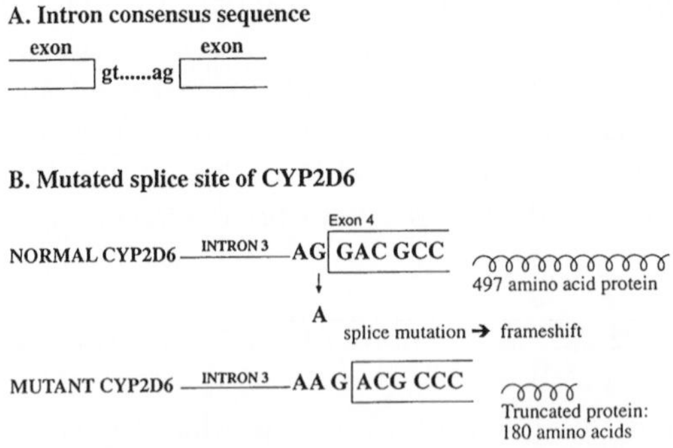

A. Intron consensus sequence

B. Mutated splice site of CYP2D6

Figure 11.5. Mutation of intron splice sites involved in formation of a mutant allele of CYP2D6*. (A) Normal intron consensus sequences of mammalian gene. (B) Mutation of the CYP2D6 splice site that moves the splice site downstream to the next 3′ consensus sequence, altering the reading frame of the protein and resulting in a premature stop codon (CYP2D6*4) and a truncated, inactive protein.

null mutation in which *CYP2D6* is completely deleted. There are also ultrarapid metabolizers of CYP2D6 substrates who contain more than two copies of the *CYP2D6* gene. These individuals produce more CYP2D6 protein, which decreases the effectiveness of commonly used doses of compounds metabolized by CYP2D6 in these individuals. The most frequent defects responsible for PMs of drugs metabolized by CYP2C19 are (a) single base mutation that causes a new splice site in exon 5 and (b) a single base substitution that produces a stop codon. Mutations that cause amino acid changes that alter activity are also responsible for additional PMs of CYP2C19 drugs. A single amino acid change in *CYP2C9* (*CYP2C9*3*) is responsible for PMs of the antidiabetic drug tolbutamide, the anticoagulant warfarin, and the anticonvulsant phenytoin. This change alters the affinity of the protein for various drugs (see Table 11.1 for examples).

11.10   POLYMORPHISMS AND DRUG METABOLISM

Polymorphisms in a drug metabolizing enzyme commonly alter the disposition and pharmacokinetics of the drug in question. For drugs with a narrow therapeutic index, this may increase toxicity. For example, PMs of CYP2D6 substrates exhibit exaggerated hypotension with debrisoquine, and several drugs are no longer used because of effects due to polymorphisms. These include: the oxytocic drug sparteine, which was taken off the market because of fetal deaths in PMs; perhexiline, which is no longer used because of increased neuropathy in PMs; and the antiarrythmic drug encainamide, which was also removed from the market because of toxicity in PMs (see Table 11.1 for examples). The polymorphism in cholinesterase can produce apnea and prolonged paralysis after administration of the muscle relaxant succinyl-choline. There is increased toxicity in CYP2C9 PMs treated with the anticonvulsant phenytoin, and much lower doses of the anticoagulant warfarin are required in CYP2C9 PMs. Some substrates are activated by drug metabolizing enzymes; and, for example, codeine is ineffective in PMs of CYP2D6.

In some cases a polymorphism may confer an advantage to PMs. For example, omeprazole is more effective in treatment of ulcers in PMs of CYP2C19, due to greater efficacy in the eradication of Helicobacteria. Moreover, in rapid metabolizers of CYP2D6 (containing multiple copies of the gene), CYP2D6 substrates, such as the anti-nausea drug Dolasteron, are less effective at any given dose.

Because of these differences in efficacy and toxicity, polymorphisms in drug metabolizing enzymes are of considerable importance to the pharmaceutical industry. Because of the possibility of increased drug toxicity or decreased efficacy in a proportion of the population, the development of a new drug may be impacted negatively if its metabolism is mediated via a pathway that is known to be polymorphic. This possibility is now always taken into account in the planning of metabolic studies and clinical trials.

11.11   METHODS USED FOR THE STUDY OF POLYMORPHISMS

Polymorphisms are discovered by resequencing genes in known PMs or by resequencing a gene in a random set of the population. Verification of a polymorphism involves developing a genetic test for a mutation, and testing a population that has been phenotyped with a drug in vivo. If the polymorphism is inherited recessively, PMs should have two mutant (m1-n) alleles, and EMs should have at least one wild-type allele. Expression of the wild-type cDNA versus the mutated cDNA in a recombinant system such as *Escherichia coli* or bacculovirus permits the investigation of the effect of that mutation on catalytic activity of the two recombinant proteins, and expression in a mammalian cell line (e.g., Cos-1 cells) is useful for determining whether the mutation affects the level of expression or stability of the protein in mammalian cells.

Phenotyping of individuals involves (a) administering the test drug and (b) collecting timed urine or blood samples for analysis of parent drug and/or metabolites. This technique has the advantage of detecting unknown genetic mutations, but it is labor-intensive and invasive to the patient or volunteer and, consequently, cannot be used on large populations.

Genotyping involves detecting known defective alleles. This can be done rapidly from a single drop of blood or a small buccal cell sample; but, to be effective, most of the known defective alleles must be known. A common type of genetic test for polymorphisms involves amplification of the affected region of the gene by the polymerase chain reaction (PCR). Mutations often affect the way that specific restriction enzymes such as *Sma*I cut the DNA. If a mutation affects a restriction enzyme site, PCR of the region can be followed by treatment with a restriction enzyme. The DNA product is then run on a gel and the cutting pattern will reveal the genotype (Figure 11.6). This is known as PCR-RFLP (PCR-restriction length polymorphism). Alternatively, if no restriction site is produced, a restriction site can often formed by placing a mismatch in one of the PCR primers so that only the normal or mutant DNA will be cut (mismatch PCR). Allele-specific *PCR* is the use of two sets of primers, the 3' end of one of the primers matches the wild-type DNA, and the 3' end of the other primer matches only the mutant DNA. With careful control of conditions, only the DNA will be amplified that is complementary to the

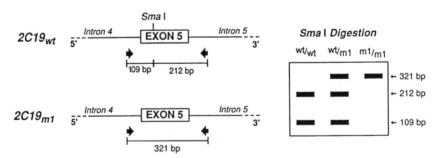

Figure 11.6. A typical restriction length polymorphism–polymerase chain reaction (RFLP–PCR) test for the *CYP2C19*2* mutation. Complementary primers that are specific for CYP2C19 (heavy arrows) are used to amplify a 312-bp fragment. The DNA is digested with a restriction enzyme *Sma*I and electrophoresced on an agarose gel. The wild-type DNA is cut by *Sma*I into two pieces of 109 bp and 212 bp, whereas the homozygous mutant allele is not cut. Individuals heterozygous for the two alleles show all three bands.

correct primer. Southern blotting is still used to identify gene deletions and duplications.

Rapid and high-throughput procedures have been developed for studying polymorphisms. Microarray procedures involve the binding of short oligonucleotides (up to 25 bp) on a solid surface such as nylon or silica. This is referred to as microarray (or chip)-based hybridizational analysis. Probes are designed so that the questionable base is in the center position of the probe. Arrays can be produced with a density of 10,000 probes per square micrometer. Over five million probes will currently fit on a 2-cm × 2-cm array. Scanning techniques such as laser confocal fluorescence can be used to analyze the results. For further information on this subject, see *Nature Genetics* **37**(6, Suppl.), June 2005. Low-density microarrays that analyze for the more common CYP2D6 and CYP2C19 methods have been developed for commercial use, and microarrays are also available to detect thousands of polymorphisms in many different xenobiotic-metabolizing enzymes.

11.12 EPIDEMIOLOGY

Epidemiology techniques, odds ratios, and so on, are used to estimate relative risk and are discussed in detail in Chapter 26. This estimates the likelihood of an occurrence of a particular disease (e.g., bladder cancer or colon cancer) if a particular risk factor is present. For example, if an individual is homozygous for a defective genotype, how likely is that individual to have the disease compared to one not in the risk group (e.g., an EM with one wild-type copy of gene)? Odds ratios significantly greater than 1.0 indicate that the PM genotype is at risk, odds ratios of 1.0 indicate no effect of the PM genotype, and odds ratios significantly less than 1.0 indicate reduced risk.

The effects of various CYP polymorphisms, as well as polymorphisms in other XMEs such as the acetylation polymorphisms and GST polymorphisms, on risk for various types of cancer is an area of considerable interest. For example, the *CYP2D6* splice site mutation has been associated with altered risk to meningioma and astrocytoma. A polymorphism in a *Msp*1 restriction site in the 3′ region of *CYP1A1* has been widely studied with respect to different cancers and has been associated with increased lung cancer in Japanese smokers. The functional relevance of the 3′ polymorphism itself seems doubtful, but it has been associated with increased inducibility of the *CYP1A1* gene, which is one possible explanation of the association with lung cancer in Japanese studies. However, the importance of the *Msp*1 polymorphism to lung cancer in Caucasians has been debated, at least partly because the low incidence of this polymorphism in Caucasians lowers the power of statistical testing. CYP2E1 metabolizes a variety of tobacco-derived chemicals including nitrosamines. A polymorphism in a *Rsa*1 restriction site in the upstream region of *CYP2E1* has been associated with increased expression of the promotor constructs in vitro. The *Rsa*1 polymorphism and a *Dra*1 restriction site polymorphism have been widely studied with respect to different cancers. The *Dra*1 restriction site has been associated with increased risk to lung cancer in Japanese smokers. Recent preliminary studies have suggested an increased risk for lung cancer and esophageal cancer in Japanese PMs of CYP2C19.

SUGGESTED READING

The chipping forecast III. *Nat. Genet.* **37**(6, Suppl.), June 2005.

Eichelbaum, M., and Gross, A. S. The genetic polymorphism of debrisoquine/sparteine metabolism—-clinical aspects. *Pharmacol. Ther.* **46**, 377–394, 1990.

Furlong, C. E., Cole, T. B., et al., Role of paraoxonase (PON1) status in pesticide sensitivity: Genetic and temporal determinants. *Neurotoxicology* **26**, 651–659, 2005.

Goldstein, J. A., and Blaisdell, J. Genetic tests which identify the principle defects in CYP2C19 responsible for the polymorphism in mephenytoin metabolism. In: Johnson, E.F., and Waterman, M. R. (Eds.). *Methods in Enzymology*, Vol. 272, *Cytochrome P450*, Part B, Academic Press, San Diego, CA, 1996, pp. 210–217.

Gonzalez, F. J. The molecular biology of cytochrome P450s. *Pharmacol. Rev.* **40**, 243–288, 1990.

Lewin, B. *Genes*, 9th ed., James and Bartlett, Boston, 2007.

Nebert, D. W. Polymorphisms in drug metabolizing enzymes: What is their clinical relevance and why do they exist? *Am. J. Hum. Genet.* **60**, 265–271, 1997.

Rendic, S. Summary of information on human CYP enzymes: Human P450 metabolism data. *Drug Metabol. Rev.* **34**, 83–448, 2002.

Richter, R. J., and Furlong, C. E. Determination of paraoxonase (PON1) status requires more than genotyping. *Pharmacogenetics.* **9**, 745–753, 1999.

Website: http://www.imm.ki.se/CYPalleles/ and references therein.

CHAPTER 12

Phase II—Conjugation of Toxicants

GERALD A. LeBLANC

Department of Environmental and Molecular Toxicology, North Carolina State University, Raleigh, North Carolina 27695

12.1   INTRODUCTION

Lipophilic toxicants can passively diffuse across the lipid-rich surface membranes of cells. Upon entry in into the cell, the toxicant is available to interact with molecular target sites of toxicity. Phase I biotransformation processes can reduce the reactivity or toxicity of the chemical through processes such as hydroxylation (Chapter 10). Toxicants can also undergo metabolic conjugation, either directly or subsequent to Phase I biotransformation. The conjugation process or Phase II detoxification involves coupling of the toxicant to small, endogenous molecules that are present within the cell. Conjugation typically reduces the reactivity of the toxicant and, hence, reduces its toxicity to the cell. Conjugation also generally facilitates the elimination of toxicants from the body through aqueous routes (e.g., urine) by increasing the aqueous solubility of the molecule.

Since both Phase I and II biotransformation processes can increase the polarity and, accordingly, the aqueous solubility of the toxicant, these biotransformations can essentially trap the toxicant in the cell by compromising its ability to passively diffuse across the surface membrane of the cell. The cellular elimination of toxicants is facilitated by membrane proteins that actively transport Phase I and II biotransformation products out of the cell and make them available for elimination from the body. The active cellular elimination processes are often referred to as Phase III detoxification/elimination processes.

12.2   CONJUGATION REACTIONS

Conjugation reactions may be divided into two types of mechanistic reactions. The first involves the formation of a conjugate in which the xenobiotic reacts with a high-energy or reactive endogenous ligand.

Type I: Xenobiotic + reactive conjugating ligand = conjugated product

Molecular and Biochemical Toxicology, Fourth Edition, edited by Robert C. Smart and Ernest Hodgson
Copyright © 2008 John Wiley & Sons, Inc.

The second type of conjugation reaction involves the coupling of the endogenous conjugating ligand with a high-energy or reactive xenobiotic. The reactivity associated with the xenobiotic is sometimes the consequence of Phase I biotransformation (e.g., epoxidation, Chapter 10).

Type II: Reactive xenobiotic + conjugating ligand = conjugated product

Type I reactions include the formation of glycosylated, sulfated, methylated, and acetylated conjugates, whereas type II reactions include peptide and glutathione conjugation.

In order for conjugation to occur, hydroxyl, amino, carboxyl, epoxide, thiol, or halogen groups must be present in the xenobiotic molecule (Table 12.1). Conjugation reactions most commonly result in the formation of hydrophilic products. These products are generally less lipid-soluble, more polar, and more readily eliminated. However, conjugation of xenobiotics to lipophilic moieties (e.g., fatty acids) can result in the storage of the xenobiotic in lipid compartments of the organism and cause a delay in excretion. Although conjugation reactions have been considered generally to result in detoxification of the xenobiotics, a number of examples of conjugation products that are the ultimate toxicants have been reported.

TABLE 12.1. Conjugation Reactions

Reaction	Enzyme	Functional Group
Glucuronidation	UDP-glycosyltransferase	OH
		COOH
		NH$_2$
		SH
		C–C
Glucosidation	UDP-glycosyltransferase	OH
		COOH
		SH
		C–C
Sulfation	Sulfotransferase	NH$_2$
		OH
Acetylation	Acetyltransferase	NH$_2$
		SO$_2$NH$_2$
		OH
Methylation	Methyltransferase	OH
		NH$_2$
		SH
Amino acid conjugation	Acyltransferase	COOH
Glutathione conjugation	Glutathion *S*-transferase	Epoxides
		Organic halides
		Organic nitro compounds
		Unsaturated compounds
Lipophilic conjugation		COOH
		OH

12.3 GLYCOSIDES

Glycosylation is the most common form of conjugation found in the animal and plant kingdom. The reactive intermediates are all derived from the universal fuel, glucose, and the supply of this molecule is less likely to be depleted than that of other endogenous intermediates. Also, glycosylation is a major pathway because the endogenous conjugating molecule has the capacity to react with a wide range of molecular groupings.

In order for glycosylation to occur, a high-energy endogenous molecule must be available to donate the conjugating group (sugar) to the aglycone (targeted xeno-biotic). Glucuronic acid and glucose are the two primary sugars used in glycosyl-ation. Other sugars (e.g., galactose) can be conjugated to an aglycone. These conjugations are typically minor contributors to xenobiotics metabolism and are largely the result of catalysis by enzymes that are distinct from the glucuronosyl/glucosyl transferases. Uridine diphosphate glucuronic acid (UDPGA) and uridine diphosphate glucose (UDPG) provide conjugating groups for glucuronidation and glucosidation reactions, respectively. Glucuronidation is a major conjugative pathway in vertebrates, whereas glucosidation generally is associated with plants and inver-tebrates. The multistep reaction for glycoside formation involves the two reactions presented below.

1. $UDPGA + ROH \xrightarrow{\text{UDP glucuronosyl transferase}} RO\text{-}\beta\text{-}D\text{-glucuronide} + UDP + H_2O$

2. $UDPG + ROH \xrightarrow{\text{UDP glucosyl transferase}} RO\text{-}\beta\text{-}D\text{-glucoside} + UDP + H_2O$

The first reaction depicts the formation of glucuronides, while the second describes glucoside formation. These reactions result in the nucleophilic displacement of the UDP moiety from the sugar molecule. The acceptor group attacks the C_1 of the pyranose ring to which UDP is attached in an α-glycosidic bond, and a Walden inversion occurs. The resulting glycoside has a β-configuration. In order for a high rate of glycosylation to occur, the reactive group of the xenobiotic must be suffi-ciently nucleophilic. Any molecule that is nucleophilic or capable of becoming nucleophilic is, potentially, a candidate for glycoside bond formation. The reactivity of the aglycone will depend on both steric and electronic factors.

12.3.1 Glucuronides

A wide variety of compounds can be substrates for glucuronidation. Four general categories of *O*-, *S*-, *N*-, and *C*-glycosides have been reported (Table 12.2). Both glucosides and glucuronides react to form phenolic, enolic, and ester glycosides. Similarly, thiolic and dithioic glycosides have been reported for the *S*-glycosides. Several types of *N*-glucuronides have been identified when the glucuronosyl moiety is attached to the nitrogen. They include the following functional groups: aromatic amino, sulfonamide, carbamyl, and heterocyclic nitrogen. Many factors affect gluc-uronidation such as age, diet, sex, species and strain differences, genetic factors, and diseases (see Chapters 13 and 14).

Xenobiotic glucuronidation occurs abundantly in the liver, the intestinal mucosa, and the kidney. Almost all other organs and tissue possess some glucuronidation

TABLE 12.2. Types of Glucuronides Formed by UGT Enzymes

O-Glucuronides

R—⟨benzene⟩—OH + UDPGA ⟶ R—⟨benzene⟩—O • $C_6H_9O_6$
Phenols

RCH_2OH + UDPGA ⟶ $RCH_2O • C_6H_9O_6$
Alcohols

$RCOOH$ + UDPGA ⟶ $RC(O)O • C_6H_9O_6$
Carboxylic acids

$R-NH-OH$ + UDPGA ⟶ $RNHO • C_6H_9O_6$
Hydroxylamines

$RC(O)NH-OH$ + UDPGA ⟶ $RC(O)NHO • C_6H_9O_6$
Hydroxamic acids

S-Glucuronides

R—⟨benzene⟩—SH + UDPGA ⟶ R—⟨benzene⟩—S • $C_6H_9O_6$
Thiophenols

⟨thiazoline⟩—SH + UDPGA ⟶ ⟨thiazoline⟩—S—$C_6H_9O_6$
Thiols

$R_2N-C(S)SH$ + UDPGA ⟶ $R_2N-C(S)S • C_6H_9O_2$
Thiocarbamic acids

N-Glucuronides

R—⟨benzene⟩—NH_2 + UDPGA ⟶ R—⟨benzene⟩—NH • $C_6H_9O_6$
Aromatic amines

$R-NHOH$ + UDPGA ⟶ $RN(OH) • C_6H_9O_6$
Hydroxylamines

$RCH_2OC(O)NH_2$ + UDPGA ⟶ $RCH_2OC(O)NH • C_6H_9O_6$
Carbamates

RSO_2-NH-R + UDPGA ⟶ $RSO_2-NR • C_6H_9O_6$
Sulfonamides

C-Glucuronides

$$R_1-\overset{\overset{O}{\|}}{C} \diagdown \\ R-\underset{\underset{O}{\|}}{C} \diagup CH + UDPGA \longrightarrow R_1-\overset{\overset{O}{\|}}{C} \diagdown \\ R-\underset{\underset{O}{\|}}{C} \diagup CH • C_6H_9O_6$$

1,3-Dicarbonyl
compounds

activity but at very low levels. Normally, the glucuronide conjugates are excreted in the urine and bile unless they are involved in enterohepatic circulation. The conjugates may be hydrolyzed by β-glucuronidases present in the various tissues or organs or upon biliary excretion by bacterial β-glucuronidases.

Glucuronidation of many substrates is low or undetectable in fetal mammalian tissues but increases with age. The rate of development is dependent upon the species, tissue, and substrate. The inability of newborns of most mammals, except the rat, to form glucuronides is associated with deficiencies in glucuronosyl transferase activity and the cofactor, UDPGA. The blood serum of newborn babies may contain pregnandiol, which is an inhibitor of glucuronide formation. Suppressed glucuroniation of bilirubin in newborns can result in neonatal jaundice.

12.3.2 The UGT Superfamily

UDP-glycosyl transferase (UGT) enzymes are the products of a multigene superfamily. Human UGTs identified to date consist of three gene families UGT1, UGT2, and UGT8, and members of the families UGT1 and UGT2 contribute to xenobiotics metabolism. The human UGT1 family consists of at least 13 enzymes that are the products of alternate splicing of a single gene. This gene contains 13 known promoters/first exons that can splice to the common exons 2 through 5, producing 13 different products. Some UGT1 family members are inducible by aryl hydrocarbon receptor (AhR) ligands [e.g., 2,3,7,8-tetrachlorodibenzo-p-dioxin (TCDD), 3-methylcholanthrene], constitutive androstane receptor (CAR) receptor ligands (e.g., phenobarbital), and pregnane X receptor (PXR) receptor ligands [e.g., pregnenolone-16α-carbonitrile (PCN)]. UGT1 family members are responsible for bilirubin and thyroxine conjugation as well as the conjugation of some xenobiotics (e.g., opioids). Humans afflicted with Crigler–Najjar syndrome type 1 or Gilbert's syndrome, as well as the Gunn strain of rat, are deficient in the ability to form glucuronide conjugates of bilirubin. This deficiency is due to a mutation in a UGT1 gene and results in the accumulation of unconjugated bilirubin (unconjugated hyperbilirubinemia).

Thyroid hyperplasia, hypertrophy, or tumorigenesis has been associated with exposure of rodents to xenobiotics such as TCDD, diniconazole, and thiazopyr. This toxicity has been attributed to the inductive effects of these compounds on hepatic UGT1. Induction of UGT1 causes increased conjugation and clearance of thyroxine and decreased circulating thyroxine levels. In response, thyroxine stimulating hormone (TSH) secretion by the pituitary increases and the thyroid gland is stimulated to produce and secrete more thyroxine. Overstimulation of the thyroid gland by TSH may account for hyperplasia, hypertrophy, and ultimately tumorigenesis in this tissue. Thus, thyroid toxicity associated with these chemicals may be secondary to effects of the chemicals on conjugative processes in the liver.

Enzyme products of the UGT2 gene family share less than 50% amino acid identity with the UGT1 family members. UGT2 enzymes appear to all be products of individual genes. Human UGT2 family consists of at least five members that are all within the UGT2B subfamily. Some of the UGT2 family members are susceptible to induction by CAR receptor ligands (e.g., phenobarbital). The UGT2 enzymes are largely responsible for glucuronide conjugation of steroid hormones and some xenobiotics (e.g., 1-naphthol, 4-nitrophenol, 4-hydroxybiphenyl).

12.4 SULFATION

Sulfate conjugation is important in the biotransformation of xenobiotics as well as many endogenous compounds such as thyroid hormones, steroid hormones, certain proteins, and some peptides. Sulfation has been demonstrated in vertebrates, invertebrates, plants, fungi, and bacteria. Sulfate ester formation is known to occur with primary, secondary, and tertiary alcohols, phenols, and arylamines. Sulfate esters are completely ionized, very water-soluble, and quickly eliminated from the organism. Sulfation contributes to the metabolic detoxification and elimination of many xenobiotics. However, some compounds are converted into a highly labile sulfate conjugates that form reactive intermediates that have been implicated in carcinogenesis and tissue damage (see Section 12.10).

The sulfate molecule used in conjugation reactions is derived from 3-phosphoadenosine-5′-phosphosulfate (PAPS). Efficient synthesis of PAPS requires an adequate amount of inorganic sulfate with L-cysteine, D-cysteine, or L-methionine serving as precursors for the inorganic sulfate. PAPS is likely synthesized in every vertebrate cell, with high concentrations produced in the liver.

The most common acceptor for sulfation is the hydroxyl group of phenols, alcohols, and N-substituted hydroxylamines (Table 12.3). In addition, sulfation of thiols and amines forming thiosulfates and sulfamates have been reported. In most species, both the sulfation and glucuronidation of the same phenolic substrates occur. Glucuronidation often provides for low-affinity, high-capacity catalysis resulting in the rapid biotransformation of substrates when present at high concentrations. Sulfation often complements glycosidation by providing high-affinity, low-capacity catalysis for efficient substrate conjugation at low substrate concentrations. Sulfate conjugates are excreted predominately in the urine and to a lesser degree in the bile.

TABLE 12.3. Examples of Sulfate Conjugation to Alcohols, Phenols, and Aromatic Amines

Alcohol Conjugation

$$C_2H_5OH + PAPS \rightleftharpoons C_2H_5OSO_3H + ADP$$

Phenol Conjugation

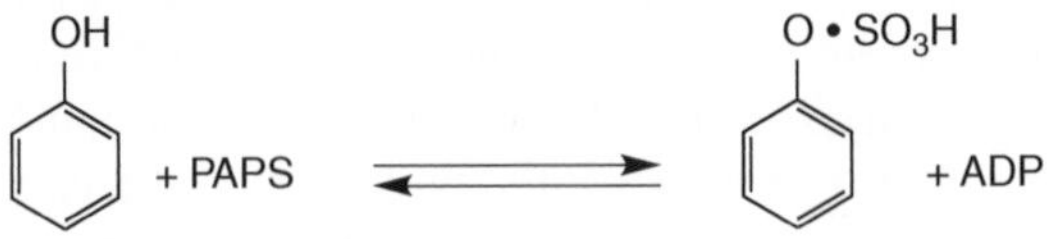

Aromatic Amine Conjugation

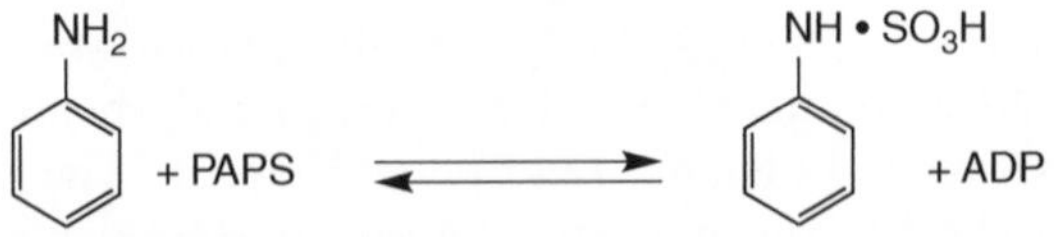

12.4.1   The SULT Superfamily

The sulfotransferase enzymes comprise a superfamily of enzymes. These enzymes can either be membrane-bound or cytosolic, though the enzymes involved in xenobiotic conjugation are primarily cytosolic. Several members of the sulfotransferase superfamily that are significant contributors to the sulfation of xenobiotics have been identified in humans. Provided below is a brief description of the known human SULTs.

Several SULT 1A subfamily members contribute to the sulfate conjugation of endogenous substrates such as thyroid hormone, catecholamines, and 17β-estradiol as well as exogenous compounds including drugs such as acetaminophen and environmental chemicals such as isoflavones. Each of these subfamily members is the product of a distinct gene. Within this subfamily, SULT 1A1 and SULT 1A2 seem to have greatest involvement in xenobiotics biotransformation. The SULT 1A enzymes are highly expressed in the brain, breast, intestine, endometrium, kidney, lung, and platelets.

SULT 1B1 functions in the sulfation of diverse endogenous substrates include cholesterol, dehydroepiandrosterone (DHEA), thyroid hormone, and dopamine. The enzyme also contributes to the metabolism of phenolic xenobiotics such as those containing 2-naphthol. It is highly expressed in the liver, colon, small intestine, and blood leukocytes.

Little is known of the substrate specificity of SULT 1C subfamily members, though they are known to contribute to the sulfation of the procarcinogen N-hydroxy-2-acetylaminofluorene.

SULT1E1 is a major contributor to the sulfation of estrogens including 17β-estradiol, esterone, and 17α-ethinylestradiol. The K_m values for these enzymes toward estrogens are several orders of magnitude below those of SULT 1A subfamily members, suggesting that these two subfamilies may function in concert for the rapid and efficient inactivation of estrogens. Some hydroxylated polychlorinated biphenyls (PCBs) are potent inhibitors of SULT 1E1. Some of the endocrine-disrupting effects of PCBs may be due to interference with normal estrogen homeostasis. Crystalographic studies have revealed that the hydroxylated PCBs bind to the ligand-binding pocket of SULT 1E1 in a manner highly similar to estradiol, suggesting that these environmental chemicals are competitive inhibitors of estradiol metabolism.

SULT 2A and 2B sulfotransferase subfamily members sulfate the 3β-hydroxyl group of a variety of steroid hormones. Dehydroepiandrosterone (DHEA) is the prototypical substrate for the SULT 2 enzymes. However, other hydroxysteroids such as testosterone and its phase I hydroxylated derivatives are substrates for these enzymes. The SULT 2 sulfotransferases also are responsible for the sulfate conjugation of a variety of alcohols and xenobiotics that have undergone phase I hydroxylation, including the polycyclic aromatic hydrocarbons (PAHs). The SULT 2 enzymes exhibit different patterns of tissue expression. SULT 2A1 is expressed primarily in the adrenal cortex, brain, liver, and intestine, while SULT 2B1 is expressed in the prostate, placenta, and trachea.

SULT 4A1 is expressed in the human brain, and little is known of its function. The enzyme appears to have little activity toward typical SULT substrates and is not likely an important contributor to xenobiotic conjugation. Human SULT 4A1 shares ~97% amino acid sequence identity with mouse and rat homologs, suggesting that this enzyme performs some highly important and conserved function.

12.4.2 SULT Regulation

The SULT enzymes are not known to be regulated by the AhR. In fact, SULT1A and SULT 2A are suppressed by AhR ligands such as TCDD, 2-acetylaminofluorene, 3-methylchlolanthrene, and β-napthoflavone. Some SULT 2A enzymes are responsive to ligands of CAR (e.g., phenobarbital, some PCBs) and a functional CAR response element has been associated with this responsiveness. Evidence for the responsiveness of human SULT 2A enzymes to CAR ligands is conflicting. Some SULT 2A subfamily members also are responsive to ligands of the glucocorticoid receptor (e.g., dexamethasone), PXR (e.g., PCN), and peroxisome proliferator activated receptor (PPAR) alpha (e.g., clofibrate). Bile acids regulate the expression of some SULT 2 family members via the farnesoid X receptor (FXR).

12.5 METHYLATION

Methylation is a common biochemical reaction involving the transfer of methyl groups from one of two methyl donor substrates (*S*-adenosylmethione or N^5-methyltetrahydrofolic acid). The transfer occurs with a wide variety of methyl-acceptor substrates (proteins, nucleic acids, phospholipids, and other molecules of diverse structures). Of the cofactors, *S*-adenosylmethionine (SAM) is the most important in the methylation of xenobiotics containing oxygen, nitrogen, or sulfur nucleophiles. SAM is biosynthesized from L-methionine and ATP by ATP:L-methionine *S*-adenosyltransferase. Methyl conjugates generally are less water-soluble than the parent compound, except for the tertiary amines. Despite the decrease in water-solubility, methylation generally is considered a detoxification reaction. Methylation reactions are classified according to substrates as follows.

12.5.1 *N*-Methylation

N-methylation is an important reaction by which primary, secondary, and tertiary amines are substrates of methylation. Most tissues catalyze the methylation of a large variety of amines. The source of the methyl group that is transferred in each instance is SAM, and the products are secondary, tertiary or quaternary *N*-methylamines as well as *S*-adenosyl-L-homocysteine (SAH). The reaction shown below is with a primary amine as substrate and is catalyzed by an amine *N*-methyltransferase.

$$RCH_2NH_2 + SAM \rightarrow RCH_2NHCH_3 + SAH$$

Humans express at least three forms of *N*-methyltransferase that are catagorized base upon substrate specificity. Nicotinamide *N*-methyltransferase methylates compounds that have an indole ring such as serotonin and tryptophan or a pyridine ring such as nicotinamide and nicotine. Histamine methyltransferase is a cytoplasmic enzyme characterized by its ability to methylate histamine to form 3-*N*-methylhistamine. The transferase is expressed in several tissues including the kidney, gastric mucosa, erythrocytes, brain, and skin. Phenylethanolamine *N*-methyltransferase, also known as noradrenaline methyltransferase, catalyzes the methylation of norepinephrine to epinephrine; the final biosynthetic step is formation of the hormone. The enzyme is located in the soluble fraction of the adrenal gland, retina, heart, and brain. Multiple forms of phenylethanolamine *N*-methyltransferase have been reported to be present in several species. At least five forms have been described in cow and rabbit, two in dog, one in rat, and one in human. A variety of phenylethanolamines, but not phenylethylamine, serve as methyl acceptor substrates when SAM is the methyl donor. Examples of natural substrates are norepinephrine, epinephrine, octopamine, normethaneprine, metanephrine, and synephrine. In general, primary amines are better substrates than secondary amines.

12.5.2 *O*-Methylation

Catechol-*O*-methyltransferase (COMT) is widely distributed throughout the animal and plant kingdom and is primarily associated with the cytosolic fraction of rat liver, skin, kidney, glandular tissue, heart, and brain. In humans, a membrane-bound COMT also has been characterized which is encoded by the same gene as the cytosolic COMT, but with a different transcription start site. Multiple polymorphic forms of COMT may be expressed within a species. COMT catalyzes the transfer of a methyl group from SAM to one of the phenolic groups of a catechol in the presence of Mg^{2+} ions. With only a few exceptions, all of the methyl acceptor substrates of COMT require the catechol moiety.

3,4-Dihydroxybenzoic acid   3-Methoxy-4-hydroxybenzoic acid

Substrates of COMT include xenobiotics catechols, catecholamines, and catechol estrogens. Three functional classes of chemicals are known to inhibit COMT. *S*-Adenosyl-I-homocysteine (SAH) is a potent inhibitor of COMT as well as the other SAM-dependent methyltransferases. Inhibition results from SAH binding to the SAM binding site on the enzyme. Certain divalent ions such as Ca^{+2} and trivalent metal ions such as the salts of lanthanides, neodymium, and europium are excellent inhibitors of COMT. A number of catechol-type substrates such as pyrogallol, flavonoids, pyrones, pyridenes, hydroxyquiolines, 3-mercaptotyramine, and tropolones are irreversible inhibitors of COMT.

Phenol *O*-methyltransferase (POMT) is a membrane-bound methyltransferase that is responsible for the methylation of phenolic compounds that are more commonly recognized as substrates for glucuronic acid or sulfate conjugation (e.g., acetaminophen). Phenol *O*-methylation is typically regarded as a minor catalytic processes with respect to xenobiotic detoxification.

12.5.3 *S*-Methylation

Methyl groups from SAM are also transferred to thiol-containing drugs and xeno-biotics. It appears that the only structural requirements are the presence of a free thiol and that the compound is not too hydrophilic in nature. *S*-Methylation has been demonstrated with mercaptoethanol, thioacetanilide, and 6-mercaptopurine, as well as with a variety of thiopurine and thiopyrimidine drugs. In general, aromatic thiols are the best substrates, while simple aliphatic thiols are less active.

$$\text{HSCH}_2\text{CH}_2\text{OH} + \text{SAM} \xrightarrow{\text{Thiol } S\text{-methyltransferase}} \text{CH}_3\text{SCH}_2\text{CH}_2\text{OH} + \text{SAH}$$

Mercaptoethanol   *S*-Methylthioethanol

Thiol methyltransferase has been detected in erythrocytes, lymphocytes, lungs, cecal, and colonic mucosae. The nature and number of thiol methyltransferases is not clear at the present time. A cytosolic enzyme and a microsomal enzyme have been reported, with the microsomal enzyme being dissociated from membrane relatively easily. The microsomal enzyme in rat liver has been purified to homogeneity. The enzyme is a 28,000-dalton monomer with an isoelectric point of 6.2. A wide variety of xenobiotic thiols are methylated, but cysteine and glutathione are not substrates. *S*-Methylation is an important component in the "thiomethyl shunt." Thiomethyl conjugates are metabolized to the methylsulfoxides by oxidation (see Chapter 10) and reenter the mercapturic acid pathway as substrates for glutathione *S*-transferase.

12.6 ACYLATION

Foreign carboxylic acids and amines undergo biological acylation to form amide conjugates. Acylation reactions are of two types. The first involves an activated conjugating intermediate, acetyl CoA, and the xenobiotic. The reaction is referred to as acetylation.

$$\text{CH}_3\text{C(O)SCoA} + \text{RNH}_2 \rightarrow \text{RNHCOCH}_3 + \text{CoASH}$$

Acetyl-CoA   Amine   Acetyl   CoA
                                conjugate

The second type involves the activation of the xenobiotic to form an acyl CoA derivative, which then reacts with an amino acid to form an amino acid conjugate.

$$\text{RC(O)SCoA} + \text{NH}_2\text{CHCOOH} \rightarrow \text{RC(O)NHCH}_2\text{COOH} + \text{CoASH}$$

Acyl-CoA   Glycine   Amino acid conjugate   CoA

12.6.1 Acetylation

Compounds containing amino or hydroxyl groups are quite readily acetylated. The only known example of acetylation of a sulfhydryl group occurs with the formation of acetyl-CoA.

The acetylation of amino groups is quite common. Five types of amino groups (arylamino, aliphatic amino, α-amino, hydrazino, and sulfonamido) can be biotrans-formed by acetylation. Besides direct acetyl transfer, inter- and intramolecular transfers of the acetyl group from the nitrogen to the oxygen of arylhydroxamic acids have been reported. Direct *O*-acetyl transfer has been shown to form carcinogenic acetoxyarylamines. Acetylation typically results in the masking of the amine group with a nonionizable acetyl moiety. As a result, the acetylated derivatives are generally less water-soluble than the parent compound.

There is a large species variation in *N*-acetylation. The hamster and rabbit possess relatively high *N*-acetylating capacities, while the dog has little or no acetylating capacity. Other species such as rats and mice are intermediate. *N*-Acetyltransferase activity has been demonstrated in liver, intestinal mucosa, colon, kidney, thymus, lung, pancreas, brain, erythrocytes, and bone marrow. *N*-Ethylmaleimide, iodoacetate, and *p*-chloromercuribenzoate are irreversible inhibitors of the *N*-acetyltransferase. A variety of divalent cations such as Cu^{2+}, Zn^{2+}, Mn^{2+}, and Ni^{2+} are also inhibitory.

Two cytoplasmic *N*-acetyltransferases, NAT1 and NAT2, have been identified in humans. NAT1 is rather ubiquitous in its expression, whereas NAT2 is associated with the liver and the gastrointestinal tract. These enzymes have different but overlapping substrate preferences. A third enzyme, NAT3, has been identified in mouse.

Polymorphisms among the *N*-acetyltransferases have been characterized in humans and other species. The human population is segregated into slow acetylators and fast acetylators based upon rates of acetylation of the drug isoniazid. The incidence of slow acetylators is high among Middle Eastern populations, intermediate in Caucasian populations, and low in Asian populations. The slow acetylator phenotype is the result of mutations in the NAT2 gene that compromises the activity of the enzyme. Slow acetylators are predisposed to toxicity of drugs that are inactivated by acetylation such as isoniazid, dapsone, sulfamethazine, procainamide, and hydralazine. This enzyme also acetylates aromatic amine dyes to which workers have been exposed industrially such as benzidine dyes, 4-aminobiphenyl, 4,4′-methylbis(2-chloroaniline)naphthylamine, and *o*-toluidine. Workers in the arylamine dye industry who are slow acetylators have been shown to have an increased risk for bladder cancer. The low activity of NAT2 in the liver of slow acetylators may make the aromatic amines more available for hydroxylation by CYP1A2. The resulting hydroxylamines then accumulate in the bladder where they are acetylated by NAT1.

12.6.2 Amino Acid Conjugation

The conjugation of carboxylic acid xenobiotics with amino acids occurs in both liver and kidney and is catalyzed by an enzyme system located in the mitochondria. Conjugation requires initial activation of the xenobiotic to a CoA derivative in a reaction catalyzed by acyl CoA ligase. The acyl CoA subsequently reacts with an amino acid, giving rise to acylated amino acid conjugate and CoA.

The reactions are catalyzed by acyl-CoA:amino acid *N*-acyltransferase, of which two distinct *N*-acyltransferases exist in mammalian mitochondria. The predominant transferase conjugates medium-chained fatty acyl CoA and substituted benzoic acid derivatives with glycine and is termed an aralkyl-CoA:glycine *N*-acyltransferase, while the other enzyme conjugates arylacetic acid derivatives with glycine, glutamine, or arginine and is an arylacetyl-CoA:amino acid *N*-transferase.

Amino acid conjugation has been reported to occur with glycine, glutamine, arginine, and taurine in mammals and certain primates. In other organisms, different amino acid acceptors are utilized in peptide conjugation. Ornithine is utilized by reptiles and some birds, while ticks utilize arginine and glutamine, insects use alanine, glycine, serine, and glutamine, and fish use taurine. There is more interspecies variation with this type of conjugation than in any other Phase II reaction. The specific amino acids employed in peptide conjugation within a class of animals generally depend on the bioavailability of the amino acids from endogenous and dietary sources and are those amino acids that are normally nonessential. The extent of peptide conjugation with glycine in mammals follows the order herbivores > omnivores > carnivores.

Amino acid conjugation typically results in detoxification and elimination of xenobiotics. Amino acid conjugation is an alternative conjugation processes for carboxylic acid-containing xenobiotics such as ibuprofen. Glucuronidation of carboxylic acid-containing xenobiotics can result in the generation of toxic acylglucuronides. Amino acid conjugation of these xenobiotics limits the production of these toxic metabolites. Amino acid conjugates of xenobiotics are generally cleared from the body by urinary elimination.

12.7 GLUTATHIONE *S*-TRANFERASES

The glutathione *S*-transferases (GST) are a family of isoenzymes involved in the conjugation of reduced glutathione with electrophilic compounds. GSTs mediate the initial reaction in the biosynthesis of mercapturic acids (Figure 12.1). The biosynthesis of mercapturic acids occurs in several steps, initially involving the formation of a glutathione conjugate and then involving formation of cysteinyl glycine conjugate as the result of removal of glutamic acid by the γ-glutamyltranspeptidase. This is followed by removal of glycine by cysteinyl *S*-conjugate—that is the premercapturic acid. Acetylation of the cysteine *S*-conjugate by *N*-acetyltransferase results in the formation of the *N*-acetyl derivative, a mercapturic acid.

The GSTs are distributed quite widely throughout the biota, having been characterized in vertebrates, invertebrates, plants, and bacteria. GST activity is ubiquitous in mammalian tissues with high levels found in tissues associated with chemical detoxification and elimination. Soluble GSTs are localized to the cytoplasm and mitochondria of cells, whereas membrane-bound GSTs are associated with the endoplasmic reticulum. GSTs also have been shown to compartmentalize within interchromatinic regions of the cell's nucleus. The cytoplasmic GSTS are the most abundant forms of the enzyme and are the forms primarily associated with xenobiotic detoxification.

GSTs are involved in the conjugative metabolism of a wide variety of electrophilic substrates such as epoxides, haloalkanes, nitroalkanes, alkenes, methyl sulfoxide derivatives, organophosphates, and aromatic halo- and nitro-compounds. Functional classes of compounds that are catalytically conjugated to glutathione include antibiotics, vasodilators, herbicides, insecticides, analgesics, anticancer agents, and carcinogens. The reaction involves nucleophilic displacement, Michael addition, and nucleophilic attack on strained oxirane rings. Each of these reactions is illustrated in Table 12.4.

The cytoplasmic GSTs comprise a gene superfamily of homo- and heterodimeric enzymes whose individual members are composed of various combinations of different monomers having molecular weights ranging from ~20,000 to 25,000 daltons.

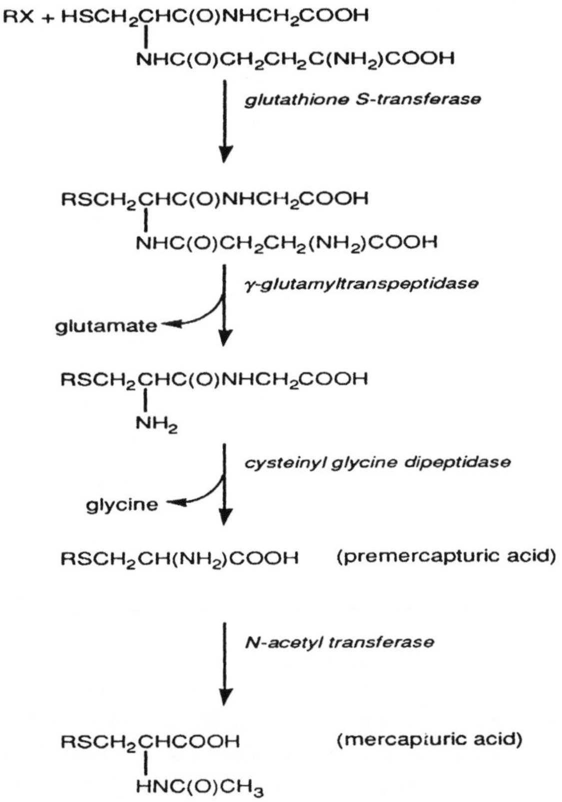

RX + HSCH$_2$CHC(O)NHCH$_2$COOH
            |
         NHC(O)CH$_2$CH$_2$C(NH$_2$)COOH

↓ *glutathione S-transferase*

RSCH$_2$CHC(O)NHCH$_2$COOH
           |
        NHC(O)CH$_2$CH$_2$(NH$_2$)COOH

↓ *γ-glutamyltranspeptidase*

glutamate ←

RSCH$_2$CHC(O)NHCH$_2$COOH
           |
          NH$_2$

↓ *cysteinyl glycine dipeptidase*

glycine ←

RSCH$_2$CH(NH$_2$)COOH   (premercapturic acid)

↓ *N-acetyl transferase*

RSCH$_2$CHCOOH   (mercapturic acid)
           |
        HNC(O)CH$_3$

Figure 12.1. Mercapturic acid biosynthesis.

TABLE 12.4. Glutathione S-Transferase-Mediated Conjugations Involving Nucleophilic Displacement, Michael Addition, and Nucleophilic Attack on Strained Oxirange Rings

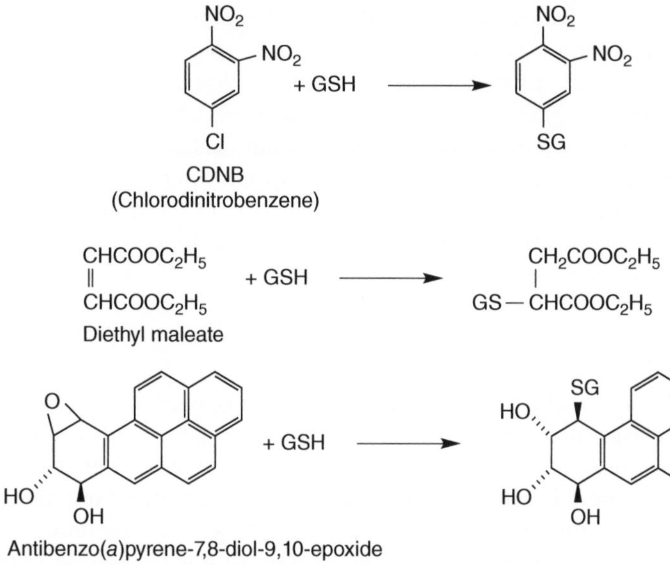

CDNB
(Chlorodinitrobenzene)

CHCOOC$_2$H$_5$
‖   + GSH ⟶
CHCOOC$_2$H$_5$
Diethyl maleate

CH$_2$COOC$_2$H$_5$
       |
GS — CHCOOC$_2$H$_5$

Antibenzo(a)pyrene-7,8-diol-9,10-epoxide

TABLE 12.5. Human Cytosolic Glutathione S-Transferases

Class	Gene Designation
Alpha	hGSTA1
	hGSTA2
	hGSTA3
	hGSTA4
	hGSTA5
Mu	hGSTM1
	hGSTM2
	hGSTM3
	hGSTM4
	hGSTM5
Omega	hGSTO1
	hGSTO2
Pi	hGSTP1
Theta	hGSTT1
	hGSTT2
Zeta	hGSTZ1

The monomeric proteins all appear to be products of individual genes. Mammalian cytosolic GSTs have been segregated in seven classes—alpha, mu, omega, pi, sigma, theta, and zeta—based upon amino acid sequence similarity. Subunits within a class typically share at least 60% homology, whereas subunits between classes are typically ~30 identical. Heterodimeric combinations occur only among subunits within the same class. The class alpha, mu, and pi GSTs are responsible for most of the catalytic activity associated with organs of detoxification such as the liver.

Various nomenclatures have been used to identify GSTs. Presently, individual GSTs are most commonly identified using a lowercase letter to designate species, an uppercase letter to designate class, and an Arabic number to designate the specific protein subunit within the class. Thus, rGSTM1-2 refers to the rat class mu GST consisting of a heterodimeric combination of subunits 1 and 2. Known human GSTs genes are presented in Table 12.5. The GSTs typically exhibit overlapping substrate specificities, although enzymes within a class sometimes exhibit substrate preferences. For example, class alpha GSTs exhibit high organic peroxidase activity toward cumene hydroperoxide relative to members of the other GST classes. Class mu and pi GSTs exhibit higher relative glutathione conjugating activity toward 1,2-dichloro-4-nitrobenzene and ethacrynic acid, respectively.

12.7.1 Allosteric Regulation

In addition to the substrate and the glutathione-binding sites, which are involved in catalysis, some class alpha GSTs contain noncatalytic ligand-binding sites. Binding of ligands to some of these sites can increase enzymatic activity. For example, interaction of class alpha GSTs with the herbicide 2,4,5-T increases catalytic activity of the GST toward 1-chloro-2,4-dinitrobenzene. Binding of the ligand to the GST appears to change the conformation of the protein in a manner that enhances its catalytic activity. Therefore, exposure to some xenobiotics may increase GST activity without affecting GST protein levels.

12.7.2   Regulation of Cell Signaling

Some GSTs associate with kinases that are involved in controlling stress responses and apoptosis. This association maintains the kinase in an inactive form. Upon release from the GST, the kinase becomes catalytically active. For example, GST pi has been shown to bind Jun kinase and inhibit Jun signaling in nonstressed cells. In response to oxidative or chemical stress, GST pi releases Jun kinase, the signaling pathway is activated, and the cells undergo apoptosis. Similarly, some GST mu proteins bind and inhibit the ASK1 kinase. In response to cellular stress (e.g., oxidative, chemical, heat), ASK1 is released, cell signaling occurs, and the cells under apoptosis in response to the stress. This role for GST in cellular responses to stress may prove to be as important in chemical toxicity as the catalytic activities associated with these proteins.

12.7.3   GST Regulation

Following treatment with xenobiotics, such as some planar aromatic hydrocarbons, phenols, and quinones, transcription of some class alpha and mu GSTs is significantly elevated. Induction by planar aromatic hydrocarbons such as TCDD, some PCBs, and some polycyclic aromatic hydrocarbons (PAHs) is dependent upon the Ah receptor and CYP 1A1 activity and appears to be mediated by two distinct regulatory pathways. The response element known as the dioxin response element (DRE), also known as the xenobiotic response element (XRE), has been identified in some class alpha GSTs. These GSTs appear to be induced in direct response to the interaction of the Ah receptor and the DRE (see Chapter 13 for more detail of this induction mechanism). Planar aromatic hydrocarbons can also induce levels of CYP 1A1 in the cell via interaction with the Ah receptor. This increase in CYP1A1 results in increased oxidative metabolism of the xenobiotic to oxidation–reduction labile metabolites, such as diphenols, aminophenols, and quinones. These reactive metabolites generate a redox signal in the cell that initiates a GST induction pathway. The antioxidant response element (ARE), also known as the electrophile response element (EpRE), has been identified in some class alpha GSTs that may respond to such an induction pathway. According to this model, a redox signal in the cell stimulates a signal transduction pathway resulting in the generation of cfos/cjun heterodimers. These heterodimers then interact with the ARE, which is a variant of the AP-1 binding site, to stimulate gene transcription. AP-1 sites are known to regulate a variety of genes involved in cell proliferation and other processes through interaction with fos/jun heterodimers. Xenobiotics such as some quinones and phenols may directly generate a redox signal (i.e., by generating oxidative stress), causing GST gene induction via the same redox signal.

Many class alpha and mu GSTs are induced by CAR receptor ligands such as phenobarbital. The glucocorticoid response element also has been identified in some class alpha GSTs. This element may mediate the induction of some GSTs by the synthetic glucocorticoid dexamethasone. PXR ligands such as 17α-hydroxypregnenolone have been shown to induce the expression of some GSTs; however, this induction appears to be independent of the PXR response element and may involve cross-talk with some other induction pathway.

GST pi expression is generally not considered to be induced by xenobiotics. However, hepatic GST pi expression is elevated 50- to 100-fold in hepatic preneoplastic tissue and in hepatocarcinomas induced by chemical carcinogenesis. Due to this high level of induction, GST pi is regarded as a marker of hepatic preneoplastic cells. Furthermore, GST pi is expressed at high levels in tumor cells from various organs that have developed resistance to the toxicity of anticancer drugs. GST pi gene expression is under the regulatory control of both enhancer and silencer elements. Deletion of the silencer region elevates gene transcription, while deletion of the enhancer region inhibits gene transcription. The enhancer region called GPE1 contains two response elements that, like the ARE, are capable of binding cfos/cjun heterodimers resulting in gene activation. Therefore, GPE1 could be responsible for the induction of GST pi during carcinogenesis. Some silencer regions of the GST pi gene contain the recognition sequence for the transcriptional regulator protein $NF_{\kappa}B$. As long as the silencer region of the gene is occupied by the appropriate transcriptional regulator protein, the enhancer region is nonfunctional and gene expression is suppressed.

GSTs also play an important role in resistance to chemical toxicity. GSTs have been implicated in resistance to the effects of a variety of chemicals including antibiotics, anticancer agents, analgesics, herbicides, insecticides, and vasodilators. Preneoplastic lesions, tumors, and cultured tumor cell lines commonly overexpress GSTs of the classes alpha, mu, and pi. This overexpression of GSTs is considered partly responsible for the resistance often associated with these cells to cancer chemotherapeutics.

12.8 CYSTEINE *S*-CONJUGATE β-LYASE

Cysteine *S*-conjugate β-lyase is responsible for converting a number of cysteine *S*-conjugates into pyruvate, ammonia, and thiols.

$$RSCH_2CH(NH_2)COOH + H_2O \rightarrow RSH + NH_3 + CH_3C(O)C(O)OH$$
Premercapturic acid   Pyruvic acid

This activity has been demonstrated in mammalian liver and kidney, as well as in intestinal bacteria. Glutathione *S*-transferases catalyze the formation of glutathione conjugates, which are processed via the mercapturic biosynthetic pathway, to acetylated cysteine *S*-conjugates. The unacetylated premercapturic acids are substrates for cysteine *S*-conjugate β-lyase, whereas the acetylated cysteine *S*-conjugates, the mercapturic acids, do not function as substrates for the enzymes.

Cysteine *S*-conjugate β-lyase activity has been implicated in the bioactivation of certain halogenated alkenes. This bioactivation results from the generation of reactive thiols. The thiol produced by the β-lyase-dependent metabolism of *S*(1,2-dichlorovinyl)-L-cysteine is an unstable electrophile that binds covalently to cellular macromolecules and leads to nephrotoxicity and genotoxicity. Alternatively, certain other cysteine *S*-conjugates form chemically stable thiols that exhibit no toxicity. The chemically stable thiols may be excreted intact or further metabolized to *S*-glucuronides or methylated to yield methylthio derivatives.

12.9 LIPOPHILIC CONJUGATION

Some metabolic conjugative processes result in the production of lipophilic derivatives that are not readily eliminated. While their physical properties are different from the classical hydrophilic conjugates, the mechanism of formation clearly defines them as conjugates (i.e., they are formed as the result of a union of xenobiotic metabolites with endogenous molecules). The reactions involve the coupling of xenobiotic acids and alcohols with endogenous intermediates of lipid synthesis (i. e., the acids with glycerol and cholesterol and the alcohols with the fatty acids).

With lipophilic conjugation, the xenobiotic metabolites appear to be incorporated in the lipid biosynthetic pathway similar to the normal constituents. One would expect the conjugates to have turnover times similar to that of their natural counterparts. Whether the lipophilic conjugates have any deleterious effect on the organism would depend upon the type and amount of bioactivity retained by the metabolites before and after conjugation.

12.10 PHASE II ACTIVATION

The biological activity of xenobiotics is generally decreased by conjugation. Phase II reactions normally increase the rate of elimination of most lipid-soluble xenobiotics, thereby terminating their possible effect in a biological system. However, the number of xenobiotics that are metabolized to penultimate or ultimate toxicants as the result of Phase II reactions is ever-expanding. Therefore, conjugation can no longer be considered strictly a detoxification process. Additional details on metabolism to reactive metabolites can be found in Chapter 20.

A number of arylamines are potent bladder carcinogens such as 4-aminobiphenyl, 1-naphthylamine, and benzidine. Metabolic activation of these carcinogens requires the action of UDP-glucuronosyl transferase on *N*-hydroxyarylamines to form *N*-glucuronides.

$$\text{Ar-NHO glucuronide} \xrightarrow{\text{pH}<7} \text{Ar-NHOH} \Longleftrightarrow \underset{\text{Arylnitrenium ion}}{\overset{+}{\text{Ar-N-OH}}}$$

The *N*-glucuronides of the *N*-hydroxyarylamines are transported to the bladder, where, upon hydrolysis by β-glucuronidase and in the presence of acidic urine, *N*-hydroxylarylamines are formed. Spontaneously, the electrophilic arylnitrenium ion forms, which can then react with nucleophilic centers of the epithelium in the bladder to initiate tumor formation.

Metabolic activation occurs with the sulfation of 2-acetylaminofluorine (2-AAF). *N*-Hydroxylation of the amino nitrogen by the monooxygenases followed by sulfation results in an unstable N–O–sulfate ester. The ester decomposes to an electrophilic nitrenium ion–carbonium ion, which can form covalent adducts with macromolecules. Several other xenobiotics such as mono- and dinitrotoluene, *N*-hydroxyphenacetin, hydroxysafrole, and other N-hydroxyarylamides are in part activated by this mechanism.

Certain halogenated alkanes and alkenes are bioactivated via glutathione conjugation to form potent nephrotoxicants. Glutathione conjugates are metabolized by the mercapturic acid pathway to the cysteine conjugates. Two different mechanisms have been proposed by which cysteine *S*-conjugates may produce nephrotoxicity. One is the spontaneous nonenzymatic formation of electrophilic episulfonium ions, which may alkylate essential renal macromolecules. The other mechanism is dependent upon β-lyase activation, forming unstable reactive thiols that react with biological macromolecules. Both mechanisms appear to be responsible for the formation of mutagenic and DNA-damaging intermediates that are responsible in the initiation of neoplastic transformations.

12.11   PHASE III ELIMINATION

Conjugation of lipophilic xenobiotics to polar cellular constituents renders the xenobiotic more water-soluble. While the lipophilic parent xenobiotics could readily diffuse into the cells, the increase in polarity associated with conjugation greatly reduces the ability of the compound to diffuse across the lipid bilayer of the cell membrane thus trapping the compound within the cell. The polar conjugates must therefore rely upon active transport processes to facilitate efflux from the cell. Hepatocytes, as well as other cells involved in chemical detoxification, are rich with members of the ATP-binding cassette superfamily of active transport proteins (ABC transporters). Cellular efflux of xenobiotics by these transporters is often referred to as Phase III elimination because Phase I or II detoxification processes often precede and are a requirement of Phase III elimination. A detailed description and discussion of elimination and transporters is presented in Chapter 15.

12.12   CONCLUSIONS

Conjugation and active transport processes provide efficient means for both the inactivation and elimination of xenobiotics. Xenobiotics may possess the molecular properties that render them suitable for active cellular efflux with no need for biotransformation (Figure 12.2). Alternatively, xenobiotics may first undergo Phase I hydroxylation, Phase II conjugation, or sequential hydroxylation and conjugation (Figure 12.2). The resulting modifications can render the xenobiotic susceptible to active efflux from the cell and, ultimately, elimination from the body. Through these processes of inactivation and efflux, conjugation and active elimination processes confer a high degree of protection against the toxicity of chemicals. The overexpression of the proteins responsible for these processes is a common mechanism of resistance to chemical toxicity. Conversely, the inhibition of these proteins or the suppression of their expression can increase susceptibility to some toxicants.

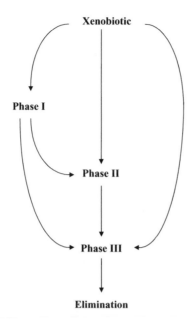

Figure 12.2. Interrelationships among Phase I (hydroxylation), Phase II (glucuronic acid, sulfate, and glutathione conjugation), and Phase III (ABC transporter-mediated efflux) detoxification processes leading to the inactivation and elimination of xenobiotics.

SUGGESTED READING

Blanchard, R. L., Freimuth, R. R., Buck, J., Weinshilboum, R. M., and Coughtrie, M. W. H. A proposed nomenclature system for the cytosolic sulfotransferase (SULT) superfamily. *Pharmacogenetics* **14**, 199–211, 2004.

Bock, K. W. Vertebrate UDP-glucuronosyltransferases: Functional and evolutionary aspects. *Biochem. Pharm.* **66**, 691–696, 2003.

Eaton, D. L., and Bammler, T. K. Concise review of the glutathione *S*-transferases and their significance to toxicology. *Toxicol. Sci.* **49**, 156–164, 1999.

Gamage, N., Barnett, A., Hempel, N., Duggleby, R. G., Windmill, K. F., Martin, J. L., and McManus, M. E. Human sulfotransferases and their role in chemical metabolism. *Toxicol. Sci.* **90**, 5–22, 2005.

Mackenzie, P. I., Bock, K. W., Burchell, B., Guillemette, C., Ikushiro, S., Iyanagi, T., Miners, J. O., Owens, I. S., and Nebert, D. W. Nomenclature update for the mammalian UDP glycosyltransferase (UGT) gene superfamily. *Pharmacogen. Genom.* **15**, 677–685, 2005.

McIlwan, C. C., Townsend, D. M., and Tew, K. D. Glutathione *S*-transferase polymorphisms: Cancer incidence and therapy. *Oncogene* **25**, 1639–1648, 2006.

Zamek-Gliszczynski, M. J., Hoffmaster, K. A., Nezasa, K., Tallman, M. N., and Brouwer, K. L. R. Integration of hepatic drug transporters and phase II metabolizing enzymes: Mechanisms of hepatic excretion of sulfate, glucuronide, and glutathione metabolites. *Eur. J. Pharm. Sci.* **27**, 447–486, 2006.

CHAPTER 13

Regulation and Polymorphisms in Phase II Genes

YOSHIAKI TSUJI

Department of Environmental and Molecular Toxicology, North Carolina State University, Raleigh, North Carolina 27695

13.1 INTRODUCTION

Two broadly classified groups of enzymes, termed phase I and phase II enzymes, metabolize and detoxify xenobiotics (industrial, environmental, dietary and medicinal chemicals). Xenobiotic metabolizing phase I enzymes, as represented by cytochrome P450 monooxygenases, are involved in the metabolism of a broad spectrum of xenobiotics; this initial metabolism, usually oxidation reactions, introduces one or more polar groups that serve as substrates for phase II detoxication reactions. This metabolic transformation sequence is a necessary step for the efficient elimination of xenobiotics. Xenobiotic metabolizing phase II enzymes are a group of enzymes that detoxify phase I reactive metabolites either via conjugation of their active functional groups with endogenous ligands (such as glutathione) or via destruction of the active centers of reactive metabolites by reduction or hydrolysis. Phase II reactions generally increase the water-solubility of compounds and thereby facilitate their excretion from the body. The emerging importance of the comprehensive study of gene–environment interactions, along with advances in toxicogenomic and pharmacogenetic approaches, has revealed that genetic variations and single nucleotide polymorphisms in phase II genes may determine individual susceptibility to various xenobiotics, ultimately having a significant impact on chemical carcinogenesis and drug metabolism. In addition, it has been shown that expression of many phase II genes is induced by phase I reactive metabolites and reactive oxygen species via a defined cis-acting enhancer element. By taking advantage of our knowledge of molecular mechanisms of phase II gene expression, the effective use of pharmacological drugs and dietary phytochemicals to induce phase II gene expression has been investigated to prevent or reverse carcinogenesis and neurodenegerative diseases.

Molecular and Biochemical Toxicology, Fourth Edition, edited by Robert C. Smart and Ernest Hodgson

13.2   ROLES OF PHASE II GENES AND POLYMORPHISMS

13.2.1   UDP-Glucuronosyl Transferases

Glucuronidation is a major detoxification system in vertebrates for biotransformation of lipophilic compounds into more hydrophilic, water-soluble compounds through conjugation with glucuronic acid (see Chapter 12). The enzymes catalyzing the glucuronidation are UDP-glucuronosyltransferases (UGTs). UGTs utilize UDP-glucuronic acid to transfer the glucuronic acid to the substrates containing a variety of functional groups including hydroxyl (–OH; phenolic, alcoholic), carboxyl (–COOH), thiophenols (–SH), and amines (–NH$_2$). Substrates of UGTs include (a) a variety of xenobiotics, including drugs, environmental agents, pesticides, and carcinogens, and (b) endobiotics such as bile acids and steroid and thyroid hormones. More than 350 xenobiotics and endobiotics are known to be UGT substrates. More than 15 human genes of the UGT superfamily, along with the corresponding cDNAs and isozymes, have so far been identified.

The C-terminal half of UGT1A proteins have identical amino acid sequences containing the ER membrane anchoring domain, while the N-terminal regions of the UGT1As have diversity within their amino acids sequences (Figure 13.1). This divergent N-terminal region serves as the binding domain for various substrates and determines the substrate specificity for each UGT1A isozyme. The diversity of the N-terminus half of UGT1As can be explained by the organization of the UGT1A genes with a unique exon 1 and common exons 2–5 (Figure 13.1). Transcription of each exon 1 cassette for a UGT1A isozyme is regulated by each promoter/enhancer

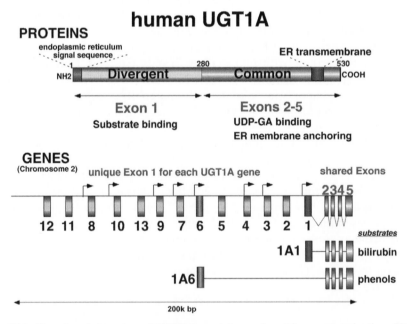

Figure 13.1. Functional domains of UGT1A proteins and multigene organization. UGT1A family proteins contain ER signal sequence at the N-terminus, followed by a divergent region and a common region at the C-terminus.

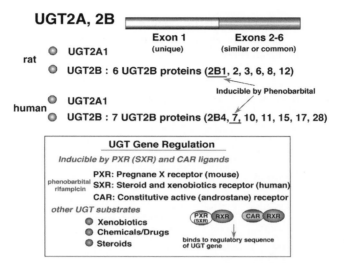

Figure 13.2. Rat and human UGT2A and 2B proteins have a structure similar to that of UGT1A, consisting of divergent and common (or conserved) regions. Some of UGT genes are inducible by phenobarbital or rifampicin through activation of SXR or CAR, heterodimer formation with RXR and bind to regulatory elements in the UGT genes.

regulatory region, which allows the unique exon 1 to be utilized and then spliced to the commonly shared exons 2–5. UGT2B subfamily members are also highly homologous proteins, which are encoded by independent homologous genes consisting of 6 exons in the clustered region of human chromosome 4 (Figure 13.2).

UGTs glucuronidate many xenobiotics and endobiotics through overlapping substrate specificities to many different classes of chemicals. Among these, phenols such as acetaminophen and 1-naphthol are preferentially glucuronidated by 1A9, 1A1, and 1A6, while bilirubin is exclusively metabolized by UGT1A1. Bilirubin is the major breakdown product of the heme-containing protein hemoglobin. Bilirubin is hydrophobic and bound to albumin. UGT1A1 plays a critical role in the conjugation and elimination of excess bilirubin, converting it to the more water-soluble bilirubin glucuronide in the liver, the key site for the detoxication and elimination of bulirubin. Insufficient or deficient UGT1A1 activity due to genetic polymorphisms therefore leads to hyperbilirubinemia, ultimately causing accumulation of bilirubin in brain tissues and neuronal toxicities. Crigler–Najjar syndrome is an inheritable unconjugated hyperbilirubinemia disease with no UGT1A1 activity (type I) or detectable but insufficient UGT1A1 activity (type II) because of various UGT1A1 polymorphisms (Figure 13.3). Most type I patients (type IA) have a mutation in the common C-terminal half encoded by the exon 2–5, and they have deficiencies in conjugating not only bilirubin but also xenobiotics normally metabolized by UGT1A1. A smaller population of type I patients (type IB) have mutations in the N-terminal divergent region determining the specificity of substrate binding (encoded by the exon 1), in which the defect in glucuronidation is mostly limited to bilirubin. In type II patients, UGT1A1 activity is typically about 10% of normal levels. In addition to polymorphisms in UGT1A1 exons, UGT promoter polymorphisms—in particular, an increased numbers of TA repeats causing decreased

Crigler-Najjar Syndrome Mutations in the UGT1A1 gene
Type I no UGT1A1 activity ➤ severe — blood exchange, transfusion or liver transplantation
Type II 10% of normal UGT1A1 activity

	Allele	Nucleotide Changes	Protein Changes	Type	Exon	Disease
	UGT1A1*1	Wild type				
	UGT1A1*2	879 del 13	Truncation	Deletion	2	CN1 (Crigler-Najjar Type 1)
	UGT1A1*3	1124 CT	S375F	Missense	4	CN1
C to T	UGT1A1*4	1069 CT	Q357X	Nonsense	3	CN1
	UGT1A1*5	991 CT	Q331 del 44	132 nt del	2	CN1
	UGT1A1*6	221 GA	G71R	Missense	1	Gilbert
	UGT1A1*7	145 TG	Y486D	Missense	5	CN2 (Crigler-Najjar Type 2)
	UGT1A1*8	625 CT	R209W	Missense	1	CN2
	UGT1A1*9	992 AG	Q331R	Missense	2	CN2
	UGT1A1*10	1021 CT	R341X	Nonsense	3	CN1
	UGT1A1*11	923 GA	G308E	Missense	2	CN1
	UGT1A1*12	524 TA	L175Q	Missense	1	CN2
	UGT1A1*13	508 del 3	F170del	Deletion	1	CN1
	UGT1A1*14	826 GC	G276R	Missense	1	CN1
	UGT1A1*15	529 TC	C177R	Missense	1	CN2
	UGT1A1*16	1070 AG	O357R	Missense	3	CN1
	UGT1A1*17	1143 CG	S381R	Missense	4	CN1
	UGT1A1*18	1201 GC	A401P	Missense	4	CN1
	UGT1A1*19	1005 GA	W335X	Missense	3	CN1
	UGT1A1*20	1102 GA	A368T	Missense	4	CN1
	UGT1A1*21	1223 ins G	Frameshift	Frameshift	4	CN2
	UGT1A1*22	875 CT	A292V	Missense	2	CN1
	UGT1A1*23	1282 AG	K426E	Missense	4	CN1
	UGT1A1*24	1309 AT	K437X	Missense	5	CN1
	UGT1A1*25	840 CA	C280X	Missense	1	CN1
	UGT1A1*26	973 del G	Frameshift	Frameshift	2	CN2
	UGT1A1*27	686 CA	P229Q	Missense	1	Gilbert
	UGT1A1*28	TAATA7	Transcription	Insertion	Promoter	Gilbert
	UGT1A1*29	1099 CG	R367G	Missense	4	Gilbert
	UGT1A1*30	44 TG	L15R	Missense	1	CN2
	UGT1A1*31	11609 CCGT	P387R	Missense	3	CN1
CC to GT	UGT1A1*32	1006 CT	R336W	Missense	3	CN1
	UGT1A1*33	881 TC	I294T	Missense	2	CN2

Figure 13.3. Allelic polymorphisms of the human UGT1A1 gene and association with unconjugated hyperbilirubinemia. (Adapted from *Annu. Rev. Pharmacol. Toxicol.* **40**, 581–616, 2000.)

UGT1A1 expression—are also associated with hyperbilirubinemia. Since SXR (steroid and xenobiotics receptor) and CAR (constitutive active androstane receptor) as well as a basic-leucine zipper transcription factor Nrf2 (NF-E2 related factor 2) are involved in transcriptional activation of some UGT1A genes including UGT1A1 (Figure 13.2), pharmacological activators of SXR (such as phenobarbital, rifampicin) and many substrates of UGTs are capable of inducing expression of UGT genes. In fact, induction therapy with phenobarbital is effective in type II Crigler–Najjar syndrome patients to induce UGT1A1 expression via activation of SXR and CAR (Figure 13.2).

Glucuronidation is generally a detoxification pathway for xenobiotics and endobiotics; however, it has been suggested that induction of UGT activity by xenobiotics may induce formation of thyroid tumors in rodents. As shown in Figure 13.4, thyroid-stimulating hormone (TSH), secreted by the pituitary grand, stimulates the production and secretion of T4 (thyroxine) in the thyroid gland. The secretion of TSH is regulated by thyroid-releasing hormone in the hypothalamus. In rodents, thyroid hormone T4 is subject to glucuronidation by UGT followed by biliary excretion. Chronic exposure to UGT-inducing xenobiotics causes more T4-glucuronide formation and biliary excretion, which decreases T4 levels in the serum (Figure 13.4). The

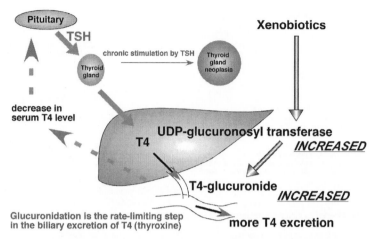

Figure 13.4. Induction UGT-related thyroid tumors.

decrease in serum T4 levels stimulates the pituitary for more TSH secretion in order to produce more T4. In this negative feedback regulation initiated by xenobiotic-mediated UGT induction, chronic stimulation of the thyroid gland by TSH enhances development of thyroid tumors.

13.2.2 Glutathione Synthesis and Glutathione S-Transferases

The synthesis and metabolism of glutathione (L-γ-glutamyl-L-cysteinylglycine, GSH) is discussed in detail in Chapters 12 and 18. In summary, GSH is a tripeptide thiol, involved in cellular antioxidant defense against oxidative stress in mammalian cells and, in addition to serving as an antioxidant, is also involved in the conjugation and detoxification of reactive metabolites by glutathione S-transferases. Glutathione (1–10 mM) is found in the cytoplasm and nucleus, primarily in the reduced form (GSH). Under normal conditions, over 90% of glutathione remains reduced. Oxidized glutathione (GSSH) can be reduced to GSH by glutathione reductase, maintaining the normal ratio of GSH:GSSG.

Glutathione is synthesized by two sequential ATP-dependent actions of γ-glutamylcysteine synthetase (γ-GCS, the rate-limiting reaction) and glutathione synthetase (GSS). γ-GCS catalyzes the formation of γ-glutamyl L-cysteine (the gamma linkage instead of the typical alpha linkage makes glutathione more resistant to peptidases), to which glycine is added by GSS. γ-GCS is a heterodimer of a catalytic heavy subunit containing a substrate binding site and a regulatory light subunit (Figure 13.5). These two subunits are encoded by different genes. In humans, the 73-kD catalytic subunit gene is on chromosome 6p12, and polymorphic 7–9 trinucleotide repeats located 10–30 bp upstream from the translation start site may affect translation of the gene. The regulatory subunit gene is on chromosome 1p22, and loss of the 1p21–22 locus appears to be associated with some cancers such as breast cancer, neuroblastoma, and pheochromocytoma, although the linkage between the loss of γ-GCS regulatory subunit and these cancers remains to be determined. Both catalytic and regulatory subunit genes are induced by various stressors and chemicals (Figure 13.5) including electrophiles/electrophilic metabolites (molecules

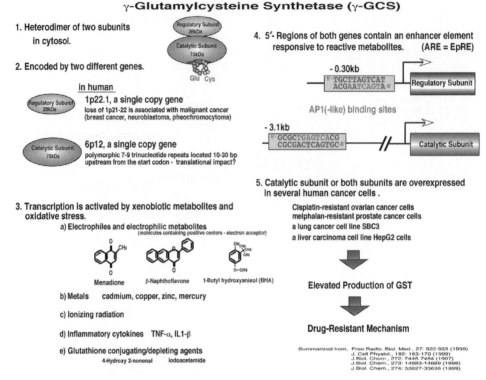

Figure 13.5. γ-Glutamylcysteine synthetase (γ-GCS), a rate-limiting and key enzyme in glutathione synthesis.

containing positive centers serving as electron acceptors), metals (cadmium, copper, mercury, zinc), ionizing radiation, inflammatory cytokines (TNF-α and IL1-β), and glutathione conjugating/depleting agents (iodoacetamide, 4-hydroxy 2-nonenal). Some of these inducing agents—in particular, electrophiles/electrophilic metabolites—activate transcription of both subunit genes via an enhancer element responsive to oxidative stress, termed an antioxidant responsive element (ARE, Figure 13.5). Either the catalytic subunit or both subunits are overexpressed in several drug-resistant human cancer cells such as cisplatin-resistant ovarian cancer cells and melphalan-resistant prostate cancer cells, suggesting that elevated production of glutathione may induce drug-resistance mechanisms in these cancer cells (Figure 13.5). Glutathione synthetase (GSS), in mammalian cells, is a homodimer of 52-kD protein. In humans, the GSS gene contains 13 exons (encoding 474 amino acids) on chromosome 20q11.2. There are rare cases of inherited GSS deficiency as a result of missense mutations and splice mutations. GSS deficiency patients have severe acidosis and hemolytic anemia.

Conjugation of glutathione to a wide variety of electrophilic chemicals is a vital detoxification reaction such as xenobiotics and reactive metabolites. Glutathione *S*-transferases (GSTs) are involved in this conjugation reaction by catalyzing the nucleophilic attack of glutathione on electrophilic moieties of substrates. GSTs are a multigene family of phase II detoxification enzymes localized in the cytoplasm

The Human GST Multigene Family

genes	M.W.	chromosomal location	substrates
alpha (GSTA1-5)	26kDa	6p12	lipid peroxidation products (anticancer drugs)
mu (GSTM1-5)	26kDa	1p13	polycyclic aromatic hydrocarbons
omega (GSTO1,2)	28-31kDa	10q24	dehydroascorbate (reduction) monomethylarsonate (reduction)
pi (GSTP1)	23kDa	11q13	polycyclic aromatic hydrocarbons (anticancer drugs)
theta (GSTT1, 2)	27kDa	22q11	halomethanes polycyclic aromatic hydrocarbons
zeta (GSTZ1)	24kDa	14q24	dihaloacetic acids, fluoroacetic acids

↓

homodimer or heterodimer within the same class

Figure 13.6. Glutathione *S*-transferases (GST), chromosomal location, and their substrates. GSTα, GSTμ, GSTω, and GSTθ classes have multiple genes. GST proteins within the same class have approximately 50% identical amino acid sequences.

(also a small fraction in mitochondria) consisting of alpha (α), mu (μ), omega (ω), pi (π), theta (θ), and zeta (ζ) in mammalian cells (Figure 13.6). GSTα, GSTμ, GSTω, and GSTθ have multiple genes to encode subclasses of GSTA1-A5, GSTM1-M5, GSTO1-O2, and GSTT1-T2, respectively. GST proteins in the same class (e.g., GSTA1-A5) are approximately 50% identical in their sequences, and they are less than 30% identical to the GST sequences in other classes. The cytoplasmic GSTs form homo- or heterodimers in the same class and catalyze conjugation of glutathione to a wide variety of reactive metabolites for detoxification (environmental carcinogens such as aflatoxin B1 and benzo(a)pyrene, chemical drugs such as acetaminophen and cyclophosphamide, pesticides such as atrazine and alachlor, and endobiotics such as 4-hydroxy-2-nonenal), in which some substrates are preferentially detoxified by a particular GST class (Figure 13.7).

Among GST superfamily members, the rat GST A2 gene has been well-characterized for the mechanisms of gene regulation in response to xenobiotics. The transcription of the rat GST A2 gene is induced by polycyclic aromatic hydrocarbons (PAH such as dioxin, benzo[a]pyrene, β-naphthoflavone, and 3-methylcholanthrene) via a cis-acting enhancer element, termed the xenobiotic responsive element (XRE), to which a PAH-activated (PAH-bound) Ah-receptor binds and activates GST A2 transcription. Phenolic antioxidants such as butylated hydroxyanisole (BHA) and its active metabolite *t*-butylhydroquinone (*t*-BHQ) also induce GST A2 gene transcription via an antioxidant response element (ARE), to which members of the basic-leucine zipper transcription factor family including Nrf2 and Maf bind (see Section 13.3 and Figure 13.12). In addition to XRE- and ARE-mediated transcriptional regulation of GST A2, glucocorticoids such as dexamethasone ($>10^{-7}$M) induce GST A2 mRNA at the transcriptional level through activation of the pregnane X receptor (PXR), while lower concentrations of dexamethasone (below

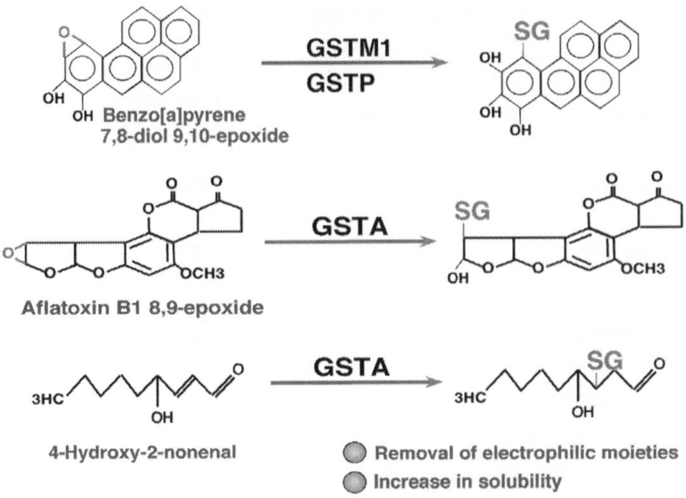

Figure 13.7. Glutathione conjugation to reactive metabolites by GSTs. Ultimate carcinogens such as benzo[*a*]pyrene 7,8-diol 9,10-epoxide and aflatoxin B1 8,9-epoxide are glutathione-conjugated for detoxification by GSTM1, GSTp, and GSTA, respectively. Lipid peroxidation product 4-hydroxy-2-nonenal is preferentially detoxified by GSTA.

10^{-7} M) repress GST A2 transcription through activation of the glucocorticoid receptor and subsequent binding to a glucocorticoid responsive element (GRE). The regulation of GST pi expression has also been well-studied since many different cancer cells have been shown to overexpress GST pi mRNA and protein without gene amplification. Elevated levels of GST pi are associated with tumor development of solid tumors (such as ovarian, testis, renal, liver, and colorectal cancers) in conjunction with cellular resistance mechanisms to chemotherapeutic drugs. In the rat and human GST pi gene, an ARE (antioxidant responsive element) located at 2500 bp and 70 bp upstream from the transcription initiation sites, respectively, plays a pivotal role in elevated expression of GST pi during hepatocarcinogenesis and basal expression (see Section 13.3 for ARE regulation).

GST polymorphisms have been identified in all six subclasses of the human GST genes; however, it remains largely inconclusive as to the influence of a particular single nucleotide polymorphic variation of the human GST gene on xenobiotic metabolism and susceptibility to cancer. However, it should be noted that homozygous deletion of GST-M1 was identified in 40–60% of the human population (approximately 50% of the Caucasian populations, 60% of Australians). The lack of GST M1 enzyme is associated with a moderate increase in the risk of lung, bladder, and colon cancers. Similar to the GST M1 gene deletion, homozygous deletion of the GST-T1 gene was identified in about 60% of Chinese and Korean populations and in about 20% in Caucasians and African-Americans. Substrates for the theta class of GSTs include dichloromethane and other halogenated compounds (Figure 13.6) such as ethylene dibromide and *p*-nitrobenzyl chloride, as well as various epoxides. Since some glutathione conjugates such as ethylene dibromide are more mutagenic (Figure 13.8), association between the GST T1 null phenotype and the risk of cancer appears to be complicated. Namely, the GST T1 null phenotype is

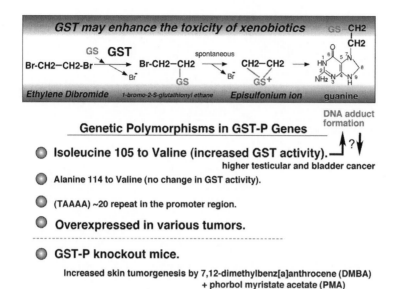

Figure 13.8. GST may enhance the toxicity of xenobiotics. Ethylene dibromide is a substrate of GST (theta and others) to be converted to 1-bromo-2-*S*-glutathionyl ethane, which facilitates spontaneous formation of the episulfonium ion. The episulfonium ion attacks the N7 position of guanine and causes DNA adduct formation. GST pi Ile105 to Val polymorphism, causing increased GSTpi activity, appears to correlate with higher testicular and bladder cancer. This may also be involved in GSTpi-mediated activation of unidentified xenobiotics. (Adapted from Henderson et al. *Proc. Natl. Acad. Sci. USA* **95**, 5275–5280, 1998.)

associated with an increased risk of some types of cancers (liver, head, and neck); but, on the other hand, the GST T1 positive phenotype may be related to an increased risk of cancer under conditions of chronic exposure to particular halogenated compounds. In addition to these GST T1 and M1 null phenotypes, a GST pi single nucleotide polymorphism for isoleucine 105 (105Ile) to valine substitution (105Val) shows higher susceptibility to testicular and bladder cancer; this may be due to increased GST activity with 105Val (Figure 13.8), although several published reports for the catalytic activity of GST pi between 105 Ile and 105Val appears to be controversial.

13.2.3 *N*-Acetyltransferases

N-Acetyltransferases (NATs) are phase II detoxification enzymes that catalyze acetyl CoA-dependent conjugation of an acetyl group ($-COCH_3$) to an amino group (*N*-acetylation) of aromatic amines ($R-NH_2$) and hydrazine compounds ($R-NH-NH_2$). NATs also catalyze *O*-acetylation of *N*-hydroxy aromatic amines produced by CYP-mediated monooxygenation (Figure 13.9). There are two functional NAT genes (NAT1 and NAT2) in humans, hamsters, and rabbits. Human NAT1 and NAT2 are encoded by two independent genes, both on chromosome 8p22, with 0.87-kb intronless coding regions, encoding cytosolic enzymes of 290 amino acids. Human NAT1 and NAT2 share high homology in their nucleotide sequences (about 90%

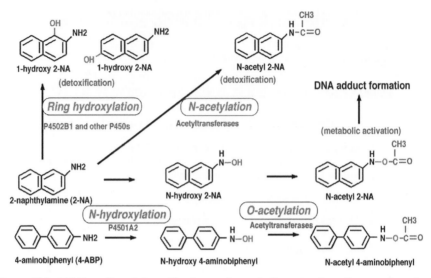

Figure 13.9. NAT-mediated detoxification and metabolic activation of aromatic amines.

identical) and amino acid sequences (about 80% identical). Expression levels of the NAT1 and NAT2 genes are regulated independently, resulting in different tissue distributions; NAT1 is ubiquitously expressed in various tissues, while NAT2 is predominantly expressed in liver and intestine. Information on molecular mechanisms of tissue-specific expression and xenobiotics-mediated regulation of NATs has been growing but is still limited. It was reported that the expression of human NAT1 is regulated by an AP1 binding sequence, to which AP1 (Jun and Fos) and YY1 transcription factors bind. Mouse NAT2 transcription is regulated by SP1 family transcription factors.

Genetic polymorphisms in NAT1 and NAT2 have been investigated relative to their associations with cancer risk since various environmental and industrial aromatic amine carcinogens are subject to NAT-mediated acetylation. In particular, the relationship between arylamine exposure to dye industry workers and a higher risk of bladder cancer has been recognized and investigated in conjunction with NAT2 genetic polymorphisms. More than two dozen human NAT1 and NAT2 polymorphisms have been identified, and the updated listing of NAT polymorphisms is available at http://www.louisville.edu/medschool/pharmacology/NAT.html. In particular, the association between NAT2 genetic polymorphisms and the risk of bladder cancer has been investigated because some urinary bladder carcinogens, including 2-naphtylamine (2-NA) and 4-aminobiphenyl (4-ABP), are preferred NAT2 substrates. Human NAT2*4 was defined as the wild-type NAT2 allele because of its fast acetylator phenotype and its frequency as one of the most common alleles. The allelic frequencies of NAT2*4 allele vary in different ethnic groups, in which Asians are highest (about 60% of Korean, Chinese, and Japanese) followed by 20–30% of Caucasians and 10–15% of Africans. Variant NAT2 alleles showing slow acetylator phenotype due to decreased catalytic activity and/or decreased protein stability are NAT2*5 (341T to C, Ile114 to Thr, 40–50% of Caucasians, 30–40% Africans), NAT2*6 (590G to A, Arg197 to Gln, 20–30% in almost all ethnic groups), NAT2*7

(857G to A, Gly286 to Glu, less than 10% in all ethnic groups), and NAT2*14 (191G to A, Arg64 to Gln, less than 10% in all ethnic groups).

NAT1 and NAT2 detoxify bicyclic aromatic amines (via *N*-acetylation) but not heterocyclic aromatic amines. On the other hand, NATs also produce aromatic amine reactive metabolites when *N*-hydroxyaromatic amines are first produced by cytochrome P450s. As shown in Figure 13.9, CYP1A2-mediated *N*-hydroxylation of 2-NA or 4-ABP and subsequent NAT2-mediated *O*-acetylation produce acetoxy intermediates, which bind to DNA and ultimately cause DNA mutations. The *O*-acetylation of these *N*-hydroxy bicyclic aromatic amines is catalyzed by both NAT1 and NAT2, while the *O*-acetylation of *N*-hydroxy hetrocyclic aromatic amines (such as 2-amino-1-methyl-6-phenylimidazo[4,5-b] pyridine generated from high-temperature cooking of red meats) are catalyzed by NAT2. Therefore, the susceptibility to aromatic amine-mediated carcinogenesis is determined by complex factors including type of aromatic amines and type of tissues expressing NAT1 or NAT2 in addition to *N*-hydroxylation activities by CYP1A2 and other monooxygenases. Even taking account of these complexities, individuals with NAT slow acetylator alleles such as NAT2*5 have been shown to be at higher risk of bladder cancer induced by aromatic amines because *N*-acetylation by NAT2 (also by NAT1) is a detoxification process of aromatic amines. Individuals with NAT2 fast acetylator alleles, on the other hand, appear to be at higher risk of colon cancer induced by heterocyclic aromatic amines because NATs do not efficiently *N*-acetylate heterocyclic aromatic amines for detoxification, but higher NAT2 activity induces *O*-acetylation of *N*-hydroxy heterocyclic aromatic amines, causing more DNA adduct formation.

13.2.4 Sulfotransferases

Sulfotransferases (SULTs) are cytosolic phase II detoxification enzymes involved in sulfonation of various xenobiotics and endobiotics. There are also membrane-bound SULTs that are not involved in phase II metabolism but are involved in the sulfonation of proteins and polysaccharides. Substrates of cytosolic SULTs include alcohols (ethanol, 2-butanol, cholesterol, bile acids), phenols (phenol, naphthol, acetaminophen), aromatic amines and hydroxyamines (2-naphthylamine, *N*-hydroxy 2-naphthylamine). SULTs transfer sulfonate (SO_3^-) to a hydroxy or amino group of a substrates from the cofactor 3'-phosphoadenosine-5'-phosphosulfate (PAPS), generating highly water-soluble metabolites for elimination through the kidney and liver.

A multigene family (SULT1–SULT5) of cytosolic SULTs has been identified in mammalian species, in which SULT1, SULT2, and SULT4 families have been identified in humans (see also Chapter 12). The human SULT1 family is composed of four subfamily members of SULT1A (further divided into 1A1, 1A2, 1A3, and 1A4, all on chromosome 16p11–12), SULT1B (1B1 on chromosome 4q11–13), SULT1C (divided into 1C2 and 1C4, both on chromosome 2q11), and SULT1E (1E1 on chromosome 4q13). The human SULT2 and SULT4 have SULT2A1 and SULT2B1 (both on chromosome 19q13) and SULT4A (on chromosome 22q13), respectively. SULT1 and SULT2, the major human SULTs, show functional differences; SULT1 enzymes are the phenol SULTs, whereas SULT2 enzymes are the alcohol SULTs. Among those SULT enzymes, SULT1A1 is highly expressed in human liver and

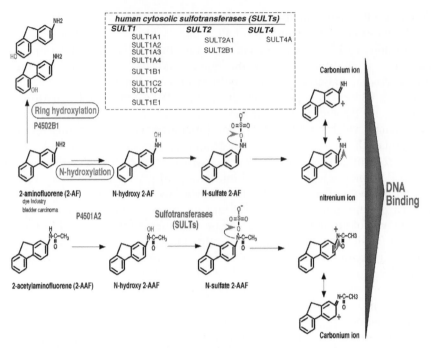

Figure 13.10. NAT-mediated metabolic activation of heterocyclic aromatic amines. 2-AF is not a NAT substrate for *N*-acetylation (compare with 2-NA, 4-ABP *N*-acetylation in Figure 13.9). 2-AF is detoxified by P4502B1 ring hydroxylation. 2-AF is also metabolized via P4501A2-mediated *N*-hydroxylation. The resultant *N*-hydroxy 2-AF is a good substrate of conjugation reactions by NATs as well as by UGTs and SULTs and is converted to reactive metabolites for DNA adduct formation.

catalyzes sulfonation of a variety of phenolic xeno- and endobiotics such as acetaminophen, 4-hydroxytamoxifen, iodothyronines, and 2-methoxyestradiol for elimination. In the manner similar to that of UGTs and NATs, SULT1A1 and other SULT subfamily members mediate bioactivation of *N*-hydroxylated aromatic and heterocyclic amines (Figure 13.10). Therefore, the influence of individual variations in SULT enzyme activities on the risk of particular cancers and susceptibility to particular xenobiotics is an important subject which needs to be examined. It was reported that human SULT1A1 transcription is regulated by SP1 and Ets transcription factors through proximal enhancer sequences to the transcription initiation site. Although polymorphisms in this enhancer region altering SULT1A1 transcription have not been identified, several polymorphisms in coding and noncoding regions that determine ethnic and inherited individual variations in SULT1A1 activity have been reported. In particular, the influence of single amino acid alteration between SULT1A1*1 (Arg213, 2/3 of Caucasians in the United States) and SULT1A1*2 (His213, 1/3 of Caucasians in the United States) alleles on the efficacy of tamoxifen treatment in breast cancer patients has been elucidated. The SULT1A1*1 (Arg213) enzyme generally shows much higher SULT activity than does SULT1A1*2 (His213) on various substrates. Tamoxifen is a competitive ligand of estradiol for the estrogen receptor and is activated to 4-hydroxytamoxifen by cytochrome P450.

4-Hydroxytamoxifen is further metabolized for detoxification via SULT1A1-mediated sulfonation; therefore, higher SULT 1A1 activity would result in rapid clearance of 4-hydroxytamoxifen, which might cause less antitumor activity of tamoxifen in patients with the SULT1A1*1 allele. Results reported so far, however, do not support this prediction but rather that patients with SULT1A1*1 allele (Arg213) show better breast cancer survival after tamoxifen treatment than do those with SULT1A1*2 (His213). Although the mechanisms behind this outcome are not clear, it might suggest that sulfonated 4-hydroxytamoxifen is another activated form of tamoxifen that may exhibit cytotoxicity to breast cancer cells. Several human SULT1A2 genomic variants have also been identified, in which the most prominent allele showing decreased SULT1A2 catalytic activity is SULT1A2*2 with two amino acid alterations from Ile to Thr (at the 7th amino acid) and from Asn to Thr (at the 235th amino acid). The SULT1A2*1 (Ile at 7th, Asn at 235th) allele was identified in approximately 50% of Caucasians in the United States, while the SULT1A2*2 allele was seen in about 30% of Caucasians in the Untied States. The bioactivation of N-hydroxylated xenobiotics such as N-hydroxy-2-acetylaminofluorene by SULT1A2*1 is 5- to 10-fold higher than SULT1A2*2 (the mutagenic activity of these enzymes was measured when they were expressed in *Salmonella typhimurium*). The relationship between SULT1A2 allelic variations and risk of cancer remains to be evaluated.

The SULT2 family are the alcohol sulfotransferases, among which SULT2A1 is the major enzyme to catalyze sulfonation of bile acids and steroids. Human SULT2A1 single nucleotide polymorphisms were identified, though rare in G187C and G781A, causing amino acid alterations from Ala 63 to Pro and from Ala261 to Thr, respectively. These changes of amino acids decrease SULT2A1 catalytic activity. In addition to these, SULT2A1 Met57 to Thr and Glu186 to Val have also been identified; however, the influence of 2A1 polymorphisms in steroid metabolism and cancer risks has not been fully elucidated.

13.3 THE ANTIOXIDANT RESPONSIVE ELEMENT AND PHASE II GENE REGULATION

The induction of antioxidant/detoxification proteins in response to chemical carcinogens, xenobiotics, and oxidants is a vital step for elimination of harmful chemicals from the body. Phase II enzymes are a group of detoxication proteins, including UDP-glucuronosyltransferase (UGT), N-acetyltransferase (NAT), sulfotransferase (SULT), glutathione S-transferases (GST), quinone oxidoreductases (NQO), and antioxidative proteins such as γ-glutamylcysteine synthetase, heme oxygenase, metallothionein, and ferritin. These proteins detoxify reactive phase I metabolites either by catalyzing the conjugation of their active functional groups with endogenous ligands, thereby facilitating their excretion, or by destruction of active centers of the compounds by chemical reactions such as reduction and hydrolysis. Thus, induction of phase II enzymes is an essential cellular defense mechanism against carcinogens and chemical stress. This is supported by the evidence that expression levels of these phase II gene products determine cellular sensitivity to oxidative stress and several chemotherapeutic drugs as well. For instance, increased GST-pi expression was observed in various types of human tumors, which is frequently associated with more aggressive cell growth, resistance to chemotherapeutic drugs (such as the DNA alkylating agent chlorambucil), and patient's poorer prognosis. Another important issue to be noted is that chemoprevention, a strategy to retard or reverse carcinogenesis (and other diseases caused by chemical and oxidative stress such as cardiovascular and neurodegenerative diseases), has been investigated to develop an approach of utilizing induction of phase II antioxidant/detoxification genes by dietary or pharmacological agents. For this goal, it is essential to understand the molecular mechanisms by which the expression of a battery of phase II genes are regulated.

The regulation of many phase II enzymes, induced at the transcriptional level by xenobiotics and oxidative stress conditions, is distinctly regulated from phase I enzymes. They are induced by either (a) monofunctional compounds (which induce only phase II enzymes) such as *tert*-butylhydroquinone (t-BHQ) and H_2O_2 or (b) bifunctional compounds (which induce both phase I and II enzymes) such as β-naphthoflavone (β-NF). A cis-acting element responsible for transcriptional activation of phase II enzymes by these compounds has been identified in the promoter regions of the genes encoding NAD(P)H quinone reductase, glutathione S-transferases (GST), γ-glutamylcysteine synthetase, and heme oxygenase (Figure 13.11). In addition, genes encoding such metal-binding proteins as metallothionein and the H and L subunits of the major intracellular iron storage protein, ferritin, have been found to contain the similar cis-acting elements (Figure 13.11). The cis-acting element is termed the antioxidant responsive element (ARE) or electrophile responsive

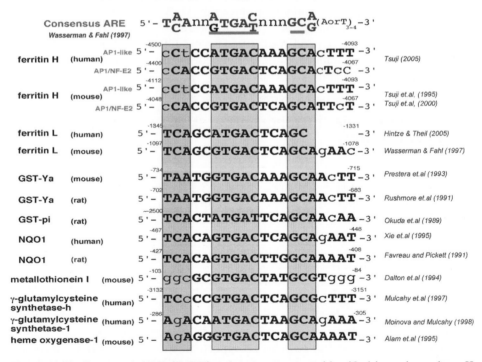

Figure 13.11. Conserved ARE (EpRE) enhancer sequences identified in various phase II genes. The negative numbers shown on the 5′ and 3′ positions of each sequence are nucleotide positions from a transcription initiation site of each gene.

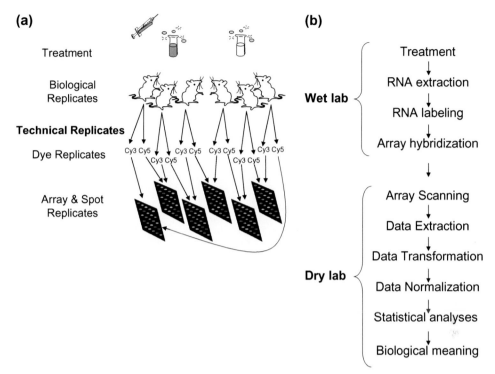

Figure 3.3. Schematic of a microarray experiment. (A) Experimental design incorporating both biological and technical replication. There are three treated mice and three control mice, providing biological replication. RNA from each mouse is labeled with each dye and hybridized more than once, providing technical replication. The hybridization design uses a loop design as shown in Figure 3.2. (B) Outline of the wet and dry lab steps involved in a microarray experiment.

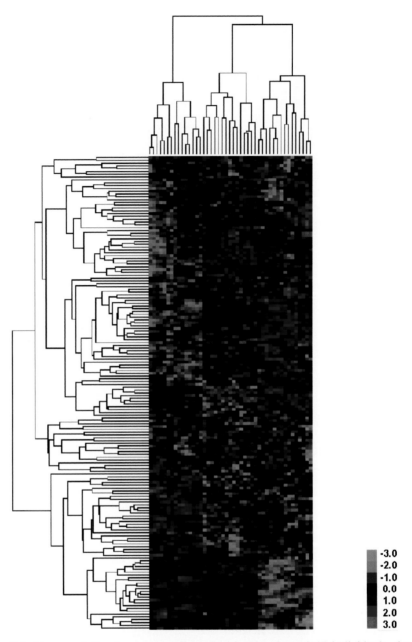

Figure 3.5. Representative heat map. Hierarchical clustering of 45 individuals clustered across the top and 169 metabolic genes clustered along the side. Notice that the 45 individuals form three groups of 15 individuals each. Each group of 15 individuals shares similar patterns of gene expression that are different among the three groups. It is the shared patterns that can be used to diagnose exposure and disease or classify cancers. The relative expression levels range from −3 to +3 as shown by the scale bar.

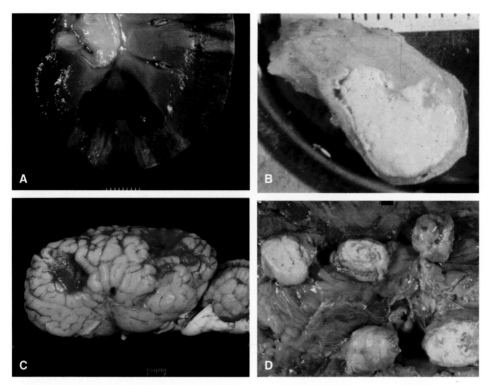

Figure 16.6. The four primary types of necrosis. (A) Coagulative. On the cut surface of this canine kidney, there is a dark red, wedge-shaped area of coagulative necrosis caused by blockage of the blood supply to the area (infarction). The pale areas are older infarcts. (B) Caseous. This bovine lymph node contains a large whitish, cheese-like area of caseous necrosis characterized by loss of the normal tissue architecture. (C) Liquefactive necrosis, equine brain. There are two large areas of liquefaction with extensive loss of brain tissue. (D) Fat necrosis, bovine abdominal fat. Necrotic fat is firm and chalky white and often becomes mineralized.

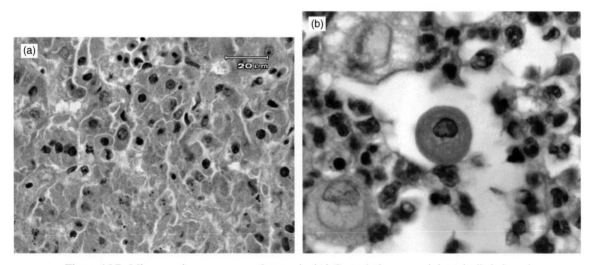

Figure 16.7. Microscopic appearance of necrosis. (A) Coagulative necrosis in a virally infected avian liver. Hepatocytes in the lower half of the photo are in various stages of necrosis, with small, pyknotic or fragmented nuclei and increased cytoplasmic eosinophilia. (B) Necrotic cells in immune-mediated skin disease, canine. The central cell has a pyknotic nucleus and intensely eosinophilic cytoplasm, while the cells at lower left and upper left are injured and swollen. The smaller cells are neutrophils.

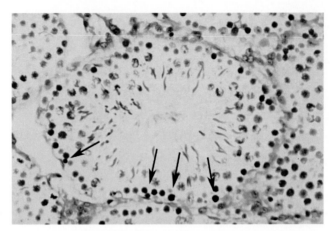

Figure 16.12. TUNEL assay used to detect apoptotic cells within a seminiferous tubule. There are many ways of detecting apoptosis at different stages on histological sections. The most commonly used method is called TUNEL (terminal deoxynucleotidyl transferase biotin-dUTP nick end-labeling). One of the major characteristics of apoptosis is the degradation of DNA after the activation of Ca/Mg-dependent endonucleases. This DNA cleavage leads to strand breaks within the DNA. The TUNEL method identifies apoptotic cells in situ by using terminal deoxynucleotidyl transferase (TdT) to transfer biotin-dUTP to these strand breaks of cleaved DNA. The biotin-labeled cleavage sites are then detected by reaction with HRP (horseradish peroxidase) conjugated streptavidin and visualized by DAB (diaminobenzidine), which gives a brown color. The tissue is counterstained with Toluidine blue to allow evaluation of tissue architecture. Figure 16.12. Photomicrograph of a section of testes from a B6C3F1 mouse that was in a National Toxicology Program bioassay. The TUNEL assay was used on this section of tissue to detect apoptotic cells. The seminiferous tubule is cut in cross section and has sperm, spermatogonia, and spermatocytes in various stages of development. The arrows point to labeled apoptotic cells in the basal layer of the seminiferous epithelium.

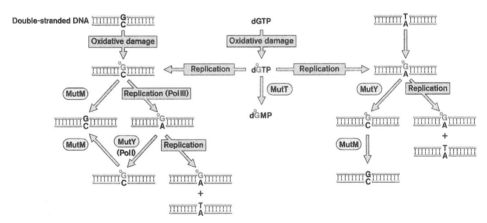

Figure 23.6. The GO system protects cells from the mutagenic effects of 8-oxoG. Exposure to oxidative damaging agents can produce 8-oxoG in DNA. This is removed by the glycosylase activity of MutM (Fpg). If 8-oxoG is present in DNA and encountered by the DNA replication machinery, A can be misincorporated opposite the 8-oxoG, creating an 8-oxoG:A mispair. This mispair can be acted upon by MutY, which removes the A in the 8-oxoG:A mispair which is then usually replaced with a C during base excision repair. This provides MutM with another opportunity to detect an 8-oxoG:C pair and remove the 8-oxoG, which is then replaced with normal guanine. A third enzyme in this pathway is MutT, which serves to sanitize the deoxynucleotide triphosphate pools by converting 8-oxoGTP to 8-oxoGMP. This prevents the direct incorporation of 8-oxoG into DNA during DNA synthesis. (Reproduced with permission from Cunningham, R. P. DNA repair: Caretakers of the genome? *Curr. Biol.* **7**(9), R576–R579, 1997.)

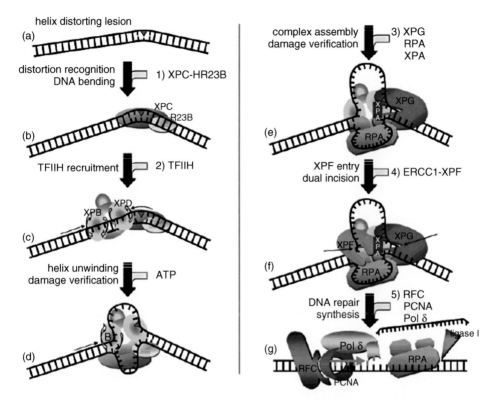

Figure 23.9. The nucleotide excision repair pathway in mammalian cells. A lesion is formed in DNA that usually creates a significant distortion. XPC-RAD23B detects the distortion and binds. This complex loads TFIIH which translocates and unwinds the helix using its helicase activities until one of the subunits encounters a chemically modified base. The other subunit continues unwinding the DNA to create a larger opened bubble structure. XPG, RPA and XPA are then recruited and the presence of a *bona fide* lesion is verified. Upon binding of ERCC1-XPF, dual incisions are made by the junction-specific endonucleases. XPG catalyzes the 3′ incision and ERCC1-XPF catalyzes the 5′ incision. RPA remains bound to the single stranded DNA and facilitates the transition to repair synthesis and ligation. (Reproduced with permission from Gillet, L. C. J., and Scharer, O. D. Molecular mechanisms of mammalian global genome nucleotide excision repair. *Chem. Rev.* **106**, 253–276, 2006.)

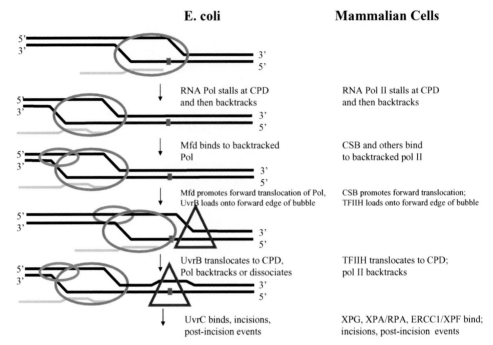

Figure 23.10. A model for TC-NER of CPDs. *E. coli*: Elongating RNA polymerase complex (orange oval) stalls at a CPD (small red square) in the transcribed strand. The polymerase complex, transcription bubble, and nascent RNA (blue line) translocate backwards. Mfd (lavender oval) binds backtracked polymerase and DNA upstream of the bubble. Mfd promotes the forward translocation of the polymerase complex. UvrB (blue triangle) binds 5′ (relative to the damaged strand), loads onto the forward edge of the bubble, and translocates to the lesion. The polymerase complex backtracks or is dissociated by Mfd. Subsequent NER processing events continue as they would in nontranscribed DNA. Mammalian cells: Same as for *E. coli* except for the following. CSB (lavender oval) binds the backtracked polymerase complex and promotes forward translocation. TFIIH (blue triangle) binds 5′ to the damage and loads onto the forward edge of the bubble. Subsequent NER events continue as they would in nontranscribed DNA. (Reproduced with permission from Mellon, I. Transcription-coupled repair: A complex affair. *Mutation Res.* **577**, 155–161, 2005.)

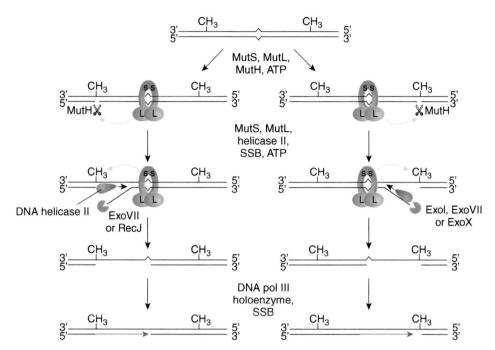

Figure 23.12. Bidirectional mismatch repair in *E. coli*. MutS binds to mismatches formed by errors made during replication. MutL binds to the MutS-bound complex. MutH incises the newly synthesized strand next to the unmethylated GATC sequence. The green arrows indicate signally between the two sites of activity. DNA helicase II unwinds the DNA and exonucleases excise the DNA from the point of incision up to and including the incorrect base. DNA Pol III synthesizes new DNA to replace the excised DNA. (Reproduced with permission from Iyer, R. R., Pluciennik, A., Burdett, V., and Modrich, P. L. DNA mismatch repair: Functions and mechanism. *Chem. Rev.* **106**, 302–323, 2006.)

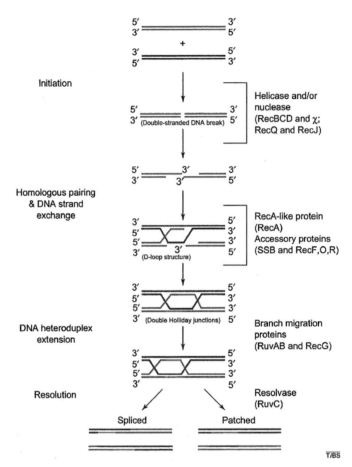

Figure 23.14. A general model for double-strand break repair using homologous recombination. The text in parentheses to the right indicates proteins that participate in each step in *E. coli*. The light blue lines indicate newly synthesized DNA. (Reproduced with permission from Kowalczykowski, S.C. Initiation of genetic recombination and recombination-dependent replication. *Trends Biochem. Sci.* **25**:156–165, 2000.)

element (EpRE). Rushmore-Pickett's group [T.H. Rushmore, M.R. Morton and CB Pickett J. Biol. Chem., *266*, 11632–11639 (1991)] and Wasserman-Fahl's group [W.W. Wasserman and W.E. Fahl Proc. Natl. Acad. Sci. USA *94*, 5361–5366 (1997)] carried out functional analysis of the ARE sequence by introducing various point mutations and identified a minimum core sequence (<u>TGACnnnGC</u>) and a consensus sequence (TA/CAnnA/G<u>TGAC</u>/TnnnGCA/G), respectively (Figure 13.11). This sequence is similar but not identical to an AP1 binding sequence (TGAC/GTCA).

Since the ARE sequence has a core AP1-like motif, it is likely that members of the Jun and Fos transcription factor families (the AP1 family) may bind to the ARE and activate the transcription of phase II genes in response to xenobiotics. However, initial studies showed that the increase in protein binding to the ARE sequences following xenobiotic treatment was not consistently observed and that transcription factors distinct from Jun and Fos family members have been demonstrated to be involved in activation of ARE. Jaiswal and his colleague [R. Venugopal and A.K. Jaiswal Proc. Natl. Acad Sci. USA *93*, 14960–14965 (1996)] initially demonstrated that Nrf1 and Nrf2 (NFE2-related factor 1 and 2) regulate ARE-mediated regulation of the NAD(P)H:quinone reductase gene. Recent studies, including analysis of knockout mice conducted by Yamamoto and his colleagues [K. Itoh et al. Biochem. Biophys. Res. Commun *236*, 313–322 (1997), K Itoh et al. Genes & Dev. *93*, 76–86 (1999)] indicated that Nrf2 is the major transcription factor responsible for transcriptional activation of many ARE-regulated phase II genes in response to xenobiotics. Namely, homozygous Nrf2 knockout mice showed significantly impaired expression of phase II genes (such as GST and NQO1) in response to phenolic antioxidants and electrophilic chemicals. These mice also showed enhanced susceptibility to xenobiotics due to the lack of Nrf2-mediated induction of phase II detoxification genes. Nrf2 is a member of the cap "n" collar (CNC) family of b-zip transcription factors that also includes Nrf1, Nrf3, and NF-E2 (Figure 13.12). Nrf2

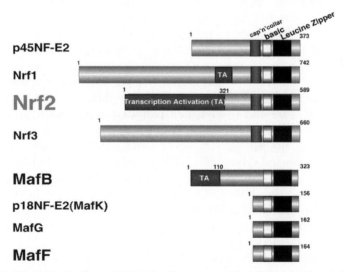

Figure 13.12. The Nrf family and Maf family transcription factors. The basic domain is involved in DNA binding and the leucine zipper domain is a dimerization module with other basic-leucine zipper transcription factors. Small Maf proteins such as MafK and MafG heterodimerize with Nrf2 for binding and activation of the ARE.

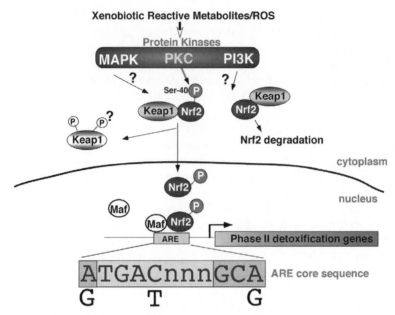

Figure 13.13. ARE activation by Nrf2. Nrf2 associates with Keap1, a repressor protein of Nrf2, in the cytoplasm. Upon xenobiotic exposure to cells, upstream kinase pathways are activated and induce dissociation of Nrf2 from Keap1. The liberated Nrf2 translocates into nucleus and dimerize with a small Maf protein and bind to an ARE enhancer sequence identified in various phase II genes.

and other CNC transcription factors form a heterodimer with a small Maf protein such as MafG or MafK, another b-zip transcription factor member, for activation of an ARE. The activation mechanisms of an ARE via Nrf2 have not been fully elucidated; however, Nrf2 appears to stay in the cytoplasm or is degraded by forming a complex with the actin binding protein Keap1 prior to xenobiotic exposure (Figure 13.13). Perhaps via a similar regulatory mechanism of NFkB activation, xenobiotic inducers of phase II genes cause dissociation of Nrf2 from Keap1, leading to nuclear localization of free Nrf2 and Nrf2/small Maf heterodimer binding to AREs of the phase II genes (Figure 13.13). Signaling pathways responsible for activation of ARE-mediated transcription of phase II genes appear to be complex and to be specific to each xenobiotic inducer; however, t-BHQ-mediated induction of the Nrf2 nuclear translocation and subsequent ARE activation utilize PI3 kinase and MAP kinase pathways, although molecular mechanisms to link these pathways to Nrf2 activation are elusive. In addition, it was demonstrated that protein kinase C-mediated phosphorylation of Nrf2 at Ser-40 in response to t-BHQ facilitates dissociation of Nrf2 from Keap1, which is important for transcriptional activation of the phase II gene via an ARE (Figure 13.13).

Chemoprevention is a strategy used to block or suppress carcinogenesis and degenerative disease using dietary or pharmacological compounds. These chemopreventive compounds include phenols [such as *tert*-butyl 4-hydroxyanisole (BHA), t-BHQ (active metabolite of BHA), ethoxyquine, curcumin (from turmeric) and resveratrol (from grapes)] and sulfur compounds [such as oltiplaz, sulforaphane

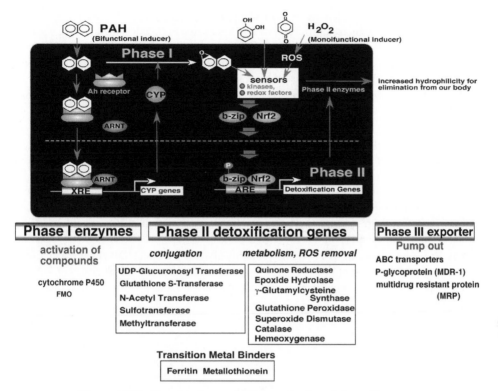

Figure 13.14. Xenobiotic metabolism and roles of phase II genes.

(from broccoli), diallyl sulfide (from garlic)], many of which induce expression of a battery of phase II detoxification genes via an ARE through activation of Nrf2 (or inactivation of the cytoplasmic inhibitor Keap1) and related family members of transcription factors. In fact, it was demonstrated that Nrf2-deficient mice showed (1) decreased expression of constitutive GST and NQO1 levels, (2) impaired inducibility of these phase II genes by a chemopreventive compound, oltiplaz, and (3) increased carcinogenesis (gastric neoplasia formation) after benzo(*a*)pyrene treatment compared with wild-type mice. Furthermore, oltiplaz was chemopreventive against benzo(*a*)pyrene-mediated carcinogenesis in wild-type mice but not in Nrf2-deficient mice. These results suggest that Nrf2 is a crucial transcription factor for suppression of chemical carcinogenesis as well as chemoprevention.

13.4 SUMMARY

Figure 13.14 is a brief summary of the material covered in this chapter.

SUGGESTED READING

Boukouvala, S., and Fakis, G. Arylamine *N*-acetyltransferases: What we learn from genes and genomes. *Drug Metab. Rev.* **37**, 511–564, 2005.

Gamage, N., Barnett, A., Hempel, N., Duggleby, R. G., Windmill, K. F., Martin, J. L., and McManus, M. E. Human sulfotransferases and their role in chemical metabolism. *Toxicol. Sci.* **90**, 5–22, 2006.

Hein, D. W. *N*-Acetyltransferase 2 genetic polymorphism: Effects of carcinogen and haplotype on urinary bladder cancer risk. *Oncogene* **25**, 1649–1658, 2006.

McIlwain, C. C., Townsend, D. M., and Tew, K. D. Glutathione *S*-transferase polymorphisms: cancer incidence and therapy. *Oncogene* **25**, 1639–1648, 2006.

Motohashi, H., and Yamamoto, M. Nrf2-Keap1 defines a physiologically important stress response mechanism. *Trends Mol. Med.* **10**, 549–557, 2004.

Nagar, S., and Remmel, R. P. Uridine diphosphoglucuronosyltransferase pharmacogenetics and cancer. *Oncogene* **25**, 1659–1672, 2006.

Nowell, S., and Falany, C. N. Pharmacogenetics of human cytosolic sulfotransferases. *Oncogene* **25**, 1673–1678, 2006.

Tukey, R. H., and Strassburg, C. P. Human UDP-glucuronosyltransferases: Metabolism, expression, and disease. *Annu. Rev. Pharmacol. Toxicol.* **40**, 581–616, 2000.

Tukey, R. H., and Strassburg, C. P. Genetic multiplicity of the human UDP-glucuronosyltransferases and regulation in the gastrointestinal tract. *Mol. Pharmacol.* **59**, 405–414, 2001.

Wolf, C. R. Chemoprevention: Increased potential to bear fruit. *Proc. Natl. Acad. Sci. USA* **98**, 2941–2943, 2001.

Developmental Effects on Xenobiotic Metabolism

MARTIN J. J. RONIS

Department of Pharmacology and Toxicology, Arkansas Children's Nutrition Center, University of Arkansas for Medical Sciences, Little Rock, Arkansas, 72202

HELEN C. CUNNY

National Institute of Environmental Health Sciences, Research Triangle Park, North Carolina 27709

14.1 INTRODUCTION

Xenobiotics are biotransformed by phase I enzymes and phase II conjugation reactions to form a variety of metabolites that are generally more water-soluble and less toxic than the parent compound. Occasionally, the enzymic action of phase I or II systems leads to the formation of unstable intermediates or reactive metabolites that are toxic or carcinogenic. Many physiological factors influence the rate of xenobiotic metabolism and the relative importance of different pathways of metabolic activation or detoxication.

Phase I and II enzyme systems exist as large gene families of enzymes, many of which have some degree of overlapping substrate specificity. It has become increasingly clear in recent years that in addition to the disposal of endogenous waste products such as bilirubin and detoxication of dietary and environmental xenobiotics, these enzymes originally evolved to regulate endogenous processes. These include steroid biosynthesis and degradation; fatty acid and vitamin metabolism, which control ligand availability to a large family of nuclear receptors and other transcription factors. Activation of these transcription factors, in turn, control patterns of gene transcription important in development, in homeostatic control of energy balance, in intermediary metabolism, and in other physiologic and biochemical processes in the adult. Therefore, it is not too surprising that these enzymes are under close temporal and spacial regulation during development in addition to being inducible by exogenous chemical agents.

In vivo metabolism and clearance of xenobiotics is the sum of many different processes including (1) the action of many different phase I and II enzymes in the

Molecular and Biochemical Toxicology, Fourth Edition, edited by Robert C. Smart and Ernest Hodgson
Copyright © 2008 John Wiley & Sons, Inc.

liver and extrahepatic tissues and (2) xenobiotic transport systems leading to excretion in urine or bile (often termed phase III). Modern biochemical and molecular biological techniques have revealed striking differences in developmental and hormonal regulation of phase I and II enzymes and phase III transporters in different tissues. In addition, xenobiotic metabolism is affected by alterations in physiology and pathology (such as found in pregnancy and disease states) and undergoes circadian, seasonal, and other types of cycling. In addition to changes in protein expression of phase I and II enzymes, rates of xenobiotic metabolism are also affected by the levels of essential cofactors such as NADPH, GSH, PAPS, and UDPGA, which may also be altered during development and by pathological conditions.

In the sections below, the ontogeny of xenobiotic metabolism is considered. This is followed by sections on the effects of pathology associated with aging, on the effects of pregnancy, and on physiological cycles. Finally, information is included on development of xenobiotic metabolism in nonmammalian species.

14.2 XENOBIOTIC METABOLISM DURING DEVELOPMENT

In general, the ability to metabolize xenobiotics is low or absent in the fetus and neonate, develops rapidly after birth, is at its highest in early adulthood, and declines in old age. The development of individual phase I and II enzymes is under complex ontogenic control. Most xenobiotic-metabolizing enzymes do not appear gradually during development. Instead clusters of different enzymes appear to develop during critical periods in an organism's life such as birth, weaning and puberty. The timely expression of these enzymes complements the physiological needs of the developing organism in its changing environment (see Table 14.1).

TABLE 14.1. Ontogeny of Human Hepatic Phase I Enzymes

Gene	Prenatal Trimester 1	2	3	Neonate	1 mo. to 1 yr	1–18 yr	Adult	Old Age
CYP1A1	++	+	+/–	–	–	–	–	–
CYP1B1	+/–	+/–	?	?	–	–	–	–
CYP1A2	–	–	–	–	+/–	++	++	?
CYP2A	–	–	–	–	+	+	+	?
CYP2B6	–	–	–	?	+	+	+	?
CYP2C9	–	–	–	+/–	+	++	++	?
CYP2E1	–	+/–	+/–	+	+	++	++	?
CYP2R1	+	–	–	?	?	?	++	?
CYP2S1	+/–	+/–	+/–	?	?	?	+/–	?
CYP3A7	++	++	++	+	–	–	–	–
CYP3A4/5	–	–	–	+	++	++	++	+++ female, + male
FMO1	++	++	+	–	–	–	–	–
FMO3	–	–	–	+/–	+	++	++	?
ADH1	+	+	+	+	+	+	+	?
ADH2	–	+	+	+	+	+	+	?
ADH3	–	–	+	+	+	+	+	?

+/– Activity or protein or mRNA detectible in a proportion of samples; +/++/+++ increasing levels of activity, protein, and mRNA; – undetectable; ? not known with certainty.

14.2.1 Embryogenesis

Before birth, metabolic capacities for handling xenobiotics are low, largely because the mother's metabolism provides fetal detoxication reactions. However, with the exception of CYP1A2, orthologs of all cytochrome P450 enzymes in gene families 1–4 have been detected at low levels in human and rodent fetal tissues and demonstrate striking differences in both tissue distribution and developmental profile during embryogenesis. Cytochrome P450-dependent monooxygenase activities appear to be essential for normal embryonic development since cytochrome P450 reductase knockout mice embryos are not viable and die on gestational day 13.5 (E13.5). When mouse embryogenesis is split into four stages—gastrulation (E7), neural patterning and somitogenesis (E11), organogenesis (E15), and the fetal period (E17)—overall P450 expression falls into four different categories. Some P450 enzymes are only highly expressed at one embryonic stage. These include CYP2R1, which is only expressed very early (E7), and CYP2E1, which is only detected late in embryogenesis (E17). Some, such as CYP1B1 and CYP4B1, appear at E11 and are expressed at similar levels in the fetus thereafter. Others, such as CYP2S1, are present at similar levels at all embryonic stages, while CYP1A2 is not present at any stage. The precise temporal patterns of individual P450 enzyme expression during embryogenesis is highly suggestive that they play specific roles in normal embryogenesis unrelated to their ability to metabolize xenobiotics. The major human embryonic hepatic cytochrome P450 is CYP3A7, which probably has an important physiological role in steroid degradation within the fetus. Other toxicologically important cytochrome P450s expressed in the fetus are CYP1A1, CYP1B1, and CYP2E1. These are involved in pro-carcinogen metabolism and clearance and in the case of CYP2E1 may play a role in fetal alcohol effects. In contrast to what occurs in adult tissues, mouse and human CYP1A1 are constitutively expressed in early stages of embryonic development. In the mouse, expression occurs in the ectoderm as early as E7 and during organogenesis is strongly expressed in heart and hindbrain, kidney, and liver. However, constitutive expression is then repressed during later development. CYP1A1 has been reported to be readily inducible in both the placenta and in those fetal tissues in which it is expressed following treatment with AhR ligands such as TCDD, PCBs, and polycyclic aromatic hydrocarbons. However, whereas the ability of xenobiotics to induce CYP1A1 expression has been reported to be reduced in embryos from mice expressing a lower-affinity form of this receptor, constitutive expression was unaffected, suggesting that developmental regulation of this enzyme is, at least to some degree, AhR-independent.

CYP1B1 is also expressed in the fetus in multiple tissues, particularly in thymus, spleen, kidney, and adrenal. A null CYP1B1 phenotype in humans has been associated with appearance of primary congenital glaucoma. Consistent with an important endogenous role for this enzyme in development of the eye, CYP1B1—which, in addition to biotransformation of xenobiotics, is also capable of metabolizing retinoid and sex steroids—is expressed in embryonic ocular tissues.

CYP2E1 has been detected at low levels in a proportion of human fetal livers late during gestation in the second and third trimesters and has been reported at somewhat higher levels in the fetal mouse brain. Other phase I enzymes are also expressed in the embryo. In humans, the flavin containing monoxygenase FMO1 is expressed at high levels during 8–15 weeks of gestation and substantially declines during later periods of fetal development. In contrast, in another important family of phase I enzymes, alcohol dehydrogenase I appears early in fetal liver development with the appearance of ADH 2 and 3 later in gestation.

Less is known about expression of phase II enzymes during fetal development. However, similar to the cytochrome P450 enzymes, it appears that their tissue-specificity and temporal expression are also precisely regulated during embryogenesis. The sulfotransferases SULT1A1 and SULT2A1, which sulfate phenols and hydroxylated steroids, respectively, are readily detectible in human fetal livers. Whereas SULT1A1 expression remains fairly constant, SULT2A1 expression increases significantly during the third trimester. In contrast, SULT1E1, the major enzyme responsible for estrogen metabolism/inactivation, is expressed at the highest levels during early gestation in male liver (possibly to protect the developing male fetus against the feminizing effects of estrogens).

Glutathione S-transferases (GSTs) of at least three classes A, M, and P have been detected in human fetal liver. GSTA expression is similar to adults, GSTM is expressed at low levels prior to birth, and GSTP is expressed at its highest levels in the early fetal period. Several UGT enzymes are also expressed in human fetal liver, including UGT1A3, UGT1A6 (which is the principal catalyst of acetaminophen glucuronidation), and UGT2B7 (which glucuronidates morphine). However, other important UGTs such as UGT1A1, the most active enzyme toward bilirubin, are absent from fetal liver. In contrast, other phase II enzymes such as epoxide hydrolases and N-acetyltransferases appear to be expressed early in fetal liver and increase slowly with age.

14.2.2 Xenobiotic Metabolism in the Neonate

Parturition represents a major physiological shift during development. The organism moves from nutrition via the placenta and ingestion of amniotic fluid to oral feeding of colostrum followed by breast milk or formula with consequent gastrointestinal maturation in response to diet. In addition, lung maturation and the shift to breathing represent another major source of exposure to environmental influences. Parturition also results in a major biochemical shift in the way the body handles xenobiotics to adapt to this new physiological environment. A major suite of P450 enzymes, particularly in family 2, are expressed at much higher levels immediately after birth than in the fetus. In the rat, overall P450 concentration in neonatal liver increases from less than 10% to 50% of adult levels within a few days of birth. One of the P450 enzymes that is regulated in this fashion is CYP2E1, and it is one of the few for which this process has been studied at the molecular level. Increased hepatic CYP2E1 expression at birth involves gene transcription. This appears to be associated with the binding of a liver-specific transcription factor HNF1α to a response element between -100 and -120 base pairs (bp) in the promoter region of the CYP2E1 gene upstream of the transcription start site. The process that initiates this response is still unknown, but appears to involve alterations in methylation status of the CYP2E1 promoter. In addition to P450 enzymes, many phase II enzymes are also expressed at markedly higher levels soon after birth. These include (a) the glutathione S-transferase M class with expression increased five-fold to adult levels and (b) the UDPGT enzyme UGT1A1 responsible for the glucuronidation of bilirubin. This latter process is important clinically for protection against the development of neonatal jaundice. Other UDGTs such as UGT2B17, which glucuronidates androgenic steroids, are also turned on at parturition. In contrast to increased expression, a number of enzymes expressed in human fetal tissues disappear in late gestation or immediately after birth. These include hepatic CYP1A1, CYP3A7, FMO1, and GSTP. In some cases such as CYP1A1 and GSTP, catalytic function is not replaced, but in the case of CYP3A7 and FMO1 a temporal switch occurs in the postnatal period. At birth, CYP3A7 rapidly disappears and is replaced by rapid increases in the major adult CYP3A enzyme CYP3A4 such that overall CYP3A expression remains relatively constant. Similarly, FMO1 disappears and is replaced over a somewhat longer period with the adult liver form FMO3. During the neonatal period and in early childhood, human infants are FMO-deficient, and regulation of FMO1 does not appear to be linked to that of FMO3. Given the differing ability of FMO1 and FMO3 to metabolize trimethylamine to its N-oxide, these data may explain isolated case reports of trimethylaminuria in neonates which resolve with age. Different sets of transcription factors have been implicated in control of the CYP3A7/CYP3A4, and FMO1/FMO3 temporal switches. The CCAAT/enhancer binding protein C/EBPα and D-element binding protein (DBP) have been suggested to regulate CYP3A7/CYP3A4, while it has been shown that HNF1α and HNF4α regulate FMO1 expression.

14.2.2a   *Effects of Infant Diet (Breast Versus Formula Feeding).* Xenobiotic metabolism is highly variable during the first year in human neonates. Part of this represents differing genetic backgrounds, different rates of development, and different environments. However, it is also possible that part of the variability may be related to differences in neonatal diets between babies who are breast-fed and those fed infant formulas of differing composition. A number of hepatic phase I and II enzymes are expressed at high levels during lactational exposure in neonatal rodents and then decline following weaning. These include CYP2E1 and CYP2B family members and a number of steroid sulfotransferases such as hydroxysteroid sulfotransferase SULT2A1. This presumably is the result of important physiological roles during this period of high fat intake and estrogen imprinting of the brain and reproductive physiology. It is not clear if lactation-specific expression of particular phase I and II enzymes occurs in human neonates. In addition, studies in which rats have been fed soy protein isolate, the major protein source in soy-containing infant formula, during early development have demonstrated significant increases in expression and in vitro microsomal activity of hepatic and intestinal CYP3A enzymes. Since the CYP3As are responsible for the rate-limiting metabolism and clearance of many pediatric medications, it is possible that similar effects may occur in infants fed soy formula compared to those fed dairy formula or breast-fed. This has not been examined to date in human clinical studies, but preliminary data from formula-fed neonatal pigs support increased CYP3A expression following soy formula feeding. In addition, in vitro culture studies with human hepatocytes exposed to phytochemical extracts from soy have reported increases in expression of the major human CYP3A, CYP3A4.

14.2.2b   *Weaning to Solid Foods.* A further group of hepatic and intestinal phase I and II enzymes are expressed at around weaning which are involved in the metabolism/detoxication of dietary xenobiotics such as flavinoids, alkaloids, and terpenes. These include CYP1A2 in humans and CYP1A1, 1A2, and CYP3A6 in the rabbit. Increased expression of CYP1 enzymes may be the result of induction by dietary factors such as aromatic amines and aromatic hydrocarbons produced during cooking and natural plant associated Ah receptor ligands such as indole 3-carbinol in cruciferous vegetables. Artificial alteration of the age of weaning by controlling food availability to the pups also had no effect on CYP3A6 expression in the rabbit, and this suggests endogenous physiological regulation of this enzyme by as yet unknown factors rather than the influence of dietary components.

14.2.3 Xenobiotic Metabolism in Childhood and Preadolescence

In general, by 1 year, many xenobiotic metabolizing enzymes in humans are expressed at levels similar to those observed in adults. However, there are some exceptions. Even at 1 year, CYP1A2 levels are only 50% of those seen in adults. Moreover, CYP2 enzymes such as CYP2C9 and CYP2E1 are expressed at 30% of adult levels at age 1 year and increase to adult levels by age 10. FMO3 expression is detectible by 1–2 years of age, increases to intermediate levels by age 11, and increases to adult levels by age 18. The molecular regulation of CYP and FMO ontogeny during this developmental period remains largely unexplored. Less is known regarding expression of human phase II enzymes during this period of development, but at least two UGTs, UGT1A6 and UGT2B17, involved in acetaminophen and androgen glucuronidation also appear not to achieve adult expression until around puberty.

14.2.4 Puberty and Adulthood

Major alterations in expression of hepatic phase I and II enzymes occur in rodents at puberty. This is largely the result of the appearance of sexual dimorphism of specific enzymes in these species which persist throughout adulthood. Male-specific rat enzymes such as CYP2C11 (testosterone 16 and 2α-hydroxylase) and sulfotransferase SULT1C1, which metabolizes N-hydroxy-2-aminofluorene (N-OH-2AAF), appear at puberty. So do female-specific rat enzymes such as CYP2C12 and SULT2A1. In contrast, other P450 enzymes such as CYP3A2, which is expressed at high levels prior to puberty in both sexes of rat, are selectively suppressed in female animals. Glutathione S-transferases (GSTs) increase in activity and become sexually dimorphic at puberty. In addition, UDPGTs also become sexually dimorphic at puberty. It appears that these differences are either directly due to differences in sex steroid concentrations in male and female rodents or the indirect effect of sex steroid imprinting of sexually dimorphic patterns of growth hormone secretion (GH) by the pituitary and that GH is the endocrine mediator that maintains sexually dimorphic enzyme expression (Figure 14.1).

In adult rats, growth hormone (GH) is secreted in a pulsatile fashion. In males, low-frequency, high-amplitude pulses are observed with nadirs between pulses in which no GH is secreted. In contrast, in females GH is released as high-frequency, low-amplitude pulses superimposed on a baseline of continuous GH secretion. The sexually dimorphic pattern of GH secretion by the pituitary appears at puberty and is determined by pulsatile secretion of the hypothalamic peptides GHRH and somatostatin 180 degrees out of sync with each other. The release patterns of GHRH, somatostatin, and GH are themselves imprinted and under feedback regulation by the sex steroids. Prenatal and neonatal androgen surges appear to imprint the mas-

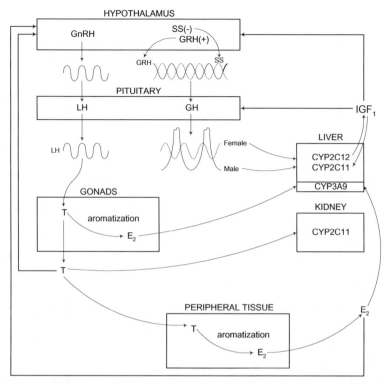

Figure 14.1. Regulation of sex-specific cytochrome P450 expression in adult rat liver and kidney by the hypothalmic–pituitary–gonadal axis. GnRH, gonadotropin releasing hormone; SS, somatostatin; GRH, growth hormone releasing hormone; GH, growth hormone; LH, lutenizing hormone; T, testosterone; E2, 17ß-estradiol; IGF1, insulin-like growth factor; CYP2C12, female-specific cytochrome P450 2C12; CYP2C11, male-specific cytochrome P450 2C11; CYP3A9, female-predominant cytochrome P450.

culine pattern of GH secretion. It is thought that aromatization of androgens to estrogens within the neonatal hypothalamus is required for the imprinting of GH since nonaromatizable androgens such as dihydrotestosterone are unable to reverse the effects of neonatal castration.

In some cases, testosterone and estrogens directly stimulate expression of xenobiotic metabolizing enzymes. An example of this is in the mouse kidney where many cytochrome P450 enzymes including CYP2E1 are under direct androgen regulation. This accounts for the unique susceptibility of the female mouse kidney to acetaminophen nephrotoxicity compared to the male mouse kidney. In the rat liver, expression of CYP3A9 has recently been shown to be positively regulated by estrogens. In the rat kidney, another male-specific cytochrome P450 enzyme, CYP2C11, has been shown to be under direct androgen regulation. Castration and hypophysectomy (Hx, removal of the pituitary), which remove testosterone either directly or indirectly via loss of luteinizing hormone (LH) signaling, both significantly reduce renal CYP2C11 expression. However, whereas testosterone replacement completely restores this enzyme, GH replacement seems to have no effect. These results are in direct contrast to the expression of CYP2C11 in male rat liver. In this case, although both castration and Hx reduce hepatic CYP2C11 expression 50% in male animals, testosterone replacement has no effect in Hx animals and intermittent administration of GH mimicking the male pattern restores levels to normal. Moreover, continuous administration of GH mimicking the female pattern completely suppresses CYP2C11 expression. In addition, passive immunization against GH in vivo also significantly suppresses hepatic CYP2C11 expression in male rats [see Figure 1, Ronis et al. (2006) in Suggested Reading]. Regulation of the female-specific rat cytochrome P4502C12 is exactly the reciprocal of CYP2C11: The female GH pattern stimulates expression and the male pattern suppresses it. These enzymes appear at puberty in the rat, and their appearance coincides with the development of the sexually dimorphic patterns of GH secretion. Neonatal castration and steroid replacement experiments have demonstrated-that both GH patterns and these hepatic P450 enzymes are imprinted by neonatal androgen exposure. In conditions such as lead poisoning where androgen surges at birth and puberty are suppressed and pubertal development is delayed, developmental expression of these enzymes is also delayed and sexual dimorphism of xenobiotic metabolism is suppressed. In addition to CYP2C11 and CYP2C12, many other rat liver, P450 enzymes are sexually dimorphic to some degree. In the male liver, CYP2C13, CYP3A2, and CYP2A2 are the major forms. In female rat liver, CYP2A1 and CYP2C7 are predominant. Epoxide hydrolase, GSTs, and some UDPGT enzymes are also expressed at a higher level in male rat liver. In contrast, many FMOs are found at higher levels in the livers of female rodents. Testosterone has been shown to significantly suppress FMO1 and FMO3 in female mice, but it is not clear if this is a direct steroid effect or mediated via alterations in GH secretion.

Sulfotransferase enzymes are also sexually dimorphic in rat liver. SULT1A1, SULT1C1, and SULT1E1, which metabolize phenol, N-OH-2AAF, and estrone, respectively, are male-dominant, whereas SULTs, which metabolize glucocorticoids, hydroxysteroids, and bile acids (SULT2As), are female-dominant. As with the P450 enzymes, the primary regulator of gender-specific SULT expression appears to be GH rather than the sex steroids directly.

A considerable amount of research has been conducted to examine which aspects of the sexually dimorphic GH pulse are responsible for sexually dimorphic enzyme expression at puberty. With CYP2C11 and 2C12, plasma GH concentrations and pulse amplitude appear unimportant for regulation. Administration of GH to Hx rats in different patterns and concentrations has revealed that 2C11 is only expressed if GH is completely absent for 1.5–2 hr between pulses. In contrast, CYP2C11 is completely suppressed and CYP2C12 is expressed at normal levels in the presence of continuous GH at concentrations as low as 3% of physiological levels. Other sexually dimorphic P450 enzymes are less dependent on GH pulse pattern. Male-rat-specific CYP3A2 is markedly elevated by Hx, and the enzyme is suppressed by GH in a pattern-independent fashion.

Recently, the molecular events underlying GH regulation of rat CYP2C11 and CYP2C12 have been examined. The GH receptor is present in the cell membrane and, upon GH binding, undergoes dimerization and a conformational change that allows the binding and activation of a member of the Janus kinase family of protein kinases known as JAK2. This kinase has been shown to autophosphorylate itself and to phosphorylate the GH receptor at specific tyrosine and serine residues. It appears that these phosphorylation events in turn allow the binding and subsequent phosphorylation of the transcription factor STAT5b. STAT5b phosphorylation releases the transcription factor for the GH receptor complex and undergoes homodimerization with a second phosphorylated STAT5b and translocation into the nucleus, where it binds its response element on the promoter of the CYP2C11 gene and initiates transcription. The reason that this signaling pathway appears to be sensitive to the pattern of GH secretion is that if GH is not removed, the GH receptor/JAK2 complex remains active and other signals are initiated which result in the permanent loss of STAT5b phosphorylation and degradation of the transcription factor. It is not yet clear what these other pathways are, but they may involve activation of other GH-dependent transcription factors such as STAT1, STAT3, the Ras/Raf kinase pathway, or hepatocyte nuclear factor 6, which in female-rats exposed to a more constant pattern of GH secretion is required for expression of the female-specific P450 enzyme CYP2C12. If GH is removed from the male for a minimum of approximately 3 hr between pulses, STAT5b in the nucleus is dephosphorylated possibly via the protein phosophatase SHP1 and then recycles into the cytosol, and the GH receptor/JAK2 complex also becomes dephosphorylated and inactivated. A second GH pulse initiates the signaling cycle all over again and CYP2C11 expression is maintained (Figure 14.2).

In humans, GH secretion patterns are much less sexually dimorphic than in rodents and there are no human orthologs of the sexually dimorphic rat liver P450s CYP2C11, 2C12, 2C13, or CYP3A2. As such, genetic polymorphism and environmental factors such as diet probably play a more important role in interindividual variations in drug and xenobiotic metabolism in adults than do gender differences. However, the major human hepatic P450 enzyme CYP3A4 does appear to be expressed overall at higher levels in women than in men. This also appears to be due to GH regulation of this enzyme. Administration of male (pulsatile) patterns of GH to human hepatocytes suppresses glucocorticoid-induced expression of

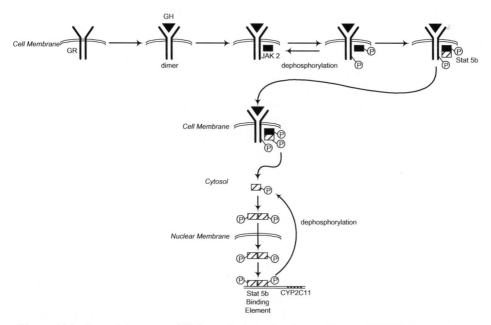

Figure 14.2. Growth hormone (GH) regulation of male-specific rat CYP2C11 expression.

CYP3A4, whereas female (continuous) patterns of GH appear to act permissively allowing expression of CYP3A4 to greater or equal levels than glucocorticoid induction alone. Like the human, little sexual dimorphism in xenobiotic metabolism is observed in the guinnea pig, dog, pig, or rhesus monkey. These species differences in gender-specific metabolism must be taken into account when extrapolating from studies in laboratory animals to humans.

14.2.5 Effects of Aging

The effects of old age on xenobiotic metabolism continues to receive much less attention than earlier developmental events. In general, drug and xenobiotic metabolism rates and clearance appear to decline to some degree, although much of this may be due to changes in liver blood flow and reduced liver size rather than reduction in expression of particular phase I or II enzymes. In aged male rats, male-specific hepatic P450 enzymes such as CYP2C11 and CYP3A2 decline while female-specific forms such as CYP2C12 begin to appear. This may reflect loss of sex steroids and sexually dimorphic patterns of growth hormone secretion. Similarly, decreases in CYP3A4 expression have been described in older men, possibly as a result of decreased testosterone levels, while CYP3A4 activities in women appear to be largely maintained or elevated. In addition, whereas glucuronidation declines with old age, other enzymes such as GST are maintained and others such as monoamine oxidases have been shown to increase.

14.2.6 Effects of Disease

Aging is associated with increased susceptibility to infection and a host of chronic pathologies including diabetes, cardiovascular disease, and cancer. Diseases cause many alterations in xenobiotoic metabolism. Conditions such as acute and chronic hepatitis, cirrhosis, and hepatic porphyria, which directly lead to liver damage, obviously result in major impairments in metabolism and clearance. However, diseases of the kidney and cardiac system also have major effects on xenobiotic clearance through changes in pharmacokinetics. Impaired kidney function results in a buildup of conjugated metabolic products which may be hydrolyzed and further metabolized prior to elimination. Circulatory disfunction results in impaired delivery of xenobiotics to the liver and other sites of metabolism.

Infection and other inflammatory stimuli have been shown to cause changes in the activity and expression of many different cytochrome P450 forms in the liver, kidney, and brain of both humans and experimental animals. This is of particular clinical importance since alterations in phase I metabolism have the potential to adversely or favorably affect the theraputic or toxic effects of drugs. In general, hepatic metabolism is compromised and the possibility of toxic drug side effects is increased. Thus infection can itself influence the potential for toxic side effects of the drugs being used to treat it. Moreover, infections can produce toxic side effects in patients on drug therapy for chronic conditions such as epilepsy, high blood pressure, or asthma, especially in those patients on drugs with a low theraputic index. In the last few years a large number of clinical and basic studies have investigated the effects of inflammation on P450 expression and there has been a tendency to lump all of these studies together. However, whereas it is clear that P450 enzymes are suppressed by various interferons (IFNs) and inflammatory cytokines such as TNFα, IL1-β, IL-6, and TGFβ, the patterns of IFNs and cytokines produced vary by the type of infection or inflammation, and this in turn results in different patterns of effects on P450 expression. This probably underlies the tremendous variability of phase I responses and drug clearance studies reported clinically in diseased patients. Animal models also differ. Administration of IFNs is probably a good model for viral infections, while endotoxin administration better elicits the effects of bacterial infections. The most common findings in humans in vivo are suppression of CYP1A2 and sometimes CYP3A4-dependent activities. In animals, IFNγ, endotoxin, IL-1β, IL-6, and TNFα have most commonly been reported to suppress CYP3As, CYP1A2, CYP2C11, and CYP2C12 expression. However, not all cytokine-mediated effects on hepatic P450 enzymes result in suppression. In human hepatocytes, TNFα, IL-1β, and IL-6 all suppress expression of CYP1A2, CYP2C forms, CYP2E1, and CYP3A enzymes, but IL-4 has been reported to increase CYP2E1 expression. In extrahepatic tissues, infections and interferon have been shown to suppress expression of pulmonary and renal P450s. However, endotoxin has been reported to increase CYP4A expression in the kidney, and inflammation is the only thing known to induce CYP2E1 in the brain. Many signaling pathways have been implicated in the effects of inflammation, IFNs, and cytokines on cytochrome P450 expression. Induction of nitric oxide synthetase (iNOS) has been suggested to mediate the suppression observed in phenobarbital-inducible CYP2B1 expression in rats treated with endotoxin since iNOS inhibitors reverse this effect. In addition, iNOS inhibitors have been shown to reverse the suppression of CYP2C11 and CYP3A2 produced in rat hepatocyte cultures by TNFα and IL-1β treatment. Cytokine-induced breakdown of cell membrane lipids has also been suggested to play a role in supression of some rat hepatic P450 enzymes such as CYP2C11. Sphingolipid backbones such as ceramide and sphingosine have been suggested to act as secondary messengers for cytokines such as TNFα and IL-1β. It has been shown that sphinogomyelinase and ceramidase are induced following treatment of hepatocytes with low concentrations of IL-1β. This results in elevations in the ceramide metabolite sphingosine, which has been shown to be a more potent inhibitor of CYP2C11 expression than ceramide in the same system. The steps leading to transcriptional downregulation of CYP2C11 by sphingosine or its metabolites are yet to be elucidated.

Carcinomas and preneoplastic lesions appear to have altered xenobiotic metabolism relative to the untransformed tissue surrounding them. In hyperplastic liver nodules induced by dietary N-OH-2-AAF, microsomal monooxygenase activities are depressed, whereas epoxide hydrolase and some phase II enzymes such UDPGTs and GSTs are increased. In preneoplastic liver foci produced by treatment with *N*-nitrosomorpholine, a "permanent" induction of UDPGT activity of up to five-fold has been observed even 330 days after dosing with the initiator. The UDPGT enzyme affected is that involved in the conjugation of planar phenols which is inducible by polycyclic aromatic hydrocarbons. A radical alteration in the pattern of GST enzyme expression is also observed with significant expression of P class GSTs which may be involved in resistance to chemotheraputic drugs. It is interesting to note that the phenolic UDPGT and GSTP are normally expressed in fetal tissues. GSTP expression is now being utilized as a marker in the detection of some preneoplastic lesions.

Diabetes also produces significant alterations in xenobiotic metabolism and can increase susceptibility to hepatotoxicity as the result of impaired insulin signaling and hyperglycemia. In rats and humans the major cytochrome P450 induced by type I diabetes is CYP2E1, which has been implicated in hepatoxicity of acetaminophen, alcohol, and industrial solvents and activation of nitrosamine carcinogens. However, many other phase I and II enzymes are also affected. In rats, drug metabolism is demasculinized, and effects on both CYP2E1 and other enzymes are reversed by insulin treatment. Fewer effects on cytochrome P450 expression have been reported to be associated with type II diabetes either in humans or in animal models, but hepatic alcohol dehydrogenase class I expression has been shown to be induced significantly as the result of disruption of insulin signaling via the Akt (protein kinase B) pathway.

14.3 EFFECTS OF PREGNANCY ON MATERNAL XENOBIOTIC METABOLISM

Pregnancy results in major physiological and biochemical changes in female animals which often affect their ability to metabolize xenobiotics. It is well known that a large number of maternal enzymes have decreased activity during pregnancy. For example, both monoamine oxidase and catechol-O-methyltransferase activity are significantly reduced. Many of the changes in xenobiotic metabolism during this period appear to be associated with altered hormone levels such as the rise in progesterone and pregnane-diol late in gestation and the release of placental hormones such as chorionic gonadotropin into the maternal circulation.

Pregnancy has differing effects on hepatic expression of cytochrome P450 enzymes depending on the species. In the rat there is an overall decrease in P450 content, and some enzymes such as CYP2E1 have been shown to be significantly suppressed. In addition, inducibility of CYP2B1 by phenobarbital is significantly reduced. In contrast, in the pregnant rabbit liver, little effect on monooxygenase activities are observed. In the rabbit lung, pregnancy selectively induces a member of P450 family 4 by more than 100-fold between days 25 and 28 of gestation. This enzyme is involved in the omega hydroxylation of prostaglandins E1, F2a, and A, and it is not clear if it plays a significant role in xenobiotic metabolism. Evidence also exists for induction of flavin-containing monooxygenase during mid- to late gestation in the rabbit. FMO2 expression increases in rabbit lung and kidney during late gestation and peaks at parturition and FMO1 is similarly increased during mid- and late gestation in maternal rabbit liver. It has been suggested that these effects are associated with rises in plasma progesterone and glucocorticoids during this period. It has also been reported that expression of gastric alcohol dehydrogenase IV is increased significantly in pregnancy and that this may contribute to increased first pass clearance of alcohol in pregnant rats.

Pregnancy also results in significant changes in conjugating enzyme systems. Increasing levels of progesterone and pregnane-diol in late pregnancy may be responsible for observed decreases in glucuronide conjugation since both steroids have been shown to be inhibitors of UDPGT in vitro. The presence of pregnane-2α-20β-diol in the milk of lactating mothers has been reported to result in jaundice in some breast-feeding infants as a consequence of inhibition of neonatal bilirubin conjugation by UDPGT. Decreases in sulfotransferase activity have also been reported in pregnant rats and rabbits.

14.4   DEVELOPMENTAL EFFECTS ON XENOBIOTIC METABOLISM IN NONMAMMALIAN SPECIES

Much less is known regarding developmental regulation of xenobiotic metabolism in species other than mammals. However, the same type of temporal and special regulation of expression appears to occur in both mammals and nonmammalian species. In fish, expression of CYP1 enzymes occurs in specific tissues during early embryogenesis similar to mammals. These enzymes may be responsible for susceptibility of fish embryos to carcinogenic agents but may also protect against teratogenic effects of some environmental chemicals such as polycyclic aromatic hydrocarbons by increasing clearance. Some developmental and gender-specific changes in phase I and II enzymes seen in nonmammalian vertebrates and invertebrates appear to be related to changes in hormonal status. In birds, glucocorticoids have been shown to induce UDPGTs in the chick embryo, and some cytochrome P450 activities such as ethylmorphine *N*-demethylase are induced by dexamethasone in the adult bird. In contrast, fish appear relatively insensitive to glucocorticoid induction even though they possess CYP3A enzymes related to the forms found in mammals. However, like rodents, many fish species demonstrate sexual dimorphism in xenobiotic metabolizing enzymes which develops during sexual maturity. For example, in mature rainbow trout a significantly higher content of cytochrome P450 is observed in the livers of male fish, and the gender difference is as much as 20 to 30-fold in the male compared to the female kidney. These gender differences appear to be due to a major suppressive effect of estrogens on some male-specific fish P450 enzymes. A similar downregulation of FMO enzymes by estrogens has been reported in fish liver during sexual maturation.

In insects, hormonal status is known to control xenobiotic metabolism during development and may have important consequences for resistance and susceptibility to pesticides. In the housefly, both hormones which control moulting—ecdysone and juvenile hormone—cause increases in P450-dependent monooxygenase activity as measured by heptachlor epoxidation. In the fruit fly, *Drosophila*, the CYP18 family of P450 enzymes which is most closely related structurally to the mammalian xenobiotic-metabolizing P450s of the CYP2 family are tightly developmentally regulated in the body wall in correlation with ecdysteroid pulses in the first, second, and third larval instars, at the time of pupation and in the pupae. It is thought that both ecdysone and juvenile hormone act via increases in gene transcription.

14.5   CYCLES IN DEVELOPMENT

In many species, xenobiotic metabolism exhibits circadian and seasonal cycles. In wild populations, these cycles may be related to many factors including day length, diet, breeding cycles, and temperature. Since most laboratory animals are kept under controlled environmental conditions, they do not display seasonal changes in metabolism. However, they do exhibit circadian rhythms in activity and endocrine function related to feeding and the light–dark cycle. In rodents, monooxygenase activities display circadian rhythms that appear to be partly related to changes in cytochrome P450 reductase expression. However, many P450 enzymes are also expressed in a circadian fashion. One P450 enzyme that demonstrates large circadian rhythms is CYP7A, which is involved in the 7α-hydroxylation of cholesterol and is active in bile acid synthesis. Others are CYP2C and CYP4A family members and xenobiotic metabolizing enzymes such as CYP2B, SULT1D1, UGT1A1, and ALDH1A1, which are under regulation by the transcription factor CAR. It appears that reductases such as P450 reductase and steroid 5α reductase, some CYPs, and CAR expression are regulated though a family of PAR-domain basic leucine zipper (PAR bZIP) transcription factors, which are cell-autonomous circadian oscillators expressed in the liver and extrahepartic tissues such as the kidney which are synchronized by a master pacemaker in the suprachiasmic nucleus of the brain hypothalamus via core oscillator components known as clock genes. Mice with mutated or knocked out clock and PAR bZip proteins are highly susceptible to drug toxicity as a result of impaired phase I and II enzyme expression and display disrupted circadian expression of these enzymes. Nonmammalian species such as some birds also display diurnal variations in xenobiotic enzymes such as B-esterase activities with daily highs and nightly lows. This has been suggested to be related to feeding patterns.

In addition to daily circadian cycles, a number of longer-term cycles have been described which affect xenobiotic metabolism. One of these is the estrus cycle in females in which significant shifts in estrogen and progesterone concentrations regulate ovulation. Expression of at least one cytochrome P450, CYP1B1, is altered in the ovary during the rat estrus cycle being expressed at high levels on the evening of pro-estrus and dramatically decreased on the morning of estrus.

An unusual long-term cycle in drug metabolism involves ethanol. In models where alcohol is continuously infused intragastrically into rats, blood and urine ethanol concentrations cycle from high peaks of 500–600 mg/dl to nadirs of nearly zero with a periodicity of 6–7 days. It is unclear whether this is an example of substrate-driven time-dependent pharmacokinetics or if an underlying endocrine-driven cycle in ethanol metabolism is being revealed. Cyclic changes in ethanol intake with the same periodicity have been observed in pigs fed beer *ad libitum* and in alcohol-addicted monkeys. It is possible that these cycles in ethanol metabolism and consumption have a relationship to binge drinking in humans. The biochemical mechanisms underlying increased ethanol metabolism appear to involve increases in hepatic expression of alcohol dehydrogenase I related to disruption of insulin signaling via Akt at high ethanol concentrations. Normally, insulin suppresses ADH I by Akt-dependent activation of the transcription factor SREBP-1, which, in turn, acts as a transcriptional repressor at the ADH I promoter. Rising blood ethanol concentrations block Akt signaling via induction of a protein suppressor TRB3 and thus de-represses ADH I expression, resulting in increases in ethanol clearance [see He et al. (2006) in Suggested Reading].

Seasonal variations in phase I and II enzymes are often seen in conjunction with breeding cycles. This is particularly true in amphibians, fish, and birds and probably reflect underlying endocrine changes associated with the establishment of reproductive competence. For example, in the razorbill, elevated metabolism of organochlorine insecticides such as aldrin has been reported in females collected in April–May. This correlated with increased ovarian size and may be related to increases in circulating estradiol and/or progesterone. Similarly, increases in lauric acid hydroxylase, benzo[*a*]pyrene hydroxylase, and ethoxycoumarin *O*-deethylase have been reported in spawning frog populations in the months April–July. In addition to reproductive state, seasonal variations in metabolism observed in marine species such as fish and molluscs may be associated with adaptations to changes in water temperature.

SUGGESTED READING

Agrawal, A. K., and Shapiro, B. H. Gender, age and dose effects of neonatally administered aspartate on the sexually dimorphic plasma growth hormone profiles regulating expression of the rat sex-dependent hepatic CYP isoforms. *Drug Metab. Disp.* **25**, 1249–1256, 1997.

Badger, T. M., Hoog, J.-O, Ronis, M. J. J., and Ingelman-Sundberg, M. Cyclic variation of class I alcohol dehydrogenase in male rats treated with ethanol. *Biochem. Biophys. Res. Commun.* **274**, 684–688, 2000.

Cambell, S. J., Henderson, C. J., Anthony, D. C., Davidson, D., Clark, A. J., and Wolf, C. R. The murine CYP1A1 gene is expressed in a restricted special and temporal pattern during embryonic development. *J. Biol. Chem.* **280**, 5828–5835, 2005.

Chen, G.-F., Ronis, M. J. J., Thomas, P. E., Flint, D. J., and Badger, T. M. Hormonal regulation of microsomal cytochrone P450 2C11 in rat liver and kidney. *J. Pharmacol. Exp. Ther.* **283**, 1486–1494, 1997.

Choudhary, D., Jansson, I., Sarfarazi, M., and Schenkman, J. B. Xenobiotic-metabolizing cytochromes P450 in ontogeny: Evolving prespective. *Drug Metab. Rev.* **36**, 549–568, 2004.

Dhir, R. N., Dworakowski, W., Thangavel, C., and Shapiro, B. H. Sexually dimorphic regulation of hepatic isoforms of human cytochrome P450 by growth hormone. *J. Pharmacol. Exp. Ther.* **316**, 87–94, 2006.

Duanmu, Z., Weckle, A., Koukouritaki, S. B., Hines, R. N., Falany, J. L., Falanay, C. N., Korcarek, T. A., and Runge-Morris, M. Developmental expression of aryl, estrogn and hydroxysteroid sulfotransferases in pre- and postnatal human liver. *J. Pharmacol. Exp. Ther.* **316**, 1310–1317, 2006.

Gachon, F., Olela, F. F., Schaad, O., Descombres, P., and Schibler, U. The circadian PAR-domain basic leucine zipper transcription factors DBP, TEF and HLF modulate basal and inducible xenobiotic detoxification. *Cell Metab.* **4**, 25–36, 2006.

Gebert, C. A., Park, S.-H., and Waxman, D. J. Regulation of signal transducer and activator of transcription (STST) 5b activation by the temporal pattern of growth hormone stimulation. *Mol. Endocrinol.* **11**, 400–414, 1997.

Gebert, C. A., Park, S.-H., and Waxman. D. J. Termination of growth hormone pulse-induced STAT5b signaling. *Mol. Endocrinol.* **13**, 38–56, 1999.

Hakkola, J., Pelkonen, O., Pasanen, M., and Raunio, H. Xenobiotic-metabolizing enzymes in the human feto-placental unit: Role in intrauterine toxicity. *Crit. Rev. Toxicol.* **28**, 35–72, 1998.

He, L., Simmen, F. A., Ronis, M. J. J., and Badger, T. M. Chronic ethanol feeding impairs insulin signaling in rats through disrupting Akt/PKB association with the cell membrane: Role of TRB3 in inhibition of Akt/PKB activation. *J. Biol. Chem.* **281**, 11126–11134, 2006.

Hines, R. N., and McCarver, D. G. The ontogeny of human drug-metabolizing enzymes: Phase I oxidative enzymes. *J. Pharmacol. Exp. Ther.* **300**, 355–360, 2002.

Klaassen, C. D., Liu, L., and Dunn, R. T., 2nd. Regulation of sulfotransferase mRNA expression in male and female rats of various ages. *Chem–Biol. Interact.* **109**, 299–313, 1998.

Koukouritaki, S. B., Simpson. P., Yeung, C. K., Rettie, A. E., and Hines, R. N. Human hepatic flavin-containing monooxygenases 1 (FMO1) and 3 (FMO3) developmental expression. *Pediatric Res.* **51**, 236–243, 2002.

Lacroix, D., Sonnier, M., Moncion, A., Cheron, G., and Cresteil, T. Expression of CYP3A in the human liver—Evidence that the shift between CYP3A7 and CYP3A4 occurs immediately after birth. *Eur. J. Biochem.* **247**, 625–634, 1997.

McCarver, D. G., and Hines, R. N. The ontogeny of human drug-metabolizing enzymes: Phase II conjugation enzymes and regulatory mechanisms. *J. Pharmacol. Exp. Ther.* **300**, 361–366, 2002.

Morgan, E. T. Regulation of cytochromes P450 during inflamation and infection. *Drug Metab. Rev.* **29**, 1129–1188, 1997.

Mugford, C. A., and Kedderis, G. L. Sex-dependent metabolism of xenobiotics. *Drug Metab. Rev.* **30**, 441–498, 1998.

Okita, R. T., and Okita, J. R. Prostaglandin-metabolizing enzymes during pregnancy: Characterization of NAD-dependent prostaglandin dehydrogenase, carbonyl reductase and P450-dependent prostaglandin omega-hydroxylase. *Crit. Rev. Biochem. Mol. Biol.* **31**, 101–126, 1996.

Ronis, M. J. J., Chen, Y., Badeaux, J., Laurenzana, E., and Badger, T. M. Induction of rat hepatic CYP3A1 and CYP3A2 by soy protein isolate or isoflavones fed during early development. *Exp. Biol. Med.* **231**, 60–69, 2006.

Stoilov, I., Rezaie, T., Jansson, I., Schenkman, J. B., and Sarfarazi, M. Expression of CYP1B1 during early murine development. *Mol. Vis.* **10**, 629–636, 2004.

Tanaka, E. In vivo age-related changes in hepatic drug-oxidizing capacity in humans. *J. Clin. Pharmacy Ther.* **23**, 247–255, 1998.

Viera, I., Sonnier, M., and Cresteil, T. Developmental expression of CYP2E1 in the human liver. Hypermethylation control of gene expression during the neonatal period. *Eur. J. Biochem.* **238**, 476–483, 1996.

Waxman, D. J., Pampori, N. A., Ram, P. A., Agrawal, A. K., and Shapiro, B. H. Interpulse interval in circulating growth hormone patterns regulates sexually dimorphic expression of hepatic cytochrome P450. *Proc. Natl. Acad. Sci. USA* **88**, 6868–6872, 1991.

CHAPTER 15

Cellular Transport and Elimination

DAVID S. MILLER

National Institute of Environmental Health Sciences, Research Triangle Park, North Carolina 27709

15.1 TRANSPORT AS A DETERMINANT OF XENOBIOTIC ACTION

This chapter focuses on the contributions of transporters and transport processes to biochemical toxicology. This is done through a consideration of the nature of biological membranes and the transport proteins they contain. Throughout the chapter, the emphasis is on xenobiotic transporters. These membrane proteins have an existence similar to that of Dr. Jekyll and Mr. Hyde. Like xenobiotic metabolizing enzymes, they protect us from toxicants and their metabolites by both driving their elimination and limiting their access to discrete body compartments (e.g., the brain). On the other hand, both enzymes and transporters impede pharmacotherapy, removing drugs from physiologically important sites of action and providing a means through which drug–drug and drug–toxicant interactions can occur.

The sixteenth-century physician-alchemist Philippus Aurelolus Theophrastus Bombastus von Hohenheim-Paracelsus is often credited with being the father of the science of toxicology. Paracelsus is perhaps best remembered for arguing that chemicals can have both therapeutic and toxic properties and that these are distinguished by dose; that is, the dose makes the poison. We now recognize that for a chemical to have an effect, therapeutic or toxic, it must access physiologically significant sites of action within the body at a high enough concentration and for a long enough time. Thus, multiple factors—that is, distribution (at the organism, tissue, and cell levels) concentration and time—combine to determine dose–response relationships. These factors, in turn, are a function of four processes: uptake, distribution, metabolism, and excretion. Three of these—uptake, distribution, and excretion—are directly affected by xenobiotic transporters; the fourth, metabolism, can be thought of as part of a multicomponent pipeline that often feeds those transporters.

A dramatic example of how transporters can alter the efficacy and toxicity of a commonly prescribed drug is provided by the following scenario. Three 70-kg males with identical symptoms see a physician and are diagnosed with mild cardiac insufficiency. For all a standard treatment with digoxin, a cardiac glycoside that increases the strength of heart muscle contractions, is recommended and all follow the

Molecular and Biochemical Toxicology, Fourth Edition, edited by Robert C. Smart and Ernest Hodgson
Copyright © 2008 John Wiley & Sons, Inc.

physician's orders to the letter. However, the prescribed drug has a different effect on each patient. The first patient complains that the drug appears to be ineffective; when measured, his plasma digoxin level is below the therapeutic range. The second patient complains of headache and nausea, and his plasma digoxin level is well above the therapeutic range. Finally, the third patient reports that the drug works as advertised; not surprisingly, his plasma digoxin level is within the therapeutic range. What happened here?

Digoxin is a drug with a narrow therapeutic index and thus a narrow range of plasma concentrations over which it is effective. Below this range, the drug is ineffective (first patient); above this range, it is toxic (second patient). Metabolism of digoxin is minimal, and the major determinants of plasma level are absorption at the gut and excretion in liver and kidney. These integrated processes, in turn, are influenced in large part by the expression and function of specific plasma membrane transporters, including p-glycoprotein and at least one member of the organic anion transport polypeptide (OATP) family. Expression of these transporters can be altered by various drugs and by disease processes (e.g., inflammation). Thus, depending on the tissue involved, increased transporter expression can reduce digoxin absorption (p-glycoprotein in the gut) or increase its excretion (p-glycoprotein in liver and kidney), while decreased expression can have the opposite effects. From this perspective, increased expression of the efflux pump could underlie the first patient's problem, and competition for transport with a second p-glycoprotein substrate (drug, metabolite, or herbal remedy) could underlie the second patient's problem.

15.2 FACTORS AFFECTING MEMBRANE/TISSUE PERMEABILITY

The physiological importance of membrane-bound transporters and ion channels is clearly evident at both the cellular and tissue levels of organization. By regulating ionic composition and cell volume and by mediating uptake of essential nutrients and export of metabolic wastes, channels and transporters play essential roles in maintaining homeostasis in all cells. Transport function underlies unique activities of many specialized cells; for example, ion channels and ion pumps are responsible for neuronal action potentials and myocyte contraction. Transporters also play important roles in barrier and excretory tissue function. A tissue's ability to drive net movements of solutes and generate substantial trans-tissue gradients is a function of the polarized (i.e., unsymmetrical) expression of selective channels and transporters. In this regard, multiple transporters and ion channels are responsible for urine and bile formation.

Clearly, one key contribution of transporters to mechanisms of xenobiotic toxicity arises from their important roles as determinants of where a foreign chemical acts, at what levels, and with what results (therapeutic effects versus toxicity). A second important consideration is that those proteins themselves and the cell membranes that contain them can be primary targets of toxicant action. Indeed, membranes and their proteins are the first cellular structures encountered by toxicants. They thus provide functionally important targets of opportunity for chemicals that have the potential to react with residues affecting specificity, coupling to cellular metabolism, regulation, and mode of action. Changes can range from subtle alterations of solute gradients to substantial disruption of cell and tissue function.

15.2.1   Cellular Level Transport

The simplest picture of a biological membrane consists of a fluid, phospholipid bilayer in which proteins are embedded (Figure 15.1). If only the lipid bilayer were considered, one would expect diffusion of solutes across these membranes to be positively correlated with lipophilicity and negatively correlated with polarity. Although this is certainly the case for model membrane systems and for some classes of solutes in actual biological membranes, the large number of exceptions to this rule define the contributions of proteins to membrane permeability. In essence, embedded proteins add three elements to the transport characteristics of the bilayer. They facilitate the movement of (polar) solutes across the bilayer, they do this selectively, and they provide a means to energetically couple such transport to cellular metabolism. The end result is the creation and maintenance of the steady-state transmembrane concentration gradients that define living cells. As a result (relative

(a)

Process	Kinetics	Driving Force
Simple Diffusion	Linear/No Competition	Concentration Gradient
Facilitated Diffusion	Saturation/Competition	Concentration Gradient
Primary Active Transport	Saturation/Competition	ATP Hydrolysis
Secondary Active Transport	Saturation/Competition	Ion/Metabolite Gradient

(b)

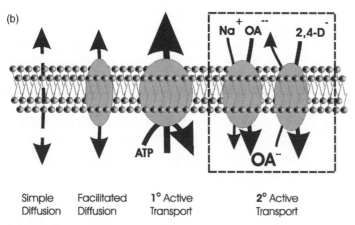

| Simple Diffusion | Facilitated Diffusion | 1° Active Transport | 2° Active Transport |

Figure 15.1. The language of membrane transport. (A) Defining characteristics of transport processes. (B) Example of each process. The arrows indicate directions of substrate movement. Cytoplasm is below the membrane. Except for simple diffusion, all processes require participation of a selective transport protein. For secondary active transport, two energetically coupled events are shown. (1) The Na gradient (out > in, established by Na,K-ATPase) drives cellular accumulation of a divalent organic anion (OA^{--}), and (2) this organic anion exchanges for the herbicide, 2,4-dichlorophenoxyacetic acid (2,4-D). The net result is OA^{--} recycling and Na-gradient-driven 2,4-D accumulation. This is the actual mechanism by which OAT1 and OAT3 mediate the concentrative uptake of organic anions from blood in renal proximal tubule cells.

to extracellular fluid), animal cell cytoplasm is high in K, low in Na, low in Ca, high in many nutrients, and low in metabolic wastes.

Membrane proteins involved in transport are classified as channels or transporters. When open, channels are variably ion-specific and have a huge capacity to transport primarily inorganic ions down their respective concentration and electrical (electrochemical) gradients. Transporters are different in several respects. They have much more restricted transport capacities than channels and they are certainly more diverse in function in that they facilitate both downhill and concentrative transport (Figure 15.1). In addition, the range of specificities found for individual transport proteins is impressive. Some only handle solutes with a narrowly restricted range of structures (e.g., certain highly specific hormone, hexose, or amino acid transporters). Others accept an extremely wide range of solutes (e.g., many multi-specific xenobiotic transporters).

For transporters, relatively low protein expression level and limited transport capacity makes for nonlinear, enzyme-like transport kinetics; that is, the transport rate saturates with increasing substrate concentration. This phenomenon is the basis for the competitive interactions generally found for chemicals that are handled by one or more common transporters; this is usually manifest as inhibition of the transport of one chemical by a structural analog. The extent to which these competitive interactions are important depends on the concentrations of the chemicals involved, their relative affinities for the common transporter, and their pharmacological/toxicological profiles (effects, effective concentrations, therapeutic index). Competition for transport is discussed below in the context of drug–drug interactions.

Without the input of potential energy, transporters, like enzymes, can only accelerate the approach to an equilibrium state; for nonelectrolytes this would mean equal concentrations on both sides of a membrane. In biology as in physics, there is no free lunch (First Law of Thermodynamics). By coupling transport to metabolic energy, transporters create steady states with substantial transmembrane concentration gradients. As shown in Figure 15.1, xenobiotic transporters do this by utilizing potential energy stored in ATP (ATPases like p-glycoprotein), in the electrical potential gradient across plasma membranes (OCTs and some OATs), or in the transmembrane gradients of ions (Na for CNTs, protons for PEPTs) and metabolites (OAT1, OAT3).

15.2.2 Tissue Level Transport

Solute transport across excretory or barrier tissues, such as renal proximal tubule epithelium or brain capillary endothelium, involves three steps arranged in series. Solutes enter the cell at one side, traverse the cytoplasm, and exit at the other side of the cell. The first and last steps involve transport across functionally different regions of the plasma membrane. They may be accomplished by simple diffusion, but solutes often receive help from selective channels or transporters. Solutes may simply diffuse through the aqueous cytoplasm, but there is often the potential for significant detours caused by binding to cytoplasmic proteins or organelle surfaces, by sequestration within organelles, or by biotransformation to metabolites with altered transport properties. The concentration of solute in the fluid bathing each surface of the epithelium and in the cytoplasm along with the weighted contribution

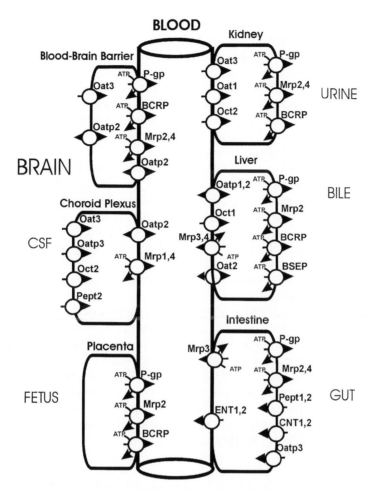

Figure 15.2. Location of xenobiotic transporters in selected barrier and excretory tissues. For simplicity, the tissues are arranged along a structure representing the vascular space. Arrows indicated direction of transport under normal conditions. This figure is not meant to be comprehensive; not all transporters expressed in a tissue are shown. Transporters driven by ATP pump substrates out of cells (efflux). Other transporters are capable of supporting substrate uptake or efflux. Which of these processes predominates depends on available driving forces—for example, substrate concentration gradient and the capability to couple transport to sources of potential energy.

of each of the biological processes (transport at each limiting membrane and through the cytoplasm) together determine individual unidirectional trans-tissue fluxes as well as the resultant net flux. Given this complex picture, it is clearly the polarized arrangement of transporters in these tissues (Figure 15.2), their differential selectivities, and their underlying energetics that determine to what extent and in which direction a given solute will be transported.

Consider two examples. First, in the brain capillary endothelium, luminal expression of the ATP-driven drug efflux pump, *p*-glycoprotein, provides a formidable barrier to drug penetration from blood to brain (Figure 15.2). This is best seen in experiments with animals in which p-glycoprotein activity has been abolished either genetically (knockout) or with specific inhibitors. In these animals, brain concentrations of drugs that are p-glycoprotein substrates are substantially increased, but increases in plasma concentrations are small. These results reflect (1) the important influence of p-glycoprotein as a gatekeeper for the CNS and (2) the more modest contribution of p-glycoprotein to excretion of the drugs in liver and kidney. These findings emphasize the power of xenobiotic efflux pumps to greatly influence the distribution of substrates in the body. They also show that transporter function can be tissue-specific as well as substrate-specific.

Second, anionic drugs or anionic products of Phase II drug metabolism produced in liver are strongly excreted into the urine in the proximal segment of the nephron. Here transport from blood to urine is by a two-step mechanism (Figure 15.2). The first step, uptake into the cell at the basolateral membrane, is mediated by two members of the OAT family of transporters, OAT1 and OAT3. For these transporters, concentrative uptake is indirectly powered by the potential energy stored in the Na gradient (Figure 15.1B). At the other side of the cell, several transporters are available to mediate efflux across the luminal plasma membrane into the urinary space. Because of the transmembrane electrical potential (inside of the cell negative relative to outside), this step is energetically downhill for organic anions. It may be facilitated by different OAT (OAT2) family members or by members of the OATP (OAT-K1,2) family. In addition, these organic anions may be pumped out of the cells by MRP2 and MRP4, two ATP-driven transporters that are localized to the luminal membrane (Figure 15.2). Thus, at each side of the epithelial cells, plasma membranes contain multiple transporters that overlap in function. This multiplicity of transporters results in increased transport capacity of each membrane and, through redundancy, lessens the consequences when function of a single transporter is lost through mutation or inhibition.

15.3 XENOBIOTIC TRANSPORTERS

15.3.1 Transporter Families

Historically, studies of xenobiotic transport have progressed through three periods: descriptive, mechanistic, and molecular. In the earliest, descriptive period, it was recognized that certain solutes were preferentially excluded from or concentrated in cells or specific fluid compartments. These studies led to the concepts of urine and bile as excretory pathways for xenobiotics and brain capillaries, testes, placenta, and intestine as barriers to xenobiotic movement. The second period has been concerned with the cellular mechanisms that were the basis for selective excretory and barrier function. Through characterization of the tissue distribution, specificities, energetics, and differential regulation of transport processes, it was established that multiple uptake and efflux transport proteins were involved and that these could be conveniently grouped into families. In the third period, specific transporters were identified and characterized at the molecular level, the molecular basis for transporter families was established, and atomic level information on transporter function (specificity and energetics) were obtained from studies of protein structure and computer simulations of protein dynamics. All three periods are still progressing in parallel to provide molecular/atomic level models that can be used to (1) describe transporter function and coupling to metabolism, (2) understand regulation of function, and (3) build complex, integrative, pharmacokinetic-based models of xenobiotic distribution and action.

Over aproximately the past 20 years, xenobiotic transport proteins have been classified into multiple superfamilies and subfamilies based on sequence homology. In mammals, most xenobiotic transporters fall within the ABC transporter or major facilitator (SLC) superfamilies (Figure 15.3). Within each of these are subfamilies of transporters with similar amino acid sequences and also roughly similar specificities. Subfamily specificities can be narrow, as for the peptide and nucleoside transporters, or more broad, as for the OATs, OATPs, and MRPs (which handle primarily anionic xenobiotics and metabolites) and the OCTs (which handle mostly cationic xenobiotics). In addition, there are two highly multispecific xenobiotic transporters, *p*-glycoprotein (ABCB1) and BCRP (ABCG2). These are the only xenobiotic transporters within their respective ABC transporter subfamilies, and each handles a truly remarkable mix of negatively charged, positively charged, and uncharged xenobiotics with molecular masses from about 300 to several thousand daltons (Figure 15.3).

Major Facilitator Superfamily (SLC)

Organic Anion Transporters (OATs; SLC22A): Multispecific, organic anion transporters/exchangers; Oat1 and Oat3 are indirectly coupled to Na
Substrates: NSAIDS, fluoroquinalones, -lactam antibiotics, nucleoside phosphonates (Oat1), PAH (Oat1 and Oat3), AZT, prostaglandins, statins, cimetidine (Oat3), MTX, thyroxine, GSH, glucuronide and sulfate conjugates

Organic Anion Transporting Polypeptides (OATPs; SLC21): Multispecific, organic anion transporters/exchangers; not Na-coupled
Substrates: MTX, thyroxines, digoxin (Oatp2), bile salts, peptides, eicosanoids, statins, GSH, glucuronide and sulfate conjugates of hormones

Organic Cation Transporters (OCTs; SLC22A): Multispecific, organic cation transporters/exchangers.
Oct1,2,3 substrates: neurotransmitter derivatives, cimetidine, amantidine, Ca channel blockers, quinacrine, quinine
OctN1,N2 substrates: carnitine and some tetraalkylammoniums

Peptide transporters (PEPT1-2; SLC15A): Proton coupled cotransporters
Substrates: di- and tripeptides, -lactam antibiotics, valacyclovir, ACE inhibitors, bestatin

Nucleoside transporters (ENT1-2, SLC28A; CNT1-3, SLC29A): Equilibrative (ENTs) and Na-coupled (CNTs) transporters
Substrates: AZT, fluorouridine, 5dFU, gemcitabine

ABC Transporters: Highly multispecific, ATP-driven, drug export pumps

p-Glycoprotein (MDR, p-gp; ABCB1) transports organic cations, weak organic bases, some uncharged drugs and some polypeptide derivatives chemotherapeutics, Ca channel blockers, digoxins, cyclosporins, HIV protease inhibitors

Multidrug Resistance-Associated Proteins (MRP1-9; ABCC sub-family) transport organic anions, weak organic acids, some uncharged drugs, some polypeptide derivatives and products of Phase II drug metabolism MTX, statins, endothelin, cyclic nucleotides (AZT, PMEA; Mrp4, Mrp5), cephalosporins, GSH, glucuronide and sulfate conjugates

BCRP (ABCG2) transports hydrophobic compounds, weak bases, organic anions, and glucuronide-, sulfate-, glutamylate- and glutathione-conjugates of many endogenous and exogenous molecules (drugs, carcinogens, etc.)

Figure 15.3. Substrate selectivity and energetics of major xenobiotic transporters.

15.3.1a   ABC Transporters. Primary active xenobiotic transporters (ATPases) are of special interest, because of their profound influence on the uptake, distribution, and excretion of drugs. For the most part, these transporters belong to the ATP-binding cassette (ABC) superfamily. Most ABC transport proteins contain two intracellular nucleotide-binding domains and two transmembrane regions consisting of a 12-transmembrane segment. Some ABC transporters may have additional transmembrane domains (MRPs) and a few are half-transporters, with a single nucleotide-binding domain (BCRP).

The best-known and most widely studied representative of the ABCB subfamily is *p*-glycoprotein (ABCB1), a phosphorylated glycoprotein with an apparent molecular weight of 170 kD. *p*-Glycoprotein was discovered over 20 years ago as a gene that was overexpressed in certain multidrug-resistant tumor cells. The gene was found to code for an ATP-driven, plasma membrane efflux transporter that accepted a remarkably wide range of substrates (MDR1; Figure 15.3). Many of the drugs that were found to be ineffective in these cells were highly lipophilic (although cationic) and thus expected to readily diffuse through membranes. Nevertheless, high expression of *p*-glycoprotein essentially abolished drug accumulation in resistant cells. Drug resistance roughly correlated with transporter expression and both resistance, and expression could be induced by exposing cells to chemotherapeutics. Inhibition of the transporter increased the sensitivity of the cells to chemotherapeutics, and extensive efforts have been under way to identify potent and specific inhibitors that can be used in the clinic to reverse *p*-glycoprotein-based multidrug resistance.

Several years after the discovery and characterization of *p*-glycoprotein, a second transporter was cloned from multidrug-resistant tumor cells that did not express *p*-glycoprotein. Although ATP-driven, this transporter had a specificity profile that was different from that of *p*-glycoprotein. It tended to prefer anionic substrates and clearly could transport not only drugs but also drug metabolites (Figure 15.3). As with *p*-glycoprotein, overexpression of the second transporter greatly limited cellular accumulation of certain chemotherapeutics. This transporter was named multidrug-resistance-associated protein (MRP).

At present, nine members of the MRP family of transporters (ABCC subfamily) have been cloned. Of these, the tissue distributions and functions of MRP1-5 are well-characterized. MRP1-3 appear to be highly multispecific, handling a large number of anionic xeniobiotics and their metabolites, especially Phase II (conjugation) products. In this regard, recent evidence indicates colocalization of some Phase II isozymes like GST-π with MRPs, suggesting especially close coupling between xenobiotic metabolism and metabolite elimination. In addition, these MRPs can transport glutathione (GSH), either by itself or in concert with certain chemotherapeutics. In hepatocytes, MRP2 has been implicated in the efflux of GSH into bile, a process that drives a significant fraction of bile flow. In some cells upregulation of MRP1 and MRP2 has been implicated in the depletion of GSH and increased sensitivity to toxicants.

MRP4,5 have more restricted specificity, with substrates including nucleotide/nucleoside-based drugs (e.g., adefovir) and endogenous organic anions (e.g., cAMP and cGMP) (Figure 15.3). Expression in smooth muscle cells (MRP5), platelet-dense granules (MRP4), and nerve cells (MRP4 and MRP5) suggest physiological function involving signaling.

BCRP (ABCG2) was first cloned from placenta and from multidrug-resistant tumor cells that did not express *p*-glycoprotein or MRP1. BCRP is a half-ABC transporter, because it contains only a single nucleotide-binding domain. It appears to be functional as a dimer or higher oligomer. BCRP transports a wide range of drugs including large, hydrophobic, organic anions and cations (Figure 15.3). Thus, the range of substrates handled overlaps with both *p*-glycoprotein and the MRPs. Low-oxygen conditions upregulate BCRP expression, and high transporter expression is found in tissues exposed to low-oxygen environments. BCRP interacts with heme and other porphyrins and is protective against protoporphyrin accumulation during hypoxia. BCRP is also expressed in a variety of stem cells and may act to protect them from toxicants.

Multidrug-resistance transporters—*p*-glycoprotein, the MRPs, and BCRP—are expressed in normal cells, especially in the cells of certain barrier and excretory tissues (Figure 15.2). In some tissues, expression of specific ABC transporters is especially high—for example, *p*-glycoprotein in the brain capillaries (blood–brain barrier), MRP2 and BCRP in renal proximal tubule, and MRP2 and *p*-glycoprotein in small intestine. In each of these tissues the transporters indicated are localized to the luminal plasma membrane, the correct location to provide a barrier to uptake (brain and intestine), or the final step in excretion (renal tubule).

15.3.1b Organic Anion Transporters (OATs; SLC22A).
Six members of the OAT family of transporters have been cloned, and the first four members have been extensively characterized. Although there are differences in specificity among family members, all handle a wide range of organic anions from waste products of cellular metabolism to drugs and drug metabolites to environmental pollutants (Figure 15.3). OAT1 and OAT3 have received particular attention since they drive potent, Na-dependent excretion of organic anions in renal proximal tubule, a process of historic importance and mechanistic interest (indirect coupling to Na; see Figure 15.1B). OAT3 is the only member of this family that is extensively expressed in brain (blood–brain barrier and choroid plexus), suggesting at a minimum a role in transport of neurotransmitter metabolites out of the CNS.

15.3.1c Organic Cation Transporters (OCTs; SLC22A).
Roughly two-thirds of commonly prescribed drugs are organic cations or weak organic bases. OCT1,2 are multispecific organic cation transporters that mediate potential-driven uptake from blood to cell in excretory epithelia (e.g., kidney and liver). OCT3 also handles some neurotransmitters and is expressed in brain as well as other tissues. OCTN1 and OCTN2 transport the essential nutrient carnitine into cells through Na/carnitine cotransport. They also handle a restricted list of other organic cations, suggesting the potential for drug–nutrient interactions at these transporters.

15.3.1d Organic Anion Transporting Polypeptides (OATPs; SLC21A).
More than 50 OATPs have been cloned. These transporters can be roughly divided into two groups: (1) transporters that show broad substrate specificity and are expressed principally in liver, kidney, and barrier/excretory tissues and (2) transporters that handle a more limited set of substrates [e.g., the prostaglandin transporter, hPGT (SCL21A2 or OATP2A1)]; these appear to have specific functions in certain cells. Unlike other transporter families (e.g., ABC transporters and OATs), OATPs do

not exhibit one-to-one correspondence between the human and rodent forms, making extrapolation of results from model species to man difficult and complicating the systematic naming of individual family members. Obviously, with so many transporters and multiple functions, it is difficult to generalize about common OATP family specificity characteristics, other than to say that all of them transport organic anions. Although many OATps appear to be capable of concentrative transport, it is not clear how such transport is coupled to cell metabolism or ion gradients.

15.3.1e *Nucleoside and Nucleotide Transporters (SCL28 and SLC29).* There are three members of the concentrative nucleoside transporter family, CNT1-3 (SCL28A1-3), and four members of the equilibrative nucleoside transporter family, ENT1-4 (SLC29A1-4). As implied by their names, the CNTs are capable of sodium-driven, concentrative cellular uptake of purine and pyrimidine nucleosides and their analogs. CNT1 prefers pyrimidines, CNT2 transports purines, and CNT3 handles both. In polarized epithelia, the CNTs are apically expressed. On the other hand, the ENTs mediate facilitative diffusion and will not support energetically uphill transport. All four ENTs are widely distributed in various cell types, and ENT1 and 2 are known to be basolateral in the polarized epithelium of the kidney. Thus, it appears that in combination the apical CNTs and the basolateral ENTs are organized to mediate absorptive transport (toward the blood) of nucleobases, nucleosides, and their therapeutic analogs.

15.3.1f *Peptide Transporters (SLC15).* All four PEPT family members are proton-coupled cotransporters that handle di- and tripeptides in addition to peptidomimetric drugs and prodrugs. PEPT1 and PEPT2 (SLC15A1 and A2) are predominantly expressed in polarized epithelia, where they mediate the apical uptake step in peptide absorption or reabsorption. For most substrates, PEPT1 is a high-capacity, low-affinity system. PEPT2 generally has a lower capacity and higher affinity. In intestine and kidney, these transporters effectively remove peptide substrates from the lumen to the blood.

15.3.2 Determining the Molecular Basis for Transport in Cells and Tissues

How does one determine which transporters are responsible for the uptake or efflux of a given chemical in a specific cell or tissue? Several pieces of information are needed. It is critical to know enough about the transport of the chemical in the native system to define the specificities and energetics involved. One must also determine through inhibition and kinetic studies whether such transport involves a single transporter or multiple processes. From these data, one can narrow the field of possibilities and then establish which candidate transport proteins are actually expressed in the cells and where they are localized. At this point, one can attempt to match the transport characteristics (specificity, energetics) of the chemical in the cells with those of the chemical in heterologous systems that are engineered to overexpress single, cloned transporters. Finally, one can use transporter knockouts or knockdowns to confirm the identity of the transporter(s) involved. This grand scheme assumes that (1) the cells express a limited number of candidate transporters, (2) transport of the chemical is remarkable enough to eliminate all but a few choices, (3) substantial information is available about transport mediated by the cloned transporters to eliminate additional possibilities, and (4) the molecular tools needed to modify transporter expression are both available and effective. The scheme fails if the transport characteristics determined in the cells are not unique to a single transporter—that is, if there are multiple matches or none.

15.4   ALTERED XENOBIOTIC TRANSPORT

15.4.1   Competition for Transport

Several notable features of xenobiotic transporters and their regulation bear importantly on mechanisms of toxicity. Many of these transporters (e.g., p-glycoprotein, MRPs, OATPs, OATs, and OCTs), are multispecific and transport a large number of substrates with a wide range of chemical structures. For some, the range extends over multiple classes of drugs (Figure 15.3). Since polypharmacy is a fact of life in the clinic, competition for transport can be a complicating factor in pharmacotherapy. That is, one can envision competition for transport not only between compounds with similar structures, but also between compounds that are structurally unrelated but are still handled by a common transporter. Indeed, since patients often receive multiple drugs, competitive inhibition of the excretion of one drug by another has become a well-documented phenomenon. Consider the cardiac glycoside, digoxin, discussed earlier. Digoxin renal excretion is decreased and plasma levels are increased when the drug is given with a number of other p-glycoprotein substrates, including verapamil, quinidine, and statins. For the opiate loperamide, used as a locally acting antidiarrheal, co-administration with quinidine leads to significant central opioid effects. In the absence of quinidine, p-glycoprotein at the blood–brain barrier limits loperamide entry into the CNS and there are no CNS effects. The quinidine effect suggests competition between loperamide and quinidine for transport by p-glycoprotein at the blood–brain barrier.

Knowledge of such competition can be put to good use in the clinic. Two examples involving Oat family transporters are particularly instructive. First, when first widely used, the antibiotic penicillin was given in high dose because of rapid renal excretion. This problem was exacerbated by short supply of the new wonder drug. It was subsequently found that the plasma half-life of the drug could be substantially prolonged if it was given with probenecid, a potent competitive inhibitor of renal organic anion transporters (especially OAT1 and OAT3) that are responsible for the first step in penicillin excretion, uptake into renal cells. Thus, through inhibition of renal excretion, lower doses of a drug that was in critically short supply could be used with no loss in effectiveness.

Second, certain nucleotide phosphonates (e.g., adefovir and cidofovir) are effective antivirals, but their use in the clinic is limited by renal toxicity. This is believed to be caused by avid uptake at the basolateral membrane of renal proximal tubule cells followed by slow transport into the urine at the apical membrane, a sequence of events that results in intracellular drug accumulation and thus toxicity. As with penicillin, the OAT family of transporters has been implicated in cidofovir uptake. Co-administration of probenecid with cidofovir has been shown to decrease renal clearance of the antiviral and reduce its nephrotoxicity, presumably through competition for uptake from blood to renal proximal tubule cell by OAT1. Thus, an understanding some aspects of renal excretory transport provided a tool that could be used on the one hand to increase the effectiveness of a given dose antibiotic by blocking excretion and on the other hand to reduce drug renal toxicity by blocking accumulation in the specific renal cells that are the sites of toxicity.

Since xenobiotic transporters handle xenobiotics and potentially toxic endogenous metabolites, there is the potential for transporter-based interactions between xenobiotics and metabolites and thus for secondary disruptive effects on both metabolite clearance and metabolic pathways. For example, the bile salt export pump (BSEP; ABCB11) is an ABC transporter that is localized to the canalicular plasma membrane hepatocytes. BSEP is of primary importance in the biliary excretion of conjugated bile acids. It also transports some drugs—for example, the cholesterol-lowering drugs, pravastatin and troglitazone. The latter drug has been withdrawn from the market largely because of hepatotoxicity likely mediated through BSEP.

15.4.2 Specific Regulation

15.4.2a Transporter Expression. Changes in transporter expression and ultimately activity can be responsible for alterations in drug effects. It is now clear that xenobiotic transporters are transcriptionally regulated through multiple pathways. Some of these are activated by stress signals (e.g., inflammation, hypoxia, toxicants, and buildup of metabolic waste products), and some are activated by xenobiotics. In many cases, xenobiotics actually cause transcriptional upregulation of transporters for which they are substrates (see below). The best-documented targets for these xenobiotics are the ligand-activated nuclear receptors that coordinately regulate transcription of drug metabolizing enzymes and the transporters responsible for excretory transport of xenobiotics and their metabolites. The pregnane-X receptor (PXR) is one such receptor expressed in barrier and excretory tissues. PXR regulates expression of multiple phase I and phase II metabolizing enzymes as well as drug efflux transporters, including *p*-glycoprotein, MRP2, and several OATPs. Ligands that activate PXR include endogenous metabolites (bile salts) as well as a large number of commonly prescribed drugs, including steroids, glucocorticoids, Ca channel blockers, chemotherapeutics, antidiabetics, and statins. Note that many of these classes of drugs contain substrates for transport by *p*-glycoprotein and MRPs and that many drug conjugates are substrates for MRPs.

Activation of PXR increases expression of multiple xenobiotic transporters in barrier and excretory tissues, thereby decreasing drug access to protected compartments (e.g., brain) and increasing renal, hepatic, and gastrointestinal excretion and thus plasma levels. The overall effect of PXR-driven *p*-glycoprotein induction would be reduced drug efficacy as a result of further tightening of specific barriers and increased excretion into bile and urine. This has been observed in the clinic where plasma concentrations of the antibiotic, rifampin, and the immunosuppressive, cyclosporine A, were reduced to below effective levels when patients were chronically treated with drugs that were PXR ligands.

15.4.2b Transporter Function. Function of certain ABC transporters appears to be regulated through mechanisms that affect their insertion into and retrieval from the plasma membrane. This is best documented in hepatocytes, where recent studies show regulated trafficking of BSEP, *p*-glycoprotein, and Mrp2 between the canalicular plasma membrane and intracellular vesicular compartments. For the hepatocytes, this provides a means of rapidly responding to changing conditions by altering the efflux transport capacity at the pole of the cell facing the biliary space. BSEP and MRP2 are important contributors to bile formation and defects in these transporters that abolish or reduce function lead to cholestasis.

15.4.3 Genetic Heterogeneity

Alterations in genes coding transporters can result in physiologically significant effects through changes in transporter expression and function. At the molecular level, effects of altered transport proteins can range from profound (no protein expression, mistargeting, loss of function) to subtle (altered transport capacities or affinities). Thus, as with drug metabolizing enzymes, these changes have the potential to contribute substantially to patient-to-patient variability in response to drugs and toxicants. They form one basis for the emerging field of personalized medicine.

Mutations of genes within the ABC transporter family causes or contributes to several human disorders, including retinal degeneration, cystic fibrosis, neurological disease, cholesterol and bile transport defects (BSEP and MRP2), anemia, and inflammatory bowel disease (BCRP). In addition, a number of single nucleotide polymorphisms (SNPs) have been identified in the coding regions of MDR1, MRP1, MRP2, and BCRP. They have been evaluated in expression systems where a number of SNPs have yielded evidence of altered expression or function. Frequencies of occurrence have been determined for a number of these SNPs in several populations, and some studies have begun to evaluate how they might affect drug distribution. At present, there are little data available and substantial controversy as to the nature of clinically relevant effects.

SUGGESTED READING

Anzai, N., Kanai, Y., and Endou, H. Organic anion transporter family: Current knowledge. *J. Pharmacol. Sci.* **100**, 411–426, 2006.

Breedveld, P., Beijnen, J. H., and Schellens, J. H. Use of *p*-glycoprotein and BCRP inhibitors to improve oral bioavailability and CNS penetration of anticancer drugs. *Trends Pharmacol Sci.* **27**, 17–24, 2006.

Choudhuri, S., and Klaassen, C. D. Structure, function, expression, genomic organization, and single nucleotide polymorphisms of human ABCB1 (MDR1), ABCC (MRP), and ABCG2 (BCRP) efflux transporters. *Int. J. Toxicol.* **25**, 231–259, 2006.

Deeley, R. G., Westlake, C., and Cole, S. P. Transmembrane transport of endo- and xenobiotics by mammalian ATP-binding cassette multidrug resistance proteins. *Physiol Rev.* **86**, 849–899, 2006.

Endres, C. J., Hsiao, P., Chung, F. S., and Unadkat, J. D. The role of transporters in drug interactions. *Eur. J. Pharm. Sci.* **27**, 501–517, 2006.

Ho, R. H., and Kim, R. B. Transporters and drug therapy: Implications for drug disposition and disease. *Clin. Pharmacol. Ther.* **78**, 260–277, 2005.

Shitara, Y., Horie, T., and Sugiyama, Y. Transporters as a determinant of drug clearance and tissue distribution. *Eur. J. Pharm. Sci.* **27**, 425–446, 2006.

Various Authors. *Toxicol. Appl. Pharmacol.* **204**, 197–360, 2005.

CHAPTER 16

Mechanisms of Cell Death

MAC LAW

Department of Population Health and Pathobiology, College of Veterinary Medicine, North Carolina State University, Raleigh NC 27695

SUSAN ELMORE

National Institute of Environmental Health Sciences, Research Triangle Park, North Carolina 27709

16.1   INTRODUCTION

One of the most fundamental concepts in toxicology is that the actual harm done by toxic compounds occurs in living flesh and blood, ultimately at the level of the *cell* and its component parts. This concept of the cell as the common denominator in tissue/organ injury was first put forth in the nineteenth century by Rudolf Virchow, generally credited as the "father" of modern pathology. This chapter will cover the nature of cell injury, followed by the molecular and biochemical mechanisms of cell death. While we will focus mainly on the cell, keep in mind that cells of course do not live in isolation but rather constantly interact with their surrounding matrix components and the total milieu of extracellular fluids, including many biochemical mediators. It is possible to have disease states that involve derangements in plasma biochemistry without remarkable morphologic changes noted in cells by routine methods (an example would be acute hypoglycemia). In addition, molecular analyses are beginning to reveal gene changes that may have profound effects on cell function without always causing marked changes in cell morphology. However, most currently understood diseases are associated with specific, detectable changes in cell and, thus, tissue morphology.

Cell death occurs in normal development and maturation but can also be a response to xenobiotics, microorganisms, and physical agents (i.e., trauma, radiation, temperature extremes) or to endogenous changes such as inflammation, genetic derangements, or altered blood supply. Cell death occurs by two primary alternative modes: apoptosis and necrosis. An understanding of the physiologic conditions and biochemical changes under which cell injury and death occur continues to be an important focus of biomedical research.

Molecular and Biochemical Toxicology, Fourth Edition, edited by Robert C. Smart and Ernest Hodgson
Copyright © 2008 John Wiley & Sons, Inc.

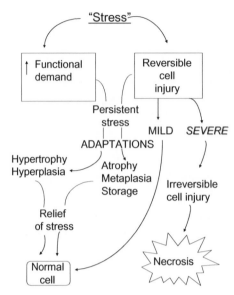

Figure 16.1. Cell stress and adaptations.

16.2   HOW CELLS/TISSUES REACT TO "STRESS"

16.2.1   Levels of Adaptation: Physiology Versus Pathology

Cells can adapt to alterations in their environment and assume a number of altered steady states (Figure 16.1). Some of these alterations could be looked upon as minor injuries with which the cell is able to cope by assuming an altered biochemical composition and morphological appearance. This process is called *adaptation*. Adaptation can be successful or unsuccessful (i.e., retrograde). More serious injury to the cell will exceed the cell's adaptive capabilities and result in more serious biochemical derangements. These are heralded by drastic changes in the cell's morphological appearance culminating in *cellular degeneration, cell death,* and finally *necrosis.* The final outcome, of course, depends on the type of cell, nature of the injury, and the magnitude and duration of the injurious influences.

16.2.2   Hypertrophy

Hypertrophy is an increase in the *size* of the cells due to synthesis of more subcellular components. An increase in the size of the organ results. Take, for example, cardiac hypertrophy in response to increased workload. The increased load is often a result of some pathological condition such as heart valve incompetence or stenosis of an outflow path. Hypertrophy of the muscles of a body builder is another good example, in which the increased workload is due to continued weight training.

16.2.3   Hyperplasia

Hyperplasia is an increase in the *number* of cells in a given organ or tissue, usually resulting in an increase in the size of the organ as well. Only certain differentiated tissue types have the capability to synthesize DNA, permitting mitotic division and, thus, hyperplasia. Although hypertrophy and hyperplasia are two distinct processes, they often occur together and may be triggered by the same factors. Hormone-induced growth of tissues such as the uterus involves both an increase in the size of the uterine smooth muscle and epithelial cells and an increase in their numbers.

16.2.4 Atrophy

Atrophy is a decrease in the size of a cell due to loss of subcellular components. Shrinkage of the involved tissue or organ results. Decreased cell size is accomplished by autophagocytosis of those cell organelles not essential for cell function at the new steady state. Smaller cells containing residual bodies composed in part of lipo-fuscin (a pigment representing the indigestible residue of cell membranes) result. The causes of atrophy include diminished blood supply, disuse or diminished work-load, inadequate nutrition, denervation, loss of hormonal stimulation, pressure (somewhat related to diminished blood supply), and aging.

Atrophic organs present a distinctive appearance. There is an overall decrease in size. Often the capsule is undulating or wrinkled. A relative increase in connective tissue versus normal tissue (parenchyma) may be apparent. Replacement of paren-chyma by fat (fatty change) is also possible. Histologically, parenchymal cells will be smaller and contain fewer organelles (fewer mitochondria and filaments, less endoplasmic reticulum) and perhaps a fine granular brownish pigment in the cyto-plasm (lipofuscin). Atrophy can progress to cell death and replacement of the lost parenchymal cells by fibrosis (scarring) if, for instance, blood supply is inadequate even to support the needs of atrophic cells. Hence, degeneration, apoptosis, and necrosis may coexist with atrophy in some instances.

16.2.5 Metaplasia

In some cases, cells in a persistently stressful environment may be replaced by a cell type that is better able to withstand the harsh environment. Replacement of an adult cell type by another adult cell type is known as metaplasia and is usually a reversible change. Perhaps the best-known example of metaplasia is the replacement of the normally tall (columnar), ciliated epithelial cells in the respiratory tract of cigarette smokers by layers of relatively flattened (squamous) epithelial cells. While the squa-mous epithelial cells are somewhat tougher than the columnar cells, important functions such as mucus secretion are lost. In addition, the continued stresses that induce metaplasia may lead to neoplastic transformation.

16.2.6 Dysplasia

Dysplasia is a special type of cell adaptation that literally means disordered growth, and it is usually associated with neoplasia or preneoplastic changes. Dysplasia implies a loss in the uniformity of the individual cells and a loss in the normal tissue growth pattern. For example, a dysplastic area involving the epidermal layers of the skin might contain jumbled, atypical epithelial cells with loss of their normal polarity and orderly differentiation.

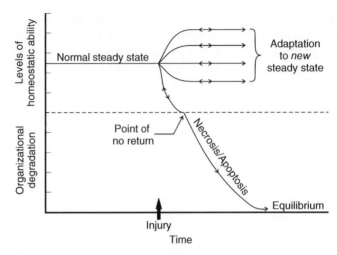

Figure 16.2. Reversible versus irreversible cell injury. Depending on the cell and/or tissue type and the relative "health" of the individual, cells may adapt to stress within a limited range of homeostatic ability. If cell injury exceeds this range, the "point of no return" is reached (irreversible injury) and the cell dies. Note that cell death and the degradative changes associated with necrosis are at different time points.

16.3   CELL INJURY AND CELL DEATH

In the previous sections, we discussed how cells are able to adapt to minor stresses in their environment within a somewhat narrow range. In cell injury, the adverse influence exceeds the cell's adaptive homeostatic abilities. The outcome depends on the magnitude and duration of the injury. Either a series of reversible biochemical and morphological changes occur or, once a point of no return is reached, irreversible biochemical and morphological changes supervene, culminating in cell death and necrosis (Figure 16.2). There is a fine line between reversible and irreversible cell injury. This line ("the point of no return") is very difficult to establish biochemically and morphologically.

At this point, a few terms need to be defined:

1. *Injury.* An adverse influence which exceeds the cell's ability to achieve a steady normal or adaptive homeostatic state.
2. *Degeneration.* An older term that implies reversible changes in injured cells. Degenerations are, then, the morphological manifestations of reversible cell injury.
3. *Cell Death.* The result of irreversible cell injury. Cell death is not synonymous with necrosis, since fixed, normal tissues are no longer alive but have not undergone the changes of necrosis.
4. *Necrosis.* The series of morphological changes following cell death in a living organism.
5. *Autolysis.* The series of morphological changes in cells following *somatic death* (i.e., death of the individual). The processes occurring are in many respects the same as those in necrosis except that the normal body responses to necrosis

(e.g., inflammation; cell proliferation, etc.) are incapable of occurring. Auto-lytic changes (= decay or rot) after death make proper examination of tissues for "real" lesions difficult, since they can mask many changes. Autolysis and other artifacts can be avoided or at least minimized by rapid cooling (but not freezing!) of the tissues, careful tissue handling, and proper fixation methods.

6. *Putrefaction.* Results from the action of saprophytic bacteria which are either normal flora of the tissue concerned or invade from the intestinal tract at the time of or after death. Putrefaction is a common accompaniment of autolysis and also occurs occasionally as a complication of necrosis.

16.3.1 Causes of Cell Injury and Cell Death

16.3.1a Hypoxia. Hypoxia, a deficiency of oxygen, may result from loss of blood supply (ischemia), decreased cardiopulmonary function that leads to reduced oxy-genation of blood, and decreased oxygen-carrying capacity of blood as in anemia. Lack of oxygen causes cell injury by reducing aerobic oxidative respiration (discussed below). It should be noted that ischemia deprives cells not only of oxygen, but also of critical metabolic substrates normally supplied by the blood such as glucose. Thus, cell damage ensues more rapidly and severely than with hypoxia alone.

16.3.1b Physical Agents. Physical agents include mechanical trauma, extremes of temperature and radiation, electric shock, and so on, which may damage tissues directly or indirectly via ischemia or secondary inflammation.

16.3.1c Chemicals. Simple chemicals such as high concentrations of glucose or salt may cause cell injury directly or by creating osmotic disturbances in cells. Caustic agents exert their direct effects by coagulation of proteins or corrosive action. Concentration is important. Poisons exert their effects at very low concentra-tions. For example, some poisons specifically deactivate vital enzyme systems: Cyanide attacks cytochrome oxidase; sodium fluoroacetate deactivates aconitase. Others act more generally through formation of reactive chemicals with unpaired "free radical" electrons that react with (a) unsaturated lipids in cell membranes, (b) specific components of amino acids in proteins, or (c) DNA. The site of entry into the body is not always the site of injury. The toxic principal of some poisons is produced by metabolism of the parent compound, and hence sites of metabolic detoxification and excretion are susceptible. For example, carbon tetrachloride (CCl_4) given orally causes degeneration and necrosis in the liver.

16.3.1d Microbial Agents. Infectious agents include viruses, bacteria, fungi, mycoplasma, rickettsia, protozoa, nematodes, and other parasites, each of which has its own unique pathogenesis. They may injure cells directly, but often the host immune response to the microbe causes much of the tissue damage. For example, it was recently demonstrated that the deadly 1918 flu virus likely killed even healthy young adults not due to direct viral effects, but because it triggered an overwhelm-ing and self-destructive immune response in the lungs. Perhaps the newest infec-tious agents on the biomedical scene are the prion proteins (PrP), which appear to damage the central nervous system by a poorly understood mechanism of protein conformational change.

16.3.1e Other Causes. Autoimmune diseases, nutritional imbalances, and genetic derangements may also cause cell injury and death. It should be noted that individual variations in genetic makeup also influence the ability of cells to withstand chemicals and other environmental insults.

16.3.2 Pathogenesis of Reversible Versus Irreversible Cell Injury

We will use ischemic cell injury as a model (Figure 16.3). Hypoxia rapidly leads to cessation of oxidative phosphorylation in mitochondria and the replenishment of ATP slows or stops. The cell quickly switches over to anaerobic glycolysis to generate ATP. Thus, tissues that have greater glycolytic capacity, such as liver, can withstand such insults longer than other tissues. Glycogen is quickly depleted and lactic acid accumulates, resulting in a decrease in intracellular pH, which is associated with clumping of nuclear chromatin as seen by electron microscopy (EM).

The normal cell possesses lower intracellular concentrations of sodium, calcium, and so on, than extracellular fluid (to balance the high intracellular osmotic colloidal pressure exerted by higher protein concentration inside the cell than outside). Conversely, the intracellular potassium concentration is substantially higher than the extracellular concentration. This balance is maintained by an ATP-driven enzymatic sodium pump (Na^+,K^+-ATPase) located in the cell membrane. The diminished ATP concentration in hypoxia leads to failure of this active transport system and sodium accumulates intracellularly while potassium diffuses out of the cell. There is a *net gain of solute* within the cell which is accompanied by an isosmotic gain of water, causing acute cell swelling. A second mechanism for cell swelling is increase in intracellular osmotic load caused by accumulation of catabolites such as inorganic phosphates, lactate, and purine nucleosides in certain disease states. In

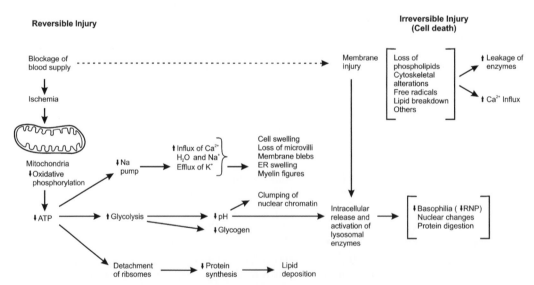

Figure 16.3. Mechanisms of cell injury following an ischemic event, such as blockage of a blood vessel by a blood clot. ER, endoplasmic reticulum; RNP, ribonucleoprotein. [Adapted from Kumar, Abbas, and Fausto (Eds.) *Robbins and Cotran Pathologic Basis of Disease*, 7th ed., Elsevier Saunders, Philadelphia, 2005, Fig. 1-22, p. 24.]

addition, failure of the Ca^{2+} pump leads to an influx of Ca^{2+}, with numerous effects on cell organelles (discussed below).

Cell swelling first manifests as dilation of the endoplasmic reticulum (ER). Detachment of ribosomes from the rough ER and dissociation of polysomes into monosomes occurs next because of diminished ATP generation, causing reduced protein synthesis. If hypoxia persists, further membrane dysfunction, manifested by more severe increase in permeability, occurs. Cell surface blebs and loss of microvilli in cells so endowed are ultrastructural correlates of more severe membrane damage (Figure 16.4A,B). The point of irreversibility has not yet been reached if oxygen is restored. Indeed, it is very difficult to define precisely when the cell is irreversibly injured.

If ischemia persists, irreversible injury follows (Figure 16.4C). It seems to coincide with (a) marked swelling of mitochondria and their cristae and (b) the appearance of amorphous mitochondrial densities and high concentrations of calcium. Mitochondria eventually rupture. Severe lysosomal membrane dysfunction, along with release and activation of lysosomal enzymes (by low pH), is a very late event occurring long after the stage of irreversibility has been reached. Lysosomal enzymes with their degradative abilities do, however, contribute much to the morphological appearance of necrotic cells. Indeed, the leakage of lysosomal and cytosolic enzymes across the abnormally permeable plasma membrane provide important clues to cell injury and cell death. These enzymes enter the blood and can be readily assayed. By choosing to assay appropriate, relatively organ-specific enzymes, a clinician/toxicologist can be appraised of cell injury and cell death in a particular organ or tissue (see Table 16.1). It should be stressed that the serum concentration of an organ-specific enzyme can't be used to predict the severity or extent of cell injury in a particular organ ordinarily. The same elevations will occur in a *severe local* versus *a mild diffuse* injury. The functional reserve capability of the organ is not evaluated in tests that measure leakage of enzymes from *individual* necrotic cells.

16.3.3 Mechanisms of Irreversibility in Cell Injury

Currently, the precise biochemical events initiating irreversible cell injury are unknown. At what stage did the cell actually die? What is the critical biochemical event responsible for the "point of no return"? There is no universally accepted biochemical explanation for the transition from reversible injury to cell death. The duration of hypoxia necessary to induce irreversible cell injury varies according to cell type and its nutritional and hormonal status:

- *Liver:* Between 1 and 2 hr, liver cells of normally fed rats have abundant glycogen and, thus, survive longer than liver cells of starved rats.
- *Brain:* Irreversible damage to neurons occurs after 3–5 min of hypoxia!

Irreversibility is consistently characterized by two phenomena:

1. The inability to reverse mitochondrial dysfunction upon reperfusion, causing ATP depletion.
2. Development of profound disturbances in membrane function.

ATP depletion is certainly an important player in ischemic damage. But, drastic lowering of cellular ATP levels by the use of metabolic inhibitors has shown that this single event does not *induce* irreversible cell injury. Likewise, as discussed above, activation and release of lysosomal enzymes, subsequent to decreased intracellular

(a)

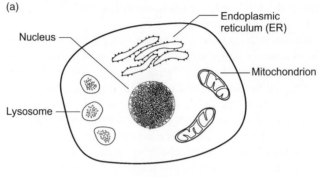

Normal Cell

(b)

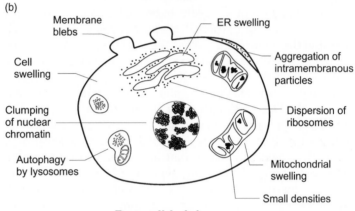

Reversible Injury

(c)

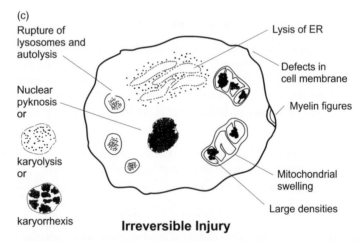

Irreversible Injury

Figure 16.4. Morphological changes associated with cell injury. In these diagrams, (A) a normal cell is compared with the changes typically seen via electron microscopy in (B) reversible and (C) irreversible cell injury. [Adapted from Cotran, Kumar, and Collins (Eds.). *Robbins Pathologic Basis of Disease*, 6th ed., Elsevier Saunders, Philadelphia, 1999, Fig. 1-6, p. 9.]

TABLE 16.1. Organ-Specific Leakage Enzymes Detectable in Serum/Plasma

Organ or Tissue	Enzyme Test	Comment
Liver	ALT	Alanine aminotransferase is fairly liver-specific in some species.
	AST	Aspartate aminotransferase is not as liver-specific; it is also found in high concentrations in skeletal muscle and myocardium.
	ALP	Alkaline phosphatase, found in liver, intestine, bone, placenta, others. ALP is **not** elevated in liver cell injury unless cholestasis results.
Skeletal muscle	CPK or CK	Creatine phosphokinase is fairly muscle-specific, although high concentrations exist in CNS.
	AST	AST is also found in the liver.
Pancreas	Lipase, Amylase	Both are cleared by the kidney and hence are elevated somewhat in renal failure as well. Lipase seems to have better predictive value.
Myocardium	CPK AST	
Kidney	—	Currently, there are no kidney-specific leakage enzymes assayed in serum. Renal function is evaluated by assaying serum levels of nitrogenous wastes—blood urea nitrogen (BUN) and creatinine—and monitoring protein levels in the urine (indicator of glomerular damage).

pH, occurs after the cell is irreversibly injured (i.e., after cell death). This latter mechanism was central to the old "suicide-bag" theory of cell death.

Recent evidence suggests that cell membrane damage is a central factor in the pathogenesis of cell injury:

(i) *Mitochondrial Dysfunction.* Increased uptake of calcium (because of ATP depletion) by mitochondria activates phospholipases, resulting in accumulation of free fatty acids. These cause changes in the permeability of mitochondrial membranes, such as the *mitochondrial permeability transition.*

(ii) *Progressive Loss of Phospholipids.* Increased degradation by endogenous phospholipases and inability of the cell to keep up with synthesis of new phospholipids (reacylation, an ATP-dependent process).

(iii) *Cytoskeletal Abnormalities.* Activated proteases lyse cytoskeletal elements and cell swelling causes detachment of cell membrane from cytoskeleton; stretching of the cell membrane results in increased membrane damage.

(iv) *Reactive Oxygen Species.* Produced within the cell and by infiltrating neutrophils and macrophages, especially after restoration of blood flow to an area (reperfusion injury). Cell injury triggers release of a number of inflammatory cytokines and chemokines which amplify the host immune response and attract neutrophils to the site.

(v) *Lipid Breakdown Products.* Unesterified free fatty acids, acyl carnitine, and lysophospholipids. These have a detergent effect on membranes and may exchange with membrane phospholipids, causing permeability changes.

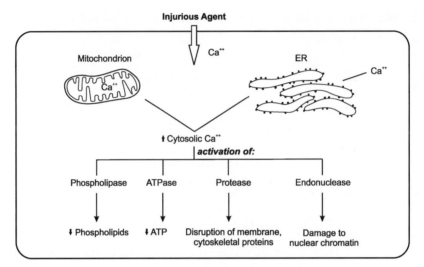

Figure 16.5. Mechanisms of membrane damage due to increased cytosolic calcium. [Adapted from Cotran, Kumar, and Collins (Eds.). *Robbins Pathologic Basis of Disease*, 6th ed., Elsevier Saunders, Philadelphia, 1999, Fig. 1-3, p. 6.]

Increased calcium flux into the cell through the damaged membranes seems to be the final common pathway mediating cell death (Figure 16.5). In fact, inhibition of calcium influx by chlorpromazine will prevent cell death in otherwise lethally injured cells in experimental models of cell injury. Similarly, causes of cell injury other than hypoxia have been postulated to directly or indirectly lead to membrane damage with subsequent increased permeability followed by increased calcium influx. The mechanisms of membrane damage range from simple complement-mediated damage to complex free radical-mediated lipid peroxidation reactions.

16.4 MORPHOLOGY OF CELL INJURY AND CELL DEATH

16.4.1 Cell Swelling (Hydropic Degeneration)

In this section we will deal with the gross and microscopic appearance of reversible and irreversible cell injury. *Reversible cell injury* has two major morphological manifestations: *cell swelling,* which is often called hydropic change or hydropic degeneration, and *fatty change*. The latter is usually only encountered in cells involved in lipid metabolism such as the hepatocyte. *Irreversible cell injury* usually manifests initially as coagulative necrosis. Depending on the cause of necrosis, the magnitude and duration of its action, and to some extent the tissue involved, other morphological types of necrosis are possible. As outlined previously, cell swelling is one of the earliest manifestations of cell injury and is due to the isosmotic uptake of sodium and water from the extracellular fluid. Cell swelling is reversible, but it may precede necrosis if the causal injury persists.

16.4.2 Types of Necrosis

Necrosis requires time to develop after cell death and depends on the continued activity of cellular enzymes. *Necrosis then is the morphological evidence of cell death* which appears some time after irreversible loss of biochemical function. Various morphologic manifestations of necrosis are recognized. The particular type that develops depends on a number of factors: the tissue and nature of the pathological process, the etiologic agent, and the time interval between cell death and sampling of the tissue. The importance in recognizing the different patterns of necrosis resides in their predictive value in determining the cause in some instances. Four major types of necrosis are shown in Figure 16.6.

16.4.2a Coagulative Necrosis. In coagulative necrosis, the gross and microscopic architecture of the necrotic tissue are still recognizable (at least to an extent). It should soon become clear that coagulative necrosis often precedes the other types

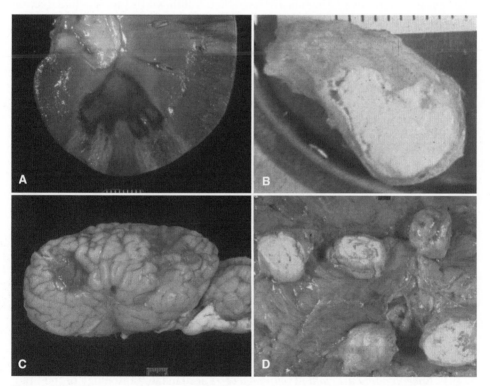

Figure 16.6. The four primary types of necrosis. (A) Coagulative. On the cut surface of this canine kidney, there is a dark red, wedge-shaped area of coagulative necrosis caused by blockage of the blood supply to the area (infarction). The pale areas are older infarcts. (B) Caseous. This bovine lymph node contains a large whitish, cheese-like area of caseous necrosis characterized by loss of the normal tissue architecture. (C) Liquefactive necrosis, equine brain. There are two large areas of liquefaction with extensive loss of brain tissue. (D) Fat necrosis, bovine abdominal fat. Necrotic fat is firm and chalky white and often becomes mineralized. See color insert.

of necrosis which need more time and additional influences to be manifest. Coagulative necrosis is usually caused by hypoxia, ischemia, or acute toxicity.

Specific examples of coagulative necrosis:

(i) *Infarction.* An infarct is a local area of coagulative necrosis due to ischemia. Thrombosis of vessels is the most common cause.

(ii) *Zenker's Necrosis.* This is the name given to coagulative necrosis in skeletal muscle.

16.4.2b   Caseous Necrosis. Caseous necrosis is characterized by loss of recognizable tissue architecture and cellular detail. The necrotic tissue is reminiscent of crumbly cheese (hence the term caseous). Certain bacterial agents (notably *Mycobacterium tuberculosis, Mycobacterium bovis*, and *Corynebacterium ovis*) are capable of inducing caseous necrosis within areas of granulomatous inflammation. This type of inflammation is composed predominantly of macrophage infiltration into tissue. The causal bacteria live inside macrophages and kill them. Incomplete liquefaction of the necrotic debris creates the characteristic appearance of the necrotic tissue (the precise reasons for this are unclear, but agent and host factors are involved).

16.4.2c   Fat Necrosis. Necrosis of adipose tissue often presents distinctive features useful in identifying the cause. Enzymatic fat necrosis results from the action of lipases released from pancreatic acinar cells during an episode of acute pancreatic necrosis and pancreatitis. There is another form of abdominal fat necrosis in some species that is not related causally to pancreatitis.

16.4.2d   Liquefactive Necrosis. Coagulative necrosis of tissue may progress to liquefactive necrosis due to the action of bacterial and/or neutrophil lytic enzymes. Coagulative necrosis in nervous tissue usually proceeds to liquefactive necrosis because of breakdown of complex lipids. Example of liquefactive necrosis are described below.

(i) *Pyogenic Abscess.* A localized bacterial infection that attracts large numbers of neutrophils. The center has a watery to creamy paste-like appearance (pus) and is white-yellow to red (particularly if blood breakdown pigments are present). The liquid center contains bacteria and large numbers of neutrophils, many of which are necrotic and fragmented. The original tissue is usually no longer recognizable in the center of the lesion. Healing usually results in scar formation (due to destruction of the tissue framework).

(ii) *Encephalomalacia and Myelomalacia.* These mean "softening" of the brain and spinal cord, respectively. In the central nervous system, coagulative necrosis progresses to liquefaction without the necessity for the lytic enzymes from inflammatory cells.

(iii) *Gangrenous Necrosis.* Gangrene results from invasion and putrefaction of necrotic tissue by saprophytic bacteria. There are two basic patterns: wet gangrene and dry gangrene. Dry gangrene typically develops in extremities that have undergone coagulative necrosis due to ischemia. Bacterial invasion occurs, but bacterial multiplication is limited by the coolness and desiccation of the necrotic tissue. Coagulative necrosis predominates over liquefaction. The necrotic extremity may eventually slough. Wet gangrene typically develops in organs and tissues that retain moisture and warmth after they become necrotic. Hence, optimal conditions are provided for bacterial growth. Liquefaction due to bacterial toxins and leukocyte enzymes complicates initial coagulative necrosis due usually to ischemia. Aspiration pneumonia is an example.

16.4.3   Microscopic Appearance of Necrosis

Histologically, the cytoplasm of necrotic cells usually shows increased eosinophilia— the red stain component of the standard hematoxylin and eosin (H&E) histological staining method—due partly to the loss of RNA (i.e., less basophilia or blueness) and partly to increased eosin binding by denatured proteins. The cytoplasm also looks more homogeneous and glassy (i.e., hyalinized). In skeletal myocytes there is a loss of cross-striations. Morphologic changes indicating cell death (necrosis) appear some 2–3 hr *after* biochemical irreversibility has occurred. This is why the dead cells in histology preparations appear morphologically normal. The first indications of necrosis are seen with the electron microscope. These include blebbing of the plasma membrane; dilated endoplasmic reticulum; dispersion of ribosomes; mitochondrial swelling, small densities, and calcific precipitates; and condensed and shrunken nuclei (Figure 16.4). By light microscopy, the denaturation of proteins and dispersion of ribosomes in the cytoplasm is reflected in increased eosinophilic staining of the cytoplasm ("dead red" cells, Figure 16.7).

Nuclear changes in dead cells are another good indicator of necrosis under light microscopy, and they appear in one of three patterns (Figure 16.7):

(i) *Pyknosis.* Characterized by nuclear shrinkage and increased basophilia; here, the DNA condenses into a solid, darkly staining mass.

(ii) *Karyorrhexis.* As the structure of the nucleus degenerates, the nucleus becomes irregularly shaped/fragmented.

(iii) *Karyolysis.* The basophilia of the chromatin may fade, presumably due to DNAse activity, leaving only tiny fragments of chromatin debris in the center of the degenerating cell. Often, residual fragments of nuclei can be seen as free debris in regions of necrosis.

16.5   APOPTOSIS

Apoptosis is an energy-dependent and carefully regulated process with specific morphological and biological features, whereas necrosis is described as an "unordered" and "accidental" form of cell death. Although the mechanisms of apoptosis and necrosis differ, there is overlap between these two processes. Evidence indicates that necrosis and apoptosis represent morphologic expressions of a shared biochemical network. For example, two factors that will convert an ongoing apoptotic process into a necrotic process include a decrease in the availability of caspases and intracellular ATP. In fact, there are enough similarities so that any single biochemical marker of apoptosis cannot be relied upon to distinguish apoptosis from necrosis.

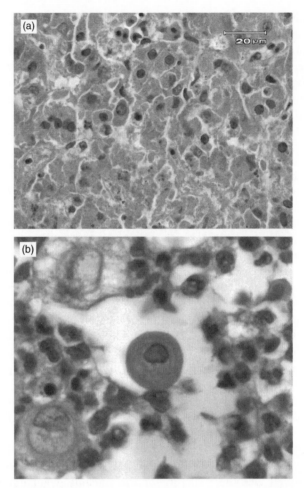

Figure 16.7. Microscopic appearance of necrosis. (A) Coagulative necrosis in a virally infected avian liver. Hepatocytes in the lower half of the photo are in various stages of necrosis, with small, pyknotic or fragmented nuclei and increased cytoplasmic eosinophilia. (B) Necrotic cells in immune-mediated skin disease, canine. The central cell has a pyknotic nucleus and intensely eosinophilic cytoplasm, while the cells at lower left and upper left are injured and swollen. The smaller cells are neutrophils. See color insert.

Whether a cell dies by necrosis or apoptosis depends in part on the nature of the cell death signal, the developmental stage, the tissue type, and the physiologic milieu. A description of the morphology and mechanisms of apoptosis follows.

16.5.1 Morphology of Apoptosis

Light and electron microscopy have identified the various morphological changes that occur during apoptosis. During the early process of apoptosis, cell shrinkage and pyknosis are visible by light microscopy. With cell shrinkage, the cells are smaller in size, the cytoplasm is dense, and the organelles are more tightly packed. Pyknosis is the result of chromatin condensation, and this is the most characteristic feature

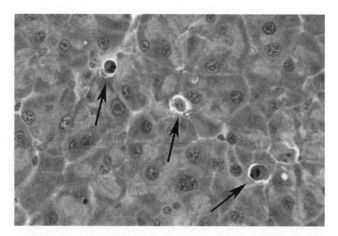

Figure 16.8. Apoptotic acinar cells in the exocrine pancreas. Photomicrograph of a section of exocrine pancreas from a B6C3F1 mouse. The arrows indicate apoptotic cells that are shrunken with condensed cytoplasm. The nuclei are pyknotic and fragmented. Note the lack of inflammation. H&E.

of apoptosis. On histologic examination with hematoxylin and eosin stain, apoptosis involves single cells or small clusters of cells. The apoptotic cell appears as a round or oval mass with dark eosinophilic cytoplasm and dense purple nuclear chromatin fragments (Figures 16.8 and 16.9). Electron microscopy can better define the subcellular changes. Early during the chromatin condensation phase, the electron-dense nuclear material characteristically aggregates peripherally under the nuclear membrane, although there can also be uniformly dense nuclei (Figure 16.10B). Extensive plasma membrane blebbing occurs followed by karyorrhexis and separation of cell fragments into apoptotic bodies during a process called "budding" (Figure 16.10C). Apoptotic bodies consist of cytoplasm with tightly packed organelles with or without a nuclear fragment. The organelle integrity is still maintained, and all of this is enclosed within an intact plasma membrane. These bodies are subsequently phagocytosed by macrophages, parenchymal cells, or neoplastic cells and degraded within phagolysosomes (Figure 16.10D). Macrophages that engulf and digest apoptotic cells are called "tingible body macrophages" and are frequently found within the reactive germinal centers of lymphoid follicles or occasionally within the thymic cortex. The tingible bodies are the bits of nuclear debris from the apoptotic cells. Because apoptotic cells do not release their cellular constituents into the surrounding interstitial tissue, there is essentially no inflammatory reaction.

Using conventional histology, it is not always easy to distinguish apoptosis from necrosis; they can occur simultaneously, depending on the intensity and duration of the stimulus, the extent of ATP depletion, and so on. Table 16.2 compares some of the major morphological features of apoptosis and necrosis.

16.5.2 Mechanisms of Apoptosis

The mechanisms of apoptosis are highly complex and sophisticated, involving an energy-dependent cascade of molecular events (Figure 16.11). There are two main apoptotic pathways: the extrinsic or death receptor pathway and the intrinsic or

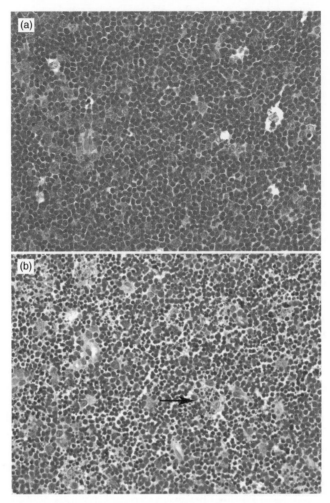

Figure 16.9. Normal thymus cortical tissue and dexamethasone-induced lymphocyte apoptosis. (A) Photomicrograph of normal thymus tissue from a control Sprague–Dawley rat. (B) Sheets of apoptotic cells in the thymus from a rat that was treated with dexamethasone to induce lymphocyte apoptosis. Under physiological conditions, apoptosis typically affects single cells or small clusters of cells. However, the degree of apoptosis in this treated rat is more severe due to the amount of dexamethasone administered (1 mg/kg bodyweight) and the time post-treatment (12 hr). The majority of lymphocytes are apoptotic, although there are a few interspersed cells that are morphologically normal and most likely represent lymphoblasts or macrophages. The apoptotic lymphocytes are small and deeply basophilic with pyknotic and often-fragmented nuclei. Macrophages are present with engulfed cytoplasmic apoptotic bodies (arrow). H&E.

mitochondrial pathway. However, there is now evidence that the two pathways are linked and that molecules in one pathway can influence the other. There is an additional pathway that involves T-cell-mediated cytotoxicity and perforin-granzyme-dependent killing of the cell. The perforin/granzyme pathway can induce apoptosis via either granzyme B or granzyme A. The extrinsic, intrinsic, and Granzyme B

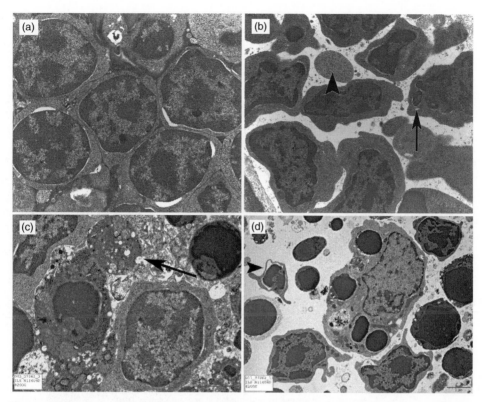

Figure 16.10. Transmission electron micrographs of normal and apoptotic thymus thymocytes. (A) Transmission electron micrograph (TEM) of the normal thymus tissue. The lymphocytes are closely packed and have large nuclei and scant cytoplasm. (B) TEM of apoptotic thymic lymphocytes in an early phase of apoptosis with condensed and peripheralized chromatin. The cytoplasm is beginning to condense and the cell outlines are irregular. The cells have separated due to loss of cell-to-cell adhesion. The arrow indicates a fragmented section of nucleus and the arrowhead most likely indicates an apoptotic body that seems to contain predominantly cytoplasm without organelles or nuclear material. (C) TEM of an apoptotic lymphocyte in the process of "budding" or extrusion of membrane-bound cytoplasm containing organelles (arrow). Once the budding has occurred, this extruded fragment will be an "apoptotic body." These apoptotic bodies are membrane-bound and thus do not release cytoplasmic contents into the interstitium. Macrophages or other adjacent healthy cells subsequently engulf the apoptotic bodies. For these reasons, apoptosis does not incite an inflammatory reaction. (D) TEM of a section of thymus with lymphocytes in various stages of apoptosis. The large cell in the center of the photomicrograph is a macrophage with engulfed intracytoplasmic apoptotic bodies. This macrophage is also called a "tingible body macrophage." The arrowhead indicates a lymphocyte in an advanced stage of apoptosis with nuclear fragmentation.

pathways converge on the same terminal, or execution pathway. This pathway is initiated by the cleavage of caspase 3 and results in DNA fragmentation, degradation of cytoskeletal and nuclear proteins, crosslinking of proteins, formation of apoptotic bodies, expression of ligands for phagocytic cell receptors, and finally uptake by phagocytic cells. The Granzyme A pathway activates a parallel, caspase-independent cell death pathway via single-stranded DNA damage.

TABLE 16.2. Comparison of Morphological Features of Apoptosis and Necrosis

Apoptosis	Necrosis
Single cells or small clusters of cells	Often contiguous cells
Cell shrinkage and convolution	Cell swelling
Pyknosis (chromatin condensation)	Pyknosis
Karyorrhexis (nuclear fragmentation)	Karyorrhexis
Karyolysis (nuclear lysis) not present	Karyolysis
Intact cell membrane	Disrupted cell membrane
Cytoplasm retained in apoptotic bodies	Cytoplasm released
No inflammation	Inflammation usually present

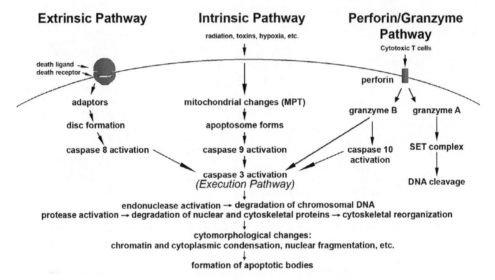

Figure 16.11. Schematic representation of apoptotic events. The two main pathways of apoptosis are extrinsic and intrinsic as well as a perforin/granzyme pathway. Each requires specific triggering signals to begin an energy-dependent cascade of molecular events. Each pathway activates its own initiator caspase (8, 9, 10), which in turn will activate the executioner caspase 3. However, granzyme A works in a caspase-independent fashion. The execution pathway results in characteristic cytomorphological features including cell shrinkage, chromatin condensation, formation of cytoplasmic blebs and apoptotic bodies, and finally phagocytosis of the apoptotic bodies by adjacent parenchymal cells, neoplastic cells, or macrophages.

16.5.2a Biochemical Features. Apoptotic cells exhibit several biochemical modifications such as protein cleavage, protein crosslinking, DNA breakdown, and phagocytic recognition that together result in the distinctive structural pathology described previously. To date, 14 caspases have now been identified and are broadly categorized into initiators, effectors (executioners), and inflammatory caspases. Caspases are widely expressed in an inactive proenzyme form in most cells and, once activated, can often activate other pro-caspases, allowing initiation of a protease cascade. Some pro-caspases can also aggregate and autoactivate. This proteolytic cascade, in which one caspase can activate other caspases, amplifies the apoptotic signaling pathway and thus leads to rapid cell death. Caspases have proteolytic activity and are able to cleave proteins at aspartic acid residues, although different caspases have different specificities involving recognition of neighboring amino acids. Once caspases are initially activated, there seems to be an irreversible commitment toward cell death. Extensive protein crosslinking is another characteristic of apoptotic cells and is achieved through the expression and activation of tissue transglutaminase. DNA breakdown by endonucleases also occurs, resulting in DNA fragments of 180–200 base pairs. A characteristic "DNA ladder" can be visualized by agarose gel electrophoresis with an ethidium bromide stain and ultraviolet illumination. Another biochemical feature is the expression of cell surface markers that result in the early phagocytic recognition of apoptotic cells by adjacent cells, permitting quick phagocytosis with minimal compromise to the surrounding tissue. This is achieved by the movement of the normal inward-facing phosphatidylserine of the cell's lipid bilayer to expression on the outer layers of the plasma membrane. Although externalization of phosphatidylserine is a well-known recognition ligand for phagocytes on the surface of the apoptotic cell, recent studies have suggested that other proteins may also be exposed on the cell surface during apoptotic cell clearance. These include (a) annexin I, which is a phosphatidylserine-binding protein, and (b) calreticulin, which is a protein that binds to an LDL-receptor-related protein on the engulfing cell and is suggested to cooperate with phosphatidylserine as a recognition signal. The adhesive glycoprotein, thrombospondin-1, can also be expressed on the outer surface of activated microvascular endothelial cells and induce receptor-mediated apoptosis.

16.5.2b *Extrinsic Pathway.*

The extrinsic signaling pathways that initiate apoptosis involve transmembrane receptor-mediated interactions. These involve death receptors that are members of the tumor necrosis factor (TNF) receptor gene superfamily. Members of the TNF receptor family share similar cysteine-rich extracellular domains and have a cytoplasmic domain of about 80 amino acids called the "death domain." This death domain plays a critical role in transmitting the death signal from the cell surface to the intracellular signaling pathways. To date, the best-characterized ligands and corresponding death receptors include FasL/FasR, TNF-α/TNFR1, Apo3L/DR3, Apo2L/DR4, and Apo2L/DR5. The sequence of events that define the extrinsic phase of apoptosis are best characterized with the FasL/FasR and TNF-α/TNFR1 models. In these models, there is clustering of receptors and binding with the homologous trimeric ligand. Upon ligand binding, cytplasmic adapter proteins are recruited which exhibit corresponding death domains that bind with the receptors. The binding of Fas ligand to Fas receptor results in the binding of the adapter protein FADD, and the binding of TNF ligand to TNF receptor results in the binding of the adapter protein TRADD with recruitment of FADD and RIP. FADD then associates with procaspase 8 via dimerization of the death effector domain. At this point, a death-inducing signaling complex (DISC) is formed, resulting in the autocatalytic activation of procaspase 8. Once caspase 8 is activated, the execution phase of apoptosis is triggered. Death-receptor-mediated apoptosis can be inhibited by a protein called c-FLIP, which will bind to FADD and caspase 8, rendering them ineffective. Another point of potential apoptosis regulation involves a protein called Toso, which has been shown to block Fas-induced apoptosis in T cells via inhibition of caspase 8 processing. See Table 16.3 for nomenclature of the major proteins involved in the extrinsic pathway.

TABLE 16.3. Extrinsic Pathway Proteins, Abbreviations, and Alternate Nomenclature

Abbreviation	Protein Name	Select Alternate Nomenclature
TNF-α	Tumor necrosis factor alpha	TNF ligand, TNFA, cachectin
TNFR1	Tumor necrosis factor receptor 1	TNF receptor, TNFRSF1A, p55 TNFR, CD120a
FasL	Fatty acid synthetase ligand	Fas ligand, TNFSF6, Apo1, apoptosis antigen ligand 1, CD95L, CD178, APT1LG1
FasR	Fatty acid synthetase receptor	Fas receptor, TNFRSF6, APT1, CD95
Apo3L	Apo3 ligand	TNFSF12, Apo3 ligand, TWEAK, DR3LG
DR3	Death receptor 3	TNFRSF12, Apo3, WSL-1, TRAMP, LARD, DDR3
Apo2L	Apo2 ligand	TNFSF10, TRAIL, TNF-related apoptosis-inducing ligand
DR4	Death receptor 4	TNFRSF10A, TRAILR1, APO2
DR5	Death receptor 5	TNFRS10B, TRAIL-R2, TRICK2, KILLER, ZTNFR9
FADD	Fas-associated death domain	MORT1
TRADD	TNF-receptor-associated death domain	TNFRSF1A associated via death domain
RIP	Receptor-interacting protein	RIPK1
DED	Death effector domain	Apoptosis antagonizing transcription factor, CHE1
Caspase 8	Cysteinyl aspartic acid-protease 8	FLICE, FADD-like Ice, Mach-1, Mch5
c-FLIP	FLICE-inhibitory protein	Casper, I-FLICE, FLAME-1, CASH, CLARP, MRIT

References:
1. Human Protein Reference Database http://www.hprd.org/.
2. ExPASy Proteomics Server http://ca.expasy.org/.

16.5.2c Perforin/Granzyme Pathway.

T-cell-mediated cytotoxicity is a variant of type IV hypersensitivity where sensitized CD8+ cells kill antigen-bearing cells. These cytotoxic T lymphocytes (CTLs) are able to kill target cells via the extrinsic pathway, and the FasL/FasR interaction is the predominant method of CTL-induced apoptosis. However, they are also able to exert their cytotoxic effects on tumor cells and virus-infected cells via a novel pathway that involves secretion of the transmembrane pore-forming molecule perforin with a subsequent exophytic release of cytoplasmic granules through the pore and into the target cell. The serine proteases granzyme A and granzyme B are the most important component within the granules.

Granzyme B will cleave proteins at aspartate residues and will therefore activate pro-caspase 10 and can cleave factors like ICAD (inhibitor of caspase-activated DNase). However, Granzyme B can also directly activate caspase 3. In this way, the upstream signaling pathways are bypassed and there is direct induction of the execution phase of apoptosis. Recent findings indicate that this method of granzyme B cytotoxicity is critical as a control mechanism for T-cell expansion of type 2 helper T (Th2) cells. Moreover, findings indicate that neither death receptors nor caspases are involved with the T-cell-receptor-induced apoptosis of activated Th2 cells because blocking their ligands has no effect on apoptosis. On the other hand, Fas–Fas ligand interaction, adapter proteins with death domains, and caspases are all involved in the apoptosis and regulation of cytotoxic Type 1 helper cells whereas granzyme B has no effect.

Granzyme A is also important in cytotoxic T-cell-induced apoptosis and activates caspase-independent pathways. Once in the cell, Granzyme A activates DNA nicking via DNAse NM23-H1, a tumor suppressor gene product. This DNase has an important role in immune surveillance to prevent cancer through the induction of tumor cell apoptosis. The nucleosome assembly protein SET normally inhibits the NM23-H1 gene. Granzyme A protease cleaves the SET complex, thus releasing inhibition of NM23-H1 and resulting in apoptotic DNA degradation. In addition to inhibiting NM23-H1, the SET complex has important functions in chromatin structure and DNA repair. The proteins that make up this complex (SET, Ape1, pp32, and HMG2) seem to work together to protect chromatin and DNA structure. Therefore, inactivation of this complex by Granzyme A most likely also contributes to apoptosis by blocking the maintenance of DNA and chromatin structure integrity.

16.5.2d *Intrinsic Pathway.* The intrinsic signaling pathways that initiate apoptosis involve a diverse array of non-receptor-mediated stimuli that produce intracellular signals that act directly on targets within the cell and are mitochondrial-initiated events. The stimuli that initiate the intrinsic pathway produce intracellular signals that may act in either a positive or negative fashion. Negative signals involve the absence of certain growth factors, hormones, and cytokines that can lead to failure of suppression of death programs, thereby triggering apoptosis. In other words, there is the withdrawal of factors, loss of apoptotic suppression, and subsequent activation of apoptosis. Other stimuli that act in a positive fashion include, but are not limited to, radiation, toxins, hypoxia, hyperthermia, viral infections, and free radicals.

All of these stimuli cause changes in the inner mitochondrial membrane that results in an opening of the mitochondrial permeability transition (MPT) pore and release of two main groups of normally sequestered pro-apoptotic proteins from the intermembrane space into the cytosol. The first group consists of cytochrome c, Smac/DIABLO, and the serine protease HtrA2/Omi. These proteins activate the caspase-dependent mitochondrial pathway. Cytochrome c binds and activates Apaf-1 as well as procaspase 9, forming an "apoptosome." The clustering of procaspase 9 in this manner leads to caspase 9 activation. Smac/DIABLO and HtrA2/Omi are reported to promote apoptosis by inhibiting IAP (inhibitors of apoptosis proteins) activity. Additional mitochondrial proteins have also been identified that interact with and suppress the action of IAP; however, gene knockout experiments suggest that binding to IAP alone may not be enough evidence to label a mitochondrial protein as "pro-apoptotic." The second group of pro-apoptotic proteins—endonuclease G, AIF, and CAD—are released from the mitochondria during apoptosis, but this is a late event that occurs after the cell has committed to die. AIF translocates to the nucleus and causes DNA fragmentation into ~50- to 300-kb pieces and condensation of peripheral nuclear chromatin. This early form of nuclear condensation is referred to as "stage I" condensation. Endonuclease G also translocates to the nucleus where it cleaves nuclear chromatin to produce oligonucleosomal DNA fragments. AIF and endonuclease G both function in a caspase-independent manner.

CAD is subsequently released from the mitochondria and translocates to the nucleus, where, after cleavage by caspase 3, it leads to oligonucleosomal DNA fragmentation and a more pronounced and advanced chromatin condensation. This later and more pronounced chromatin condensation is referred to as "stage II" condensation.

The control and regulation of these apoptotic mitochondrial events occurs through the members of the Bcl-2 family of proteins. These proteins govern mitochondrial membrane permeability and can be either pro-apoptotic or anti-apoptotic. To date, there are a total of 25 genes identified in the Bcl-2 family. Some of the anti-apoptotic proteins include Bcl-2, Bcl-x, Bcl-XL, Bcl-XS, Bcl-w, and BAG, and some of the pro-apoptotic proteins include Bcl-10, Bax, Bak, Bid, Bad, Bim, Bik, and Blk. These proteins have special significance since they can determine if the cell commits to apoptosis or aborts the process. It is thought that the main mechanism of action of the Bcl-2 family of proteins is the regulation of cytochrome C release from the mitochondria via alteration of mitochondrial membrane permeability. A few possible mechanisms have been studied, but none have been proven definitively. Mitochondrial damage in the Fas pathway of apoptosis is mediated by the caspase 8 cleavage of Bid. Serine phosphorylation of Bad is associated with 14-3-3, a member of a family of multifunctional phosphoserine binding molecules. When Bad is phosphorylated, it is trapped by 14-3-3 and sequestered in the cytosol; but once Bad is unphosphorylated, it will translocate to the mitochondria to release cytochrome C. Bad can also heterodimerize with Bcl-Xl or Bcl-2, neutralizing their protective effect and promoting cell death. When not sequestered by Bad, both Bcl-2 and Bcl-Xl inhibit the release of cytochrome C from the mitochondria, although the mechanism is not well understood. Bcl-2 and Bcl-Xl will also bind to Apaf-1, thus inhibiting the association of procaspase 9 with Apaf-1 and preventing formation of the apoptosome. An additional protein designated "Aven" appears to bind both Bcl-Xl and Apaf-1, thereby preventing activation of procaspase 9. There is evidence that overexpression of either Bcl-2 or Bcl-Xl will downregulate the other, indicating a reciprocal regulation between these two proteins.

Puma and Noxa are two members of the Bcl2 family that are also involved in pro-apoptosis. Puma plays an important role in p53-mediated apoptosis. It was shown that, in vitro, overexpression of Puma is accompanied by increased BAX expression, BAX conformational change, translocation to the mitochondria, cytochrome C release, and reduction in the mitochondrial membrane potential. Noxa is also a candidate mediator of p53-induced apoptosis. Studies show that this protein can localize to the mitochondria and interact with anti-apoptotic Bcl-2 family members, resulting in the activation of caspase-9. Since both Puma and Noxa are induced by p53, they might mediate the apoptosis that is elicited by genotoxic damage or oncogene activation. Further elucidation of these pathways should have important implications for tumorigenesis and therapy. See Table 16.4 for nomenclature of the proteins involve in the intrinsic pathway.

16.5.2e Execution Pathway. The extrinsic and intrinsic pathways both end at the point of the execution phase, considered the final pathway of apoptosis. It is the activation of the execution caspases that begins this phase of apoptosis. Execution caspases activate cytoplasmic endonuclease, which degrades nuclear material; they also activate proteases, which degrade the nuclear and cytoskeletal proteins. Caspase

TABLE 16.4. Intrinsic Pathway Proteins, Abbreviations, and Alternate Nomenclature

Abbreviation	Protein Name	Select Alternate Nomenclature
Smac/DIABLO	Second mitochondrial activator of caspases/direct IAP-binding protein with low PI	None
HtrA2/Omi	High-temperature requirement	Omi stress regulated endoprotease, serine protease Omi protein A2
IAP	Inhibitor of apoptosis proteins API1, complex protein	XIAP, API3, ILP, HILP, HIAP2, cIAP1, MIHB, NFR2-TRAF signaling
Apaf-1	Apoptotic protease-activating factor	APAF1
Caspase 9	cysteinyl aspartic acid-protease-9	ICE-LAP6, Mch6, Apaf-3
AIF	Apoptosis-inducing factor	Programmed cell death protein 8, mitochondrial
CAD	Caspase-activated DNase	CAD/CPAN/DFF40
Bcl-2	B-cell lymphoma protein 2	Apoptosis regulator Bcl-2
Bcl-x	BCL2-like 1 protein	BCL2-related protein
Bcl-XL	BCL2-related protein, long isoform	BCL2L protein, long form of Bcl-x
Bcl-XS	BCL2-related protein, short isoform	
Bcl-w	BCL2-like 2 protein	Apoptosis regulator BclW
BAG	BCL2-associated athanogene	BAG family molecular chaperone regulator
Bcl-10	B-cell lymphoma protein 10	mE10, CARMEN, CLAP, CIPER
BAX	BCL2-associated X protein	apoptosis regulator BAX
BAK	BCL2 antagonist killer 1	BCL2L7, cell death inhibitor 1
BID	BH3 interacting domain death agonist	p22 BID
BAD	BCL2 antagonist of cell death	BCL2-binding protein, BCL2L8, BCL2-binding component 6, BBC6, Bcl-XL/Bcl-2 associated death promoter
BIM	BCL2 interacting protein BIM	BCL2-like 11
BIK	BCL2 interacting killer	NBK, BP4, BIP1, apoptosis-inducing NBK
Blk	Bik-like killer protein	B lymphoid tyrosine kinase, p55-BLK, MGC10442
Puma	BCL2-binding component 3	JFY1, PUMA/JFY1, p53 upregulated modulator of apoptosis
Noxa	Phorbol-12-myristate-13-acetate-induced protein 1	PMA-induced protein 1, APR
14-3-3	Tyrosine 3-monooxygenase/tryptophan 5-Monooxygenase activation protein	14-3-3 eta, theta, zeta, beta, epsilon, sigma Gamma
Aven	Cell death regulator Aven	None

References:
1. Human Protein Reference Database http://www.hprd.org/.
2. ExPASy Proteomics Server http://ca.expasy.org/.

TABLE 16.5. Execution Pathway Proteins, Abbreviations and Alternate Nomenclature

Abbreviation	Protein Name	Select Alternate Nomenclature
Caspase 3	Cysteinyl aspartic acid-protease-3	CPP32, Yama, Apopain, SCA-1, LICE
Caspase 6	Cysteinyl aspartic acid-protease-6	Mch-2
Caspase 7	Cysteinyl aspartic acid-protease-7	Mch-3, ICE-LAP-3, CMH-1
Caspase 10	Cysteinyl aspartic acid-protease-10	Mch4, FLICE-2
PARP	Poly(ADP-ribose) polymerase	ADP ribosyl transferase, ADPRT1, PPOL
Alpha fodrin	Spectrin alpha chain	Alpha-II spectrin, fodrin alpha chain
NuMA	Nuclear mitotic apparatus protein	SP-H antigen
CAD	Caspase-activated DNase	DNA fragmentation factor subunit beta, DFF-40, caspase-activated nuclease, CPAN
ICAD	Inhibitor of CAD	DNA fragmentation factor subunit alpha, DFF-45

References:
1. Human Protein Reference Database http://www.hprd.org/.
2. ExPASy Proteomics Server http://ca.expasy.org/.

3, caspase 6, and caspase 7 function as effector or "executioner" caspases, cleaving various substrates including cytokeratins, PARP, the plasma membrane cytoskeletal protein alpha fodrin, the nuclear protein NuMA, and others, which ultimately cause the morphological and biochemical changes seen in apoptotic cells. Caspase 3 is considered to be the most important of the executioner caspases and is activated by any of the initiator caspases (caspase 8, caspase 9, or caspase 10). Caspase 3 specifically activates the endonuclease CAD. In proliferating cells, CAD is complexed with its inhibitor, ICAD. In apoptotic cells, activated caspase 3 cleaves ICAD to release CAD. CAD then degrades chromosomal DNA within the nuclei and causes chromatin condensation. Caspase 3 also induces cytoskeletal reorganization and disintegration of the cell into apoptotic bodies. Gelsolin, an actin binding protein, has been identified as one of the key substrates of activated caspase 3. Gelsolin will typically nucleate actin polymerization and also binds phosphatidylinositol biphosphate, linking actin organization and signal transduction. Caspase 3 will cleave gelsolin, and the cleaved fragments of gelsolin in turn cleave actin filaments in a calcium independent manner. This results in disruption of the cytoskeleton, intracellular transport, cell division, and signal transduction. Phagocytic uptake of apoptotic cells is the last component of apoptosis. Caspase 3 activity also leads to the externalization of phosphatidylserine on apoptotic cells and their fragments. This facilitates the early recognition of apoptotic cells by phagocytes, allowing for their early uptake and disposal. This process of early and efficient uptake with no release of cellular constituents results in essentially no inflammatory response. See Table 16.5 for nomenclature of the proteins that are involved in the execution pathway.

16.5.3 Physiologic Apoptosis

The role of apoptosis in normal physiology is as significant as that of its counterpart, mitosis. It demonstrates a complementary but opposite role to mitosis and cell proliferation in the regulation of various cell populations. It is estimated that to maintain homeostasis in the adult human body, around 10 billion cells are made each day just to balance those dying by apoptosis. And that number can increase significantly when there is increased apoptosis during normal development and aging or during disease. Apoptosis is critically important during various developmental processes. As examples, both the nervous system and the immune system arise through overproduction of cells. This initial overproduction is then followed by the death of those cells that fail to establish functional synaptic connections or productive antigen specificities, respectively. Apoptosis is also necessary to rid the body of pathogen-invaded cells and is a vital component of wound healing in that it is involved in the removal of inflammatory cells and the evolution of granulation tissue into scar tissue. Apoptosis is also needed to eliminate activated or auto-aggressive immune cells either during maturation in the central lymphoid organs (bone marrow and thymus) or in peripheral tissues. Additionally, apoptosis is central to remodeling in the adult, such as the follicular atresia of the post-ovulatory follicle and post-weaning mammary gland involution, to name a couple of examples. Furthermore, as organisms grow older, cells begin to deteriorate and are eliminated via apoptosis. It is clear that apoptosis has to be tightly regulated since too little or too much cell death may lead to pathology, including developmental defects, autoimmune diseases, neurodegeneration, or cancer.

16.5.4 Pathologic Apoptosis

Abnormalities in cell death regulation can be a significant component of diseases such as cancer, autoimmune lymphoproliferative syndrome, AIDS, ischemia, and neurodegenerative diseases such as Parkinson's disease, Alzheimer's disease, Huntington's disease, and amyotrophic lateral sclerosis. Some conditions feature insufficient apoptosis, whereas others feature excessive apoptosis.

Cancer is an example where the normal mechanisms of cell cycle regulation are dysfunctional, with an overproliferation of cells and/or decreased removal of cells. In fact, suppression of apoptosis during carcinogenesis is thought to play a central role in the development and progression of some cancers. There are a variety of molecular mechanisms that tumor cells use to suppress apoptosis.

Tumor cells can acquire resistance to apoptosis by the expression of anti-apoptotic proteins such as Bcl-2 or by the downregulation or mutation of pro-apoptotic proteins such as Bax. Certain forms of human B-cell lymphoma have overexpression of Bcl-2, and this is one of the first and strongest lines of evidence that failure of cell death contributes to cancer. Another method of apoptosis suppression in cancer involves evasion of immune surveillance. Certain immune cells (T cells and natural killer cells) normally destroy tumor cells via the perforin/Granzyme B pathway or the death-receptor pathway. In order to evade immune destruction, some tumor cells will diminish the response of the death receptor pathway to FasL produced by T cells. This has been shown to occur in a variety of ways including downregulation of the Fas receptor on tumor cells. Other mechanisms include secretion of high levels of a soluble form of the Fas receptor that will sequester the Fas ligand or expression of Fas ligand on the surface of tumor cells.

Alterations of various cell signaling pathways can result in dysregulation of apoptosis and lead to cancer. The p53 tumor suppressor gene is a transcription factor that regulates the cell cycle and is the most widely mutated gene in human tumorigenesis. The critical role of p53 is evident by the fact that it is mutated in over 50% of all human cancers. p53 can activate DNA repair proteins when DNA has sustained damage, can hold the cell cycle at the G_1/S regulation point on DNA damage recognition, and can initiate apoptosis if the DNA damage proves to be irreparable. Tumorigenesis can occur if this system goes awry. If the p53 gene is damaged, then tumor suppression is severely reduced. The p53 gene can be damaged by radiation, various chemicals, and viruses such as the human papillomavirus (HPV). People who inherit only one functional copy of this gene will most likely develop Li–Fraumeni syndrome, which is characterized by the development of tumors in early adulthood.

The ataxia telangiectasia-mutated gene (ATM) has also been shown to be involved in tumorigenesis via the ATM/p53 signaling pathway. The ATM gene encodes a protein kinase that acts as a tumor suppressor. ATM activation, via ionizing radiation damage to DNA, stimulates DNA repair and blocks cell cycle progression. One mechanism through which this occurs is ATM-dependent phosphorylation of p53. As mentioned previously, p53 then signals growth arrest of the cell at a checkpoint to allow for DNA damage repair or can cause the cell to undergo apoptosis if the damage cannot be repaired. This system can also be inactivated by a number of mechanisms including somatic genetic/epigenetic alterations and expression of oncogenic viral proteins such as the HPV, leading to tumorigenesis. Other cell signaling pathways can also be involved in tumor development. For example, upregulation of the phosphatidylinositol 3-kinase/AKT pathway in tumor cells renders them independent of survival signals. In addition to regulation of apoptosis, this pathway regulates other cellular processes, such as proliferation, growth, and cytoskeletal rearrangement.

In addition to cancer, too little apoptosis can also result in diseases such as autoimmune lymphoproliferative syndrome (ALPS). This occurs when there is insufficient apoptosis of auto-aggressive T cells, resulting in multiple autoimmune diseases. An overproliferation of B cells occurs as well, resulting in excess immunoglobulin production, leading to autoimmunity. Some of the common diseases of ALPS include hemolytic anemia, immune-mediated thrombocytopenia, and autoimmune neutropenia. The different types of this condition are caused by different mutations. Type 1A results from a mutation in the death domain of the Fas receptor, Type 1B results from a mutation in Fas ligand, and Type 2 results from a mutation in caspase 10, reducing its activity.

Excessive apoptosis may also be a feature of some conditions such as autoimmune diseases, neurodegenerative diseases, and ischemia-associated injury. Acquired immune deficiency syndrome (AIDS) is an example of an autoimmune disease that results from infection with the human immunodeficiency virus (HIV). This virus infects CD4$^+$ T cells by binding to the CD4 receptor. The virus is subsequently internalized into the T cell, where the HIV Tat protein is thought to increase the expression of the Fas receptor, resulting in excessive apoptosis of T cells. Alzheimer's disease is a neurodegenerative condition that is thought to be caused by mutations in certain proteins such as APP (amyloid precursor protein) and presenilins. Presenilins are thought to be involved in the processing of APP to amyloid β. This condition is associated with the deposition of amyloid β in extracellular deposits known as plaques, and amyloid β is thought to be neurotoxic when found in aggregated plaque form. It is thought to induce apoptosis by causing oxidative stress or signaling through death receptors. It may also activate microglia, which would result in TNFα secretion and activation of the TNF-R1, leading to apoptosis. Excessive apoptosis is also thought to play an important role in various ischemia-associated injuries. One example is myocardial ischemia caused by an insufficient blood supply, leading to a decrease in oxygen delivery to, and subsequent death of, the cardiomyocytes. Although necrosis does occur, overexpression of BAX has been detected in ischemic myocardial tissue, and therapy aimed at reducing apoptosis has shown some success in reducing the degree of tissue damage. One hypothesis is that the damage produced by ischemia is capable of initiating apoptosis; but if ischemia is prolonged, necrosis occurs. If energy production is restored, as with reperfusion, the apoptotic cascade that was initiated by ischemia may proceed. Although the extent to which apoptosis is involved in myocardial ischemia remains to be clarified, there is clear evidence that supports a role for this mode of cell death.

16.5.5 Assays for Apoptosis

Since apoptosis occurs via a complex signaling cascade that is tightly regulated at multiple points, there are many opportunities to evaluate the activity of the proteins involved. As the activators, effectors, and regulators of this cascade continue to be elucidated, a large number of apoptosis assays are devised to detect and count apoptotic cells. Although most of the commonly used apoptosis assays detect later events that occur in the terminal phase of apoptosis, some laboratories employ two or more distinct assays to confirm that cell death is occurring via apoptosis. One assay will detect early (initiation) apoptotic events, and a different assay will target a later (execution) event. The second assay is generally based on a different principle and is used to confirm apoptosis.

Understanding the kinetics of cell death in your model system is critical. Some proteins, such as caspases, are expressed only transiently. Cultured cells undergoing apoptosis in vitro will eventually undergo secondary necrosis.

Apoptotic cells in any system can die and disappear relatively quickly. The time from initiation of apoptosis to completion can occur as quickly as 2–3 hr. Therefore a false negative can occur if the assay is done too early or too late. Moreover, apoptosis can occur at low frequency or in specific sites within organs, tissues, and cultures. In such cases, the ability to rapidly survey large areas could be useful. In general, if detailed information on the mechanism of cell death is desired, the duration of toxin exposure, the concentration of the test compound, and the choice of assay endpoint become critical.

A detailed description of all methodologies and assays for detecting apoptosis is beyond the scope of this chapter. However, some of the most commonly employed assays are mentioned and briefly described. Apoptosis assays, based on methodology, can be classified into six major groups, and a subset of the available assays in each group is indicated and briefly discussed:

1. Cytomorphological alterations
2. DNA fragmentation
3. Detection of caspases, cleaved substrates, regulators and inhibitors
4. Membrane alterations
5. Detection of apoptosis in whole mounts
6. Mitochondrial assays

16.5.5a Cytomorphological Alterations. The evaluation of hematoxylin and eosin-stained tissue sections with light microscopy does allow the visualization of apoptotic cells. Apoptosis involves single cells or small clusters of cells, and these cells will appear smaller in size with dense cytoplasm. Chromatin aggregation results in deeply basophilic pyknotic cells. Phagocytized apoptotic bodies can also be visualized with this method; and because there is no release of cytoplasmic contents, there is generally no inflammatory reaction. Although a single apoptotic cell can be visualized with this method, confirmation with other methods may be necessary. Because the morphological events of apoptosis are rapid and the fragments are quickly phagocytized, considerable apoptosis may occur in some tissues before it is histologically apparent. Additionally, this method detects the later events of apoptosis, so cells in the early phase of apoptosis will not be detected.

Resin-embedded tissues can be stained with toluidine blue or methylene blue to reveal intensely stained apoptotic cells when evaluated by standard light microscopy. This methodology depends on the nuclear and cytoplasmic condensation that occurs during apoptosis. The tissue and cellular details are preserved with this technique, and surveys of large tissue regions are distinct advantages. However, smaller apoptotic bodies will not be detected, and healthy cells with large dense intracellular granules can be mistaken for apoptotic cells or debris. Additionally, there is loss of antigenicity during processing so that immunohistological or enzyme assays cannot be performed on the same tissue. However, this tissue may be used for transmission electron microscopy (TEM).

TEM is considered the gold standard to confirm apoptosis. This is because categorization of an apoptotic cell is irrefutable if the cell contains certain ultrastructural morphological characteristics. These characteristics are (1) electron-dense nucleus (marginalization in the early phase), (2) nuclear fragmentation, (3) intact cell membrane even late in the cell disintegration phase, (4) disorganized cytoplasmic organization, (5) large clear vacuoles, and (6) blebs at the cell surface. As apoptosis progresses, these cells will lose the cell-to-cell adhesions and will separate from neighboring cells. During the later phase of apoptosis, the cell will fragment into apoptotic bodies with intact cell membranes and will contain cytoplasmic organelles with or without nuclear fragments. Phagocytosis of apoptotic bodies can also be appreciated with TEM. The main disadvantages of TEM are the cost, time expenditure, and the ability to only assay a small region at a time.

16.5.5b DNA Fragmentation. The DNA laddering technique is used to visualize the endonuclease cleavage products of apoptosis. This assay involves extraction of DNA from a lysed cell homogenate followed by agarose gel electrophoresis. This results in a characteristic "DNA ladder" with each band in the ladder separated in size by approximately 200 base pairs. This methodology is easy to perform, has a sensitivity of 1×10^6 cells (i.e., level of detection is as few as 1,000,000 cells), and is useful for tissues and cell cultures with high numbers of apoptotic cells per tissue mass or volume, respectively. On the other hand, it is not recommended in cases with low numbers of apoptotic cells. There are other disadvantages to this assay. Since DNA fragmentation occurs in the later phase of apoptosis, the absence of a DNA ladder does not eliminate the potential that cells are undergoing early apoptosis. Additionally, DNA fragmentation can occur during preparation, making it difficult to produce a nucleosome ladder and necrotic cells can also generate DNA fragments.

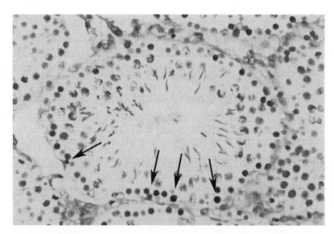

Figure 16.12. TUNEL assay used to detect apoptotic cells within a seminiferous tubule. There are many ways of detecting apoptosis at different stages on histological sections. The most commonly used method is called TUNEL (terminal deoxynucleotidyl transferase biotin-dUTP nick end-labeling). One of the major characteristics of apoptosis is the degradation of DNA after the activation of Ca/Mg-dependent endonucleases. This DNA cleavage leads to strand breaks within the DNA. The TUNEL method identifies apoptotic cells in situ by using terminal deoxynucleotidyl transferase (TdT) to transfer biotin-dUTP to these strand breaks of cleaved DNA. The biotin-labeled cleavage sites are then detected by reaction with HRP (horseradish peroxidase) conjugated streptavidin and visualized by DAB (diaminobenzidine), which gives a brown color. The tissue is counterstained with Toluidine blue to allow evaluation of tissue architecture. Figure 16.12. Photomicrograph of a section of testes from a B6C3F1 mouse that was in a National Toxicology Program bioassay. The TUNEL assay was used on this section of tissue to detect apoptotic cells. The seminiferous tubule is cut in cross section and has sperm, spermatogonia, and spermatocytes in various stages of development. The arrows point to labeled apoptotic cells in the basal layer of the seminiferous epithelium. See color insert.

The TUNEL (terminal dUTP nick end-labeling) method is used to assay the endonuclease cleavage products by enzymatically end-labeling the DNA strand breaks. Terminal transferase is used to add labeled UTP to the 3′ end of the DNA fragments. The dUTP can then be labeled with a variety of probes to allow detection by light microscopy, fluorescence microscopy, or flow cytometry (Figure 16.12). The assays are available as kits and can be acquired from a variety of companies. This assay is also very sensitive, allowing detection of a single cell via fluorescence microscopy or as few as ~100 cells via flow cytometry. It is also a fast technique and can be completed within 3 hr. The disadvantages are cost and the unknown parameter of how many breaks are necessary for detection by this method. This method is also subject to false positives from necrotic cells and cells in the process of DNA repair and gene transcription. For these reasons, it is usually paired with another assay.

16.5.5c Detection of Caspases, Cleaved Substrates, Regulators, and Inhibitors. There are more than 13 known caspases (procaspases or active cysteine caspases) that can be detected using various types of caspase activity assays. There are also immunohistochemistry assays that can detect cleaved substrates such as

PARP and fluorescently conjugated caspase inhibitors that can label active caspases within cells. Caspase activation can be detected in a variety of ways including Western blot, immunoprecipitation, and immunohistochemistry. Both polyclonal and monoclonal antibodies are available to proactive and active caspases. One method of caspase detection requires (a) cell lysis in order to release the enzymes into the solution, (b) coating of microwells with anti-caspases, and (c) detection with a fluorescent-labeled substrate. Detection of caspase activity by this method usually requires 1×10^5 cells. This technique allows you to select for individual initiator or execution caspases. It also allows for rapid and consistent quantification of apoptotic cells. The major disadvantage is that the integrity of the tissue is destroyed, thereby eliminating the possibility of localizing the apoptotic event within the tissue or determining the type of cell that is undergoing apoptosis. Another disadvantage is that caspase activation does not necessarily indicate that apoptosis will occur. Moreover, there is tremendous overlap in the substrate preferences of the members of the caspase family, affecting the specificity of the assay.

Apoptosis PCR microarray is a relatively new methodology that uses real-time PCR to profile the expression of at least 112 key genes involved in apoptosis. The array includes: the TNF ligands and their receptors; members of the bcl-2, caspase, IAP, TRAF (TNF receptor associated factor), CARD (caspase activation and recruitment domain), death domain, death effector domain, and CIDE (cell death-inducing DFFA-like effector) families; and genes involved in the p53 and ATM pathways. Also included in the evaluation are genes involved in anti-apoptosis. This assay allows for the evaluation of the expression of a focused panel of genes related to apoptosis.

16.5.5d Membrane Alterations. Externalization of phosphatidylserine residues on the outer plasma membrane of apoptotic cells allows detection via Annexin V in tissues, embryos, or cultured cells. Once the apoptotic cells are bound with FITC-labeled annexin V, they can be visualized with fluorescent microscopy. The advantages are sensitivity (can detect a single apoptotic cell) and the ability to confirm the activity of initiator caspases. The disadvantage is that the membranes of necrotic cells are labeled as well. The transfer of phosphatidylserine to the outside of the cell membrane will also permit the transport of certain dyes into the cell in a unidirectional manor. As the cell accumulates dye and shrinks in volume, the cell dye content becomes more concentrated and can be visualized with light microscopy. This dye-uptake bioassay works on cell culture, does not label necrotic cells, and has a high level of sensitivity (can detect a single apoptotic cell).

16.5.5e Detection of Apoptosis in Whole Mounts. Apoptosis can also be visualized in whole mounts of embryos or tissues using dyes such as acridine orange (AO), Nile blue sulfate (NBS), and neutral red (NR). Since these dyes are acidophilic, they are concentrated in areas of high lysosomal and phagocytotic activity. The results would need to be validated with other apoptosis assays because these dyes cannot distinguish between lysosomes degrading apoptotic debris from degradation of other debris such as microorganisms. Although all of these dyes are fast and inexpensive, they have certain disadvantages. AO is toxic and mutagenic and quenches rapidly under standard conditions, whereas NBS and NR do not penetrate thick tissues and can be lost during preparation for sectioning. LysoTracker Red is another dye that acts in a similar way; however, this dye can be used with laser confocal microscopy to provide three-dimensional imaging of apoptotic cells. This dye is stable during processing, penetrates thick tissues, and is resistant to quenching. This dye can be used for cell culture as well as for whole mounts of embryos, tissues, or organs.

16.5.5f *Mitochondrial Assays.* Mitochondrial assays and cytochrome C release allow the detection of changes in the early phase of the intrinsic pathway. Laser scanning confocal microscopy (LSCM) creates submicron thin optical slices through living cells that can be used to monitor several mitochondrial events in intact single cells over time. Mitochondrial permeability transition (MPT), mitochondrial depolarization, Ca^{2+} fluxes, mitochondrial redox status, and reactive oxygen species can all be monitored with this methodology. The main disadvantage is that the mitochondrial parameters that this methodology monitors can also occur during necrosis. The electrochemical gradient across the mitochondrial outer membrane (MOM) collapses during apoptosis, allowing detection with a fluorescent cationic dye. In healthy cells, this lipophilic dye accumulates in the mitochondria, forming aggregates that emit a specific fluorescence. In apoptotic cells the MOM does not maintain the electrochemical gradient and the cationic dye diffuses into the cytoplasm, where it emits a fluorescence that is different from the aggregated form. Other mitochondrial dyes can be used that measure the redox potential or metabolic activity of the mitochondria in cells. However, these dyes do not address the mechanism of cell death and are usually used in conjunction with other apoptosis detection methods such as a caspase assay. Cytochrome C release from the mitochondria can also be assayed using fluorescence and electron microscopy in living or fixed cells. However, cytochrome C becomes unstable once it is released into the cytoplasm. Therefore a non-apoptotic control should be used to ensure that the staining conditions used are able to detect any available cytochrome C. Apoptotic or anti-apoptotic regulator proteins such as Bax, Bid, and Bcl-2 can also be detected using fluorescence and confocal microscopy. However, the fluorescent protein tag may alter the interaction of the native protein with other proteins. Therefore, other apoptosis assays should be used to confirm the results.

ACKNOWLEDGMENTS

We gratefully acknowledge Roy Pool, DVM, PhD, whose course notes on General Pathology provided an excellent framework for the sections on cell injury and necrosis. We also acknowledge Greg Miller, NCSU Communications Services, and Elizabeth Ney, NIEHS, for graphic design of the figures. Support for this chapter was provided in part by the Intramural Research Program of the NIH, National Institute of Environmental Health Sciences.

SUGGESTED READING

Ashkenazi, A., and Dixit, V. M. Death receptors: Signaling and modulation. *Science* **281**, 1305–1308, 1998.

Cohen, G. M. Caspases: The executioners of apoptosis. *Biochem. J.* **326**(Pt 1), 1–16, 1997.

Fulda, S., and Debatin, K. M. Extrinsic versus intrinsic apoptosis pathways in anticancer chemotherapy. *Oncogene* **25**, 4798–4811, 2006.

Gozuacik, D., and Kimchi, A. Autophagy as a cell death and tumor suppressor mechanism. *Oncogene* **23**, 2891–2906, 2004.

Hengartner, M. O. The biochemistry of apoptosis. *Nature* **407**, 770–776, 2000.

Hirsch, T., Marchetti, P., Susin, S. A., Dallaporta, B., Zamzami, N., Marzo, I., Geuskens, M., and Kroemer, G. The apoptosis-necrosis paradox. Apoptogenic proteases activated after mitochondrial permeability transition determine the mode of cell death. *Oncogene* **15**, 1573–1581, 1997.

Igney, F. H., and Krammer, P. H. Death and anti-death: Tumour resistance to apoptosis. *Nat. Rev. Cancer* **2**, 277–288, 2002.

Kerr, J. F. History of the events leading to the formulation of the apoptosis concept. *Toxicology* **181–182**, 471–474, 2002.

King, K. L., and Cidlowski, J. A. Cell cycle regulation and apoptosis. *Annu. Rev. Physiol.* **60**, 601–617.

Klionsky, D. J., and Emr, S. D. Autophagy as a regulated pathway of cellular degradation. *Science* **290**, 1717–1721, 2000.

Kumar, V., Abbas, A. K., and Fausto, N. Cellular adaptations, cell injury, and cell death. In: *Robbins and Cotran Pathologic Basis of Disease*, 7th ed. Elsevier Saunders, Philadelphia, 2005, pp. 3–46.

Newmeyer, D. D., and Ferguson-Miller, S. Mitochondria: Releasing power for life and unleashing the machineries of death. *Cell* **112**, 481–490, 2003.

Nicholson, D. W. From bench to clinic with apoptosis-based therapeutic agents. *Nature* **407**, 810–816, 2000.

Norbury, C. J., and Hickson, I. D. Cellular responses to DNA damage. *Annu. Rev. Pharmacol. Toxicol.* **41**, 367–401, 2001.

Pardo, J., Bosque, A., Brehm, R., Wallich, R., Naval, J., Mullbacher, A., Anel, A., and Simon, M. M. Apoptotic pathways are selectively activated by granzyme A and/or granzyme B in CTL-mediated target cell lysis. *J. Cell Biol.* **167**, 457–468, 2004.

Pietenpol, J. A., and Stewart, Z. A. Cell cycle checkpoint signaling: Cell cycle arrest versus apoptosis. *Toxicology* **181–182**, 475–481, 2002.

Rai, N. K., Tripathi, K., Sharma, D., and Shukla, V. K. Apoptosis: A basic physiologic process in wound healing. *Int. J. Low. Extrem. Wounds* **4**, 138–144, 2005.

Schwartz, L. M., Smith, S. W., Jones, M. E., and Osborne, B. A. Do all programmed cell deaths occur via apoptosis? *Proc. Natl. Acad. Sci. USA* **90**, 980–984, 1993.

Strasser, A., O'Connor, L., and Dixit, V. M. Apoptosis signaling. *Annu. Rev. Biochem.* **69**, 217–245, 2000.

Susin, S. A., Daugas, E., Ravagnan, L., Samejima, K., Zamzami, N., Loeffler, M., Costantini, P., Ferri, K. F., Irinopoulou, T., Prevost, M. C., Brothers, G., Mak, T. W., Penninger, J., Earnshaw, W. C., and Kroemer, G. Two distinct pathways leading to nuclear apoptosis. *J. Exp. Med.* **192**, 571–580, 2000.

Trump, B. F., Berezesky, I. K., Chang, S. H., and Phelps, P. C. The pathways of cell death: Oncosis, apoptosis, and necrosis. *Toxicol. Pathol.* **25**, 82–88, 1997.

Vaux, D. L. Apoptosis and toxicology—What relevance? *Toxicology* **181–182**, 3–7, 2002.

Zeiss, C. J. The apoptosis-necrosis continuum: insights from genetically altered mice. *Vet. Pathol.* **40**, 481–495, 2003.

CHAPTER 17

Mitochondrial Dysfunction

JUN NINOMIYA-TSUJI

Department of Environmental and Molecular Toxicology, North Carolina State University, Raleigh, North Carolina 27695

17.1 INTRODUCTION

Mitochondria, the energy powerhouses of cells, generate most ATP through energy-yielding oxidative reactions, namely those of the citric acid cycle [also known as the tricarboxylic acid (TCA) cycle or the Krebs cycle], fatty acid oxidation, and the respiratory chain (also known as the electron transport chain) by a process known as oxidative phosphorylation. Therefore, mitochondrial dysfunction causes a significant reduction of ATP and impairs almost all biological processes in cells potentially leading to necrotic cell death. As discussed in the previous chapter, necrotic cells cause undesired responses such as inflammation. To avoid necrotic cell death, dysfunctional mitochondria can induce apoptosis if the damage is not extreme. The mitochondrial apoptotic pathway is induced not only by dysfunction of mitochondria but also by extramitochondrial cellular damage or environmental stress to the cells. It is the major pathway of apoptosis, which is associated with xenobiotic-mediated cell toxicity and ultimately cardiac toxicity, hepatotoxicity, and neurotoxicity. In addition, the primary targets of many pesticides, herbicides, and environmental toxicants are the energy-yielding oxidative reactions in mitochondria, and they induce dysfunction of mitochondria. In this chapter, we introduce mitochondrial functions and discuss the mechanisms by which mitochondrial dysfunction activates the apoptotic and necrotic pathway. We also discuss the mechanism of how extramitochondrial stress causes mitochondrial apoptosis.

17.2 MITOCHONDRIAL FUNCTION

17.2.1 Structure: Outer and Inner Membranes

Mitochondria are the energy powerhouses of cells. Mitochondrial membranes play an essential and unique function to generate energy. Mitochondria have two specialized membranes (outer and inner membranes), which are very different. Inside the inner membrane is the matrix, and the space between the outer and inner mem-

Molecular and Biochemical Toxicology, Fourth Edition, edited by Robert C. Smart and Ernest Hodgson
Copyright © 2008 John Wiley & Sons, Inc.

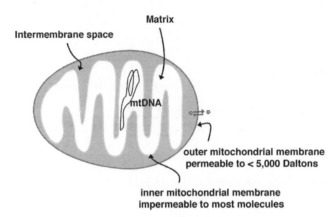

Figure 17.1. Mitochondria have two specialized membranes. mtDNA, stands for mitochondrial DNA.

branes is the intermembrane space (Figure 17.1). The outer membrane contains an abundance of large pore-forming proteins (permeability transition pore, PTP) as described in Section 17.2.4. It is permeable to molecules less than 5000 daltons, including all metabolites of energy-generating reactions, namely ATP, ADP, fatty acid, and pyruvate. In contrast, the inner membrane is impermeable to most molecules. The inner membrane contains a high proportion of a unique dimeric phospholipids, cardiolipin, which makes the membrane especially impermeable to ions. This membrane contains many transport proteins to selectively incorporate the metabolites into the matrix and export the generated ATP to be available to biological processes in cells. The energy-yielding oxidative reactions, the citric acid cycle, the fatty acid oxidation, and the respiratory chain take place in the matrix. The matrix contains mitochondrial DNA (mtDNA), which encodes 13 proteins of the respiratory chain. mtDNA is unique; it contains no intron and very little noncoding DNA, has a genetic code similar to that of bacteria, and is maternally inherited. mtDNA genes are transcribed and translated in the matrix, and the products function on the inner membrane.

17.2.2 Generating High-Energy Electrons (TCA Cycle, Fatty Acid Oxidation, FADH$_2$, and NADH)

The energy-yielding reactions in the mitochondrial matrix are initiated from pyruvate (a metabolite of glycolysis in the cytoplasm) and fatty acids, which are derived from carbohydrates, fats, and other nutrients. Pyruvate and fatty acids are transported into the matrix through specific transporters, and they are utilized to generate high-energy electrons carried by the carrier molecules, NADH and FADH$_2$, through the fatty acid oxidation and the citric acid cycles. Acetyl CoA is generated from pyruvate and from fat through fatty acid oxidation. Acetyl CoA enters the citric acid cycle; and from the two carbon atoms of acetyl CoA, two carbon atoms leave the cycle as CO_2. In the four oxidation–reduction reactions in the cycle, three pairs of electrons are transferred to NAD$^+$ and one pair to FAD.

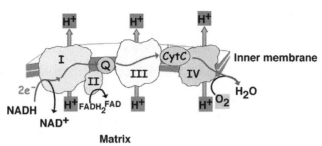

Figure 17.2. Respiratory chain. Electrons flow through Complex I (I)/Complex II (II), ubiquinone (Q), Complex III (III), cytochrome c (CytC), and Complex IV (IV) and protons are pumped from the matrix to the intermembrane space.

17.2.3 Respiratory Chain (Complex I, II, III, and IV, Ubiquinone, Cytochrome c, Proton Pump, Membrane Potential, Proton Motive Force)

The energy saved in the electron carriers NADH and FADH2 is transferred into the respiratory chain. The enzyme complexes of the respiratory chain are physically located in the inner membrane (Figure 17.2). The unique characteristic (impermeability to almost all molecules) of the inner membrane is essential for this reaction as described below. The reaction in the respiratory chain is very simple.

$$H_2 + \frac{1}{2}O_2 \rightarrow H_2O$$

This is an energetically favorable reaction, and the reaction is processed through many small steps so that the energy can be saved without loss as heat. During the small steps in the respiratory chain, the energy from electron transfers is used to pump the protons from the matrix to the intermembrane space. In this section, we describe the reactions and the mechanisms by which ATP is generated. H_2 is separated to protons (H^+) and electrons in previous reactions in the citric acid cycle. The electrons carried by NADH and $FADH_2$ are transferred into four enzyme complexes, Complex I, II, III, and IV. In the final step, the electrons are donated to O_2 resulting in H_2O.

Step 1: Complex I. Complex I is the NADH dehydrogenase complex consisting of more than 40 polypeptides. It accepts electrons from NADH and passes them through flavin and iron–sulfur centers to the electron carrier ubiquinone (Q). By using one NADH (= 2 electrons), this reaction pumps the 4 protons (2 protons/ electron) from the matrix to the intermembrane space.

$$NADH + Q + 5H^+_{matrix} \rightarrow NAD^+ + QH_2 + 4H^+_{intermembrane\ space}$$

Ubiquinone. An electron carrier in the respiratory chain, ubiquinone is a small hydrophobic molecule that can move freely in inner membrane. It is also known as coenzyme Q.

Step 1': Complex II. When $FADH_2$ is utilized as an electron donor, the succinate-Q reductase complex (Complex II) transfers the electrons from $FADH_2$ to ubiquinone through iron–sulfur centers. This reaction does not have a proton pumping function.

Step 2: Complex III. The cytochrome b-c 1 complex (Complex III) consists of at least 11 polypeptides and functions as a dimer. It accepts electrons from ubiquinone and transfers them to the next carrier, cytochrome c (CytC). This reaction pumps four protons (2 protons/electron) to the intermembrane space.

$$QH_2 + 2Cyt_{coxidized} + 2H^+_{matrix} \rightarrow Q + 2CtyC_{reduced} + 4H^+_{intermembrane\ space}$$

Cytochrome c. An electron carrier in the respiratory chain, cytochrome c consists of a small protein (114 amino acids in human) and a bound heme group. The iron atom in the heme group can accept an electron. It has a pink color with absorption at 415, 520, and 549 nm that is derived from the heme group.

Step 3: Complex IV. The cytochrome oxidase complex consists of at least 13 polypeptides and functions as a dimer. It accepts electrons form cytochrome c and transfers them to the final acceptor oxygen. From two molecules of reduced cytochrome c, this reaction pumps two protons (one proton/electron) to the intermembrane space.

$$4CytC_{reduced} + 8H^+_{matrix} + O_2 \rightarrow 4CtyC_{oxidized} + 2H_2O + 4H^+_{intermembrane\ space}$$

Step 4: ATP Synthase. The respiratory chain pumps protons toward to the intermembrane space, which results in creating a pH gradient = membrane potential ($\Delta\Psi m$). This produces proton motive force from the intermembrane space back to the matrix. ATP synthase is a big multisubunit protein (>500 kD) in which protons flow from the intermembrane space to the matrix. The proton flow energy utilizes to phosphorylate ADP resulting in ATP.

$$ADP + Pi \rightarrow ATP$$

It is known that three or four protons are required to flow through the ATP synthase to generate one molecule of ATP.

17.2.4 Permeability Transition Pore (VDAC. ANT)

ATP generated in the matrix is transported to the intermembrane space by adenine nucleotide translocase (ANT) that exchanges ADP in the intermembrane space with ATP in the matrix. ATP then goes to the cytosol, where it is utilized in a variety of biological processes. Molecules such as NADH, FADH2, ATP, ADP, and small proteins (<5000 daltons) can freely cross the mitochondrial outer membrane through a channel, the voltage-dependent anion channel (VDAC). VDAC in the outer membrane forms a complex with ANT in the inner membrane (Figure 17.3). This complex is called the "permeability transition pore" (PTP) and consists of VDAC, ANT, and several VDAC/ANT-binding proteins such as cyclophilin D (Cyp D) and the periph-

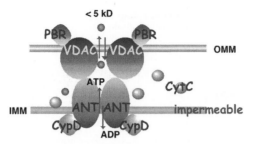

Figure 17.3. Permeability transition pore (PTP). The PTP consists of voltage-dependent anion channel (VDAC), adenine nucleotide translocase (ANT) and several associated molecules including cyclophilin D (CypD) and peripheral benzodiazepine receptor (PBR). IMM, inner mitochondrial membrane; OMM, outer mitochondrial membrane; CytC, cytochrome c.

eral benzodiazepine receptor (PBR) (Figure 17.3). PTP is very abundant in mitochondria. As discussed above, PTP normally functions to exchange the substrates and products of the energy-yielding reactions in mitochondria. PTP opening, the state that unregulated exchange (leak) of mitochondrial components occur, is coupled with apoptosis and necrosis.

17.3 MITOCHONDRIAL APOTOSIS/NECROSIS

17.3.1 Mitochondrial Permeability Transition

Besides energy generation, one of the most important functions of mitochondria is inducing apoptosis (mitochondrial apoptosis), which is essential for eliminating dysfunctional cells safely from the body. Importantly, every apoptosis induced by a variety of stimuli or stress is always coupled with mitochondrial membrane permeabilization [also called mitochondrial permeability transition (MPT)]. There are two types of mitochondrial membrane permeabilization: One type is mitochondrial outer-membrane permeabilization in which relatively bigger proteins are leaked from intermembrane space through Bax/Bak pores as described in Section 17.3.4; the other type is the PTP opening that disrupts impermeability of the inner membrane, resulting in influx of water/ions into the matrix and leading to loss of membrane potential, swelling, and loss of ATP. From recent studies utilizing genetically engineered animal models and siRNA methods, it is generally accepted that the Bax/Bak pore is the major inducer of apoptosis and that PTP opening is instead associated with necrosis. Mitochondrial membrane permeabilization causes leakage of mitochondrial proteins into the cytosol resulting in activation of caspases. The mechanism of caspase activation is discussed in Sections 17.3.2 and 17.3.3.

17.3.2 Caspase Activation: Effector and Initiator Caspases

Caspases belong to the family of cystein proteases that cleave adjacent to aspartate residues in the substrate. Fourteen mammalian caspases have been identified and divided into three classes. The first class is the inflammatory caspases, caspase 1, 4,

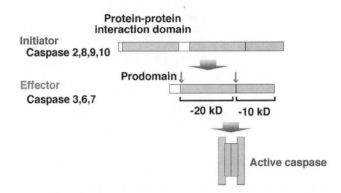

Figure 17.4. Initiator and effector caspases. Initiator caspases cleave and activate effector caspases. Cleavage sites are shown.

5, and 13. They are activated in response to an inflammatory signal and are believed not to be involved in apoptosis. Caspase 1 and 5 participate in proteolytic maturation of a proinflammatory cytokine interleukin 1. The second class is the initiator caspases, which include caspases 2, 8, 9 and 10, which function to activate the third class, effector caspases, caspases 3, 6, and 7. An apoptotic signal activates the initiator caspase, which, in turn, cleaves the precursor of the effector caspase resulting in formation of active effector caspase (Figure 17.4). For example, caspase 9 cleaves 2 sites of caspase 3 and generate 3 small domains. The N-terminal pro-domain is degraded, and the remaining polypeptides form a catalytically active enzyme.

How Are the Initiator Caspases Activated? The initiator caspases are relatively bigger than the effector caspases and have domains such as CARD (caspase recruitment domain) and DED (death effector domain). Their activity is regulated through protein–protein interactions. Caspase 8 has a DED domain that is associated with the death receptor family complexes such as tumor necrosis factor receptor and CD95 upon stimulation. The association triggers activation by auto-cleavage. The CARD domain of caspase 9 functions in its oligomerization. When apoptotic stimuli induce the cytochrome c release (described in detail in Section 17.3.3), caspase 9 forms a large oligomeric complex so-called "apoptosome" that has seven spokes. Apoptosome comprises seven monomers of cytochrome c, caspase 9, apoptosis protease activation factor-1 (Apaf1), and ATP (Figure 17.5). The apoptosome is the active caspase 9 and cleaves caspase 3 as described above.

17.3.3 Mitochondria Are "Poison Cabinets"

It had long been known that apoptosis and activation of caspase are coupled with mitochondrial permeabilization. However, until 1996, the molecular link between mitochondria and caspase activation had not been known. Dr. Wang's group at the University of Texas isolated activators of caspase 3 by using classical fractionation methods in 1996–1997. One protein that was required for activation of caspase 3 was a 15-kD protein with pink color, which was cytochrome c. This finding is rather surprising because cytochrome c is localized in mitochondria and has a well-established role as an electron carrier. Cytochrome c was later shown to be released

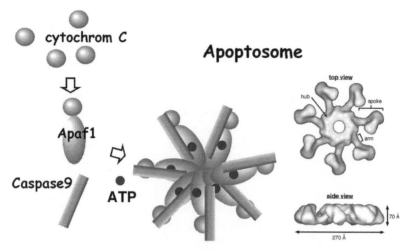

Figure 17.5. Apoptosome. (Adapted from Acehan, D., Jiang, X., Morgan, D. G., Heuser, J. E., Wang, X., and Akey, C. W. Three-dimensional structure of the apoptosome: Implications for assembly, procaspase-9 binding, and activation. *Mol. Cell.* **9**, 423–432, 2002.)

TABLE 17.1. Mitochondrial Pro-apoptotic Proteins

	Approximate Size	Function in Mitochondria	Function in Apoptosis
Cytochrome c	15 kD	Electron carrier	Component of apoptosome (activation of caspase 9)
Diablo (also known as Smac)	28 kD	Unknown	Inhibition of IAPs
Omi (also known as Htra2)	38 and 40 kD	Unknown	Inhibition of IAPs
AIF	57 kD	Oxidoreductase	Chromatin condensation
Endonuclase G	35 kD	Unknown (mtDNA repair?)	DNA fragmentation

from mitochondria when permeabilized, and it is an essential component of apoptosome (Figure 17.5). Subsequently, Apaf1 and caspase 9 were identified. Cytochrome c is a protein conserved through evolution from bacteria to mammals; however, it gains its apoptotic role only in vertebrates. It is known that cytochrome c in worms and fruit flies does not function in apoptosis. It should be noted here that because the apoptosome requires ATP as an essential component, apoptosis is an ATP-dependent process.

Mitochondria release not only cytochrome c but also many pro-apoptotic factors (Table 17.1). They are normally localized in the intermembrane space of mitochondria. However, except for cytochrome c and the apoptosis inducing factor (AIF), their functions in the mitochondria have not been determined or they may have no function under normal conditions. Because they are larger than 5 kD, they remain inside the mitochondrion. Once the mitochondrial outer membrane is permeabilized, cytochrome c, Diablo (also known as Smac), Omi, AIF, and endonuclease G (EndoG) are released to the cytosol. Cytochrome c participates in apoptosome formation. Diablo/Smac and Omi inhibit inhibitors of apoptosis (IAP), which act as endogenous inhibitor of caspases. IAPs associate with caspases and directly inhibit their catalytic activity, which functions to downregulate caspase activation. When Diablo/Smac and Omi bind to IAPs, interaction of IAPs with caspases is disrupted, resulting in activation of caspases. Therefore, release of Diablo/Smac and Omi facilitates the caspase-induced apoptosis pathway. In contrast, AIF and EndoG directly induce DNA fragmentation and chromatin condensation (hallmarks of nuclear apoptosis), which is independent of caspase activation. AIF shares significant homology with bacterial NADH oxidoreductases. Indeed, it has NADH oxidoreductase activity, and AIF-deficient cells show reduced activity of complex I and III in the respiratory chain. AIF is inserted into the inner mitochondrial membrane via its N-terminal domain and functions to optimize the respiratory chain reactions. Upon apoptotic stimulation, the membrane insertion domain of AIF is cleaved and C-terminal AIF is released from the mitochondria. AIF goes to the nucleus and binds directly to DNA. AIF-bound DNA clusters and undergoes chromatin condensation. AIF knockout mice do not show any defect in apoptosis during development but are lethal because of impaired respiratory chain function. Therefore, AIF may participate in some specific apoptotic signaling pathways but not those during development. EndoG induces DNA fragmentation. However, EndoG knockout mice are normal and EndoG-deficient cells are also normal in apoptotic responses. EndoG is likely to facilitate apoptotic DNA fragmentation but is not essential.

17.3.4   Interplay of Bcl2 Family (BH3-only, Bcl2, Bax/Bak)

Bcl-2 family proteins are major regulators of mitochondrial outer membrane permeability (MOMP). Bcl-2 family proteins have conserved regions, Bcl-2 homology regions 1–3 (BH1, BH2, and BH3). Bcl-2 family proteins are divided into three subfamilies (Figure 17.6), Bcl-2, Bax/Bak, and BH3-only subfamilies. Bcl-2 subfamily proteins have BH1, 2, and 3, and promote cell survival (anti-apoptotic Bcl-2 family). Bax/Bak subfamily proteins have BH1, 2, and 3 and are pro-apoptotic. Bax/Bak are the executioners of MOMP by forming pores in the mitochondrial outer membrane. BH3-only subfamily proteins are also pro-apoptotic and function as inhibitors of anti-apoptotic Bcl-2 family proteins or activators of Bax/Bak.

BH3-only Proteins Are Sensors of Apoptosis.   A number of BH3-only proteins have been identified. Bim is one of potent pro-apoptotic BH3-only proteins and is transcriptionally or posttranscriptionally upregulated by growth factor depletion, Ca^{2+} influx, or other stresses. Bim binds to and inhibits all types of anti-apoptotic Bcl-2 family proteins as described blow. Bim can directly activate Bax. Bid is another potent pro-apoptotic BH3-only protein and is activated by cleavage of N-terminal region by caspase 8 upon tumor necrosis factor stimulation. The 15-kD C-terminal fragment tBid is translocated into the mitochondrial outer membrane and binds to and inhibits all types of anti-apoptotic Bcl-2 family proteins. tBid can also directly activate Bax. Puma and Noxa are transcriptionally regulated by p53. p53-activating stimuli such as DNA damage induce an increase of p53 that directly binds to and activates the promoter region of Puma and Noxa genes. Puma binds to and inhibits

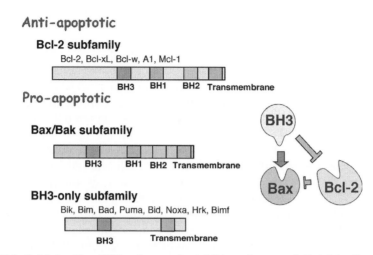

Figure 17.6. Bcl-2 families. BH3-only proteins inhibit anti-apoptotic Bcl-2 family proteins or directly activate Bax/Bak. Bcl-2 family proteins serve as inhibitors of Bax/Bak. BH, Bcl-2 homology domain.

TABLE 17.2. Role of BH3-only Family Proteins

	Activator	Target	Function
Bim	Ca2+ flux, Taxol, UV, other stress	Bcl-2, Bcl-xL, Bcl-w, A1, Mcl-1, Bax	Inhibition of Bcl-2 family/ activation of Bax/Bak
Bid	Tumor necrosis factor, CD95 (FASL), granzyme B[a]	Bcl-2, Bcl-xL, Bcl-w, A1, Mcl-1, Bax	Inhibition of Bcl-2 family/ activation of Bax/Bak
Puma	p53	Bcl-2, Bcl-xL, Bcl-w, A1, Mcl-1	Inhibition of Bcl-2 family
Noxa	p53	A1, Mcl-1	Inhibition of Bcl-2 family
Bad	Growth factor deprivation	Bcl-2, Bcl-xL, Bcl-w	Inhibition of Bcl-2 family
Bimf	Anoikis,[b] UV	Bcl-2, Bcl-xL, Bcl-w	Inhibition of Bcl-2 family
Bik	p53 and other stress	Bcl-2, Bcl-xL, Bcl-w	Inhibition of Bcl-2 family
Hrk	Stress (not well-studied)	Bcl-2, Bcl-xL, Bcl-w	Inhibition of Bcl-2 family

[a]Granzyme B, a serine proteinase, is produced in cytotoxic T lymphocytes and functions to kill the virus infected or tumor cells.
[b]Anoikis is epithelial cell death induced by detachment from basement membrane (extracellular matrix).

all types of anti-apoptotic Bcl-2 family proteins, whereas Noxa targets only A1 and Mcl-1. We summarize the activators and targets of BH3-only proteins in Table 17.2.

Anti-apoptotic Bcl-2 Family Proteins Are Targets of BH3-only Proteins and Inhibitors of Bax/Bak. Bcl-2, Bcl-xL, Bcl-w, A1, and Mcl-1 are anti-apoptotic Bcl-2 family proteins and are localized in the cytosol and on the cytoplasmic face of the mitochondrial outer membrane. In healthy cells they sequester Bax/Bak proteins, presumably through direct association. When BH3-only proteins bind to Bcl-2 family proteins, they release Bax/Bak, resulting in activation of Bax/Bak pore formation.

Bax/Bak Forms a Pore on Mitochondrial Outer Membrane. Bax/Bak is regulated by three different mechanisms: transcriptional activation by p53, binding of Bim and tBid (activation of Bax/Bak), and release from Bcl-2 family proteins (free Bax/Bak). These result in increase of free (or activated) Bax/Bak on the mitochondrial outer membrane, which induces Bax/Bak oligomerization and pore formation, which allows >5 kD proteins (such as cytochrome c) to leak from the mitochondrial inter-membrane space. Bax and Bak are essential for apoptotic cell death. The double knockout Bax-/- Bak-/- mice exhibit impaired apoptosis during development. Although most of double mutant mice are lethal during embryogenesis, some mice reach adulthood and display interdigital webs that should undergo apoptosis during development. Bax-/- Bak-/- cells are resistant to many apoptotic stimuli including DNA damage, growth factor deprivation, and ER stress. Bax-/- Bak-/- cells undergo necrotic or autophagic (see the note below) cell death when cells are excessively damaged.

Autophagy. Autophagy is a mechanism of cell survival under starving condition. Cells degrade their own cytoplasmic components and re-use them as energy sources. The cellular components are incorporated into double-membrane autophagic vacuoles, also known as autophagosomes, which transport them to the lysosome for degradation. Recent studies have revealed that this pathway participates in not only cell survival but also cell death. When cells are stressed, adaptive pathways including autophagy are activated to protect cells from death. If the stress is excessive, cells die normally through the Bax/Bak-caspase-dependent apoptosis. However, when the Bax/Bak-caspase pathway is somehow not activated, the autophagy pathway is activated and cells undergo autophagic cell death, which is characterized by an accumulation of autophagic vacuoles in the cytoplasm.

17.3.5 Permeability Transition Pore Opening

The permeability transition pore (PTP) consists of a voltage-dependent anion channel (VDAC), adenine nucleotide translocase (ANT), and several associated molecules including cyclophilin D (Cyp D) and the peripheral benzodiazepine receptor (PBR) described in Section 17.2.4 (Figure 17.3). PTP normally functions to exchange metabolites through the mitochondrial outer and inner membranes. Ca^{2+} is a prominent modulator of PTP. Increase of cytosolic Ca^{2+} concentration triggers a conformational change of ANT, which leads to "pore opening." The PTP opening destroys the impermeability of the mitochondrial inner membrane. Because the matrix normally keeps high osmolarity by the impermeable membrane, pore opening causes influx of solutes and water into the matrix resulting in swelling. The mitochondrial swelling ruptures the outer membrane, causing release of mitochondrial pro-apoptotic proteins such as cytochrome c. Concurrently, function of the respiratory chain is disrupted by permeabilized inner membrane resulting in loss of membrane potential, which ultimately leads to ATP loss. Cytochrome c release activates caspase-dependent apoptosis; however, the ATP loss prevents apoptotic cell death and induces necrotic or autophagic cell death. It is generally accepted

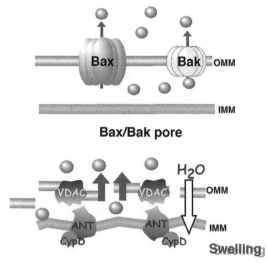

Figure 17.7. Two types of mitochondrial membrane permeabilization. Upper: Bax/Bak pore leads to release of intermembrane space proteins, but the inner membrane is intact. Lower: PTP (permeability transition pore) opening destroys the impermeability of the inner mitochondrial membrane (IMM). The pore opening causes influx of solutes and water into the matrix resulting in swelling. The mitochondrial swelling ruptures outer mitochondrial membrane (OMM). CypD, cyclophilin D.

that "strong PTP modulators" lead to necrotic cell death due to irreversible mitochondrial damage and significant loss of ATP, but "mild PTP modulators" lead to apoptosis because a subpopulation of mitochondria is undamaged and produces enough ATP.

Ca^{2+} and oxidative stress are well-studied modulators of PTP that leads to "pore opening." A component of PTP, cyclophilin D, functions to facilitate PTP opening. The immunosuppressive drug cyclosporin A binds to and inhibits cyclophilin D. Studies using cyclophilin D knockout mice and cyclosporin A have revealed that PTP opening is important for Ca^{2+}- and oxidative stress-induced cell death but dispensable for the Bax/Bak-caspase-dependent apoptosis. Thus, there are two independent mitochondrial permeabilization pores: One is Bax/Bak pore that is formed upon stimulation; the other is PTP that has normal functions but causes leakage upon challenge by modulators (Figure 17.7). Many BH3-only activating death stimuli such as DNA damage primarily utilize the Bax/Bak pore but not PTP opening to release cytochrome c. This is anticipated, because the Bax/Bak pore only affect the intermembrane space but not the energy yielding reactions in mitochondria, in which cells undergo apoptosis and avoid necrotic cell death.

17.3.6   Cross-Talk of Bax/Bak and PTP Opening Pathways

Dying cells usually show characteristics of both apoptotic and necrotic cell death. This is because the Bax/Bak and PTP opening pathways have cross-talk at several stages. Indeed, loss of mitochondrial membrane potential is a hallmark of apoptosis even though it is coupled with loss of ATP and necrotic cell death. Cross-talk happens at the stage of activation of caspases. Caspases cleave many substrates including the p75 subunit of Complex I of the respiratory chain, resulting in ablation of the respiratory chain and loss of membrane potential. The other major cross-talk happens at the stage of generation of reactive oxygen species (ROS). ROS are constitutively produced at some levels as byproducts of the respiratory chain and are reduced and detoxified by several mechanisms. Inhibition of the respiratory chain causes accumulation of electrons in the highly reactive electron carriers, which induces excessive amount of ROS production. ROS generated by inhibition of the respiratory chain is sensed by BH3-only proteins and also modulate PTP, resulting in PTP opening. The typical pathways leading to apoptotic and necrotic cell death are summarized below.

1. *Stress Is Sensed by BH3-only Proteins.* → BH3-only proteins inhibit anti-apoptotic Bcl-2 proteins. → Bax/Bak pore is formed. → Cytochrome c and other apoptotic proteins are released from mitochondria. → They activate caspases. → Caspases cleave the complex I subunit of the respiratory chain. → It causes loss of membrane potential and generates ROS. → ROS modulates and open PTP and activates BH3-only proteins. →

2. *PTP Is Modulated by ROS or by Ca^2+.* → It releases cytochrome c and other apoptotic proteins and causes loss of membrane potential. → More ROS is generated and caspases are activated. → ROS activates BH3-only proteins. →

3. *Respiratory Chain Is Blocked or Damaged by Toxicants.* → Membrane potential is declined and ROS is generated. → ROS causes PTP opening and activates BH3-only proteins. → Apoptotic proteins are released from mitochondria. → Cytochrome c activates caspases.

17.4 TOXICANT-INDUCED MITOCHONDRIAL APOPTOSIS/NECROSIS

17.4.1 Electron Transport Inhibitors

Because the respiratory chain is a conserved mechanism to produce energy and energy is critical for cell survival, many pesticides and herbicides target the respiratory chain (Figure 17.8). In addition, toxins and environmental stress such as hypoxia also affect the respiratory chain. Rotenone is a widely used pesticide and fish toxin that inhibits Complex I. Antimycin inhibits Complex III. Carbon monoxide, azide, and cyanide bind to heme a3 of Complex IV at high affinity, thereby inhibiting the oxidase activity. Temporary lack of oxygen (ischemia and hypoxia) also inhibits electron flow due to the lack of electron acceptor oxygen. Mitochondrial DNA (mtDNA) encodes 13 proteins of the respiratory chain. Because mtDNA has almost no protective proteins, UV and reactive oxygen species (ROS) easily damage mtDNA. Mutation of mtDNA results in production of dysfunctional proteins essential for the respiratory chain causing inhibition of electron flow. Inhibition of electron flow leads to accumulation of reduced ubiquinone (highly reactive) and reduced

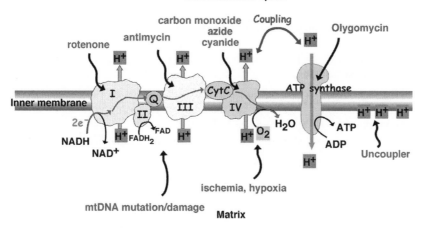

Figure 17.8. Toxicants target the respiratory chain.

cytochrome c. The reduced ubiquinone, Complex I, and Complex III donate electrons directly to oxygen and generate superoxide anions $O_2^{-\bullet}$. ROS modulates PTP and activates BH3-only proteins, which results in activation of the caspase pathway. However, inhibition of the respiratory chain reduces ATP production, which prevents ATP-dependent apoptosis and low ATP condition (metabolic stress) activates autophagy. Therefore, these toxicants more likely induce necrotic and autophagic cell death rather than apoptosis.

17.4.2   Energy Transfer Inhibitors

Inhibition of ATP synthase (energy transfer) reduces proton flow from the intermembrane space to the matrix, which inhibits electron flow in the respiratory chain. Oligomycin, a macrolide antibiotic, prevents phosphoryl group transfer of ATP synthase. Dicyclohexylcarbodimide (DCCD) binds to and inhibits ATP synthase. Similar to the inhibitors of Complexes I, III, and IV, energy transfer inhibitors cause accumulation of reactive electrons and generate ROS.

17.4.3   Uncouplers

Several compounds, such as 2,4-dinitrophenol (DNP) and pentachlorophenol (PCP), are known to act as uncoupling agents. "Uncoupling" is the state in which ATP synthesis is inhibited by disruption of pH gradient. DNP is a small, lipid-soluble compound and carries protons across the inner membrane. The dissociable protons increase protons in the matrix, resulting in abolishment of pH gradient/proton motive force that is required for ATP synthesis. Uncouplers reduce production of ATP, which generate a condition of "metabolic stress." Metabolic stress induces autophagy and other stress responses including activation of p38 and c-Jun N-terminal kinase (JNK). These pathways lead to apoptotic, necrotic, and autophagic cell death.

SUGGESTED READING

Danial, N. N., and Korsmeyer, S. J. Cell death: Critical control points. *Cell* **116**, 205–219, 2004.

Debnath, J., Baehrecke, E. H., and Kroemer, G. Does autophagy contribute to cell death? *Autophagy* **1**, 66–74, 2005.

Li, P., Nijhawan, D., and Wang, X. Mitochondrial activation of apoptosis. *Cell* **116**, S57–S59.

Newmeyer, D. D., and Ferguson-Miller, S. Mitochondria: Releasing power for life and unleashing the machineries of death. *Cell* **112**, 481–490, 2003.

Willis, S. N., and Adams, J. M. Life in the balance: How BH3-only proteins induce apoptosis. *Curr. Opin. Cell Biol.* **17**, 617–625, 2005.

CHAPTER 18

Glutathione-Dependent Mechanisms in Chemically Induced Cell Injury and Cellular Protection Mechanisms

DONALD J. REED

Department of Biochemistry and Biophysics, Oregon State University, Corvallis, Oregon 97331

18.1 INTRODUCTION

In this chapter, glutathione (GSH)-dependent mechanisms of cell injury and death are discussed along with the cellular protective systems that depend on glutathione. These protective systems are thought to be highly important in the detoxication of products from the bioactivation of chemicals and oxygen metabolism. However, certain chemicals form conjugates with GSH that lead to reactive intermediates that can cause cell injury and death rather than cellular protection. Cells produce a constant supply of reactive oxygen species mainly within their mitochondria; without protection, chemically induced injury and death of cells and tissues with concomitant loss of organ functions would be greater as a result of chemical exposure.

The GSH-dependent systems that protect against cell injury and death are important components of biochemical toxicology. Although several hypotheses exist as the basis for mechanisms of chemical effects on cells, two will be described in this chapter: receptor-mediated and chemical-mediated effects to illustrate GSH-dependent protection. One hypothesis suggests that chemically induced cell injury is the result of covalent binding of biological reactive intermediates to cellular molecules, including macromolecules. Although this is a predominant mechanism for loss of cellular functions and expression of cell injury, oxidative stress, as a byproduct of the metabolism of molecular oxygen, can be coupled to the effects of covalent binding and, in some instances, have a major role in cell injury and death. Oxygen is metabolized to oxygen-containing reactive intermediates, many of which can undergo a cycling process between oxidation and reduction. This process, known as redox cycling, can cause oxidative stress in cells. Such oxidative stress can result in (a) oxidation of cellular constituents that are protective, such as low-molecular-weight thiol-containing compounds including GSH, protein thiols, and other functional groups on macromolecules, and (b) peroxidation of lipids

Molecular and Biochemical Toxicology, Fourth Edition, edited by Robert C. Smart and Ernest Hodgson
Copyright © 2008 John Wiley & Sons, Inc.

and other susceptible cellular constituents. It is known that many chemicals can (i) undergo bioactivation to form biological reactive intermediates that bind to macromolecules and (ii) enhance the formation of oxygen radicals with a concomitant oxidative stress.

Receptor-mediated effects of chemicals are important features of cell injury and death and have led to a new area of inquiry, namely, signal transduction pathways that are dependent on glutathione. These pathways are being elucidated at a rapid pace and have been a major component in development of new concepts in cellular signal transduction.

18.2 GLUTATHIONE-DEPENDENT CONJUGATION OF CHEMICALS

A major cellular function of GSH is the detoxification of chemicals designated as xenobiotics or their metabolites (this aspect is covered in Chapter 12).

In reactions catalyzed by glutathione *S*-transferases (GSTs), electrophiles are conjugated with GSH and excreted by cells or in the case of hepatocytes excreted into the bile. Some electrophiles are so reactive toward GSH that spontaneous reactions occur simultaneously along with catalyzed reactions to form conjugates. Once outside the cell, the conjugates are subject to degradation, leading to the formation of mercapturic acids first by the loss of the gamma-glutamyl moiety by gamma-glutamyl transpeptidase (GGT) activity to form cysteinyl-glycine conjugates that are imported into the inside of cells again where dipeptidase activity cleaves the cysteinyl-glycine bond to form cysteine conjugates. To limit cellular and organ retention of the cysteine conjugates, they are converted to *N*-acetylcysteine conjugates known as mercapturic acids; these are excreted into the urine, bile, or both, depending on several factors including molecular weight of the mercapturic acids (Chapter 12).

GSTs are a family of isoenzymes that conjugate GSH with electrophilic compounds (see Chapter 12). They not only serve for the detoxification of xenobiotics but also play a role in bioactivation of various chemicals including certain halogenated compounds. GSTs also have physiological catalytic functions. For example, the isomerization of 3-keto steroids and biosynthesis of leukotriene A4 and eicosanoids involve certain GSTs. Some GSTs have peroxidase activity which assists in the prevention of oxidative stress.

GSTs are present as soluble and membrane-bound forms. There are six species-independent classes of cytosolic GSTs—alpha, mu, pi, sigma, theta, and zeta—and one class of microsomal GSTs. Cytosolic GSTs are dimeric proteins composed of two identical or nonidentical subunits. Classes pi and theta contain homodimers, whereas classes of alpha and mu are more complex and display multiplicity of homodimeric and heterodimeric isoenzyme forms. These subunits have a catalytic center with two binding sites: a highly specific glutathione binding site and a hydrophobic side where acceptor substrates are accommodated. The zeta form is identical with maleylacetoacetate isomerase.

Microsomal GSTs can account for as much as 3% of endoplasmic reticulum protein. Microsomal GSTs are involved in protection against lipid peroxidation because they have been shown to reduce fatty acid hydroperoxides. GSTs are found in the cytosol, mitochondria, nucleus, and the nucleolus. A GSH-dependent peroxidase in rat liver nuclear membranes has been shown to be capable of inhibiting lipid peroxidation. Since the nuclear membrane regulates transport mRNA into the cytosol, this GST is thought to aid in the protection of nuclear division from oxidative damage. It is known that DNA itself is associated with certain regions of the nuclear membrane and may benefit from such peroxidase protection.

18.3 GSH-DEPENDENT BIOACTIVATION OF CHEMICALS

The formation of GSH conjugates with electrophiles generally cause the electrophile to be both less toxic and more readily excreted. However, there is a growing group of chemicals, many being halogenated compounds, that increase in toxicity due to being conjugated with GSH. GSH conjugates that cause toxicity have been divided into three groups: (1) reactive, direct-acting GSH conjugates, (2) glutathione conjugates in which GSH functions as a transporter molecule by releasing reversibly bound electrophilic compounds, and (3) GSH conjugates that require further bioactivation, either by catabolism of the GSH moiety or by activation of the chemically derived moiety.

18.3.1 Enzymes Involved in Bioactivation

GSTs participate in bioactivation of xenobiotics by (1) forming direct-acting glutathione conjugates without further enzymic action, (2) functioning as a transporter molecule that utilizes membrane transport proteins to release reversibly bound electrophiles at target tissues, (3) forming GSH conjugates that are bioactivated by subsequent metabolism of the GSH moiety by the action of gamma-glutamyl transpeptidase (GGT) and peptidases, and (4) reductive enzymes performing reductive bioactivation. For example, selenite activation by reduction with GSH to form selenodiglutathione (GS-Se-SG), a potent tumor inhibitor.

18.3.2 GSH-Dependent Biologically Reactive Intermediates

The bioactivation and metabolism of many classes of xenobiotics have been well-documented and reviewed. Formation of direct-acting GSH conjugates include the bioactivation of haloalkanes and haloalkenes. GSH conjugates form as the initial step, but are unstable and give rise to toxic metabolites. With dihalomethanes, formaldehyde is formed. Vicinal dihaloethanes, including 1,2-dichloroethane and 1,2-dibromoethane, have been shown to be mutagenic and carcinogenic. Their bioactivation leads to the formation of half-sulfur mustards that can undergo an intramolecular cyclization to give a highly reactive episulfonium ion that may react with cellular nucleophiles (Figure 18.1).

Examples for GSH as a transporter of reversibly bound electrophiles are conjugates of isocyanates, isothiocyanates, alpha and beta unsaturated aldehydes, and aldehydes. These compounds form labile conjugates that may again disassociate to the parent electrophile and glutathione. The electrophile can then react with endogenous nucleophiles to form more thermodynamically favored adducts.

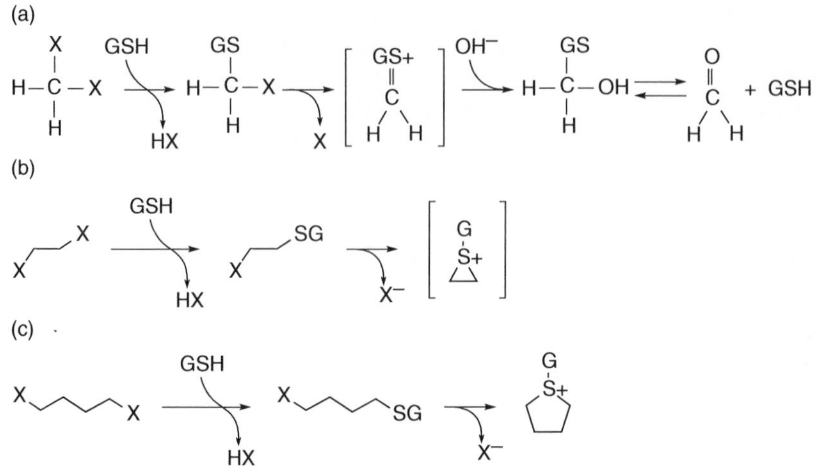

(a)

(b)

(c)

Figure 18.1. Bioactivation of dihalogenated alkanes by GSH conjugation. (A) Dihalomethanes, (B) 1,2-dihaloethanes, and (C) 1,4-dihalobutanes.

18.4 OXIDATIVE STRESS

18.4.1 Introduction

A major concept is that a mechanism for chemical-induced cell death could be based upon oxidative stress. This advance in our knowledge is in part due to the many new findings that have been reported about reactive oxygen species (ROSs) and cell defenses. As targets of oxidants have been identified and intracellular events have been described in detail, many of the defense systems are dependent upon intracellular GSH. Inflammation that is based on cytokines and stress responses has added additional dimensions to various types of liver injury, nonhepatic diseases, immune function, and HIV. A major discovery has been the oxidative events associated with ischemia and reperfusion. Liver diseases including alcoholic liver disease, metal storage diseases, and chloestatic liver disease have been described in terms of an oxidative stress component. The occurrence of reactive oxygen species in normal metabolism by mammalian organs was first demonstrated in the liver. Rates of hydrogen peroxide production of 50–100 nmol/min/g liver under normal conditions could be enhanced up to sevenfold in the presence of substrates for peroxisomal oxidases. Subsequently, it was shown that oxidative stress-induced hepatocyte injury may accompany metabolism of chemicals to biological reactive intermediates. From these studies came the concept that a imbalance in the prooxidant/antioxidant steady state, potentially leading to damage, is termed "oxidative stress."

The loss of control of endogenous oxidative events in the use of molecular oxygen by the cell is the major factor in oxidative stress injury. Such a process, which is known as chemical-induced oxidative stress, may occur to an extent that ranges from a minor to a major contribution to overall toxicity. For example, chemicals that are known to undergo redox cycling cause exogenous oxidative stress to such a degree that they play a major role in chemically induced cell injury. In addition, some of these chemicals are known to form adducts with cellular constituents, particularly proteins that are necessary for protection against oxygen toxicity. However, as indicated above, all tissues and cells contain systems for detoxification of biological reactive intermediates and antioxidants and antioxidant enzyme systems to prevent or limit cellular damage due to oxidative stress events.

Normal fluxes of proxidants serve useful functions at the cellular and whole organism levels and therefore need to be balanced with sufficient antioxidant defense to maintain a biological steady state. For example, the continual formation of reactive oxygen species is a physiological necessity and an unavoidable consequence of oxygen metabolism. When reactive oxygen species are generated in excess, they can be toxic, particularly in the presence of transition metal ions such as iron or copper. However, defense systems are present and functioning under normal conditions. Therefore, endogenous free radicals do not necessarily place biological tissues and cells at risk. Chemical exposures can cause these defense systems to be overwhelmed during various pathological conditions including anoxia, radiation, and loss of control of intracellular and extracellular constituents such as calcium.

It is estimated that nearly 90% of the total O_2 consumed by mammalian species is delivered to mitochondria where a four-electron reduction to H_2O by the respiratory chain is coupled to ATP synthesis. Nearly 4% of mitochondrial O_2 is incompletely reduced by leakage of electrons along the respiratory chain, especially at ubiquinone, forming ROS such as superoxide (O_2^-), hydrogen peroxide (H_2O_2), singlet oxygen, and hydroxyl radical ($HO^\bullet$). It has been calculated that during normal metabolism, one rat liver mitochondrion produces 3×10^7 superoxide radicals per day. It is estimated that superoxide and hydrogen peroxide steady-state concentrations are in the picomolar and nanomolar range, respectively. However, an estimated hepatocyte steady-state H_2O_2 concentration may be up to 25 μM. These events are thought to contribute over 85% of the free radical production in mammalian species.

Aerobic Metabolism: Endogenous Oxidative Stress. If endogenously formed H_2O_2 is not detoxified to H_2O, formation of the $HO^\bullet$ radical by a metal (iron)-catalyzed Haber–Weiss or Fenton reaction can occur [Eqs. (18.1a) and (18.1b)]. The $HO^\bullet$ species is one of the most reactive and short-lived biological radicals and has the potential to initiate lipid peroxidation of biological membranes, although not as effectively as other radicals including the $ROO^\bullet$ radical. Unless termination reactions occur, the process of lipid peroxidation will propagate, resulting in potentially high levels of oxidative stress. Therefore, detoxification of endogenously produced H_2O_2 is critical for redox maintenance of mitochondrial as well as cellular homeostasis.

$$\text{Haber–Weiss Reaction:} \quad O_2^- + H_2O_2 + \rightarrow + O_2 + OH^- + OH^\bullet \quad (18.1a)$$

$$\text{Fenton Reaction:} \quad H_2O_2 + Fe^{2+} \rightarrow Fe^{3+} + OH^- + OH^\bullet \quad (18.1b)$$

Other reactive oxygen species of biological relevance include singlet oxygen ($^1O_2$), hypochlorous acid (HOCl), nitric oxide (NO), and peroxynitrite. Thus, several reactive oxygen species can be formed by one-electron step reductions of molecular oxygen and spontaneous reactions with nitric oxide including the formation of the oxidant, peroxynitrite (Figure 18.2).

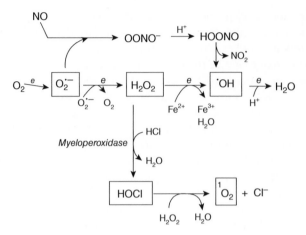

Figure 18.2. Reactive oxygen species formed by one-electron reductions of molecular oxygen.

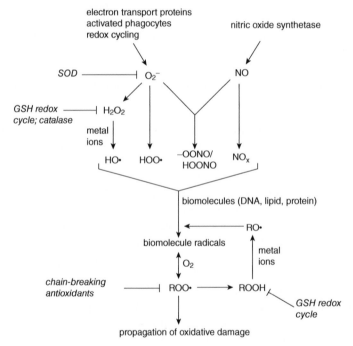

Figure 18.3. Origin of reactive oxygen and nitrogen species and sites of blocking their oxidant challenges by antioxidant defenses.

18.4.2 The Organization of Antioxidant Defense Against Oxidative Stress

The production of oxygen free radicals was not considered to be of biological consequence until 1969, when McCord and Fridovich discovered that superoxide dismutase exists in essentially every mammalian cell, suggesting ubiquitous superoxide formation in vivo (Figure 18.3).

Later, oxygen free radicals were linked to inflammatory disease states, and the imbalance between proxidants and antioxidants was defined as oxidative stress. With the discovery of the second messenger functions of nitric oxide and the role of peroxynitrite in both toxicity and signaling, the physiological and pathological roles of nitric oxide and superoxide are being investigated vigorously as well as the defense systems that must balance between protection and providing a suitable environment for signal transduction and other biological functions based on these molecules. The evolution of bioactivation processes that form biological reactive intermediates is thought to have necessitated the concomitant evolution of cellular protection systems for cell survival in an oxygen-containing atmosphere. All tissues and cells contain systems for detoxification of biological reactive intermediates and to prevent or limit cellular damage due to oxidative stress events.

18.4.3 Superoxide Dismutase (SOD) Defense

The importance of the defense enzymes Se-glutathione peroxidase, catalase, Cu/Zn-SOD, and Mn-SOD for cell survival against oxidative stress is well-established, with each enzyme having specific functions (Figure 18.4).

In part, the function of these enzymes is based on studies that have utilized specific chemical enzyme inhibitors including aminotriazole for catalase, diethyldithiocarbamate for Cu/Zn-SOD, mercaptosuccinate for SeGpx, BCNU for glutathione reductase, and BSO for GSH synthesis. A part of the basal defense that is provided by other antioxidant enzymes includes PH Gpx, Mn-SOD, thioredoxin, and peroxiredoxins. In transfection studies in which overexpression of a specific enzyme is studied, the physiological responses indicate that the metabolism of reactive oxygen species may have a critical balance. For example, Mn-SOD transfected mouse cells overexpressing SOD activity are more resistant to hyperoxia and paraquat. Yet, SOD-enriched bacteria display increased sensitivity to hyperoxia and paraquat. SOD has an important role in defense against degenerative disease, but much remains to be understood concerning effects of manipulation of SOD expression as

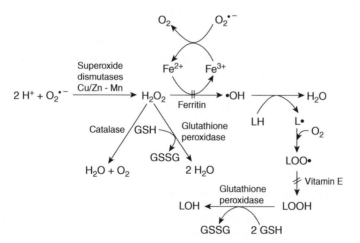

Figure 18.4. Cellular defense systems for the metabolism and inactivation of reactive oxygen species.

an intervention. An explanation for hydrogen peroxide toxicity and adaptive defense by exposure to low levels of hydrogen peroxide is that the toxicity is maximized by optimizing the concentrations of superoxide and hydrogen peroxide for additional free radical formation. Transfection experiments could therefore cause increased or decreased toxicity, depending on concentrations of hydrogen peroxide and superoxide. In addition, iron (or copper) has the potential for a major influence by being optimized in its redox ratio (Fe^{2+}/Fe^{3+}) for increasing the reactivity of hydrogen peroxide (see Recommended Reading section).

18.4.4 Role of GSH Protection Against Reactive Oxygen Species

All living organisms have evolved protective systems to minimize injurious events that result from bioactivation of chemicals including xenobiotics and oxidative products of cellular metabolism of molecular oxygen (Figures 18.3 and 18.4). The major protective system is dependent upon GSH. GSH acts both as a nucleophilic "scavenger" of numerous compounds and their metabolites, via enzymatic and chemical mechanisms, converting electrophilic centers to thioether bonds, and as a cofactor in the GSH peroxidase-mediated destruction of hydroperoxides. GSH depletion to about 15–20% of total glutathione levels can impair the cell's defense against the toxic actions of such compounds and may lead to cell injury and death. Oxidative stress as a mechanism for chemical-induced hepatocyte death is based on understanding that the consumption of oxygen even during normal environmental conditions requires a considerable amount of cellular resources on a constant basis to detoxify toxic oxygen metabolites. If not reduced, these metabolites can lead to the formation of very reactive radicals including the hydroxyl radical and cause the formation of lipid hydroperoxides that can damage membranes, nucleic acids, proteins, and their functions. Failure to provide or maintain the cellular protective systems is now known to cause serious human diseases that can be exacerbated by exposures to toxic chemicals.

Depending on the cell type, the intracellular concentration of GSH is maintained in the range of 0.5–10 mM. Concentrations in the liver are 4–8 mM. Nearly all the glutathione is present as GSH. Less than 5% of the total is present as glutathione disulfide (GSSG). This is so because of the redox status of glutathione that is maintained by intracellular glutathione reductase and NADPH (Figure 18.5).

Continual endogenous production of reduced oxygen species, including hydrogen peroxide and lipid hydroperoxides, causes constant production of some GSSG, however. The GSH content of various organs and tissues represents at least 90% of the total nonprotein, low-molecular-weight thiols. The liver content of GSH is nearly twice that found in kidney and testes and over threefold greater than in the lung. The importance of hepatic GSH for protection against reactive intermediates has been reviewed extensively.

18.4.5 Organelle Glutathione Protection Against Oxidative Stress

Compartmentation of glutathione has been demonstrated in that a separate pool of glutathione exist in the cytoplasm from that in the mitochondria (Figure 18.6). The cytosolic pool of glutathione has been characterized in terms of cellular protection (Table 18.1).

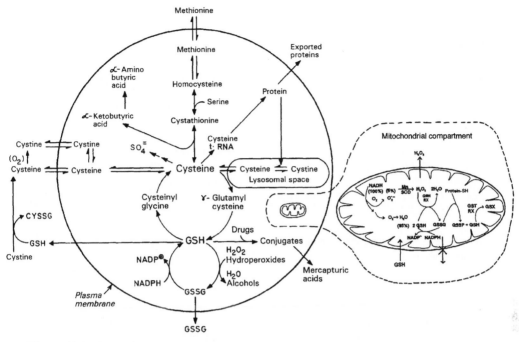

Figure 18.5. GSH biosynthesis and the cystathionine pathway for cysteine biosynthesis in liver and the protective functions of GSH.

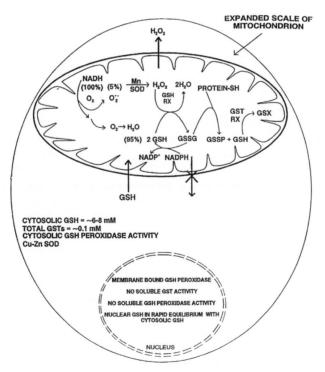

Figure 18.6. Cellular defense systems for the metabolism and inactivation of reactive oxygen species.

TABLE 18.1. Features of Cytosolic Glutathione that Are Important in Protection Against Chemical Toxicity

- High intracellular concentration (1–8 mM)
- Rapid synthesis by cytosolic enzymes in response to chemical depletion
- GSH, glutathione conjugate, and GSSG transport out of most cell types via plasma membrane transporters
- Very dynamic glutathione redox cycle activity to maintain GSH/GSSG ratio of about 50–100
- Protein thiol/GSH ratio is about 2–4:1
- High concentration of glutathione transferases (about 0.1 mM)
- Stable pool of GSH as degradation is initiated by extracellular gamma-glutamyl transpeptidase
- Signal transduction based on glutathione interactions including glutathionylation

TABLE 18.2. Features of Mitochondrial Glutathione that Are Important in Protection Against Chemical Toxicity

- Separate physiological pool of glutathione from cytosolic glutathione
- High concentration in the matrix (estimated at 10–12 mM)
- Only extramitochondrial synthesis of glutathione
- GSH transport; lack of GSSG transport in and out mitochondrial inner membrane
- Active glutathione redox cycle that maintains a GSH/GSSG ratio of about 25–50
- Protein thiol/GSH ratio is about 10
- Glutathione transferase activity
- High levels of glutathionylation of mitochondrial protein thiols

A separate pool has been proposed for the nucleus; however, that finding remains in dispute. Mitochondrial glutathione is a separate physiological pool of glutathione in agreement with the observation that the liver has two pools of GSH. One has a fast (2-hr) and the other a slow (30-hr) turnover. In freshly isolated rat hepatocytes, the mitochondrial pool of GSH (about 10% of the total cellular pool) has a half-life of 30 hr while the half-life of the cytoplasmic pool is 2 hr. It has been concluded that the mitochondrial pool represents the stable pool of GSH observed in whole animals.

Mitochondrial GSH functions as a discrete pool separate from cytosolic GSH. Mitochondrial GSH is mostly retained and therefore largely impermeable to the inner membrane following isolation of mitochondria, where the concentration of mitochondrial GSH (10 mM) is higher than cytosolic GSH (7 mM). The mitochondrial pool of glutathione has features that are very important for the protection of mitochondrial functions (Table 18.2). Because the main source of ROS in cells are the mitochondria, it is critical for the cell to maintain an adequate level of GSH as well as an adequate GSH:GSSG ratio. The ratio of GSH:GSSG in mitochondria is approximately 10:1 under normal (untreated) conditions. Unlike cytosolic GSSG, mitochondrial GSSG is not effluxed from the soluble compartment. During oxidative stress induced with *t*-butyl hydroperoxide, GSSG is accumulated in the mitochondrial matrix and eventually reduced back to GSH. However, as the redox state of the mitochondria increased, an increase in protein mixed disulfides by glutathionylation is also observed. Thus, mitochondria are more sensitive to redox changes in GSH:GSSG than the cytosol, and therefore mitochondria may be more susceptible to the damaging effects of oxidative stress. These findings suggest that under certain experimental conditions, irreversible cell injury due to oxidative challenge may result from irreversible changes in mitochondrial function mainly from loss of critical protein thiol functions.

Rat liver mitochondria contain the enzymes and cofactors necessary for the GSH/GSSG redox cycle (Figures 18.5 and 18.6) but do not contain catalase. Catalase is a major enzyme for the detoxifying of hydrogen peroxide in peroxisomes and is found in heart mitochondria, but not in other tissues including skeletal muscle. A primary function of mitochondrial glutathione (GSH) is for the detoxification of endogenously produced H_2O_2. This redox cycle requires the enzymes GSH peroxidase (selenium containing) and glutathione reductase along with the cofactors GSH and NADPH. In addition to detoxifying H_2O_2, GSH also protects protein thiols from oxidation.

18.5 GLUTATHIONE-DEPENDENT CELLULAR DEFENSE SYSTEMS

18.5.1 Glutathione-Dependent Protection

We review here the pathways by which chemicals are metabolized to proximate toxins, the mechanisms by which biological reactive intermediates are toxic to and damage cells, and the mechanism by which cells are protected by preventing or limiting cellular damage.

Certain chemicals, including some cancer chemotherapeutic agents such as nitrogen mustards, are so chemically reactive that they are direct-acting electrophiles and therefore do not require bioactivation for either pharmacologic action or toxicity. Of course, because these agents are so very reactive chemically, they have the ability to challenge and defeat the cellular protective systems, especially the glutathione-dependent systems. Depletion of protective constituents of the cell can result in extensive damage to both normal and cancer cells. Both the extent and the molecular targets of covalent binding by biological reactive intermediates vary with the properties of the intermediates, especially their chemical reactivity and, to some extent, sites of formation. The liver, because it is a primary site of xenobiotic metabolism, is also a major source of biological reactive intermediates derived from xenobiotics. The half-life of a biological reactive intermediate determines, in part, the site of the molecular targets with which it can react. Some biological reactive intermediates possess sufficient stability to be transported throughout the body. Also, there are threshold doses for toxicity of certain chemicals that relate to the rate of formation of biological reactive intermediates. Above such a threshold level, covalent binding of a biological reactive intermediate generally increases greatly for relatively small increases in dose of the chemical. The concept of a threshold dose, which is very important for safe use of drugs, has been considered in detail for many drugs and is now being considered for limitation of exposures to many xenobiotics, such as dioxin, a potent inducer of drug metabolizing enzymes and a carcinogen.

Bromobenzene, displays a dose response for toxicity that illustrates the threshold for an observable toxic effect. The rate of bromobenzene metabolism is dependent on the level of specific cytochrome P450s (Figure 18.7). The concentration of GSH

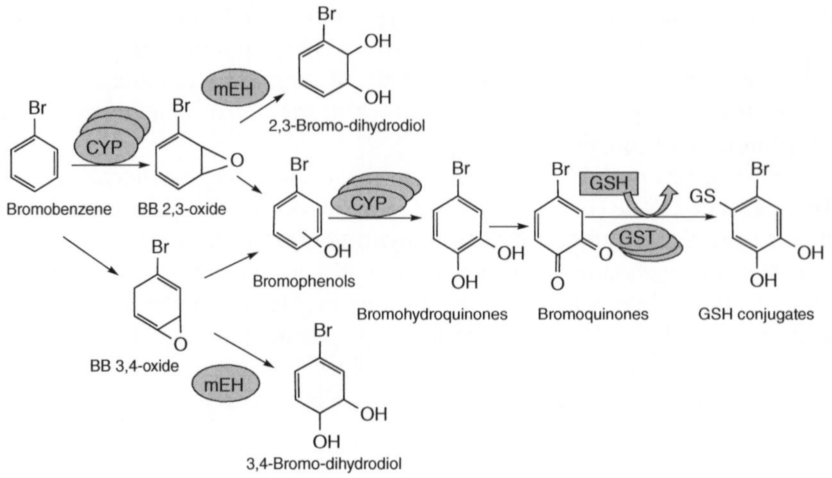

Figure 18.7. Mechanism of biotransformation of bromobenzene in rat liver. [Adapted from Heijne et al. (2004).]

in liver markedly decreases after the administration of toxic doses of bromobenzene, especially after induction of cytochrome P450s by prior administration of pheno-barbital. The basis for this depletion is the conjugation of a biological reactive intermediate, an epoxide of bromobenzene, with GSH. Eventually, the rate of for-mation of the conjugate of the reactive epoxide becomes limited by the availability of GSH, which, in turn, is limited by its rate of synthesis. Therefore, the rate of reac-tion between biological reactive intermediates of bromobenzene and macromole-cules in the liver increases markedly to a level that causes extensive GSH depletion. Even the rate of biosynthesis of GSH is not adequate for conjugation and cellular protection. Obviously, many factors can contribute to the rate of biosynthesis of GSH, including fasting or starvation, and can make a substantial difference in the tolerance to certain chemicals. Thus, both the rate of replenishment of GSH and the rate of depletion are critical factors in acute hepatotoxicity by bromobenzene. The enhancement of hepatotoxicity by depletion of GSH with diethyl maleate has been noted during the metabolism of bromobenzene. When hepatocytes were pre-treated with diethyl maleate, which reduced intracellular levels of GSH by about 70%, added bromobenzene caused levels of GSH to fall to 5% of initial levels, and cell death (75% by 5 hr) was noted. Addition of cysteine, methionine, or N-acetyl-cysteine prevented bromobenzene-induced toxicity, but did not prevent depletion of GSH. Bromobenzene, in the presence of a cysteine source, reduced initial levels of GSH to about 40% of control. The presence of metyrapone, an inhibitor of cyto-chrome P450-dependent monooxygenase reactions, eliminated bromobenzene-induced toxicity. These data are consistent with a requirement for bromobenzene activation prior to GSH conjugation. Redox cycling of a metabolite of bromoben-zene derived from GSH conjugation has been shown to cause kidney toxicity.

Acetaminophen (paracetamol, 4-hydroxyacetanilide, APAP), a commonly used analgesic drug, causes centrilobular hepatic necrosis upon overdosage. APAP dis-plays toxicity characteristics that demonstrate, very clearly, dependence upon GSH for protection. Hepatotoxicity, including liver failure, often occurs when APAP is

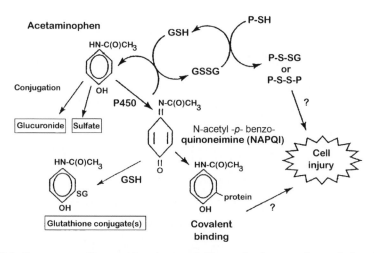

Figure 18.8. Structures of acetaminophen and *N*-acetyl-*p*-benzoquinone imine (NAPQI) metabolites and GSH defense.

overdosed. The drug mainly undergoes sulfation and glucuronidation in most species at normal dose levels (Figure 18.8).

At dose levels that begin to saturate the pathways of sulfation and glucuronidation, however, bioactivation occurs and a small fraction of the drug is metabolized by cytochrome P450 into the reactive intermediate *N*-acetyl-*p*-benzoquinoneimine (NAPQI) (Figure 18.8). As the normal conjugation pathways of sulfation and glucuronidation become saturated due to the overdose, the hepatic cytochrome P450 systems catalyze the bioactivation of APAP to an increasing extent. NAPQI is a highly reactive compound. It reacts and covalently binds with cell components including macromolecules if not detoxified by conjugation with GSH, which depletes GSH, increases oxidative stress, and diminishes the efficiency of cytoprotective mechanisms. Thus, overdosage with acetaminophen causes depletion of cellular glutathione followed by oxidation and/or alkylation (covalent binding) of cysteinyl residues in cellular proteins. Although many cellular events occur resulting in cytotoxicity, GSH depletion in liver is generally accepted as the critical initial event in APAP-induced hepatoxicity. NAPQI is chemically reactive and can undergo a nonenzymatic two-electron reduction in the presence of glutathione to yield stoichiometric amounts of APAP and GSSG. As GSH is depleted, cellular macromolecules become a greater proportion of the target molecules for covalent alkylation.

Peroxynitrite is considered to be a critical mediator of APAP hepatotoxicity with protection by GSH (Figure 18.9). Alkylation of proteins, especially mitochondrial proteins by reactive metabolite formation, involves mitochondrial dysfunction leading to the formation of reactive oxygen and peroxynitrite, which trigger the membrane permeability transition and the collapse of the mitochondrial membrane potential. Diminished capacity to synthesize ATP is accompanied by the release of several mitochondrial proteins including certain caspases and endonucleases, cytochrome c, and other proteins leading to DNA fragmentation. Transgenic animal studies have provided valuable information supporting these events. Mice deficient in Cu,Zn-superoxide dismutase are resistant to APAP toxicity. Knockout of SOD

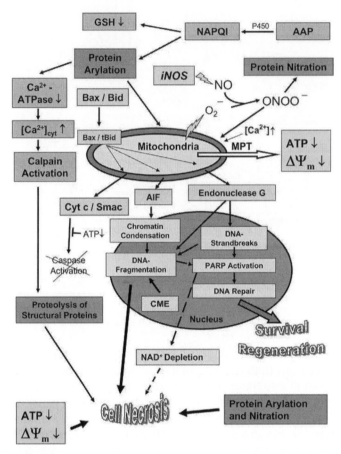

Figure 18.9. Proposed scheme for acetaminophen toxicity and the role of mitochondria and peroxnitrite. [Adapted from Jaeschke and Bajt (2006).]

is associated with attenuated hepatic GSH depletion, modest shifts in APAP metabolism, decreased levels of the APAP metabolizing enzyme, CYP2E1, and blocked APAP-mediated protein nitration. Double null of selenium-GSH peroxidase-1 and Cu,Zn-superoxide dismutase enhances resistance of mouse primary hepatocytes to APAP toxicity.

The double knockout attenuates the peroxynitrite-mediated protein nitration in primary hepatrocyte studies. However, the double knockout primary hepatocytes have enhanced resistance to cytotoxicity by APAP but not by NAPQI. Figure 18.10 is a schematic illustration of impacts of singular or double knockouts of GPX1 and SOD1 on APAP- or NAPQI-induced cell death and peroxynitrite-mediated nitro-tyrosine formation. As glutathione synthesis is rate-limited by the first step in GSH synthesis, blocking of GSH depletion is a key factor in the attenuation of APAP toxicity. Other evidence that the rate of GSH synthesis and in turn the intracellular level of GSH is crucial during an overdose of APAP is the finding that attenuation of APAP-induced liver injury occurs with enhanced GSH synthesis. Conditionally, overexpression of glutamate cysteine ligase in transgenic mice has shown that enhancement of the levels of glutamate cysteine ligase provides higher levels of

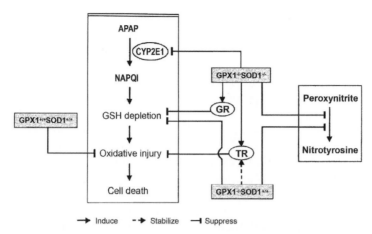

Figure 18.10. Schematic illustration of impacts of singular or double knockouts on acetaminophen (APAP) toxicity. [Adapted from Zhu and Lei (2006).]

GSH protection from APAP toxicity. Mitochondrial GSH depletion, peroxynitrite formation, and mitochondrial permeability transition appear to be critical for APAP hepatocellular necrosis.

BCNU (1,3-bis(2-chloroethyl)-1-nitrosourea)-induced inactivation of glutathione reductase prevents the reduction of NAPQI-generated GSSG and increases cytotoxicity but has no effect on the covalent binding of NAPQI to cellular proteins. A competing reaction at physiological pH is the formation of an APAP-GSH conjugate. Incubation of NAPQI with hepatocytes yields the same reaction products in control and in BCNU-treated hepatocytes. The thiol, dithiothreitol, protects against cytotoxicity even after covalent binding has occurred. It has been speculated that the toxicity of NAPQI for isolated hepatocytes may result primarily from its oxidative effects on cellular proteins rather than the formation of APAP–GSH conjugate. Thus, another possible mechanism contributing to APAP-induced cell injury is oxidative stress. NAPQI-induced cytotoxicity is accompanied by oxidation of thiol groups in proteins, and cytotoxicity of NAPQI can be prevented by a reducing agent for disulfides [e.g., dithiothreitol (DTT)]. There is uncertainty, however, about the relative contributions of oxidative stress and injury mediated by covalent binding of NAPQI during overdosage of APAP (500 mg/kg i.p.). This dose increased the mitochondrial GSSG content from 2% in controls to greater than 20% in treated animals. It is known that GSSG does not efflux from mitochondria and must be reduced back to GSH by mitochondrial-derived reducing equivalents. Therefore, if levels of GSSG approach 20% in treated animals, then the ability to maintain mitochondrial integrity is in question as loss of calcium regulation potentiates the mitochondria for undergoing MPT. BCNU (1,3-bis(2-chloroethyl)-1-nitrosourea) increases the susceptibility of hepatocytes to oxidative stress and increases both the formation of GSSG and cell killing produced by APAP. Interestingly, APAP administered by repeat exposures of incremental doses provides protection against APAP-induced lethality in mice. Such autoprotection is thought to be dependent upon the ability to sustain repair processes associated with cell proliferation. APAP exposure affects cell replacement and organ repair by the modulation of signal transduction and inhibition of entrance of cells into the cell cycle. APAP inhibition of passage of cells through G1 and S phases has been proposed to interfere with organ regeneration and to exacerbate acute liver injury damage caused by APAP bioactivation products including NAPQI.

Large doses of morphine in rats deplete hepatic GSH and are associated with threefold elevation in activities of glutamic oxaloacetic transaminase (SGOT) and glutamic pyruvic transaminase (SGPT) in serum. Prior treatment of the rats with phenobarbital, which increases production of certain cytochrome P450 enzymes, enhances the morphine-elicited rise in SGOT and SGPT. The bioactivation of morphine involves a cytochrome P450-dependent oxidation at the benzylic C-10 position to form an electrophilic species that can react with nucleophilic thiols such as glutathione. Thus, metabolism of morphine can deplete thiols and cause liver damage. Overdosage or use of morphine in combination with other drugs that require glutathione for detoxification will therefore increase the risk of toxicity due to either agent. Alkylation of critical cellular targets by the metabolic intermediate appears to be important in the hepatotoxicity of morphine.

Because of their chemical reactivity, biological reactive intermediates can bind covalently to cellular proteins, including those present in the cytoplasm, endoplasmic reticulum, nucleus, and nucleic acids. Such covalent binding to DNA is used as a quantitative indicator of genotoxicity of many xenobiotics. Because there is only a general understanding of how adduct formation alters the function of specific proteins, the relationship between covalent binding to proteins and the cytotoxic properties of xenobiotics remains uncertain, Also, little attention has been given to the significance of damage to either proteins or DNA as related to acute cellular toxicity.

18.5.2 Depletion of Glutathione by Chemicals and Fasting

The concentration of GSH in the liver is altered in rats by diurnal or circadian variations, as well as by starvation. The diurnal variation in hepatic GSH results in the highest levels of GSH at night and early morning and the lowest levels in the late afternoon. The maximum variation is as much as 25–30%. Starvation limits the availability of methionine for synthesis of GSH in the liver, and it decreases the concentration of GSH by about 50% of the level in fed animals. Assuming that GSH is a physiological reservoir for plasma cysteine, efflux of GSH from liver will continue during starvation, and the released cysteine will help maintain levels of GSH in other organs including the kidney.

In vivo treatment of rats with an inhibitor of γ-glutamyl transpeptidase AT-125 (L-alphaS,5S)-alpha-amino-3-chloro-4,5-dihydro-5-isoxazoleacetic acid) prevents degradation of GSH in plasma, leading to massive urinary excretion of GSH. This treatment also lowers the hepatic content of GSH because it inhibits recycling of cysteine to the liver. A physiologic decrease in interorgan recycling of cysteine to the liver for synthesis of GSH also may account in part for the decrease of hepatic GSH during starvation and for the marked diurnal variation in concentration of GSH in liver. As mentioned, the nadir occurs in the late afternoon, whereas the early morning peak occurs shortly after the animals are fed.

The efflux of liver GSH and metabolism of the resulting plasma GSH and GSSG appears to help ensure a continuous supply of plasma cysteine. Interorgan GSH has a half-life of 1–2 min, with the brush border membrane of the kidneys providing gamma-glutamyl transpeptidase for the initial step in the breakdown of GSH to cysteine. This cysteine pool should, in turn, minimize the degree of fluctuation of GSH concentrations within the various body organs and cell types that require cysteine or cystine, or both, rather than methionine for synthesis of GSH.

Depletion of cellular GSH has been widely studied with hundreds of chemicals including APAP and bromobenzene. These studies demonstrated very clearly that bioactivation followed by GSH adduct formation causes depletion of cytosolic glutathione and oxidative stress as indicated by indicators including enhanced levels of GSSG, lipid peroxidation, and loss of membrane integrity.

Fasting enhances the toxicity of many chemicals. One of the earliest studies of this phenomenon compared the effects of fasting and various diets on chloroform-induced hepatotoxicity. Increased hepatotoxicity in association with fasting occurs with chemicals that are capable of depleting GSH, including carbon tetrachloride, 1,1-dichloroethylene, APAP, bromobenzene, and many others. Because fasting decreases the hepatic concentration of GSH in mice and rats, such a decrease could account for the enhanced toxicity of many of these chemicals in fasted animals. In several instances, as a result of a depletion of GSH in the liver after pretreatment with diethyl maleate, APAP, bromobenzene, carbon tetrachloride, and anthracyclines, showed increased hepatotoxicity.

18.5.3 Glutathione Compartmentation and Chemical-Induced Injury

Many studies have been conducted that address mitochondria as target organelles of certain types of irreversible cell injury, as related to chemically induced effects, oxidative stress, and disrupted Ca^{2+} homeostasis. GSH-dependent protection against lipid peroxidation has been demonstrated in mitochondria, nuclei, microsomes, and cytosol of rat liver. Lipid peroxidation induced in mitochondria is inhibited by respiratory substrates such as succinate, which leads indirectly to reduction of ubiquinone to ubiquinol. The latter is a potent antioxidant. The essential factor in preventing accumulation of lipid hydroperoxides and lysis of membranes in mitochondria, however, is glutathione peroxidase. Although prevention of free radical attack on membrane lipids may occur by an electron shuttle that utilizes vitamin E and GSH in microsomes, similar activity may be limited in mitochondria. Instead, mitochondrial GSH transferase(s) may prevent lipid peroxidation in mitochondria by a nonselenium GSH-dependent peroxidase activity. Three GSH transferases have been isolated from the mitochondrial matrix, and nearly 5% of the mitochondrial outer-membrane protein consists of microsomal glutathione transferase. GSH transferase in the outer mitochondrial membrane could provide the GSH-dependent protection of mitochondria by scavenging lipid radicals by a mechanism that requires vitamin E and is abolished by bromosulfophthalein. Mitochondrial phospholipid hydroperoxide glutathione peroxidase (PHGPx) has been suggested as having a major role in preventing oxidative injury to cells. PHGPx is synthesized in the cytoplasm as a long form (23 kD) and a short form (20 kD). The long form contains a leader sequence that is utilized for transport to mitochondria. PHGPx in the mitochondria protects against the lipid hydroperoxides generated in the mitochondria and is thought to participate in the regulation of signal transduction pathways that are triggered by reactive oxygen species in mitochondria. Phospholipid monohydroperoxide has been shown to cause dissipation of mitochondrial membrane potential.

Protection against this effect was achieved by transfection of cells with PHGPx gene. Overexpression of PHGPx has been shown to suppress cell death that is due to oxidative damage.

Mitochondrial oxidative stress and mitochondrial GSH defense affects transcription factor activation. Oxidant stress in mitochondria not only can promote the loss of mitochondrial GSH and mitochondrial functions, but also can promote extramitochondrial activation of NF-κB and therefore may affect nuclear gene expression. Mitochondria are targets of cytokines leading to the overproduction of reactive oxygen species induced by ceramide, a lipid intermediate of cytokine action and closely associated with apoptosis. Chronic ethanol intake depletes liver mitochondrial glutathione due to an ethanol-induced defect in the transport of GSH from cytosol into the mitochondrial matirix. This sensitizes liver cells to the prooxidant effects of cytokines and prooxidants generated by the oxidative metabolism of ethanol.

The mitochondrial pool of GSH, as stated already, has a half-life of about 30 hr. It is expected, therefore, that fasting will not deplete this pool of GSH. Fasting, in fact, does not increase the spontaneous rates or carbon tetrachloride-induced rates of lipid peroxidation. Hence, it may be that lipid peroxidation events are related to the size of the mitochondrial pool of GSH in liver. Moreover, perhaps certain "antioxidant" proteins in the mitochondria participate with GSH in preventing lipid peroxidation.

The oxidation–reduction state of vicinal thiols has a role in the opening of the membrane permeability transition pore. Oxidants including menadione, diamide, arsenite, and *tert*-butyl hydroperoxide can cause cell death by one or more mechanisms that are responsive to the effects of cyclosporine A to prevent such cell death presumably via mitochondrial membrane permeability transition. Oxidants cause vicinal thiols to form a protein disulfide to enhance pore opening, which can be reversed by thiols such as DTT. The recruitment of mitochondrial cyclophilin to the mitochondrial inner membrane under conditions of oxidative stress that enhance the opening of a Ca^{2+}-sensitive nonspecific channel provides strong evidence that the adenine nucleotide translocase is a component of the permeability pore. Evidence for the participation of MPT in loss of cell viability during exposure to toxic agents is abundant because the collapse of the inner mitochondrial membrane potential causes the loss of ATP generation.

One of the main questions is whether this sudden collapse is closely associated with the loss of mitochondria structure and function, loss of energy production, protein synthesis, and the status of cytoplasmic and/or mitochondrial GSH.

18.5.4 Glutathione Redox Cycle

It is apparent that a major protective role against the reactive chemical intermediates, which are generated by bioreduction and cause oxidative stress by redox cycling, is provided by the ubiquitous glutathione redox cycle. This cycle utilizes NADPH- and, indirectly, NADH-reducing equivalents in the mitochondrial matrix as well as the cytoplasm to provide GSH by the glutathione reductase-catalyzed reduction of GSSG (Figure 18.11).

The rates of NADPH consumption in liver by the various NADPH-dependent enzymes indicate that GSH reductase has by far the highest rate to detoxify the reactive oxygen species generated by the various sources.

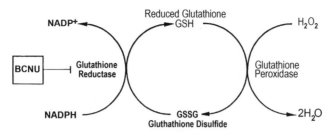

Figure 18.11. Cellular defense system by the glutathione redox cycle and the modulation of GSH.

With a rat liver perfusion system, extra production of 70 nmol/min/g liver of GSSG can occur before GSSG is excreted into the bile as a result of t-butylhydroperoxide in the perfusate. These findings suggest that liver GSH could be oxidized to GSSG and reduced back to GSH over 10 times per minute prior to exceeding the reducing capacity of the liver. The liver possesses high resistance to intracellular reactive oxygen formation that is dependent upon the GSH redox cycle system. Therefore, when the glutathione redox cycle is functioning at maximum capacity to eliminate hydrogen peroxide, a major regulatory effect is imposed on other NADPH-dependent pathways. The ability of the glutathione redox cycle to consume major quantities of reducing equivalents (NADPH) is further evidence that oxidative stress is a mechanism for injury and cell death since loss of this redox cycle potentiates hepatocyte death.

The mitochondrial glutathione redox cycle has a role in regulating mitochondrial oxidations in liver. Various oxidants decrease O_2 uptake by isolated mitochondria and cause a complete turnover of GSH via glutathione peroxidase every 10 min. It appears that a continuous flow of reducing equivalents through the glutathione redox cycle is balanced by a continuous formation of mitochondrial NADPH, which is needed for glutathione reductase activity. In addition, metabolism of hydrogen peroxide in mitochondria poses a regulatory function in regard to the oxidation of substrates by lipoamide-dependent ketoacid oxidases, which generate NADPH-reducing equivalents. The entire NADPH : NADP$^+$ pool may turn over at least once every minute during a maximum oxidant challenge.

Examples of chemicals that are associated with injury involving redox cycling include bromobenzene, carbon tetrachloride, ethanol, paraquat, diquat, menadione, and iron loading. Drugs that can undergo redox cycling include anthracyclines like adriamycin, mitomycin C, bleomycin, β-lapachone, alloxan, APAP, and the radiosensitizers (e.g., misonidazole and metronidazole). It is now known that activation of oxygen by reduction can play an important role in the toxicity of these drugs and chemicals. In general, they are enzymatically reduced or oxidized by a one-electron transfer. After a one-electron redox reaction, the intermediate formed then transfers the extra one electron to molecular oxygen to yield the superoxide anion radical and the parent drug or chemical which can undergo repeated one electron reduction and/or oxidation to then provide one electron to molecular oxygen. This process of generation of reactive oxygen species, which is termed redox cycling, is involved in the toxicity of many hydroquinones, quinones, metal chelates, nitro compounds, amines, and azo compounds (Figure 18.12).

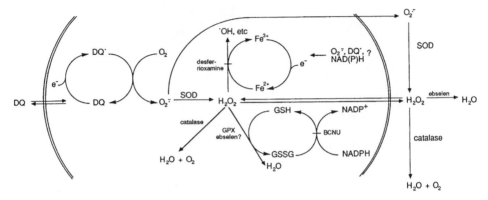

Figure 18.12. Redox cycling by diquat (DQ).

In general, chemical toxicity compromises the ability of cellular processes to detoxify endogenous oxidative events related to oxygen metabolism. Furthermore, the inherent toxicity of many chemicals includes the ability to produce additional oxidative stress, which can then become a dominant component of the overall toxicity. As mentioned above, all tissues and cells contain defense systems for detoxification of biological reactive intermediates of oxygen to prevent or limit cellular damage Whether certain chemicals may cause both reactive intermediate effects such as covalent binding and redox cycling stress is being examined.

Diquat (DQ), a bipyridyl herbicide, is a model compound for redox cycling (Figure 18.12). It generates large amounts of superoxide and hydrogen peroxide within cells. Isolated, perfused rat liver when treated with diquat (200 μM) in the perfusate can generate 1 μmol/min/g liver of superoxide. Only after depletion of GSH (10% of control) with phorone pretreatment (200 mg/kg body weight) and either ferrous sulfate (100 mg/kg) or BCNU (40 mg/kg) is it possible to significantly increase the diquat-induced liver injury. These findings support other evidence that the protective role of the glutathione redox cycle in the toxicity of diquat in vivo and in vitro. Since diquat toxicity is mediated by redox cycling, it is greatly enhanced by prior treatment with BCNU, an inhibitor of glutathione reductase. Ebselen, a synthetic compound possessing GSH peroxidase-like activity, protects against diquat cytotoxicity when extracellular GSH is present in the medium. Since superoxide can reduce ferric iron to ferrous iron, desferrioxamine, which chelates intracellular iron in the ferric state with an affinity constant of 10^{31}, provides considerable protection against diquat-induced toxicity. Therefore, hydrogen peroxide and transition metals have been suggested as major contributors to diquat toxicity. Even though the hydroxyl radical or a related species seems the most likely ultimate toxic product of the hydrogen peroxide/ferrous iron interaction, scavengers of hydroxyl radical afford only minimal protection. The high degree of reactivity of hydroxyl radicals ensures, however, that the site of interaction with cellular components is within close proximity (a few angstroms) of the site of generation of the radical. *t*-Butylhydroperoxide (TBH) is widely utilized to investigate the mechanisms of cell injury initiated by oxidative stress. Although the mechanism remains unknown, lipid peroxidation and altered thiol status are implicated as components. Evidence has been presented that supports the hypothesis that the mitochondrial MPT can occur under conditions of oxidative stress and contribute to the mechanism of cell death. Importantly, CsA, a high-activity inhibitor of the transition, prolongs survival of hepatocytes treated with TBH under conditions where extensive lipid peroxidation is prevented. During oxidative stress induced by TBH treatment, cultured hepatocytes display the characteristics of MPT which quickly lead to mitochondrial depolarization and cell death.

18.5.5 Regulation of Glutathione Biosynthesis

Through the action of γ-glutamylcysteine synthetase (GLC) and glutathione synthetase (GS), GSH can be synthesized in the cytosol of most mammalian cells.

$$\text{L-Glu} + \text{L-cys} + \text{ATP} \xrightarrow{\text{GCL}} \gamma\text{-glutamyl-L-cys} + \text{ADP} + \text{Pi}$$

$$\gamma\text{-L-glutamyl-L-cys} + \text{Gly} + \text{ATP} \xrightarrow{\text{GS}} \gamma\text{-L-Glutamyl-L-cysteine} + \text{ADP} + \text{Pi}$$

ATP is required for both reactions, and the product GSH exerts control by feedback inhibition. The availability of cysteine is important because it is thought to be the rate-limiting step in GSH biosynthesis. Constitutive and regulated synthesis of GSH maintains a high cytosolic level of this unique tripeptide. GSH has a glutamyl moiety attached to a cysteinyl glycine dipeptide by an amide rather than a second peptide linkage, which makes it resistant to intracellular degradation. The level of cysteine is regulated at about 0.04 mM in order to keep the cellular toxicity of cysteine to a minimum. Regulation of synthesis of cysteine from the sulfur of methionine and the carbon of serine via the cystathionine pathway is important at times of low levels of GSH. At high cysteine levels, hormonal control of cysteine dioxygenase is involved in the rapid depletion of cellular cysteine to form ammonia, pyruvate, and sulfate. The unidirectional process of trans-sulfuration in which methionine sulfur and serine carbon are utilized in cysteine biosynthesis via the cystathionine pathway is essential for the maintenance of GSH stores in the body (Figure 18.5). Thus, the cystathionine pathway is of major importance to pathways of drug metabolism that involve GSH or cysteine, or both. Depletion of GSH by rapid conjugation can increase synthesis of GSH to rates as high as 2–3 μmol/hr/g wet liver tissue. The cysteine pool in the liver, which is about 0.2 μmol/g, has an estimated half-life of 2–3 min at such high rates of synthesis of GSH. Although the cystathionine pathway appears to be highly responsive to the need for cysteine biosynthesis in the liver, the organ distribution of the pathway is limited. Evidence indicates that in mammals, such as rats, the liver is the main site of cysteine biosynthesis, which occurs via the cystathionine pathway. Maintenance of high concentrations of GSH in the liver, in association with high rates of secretion into plasma and extensive extracellular degradation of GSH and GSSG, supports the concept that liver GSH is a physiological reservoir of cysteine.

Oxidants are known to cause the induction of GSH biosynthesis. A remarkably large number of compounds including those forming GSH conjugates also increase transcription of the gamma-glutamylcysteine synthetase genes. Compounds that can increase the rate of GSH synthesis include (a) 4-hydroxy-2-nonenal, menadione, and related quinones and (b), reactive oxygen and nitrogen species, including hydrogen peroxide and nitric oxide and heavy metals. The balance between depletion and synthesis of GSH is maintained largely by *de novo* synthesis responding to depletion as a result of GSH degradation and oxidation to GSSG and mixed disulfides of GSH as well as conjugation of GSH.

18.6   GLUTATHIONE/GLUTATHIONYLATION DEPENDENT SIGNALING SYSTEMS AND ANTIOXIDANT DEFENSE

Protein thiols can interact with glutathione by several mechanisms that permit glutathionylation to participate in the regulation and antioxidant protection of specific protein thiols. These reactions are involved in the consequences of oxidative stress caused by reactive species of oxygen and nitrogen and include the formation of protein sulfenic acids and their participation in regulatory processes.

18.6.1   Formation of GSH-Protein Mixed Disulfides—Glutathionylation

Protein glutathionylation, or glutathiolation, is the formation of a mixed disulfide between a protein thiol and the thiol of GSH (Figure 18.13). The reversible glutathionylation of proteins is important in the response of cells to oxidative damage and may have an important role in redox signaling. This section describes (a) the mechanisms by which protein thiols can be glutathionylated and (b) the basis for the selectivity of which protein thiols participate in this important function in biology. Topics included are thiol–disulfide exchange, thiol oxidation, glutathionylation induced by nitric oxide, fate of glutathionylated proteins, glutaredoxin, thioredoxin, and physiological roles of protein glutathionylation. Protein thiols may become glutathionylated independent of the redox state of the glutathione pool. Examples include fate of protein thiols following *S*-nitrosylation or sulfenic acid formation.

18.6.2   Role of Protein Sulfenic Acids

Sulfenic acids in proteins are derived from a pool of protein thiols that possess a high susceptibility to oxidation. Sulfenic acids are inherently unstable and highly reactive. They can be stabilized within the protein environment as they are formed. The oxidation of specific protein thiols that possess such high chemical reactivity can be oxidized by even very low concentrations of oxidant including hydrogen peroxide. It has been suggested that an important role for protein sulfenic acids is their ability regulate activity in enzymes for which they are not a part of the catalytic mechanisms. Many enzymes have been shown to be susceptible to this modification and inactivated by it when an essential cysteinyl residue at their active site is oxidized to a sulfenic acid residue. Such oxidations can be reversed by a chemical reductant including GSH. If the enzyme is glutathionylated as part of their reversible inactivation, this may occur by first being oxidized to a sulfenic acid and then condensed with GSH to form a protein glutathione mixed disulfide. Thus, redox regulation of protein catalytic activity may occur via oxidation of protein thiol groups to generate reversibly oxidized cysteinyl residue by reaction with peroxides, nitric oxide, peroxynitrite, and other reactive oxygen or nitrogen species. Redox regulation includes a wide variety of redox sensors that are capable of modulation

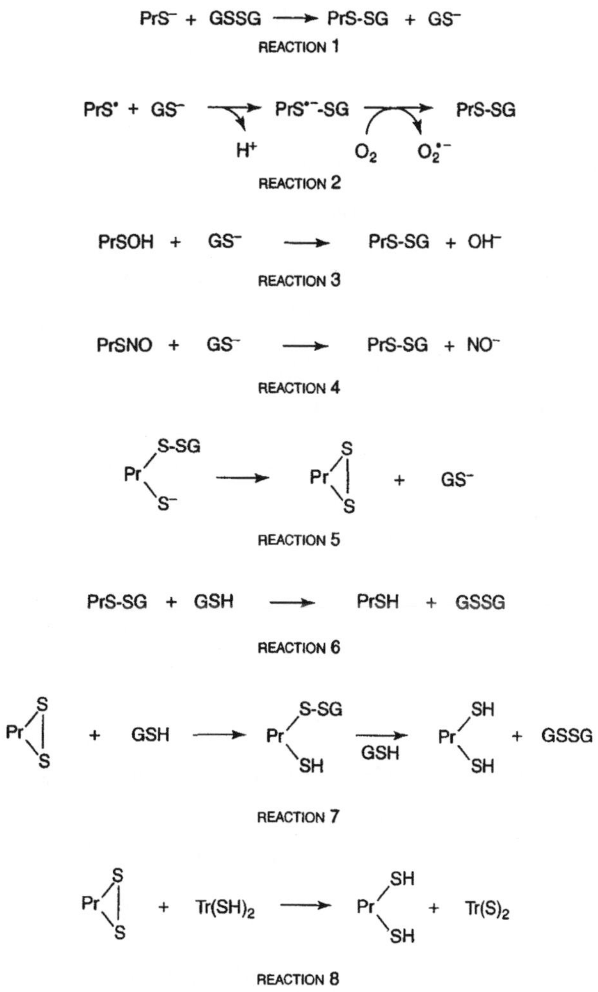
PrS⁻ + GSSG → PrS-SG + GS⁻
REACTION 1
PrS• + GS⁻ → PrS•⁻-SG → PrS-SG
H⁺
O2
O2•⁻
REACTION 2
PrSOH + GS⁻ → PrS-SG + OH⁻
REACTION 3
PrSNO + GS⁻ → PrS-SG + NO⁻
REACTION 4
S-SG
Pr
S⁻
+ GS⁻
REACTION 5
PrS-SG + GSH → PrSH + GSSG
REACTION 6
+ GSH
GSH
+ GSSG
REACTION 7
+ Tr(SH)2
+ Tr(S)2
REACTION 8

Figure 18.13. Reaction schemes associated with glutathionylation. [Adapted from Filipovska and Murphy (2006).]

of protein function in response to changing environmental and/or intracellular signals. Hydrogen peroxide and NO and other reactive oxygen species and nitrogen species are now known to be involved in redox signaling pathways that may contribute to normal cell function as well as disease progression.

18.6.3 Glutathionylation Induced by Nitric Oxide

Glutathionylated proteins can be formed by the interaction of nitric oxide and protein thiols. For example, exposure of mitochondria to NO can lead to the formation of peroxynitrite, an oxidant, that can cause protein glutathionylation. Protein glutathionylation may also occur via the formation of a nitroso thiol protein (PrSNO) followed by the glutathionylate anion displacement of the nitroxyl anion (NO–) by GSH to form protein glutathionylation as shown in Figure 18.13.

Also, formation of GSNO may lead to either the breakdown of GSNO to form GSSG or reaction with protein thiols to displace NO– and generate a glutathionylated protein.

18.6.4   Fate of Glutathionylated Proteins

Members of one class of protein thiols undergo reversible glutathionylation. They appear to have two major biological functions: (1) participation in cellular antioxidant defense mechanisms and (2) a component of a redox signaling pathway. These glutathionylated proteins have characteristics suitable for both of these functions. The disulfide bond between the protein and glutathione is readily reversible under reducing conditions but can be maintained indefinitely as a glutathionylated protein under oxidizing conditions. Members of a second class of protein thiols are persistently glutathionylated, and they represent a smaller proportion than those protein thiols that form intraprotein disulfides. This may be due to glutathionylation normally occurring on protein thiols that are adjacent to a second thiol that rapidly displaces GSH to form an internal disulfide. Glutaredoxin (GR) and thioredoxin (TR) are vital for a rapid response of the protein thiol status to the GSH:GSSG ratio and the reversal of glutathionylation by thiol-disulife exchange. The GSH:GSSG ratio is maintained by the action of glutathione reductase (GR), which, in turn, is dependent upon the redox status of the pyridine nucleotides NADPH:NADP.

18.6.5   Protein Glutathionylation as an Antioxidant Defense

Interaction of the glutathione pool with protein thiols can occur in several ways to be an antioxidant defense. The concentration of GSH in mitochondria is very high (>6–8mM), yet it is exceeded by the concentration of exposed reactive protein thiols. Thus, the direct reaction of protein thiols with ROS is thought to be an important antioxidant defense. Protein thiols are converted to thiyl radicals or sulfenic acids by reaction with ROS. The rapid reaction of GSH with thiyl radicals and sulfenic acids has a major role in limiting the formation of higher oxidation states of oxidized protein thiols. In addition, the reaction of GSH with glutathionylated proteins provides an anioxidant defense that extends that of GSH and does not require the maintenance of a high GSH:GSSG ratio as does the action of glutathione peroxidases and glutathione-S-transferases.

18.6.6   Protein Glutathionylation as a Regulatory Response

There is growing evidence of a role for protein glutathionylation in redox sensing and signaling. The activity of enzymes, transcription factors, and transporters can be dramatically affected by glutathionylation and the formation intraprotein disulides. Glutathionylation enables reversible response to the ambient GSH:GSSG ratio, as does direct reaction of NO, hydrogen peroxide, or peroxynitrite with protein thiols, followed by glutathionylation. The nature of these interactions supports an important physiological role for protein glutathionylation in regulation of cellular processes.

18.7 CONCLUSIONS

GSH has a crucial role in the protection and prevention of cell injury and death. This role is illustrated in the multitude of processes that involve GSH in prevention of limiting the damaging effects of chemical exposure. As a primary protecting agent, the GSH content of cells and subcellular compartments is important as is the ability to maintain the rate of replenishment during depleting conditions including oxidation and conjugation. Some GSH conjugation reactions lead to the formation of reactive intermediates rather that detoxication. Such GSH-dependent bioactivation occurs with some halogenated alkanes and alkenes.

Mitochondrial membrane permeability transition and its role in the mechanism of toxicity of many chemicals is a major area of research focus. The protein thiol status of mitochondria is crucial to the control of the inner-membrane potential for ATP synthesis and the membrane permeability transition as well as cellular integrity. Protein thiols appear to have a major role in signal transduction and are important in the responses to chemical exposure.

ACKNOWLEDGEMENT

Support for this chapter was from NIH grant ES-00210.

SUGGESTED READING

Babson, J. R., and Reed, D. J. Inactivation of glutathione reductase by 2-chloroethyl nitrosourea-derived isocyanates. *Biochem. Biophys. Acta.* **83**: 754–762, 1978.

Babson, J. R., Abell, N. S., and Reed, D. J. Protective role of the glutathione redox cycle against adriamycin-mediated toxicity in isolated hepatocytes. *Biochem. Pharmacol.* **30**: 2299–2304, 1981.

Beckman, J. S., and Crow, J. P. Pathological implications of nitric oxide, superoxide, and peroxynitrite formation. *Biochem. Soc. Transact.* **21**: 330–334, 1993.

Botta, D., Shi, S., White, C. C., Dabrowski, M. J., Keener, C. L., Srinouanprachanh, S. L., Farin, F. M., Ware, C. B., Ladiges, W. C., Pierce, R. H., Fausto, N., and Kavanagh, T. J. Acetaminophen-induced liver injury attenuated in male glutamate cysteine ligase transgenic mice. *J. Biol. Chem.* **281**: 28865–28875, 2006.

Commandeur, J. N. M., Stijntjes, G. J., and Vermeulen, N. P. E. Enzymes and transport systems involved in the formation and disposition of glutathione s-conjugates. *Pharmacol. Rev.* **47**: 271–330, 1995.

Dickinson, D. A., and Forman, H. J. Glutathione in defense and signaling lessons from a small thiol. *Ann. N. Y. Acad. Sci.* **973**: 488–504, 2002.

Filipovska, A., and Murphy, M. P. Overview of protein glutathionylation. In: Bus, J. S., Costa, L. G., Hodgson, E., Lawrence, D. A., and Reed, D. J., (Eds.). *Current Protocols in Toxicology*, John Wiley & Sons, Hoboken, NJ, 2006, pp. 6.10.1–6.10.8.

Fridovich, I. Superoxide dismutases. *Adv. Enzymol. Relat. Areas Mol. Biol.* **41**: 35–97, 1974.

Heijne, W. H. M., Slitt, A. L., van Bladeren, P. J., Groten, J. P., Klaassen, C. D., Stierum, R. H., and Van Ommen, B. Bromobenzene-induced hepatotoxicity at the transcriptome level. *Toxicol. Sci.* **79**: 411–422, 2004.

Jaeschke, H., and Bajt, M. L. Intracellular signaling mechanisms of acetaminophen-induced liver cell death. *Toxicol. Sci.* **89**: 31–41, 2006.

Kosower, N. S., and Kosower, E. M. Glutathione status of cells. *Int. Rev. Cytol.* **54**: 109–160, 1978.

Meredith, M. J., and Reed, D. J. Status of the mitochondrial pool of glutathione in the isolated hepatocyte. *J. Biol. Chem.* **257**: 3747–3753, 1982.

Poole, L. Overview of protein sulfenic acids. In: Bus, J. S., Costa, L. G., Hodgson, E., Lawrence, D. A., and Reed, D. J. (Eds.). *Current Protocols in Toxicology* John Wiley & Sons, Hoboken, NJ, 2006; pp. 17.1.1–17.1.15.

Reed, D. J., and Beatty, P. W. Biosynthesis and regulation of glutathione. Toxicological implications. In: Hodgson, E., Bend, J. R., and Philpot, R. M. (Eds.). Reviews in Biochemical Toxicology, Elsevier Press, New York, 1980, pp. 213–241.

Reed, D. J. Glutathione: Toxicological implications. *Annu. Rev. Pharmacol. Toxicol.* **30**: 603–631, 1990.

Reed, D. J. Cellular defense mechanisms against reactive metabolites. In: Anders, M. W. (Ed.). *Bioactivation of Foreign Compounds*, Academic Press, Orlando, FL, 1985, pp. 71–108.

Reed, D. J. Toxicity of oxygen. In: De Matteis, F., and Smith, L. L. (Eds.). *Molecular and Cellular Mechanisms of Toxicity*, CRC Press, Boca Raton, FL, 1995, pp. 35–68.

Reed, D. J. Regulation of reductive processes by glutathione. *Biochem. Pharmacol.* **35**: 7–13, 1986.

Reed, D. J. Mitochondrial glutathione and chemically induced stress including ethanol. *Drug Metab. Rev.* **36**: 569–582, 2004.

Rinaldi, R., Eliasson, E., Swedmark, S., and Morgenstern, R. Reactive intermediates and the dynamics of glutathione transferases. *Drug Metab. Dispos.* **30**: 1053–1058, 2002.

St. Clair, D. K., Oberley, T. D., Ho, Y.-S., and Wheeler, K. T. Overproduction of human MnSOD modulates paraquat-mediated toxicity in mammalian cells. *FEBS Lett.* **293**: 199–203, 1991.

Turrens, J. F. Mitochondrial formation of reactive oxygen species. *J. Physiol.* **552**: 335–344, 2003.

Warner, D. S., Sheng, H., and Batinic-Haberle, I. Oxidants, antioxidants and the ischemic brain. *J. Exp. Biol.* **207**: 3221–3231, 2004.

Zhu, J.-H., and Lei, X. G. Double null of selenium-glutathione peroxidase-1 and copper,zinc-superoxide dismutase enhances resistance of mouse primary hepatocytes to acetaminophen toxicity. *Exp. Biol. Med.* **231**: 545–552, 2006.

CHAPTER 19

Toxicant–Receptor Interactions: Fundamental Principles

RICHARD B. MAILMAN

Departments of Psychiatry, Pharmacology, Neurology and Medicinal Chemistry, University of North Carolina School of Medicine, Chapel Hill, North Carolina 27599-7160

19.1 DEFINITION OF A RECEPTOR

Although the concept of a receptor as a selective locus of action was pioneered more than a century ago, receptor theory remains a critical concept in toxicology and in many other branches of biology. In one sense, receptors may be considered as macromolecules that bind small molecules (commonly termed ligands) with high affinity and thereby initiate a characteristic biochemical effect. In some fields of biology, the term receptor has specific constraints. For example, in a cell biological context, the term refers to macromolecules (intracellular or cell surface) that recognize endogenous ligands; these ligands may be small (e.g., neurotransmitters, hormones, autacoids) or large (e.g., proteins involved in protein sorting or in intracellular scaffolding). In toxicology and pharmacology, the term "receptor" is often used to refer to the high-affinity binding site that initiates the functional change(s) induced by a xenobiotic (e.g., toxicity). It is important to note that the interaction of two macromolecules can be studied in much the same way as a ligand and a single macromolecule, and such macromolecular interactions have great importance to (a) biological systems in general and (b) toxicology in particular. The use of fundamental structural precepts (both steric and electrostatic) to understand the "fit" between receptor and ligand has become more feasible during the past few years based on advances in molecular biology and in computational chemistry and molecular modeling.

19.1.1 A Brief History of the Concept of Receptors

The concept of receptors is credited to the independent work of Paul Ehrlich (1845–1915) and J. N. Langley (1852–1926). With Ehrlich, the concept appeared to originate from his immunochemical studies on antibody–antigen interactions. Based on the high degree of specificity of antibodies for antigens, Ehrlich postulated the existence of stereospecific, complementary sites on the two molecules. Similar

Molecular and Biochemical Toxicology, Fourth Edition, edited by Robert C. Smart and Ernest Hodgson
Copyright © 2008 John Wiley & Sons, Inc.

reasoning led Emil Fisher to formulate the idea of a lock-and-key fit between enzymes and their substrates. In later studies with arsenicals to be used against syphilis and trypanosomes, Ehrlich observed that slight modification in chemical structure could dramatically affect the potency of a compound. This high degree of stereospecificity suggested to Ehrlich that his receptor theory was also relevant to drugs, and it also led to the concept of structure–activity relationships. Although this idea seems straightforward today, it was decades before physical evidence for this idea could be gleaned.

The physiological importance of receptors was shown by Langley. Studying the denervated frog neuromuscular junction, Langley observed that muscle contraction could be elicited when nicotine was applied to the denervated muscle and that this effect could be blocked by curare. He postulated the existence of "the receptive substance" that could bind both nicotine and curare; this was the first formulation of what is now called the neurotransmitter receptor. Langley also speculated that the formation of complexes between drug and receptive substance could be described lawfully through consideration of the relative concentration of a drug and its affinity for the receptive substance. This latter speculation predated the formal mathematical description of drug–receptor interactions according to mass action principles.

19.1.2   Toxicants and Drugs as Ligands for Receptors

The rationale for applying principles derived from the study of drug–receptor interactions to questions concerning possible toxicant–receptor interactions becomes clear when one considers that a drug is defined broadly as any chemical agent that affects living organisms; thus, a poison (toxicant) can be considered as a drug that has been administered at a dose level that results in detrimental effects, even death. Indeed, most clinically useful drugs have beneficial effects at one dose, yet become toxic (or even lethal) at higher doses. Clinical pharmacologists and toxicologists have quantified these observations in the term therapeutic index that can be defined:

$$\text{Therapeutic index} = \frac{\text{TD50}}{\text{ED50}}$$

The TD50 is the dose of a drug that would be toxic (or lethal) to one-half of a given population; the ED50 (ED for effective dose) is the dose that produces the beneficial effect in one-half of a comparable population. The therapeutic index is large for drugs that cause toxicity only at doses much higher than needed for therapy, and it is small for drugs that have a narrow margin of safety.

Although drugs are of considerable interest to toxicologists, other chemical agents are of at least equal importance. These include agricultural and industrial chemicals (e.g., herbicides, insecticides, chemical wastes, industrial solvents, etc.), as well as environmental contaminants. Whatever the source of a xenobiotic—drug or pollutant—its interaction with receptors can be described by the same general principles.

19.1.3   Is There a Normal Function for Toxicant Receptors?

As noted, a toxicant or drug receptor may be defined loosely as "a macromolecule with which a drug (or toxicant) interacts with high affinity to produce its characteristic effect." Inherent in this is the fact that binding of the toxicant (or drug) to the receptor causes a specific and predictable biological response. The obvious question, then, is whether the receptor of interest is one to which endogenous ligands bind or, rather, an opportunistic site for which the toxicant has affinity. Since this distinction is of general importance, several examples may be useful.

The pharmacological and toxicological properties of opium alkaloids (e.g., morphine and codeine) were known for centuries, and the framework provided by Langley, Ehrlich, Dale, and others led to the hypothesis that a specific receptor existed for these drugs. By the early 1970s, the approaches outlined in this chapter first were used to identify the "morphine receptor" (i.e., the target for the actions of morphine and other opiates). This was followed closely by evidence for three related opioid receptors (i.e., μ-, δ-, and κ-opioid receptors) that were the targets for endogenous neurotransmitter opioid peptides and for xenobiotics like morphine. Although the opioid system was one of the first systems in which direct evidence was presented for the binding of a foreign compound to a receptor, receptor methods are now widely used to study the mechanism of action of drugs and toxicants. Another similar example relates to the alkaloids derived from the "deadly nightshade" (*Belladonna* sp.). It is now known that atropine and related compounds bind to a subtype of receptor for the endogenous neurotransmitter acetylcholine. These are only two early and well-studied examples.

When an endogenous ligand is not known for a toxicant receptor, it raises the question whether the toxicant has found an adventitious site. An interesting historic example is the GABA$_A$ receptor, a schematic of which is shown in Figure 19.1. The GABA$_A$ receptor is one of the two major subclasses of recognition sites for the

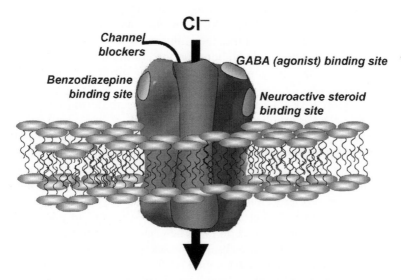

Figure 19.1. Schematic representation of GABA$_A$ receptor. This polymeric receptor regulates the influx of chloride, thus providing an inhibitory neuronal influence. It consists of five subunits of three major types (α, β, and γ), each of which has several subforms. There are differences in the subunit composition of the receptor, depending on where it is expressed. There are many sites at which xenobiotics may bind, thereby affecting the function of this receptor complex.

inhibitory neurotransmitter GABA (γ-aminobutyric acid). The $GABA_A$ receptor belongs to, and is typical of, a superfamily of ligand-gated ion channels called ionotropic receptors (see Section 19.2). It is a polymer (in this case of five protein subunits each of ~50 kD) arranged in a circle to form a channel. Each protein subunit is actually a string of amino acids which passes in and out of the cell membrane four times. The $GABA_A$ receptor functions as an ion channel for the chloride anion. This channel remains closed until GABA binds to the recognition site.

The same portion of the receptor that recognizes GABA also binds certain toxicants such as the mushroom alkaloid muscimol. There is another site on the receptor that can bind a class of neuromodulators called neuroactive steroids. Several toxins, including picrotoxin, bind to the pore of this receptor. There is a third and distinct site on the $GABA_A$ complex that was known for many years to bind a class of drugs called benzodiazepines (e.g., diazepam or Valium[R]). For many years, the function of this second site was unknown, but it is now thought that a peptide called DBI (diazepam-binding inhibitor) is the endogenous ligand. As molecular genetics has provided information on the structure and function of receptors in the genome, it has resulted in better understanding of the receptor mechanisms of action of many toxicants.

19.1.4 Functional Biochemistry of Toxicant Receptors

From the viewpoint of biochemistry and neuroscience, "receptor" refers to a distinct class of molecules that have unique and characteristic functions and modes of action that involve endogenous ligands. Enzymes are, in one sense, receptors with respect to their substrates (and inhibitors, too). As will be demonstrated later in this chapter, the interaction of receptors with their ligands is similar conceptually to the interaction of enzymes and substrates or inhibitors. In practice, however, the interaction of ligands with receptors often differs from enzyme–substrate interactions in two qualitative, but important, ways. First, the affinity of ligands for receptors is usually two or more orders of magnitude higher than the affinity of enzymes for their substrates; second, enzymes act by biochemically altering substrate molecules, whereas receptors do not.

As noted above, toxicant receptors, per se, may serve a physiological role as receptors for a diverse group of endogenous ligands such as neurotransmitters, hormones, and autacoids. A common characteristic of these receptors is that they transduce chemical information of one type (the concentration of the ligand) into a variety of other secondary chemical events (including synthesis of second messengers, changes in ion flux or transport, and the like). For example, when the receptors are present on the extracellular surface of neurons, the ultimate effect of ligand binding is a change in the electrical excitability of the cell. In addition to postsynaptic receptors, cell surface receptors on neurons may be present on either presynaptic membranes or cell bodies. In these cases, they are termed auto-receptors and play a self-regulatory role (see Chapter 31 for specific examples).

Some ligands do not interact principally with cell surface receptors, but diffuse into cells and bind to intracellular receptors in the cytoplasm. For example, ligand binding to cytoplasmic steroid receptors initiates a process that is not well understood but that involves the movement of steroid-bound receptor into the cell nucleus, where the receptor molecule interacts with genomic material, resulting in alterations in gene expression and protein synthesis.

As will become clear, this chapter is focused on toxicants for which the receptor is a high-affinity recognition site of the type discussed in the previous paragraphs. It should be noted explicitly that other toxicants have "receptors," but fall into more complex situations not appropriate for this chapter. For example, some toxicants inhibit enzymes, or are themselves enzymes. Such interesting compounds include the organophosphate and carbamate insecticides (acetylcholine esterase inhibitors) and diphtheria toxin (an enzyme).

19.1.5 Types of Interactions Between Toxicants and Receptors

Toxicant molecules that gain access to receptor sites in various tissues may then interact with those receptors in several ways. First, the interaction may mimic the endogenous ligand and cause agonist-like actions. Second, the molecule may bind to the receptor without causing resultant activation, thereby blocking access of an endogenous ligand (antagonist actions). Third, the toxicant may produce allosteric effects on receptors; for example, it has been shown that some toxicants, rather than binding to the same site as endogenous ligands, bind instead to an adjacent part of the macromolecule (see Section 19.2.2 and Figure 19.1 for examples). This interaction causes allosteric changes that affect the function of the complex, and sometimes the binding of the neurotransmitter itself. Finally, some macromolecules may have no endogenous ligands, yet bind specific toxicants that cause physiological changes via this interaction.

The four mechanisms listed above involve direct interaction of a toxicant with a receptor; in such cases, the toxicant–receptor interaction is likely to be involved in the mechanism of action. In many cases, toxicants may affect receptor function indirectly. For example, in the nervous system, decreases in synaptic transmission (by receptor blockade or damage to a neuron) may lead to increases in the number of receptors on the target neuron (so-called upregulation). This is often felt to be one of the compensatory mechanisms by which the nervous system responds to such perturbation. Conversely, increases in synaptic transmission (e.g., by long-term receptor activation) may lead to compensatory decreases (downregulation) in receptor number. The techniques by which such mechanisms are studied are described later in this chapter. It should be noted that the availability of molecular probes now permits evaluation not only of the characteristics of the binding sites, but also of the expression of the mRNA for the receptor(s) under study.

19.1.6 Goals and Definitions

The object of this chapter is to provide the foundation necessary for understanding the interactions of small molecules (ligands, be they toxicants or drugs) with receptors, and also the fundamentals of the techniques commonly employed for such analysis. It should be noted that the same theory is also useful for understanding the interaction of two large molecules, and the application of similar theory to such problems has increased as the importance of macromolecular interactions has been realized in recent years.

19.1.6a Definitions. There are several terms in common use that should be defined.

- *Affinity:* This is related to the "tenacity" by which a drug binds to its receptor, and it reflects the difference between the rates of association and dissociation of the ligand. Thus, a ligand that binds tightly to a receptor has a small equilibrium dissociation constant, or a large equilibrium association constant. A toxicant with high affinity for a receptor will bind to that receptor even at low concentrations, whereas binding of a low affinity toxicant will be evident only at higher concentrations.

- *Potency:* This term refers to the ability of a toxicant to cause a measured biological change (any one of interest). If the change occurs at low concentrations or doses, the compound is said to have high potency. If high concentrations or doses are required, the compound is said to have low potency.

- *Intrinsic Activity (Efficacy):* The maximal response caused by a compound (i.e., toxicant or drug) in any given test preparation. Intrinsic activity (efficacy) is always defined relative to a standard compound such as the endogenous ligand for a receptor or the prototypical toxicant for a specific system. For drugs, a *full agonist* causes a maximal effect equal to that of the endogenous ligand or reference compound; a *partial agonist* causes less than a maximal response. An *antagonist* binds to the same site but causes no functional response. More recently, ligands have also been classified as inverse agonists (decrease constitutive activity in the opposite direction as agonists). These issues will be discussed later in Section 19.4.

19.2   RECEPTOR SUPERFAMILIES

Receptors can be divided into several distinct classes based on the effector mechanisms evoked by ligand binding (Figure 19.2). Three major classes of receptors that have endogenous ligands and also bind toxicants include the G protein-coupled receptors, the ionotropic receptors (also known as ligand-gated ion channels), and intracellular steroid receptors. In addition, the members of a superfamily of enzyme-linked receptors bind a variety of polypeptide hormones and growth factors, and they exert pleiotropic effects on cell physiology. Although these receptors are not known to bind xenobiotic toxicants, their central role in controlling cell proliferation and cell death makes it likely that they participate ultimately in the perturbations and adaptations induced by exposure to a variety of toxicants. Finally, voltage-gated ion channels are macromolecules that span the cell membrane and provide a pathway for particular ions. Although no endogenous ligands have been identified, a variety of xenobiotic toxicants act on these sites. Recent progress in the molecular cloning of receptors has demonstrated considerable amino acid sequence homology among receptors that share a common effector mode; such observations provide an additional rational basis for classification, but more importantly, they enable a search for characteristics (i.e., structural motifs) that must ultimately confer specificity of receptor–effector coupling.

19.2.1   G Protein-Coupled Receptors (GPCRs)

After binding of ligands, many neurotransmitter receptors (e.g., the dopamine, adrenergic, $GABA_B$, metabotropic glutamate, and muscarinic cholinergic receptors)

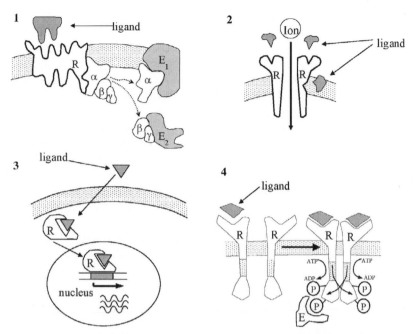

Figure 19.2. Schematic representation of some receptor superfamilies. This figure represents the common mechanisms of the four major receptor families discussed in Section 19.2. Example 1 is of the G protein-coupled receptor. Example 2 represents the ionotropic receptor (ligand-gated ion channel) superfamily, on which multiple binding sites are known to exist (see Figure 19.1). Example 3 represents the steroid receptor superfamily. Example 4 depicts the enzyme-linked receptor superfamily.

use large guanine nucleotide-binding proteins (G proteins) as their transduction systems. Thus, binding of the ligand to the receptor ultimately changes the rate of synthesis or degradation of various cytoplasmic effectors that are modulated by these G proteins (see **1** in Figure 19.2). The G protein-coupled receptors contain a single polypeptide chain with seven transmembrane spanning regions. There is an extracellular/intramembrane domain for ligand recognition, and cytoplasmic domains involved in coupling to G proteins. Interaction of the ligand with the receptor initiates a cascade of secondary events that are initiated by G protein activation. This, in turn, results in the activation or inhibition of specific enzymes (e.g., adenylate cyclase or phospholipase C), with subsequent changes in the turnover of intracellular "second messengers" such as cAMP, diacylglycerol, phosphoinositols, the direct regulation of ion channels, and so on. Together, these biochemical events determine the ultimate cellular consequences of ligand binding.

The best-characterized model for receptor–G protein interaction is derived from studies of receptors linked to the stimulation of adenylate cyclase (e.g., β_2-adrenergic receptors). In the absence of ligand, the receptor is presumed to exist in a high-affinity state stabilized by the heterotrimer G protein that has GDP tightly bound to the α-subunit of the G protein. Agonist binding to the receptor induces a conformational change in the complex of receptor and G protein, promoting the exchange of GDP for GTP. This results in the subsequent dissociation of the

G protein subunits into α and $\beta\gamma$ subunits. Both the α and $\beta\gamma$ subunits may elicit secondary events such as the activation or inhibition of the catalytic unit of adenylate cyclase.

Adenylate cyclase is considered as a second messenger that catalyzes the formation of cAMP (cyclic adenosine monophosphate) from ATP; this results in alterations in intracellular cAMP levels that change the activity of certain enzymes—that is, enzymes that ultimately mediate many of the changes caused by the neurotransmitter. For example, there are protein kinases in the brain whose activity is dependent upon these cyclic nucleotides; the presence or absence of cAMP alters the rate at which these kinases phosphorylate other proteins (using ATP as substrate). The phosphorylated products of these protein kinases are enzymes whose activity to effect certain reactions is thereby altered. One example of a reaction that is altered is the transport of cations (e.g., Na^+, K^+) by the enzyme adenosine triphosphatase (ATPase).

These processes are all possible loci for biochemical attack by toxicants. For instance, compounds like ephedrine or mescaline are alkaloids that act as ligands at GPCRs. Many drugs have been made to act at these sites. Another possible site of attack is at the G protein level. Diphtheria and cholera toxin act on guanine nucleotide-binding proteins linked to the stimulation (G_s) or inhibition (G_i) of adenylate cyclase. Both toxins act by enzymatically transferring an ADP-ribosyl group to a G protein subunit. In the case of G_s, the result is an enduring activation of adenylate cyclase, while in the case of G_i, ribosylation disrupts hormonal inhibition of adenylate cyclase.

19.2.2   Ionotropic Receptors (Ligand-Gated Ion Channels)

A number of receptors are transmembrane proteins whose structure incorporates an ion channel through the cell membrane (see **2** in Figure 19.2). Ligand binding presumably induces conformational changes in the receptor protein that open the channel and allow ions of particular size and charge to pass through, thus altering the ionic concentrations (Na^+, K^+, Cl^-, and/or Ca^{2+}) across the membrane. In neuronal cells, the resultant shifts in membrane potential are either depolarizing or hyperpolarizing, making the cell more or less excitable, respectively. Two prototypical examples of ionotropic receptors are the nicotinic acetylcholine receptor and the GABA$_A$ receptor (see Figure 19.1 for the latter). In both cases, molecular cloning indicates that these receptors are formed from multiple subunits that are transcribed separately and then assembled after translation. For example, the GABA$_A$ receptor, one of the two major subclasses of receptors for the neurotransmitter GABA (γ-aminobutyric acid), is known to be a polymer with four or five subunits which together form a chloride channel. Besides the loci to which GABA binds, there are several other binding sites that have been characterized and are of interest to toxicologists (see Figure 19.1).

Several receptors that recognize the excitatory amino acid glutamate (e.g., the NMDA, kainate, and quisqualate classes) represent another group of ligand-gated ion channels. The NMDA receptor has been the focus of intense interest in recent years. This receptor appears to be activated primarily by membrane depolarizations that occur during periods of heightened neuronal activity. The activated NMDA receptor channel allows both Na^+ and Ca^{2+} to flow into the cell. The detrimental effects of massive Ca^{2+} influx have been implicated in the selective cytotoxicity that occurs in brain regions rich in NMDA receptor (e.g., hippocampus) during episodes of disinhibited neuronal activity, such as during seizures. Some years ago, a major poisoning episode in Canada involved domoic acid, a marine toxin that has sometimes contaminated commercial seafood. Domoic acid works as an ionotropic glutamate receptor ligand, apparently causing toxicity by permitting calcium influx through some members of these receptors.

19.2.3 Intracellular Steroid Receptors

In contrast to the preceding three categories of cell surface receptors, binding sites for diffusible steroids can be found intracellularly, although the exact anatomical locus (cytoplasm versus nucleus) remains controversial (see **3** in Figure 19.2). Cloning of several of these intracellular receptors has verified the presence of distinct steroid-binding and DNA-binding subunits. Endogenous steroid ligands include the estrogens, androgens, progestins, glucocorticoids, mineralocorticoids, and vitamin D, as well as some active metabolites of these compounds. The proximal effect of ligand occupation of steroid receptors is the activation (or disinhibition) of a DNA-binding domain, enabling the interaction of this domain with regulatory DNA sequences (promoter regions) that influence the rate of transcription of particular genes. Compared to the previously described receptor-mediated events, the changes in gene expression that are mediated by steroid binding are both slower and more enduring. One fascinating line of recent research has examined the role of glucocorticoids in promoting cellular damage during episodes of heightened neural excitability (e.g., during ischemia).

19.2.4 Enzyme-Linked Receptors

A large and heterogeneous superfamily of cell surface receptors is defined by the presence of a single transmembrane segment and a direct or indirect association with enzyme activity (see **4** in Figure 19.2). Ligands for these receptors include cytokines, interferons, and many growth factors. Some members of this receptor superfamily possess intrinsic catalytic activity within their cytoplasmic domain (e.g., tyrosine kinase, serine-threonine kinase, guanylate cyclase, and phosphotyrosine phosphatase). In other cases, receptors are devoid of intrinsic enzyme activity but associate directly with cytoplasmic enzymes. The intracellular "signal" generated by ligand binding of enzyme-linked receptors is conveyed by phosphorylation/dephosphorylation of various intracellular targets, or by generation of cGMP.

One important subclass of enzyme-linked receptors whose signaling pathway has been delineated recently is receptor tyrosine kinases (RTK). Two of the prototypical RTKs bind insulin and epidermal growth factor. RTK receptor structure is defined by three elements: an extracellular domain with a ligand-binding site, a single hydrophobic alpha helix that spans the membrane, and an intracellular domain with intrinsic protein tyrosine kinase activity. In most cases, RTK receptors exist as monomers in the ligand-unbound state. The binding of soluble or membrane-bound peptide or protein ligands promotes the formation of dimers. This ligand-triggered dimerization elicits autophosphorylation of specific intracellular tyrosine residues. These phosphotyrosines serve as binding sites for distinct cytosolic proteins that contain a polypeptide domain called the Src homology 2 (SH2) domain. Slight differences in the amino acid sequence among SH2-containing proteins confer receptor specificity by enabling recognition of distinct amino acid sequences surrounding a phosphotyrosine residue in the receptor. SH2-containing proteins include a number of cytosolic enzymes and specific adapter proteins that link the RTK receptor to other signaling molecules. Ras proteins have been shown to be involved in the signal transduction pathway elicited by ligand binding to a number of RTKs, whose ultimate effects include regulation of cell growth, differentiation, and metabolism. The demonstration of RTK and Ras involvement in human cancers has sparked great interest in further elucidation of this important signaling cascade.

19.2.5   Additional Cell Surface Targets for Toxicants

Electrical potentials in excitable cells arise from the differential localization of a number of ions (e.g., Na^+, K^+, Ca^{2+}, and Cl^-) across the semipermeable plasma membrane. Selective changes in the permeability of the membrane to certain ions produce alterations in the membrane potential that ultimately are responsible for the propagation of electrical signals between cells. Voltage-gated ion channels are macromolecules that span the cell membrane and provide a pathway for particular ions (similar to **2** in Figure 19.2). The mechanism for voltage dependency of these channels currently is unclear, although it probably involves a conformational change induced by altered interactions among charged amino acid residues of the channel protein. The best-characterized voltage-gated conductance channel is the Na^+ channel. It is a heavily glycosylated protein formed of a membrane spanning α-subunit and one or two polypeptide ß-units that are situated on the extracellular membrane face. There are a number of binding sites present on voltage-gated channels that alter their function. Although no endogenous ligands for these sites have been identified, a variety of xenobiotic toxicants act on these sites. The puffer fish toxin tetrodotoxin and the paralytic shellfish toxin saxitoxin bind to the same site near the extracellular opening of the Na^+ channel, thereby blocking Na^+ transport through the channel. Several lipid-soluble anesthetics (e.g., lidocaine) appear to bind to a hydrophobic site on the protein. Other toxicants act to increase spontaneous channel opening (e.g., aconitine) or channel closing (e.g., scorpion toxins).

19.2.6   The Importance of Understanding Toxicant–Receptor Interactions

The approaches outlined in this chapter are important for several distinct reasons. First, they can be applied to answer mechanistic questions about the proximal actions of a toxicant. The methods and strategies of such approaches, along with appropriate examples, will be provided later in this chapter. A second distinct rationale also is of importance. Toxic insult (from nervous system damage or chronic drug administration) may cause compensatory changes that include alterations in receptor characteristics. Such changes may be essential to understanding the toxicant-induced physiological changes and, possibly, essential to the chain of events initiated by the toxicant. Finally, finding specific receptors for toxicants often provides important information that can be useful in understanding not only the toxicant, but also cellular function. Consider that finding a receptor for morphine certainly provided information about its mode of action; but more importantly, it opened a new horizon in understanding how ligands acting at that family of receptors affected nervous system function.

It will become apparent in the material that follows that much of the present discussion bears a great similarity to elementary enzyme kinetics. In fact, one can often understand a great deal about relevant toxicological phenomena by applying simple mass action considerations. The emphasis in this chapter shall be on the recognition of a high-affinity toxicant with a macromolecule, under the assumption that one is dealing with a simple reversible phenomenon. In fact, this is often not the case. With enzyme substrates, the reaction is often not just bimolecular, and a variety of mechanisms can be invoked to yield product. Similarly, in many cases where the reaction is between a neurotransmitter receptor and a ligand, various phenomena such as internalization or trapping can complicate this picture.

19.3   THE STUDY OF RECEPTOR–TOXICANT INTERACTIONS

19.3.1   Development of Radioligand Binding Assays

As noted above, many neurotoxicants bind specifically to sites on neurons or glia; consequently, receptor binding assays have become a useful method to study toxicants that have high-affinity interactions with a binding site. Prior to the 1970s, the study of receptors was limited largely to indirect inferences from measurement of physiological responses suspected of being mediated by the receptor of interest. The availability of radiolabeled drugs with high affinity for specific receptors enabled the development of binding assays to assess receptor–ligand interactions in isolated tissue preparations. In the two decades since these methods were introduced, in vitro binding assays have become routine. Advances in molecular biology have paved the way for new generations of techniques toward this same end, yet the radioreceptor methods described below are still used widely and remain a powerful tool when one wishes to elucidate the mechanism of action of small molecules.

The essential ingredient in this technique is the availability of radiolabeled ligand; this radioligand is a molecule in which one or more radioactive atoms has been incorporated into the structure without significantly changing its receptor recognition characteristics. Often, scientists can choose among several radiolabeled agents specific for a particular receptor type. The choice of the type of radioactive isotope is governed by the half-life of the isotope, the amount of receptor in a tissue preparation, and the affinity of the receptor for the radioligand. In practice, specific activities of at least 10 Ci/mmol are required, meaning that ^3H, ^{125}I, ^{35}S, and ^{32}P can be used, whereas ^{14}C cannot. As an example, a tritiated ligand has a much lower molar specific activity (~30 Ci/mmol) compared to a similar ligand with a single ^{125}I atom (~2200 Ci/mmol); the former thus affords lower sensitivity. On the other hand, the replacement of hydrogen with tritium is unlikely to alter receptor recognition characteristics, whereas addition of a bulky iodine atom may have profound effects. Nonetheless, whereas ligands using ^3H are used commonly, ^{125}I, ^{35}S, and ^{32}P containing molecules also are sometimes employed.

Radioligand binding assays are performed typically in solutions of extensively washed membranes prepared from tissue homogenates or cultured cells. Assays of receptor binding to slide-mounted thin tissue sections (quantitative receptor autoradiography) also have become popular, and they are useful particularly for the study of the anatomical localization of binding sites. In both cases, tissue is incubated with radioligand for a specific duration, followed by measurement of the amount of ligand–receptor complex. Accurate assessment of the amount of radioactivity bound to receptors requires a method to separate bound ligand from unbound and a nonbiased estimate of the amount of ligand bound specifically to the receptor of interest.

There are three criteria that are important to determine if a suspected toxicant binding site is biologically meaningful, and also whether it may be a receptor with a known physiological function. (These issues are discussed in greater detail later in the chapter.) There should be a finite number of receptors in a given amount of tissue; thus, the binding should be saturable. As increasing amounts of the radioligand are added to the tissue, the amount of specific binding should plateau [see Figure 19.4 (top) and discussion in Section 19.3.2b]. If the binding site is a known or suspected physiological receptor, other drugs or toxicants that bind to the same receptor should compete with the radioligand for receptor occupancy. The selectivity and potency of this binding should parallel the way they enhance or block the physiological effects of the toxicant believed to be mediated through the receptor. This is usually ascertained by use of competition assays (described in detail in Section 19.3.4). Finally, the kinetics of binding often will be consistent with the time course of the biological effect elicited by the toxicant.

19.3.2 Equilibrium Determination of Affinity (Equilibrium Constants)

19.3.2a Law of Mass Action and Fractional Occupancy. Although toxicant–receptor interactions may involve very complex models, they often follow an elementary mass-action model that results in algebraic expressions similar to those seen in Michaelis–Menten enzyme kinetics. Essentially, this model (Figure 19.3) is the reversible interaction of a toxicant with a binding site (i.e., "receptor"). Although the interaction may generate physiological responses, for receptor analyses it can be modeled without making some of the assumptions needed in enzyme kinetics (e.g., initial velocity conditions). Thus, we can illustrate this model as shown in Figure 19.3.

Thus, for a toxicant (ligand) interacting with a single (or homogeneous population) of receptor(s), mass-action principles require:

$$\text{Ligand} + \text{Receptor} \underset{k_{off}}{\overset{k_{on}}{\rightleftharpoons}} \text{Ligand} - \text{Receptor} \tag{19.1}$$

For any given system, there is a tendency for ligand and receptor to remain associated (the association rate constant k_{on}) and for the ligand–receptor complex to

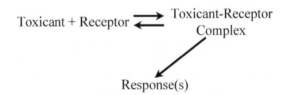

Figure 19.3. Simplest model of how toxicant–receptor interactions can lead to functional consequences.

dissociate (the dissociation rate constant k_{off}). Equilibrium is reached when the rate of formation of new ligand–receptor complexes equals the rate at which existing the ligand–receptor complexes dissociate. Thus, equilibrium is reached when the rate of complex formation equals the rate of complex dissociation, as shown algebraically in Eq. (19.2).

$$[\text{LIGAND}] \cdot [\text{RECEPTOR}] \cdot k_{ON} = [\text{LINGAND-RECEPTOR}] \cdot k_{OFF} \tag{19.2}$$

From basic principles of mass action, we know the relationship between these rate constants and the association (K_A) and dissociation (K_D) equilibrium constants:

$$\frac{1}{K_A} = \frac{k_{off}}{k_{on}} = K_D \tag{19.3}$$

Thus, we can rearrange Eq. (19.2) to define the equilibrium dissociation constant K_D:

$$\frac{[\text{Ligand}] \cdot [\text{Receptor}]}{[\text{Ligand} \cdot \text{Receptor}]} = \frac{k_{off}}{k_{on}} = K_D \tag{19.4}$$

From this, it is clear that the K_D is the concentration of ligand that, at equilibrium, will cause binding to half the receptors. The K_D is the equilibrium dissociation constant, whereas the k_{off} is the dissociation rate constant. A summary of these definitions (and their chemical units) is shown in Table 19.1.

A term that is sometimes useful for biological scientists is *fractional occupancy*. Based on the law of mass action, this term describes (at equilibrium) receptor occupancy as a function of ligand concentration. Specifically:

$$\text{Fractional occupancy} = \frac{[\text{Ligand} \cdot \text{Receptor}]}{[\text{Total receptor}]} \tag{19.5}$$

Since [Total receptor] = [Receptor] + [Ligand · Receptor], then

$$\text{Fractional occupancy} = \frac{[\text{Ligand} \cdot \text{Receptor}]}{[\text{Receptor}] + [\text{Ligand} \cdot \text{Receptor}]} \tag{19.5a}$$

From the equation for K_D derived above, it is seen that

$$[\text{Receptor}] = \frac{K_D \cdot [\text{Ligand} \cdot \text{Receptor}]}{[\text{Ligand}]} \tag{19.6}$$

TABLE 19.1. Chemical Constants and Their Units

Variable	Name	Units
k_{on}	Association rate constant or "on" rate constant	$M^{-1}min^{-1}$
k_{off}	Dissociation rate constant or "off" rate constant	min^{-1}
K_D	Equilibrium dissociation constant	M

TABLE 19.2. Values for Fractional Occupancy (FO) Versus Concentration or log Concentration for the Bimolecular Model

[ligand] (nM)	log[ligand]	FO
0.01	−2	0.009901
0.03	−1.52288	0.029126
0.1	−1	0.090909
0.3	−0.52288	0.230769
1	0	0.5
3	0.477121	0.75
10	1	0.909091
30	1.477121	0.967742
100	2	0.990099

One can substitute this value for [Receptor] in both the numerator and the denominator of the equality for fractional occupancy and, by simplifying, obtain the following. (Try to derive this yourself on a piece of paper.)

$$\text{Fractional occupancy} = \frac{[\text{Ligand}]}{[\text{Ligand}] + K_D} \tag{19.7}$$

Equation (19.7) assumes that the system is at equilibrium. To make sense of it, think about a few different values for [ligand]. When [ligand] = 0, the fractional occupancy equals zero. When [ligand] is very, very high (i.e., many times the K_D), the fractional occupancy approaches 100%. When [ligand] = K_D, fractional occupancy is 50% (just as the Michaelis–Menten constant K_m describes the concentration of enzyme substrate that gives half-maximal velocity).

Fractional Occupancy Is an Important Concept for Several Reasons. For example, if one knows the K_D of a ligand for a target site, one can estimate the occupancy, and hence the effects, that might be caused by a given concentration of the toxicant. Similarly, in experimental toxicology, it permits one to choose concentrations of a test compound that are likely to cause the desired effects without being unnecessarily high and causing undesired secondary effects. Finally, this equation makes clear why semilog dose response curves are so useful in toxicology. Table 19.2 illustrates the fractional occupancy (FO) values calculated for a hypothetical toxicant with a $K_D = 1\,nM$ (an arbitrary value chosen as an easy example). (Be sure that you can calculate the FO values as shown in this table.)

One can then plot these data as shown in Figure 19.4.

19.3.2b The Theoretical Basis for Characterizing Receptors Using Saturation Radioligand Assays.
Radioreceptor assays were first developed in the early 1970s. They were based on two simple, but very elegant, concepts.

1. If a ligand had high affinity for a macromolecular target (as had been shown by classical pharmacological studies over many decades), it should be thermodynamically possible to measure the binding of the ligand to the receptor without

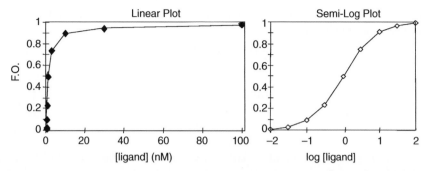

Figure 19.4. Plot of fractional occupancy (FO) versus the [ligand] and the log [ligand]. This provides the theoretical basis for the utility of semilog plots; a sigmoidal curve results that allows maximal visual extrapolation of information from the most biologically meaningful part of the dose–response curve.

the need to perform equilibrium dialysis (the only method then used) as long as one could separate the ligand–receptor complex from the free ligand.

2. By labeling ligands with appropriate radioactive atoms, one could detect the ligand–receptor sensitively and rapidly. (This was the key point, since chemical methods were neither sufficiently sensitive, nor inexpensive, for this use.)

We begin with the simple law of mass action shown in Eq. (19.1):

$$\text{Ligand} + \text{Receptor} \underset{k_{\text{off}}}{\overset{k_{\text{on}}}{\rightleftharpoons}} \text{Ligand} - \text{Receptor} \tag{19.1}$$

We developed the following equation:

$$\frac{[\text{Ligand}]\cdot[\text{Receptor}]}{[\text{Ligand}\cdot\text{Receptor}]} = \frac{k_{\text{off}}}{k_{\text{on}}} = K_{\text{D}} \tag{19.4}$$

It is useful to replace some of these terms with equivalent ones that are common lab jargon in the field and that reflect the measurements that are actually made in characterizing receptors. These names are as follows:

The [Ligand–Receptor] complex is called "Bound" or "B".
The unbound ligand is called "Free" or "F".

As noted, both B and F are measurable experimentally. We can then substitute in the above equation [originally Eq. (19.4)] with these terms and obtain the following:

$$\frac{F\cdot[\text{Receptor}]}{B} = K_{\text{D}} \tag{19.8}$$

Since F and B are independent variables, and we wish to solve for K_{D}, it is necessary to be able to quantify the unbound receptor. Yet this is almost always technically impossible or impractical. On the other hand we know the following:

$$[\text{Total receptor}] = [\text{Receptor}] + [\text{Ligand} - \text{Receptor}] \tag{19.9}$$

The [Total receptor] present is commonly termed the "B_{max}" (i.e., the maximal number of binding sites). Thus,

$$B_{max} = [\text{Receptor}] + B \tag{19.10}$$

Equation (19.10) can be rearranged:

$$[\text{Receptor}] = B_{max} - B \tag{19.11}$$

We now take Eq. (19.11) and substitute the right-hand side of Eq. (19.11) for the [Receptor] in Eq. (19.8) as follows:

$$\frac{F \cdot [\text{Receptor}]}{B} = K_D \tag{19.8}$$

$$\frac{F \cdot (B_{max} - B)}{B} = K_D \tag{19.12}$$

We can simplify this by multiplying both sides by B and expanding the left-hand parenthetical expression to yield

$$F \cdot B_{max} - F \cdot B = K_D * B \tag{19.13}$$

With simple rearrangement, we get

$$K_D \cdot B + F \cdot B = F \cdot B_{max}$$

This can be factored to

$$B(K_D + F) = F \cdot B_{max}$$

resulting in the following equation:

$$B = \frac{F \cdot B_{max}}{K_D + F} \tag{19.14}$$

This equation should look familiar, because it is functionally identical to the Michaelis–Menten equation of enzyme kinetics. This equation also should make clear the experimental design to be used in determining K_D and B_{max} using saturation isotherms. We have as the independent variable $[F]$ and as dependent variable B. A successful experiment should permit the estimation of the two biologically meaningful constants: K_D and B_{max}.

The K_D and B_{max} are of interest to toxicologists and other biological scientists because of the information they convey. The B_{max} defines the number of available binding sites. This can provide clues about whether or not a particular tissue may bind the toxicant, or whether a toxicant alters the available number of binding sites (e.g., through killing cells or causing compensatory responses). The K_D defines the affinity of the toxicant for the binding site. This provides clues about the likelihood of the toxicant working through the mechanism being tested, and also clues about whether intoxication alters the binding site through some compensatory or toxic response.

19.3.2c   Experimental Forms of Data from Saturation Radioligand Assays.

As we did with fractional occupancy, we can predict the data that would result from a toxicant-receptor interaction that was modeled by Eq. (19.14). Let's see what theoretical data would look like if we assume (for the sake of easy calculations) that the $B_{max} = 100$ and the $K_D = 1\,nM$. (This is a similar exercise to what we did with fractional occupancy.) Picking arbitrary $[F]$ concentrations, Table 19.3 shows the derived values [try this yourself using Eq. (19.14)].

If we plot these data in either a linear or semilogarithmic format, a nonlinear plot results as shown in Figure 19.5 (panel A). In the era when computers were not readily available, it was recognized that one could linearize Eq. (19.14) by several simple algebraic rearrangements. Some of these (and their names) are shown in Figure 19.5 with both their common name and how a typical plot of the linearized data would appear. If data are near-ideal (e.g., the calculated data in Table 19.3), then all of these plots will yield *identical* results for K_D and B_{max} when their slope and intercept(s) are calculated. In practice, they each are affected differently by sources of experimental variance. In any event, direct calculations should now be done using computerized nonlinear regression [e.g., based on Eq. (19.14)]. This can be done with specialized programs such as Prism (GraphPad, Inc), or with general-purpose scientific plotting software (e.g., SigmaPlot has specific radioligand binding modules).

One of the reasons, however, that Scatchard plots still are used (or at least often expected for publication) is that they permit quick visual estimates of the B_{max} (i.e., the B_{max} is the X-intercept of the extrapolated line) and of relative toxicant affinities (reflected as the negative reciprocal of the slope). The steeper the slope, the smaller the K_D, and the greater the affinity of the toxicant for the receptor.

Figure 19.6 (bottom) shows the Scatchard transformation of the saturation data shown in Figure 19.6 (top). In this case, a linear regression analysis adequately fits these data. As noted above, the slope of the regressed line is the negative reciprocal of the K_D, and the x-intercept is the B_{max}. Curvilinear Scatchard plots can indicate the presence of multiple populations of receptors with differing affinities for the

TABLE 19.3. Derived Values for Theoretical Saturation Plot

F (nM)	$\log [F]$	B
0.01	−2	0.99
0.03	−1.52	2.91
0.1	−1	9.09
0.3	−0.52	23.08
1	0	50.00
3	0.48	75.00
10	1	90.91
30	1.48	96.77
100	2	99.01

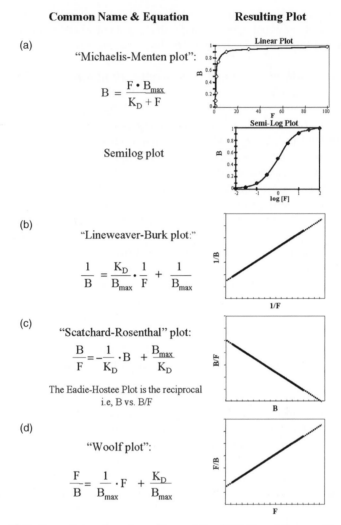

Common Name & Equation   **Resulting Plot**

(a)

"Michaelis-Menten plot":

$$B = \frac{F \cdot B_{max}}{K_D + F}$$

Semilog plot

(b)

"Lineweaver-Burk plot:"

$$\frac{1}{B} = \frac{K_D}{B_{max}} \cdot \frac{1}{F} + \frac{1}{B_{max}}$$

(c)

"Scatchard-Rosenthal" plot:

$$\frac{B}{F} = -\frac{1}{K_D} \cdot B + \frac{B_{max}}{K_D}$$

The Eadie-Hostee Plot is the reciprocal
i.e, B vs. B/F

(d)

"Woolf plot":

$$\frac{F}{B} = \frac{1}{B_{max}} \cdot F + \frac{K_D}{B_{max}}$$

Figure 19.5. Some linear transformations of Eq. (19.14) and the resulting plots.

toxicant. Alternatively, they may indicate positive or negative cooperativity (producing upwardly or downwardly "concave" plots), where the binding of one toxicant molecule influences the binding of subsequent molecules.

A potential problem associated with Scatchard analysis occurs when the range of radiolabeled toxicant concentrations used is too narrow, such that the highest concentration does not equal or exceed the K_D (that concentration of toxicant which results in the occupation of one half of the total binding sites). Although not apparent from a Scatchard plot of such data, estimates of B_{max} under such conditions are prone to significant error.

Two other possible transformations of saturation binding data mentioned earlier are the Hofstee (B versus B/F) and Woolf Plots (F/B versus F). As with the Scatchard equation, these equations can be derived from algebraic manipulations of the equations listed above.

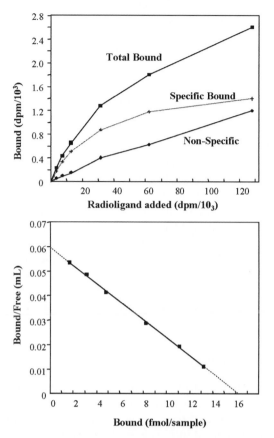

Figure 19.6. Characterization of a receptor using a saturation isotherm with a radioligand. (Top) In this experiment, increasing amounts of radioligand are added to tubes containing a constant amount of tissue. A duplicate series of tubes also includes an excess (>100-fold) of a known competitor for the receptor being studied. The solid lines indicate the total amount of radioligand bound to the receptor (Total Bound) and the nonspecific binding (Nonspecific). The nonspecific binding is due to physicochemical interactions of the radioligand with proteins or lipid (e.g., dissolving into membrane lipids). The specific binding (the term of interest) is obtained by subtracting the Nonspecific from the Total Bound. As predicted by Eq. (19.14), a plot of specific binding versus ligand added yields a rectangular hyperbola. (Bottom) A Scatchard plot of these data (based on equation shown in Figure 19.5, panel B). With these data, the extrapolated density of receptors is approximately 16 fmol. The KD could be calculated based on −1/slope.

19.3.3 Kinetic Determinations of Equilibrium Constants

Although equilibrium methods are used most commonly to estimate parameters of receptor binding, separate kinetic analyses should be employed during the initial characterization of toxicant binding—for example, when a novel toxicant radioligand becomes available or an established radioligand is planned for use in a new tissue type or species. Several of the general references cited at the end of this chapter provide an excellent detailed discussion of kinetic binding methods and the derivation of the associated equations. Briefly, kinetic experiments to determine association and dissociation rate constants are conducted separately. Association rate is estimated by measuring the amount of specific toxicant bound as a function of time; assuming that the amount of free radioligand does not change appreciably over time, then a pseudo-first-order rate equation is used to estimate k_1. The rate of dissociation of the toxicant is determined by incubating tissue with toxicant, allowing the reaction to proceed to equilibrium, and then stopping it either by infinite dilution or by the addition of an excess of nonradiolabeled ligand. This stops the forward association reaction, so that measurement of the amount that remains bound as a function of time can be attributed solely to the dissociation between receptor and toxicant. The rate of change of the receptor–igand complex over time is used to estimate K_{-1}, the dissociation rate constant.

In theory, K_D (i.e., k_{-1}/k_1) should be the same whether determined by kinetic or equilibrium approaches. In practice, however, moderate differences arise that are often attributed to technical problems associated with separating bound from free rapidly without losing a significant proportion of receptor–toxicant complex. This problem is troublesome, particularly when estimating the amount bound at early time points in association or dissociation experiments, when the amount of bound ligand is changing rapidly. Large differences between the K_D as determined in saturation and kinetic experiments, however, may indicate that the reaction is more complex than a simple bimolecular reversible reaction.

19.3.4 Competition Binding Assays

Competition binding assays, also referred to as indirect binding studies, involve measuring the competition between a radioligand and an unlabeled drug for a specific receptor site. The concentration of the radioligand is held constant, while the concentration of the unlabeled drug is varied over a wide range of concentrations. If both drugs compete for the same receptor site, then the amount of radioligand bound will decrease systematically as the concentration of the unlabeled drug increases (i.e., as more of the unlabeled drug binds). Competition studies are particularly attractive because they allow the study of many drugs that are not suitable for radiolabeling and direct binding measurements due to their low affinity. Applications of this technique include determining the binding specificity of a radiolabeled toxicant, comparing the potencies of a series of unlabeled toxicants for the radiolabeled receptor site, and providing a reliable estimate of nonspecific binding.

Data from competition binding experiments are plotted routinely as the percent inhibition of specific radioligand binding (total – nonspecific) as a function of $\log_{10}$ of the concentration of competitor. The 100% binding value is defined as the amount of specific radioligand binding in the absence of competitor. When comparing the potency of several compounds, it is useful to plot several competition ("displacement") curves together because the relative position of the competition curves for various ligands will be related to their potency in displacing the radioligand, with the leftmost curves representing the most potent competitors.

Figure 19.7 (top) illustrates competition binding data of two compounds that can compete for radioligands that label the receptor for the neurotransmitter dopamine (specifically, the D_1 receptor subclass). It can be seen that Compound A can eliminate specific binding at concentrations about 100-fold lower than required for Compound B. In toxicological (and pharmacological) terms, one would say that

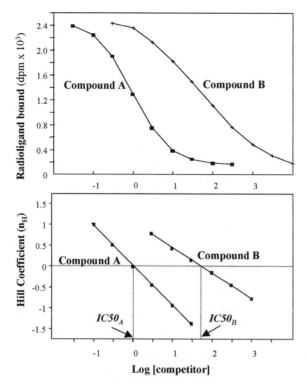

Figure 19.7. Competition curves for two compounds versus a known radioligand. (Top) These data represent the competition of two compounds with a known radioligand (in this case a radioligand that labels the dopamine D_1 receptor, a member of the G protein-coupled super-family). It is important to note not only the left–right difference between Compound A and Compound B, but also the difference in the shape of their competition curves. (Bottom) A Hill plot [based on Eq. (19.20)] of the competition curves shown in the top figure provides two pieces of data. First, the slopes of the lines are different (Compound A = ~1.0; Compound B = ~0.6), which has important mechanistic meaning that is discussed in Section 19.3.4b. Hill plots also allow more precise estimation of IC50s. By definition, at 50% inhibition, the Hill coefficient is 0. As shown, one can estimate the IC50 for each compound from this plot.

Compound A is more potent than Compound B in this assay (or more rigorously, A has higher affinity). Specifically, higher concentrations of Compound B are required to cause 50% inhibition of binding than are required of Compound B. One other point should be obvious in these data in Figure 19.7. The slopes in the top curves are clearly not parallel for Compound A and Compound B, with the latter being more shallow. The shape of the competition curves also provides important information concerning the nature of the interaction between receptor and ligand, as will be described in Sections 19.3.4b–19.3.4d.

The relevant kinetic model for competition experiments with a radiolabeled drug [D] and an unlabeled competitor [I] is shown in the two equations in (19.15). When both sets of reactions have proceeded to equilibrium, the net rate of formation of both (DR) and (IR) are zero, and the following Eq. (19.16) can be derived from mass-action principles.

$$D + R \underset{k_{off}}{\overset{k_{on}}{\rightleftharpoons}} D \cdot R$$

$$I + R \underset{k_{off}}{\overset{k_{on}}{\rightleftharpoons}} I \cdot R \tag{19.15}$$

$$I \cdot R = \frac{B_{max} * I}{I + K_I * \left(1 + \dfrac{D}{K_D}\right)} \tag{19.16}$$

In Eq. (19.16), K_I is the equilibrium dissociation constant of the unlabeled competitor, while K_D is the dissociation constant of the radioligand. K_I values of different toxicants can be compared to determine the rank order of potency of toxicants in competing for the radiolabeled receptor site, where a low K_I indicates a high potency. K_I values are routinely determined from experimentally derived IC50 values (the concentration of inhibitor that produces a 50% reduction in total specific binding of the radiolabeled drug). Although IC50 values may be estimated (with considerable error) from visual inspection of plots such as in Figure 19.7, more rigorous techniques are applied typically, including iterative nonlinear regression analysis and indirect Hill plots (see Section 19.3.4b).

19.3.4a   The Cheng–Prusoff Equation. For purely competitive interactions (that fit the model shown in Figure 19.1), the relation between the IC50 and the K_I is described by the Cheng–Prusoff equation:

$$K_I = \frac{IC50}{1 + \dfrac{[L]}{K_D}} \tag{19.17}$$

where L is the concentration of the radioligand having affinity K_D. If the concentration of the radioligand is equal to its K_D, then the denominator becomes 2 and the K_I of the competitor is equal to one-half of the IC50. In contrast, if the concentration of radioligand is extremely low, then the denominator becomes unity, and the $K_I \Rightarrow IC50$.

It is important to note that the application of the Cheng–Prusoff equation requires several assumptions: (1) Both radioactive and competitor ligands are interacting reversibly with a single population of sites (bimolecular reactions); (2) the reaction is at steady state; (3) nonspecific binding is accurately estimated; and (4) the concentration of receptor is much less than the K_D for the drug or the competitor. It is important to recognize that there are really many Cheng and Prusoff equations as described in their oft-cited paper 1973 paper. The one commonly cited is only applicable to competitive bimolecular reactions following the model shown in Eq. (19.1).

19.3.4b   The Hill Plot. The Hill plot was used frequently to determine whether obtained binding data deviate significantly from what would be expected from a simple reversible bimolecular reaction. The Hill transformation was developed originally to describe the cooperativity exhibited when O_2 bound to hemoglobin. Hill plots can be direct or indirect, depending on whether the data are obtained from saturation or competition experiments, respectively. The Hill equation can be derived from elementary enzyme kinetic principles.

The more generalized form of the Michaelis–Menten equation is shown in Eq. (19.18):

$$B = \frac{B_{max} \cdot [D]^n}{K_D + [D]^n} \tag{19.18}$$

where n reflects whether or not the binding sites behave as a single homogeneous population rather than multiple populations or interactive sites.

If we do a series of algebraic transformations, we can arrive at the commonly used Hill equation:

$$\log\left[\frac{B}{B_{max} - B}\right] = n \cdot \log[D] - \log K_D \tag{19.19}$$

As can be seen by inspection, the Hill equation (19.19) can be plotted as a straight line in the form $y = mx + b$, where y is the "pseudo-logit" term on the left-hand side of Eq. (19.19), and $x = \log[D]$). In such a plot, the slope is equal to n (often called n_H; the Hill number) and the y intercept is equal to $-\log K_D$. In the case of saturation binding data, $K_{D'}$ is the apparent equilibrium dissociation constant whose value reflects both the intrinsic K_D and the factors that determine how the binding of one ligand molecule affects the binding of subsequent molecules. If such sequential binding effects are not present, the $n_H = 1$, and $K_{D'}$ will equal the K_D. Alternatively, when n_H is significantly different from unity, then $K_{D'}$ is not equivalent to K_D. An equation of similar form (but opposite slope) can be derived for competition experiments (see below).

19.3.4c *Visual Inspection of Binding Data.*

There are several important inferences that scientists make from examination of Hill plots. If the interaction of a ligand and its receptor follows the simple bimolecular reaction shown in Eq. (19.1), the value of the Hill coefficient n (or more commonly often n_H) will equal 1. (Please note, however, that it is usually impossible experimentally to distinguish between a single population of receptors versus several receptors having nearly similar or identical K_D's for the ligand.) A Hill coefficient significantly less than 1 suggests a more complex situation. These possibilities include interaction of the ligand with multiple populations of binding sites (e.g., different receptors), a single class of receptor that exists in different affinity states, or even negative cooperativity among the same receptors. Alternatively, a Hill coefficient significantly greater than 1 might be obtained when the binding sites exhibit positive cooperativity.

In toxicology, an even more common use of Hill plots is for competition assays. In contrast to the direct Hill plot [from Eq. (19.19)], competition assays use indirect Hill plots. These have essentially the same form and interpretation, but have negative rather than positive slope. Again, an indirect Hill coefficient (n_H) of -1 is consistent with (although it does not prove) the notion that both competitor and radioligand are interacting with one population of receptors via a simple reversible bimolecular reaction. Similarly, an n_H significantly less in magnitude than -1 (e.g., -0.6) is consistent with multiple receptors or negatively cooperative receptors.

Figure 19.7 (bottom) shows the Hill plots from two of the competition curves illustrated in Figure 19.7 (top). In one case, $n_H = -1$, whereas in the second case it is significantly lower (approximately –0.6). The most parsimonious explanation for these data is that the binding of Compound A is occurring to a homogeneous population of binding sites, whereas that of Compound B is to multiple or negatively cooperative binding sites.

Practicing scientists are able to make reliable estimations by visual inspecting competition curves and looking at the steepness of those curves. Such a visual inspection is based on the following line of reasoning between the apparent steepness of a competition curve and the derived Hill coefficient. First, consider that the "indirect" Hill equation can take the following form:

$$\log\left[\frac{B}{100-B}\right] = n \cdot \log[D] - n \cdot \log \mathrm{IC50} \qquad (19.20)$$

$$y \text{ value} \quad \text{slope} \quad x \text{ value} \quad \text{intercept}$$

where 100 represents the total binding (i.e., 100%) of the radioligand in the absence of competitor and B represents the percent of total binding found at each concentration of competitor $[D]$. For our purposes, we shall ignore the intercept term in this equation.

From this equation, we can construct an example that is extremely useful in visually examining receptor data, indeed any data fitting the simple mass action model of Eq. (19.1). This means that a toxicant under study is competing with the radioligand for one, and only one, population of sites. By definition, the Hill slope (often called n_H) must equal 1. It turns out that by memorizing the numbers "9" and "91," one can do a very useful preliminary analysis of data fit to such a mass-action model. The reason for this is as follows:

At 91% of total binding, the value for the left-hand $\left(\text{i.e., } \log\left[\frac{B}{100-B}\right]\right)$ part of Eq. (19.20) can be calculated as follows:

$$\log\left(\frac{91}{100-91}\right) = \log\left(\frac{91}{9}\right) = \log(10.111) = 1.005$$

One can perform the same calculations for 9% of total binding:

$$\log\left(\frac{91}{100-9}\right) = \log\left(\frac{9}{91}\right) = \log(0.989) = -1.005$$

This means that the Δ (change) in the left-hand side of the equation when going from 91% to 9% of total binding will be $1.005 - (-1.005) = 2.01$ Hill units. Remember, as shown in Eq. (19.20), the Hill number (n_H) is a slope function, in which the denominator is the log of the concentration range. If $n_H = 1$ [that is, a system that meets the constraints of the model of Eq. (19.1)], then the log of the concentration range to cause this change from 91% to 9% total binding must also have a value of 2.01. Of course, the antilog of 2.01 is ~100. *Thus, if a curve has "normal" steepness [i.e., it fits Eq. (19.1)], then a change from 91% to 9% of binding must occur with a change in concentration of 100-fold.* This means that one can visually inspect dose–response data to estimate whether the 91% to 9% range spans a two log order (i.e.,

100-fold) concentration range. This is a very useful way to estimate whether data are consistent with the bimolecular model.

If the change from 9% to 91% requires more than a 100-fold concentration change, then the curve is "shallow" and cannot be explained by the model in Eq. (19.1). If the change from 9% to 91% requires less than a 100-fold concentration change, than the curve is "steep" and also does not meet the model of Eq. (19.1). Of these two situations, the former is more common. One common example of shallow competition curves involves the binding of ligands to G protein-coupled receptors (GPCRs; see 19.1.3). In such cases, the shallow steepness of the curve often is attributed to the existence of a single receptor population that exists in two interconvertible affinity states. According to this model, GPCRs exist in a complex with G proteins and have GDP bound to the α-subunit of the G protein. These receptors are in a high-affinity state for agonists. Some of the receptors are not associated with G proteins, and they exist in a low-affinity state for agonists. Thus, normally an agonist competition curve looks shallow because it detects two populations of binding sites having very different affinity.

With these GPCRs, the addition of GTP (or related guanine nucleotide analogs) to an assay system can convert those receptors in a high-affinity to a low-affinity state by displacing the GDP. Thus, one goes from having two sites (one with high and one with low affinity for agonists) to one site (all low affinity). Experimentally, the competition curve changes from a shallow curve (due to two different populations of binding sites) to a curve of normal steepness (representing only the low-affinity site). Moreover, the curve shifts rightward. Thus, one can, even by visual inspection, test hypotheses about the underlying model. Since dose–response relationships are at the heart of toxicology, it is important to be able to make such estimations facilely.

Finally, two practical points relative to Hill plots also should be made. As can be seen in Eq. (19.20), at the IC50 of a competitor, the left-hand term will have a value of 0. Hill plots are used frequently to calculate the IC50 graphically; this value can then be substituted into the Cheng–Prusoff equation to determine a K_I value (see Section 19.3.4a). Experimentally, those data representing less than 5%, or more than 95%, radioligand occupancy are of little use since at these extremes, systematic or random sources of experimental error (e.g., counting error, pipeting error) make use of these points problematic.

19.3.4d Complex Binding Phenomena.
In many cases, the assumption that a toxicant is interacting with a single population of binding sites is not tenable, as indicated by curvilinear Scatchard plots or Hill coefficients different from unity. A lengthy discussion of the alternative mathematical models used to describe possible receptor–toxicant interactions in such cases is beyond the scope of this chapter. Briefly, such analyses rely on powerful statistical computer programs (i.e., Prism or LIGAND) that can determine whether binding data can be modeled with greater precision by assuming the existence of two (or more) receptor sites, rather than one. If so, separate parameter estimates can be obtained for the two sites. Although two-site analysis is used frequently, the ability to resolve more than two receptor sites currently is limited. In all cases, it is important to emphasize that such analyses should be used to generate testable hypotheses about the multiple sites, rather than as proof for their existence. Finally, it should be recognized that a number of

technical problems (failure to reach equilibrium at low drug concentrations) could produce data consistent with multiple sites; thus, a careful consideration of potential artifacts is wise prior to entertaining the idea of multiple receptor populations.

19.3.5 How to Conduct a Radioreceptor Assay

19.3.5a Methods to Separate Bound from Free Radioligand. Three methods have been used traditionally to separate bound from free ligand in radioreceptor assays. Dialysis, originally widely used to study enzyme–substrate interactions, has several technical problems that prevent its wider use. These include degradation or sticking of receptor or ligand, the cumbersome nature of assays when large numbers of samples are needed, and the long time to obtain equilibrium. Centrifugation to separate ligand bound to tissue receptors from free ligand in solution is limited because the ligand dissociates from the receptor during pelleting and a large amount of drug receptor complex is therefore lost. There is an inverse relationship between the affinity of the ligand for the receptor (i.e., $1/K_D$) and the time allowable for centrifugation. One can calculate this time based on the K_D of the radioligand and the derived rate constants; the relationship will be logarithmic. For example, assuming that separation must be complete in $0.15t_{1/2}$ to avoid losing more than 10% of ligand–receptor complex, then centrifugation must be accomplished in 10 sec for a ligand with a K_D of 10 nM and in 1000 sec for a ligand with a K_D of 0.1 nM.

The most widely used method to separate bound from free radioactivity is vacuum filtration. The suspension of tissue homogenate is diluted and is rapidly aspirated through filters that retain the tissue membranes but not the solvent. With a sufficiently high ligand affinity for the receptor, several washes of the filter can be used to achieve low levels of nonspecific binding. The use of ice-cold buffer for these washes decreases the dissociation rate of the receptor–ligand complex, permitting more extensive washing with lower nonspecific binding. The ability to prepare large numbers of samples simultaneously through the use of commercially available filtration systems is a major advantage of this method. An excellent summary of the factors that should be considered in choosing a separation method is provided by Limbird (2004); see Suggested Reading.

19.3.5b Methods to Estimate Nonspecifically Bound Radioligand. It is important to recognize that the total amount of radioligand bound to tissue represents ligand not only bound to high-affinity receptors, but also bound "nonspecifically" to other tissue components as well as to assay materials (filter paper, glass or plastic tubes, etc.). In contrast to receptor-bound ligand, nonspecifically bound ligand is assumed to be nonsaturable; nonspecific binding continues to increase as radioligand concentration increases. Most nonspecific binding often is believed to be due to hydrophobic interactions between the radioligand and lipophilic tissue constituents, such as membrane phospholipids. One may conceptualize this process as a sort of "dissolving" of the radioligand into biological membranes. Thus, while this process will be slow ("low affinity"), in an aqueous environment it will be both essentially irreversible and of high capacity relative to the limited number of specific receptors.

A method used commonly to estimate the nonspecific binding component involves measuring the amount of radioligand bound in the presence of a large excess of unlabeled ligand ($>100 * K_D$). This large excess should compete for receptor sites, preventing almost all binding of radiolabeled ligand. Conversely, the model discussed in the previous paragraph would predict that radioligand bound to nonspecific components should be unaffected; a nearly infinite number of sites for nonspecific binding exist relative to specific receptor(s). Thus, the amount of radioactivity remaining under these conditions is taken as a measure of nonspecific binding. It is calculated as follows:

$$\text{Specific binding} = \text{Total binding} - \text{Nonspecific binding}$$

where "Total binding" is the amount of radioligand bound in the absence of a competitor, and "Nonspecific binding" is that bound in the presence of an excess of competitor. The theory that is applied to the analyses of such data was discussed earlier.

19.3.6   The Relation Between Receptor Occupancy and Biological Response

The above discussion has focused on methods for describing the proximal events in receptor-mediated responses to toxicants, namely the binding of a toxicant ligand to a biochemically defined receptor. A critical assumption is that such an approach will elucidate the mechanisms by which receptor–toxicant interactions produce their ultimate toxic effects. The validity of this assumption rests on demonstrating that the binding of a ligand is related lawfully to some biological response. For instance, a straightforward prediction might be that the binding affinity of a toxicant for a receptor should correlate with its propensity to either stimulate or block a functional response. More often than not, however, this relation is complex rather than simple.

Stephenson is credited with introducing the concept of a complex relation between receptor occupancy and tissue response. His formulations were designed to explain prior anomalous data that suggested that drug effects were often not linearly related to receptor occupancy, as shown by the frequent finding that the maximal drug effect achieved was not linearly related to the concentration of drug applied. Stephenson introduced three postulates to explain such aberrant data. First, he stated that some agonists could produce a maximal response by occupying a small fraction of a receptor population. Second, he stated formally that the tissue response is not related linearly to receptor occupation. Finally, he introduced the concept of differing efficacies of agonists, such that an agonist with high efficacy may elicit a maximal effect when only a small percentage of receptors are occupied, while an agonist with low efficacy may require occupation of a substantially greater percentage of receptors. A partial agonist is a special case of a low efficacy agonist. That is, the efficacy of a partial agonist is so low that it fails to elicit a maximal response even at saturating concentrations, when all receptors presumably are occupied. The potential antagonistic effects of partial agonists present at high concentrations can now be appreciated; partial agonists can compete effectively for receptor binding sites with more efficacious ligands, although partial agonists have only minimal biological effects. Figure 19.8 provides an example of the functional response produced by a partial agonist and a full agonist. It should be noted that the com-

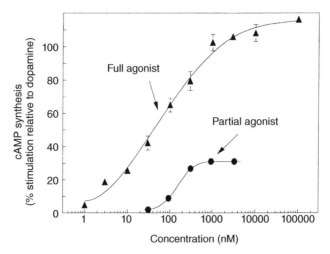

Figure 19.8. Dose–response plot showing functional differences between partial and full agonist. These data illustrate that two compounds of similar potency may differ in functional efficacy. In this case, the full agonist causes the same degree of stimulation as does the endogenous ligand. Conversely, the partial agonist, although of similar potency, is only partially efficacious. These data are modified from actual published studies from the authors' laboratory (Brewster et al. *J. Med. Chem.* 330: 1756–1764, 1990).

pounds in Figure 19.8 can completely compete for the receptor believed to mediate this functional response. Several distinct mechanisms may explain partial functional efficacy, and it is not always clear which are operative. It is important to remember that functional efficacy and affinity for the receptor (potency) need not correlate.

The phenomenon of partial agonism is but one example of the complexities encountered when relating binding data to biological response. Another noteworthy example is provided by the finding that many receptor populations appear to be larger than would be required to produce a maximal response. This phenomenon is referred to as "spare receptors" or, alternately, as "receptor reserve." The strongest empirical data supporting this notion comes from studies using irreversible receptor inactivation which demonstrate that the maximum tissue response is not depressed until a large percentage of the receptor binding sites are eliminated.

Finally, it should be noted that a comparison of receptor binding data with functional response is inherently difficult. In the case of binding assays prepared from tissue homogenates, receptors are isolated from components that are involved in the ultimate cellular response; thus, the assays are conducted in artificially pure preparations. In contrast, functional assays typically require greater preservation of cellular integrity. This also means that receptor regulatory processes are more likely to be evident during functional assays. For instance, a common finding is that the response of an agonist decreases as a function of prior exposure to the same or a similar agonist; this phenomenon is referred to as homologous or heterologous desensitization, respectively. In some model systems (e.g., β-adrenergic receptor), one mechanism for desensitization involves internalization and sequestration of receptor molecules. Additionally, an intracellular enzyme (β-adrenergic receptor kinase; BARK), has been found to be activated by agonist stimulation; this kinase phosphorylates particular amino acid residues on the receptor which are involved in some aspects of the desensitized response. Clearly, such dynamic receptor regulation is a critical aspect of receptor–toxicant interactions, yet such processes are not likely to be captured in receptor binding assays using extensively washed membranes.

This preceding discussion was not intended to convince one that the study of receptor binding is of limited utility, but rather that it should be accompanied by an examination of functional endpoints, with a realization of the potential challenges (and rewards) involved in relating the two.

19.4 RELATIONSHIP OF RECEPTOR OCCUPANCY TO FUNCTIONAL EFFECTS

19.4.1 Classical Definitions of Receptor-Mediated Functional Effects

A common tenet of pharmacology is that ligands can be characterized by the nature of the functional effects elicited by their interaction with their target receptor. In a given assay, a ligand can be classified as a full agonist, partial agonist, neutral antagonist, or inverse agonist. This led to the theory of "intrinsic efficacy," originally proposed as a measure of the stimulus per receptor molecule produced by a ligand. According to this notion, full agonists possess sufficiently high intrinsic efficacy such that they maximally stimulate all cellular responses linked to a given receptor. Partial agonists possess lower degrees of intrinsic efficacy (leading to submaximal responses), whereas inverse agonists reduce constitutive (ligand-independent, basal) receptor signaling. Neutral antagonists possess no intrinsic efficacy, but occupy the receptor to block the effects of full, partial, or inverse agonists. Thus, "intrinsic efficacy" referred to an invariant characteristic of a ligand—as opposed to the operational term "intrinsic activity," which simply refers to the maximal effect (E_{max}) of a ligand relative to a reference agonist in any given experimental system.

19.4.2 Evolving Concepts of Receptor-Induced Functional Changes

With the GPCRs in particular, it is now clear that a single receptor can signal through many signaling pathways, depending on the complement of G proteins and other signaling molecules associated with the receptor in a given cellular locale. As discussed in the previous section, the traditional view has been that a ligand will have the same functional characteristics at every function modulated by a single receptor. Thus, a full agonist would be expected to activate all of the signaling pathways linked to a receptor to the same degree, and a ligand that antagonizes one signaling pathway via a specific receptor should antagonize every pathway coupled to that receptor to the same extent.

It now seems clear that this premise is frequently incorrect. There has been a rash of data emerging within the past decade showing that certain ligands have quite diverse functional consequences mediated via a single receptor. Thus, it is clear that the classical concept of "intrinsic efficacy" as a system-independent constant, although once having conceptual utility, is probably not correct. Indeed, evidence that this was of importance is suggested by the fact that different research groups coined unique names for essentially the same phenomenon (e.g., "functional selectivity," "agonist-directed trafficking of receptor stimulus," "biased agonism," "protean agonism," "differential engagement," "stimulus trafficking," etc.).

There are three reasons why this new concept [see Urban et al. (2006) for review] is of importance to molecular toxicology. In the narrowest sense, it may be critical to understanding the exact mechanism of some toxicants. In a broader sense, it certainly affects clinical toxicology and toxicological aspects of drug development. As an example, it has been shown recently that agonists for the serotonin 5-HT$_{2B}$ receptor can promote heart valvuopathies (e.g., as are caused by fenfluramine). Until the specific signaling pathway responsible for this effect is known, one sometimes may generate misleading hypotheses if a drug candidate is screened at a single functional pathway that is not the one mediating toxicity in the heart. By similar reasoning, this mechanism suggests that one may be able to do drug optimization by selecting compounds with differential effects at two signaling pathways mediated by the same receptor. Finally, for ligand-selective toxicants, risk assessment estimates can often be more precise if we know the affinity of a ligand for its receptor. Awareness of the target receptor's key functional pathways, as well as the effects of the toxicant at each, should allow even greater predictive accuracy.

SUGGESTED READING

Cheng, Y. C., and Prusoff, W. H. Relationship between the inhibition constant (K_i) and the concentration of inhibitor which causes 50 percent inhibition (I50) of an enzymatic reaction. *Biochem. Pharmacol.* **22**, 3099–3108, 1973. *This is one of the most commonly cited, and least read, papers in the field.*

Foreman, J. C., and Johansen, T. (Eds.). *Textbook of Receptor Pharmacology*, 2nd ed., CRC Press, Boca Raton, FL, 2003.

Kenakin, T. *A Pharmacology Primer: Theory, Application and Methods*. Elsevier, San Diego. 2004.

Limbird, L. E. *Cell Surface Receptors: A Short Course on Theory and Methods*, 3rd ed., Springer, New York, 2004.

Urban, J. D., Clarke, W. P., von Zastrow, M., Nichols, D. E., Kobilka, B. K., Weinstein, H., Javitch, J. A., Roth, B. L., Christopoulos, A., Sexton, P., Miller, K., Spedding, M., and Mailman, R. B. Functional selectivity and classical concepts of quantitative pharmacology. *J. Pharmacol. Exp. Ther.* **320**, 1–13, 2007.

www.graphpad.com. This is the website of the company that markets Prism, the most widely used software for receptor-ligand analyses. There is a wealth of background information offered free by this company (see Resource Library on home page) that can help even experienced scientists avoid common mistakes.

CHAPTER 20

Reactive Oxygen/Reactive Metabolites and Toxicity

ELIZABETH L. MACKENZIE

Education and Training Systems International, Chapel Hill, North Carolina 27514

20.1 INTRODUCTION

During the detoxication of xenobiotics, the parent compound is enzymatically converted to a more polar form by phase I enzymes, including cytochrome P450s (CYPs), prostaglandin synthetase, flavin monooxygenase (FMO), and alcohol or aldehyde dehydrogenase. In most cases these modifications create a polar compound, ready to be conjugated to an endogenous metabolite by phase II enzymes and subsequently eliminated; however, phase I modifications can cause the activation of certain xenobiotics to more toxic forms that deleteriously react with critical cellular components, including nucleic acids, proteins, and lipids. Although phase II reactions generally result in further detoxication, conjugation may (in some cases) also cause activation.

Typical reactive metabolite categories are shown in Figure 20.1. Most are electrophilic in nature (from the Greek words for "electron loving," containing a positive center). As a result, they covalently bind to nucleophilic molecules ("lover of nuclei," donates electrons, negative center), including proteins and nucleic acids, as well as glutathione. Instead of being electrophilic, reactive metabolites may be free radicals, which contain an unpaired electron, or they may also be capable of producing radicals. Through interaction between these types of reactive intermediates and oxygen-containing molecules, reactive oxygen species may form and can also damage important cellular components.

Our current understanding of metabolic activation and resulting cellular damage stems from studies regarding toxic bioactivation in chemical carcinogenesis. In the mid-1900s, James and Elizabeth Miller's studies of the aminoazo dye and rat hepatocarcinogen, *N,N*-dimethyl-4-aminoazobenzene (DAB), demonstrated that chemical carcinogens bind and modify protein and nucleic acid. Later they discovered that metabolic conversion of 2-acetylaminofluorene (2-AAF) to *N*-hydroxyl-AAF was essential for carcinogenesis.

Such revelations led to further study of xenobiotic toxicity and carcinogenesis, illuminating mechanisms involved. In recent years, numerous studies have revealed

Molecular and Biochemical Toxicology, Fourth Edition, edited by Robert C. Smart and Ernest Hodgson
Copyright © 2008 John Wiley & Sons, Inc.

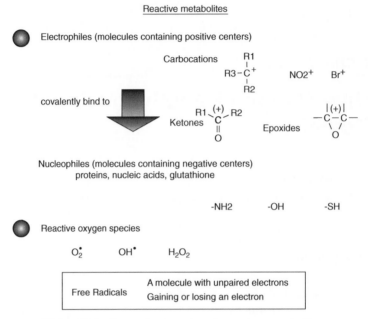

Figure 20.1. General categories of reactive metabolites.

the prevalence of such metabolic activations in toxicity as well as the specific enzymes and reactions involved. At least six carcinogenic components of tobacco smoke, including benzo(*a*)pyrene (BaP) and *N*-nitrosodimethylamine (NDMA), have been shown to form DNA or protein adducts in the exposed lung. BaP, for example, is metabolized to an epoxide by Cyp1A1, and epoxide hydrolase then converts it to a diol. Further oxidation leads to the generation of a diol epoxide. It is these diol hypoxides that react with DNA to form BaP diol-epoxide deoxygaunosine adducts. A strong correlation has been established between these adduct biomarkers, tobacco product use, and cancer. Along the same lines, studies have also shown that antioxidant compounds, such as flavanoids, are cytoprotective against such adduct formation, while the lowered enzyme activity associated with the GSTM1 polymorphism is associated with increased bulky adduct formation. Thus, toxicity is not merely a consequence of the chemical character of the compound, but also of the enzymatic composition and activity. In addition to genetic determinants such as the GSTM1 polymorphism, gender, nutrition, disease, and exposure to other chemicals also influence individual enzymatic status.

20.2   ENZYMES INVOLVED IN BIOACTIVATION

20.2.1   Phase I Oxidations

20.2.1a   Cytochrome P450. A majority of enzymes involved in xenobiotic detoxication can also catalyze the formation of reactive metabolic intermediates; however, oxidation reactions have the highest propensity to produce reactive metabolites. The CYPs are the major enzyme family responsible for the oxidation of xenobiotics (see Chapters 9 and 10). This is in part due to their ubiquitous presence—especially in the liver, but also in the lung, kidney, and skin—and also because of the large number of isoforms and the diversity of substrate specificity, as well as the inducibility of many of the isoforms.

A number of families consisting of one or more distinct proteins comprise this enzyme group. Many xenobiotics are preferentially metabolized by a particular CYP or CYP family; however, they may also be substrates for other CYPs. In these instances, the differences in enzyme activity are in the rate at which the oxidation occurs, the site of the modification on the parent compound (regioselectivity), and the steric configuration of the product (stereoselectivity).

A classic example is the activation of the carcinogen benzo(*a*)pyrene to (+)-benzo[*a*]pyrene 7,8 diol-9,10-epoxide by CYP1A1, which was already mentioned. CYP1A1 and CYP1A2 are considered the most important CYPs in the activation of chemical carcinogens, including polycyclic aromatic hydrocarbons as well as aromatic and heterocyclic amines. CYP1A1 is the predominant enzyme responsible for converting 2-AF to *N*-hydroxy 2-AF; however, the flavin-containing monooxygenase and prostaglandin synthetase have also been indicated in the production of the reactive intermediates (see Section 20.3). In addition to the metabolism of amines and hydrocarbon molecules, CYPs are also involved in the metabolism of other chemicals, including acetaminophen and aflatoxin B (see Table 20.1 for examples of CYP substrates).

Despite the prevalence of CYP1A1 and CYP1A2 in bioactivation reactions, all CYPs have the potential to produce reactive metabolites. CYP2E1 also has been indicated in chemical toxicity through the metabolism of low-molecular-weight chemicals as well as in the production of oxidative stress. A well-characterized example of bioactivation by CYP2E1 is that of the hepatotoxicity of ethanol and acetaminophen metabolism in concert. CYP2E1 is the primary CYP for ethanol oxidation and, furthermore, ethanol induces CYP2E1, causing increased metabolism of acetaminophen by the same CYP. This particular pathway, in contrast to other

TABLE 20.1. Substrates for Bioactivation Reactions

Cytochrome P450	Flavin-Containing Monooxygenase	Prostaglandin Synthetase
Acetaminofluorene	2-Aminofluorene	Acetaminophen
Acetaminophen	3,3′-Iminodipropionitrile	2-Amino-naphthalene
Aflatoxin B$_1$	Ketaconozole	Amphetamines
Benzo[*a*]pyrene	2-Mercaptoimidazoles	BHT (in the presence of BHA)
Bromobenzene	Thioamides	5-Nitrofurans
Chloroform	Thiocarbamates	Phenylbutazone
N,N-Dimethyl-4-aminoazobenzene	Thiocarbamides	Thalidomide
7,12-Dimethylbenzanthracene		
Ethanol		
Halothane		
N-Nitrosodimethylamine		
Vinyl chloride		

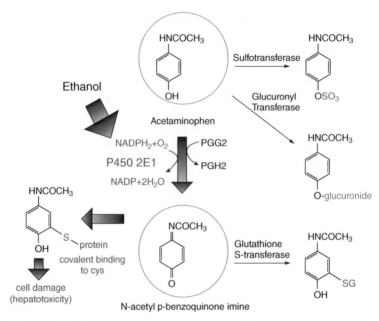

Figure 20.2. Mechanism of detoxication and bioactivation of ethanol.

possible phase II metabolic conjugation pathways, leads to the production of the highly reactive quinone, N-acetyl p-benzoquinone imine (NAPQI), which is capable of covalently binding to protein (Figure 20.2). Subsequent conjugation of NAPQI with glutathione can inactivate it. Therefore, available levels of glutathione and activity of GSTs are critically important to reducing toxicity. Cyp2E1 knockout mice are resistant to acetominophen hepatotoxicity. CYP2E1 also plays an important role in the production of oxidative stress. It functions to oxidize ethanol to acetaldehyde and 1-hydroxyethyl radical and thereby produces oxidative stress. In addition to the metabolic products of oxidation, the mechanism of action of CYPs produces super-oxide anion and hydrogen peroxide. During the catalytic cycle they use NADPH to reduce O_2, thereby causing reactive oxygen species production. Alcohol steatohepa-titis, the precursor to cirrhosis, has been linked to oxidative stress-induced damage.

20.2.1b   Flavin-Containing Monooxygenase. FMOs are responsible for the oxygenation of xenobiotics that contain sulfur, nitrogen, and, to a lesser extent, selenium, iodine, boron, phosphorus, and carbon, although oxidation at carbon atoms is rare. The soft-nucleophile substrates of various FMOs are indicated in Table 20.1. Its mechanism is similar to that of CYP, because it utilizes NADPH to reduce molecular oxygen to water, allowing the other oxygen atom to oxidize the substrate; however, unlike CYP, this process is reductase-independent. In addition, FMO does not catalyze carbon oxidations and hydroxylations as CYPs do, despite some overlapping substrate specificities. For example, CYPs activate the breast cancer therapeutic, tamoxifen, through an α-hydroxylation and subsequent sulfonation, whereas FMO catalyzes N-oxygenation, leading to detoxication (see Chapter 10). For the most part, metabolism by FMO leads to detoxication; however, in some

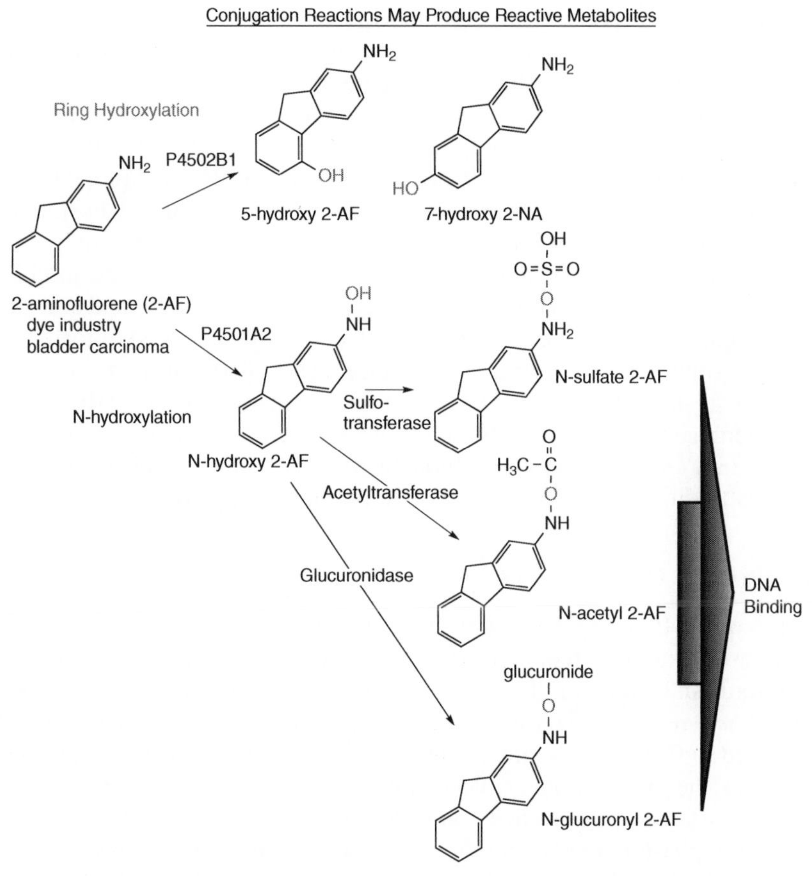

Figure 20.3. Pathways of metabolism of 2-aminofluorene.

instances it may produce a reactive intermediate. For example, FMO catalyzes the *N*-hydroxylation of 2-aminofluorene (Figure 20.3) and is involved in the conversion of 3,3′-iminodipropionitrile to its neurotoxic metabolite, N-hydroxy-3,3′-iminodipropionitrile. FMO also is responsible for the activation of the antifungal agent ketaconazole; its activity correlates with covalent binding of the reactive intermediate to protein and subsequent hepatotoxicity.

20.2.1c Prostaglandin Synthetase. Bioactivation reactions also occur as a result of prostaglandin synthesis. The two-step biosynthesis reaction is catalyzed by the enzyme system prostaglandin synthetase (PGS) (see Chapter 9). Initially, arachidonic acid is oxidized by cyclooxygenase, forming prostaglandin G2 (PGG2), a hydroperoxide. PGH synthetase then catalyzes the reduction of PGG2 to PGH2. During this reduction, xenobiotics may be cooxidized. Because the cooxidation is a consequence of the metabolism of an exogenous compound, substrate-specificity is broad in that any xenobiotic that can accept the direct transfer of the peroxide oxygen can be oxidized in this manner. Acetaminophen, which is activated by CYP2E1, can also be oxidized by PGS to form the reactive quinone, NAPQI. Table 20.1 provides examples of compounds that are activated by PGS. Bioactivation of amphetamines that produce free radical intermediates and cause DNA damage in the CNS have recently been shown to be a result of PGS cooxidation. Inhibition of PGS blocked amphetamine-induced neurodegeneration. Thus, rates of prostaglandin synthesis are physiologically related to the incidence of bioactivation of other chemicals. Therefore, the availability of arachadonic acid, as well as the expression and activity of PHS, influences the incidence of cooxidation reactions. The amount of free arachadonic is mediated by a number of factors including hormones, cell cycle progression, phospholipase, and xenobiotics like TPA. PGS activity is stimulated by an increase in arachadonic acid, however, it may also be inhibited by endogenous and exogenous compounds, including acetylsalicyclic acid and steroids. PGS is thought to convert the sedative, thalidomide, which gained notoriety after a rash of embryopathies revealed it as a developmental toxicant, to a reactive intermediate that results in the production of reactive oxygen species (ROS), causing oxidative damage to DNA and other biomolecules. One piece of evidence supporting a role for PGS in thalidomide bioactivation is that treatment with the PHS inhibitor acetylsalicylic acid decreases thalidomide teratogenicity.

20.2.2 Phase II Conjugations

In general, phase II conjugations have less propensity to produce toxic metabolites than do phase I oxidation reactions. Phase II enzymes typically involve the conjugation of glutathione, glucuronide, sulfate, or methyl groups to the intermediates produced by phase I reactions. Phase II reactions, enzyme regulation, and polymorphisms are described in Chapters 12 and 13. The nature of conjugation generally does not lend itself to the production of electrophilic molecules. In fact, in the previous example of acetaminophen toxicity, glutathione conjugation detoxifies NAPQI; as a general rule, phase II reactions reduce toxicity. In some cases, however, conjugation leads to the production of a reactive intermediate.

Glutathione is perhaps the most important nonenzymatic reducing agent in terms of detoxifying reactive intermediates. It acts as a competing nucleophile, substituting for targets in DNA and protein. In a number of examples of activation reactions by oxidative metabolism, glutathione conjugation is the major mechanism of detoxication; however, there are some instances in which glutathione activates a xenobiotic. One example of glutathione conjugation-induced activation occurs during the metabolism of ethylene dibromide (1,2-dibromoethane). Both oxidative metabolism and glutathione conjugation can cause the formation of electrophilic species. Glutathione addition in particular forms the half-sulfur mustard, 1-bromo-2-S-glutathionyl ethane, which rearranges to the episulfonium ion and ultimately leads to the creation of the DNA adduct S-[2-(N^7-guanyl)ethyl]glutathione. (For more information on glutathione, see Chapter 18).

Sulfate conjugation, or sulfation, has been indicated in bioactivation because sulfate is an excellent leaving group; the removal of sulfate results in the formation of an electrophile, which can bind to DNA and cause mutation and carcinogenesis. Originally, the sulfate-conjugated metabolite of 2-AAF was recognized as necessary for reactivity with DNA. Later, sulfate conjugation was shown to be involved in the reactivity of 7,12 dimethylbenz[*a*]anthrycene and the sassafrass derivitive, safrole. Recently, concern arose regarding bioactivation of tamoxifen, an anti-estrogen utilized in breast cancer therapy, due to its hepatocarcinogenicity in rats. Tamoxifen is

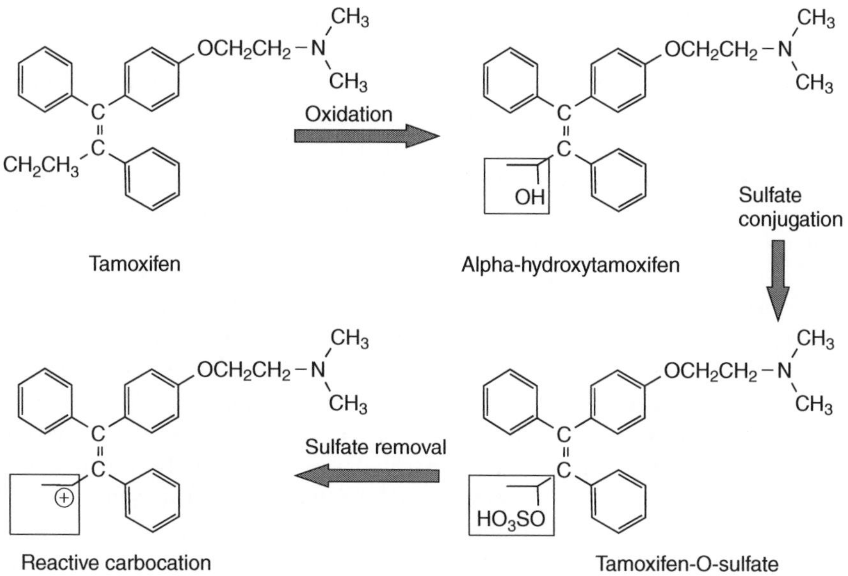

Figure 20.4. Mechanism of bioactivation of tamoxifen.

activated via an initial α-hydroxylation event, along with subsequent sulfonation to a sulfate ester formation, and finally this ester collapses to cause the formation of a carbon radical. If tamoxifen is glucuronidated instead, detoxication occurs (Figure 20.4). Fortunately, human metabolism favors the glucuronidation reaction as opposed to sulfation, which is the primary metabolic conjugation reaction in rats; thus, the potential for hepatocarcinogenicity is extremely low in humans.

Glucuronide conjugation in the case of tamoxifen metabolism is a detoxication reaction. Glucuronidation typically results in the formation of more stable intermediates compared to sulfate conjugations. While sulfation products are labile at pH 7.4, glucuronides are unstable only at extreme pHs. This explains why the N-hydroxy-AAF that results from the glucuronidation reaction is reactive only at higher pHs. Because of this stability, they tend to be rather long-lived and may be capable of being transported to another target site where conditions favor their reactivity. This is the case of N-hydroxy derivatives of aromatic amines formed by an initial oxidation by CYP and subsequent N-glucuronidation. This metabolite is transported from the neutral liver to the acidic bladder, where they are hydrolyzed and converted to nitrenium ions, which are extremely reactive.

20.2.3 Intestinal Microflora

Intestinal microflora are capable of impacting xenobiotic metabolism by causing enterohepatic circulation and delayed excretion and by catalyzing many of the reactions that also occur as a result of detoxication and bioactivation reactions by phase I and II enzymes. The carbohydrate amygdalin, which contains a cyanide substituent, is found in the kernels of various fruits including plum, cherry, peach, and apricot as well as in almonds. Hydrolysis by the β-glucosidases in intestinal bacteria yields reactive intermediates capable of releasing cyanide.

Because individual microbiotic composition varies, metabolic effects due to intestinal microflora vary and may explain differences in individual toxicity. Intestinal microflora is also an important consideration in terms of differences in metabolism as a result of oral exposure versus other routes of administration. In a recent study, germ-free mice had decreased sensitivity to the intestinal toxicity of the chemotherapeutic agent, irinotecan, suggesting that intestinal microflora are significantly involved in the potentiation of the deleterious side effect. Thus, any factors that may affect microflora composition, such as age, diet, GI disease, xenobiotics, and antibiotics, can influence enzymatic metabolism. Glucuronides are predominantly eliminated through biliary excretion; therefore, additional metabolism by intestinal microflora, specifically hydrolysis by the glucuronidases can yield ultimate carcinogens that may target intestinal tissues. Cycasin is a glucoside found in cycad nuts that is activated by this pathway to the ultimate carcinogen, methylazoxymethanol. In addition, recent studies examining infection with the causative bacterium in stomach ulcer, *Heliobacter pylori*, demonstrated a correlation between infection and the incidence of *N*-methyl-*N*-nitrosourea (MNU)-induced gastric cancer. Such bioactivation potential of intestinal microflora, along with the potential toxicity of xenobiotics and pharmaceutics, needs to be an important consideration in the pathogenesis of colon and other GI cancers.

20.3   STABILITY OF REACTIVE METABOLITES

Reactive metabolites typically fall into several catergories based on their half-lives. The stability of a reactive intermediate is an important consideration in terms of its ability to induce damage. Also of significance to the consequence of bioactivation is the distance that the reactive intermediate may be transported. Due to their high degree of reactivity, combined with a resulting propensity to participate in further reactions, most reactive species tend to be somewhat short-lived. We can classify reactive metabolites into three groups based upon their stability and diffusion rate or ability to be transported to other tissues: ultra-short-lived, short-lived, and long-lived.

Ultra-short-lived metabolites typically fall into two categories: reactive oxygen species (ROS) and suicide substrates. The extremely reactive hydroxyl radical (see next section) reacts immediately with adjacent nucleophiles including DNA, protein, and lipid. The diffusion-limited rate of reaction for the hydroxyl radical is between 10^7 and 10^{10} $M^{-1}s^{-1}$. The diffusion distance of the hydroxyl radical is estimated to be 3 nM. Since this is approximately the diameter of many proteins, this dictates that the hydroxyl radical essentially can only react with biomolecules where it was formed. Suicide substrates are different from ROS, because they are produced by the enzyme that metabolizes them, and then they immediately react with the active site of the same enzyme. Olefins and acetylenes are two chemical classes that act as suicide substrates for CYPs. They bind irreversibly to a nitrogen within the heme of CYP, destroying this moiety by causing the release of iron. Heme is the most significant target of binding for suicide substrates because destruction or alteration of this site destroys the activity of the enzyme. There are many examples of enzyme inhibitors binding to the heme of CYP: The reactive intermediate of 1-aminobenzotriazole (1-ABT) inhibits CYP activity by binding heme. 1-ABT is often employed experimentally because of the efficacy of its inhibitory effect. Methapyrilene, an ethylenediamine H1 histamine receptor antagonist that was utilized as an over-the-counter sedative and rhinitis remedy until its use was discontinued due to hepatotoxicity and hepatocarcinogenicity in rats, elicits its damage by binding and inactivating CYP2C11. Methylenedioxyphenyl compounds such as the safrole and isosafrole carcinogens, as well as the well-known insectide synergist piperonyl butoxide, also form a reactive intermediate that irreversibly binds to the heme iron. This effect by piperonyl butoxide helps explain its synergy with insecticides, since insecticides that are metabolized through detoxication reactions catalyzed by the CYP targets of piperonyl butoxide will not be detoxified because of the inhibition. The sulfur released during the CYP-catalyzed oxidation of parathion to paraoxon can bind to the enzyme and destroy its function. Suicide substrates are currently in focus as a potential avenue for addressing the growing problem of resistance by developing drugs that can also destroy the function of the enzymes produced to detoxify them in resistant strains.

Short-lived intermediates do not exhibit widespread transport or diffusivity. The deleterious effects caused by their binding are limited to their point of origin or adjacent cells. In the consideration of damage of a particular target organ, it is logical that potential for damage relates to (a) the amount of enzyme responsible for catalysis of the reactive metabolite available in that tissue and (b) the concentration of the parent compound. This explains why many xenobiotics exert deleterious effects in the liver. The concentration of CYPs in the liver is extremely high compared to other tissues. In addition, a commonly damaged area in hepatotoxicity is the centrolobular region, the region of the liver with the highest concentration of CYPs. For example, acetaminophen-induced necrosis occurs for the most part in the centrolobular region. In this case, not only is the concentration of P450 high, but there is a distinct discrepancy between CYP levels and the amount of glutathione available for conjugation of NAPQI. Peroxynitrite is a relatively long-lived intermediate that has been linked to tumor formation and progression. It forms from reaction of the radical species NO and O_2^- that are enzymatically formed during inflammation. Its half-life is approximately 1 sec, and its diffusion distance is estimated to be 10,000 times that of the hydroxyl radical. There are other reactive intermediates that have even longer half-lives that are transported to target tissues where they cause damage. As a result, some xenobiotics that are bioactivated in the liver cause damage in a totally different tissue. Thus, in some instances of chemical carcinogenesis due to a bioactivation event, the liver is not the site of the cancer. An excellent example of this is benzo[*a*]pyrene. Benzo[*a*]pyrene exposure correlates with the formation of lung tumors despite the fact that its metabolism predominantly occurs in the liver. Studies in which the reactive metabolite, 7,8,-diol-9,10-epoxide, was administered into the intraperitoneal space resulted in the formation of lung tumors, suggesting that it may be transported via the blood to target site. Another potential cause for this downstream target tissue damage may result from differences in the physiology of the different tissues. Fairly stable metabolic products may be formed, however, as they may be transported to tissues with conditions that favor transformation into a different reactive intermediate. The bladder carcinogens, aromatic amines, are *N*-hydroxylated and glucuronidated before being transported to the kidney. In the bladder, acidic conditions cause the release of the glucuronide, and the reactive *N*-hydroxy compound is then available to covalently bind biomolecules.

20.4   FACTORS AFFECTING ACTIVATION AND TOXICITY

Typically, metabolic processes leading to detoxication of the parent compound are favored and activation pathways are minor; however, some conditions cause a shift toward bioactivation events becoming predominant. When an organism is exposed to a large quantity of a xenobiotic, it may saturate or activate certain enzymes, thereby altering the resultant detoxication pathway. A number of physiological and genetic factors, as well as metabolic and chemical interactions, may also favor the activation of a parent compound as opposed to its detoxication.

20.4.1   Saturation of Detoxication Pathways

The intricate and effective system of xenobiotic detoxication requires enzymes and cofactors to chemically alter parent compounds. In some instances the nature of the xenobiotic insult is such that the pathway may become saturated when there is either not enough of the enzyme or its mandatory cofactor available to detoxify all of the parent compound. In this case there may be a shift toward more activation events through minor pathways. Conditions that could lead to saturation include an extremely high level of exposure or exposure to another compound that utilizes the same detoxication system.

An example of enzyme saturation leading to reactive intermediate formation is acetaminophen. Acetaminophen at therapeutic doses is readily conjugated with glucuronide and sulfate, with only a small fraction being metabolized by CYPs. When levels of acetaminophen exceed the capacity of the conjugation system, levels of PAPS and UDGPA are exhausted, resulting in an excess of the toxic product of CYP oxidation, the quinoneimine. Conjugation of glutathione to quinoneimine effectively detoxifies the reactive intermediate leading to its excretion.

20.4.2   Enzyme Induction

As previously mentioned, many of the enzymes involved in xenobiotic metabolism are inducible. Inducibility allows for more enzymatic activity, thereby ensuring an adequate detoxication response; however, it also provides a mechanism whereby an activation pathway may be increased. This occurs in the example given earlier of the combined effects of ethanol and acetaminophen. When CYP2E1 is induced by ethanol prior to administration of acetaminophen, subsequent activation of acetaminophen to NAPQI is prevalent; however, without induction by ethanol, CYP2E1 is not the predominant enzyme for metabolizing acetaminophen, and detoxication is favored. Interestingly, simultaneous administration decreases the toxicity of acetaminophen because both are substrates for 2E1; ethanol acts as a competitive inhibitor, thereby blocking the activation of acetaminophen.

20.4.3   Genetic and Physiological Factors

In addition to enzyme induction or inhibition by various substrates, genetic and physiological factors have a significant impact on xenobiotic metabolism. In most fish species, for example, the CYP1A subfamily consists of one ancestral gene, with two paralogs resulting from a gene duplication event and subsequent sequence conversion events in mammals and birds. In terms of CYP composition, major differences exist between species. Such differences probably result from positive selection pressure. Large variations in expression levels and composition of detoxication enzyme complement exist within a population, and there are also differences in expression from tissue to tissue in individuals. Differences in enzyme composition between species explain why insects are susceptible to the pesticide malathion, while mammals exhibit extremely low sensitivity. Both species catalyze CYP-mediated oxidation to the toxic acetylcholinesterase inhibitor, malaoxon; however, in mammals the predominant route of metabolism is not bioactivation, but rather the hydrolysis of malathion to the malathion monoacid by carboxylesterase. Since insects contain virtually no carboxylesterase, the bioactivation reaction occurs almost exclusively. Another example of such species differences is the resistance of guinea pigs to 2-AAF. Because guinea pigs do not catalyze the hydroxylation, 2-AAF cannot be converted to its carcinogenic form. Even within the same species, there is evidence of altered expression levels of metabolic enzymes due to selection pressure from polluted environments.

Expression levels are controlled by a number of factors in addition to xenobiotic exposures and species differences; diet, physiological factors, and genetic polymorphisms also dramatically impact enzyme composition and function. Endogenous substrates such as hormones, cytokines, and steroids may lead to the induction of enzymes either by activating transcription of the target genes or by enhancing protein synthesis. In addition to the action of an exogenous or endogenous compound in terms of altering enzyme expression, genetic polymorphisms can also substantially alter expression levels as well as enzymatic activity (see Chapters 11 and 13). A number of polymorphisms in CYPs have been observed in the population and have been correlated with poor metabolizer and rapid metabolizer phenotypes, which can dramatically affect the toxicity and efficacy of xenobiotics and pharmaceuticals.

20.4.4 Metabolic Interactions

Metabolic interactions resulting from concurrent detoxication of xenobiotics may modulate the activity of enzymes and thereby result in bioactivation reactions or a decrease in efficacy of a drug. The previous example of induction of CYP2E1 by ethanol and subsequent activation of acetaminophen to NAPQI is a case of metabolic interaction as well, where the mechanism is enzyme induction.

In another example, the nuclear orphan receptor, steroid X receptor (SXR), regulates the induction of CYP3A. Exogenous compounds such as dexamethasone, progesterone, clotrimazole, and phenobarbitol all seem to induce CYP3A through SXR, and CYP3A4 is responsible for the metabolism of aflatoxin B, erythomycin, tamoxifen, and a number of important pharmaceuticals and other xenobiotics. This induction enhances the metabolism of the substrate, thus changing the pharmacokinetics of the therapeutic agent (e.g., erythromycin), but it may also increase potential activations catalyzed by CYP3A4.

20.4.5 DNA Adduct Formation

DNA adducts can be formed as a result of covalent binding of hard electrophiles to DNA during bioactivation. For example, DNA adduct formation has been linked to the carcinogenesis of chemicals found in cooked meats and cigarette smoke. It is believed that covalent binding of the chemical to DNA, thereby creating DNA adducts, may be the critical initiation step in chemical carcinogenesis.

The consequences of covalent modification of DNA bases include alteration of the structure and subsequent processes of replication, transcription, and repair. If the proper DNA conformation is not restored by repair mechanisms (see Chapter 23) and adducts persist, these alterations may cause mutations and ultimately result in cancer development. This becomes especially critical if the deleterious modification is located in a tumor suppressor gene or oncogene. 2-Amino-1-methyl-6-phenylimidazo[4,5-*b*]pyridine (PhIP) is the most mass-abundant heterocyclic amine (food mutagen) found in cooked meats. C8-dG pairs with dC PhIP adduct duplex, disrupting hydrogen bonding of the modified dG.dC base pair. This causes dG and dC bases to be shifted into the major groove. Due to this intercalation, the DNA becomes deformed, with an expanded minor groove and a compacted major groove.

In smokers many adducts form as a result of bioactivation of many of the pro-carcinogenic compounds in tobacco smoke. For example, benzo(*a*)pyrene (BaP) is converted to an epoxide by CYPs with successive conversion to a diol-epoxide by epoxide hydrolase and further oxidation, which confers DNA reactivity (Figure 20.5). 4-(Methylnitrosamino)-1-(3-pyridyl)-1-butanone (NNK) is oxized by P450s, resulting in the formation of alpha-hydroxyNNK. These extremely unstable intermediates are immediately converted into aldehydes and diazonium ions, of which the former binds covalently to DNA. Certain types of adducts can be correlated with cancer incidence, and their detection may serve as a biomarkers of exposure and damage and provide insights into mechanisms of carcinogenic compounds.

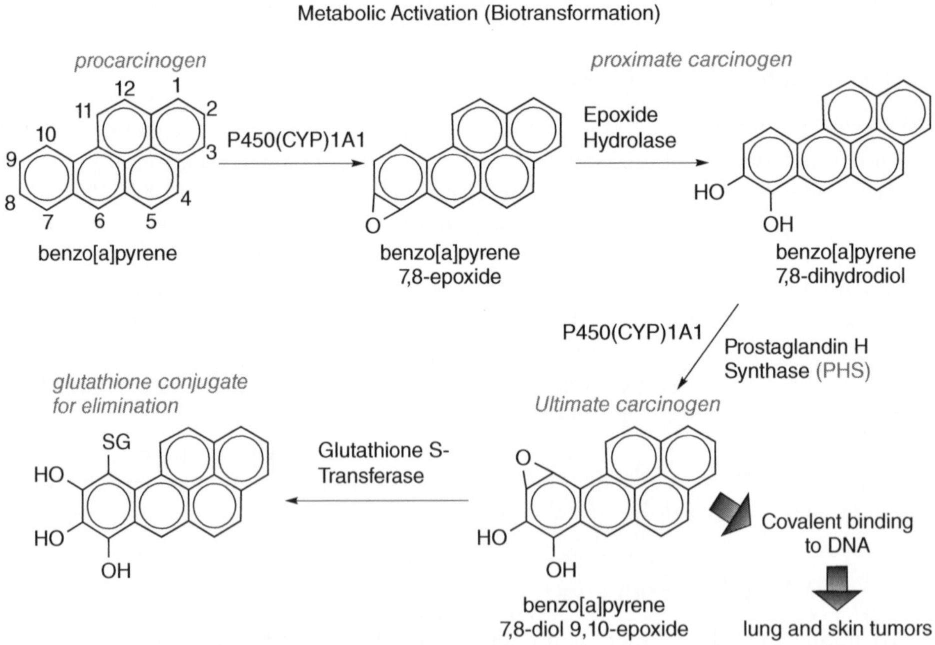

Figure 20.5. Mechanism of bioactivation of benzo(*a*)pyrene.

In addition to DNA adducts that occur as a result of covalent binding of reactive intermediates generated by oxidation or conjugation of parent compounds to DNA, reactive oxygen species produced during xenobiotic metabolism can also react with nucleophilic biomolecules.

20.4.6 Redox Cycling

Redox cycling describes the reversible transfer of electrons between two substances. This transfer results in an increase in oxidation number of the electron donor, also known as the reducer. Simultaneously, the oxidant gains electrons, thereby becoming reduced. Oxidants typically contain elements with high oxidation numbers or electronegativity (e.g., H_2O_2, O, F, Cl, Br), which are capable of gaining electrons. Conversely, the reducer, or electron donor, typically is more nucleophilic. Nucleophilic strength indicates the propensity of a compound to donate an electron pair, thus, reductants are varied and include electropositive metals, including Fe^{2+} and Cu^+. Some predictions of the potential role of a compound in redox reactions can be made if we consider the constituent elements and their location on the periodic table. In general, nucleophilicity decreases and electronegativity increases as we move left to right down a row on the periodic table. Also, because size increases as we move from the top of a column to the bottom, nucleophilicity increases because electrons are held more loosely due to the radius increase.

Redox reactions are ubiquitous and occur in numerous biological processes. In mitochondrial respiration, ATP production is coupled to electron transport. Electrons are transferred from complex I, II, III, and IV, through a series of redox reactions that create a proton gradient outside of the mitochondrial membrane allowing for the production of ATP. Redox reactions are also extremely important in metabolism. CYPs and FMOs rely on redox reactions for their catalytic function (see Chapter 10).

The metabolism of quinone-containing chemotherapeutic drugs, such as adriamycin, involves enzymatic reduction by CYP reductase of the quinone by one electron (Figure 20.6). This reduction results in the formation of the adriamycin semiquinone radical. The electron transfer from semiquinone to oxygen reforms the original quinone. This reduction by a reductase, along with subsequent oxidation by molecular oxygen, comprises redox cycling.

20.4.7 Target Organ Toxicity

Reactive metabolites may cause toxicity in target organs due to transport, or also because of their chemical character and the conditions in a particular microenvironment that may maximize reactivity, as well as the susceptibility of the site to damage. Specific examples of target organ toxicity are described in detail in Chapters 27–35.

20.5 REACTIVE OXYGEN SPECIES AND TOXICITY

As discussed in Section 20.4.6, many bioactivation reactions can lead to the generation of reactive oxygen species (ROS) through reactions with molecular oxygen. Like

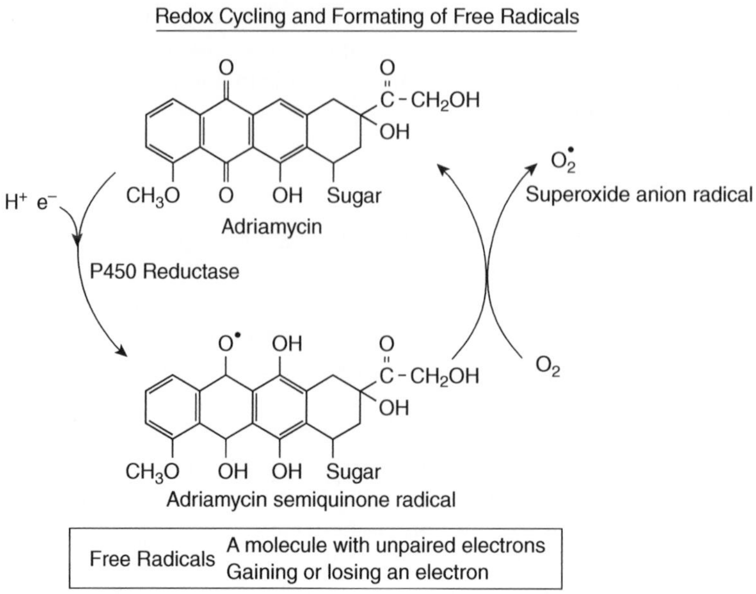

Figure 20.6. Redox cycling of adriamycin and free radical formation.

other electrophilic reactive intermediates, ROS also damage lipids, DNA, and protein. Because they are ubiquitous and potentially harmful, efficient detoxication systems have evolved to maintain a healthy cellular oxidative status. ROS in excess of the capabilities of detoxication systems causes oxidative stress in the cell. Such oxidative stress has been linked in recent years to the pathogenesis of a number of neurodegenerative diseases and to cancer, as well as to the aging process in general.

20.5.1 Generation and Detoxication of Reactive Oxygen Species

Free radicals are chemicals with unpaired electrons, which are extremely reactive, making them likely to participate in chemical reactions. Their reactions and covalent binding to DNA, lipids, and protein lead to deleterious effects. The formation of reactive oxygen species, namely superoxide anion, hydrogen peroxide, singlet oxygen, nitric oxide, and the hydroxyl radical, are either necessary for, or byproducts of, aerobic metabolism. The primary event that converts oxygen from an inert ground state is a change in "electron spin pairing." Normally, molecular oxygen is in a triplet state, $3O_2$, with parallel spins of each unpaired electron on the oxygen atom; this conveys nonreactivity. Many chemical mechanisms can cause a change in spin pairing, namely an increase in one electron to a higher energy level, creating antiparallel spins in unpaired electrons. This may be caused by UV radiation, or endoperoxides, and can produce singlet oxygen; a one electron reduction of an atom of oxygen by metals and nucleophiles can produce superoxide anion (O_2^-); oxygen can be enzymatically activated as well; chemical reactions resulting in the removal of an electron or hydrogen can produce peroxyl radicals (Figure 20.7).

Metabolic pathways such as the electron transport chain of mitochondrial respiration and biochemical reduction of oxygen by enzymes of xenobiotic metabolism,

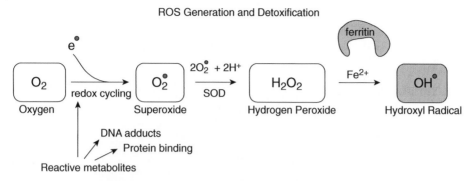

Figure 20.7. Mechanism of reactive oxygen species generation and detoxification.

TABLE 20.2. Sources of Reactive Oxygen Species

Enzymatic catalysis (oxidation, peroxidation)
Growth factors
Insulin
Interleukins
Metal catalysis (e.g., Fenton reagent)
Oxidative burst in immune response
Redox cycling (e.g., paraquat, dopamine, quinines)
TNFα
Ultraviolet light

as well as inflammation and phagocytosis, can lead to the production of ROS (see Table 20.2 for sources of ROS). A major source of reactive oxygen species is mitochondrial electron transport. In this process, NADH and flavoprotein-linked dehydrogenases donate electrons to reduce oxygen to water during the production of ATP. The reduction of oxygen to water is a four-electron transfer by cytochrome c oxidase, and it occurs following serial electron transfers along the inner mitochondrial membrane.

Signaling pathways, likewise, play a role in the production of ROS: Inflammatory cytokines such as some interleukins and tumor necrosis factor alpha, as well as growth factors including epidermal growth factor, plasma-derived growth factor, and insulin, also lead to increased amounts of ROS in the cell via flavin-containing enzymes such as NADPH oxidase. Finally, exogenous stressors such as oxidants (e.g., peroxide), redox cycling compounds such as paraquat or quinones, UV radiation, and hyperoxia can produce excess ROS.

There are several different species of ROS that have various abilities to produce deleterious effects in the cell, including the hydroxyl radical and peroxynitrite. The superoxide anion, O_2^{-}, is capable of producing H_2O_2, which can cause damage or lead to the production of the hydroxyl radical, $OH^{\bullet}$, which is the most toxic ROS. Metals serve a catalytic role in the production of the hydroxyl radical. For example, iron is able to increase the reactive potential of H_2O_2 by allowing for Fenton or Haber–Weiss reactions to occur, resulting in (a) an increase in oxidation of the metal and (b) a reduction of the H_2 to form the hydroxyl radical (Figure 20.8). Because

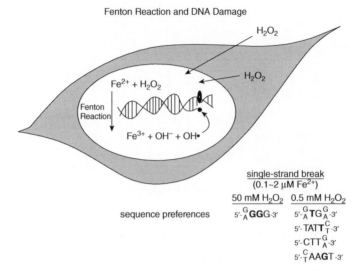

Figure 20.8. Iron-mediated generation of reactive oxygen species and generation of single-strand DNA breaks. (Adapted from Henle et al. *J. Biol. Chem.* **274**, 962–971, 1999).

of the propensity of ROS to react with and damage important biomolecules, cells have evolved biochemical systems for the detoxication of such reactive species. Antioxidant compounds and enzymes capable of reducing such molecules protect the cells and help maintain the redox balance.

Antioxidant compounds are an important defense for immediate detoxication of highly reactive intermediates. They act as competing nucleophiles and can bind to the intermediate, forming a less reactive species. Glutathione is one antioxidant molecule that can directly interact with free radicals. Other chemical antioxidants include vitamins A (retinol), C (ascorbic acid), and E. Ascorbic acid, for example, may react directly with reactive intermediates by hydrogen abstraction, resulting in the formation of dehydroascorbic acid.

Enzymatic antioxidants are also critically important to limit intracellular ROS levels. In the case of the superoxide anion, superoxide dismutases (SOD) convert the anion into H_2O_2, which is then altered by catalase or glutathione peroxidase to produce water. The conversion of the superoxide anion to H_2O_2 is necessary for the detoxication process; however, H_2O_2 can be harmful since, as we mentioned earlier, it can be converted to the more hydroxyl radical, as well as causing damage by itself. Mammals have several forms of SOD including copper–zinc SOD, SOD3 (extracellular), SOD1 (cytosolic), and manganese SOD (mitochondrial intermembrane).

Damage by free radicals or ROS occurs under conditions of oxidative stress. Normally, antioxidant defenses are able to counter ROS production and thereby block damage; however, oxidative stress can occur when (a) there is an imbalance between the detoxication pathway, which utilizes enzymes and antioxidant compounds to convert ROS to benign molecules, (b) there is either malfunctioning or deficient in some way, or (c) the amount of ROS overloads the system (Figure 20.9).

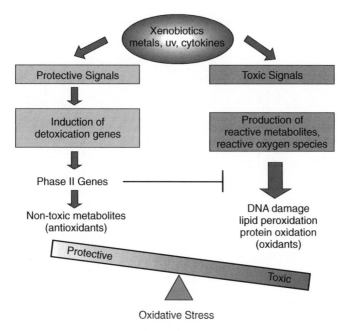

Figure 20.9. Protective and toxic signal balance in xenobiotic metabolism and generation of oxidative stress.

20.5.2   Oxidative DNA Damage

Among the damage caused by ROS in conditions of oxidative stress is oxidative modification of DNA, either by binding and chemically altering bases or by inducing conformation changes in DNA that cause strand breakage (see Chapter 22). Such alterations can lead to mutations and problems with DNA replication, unless they are fixed by DNA repair mechanisms (see Chapter 23). Oxidative base modifications are indicated in Figure 20.10.

The hydroxyl radical reacts with DNA, causing single-strand breaks and base alterations, including DNA adducts and mutations. It also causes damage to membranes and proteins through lipid peroxidation and DNA protein crosslinks (see Section 20.5.3). Reactions of the hydroxyl radical with DNA occur by removal of a hydrogen from any of the five carbons of deoxyribose, leading to the formation of a carbon radical. A number of subsequent reactions involving the carbon radical are then possible: Oxygen may react with it, forming peroxyl radicals. These peroxyl radicals result in the destruction of a carbon–carbon bond, causing a sensitive site for strand breaks. Conversion to an oxyl radical at C5 causes strand breakage and loss of a base. In addition, a hydroxyl radical also binds directly to heterocyclic DNA bases. It can add to the C5–C6 double bond in pyrimidines, thereby creating two base radicals, including 5-hydroxy-6-yl and 6-hydroxy-5-yl. In purines, it can bind to C4, C5, and C8 positions, producing similar adducts, and in the case of C8-hydroxyl, radicals of purines also disrupt the imidiazole ring. This is one pathway by which 8-oxo-deoxyguanosine arises. This DNA adduct also forms as a result of generation of free radical intermediates of amphetamines.

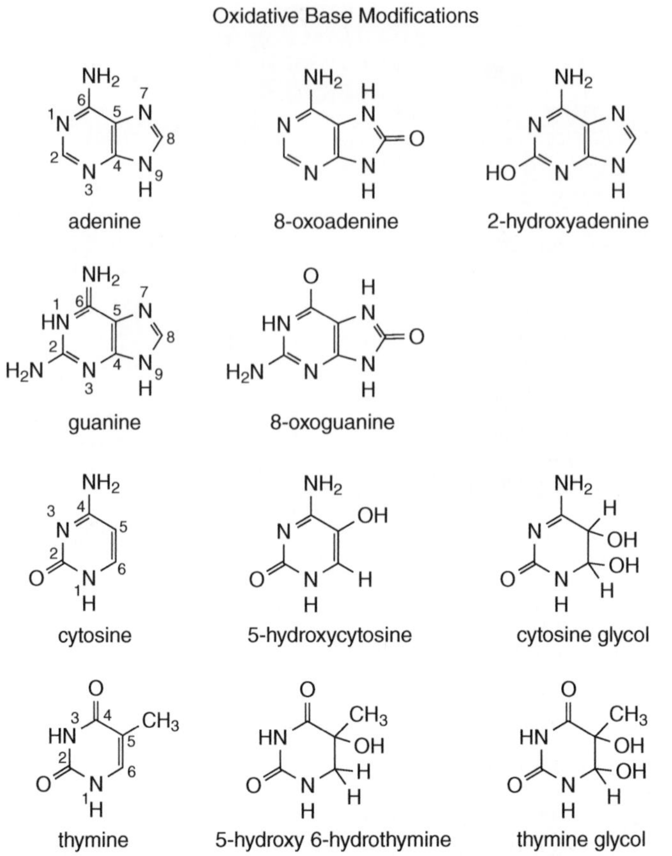

Oxidative Base Modifications

Figure 20.10. Summary of oxidative base modifications.

In contrast to the hydroxyl radical, which has no preference for DNA binding and interacts with all bases and deoxyribose, peroxynitrite preferentially reacts with guanine and therefore primarily causes the formation of 8-oxo-deoxygaunosine adducts. Interestingly, this particular DNA adduct is also a target for further oxidation by peroxinitrite, so the presence of such adducts creates the predisposition for accumulation of adducts. Furthermore, this adduct will pair with deoxyadenine, causing a G-to-T transversion during replication (Figure 20.11).

Free iron and H_2O_2 can induce Fenton reaction-mediated single-strand breaks. Interestingly, there appears to be a sequence preference for the site of the DNA damage (Figure 20.8). These include RTGR, TATT, CTTR, and AAGT, and sequences that are found within the human telomere (5'-TTAGGG-3'). This implies that there may be some connection between iron and oxidative damage in aging.

Another consequence of DNA oxidation is the formation of DNA protein cross-links between carbon radicals and the carbon chains of amino acid residues. This is problematic for two reasons: First, it leads to errors in DNA replication, and second, it disrupts the structure and function of the protein.

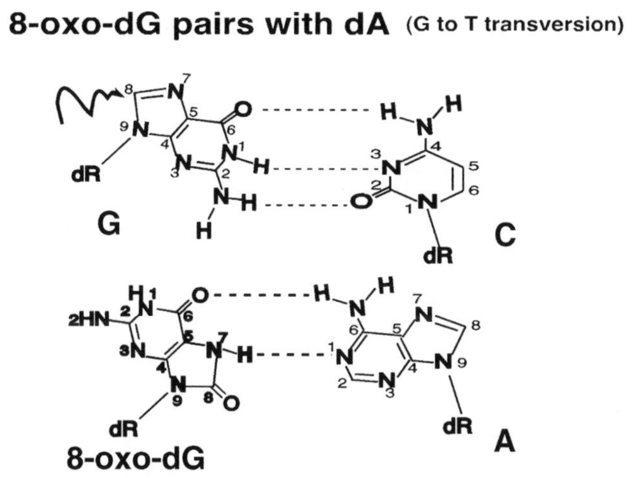

8-oxo-dG pairs with dA (G to T transversion)

Figure 20.11. Specific mechanism of oxidative DNA damage by 8-oxo-dG binding to dA.

20.5.3 Protein Oxidation

In addition to DNA protein crosslinking, protein can also be the direct target of oxidative damage. There are many deleterious modifications of protein that can be classified as several types of reactions. The hydroxyl radical can abstract a hydrogen from any of the carbons in the amino acid, forming a carbon radical that may subsequently be converted into an alkyl-peroxyl radical. This radical is susceptible to other reactions that eventually lead to cleavage of peptide bonds. Peptide bond cleavage may occur through acid hydrolysis, or through oxidation of glutamyl side chains. The formation of carbon radicals can also lead to protein crosslinking if two are present. Oxidation of sulfhydryl groups can also lead to protein crosslinking (Figure 20.12). One of the consequences of protein crosslinking is that the crosslinked protein is resistant to proteosomal degradation. Its activity may also be inhibited. This mechanism of proteosomal inhibition has been implicated in many age/oxidative stress-related diseases, and accumulation of oxidatively damaged proteins have been noted in many disorders including, but not limited to, Parkinson's disease, Alzheimer's disease, progeria, rheumatoid arthritis, and amyotrophic lateral sclerosis.

Protein oxidation was once thought to be simply a consequential damage to cellular components following the accidental production of oxidative stress. However, examples of functional generation of ROS have also been identified. For instance, ROS is produced in the oxidative burst of phagocytes. The phagocyte NADPH oxidase is inactive until infection or inflammation activates it, and activation causes production of ROS. Recently, enzymes that catalyze the formation of ROS have been identified in nonphagocytic cells, including NADH oxidase (NOX) and dual oxidase (DUOX), which are both capable of producing ROS in a controlled manner. Thus, it seems that in addition to causing cell damage, ROS may also be involved in some biological functions. It appears that some oxidative modifications of protein may serve a signaling function and may be a reversible modifier of enzymatic activity. Cysteine residues are substrates for different forms of oxidation, including some

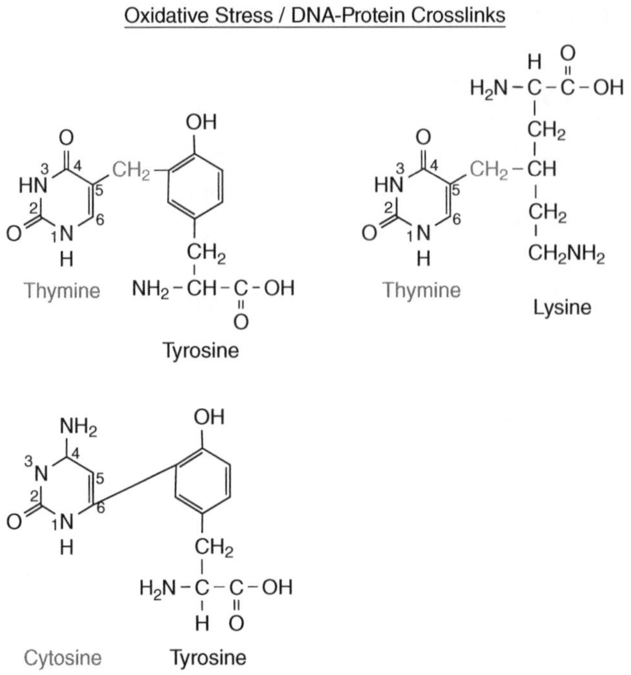

Oxidative Stress / DNA-Protein Crosslinks

Figure 20.12. Formation of DNA protein crosslinks in oxidative stress. (Adapted from Williams, G. M., and Jeffrey, A. M. *Reg. Tox. Pharm.* **32**(3), 283–282, 2002).

modifications that are reversible, such as disulfide formation, glutathionylation, and *S*-nitrosylation.

While oxidoreduction of proteins was previously viewed as damage under conditions of oxidative stress, recently there is mounting evidence that redox modifications of protein thiols/disulfides may regulate the activity of the protein. DNA binding of the AP-1 family transcription factor, Fos-Jun, was altered by a redox change caused by a single conserved cysteine residue in the DNA-binding domains of the two proteins. The oxyR gene positively regulates oxidative stress-responsive genes in bacteria, and it is also a sensor and regulator of oxidative stress, such that only OxyR with oxidized cysteine activates transcription of antioxidant genes. Recently, another sensor of oxidative stress has been identified in *Bacillus subtilus*. The PerR transcription factor is activated by the addition of an oxygen to a histidine residue. This reaction is catalyzed by ferrous iron and is a sensing mechanism for H_2O_2. Thus, protein oxidations that occur as a result of oxidative stress may serve an important role in activating antioxidant systems, however, the accidental modification of other proteins causes damage that may be involved in aging and the pathogenesis of certain disease states.

20.5.4 Lipid Peroxidation

Lipids do not escape damage by reactive intermediates. Lipids are oxidized in number of ways. They may be oxidized enzymatically, or via free-radical interaction (enzyme independent), or they can also be oxidized in a free-radical-independent,

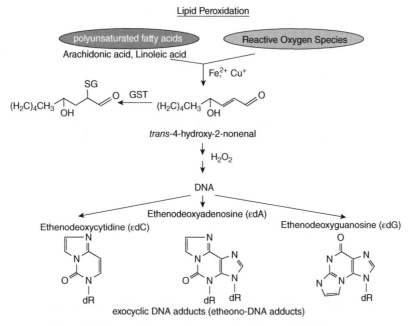

Figure 20.13. Lipid peroxidation and oxidative DNA damage.

nonenzymatic manner. Lipid oxidation may be catalyzed by lipoxygenases, cyclo-oxygenases, and CYPs. Several CYP families have demonstrated the ability to catalyze formation of hydroxycholesterol and side-chain oxidation products. Free-radical-mediated peroxidation of polyunsaturated fatty acids (PUFA) can take place through distinct mechanisms (Figure 20.13). In the first, hydrogen is abstracted from PUFA to the reactive intermediate, creating a carbon-centered radical. This lipid carbon radical may then react with oxygen, giving rise to the peroxyl radical. This reaction may also be reversed, releasing oxygen. If the PUFA contains more than three double bonds, then it may be able to undergo cyclization of the peroxyl radical, since it is capable of rearrangement. Antioxidant treatment typically reduces the formation of lipid peroxidation products. It is thought that oxidative modification of low-density lipoprotein ("bad cholesterol") may play a role in the development of artherosclerosis, but lipid oxidation also has other deleterious effects, including (a) damage of biological membranes and binding to DNA, creating etheno-DNA adducts, and (b) ability to cause the production of more ROS and resulting oxidative damage to other macromolecules.

Damage to critical biomolecules caused by ROS accumulates as the reactive oxygen increases and antioxidant defenses are overwhelmed, resulting in oxidative stress. The oxidative stress theory of aging suggests that ROS induces increasing damage that leads to the functional degeneration associated with aging (Figure 20.14). The level of superoxide negatively influences life spans of *Drosophila* and *Caenorhabditis elegans*. Overexpression of the antioxidant enzymes SOD and cata-lase, as well as treatment with the glutathione precursor *N*-acetlycysteine and the antioxidant melatonin, increased *Drosophila* life span. Generation of increased amounts of reactive oxygen from mitochondrial respiration, especially superoxide

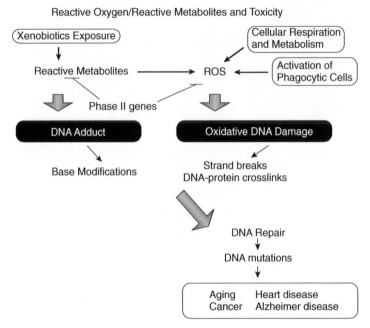

Figure 20.14. Consequences of oxidative damage.

anion and H_2O_2, correlates with advanced age. Targets of damage caused by increased levels of ROS may mediate the age-associated decline in functionality. Protein oxidation that damages the proteosomal degradation system may lead to the accumulation of misfolded proteins that is observed in some neurodegenerative conditions, such as Parkinson's disease and Alzheimer's disease. Oxidized proteins lose catalytic function and often cannot be degraded. Mutations that result from DNA adducts may alter the genes coding for important regulatory proteins, including tumor supressors and oncogenes, thereby playing a causative role in carcinogenesis. Furthermore, mutations may occur in the coding regions of antioxidant enzymes as well, increasing the potential for oxidative stress. The shortening of telomeres that occurs during aging may be partially a result of Fenton reaction-mediated strand breaks. Lipid oxidation products are also associated with cardiovascular disease and have the ability to catalyze more ROS-mediated damage and apoptosis. Thus, it seems that oxidative stress may be an important generalized mechanism of damage over time, which leads to aging as well as to the increased prevalence of cancer, neurodegeneration, and cardiovascular ailments in advanced stages of life. Future studies that elucidate clear mechanisms of controlling the redox state of cells may provide new strategies for preventing or delaying some of this macromolecular damage and its resulting deleterious effects.

SUGGESTED READING

Aust, A., and Eveleigh, J. Mechanisms of DNA oxidation. *Proc. Soc. Exp. Biol. Med.* **222**, 246–252, 1999.

Biswas, S., Chida, A., and Rahman, I. Redox modifications of protein-thiols: Emerging roles in cell signaling. *Biochem. Pharmacol.* **71**, 551–564, 2006.

Gonzalez, F. Role of cytochromes P450 in chemical toxicity and oxidative stress: Studies with CYP2E1. *Mutat. Res.* **569**, 101–110, 2005.

Hecht, S. Tobacco carcinogens, their biomarkers and tobacco-induced cancer. *Nat. Rev. Cancer* **3**, 733–745, 2003.

Krueger, S., and Williams, D. Mammalian flavin-containing monooxygenases: structure/function, genetic polymorphisms and role in drug metabolism. *Pharmacol. Ther.* **106**, 357–387, 2005.

Lambeth, J. Nox enzymes and the biology of reactive oxygen. *Nat. Rev. Immunol.* **4**, 181–189, 2004.

Niki, E., Yoshida, Y., Saito, Y., and Noguchi, N. Lipid peroxidation: Mechanisms, inhibition, and biological effects. *Biochem. Biophys. Res. Commun.* **338**, 668–676, 2005.

Park, B., Kitteringham, N., Maggs, J., Pirohamed, M., and Williams, D. The role of metabolic activation in drug-induced hepatotoxicity. *Annu. Rev. Pharmacol. Toxicol.* **45**, 177–202, 2005.

Sohal, R. Role of oxidative stress and protein oxidation in the aging process. *Free Radi. Biol. Med.* **33**(1), 37–44, 2002.

Stadtman, E. Protein oxidation in aging and age related diseases. *Ann. NY Acad. Sci.* **928**, 22–38, 2001.

CHAPTER 21

Metals

DAVID B. BUCHWALTER

Department of Environmental and Molecular Toxicology, North Carolina State University, Raleigh, North Carolina 27695

21.1   INTRODUCTION

Eighty of the 105 normally occurring elements in the periodic table are metals (Figure 21.1). As living cells evolved, metal ions became inexorably involved in a wide range of physiological processes. Metals can be important as bulk nutrients (Ca, Na, K, and Mg) or as trace nutrients (As, Cr, Co, Cu, Fe, Mn, Mo, Ni, Se, Si, Sn, V, and Zn). Metals also play fundamental roles in the function of several proteins. Zinc finger proteins and the iron associated with heme-containing proteins (e.g., cytochrome P450s, hemoglobin) are two prominent examples. Pathologies can develop if metals essential to biological functions are not appropriately regulated. Both deficiencies and excesses in essential metal concentrations in cells can have physiological consequences. Other metals such as lead and mercury have no physiological functions, and they can interfere with a host of processes via associations with biological macromolecules or by mimicking or otherwise affecting essential metals. As elements, metals are neither created nor destroyed by humans. However, changes in the concentrations of lead in arctic ice cores, along with global mercury transport and deposition, clearly show how human uses of metals have increased their concentrations in our immediate environment. The widespread use of metals in industrial processes, consumer goods, and even medical treatments makes exposure to metals unavoidable.

There are few broad generalizations that can be safely made about the toxicology of metals. However, metals do tend to be highly reactive in biological systems and interact with numerous macromolecules. In biological systems, the more toxic metals typically only fleetingly exist in their free ionic forms and are usually complexed with proteins and other macromolecules. Sulfhydryl, carboxyl, and phosphate groups are common targets of metal binding, as are amines and various anions including chloride anions. This characteristic leads to their potential effects in several different organ systems and biochemical pathways. Because essential metals are so prevalent in biological systems, and nonessential metals can act as mimics of essential metals, the array of potentially toxic interactions that metals can participate in is staggering. A comprehensive treatment of the biochemical toxicology of

Molecular and Biochemical Toxicology, Fourth Edition, edited by Robert C. Smart and Ernest Hodgson
Copyright © 2008 John Wiley & Sons, Inc.

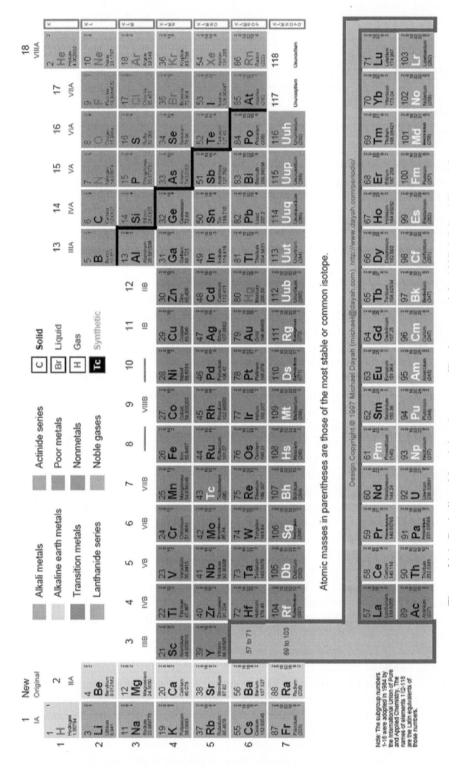

Figure 21.1. Periodic table of the elements. (Design copyright 1997, Michael Dayah.)

metals would fill several volumes. Writing a short chapter on the toxicology of metals is a daunting task due to the massive amount of important information that must be omitted. This chapter is intended to serve as a basic introduction to the biochemical toxicology of metals and is not intended to be comprehensive.

21.1.1 Importance of Essential Metals in Physiological Systems

To highlight the importance of trace metals in biological systems, we'll first move across the periodic table of the elements, providing examples of essential metal functions. This is intended to provide a brief overview into the variety of functions that metals can play. In part, this wide range of critical essential metal functions provides the basis for the wide range of toxicological interactions that metals can have with biological systems.

The alkali metals (groups 1A of the periodic table) include Na and K, metals that are common both in the environment and in biological systems. Because group 1A metals only contain one electron in their outer shell, they can only form ionic bonds. Na^+ and K^+ are major cellular electrolytes that regulate metabolic and signaling processes via concentration gradients across cell membranes. In mammalian cells, the intracellular concentration of Na^+ is typically 10 mM, whereas $[K^+]$ is typically 145 mM. Conversely, extracellular $[Na^+]$ is typically 145 mM, whereas $[K^+]$ is typically 5 mM. Aside from its importance in maintaining concentration gradients across membranes, potassium also plays important roles as a cofactor in several enzyme systems, including pyruvate and aspartate kinases. Potassium appears to bind to oxygen in the activation of these enzymes.

The alkaline earth metals (group 2A of the periodic table) include Mg and Ca, which play both structural and physiological roles. Aside from its structural importance in bones and teeth, calcium is critical in processes ranging from vascular tone, nerve impulse transmission, muscle contraction, blood clot formation, the secretion of hormones such as insulin, and cell signaling. Calcium levels in cells, blood, and extracellular fluid are very tightly controlled. If calcium intake is insufficient, calcium is liberated from bones in order to support these physiological functions.

There are no known biological functions for any of the transition metals in groups 3B and 4B. In group 5B, vanadium is the only element with known biological function, although these functions are not well-characterized because its total mass in the human body is trace (~2 mg V) relative to more common elements (~250 g K). Vanadium can exist in three oxidation states: +2, +4, and +5. Vanadium deficiency in mammals is associated with poor reproductive performance and impaired bone development. Recent work has demonstrated that vanadium has insulin-like properties. The mechanism responsible for the insulin-like effects of vanadium compounds involve the activation of insulin-signaling pathways including (a) the mitogen-activated protein kinases (MAPKs) extracellular signal-regulated kinase 1/2 (ERK1/2) and p38MAPK and (b) phosphatidylinositol 3-kinase (PI3-K)/protein kinase B (PKB).

Group 6B includes both chromium and molybdenum as essential trace elements. The oxidation state of Cr is of critical importance to its biological activity. The most common oxidation state is Cr^{3+} or Cr(III). Chromium(VI) is used in electroplating and in other industrial applications and is recognized as a carcinogen when inhaled. The essentiality of Cr was discovered when weaning rats were fed Cr-deficient diets, and subsequently developed glucose intolerance. It is still unclear what the precise structure of biologically active Cr is; however, it is suggested that a low-molecular-weight chromium-binding substance is involved in insulin–insulin receptor interactions.

.Of the group 7B metals, only manganese is recognized as an essential element. Mn can also be toxic in biological systems. Mn can exist in several oxidation states, but appears to most commonly occur in biological systems as Mn(II), where it is involved as a constituent of some enzyme systems such as manganese superoxide dismutase (MnSOD)—the most important antioxidant in mitochondria. Interestingly, Mn(II) is similar in size and charge density to Mg(II) and Zn(II), and it can replace these metals in some enzyme systems with little effect on enzyme activities. Mn also activates many enzymes involved in the metabolism of carbohydrates, amino acids, and cholesterol, and it is involved in cartilage and bone development as well as wound healing. Mn toxicity is associated with neurological problems, with inhalation as the primary cause of these problems. This is because inhaled Mn is transported directly to the brain before the liver has an opportunity to process it. Dietary Mn toxicity has been observed in humans who drank contaminated water, or via dietary supplements.

Columns 8–10 of the periodic table contain the group 8B elements. Iron is a group 8B metal with several essential functions. Iron occurs in biological systems in two oxidation states, Fe(II) and Fe(III), making it ideally suited as a cofactor in oxidation–reduction reactions. Iron forms the central structural component of heme, a compound found in several important biomolecules including hemoglobin and cytochromes (Figure 21.2). Approximately two-thirds of the body's iron is found in hemoglobin. The main function of heme is the retention and delivery of O_2 for enzymatic reactions. Iron is absorbed in the intestine as +2, and it is oxidized to the +3 state by apoferritin. Ferritin is the major iron storage protein in mammals.

Group 11 (or 1B) contains copper, which is the third most common transition metal found in biological systems. Copper in solution has two stable oxidation states, cuprous (Cu^{1+}) and cupric (Cu^{2+}) ion. The ability of copper to easily accept and donate *electrons* explains its important role in oxidation–reduction reactions and

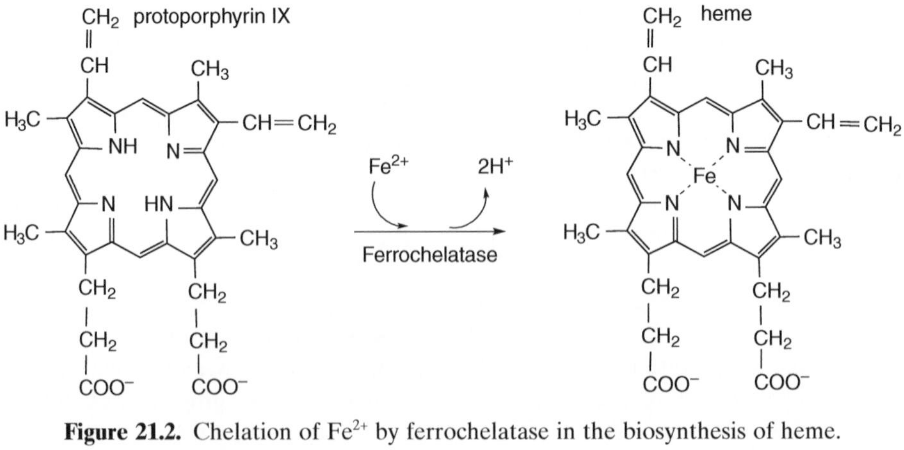

Figure 21.2. Chelation of Fe^{2+} by ferrochelatase in the biosynthesis of heme.

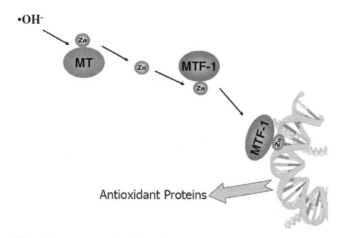

Figure 21.3. Zinc as an antioxidant. (Image courtesy of Christer Hogstrand.)

the *scavenging* of *free radicals*. Important cuproenzymes include cytochrome c oxidase (energetics), lysyl oxidase (connective tissue), ferroxidase I and II (iron metabolism), tyrosinase (eyes, skin, hair), and dopamine-b-hydroxylase (brain). Copper can be toxic to cells if present at elevated concentrations. Excess copper must be sequestered by proteins such as metallothioneins (see below) to prevent toxicity from occurring.

Group 12 (or 2B) contains zinc, which, after iron, is the second most common transition metal found in biological systems. Zinc function can be structural, regulatory, or catalytic. Zinc use in physiological systems is limited to the Zn(II) state. Therefore, we don't find zinc itself involved in redox reactions as we do with copper or iron. Zinc-containing enzymes such as alcohol dehydrogenase and superoxide dismutase do perform catalytic oxidations of their respective substrates, but Zn(II) is not a redox cycling cofactor. Zinc finger proteins are common transcription factors, with over 4500 described in humans to date. Zinc may also play a role as a mediator of antioxidant responses in cells (Figure 21.3). Zinc also plays a pivotal role in apotosis (Figure 21.4) via interactions with caspases. Zinc also is involved in hormone release and cell signaling. The regulation of Zn concentrations in cells is, in part, regulated by metallothioneins, an important family of proteins that will be discussed in further detail below.

21.1.2 Examples and Causes of Essential Metal Deficiency

Given the broad spectrum of important functions that essential metals perform in biological systems, it should not be surprising that essential metal deficiency can cause significant problems. Deficiencies can arise if the supply of essential metals are lacking in the diet, as a result of impaired regulation or as result of exposure to nonessential toxic metals.

The most common essential metal deficiency in the United States is that of iron. Iron-deficiency anemia is the result of inadequate iron availability, leading to decreased hemoglobin production. Clinically, this is recognized by smaller red blood cells with lower hemoglobin content than normal red blood cells. Other nutritional

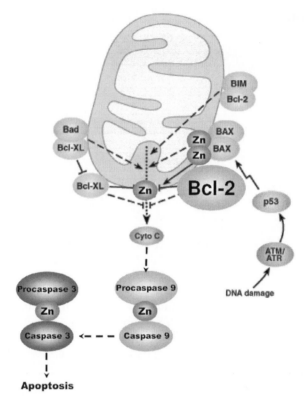

Figure 21.4. Zinc and apoptosis. (Image courtesy of Christer Hogstrand.)

factors such as folic acid and vitamin B12 can play a role in iron-deficiency anemia. Typically, essential metal deficiencies are generally not due to a lack of dietary supply as is the case with iron, but instead involve more complex physiological pathways. For example, calcium deficiency is usually a result of parathyroid function rather than dietary levels of intake.

Some of the pathologies associated with trace metal deficiencies have been understood via the study of genetic disorders involving trace metal regulation. *Menke's disease* is an X-linked recessive disorder associated with copper deficiency. More specifically, the disease is the result of mutations to the gene ATP7A, which encodes a copper transporting P-type ATPase. Mutations in this gene result in the accumulation of Cu in the intestine and kidneys, because the transport function of the protein is either fully or partially abolished, depending on the severity of the mutation. This failure to transport Cu leads to systemic Cu deficiency. Approximately one in 200,000 males is born with this disorder, which can result in severe mental retardation and connective tissue abnormalities. Unusually kinky white hair is one clinical sign that is used for diagnosis. Few baby boys survive past three years of age in severe cases, which are the result of frameshift mutations that render this copper transporting protein completely ineffective. Other point mutations result in a condition known as occipital horn syndrome, which also produces similar symptoms but are not as rapidly lethal.

A classic example of essential metal deficiency resulting from nonessential metal exposure is *Itai itai* disease. Cadmium pollution in the Jinzu River basin in Japan resulted in severe nephrotoxicity in approximately 184 people. Renal tubule damage caused excessive loss of electrolytes and small proteins from the urine. In severe cases, urinary Ca loss was so severe that bone Ca was mobilized, resulting in osteomalacia. Renal tubular defects persisted for life and induced hypophosphatemia, hyperuricemia, and hyperchloremia, which are characteristic biochemical features of Itai-itai disease (see Section 21.6.1).

21.1.3   Examples and Causes of Essential Metal Excess

Copper toxicity due to dietary consumption is rare in the general population. Acute copper poisoning has occurred through the contamination of beverages by storage in copper-containing containers as well as from contaminated water supplies. However, a genetic disorder, *Wilson's disease*, results in excessive copper in the liver, kidney, and brain (and, to a lesser extent, other organs). This autosomal recessive disorder is the result of mutations to the ATP7B gene, which is similar (56%) to the ATP7A gene associated with Menke's disease. Both code for copper transporting ATPases. The major difference between these disorders lies in the tissue-specific expression of the proteins. In the case of Wilson's disease, mutations in the ATP7B gene result in impaired biliary excretion of copper, which is the major pathway for excess Cu elimination.

Iron is another essential metal that can overload the body as a result of genetic disorders. Hereditary hemochromatosis and sub-Saharan African hemochromatosis are two examples. These two disorders differ in that hereditary hemochromatosis results in excessive iron when iron intake levels are normal, while sub-Saharan African hemochromatosis requires excessive intake of Fe coupled with a genetic predisposition to poorly regulate iron. Generally speaking, toxicity associated with excess essential metals tends to be rare, and it most frequently occurs in people who inappropriately consume dietary supplements.

21.1.4   Toxicity Associated with Nonessential Metals

The majority of this chapter focuses on the toxicology of nonessential metals. Cadmium, lead, and mercury are three nonessential metals that have been investigated in great detail over the years, and they will be highlighted below. Other nonessential metals such as aluminum, beryllium, and nickel have not received as much attention but can pose toxicity issues. Arsenic and selenium are technically not metals, but are often included in discussions of metal toxicology. These elements will not be discussed here.

21.2   UNDERSTANDING METAL ION REACTIONS IN BIOLOGICAL SYSTEMS

Metals tend to be highly reactive in biological systems. Unfortunately, the behavior of metals usually cannot be predicted based on their ionic chemistry alone.

21.2.1 Metal Complexation with Biological Molecules

One of the major ways in which metals interact with biological molecules is via complexation. The formation of metal complexes can be described by this simple equilibrium equation:

$$M + L \leftrightarrow ML$$

where M is the metal and L is the biological ligand. Common biological ligands are nitrogen, oxygen, and sulfur in proteins. The strength of the complex is naturally dependent on the nature of the metal ion and the nature of the ligand. The stability constant of the metal–ligand complex (K_{ML}) can be described as follows:

$$K_{ML} = \frac{[ML]}{[M][L]}$$

How Does the Nature of the Metal Ion Influence Complexation? If the metal–ligand complex is the result of electrostatic attraction between the metal ion and the ligand, then both charge and metal ion size are influential. For example, the stability constants (K_{ML}) for phosphate complexes are as follows:

$$Fe^{3+} > Mg^{2+} > Li^+$$

Similarly,

$$Mg^{2+} > Ca^{2+} > Sr^{2+} > Ba^{2+}$$

Here, the charge density of the metal ion plays a pivotal role in the stability constants. The first series contains metal ions of similar size but different charges, while the ions in the second series increase with size from Mg^{2+} to Ba^{2+}. In both cases, ions with higher charge densities form stronger complexes. Sometimes ions with similar sizes and charge densities can be transported by the same transport proteins. For example, Cd^{2+} (0.97 Å) can be transported via Ca^{2+} (0.99 Å) channels.

Not all complexes are purely electrostatic. In fact, many metal complexes in biological systems have covalent interactions as well. In these cases, the ligand donates a pair of electrons (acting as a Lewis base) to the metal, which functions as a Lewis acid. Therefore, metals can be evaluated based on their abilities to accept electron pairs. Alkali metal ions (Na^+, K^+) and alkaline earth metals (Mg^{2+}, Ca^{2+}) tend to not form stable complexes with Lewis base ligands. Transition metal ions, particularly those with vacant d-orbitals, will form more stable complexes with Lewis base-acting ligands.

How Does the Nature of the Ligand Influence Complexation? Several factors come into play in determining the stability of metal–ligand complexes. If the ligand has only one coordination site, then the stability of the complex is directly related to the basicity of the ligand. However, many biological ligands have more than one coordination site or denticity. Monodentate ligands have one coordination site. Bidentate ligands have two coordination sites, and so on. More coordination sites

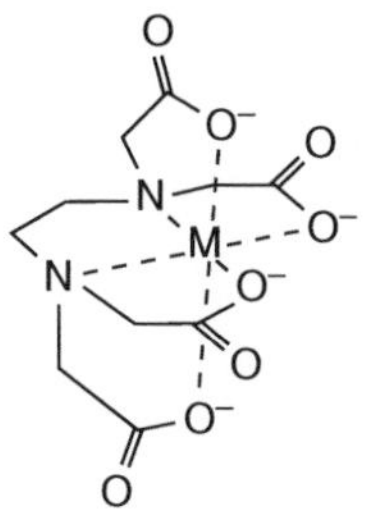

Figure 21.5. Chelation of a metal ion by EDTA.

result in more stable complexes. Chelation is a term derived from the Greek word for claw (chele), which describes a special kind of metal–ligand complex in which bi- or multidentate ligands form a ring structure that tightly binds the metal ion. EDTA (Figure 21.5) is a classic metal chelator. Chelation is an important process in biological systems. Proteins such as metallothioneins can chelate several metal ions including zinc, mercury, cadmium, and copper. [Also see structure of heme (Figure 21.2).] Several other factors come into play in determining the ligand's influence on complexation and stability constants, including charge, ring size, and degree of substitution.

A Note about Chelation Therapies. Chelation therapies are used to prevent or treat metal-induced toxicities. They are often used in acute poisoning scenarios, but can also be used to assess exposure. One of the major challenges in the management of chelation therapies is the tendency for chelating agents to interact with essential metals, particularly calcium and zinc. Chelation therapies should only be administered by a physician due to the potential to disrupt essential metal functions. The Food and Drug Administration does not regulate dietary supplements, and several "do it yourself" chelation therapies are available. These are not advisable.

21.2.2   Nieboer and Richardson's Metal Classification Scheme

In a now classic paper from 1980, Nieboer and Richardson devised a metal ion classification scheme based on the binding preferences of metal ions with components of biological systems. Using X-ray chrystallographic data, they separated metals into three categories based on their affinities for O-, N-, or S-containing ligands. Their classification scheme was derived from earlier work that recognized that metal ions could function as "hard acids" and "soft acids." Nieboer and Richardson divided metal ions into three categories: class A (previously termed hard acids), class B (previously termed soft acids), and borderline.

Class A metal ions have the following preferences for ligands:

$$F^- > Cl^- > Br^- > I^-$$

and for metal-binding donor atoms in ligands:

$$O > P > N > S$$

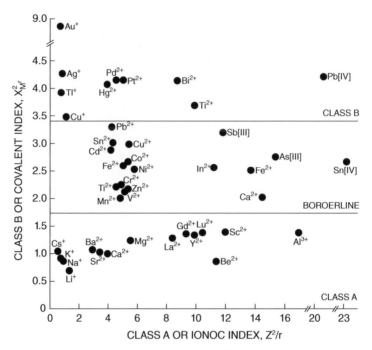

Figure 21.6. Metal classification.

Class B metal ions exhibit the opposite preference for ligands:

$$I^- > Br^- > Cl^- > F^-$$

and for metal-binding donor atoms in ligands:

$$S > N > O$$

Borderline metal ions form an intermediate group that does not exhibit strong preferences for some ligands over others.

Class A metals have a tendency to form ionic bonds. Class B metals have a tendency to form covalent bonds. A class A or ionic index based on the charge and radius of the ion is plotted against a class B or covalent index in Figure 21.6. The covalent index is based on the energy of the empty valence orbital of the metal ion, and it can be used to describe the ion's ability to accept electrons and form covalent bonds. Note that metal ions that serve as electrolytes and macronutrients in biological systems tend to be class A metal ions. Metals that have stronger interactions with, or are structurally associated with, proteins tend to be borderline or class B metal ions. Metal ions that are of highest concern for causing toxicity tend to be class B and borderline metals.

21.3   MODES OF METAL TOXICITY

At this point, it should be apparent that metals play such important and varied roles in physiological systems and that the biochemistry of metal ions is tremendously complex. It should not be surprising that it is very difficult to make broad generalizations about metal toxicity. Perhaps the only common feature aspect of metal toxicity is the ability of a single metal to interact with several macromolecular targets. For example, Cd binds or interacts with glutathione, alkaline phosphatases, malate dehydrogenase, carboxypeptidase, and RNA polymerase. Similarly, a single target or process can be affected by several different metals. Na^+/K^+-ATPase can be disrupted by lead, cadmium, copper, and zinc. Historically, metals have been thought to elicit toxic responses by interacting with biological molecules in three major ways:

1. *Blocking the Essential Biological Functional Groups of Biomolecules.* Class B and borderline metals are the most likely group of metals to abolish the activity of proteins via interactions with functional centers. Cysteine groups are often catalytic centers in proteins. As indicated above, Class B and borderline metals can form strong bonds with sulfur-containing molecules including cysteine. Mercury has been shown to cause deactivation by binding to cysteine residues in alkaline phosphatase, lactate dehydrogenase, and glucose-6-phosphatase.

2. *Displacing the Essential Metal Ion in Biomolecules.* It is estimated that approximately one third of all enzymes require metal as a cofactor or as a structural component. Those that involve metals as a structural component do so either for catalytic capability, for redox potential, or to confer steric arrangements necessary to protein function. Metals can cause toxicity via substitution reactions in which the native, essential metal is displaced/replaced by another metal. In some cases, the enzyme can still function after such a displacement reaction. More often, however, enzyme function is diminished or completely abolished. For example, Cd can substitute for Zn in the protein farnesyl protein transferase, an important enzyme in adding farnesyl groups to proteins such as Ras. In this case, Cd diminishes the activity of the protein by 50%. Pb can substitute for Zn in δ-aminolevulinic acid dehydratase (ALAD), and it causes inhibition *in vivo* and *in vitro*. ALAD contains eight subunits, each of which requires Zn. Another classic example of metal ions substituting for other metal ions is Pb substitution for Ca in bones.

3. *Modifying the Active Conformation of Biomolecules.* Class B metals such as Hg have been shown to alter the steric conformation of proteins via interactions with sulfur atoms, particularly disulphide bonds. For example, Hg insertion between the two sulfurs involved in a disulfide bond can significantly alter the shape of the protein and reduce or abolish its activity. This has been demonstrated with ribonuclease, for example.

21.4 METALS AND OXIDATIVE STRESS

The generation of reactive oxygen species (ROS) via normal metabolic processes or exposure to xenobiotics is a challenge that cells need to counteract (see Chapter 20). Metals play prominent roles in the biochemistry of reactive oxygen species as both causal agents and as essential players in enzyme systems that convert ROS to more benign compounds (Figure 21.7). The inability to effectively reduce ROS can lead to lipid peroxidation and eventually cell death.

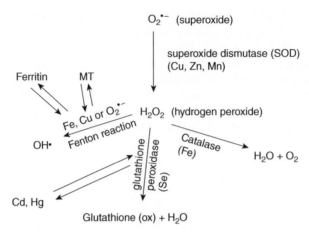

Figure 21.7. Metals and reactive oxygen species (ROS).

Metals and Protein Systems that Counteract ROS. Three major enzymes that play prominent roles in ROS biochemistry are superoxide dismutase, catalase, and glutathione peroxidase (Figure 21.7). Each of these proteins requires metals to perform their respective functions. Superoxide dismutase (SOD) exists in three forms in humans: cytoplasmic SOD1, mitochondrial SOD2, and extracellular SOD3. SOD1 and SOD3 require both Cu and Zn, while SOD2 requires Mn. SOD performs the conversion of superoxide to hydrogen peroxide. Catalase is a tetramer containing four heme groups. It performs the conversion of hydrogen peroxide to water and oxygen. Glutathione peroxidase contains four selenocysteine amino acid residues, which are essential in its ability to convert hydrogen peroxide to water.

Other proteins play important roles in controlling free metal concentrations in the cytosol. Of particular importance are ferritin and metallothionein. Ferritin is the major iron storage protein in the cell. It is critical that cytosolic iron is kept at low levels, because iron can catalyze the Fenton reaction, which generates the most toxic of the ROS—the hydroxyl radical. Copper and superoxide can also participate in the fenton reaction. Metallothioneins are another important family of proteins that helps control cytosolic concentrations of metals such as Cu and Cd. Glutathione is another peptide that controls free Hg and Cd levels in the cell (Figure 21.7).

21.5 METALLOTHIONEINS

21.5.1 Introduction

Metallothioneins (MTs) are a superfamily of low-molecular-weight (<7000-dalton) intracellular metal-binding proteins, which, in many species, play a critical role in (a) the detoxification of nonessential metals such as Cd^{2+} and Hg^{2+} and (b) the regulation of intracellular concentrations of essential metals such as Zn^{2+} and Cu^+. In 1957, Kagi and Vallee first purified and characterized MT as a cadmium-binding protein in equine kidney.

21.5.2 Classification and Structure

Metallothioneins are evolutionarily conserved in that they contain a high cysteine content and lack of aromatic amino acids. However, few invertebrate MTs have been characterized, and these can exhibit wide variation in noncysteine amino acid residues. Initially, MTs were classified according to their structural characteristics. Class I MTs consist of polypeptides with highly conserved cysteine residue sequences and closely resemble the equine renal MT. Mammalian MTs consist of 61–68 amino acids residues and the sequence is highly conserved with respect to the position of the cysteine residues (e.g., cys-x-cys, cys-x-y-cys, and cys-cys sequences, where x and y are noncysteine, non-aromatic amino acids). Class II MTs have less conserved cysteine residues and are distantly related to mammalian MTs. Class III MTs are defined as atypical and consist of enzymatically synthesized peptides such as phytochelatins and cadystins. This former classification scheme has been replaced by a more complex system to include the increasing number of identified isoforms.

Spectroscopic analysis of MT indicates tetrahedral–thiolate clusters in which divalent metals are bound by sulfur atoms. However, apo-MT (metal-free thionein) exists as a randomly coiled polypeptide, and the structure of the holoprotein is determined by coordinated metal ions. M-MT consists of two subunits as identified by Winge and co-workers using two-dimensional NMR. The two-cluster structure of MT has been more recently confirmed by Stout et al. using X-ray crystallography. Figure 21.8 depicts the more stable α-domain, also known as the C-terminal, which binds four divalent metals while the β-domain, also known as the N-terminal, is capable of binding three divalent metals. The stoichiometry of MT dictates that up to 7 divalent metals or 12 monovalent metals can form thiolate complexes with MT. MT lacks cellular specificity in binding metal ions; however, it has the greatest propensity for incorporating group Ib and IIb transition metals. The binding affinity of different metals for MT varies accordingly: $Zn^{2+} < Cd^{2+} < Cu^{2+} < Ag^+ = Hg^{2+} = Bi^{3+}$, therefore Zn can be readily displaced by other metal ions.

21.5.3 Function

Researchers have proposed that metallothioneins possess multiple physiological functions including protection against metal toxicity, zinc homeostasis, and defense

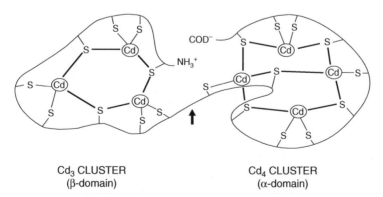

Cd₃ CLUSTER
(β-domain)

Cd₄ CLUSTER
(α-domain)

Figure 21.8. Metallothionein- and Cd-binding clusters.

against reactive oxygen species (ROS). Yet, recent work with transgenic and knock-out mice has questioned the validity of the later two functions, while confirming that MT does indeed play an integral role in protection against cadmium toxicity. For instance, Klaassen and co-workers observed that in the absence of MT-1 and MT-2, incidence of CdCl$_2$-induced lethality and hepatotoxicity increased, whereas MT overexpression proved to be protective. Using the MT -/- mouse model, Cd exposure produced considerably variable toxicity, leading researchers to believe that there are factors other than MT determining Cd toxicity. Despite the important physiological functions of MT, it should be noted that MTs, due to their redox properties, are not necessarily a sink for metals entering cells. On the contrary, metal exchange reactions as well as nucleophilic reactions carried out by its numerous sulfhydryl groups make MT a rather reactive protein. Consequently, there is much work to be done in order to further elucidate the physiological functions of metallothionein.

21.5.4 Degradation

The half-life of M-MT is dependent on the binding affinity of thionein for different metal ions. For instance, upon oxidation, Cu-MT forms insoluble polymers which are biologically unavailable and are eventually eliminated via biliary secretion. In contrast, thionein has lower affinity for Zn, making it more easily released from the protein and rendering the ion available for cellular processes. Furthermore, the rate of degradation may be influenced by differences in metal distribution between MT isoforms. It has been determined that MT degradation can occur in lysosomal and nonlysosomal (cytosolic) compartments.

21.5.5 Isoforms

All research performed thus far on mammals indicates that within a single species, multiple MT genes exist that code for a family of isoforms, known as isometallo-thioneins. At least four subgroups of mammalian MT genes exist and are named MT-1, MT-2, MT-3, and MT-4. The best-studied MT-1 and MT-2 isoforms are the most widely expressed in various tissues and are distinguishable by a single negative charge. Within the MT-1 and MT-2 subgroups, different structures have been observed and these isoforms are designated by lowercase letters. In humans, the most highly expressed isoform is MT-2a which accounts for roughly half of all MT expression. Most of the existing MT research has been performed on rodents, more specifically mice. There are four known murine MT genes, and they are located on chromosome 8 and encode MT-1 through MT-4 proteins. In humans, 17 MT genes are located on chromosome 16, of which 10 are functional and code for multiple isoforms of hMT-1 and one isoform of hMT-2. MT-3 is found primarily in the brain, although recently, it has been identified in maternal deciduum, reproductive tissues, stomach, tongue, kidney, and heart. Little research has been conducted on MT-4 however, the gene has also been found in the placenta and previously only in differentiating stratified squamous epithelium. It is unclear as to whether or not different MT isoforms have discrete functions, however recent data indicate that MT-1 and MT-3 may possess distinct properties and physiological roles. Posttranslational modifications are responsible for heterogeneity among MTs.

21.5.6   Induction and Regulation of MT Gene Expression

MT gene expression is largely regulated at the transcriptional level. Enhanced MT expression may also be attributed to MT gene amplification. However, the specific mechanism by which MT gene expression is regulated in vivo in response to metal exposure is as yet unclear. In addition to metals, a number of other stimuli are known to induce MT, namely glucocorticoids, cytokines, oxidative stress, and a variety of chemicals.

The 5' end of MT-1 and MT-2 genes possess a TATA box, or core promoter element and numerous response elements that confer metal inducibility on the MT gene promoter (Figure 21.8). Some of these response elements such as AP1 and AP2 in humans and in mouse, the antioxidant response element (ARE) and upstream stimulatory factor (USF) provide putative binding sites for MT transcription factors. The most common of these cis-acting proximal elements are the metal responsive elements (MREs), motifs that are conserved across vertebrate and invertebrate species. Multiple copies of MREs exist in the MT promoter region and act synergistically to enhance activity.

The constitutively active metal-responsive transcription factor 1 (MTF-1) specifically binds to these MREs, resulting in transcriptional activation of MT gene expression. MTF-1 is a zinc finger transcription factor under the control of MTI (metallothionein transcription inhibitor). The DNA-binding ability of MTF-1 as well as its zinc sensory function is attributed to the zinc finger domain. However, recent evidence indicates that interactions between the zinc finger domain and other portions of MTF-1 may result in the zinc response of MTF-1. Figure 21.9 illustrates that in the absence of Zn, MTF-1 complexes with MTI, thereby inhibiting MTF-1 interaction with the MRE and preventing MT transactivation. Conversely, in the presence of Zn, MTF-1 is unbound by MTI and is free to bind to MREs whereupon MT gene transcription is activated.

In addition to Zn, evidence exists that oxidative stress can also activate MTF-1, which enhances binding to MREs; as a consequence, MT gene transcription is activated (Figure 21.10). Cadmium, although a more powerful MT inducer than Zn, is not capable of activating MTF-1 because it has little effect on MTF-1 DNA-binding activity; therefore, Cd must be acting through some other mechanism to upregulate MT expression. Further research has confirmed that Cd and Zn can in fact activate the MT promoter via distinct mechanisms and that Cd-induced activation of MRE is independent of Zn and MTF-1. Interestingly, mice lacking both alleles for the

Human MT-IIA Proximal Promoter

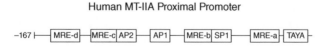

Figure 21.9. Human MT promoter. (Adapted from Murphy, B. J., Andrews, G. K. Activation of metallothionein gene expression by hypoxia involves metal response elements and metal transcription factor-1. *Cancer Res.* **59**, 1315–1322, 1999.)

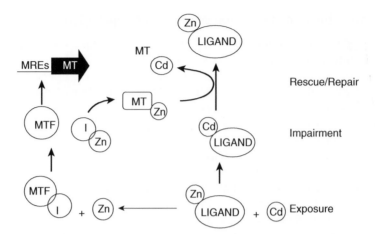

Figure 21.10. MT detoxification of Cd. (From Amiard et al. *Aquat. Toxicol.* **76**, 160–202, 2006.)

MTF-1 gene suffer embryonic lethality as a result of liver failure, while mice lacking both alleles for the MT gene develop normally despite an increase in cadmium sensitivity. This indicates that MTF-1 not only is involved in MT gene transcription, but also has been shown to be involved in regulating other genes that are crucial to embryonic development.

21.6 EXAMPLES OF TOXIC METALS

21.6.1 Cadmium

Cadmium is a relatively rare metal that is generally considered nonessential in mammals, although it has been shown to be essential in a specialized marine diatom carbonic anhydrase. As is the case with other toxic metals, cadmium does not have a single target, but rather can disrupt many physiological processes. At the organ level, targets for Cd include kidney, liver, intestine, lungs, and bone. Cd is also a carcinogen. Diet is the major route of exposure for the majority of the population, although inhalation is an important route of exposure in smokers and in some occupational settings. Smokers typically carry twice the Cd body burden of non-smokers. Cadmium accumulates in the body over time, primarily in the liver and kidneys, but also can affect bone and other organs.

The pharmacokinetics of Cd has received quite a bit of attention. Experiments with rodents have been performed with CdCl$_2$ and with Cd complexed with metallo-thionein (MT). Intestinal absorption of Cd administered as CdCl$_2$ is significantly higher than Cd-MT. Cd (as CdCl$_2$) is rapidly absorbed in the duodenum, although the exact mechanisms remain unclear. It appears that MT does not play a role in the uptake process, because MT-deficient mice showed similar uptake kinetics as normal mice. In other experiments, pretreatment with zinc to induce MT expression in the intestine did not alter the rates of intestinal Cd uptake. Cadmium complexed with MT is also accumulated in the intestine, although not as rapidly as free Cd^{2+}. Interestingly, iron plays a protective role in limiting Cd-MT accumulation in intestinal cells, whereas calcium and zinc are not as protective. Once inside the cell, Cd can bind to MT and other proteins. Movement across the basolateral membrane into circulation is much slower than transfer into the cell, and it appears to be slowed by complexation with MT. Cd remains in the blood for a relatively short duration, bound to red blood cells (65%), albumin, and other proteins. Liver Cd concentrations plateau in the first 4–8 hr after an initial single oral dose, while kidney Cd levels rise over the course of several days. It is believed that Cd bound to MT is the predominant form transported to the kidney. In feeding studies with Cd-MT, kidney is the preferred deposition site rather than liver. However, studies with MT knockout mice demonstrate that intestine to kidney Cd transport occurs in the absence of MT. Perhaps other cysteine-rich intestinal proteins play a role in this transfer.

Kidney. Several lines of evidence suggest that sites of renal Cd-MT uptake are the S1 and S2 segments of the proximal tubule and that uptake is mediated by a low-molecular-weight protein transporter. The protein βμG can compete with Cd-MT for uptake; and at high Cd-MT doses, βμG is lost in the urine. Urinary βμG is sometimes used to assess renal damage and has been used to indicate Cd-induced kidney damage in occupationally exposed individuals. It appears that once inside the proximal tubule cell, Cd is rapidly dissociated from MT by endo-lysosomes and is released into the cytosol where it induces oxidative stress (Figure 21.11). Reactive oxygen species are thought to disrupt Na$^+$/K$^+$-ATPases and, if unabated, can induce apoptosis. Loss of the apical brush border and reduced infolding of the basolateral membrane have also been observed. The fragmentation and depolymerization of microtubules also occur following Cd treatment of cells. Glutathione (GSH) and multidrug resistant *p*-glycoproteins are thought to play a protective role in the kidney. Cd can also be complexed with endogenous MT. Cd has a long half-life in the cell, however, and has been shown to generate reactive oxygen species. Renal

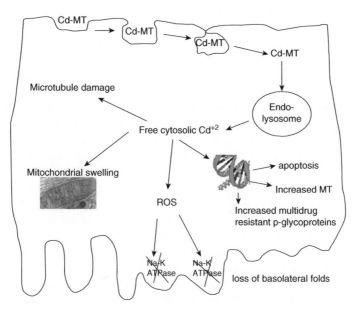

Figure 21.11. Cd nephrotoxicity.

toxicity is manifested by loss of proteins, amino acids, glucose, and electrolytes in the urine. Normal functioning proximal tubules resorb these substances. All indications are that nephrotoxicity is not reversible. Occupationally exposed workers continue to show renal problems years after the cessation of Cd exposure.

Liver. In humans, chronic Cd exposure does not typically result in hepatotoxicity. In laboratory animals, the liver accumulates the largest concentrations of Cd after acute or chronic exposures. In chronically exposed rats, liver injury occurs prior to renal dysfunction. Chronic Cd effects in the liver include increased plasma activities of alanine and aspartate aminotransferases, structural irregularities in hepatocytes, and decreased microsomal mixed function oxidase and CYP450 activities. Acute exposures in rats result in hepatic necrosis, particularly in parenchymal cells. Additionally, rough endoplasmic reticulum deteriorates, while smooth endoplasmic reticulum proliferates. Mitochondria are also degraded. As is the case with chronic exposure, microsomal mixed function oxidases and CYP450s are inhibited.

Cardiovascular System. Cadmium is a cardiodepressant, altering both mechanical and electrical functions. The mechanical effect is likely due, in part, to Cd antagonism of Ca function in myocardial contraction. Cd interference with Ca plays a role in altered electrical activity of the heart, prolonging conduction intervals. Cd is 100 times more potent in its electrical disruption than it is in its mechanical disruption. Endothelial cells are targets of Cd toxicity in several organ systems including testes, liver, kidney, central nervous system, ovary, uterus, and placenta. Altered endothelial cell morphology has been described in each of these tissues. Mechanisms underlying Cd-induced cytotoxicity in experiments using endothelial cell cultures have proven elusive. In canine coronary artery cells, Cd increased intercellular Ca concentrations and stimulated Ca efflux. Zn pretreatment intensified this effect. In bovine pulmonary artery endothelial cells, Cd reduced ATP concentrations and altered phospholipid synthesis. Cd-induced inflammation in endothelial cells could result from an increase in arachidonic acid. In human umbilical vein endothelial cells, Cd exposure resulted in inhibition of tissue plasminogen activator and stimulated the release of plasminogen activator inhibitor-1. Zn and Cu have been protective in some experiments with endothelial cells, and they have intensified toxicity in other studied.

Skeletal System. The skeletal system is a dynamic system with several critical functions. Aside from obvious protective, mechanical, and locomotor functions, the skeletal system also is a major player in the storage of ions and ion homeostasis, as well as acting as the sole site of bone marrow and associated hemopoieic functions. Cadmium is known to affect the skeletal system, and Cd effects on the skeletal system have been studied in vivo and in vitro. These research efforts were largely driven by the need to better understand cadmium-induced skeletal problems in the wake of a relatively large-scale environmental illness incident in Japan.

Itai-itai disease was identified in residents of the Jinzu River Basin in Japan in the mid-1940s. A lead and zinc mine polluted the river with high levels of Cd. The river water was used in the irrigation of rice fields and for drinking water. People affected by the disease suffered severe kidney dysfunction and painful skeletal symptoms. In the worst cases, bones would break from slight pressure, even from simply coughing. It is important to note that the worst cases were presented in postmenopausal women. Reduced estrogen production in these patients likely made them more susceptible to Cd-induced effects. It is also likely that malnutrition could have also increased Cd susceptibility. Nonetheless, cases of bone related problems also presented in Cd workers in both France and Sweden. In these cases, skeletal issues were also accompanied by renal problems. Treatment with high doses of vitamin D was required for these patients.

Because Cd accumulation in bones is considerably lower that concentrations found in liver and kidney, it was initially hypothesized that that skeletal effects were secondary to renal toxicity. This notion is still reported in several texts. A number of rodent studies have been conducted that clearly demonstrate Cd-induced skeletal effects at lower concentrations and/or at shorter exposure durations than is needed to induce renal toxicity. Many in vivo studies involve measuring the rates of ^{45}Ca mobilization from bone. Calcium naturally is released from bones as they are remodeled. Remodeling is the process by which bone is broken down and rebuilt in order to optimize structure and architecture. Approximately 7–8% of the skeletal system (by mass) is remodeled annually. Calcium is mobilized as small packets of bone are released. In experiments with rodents, ^{45}Ca was released from bones nearly twice as quickly as control animals after 10 days of dietary Cd exposure at 50 ppm. This Ca release preceded renal toxicity symptoms. Similar studies showed increased ^{45}Ca release from both dog and mice as early as three days post Cd exposure via the diet.

In vitro studies have shed some light on the mechanisms responsible for Cd induced osteotoxicity. In studies with cultured bones and bone cell cultures, several Cd induced effects have been observed. Consistent with in vivo studies, Cd exposure (10 nM) to cultured fetal rat limb bones in culture resulted in significant demineralization. Other work with bone cell cultures showed release of Ca and prostaglandin PGE_2 into the culture medium in response to Cd exposure. Prostaglandins play a complex, biphasic role in bone remodeling, by interacting with osteoclasts. Initially, prostaglandins inhibit osteoclast function. (Osteoclasts are related to macrophages and are primarily involved in the breakdown portion of bone remodeling while osteoblasts are involved with new bone synthesis.) The major long-term effect of prostaglandins in cultured bone tissue is increased bone resorption by stimulating the formation of new osteoclasts. The release of Ca and prostaglandins by bone cultures could be blocked by the administration of indomethacin, which is an inhibitor of prostaglandin synthesis. Other studies showed an increase in tartrate-resistant acid phosphatase in response to Cd exposure. This enzyme is found in osteoclasts, which play a role in bone degeneration in the remodeling process.

Signaling Pathways. Cadmium has also been implicated in the disruption of several signaling pathways. Recent work suggests that the glomerulus is affected by Cd exposure via the p38 MAPK/HSP25 signaling pathway. Hirano and co-workers demonstrated that Cd induces contraction in mesangial cells, which is thought to be associated with reduced glomerular filtration. Other work has demonstrated that Cd activates MAPK, which, in turn, activates c-fos proto-oncogenes in mesangial cells. Cadmium is also known to activate other transcription factors such as egr-1, c-myc, and c-jun. Due to their similar sizes and charge densities, Cd^{2+} also disrupts Ca^{2+}, requiring protein activities and signaling pathways. Inosotol triphosphate, calmodulin kinase, and protein kinase C all increase in response to cadmium, while Ca^{2+} ATPase is inhibited.

21.6.2 Lead

Lead is a nonessential metal that continues to cause toxicity issues despite wide-spread reductions of lead use in common products. The cessation of lead use in gasoline and lead soldered food containers has reduced exposure and lowered blood lead levels in the general population in the United States. Nevertheless, in urban communities, blood lead levels in children can be alarmingly high (Figure 21.12), particularly among African Americans. The Centers for Disease Control has set a guideline for lead levels in blood at $10\,\mu g/dl$ ($0.48\,\mu M$). In the early 1990s the average blood Pb concentration among African American children in urban centers was $13.9\,\mu g/dl$, and 35% of those sampled exceeded CDC guidelines. The major source of Pb in these cases is from lead-based paint formulations in older buildings, with exposure largely occurring via dust.

Lead can be toxic to several organ systems. The nervous system is particularly sensitive to lead exposure and will be the focus of this summary. At the cellular level, lead can cause apoptosis and excitotoxicity, affect mitochondria, and alter neurotransmitter dynamics. As is the case with Cd, lead can also substitute for calcium, causing numerous deleterious effects. Lead readily crosses the blood–brain barrier, probably via Ca-ATPase pumps. In experiments with various isolated cell lines, Pb^{+2} uptake into cells' is increased then the cells' Ca stores are reduced.

In neurons, lead can alter Ca^{2+} regulation, resulting in an increase in intracellular Ca^{2+} concentrations. This increase in Ca^{2+} is one potential trigger for the opening of the mitochondrial transition pore, which initiates apoptosis. Very low concentrations of Pb also can cause the mitochondria to release Ca^{2+}, which can also lead to apoptosis. The death of neuronal cells due to lead exposure is well-studied in the retina

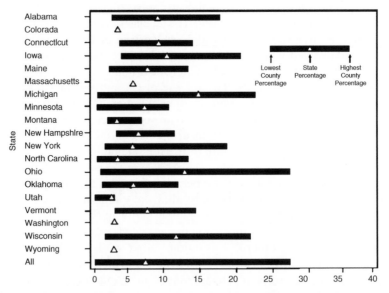

Figure 21.12. State-specific percentage of children aged <6 with blood lead levels $\geq10\,\mu g/dL$, 1998. Only counties with ≥200 children tested for BLL are included. Coloardo, Washington, and Wyoming had <2 counties with 200 children tested, and Massachusetts did not report county of residence. (From CDC website.)

and appears to be related to the opening of the mitochondrial transition pore as well. Vision problems have been observed in humans and lab animals that have had developmental lead exposure. Hearing impairment has also been observed in teenagers with blood Pb levels >20 μg/dl.

Epidemiological studies have examined the relationship between blood Pb level at different early developmental stages with psychomotor and cognitive deficits. Elevated blood Pb levels are associated with reduced IQ scores. One possible mechanism underlying this neurological effect is that lead disrupts the timing of cell-to-cell connections in the developing neuronal circuitry. As mentioned above, lead can also mimic Ca in certain situations, including regulation of protein kinase C (PKC). Lead levels can stimulate PKC at very low concentrations relative to concentrations of Ca involved in PKC upregulation. It is possible that PKC plays a role in memory and learning. Another possible mechanism is Pb interactions with Ca^{2+}-dependent releases of neurotransmitters such as acetylcholine, dopamine, and glutamate.

Another mechanism by which Pb can affect the nervous system is by affecting glial cells. Glial cells are non-neuronal cells that perform important support functions in the nervous system, including structural support, nutritional support, and myelination, and in some cases are involved in signal transmission. Two major glial cell types—oligodendroglia and astroglia—are affected by lead. Oligodendroglia form the myelin sheath around neurons. Exposure to lead causes a hypomyelination and demyelination. Astroglia or astrocytes regulate the external environment of neurons by regulation of ionic composition and recycling neurotransmitters. These cells appear to be sinks for Pb, although it is unclear whether astroglial accumulation of Pb is protective or a potential source of Pb that can subsequently be released in the brain or other neurons and cause damage.

Lead exposure can cause hematological problems, which can indirectly cause neurotoxicity. Among the best-known hematological effects of Pb exposure is inhibition of δ-aminolevulinic acid dehydratase (ALA-D) (Figure 21.13). The hematological consequences of this process are described below. There are also neurological consequences of this inhibition. ALA is known to inhibit GABA release and may also compete with GABA at the receptor site. Lead disruption of heme synthesis (see below) can result in anemia, which can also affect cognitive processes.

A second major lead-induced toxicity involves interruption of heme synthesis. Lead interacts at several steps in the heme biosynthetic pathway (Figure 21.13). As mentioned above, Pb inhibits the enzyme δ-aminolevulinic acid dehydratase (ALA-D), which catalyzes the second step of heme synthesis involving the condensation of two molecules of aminolevulinic acid (ALA) to form porphobilinogen. The result of this inhibition is the accumulation of aminolevulinic acid in the serum and increased excretion of ALA in the urine. A second major disruption of the heme biosynthetic pathway is Pb inhibition of ferrochelatase. This enzyme is responsible for the incorporation of the ferrous ion (Fe^{2+}) into protoporphirin IX to produce heme (Figure 21.2). Accumulated protoporphirin is incorporated into red blood cells and chelates zinc as the cells circulate. This zinc–protoporphirin complex is fluorescent and used to diagnose Pb poisoning.

Skeletal System. Bones are a major sink for Pb in humans. Lead is stored in bones via the same processes that regulate calcium, and Pb substitutes well for Ca in the

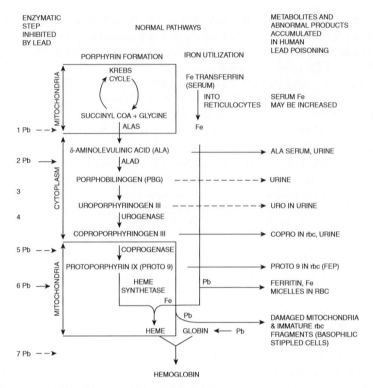

Figure 21.13. Schematic diagram of hemoglobin synthesis – – ➤ = weak evidence of Pb inhibition; —➤ = strong evidence of Pb inhibition. (Adapted from Natl. Acad. Sci.-Natl. Res. Counc. 1972. Airborne Lead in Perspective. Washington DC: Natl. Acad. Sci.)

lattice structures of bones. Despite the fact that bone preferentially accumulates Pb, it does not appear that Pb is as osteotoxic as Cd. It does appear that lead can adversely affect bone during fetal and postnatal development, however. Due to concerns about lead exposure and possible effects on other organ systems, bone provides a means for assessing Pb exposure. Because lead levels in blood are often transient, bones provide a more integrated view of lead exposure. Often, chelation treatments are used to attempt to assess Pb exposures. However, the relative contribution of excreted lead from soft tissues and skeletal tissue are not well-characterized by this approach. Direct measurement of Pb levels in bone is preferred and can be done by biopsy or by X-ray fluorescence.

21.6.3 Mercury

Mercury exists in the environment in several forms. Elemental mercury is a liquid at room temperature with a high vapor pressure. Inorganic mercury can exist as a cation in two oxidation states, mercurous (+1) and mercuric (+2). Mercury can also covalently bind to carbon, resulting in several organomercuric compounds, the most common being methyl mercury. Each of these forms has the potential to be toxic, depending on the route of exposure. Because mercury is globally distributed via atmospheric deposition, all humans are exposed to some degree. The chemistry of mercury in the atmosphere, in the environment, and in the internal environment of living organisms is complex. In the environment, the conversion of inorganic to organic mercury is performed by microorganisms in soils, aquatic, and marine environments. In aquatic environments, sulfur-reducing bacteria are the primary culprits in mercury methylation, particularly under anaerobic conditions. Methyl mercury rapidly enters food chains and biomagnifies, leading to high concentrations in top predatory fish. Several states have fish consumption advisories due to the potential health risks to humans eating contaminated fish. Other routes of human exposure include inhalation exposure to metallic mercury, dental amalgam fillings, and exposure to an ever-shrinking variety of consumer goods and health care products.

The toxicity of mercury is dependent on its form and route of exposure. For example, liquid elemental mercury is very poorly assimilated in the gastrointestinal tract and is not considered a toxicological hazard. Elemental Hg has a very high vapor pressure, however, and inhaled Hg vapors are readily absorbed and distributed systemically. The assimilation of mercuric mercury from food is relatively low (approximately 15%), while 90–95% of methyl mercury is absorbed by the gastrointestinal tract. Biotransformation can occur in the body, with elemental and methyl mercury being oxidized to divalent mercury. Divalent mercury has a high affinity for thiol groups, and can nonspecifically affect several enzymes. Similar to Cd, mercury induces oxidative stress, lipid peroxidation, and mitochondrial dysfunction, among other problems. Like Cd, mercury binds strongly to metallothionein, but interestingly, the half-life of this complex is much shorter than Cd-MT in the kidney. In the liver, glutathione plays a major role in trafficking methyl and inorganic mercury. This discussion will be limited to neurological effects of methyl mercury and renal effects of mercuric mercury.

Neurological Effects. Methyl mercury is a notorious central nervous system toxicant. Mass poisoning of humans in the Minamata Bay region of Japan in the 1950s and in Iraq in the 1970s resulted in severe teratogenesis, with neurological effects such as severe mental retardation, poor reflexes, speech pathologies, and other problems associated with in utero exposures. Other epidemiological studies have focused on several populations, including the Cree of Canada, a Peruvian population, and people from the Faroe Islands, Seychelle Islands, and New Zealand. Attempts to reconstruct exposure levels by correlating neurological evaluations with maternal Hg levels in hair and umbilical cord blood have revealed relationships between exposures to MeHg and developmental effects.

Neurons are thought to be poor in glutathione activity, making them particularly susceptible to mercury insult. An increase in reactive oxygen species has been observed in different portions of the brain in rodents after exposure to methyl mercury. Other studies have demonstrated that cytotoxicity occurred in the absence of lipid peroxidation, with α-tocopherol being used to control peroxidative processes. One hypothesis for the neurological effects of Hg is that Ca homeostasis is disrupted, leading to a sequence of events culminating in oxidative stress. Patch-clamp studies show that both methyl and divalent mercury retard current transduction in dorsal root ganglion cells. This effect does not appear to be mediated by mercury interaction with sodium or calcium channels; however, increasing the intracellular calcium concentration is stimulatory. Different forms of mercury affect different parts of the neuron. The high lipophilicity of methyl mercury may lead to its preferential accumulation in myelin, resulting in decreased excitability. Divalent mercury may interfere with calcium signaling in cholinergic processes.

Of particular concern are the potential effects of mercury exposure to the developing nervous system. In utero exposure to mercury results in the disruption of neural networks by altering neuronal migration and affecting neuronal organization patterns in the cerebral cortex. Several studies have focused on mercury interactions with microtubules to explain these observations. Specifically, MeHg binds to tubulin and can inhibit the polymerization of microtubules. Divalent mercury can have the same effect on microtubule polymerization. At higher concentrations, mercury can affect several enzyme systems in the nervous system, including inhibition of protein kinase C and cerebral alkaline phosphatase.

Nephrotoxicity. Diet is the major route of exposure to mercuric mercury, with assimilation ranging from 8% to 25% of the ingested dose. These data were obtained from human volunteers. Other aspects of the toxicity of mercuric chloride have been obtained from people using mercuric chloride, previously used as an antiseptic, to commit suicide. The acute cause of death is severe gastrointestinal damage, cardiovascular shock, and renal failure. In chronically exposed animals, renal damage is the major toxic endpoint. In animal studies, approximately 30% of the body burden is found in the kidneys. Little mercuric mercury crosses the blood–brain barrier relative to mercury vapor or methyl mercury. At the kidney, it appears that there are two mechanisms for the uptake of inorganic mercury at the proximal tubule. Luminal uptake is linked to the activity of γ-glutamyltransferase, which functions to catalytically cleave the γ-glutamylcysteine bond in glutathione. Basal transport likely occurs via an anion transporter.

There are two distinctive mechanisms by which mercuric mercury affects the kidney. This form of mercury stimulates the immune system, leading to excessive antibody concentrations at the glomerular membrane. This strongly affects the selectivity of the glomerulus, allowing albumin and other proteins to be filtered and lost in the urine. This also results in edema. The second mechanism of mercuric mercury-induced renal damage is via lipid peroxidation and oxidative stress. This may occur as a result of mercury inhibiting the activities of superoxide dismutase, catalase, glutathione peroxidase, and glutathione disulfide reductase. Other work has demonstrated that mercury exposure increases hydrogen peroxide in the mitochondria of renal epithelial cells.

21.7 METALS AND CANCER

The current paradigm of chemical carcinogenesis involves three stages: initiation, promotion, and progression (see Chapter 24). It appears that metals can play roles in each of these steps. Unlike other organic toxicants, metals are generally not bioactivated or metabolized to reactive intermediates, nor are they broken down to less toxic subunits. There are, of course, exceptions involving addition or subtraction of alkyl groups (demethylation of methyl Hg and the methylation of As).

While several metals show at least some evidence of carcinogenicity (see Table 21.1), five metals are classified by the International Agency for Research on Cancer (IARC) as carcinogenic to humans. Arsenic and arsenic compounds, beryllium and

TABLE 21.1. International Agency for Research on Cancer (IARC) Summary of Metal Carcinogenic Risk to Humans

Metal (or Process Involving Metal Exposure)	Evidence for Carcinogenicity in Humans	Evidence for Carcinogenicity in Animals	IARC Rating[a]
Arsenic[b]	Sufficient	Limited	1
Beryllium[b]	Sufficient	Sufficient	1
Cadmium[b]	Sufficient	Sufficient	1
Hexavalent chromium[b]	Sufficient	Sufficient	1
Nickel (metallic)	Insufficient	Sufficient	2B
Nickel compounds	Sufficient	Sufficient	1
Underground hematite mining with radon exposure	Sufficient	Sufficient	1
Iron and steel founding	Sufficient		1
Cisplatin	Insufficient	Sufficient	2A
Cobalt[b]	Insufficient	Sufficient	2B
Iron dextran	Insufficient	Sufficient	2B
Inorganic lead compounds	Insufficient	Sufficient	2B
Methylmercury compounds	Inadequate	Adequate	2B
Implanted foreign bodies of metallic Co, metallic Ni, Ni/Cr/Fe alloys	Inadequate	Adequate	2B
Welding fumes and/or gases	Limited	Inadequate	2B

[a]1, Carcinogenic to humans; 2A, probably carcinogenic to humans; 2B, possibly carcinogenic to humans.
[b]includes elemental and metal-containing compounds.
Source: Michael P. Waalks. Metal carcinogenesis. In: Sarkar, B. (Eds.). *Heavy Metals in the Environment*, Marcel Dekker, New York, 2002.

beryllium compounds, cadmium and cadmium compounds, hexavalent chromium compounds, and nickel and nickel compounds have been rated as human carcinogens, largely based on epidemiological studies and corroborating evidence of carcinogenicity in laboratory animals. Arsenic and arsenic compounds present an interesting case because evidence of carcinogenicity in humans is unequivocal, whereas arsenic and arsenic compounds have not been strongly carcinogenic in animal models. This raises the unsettling likelihood that humans are considerably more susceptible to cancer risks posed by arsenic and arsenic compounds. Epidemiological and laboratory studies with the other metals listed above produce strong evidence for their carcinogenicity.

The mechanistic basis for metal carcinogenicity has proven to be elusive, although several possible mechanisms have been proposed. The difficulties in defining mechanisms are numerous. First, the reactivity of metals with several biological processes is dose-dependent and is often affected by the localized concentrations of other essential metals. Many in vitro studies use exceptionally high metal concentrations that are well beyond exposure levels that could occur in vivo without causing acute toxicity and lethality. Other studies that have produced cancers in lab animals have done so by direct injection of metal into the organ of study (i.e., testes, bone, etc.). Often, in these studies, cancers form at the site of injection and cannot be replicated via oral or inhalation routes of exposure. Another difficulty in studying metals is that metal ions often form ionic bonds with macromolecules. Ionic bonds are very easy to disrupt, severely handicapping the researcher's ability to determine the precise location of metal in the cell, particularly if the cell must be disturbed in running a particular assay. Immunohistochemical approaches and other localization techniques have not been very helpful in determining sites of action in live or fixed cells. Finally, as is the case with all carcinogens, the time lag between initiation and tumor formation makes the unequivocal determination of carcinogens difficult.

Despite these difficulties, several possible mechanisms of metal carcinogenicity have been proposed. Genotoxic mechanisms of metal carcinogenesis include direct DNA damage:

- Direct single or double DNA strand breaks
- DNA binding and/or crosslinking by a metal
- Altered DNA methylation (nickel is a good example)
- Free radical attack of DNA (reduction of hexavalent chromium creates free radicals)
- Metal binding to DNA results in reduced/impaired DNA repair mechanisms, leading to mutagenesis

Indirect events leading to possible metal carcinogenesis include:

- Generation of DNA-attacking free radicals (Ni, Cr, Fe, Cu)
- Replacement of Zn in zinc finger DNA-binding proteins, resulting in altered gene expression
- Inhibition of DNA repair (leading to increased mutation rates)
- Interactions with DNA polymerase (reducing accuracy and leading to mutation)
- Reduction of cellular methyl group stores, resulting in demethylation of DNA and altered gene expression (arsenic)
- Metal binding to Zn finger proteins or other transcription factors, altering protein conformation and altering gene expression patterns

ACKNOWLEDGMENTS

C.M. Martin contributed to the compilation of this chapter, and L. Xie and T. Pandolfo provided editorial assistance.

SUGGESTED READING

Andrews, G. K. Regulation of metallothionein gene expression by oxidative stress and metal ions. *Biochemi. Pharmacol.* **59**, 95–104, 2000.

Bhattacharya, P. K. *Metal Ions in Biochemistry*, Alpha Science International, Harrow, UK, 2005, 217 pp.

Klassen, C. D. (Ed.). *Casarett & Doull's Toxicology. The Basic Science of Poisons*, 6th ed. McGraw-Hill, New York, 2001, 1236 pp.

Coyle, P., Philcox, J. C., Carey, L. C., and Rofe, A. M. Metallothionein: The multipurpose protein. *Cell. Mol. Life Sci.* **59**, 627–647.

Chu, W. A., Moehlenkamps, J. D., Bittel, D., Andrews, G. K., and Johnson, J. A. Cadmium-mediated activation of the metal response element in human neuroblastoma cells lacking functional metal response element-binding transcription factor-1. *J. Bio. Chem.* **274**(9), 5279–5284, 1999.

Goyer, R. A., Klassen, C. D., and Waalkes, M. P. (Eds.). *Metal Toxicology*, Academic Press, San Diego, 1995, 525 pp.

Hirano, S., Sun, X. K., DeGuzman, C. A., Ransom, R. F., McLeish, K. R., Smoyer, W. E., Shelden, E. A., Welsh, M. J., and Benndorf, R. p38 MAPK/HSP25 signaling mediates cadmium-induced contraction of mesangial cells and renal glomeruli. *Am. J. Renal Physiol.* **288**, F1133–F1143, 2005.

Klaassen, C. D., Liu, J., and Choudhuri, S. Metallothionein: An intracellular protein to protect against cadmium toxicity. *Annu. Rev. Pharmacol. Toxicol.* **39**, 267–294, 1999.

Lidsky, T. I., and Scneider, J. S. Lead neurotoxicity in children: Basic mechanisms and clinical correlates. *Brain* **126**, 5–16, 2003.

Miles, A. T., Hawksworth, G. M., Beattie, J. H., and Rodilla, V. Induction, regulation, degradation, and biological significance of mammalian metallothioneins. *Criti. Rev. Biochem. Mol. Bio.* **35**(1), 35–70, 2000.

Zalpus, R. K., and Koropatnick, J. (Eds.). *Molecular Biology and Toxicology of Metals*, Taylor and Francis, London, 2000, 596 pp.

Nieboer, E., and Richardson, D. H. S. The replacement of the nondescript term "heavy metals" by a biologically and chemically significant classification of metal ions. *Environ. Pollut. (Series B)*. **1**, 3–26, 1980.

Sarkar, B. (Ed.). *Heavy Metals in the Environment*, Marcel Dekker, New York, 2002, 725 pp.

CHAPTER 22

DNA Damage and Mutagenesis

ZHIGANG WANG

Graduate Center for Toxicology, University of Kentucky, Lexington, Kentucky 40536

22.1 INTRODUCTION

DNA is chemically highly reactive. Bases in DNA are particularly susceptible to chemical reactions because they contain nucleophilic centers and unstable double bonds. Consequently, DNA is frequently subject to chemical modifications. DNA damage often alters the coding property of the bases and/or results in profound effects on the DNA structure. Therefore, DNA damage can lead to cytotoxicity (cell death) and genotoxicity (mutagenesis). While cytotoxicity is an acute biological effect of DNA damage, genotoxicity is a chronological effect that may take many years to manifest its physiological consequences on the organism, as in the case of cancer. In response, cells contain complex systems to counter the deleterious effects of DNA damage. In higher eukaryotes, cellular responses to DNA damage consist of DNA repair, cell cycle checkpoint response, apoptosis, and damage tolerance.

DNA repair is the major cellular defense mechanism against DNA damage in that the lesion is physically removed by enzymatic reactions. Most repair mechanisms take the advantage of identical genetic information stored in the undamaged strand of the duplex DNA to restore the damaged sequence to its original state. That is, using the complementary and undamaged strand of DNA as the template, a small repair patch is synthesized replacing the damaged sequence. This underscores the importance of storing genetic information in the form of double-stranded DNA due to the ease of repairing DNA damage and thus ensuring genetic stability. This may well be the fundamental basis for free-living organisms to have evolved double-stranded DNA for storing the genetic information.

Based on the source of the damaging agent, DNA damage is divided into two major classes: endogenous DNA damage and environmental DNA damage. While endogenous DNA damage is produced by agents that are normally present inside the cell or as a result of normal cellular functions, environmental DNA damage is produced by environmental agents such as radiation and many chemicals. Some DNA lesions are produced exclusively by environmental agents, such as the case of the major DNA damage produced by UV radiation. Many types of DNA damage, however, can be generated both spontaneously in cells and through induction by environmental agents, as in the case of oxidative damage and base loss by depurination.

22.2   ENDOGENOUS DNA DAMAGE

Endogenous DNA damage occurs in the absence of environmental agents. Therefore, this class of DNA damage is unavoidable. These DNA alterations likely make significant contributions to the development of sporadic cancers and hereditary diseases.

22.2.1   DNA Base Mismatches

In mismatches, each base is chemically intact. However, instead of forming the Watson–Crick base pairs (A:T, C:G), the base is incorrectly paired in the duplex DNA. If not repaired, mismatches in DNA will lead to mutations in the next round of replication. The major source of DNA base mismatches is replication. When a template base is copied to the daughter strand during replication, the correct base is normally selected due to hydrogen bonding (base-pairing energetics) and geometric fitting at the active site of the polymerase. This yields an error rate of 10^{-4}–10^{-5} per base pair per replication at the nucleotide insertion step of DNA synthesis. The $3' \rightarrow 5'$ exonuclease activity of the replicative polymerases δ and ε provides the first line of defense against mismatches during replication. The misincorporated nucleotide can be readily removed by this exonuclease activity of the polymerase and is therefore called proofreading. The proofreading exonuclease activity improves the fidelity of DNA synthesis to an error rate of ~10^{-7} per base pair per replication. Mispaired bases that have escaped removal by the proofreading exonuclease are subject to the second line of defense, mismatch repair. This repair mechanism further increases the fidelity of DNA synthesis by 50- to 1000-fold to an overall error rate of ~10^{-9} per base pair per replication. This level of accuracy is extraordinary. Imagine the analogy where only one typo is allowed when one is requested to type about 1 million single-spaced pages, or about 1000 books. Nevertheless, about 3×10^9 base pairs are copied each time a human cell replicates its genome. That means that every time a human cell replicates itself, a few mismatches, thus a few mutations, will be generated.

22.2.2   Base Deamination

Cytosine (C), adenine (A), and guanine (G) contain an exocyclic amino group, which can deaminate to become uracil (U), hypoxanthine, and xanthine, respectively (Figure 22.1). Deamination of cytosine occurs at a much higher rate than that of adenine or guanine and is thus of greater biological importance. Spontaneous cytosine deamination occurs about 100-fold faster in single-stranded DNA than in double-stranded DNA. Based on deamination rate constants measured by in vitro experiments, it is estimated that about 500 cytosines in DNA are deaminated within a human cell every day. Since uracil is a base component of RNA, but not DNA, uracil in DNA is called an inappropriate base. When cytosine is deaminated to uracil in DNA, the original C:G base pairing would be altered to U:A base pairing in the

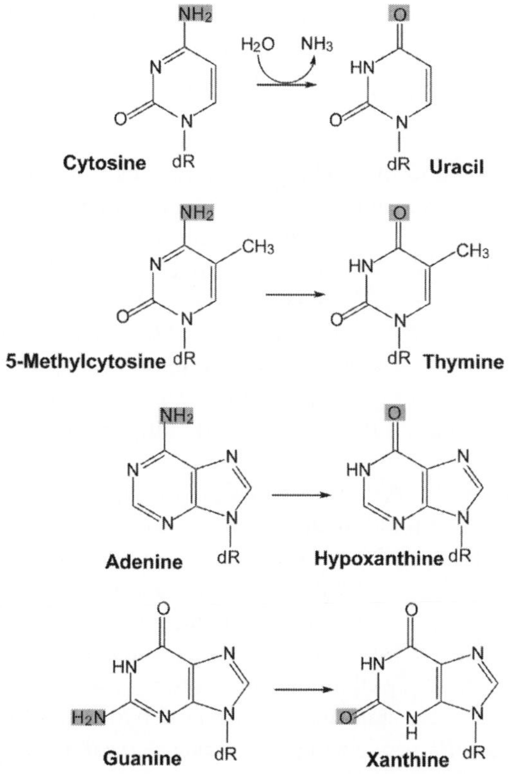

Figure 22.1. Deamination of DNA bases and the products produced. The deoxyribose moiety is designated by dR.

next round of replication, resulting in C → T (thymine) mutation. It is believed that deamination of cytosine makes an important contribution to spontaneous mutations in cells. Not surprisingly, removal of uracil from DNA constitutes a highly efficient repair mechanism in cells. Thymine, instead of uracil, evolved as a base component in DNA. With respect to genomic stability, thymine is advantageous in that the cell is able to recognize the deamination product of cytosine in DNA as an inappropriate base and remove it accordingly through DNA repair.

Base deamination is accelerated by nitrous acid and sodium bisulfite. Nitrous acid nonspecifically promotes base deamination in both double-stranded and single-stranded DNA. In contrast, sodium bisulfite specifically promotes cytosine deamination in single-stranded DNA. Some single-stranded regions do exist in cellular DNA, at least temporarily—for example, during transcription. In some cases, sodium bisulfite is used as a food additive. It is also used in the wine industry.

In mammalian genomes, some cytosine residues of the CpG (cytosines adjacent to guanines) sequences in DNA are methylated, forming 5-methylcytosine. Deamination of 5-methylcytosine, however, yields thymine, a normal base component of DNA. In single-stranded DNA, this is a challenging problem as cells are not able to determine that this thymine is "abnormal." In double-stranded DNA, however, deamination of the 5-methylcytosine in a methylated C:G base pair yields a T:G mismatch. Cells are therefore able to distinguish the thymine in a T:G mismatch as "abnormal" and thus remove it by a glycosylase through the base excision repair pathway.

22.2.3 AP Site

In DNA, bases are linked to the sugar-phosphate backbone by the *N*-glycosyl bond. This bond is subject to cleavage by hydrolysis. The result is base loss, leaving an abasic site in the DNA. The abasic site is often referred to as apurinic/apyrimidinic site, or AP site for short. It is sometimes called a noncoding lesion, referring to the fact that the lesion does not contain a coding base. Depurination (loss of G and A) occurs much faster than depyrimidination (loss of C and T). Base loss is accelerated at acidic pH. In the test tube, bases are readily released from DNA by acid hydrolysis at a high temperature. Based on depurination rate constants measured by in vitro experiments, it was estimated that about 10,000 purines are released from DNA in a human cell every day.

The deoxyribose at the AP site exists in equilibrium between a closed-chain furanose and an open-chain aldehyde (Figure 22.2). Due to the aldehyde form, AP sites in DNA are unstable. AP sites are prone to undergo a β-elimination reaction, yielding a DNA strand break at the 3′ side of the AP site (Figure 22.2). Alkaline pH and high temperatures promote β-elimination. Thus, AP sites in DNA can be conveniently measured by a DNA strand cleavage assay following alkaline treatment or heating. After β-elimination reaction, a second elimination called δ-elimination can occur, releasing the AP site as 4-hydroxy-pent-2,4-dienal (Figure 22.2). In the presence of a reducing agent such as sodium borohydride, the pentose aldehyde is reduced to an alcohol, thereby preventing β-elimination to occur. Enzymatic cleavage by AP endonucleases occurs at the 5′ side of the AP site. These different products of AP site cleavage within an oligonucleotide can be separated and identified by electrophoresis on a denaturing polyacrylamide gel (Figure 22.3).

Because of its chemical instability, the AP site is often difficult to study directly. Therefore, tetrahydrofuran (Figure 22.4) has been widely used in studies to produce a stable AP site analog. Lacking the aldehyde group, this AP site analog is unable to undergo the β-elimination reaction and is thus highly stable. Furthermore, this analog can be incorporated into an oligonucleotide at a specific site through automated DNA synthesis via phosphoramidite methods. DNA containing the AP site analog is a good substrate of AP endonucleases, because the endonuclease cleaves DNA at the 5′ side of the AP analog. Such DNA, however, is resistant to AP lyases, since the lyase cleaves DNA at the 3′ side of a natural AP site through β-elimination.

22.2.4 Oxidative DNA Damage

Oxygen consumption is an efficient way of generating the bioenergy, ATP. However, a consequence is the formation of undesired byproducts: reactive oxygen species (ROS), which can damage DNA. Therefore, for aerobic organisms, oxidative damage constitutes a major source of spontaneous DNA damage. The major ROS in cells include superoxide radical ($^{\cdot}O_2^-$), hydrogen peroxide (H_2O_2), and hydroxyl radical ($^{\cdot}OH$). Superoxide radical is formed by oxygen accepting an electron. Normally,

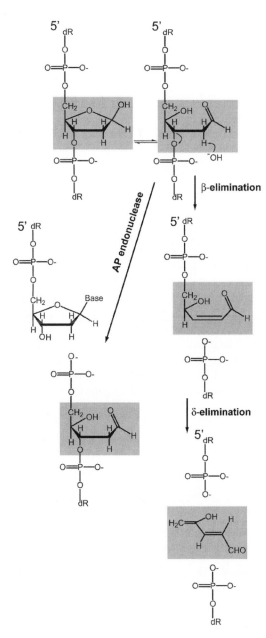

Figure 22.2. Enzymatic and chemical cleavage of DNA strand at the AP site. AP sites are chemically unstable, undergoing spontaneous β-elimination leading to DNA strand break at the 3′ side of the AP site. Only the AP endonuclease generates a 3′ terminus at the strand break that can be used by a DNA polymerase for DNA synthesis. The β-elimination reaction can also be catalyzed by AP lyases. The fate of the baseless deoxyribose is highlighted.

(a)

(b)

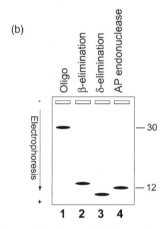

Figure 22.3. Separation and identification of different cleavage products at the AP site. (A) Duplex oligonucleotide containing an AP site. The damaged strand is labeled by ^{32}P at its 5′ end, as indicated by an asterisk. The AP site is designated in the DNA sequence by the X. (B) Schematic illustration of a gel showing the different mobility of the various cleavage products at the AP site. The various products contain a different 3′ end on the labeled 5′ DNA fragment. These products are separated by electrophoresis on a 20% denaturing poly-acrylamide gel, and the products are visualized by autoradiography.

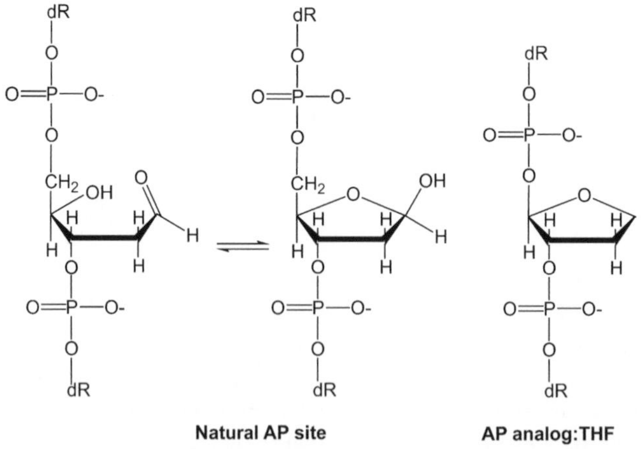

Natural AP site   **AP analog:THF**

Figure 22.4. Structural comparison between a natural AP site and its analog THF (tetrahydrofuran). The deoxyribose at the natural AP site exists in equilibrium between an open-chain aldehyde and a closed-chain furanose.

superoxide radical is eliminated by a cellular defense mechanism involving sequential actions of superoxide dismutase (SOD) and catalase, respectively, as follows.

$$2^{\cdot}O_2^- + 2H^+ \xrightarrow{\text{SOD}} H_2O_2 + O_2$$

$$2H_2O_2 \xrightarrow{\text{catalase}} 2H_2O + O_2$$

In the presence of ferrous ions (Fe^{2+}), H_2O_2 is converted to hydroxyl radical through the Fenton reaction:

$$Fe^{2+} + H_2O_2 \rightarrow Fe^{3+} + {}^{\cdot}OH + OH^-$$

Hydroxyl radical is the main cause of spontaneous oxidative DNA damage in cells. It is extremely reactive and is likely to be short-lived due to reactions with other organic molecules including DNA. As a result, hydroxyl radical is unlikely to diffuse far away from its production site. Thus, DNA damage by hydroxyl radical is thought to occur at sites of Fe^{2+}-complexed DNA. One important mechanism to limit cellular exposure to free ferrous ion is by storing these ions in the form of ferritin (iron storage protein), which is abundant in the heart and liver. Other transition metal ions such as copper can also catalyze the Fenton reaction, thus promoting oxidative stress and oxidative DNA damage.

Hydroxyl radical can react with the deoxyribose component, resulting in DNA single-strand break, or can react with DNA bases, yielding oxidative base damage. Oxidative DNA base damage is complex, with over 80 different lesions known. Nevertheless, 8-oxogaunine (also known as 8-hydroxyguanine), thymine glycol, cytosine hydrate, and 2,6-diamino-4-hydroxy-5-formamidopyrimidine (FaPy) (a form of imidazole ring-opened derivative of guanine) are among the major forms of oxidative base damage (Figure 22.5). It has been estimated that about 1000–2000 8-oxoguanine lesions and 2000 thymine glycols and cytosine hydrates are formed in the DNA of a human cell every day. Among the various types of oxidative base damage, 8-oxoguanine is highly mutagenic, because it can base pair with either C or A. A simple oxidation at its C8 position greatly alters the coding property of guanine. This lesion is called miscoding, and it frequently results in G → T transversion mutations.

Another mechanism of spontaneous oxidative DNA damage is mediated by peroxynitrite, a nitric oxide derivative. Nitric oxide (NO${}^{\cdot}$) is a key molecule in

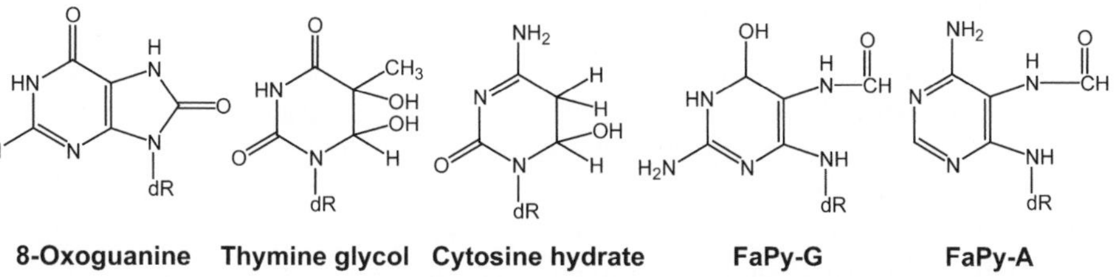

| 8-Oxoguanine | Thymine glycol | Cytosine hydrate | FaPy-G | FaPy-A |

Figure 22.5. Representatives of the major oxidative base damage. The 8-oxoguanine is also known as 8-hydroxyguanine due to isomerization. FaPy-G, 2,6-diamino-4-hydroxy-5-formamidopyrimidine; FaPy-A, 2,6-diamino-5-formamidopyrimidine.

biology. It functions in multiple physiological pathways such as neurotransmission, vasodilation, and immune response. Due to its high chemical reactivity, nitric oxide is relatively unstable and can be toxic. Nitric oxide reacts with superoxide radical in an extremely fast manner, forming peroxynitrite:

$$NO^{\boldsymbol{\cdot}} + {}^{\boldsymbol{\cdot}}O_2^- \rightarrow ONOO^-$$

The rate constant of this reaction is approximately $6.7 \times 10^9 \, M^{-1} s^{-1}$, which is near the diffusion limit. In comparison, the SOD-catalyzed fast decomposition of super-oxide radical is about 3.5-fold slower. In contrast to hydroxyl radical, peroxynitrite is able to diffuse in cells. Peroxynitrite reacts with DNA through oxidation and nitration. The reaction products are complex. Nevertheless, the major lesions include DNA single-strand breaks, 8-oxoguanine, fragmented guanine derivatives, 8-nitroguanine, and AP sites. Most single-strand breaks likely result from oxidation of the deoxyribose component of DNA. As for the base, peroxynitrite reacts mainly with guanine. Once 8-oxoguanine is formed in DNA, peroxynitrite can react with it faster than with guanine, yielding multiple secondary oxidation products. Nitration of guanine in DNA by peroxynitrite generates 8-nitroguanine. This lesion is relatively unstable, with a half-life of about 4 hr at $37\,^{\circ}C$. It depurinates, leaving an AP site in DNA.

Reaction of nitric oxide with oxygen forms another DNA damaging agent, N_2O_3. It is a powerful nitrosating agent, targeting the primary amines of DNA bases. Nitrosated C, 5-methyl-C, A, and G are unstable, leading to deamination and yielding U, T, hypoxanthine, and xanthine, respectively. Xanthine in DNA is unstable. It depurinates leaving an AP site in DNA. Other minor lesions produced by N_2O_3 include DNA single-strand breaks and DNA crosslinks.

22.2.5 DNA Adducts Formed from Lipid Peroxidation Products

Polyunsaturated fatty acids are a major type of biomolecules commonly found in cellular membranes and in our diet. The methylene groups positioned between cis double bonds in a polyunsaturated fatty acid are highly reactive towards oxidizing agents, such as $^{\boldsymbol{\cdot}}OH$. Oxygen radicals readily abstract electrons from such methylene groups, forming carbon radicals (Figure 22.6). Carbon radicals react with O_2 with an extremely fast rate (limited only by diffusion), forming peroxyl radicals, which represent the initial products of lipid peroxidation. This initiates a free radical chain reaction. Complex chemicals are generated as the end products of lipid peroxidation. Some reactions involve metal ions such as ferrous ions. Many of these end products are reactive electrophiles that can damage DNA. With respect to DNA damage, the important lipid peroxidation products are aldehydes, including malondialdehyde, acrolein, crotonaldehyde, and 4-hydroxy-2-nonenal (Figure 22.6). DNA damage by malondialdehyde yields the major adducts M_1G, M_1A, and M_1C (Figure 22.7A). In vitro reactions of malondialdehyde with DNA indicate that the amount of M_1G is about 5 times of that of M_1A. Only trace amounts of M_1C are formed. Reactions of acrolein, crotonaldehyde, and 4-hydroxy-2-nonenal with DNA yield propano-DNA adducts such as $1,N^2$-acrolein-dG, $1,N^2$-crotonaldehyde-dG, and $1,N^2$-HNE-dG, respectively (Figure 22.7B). The 4-hydroxy-2-nonenal can further epoxidize to 2,3-epoxy-4-hydroxynonenal (Figure 22.6), which reacts with DNA to

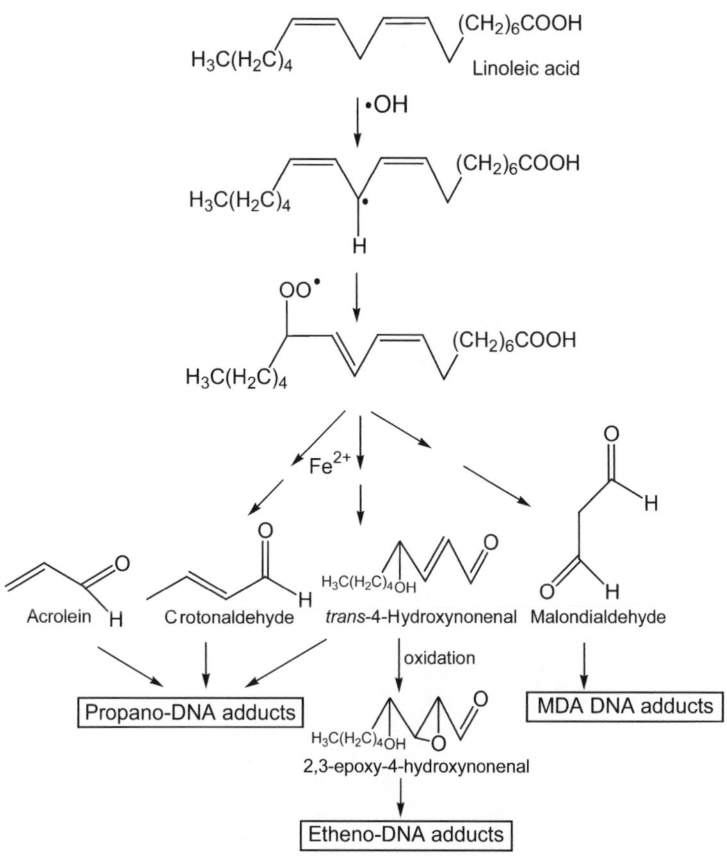

Figure 22.6. Major lipid peroxidation products that can damage DNA. Linoleic acid is a ω-6 polyunsaturated fatty acid. It is used as an example to illustrate peroxidation of polyunsaturated fatty acids. Many different products are formed following peroxidation of the polyunsaturated fatty acids. Only four major products are shown that are known to damage DNA. MDA, malondialdehyde.

form etheno DNA adducts, $1,N^2$-etheno-dG, $N^2,3$-etheno-dG, $1,N^6$-etheno-dA, and $3,N^4$-etheno-dC (Figure 22.7C).

It has been estimated that about 1000 DNA lesions are produced by lipid peroxidation products in a human cell every day. In metal storage diseases such as Wilson's disease and hemochromatosis, DNA damage by lipid peroxidation products is increased. Wilson's disease is a rare autosomal recessive disorder affecting copper metabolism. The transport of copper by the copper-transporting P-type ATPase is defective in Wilson's disease due to mutations in the *ATP7B* gene, resulting in excessive deposition of copper in the liver. Hemochromatosis is the most common form of excessive iron disease. In hemochromatosis patients, excessive iron is absorbed from food intake, leading to higher levels of iron stored in tissues, especially the liver, heart, and pancreas. There are indications that dietary intake of ω-6 polyunsaturated fatty acids also leads to higher levels of DNA damage by lipid peroxidation products.

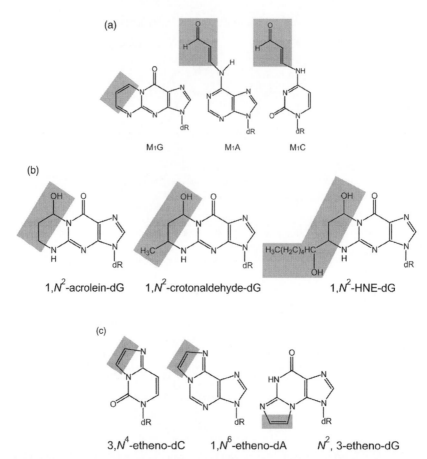

Figure 22.7. The major DNA lesions of the lipid peroxidation products. (A) DNA lesions produced by malondialdehyde. M_1 denotes the monomeric form of malondialdehyde. Malondialdehyde can polymerize to form dimers and trimers that can also react with DNA. The resulting lesions are designated as M_2 and M_3, respectively (e.g., M_2G). These lesions, however, may not be significant in cells as polymerization of malondialdehyde is relatively slow at neutral pH. (B) The $1,N^2$-propano-dG DNA adducts produced by acrolein, crotonaldehyde, and 4-hydroxy-2-nonenal (HNE). Stereochemistry is not shown. The $1,N^2$-acrolein-dG consists of three isomers. The $1,N^2$-crotonaldehyde-dG consists of two isomers. The $1,N^2$-HNE-dG consists of four isomers. (C) Etheno DNA adducts produced by 2,3-epoxy-4-hydroxynonenal. Further oxidation of 4-hydroxynonenal produces 2,3-epoxy-4-hydroxynonenal, which reacts with DNA to form the exocyclic etheno adducts.

22.2.6 DNA Methylation

In the mammalian genome, CpG sequences (cytosines followed by guanines) are often methylated at the C5 position of the cytosine by DNA methyltransferases. This physiological methylation plays a critical role in development and is the key mechanism of epigenetic gene silencing. However, methylation at other positions of DNA, which normally occurs through nonenzymatic chemical reactions, forms an important class of DNA damage. *S*-Adenosylmethionine (SAM) is the major donor of methyl groups for various biosynthesis in cells. In vitro, SAM acts as a weak DNA methylating agent through nonenzymatic reactions, yielding 7-methylguanine and 3-methyladenine. Therefore, SAM is believed to be a major source of spontaneous DNA methylation damage in cells. It is estimated that SAM alone produces about 6000 7-methylguanines and 1200 3-methyladenines in a human cell every day.

22.2.7 Incorporation of Inappropriate and Damaged dNTP into DNA During Replication

Like DNA, the bases of dNTPs are nucleophilic and highly reactive. Therefore, the dNTP pools for DNA synthesis are also subject to damage. While the vast majority of damaged dNTPs cannot be effectively recognized by DNA polymerases, dUTP and 8-oxo-dGTP (an oxidized dGTP) are efficiently utilized by the polymerases during replication. The result is incorporation of U and 8-oxoguanine, respectively, into DNA.

22.2.7a Incorporation of Uracil into DNA. In cells, dUMP is used for the synthesis of dTMP, which is further phosphorylated to dTTP for DNA synthesis. Therefore, a cellular pool of dUTP exists via phosphorylation of dUMP. Due to the presence of dUTPase, which hydrolysizes dUTP to dUMP, the steady-state concentration of dUTP is low relative to that of dTTP. Nevertheless, uracil is occasionally incorporated into DNA by a polymerase during replication as a result of the small dUTP pool in cells. The coding property of U is equivalent to T. Thus, incorporation of U in place of T results in U : A base pairing, which retains the same genetic coding of the DNA. This is in contrast to uracil in DNA derived from C deamination, which is mutagenic.

Misincorporated uracil in the U : A base pair is subject to efficient repair by uracil-DNA glycosylase via the base excision repair pathway. Consequently, uracil is not detectable in DNA under normal growth conditions. In *E. coli*, uracil becomes detectable in DNA when uracil-DNA glycosylase is inactivated (the *ung* mutant). Uracil content in DNA can reach up to 0.5% of all bases when the dUTPase is additionally inactivated (the *ung dut* double mutant) to increase the cellular dUTP pool. Inhibition of dihydrofolate reductase by methotrexate (an anticancer drug) leads to inhibition of dTMP synthesis from dUMP by thymidylate synthase. Thus, after treatment of cultured human cells with this drug, the intracellular dUTP concentration is greatly increased and the dTTP concentration is reduced, leading to readily detectable levels of uracil in DNA. Dietary deficiency in folate also leads to detectable levels of uracil in human DNA. The folate derivative N^5,N^{10}-methylenetetrahydrofolate is required for dTMP synthesis from dUMP. Insufficient folate in cells would result in decreased levels of dTTP and increased levels of dUTP, thus promoting uracil incorporation into genomic DNA.

When uracil is extensively incorporated into DNA, attempted repair by uracil-DNA glycosylase followed by AP endonuclease cleavage of DNA yields extensive DNA strand breaks located in close proximity that overwhelm the downstream repair reactions. The result is extensive DNA fragmentation at sites corresponding to the uracil sites—that is, DNA breakdown into small pieces. Therefore, DNA containing extensively incorporated uracil is incompatible with uracil-DNA glycosylase. When bacteriophage M13 is propagated in the *E. coli ung dut* cells, the phage contains extensive U in place of T in its DNA genome. Such phage is unable to survive in wild-type *E. coli* cells due to DNA fragmentation initiated by uracil-DNA glycosylase. In the *ung* mutant cells where the uracil-DNA glycosylase gene has been deleted, the uracil-containing M13 becomes viable.

22.2.7b   Genomic DNA of the B. subtilis Phage PBS2 Naturally Contains Uracil Instead of Thymine.

DNA contains T whereas RNA contains U as a coding base. An exception to this rule is found in the bateriophage PBS2. This *B. subtilits* phage naturally contains uracil instead of thymine in its DNA genome. PBS2 is among the largest phages known with a 252-kb genome. The host *B. subtilis* contains two major enzymes to prevent incorporation of uracil into DNA: dUTPase and uracil-DNA glycosylase. Therefore, bacteriophage PBS2 has to somehow counteract the host enzymatic mechanisms for eliminating uracil from DNA so that the progeny phage genome can be synthesized in favor of incorporating dUMP rather than dTMP. In addition, the phage DNA has to be protected from degradation by the host uracil-excision repair system initiated by uracil-DNA glycosylase. The strategy that PBS2 phage utilizes is to induce several proteins and enzymes that act in this regard during infection. These are uracil-DNA glycosylase inhibitor, dUTPase inhibitor, dTTPase, dTMPase, dUMP kinase, and dCTP deaminase. After infection of *B. subtilis* by PBS2, these phage-induced activities function collectively, resulting in a dramatic alteration of the intracellular dTTP and dUTP pools. The size of dTTP pool is greatly reduced, whereas the size of dUTP pool is dramatically increased. Consequently, dUMP rather than dTMP is incorporated into the newly synthesized PBS2 DNA during the phage replication. The host uracil repair pathway is inactivated by the uracil-DNA glycosylase inhibitor such that U is stably contained in the PBS2 DNA.

PBS2 codes its own DNA polymerase. The PBS2 polymerase utilizes dUTP and dTTP equally well as substrates in vitro. The apparent K_m values for dUTP and dTTP differ by less than twofold. Thus, the PBS2 phage-induced DNA polymerase has little selectivity for dUTP versus dTTP. The in vivo concentrations of dTTP in uninfected cells and of dUTP in infected cells are approximately 30–100 μM. Thus, the replacement of dTTP pool by dUTP after infection may largely contribute to the synthesis of uracil-containing progeny phage DNA. The strategy used by PBS2 phage to synthesize uracil-containing DNA is to dramatically increase the dUTP pool size and deplete the dTTP pool. This is reminiscent of the cellular mechanism in controlling the synthesis of thymine-containing DNA in most organisms. That is, the selective incorporation of dTMP or dUMP is largely dependent on the relative intracellular substrate concentrations, rather than altering the substrate specificity of DNA polymerases. The biological significance, if any, of employing uracil instead of thymine by PBS2 as a base component in its DNA is not understood.

22.2.7c   Incorporation of 8-Oxoguanine into DNA.

The oxidized dGTP, 8-oxo-dGTP, can be efficiently utilized by replicative polymerases for DNA synthesis. The 8-oxoguanine is miscoding, which is able to pair with either template C or A during replication. If not repaired, the incorporated 8-oxoguanine can lead to A → C and G → T transversion mutations during the next round of DNA synthesis. Therefore, 8-oxo-dGTP is a potent mutagenic precursor of DNA synthesis.

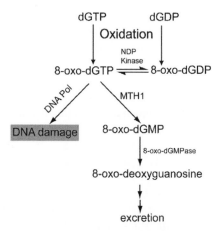

Figure 22.8. Formation and metabolism of 8-oxo-dGTP. NDP kinase, nucleoside diphosphate kinase; MTH1, 8-oxo-dGTPase. DNA damage is produced through incorporation of 8-oxogaunine into DNA by a DNA polymerase during replication. Pol, polymerase.

In addition to direct oxidation of dGTP, 8-oxo-dGTP may also be derived from oxidized dGDP, 8-oxo-dGDP. The human nucleoside diphosphate kinase that converts nucleoside diphosphates to the corresponding triphosphates is able to convert 8-oxo-dGDP to 8-oxo-dGTP, albeit at about 1/3 efficiency. As an effective defense mechanism against 8-oxo-guanine incorporation into DNA, cells contain 8-oxo-dGTPase that hydrolyzes 8-oxo-dGTP to 8-oxo-dGMP, thus reducing the 8-oxo-dGTP pool to a minimum. The 8-oxo-dGTPase is coded by the *mutT* gene in *E. coli* and *MTH1* (*mutT* homolog) gene in humans (Figure 22.8). Once hydrolyzed, 8-oxo-dGMP cannot be converted back to 8-oxo-dGTP, because the human guanylate kinase that phosphorylates both GMP and dGMP to GDP and dGDP, respectively, is inactive on 8-oxo-dGMP. The 8-oxo-dGMP is further dephosphorylated to 8-oxo-deoxyguanosine that can be readily transported across the cell membrane for excretion through the urine (Figure 22.8).

22.3   ENVIRONMENTAL DNA DAMAGE

Environmental DNA damage occurs as a result of cellular exposure to environmental agents that react with DNA. Therefore, this class of DNA damage is avoidable or manageable by eliminating or reducing the exposure of the organism to DNA damaging agents. There are two types of environmental DNA damaging agents: (a) physical agents such as radiation and (b) chemical agents such as many chemical carcinogens. While a few DNA damaging chemicals are of biological origin, as in the case of mycotoxins, the majority of these environmental chemicals result from human production/activities.

22.3.1   Mechanisms of Ionizing Radiation-Induced DNA Damage

Ionization radiation is common in our environment. Radiation sources for human exposure include: radon, rocks, and soil in some areas, cosmic radiation; and medical/

dental X-rays. The radiation energy is deposited as excitation and ionization. It is the ionization that causes DNA damage. All cellular molecules, including DNA, can be ionized by radiation, forming cation radicals.

$$\textit{Ionization:} \quad R \rightarrow {}^{\bullet}R^{+} + e^{-}$$

The cation radical can react with another molecule such as DNA, initiating a chain reaction.

$$^{\bullet}R^{+} + X \rightarrow R + {}^{\bullet}X^{+}$$

DNA suffers radiation damage by both a direct mechanism and an indirect mechanism. The direct damage results from radiation-induced ionization of DNA itself. The indirect damage results from attack on DNA by other free radicals. Since water is the predominant molecule in cells, the major source of indirect radiation DNA damage is from water radiolysis.

$$\textit{Ionization:} \quad H_2O \rightarrow H_2O^{\bullet+} + e^{-}$$

The $H_2O^{\bullet+}$ rapidly decomposes to a proton and a hydroxyl radical.

$$H_2O^{\bullet+} \rightarrow H^{+} + {}^{\bullet}OH$$

In the presence of oxygen, the released electron reacts quickly with it to form a superoxide radical.

$$O_2 + e^{-} \rightarrow {}^{\bullet}O_2^{-}$$

As discussed in Section 22.2.4, superoxide radicals can lead to the formation of hydroxyl radicals, the major DNA damaging ROS. It is estimated that ~65% of radiation DNA damage is due to ${}^{\bullet}OH$ radical attack, and ~35% is generated by direct DNA ionization. Hence, ionizing radiation also constitutes a major environmental agent of oxidative DNA damage. Together, the direct and the indirect radiation mechanisms produce a highly complex spectrum of various types of DNA lesions. These lesions consist of base damage, single-strand breaks, multiply damaged sites, double-strand breaks, deoxyribose lesions, and DNA–protein crosslinks.

22.3.1a Base Damage and Single-Strand Breaks.

Both oxidative DNA damage and indirect radiation DNA damage are mediated predominantly by hydroxyl radicals. Hence, base damage induced by ROS and radiation are very similar. Radiation adds another level of complexity as some lesions are produced by direct ionization of the bases. Similar to oxidative base damage by ROS, the major radiation-induced base damage includes 8-oxogaunine, thymine glycol, cytosine glycol, and formamidopyrimidines (FaPy-guanine and FaPy-adenine) (Figure 22.5). Many other types of base damage have been identified as minor lesions.

Single-strand breaks are generated by radiation. However, these breaks cannot be repaired by a simple DNA ligation, which requires a 3′ OH group and a 5′ phosphate group at the nick. Studies indicate that, at these breaks, phosphate groups are

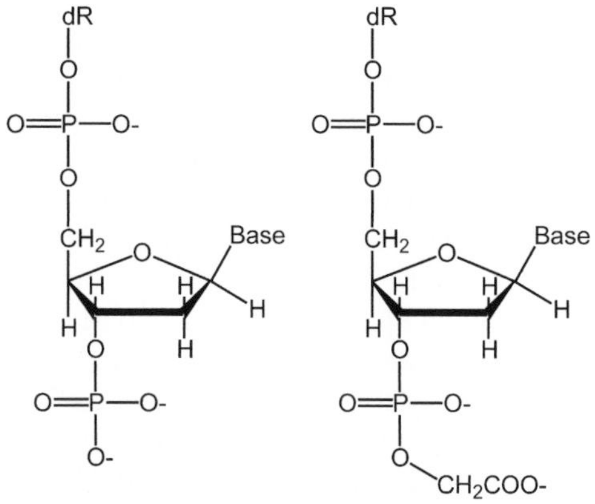

dR

O

O=P—O-

O

CH$_2$ Base

3' Phosphate

dR

O

O=P—O-

O

CH$_2$ Base

3' Phosphoglycolate

Figure 22.9. The 3' end of DNA strand breaks induced by ionization radiation often contains a phosphate or phosphoglycolate group. These groups can be removed by an AP endonuclease.

present at the 5' termini. However, 70% of the 3' termini contain a phosphate group and 30% contain a phosphoglycolate moiety (Figure 22.9). These 3' termini are subject to repair by AP endonucleases. Since strand break induced by radiation is not a clean breakage at the phosphodiester bond of DNA, many strand breaks contain a one-nucleotide gap. Nevertheless, such single-strand breaks are readily repaired in cells by the base excision repair pathway.

22.3.1b Multiply Damaged Sites and Double-Strand Breaks.
Multiply damaged sites (MDS) are closely positioned lesions on both strands of DNA. Double-strand break is an example of MDS. Formation of MDS by ionizing radiation occurs through ionizing energy deposition and the subsequent reactions of the free radicals produced. MDS is relatively unique to ionizing radiation. In contrast to singly damaged sites, MDS is highly problematic for repair in cells. Normally, the major excision repair pathways utilize the undamaged strand as the template for repair. At locations of MDS, normal excision repair is rendered ineffective by damage to the template strand. Therefore, MDS has a more profound effect on cell killing and mutagenesis than do singly damaged base lesions and single-strand breaks. Studies indicate that MDS is indeed more cytotoxic than singly damaged sites.

As a form of MDS, double-strand break is a highly potent lesion for cell killing. It is estimated that double-strand breaks measured in cells are <10% of the total MDS following ionizing radiation. Yet double-strand break contributes greatly to cell killing by radiation. In fact, double-strand break is considered the hallmark of DNA damage by ionizing radiation, due to its strong biological effect. In experimental biology, ionizing radiations such as γ-ray and X-ray have been widely used as an effective method for introducing double-strand breaks in DNA. Ionizing radiation has also been widely used in radiation therapy for cancer patients. Decreasing thiol levels and increasing oxygen levels sensitizes (enhances) ionizing radiation effects. This has been used as a therapeutic strategy to enhance the effect of radiation therapy in the clinic.

22.3.2 UV Radiation

Ultraviolet (UV) is a component of the sunlight spectrum. Sunlight is practically unavoidable. Prolonged sunlight exposure is a desirable attraction for many people in the Western culture due to its tanning effect on the skin. Therefore, UV radiation is probably the most prevalent environmental DNA damaging agent and carcinogen in human populations. The UV spectrum is divided into three segments: UVA (320–400 nm), UVB (290–320 nm), and UVC (100–290 nm). The maximum photo absorption of DNA is 260 nm, owing to its bases. Beyond 260 nm, UV absorption of DNA decreases with increasing wavelength and becomes insignificant above 300 nm. Thus, UV damage is most efficiently induced by UVC. UVB also damages DNA, albeit less efficiently. The atmospheric ozone layer effectively blocks out UVC. DNA damage on the surface of the earth is therefore mediated by UVB. This underscores the important protective role of the ozone layer against mutagenic and carcinogenic effects of the sunlight for humans.

22.3.2a The Major UV Lesions: Cyclobutane Pyrimidine Dimers and (6–4) Photoproducts.
Although UVB is more relevant to human exposure, UVC has been extensively used in the laboratory for studies on UV damage and repair. Historically, a convenient and effective UV source in the laboratory is germicidal lamps, which emits UV radiation at a maximum wavelength of 254 nm, very close to the DNA absorption peak of 260 nm. Thus, germicidal lamps are highly efficient for inducing UV damage on DNA. Similar types of UV lesions are produced by UVB and UVC. Thus, extensively documented results of UVC studies in the literature can be extrapolated qualitatively to the understanding of UVB radiation. For this reason, UV damage in this chapter is meant to be those DNA lesions induced by UVB and UVC, unless otherwise indicated.

There are two major types of DNA damage following UV radiation: cyclobutane pyrimidine dimers (CPD) and pyrimidine-pyrimidone (6–4) photoproducts that are often simply called (6–4) photoproducts. Formation of these UV lesions may be influenced by the DNA sequence. Nevertheless, in general, CPDs are more abundant than (6–4) photoproducts in UV irradiated DNA. For example, in UVC-irradiated DNA, the overall ratio of CPDs to (6–4) photoproducts is about 3:1.

22.3.2a.i Cyclobutane Pyrimidine Dimers.
CPD is formed by covalently linking two adjacent pyrimidines on the same DNA strand via a four-member ring (hence the name cyclobutane) resulting from a $(2 + 2)$ cycloaddition reaction between the C5,C6 double bonds of the two pyrimidines (Figure 22.10A). In duplex DNA, CPDs are formed predominantly in the cis–syn stereoisoform. Small amounts of trans–syn CPDs are additionally formed when single-strand DNA is irradiated with UV. In DNA, most CPDs are formed at the TT sequence (often called TT dimer for simplicity), and the least CPDs are at the sequence CC. For example, when plasmid DNA is irradiated with 254-nm UV, the ratio of various CPDs is 68:16:13:3 for TT:TC:CT:CC dimers. At 5-methyl CpG sites, the sunlight absorption is increased by 5- to 10-fold. As a result, the frequency of CPDs at T^mC and C^mC at these methylated CpG sites are expected to be higher relative to the corresponding sequences without 5-methylation.

Very high levels of TT dimers can be experimentally obtained by sensitized UV radiation in the presence of a photosensitizer such as acetophenone. The lowest

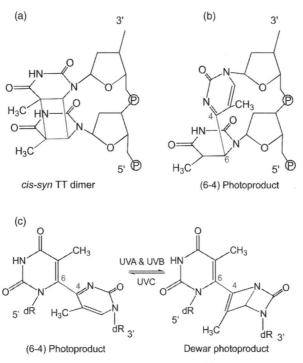

Figure 22.10. Structures of TT dimer and TT (6–4) photoproduct produced by UV radiation. (A) Structure of a cis–syn TT dimer. (B) Structure of a TT (6–4) photoproduct. The DNA backbone is drawn from 3′ to 5′. (C) Secondary photoreaction of (6–4) photoproducts to form the Dewar photoproducts. A TT (6–4) photoproduct and its corresponding Dewar photoproduct is shown.

triplet energy state of acetophenone is higher than T but lower than A, C, and G. When DNA is UV-irradiated in the presence of acetophenone, the photosensitizer is excited to a triplet state such that its energy can be transferred to T in DNA, leading to formation of TT dimers. The excited acetophenone can also transfer its energy to oxygen, leading to oxidative damage. The (6–4) photoproducts are not formed via a triplet state intermediate. Thus, UV radiation of DNA at 300 nm in the presence of acetophenone under nitrogen or argon produces high levels of TT dimers in the DNA without (6–4) photoproducts, a method that has greatly facilitated studies of TT dimers.

TT dimer is highly stable, which makes it possible to study its structure, repair, and mutagenesis. Crystal structure indicates that TT dimer is distorting to duplex DNA to some extent. The DNA helix is bent by about 30° toward the major groove and unwound by about 9° by the dimer. Despite the localized DNA distortion, the TT dimer is accommodated inside the DNA helix and some hydrogen bonding is still possible.

In contrast to the extraordinary stability of TT dimer, C-containing CPDs are prone to deamination, converting C to U or 5-methyl-C to T in the dimer. The half-life for deamination of TC and CT dimers appears to be several hours. It is very likely that CPD deamination is biologically important, as it occurs well within the time scale of cell division cycle of ~24 hr. Thus, some unrepaired CPDs may have been deaminated when such lesion sites are copied by translesion synthesis during replication. Accordingly, John-Stephen Taylor and colleagues postulated that deamination is the underlying mechanism for mutagenesis of the C-containing CPDs. Although C-containing dimers account for a smaller CPD fraction compared to TT dimers, they are much more mutagenic in cells than TT dimers.

22.3.2a.ii (6–4) Photoproducts. Pyrimidine-pyrimidone (6–4) photoproduct is formed by a covalent bond linking two adjacent pyrimidines on the same DNA strand. In the (6–4) photoproduct, the 5′ component remains a pyrimidine ring, while the 3′ component becomes a pyrimidone ring. The two components are linked together between the C6 of the 5′ pyrimidine ring and the C4 of the 3′ pyrimidone ring, hence the name (6–4) (Figure 22.10B). Only one stereoisomer of (6–4) photoproduct is formed in DNA. The (6–4) photoproducts are formed more frequently at TC or CC sequences than TT. Little is formed at the CT sequences. Formation of (6–4) photoproduct is inhibited by 5-methyl-cytosine. The (6–4) photoproducts are generally stable under neutral pH. However, the 3′ pyrimidone ring degrades to produce a DNA strand break under hot alkaline conditions (80–100 °C). In contrast to CPDs, the pyrimidone ring of the (6–4) photoproduct absorbs UV light with a maximum at ~325 nm. Therefore, further absorption of UV light (UVA and UVB) by (6–4) photoproduct leads to the secondary photoreaction, producing the Dewar photoproduct (Figure 22.10C).

In (6–4) photoproducts, the 3′ pyrimidone ring is nearly perpendicular to the 5′ pyrimidine ring and is oriented roughly parallel to the DNA helix axis. The result is a hole in DNA, which, in turn, disrupts base stacking with the base to the 3′ side of the lesion and between the two bases opposite the lesion. Based on NMR structural studies, duplex DNA containing a TT (6–4) photoproduct opposed to AA is bent by 44° toward the major groove, and unwound by 32°. When A is replaced by a G opposite the 3′ T of the TT (6–4) photoproduct, DNA bending is decreased to 27° and DNA unwinding is decreased to 2°, according to NMR analyses. This may provide a structural basis for the observation that G is frequently inserted opposite the 3′ T of the TT (6–4) photoproduct during translesion synthesis. The (6–4) photoproducts are more distorting than CPDs to the DNA structure. Consequently, (6–4) photoproducts are repaired much faster than CPDs by nucleotide excision repair. NMR studies indicate that the TT (6–4) photoproduct and its Dewar derivative cause similar levels of DNA destabilization. Consistently, both forms of the lesion are repaired at similar rates by the *E. coli* uvrABC system in vitro.

22.3.2b *Minor DNA Lesions of UV Radiation.* Several types of minor DNA damage are known following UV radiation. These include cytosine hydrate, thymine glycol, and single-stand break. Cytosine hydrate is formed by addition of a water molecule across the C5,C6 double bond of cytosine in DNA, generating 5,6-dihydro-6-hydroxy-cytosine (Figure 22.5). Thymine glycol is a significant lesion of oxidative damage and ionizing radiation.

22.3.3 DNA Damage by Chemicals

Many chemicals react with DNA, especially its bases, leading to various types of DNA damage. Increasing numbers of chemicals that can damage DNA are being synthesized as a result of industrial development. DNA damaging chemicals can be found in our environment and even in our foods. We can avoid human exposure to many DNA damaging chemicals or limit our exposure to a minimum, but we cannot practically live in an environment that is completely devoid of these chemicals. On the other hand, a few DNA damaging chemicals have been widely used in the clinic as anticancer drugs, taking advantage of the cytotoxicity associated with DNA damage. Most chemical carcinogens can damage DNA. The carcinogenic effects of these compounds are attributable to their ability to damage DNA and subsequently induce mutations. Thus, DNA damaging agents can cause cancer or can treat cancer, depending on the specific agent and how it is used. DNA damaging chemicals can be divided into two categories: chemicals that do not require bioactivation/metabolic activation (direct acting carcinogen) and chemicals that require bioactivation/metabolic activation (indirect acting carcinogen). Only a few chemicals are discussed here whose reactions with DNA are relatively well-studied. Many more are actively being studied and many others remain to be studied.

22.3.3a   DNA Damaging Chemicals without Metabolic Activation. These are chemical compounds that are reactive to DNA without further modification or metabolism in cells. Examples in this category are cisplatin compounds and simple alkylating agents.

22.3.3a.i   Induced Oxidative DNA Damage. Many chemicals can induce oxidative DNA damage. H_2O_2 and osmium tetroxide are examples of such two chemicals. Treatment of cells with H_2O_2 produces a complex spectrum of various types of oxidative DNA lesions. Osmium tetroxide (OsO_4), on the other hand, reacts specifically with single-strand DNA, producing predominantly thymine glycol as the major damage. Single-strand breaks are formed as a minor product of OsO_4 treatment. Because of its relative specificity, OsO_4-damaged DNA is often used for repair studies. OsO_4-modified DNA can be conveniently obtained in vitro by incubating purified DNA such as plasmids with the compound at 70 °C. High temperature regionally denatures double-strand DNA allowing OsO_4 to react with single-strand segments. The damaged DNA can be further purified—for example, via centrifugation in a 5–20% sucrose gradient for plasmids—to remove the small fraction of DNA containing single-strand breaks.

22.3.3a.ii   Cisplatin Compounds. Cisplatin (Figure 22.11) is widely used as an anticancer drug in the clinic. It is highly effective against testicular cancers. Cisplatin is also used in the treatment of ovary, bladder, lung, head, and neck cancers and lymphoma. Cytotoxic effect of cisplatin was first discovered by Barnett Rosenberg and colleagues in 1965 while studying the effects of electric current on bacteria growth. The growth inhibition observed turned out to be platinum complex of ammonia and chloride, which was produced in the medium from the platinum electrode. Subsequent studies by Barnett Rosenberg and colleagues demonstrated that cisplatin possesses antitumor activity. In 1978, cisplatin was approved by FDA for clinical use as an anticancer drug.

The major DNA lesions of cisplatin are intrastrand crosslinks, with cisplatin binding at the N7 position of purines. In vitro, reaction of *cis*-diamminedichloroplatinum (II) with DNA forms intrastrand crosslinks at the GG (65%), AG (25%),

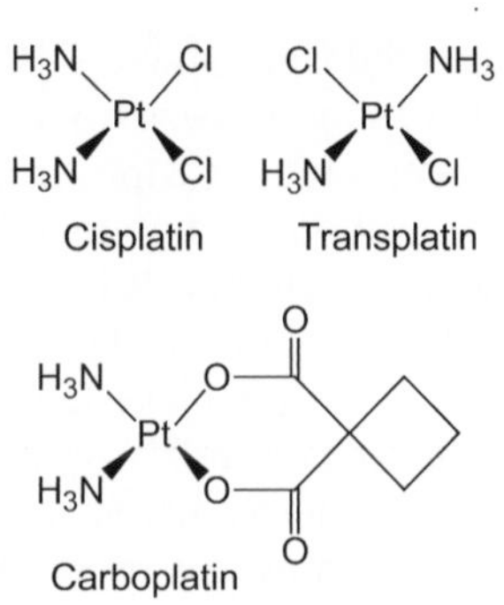

Figure 22.11. Structures of platinum compounds. Cisplatin and carboplatin are active agents used in cancer chemotherapy. Transplatin is inactive against tumors.

and GNG (6%) sequences. Minor interstand crosslinks at GC/CG (2%) are also produced. The anticancer activity of cisplatin is believed to result from cytotoxicity induced by cisplatin DNA adducts. The structure of cisplatin GG 1,2-intrastrand crosslink has been solved by X-ray crystal diffraction and NMR spectroscopy. The duplex DNA is unwound by ~21–25° and kinked ~58° toward the major groove at the damaged GG site. The minor groove is significantly perturbed. The DNA structural distortion by cisplatin 1,2-intrastrand crosslinks resembles that observed in DNA bound by the HMG-domain proteins. Thus, these proteins efficiently bind to cisplatin damaged DNA. One theory is that the cisplatin lesions bound by the HMG-domain proteins are shielded from repair by the nucleotide excision repair pathway, leading to effective killing of the cancer cells. It is not clear why cisplatin is especially effective against testicular cancers. Remarkably, transplatin (Figure 22.11) has no activity against tumors. The geometry of transplatin does not allow 1,2-intrastrand crosslinks. The major transplatin lesions are 1,3-intrastrand and interstrand crosslinks. HMG-domain proteins specifically recognize cisplatin 1,2-crosslinks. Thus, these proteins do not bind to transplatin-damaged DNA. These differences suggest that the anticancer activity of cisplatin originated from the 1,2-intrastrand crosslinks. Due to the renal and neurotoxicities of cisplatin, another cisplatin compound, carboplatin (Figure 22.11), was developed for clinical use. Carboplatin has reduced side effects and its primary side effect is hematopoietic toxicity.

22.3.3a.iii Simple Alkylating Agents. These are direct alkylating agents without the need for activation. They alkylate DNA by either the S_N1 (Substitution, Nucleophilic, unimolecular) or the S_N2 (Substitution, Nucleophilic, bimolecular) mechanism.

$$RX \rightarrow R^+ + X^- \xrightarrow{DNA} DNA\text{-}R + X^- \qquad S_N1 \text{ alkylation}$$

$$RX \xrightarrow{DNA} [\overset{\delta^-}{R}\text{-----}\overset{\delta^+}{DNA}\text{-----}\overset{\delta^-}{X}] \rightarrow DNA\text{-}R + X^- \qquad S_N2 \text{ alkylation}$$

S_N1 substitution involves an intermediate carbonium ion prior to its reaction with DNA, and proceeds with a first-order kinetics. The rate limiting step is the formation of the carbonium ion. S_N2 substitution depends on both the alkylating agent and DNA. Thus, the reaction proceeds with a second-order kinetics. Multiple sites on

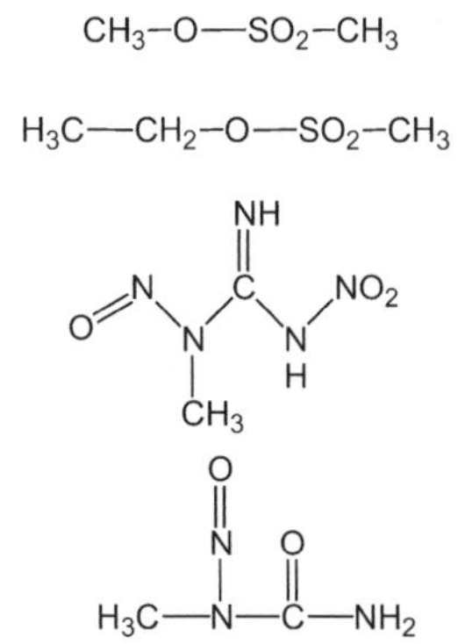

CH$_3$–O—SO$_2$–CH$_3$ MMS

H$_3$C—CH$_2$–O—SO$_2$–CH$_3$ EMS

MNNG

MNU

Figure 22.12. Structures of simple alkylating agents frequently used in the laboratory. MMS, methyl methanesulfonate; EMS, ethyl methanesulfonate; MNNG, *N*-methyl-*N'*-nitro-*N*-nitrosoguanidine; MNU, methylnitrosourea.

DNA can be alkylated. These include: N^1, N^3, N^6, and N^7 of adenine; N^1, N^2, N^3, N^7, and O^6 of guanine; N^3, N^4, and O^2 of cytosine; N^3, O^2, and O^4 of thymine; and O in the phosphodiester bond of the DNA sugar-phosphate backbone to form phosphotriesters. The exocyclic atoms are numbered in superscript. The base ring nitrogens are generally more nucleophilic than oxygens. In particular, the N^7 of guanine and the N^3 of adenine are the most reactive. Alkylation at these two positions destabilizes the *N*-glycosyl bond linking the base and the deoxyribose, leading to frequent depurination and the formation of AP sites in DNA.

Several simple alkylating agents frequently used in the laboratory include: methyl methanesulfonate (MMS), ethyl methanesulfonate (EMS), *N*-methyl-*N'*-nitro-*N*-nitrosoguanidine (MNNG), and methylnitrosourea (MNU) (Figure 22.12). MMS treatment of DNA yields predominantly N^7-methylguanine (83%) and N^3-methyladenine (11%), and a few minor products including O^6-methylguanine (0.3%). Thus, AP sites are often produced following MMS treatment of DNA. O^6-methylguanine is a miscoding lesion, which can pair with either C or T that results in G → A transition mutations. Therefore, this lesion is highly mutagenic. MNNG and EMS are much better agents than MMS in producing O^6-methylguanine in DNA.

22.3.3a.iv Psoralens are DNA Crosslinking Agents. Psoralens have been used to treat psoriasis, a human skin disease. Psoralens such as 8-methoxypsoralen has been frequently used in the laboratory as a model system for repair studies of interstrand DNA crosslinks. Due to their planar tricyclic configuration, psoralens can intercalate into DNA. Upon photoreactivation by UVA, two major types of DNA damage are formed: (1) psoralen monoadduct covalently linked to thymine as a result of addition reaction across its C5,C6 double bond and (2) interstrand DNA crosslink at TA sequences. The crosslink is produced by two independent addition reactions between the C3,C4 double bond of psoralen and thymine on one DNA strand, and between the C4',C5' double bond of psoralen and thymine on the other DNA strand.

Interstrand crosslinks pose a considerable challenge for replication, transcription, and repair. The two DNA strands are covalently linked together by the crosslinking agent, and thus cannot be separated. Crosslinks are repaired in cells, however, their repair is the least understood among all repair mechanisms. The strong cytotoxic effect of crosslinking agents has been exploited for cancer chemotherapies in the clinic.

22.3.3b DNA Damaging Chemicals that Require Metabolic Activation. These are chemical compounds that are relatively nonpolar and are thus inactive by themselves. These compounds are normally metabolized in cells (e.g., by P450 enzymes) to more polar and oxygenated products such that they can be excreted from the body. This constitutes the important cellular detoxification system. Sometimes, however, the detoxification process converts the inert parent compound into electrophiles that are reactive towards DNA, achieving metabolic activation or bioactivation into genotoxic and carcinogenic derivatives. These reactive metabolites that bind covalently to DNA are often called the ultimate carcinogen.

22.3.3b.i Chemotherapeutic Alkylating Agents. Toxic gases were used as chemical weapons in World War I. The most devastating gas was sulfur mustard (ClCH$_2$CH$_2$-S-CH$_2$CH$_2$Cl, bischloroethylsulfide). Its effects on humans led to studies on the treatment of cancer by sulfur mustard. Later, another closely related compound, nitrogen mustard of World War II, was used for clinical studies as an anti cancer agent. Now, the most widely used chemotherapeutic alkylating agents are the nitrogen mustards. Five nitrogen mustards are commonly used for cancer therapy in the clinic: mechlorethamine (the original nitrogen mustard), cyclophosphamide, ifosfamine, melphalan, and chlorambucil (Figure 22.13). Cyclophosphamide is the most widely used chemotherapeutic alkylating agent and is used for the treatment of lymphoma, some leukemia, and some solid tumors. It is also used to treat some autoimmune diseases.

The mustards are S$_N$1 agents. Nitrogen mustards react with DNA through an aziridinium intermediate. These mustards are bifunctional alkylating agents—that is, capable of forming DNA interstrand crosslinks. While sulfur mustard and mechlorethamine are direct alkylating agents without the need for activation, cyclophosphamide is a prodrug that undergoes metabolic activation in the body (Figure 22.14).

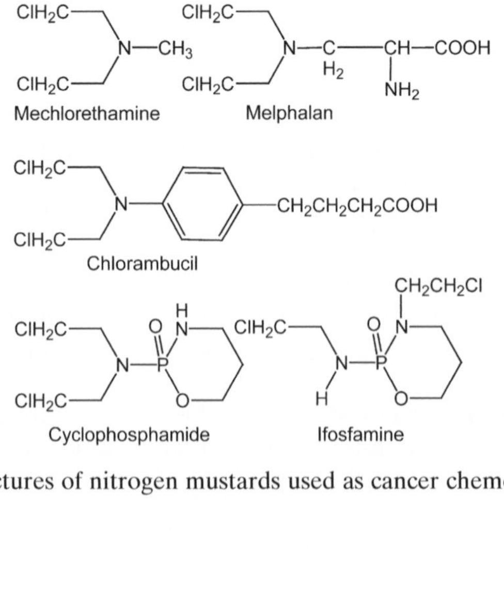

Figure 22.13. Structures of nitrogen mustards used as cancer chemotherapeutic drugs.

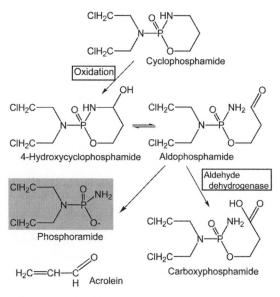

Figure 22.14. Metabolism of cyclophosphamide. The major therapeutic metabolite, phosphoramide, is highlighted. [Adapted from Colvin, O. M. In: Holland, J. F., Bast, R. C., Jr., Morton, D. L., Frei, E., III, Kufe, D. W., and Weichselbaum, R. R. (Eds.). *Cancer Medicine*, Vol. 1, 4th ed., Williams & Wilkins, Baltimore, 1997, pp. 949–975.]

Cyclophosphamide is converted to 4-hydroxycyclophosphamide in the liver by microsomal oxidation, which exists in equilibrium with its tautomer, aldophosphamide. At physiological pH, the predominant form is 4-hydroxycyclophosphamide. The 4-hydroxycyclophosphamide/aldophosphamide mixture diffuses from hepatocytes into the plasma and is distributed throughout the body. The 4-hydroxycyclophosphamide is relatively nonpolar. Thus, it readily enters cells by diffusion. In cells, aldophosphamide spontaneously decomposes to phosphoramide mustard and acrolein (Figure 22.14). Phosphoramide mustard is the direct alkylating agent that reacts with DNA and is the major metabolite responsible for the therapeutic effects of cyclophosphamide. Acrolein is also an electrophile that can damage DNA. In the presence of aldehyde dehydrogenase, cyclophosphamide is oxidized to carboxyphosphamide, which is an inactive metabolite that is excreted in the urine. The majority (~80%) of an administered dose of cyclophosphamide is eliminated through this mechanism. Hepatocytes, primitive hematopoietic cells, intestine stem cells, and mucosal absorptive cells contain high concentrations of aldehyde dehydrogenase. Therefore, cyclophosphamide is associated with less gastrointestinal and hematopoietic toxicity as compared to other chemotherapeutic alkylating agents, which makes cyclophosphamide a popular choice among its peers.

Another alkylating drug in clinical use is mitomycin C. As a natural product isolated from *Stremtomyces lavendulae*, mytomycin C is an aziridine compound, which is closely related to nitrogen mustards. This compound is best known to the DNA repair field as a DNA interstrand crosslinking agent. Mytomycin C has been used to treat breast cancer and cancers of the gastrointestinal tract.

Alkylating agents kill cells by damaging DNA. Another important consequence of DNA damage is mutagenesis. Not surprisingly, secondary tumor induction as a result of chemotherapy has been observed with alkylating agents. The rate of acute leukemia after alkylating agent therapy may be 10% or higher in some groups of patients. An increasing rate of solid tumors is also noted in association with therapeutic alkylating agents. With improving cancer therapy and increasing life span of cancer patients, addressing the issue of secondary tumor resulted from some chemotherapy is becoming more and more important.

22.3.3b.ii N-2-Acetyl-2-Aminofluorene. *N*-2-Acetyl-2-aminofluorene (AAF) is an aromatic amine. Just before World War II, AAF was considered a promising insecticide until it was shown in 1941 to be tumorigenic in rats. Nevertheless, AAF became one of the most widely studied model DNA lesions in the field of DNA repair and mutagenesis, and it is often used in liver tumorigenesis studies in animal models. In cells, AAF metabolism is initiated by the cytochrome P450 system. Among the final metabolites are *N*-acetoxy-*N*-2-acetylaminofluorene (AAAF) and *N*-acetoxyaminofluorene (Figure 22.15). Both are reactive with guanine in DNA. The major DNA lesions are the C8 adducts of guanine: AAF-dG by AAAF and AF-dG by *N*-acetoxyaminofluorene, respectively (Figure 22.15).

22.3.3b.iii Polycyclic Aromatic Hydrocarbons. Polycyclic aromatic hydrocarbons (PAH) are a group of significant environmental pollutants. They are produced by the incomplete combustion of organic materials such as wood or coal and are also commonly found in tobacco smoke, automobile exhaust, and foods charred during cooking. PAHs are nonpolar and inert compounds themselves. These compounds

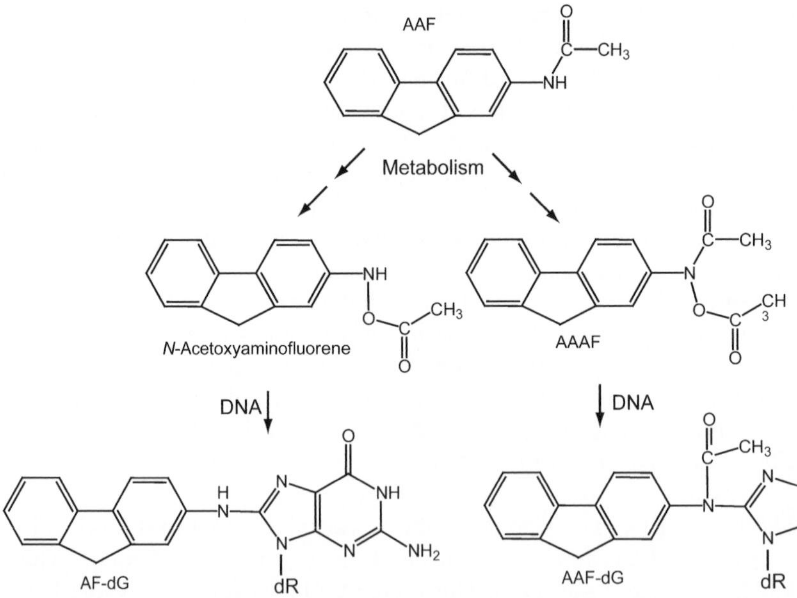

Figure 22.15. Metabolism of *N*-2-acetyl-2-aminofluorene (AAF) to form the ultimate carcinogens that damage DNA. Two metabolites are reactive toward DNA, forming AF-dG and AAF-dG as the major DNA adducts, respectively. AAAF, *N*-acetoxy-*N*-2-acetylaminofluorene.

are metabolized in cells by the P450 system for excretion. Among the metabolites are highly reactive bay region dihydrodiol epoxide derivatives of PAHs. Dihydrodiol epoxides are electrophilic and readily react with DNA, forming covalently linked bulky adducts on DNA bases. Therefore, PAHs in general are mutagenic and carcinogenic.

The carcinogenic effect of PAHs was first noticed over 200 years ago. It was observed that chimney sweeping was associated with high incidence of skin (scrotum) cancer among the English chimney sweepers then. Chimneys were used for burning coal, and thus contaminated by high levels of PAHs. Later, taking daily baths was recommended to the chimney sweepers as a preventive measure against skin cancer. In the 1930s, benzo[a]pyrene was isolated from crude coal tar and was identified as a potent carcinogen.

Benzo[a]pyrene is an extensively studied PAH compound. The mutagenic metabolites of benzo[a]pyrene are the (+)-7R,8S,9S,10R-anti-benzo[a]pyrene-7,8-dihydrodiol-9,10-epoxide, (+)-anti-BPDE, and the (−)-7R,8S,9S,10R enantiomer, (−)-anti-BPDE (Figure 22.16A,B). DNA damage occurs mainly by adduct formation between the C10 position of anti-BPDE to the N2 position of guanine. Four stereoisomeric bulky adducts are produced: 10S (+)-trans-anti-BPDE-N^2-dG, 10R (+)-cis-anti-BPDE-N^2-dG, 10R (−)-trans-anti-BPDE-N^2-dG, and 10S (−)-cis-anti-BPDE-N^2-dG. In vitro, the reaction of (+)-anti-BPDE with DNA forms predominantly the (+)-trans-anti-BPDE-N^2-dG adduct, while the reaction of (−)-anti-BPDE produces mainly the (−)-trans-anti-BPDE-N^2-dG adduct. In cells, the major benzo[a]pyrene DNA adduct is (+)-trans-anti-BPDE-N^2-dG.

Structures of the four anti-BPDE-N^2-dG DNA adducts have been solved by NMR. In a duplex DNA containing a (+)- or (−)-trans-anti-BPDE-N^2-dG adduct, the pyrenyl residues are not intercalated between adjacent base pairs. In the case of the (+)-trans-anti-BPDE-N^2-dG adduct, the aromatic pyrenyl residue stacks primarily over an adjacent sugar ring in the complementary strand in the minor groove and is oriented toward the 5′ end of the modified strand. For the (−)-trans-anti-BPDE-N^2-dG adduct, the pyrenyl residue stacks mainly over a sugar ring in the complementary strand in the minor groove, and is oriented toward the 3′ end of the modified strand. Thus, the stereoisomeric (+)- or (−)-trans-anti-BPDE-N^2-dG adducts adopt different configurations in duplex DNA as a result of the different absolute configurations of substituents about the four chiral carbon atoms.

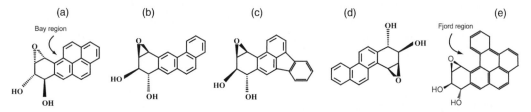

Figure 22.16. Structures of the dihydrodiol epoxides of several polycyclic aromatic hydrocarbons. (A) Benzo[a]pyrene-trans-7,8-dihydrodiol-9,10-epoxide (BPDE); (B) benz[a]anthracene-trans-8,9-dihydrodiol-10,11-epoxide (BADE); (C) benzo[b]fluoranthene-trans-9,10-dihydrodiol-11,12-epoxide (BFDE); (D) chrysene-trans-1,2-dihydrodiol-3,4-epoxide (CDE); and (E) dibenzo[a,l]pyrene-trans-11,12-dihydrodiol-13,14-epoxide (DBPDE). Only one enantiomer for each dihydrodiol epoxide is shown.

Another PAH with even more potent carcinogenicity than benzo[*a*]pyrene is dibenzo[*a,l*] pyrene (DB[*a,l*]P). DB[*a,l*]P is similar to benzo[*a*]pyrene (Figure 22.16E), but with an extra benzene ring. The addition of this ring turns the bay region of benzo[*a*]pyrene into a fjord region. A fjord-region PAH is generally more carcinogenic than its corresponding bay-region PAH. In cells, the 11,12-diol-13,14-epoxide of DB[*a,l*]P is formed. Activation of DB[*a,l*]P occurs with high stereoselectivity, producing the (−)-*anti*-(11*R*,12*S*,13*S*,14*R*)- and the (+)-*syn*-(11*S*,12*R*,13*S*,14*R*)-DB[*a,l*]PDE. The major lesions are DB[*a,l*]PDE-dA DNA adducts (>73%). The *anti*-DB[*a,l*]PDE formed ~17% guanine adducts, while the *syn*-DB[*a,l*]PDE produced less than 9% guanine adducts. DB[*a,l*]P is probably the most potent chemical carcinogen known. It is not completely understood why this PAH carcinogen is so potent. One important contributing factor is that DB[*a,l*]PDE is more reactive toward DNA than BPDE. Severalfold more DB[*a,l*]P DNA adducts are produced than benzo[*a*]pyrene DNA adducts at the same concentration of the respective diol epoxides. Several other PAH diol epoxides are shown in Figure 22.16.

22.3.3b.iv   Aflatoxin B1. Aflatoxins are natural products (i.e., mytotoxins) produced by the fungi *Aspergillus flavus* and *A. parasiticus*. They are potent liver carcinogens, especially aflatoxin B1. These fungi can grow on cereal grains. Therefore, molded rice, peanuts, and corn pose a health risk to humans if consumed. In cells, aflatoxin B1 is metabolically activated by the cytochrome P450 system to form mainly aflatoxin B1-8,9-epoxide. The major DNA damage is aflatoxin B1 adducts at the N7 position of guanine. These products are not stable. Depurination occurs leaving AP sites in DNA. Formamidopyrimidine derivatives with the imidazole ring opened are also formed (Figure 22.17).

22.3.3b.v   Oxidative DNA Damage by Quinone Redox Cycling. Quinones may exert multiple toxic effects in vivo, one of which is oxidative stress through redox cycling. Quinones can be produced in cells as metabolic products from many organic compounds. For example, phenols, hydroquinones, and catechols can be converted to quinones by monooxygenases or peroxidases; and benzene can be metabolized to *p*-benzoquinone. Some quinones are highly active for redox cycling. They can undergo enzymatic and nonenzymatic redox cycling with their corresponding semiquinones. ROS is generated with each redox cycle (Figure 22.18). Thus, each quinone molecule can generate multiple molecules of ROS until the redox cycle is terminated. When redox cycling between quinone and its semiquinone occurs in cells, the ROS byproducts lead to oxidative DNA damage via hydroxyl radicals as discussed in Section 22.2.4.

22.4   CONCEPTS OF MUTAGENESIS

22.4.1   Definition of Terms

Several terms are frequently used to describe DNA mutations and related studies. These terms are defined here to avoid potential confusions. They are divided into three categories: (a) general terms, (b) terms that are based on the phenotypic consequences, and (c) terms that are based on DNA sequence changes.

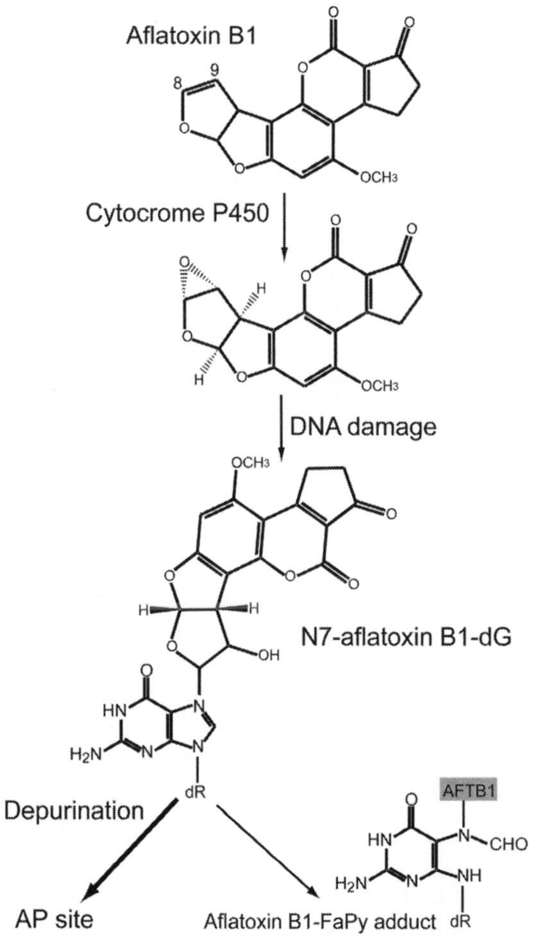

Figure 22.17. Metabolism of aflatoxin B1 (AFTB1) to form the ultimate carcinogen that damage DNA. The major aflatoxin B1 adduct, N7-aflatoxin B1-dG, is unstable. It depurinates to leave an AP site in DNA. The imidazole ring-opened derivative can also be formed from the initial adduct. [Adapted from Friedberg, E. C., Walker, G. C., Siede, W., Wood, R. D., Schultz, R. A., and Ellengerger, T. *DNA Repair and Mutagenesis*, 2nd ed., ASM Press, Washington, D.C., 2006.]

22.4.1a *General Terms*

Genome. The complete set of genetic material of an organism. For animals, the genome consists of the nuclear DNA and the mitochondrial DNA. The nuclear and the mitochondrial DNA can be referred to as the nuclear genome and the mitochondrial genome, respectively. For plants, the genome consists of nuclear DNA, the mitochondrial DNA, and the chloroplast DNA.

Genotype. The genetic information encoded in the genome of an organism. Literally, it is the type of genome.

Phenotype. The observable characteristics or traits of an organism. Literally, it is the type of phenomena. Most phenotypic traits are determined by genotype. However, not all phenotypic traits are genetic.

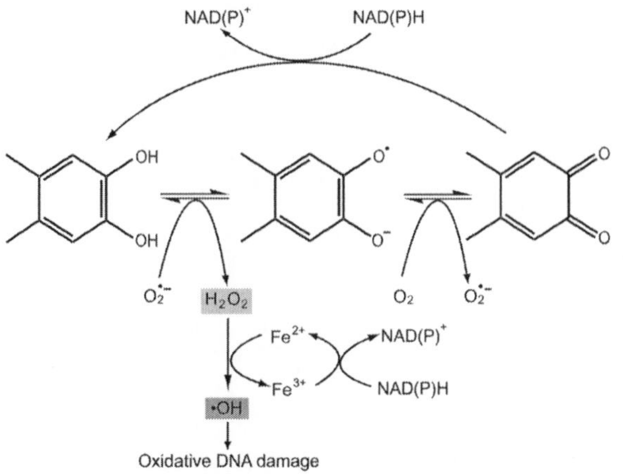

Figure 22.18. Quinone redox cycling. A substituted catechol is shown as an example. [Adapted from Squadrito, G. L., Cueto, R., Dellinger, B., and Pryor, W. A. *Free Radic. Biol. Med.* **31**, 1132–1138, 2001.]

Mutation. Heritable change in the genomic sequence of an organism. Changes of genotype result from mutation. However, not all mutations change the genotype. In fact, most mutations in human nuclear genome do not change the genotype, thus they are inconsequential. The major reason for this is that the vast majority of human nuclear genome is noncoding. Thus, most mutations, assuming random occurrence in the genome, do not hit the coding sequences.

Mutagenesis. The process that generates mutations.

Mutant. An organism that has one or more mutations in its genome.

Allele. One of the variant forms of a gene at a particular locus on a chromosome. Thus, different mutant allele means different mutations on the same gene; and different mutants of the same gene are said to be alleles of each other.

Pleiotropic Phenotype. Describes the multiple effects of a mutation on the phenotype of an organism.

Wild-type. Refers to an organism in its normal genotypic state relative to a mutant, from which the mutant is derived by mutation.

22.4.1b *Terms Based on Phenotypic Consequences*

Conditional Mutation. Phenotype of the organism changes only under a specific condition called the restrictive condition, but remains normal under other conditions called the permissive condition.

Temperature-Sensitive Mutation. It is a special case of conditional mutation in which the restrictive condition is elevated or lowered temperature. Temperature-sensitive mutants are commonly used to analyze the function of essential genes, because inactivating mutations such as deletion mutations are not possible for these genes.

Lethal Mutation. This refers to a mutation that causes cell death.

Synthetic Lethality. A mutant organism is viable when separately carrying only one of two or more mutant genes. When these mutations occur together or when these individual mutant genes are combined together in the same organism, lethality results.

Auxotroph. A mutation that causes a defect in the synthesis of an essential metabolite such that the growth of the organism is dependant on exogenous supply of this metabolite.

Forward Mutation. A mutation that occurs in the wild-type gene to yield a mutant gene.

Reverse Mutation. A mutation that reverses the mutant phenotype back to the wild-type phenotype. This does not necessarily always result from mutating the mutant gene sequence back to the wild-type sequence.

22.4.1c   Terms Based on DNA Sequence Changes

Point Mutation. A mutation occurs at a DNA base in which the base is mutated to another base, is deleted, or another base is inserted.

Base Substitution Mutation. A mutation where one base is mutated to a different base. That is, a DNA base is replaced by one of the other three bases.

Transition Mutation. Occurs when one purine is mutated to another purine or one pyrimidine is mutated to another pyrimidine, that is, $A \rightarrow G$, $G \rightarrow A$, $C \rightarrow T$, and $T \rightarrow C$.

Transversion Mutation. Occurs when a purine is mutated to a pyrimidine or vise versa, i.e., $A \rightarrow C$, $A \rightarrow T$, $G \rightarrow C$, $G \rightarrow T$, $C \rightarrow A$, $C \rightarrow G$, $T \rightarrow A$, and $T \rightarrow G$.

Frameshift Mutation. Deletion or insertion of a base or bases that alters the reading frame of the gene. In contrast to base substitution mutations where the mutation often has no consequence to the gene function or has only a partial effect on the gene function, a frameshift mutation completely changes the amino acid sequence downstream of the mutation site within a protein-coding gene. Thus, frameshift mutation is a severe type of mutation, which results in inactivation of the protein product most of the time.

Silent Mutation. A base substitution that does not change the amino acid of the protein-coding gene.

Missense Mutation. A base substitution that mutates one amino acid to another amino acid in the gene product.

Nonsense Mutation. A base substitution that mutates the codon of one amino acid to a stop codon, TAA (ochre), TAG (amber), or TGA (opal). A nonsense mutation results in protein truncation of a protein-coding gene. Thus, nonsense mutation is another type of severe mutation, which results in inactivation of the protein product most of the time.

22.4.2   Origin of Mutagenesis

Mutations mainly originate from limited fidelity of DNA replication and DNA damage. Additionally, nonhomologous end joining (NHEJ) during repair of double-strand breaks also leads to mutagenesis.

Cells contain extremely efficient replication apparatus to duplicate their nuclear genomes during cell division. Replicative DNA synthesis occurs with extraordinarily high efficiency and fidelity. As discussed in Section 22.2.1, the error rate is lowered to ~10^{-9} per base pair per replication, due to combined effects of accurate DNA synthesis by Polε and Polδ, $3' \rightarrow 5'$ proofreading exonuclease activity, and mismatch repair. However, the replication apparatus does have a tiny margin of error. When massive amounts of bases in the human nuclear genome (~3×10^9 bp) are copied during each round of replication, a few mutations will theoretically be produced. In addition to mis-incorporation, DNA synthesis at single nucleotide repeat sequences and microsatellite sequences are prone to slippage. Such replication slippage can result in deletions or insertions.

Spontaneous DNA damage is unavoidable, and some environmental DNA damage is practically unavoidable. Therefore, the human genome contains DNA damage at any given moment. For example, oxidative damage is always detected by sensitive techniques in human DNA even when the cellular repair mechanisms are fully functional in a normal subject, regardless when the blood sample is collected. When encountering DNA damage, the replicative polymerases become ineffective, because they have evolved to be so highly efficient and specific to deal with the normal DNA template. Specialized DNA polymerases are required to copy the damaged sites of the DNA template. The challenge for these specialized polymerases is not high efficiency and accuracy, but merely being able to copy the damaged sites. Consequently, these polymerases have very low fidelity for DNA synthesis and lack the $3 \rightarrow 5'$ proofreading exonuclease activity. Additionally, base damage often results in alterations of its coding property. These factors together determine that replication through a damaged site on DNA (a cellular process called translesion synthesis) is highly error-prone.

The limited replication fidelity and replication across from damaged sites on DNA make it certain that mutations are produced every time cells duplicate. According to this concept, then, agents that strongly promote cell division are carcinogenic, since they promote mutagenesis by promoting replication, although they are not direct mutagens per se.

Double-strand break is an important class of DNA damage. In cells, double-strand breaks are repaired by recombination. Whereas homologous recombination is mostly accurate, NHEJ is highly error-prone. Therefore, repair of double-strand breaks by NHEJ frequently results in mutagenesis at sites of DNA joining.

22.4.3 Biological Significance of Mutagenesis

Mutagenesis forms the fundamental basis of evolution and adaptation, human hereditary diseases, and cancer. Most mutations in the human genome are expected to be neutral—that is, having no functional consequences. Nevertheless, some mutations may be harmful to an individual. For example, mutations can cause hereditary diseases and cancer. However, in the context of the whole species, mutations are essential for evolution and adaptation. Under genotoxic stress conditions, the genome becomes more flexible in the sense that it is more prone to permanent change through mutagenesis. This would help a species to survive and subsequently proliferate in a changing environment. Hence, mutagenesis capacity is critical for the long-term survival of a species.

In higher eukaryotes, somatic hypermutation is a key step in the development of immunoglobulin genes to generate diverse antibodies. It introduces mainly point mutations into the V region of Ig genes at a rate of 10^{-3}–10^{-4}/bp/generation, which is ~10^5 to 10^6-fold higher than the spontaneous mutation rate in the rest of the genome. Somatic hypermutation may be viewed as a precisely controlled and super-fast gene evolution in B cells of the germinal center. The powerful gene diversity produced by hypermutagenesis has been employed by nature as the highly efficient means of creating antibody diversity. The possibilities of antibody diversity are essentially unlimited. For a peptide that is a mere 100 amino acids long, the theoretical diversities possible are 20^{100}, that is, ~10^{130}! Therefore, mutagenesis is a key component of the immunological defense system in higher eukaryotes.

22.5   MECHANISMS OF DNA DAMAGE-INDUCED MUTAGENESIS

In most cases, the mutagenic consequence of DNA damage is microgenomic alteration—that is, limited DNA sequence alterations such as point mutations, small deletions, and small insertions. Among the various types of DNA damage, base modifications greatly predominate. Base damage-induced mutagenesis occurs mainly through a common mechanism—that is, error-prone translesion synthesis. Double-strand breaks account for a small fraction of overall DNA damage. Mutagenesis induced by double-strand breaks is mediated mainly by a distinct mechanism—that is, nonhomologous end joining. Some DNA damage can also lead to macrogenomic alteration—that is, chromosomal aberrations. These various mechanisms of damage-induced mutagenesis are presented.

22.5.1   Chromosomal Aberrations

Chromosomal aberrations consist of numerical chromosomal alterations and structural chromosomal alterations. Numerical chromosomal alterations result in aneuploidy—that is, loss or gain of one or more chromosomes. Chromosomal breaks and rearrangements such as translocation, deletion, and gene amplification are structural chromosomal alterations. With proper staining or FISH (fluorescence in situ hybridization) painting of the metaphase chromosomes, these chromosomal aberrations are microscopically visible. Thus, they are genomic alterations occurring at the macroscale, relative to the microscale genome at the DNA sequence level.

An agent that causes chromosomal breakage is called a clastogen. Ionizing radiation and benzene are examples of clastogens, which produce chromosomal aberrations following cellular exposure to these agents. When chromosomal aberrations are induced by DNA damaging agents, double-strand break is the principal lesion leading to the abnormality. Therefore, one important consequence of ionizing radiation is chromosomal aberrations. UV radiation and the majority of chemical mutagens do not induce double-strand breaks directly. However, single-strand breaks and some labile sites like AP site can be converted into double-strand breaks during replication through the process called fork collapse. Thus, some other DNA damaging agents are also associated with clastogenic effects.

Unrepaired double-strand breaks may lead to broken chromosomes. Double-strand breaks are repaired by recombination. In higher eukaryotes, the recombination mechanisms are homologous recombination and nonhomologous end joining. Homologous recombination requires regions of extensive sequence homology. Normally, homologous recombination occurs between sister chromatids in mitotic cells. The result of such recombination is accurate repair of the double-strand break. However, when recombination occurs between homologous DNA sequences in different chromosomes (i.e., ectopic homologous recombination), reciprocal translocations, dicentric chromosome, and acentric fragments may be formed (Figure 22.19). When recombination occurs between two direct repeat sequences within the same chromosome, also called single-strand annealing, the result is deletion of one repeat unit and the intervening sequence between the repeats (Figure 22.19). Only one double-strand break is needed to initiate homologous recombination. If two double-strand breaks are formed on different chromosomes, repair by nonhomologous end joining may also produce dicentric chromosomes and acentric fragments (Figure 22.19).

During chromosomal segregation, the dicentric chromosome is pulled toward the two opposite directions due to two centromeres on the same chromosome, leading to chromosomal breakage, which in turn may lead to another translocation. In contrast, acentric fragments do not contain any centromeres. Thus, when the cell divides, some acentric fragments are excluded from the daughter nuclei, resulting in small extra nuclei within the cytoplasm, either on their own or in conjunction with other fragments. These structures are called micronuclei. Micronuclei are sometimes used as a marker or indication of chromosomal aberrations.

Chromosomal aberration is spontaneously elevated in the human chromosomal breakage syndromes, which include ataxia telangiectasia, Nijmegen breakage syndrome, ataxia telangiectasia-like disorder, Bloom syndrome, Werner syndrome, and Fanconi anemia. These are hereditary autosomal recessive diseases characterized by chromosomal instability and cancer predisposition. Ataxia telangiectasia, Nijmegen breakage syndrome, and ataxia telangiectasia-like disorder are defective in the

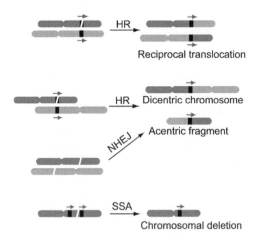

Figure 22.19. Mechanistic models for the formation of chromosomal aberrations. Regions of sequence homology (e.g., repeats and pseudogenes) are shown as black boxes with the arrow indicating orientation. HR, homologous recombination; NHEJ, nonhomologous end joining; SSA, single-strand annealing.

ATM, NBS1, and *MRE11* genes, respectively. These gene products are involved in signaling and repair of double-strand breaks. Their corresponding mutant cells are sensitive to ionizing radiation. Bloom syndrome and Werner syndrome are defective in the *BLM* and *WRN* genes, respectively. These two genes belong to the RecQ family of DNA helicases. The precise functions of *BLM* and *WRN* are not very clear. They probably act to resolve damage-stalled replication forks to prevent fork collapse. The hallmark of Bloom syndrome cells is high levels of sister chromatid exchange. Fanconi anemia is complex, involving multiple genes. One important role of the Fanconi anemia proteins is repair of DNA crosslinks. DNA damage-induced chromosomal aberrations are generally more pronounced in these mutant cells.

Chromosomal aberrations alter genomic sequence. Thus they are mutagenic. More importantly, chromosomal aberrations may result in loss of tumor suppressor genes and/or activation of oncogenes. Chromosomal aberrations occur quantitatively at a much smaller scale compared to DNA sequence mutations, their biological consequences, however, are generally more severe than most point mutations.

22.5.2 Mutagenesis Induced by Double-Strand Breaks

In addition to chromosomal aberrations that can be induced by DNA double-strand breaks, other mutations at the DNA sequence level (microgenomic alteration) can also be induced by double-strand breaks. In higher eukaryotes, double-strand break repair by homologous recombination normally occurs between sister chromatids, and the result is error-free. Essentially, the genetic information lost at the double-strand break is restored by using its intact sister chromatid as the repair template. Thus, repair by homologous recombination is restricted to the late S phase and the G2 phase of the cell cycle. Double-strand break repair by NHEJ, however, occurs throughout the cell cycle including the late S and G2. Therefore, NHEJ is the major repair mechanism for double-strand breaks in higher eukaryotes. Unlike homologous recombination, NHEJ does not rely on sister chromatid. Consequently, the genetic information that is lost at the double-strand break is permanently lost, and thus NHEJ is highly mutagenic. In contrast to base damage-induced mutagenesis, double-strand break-induced mutagenesis occurs as a result of repairing the break.

The NHEJ repair pathway proceeds in either microhomology-dependent or microhomology-independent manner. Regardless of whether or not microhomology of a few nucleotides exists at the ends of the double-strand break, most NHEJ involves DNA end processing that includes nucleotide removal from the DNA termini and gap filling by a DNA polymerase. After the DNA ends are finally ligated, most of the repaired products contain small deletions at the joining sites. Occasionally, small insertions at the joining sites are also produced.

22.5.3 DNA Damage Tolerance

DNA repair is the major cellular defense system against DNA damage, which physically removes the lesions from DNA. Cell cycle checkpoint control empowers the repair system with better efficiency in that the cell cycle is temporarily halted to allow more time for DNA repair. In multicell organisms, apoptosis is additionally employed to remove excessively damaged cells. Nevertheless, these cellular defense systems do not function with perfection, leading to some DNA damage that persists during replication. Several factors further promote persistence of DNA lesions during replication. These include (a) high levels of damage, (b) poorly repaired lesions, (c) inefficiently repaired genomic regions, and (d) damage sustained in the S phase of the cell cycle. Replicative polymerases have evolved to perform highly efficient and accurate DNA synthesis from templates of normal structure and chemical compositions. When the DNA template is damaged, the replicative polymerases become ineffective. That is, DNA lesions frequently block replication. Cells would die if replication cannot be completed. In response to the unrepaired DNA lesions during replication, cells have evolved a sophisticated system. It allows the cell to replicate its genome in the presence of DNA damage that would normally block the replicative polymerases. This system tolerates, rather than removes, DNA damage; hence it is called damage tolerance. After replication, the tolerated lesions are then subject to removal by DNA repair systems. In eukaryotes, damage tolerance consists of at least two mechanisms: (a) error-free postreplication repair, also known as template switching, and (b) translesion synthesis (Figure 22.20).

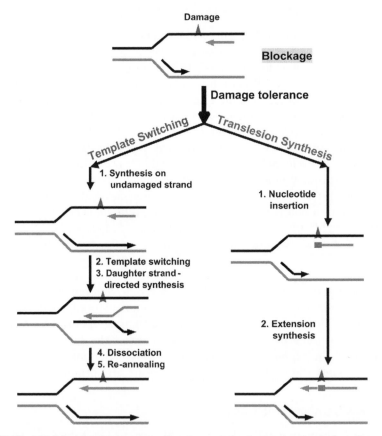

Figure 22.20. Models of two damage tolerance mechanisms. At the lesion site, template switching (the left pathway) uses the newly synthesized daughter strand as the template for DNA synthesis, thus, bypassing the lesion in an error-free manner. In contrast, translesion synthesis (the right pathway) directly copies the damaged site on the template. Consequently, mutations, shown as a square, are often generated opposite the lesion.

The term postreplication repair originated from experiments employing an alkaline sucrose gradient to examine DNA synthesis following UV radiation of the cell. Immediately after UV radiation, smaller DNA fragments are generated as detected by the alkaline sucrose gradient. With extended incubation time, these smaller DNA fragments are converted to large fragments that are normally detected without DNA damage. Now it is clear that this phenomenon reflects the action of DNA damage tolerance. Since the DNA lesion is not physically removed, postreplication "repair" may be misleading. Therefore, template switching is a preferred term to describe this mechanism of damage tolerance.

In eukaryotes, template switching requires at least Rad6, Rad18, Rad5, Mms2-Ubc13 complex, PCNA, and Polδ. The molecular details of template switching remain largely unknown in eukaryotes. Nevertheless, a conceptual model is shown in Figure 22.20. In this model, when DNA synthesis is blocked by a template lesion, synthesis on the undamaged template strand may continue to a limited extent. Then, by using the newly synthesized daughter strand as the template (template switching), the lesion-blocked DNA synthesis can proceed further downstream beyond the site corresponding to the lesion. Following dissociation of the two newly synthesized daughter strands, each is re-annealed to its original parental strand, thus bypassing the damaged site on the parental DNA strand (Figure 22.20). Template switching avoids copying the damaged site of the DNA template. Therefore, the result is error-free tolerance of the damage. In contrast, translesion synthesis directly copies the damaged site of the template. Consequently, mutations are often produced at sites of the damage (Figure 22.20). Mechanisms of translesion synthesis are presented below.

22.5.4 Error-Prone Translesion Synthesis Is the Major Mechanism of Base Damage-Induced Mutagenesis

Translesion synthesis is the cellular process that directly copies damaged sites of the template during DNA synthesis. It consists of nucleotide insertion opposite the lesion and extension synthesis from opposite the lesion (Figure 22.20). A DNA polymerase that performs the insertion step, the extension step, or both is referred to as a translesion polymerase or bypass polymerase. The term *lesion bypass* is sometimes interchangeably used with translesion synthesis. Depending on the accuracy of nucleotide insertion opposite the DNA damage, translesion synthesis is further divided into two categories: error-free and error-prone. Error-free translesion synthesis predominantly inserts the correct nucleotide opposite the lesion. Thus, it is a mutation-avoiding mechanism that suppresses DNA damage-induced mutagenesis. Error-prone translesion synthesis frequently inserts an incorrect nucleotide opposite the lesion. Thus, it is a mutation-generating mechanism that promotes DNA damage-induced mutagenesis. Error-prone translesion synthesis constitutes the major mechanism of base damage-induced mutagenesis in cells.

For a given lesion, error-free or error-prone synthesis by a bypass polymerase is often determined by in vitro translesion synthesis assays (Figure 22.21). The assay involves in vitro DNA synthesis by a purified bypass polymerase from an oligonucleotide template containing a site-specific lesion. A DNA primer labeled with ^{32}P at its 5' end is annealed to the damaged template prior to assembling the assay reactions. Following the polymerase reaction, products are separated by electrophoresis

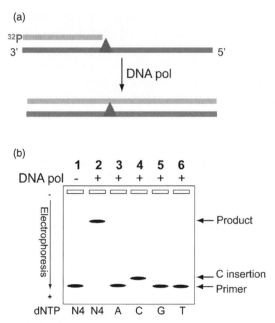

Figure 22.21. In vitro translesion synthesis assays for studying lesion bypass and base damage-induced mutagenesis. (A) Schematic illustration of the tranlesion synthesis reaction. The reaction is catalyzed by a purified bypass polymerase from an oligonucleotide template containing a site-specific lesion (triangle). The DNA primer is labeled with ^{32}P at its 5′ end and is annealed to the damaged template. (B) Following translesion synthesis reactions, products are separated by electrophoresis on a 20% denaturing polyacrylamide gel. The DNA products are visualized by autoradiography of the gel. Translesion synthesis reactions are carried out in the presence of all four dNTPs (N4) or only one deoxyribonucleoside triphosphate at a time, that is, dATP (A), dCTP (C), dGTP (G), or dTTP (T). DNA pol, DNA polymerase.

on a 20% denaturing polyacrylamide gel. The DNA product bands are visualized by autoradiography of the gel (Figure 22.21). Such analysis determines whether a DNA polymerase is capable of translesion synthesis across from a particular lesion. To further determine what nucleotide is inserted opposite the lesion (hence error-free or error-prone), the assays are performed in the presence of only one deoxyribonucleoside triphosphate at a time (i.e., dATP, dGTP, dCTP, or dTTP). These biochemical analyses are able to reveal the translesion synthesis property of a particular polymerase in response to a particular lesion.

Error-free versus error-prone is a relative description for the accuracy of translesion synthesis. Sometimes, it may not be obvious to distinguish between error-free and error-prone based on in vitro biochemical analysis of a polymerase in response to a specific lesion. The ultimate distinction between these two modes of translesion synthesis in cells can be made through genetic analysis. If the polymerase activity suppresses the lesion-induced mutagenesis, then, it is error-free. If the polymerase activity promotes the lesion-induced mutagenesis, then, it is error-prone.

22.5.5 Base Damage-Induced Mutagenesis Is A Major Component of the SOS Response in *E. coli*

In 1973, Miroslav Radman postulated the SOS response hypothesis in *E. coli*. Now, it is well understood that base damage-induced mutagenesis is a major component of the SOS response in *E. coli*. This is a tightly regulated transcriptional control system. The master controller is the LexA protein, and the master regulator is the RecA protein. Under normal growth conditions, basal levels and small amounts of LexA are expressed. LexA protein is the transcriptional repressor of over 40 genes including its own gene *lexA* and the *recA* gene. Basal levels of RecA are also expressed. LexA forms a homodimer and binds to similar operator sequences (SOS boxes) of the regulated genes or operons. When DNA is significantly damaged or replication is blocked, the co-protease activity of RecA is activated by forming RecA filaments on single-strand DNA that has been created by disrupted DNA replication. The interaction between activated RecA and LexA results in autopro-teolytic cleavage of LexA at its A84-G85 peptide bond. The cleaved LexA is inac-tivated as a transcriptional repressor, which decreases the cellular LexA repressor concentration. As a result, expression of the genes under the SOS control is induced. The induced gene products include DNA polymerase II and proteins involved in nucleotide excision repair, recombination, and translesion synthesis. Following DNA repair, translesion synthesis, recombination, and DNA synthesis, single-strand regions of the genome diminish, thus eliminating the SOS signal. RecA returns to its SOS-inactive form. Concomitantly, the induced LexA repressor accumulates to higher concentrations, leading to recovery of the SOS genes to their repressed states; that is, the SOS returns to its off state (Figure 22.22).

The *E. coli* translesion polymerases, DNA polymerases II, IV, and V, are under the control of the SOS system. While DNA polymerases II, IV are involved in translesion synthesis of a few selected types of lesions, DNA polymerase V is the

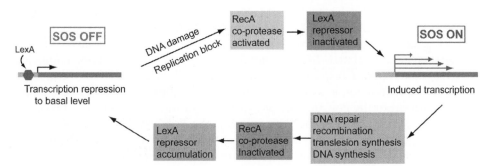

Figure 22.22. SOS response in *E. coli*. Under normal growth conditions (SOS off), genes under the SOS control are repressed by the LexA repressor. DNA damage or replication block triggers SOS response, leading to activation of the RecA co-protease and subsequent inactivation of the LexA repressor by RecA-assisted auto cleavage. This results in induced transcription of the various SOS genes (SOS on). Combined cellular activities such as DNA repair and translesion synthesis eventually removes the SOS signal. Consequently, the RecA co-protease is inactivated and the LexA repressor is accumulated in cells, returning cells to the SOS off state.

major translesion polymerase and is responsible for the majority of base damage-induced mutagenesis in *E. coli* and in prokaryotes in general. Translesion synthesis by DNA polymerase V is stimulated by RecA, SSB (single-strand DNA binding protein), the β processivity clamp, and the γ clamp loader.

DNA polymerase IV is encoded by the *dinB* gene. DNA polymerase IV additionally functions in untargeted mutagenesis. Untargeted mutagenesis refers to induced mutations that arise in undamaged regions of DNA. In *E. coli*, untargeted mutagenesis consists of mostly −1 deletions, consistent with the biochemical property of DNA polymerase IV in that this polymerase is prone to produce −1 deletions when copying DNA.

Polymerase V is formed by two gene products: the catalytic subunit UmuC and the regulatory protein UmuD. UmuD is inactive for translesion synthesis and mutagenesis until it is autocleaved at its C24-G25 peptide bond, removing the N-terminal 24 amino acids with the remaining C-terminal part designated as UmuD′. Like the autocleavage of LexA, the proteolytic cleavage of UmuD is activated by the active RecA co-protease (RecA filaments on single-strand DNA). UmuD′ forms a homodimer UmuD′$_2$, which in turn interacts with UmuC, forming a tight complex UmuD′$_2$C. This complex is DNA polymerase V and is the active polymerase for translesion synthesis and mutagenesis. Dimers of UmuD$_2$ and UmuD-UmuD′ can also be formed. However, they are inactive for translesion synthesis and mutagenesis. Thus, UmuD autocleavage forms an important posttranslational control for DNA polymerase V and, hence, for translesion synthesis and mutagenesis.

22.5.6   The Polζ Mutagenesis Pathway in Eukaryotes

Studies on DNA damage-induced mutagenesis in eukaryotes began with genetic analyses in the yeast *S. cerevisiae* at the beginning of the 1970s. Several yeast mutant strains were isolated that exhibited deficiency in UV-induced mutagenesis. That is, the corresponding wild-type genes are required for UV-induced mutagenesis. These genetically identified genes include *RAD6, RAD18, REV1* (required for reversion mutation), *REV3, REV6,* and *REV7.* Except for the *REV6* gene, these yeast mutagenesis genes and their human homologous genes have been cloned. Humans contain two closely related *RAD6* homologs, designated *HHR6A* (<u>h</u>uman <u>h</u>omolog of <u>RAD6</u>) and *HHR6B*, respectively. Genetic experiments further showed that these genes normally function in a common mutagenesis pathway called the Polζ mutagenesis pathway or the Polζ translesion synthesis pathway. More recently, two more proteins are found to act in the Polζ pathway: the Pol32 subunit of yeast Polδ (a replication polymerase) and the PCNA processivity clamp. It appears that the Polζ pathway is the major mechanism of translesion synthesis and base damage-induced mutagenesis in eukaryotes.

22.5.6a   The Rad6–Rad18 Complex Is a Ubiquitin Ligase Complex. Rad6 is a ubiquitin-conjugating enzyme and forms a stable complex with Rad18. Rad18 is a ring finger protein containing sequence motifs of a typical ubiquitin ligase. The Rad6–Rad18 complex is able to catalyze PCNA ubiquitination. Thus, this complex is expected to be a ubiquitin ligase complex. PCNA ubiquitination by the Rad6–Rad18 complex functions at an early step of the Polζ mutagenesis pathway and is an important control for mutagenesis (see Section 22.5.9).

Ubiquitin is an evolutionarily conserved and ubiquitous protein (hence the name ubiquitin) of 76 amino acids. For protein ubiquitination, ubiquitin is first activated by an E1 enzyme through adenylation by ATP and subsequent transfer of the ubiquitin to a thiol group of the enzyme. Then, ubiquitin is transferred to the thiol group of a cysteine residue on an E2 enzyme. Finally, the target protein is recognized and bound by an E3 enzyme for ubiquitin transfer from the E2 enzyme to the target, achieving mono-ubiquitination. Poly-ubiquitination can be achieved by linking multiple ubiqitin units at the Lys48 or Lys63 of the ubiquitin. Poly-ubiquitinated proteins are typically targeted for degradation by the proteasome, a large multisubunit protease complex. Mono-ubiquitinated PCNA by Rad6-Rad18, however, leads to functional modification rather than protein degradation.

22.5.6b *Rev3–Rev7 Complex Forms the Translesion Polymerase Polζ.* The

Rev3–Rev7 protein complex forms Polζ, in which Rev3 is the catalytic subunit and Rev7 is a noncatalytic subunit. Rev3 protein belongs to the B family of DNA polymerases. This family also includes the replicative polymerases α, δ, and ε. However, unlike the replicative polymerases, REV3 gene is not essential for growth in yeast cells. Therefore, Rev3 is a specialized polymerase specifically devoted for translesion synthesis.

Purified yeast Polζ is able to perform limited nucleotide insertions opposite several DNA lesions such as TT (6–4) photoproduct, AAF-dG adduct, and (+) or (−)-*trans-anti*-BPDE-N^2-dG adduct. Furthermore, Polζ also catalyzes extension synthesis from opposite many types of lesions with varying efficiencies, including an AP site, cis–syn TT dimer, (6–4) photoproduct, AAF-dG adduct, (+) or (−)-*trans-anti*-BPDE-N^2-dG adduct, and an acrolein-derived dG adduct. Therefore, it has been proposed that Polζ functions both as an insertion polymerase and an extension polymerase. It appears that the extension activity of Polζ is versatile. Thus, it is believed that Polζ is a major extension polymerase during translesion synthesis in eukaryotes.

REV3 (353 kD) is one of the largest proteins in human cells, making it highly difficult to produce the recombinant protein for in vitro biochemical studies. In addition to its C-terminal polymerase domain, human REV3 protein contains a large N-terminal region, accounting for ~3/4 of the protein. The precise role of this large N-terminal region is not known. It is possible that this region may be involved in protein–protein interactions during the recruitment of Polζ to sites of DNA damage during translesion synthesis. REV3 knockout is embryonic lethal in mice. The failed embryos exhibited increased double-strand breaks and massive apoptosis. It was postulated that DNA double-strand breaks accumulate at sites of unreplicated DNA lesions in the absence of Polζ, leading to excessive apoptosis that results in lethality. Furthermore, a Rev3$^{-/-}$ cell line could not be established from the early embryos of the Polζ knockout mice unless the p53 gene is additionally inactivated, suggesting that Polζ is essential for long-term cell survival. This is in contrast to the yeast and chicken cells, in which deletion of the *REV3* gene is compatible with cell survival under normal growth conditions.

Opposite the 3′ T of a TT dimer, Polζ is unable to insert a nucleotide in vitro, although it is active for translesion synthesis opposite the 5′ T of the dimer. Thus, other translesion polymerases are required to bypass the 3′ T of the TT dimer and other lesions for which Polζ is inactive. During translesion synthesis, the active site of the polymerase must be able to accommodate the damaged template base. Since there are various types of base lesions that differ drastically in chemistry and structure, it is not surprising that multiple translesion polymerases are devoted to copy the damaged sites of DNA, with each polymerase possessing different lesion specificities. The Y family DNA polymerases are those additional translesion polymerases.

22.5.7 The Y Family of DNA Polymerases

When the yeast *REV1* gene was cloned in 1989, it was noticed that the Rev1 protein shares some sequence homology with the *E. coli* UmuC protein. In 1996, Christopher Lawrence and colleagues discovered that the yeast Rev1 protein is a DNA template-dependent deoxycytidyl (dCMP) transferase opposite a template G and AP site. Shortly after, the yeast *RAD30* gene was cloned as a homolog of the *E. coli dinB* and *umuC*. Search for a human homolog of the yeast *RAD30* gene led to the cloning of three genes: *RAD30, RAD30B*, and *DINB1*. Since Rad30 protein shares sequence homology with Rev1, it was suspected that Rad30 might possess a similar dCMP transferase. In 1999, experiments attempting to test this prediction revealed that Rad30 protein is in fact a novel DNA polymerase, designated Polη, which is capable of error-free translesion synthesis across from a template cis–syn TT dimer, a major UV damage. At about the same time, Fumio Hanaoka and colleagues independently discovered that the human XPV (xeroderma pigmentosum variant) protein is a DNA polymerase capable of error-free translesion synthesis across from a template cis–syn TT dimer. The fact that XPV protein is encoded by the human *RAD30* gene was then quickly established in 1999. The *E. coli* homologs, DinB and UmuC (complexed with UmuD$'_2$), were separately shown to be DNA polymerases IV and V, respectively, in 1999. Subsequently, it was demonstrated that the human *DINB1* and *RAD30B* genes code for Polκ and Polι, respectively. By then, the UmuC superfamily was established that comprises a new family of DNA polymerases, which was later renamed the Y family of DNA polymerases. New gene names were also reissued as *POLH* for Polη, *POLI* for Polι, and *POLK* for Polκ. The human Y family members are REV1, Polη, Polι, and Polκ. All of the human Y family genes were reported in 1999 by the laboratories of Fumio Hanaoka, Louis Prakash, Satya Prakash (*RAD30*), Roger Woodgate (*RAD30B*), Errol Friedberg, Haruo Ohmori (*DINB1*), and Zhigang Wang (*REV1*). In the yeast *S. cerevisiae*, however, the Y family members consist of Rev1 and Polη only, while *S. pombe* contains Rev1, Polη, and Polκ. The most prominent biochemical activity of the Y family DNA polymerases is translesion synthesis. Most, perhaps all, of the Y family polymerases are involved in translesion synthesis in cells, from *E. coli* to humans.

22.5.7a Rev1 is a dCMP Transferase. Properties of REV1 protein are revealed by extensive biochemical studies on the purified yeast and human enzymes. Opposite runs of template G, REV1 behaves like a typical DNA polymerase with nucleotide incorporation and extension, synthesizing runs of C in the daughter strand of DNA. Opposite a template lesion such as AP site, uracil, 8-oxoguanine, 1,N^6-ethenoadenine, (+)-*trans-anti*-BPDE-N^2-dG, and (−)-*trans-anti*-BPDE-N^2-dG, the REV1 dCMP transferase is able to effectively insert a C. Even opposite template A, T, and C, REV1 is able to insert a C, although the efficiency is reduced when compared to a template G. Therefore, REV1 recognizes undamaged template A, C, and T and several damaged bases and preferentially inserts a C opposite each without an exception. Such template-dependent dCMP transferase is highly unusual. Rules of hydrogen bonding and geometric fitting at the polymerase active site that governs normal DNA synthesis simply do not apply to the REV1 dCMP transferase. The molecular mechanism of C insertion is now revealed by the crystal structure of the yeast Rev1 catalytic domain. Rev1 uses the Arg324 of the protein, instead of the DNA template G, as the template for choosing dCTP as the incoming base during catalysis.

22.5.7b   *DNA Polymerase η.* The hallmark of Polη is efficient error-free translesion synthesis across from UV-induced TT dimers discovered by the laboratories of Fumio Hanaoka, Louis Prakash, and Satya Prakash. Polη plays a critical role in response to UV radiation in eukaryotes. Yeast cells lacking Polη exhibit increased sensitivity to UV radiation. In humans, defects in the Polη gene result in the hereditary disease xeroderma pigmentosum variant (XPV). Thus, the Polη gene is variously designated as *XPV, RAD30,* and *POLH.* XPV patients exhibit sensitivity to the sunlight and a predisposition to skin cancer. XPV patients are proficient in nucleotide excision repair, but are deficient in error-free translesion synthesis following UV radiation. In vitro, purified Polη is able to efficiently bypass a TT dimer in an error-free manner by predominantly inserting the correct AA opposite the lesion. In XPV cells, UV lesions that are normally correctly copied by Polη are instead bypassed by other translesion polymerases with lower efficiency and in an error-prone manner. Therefore, these mutant cells are more sensitive to UV and are hypermutable by UV.

Although the molecular defect of XPV is very different from that of the other XP patients (XPA, XPB, XPC, XPD, XPE, XPF, and XPG), who are deficient in nucleotide excision repair, the clinical manifestations of the diseases are quite similar. This is not surprising because the defect in either Polη or nucleotide excision repair results in a common problem: genomic overload of TT dimers and perhaps other CPDs for error-prone translesion synthesis by other bypass polymerases during replication. The result is predictable: elevated cytotoxicity and mutagenesis induced by the UV component of the sunlight, which constitute the cellular bases of XP diseases.

In addition to TT dimers, Polη is also able to perform translesion synthesis opposite a variety of chemical-induced DNA damage with either error-free or error-prone outcome. Additional error-free translesion synthesis by Polη include lesions such as acetylaminofluorene (AAF) adducts and thymine glycols. Error-prone translesion synthesis by Polη include lesions such as AP sites, ($\pm$)-*trans-anti*-benzo[*a*]pyrene-N^2-dG adducts, and *p*-benzoquinone DNA adducts. Polη is efficient in copying 8-oxoguanine in DNA in vitro. The yeast Polη predominantly inserts the correct C opposite the lesion, while the human Polη only slightly prefers C insertion. Thus, whether Polη is error-free or error-prone in response to 8-oxoguanine in human cells remain to be determined by in vivo genetic experiments.

22.5.7c   *DNA Polymerase κ.* The hallmark of Polκ is efficient error-free bypass of ($-$)-*trans-anti*-BPDE-N^2-dG adducts discovered by Zhigang Wang and colleagues. Polκ also bypasses the ($+$)-*trans-anti*-BPDE-N^2-dG adduct with the correct C insertion opposite the lesion, although less efficiently as compared to its stereoisomer (–)-*trans-anti*-BPDE-N^2-dG adduct. These DNA adducts are formed in cells by the parent compound benzo[*a*]pyrene.

Translesion synthesis is also observed with Polκ in vitro opposite several other DNA lesions. Additional error-free translesion synthesis by Polκ include lesions such as 8-oxoguanine and thymine glycols. Error-prone translesion synthesis by Polκ include AP sites, preferring A insertion opposite the lesion, and AAF-dG adducts with frequent T mis-insertion. Unlike the *E. coli* DNA polymerase IV, which exhibits the most homology to Polκ, base substitutions, rather than –1 deletions, are the predominant errors of Polκ when copying undamaged DNA.

22.5.7d   DNA Polymerase ι.

The hallmark of human Polι is that it violates the Watson–Crick base-pairing rule opposite the template T discovered by the laboratories of Zhigang Wang, Roger Woodgate, Louis Prakash, and Satya Prakash. This polymerase incorporates G opposite the template T with 3-10-fold higher rates than A. Due to low catalytic efficiencies opposite template T and C, human Polι is able to synthesize only very short stretches of DNA. Functional implications of these striking biochemical features are not fully understood.

Polι is capable of both error-free and error-prone translesion syntheses in vitro. In response to 8-oxoguanine in DNA, human Polι predominantly incorporates C opposite the lesion. Human Polι is able to incorporate the correct C opposite the AAF-adducted guanine. However, further DNA synthesis is blocked by the lesion. Human Polι incorporates one nucleotide opposite a template AP site more efficiently than opposite a template T, with the incorporation specificity of G > T > A > C. After incorporating one nucleotide opposite the AP site, further DNA synthesis is aborted.

Surprisingly, human Polι prefers A incorporation opposite the 3′ T of a template TT (6–4) photoproduct, in contrast to the much preferred G incorporation opposite the undamaged template T. Nucleotide incorporation opposite the 5′ T of the TT (6–4) photoproduct, however, is largely blocked by the lesion. Opposite a template TT dimer, human Polι has a very limited activity, preferentially incorporating a T opposite the 3′ T of the lesion. This activity, albeit very inefficient, may contribute to TT dimer-induced mutagenesis in XPV cells that lack Polη.

22.5.7e   Common Biochemical Properties of the Y Family DNA Polymerases.

The Y family polymerases share several biochemical properties: (a) lacking 3′ → 5′ proofreading exonuclease activity; (b) synthesizing DNA in a more or less distributive manner; (c) capable of both error-free and error-prone translesion synthesis, depending on the lesion; and (d) synthesizing DNA from undamaged templates with extraordinarily low fidelity. These features are well-suited for the task of a translesion polymerase. A proofreading activity would have caused serious problems for translesion synthesis. Since DNA damage often changes the coding property of the DNA bases, incorrect nucleotides are frequently inserted opposite the lesion. A proofreading activity would have led to a futile cycle of nucleotide insertion and removal, followed by another insertion and removal, and so on. Because the Y family polymerases copy undamaged DNA with very low fidelity, it would have been disastrous for the genome if these translesion polymerases stay on the template for too long following translesion synthesis. Thus, these polymerases are distributive in that they fall off the template after synthesizing only one or a few nucleotides, minimizing the possibility of errors in the undamaged regions of DNA near the lesion sites.

The low-fidelity nature of the Y family polymerases probably reflects an inevitable consequence of their biological function in translesion synthesis. It was initially speculated that a Y family polymerase contains a loose and flexible active site such that a damaged template base can be accommodated for translesion synthesis. This concept is proved to be correct as revealed by crystal structures of yeast Polη and Dpo4, a Polκ homolog in *Sulfolobus solfataricus*. Consequently, when copying undamaged DNA, the active site of a Y family polymerase would lose the stringent geometry constraints that characterize highly accurate Watson–Crick base pairing, resulting in extraordinarily low-fidelity DNA synthesis. Such infidelity reaches to an extreme extent for Polι when copying undamaged template T, where G is preferred by the polymerase over the correct A.

22.5.8 Mechanistic Models of Translesion Synthesis

It appears that multiple mechanisms exist for translesion synthesis, due to the involvement of multiple bypass polymerases. In the simplest case, one polymerase inserts a nucleotide opposite the lesion, and then the same polymerase extends the synthesis from opposite the lesion. This constitutes the one-polymerase two-step mechanism (Figure 22.23). Examples of this mode of translesion synthesis include the bypass of a TT dimer by Polη and the bypass of a (–)-*trans-anti*-benzo[*a*]pyrene-N^2-dG by Polκ. In a more complex scheme, following nucleotide insertion opposite the lesion by one polymerase, subsequent extension synthesis is catalyzed by another polymerase. This constitutes the two-polymerase two-step mechanism (Figure 22.23). Polζ is believed to be the major extension polymerase during translesion synthesis by the two-polymerase two-step mechanism. Additionally, Polκ and Polη may also catalyze extension synthesis during the bypass of some selected lesions.

The choice of one-polymerase two-step versus two-polymerase two-step in cells most likely depends on the specific type of lesion. Apparently, efficient bypass of a

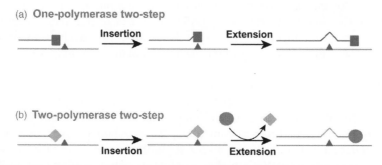

Figure 22.23. Mechanistic models of translesion synthesis. (A) The one-polymerase two-step model. The nucleotide insertion step and the extension step are catalyzed by the same translesion polymerase. (B) The two-polymerase two-step model. A polymerase inserts one nucleotide opposite the lesion at the insertion step. Subsequently, a different translesion polymerase replaces the insertion polymerase and then catalyzes the extension synthesis from opposite the lesion. The DNA lesion is shown as a triangle on the template.

lesion by a single polymerase, as TT dimer bypass by Polη, is exceptional. Thus, translesion synthesis of most types of lesions likely involves the two-polymerase two-step mechanism. It is possible that some lesions are bypassed by both mechanisms of translesion synthesis, where a fraction of the bypass involves a single polymerase while the remaining bypass requires the combination of two different polymerases. Hence, in vivo translesion synthesis would often involve the participation of multiple polymerases. Such a multiple-polymerase mode of translesion synthesis has been observed in yeast cells.

22.5.9   Control of Translesion Synthesis and Mutagenesis

In *E. coli*, DNA damage-induced mutagenesis is tightly controlled by the SOS regulatory system. Eukaryotes, on the other hand, do not contain a similar SOS response system. Nevertheless, translesion synthesis and base damage-induced mutagenesis are controlled in eukaryotic cells at two levels. First, the bypass polymerases are controlled to a low concentration in cells. Second, the extent by which translesion synthesis contributes to damage tolerance is controlled in cells.

Due to their low-fidelity nature, the Y family polymerases must be excluded from normal DNA replication in order to maintain genomic stability. Conceivably, these specialized polymerases are accessible only to the damaged sites on the template through a recruiting mechanism. Additionally, these polymerases may be regulated transcriptionally and posttranscriptionally. The Polη expression in yeast is indeed inducible by UV radiation. Polη also contains a nuclear localization sequence in its C-terminal region. Deleting this region does not affect its polymerase activity, but renders Polη biologically inactive in response to UV radiation. Thus, protein truncation at the C-terminus may represent an important control mechanism.

Translesion synthesis is often mutagenic, leading to genome instability, whereas template switching is error-free, promoting genome stability in the presence of base damage. Thus, template switching suppresses the carcinogenic effect of base damage, whereas error-prone translesion synthesis promotes the carcinogenic effect of base damage. Consequently, which damage tolerance pathway the cell commits itself to in response to unrepaired base damage during replication has a profound effect on genome stability and carcinogenesis. It has long been a puzzling mystery as to how cells control the division of the two parallel pathways of template switching and translesion synthesis (Figure 22.20). That is, what is the mechanism of cellular control that channels the base damage to template switching or translesion synthesis?

Earlier genetic studies in yeast have indicated that the Rad6–Rad18 protein complex is involved in controlling both the template switching and the translesion synthesis pathways. Most recent genetic analyses in yeast showed that ubiquitinated PCNA plays an important role in controlling the two damage tolerance pathways. Genetic studies indicate that PCNA mono-ubiquitination by Rad6–Rad18 is important for the translesion synthesis pathway, while its poly-ubiquitination by Mms2–Ubc13–Rad5 is important for the template switching pathway. A model of this control is shown in Figure 22.24. In this model, the molecular switch that controls the two damage tolerance pathways is PCNA, the polymerase sliding clamp for DNA synthesis; and the direction of the molecular switch is determined by the status of PCNA ubiquitination. Specifically, it was proposed that mono-ubiquitination of

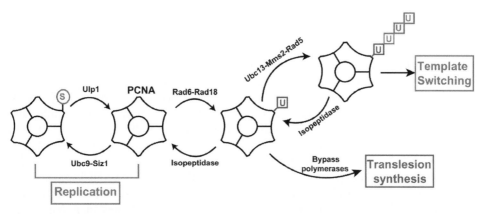

Figure 22.24. A molecular switch model for the control of the two pathways of DNA damage tolerance, template switching, and translesion synthesis. The PCNA processivity DNA clamp is shown as a homotrimer. Both SUMO-modification (S) and ubiquitination (U) occur at the K164 position of PCNA. [Adapted from Stelter, P., and Ulrich, H. D. *Nature* **425**, 188–191, 2003.]

PCNA at its K164 by the Rad6–Rad18 complex leads to the translesion synthesis pathway. Further ubiquitination (poly-ubiquitination) of the mono-ubiquitinated PCNA at the same K164 by the Mms2–Ubc13–Rad5 complex leads to the template switching pathway. At the same K164 site, SUMO-modification of PCNA also occurs in the S phase, which may promote normal replication. These three forms of modified PCNA may be reversed by isopeptidases.

22.5.10 Specificity of Base Damage-Induced Mutagenesis: Some Examples

Since the discovery of the Y family DNA polymerases, translesion synthesis has been extensively studied using in vitro biochemistry. Biochemical approach is an extremely powerful tool and has yielded detailed molecular information and models of translesion synthesis. However, biochemistry cannot precisely duplicate the in vivo condition. Therefore, biochemical results need to be validated by in vivo genetics. By combining biochemical and genetic approaches, insightful information about translesion synthesis in cells is emerging. Several examples are presented here that illustrate the specificity of lesion-induced mutagenesis and the underlying mechanisms.

22.5.10a AP Sites. AP sites are a major type of spontaneous DNA lesions, although they can also be induced by many environmental agents. AP sites are noncoding. Hence, translesion synthesis of an AP site by any polymerase is error-prone. Consequently, AP sites are highly mutagenic. In *E. coli*, translesion synthesis of AP sites results in preferential incorporation of A opposite the lesion. Such biased A insertions do not appear to be the case in mammals. Insertions of all four bases opposite AP sites have been detected in cultured mammalian cells.

In yeast cells, C is predominantly inserted opposite AP sites, which is dependent on the Rev1 and Polζ function. The Rev1 dCMP transferase is efficient in inserting a C opposite an AP site in vitro, but it cannot catalyze extension synthesis from opposite the lesion. The combination of Rev1 and Polζ, however, results in bypass of the AP site in vitro. Thus, this two-polymerase two-step action of Rev1-Polζ constitutes the major mechanism of translesion synthesis of AP sites in yeast cells. Because the human REV1 also efficiently inserts a C opposite an AP site, a similar REV1–Polζ mechanism for translesion synthesis of AP sites is most likely operational in humans as well. However, in contrast to yeast, human cells additionally contain the Y family polymerases Polι and Polκ. Human Polι efficiently inserts a G and less frequently a T opposite an AP site in vitro. Extension synthesis was not observed. Human Polκ preferentially inserts an A opposite an AP site in vitro. Efficient extension synthesis can be achieved by Polκ through a −1 deletion mechanism if the next template base is a T. It is likely that REV1, Polι, and Polκ all participate in nucleotide insertions during translesion synthesis of AP sites in human cells, resulting in insertions of C, G, A, and T opposite the lesions.

22.5.10b   UV Photoproducts. The major DNA lesions of UV radiation are cyclobutane pyrimidine dimers (CPDs) and (6–4) photoproducts. Because fewer (6–4) photoproducts are formed and more importantly they are repaired more efficiently than CPDs, mutagenesis of UV radiation mainly results from CPDs. Base substitutions predominate. The major mutation is C → T transitions. In normal cells, C-containing CPDs are highly mutagenic, whereas TT dimer is only slightly mutagenic due to error-free translesion synthesis by Polη. In XPV cells lacking Polη, mutagenesis at TT dimers are much increased, although mutations at C-containing CPDs still account for the majority.

Among the UV lesions, cis–syn TT dimers and TT (6–4) photoproducts are quite stable and have been extensively studied as a result. TT dimers are efficiently bypassed by Polη by inserting AA opposite the lesion. The 3′ T of the dimer, the first T encountered by the polymerase, completely blocks human Polκ and yeast Polζ in vitro. In contrast, Polζ is able to efficiently insert the correct A opposite the 5′ T of the dimer and subsequently extends the synthesis beyond the lesion. Polκ is also able to perform extension synthesis from a G opposite the 3′ T of the TT dimer. Opposite a template TT dimer, human Polι has a very limited activity, preferentially inserting a T and less frequently a G opposite the 3′ T of the lesion. This activity, albeit inefficient, may make a significant contribution to TT dimer-induced mutagenesis in the absence of Polη, such as the XPV cells. Extension synthesis by Polι from opposite the 3′ T of the dimer, however, is insignificant. The extension synthesis following Polι-catalyzed insertion may thus involve Polζ and Polκ. Consistent with a role of Polκ in extension synthesis of TT dimers, Polκ-knockout mouse cells are slightly sensitive to UV radiation. Polη is the predominant translesion polymerase for the bypass of TT dimers and probably other CPDs as well, which accounts for the severe phenotype of XPV cells in response to UV radiation.

In contrast to TT dimers, TT (6–4) photoproducts cannot be bypassed by Polη alone in vitro. Instead, Polη is able to insert a G opposite the 3′ T of the TT (6–4) photoproduct before aborting DNA synthesis. The resulting intermediate of translesion synthesis is a substrate for extension synthesis by Polζ. Coordination between these two polymerases could therefore achieve bypass of TT (6–4) photoproducts by the two-polymerase two-step mechanism of translesion synthesis. This indeed occurs in yeast cells and is the major mechanism of G mis-insertion opposite the 3′

T of TT (6–4) photoproducts, leading to T → C transition mutations. Polζ also possesses limited nucleotide insertion activity opposite the TT (6–4) photoproduct, frequently mis-inserting a T opposite the 3′ T but predominantly inserting the correct A opposite the 5′ T of the lesion. Thus, in yeast cells lacking Polη, the mutation spectrum at the presumed TT (6–4) photoproducts were altered from predominant T → C to predominant T → A mutations as a result of T insertion opposite the 3′ T of the lesion. In mammalian cells, TT (6–4) photoproducts also mainly induce T → C transition mutations as a result of G mis-insertion opposite the 3′ T of the lesion. Given the yeast results, a major role of Polη in T → C mutations induced by TT (6–4) photoproducts is anticipated in mammals. REV1 is catalytically inactive in response to TT dimers and TT (6–4) photoproducts. However, REV1 is required for UV-induced mutagenesis as revealed by genetic experiments. Thus, REV1 likely plays a noncatalytic role during translesion synthesis and mutagenesis of UV lesions.

With respect to UV mutagenesis, Polη plays two opposing functions: suppressing mutagenesis in response to TT dimers versus promoting mutagenesis in response to TT (6–4) photoproducts. Then, why are XPV cells lacking Polη hypermutable by UV radiation? Two key factors are probably responsible for this. First, the yield of TT (6–4) photoproducts is significantly lower than TT dimers. Second and perhaps more importantly, TT (6–4) photoproducts are rapidly removed from DNA by nucleotide excision repair. In contrast, TT dimers are poorly repaired, especially in the nontranscribed strand of an active gene and in the transcriptionally silent regions of the genome. Hence, TT dimers, rather than TT (6–4) photoproducts, are much more prevalent UV lesions left unrepaired during replication following cell exposure to the sunlight. Consequently, the error-free translesion synthesis of Polη predominates its biological function in response to UV radiation. In lower eukaryotes, CPDs can be alternatively repaired by a photolyase. Unfortunately, this enzyme was lost during evolution to mammals. This fact further underscores our dependence on Polη in protection against the cytotoxic and mutagenic effects of UV radiation. Without this single translesion polymerase, life becomes very hazardous under the sun.

22.5.10c BPDE-N²-dG Adducts. BPDE-N^2-dG adducts are the major DNA lesions derived from benzo[*a*]pyrene. The major mutations produced by these lesions are G → T transversions. Human Polκ performs error-free translesion synthesis efficiently across from the (–)-*trans-anti*-BPDE-N^2-dG adducts, and less efficiently across from the (+)-*trans-anti*-BPDE-N^2-dG adducts in vitro. Some sequence contexts dramatically affect the efficiency of Polκ-catalyzed translesion synthesis across from the (+)-*trans-anti*-BPDE-N^2-dG adduct. Unlike DNA repair that is more or less independent on the sequence context, translesion synthesis and mutagenesis is often significantly influenced by sequence contexts. Thus, sequence contexts may make a significant contribution to mutation hot spots in cells. The biochemical results predict that Polκ may function to suppress BPDE-induced mutagenesis in cells. This prediction is proved to be correct in Polκ-knockout mouse cells. Purified human REV1 mainly inserts the correct C opposite (±)-*trans-anti*-BPDE-N^2-dG adducts. Extension synthesis cannot be performed by REV1, but can be effectively catalyzed by human Polκ in vitro. Thus, Polκ may additionally function as an extension polymerase in cells during bypass of BPDE-dG adducts by multiple translesion polymerases.

Polη performs error-prone translesion synthesis opposite (+)- and (−)-*trans-anti*-BPDE-N^2-dG DNA adducts by predominantly inserting A opposite the lesion in vitro. This polymerase is more active in response to the former isomeric lesion. In yeast cells, Polη, Polζ, and Rev1 are all required for G → T transvertion mutations. The likely mechanism is A insertion opposite the lesion by Polη followed by extension synthesis by Polζ. Rev1 probably plays a noncatalytic role in such a mutagenic bypass of the BPDE lesions.

22.5.10d AAF-dG Adducts. Based on forward mutation assays, the major mutations induced by AAF DNA adducts are frameshift mutations in yeast and human cells. In yeast, the Polζ mutagenesis pathway plays a major role in error-prone translesion synthesis of AAF-dG adducts. Consistent with these in vivo results, yeast Polζ is able to perform limited translesion synthesis across from an AAF-dG adduct in vitro, mis-inserting a G opposite the lesion. Furthermore, Polζ is also capable of extension synthesis from opposite the AAF-dG adduct. The REV1 dCMP transferase, on the other hand, is essentially inactive in response to a template AAF-dG adduct. Human Polκ is capable of error-prone translesion synthesis in vitro, inserting T or C at similar frequencies and A at a lower frequency opposite the AAF-dG adduct.

Efficient error-free nucleotide insertion opposite an AAF-dG adduct can be catalyzed by Polη in vitro. The human Polη is more efficient in subsequent extension synthesis as compared to the yeast Polη. If the error-free translesion synthesis activity of Polη is utilized in cells in response to AAF-dG adducts, this polymerase would function to suppress AAF-induced mutagenesis. In one study with yeast cells, both an error-free bypass role and a frameshift mutagenesis role of Polη were reported. Hence, it is still unknown about the contribution of Polη to AAF-induced mutagenesis. Opposite a template AAF-dG adduct, human Polι is able to insert predominantly a C in vitro. Subsequent extension synthesis, however, was not observed.

22.5.10e Cisplatin. Cisplatin [*cis*-diamminedichloroplatinum(II)] is a clinically used chemotherapeutic agent in the treatment of a variety of human cancers. The anticancer activity of cisplatin is believed to result from cytotoxicity induced by cisplatin DNA adducts. Not surprisingly, cisplatin itself is a mutagen and carcinogen. G → T and A → T transversions appear to be the main mutations induced by cisplatin.

Little is known about the mechanism of cisplatin-induced mutagenesis in eukaryotic cells. Presumably, error-prone translesion synthesis is the underlying mechanism for cisplatin mutagenesis. In vitro, human Polη can effectively perform translesion synthesis across from a template cisplatin-GG intrastrand crosslink, inserting mainly the correct C opposite the damaged 3′ G. Opposite the 5′ damaged G, C, and A insertions, respectively, were reported by different studies, which may reflect a strong sequence context effect on nucleotide selection by Polη in response to the damaged 5′ G of the cisplatin damage. At some sequence contexts, deletions were also observed during in vitro translesion synthesis of cisplatin-GG intrastrand crosslink by human Polη.

22.5.11 Translesion Synthesis and Immunology: Somatic Hypermutation

In addition to facilitating cell survival in the presence of DNA damage and creating genomic variability, somatic hypermutation is another important biological function of translesion synthesis and mutagenesis in higher eukaryotes. Essentially unlimited antibody diversity is created from a limited set of immunoglobulin genes through a two-step development process. The first level of immunoglobulin gene diversity is accomplished by V(D)J recombination. The gene segments of V (variable, 200–2000 segments), J (joining, 12 segments), and sometimes D (diversity, 4 segments) are randomly combined to form a functional V(D)J sequence coding for the antigen-binding V region. Double-strand breaks are introduced at specific sites by the RAG1–RAG2 (recombination activation gene) complex and are repaired by the error-prone NHEJ. Thus, additional mutations are produced at the segment joining sites. The second level of immunoglobulin gene diversity is accomplished by somatic hypermutation in the V region. Somatic hypermutations occur mostly within three stretches of complementarity-determining regions (CDR) in the V region. Amino acid residues that contact the antigen are clustered in the CDRs (Figure 22.25).

Somatic hypermutation occurs in B cells in the germinal center of the secondary lymphoid tissues. The mutation rate is extraordinarily high ($\sim 10^{-3}$–10^{-4}/bp/generation), up to 10^6-fold higher than the spontaneous mutation rate. Cell viability is incompatible with such a high rate of mutations. Therefore, somatic hypermutation must be somehow controlled with great precision. Spreading of the somatic hypermutation to other regions in the B cell can lead to B-cell lymphoma. Somatic hypermutation occurs specifically in the V region of the immunoglobulin gene; it also requires transcription of the target sequences and the presence of the Ig enhancer, but not a specific promoter.

Somatic hypermutation is initiated by cytosine deamination in single-strand DNA by AID (activation-induced cytidine deaminase) that is specifically expressed in the B cells of the germinal center. Single-strand regions of DNA are probably created by transcription. Deamination of C yields U in DNA. Replication of U residues in DNA leads to C → T transition mutations. U residues can be removed by uracil-DNA glycosylase, producing AP sites in DNA. Replication of the AP sites requires translesion synthesis. Mutations are generated following translesion synthesis across from AP sites. Accordingly, Polζ and REV1 play a major role, and Polη

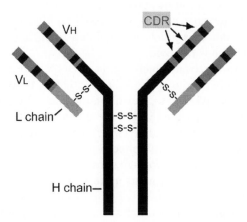

Figure 22.25. Schematic illustration of an immunoglobulin IgG. The complex consists of two heavy (H) chains and two light (L) chains. Somatic hypermutation that greatly contributes to antibody diversity occurs in the variable (V) regions of both the heavy and the light chins, especially the complementarity-determining regions (CDR).

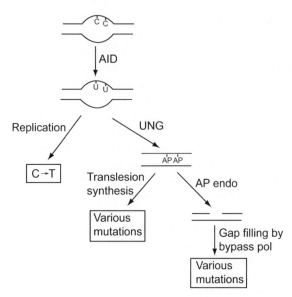

Figure 22.26. Mechanistic model of somatic hypermutation. Somatic hypermutation is initiated by AID via cytosine deamination in single-strand regions of DNA. Uracil-containing DNA leads to mutagenesis by three different mechanisms: replication, translesion synthesis, and gap filling.

and Polι appear to be additionally involved in somatic hypermutation. Furthermore, AP sites are chemically unstable, prone to β-elimination to produce DNA strand breaks. Attempted repair by AP endonucleases can also produce DNA strand breaks at the AP sites. Therefore, closely clustered AP sites in single-strand regions are prone to produce DNA gaps. Such gaps may be filled in by error-prone translesion polymerases, generating mutations in the gap filling process. This model of somatic hypermutation is illustrated in Figure 22.26. Understanding of the molecular mechanisms of somatic hypermutation has progressed a great deal in recent years as our understanding on translesion synthesis progressed rapidly. The next big question is how somatic hypermutation is precisely regulated such that only the V region of the immunoglobulin gene is targeted.

A simple chemistry of cytosine deamination in DNA leads to the evolution of an efficient, ancient, and ubiquitous repair system—that is, base excision repair initiated by uracil-DNA glycosylase—to prevent mutagenesis by cytosine deamination. This powerful mutagenesis system by cytosine deamination was later taken advantage of during the evolution of higher eukaryotes to create precisely controlled mutagenesis to achieve essentially unlimited gene diversity for producing antibodies against invasion by essentially any foreign substances. We humans complain about unbearable heat when the whether heats up to 40°C. Indeed, at 60°C, most organisms will be killed quickly. Yet, in nearly boiling water, a few organisms do thrive. Life finds a way! Mutagenesis makes such seemingly incomprehensible phenomena possible, attesting the power of genomic flexibility for great creations, the greatest of which is our human species that one day will be able to intervene on the mutagenesis or the evolution itself!

SUGGESTED READING

Friedberg, E. C., Walker, G. C., Siede, W., Wood, R. D., Schultz, R. A., and Ellengerger, T. *DNA Repair and Mutagenesis*, 2nd ed., ASM Press, Washington, D.C., 2006.

Ling, H., Boudsocq, F., Woodgate, R., and Yang, W. Crystal structure of a Y-family DNA polymerase in action: A mechanism for error-prone and lesion-bypass replication. *Cell* **107**, 91–102, 2001.

Ohmori, H., Friedberg, E. C., Fuchs, R. P., Goodman, M. F., Hanaoka, F., Hinkle, D., Kunkel, T. A., Lawrence, C. W., Livneh, Z., Nohmi, T. et al. The Y-family of DNA polymerases. *Mol. Cell* **8**, 7–8, 2001.

Stelter, P., and Ulrich, H. D. Control of spontaneous and damage-induced mutagenesis by SUMO and ubiquitin conjugation. *Nature* **425**, 188–191, 2003.

Taylor, J.-S. Structure and properties of DNA photoproducts. In: Siede, W., Kow, Y. W., and Doetsch, P. W. (Eds.). *DNA Damage Recognition*, Taylor & Francis, New York, 2006, pp. 67–94.

Wang, Z. DNA damage-induced mutagenesis: A novel target for cancer prevention. *Mol. Interv.* **1**, 269–281, 2001.

CHAPTER 23

DNA Repair

ISABEL MELLON

Graduate Center for Toxicology, University of Kentucky, Lexington, Kentucky 40536

23.1 INTRODUCTON

The preceding chapter describes the plethora of types of damage/lesions that can be introduced into the DNA of all living cells. Damage can be introduced as a consequence of (1) the inherent chemical instability of DNA, (2) reactive molecules within the cell, (3) the lack of absolute fidelity of the DNA polymerases that replicate the genome, and (4) exposure to chemical and physical agents present in the environment. DNA damage is often categorized as *spontaneously* induced when it is produced by activities that occur within the cell or *environmentally* induced when it is produced by agents that are present outside the cell. DNA damage can mean a modification or alteration of the normal structure of DNA that physically changes the phosphodiester backbone, the deoxyribose sugar, or one of the bases to an abnormal structure. However, it is important to note that DNA damage can also mean a change in the informational properties of the DNA. So, for example, a DNA polymerase error that results in the inappropriate incorporation of thymine instead of cytosine opposite a template guanine during DNA synthesis does not result in an alteration of the normal chemistry of DNA. Thymine and cytosine are normally present in DNA. However, the informational properties of the DNA may be changed when cytosine is replaced by thymine.

The persistence of DNA damage can have extremely deleterious consequences for the cell or organism. It can lead to the formation of mutations, either small-scale single nucleotide base changes or large-scale chromosomal rearrangements. It can alter gene expression by interfering with RNA polymerase. It can block replication of the genome by interfering with DNA polymerases. Alterations in any of these processes can contribute to the capacity of DNA damage to kill cells. In addition, DNA damage has been implicated in the etiology of cancer, the aging process, and a variety of other human disease states. Hence it is probably not surprising that cells have evolved a complex array of different DNA repair pathways that help preserve the integrity of DNA. The past 25 years has witnessed an explosion in the investigation and understanding of different DNA repair pathways at the genetic and biochemical level. The general strategy for many DNA repair pathways appears to have

Molecular and Biochemical Toxicology, Fourth Edition, edited by Robert C. Smart and Ernest Hodgson
Copyright © 2008 John Wiley & Sons, Inc.

TABLE 23.1. DNA Repair Pathways

Repair Pathway	Substrate	Mechanism
Direct reversal of damage	UV damage, Alkylation damage	Simplest type: Damage is removed in a single, enzyme catalyzed reaction.
Excision repair		Damage is removed ("excised") as a free base or within a stretch of nucleotides.
Base excision repair	Small "nonbulky" lesions	
Nucleotide excision repair	"Bulky" helix-distorting lesions	
Transcription-coupled nucleotide excision repair	"Bulky" lesions that block RNA polymerase elongation	
Mismatch repair	Inappropriate mispaired bases	
Recombinational repair	Double-strand breaks Interstrand crosslinks Damage-arrested DNA replication complexes	Requires sequence homology, DNA strand exchange, the formation of Holliday junctions, and resolution of Holliday junctions.
Nonhomologous end joining	Double-strand breaks	Direct end-to-end fusion that does not require large regions of sequence homology or strand exchange. Integrity of the sequence is often not completely restored.

been conserved in organisms as diverse as bacteria, yeast, and humans. However, the number and repertoire of proteins that participate in each repair pathway tends to be greater and more complex in mammals than in bacteria.

Table 23.1 lists the major DNA repair pathways that exist in many of the organisms commonly studied to date. As will become clear in the sections that describe each pathway in detail, vastly different strategies have evolved to fix or tolerate different types of DNA damage. As stated above, many moieties in DNA are subject to damage or modification, but the *direct reversal* and *excision repair* pathways generally deal with damage to the bases. *Direct reversal* is the simplest type of DNA repair where the damaged base is simply restored to its original state by a single enzyme-catalyzed reaction. *Excision repair* pathways are more complicated. Instead of the damage being simply reversed, it is *excised* and removed as free bases or nucleotides. There are three different types of excision repair: *base excision* repair, *nucleotide excision* repair, and *mismatch* repair. In general, these three types of excision repair are mechanistically and biochemically distinct. They are largely "directed" by the damaged substrates that they remove from the genome. *Base excision* repair generally removes lesions or inappropriate bases that are "nonbulky" in nature because they often do not significantly perturb the structure of the DNA helix where they are formed. In contrast, *nucleotide excision* repair generally removes damage that is described as "bulky" in nature because it usually creates a significant distortion of the DNA helix where it resides. *Transcription-coupled nucleotide excision* repair is a specialized form of nucleotide excision repair. It removes

DNA damage when the damage has blocked or arrested an RNA polymerase complex that is in the process of transcription elongation. The third type of excision repair, *mismatch repair*, usually serves to remove inappropriate or incorrect bases in DNA rather than modified bases. Mismatches or mispairs are frequently formed during DNA replication when the DNA polymerase mistakenly incorporates an incorrect nucleotide into the nascent DNA strand. For example, a DNA polymerase could mistakenly incorporate a thymine opposite a template guanine, which would result in a GT mispair. If this mispair is not repaired by the mismatch repair pathway, it would subsequently result in the production of a mutation at that site. *Recombinational* repair and *nonhomologous* end joining differ from direct reversal and excision repair pathways. Instead of removing the damage and restoring it back to its original state, they act as *tolerance* mechanisms where the damage is not always immediately removed but the cell's ability to survive is enhanced. In addition, cells have evolved a complex network of signal transduction pathways that respond to DNA damage and bring about arrest at different stages of the cell cycle. These *cell cycle checkpoints* provide the cell with more time to repair DNA damage and in doing so reduce the deleterious consequences of encounters of DNA damage with the DNA replication machinery or other transactions.

23.2 DIRECT REVERSAL OF BASE DAMAGE

Direct reversal of damage occurs in a single-step reaction that is catalyzed by a single polypeptide. Despite its simplicity, this type of repair is rare and limited to a small number of different types of DNA damage. It occurs for damage produced by UV light in bacteria and yeast, but it probably does not occur for the removal of UV damage in humans or other mammals. Direct reversal also occurs for certain types of damage produced by alkylating agents in bacteria, yeast, and humans. It has the advantage of requiring only one reaction step and is likely to be *error-free*. The teleological basis for its apparent rarity in nature when compared to the other, more complex repair pathways is unclear.

23.2.1 UV Radiation

Most organisms have to deal with the harmful effects of UV radiation and have probably had to do so since life began. As described in Chapter 22, UV radiation reacts with adjacent pyrimidines in DNA and introduces two major types of DNA damage. The most frequent is the cyclobutane pyrimidine dimer (CPD) and the less frequent is the (6–4) pyrimidine–pyrimidone photoproduct ((6–4)PP). They are formed at a ratio of roughly 4:1. CPDs in *E. coli* can be repaired by direct reversal in a light-dependent reaction that is called *photoreactivation* and is catalyzed by the enzyme *photolyase* or, more specifically, pyrimidine dimmer–DNA photolyase (PD–DNA photolyase). In fact, photoreactivation was the first DNA repair pathway that was discovered and this occurred in the late 1940s. PD–DNA photolyase binds the damaged DNA; and in a reaction dependent upon absorption of photoreactivating wavelengths of light (300–500 nm), it breaks the covalent linkage of adjacent pyrimidines and converts them back to adjacent monomers (Figure 23.1). Exposure of cells to photoreactivating wavelengths of light dramatically enhances their ability

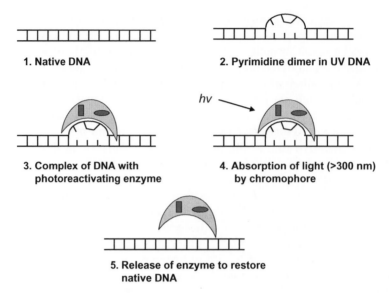

1. Native DNA

2. Pyrimidine dimer in UV DNA

3. Complex of DNA with photoreactivating enzyme

4. Absorption of light (>300 nm) by chromophore

5. Release of enzyme to restore native DNA

Figure 23.1. Schematic illustration of direct reversal of a pyrimidine dimer by the enzyme DNA photolyase. The enzyme binds to the pyrimidine dimer present in DNA. The square and triangle represent the two noncovalently bound chromophores that are present in all photolyases. The chromophores harness the energy of photoreactivating blue wavelengths of light and use them to catalyze the breakage of the pyrimidine dimer back to adjacent monomers. [Adapted from Friedberg, E. C., Walker, G. C., Siede, W., Wood, R. D., Schultz, R. A., and Ellenberger, T. (Eds.). *DNA Repair and Mutagenesis*, 2nd ed., ASM, Washington, D.C., 2006.]

to survive exposure to UV irradiation, which is how this repair pathway was initially discovered by Albert Kelner.

Photoreactivation is fairly ubiquitous in nature and it has been identified in all three kingdoms in a wide variety of organisms including bacteria, fungi, green algae, plants, fruit flies, and fish. PD–DNA photolyase activity appears to be absent from placental mammals; hence, for some reason, it appears to have been lost during very late stages of evolution. A great deal of information about the phenomenon of photoreactivation was obtained studying *E. coli*. PD–DNA photolyase activity was identified in this organism in the late 1950s by Stan Rupert and Sol Goodgal, but it was not until the arrival and implementation of recombinant DNA technologies in the late 1970s and the 1980s that it was purified to homogeneity from *E. coli* and yeast. PD–DNA photolyase is present in extremely small quantities in the cell. For reasons that are not entirely clear, many DNA repair proteins are generally present in very small quantities, and this significantly hampered the ability to purify them and characterize them at the biochemical level.

The purified *E. coli* protein has a molecular weight of 49 kD. It does not require any divalent cation for activity. It contains two different noncovalently bound *chromophores* that absorb light. One chromophore is flavin adenine dinucleotide (FADH- or FADH2). The other is 5,10-methenyltetrahydrofolyl polyglutamate (MTHF). The absorption of light by the chromophores is essential for the enzymatic reversal of the pyrimidine dimer back to the original pyrimidine monomers. However,

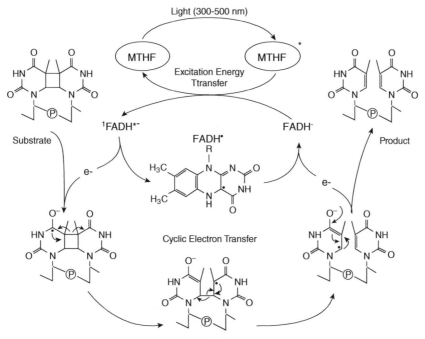

Figure 23.2. Reaction mechanism of PD–DNA photolyase. A photon of blue light is absorbed by the MTHF chromophore that acts as a photoantenna. The excited energy is transferred to the flavin chromophore (FADH⁻). The excited flavin (FADH*⁻) acts as a photocatalyst and transfers an electron to a CPD in DNA. The thymines are restored to their native state and the electron is transferred back to the flavin. (Reproduced with permission from Sancar, A. Structure and function of DNA photolyase cryptochrome blue-light photoreceptors. *Chem. Rev.* **103**, 2203–2237, 2003.)

the absorption of light is not necessary for binding of PD–DNA photolyase to pyrimidine dimers. Hence, the biochemical reaction has a light-independent or "dark" step where the enzyme first finds and binds to the CPD. This is followed by a "light" step where the chromophores absorb light at wavelengths between 300 to 500 nm, and the enzyme catalyzes the monomerization of the CPD. Binding is highly specific, and the discrimination ratio between specific and nonspecific binding is ~10⁵. The monomerization reaction is also highly specific and the PD–DNA photolyase does not photoreactivate the other major UV light-induced photoproduct, the 6–4PP. The MTHF chromophore acts as a *photoantenna* and absorbs a blue light photon (Figure 23.2). This energy is transferred to the FADH⁻ chromophore which donates an electron to the CPD converting it to a CPD anion. In doing so, the FAD chromophore acts as a *photocatalyst*. The CPD anion spontaneously rearranges to form pyrimidine monomers and the electron is transferred back to the FADH⁻ chromophore. Crystal studies of PD–DNA photolyase suggest that the CPD flips out of the DNA helix and into a cavity in the enzyme. This is one of several examples where "*base-flipping*" is an important mechanism in DNA repair.

For many years it was assumed that photoreactivation was a process that was limited to the repair of CPDs and did not occur for (6–4)PPs. This assumption was based on differences in the chemical structures of the two different photoproducts because it was thought that even if an enzyme could hydrolyze the bond between the pyrimidines comprising a (6–4)PP, it could not restore the pyrimidines back to their native state. However this turned out not to be the case and in the early 1990s an activity capable of photoreactivating (6–4)PPs was identified in *Drosophila*. A similar activity has also been identified in some vertebrates and plants. These enzymes are referred to as *(6–4) photoproduct-DNA photolyases ((6–4)PP-DNA photolyase)* to distinguish them from PD–DNA photolyases. The precise details of photoreactivation for this class of enzymes are not as well understood. They contain FAD as a chromophore and use two amino acid histidine residues present in the enzyme for the transfer of protons between the amino acids and the (6–4)PP. Electron transfer using the light-activated FAD chromophore bound to the enzyme results in restoration of the original pyrimidine monomers.

23.2.2 Alkylation Damage

Certain types of DNA damage introduced by chemical compounds collectively referred to as alkylating agents (see Chapter 22) are also repaired by direct reversal. This type of repair was initially alluded to by the discovery of a phenomenon called the *adaptive response*. When cultures of *E. coli* are treated with high doses of the alkylating agent *N*-methyl-*N'*-nitro-*N*-nitrosoguanidine (MNNG), they accumulate large numbers of mutations and many cells die. However, there is a significant reduction in the number of mutations and dead cells when cultures are first treated with a very low dose of MNNG, and this is then followed by treatment with the larger dose. This phenomenon was discovered by John Cairns and his then graduate student, Leona Samson, in the mid-1970s.

Many different bases in DNA are alkylated when cells are exposed to MNNG. One of the products formed is O^6-methylguanine and this base modification is removed by direct reversal. The protein responsible for direct reversal of this lesion in *E. coli* is O^6-alkylguanine-DNA alkyltransferase I (O^6-AGT I). It uses its methyltransferase activity to transfer the methyl group from the O6 position of O^6-methylguanine to a cysteine residue present in the enzyme thereby restoring the base to guanine (Figure 23.3A). In addition, this protein can also remove methyl groups from O^4-methylthymine and from methylphosphotriesters that are formed when the phosphodiester backbone becomes methylated (Figure 23.3B,C). It can also act on alkyl groups that are larger than methyl groups including ethyl, propyl, and butyl groups. Each molecule of the enzyme can transfer two alkyl groups from the damaged DNA. However, the cysteine residues that receive the alky groups from alkylated bases or alkylated phosphotriesters reside in different regions off the protein. The cysteine group that receives alky groups from alkylated bases resides in the C-terminal portion of the protein, Cys321, while the cysteine group that receives the alkyl group from alkylated phosphotriesters resides in the N-terminal portion of the protein, Cys38.

O^6-AGT I plays a key role in the *ada*ptive response and hence, it is sometimes also referred to as Ada and it is encoded by the *ada* gene in *E. coli*. It is unusual in that it becomes inactivated as a consequence of the transfer of the alkyl group to its cysteine residues and hence it is described as a *suicide enzyme*. O^6-AGT I has dual functions: one as a repair protein and the other as a transcriptional regulator of genes involved in the repair of alkylation damage. Levels of the protein increase

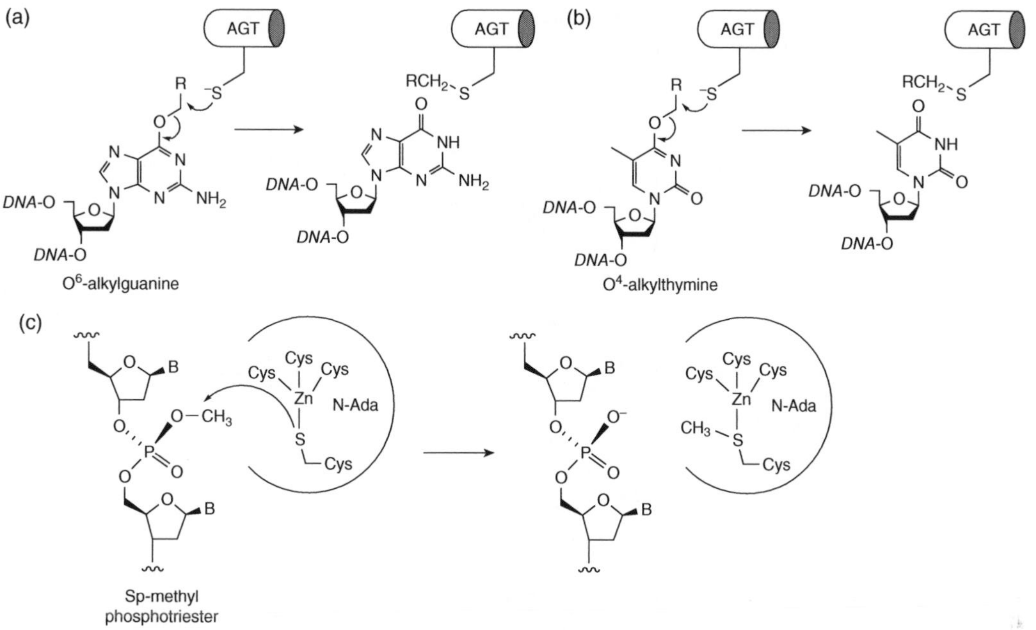

Figure 23.3. O^6-guanine alkyl transferase activities. (A) O^6-AGT I uses a cysteine residue present in the C-terminal region of the protein to remove an alkyl group from the O^6position of guanine and (B) the O^4position of thymine. (C) A zinc-bound cysteine present in the N-terminal region of the protein transfers a methyl group from a methyl phosphotriester. O^6-AGT I is also referred to as Ada because it is involved in the *ada*ptive response. All transfers to cysteine residues are irreversible. (Reproduced with permission from Mishina, Y., Duguid, E. M., and He, C. Direct reversal of DNA alkylation damage. *Chem. Rev.* **106**, 215–232, 2006.)

several hundredfold when cultures of *E. coli* are treated with an alkylating agent. This change is brought about by the role of O^6-AGT I as a transcriptional regulator. When the protein is alkylated at Cys38 after transfer of an alkyl group from an alkylated phosphotriester, it is converted to a transcriptional activator that binds to the promoter regions and upregulates many genes involved in the repair of alkylation damage including the *ada* gene (Figure 23.4). It is unclear how the addition of an alkyl group to cysteine 36 actually converts the protein to a transcriptional activator that then acquires the ability to recognize and specifically bind regulatory sequences in the DNA.

The adaptive response is observed as enhanced cell survival and reduced mutations when cells are pretreated with low doses of an alkylating agent. This is brought about by the formation of alkyl phosphotriesters by the alkylating agent. The alkyl group is transferred from the alkyl phosphotriester to the Cys38 of O^6-AGT I which converts it to a transcriptional activator. The *ada* gene, which encodes the O^6-AGT I/Ada protein, is greatly upregulated and several hundred more molecules of the protein are synthesized. Hence, when the cells are subsequently exposed to a much higher dose of the alkylating agent, they possess several hundred more molecules of the protein that can remove the alkylated bases, which, in turn, reduces the number of mutations that are formed.

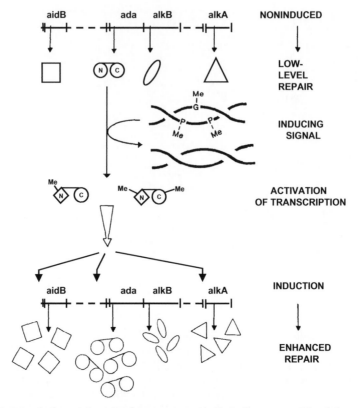

Figure 23.4. Regulation of the adaptive response to alkylating agents. The Ada regulon contains the *ada* gene, which encodes O^6-AGT I (also referred to as Ada), and several other genes involved in the response to alkylation damage. The O^6-AGT I polypeptide is represented showing its N-terminal and C-terminal regions (as circles) that contain cysteine residues. The cysteine group in the N-terminal region receives alkyl groups from alkylated phosphotriesters. The cysteine group in the C-terminal region receives alkyl groups from O^6-alkyl guanine or O^4-methyl thymine. When an alkyl group is transferred from an alkylated phosphotriester to the N-terminal cysteine, O^6-AGT I is converted to a transcriptional activator. It then binds to the promoter regions of genes in the Ada regulon and increases the transcription of the genes. This results in an increased expression of O^6-AGT I (Ada), which, in turn, increases the repair of alkylation damage. (Reproduced with permission from Lindahl, T., Sedgwick, B., Sekiguchi, M., and Nakabeppu, Y. Regulation and expression of the adaptive response to alkylating agents. *Annu. Rev. Biochem.* **57**, 133–157, 1988.)

It has been proposed that the specific transfer of alkyl groups from alkyl phosphotriesters rather than alkylated bases acts as an activation signal because the alkyl phosphotriesters are formed at much lower levels in the cell. The abundance of constitutive (uninduced) levels of protein may be sufficient to repair low levels of alkylation damage. However, when the level of damage is high enough to generate sufficient numbers of alkyl phosphotriesters (the less frequent lesion), there may be insufficient levels of protein for repair of all of the lesions and hence there would be a need to induce the *ada* gene. There are at least two possibilities for how the adaptive response may be turned off. The alkylated form of O^6-AGT I may simply be diluted by cell division. Alternatively, there may be mechanisms that specifically cleave and degrade the alkylated form of the protein.

The adaptive response is a relatively slow process. It takes approximately 1 h for the abundance of O^6-AGT I to change from only a few molecules to several thousand after cultures of *E. coli* are treated with a low dose of an alkylating agent. Since many cells replicate their DNA during this time frame, they would be at risk for accumulating mutations at sites of alkylated bases that were not repaired as a consequence of low levels of the repair protein. Presumably to prevent this from happening, *E. coli* (and other prokaryotes) possesses a second alkyl transferase called O^6-alkylguanine-DNA alkyltransferase II (O^6-AGT II) and it is encoded by the *ogt* gene. It differs from O^6-AGT I in that it is not induced by low levels of alkylating agents and it does not remove alkyl groups from alkyl phosphotriesters. Its function is to remove alkyl groups from O^6-alkyguanine and O^4-alkylthymine before or while the *ada* gene is induced during the adaptive response. Interestingly, O^6-AGT II has a preference for O^4-alkylthymine while O^6-AGT I has a preference for O^6-alkylguanine, but both proteins repair both alkylated bases.

O^6-AGT activity has been identified in a large number of organisms including other prokaryotes, yeast, mammals, and members of the *Archaea* kingdom. In humans, only one activity has been identified and it is fairly specific for the removal of O^6-methylguanine. It does not repair methyl-phosphotriesters, and O^4-methylthymine is a poor substrate. Hence, the human gene and protein are called O^6-methylguanine methyl transferase (O^6-MGMT). Like the *E. coli* protein, it is inactivated when it transfers the methyl group from O^6-methyl guanine to itself. Pretreatment of human cells with low doses of alkylating agents renders them more resistant to subsequent treatment with higher doses of the agent. However, there is no accompanying reduction in the formation of mutations. Hence, there is no evidence for an alkylation damage specific adaptive response to mutagenesis in human cells.

Interestingly, there are large variations in O^6-MGMT activity among different rodent and human cell lines. This has led to the description of mer+ or mer– phenotypes. Cells are considered mer+ if they have significant levels of the transferase activity and mer– if they have no transferase activity. Sometimes in the literature this is also referred to as mex+ or mex– phenotypes. The reduced transferase activity may be a consequence of hypermethylation of the promoter region of the O^6-MGMT gene, a more compact chromatin conformation that reduces expression of the gene or changes in regulatory proteins that regulate transcription of the gene. The "turning-off" of the O^6-MGMT gene may have clinical significance. This can impact both the cause of certain types of cancer and the therapeutic treatment of tumors. There is significant evidence that O^6-methyl guanine in DNA is mutagenic and carcinogenic. Hence, inactivation or reduction of the transferase that repairs this type of damage would likely lead to the accumulation of this type of damage, which could play a role in the etiology of certain types of cancer in humans. There is some evidence to support this in that hypermethylation of the promoter region of the O^6-MGMT gene has been found in esophageal and colorectal tumors. On the other side of the coin, alkylating agents are also used as therapeutic agents to treat certain forms of cancer because they kill cells. One approach is to try to increase the efficacy of the alkylating agents that are used in chemotherapy by inactivating or inhibiting the transferase in the tumor cells.

23.3   BASE EXCISION REPAIR

Base excision repair (BER) is a complex multistep pathway that is initiated by enzymes called DNA *glycosylases*, which recognize a damaged base and remove it by cleavage of the *N-glycosidic bond* that connects the base to the deoxyribose-sugar backbone (Figure 23.5). Hence, the name base excision repair derives from the fact that this type of repair involves removal or excision of an altered base. This is in contrast to the nucleotide excision repair pathway where a short oligonucleotide containing the damage is excised.

The removal of an altered base by the glycosylase results in the formation of an *abasic site* (*AP site*) (sometimes also referred to as an apurinic or apyrimidinic site). The AP site is acted upon by an *AP endonuclease* that makes an incision in the damaged strand by hydrolyzing the phosphodiester bond 5′ to the AP site (Figure 23.5). Alternatively, some glycosylases have an associated *lyase* activity that incises

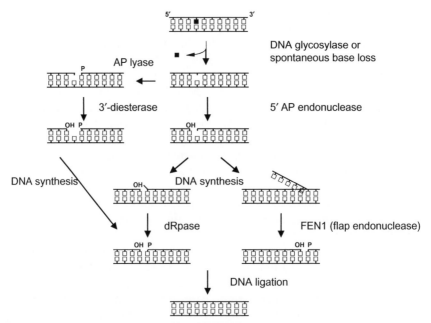

Figure 23.5. Schematic illustration of the base excision repair pathway. The activity of a DNA glycosylase or spontaneous base loss breaks the *N*-glycosydic bond between the base and the deoxyribose sugar creating an AP site. The AP endonuclease hydrolyzes the phosphodiester backbone 5′ to the AP site, creating a strand break. A one nucleotide gap is filled in by DNA polymerase. dRpase and flap endonucleases modify the DNA ends to facilitate DNA synthesis and ligation. DNA ligase seals the final nick. In the "short patch" base excision repair pathway, a single nucleotide gap is replaced. The right branch after 5′ AP endonuclease activity represents "long patch" base excision repair where the flap that is created by DNA synthesis can be processed by FEN1. The left branch after the glycosylase-mediated step represents the processing of DNA ends that occurs when the glycosylase possesses a lyase activity. (Adapted from Wyatt, M. D., Allan, J. M., Lau, A. Y., Ellenberger, T. E., and Samson, L. D. 3-Methyladenine DNA glycosylases: Structure, function, and biological importance. *Bioessays* **21**, 668–676, 1999.)

the damaged strand 3' to the AP site (see left-hand side of Figure 23.5). The structures that are formed by these activities are further processed by other enzymes including *exonucleases* and *DNA-deoxyribosphosphodiesterases (dRpase)*. In *short-patch* BER, these activities lead to the formation of a one nucleotide gap that is filled in by DNA polymerase-dependent repair synthesis. In *long-patch* BER, there is additional strand displacement resulting from the synthesis of a longer patch that results in the formation of a 5' flap that is cleaved by a *flap endonuclease* (FEN1). The pathway is complete when a *DNA ligase* creates a phosphodiester bond that seals the nick and restores integrity to the DNA. Details of each step in BER will be described in the following sections.

23.3.1 Glycosylases

There are many different types of damaged or inappropriate bases that are removed by BER. These types of base alterations are frequently formed endogenously as byproducts of cellular metabolism. Correspondingly, there is a large collection of different DNA glycosylases that are specific for each type of altered base or a subset of bases that are very similar in structure. Tables 23.2 and 23.3 list glycosylases that have been characterized in *E. coli* and humans. There are two types of glycosylases. *Monofunctional* glycosylases simply remove the base, leaving an AP site. Bifunctional glycosylases remove the base, and they cleave the phosphodiester backbone 3' to the AP site using their lyase activity.

The active site of many (but not all) glycosylases is flanked by a highly conserved double-strand DNA-binding motif called a *helix–hairpin–helix (HhH) motif*. This motif is comprised of two α-helices connected by a hairpin turn, and it is located near a deep cleft that contains the active site of the enzyme. Glycosylases that possess the HhH motif have been categorized as members of the *HhH superfamily*. Members of this family have highly divergent amino acid sequences, but their overall structure is highly conserved. The damaged or inappropriate base is forced or "flipped" out of the DNA helix and positioned into the active site of the enzyme

TABLE 23.2. DNA Glycosylases in *E. coli*

Protein	Common Name	Examples of Activities
Ung	Uracil-DNA glycosylase	Removes uracil
Mug	Mug-DNA glycosylase	Removes uracil, thymine or ethenocytosine opposite guanine
Fpg (MutM)[a]	FaPy-DNA glycosylase	Removes oxidized and ring-opened purines including 8-oxoG and formamidopyrimidine
MutY	MutY-DNA glycosylase	Removes adenine opposite 8-oxoG
Nth (endoIII)[a]	Endonuclease III	Removes ring-saturated of fragmented pyrimidines
TagA	3-Methyladenine-DNA glycosylase I	Removes 3-methyladenine and 3-ethyladenine
AlkA	3-Methyladenine-DNA glycosylase II	Removes 3-methylpurines, 7-methylpurines, 3- and 7-ethylpurines, ethenoadenine and O^2-methylpyrimidines

[a]Glycosylases with an associated lyase activity.

TABLE 23.3. DNA Glycosylases in Human Cells

Protein	Common Name	Examples of Activities
UNG	Uracil-DNA glycosylase	Removes uracil
SMUG1	SMUG DNA glycosylase	Removes uracil, 5-hydroxymethyluracil
TDG	Thymine-DNA glycosylase	Removes U, T, or etheno-C opposite G (preferably CpG sites)
OGG1[a]	8-oxoG-DNA glycosylase	Removes oxidized and ring-opened purines including 8-oxoG and formamidopyrimidine
MYH	MutY homolog DNA	Removes adenine opposite 8-oxoG, 2-OH-A opposite G
NTH1[a]	Endonuclease III	Removes ring-saturated of fragmented pyrimidines
MPG	3-Methyladenine-DNA glycosylase I	Removes 3-methylpurines, hypoxanthine, and ethenoadenine

[a]Glycosylases with an associated lyase activity.

by the process of base flipping. The amino acid residues lining the active sites of the different members of this family are variable, which contributes to the specificity of each glycosylase for different types of damaged bases.

23.3.1a Uracil. Uracil is frequently inappropriately formed in DNA (see Chapter 22). An activity that can remove uracil from DNA was first described by Tomas Lindahl studying *E. coli* in the mid-1970s. *Uracil–DNA–glycosylases (UDGs)* are a collection of families of enzymes that remove uracil from DNA, and they are present in all kingdoms. In general they are small, robust enzymes that do not require a metal cofactor but their activity can be strongly stimulated by Mg^{2+}. They are highly specific for uracil in either single-strand or double-strand DNA. There are several different families of these enzymes. One family (family 1) is represented by the enzyme Ung in *E. coli*. It is a monofunctional glycosylase that removes uracil and creates an AP site. It is encoded by the *ung* gene. The mechanism of action involves the enzyme sliding along the minor groove of DNA, scanning for the presence of an abnormal base. When uracil is detected, it kinks the phosphodiester backbone of the strand containing the abnormal base. The base is flipped into the active site of the enzyme and the *N*-glycosylic bond is cleaved, releasing the uracil base and leaving an AP site. The shape of the active site pocket and the position of specific amino acids within the pocket prevent recognition or cleavage of other bases. In mammalian cells, UNG largely functions to remove uracil that has been misincorporated at sites of DNA replication.

A second class of UDGs (family 2) recognizes uracil when it is present in mismatches. The deamination of cytosine produces uracil and consequently a G-U mispair in double-strand DNA. In eukaryotes, *thymine–DNA glycosylase (TDG)* generally removes thymine when it is mispaired with guanine. In addition, it can also remove uracil when it is present in a G–U mispair. Bacteria possess structural homologs of the eukaryotic TDGs that have been called *Mug* for mismatch-specific uracil–DNA glycosylase. However, Mugs only remove uracil, not thymine, when it is mispaired with guanine. Mug and TDG have broader substrate specificities and can also remove the alkylated bases ethenocytosine when it is opposite guanine and ethenoguanine when it is opposite cytosine. These damaged bases can be formed by lipid epoxidation and by exposure to certain chemicals (see Chapter 22). This family of enzymes does not use base flipping as a mechanism.

A third class of UDGs (family 3) was discovered and found to be very active on uracil present in single-strand DNA. Hence they were named *SMUG* for single-strand-specific monofunctional uracil–DNA glycosylase. However, their biological substrates are probably primarily double-stranded DNA. SMUG's primary role is in the removal of 5-hydroxymethyluracil that is produced by oxidative agents. There are additional families of UDGs that have been identified in *Archaea* and some bacteria. In many cases they have likely evolved to deal with the consequences of living at significantly higher temperatures on deamination rates and damage formation.

23.3.1b Alkylated Bases. In contrast to UDGs, most other glycosylases recognize a broader range of substrates. Given the extremely large number of different types of base damage, it would probably be inefficient to have one unique enzyme for the removal of each type of base damage. Hence, many glycosylases evolved to remove a wider range of bases. However, this is countered by the need to retain a high degree of damage-specificity so that normal bases are not also removed by the enzymes.

Certain alkylating agents can introduce substantial amounts of the damaged bases 7-methylguanine (7-meG) and 3-methyladenine (3-meA). These base modifications are relatively labile and exhibit a relatively high level of spontaneous depurination. However, there are specific glycosylases that also aid in their removal. In *E. coli, 3-methyladenine-DNA glycosylase I (TagA)* (Tag is for 3-methyladenine-DNA glycosylase) is highly specific for the removal of 3-methyladenine. It does not remove 7-meG or O^6-methylguanine, which are also formed by many simple alkylating agents. *E. coli* possesses a second glycosylase that was originally called *3-methyladenine-DNA glycosylase II*, and it also removes 3-methyladenine. It is encoded by the *alkA* gene and is generally now referred to as AlkA. AlkA recognizes a broader range of substrates than does TagA. It also removes 3-methylguanine, 7-meG, 7-methyladenine, and 1, N^6-ethenoadenine. The *alka* gene is induced following exposure of cells to an alkylating agent and functions as part of the adaptive response. In untreated cells, AlkA represents only 5–10% of the total 3-methyladenine–DNA glycosylase activity. In treated cells, its abundance changes to 50–70% of the total activity. Homologs of AlkA have been found in other bacteria and eukaryotes but not in mammals. Mammals possess a different 3-methyladenine-DNA glycosylase—called *MPG*, for *N*-methylpurine-DNA glycosylase—that recognizes a wide range of alkylated purines and hypoxanthine.

23.3.1c Oxidized and Fragmented Bases. Reactive oxygen species (ROS) and ionizing radiation produce a diverse array of base damage that include 7,8-dihydro-8-oxoguanosine (8-oxoG), 5-hydroxycytosine (5-OH-C), 2,6-diamino-4-hydroxy-5-*N*-methylformamidopyrimidine (FaPy-G), and 4,6-diamino-5-*N*-methylformamidopyrimidine (FaPy-A) (see Chapter 22). The latter two lesions are imidazole ring-opened forms of guanine and adenine. This diverse group of lesions is removed by the *E. coli* glycosylase called *Fpg*, for *formamidopyrimidine-DNA glycosylase*. This enzyme is also referred to as MutM. Fpg does not contain the HhH motif and is not a member of the HhH superfamily. In addition to acting as a glycosylase, it catalyzes a β,δ-elimination reaction that incises the DNA at the AP site, leaving 3′ and 5′ phosphoryl groups and a single nucleotide gap. Cells that are defective in the *fpg* gene exhibit a strong mutator phenotype.

E. coli possesses a second glycosylase, *MutY*, that removes 8-oxoG. MutY is a member of the HhH superfamily. During DNA replication, sometimes the DNA polymerase incorrectly inserts an A opposite 8-oxoG resulting in an 8-oxoG:A mispair. The MutY glycosylase removes the A from the undamaged strand, which is then usually replaced with a C by BER. This leads to the formation of an 8-oxoG:C mispair that can be acted upon by Fpg(MutM), which then removes the 8-oxoG. This has been referred to as the "GO system" (Figure 23.6). A third enzyme, *MutT*, which is not a glycosylase, also participates in this pathway by cleansing dGTP nucleotide pools that have been damaged and prevents their incorporation during DNA synthesis. MutT hydrolyzes 8-oxodGTP to the monophosphate form 8-oxo-dGMP and prevents it from being incorporated into DNA (Figure 23.6).

Inactivation of the *fpg* (*mutM*), *mutY*, or *mutT* genes in *E. coli* results in a significant mutator phenotype, which is how they were originally named. Humans possess a homolog of the *mutY* gene, called *MYH* for mutY homolog. Interestingly, mutations in the human MYH gene have been found in certain individuals that have inherited a predisposition to developing colon cancer. In addition, similar mutations in MYH have been found in colon tumors from individuals that are not known to

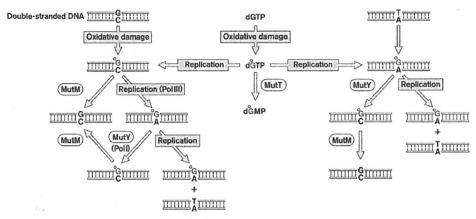

Figure 23.6. The GO system protects cells from the mutagenic effects of 8-oxoG. Exposure to oxidative damaging agents can produce 8-oxoG in DNA. This is removed by the glycosylase activity of MutM (Fpg). If 8-oxoG is present in DNA and encountered by the DNA replication machinery, A can be misincorporated opposite the 8-oxoG, creating an 8-oxoG:A mispair. This mispair can be acted upon by MutY, which removes the A in the 8-oxoG:A mispair which is then usually replaced with a C during base excision repair. This provides MutM with another opportunity to detect an 8-oxoG:C pair and remove the 8-oxoG, which is then replaced with normal guanine. A third enzyme in this pathway is MutT, which serves to sanitize the deoxynucleotide triphosphate pools by converting 8-oxoGTP to 8-oxoGMP. This prevents the direct incorporation of 8-oxoG into DNA during DNA synthesis. (Reproduced with permission from Cunningham, R. P. DNA repair: Caretakers of the genome? *Curr. Biol.* **7**(9), R576–R579, 1997.) See color insert.

be genetically predisposed. This is an example of how a deficiency in BER can pre-dispose humans to cancer.

In addition to possessing enzymes described above that remove oxidized and fragmented purines, cells also possess enzymes that remove oxidized and fragmented pyrimidines. One such enzyme is *endonuclease III* in *E. coli*. It was originally identified as an activity that nicked UV-irradiated or oxidatively damaged DNA, which is why it was first described as an endonuclease. However, it is a glycosylase with an associated lyase activity and not an endonuclease. It is encoded by the *nth* gene and is now generally referred to as Nth. Nth protein recognizes a wide variety of altered pyrimidines including thymine glycol, thymine, and cytosine hydrates and urea. It is a member of the HhH superfamily and uses base flipping to remove the damaged base. There are Nth homologs in many organisms including yeast and mammals. In addition, there are "back-up" glycosylases that also remove these types of damaged bases.

23.3.1d Pyrimidine Dimers. Certain organisms contain glycosylases that remove cyclobutane pyrimidine dimers that are formed by UV irradiation. The two enzymes that have been most extensively studied are from the bacterium *M. luteus* and the phage T4. They were initially described as endonucleases. However, it was subsequently determined that they possess glycosylase and lyase activities and not endonuclease activity. In addition, it was discovered that the glycosylase activity only cleaves the *N*-glycosidic bond connecting the 5′ member of the pyrimidine dimer to the deoxyribosephosphate backbone. This distinguishes them from other glycosylases. Because the two bases that comprise the pyrimidine dimer are connected through their pyrimidine rings and the 3′ member of the dimer remains connected to the deoxyribosephosphate backbone, this cleavage reaction does not result in the release of a free base. Instead, the pyrimidine dimer is subsequently removed as a dinucleotide or oligonucleotide during postincision degradation of DNA. *E. coli*, yeast, and mammals do not have this enzyme.

23.3.2 AP Endonucleases

AP sites are formed as part of the BER pathway. In addition, they are formed spontaneously as a consequence of inherent instability in the structure of DNA. Hence, the AP endonucleases that participate in their repair are ubiquitous in nature. The major AP endonuclease in *E. coli* is called *exonuclease III*. It was initially characterized as an enzyme with 3′ to 5′ exonuclease activity and phosphatase activity. It was subsequently characterized as also possessing a 5′ AP endonuclease activity. It is encoded by the *xthA* gene and is also referred to as XthA. It hydrolyzes the phosphodiester bond 5′ to an AP site generating a 3′ OH and a 5′ deoxyribosephosphate. Its activity is different from the lyase activity associated with some glycosylases (Figure 23.7). A similar protein is present in mammals and has been designated *APEX nuclease* (AP endonuclease/exonuclease). Mice with engineered disruptions of this gene exhibit embryonic lethality. This and other observations indicate that AP endonuclease activity is essential for viability of the cell.

E. coli possesses a second AP endonuclease, *endonuclease IV*. It is encoded by the *nfo* gene, and it is also referred to as *Nfo*. Nfo only comprises about 10% of the AP endonuclease activity in the cell. It acts in the same way that XthA does and

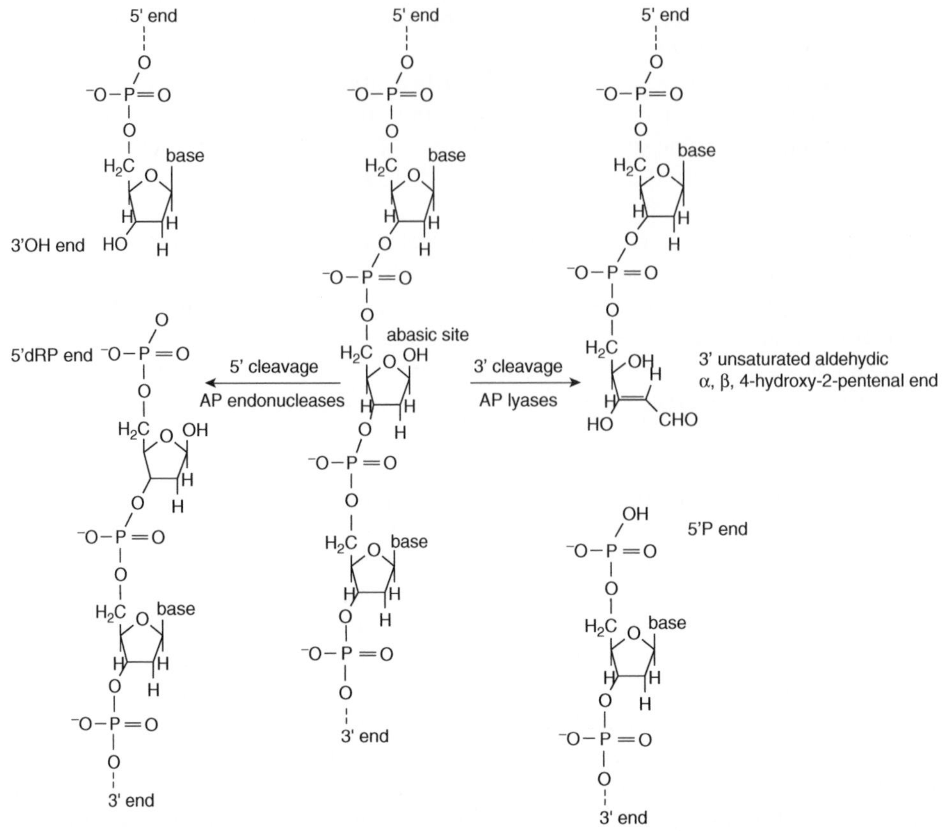

Figure 23.7. Structure of an AP site (center) and structures produced by 5′ AP endonuclease activity (left) or lyase activity (right). These are formed in double-strand DNA, but only one strand is represented in the figure. 5′ AP-endonuclease activity results in the formation of a 5′-deoxyribose-phosphate end (5′-dRP), which must be subsequently processed by a dRPase (see text). Lyase activity results in the formation of 3′-unsaturated aldehydic α, β, 4-hydroxy-2-pentenal end, which must be subsequently acted upon by a 5′ AP endonuclease. (Reproduced with permission from Boiteux, S., and Guillet, M. Abasic sites in DNA: repair and biological consequences in *Saccharomyces cerevisiae*. *DNA Repair* **3**, 1–12, 2004.)

cleaves 5′ to an AP site. It also possesses phosphodiesterase activity, but it does not possess a 3′ to 5′ exonuclease activity. In general, it acts on the same collection of substrates as XthA. However, *nfo* mutant strains, exhibit greater sensitivity to bleomycin than do *xthA* mutant strains, suggesting that Nfo may have greater specificity for some DNA lesions introduced by bleomycin. In addition, the *nfo* gene is inducible by agents that generate superoxide radicals.

23.3.3 Repair Synthesis and Ligation

The concluding steps of BER involve DNA synthesis to replace the damaged nucleotide and ligation. If the glycosylase has an associated AP lyase activity, it can generate a 3′-α, β unsaturated aldehyde (Figure 23.7). This residue must be processed by the phosphodiesterase activity of an AP endonuclease (Figure 23.5). If an AP endonuclease directly incises the DNA 5′ to the AP site, it creates a 3′ OH and a 5′ deoxyribose-phosphate (dRp). In either case, the 3′ OH serves as a primer terminus that a DNA polymerase can extend from to synthesize a repair patch of one nucleotide. However, the dRp residue, if present, must be removed by a deoxyribophosphodiesterase (dRpase) or by a flap endonuclease in long patch BER. In *E. coli* it is likely that DNA polymerase I is responsible for the gap-filling synthesis step. There are a collection of different enzymes in *E. coli* that may act as dRpases. In mammalian cells, DNA polymerase β (pol β) is the DNA polymerase that is responsible for gap filling during most BER events. In addition, pol β also has an important dRpase function. The two activities reside in two different domains of the protein.

In mammalian cells, BER can take place by either "short-patch repair," where a single nucleotide is incorporated, or "long-patch repair," where several nucleotides are incorporated. The factors that determine which pathway is used are not entirely understood. In short-patch repair, pol β incorporates a single nucleotide into the repair patch and uses its dRpase activity to remove the dRp residue. It is likely that the choice to employ long-patch repair occurs when there is difficulty processing the dRp residue. Other factors that impact the choice to use long-patch BER probably include the glycosylases that initially process the damaged base and the nature of the lesion itself. Different proteins are used. Because the patch is longer, either DNA polymerase δ or DNA polymerase ε performs the DNA synthesis step. FEN1 (flap-endonuclease 1) can either act as a 5′ to 3′ exonuclease or act as an endonuclease and incise a 5′ "flapped" structure (Figure 23.5). A flap can be created when there is strand-displacement by the DNA polymerase as it synthesizes the longer patch. PCNA (proliferating-cell nuclear antigen) stimulates the activity of FEN1.

The last step in BER is to create the final phosphodiester bond that seals the nick formed after DNA synthesis. In *E.coli* this is performed by a single DNA ligase that is encoded by the *ligA* gene. In mammalian cells, three genes have been identified that encode DNA ligases and they are designated *LIG1*, *LIG3*, and *LIG4*. Ligase I and an isoform of ligase III designate ligase IIIα can participate in BER, but studies suggest that the ligase IIIα isoform carries out most of the ligation reactions in BER.

AP sites, left unrepaired, impede the progression of DNA polymerases during DNA replication and can be mutagenic in the synthesis of both DNA and RNA. AP sites can also be converted to toxic lesions by spontaneous rearrangements that produce structures that can crosslink with proteins and lipids. They can also create toxic lesions by the activities of topoisomerase I and topoisomerase II. In addition, the processing of AP sites by the BER proteins can produce structures that are toxic. Hence, to prevent these deleterious events from occurring, the BER process is likely coordinated and each step involves a "handing off" from one enzyme in the pathway to the next.

23.4 NUCLEOTIDE EXCISION REPAIR

Nucleotide excision repair (NER) removes an assortment of different types of DNA damage. It removes chemical adducts introduced by exposure to chemical carcinogens and CPDs and (6–4) PPs produced by UV light. A major difference between the BER and NER pathways has to do with DNA damage recognition. The preceding sections describes how the BER pathway is initiated by many different glycosylases and how each one recognizes a specific type of damaged base or a subset of structurally related damaged bases. In sharp contrast, the NER pathway recognizes many different types of lesions that can be structurally dissimilar. Hence, it is likely that NER recognizes distortions of the DNA helix rather than the lesion itself. There are several steps in the process: (1) damage recognition and lesion verification, (2) unwinding of the DNA at the lesion, (3) two incisions are made, one on each side of the lesion, (4) removal (excision) of a stretch of DNA containing the lesion, (5) DNA synthesis to replace the excised DNA, and (6) ligation of the newly synthesized DNA to the parental strand. While the repertoires of proteins that orchestrate the process differ, the general strategy has been conserved in *E. coli*, yeast, and mammalian systems.

UV photoproducts have historically served as model substrates in the investigation of DNA pathways. They are easy and inexpensive to introduce by using a simple germicidal bulb, and they are extremely stable. In addition, they have clear biological relevance because most, if not all, living organisms have to contend with them. UV photoproducts are clearly mutagenic and certain human diseases are dramatic illustrations of their carcinogenicity (see Section 23.4.5). After the discovery of photoreactivation, repair associated with UV light was described as either light-dependent (photoreactivation that required visible light) or light-independent (recovery processes that did not require visible light). In the early 1960s the first direct evidence for an "excision repair" pathway was independently obtained by two different groups of investigators: Richard Setlow working with Dick Carrier and Dick Boyce working with Paul Howard-Flanders. They found that when they exposed cells to UV irradiation and then incubated them in the dark to allow them to recover, CPDs were removed from a high-molecular-weight acid-insoluble fraction of DNA and appeared in a low-molecular-weight acid-soluble fraction. They correctly concluded that the removal of CPDs from the high-molecular-weight fraction and their appearance in the low-molecular-weight fraction represented excision repair. These seminal observations were followed in the mid-1960s by the important demonstration of repair synthesis in UV-irradiated cultures of *E. coli* by Phil Hanawalt and his then graduate student Dave Pettijohn. These discoveries of a nucleotide excision repair pathway preceded the discoveries of the base excision repair pathways described in Section 23.3 by many years. Details of each step in NER will be described in the following sections.

23.4.1 DNA Damage Recognition

The NER pathway has the daunting task of being required to have the ability to recognize a wide range of structurally unrelated substrates and yet at the same time be able to distinguish damaged from undamaged DNA. How this is achieved is not completely understood. It is likely that the conformational distortion of the DNA created by the presence of the damaged base plays one key role in damage recognition. Lesions that are efficient substrates for NER are generally referred to as "bulky" in nature because they create large distortions in the helix. For example, a CPD unwinds the helix by 19.7° and produces a bend of 27°. Hence, unwinding and bending of the helix are important in damage recognition. In addition, the presence of a covalent modification to the base or bases is a second key element in damage recognition. Single base mismatches, loops, and intercalating agents that are noncovalently bound within the helix are not efficient substrates. In some cases, lesions that are not particularly distorting, such as thymine glycol, are weak substrates for NER when studied in vitro. However, they can be converted to better substrates if some of the surrounding DNA is locally unwound by the introduction of mismatched DNA. In bacterial and mammalian cells, different modifications are recognized with different efficiencies. It is likely that the degree to which a lesion creates a distortion in the helix is a significant determinant in the efficiency with which it is a substrate.

23.4.2 NER in *E. coli*

NER in *E. coli* is understood in detail and has served as a paradigm for the investigation of other organisms. Damage recognition and processing is carried out by the *UvrABC system*, which is also referred to as the *UvrABC damage-specific endonuclease* or *UvrABC exinuclease*. It is comprised of three proteins: UvrA, UvrB, and UvrC (for UV radiation). These three proteins do not function together in a complex but rather act in sequence. The elucidation of the biochemical details of this pathway were severely hampered by the low abundance of these proteins in the cell. In addition, other nucleases present in the cells that nonspecifically nick undamaged DNA confounded the problem. Hence, as was the case for many DNA repair pathways, the biochemical elucidation of the NER pathway required the advent of recombinant DNA methodologies in the 1970s and 1980s to provide the necessary tools.

A favored model for damage recognition and repair is as follows (Figure 23.8). UvrA is a *DNA-independent ATPase* and a *DNA-binding protein*. It contains two ATPase domains, which are important for ATP hydrolysis. In addition, it contains two zinc finger motifs and a *helix–turn–helix (HtH) motif*, which are important in DNA binding and conformational changes to the protein. UvrA protein forms a dimer, and this dimerization is stimulated by ATP. It then binds UvrB protein to form an UvrA$_2$B complex, and the UvrA$_2$B complex binds DNA. Although it has been controversial, it is now generally held that the UvrA$_2$B complex is responsible for DNA damage recognition. UvrB has six helical domains and has strong structural similarity to helicase proteins. The helicase activity of UvrB in the UvrA$_2$B complex may enable scanning for damage by providing the motor that powers translocation of the complex along the DNA. As it translocates along the DNA, it unwinds it to search for damaged bases. The complex may translocate along the DNA for a short distance in one direction (unidirectional). If no lesion is encountered, then it dissociates. When it encounters a damaged base, it unwinds the DNA around the lesion (~5 bp) and kinks the helix (~130°). UvrB is loaded and becomes tightly and stably bound to the damaged site. UvrA$_2$ then dissociates, leaving an unwound preincision complex containing UvrB.

It is also thought that one of the functions of UvrB is in lesion verification. A region of UvrB, a flexible β-hairpin structure, is inserted into the DNA helix to verify that the distortion represents bona fide DNA damage and to determine which strand contains the damage. Atomic force microscopy has revealed that the DNA is actually wound around the UvrB protein, and it has been suggested that the

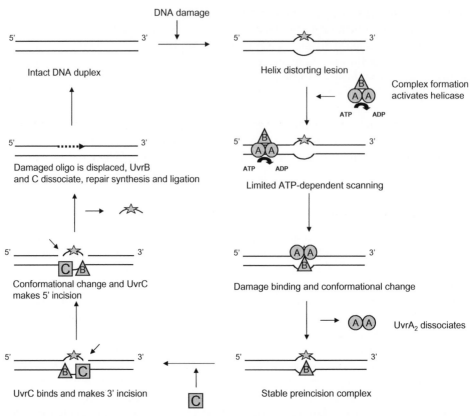

Figure 23.8. DNA damage recognition and repair by the NER machinery in *E. coli*. In solution, two molecules of UvrA bind and form a dimer. Dimerization is driven by binding of ATP. UvrB can bind UvrA$_2$ in solution or on DNA, creating a UvrA$_2$B complex. Upon binding to DNA, the UvrA$_2$B complex can translocate along the DNA for short distances, thereby hydrolyzing ATP. When DNA damage is encountered the complex undergoes a conformational change whereby the lesion that was initially in close contact with UvrA is transferred to UvrB. UvrA dissociates, leaving a stable UvrB-bound DNA complex. UvrC binds this complex and catalyzes the 3′ incision, and this is followed by the 5′ incision. The damaged oligonucleotide is displaced, UvrA and UvrB dissociate, and repair synthesis and ligation restore the intact DNA duplex. (Adapted from Batty D. P., and Wood, R. D. Damage recognition in nucleotide excision repair. *Gene* **241**, 193–204, 2000.)

damaged base is then flipped out of the helix. There is also more recent evidence that UvrB is also present as a dimer. Hence, the damage recognition complex may actually be UvrA$_2$B$_2$.

The process of DNA damage recognition produces a complex that is now "primed" for incision. The DNA damage bound UvrB complex is then recognized by the next protein required for NER, UvrC protein, which binds and produces UvrBC complex (Figure 23.8). The UvrBC complex produces an incision on each side of the lesion: The first incision is made at the fourth or fifth phosphodiester bond 3′ to the lesion, and the second incision is made at the eigth phosphodiester bond 5′ to the lesion. The catalytic sites for the 3′ and 5′ incisions are located in separate regions of the N-terminal and C-terminal regions of UvrC. However, it is controversial as to whether one or two molecules of UvrC are required for incision.

UvrD protein, also called helicase II, is another helicase involved in NER. It unwinds and releases the damaged oligonucleotide produced by the incisions. In addition it releases UvrC from the postincision complex. After displacement of the damaged oligo and displacement of UvrC, UvrB protein remains bound to the gap. DNA polymerase I displaces UvrB and synthesizes new DNA to fill in the gap. After the repair patch is synthesized, DNA ligase seals the nick. The size of the repair patch is 12 nucleotides approximately 90% of the time.

Clearly, the process of NER is very complex even in a simpler organism such as *E. coli*. While much is known about the process, there is still a great deal to be learned. In particular, it is not completely clear how the damage recognition complex discriminates the particular class of substrates it efficiently recognizes from those lesions that it does not recognize and undamaged DNA.

23.4.3 NER in Mammalian Cells

The demonstration of DNA repair synthesis in mammalian cells quickly followed the discoveries made in *E. coli*. Robert Painter in the mid-1960s developed a novel method using autoradiography to detect *unscheduled DNA synthesis (UDS)*. DNA synthesis that involves replication of the entire genome occurs during the S phase of the cell cycle. DNA synthesis that occurs outside of the S phase was described as "unscheduled," or UDS. This UDS was DNA damage-dependent and reflected synthesis of the small DNA repair patches that occur during NER. This was quickly followed by the discovery made by Jim Cleaver using similar methodology that cells from patients with the sunlight-sensitive and skin-cancer prone hereditary disease, *xeroderma pigmentosum (XP)*, are defective in UDS. This discovery led to the hypothesis that patients with XP are defective in NER as a consequence of inheriting defects in genes involved in DNA repair. This hypothesis proved to be correct, and cell lines derived from XP patients provided a critical resource for the cloning of human NER genes and characterizing the gene products at the biochemical level. In addition, many mutant rodent cell lines were generated that were identified as being defective in NER, and these cells lines also provided critical resources in the cloning of the NER genes. In fact, many NER genes initially received ERCC as part of their designation. *ERCC* stands for *excision repair cross-complementing* because the genes were identified after correcting the DNA damage sensitivity of mutant rodent cell lines by transfecting them with human DNA. After more than 20 years of work from many different investigators working all over the world, the NER pathway was reconstituted in a cell-free system by Rick Wood and Tomas Lindahl in the late 1980s.

The general strategy of NER in mammalian systems closely parallels that of *E. coli*. However, the repertoire of proteins required for mammalian NER is significantly more complex (Figure 23.9). As in *E. coli*, the distortion of the helix is important in DNA damage recognition. There is considerable evidence that XPC-RAD23B complex is involved in an early step of recognizing the distortion. The recognition of UV-induced damage also involves the proteins DDB1 and DDB2 (DDB stands for DNA damage binding protein). XPC-RAD23B binds to a region of DNA ~30 nucleotides in length and contacts both strands. TFIIH complex is then recruited

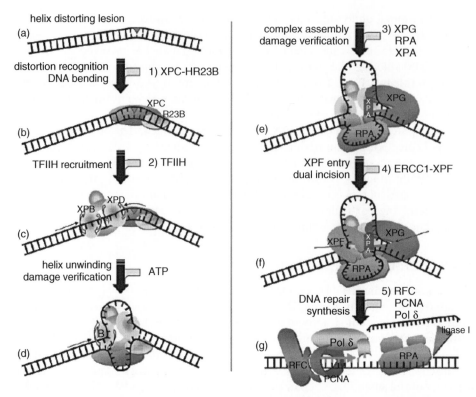

Figure 23.9. The nucleotide excision repair pathway in mammalian cells. A lesion is formed in DNA that usually creates a significant distortion. XPC-RAD23B detects the distortion and binds. This complex loads TFIIH which translocates and unwinds the helix using its helicase activities until one of the subunits encounters a chemically modified base. The other subunit continues unwinding the DNA to create a larger opened bubble structure. XPG, RPA and XPA are then recruited and the presence of a *bona fide* lesion is verified. Upon binding of ERCC1-XPF, dual incisions are made by the junction-specific endonucleases. XPG catalyzes the 3′ incision and ERCC1-XPF catalyzes the 5′ incision. RPA remains bound to the single stranded DNA and facilitates the transition to repair synthesis and ligation. (Reproduced with permission from Gillet, L. C. J., and Scharer, O. D. Molecular mechanisms of mammalian global genome nucleotide excision repair. *Chem. Rev.* **106**, 253–276, 2006.) See color insert.

which results in the unwinding of DNA near the lesion. TFIIH is involved in two metabolic pathways: It functions in NER, and it also functions in the initiation step of transcription. It is a large complex that contains 10 subunits. Two of the subunits are XPB protein and XPD protein. XPB is a helicase with 3′ to 5′ polarity (relative to the single strand of DNA with which it is bound). XPD is also a helicase but with opposite polarity. It unwinds in the 5′ to 3′ direction (the same directionality as *E. coli* UvrA$_2$B complex). The helicase activities of XPB and XPD within TFIIH create an unwound, open-helix bubble-like structure of 25–30 nucleotides around the lesion. XPG, XPA, RPA, and ERCC1-XPF assemble to form a stable *preincision complex* and in doing so identify the specific sites of incision. The complex formed by the association of XPA, and RPA may function in lesion verification. Dual incisions are carried out: the 3′ incision by XPG and the 5′ incision by ERCC1-XPF.

The catalytic activity for the 5′ incision resides in XPF. XPG and the ERCC1-XPF complex represent a specific class of endonucleases called *junction-specific endonucleases*. They cut DNA at a junction between a duplex and a single strand. XPG cuts when the single strand moves 3′ to 5′ away from the junction. ERCC1-XPF cuts when the single strand moves 5′ to 3′ away from the junction. Many of the genes described above have the designation "XP" because they are mutated in patients with the disease xeroderma pigmentosum (see Section 23.4.5).

The dual incisions results in the release of an oligonucleotide containing the damaged base that is on average about 27 nucleotides in length. The gap that is formed is filled in by either DNA polymerase δ or DNA polymerase ε in reactions dependent upon PCNA. DNA ligase I seals the final nick.

23.4.4 NER and Transcription

Transcription-coupled nucleotide excision repair (TC-NER) is generally observed as more rapid or more efficient removal of certain types of DNA damage from the transcribed strands of expressed genes compared with the nontranscribed strands. Many types of DNA damage including CPDs formed by exposure to UV light pose blocks to transcription elongation. CPDs in the transcribed strands of expressed genes pose blocks to RNA polymerase elongation, while those in the nontranscribed strand are generally bypassed. It is generally believed that the blockage of the RNA polymerase complex at the damaged site is an early event that initiates TC-NER. In general, RNA synthesis is globally reduced when cells are exposed to UV light. One of the earliest indications of the existence of TC-NER of UV damage was the key observation made by Lynn Mayne and Alan Lehmann that when mammalian cells are exposed to UV light, RNA synthesis resumes before any significant amount of UV-induced damage is removed from the bulk of the genome. This key early observation is likely explained by the subsequent observations of more rapid removal of UV-induced damage from active genes and from the transcribed strands of expressed genes compared with the nontranscribed strands and unexpressed regions of the genomes of mammalian cells. The selective repair of damage from the transcribed strands of expressed genes, which only comprises a small percentage of the mammalian genome, allows transcription to resume in the absence of significant repair in the bulk of the DNA. This "strand-specific repair" that was first described for mammalian cells by Isabel Mellon working with Phil Hanawalt in the mid-1980s was then subsequently documented in *E. coli* and in yeast.

Substrates for TC-NER include CPDs and (6–4) PPs produced by UV light and certain bulky lesions induced by chemical agents. Hence, this subpathway of NER has been conserved from bacteria to humans and operates on many different lesions. The NER pathway has also been referred to as global genome repair (GGR) because it represents repair of the nontranscribed strands of expressed genes and repair of unexpressed regions of the genome. Many of the same proteins are required for TC-NER and NER. However, the two pathways differ at the damage recognition step. For TC-NER, damage recognition is initiated by the stalling of RNA polymerase complexes at lesions in the transcribed strands of expressed genes. In addition, TC-NER requires other proteins that likely function in dealing with RNA polymerase when it becomes stalled at a lesion.

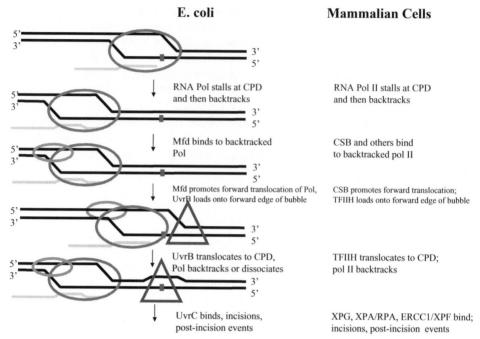

E. coli   **Mammalian Cells**

RNA Pol stalls at CPD   RNA Pol II stalls at CPD
and then backtracks   and then backtracks

Mfd binds to backtracked   CSB and others bind
Pol   to backtracked pol II

Mfd promotes forward translocation of Pol,   CSB promotes forward translocation;
UvrB loads onto forward edge of bubble   TFIIH loads onto forward edge of bubble

UvrB translocates to CPD,   TFIIH translocates to CPD;
Pol backtracks or dissociates   pol II backtracks

UvrC binds, incisions,   XPG, XPA/RPA, ERCC1/XPF bind;
post-incision events   incisions, post-incision events

Figure 23.10. A model for TC-NER of CPDs. *E. coli*: Elongating RNA polymerase complex (orange oval) stalls at a CPD (small red square) in the transcribed strand. The polymerase complex, transcription bubble, and nascent RNA (blue line) translocate backwards. Mfd (lavender oval) binds backtracked polymerase and DNA upstream of the bubble. Mfd promotes the forward translocation of the polymerase complex. UvrB (blue triangle) binds 5′ (relative to the damaged strand), loads onto the forward edge of the bubble, and translocates to the lesion. The polymerase complex backtracks or is dissociated by Mfd. Subsequent NER processing events continue as they would in nontranscribed DNA. Mammalian cells: Same as for *E. coli* except for the following. CSB (lavender oval) binds the backtracked polymerase complex and promotes forward translocation. TFIIH (blue triangle) binds 5′ to the damage and loads onto the forward edge of the bubble. Subsequent NER events continue as they would in nontranscribed DNA. (Reproduced with permission from Mellon, I. Transcription-coupled repair: A complex affair. *Mutation Res.* **577**, 155–161, 2005.) See color insert.

Genetic and biochemical studies in *E. coli* have interpreted that both TC-NER and NER require UvrA, UvrB, UvrC, and UvrD. TC-NER also requires an additional protein called Mfd and active transcription. A model for TC-NER in *E. coli* is as follows (Figure 23.10). After UV irradiation, RNA polymerase complex elongates until it encounters a CPD on the transcribed strand. The polymerase then translocates (moves) backwards. This may be an intrinsic property of the RNA polymerase complex or it may be assisted by the stimulatory factors. Mfd recognizes the backtracked complex and binds to it and to DNA upstream of the complex. Mfd then induces forward translocation of the polymerase until it re-encounters and perhaps even bypasses the lesion for a short distance (perhaps achieving a hyper-translocated conformation). When in this mode, the "jaws" of the RNA polymerase complex may be in a more open state that opens up the accessibility of downstream DNA. UvrA$_2$B or perhaps UvrB alone loads 5′ to the lesion (relative to the damaged strand). The loading of UvrB is facilitated by features of the transcription bubble brought about by the forward translocation induced by Mfd. At this point the polymerase may backtrack again or may be completely released by Mfd. UvrC then binds to the lesion-bound UvrB complex, resulting in a stable preincision complex and subsequent downstream NER events continue as they would in nontranscribed DNA. The salient point of this model is that the "coupling" of NER to transcription is mediated by the correct positioning of the transcription bubble at the lesion rather than by direct physical interactions between NER proteins and transcription factors. Mfd may serve two functions. One is to maintain the transcription bubble at the site of the lesion by reversing backtracked complexes. The other may be to ultimately displace the complex from the damaged site to allow incision and DNA synthesis.

With the exception of XPC-RAD23B, the same repertoire of proteins described for NER in mammalian cells are also required for TC-NER. It is likely that in TC-NER, the RNA polymerase complex replaces the function of XPC-RAD23B in damage recognition. Genetic studies have indicated a requirement for additional genes in TC-NER in mammalian cells, and biochemical studies have implicated the direct involvement of CSA, CSB, and XAB2 in TC-NER. As in *E. coli*, TC-NER in mammalian cells may be dependent upon the positioning of the transcription bubble at the lesion and not necessarily on direct interactions between NER proteins and transcription factors. CSA, CSB, TFIIH, and XAB2 may serve essential roles in remodelling the transcription bubble to facilitate TC-NER (Figure 23.10). The genes encoding the CSA and CSB proteins are mutated in the disease Cockayne syndrome (see Section 23.4.5).

23.4.5 Human Diseases

Several different human diseases have been found to be associated with genetic defects in DNA repair. This section will focus on those associated with defects in NER genes. As described above, one of the major discoveries in the field of NER was the finding that patients with the disease XP are deficient in the NER pathway. The disease XP is genetically heterogeneous. There are eight different *complementation groups* representing eight different genes; a defect in any one gene can create the disease XP and the associated characteristics of the disease. Of the eight different genes, seven encode proteins involved in NER and one encodes a protein involved in DNA synthesis at sites of DNA damage (see Chapter 22). The seven genes involved in NER are designates XPA, XPB, XPC, XPD, XPE, XPF, and XPG. Each product functions in a specific step of NER as described in Section in 23.4.3. XPE encodes DDB2.

Cells from different patients with XP were originally classified as belonging to different complementation groups. This was done by a stepwise cell fusion of cells from different individuals and assaying for UDS. Since cells from XP patients are deficient in NER, they exhibit very low levels of UDS when they are irradiated with UV light. However, if patients from different complementation groups are fused and then assayed for UDS following UV irradiation, UDS is restored to near "normal" levels. The reason for the "complementation" is that the cells from the different individuals have different NER gene defects. For example, if an XPA⁻ cell is fused with an XPB⁻ cell, the XPB present in the XPA⁻ cell complements that which is missing in the XPB⁻ cell and vice versa. In contrast, if an XPA⁻ cell is fused with another XPA⁻ cell, no complementation is achieved because both cells are missing XPA. This method is still used today to identify the genetic defects in newly identified XP patients.

XP patients are highly predisposed to skin cancer. They have a high risk of developing basal cell carcinoma, squamous cell carcinoma, and melanoma. This and their other skin abnormalities are all clearly related to expose to UV light. The UV light introduces UV photoproducts in their DNA. However, because the patient is unable to repair them or repairs them poorly, the damage persists in their cells. This leads to the accumulation of many mutations in their DNA at the sites of DNA damage, which, in turn, leads to alterations in cell cycle check points and greater genetic instability. These alterations are driving forces in the generation of skin tumors. Many XP patients start displaying skin problems within their first year of life and develop skin cancer within their first decade of life. This is in sharp contrast to the general population, where skin cancer generally does not appear until people are in their sixties. Thus the appearance of skin cancer in the XP population precedes the appearance of skin cancer in the general non-XP population by 50 years. XP is inherited through an autosomal recessive mode, and patients must inherit a mutant allele of the same NER gene from both of their parents. Hence, it is an extremely rare disease. However, people who carry only one mutant allele are heterozygotes or "carriers" and they are much more frequent in the population: ~1 in 300. Their risk for skin and other forms of cancer is unclear. Some XP patients also have neurological complications, but the reason for this is also unclear.

Patients with the disease *Cockayne syndrome (CS)* have inherited mutant alleles of genes that confer a deficiency in the TC-NER pathway. Patients are similar to XP in that they are hypersensitive to sunlight. However, many of the characteristics of CS are quite different. Patients appear normal at birth, but symptoms of the disease rapidly appear and the average age of death is 12 years. They develop extremely severe sunburns within the first year or two after birth, but they do not have the severe pigmentation abnormalities that XP patients have. This is followed by severe growth and neurological arrest. Patients are small in stature, display a generalized wasting, are mentally retarded, have long arms and legs relative to the rest of their body, have sunken eyes, have large ears and nose, and develop cataracts. There are two genes associated with Cockayne syndrome, CSA and CSB. Cells from CS patients are specifically deficient in the TC-NER pathway. They are unable to selectively remove UV damage from the transcribed strands of active genes. However, with the exception of the UV-photosensitivity, it is extremely unlikely that UV plays any role in the cause of the other major symptoms associated with the disease. It has been suggested that instead the major symptoms of the disease may be caused by a defect in transcription-coupled repair of other types of damage that occur spontaneously in the cell or by a subtle defect in the transcription process itself. Like XP, CS is inherited by an autosomal recessive mode and it, too, is an extremely rare disease. CS patients do not show any predisposition to skin cancer. There are some individuals that display symptoms of both XP and CS (XP-CS). They possess mutations in XPB, XPD, or XPG. The mechanisms underlying this are not completely understood.

Trichothiodystrophy (TTD) is a disease that is characterized by sulfur-deficient brittle hair, ichthyosis (fish-like scales on the skin), mental retardation, and grown defects. About half of TTD patients exhibit UV-sensitivity. TTD patients are deficient in the XPD gene, the XPB gene, or the TTDA gene. Recently, it was determined that the TTDA gene encoded a new protein found to be associated with the TFIIH complex. Hence, all three proteins—XPB, XPD, and TTDA—are components of TFIIH. It is likely that with the exception of the sensitivity to UV light, many of the clinical symptoms of TTD are caused by a defect in transcription rather than a defect in DNA repair.

One striking question relates to how defects in the same gene—for example, XPD—produce three different disease states: XP, XP-CS, and TTD. Mutations that cause the different diseases, in general, reside in different regions of the XPD gene. A more in-depth answer is complex and outside the scope of this chapter. Readers are referred to the Friedberg et al. book listed at the end of this chapter for a more detailed discussion of this fascinating question.

23.5 MISMATCH REPAIR

The NER pathway is largely responsible for removing modified bases. The *mismatch repair (MMR) pathway* is different in that it is largely responsible for removing inappropriate bases. As described in the introduction to this chapter, the frequent causes for the introduction of inappropriate bases in DNA are the DNA polymerases that replicate the genome. When DNA polymerases copy *parental DNA* during synthesis of new DNA, they occasionally make errors and incorporate incorrect deoxynucleotides into the nascent strand. This produces *mismatched base pairs* (Figure 23.11). In addition, there can be slippage events during replication where either the new strand or the template parental strand slips out of frame and creates a small loop in the DNA. These errors are corrected by the MMR pathway. The challenge for this type of repair is in determining which base in the mispair is the incorrect one. In NER, there is a modified base the repair machinery recognizes. In contrast, a mispair contains two normal bases. Hence the MMR machinery has to determine which of the two bases is incorrect and remove it. This occurs by strand discrimination. The new strand would likely contain the incorrect base because it was misincorporated during synthesis of the new strand. Hence, the MMR pathway determines which strand of DNA is the newly synthesized one. How this *strand-discrimination* step occurs in *E. coli* is well understood. In mammalian cells, this step is not completely understood.

23.5.1 MMR in *E. Coli*

The existence of a MMR repair pathway was first put forth in the mid-1960s by Robin Holliday based on observations made studying recombination and by Evelyn Witkin based on observations made studying mutagenesis. A decade later in the mid-1970s, Robert Wagner and Matthew Meselson put forth the Wagner–Meselson model for MMR (Figure 23.11). In the years that followed, a large number of genetic and biochemical studies from many different groups have supported different aspects of this model for MMR in *E. coli*.

Escherichia coli contains the palindromic sequence, *GATC*, scattered through out its genome. The A in the GATC is methylated on both strands (Figure 23.11), and

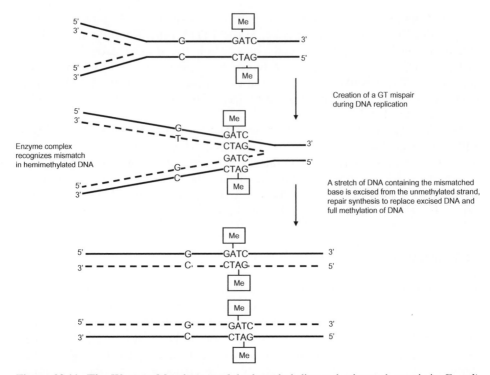

Figure 23.11. The Wagner–Meselson model of methyl-directed mismatch repair in *E. coli.* GATC sequences are repeated many times and scattered through out the genome of *E. coli.* They are methylated at the 6 position of adenine. During semiconservative replication, a T has been misincorporated opposite a template G. The new strands are shorter than the template strands. The decision to incise the strand containing the G and ultimately remove it is determined by the methylation status of the adjacent GATC sequence. For a short time following replication, the GATC sequence in the newly synthesized strand remains unmethylated. An enzyme involved in mismatch repair incises the newly synthesized strand near the unmethylated GATC sequence, and the DNA up to an including the incorrect T is removed. Repair synthesis replaces the excised DNA. Ligase seals the final nick, and methylation of the newly synthesized GATC sequence restores the DNA to its original state. (Adapted from Wagner, R., and Meselson, M. Repair tracts in mismatched DNA heteroduplexes. *Proc. Natl. Acad. Sci. USA* **73**, 4135–4139, 1976.)

methylation of the A in each strand occurs by *DAM methylase*. The methylation status of the GATC sequence serves as the basis for strand discrimination. After the GATC sequence is replicated, the A in the newly synthesized strand is unmethylated. There is a lag period between when the GATC sequence is synthesized and when it becomes methylated by DAM methylase. It is during this window that strand discrimination can take place. *Escherichia coli* possess three proteins involved in different stages of MMR: *MutH, MutS, and MutL* (Figure 23.12). The corresponding genes were identified by the mutator phenotype they conveyed when they were disrupted. Hence they were designated Mut for mutator.

MutS recognizes and binds to single base mismatches and small insertion deletion mispairs (or loops) (Figure 23.12). It can bind mismatches as a dimer or tetramer,

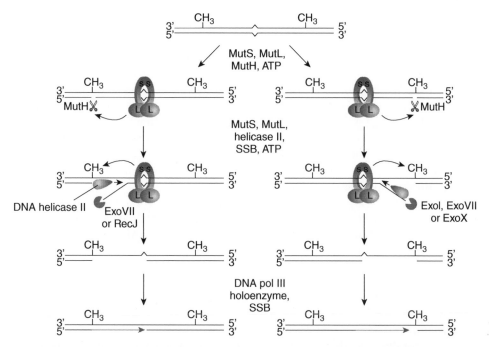

Figure 23.12. Bidirectional mismatch repair in *E. coli.* MutS binds to mismatches formed by errors made during replication. MutL binds to the MutS-bound complex. MutH incises the newly synthesized strand next to the unmethylated GATC sequence. The green arrows indicate signally between the two sites of activity. DNA helicase II unwinds the DNA and exonucleases excise the DNA from the point of incision up to and including the incorrect base. DNA Pol III synthesizes new DNA to replace the excised DNA. (Reproduced with permission from Iyer, R. R., Pluciennik, A., Burdett, V., and Modrich, P. L. DNA mismatch repair: Functions and mechanism. *Chem. Rev.* **106**, 302–323, 2006.)

and it possesses a weak ATPase activity that is essential for MMR. The binding of MutS to a mismatch initiates mismatch repair. MutL does not bind mismatches directly. It exists as a dimer, and it binds to the MutS-mismatch complex. It, too, possesses a weak ATPase activity that is required for MMR. MutH recognizes and binds hemimethylated GATC sequences that are formed after DNA replication. In a reaction stimulated by MutS and MutL, MutH incises the unmethylated strand (new strand) immediately 5′ to a GATC sequence. The incised sequence can be on either the 5′ or 3′ side of the mismatch (Figure 23.12). Hence, MMR is *bidirectional.* After the incision, helicase II is loaded and it displaces the nicked strand by unwinding the DNA between the GATC sequence and the mismatch that can be as far as 1000 bases away. As the DNA strand is unwound, it is degraded by a collection of nucleases that are determined by whether the incision occurred 5′ to the mismatch and requires a 5′ to 3′ exonuclease or whether it occurred 3′ to the mismatch and requires a 3′ to 5′ exonuclease. DNA polymerase III synthesizes new DNA to replace the excised DNA containing the incorrect nucleotide and ligase seals the final nick. The *E. coli* MMR pathway was reconstituted in a cell-free system by Paul Modrich and colleagues in the late 1980s.

23.5.2 MMR in Mammalian Cells

Mammalian cells possess a functionally similar yet more complex MMR system (Figure 23.13). However, mammalian cells, like many prokaryotes other than *E. coli*, do not possess a dam-dependent methylation pathway and hence must rely on a different mechanism for strand discrimination. Mammalian cells possess multiple homologs of the bacterial MutS and MutL proteins. Three MutS homologs are called MSH2, MSH3, and MSH6 (for MutS homolog). They function as heterodimers. MSH2 can form a complex with either MSH6 or MSH3 to form MSH2–MSH6 and MSH2–MSH3 heterodimers. These are also referred to as MutSα and MutSβ, respectively. MutSα binds to all combinations of single base mispairs and small loops. MutSβ recognizes larger loops.

Mammalian cells also contain several MutL homologs. Three major MutL homologs are called MLH1, PMS2, and PMS1. MLH1 forms a complex with either PMS2 or PMS1 to form MLH1-PMS2 (MutLα) and MLH1-PMS1 (MutLβ). MutLα forms a complex with MutSα to repair single base mispairs and small loops. MutLα forms a complex with MutSβ to repair larger loops. MMR in mammals is also bi-directional; and after binding of the MutS and MutL homolog proteins, helicases are recruited to displace the stand and exonucleases are recruited to degrade it. ExoI has a predominant 5′ to 3′ exonuclease activity, but it can also act as a 3′ to 5′ exonuclease and, hence, may be the only exonuclease required.

It is generally held that it is likely that the signal for strand discrimination in mammalian MMR and perhaps many other organisms as well is provided by the DNA termini that are present in the strands of nascent DNA in the replication fork. The terminus that the DNA polymerase is in the process of extending also provides an entry site for the helicase to displace the strand and ExoI to degrade the strand. What is unclear is how communication is achieved between the site of the mismatch and the terminus that is being extended by the DNA polymerase because these may be several thousand base pairs away from each other. There are several models currently under hot debate. One model proposed by Rick Fishel is called the

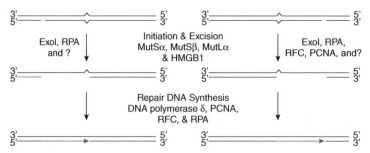

Figure 23.13. Bidirectional mismatch repair in mammals studied in vitro. Mismatch repair in vitro can be directed by a strand break located either 5′ or 3′ to the mismatch. It is likely that in vivo the strand break is provided by the termini of DNA strands that are being synthesized. MutSα, MutSβ, and MutLα are involved in mismatch recognition and processing. Exonucleases and other activities are involved in excision. Repair synthesis is carried out by DNA polymerase δ or ε. (Reproduced with permission from Iyer, R. R., Pluciennik, A., Burdett, V., and Modrich, P. L. DNA mismatch repair: Functions and mechanism. *Chem. Rev.* **106**, 302–323, 2006.)

molecular switch model. As an alternative, there is recent evidence from Paul Modrich and colleagues that MutLα actually contains an endonuclease activity. This would be an alternative way to define the sites of DNA removed by the MMR pathway. Future studies are required to confirm this or other models.

23.5.3 MMR and Cancer

Interest in the field of MMR exploded when in the 1990s it was discovered that mutations in MMR genes are associated with the disease *hereditary nonpolyposis colon cancer (HNPCC)*. HNPCC is one of the most common diseases where people are predisposed to developing cancer. All HNPCC patients are predisposed to specifically developing colon cancer. Some patients also develop other forms of cancer including endometrial, ovarian, gastric, and pancreatic. There are rigorous criteria that are used for diagnosis. One criterion is that HNPCC kindreds (families) must have at least three relatives within two generations that have colorectal cancer and one of the three must have been diagnosed when they were younger than 50 years of age. Hence, HNPCC patients develop colon cancer, they develop it at a much younger age than the general population, and it appears to be familial (heritable).

Clues to the association between defects in MMR genes and HNPCC were provided by the investigation of specific forms of genetic instability in *E. coli* and *S. cerevisiae*. These and other organisms including mammals contain simple repeat sequences that can be runs of repeating mononucleotides, dinucleotides, trinucleotides, and so on. Investigation of the simple dinucleotide repeat poly(GT)$_n$ in *E. coli* in the late 1980s resulted in the observation that this nucleotide repeat was unstable in strains with mutations in MMR genes. This was followed by the important observation in the early 1990s that a similar kind of repeat instability was observed in cells derived from patients with HNPCC. In addition, very soon after, it was demonstrated that a similar kind of repeat instability was conferred by inactivation of MMR genes in yeast. All three observations led to the prediction that HNPCC was associated with defects in MMR genes. This prediction was in fact rapidly demonstrated by several groups of investigators including Bert Vogelstein, Richard Kolodner, Rick Fishel, Paul Modrich, Tom Kunkel, and many of their colleagues. Among all of the MMR genes, the MSH2 and MLH1 genes are the most frequently altered in HNPCC.

HNPCC is different from many of the other human diseases associated with deficiencies in DNA repair. Individuals are heterozygous and contain one mutant allele of a MMR gene and one wild-type ("good") copy. Hence, it is much more frequent in the population because only one mutant allele is required to cause the disease. Most studies indicate that the heterozygous state does not significantly reduce a cell's capacity to carry out MMR so most cells from HNPCC patients are probably relatively "normal" in regards to MMR. However, it is likely that a second event occurs during the lifetime of the patient that results in the inactivation of the second "good" allele. Once this happens, the cell's ability to carry out MMR is severely compromised. This loss of MMR creates widespread spontaneous mutagenesis and genetic instability and is likely a driving force in tumorigenesis.

The instability of simple repeat sequences is called *microsatellite instability (MSI)*, and it is used as a tool for the diagnosis of HNPCC. It is thought to be caused by slippage events that occur during DNA synthesis. Since the sequence is repeated, one or more of the repeating units can misalign and slip out of the duplex during DNA replication, creating a small loop in either the new strand or the template strand. If these slippage events are not detected and repaired by MMR, they lead to the insertion or deletion of one or more of the repeat units during the next round of replication. Hence, the number of repeat units changes as the cells replicate and divide.

Importantly, defects in MMR have been found in colon tumors and other types of tumors in patients that have no known hereditary predisposition to cancer. These are referred to as *sporadic* forms of cancer. It is thought that in these cases, patients are born with both alleles of a MMR gene intact. However, during there lifetime they acquire mutations in both alleles. This, at least in part, may explain differences in the age of onset of colon cancer in HNPCC patients: It usually occurs when patients are in their forties and fifties, but sporadic cases tend to occur after people reach the age of 70.

23.6   RECOMBINATIONAL REPAIR

Recombinational repair can be brought into play in response to a variety of different insults and as a consequence of the interaction of the replication machinery with different types of DNA damage. The damaged DNA is not always restored to its original state. This distinguishes it from each of the repair mechanisms described above. Double-strand breaks (DSBs) can be formed several different ways. They can be directly formed by agents such as ionizing radiation. They can also be formed when the replication machinery encounter a nick in the template for the leading or lagging strand. Other alterations such as single-strand gaps can be formed when the replication machinery encounters damage in the template of the leading or lagging strand. When damage blocks the synthesis of both strands, it can lead to the formation of structures that resemble a chicken foot and, hence, are called chicken foot structures. This collection of lesions or structures oftentimes invokes recombinational repair mechanisms. Recombinational repair acts as a tolerance mechanism that enables the cell to continue dividing without necessarily restoring the DNA to its original state.

23.6.1   Double-Strand Break Repair in *E. coli*

Homologous recombination was first described in *E. coli* by Joshua Lederberg in the mid-1940s. It involves the exchange of DNA between two molecules that have nearly identical sequence. Different aspects of the homologous recombination pathway are brought into play for the repair of DSBs, and this is called DSB repair (Figure 23.14). A problem for the repair of double-strand breaks is that after or during their formation, the ends become broken and nucleotides are lost. Hence, they cannot simply be religated without losing genetic material. Instead, their repair requires the participation of another DNA duplex with the same base sequence. DSB repair is highly dependent on *RecA* protein that plays a central role. RecA coats the DNA molecules to form nucleoprotein filaments that are important in pairing the homologous DNA molecules and promoting strand exchange. The

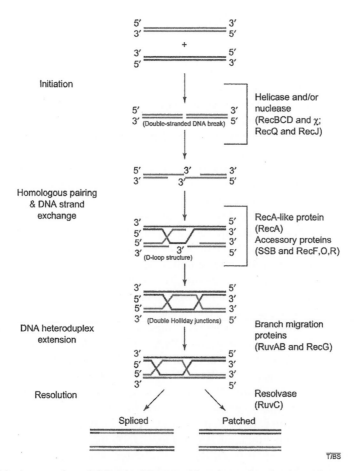

Figure 23.14. A general model for double-strand break repair using homologous recombination. The text in parentheses to the right indicates proteins that participate in each step in *E. coli*. The light blue lines indicate newly synthesized DNA. (Reproduced with permission from Kowalczykowski, S.C. Initiation of genetic recombination and recombination-dependent replication. *Trends Biochem. Sci.* **25**:156–165, 2000.) See color insert.

double-strand break is processed by helicases or nucleases that function to expose 3′ overhangs that are then used to invade the intact homologous duplex. Repair synthesis from the 3′ end of the strand that invaded the undamaged duplex uses the complementary undamaged strand as a template to fill in the nucleotides lost from the broken ends. Strand invasion and repair synthesis result in the formation of a double *Holliday junction*. Once a double Holliday junction is formed, other proteins can promote branch migration which extends the heteroduplex. Different strands can be incised to resolve the Holliday junctions. This choice determines whether the resulting molecules are nonrecombinant and contain patches or whether they are recombinant and have exchanged flanking DNA. Similar recombinational mechanisms repair daughter strand gaps that can be formed when the replication machinery encounters a lesion in the template.

23.6.2 Double-Strand Break Repair in Eukaryotic Cells

Eukaryotic cells possess two pathways for the repair of DSBs. One pathway is homologous recombination and appears to be very similar to the one described above in *E. coli*. However, much less is known about the details of this pathway in mammalian cells. In yeast, homologous recombination is the major pathway for the repair of DSBs. In mammalian cells it is restricted to the late S and G_2 phases of the cell cycle. The second pathway is called *nonhomologous end joining (NHEJ)*, and it is thought to be the major pathway for the repair of DSBs in mammalian cells. The two pathways differ in the proteins that participate in them, and they also significantly differ in the outcomes they produce. Homologous recombination produces few errors, whereas NHEJ is error-prone.

NHEJ is required for the V(D)J recombination pathway that occurs in the vertebrate immune system and is required to generate the repertoire of antigen receptors in lymphocytes. Mice that are deficient in this pathway are immunodeficient. They are also sensitive to ionizing radiation, and this observation provided clues that this pathway is also involved in the processing of DSBs produced by ionizing radiation. Humans that are deficient in this repair pathway are also immunodeficient and sensitive to ionizing radiation. A model for the NHEJ pathway is as follows (Figure 23.15). An early event involves the binding of *Ku protein* to each of the ends formed by the DSB. Ku threads onto the DNA to help prevent any additional breakage of the ends and helps align them for ultimate rejoining. This is followed by Ku-dependent recruitment and activation of *DNA-dependent protein kinase* (DNA-PKcs), which, in turn, recruits *Artemis*. Artemis acts as a nuclease to trim the ends of the DNA to remove 5′ overhangs and trim 3′ overhangs. DNA-PKcs may help synapse the two ends. If there are gaps present, they are filled in by DNA polymerases. Ligation is performed by a complex of *XRCC4* and DNA ligase IV. NHEJ is the major repair pathway for the repair of DSBs in vertebrates, whereas homologous recombination appears to be repair mechanism of choice in yeast. It may be that the much larger and highly repetitive genomes of vertebrates creates too great a challenge in finding the correct homologous partner for homologous recombination to be an efficient mode of repair.

The repair of DSBs by homologous recombination in mammalian cells is not understood in detail. As in *E. coli*, it involves nucleases to process the ends of the DSBs and produce DNA tails with 3′ overhangs. There is evidence that a complex of proteins called *MRN (for Mre11/Rad50/NBS)* acts as a nuclease in this step, but this is not entirely clear. A key protein is *Rad51*, which acts in the same way as RecA to coat the DNA and form nucleoprotein filaments. It participates in pairing and strand invasion of the damaged strand into the undamaged duplex. These steps also involve the *single-strand DNA-binding protein replication protein A (RPA), Rad52* and *Rad54*. There are several paralogs of Rad51 that may also be involved in these steps as well. DNA polymerases replace the missing DNA and ligase seals the nicks. During this process, Holliday junctions are formed, there is branch migration, and ultimately nucleases act to resolve the Holliday junctions to produce two intact duplexes.

The homologous recombination repair pathway is highly complex, in part perhaps as a consequence of the diverse collection of structures on which it acts. Many proteins have been implicated in this pathway, but their real specific roles in the

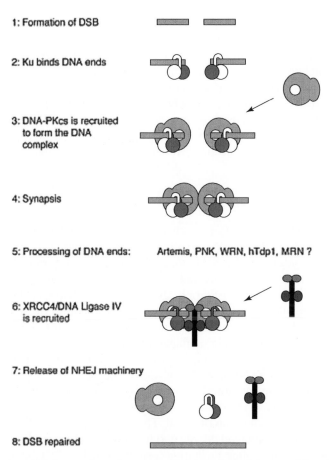

1: Formation of DSB

2: Ku binds DNA ends

3: DNA-PKcs is recruited
   to form the DNA
   complex

4: Synapsis

5: Processing of DNA ends: Artemis, PNK, WRN, hTdp1, MRN ?

6: XRCC4/DNA Ligase IV
   is recruited

7: Release of NHEJ machinery

8: DSB repaired

Figure 23.15. A model for nonhomologous end joining. A double-strand break can be introduced by either a DNA damaging agent or during the encounter of the replication machinery with DNA damage. Ku heterodimer binds to the ends. It may translocate away from the ends and recruit DNA-PKcs to the ends. The binding of Ku to DNA-PKcs forms an active DNA-PK protein kinase. Artemis and additional proteins may assemble to process the ends of DNA. This is necessary because certain DNA damaging agents generate DNA ends that are either blocked or are not directly ligatable. The two broken DNA ends are brought together by the two DNA-PK molecules present on each end. XRCC4/DNA ligase IV may also be involved in processing the ends before they join them. (Reproduced with permission from Lees-Miller, S. P., and Meek, K. Repair of DNA double-strand breaks by non-homologous end joining. *Biochimie* **85**, 1161–1173, 2003.)

cell are unclear. Alterations in the *BRCA1, and BRCA2 genes* predispose individuals to breast and ovarian cancer. Both genes have been implicated in DSB repair. BRCA2 has been shown to directly interact with Rad51 and may play a role in mediating Rad51-dependent strand exchange by recruiting it to the DSB. The interaction of BRCA1 with Rad51 appears to be indirect. Targeted disruptions of Rad51, BRCA1, and BRCA2 in mice lead to embryonic lethality. In addition, human homologs of the *RecQ family of helicases* in *E. coli* have been implicated in homologous recombination. The human helicases are *Bloom (BLM), Werner (WRN), and*

Rothmund–Thomson (Recql 4) proteins. The genes that encode these proteins are altered in human syndromes of the same name. Patients with these diseases are predisposed to different forms of cancer. Cell lines derived from patients with these diseases exhibit chromosome instability and accumulate abnormal replication intermediates. When these helicase proteins are studied in vitro, they can unwind Holliday junction-like structures. A characteristic of many proteins involved in homologous recombination is that they can be visualized as forming subnuclear structures in cells in response to DNA damage. This can be directly visualized using immunofluorescence and the structures that are formed are called foci (see Section 23.8.2).

23.6.3 Interstrand Crosslink Repair

Chapter 22 describes how different chemical agents react with DNA and form *interstrand crosslinks (ICLs)* that covalently link the two strands together. This poses a special problem for the cell in that the replication and transcription machineries are unable to progress or unwind the helix at the sites where the two strands are covalently linked to each other. It also presents a challenge for DNA repair since the damaged sites occur in close proximity to each other on opposing strands. Hence it is difficult for one strand to serve as a template for repair of the other strand because they both contain damaged DNA.

The repair of ICLs is best understood in *E. coli*. The major pathway for ICL repair makes use of the NER pathway and recombination (Figure 23.16). The UvrABC complex forms incisions on each side of the damage on one strand. A gap is formed on the 3′ side of the crosslink. This could be achieved by the 5′ to 3′ exonuclease activity of Pol I or UvrABC itself. The gap is then invaded by a DNA strand from a homologous undamaged duplex and this pairing and strand invasion is mediated by RecA. The small oligonucleotide formed after UvrABC-mediated incision is attached to the opposite strand by the crosslink. This is displaced by the invading strand from the undamaged duplex. The opposite side of the crosslink is then incised by UvrABC, which releases the small region of duplex oligonucleotide containing the crosslink. The gap is filled in by repair synthesis and ligation.

The repair of ICLs in mammalian cells is more complex and occurs by multiple different pathways. One pathway likely involves a combined action of the NER and recombination pathways similar to *E. coli*. Mammalian cells with mutations in NER genes are sensitive to crosslinking agents. However, mutations in ERCC1 or XPF confer a greater sensitivity to these agents and a model has been proposed for ICL repair that utilizes the junction-specific endonuclease activity of ERCC1-XPF in early events of ICL processing. In this model, the actions of a helicase or encounters with the replication fork unwind the duplex immediately 3′ to the ICL. This creates a Y structure that can be cleaved at the 3′ junction between double-strand and single-strand DNA by ERCC1-XPF. Then ERCC1-XPF incises the same strand 5′ to the ICL. Either homologous recombination or the collapse of the replication fork fills in the gap. NER can then act to incise the DNA on the opposite arm of the crosslink and complete repair. This processing of ICLs is relatively error-free. Other pathways exist that create DSBs near the ICLs and invoke a DSB mode of repair that can employ specialized translesion DNA synthesis polymerases (see Chapter 22). Some of these are likely to be more error prone.

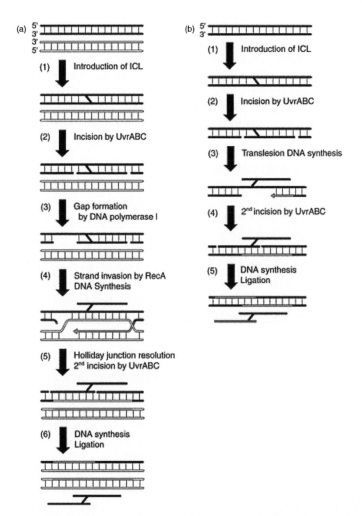

Figure 23.16. The repair of interstrand crosslinks in *E. coli*. (A) The two sister chromatids are represented by black and white duplexes. The ICL is represented by a thick black line connecting the two strands in the black duplex. The ICL is incised by the UvrABC system 3′ and 5′ to the ICL on one strand. The exonuclease activity of DNA polymerase I produces a gap on the 3′ side of incised ICL. RecA coates the single-strand region and the RecA filament pairs with the homologous duplex and performs strand exchange past the crosslink. This promotes DNA synthesis in the homologous duplex. The holiday junctions are resolved to produce two separate duplex molecules. The UvrABC system incises the crosslinked oligonucleotide. DNA synthesis and ligation restore the integrity of the DNA. (B) Repair of the ICL by the UvrABC system and translesion synthesis. (Reproduced with permission from Dronkert, M. L. G., and Kanaar, R. Repair of DNA Interstrand cross-links. *Mutati. Res.* **486**, 217–247, 2001.)

Fanconi anemia (FA) is a rare genetic disease. Patients are highly predisposed to developing a wide variety of different types of cancer. Cells from FA patients are specifically hypersensitive to crosslinking agents. These observations lead to the prediction that FA cells are deficient in the repair of ICLs. While evidence for this is conflicting, a large number of studies in this field have determined that FA represents a novel DNA damage response pathway that confers resistance to treatment with crosslinking agents. Unraveling this pathway has proven to be difficult and complicated by the identification of at least 12 complementation groups in this disease. Eleven FA genes have been cloned; and, interestingly, one of the genes is BRCA2. The function of the FA pathway may be to coordinate the multiple DNA repair pathways that function in ICL processing and response to ICL damage.

23.7   DNA REPAIR AND CHROMATIN STRUCTURE

Many biochemical studies that have played crucial roles in determining the mechanistic details of the different DNA repair pathways have been largely conducted using naked DNA as a damaged substrate. However, in the cell, the DNA is complexed with a large variety of histone and nonhistone proteins. The purpose of this is to package the DNA in what is likely an ordered way so that it can fit inside the nucleus and yet be available for many metabolic transactions. How the packaging of DNA into chromatin influences the various repair pathways that must first detect and then gain access to the damage in DNA is not well understood.

23.7.1   Chromatin Structure

If the genomic DNA in a mammalian cell were stretched out, it would be 2 meters in length. Yet it must fit into the nucleus that has a radius of about 10^{-5} meters. This amazing achievement is accomplished by several different levels of compaction. The first level of compaction occurs at the level of the nucleosome. Each nucleosome is comprised of 170–240 bp of DNA wrapped around a "histone octomer or core" containing two molecules each of histone H2A, H2B, H3, and H4 and a linker region that binds to histone H1. The DNA is tightly wrapped 1.5 times around the histone core, and the linker region connects each core to the next; this creates what is generally visualized as "beads on a string." This results in ~7-fold compaction of the DNA. An additional 50-fold level of compaction is achieved by the formation of what are generally called 30-nm fibers, and an additional 250 fold level of compaction results in the formation of a metaphase chromosome. Within chromatin, loops of 30–100 kb of DNA are formed that are referred to as domains. The domains are anchored by a support structure called the nuclear matrix. The nuclear matrix is an important docking place for the metabolic enzymes that carry out DNA replication, transcription, and repair.

23.7.2   Chromatin Remodeling in DNA Repair

The nucleotide excision repair pathway recognizes specific distortions in DNA to find the damage and repair it. It is difficult to see how this is achieved when the damage is imbedded in DNA compacted with chromatin. Studies performed in vitro and in vivo indicate that DNA damage present in chromatin or within the nucleosome is actually shielded from repair enzymes. Models have been proposed that invoke chromatin remodeling enzymes to help the NER proteins gain access to the damage, and there is experimental evidence to support them. Several different complexes have been identified that alter the association of histones with DNA. These proteins contain an ATPase subunit that belongs to the *SNF2 superfamily.* One such complex is *SWI/SNF* (for switch/sucrose nonfermenting). It increases accessibility of nucleosomal DNA to transcription factors and restriction enzymes, and it induces sliding of the histone octomer. There is evidence that the addition of SWI/SNF can help the enzyme photolyase gain access to UV damage. There is genetic evidence that these complexes may help NER proteins gain access to UV damage. After repair takes place, it is likely that the appropriate chromatin structure must be restored. It is not clear how this takes place. It has been proposed that histone chaperones participate in this process, and *the chromatin assembly factor CAF1* has been found to promote chromatin assembly associated with NER. There is recent evidence that during this process, new histone proteins are deposited instead of simply recycling the old ones.

Chromatin structure can also influence damage formation. CPDs are probably formed uniformly within the nucleosome core and linker regions and in chromatin in general. However, there is evidence that 6–4(PPs) are distributed in the linker region of the nucleosome. It is unclear if they are preferentially formed there or if their formation in nucleosome cores causes the nucleosomes to reposition so that the 6–4(PPs) are moved into the linker region where they are somehow better accommodated. There is evidence that chromatin structure can significantly influence the distribution of some chemical modifications to DNA. The chromatin structure of genes that are actively transcribed is generally thought to be in a more open or less compact state. In contrast, certain repetitive sequences in the genome are in a more compact state. This likely influences DNA damage formation and DNA damage removal. The acetylation of histones plays a significant role in opening-up or decondensing chromatin. There is evidence that hyperacetylation of histones enhances the removal of DNA damage by the NER pathway.

The ability of BER proteins to recognize and remove damage may also be influenced by the remodeling of chromatin. Mammalian cells contain *poly(ADP-ribose) polymerases* that add branched chains of ADP-ribose to proteins. Histones are substrates for one of these enzymes, *PARP1*, and it has been suggested that this modification of histones helps remodel chromatin for BER. PARP1 binds tightly to single-strand breaks and dRp residues, and PARP1 and PARP2 interact with several BER proteins.

23.8   DNA DAMAGE AND CELL CYCLE CHECKPOINTS

In addition to possessing mechanisms that remove damage from DNA, cells have evolved complex mechanisms to "*sense*" that damage has occurred and "*transduce signals*" with the end result of arresting the cell at different stages of the cell cycle. This *cell cycle arrest* can provide the necessary time to remove damage and prevent the formation of mutations that could be formed when the replication machinery encounters the damage before it is repaired. It provides time to remove damage and prevent potentially deleterious structures from being formed when the replication and transcription machineries encounter certain types of DNA damage. It can prevent problems that could arise should a cell enter mitosis and be unable to properly segregate chromosomes to daughter cells due to the presence of DNA

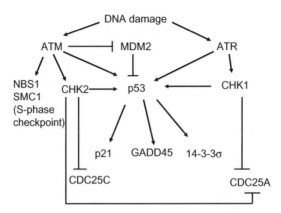

Figure 23.17. DNA damage is translated into checkpoint responses ATM and ATR are kinases activated by DNA damage and these kinases phosphorlayte and regulate proteins such as the checkpoint kinases (CHK1 and CHK2) which phosphorylate and activate p53. p53 increases the expression of p21(regulates the G1/S, S, and G2/M checkpoints), GADD45 (G2/M checkpoint), and 14-3-3σ (G2/M checkpoint). CDC25A regulates the G2/M and S-phase checkpoints, whereas CDC25C regulates G2/M checkpoint. NBS1 and SMC1 are important effectors of the S-phase checkpoint.

damage. The progression of the cell through each phase of the cell cycle—G_1, S, G_2, and M—is a highly regulated process. *Cell cycle checkpoints* are activated by signaling cascades and temporarily arrest cell cycle progression. There are checkpoints, at the G_1/S and G_2/M boundaries and in the S-phase of the cell cycle. On the top of these signaling pathways are two proteins, *ATM* and *ATR*, that are members of the *phosphatidylinositol-3-kinase-related kinases (PIKKs)* (Figure 23.17). They are induced by many types of DNA damage and by stalled replication forks and initiate the signally cascade by phosphorylating substrates that control cell cycle progression and DNA repair pathways.

23.8.1   DNA Damage Sensing: ATM and ATR

Ataxia telangiectasia (AT) is a hereditary disease; patients with AT are highly sensitive to ionizing radiation, and they are immunodeficient and predisposed to cancer. Normal cells shut down replicative DNA synthesis after treatment with ionizing radiation or chemicals that mimic the effects of ionizing radiation (radiomimetic). It has been known for many years that cells from AT patients are unable to similarly shut down DNA synthesis, and they were characterized as having "radioresistant DNA synthesis." The investigation of this phenomenon has revealed and unraveled major DNA damage response cell cycle checkpoint pathways. The gene that is defective in AT is called ATM (for ataxia telangiectasia mutated). The ATM protein senses the presence of certain types of DNA damage and becomes activated. The activated ATM then activates target molecules that function to arrest the cell in G_1/S. When the cell is blocked from entering S by this G_1/S checkpoint, it cannot carry out replicative DNA synthesis. Since AT cells are defective in the ATM gene, they do not arrest and hence they exhibit radioresistant DNA synthesis. ATM also triggers a G_2/M checkpoint.

A model for how ATM initiates a cell cycle checkpoint is a follows. In the absence of DNA damage or other sources of stress, ATM forms dimers and the "FAT" domain of one ATM molecule interacts with the kinase domain of the other molecule. In this form the kinase activity of ATM is inactive and it cannot phosphorylate its protein targets. The introduction of double-strand breaks formed by ionizing radiation causes a disruption in chromatin structure. This change in chromatin structure causes the two ATM molecules to phosphorylate each other at serine 1981. This phosphorylation causes them to dissociate from each other, and each molecule is then able to phorphorylate their target proteins that are involved in cell cycle arrest, DNA repair, and apoptosis. This response to ionizing radiation is extremely rapid and occurs within a few minutes. It is also striking that as few as two DSBs in the genome can induce the response. The activation of ATM also requires the MRN complex with is involved in DSB repair by homologous recombination. It is unclear exactly how MRN functions in the activation of ATM. It may physically interact with ATM or it may process DSBs and in doing so contribute to the activation of ATM.

ATR is another member of the PIKK superfamily that plays an important role in cell cycle checkpoint control. ATR stands for ATM and Rad3 related. Rad3 is a gene involved in cell cycle check point control in *S. pombe*. ATM functions in response to ionizing radiation or other agents or processes that generate DSBs. In contrast, ATR functions more in response to UV-induced damage and other types of DNA damage that block the replication machinery. There is some crosstalk between the two pathways. The regulation of ATR appears to be different from ATM. It does not appear to form dimmers, and it does not appear to autophosphorylate itself in vivo. Instead a DNA-binding protein called *ATRIP* partners with ATR and directs it to sites of DNA damage. RPA is a single-strand DNA-binding complex, and it interacts with ATR-ATRIP. ATR-ATRIP may be recruited to sites of damage or replication intermediates that contain single-strand DNA through interactions with RPA. In addition, this pathway requires *RAD17-RFC2-5* and a complex related to PCNA that is composed of RAD9, HUS1, and RAD1 (*9–1–1*). It has been proposed that RAD17-RFC2-5 and 9–1–1 are loaded at blocked replication sites and this is required for checkpoint signaling. It regulates G1/S, mid-S, and G2/M checkpoints.

23.8.2 Mediators and Adaptors

Mediators and adaptors function to promote interactions between ATM or ATR and their effector targets. They convert sensor input into kinase modifications, and they may serve as signal amplifiers. As mentioned in Section 23.6.3, the introduction of DSBs by ionizing radiation results in a dramatic colocalization of certain DNA damage response proteins to subnuclear structures that are discernible as foci by fluorescence microscopy. These are often referred to as IR-induced nuclear foci (*IRIF*). IRIFs are generally thought to represent sites or clusters of sites of double-strand break repair. Some of the proteins that colocalize to IRIFs contain *BRCA1 carboxy-terminal (BRCT)* domains. It is thought that these proteins interact with regions of chromatin that flank the sites of the DSBs. A variant of histone H2A, *H2AX*, is phosphorylated by PIKKs in response to DSBs to form γ*H2AX*. The phophorylation of H2AX is a key regulator of the formation of IRIFs. ATR is the primary kinase that phosphorylates H2AX in response to UV irradiation or many other agents that block replication. ATM is the primary kinase in response to ionizing radiation.

The phospho-epitope in γH2AX is recognized by several proteins, and these interactions play critical roles in signal transduction. Certain proteins possess *forkhead-associated (FHA) domains* that can interact with proteins containing phosphorylated threonines when they are present within specific amino acid sequences. In addition, proteins containing tandem BRCT domains can bind specific phosporylated peptide domains. Among these proteins, *MDC1* has been determined to be the likely binding partner for γH2AX and binds through its tandem BRCT domain. MDC1 may then act as a scaffold to recruit other proteins to the foci, including MRN. Other proteins that are recruited to DSBs include *53BP1* and BRCA1. The formation of IRIFs has two roles. One is in the repair of DSBs. They likely serve to hold together the two ends of DNA formed by a DSB and prevent them from incorrectly synapsing with other ends of DNA formed in the nucleus. They also may serve as sites of chromatin remodeling that promote DSB repair. The other major role of IRIFs is in the activation of cell cycle check points. They may act to amplify the signal so that only a few DSBs are sufficient to generate a full cell cycle arrest.

23.8.3 Effector Targets

As mentioned in the previous sections, cells, in response to DNA damage, activate cell cycle checkpoints that temporarily arrest cell cycle progression. This pause in cell cycle progression prevents the replication of damaged DNA and allows time for DNA repair. Checkpoints occur in S phase of the cell cycle as well as at the G1/S and G2/M transitions. ATM/ATR activate checkpoint kinases (CHK1 and CHK2) and these kinases phosphorylate numerous proteins that are directly or indirectly involved in all three checkpoints (Figure 23.17). MDM2 negatively regulates p53 protein by mediating its nuclear export and ubiquitination resulting in proteosomal degradation of p53. MDM2 also decreases p53 transcriptional activity by directly binding to p53. Phosphorylation of MDM2 by ATM inactivates MDM2, and phosphorylation of p53 mediated by ATM/ATR and CHK1/CHK2 results in the stabilization and activation of p53, which positively regulates the expression of the effector, cyclin-dependent kinase inhibitor p21 (Figure 23.17) (see Chapter 24 for more p53 details). p21 blocks cell cycle progression at G1/S and G2/M transitions as well as S-phase by the inhibition of cyclin dependent kinase 1 (CDK1) and CDK2 activity (see Chapter 24 for cell cycle details). As shown in Figure 23.17, p53 also increases the expression of GADD45 (growth arrest and DNA damage inducible) and 14-3-3σ, both of which function in the G2/M checkpoint. 14-3-3α is thought to sequester cyclinB1/CDK1 in the cytoplasm where is inactive, while GADD45 is thought to interfere with the interaction between CDK1 and cyclin B1. CDC25A is protein phosphatase that removes phosphate groups located on CDK2 and CDK1, allowing for their activation. Upon DNA damage, CHK1 and CHK2 phosphorylate CDC25A and target it for proteosomal degradation, thereby preventing the dephosphorylation of CDK2 and CDK1 and producing an inhibition of cell cycle progression in S-phase and the G2/M transition of the cell cycle. CDC25C is also a protein phosphatase and it dephosphorylates CDK1. CHK2 phosphorylation of CDC25C prevents the dephosphorylation of CDK1 and results in a block in G2/M transition. As shown in Figure 23.17, NBS1 (Nijmegaen breakage syndrome) and SMC1 (structure maintenance of chromosomes 1) are important effectors of the S-phase checkpoint regulated directly by ATM activation.

23.9 SUMMARY

Clearly we have achieved a wealth of understanding of the multitude DNA repair and cell cycle damage response pathways. It is also clear that alterations in these pathways contribute to cancer, aging, and various other human disease states. Now that we have made such incredible advances in understanding the genetic and biochemical details of these pathways, it will be the ultimate challenge to apply this understanding to devise better ways to ameliorate or even cure the many human disease states that they impact.

SUGGESTED READING

Friedberg, E. C., Walker, G. C., Siede, W., Wood, R. D., Schultz, R. A., and Ellenberger, T. (Eds.). *DNA Repair and Mutagenesis*, 2nd ed. ASM, Washington, D.C., 2006.

Gillet, L. C. J., and Scharer, O. D. Molecular mechanisms of mammalian global genome nucleotide excision repair. *Chem. Rev.* **106**, 253–276, 2006.

Huffman, J. L., Sundheim, O., and Tainer, J. A. DNA base damage recognition and removal: new twists and grooves. *Mutat. Res.* **577**, 55–76, 2005.

Iyer, R. R., Pluciennik, A., Burdett, V., and Modrich, P. L. DNA mismatch repair: Functions and mechanism. *Chem. Rev.* **106**, 302–323, 2006.

Kadyrov, F. A., Dzantiev, L., Constantin, N., and Modrich, P. Endonucleolytic function of MutLalpha in human mismatch repair. *Cell* **126**, 297–308, 2006.

Ljungmna, M. (Ed.). Mechanisms of DNA repair special issue dedicated to Philip Hanawalt, *Mutat. Res.* **577**, 2005.

Mellon, I. Transcription-coupled repair: A complex affair. *Mutat. Res.* **577**, 155–161, 2005.

Sancar, A. Structure and function of DNA photolyase cryptochrome blue-light photoreceptors. *Chem Rev.* **103**, 2203–2237, 2003.

Shiloh, Y. Bridge over broken ends. The cellular response to DNA strand breaks in health and disease. Special issue of *DNA Repair.* **3**(8–9), 779–1254, 2004.

Siede, W., Wah Kow, Y., and Doetsch, P. (Eds.). *DNA Damage Recognition*, Taylor and Francis, New York, 2006.

Van Houten, B., Croteau, D. L., Della Vecchia, M. J., Wang, H., and Kisker, C. Close-fitting sleeves: DNA damage recognition by the UvrABC nuclease system. *Mutat. Res.* **577**, 92–117, 2005.

CHAPTER 24

Carcinogenesis

ROBERT C. SMART

Department of Environmental and Molecular Toxicology, North Carolina State University, Raleigh, North Carolina 27695-7633

SARAH J. EWING

Penn State Erie, The Behrend College, Erie, Pennsylvania 16563

KARI D. LOOMIS

Functional Genomics Program, North Carolina State University, Raleigh, North Carolina 27695

24.1 INTRODUCTION AND HISTORICAL PERSPECTIVE

English physicians, John Hill in 1761 and Sir Percival Pott in 1775, made two of the earliest observations linking chemical substance exposure to an increased incidence of human cancer. Hill observed an increased incidence of nasal cancer among snuff users, while Pott observed chimney sweeps had an increased incidence of skin cancer of the scrotum. Pott attributed this increase to topical exposure to soot and coal tar. It was not until nearly a century and a half later in 1915 when two Japanese scientists, Yamagiwa and Ichikawa, substantiated Pott's observation by demonstrating that multiple topical applications of coal tar to rabbit skin produced skin cancer. Thus, Yamagiwa and Ichikawa are considered the fathers of experimental chemical carcinogenesis. Their experimental results are important for two major reasons: (1) it was the first demonstration that a chemical or substance could produce cancer, and (2) it confirmed Pott's initial observation and established a relationship between human epidemiological studies and laboratory animal carcinogenicity. In the 1930s, Kennaway, Cook, and co-workers isolated several carcinogenic chemicals from coal tar, one of which was identified as benzo[*a*]pyrene, a polycyclic aromatic hydrocarbon. Benzo[*a*]pyrene results from the incomplete combustion of organic molecules and is also a component of cigarette smoke.

The concept that cancer involves an alteration in the genetic material of a somatic cell (somatic mutation theory) was first introduced by Theodor Boveri in 1914. Boveri suggested that cancer was related to chromosome abnormalities in somatic cells. With respect to chemical carcinogenesis, James and Elizabeth Miller in the 1950's and 1960s observed that a wide variety of structurally diverse chemicals could produce cancer in laboratory animals. They suggested that many of these diverse

Molecular and Biochemical Toxicology, Fourth Edition, edited by Robert C. Smart and Ernest Hodgson

chemicals require metabolic activation within the cell to produce a "common" electrophilic reactive intermediate. These electrophilic intermediates then covalently bind to nucleophilic centers on proteins, RNA or DNA. The Millers termed this the electrophilic theory of chemical carcinogenesis. Today, we know that many carcinogens, including benzo[*a*]pyrene (mentioned above), are metabolized by cytochrome P-450 to produce reactive electrophilic intermediates capable of covalently binding to DNA and producing mutations in genes. Thus, the electrophillic theory of chemical carcinogenesis is in accord with the somatic mutation theory.

Specific genes found in normal cells, termed proto-oncogenes and tumor suppressor genes, can be mutational targets for chemical carcinogens. Proto-oncogenes are frequently involved in the positive regulation of cell proliferation, and gain-of-function mutations in these proto-oncogenes result in cancer producing oncogenes. Tumor suppressor genes, on the other hand, often function as negative regulators of cell proliferation. Loss-of-function mutations in tumor suppressor genes release cells from the negative proliferation restraint imposed by the wild-type gene, thereby contributing to cancer development. DNA stability genes responsible for genome maintenance are also altered in some cancers and contribute to genomic instability and the accumulation of mutations in oncogenes and tumor suppressor genes. Cancer arises from the accumulation of sequential mutations within a single cell lineage, with each additional mutation conferring a greater proliferative advantage.

While the mechanism of action of many carcinogens is consistent with their mutagenicity, other agents can contribute to cancer development through an epigenetic process. Epigenetic mechanisms do not involve mutation of DNA, but rather involve the alteration in the expression of genes that function in important cellular processes such as cell proliferation, differentiation, and apoptosis. Agents such as tumor promoters and certain hormones contribute to cancer development through epigenetic mechanisms. Many of these agents stimulate the proliferation of "initiated" cells, which are cells that contain mutated oncogenes/tumor suppressor genes.

Lifestyle (environmental) factors including smoking, diet, cultural and sexual behavior, occupation, natural and medical radiation, as well as exposure to substances in air, water, and soil all influence the carcinogenic process in humans. Cancer susceptibility is determined by complex interactions between age, environment, and an individual's genetic make-up.

24.2 HUMAN CANCER

An important aspect of toxicology is the identification of potential human carcinogens and, thus, the protection of human health. To begin to appreciate the complexity of this subject, it is important to have some understanding of human cancer and its etiologies. Cancer is not a single disease but a large group of diseases (~110), all of which can be characterized by the uncontrolled growth of an abnormal cell to produce a population of cells that have acquired the ability to multiply and invade surrounding and distant tissues (metastasis). It is the invasive/metastatic characteristic of cancer that imparts its lethality to the host. Epidemiological studies have revealed that the incidence of most cancers increases exponentially with age (Figure 24.1). From these studies it has been suggested that over time, cells accumulate somatic mutations or "hits" and that three to seven mutations or hits are necessary

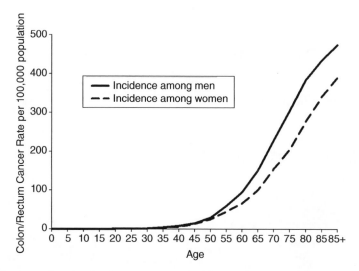

Figure 24.1. Age-related colon/rectum cancer incidence in the United States (2000–2003). Epidemiological studies have revealed that the incidence of most human cancers increase exponentially with age. (From NCI SEER. *Cancer Stati. Rev.* 1975–2003 http://seer.cancer.gov.)

for cancer development. Molecular analyses of human tumors have now confirmed the accumulation of somatic mutations in critical genes in cancer development.

Most cancers are monoclonal (derived from a single cell, likely a tissue/organ specific stem cell) in origin and arise from the accumulation of sequential mutations/hits. Each additional mutation within the cell provides a further proliferative advantage and this drives clonal proliferation/expansion and tumor development (Figure 24.2). Somatic mutations can result from numerous mechanisms, including imperfect DNA replication, imperfect DNA repair, oxidative DNA damage, and DNA damage caused by environmental carcinogens. Gain-of-function mutations in oncogenes and loss-of-function mutations in tumor suppressor genes provide a selective proliferative/survival advantage to the cell. Genes that are important in the maintenance of genome stability are also mutated in cancer. As the tumor progresses, the tumor cell genome becomes increasingly unstable, and the rate at which mutations are acquired increases greatly. A simplified view of the cancer process is shown Figure 24.3. It often requires decades for a tumor cell to accumulate the number of mutations/alterations needed to produce a clinically detectable cancer. Epigenetic changes also contribute to clonal expansion and tumor progression. For example, methylation of cytosine residues contained within CpG islands (cytosine/ guanosine) within the promoter region of a tumor suppressor gene can silence its expression. The methylation is mediated by cellular enzymes known as methyltransferases.

Cancer is a type of neoplasm or tumor. While technically a "tumor" is defined as only a tissue swelling, it is now used as a synonym for a neoplasm. A neoplasm is an abnormal mass of tissue, the growth of which exceeds and is uncoordinated with the normal tissue, and persists after cessation of the stimuli which evoked it. There are two basic types of neoplasms, termed benign and malignant. The general characteristics of these tumors are defined in Table 24.1. Cancer is the common name for a malignant neoplasm. Neoplasms are composed of two main components:

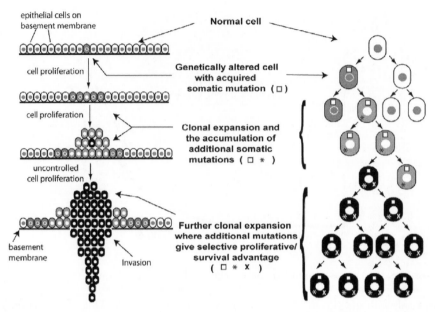

Figure 24.2. Monoclonal nature of cancer. Most cancers are monoclonal in origin and are derived from the accumulation of sequential mutations in an individual cell. Each additional mutation within the cell provides a further proliferative/survival advantage, and this process drives clonal expansion and tumor development.

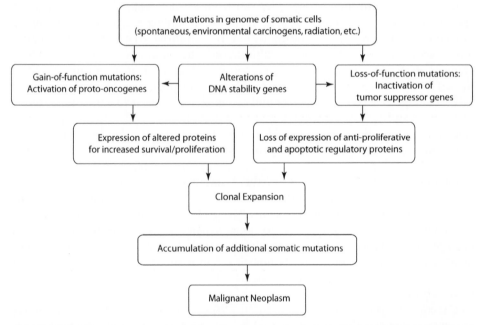

Figure 24.3. General aspects of the cancer process. Somatic mutations involving gain-of-function mutations in proto-oncogenes and loss-of-function mutations in tumor suppressor genes provide for a selective proliferative/survival advantage and are critical events in tumorigenesis. Inactivating mutations in genes involved in genomic maintenance can result in genomic instability.

TABLE 24.1. Some General Characteristics of Malignant and Benign Neoplasms

Benign	Malignant
Generally slow growing	May be slow to rapid growing
Few mitotic figures	Numerous mitotic figures
Well-differentiated and architecture resembles that of parent tissue	Some lack differentiation, disorganized; loss of parent tissue architecture
Sharply demarkated mass that does not invade surrounding tissue	Locally invasive, infiltrating into surrounding normal tissue
No metastases	Metastases

(1) the parenchyma that contains proliferating neoplastic cells and (2) the stroma that provides structural support and is composed of connective tissue, blood vessels, and immune cells. The stroma is not inert, because the endothelial, connective, and immune cells contained within the stroma contribute to the proliferation/survival of the neoplastic cells of the parenchyma.

All cancers are thought to have acquired the following capabilities: (1) self-sufficient signals for growth (no longer dependent on growth factors for proliferation), (2) the ability to evade apoptosis, (3) the capacity for extended proliferation (immortal), (4) resistance to growth-inhibiting signals, (5) the ability for tissue invasion and metastasis, and (6) the ability to induce blood-vessel formation (angiogenesis). In terms of cancer nomenclature, cancers derived from epithelial cells are termed carcinomas. The majority of human cancers (~80%) are carcinomas. For example, carcinomas are derived from the epithelial cells of the stomach, small and large intestine, skin, liver, lung, pancreas, mammary gland, ovary, prostate, urinary bladder, mouth, and esophagus. Sarcomas are rare and are derived from cells of mesenchymal tissues, including connective tissue, bone, fat, and muscle. Leukemias are derived from blood forming stem cells within the bone marrow and lymphomas from B and T lymphocytes within the lymphoid tissue. Melanoma is derived from melanocytes, whereas retinoblastoma, glioblastoma and neuroblastoma are derived from the stem cells of the retina, glia and neurons, respectively.

24.2.1 Causes, Incidence, and Mortality Rates

According to the American Cancer Society, (i) the lifetime risk for developing cancer in the United States is 1 in 3 for women and 1 in 2 for men, (ii) in 2007 about 1.4 million new cancer cases are expected to be diagnosed not including carcinoma in situ or basal or squamous carcinomas of the skin, and (iii) cancer is a leading cause of death in the United States; approximately 23% of all deaths are due to cancer. Cancer susceptibility is determined by complex interactions between age, environment, and an individual's genetic background. The incidence of specific cancer types in various geographic locations around the world can vary greatly, in some cases over 100-fold. The geographic migration of immigrant populations where differences in specific types of cancer incidence exist between the old and new communities have provided a great deal of information regarding the role of the environment in the development of certain cancers. For example, when Japanese people emigrate from Japan to the United States, they eventually aquire the specific cancer rates that are more representative of the U.S. population and less like the cancer

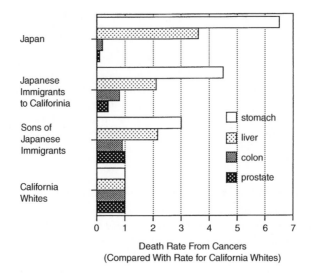

Figure 24.4. Cancer death rates in Japanese immigrants in the United States. The cancer death rate for each type of cancer is normalized to one for California white males and then compared to Japanese immigrants of a similar age. (Adapted from J. Cairns. In *Readings in Scientific American—Cancer Biology*, 1986, p. 13.)

rates of the Japanese population living in Japan (Figure 24.4). From these types of epidemiological studies, it is estimated that 65–80% of all cancers are associated with the environment in which we live and work. It should be noted that the term environment is not restricted to exposure to pollutants or man-made chemicals in the environment. The term environment instead applies to all aspects of our lifestyle, including smoking, diet, cultural and sexual behavior, occupation, natural and medical radiation, as well as exposure to substances in air, water, and soil. The major factors associated with cancer and their estimated contribution to human cancer deaths are listed in Figure 24.5. Notice that only a small percentage of total cancer occurs in individuals with a hereditary cancer syndrome/familial cancer syndromes. However, an individual's genetic background is the "stage" upon which cancer develops, and specific cancer susceptibility or cancer predisposition genes have been identified in humans. For example, women who have a first-degree relative (mother, sister, or daughter) who developed breast cancer have approximately twice the risk of developing breast cancer compared to women who do not have a family history of this cancer. Women that inherit germ-line mutant forms of BRCA1 or BRCA2 susceptibility genes have an increased risk of breast cancer. Genetic polymorphisms in enzymes responsible for the activation/deactivation of chemical carcinogens may represent another risk factor. For example, certain polymorphisms in the *N*-acetyl-transferase gene are associated with increased risk of bladder cancer. Many of these types of genetic risk factors are generally of low penetrance (low to moderate increased risk of developing the associated cancer); however, their effect can be amplified through gene-environment interactions (see Chapters 11, 13, 23, and 26). Although the values shown in Figure 24.5 are best estimates, it is clear that smoking and diet constitute the major factors associated with human cancer deaths. If one considers all of the categories that pertain to man-made chemicals, it is estimated their contribution to human cancer death is less than 10%. However, 10% of the

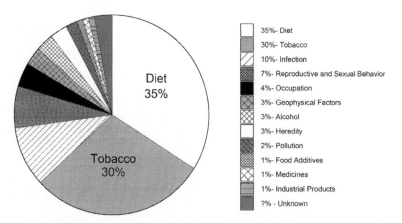

	35%- Diet
	30%- Tobacco
	10%- Infection
	7%- Reproductive and Sexual Behavior
	4%- Occupation
	3%- Geophysical Factors
	3%- Alcohol
	3%- Heredity
	2%- Pollution
	1%- Food Additives
	1%- Medicines
	1%- Industrial Products
	?% - Unknown

Figure 24.5. Proportion of cancer deaths attributed to various different factors. Values are a best estimate as determined by epidemiological studies. (Adapted from R. Doll and R. Peto. *The Causes of Cancer: Quantitative Estimates of Avoidable Risks of Cancer in the United States Today*, Oxford Medical Publications, 1981.)

total annual cancer deaths represent a large number of deaths; about 56,000 deaths/ year. The factors listed in Figure 24.5 are not mutually exclusive because there are likely to be complex interactions between these factors in the multistep process of carcinogenesis.

Cancer cases and cancer deaths by site and sex for the United States are shown in Figure 24.6. Breast, lung, and colon/rectum cancers are the major cancer site in females, whereas prostate, lung, and colon/rectum are the major cancer sites in males. A comparison of cancer deaths versus incidence for a given site provides an idea of the prognosis associated with different cancers. For example, the prognosis for lung cancer cases is poor, while that for breast or prostate cancer cases is much better. Age-adjusted cancer mortality rates (1930–2002) for selected sites in males are shown in Figure 24.7 and for females in Figure 24.8. The increase in the mortality rate associated with lung cancer in both females and males is striking and is the result of cigarette smoking. It is estimated that 87% of lung cancers are due to tobacco use. Lung cancer death rates in males began to increase in the mid-1930s, whereas lung cancer death rates for women did not increase until the mid-1960s. These time differences are explained by the use of cigarettes. Cigarette smoking among females did not become popular until the 1940s, whereas smoking among males was popular in the early 1900s. Taking into account these differences along with a 20 to 25-year lag period for the cancer to develop explains the differences in the temporal increase in lung cancer death rates in males and females. In addition to lung cancer, cigarette smoking also plays a significant role in cancer of the oral cavity, esophagus, pancreas, pharynx, larynx, bladder, kidney, stomach, and uterine cervix. Smoking accounts for at least 30% of all cancer deaths. Another disturbing statistic is that lung cancer, a theoretically preventable cancer, has surpassed breast cancer as the cancer responsible for the greatest number of cancer deaths in women. According to the American Cancer Society, the age-adjusted national overall cancer death rate has decreased slightly from 1950 to 2003. However, it is important to realize that the death rates for some types of

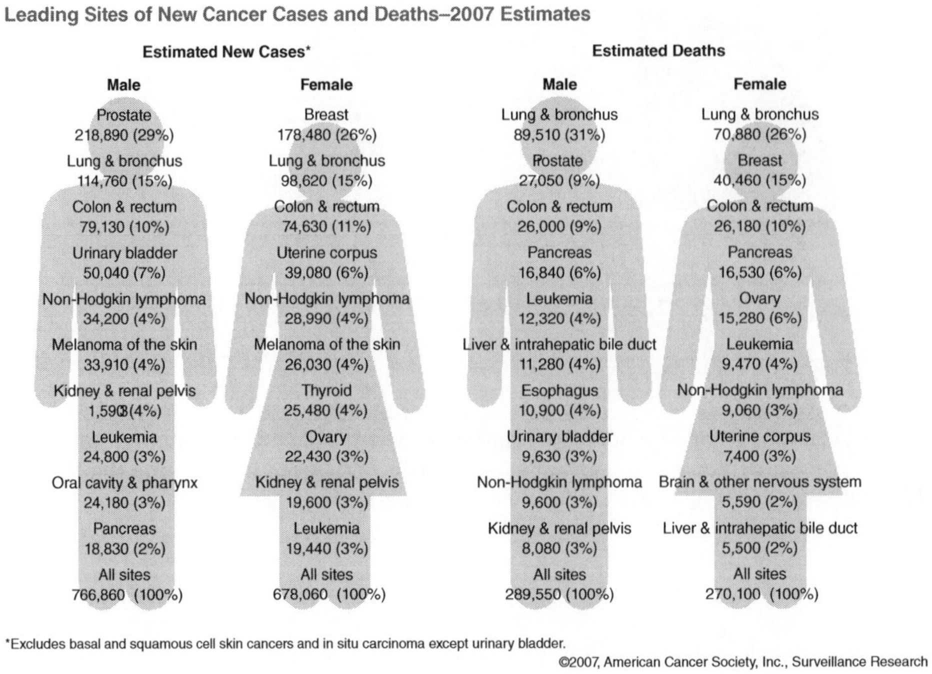

Leading Sites of New Cancer Cases and Deaths–2007 Estimates
Estimated New Cases*
Estimated Deaths
Male
Female
Male
Female
Prostate
218,890 (29%)
Lung & bronchus
114,760 (15%)
Colon & rectum
79,130 (10%)
Urinary bladder
50,040 (7%)
Non-Hodgkin lymphoma
34,200 (4%)
Melanoma of the skin
33,910 (4%)
Kidney & renal pelvis
1,593(4%)
Leukemia
24,800 (3%)
Oral cavity & pharynx
24,180 (3%)
Pancreas
18,830 (2%)
All sites
766,860 (100%)
Breast
178,480 (26%)
Lung & bronchus
98,620 (15%)
Colon & rectum
74,630 (11%)
Uterine corpus
39,080 (6%)
Non-Hodgkin lymphoma
28,990 (4%)
Melanoma of the skin
26,030 (4%)
Thyroid
25,480 (4%)
Ovary
22,430 (3%)
Kidney & renal pelvis
19,600 (3%)
Leukemia
19,440 (3%)
All sites
678,060 (100%)
Lung & bronchus
89,510 (31%)
Prostate
27,050 (9%)
Colon & rectum
26,000 (9%)
Pancreas
16,840 (6%)
Leukemia
12,320 (4%)
Liver & intrahepatic bile duct
11,280 (4%)
Esophagus
10,900 (4%)
Urinary bladder
9,630 (3%)
Non-Hodgkin lymphoma
9,600 (3%)
Kidney & renal pelvis
8,080 (3%)
All sites
289,550 (100%)
Lung & bronchus
70,880 (26%)
Breast
40,460 (15%)
Colon & rectum
26,180 (10%)
Pancreas
16,530 (6%)
Ovary
15,280 (6%)
Leukemia
9,470 (4%)
Non-Hodgkin lymphoma
9,060 (3%)
Uterine corpus
7,400 (3%)
Brain & other nervous system
5,590 (2%)
Liver & intrahepatic bile duct
5,500 (2%)
All sites
270,100 (100%)
*Excludes basal and squamous cell skin cancers and in situ carcinoma except urinary bladder.
©2007, American Cancer Society, Inc., Surveillance Research

Figure 24.6. Cancer cases and cancer deaths by sites and sex—2007 estimates. (Reprinted with permission of the American Cancer Society, *Cancer Facts and Figures 2007*, American Cancer Society, Atlanta, 2007.)

Age-Adjusted Cancer DeathRates,*Males bySite, US, 1930–2003
Rate per 100,000 male population
100
80
60
40
20
0
1930
1940
1950
1960
1970
1980
1990
2003
Lung & bronchus
Stomach
Prostate
Colon & rectum
Pancreas
Leukemia
Liver
*Per 100,000, age-adjusted to the 2000 US standard population.
Note: Due to changes in ICD coding, numerator information has changed over time. Rates for cancer of the liver, lung and bronchus, and colon and rectum are affected by these coding changes.
Source: US Mortality Public Use Data Tapes1960 to 2003, US Mortality Volumes 1930 to 1959, National Center for Health Statistics, Centers for Disease Control and Prevention, 2006.
American CancerSociety, Surveillance Research, 2007

Figure 24.7. Age-adjusted cancer death rates in males from 1930 to 2003. (Reprinted with permission of the American Cancer Society, *Cancer Facts and Figures 2007*, American Cancer Society, Atlanta, 2007.)

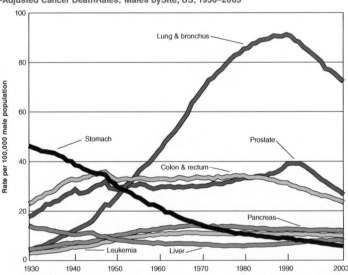

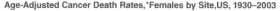

Age-Adjusted Cancer Death Rates,*Females by Site,US, 1930–2003

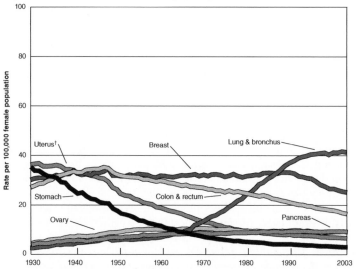

*Per 100,000, age-adjusted to the 2000 US standard population. †Uterus cancer death rates are for uterine cervix and uterine corpus combined.
Note: Due to changes in ICD coding, numerator information has changed over time. Rates for cancer of the lung and bronchus, colon and rectum, and ovary are affected by these coding changes.
Source: US Mortality Public Use Data Tapes 1960 to 2003, US Mortality Volumes 1930 to 1959, National Center for Health Statistics, Centers for Disease Control and Prevention, 2006.

American Cancer Society, Surveillance Research, 2007

Figure 24.8. Age-adjusted cancer death rates in females from 1930 to 2003. (Reprinted with permission of the American Cancer Society, *Cancer Facts and Figures*, American Cancer Society, Atlanta, 2007.)

cancers are increasing, while the rates for others are decreasing or remaining constant (Figure 24.7 and 24.8).

24.2.2 Classification of Carcinogens

Epidemiological studies have provided sufficient evidence that exposure to a certain chemicals, agents, or processes is associated with human cancer (see Chapter 26). For example, the following causal associations have emerged between exposure and the development of specific human cancers: vinyl chloride and hepatic cancer, amine dyes and bladder cancer, benzene and leukemia, and cigarette smoking and lung cancer. Naturally occurring chemicals or agents such as asbestos, aflatoxin B_1, betel nut, nickel, and certain arsenic compounds are also associated with an increased incidence of certain human cancers. Both epidemiological studies and rodent carcinogenicity laboratory studies are important in the identification of potential human carcinogens. The strongest evidence for establishing whether exposure to a given chemical is carcinogenic in humans comes from epidemiological studies. The International Agency for Research on Cancer (IARC), part of the World Health Organization, has developed a widely used system for the classification of carcinogens (Table 24.2). IARC classifies agents into four groups. Group 1 agents are classified as "known human carcinogens," and for all of these agents there is sufficient evidence from epidemiological studies to support a causal association between exposure to the agents and cancer (Table 24.3). Table 24.3 also includes information

TABLE 24.2. International Agency for Research on Cancer (IARC) Classification of Carcinogens

Group 1: The agent is *carcinogenic to humans.* This category is used when there is *sufficient evidence of carcinogenicity* in humans. Exceptionally, an agent may be placed in this category when evidence of carcinogenicity in humans is less than *sufficient*, but there is *sufficient evidence of carcinogenicity* in experimental animals and strong evidence in exposed humans that the agent acts through a relevant mechanism of carcinogenicity.

Group 2. This category includes agents for which, at one extreme, the degree of evidence of carcinogenicity in humans is almost *sufficient*, as well as those for which, at the other extreme, there are no human data but for which there is evidence of carcinogenicity in experimental animals. Agents are assigned to either Group 2A (*probably carcinogenic to humans*) or Group 2B (*possibly carcinogenic to humans*) on the basis of epidemiological and experimental evidence of carcinogenicity and mechanistic and other relevant data. The terms *probably carcinogenic* and *possibly carcinogenic* have no quantitative significance and are used simply as descriptors of different levels of evidence of human carcinogenicity, with *probably carcinogenic* signifying a higher level of evidence than *possibly carcinogenic.*

Group 2A: The agent is *probably carcinogenic to humans.* This category is used when there is *limited evidence of carcinogenicity* in humans and *sufficient evidence of carcinogenicity* in experimental animals. In some cases, an agent may be classified in this category when there is *inadequate evidence of carcinogenicity* in humans and *sufficient evidence of carcinogenicity* in experimental animals and strong evidence that the carcinogenesis is mediated by a mechanism that also operates in humans. Exceptionally, an agent may be classified in this category solely on the basis of *limited evidence of carcinogenicity* in humans. An agent may be assigned to this category if it clearly belongs, based on mechanistic considerations, to a class of agents for which one or more members have been classified in Group 1 or Group 2A.

Group 2B: The agent is *possibly carcinogenic to humans.* This category is used for agents for which there is *limited evidence of carcinogenicity* in humans and less than *sufficient evidence of carcinogenicity* in experimental animals. It may also be used when there is *inadequate evidence of carcinogenicity* in humans but there is *sufficient evidence of carcinogenicity* in experimental animals. In some instances, an agent for which there is *inadequate evidence of carcinogenicity* in humans and less than *sufficient evidence of carcinogenicity* in experimental animals together with supporting evidence from mechanistic and other relevant data may be placed in this group. An agent may be classified in this category solely on the basis of strong evidence from mechanistic and other relevant data.

Group 3: The agent is *not classifiable as to its carcinogenicity to humans.* This category is used most commonly for agents for which the evidence of carcinogenicity is *inadequate* in humans and *inadequate* or *limited* in experimental animals. Exceptionally, agents for which the evidence of carcinogenicity is *inadequate* in humans but *sufficient* in experimental animals may be placed in this category when there is strong evidence that the mechanism of carcinogenicity in experimental animals does not operate in humans. Agents that do not fall into any other group are also placed in this category. An evaluation in Group 3 is not a determination of noncarcinogenicity or overall safety. It often means that further research is needed, especially when exposures are widespread or the cancer data are consistent with differing interpretations.

Group 4: The agent is probably not carcinogenic to humans. This category is used for agents for which there is *evidence suggesting lack of carcinogenicity* in humans and in experimental animals. In some instances, agents for which there is *inadequate evidence of carcinogenicity* in humans but *evidence suggesting lack of carcinogenicity* in experimental animals, consistently and strongly supported by a broad range of mechanistic and other relevant data, may be classified in this group.

Source: IARC, http://monographs.iarc.fr/ENG/Preamble/currentb6evalrationale0706.php.

TABLE 24.3. International Agency for Research on Cancer (IARC) List of Agents, Substances, Mixtures or Exposure Circumstances Known to Be Carcinogenic to Humans

Agents and Groups of Agents

Aflatoxins (naturally occurring mixtures of)

4-Aminobiphenyl

Arsenic and arsenic compounds

(*Note:* This evaluation applies to the group of compounds as a whole and not necessarily to all individual compounds within the group.)

Asbestos

Azathioprine

Benzene

Benzidine

Benzo[a]pyrene

Beryllium and beryllium compounds

N,N-Bis(2-chloroethyl)-2-naphthylamine (Chlornaphazine)

Bis(chloromethyl)ether and chloromethyl methyl ether (technical grade)

1,4-Butanediol dimethanesulfonate (Busulfan; Myleran)

Cadmium and cadmium compounds

Chlorambucil

1-(2-Chloroethyl)-3-(4-methylcyclohexyl)-1-nitrosourea (methyl-CCNU; Semustine)

Chromium[VI] compounds

Cyclophosphamide

Cyclosporin (ciclosporin)

Diethylstilbestrol (DES)

Epstein–Barr virus

Erionite

Estrogen therapy, postmenopausal

Estrogen-progesterone postmenopausal therapy

Estrogen-progesterone oral contraceptives

Estrogens, nonsteroidal

(*Note:* This evaluation applies to the group of compounds as a whole and not necessarily to all individual compounds within the group.)

Estrogens, steroidal

(*Note:* This evaluation applies to the group of compounds as a whole and not necessarily to all individual compounds within the group.)

Ethanol in alcoholic beverages

Ethylene oxide

Etoposide in combination with cisplatin and bleomycin

Formaldehyde

Gallium arsenide

Gamma radiation

Helicobacter pylori (infection with)

Hepatitis B virus (chronic infection with)

Hepatitis C virus (chronic infection with)

Herbal remedies containing plant species of the genus *Aristolochia*

Human immunodeficiency virus type 1 (infection with)

Human papillomavirus type 16, 18, 31, 33, 39, 45, 51, 52, 56, 58, 59 and 66

Human T-cell lymphotropic virus type I

Melphalan

8-Methoxypsoralen (Methoxsalen) plus ultraviolet A radiation

MOPP and other combined chemotherapy including alkylating agents

Mustard gas (sulfur mustard)

TABLE 24.3. *Continued*

2-Naphthylamine

Neutrons

Nickel compounds

Opisthorchis viverrini (infection with)

Oral contraceptives, combined

(*Note:* There is also conclusive evidence that these agents have a protective effect against
  cancers of the ovary and endometrium.)

Oral contraceptives, sequential

Phosphorus-32, as phosphate

Plutonium-239 and its decay products (may contain plutonium-240 and other isotopes), as
  aerosols

Radioiodines, short-lived isotopes, including iodine-131, from atomic reactor accidents and
  nuclear weapons detonation (exposure during childhood)

Radionuclides, alpha-particle-emitting, internally deposited

(*Note:* Specific radionuclides for which there is sufficient evidence for carcinogenicity to
  humans are also listed individually as Group 1 agents.)

Radionuclides, beta-particle-emitting, internally deposited

(*Note:* Specific radionuclides for which there is sufficient evidence for carcinogenicity to
  humans are also listed individually as Group 1 agents.)

Radium-224 and its decay products

Radium-226 and its decay products

Radium-228 and its decay products

Radon-222 and its decay products

Schistosoma hematobium (infection with)

Silica, crystalline (inhaled in the form of quartz or cristobalite from occupational sources)

Solar radiation

Talc containing asbestiform fibers

Tamoxifen

(*Note:* There is also conclusive evidence that this agent (tamoxifen) reduces the risk of
  contralateral breast cancer.)

2,3,7,8-Tetrachlorodibenzo-para-dioxin

Thiotepa

Thorium-232 and its decay products, administered intravenously as a colloidal dispersion of
  thorium-232 dioxide

Treosulfan

Vinyl chloride

X- and gamma radiation

Mixtures

Alcoholic beverages

Analgesic mixtures containing phenacetin

Areca nut

Betel quid with tobacco

Betel quid without tobacco

Coal-tar pitches

Coal-tars

Mineral oils, untreated and mildly treated

Phenacetin, analgesic mixtures

Salted fish (Chinese-style)

Shale-oils

Soots

TABLE 24.3. *Continued*

Tobacco products, smokeless
Wood dust

Exposure Circumstances

Aluminum production
Arsenic in drinking water
Auramine, manufacture of
Boot and shoe manufacture and repair
Chimney sweeping
Coal gasification
Coke production
Furniture and cabinet making
Hematite mining (underground) with exposure to radon
Involuntary smoking
Iron and steel founding
Isopropanol manufacture (strong-acid process)
Magenta, manufacture of
Painter (occupational exposure as a)
Paving and roofing with coal far pitch
Rubber industry
Strong inorganic acid mists containing sulfuric acid (occupational exposure to)
Tobacco smoking

Source: IARC, http://monographs.iarc.fr/ENG/Classification/crthgr01.php.

on carcinogenic complex mixtures and occupations associated with increased cancer incidence.

It often takes 20–30 years after carcinogen exposure for a clinically detectable cancer to develop; therefore, epidemiological studies are complicated by the interference of a large number of confounding variables. This delay is problematic and can also result in inaccurate historical exposure information and prevent the timely identification of a putative carcinogen, resulting in unnecessary human exposure. Therefore, a bioassay using both laboratory mice and rats, was developed. This bioassay is termed the long-term rodent bioassay or the 2-year rodent carcinogenesis bioassay, and it is currently utilized to identify potential human carcinogens. Identification and classification of potential human carcinogens through the 2-year rodent carcinogenesis bioassay is complicated by species differences, use of high doses (MTD, maximum tolerated dose), short life span of the rodents, small sample size, and the need to extrapolate from high to low doses for human risk assessment. Although these problems are by no means trivial, the 2-year rodent carcinogenesis bioassay remains the "gold standard" for the classification of potential human carcinogens. Importantly, most, if not all, known human carcinogens are positive in the rodent carcinogenesis bioassay. However, it is not known whether most rodent carcinogens are human carcinogens. If human data are lacking, then a positive rodent carcinogenesis bioassay result is weighted as described in Table 24.2. Carcinogens are also classified by the National Toxicology Program (NTP), and approximately every 2 years the NTP releases the *Report on Carcinogens*, which classifies agents according to their carcinogenicity.

24.3   CATEGORIZATION OF AGENTS ASSOCIATED WITH CARCINOGENESIS

Chemical agents that contribute to cancer development can be divided into two major categories based on whether they are or are not mutagenic in in vitro mutagenicity assays (see Chapter 25). *DNA damaging agents* (genotoxic) are mutagenic in in vitro mutagenicity assays and produce permanent alterations in the DNA in vivo. *Epigenetic agents* (nongenotoxic), on the other hand, are not mutagenic in in vitro mutagenesis assays, and these agents are not thought to alter the primary sequence of DNA in vivo.

24.3.1   DNA Damaging Agents

DNA damaging agents produce three general types of genetic alterations: (i) small-scale changes, such as point mutations, involving single base-pair substitutions that can result in amino acid substitutions in the encoded protein, and frame shift mutations, involving the loss or gain of one or two base pairs resulting in an altered reading frame and gross alterations in the encoded protein, (ii) chromosome aberrations, including gross chromosomal rearrangement such as deletions, duplications, inversions and translocations, and (iii) aneuploidy (gain or loss of one or more chromosomes) (see Chapter 22).

DNA damaging agents can be divided into four major categories: direct, indirect, radiation, and inorganic. Direct-acting carcinogens are carcinogens that are intrinsically reactive compounds that do not require metabolic activation by cellular enzymes to covalently interact with DNA and produce mutations. Examples include *N*-methyl-*N*-nitrosourea and *N*-methyl-*N'*-nitro-*N*-nitrosoguanidine; the alkyl alkanesulfonates, such as methyl methanesulfonate; the lactones such as beta propiolactone; and the nitrogen and sulfur mustards. Indirect-acting carcinogens are carcinogens that require metabolic activation by cellular enzymes to form the ultimate carcinogenic species that covalently binds to DNA. Examples include dimethylnitrosamine, benzo[*a*]pyrene, 7,12-dimethylbenz[*a*]anthracene, aflatoxin B1, and 2-acetylaminofluorene (Figure 24.9). Ionizing radiation produces DNA damage through direct ionization of DNA to produce DNA strand breaks or indirectly, via the ionization of water to produce reactive oxygen species (ROS) that damage DNA. ROS are also produced by various cellular processes including respiration and lipid peroxidation. Ultraviolet radiation (UVR) from the sun is responsible for approximately 1 million new cases of human basal and squamous cell skin cancers each year. UVR causes DNA damage mostly in the form of cyclobutane pyrimidine dimers, 6–4 photoproducts, and DNA strand breaks. Inorganic agents include metals such as arsenic, chromium, and nickel. These agents are considered DNA damaging agents, although in many cases a definitive mechanism is unknown.

24.3.2   Epigenetic Agents

Epigenetic agents contribute to carcinogenesis; however, these agents do not produce mutations in DNA. Instead, they alter the expression of certain genes and/or alter pathways that influence cellular events related to proliferation, differentiation, or apoptosis. Epigenetic agents can alter gene expression through numerous mecha-

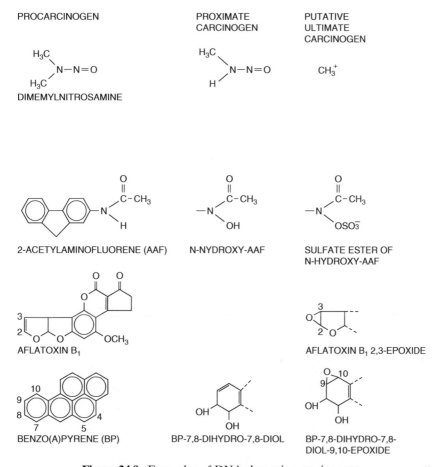

Figure 24.9. Examples of DNA-damaging carcinogens.

nisms including: (i) posttranslational modification of proteins such as transcription factors and kinases that often impinge on transcriptional regulation, (ii) methylation of CpG islands within promoter regions, (iii) histone modification (methylation/ acetylation), and (iv) receptor–ligand interactions. Some epigenetic agents can also suppress the immune system allowing cancer cells to escape detection and destruction. Epigenetic agents can be divided into four major categories: (i) *hormones* such as conjugated estrogens and diethylstilbestrol, (ii) *immunosuppressive xenobiotics* such as azathioprine and cyclosporin A, (iii) *solid-state agents* including plastic implants and asbestos, and (iv) *tumor promoters*, which include 12-*O*-tetradecanoylphorbol-13-acetate (TPA) also known as phorbol myristate acetate (PMA), peroxisome proliferators, TCDD, and phenobarbital; all of which are rodent tumor promoters (Figure 24.10). Tumor promoters favor the proliferation of cells with an altered genotype known as initiated cells (i.e., cells containing a mutated oncogene(s) and/or tumor suppressor gene(s)), thus "promoting" their selective clonal expansion. By definition, tumor promoters are not classified as carcinogens because they are considered inactive in the absence of initiated cells. However, this may be a matter of semantics as dormant initiated cells are likely to be present in human

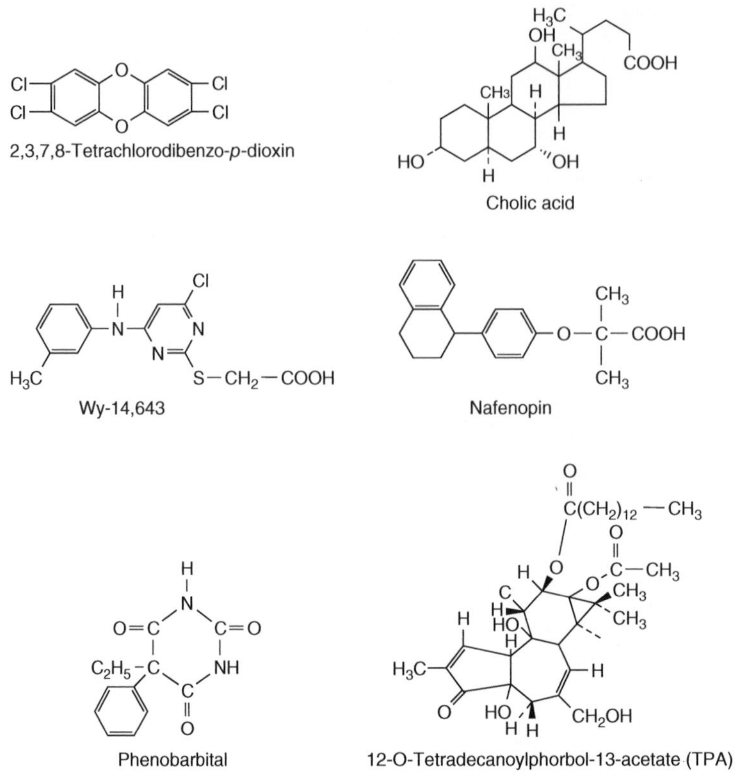

Figure 24.10. Examples of tumor promoters.

adults. In humans, diet (including caloric, fat, and protein intake), excess alcohol, and late age of pregnancy are considered to influence cancer development through a promotion mechanism. While cigarette smoking and UVR exposure produce DNA damage, they also are considered to have tumor promoting activity.

The classification of epigenetic agents can be complicated as some epigenetic agents can induce oxidative DNA damage in vivo but not in in vitro mutagenesis assays, and thus are not mutagenic in mutagenecity assays. For example, certain estrogen hormones possess this ability and this may contribute to their carcinogenicity. Thus, it is important to keep in mind that there can be overlap in the mechanisms and categorization of the agents that contribute to the cancer process.

24.4 SOMATIC MUTATION THEORY

The somatic mutation theory of carcinogenesis states that mutation of DNA within somatic cells is required for neoplasia. The major pieces of evidence supporting the somatic mutation theory of carcinogenesis are shown in Table 24.4.

24.4.1 Electrophilic Theory, Metabolic Activation, and DNA Adducts

In the 1950–1960s Elizabeth and James Miller observed that a diverse array of chemicals could produce cancer in laboratory animals. In an attempt to explain this,

TABLE 24.4. Major Evidence Supporting a Genetic Mechanism of Carcinogenesis

1. Cancer is a heritable stable change.
2. Tumors are generally clonal in nature.
3. Many carcinogens are or can be metabolized to electrophilic intermediates which can covalently bind to DNA.
4. Many carcinogens are mutagens.
5. Autosomal dominant-inherited cancer syndromes provide direct evidence for a genetic component in the origin of cancer.
6. Autosomal recessive-inherited cancer syndromes associated with chromosome fragility or decreased DNA repair predisposes affected individuals to cancer.
7. Most, if not all, cancers display chromosomal abnormalities.
8. The transformed phenotype can be transferred from a tumor cell to a non-tumor cell by DNA transfection.
9. Cells can be transformed with oncogenes.
10. Proto-oncogenes are activated by mutation in cancer cells.
11. Tumor suppressor genes are inactivated by mutation in cancer cells.

they hypothesized that many carcinogens are metabolically activated to electrophilic intermediates capable of covalent binding to nucleophilic sites in DNA, RNA, and protein. They proposed that the production/presence of a highly reactive electrophilic intermediate is common to most chemical carcinogens. The Millers termed this the *electrophilic theory of chemical carcinogenesis*. We now know that many carcinogens are metabolized by cellular enzymes (metabolic activation) to produce reactive electrophilic intermediates that are highly mutagenic and carcinogenic. From this concept of metabolic activation, the important terms *parent, proximate*, and *ultimate* carcinogen were developed. A *parent carcinogen* (also called a procarcinogen) is a compound that must be metabolized in order to have carcinogenic activity. *A proximate carcinogen* is an intermediate metabolite requiring further metabolism to produce the *ultimate carcinogen*, which is the electrophilic metabolite that covalently binds to DNA forming a DNA adduct. Examples of chemical carcinogens that require metabolic activation by cellular enzymes include benzo[*a*]pyrene, 7,12-dimethylbenz[*a*]anthracene, aflatoxin B1, and 2-acetylaminofluorene. Some of the parent structures of these compounds as well as their proximate and ultimate carcinogenic species are shown in Figure 24.9. If the DNA adduct, or any base modification, is improperly repaired or is not repaired before the cell replicates, a mutation can occur. If this mutation occurs within a critical region of a proto-oncogene or tumor suppressor gene, then this would be an important event in carcinogenesis. Therefore, misrepair or lack of repair followed by DNA replication is an important aspect of chemical carcinogenesis (see Chapter 22). Thus, the electrophilic theory of chemical carcinogenesis is in accord with the somatic mutation theory of carcinogenesis.

Metabolic activation of chemical carcinogens by cytochrome P-450 is well-documented. In addition to the Phase 1 enzymes, Phase 2 enzymes such as GSH transferases can also participate in metabolically activating or inactivating (detoxification) chemical carcinogens. Human genetic polymorphisms have been characterized in some Phase 1 and Phase 2 genes (see Chapters 11 and 13). Such polymorphisms could alter an individual's susceptibility to cancer induced by certain chemical

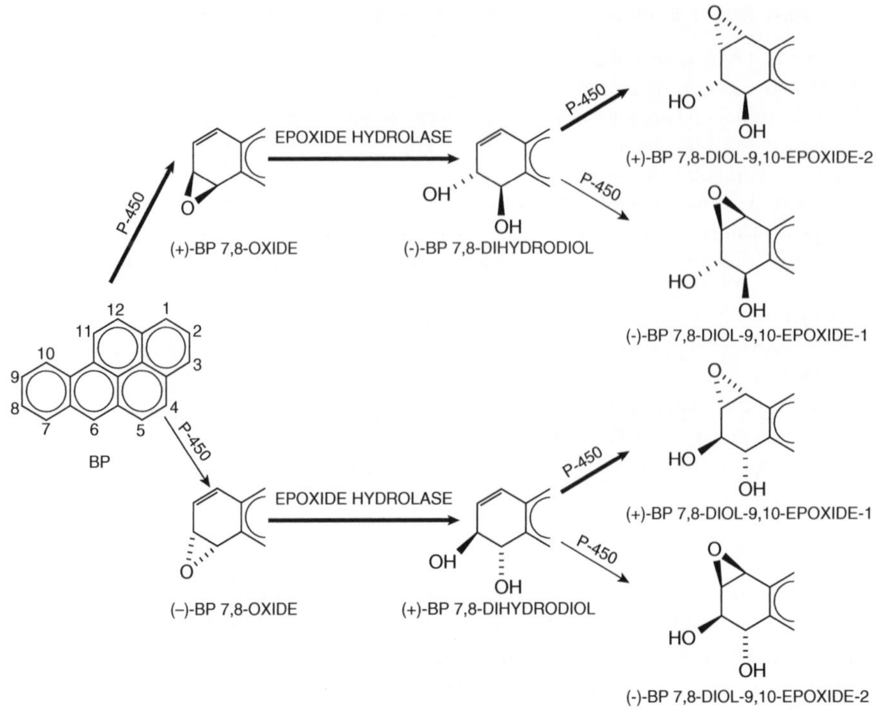

Figure 24.11. Metabolic activation of benzo[*a*]pyrene to the ultimate carcinogenic species. Benzo[*a*]pyrene is metabolized by cytochromes P-450 and epoxide hydrolase to form the ultimate carcinogen, (+)benzo[*a*]pyrene-7,8-diol-9,10 epoxide-2. (Adapted from Conney, A. H. *Cancer Res.* **42**, 4875, 1982.)

carcinogens. These genes are often termed metabolic susceptibility genes. The metabolic activation of benzo[*a*]pyrene has been extensively studied, and at least 15 major Phase 1 metabolites have been identified. Many of these Phase 1 metabolites are further metabolized by Phase 2 enzymes. Extensive research has elucidated the metabolites that are important in the carcinogenic process. As shown in Figure 24.11, the parent carcinogen, benzo[*a*]pyrene, is metabolized by cytochrome P-450 to benzo[*a*]pyrene-7,8 epoxide, which is then hydrated by epoxide hydrolase to form benzo[*a*]pyrene-7,8-diol. Benzo[*a*]pyrene-7,8-diol is considered the proximate carcinogen because it must be further metabolized by cytochrome P-450 to form the ultimate carcinogen: the bay region diol epoxide, (+)-benzo[*a*]pyrene-7,8-diol-9,10-epoxide-2. It is this reactive electrophilic intermediate that covalently binds to DNA, forming DNA adducts. (+)-Benzo[*a*]pyrene-7,8-diol-9,10-epoxide-2 binds preferentially to deoxyguanine residues, forming an N-2 adduct (Figure 24.12). (+)-Benzo[*a*]pyrene-7,8-diol-9,10-epoxide-2 is highly mutagenic in eukaroytic and prokaryotic cells and is carcinogenic in rodents. It is important to note that not only is the chemical configuration of the metabolites of many polycyclic aromatic hydrocarbons important for their carcinogenic activity, but so is their chemical conformation and stereospecificity (Figure 24.11). For example, four different stereoisomers of benzo[*a*]pyrene-7,8-diol-9,10 epoxide are formed. Each one only differs with respect to whether the epoxide or hydroxyl groups are above or below the plane of

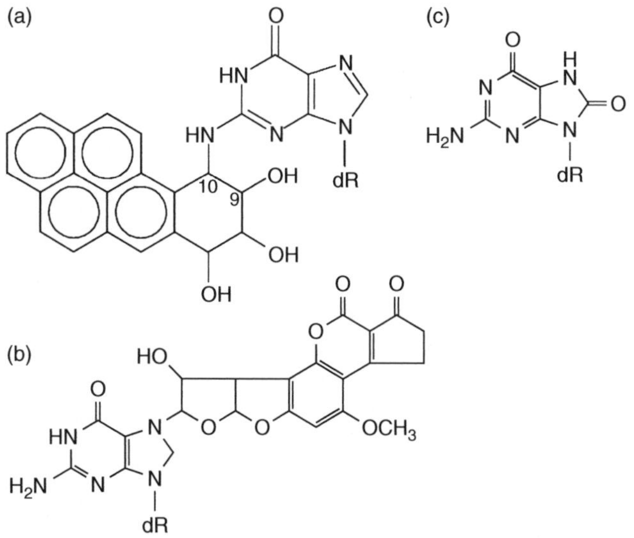

Figure 24.12. Examples of DNA adducts and oxidized guanine nucleotide. (A) Benzo[*a*]pyrene-7,8-diol-9,10-epoxide-*N*-2 guanine adduct. (B) Aflatoxin epoxide N7 guanine adduct. (C) 8-Oxoguanine is commonly produced by reactive oxygen species, dR = deoxyribose.

the flat benzo[*a*]pyrene molecule; and yet only one, (+)-benzo[*a*]pyrene-7,8-diol-9,10-epoxide-2, has significant carcinogenic potential. Many polycyclic aromatic hydrocarbons are metabolized to bay-region diol epoxides. The bay region theory suggests that the bay region diol epoxides are the ultimate carcinogenic metabolites of polycyclic aromatic hydrocarbons. Examples of bay region diol epoxides that are considered the ultimate carcinogenic species of a number of polycyclic aromatic hydrocarbons are shown in Figure 24.13.

24.4.2 DNA Damage

DNA can be altered by oxidative damage, chemical carcinogen adduction, alkylation, deamination, and hydrolysis of bases to produce an apurinic/apyrimidinic site (AP site) as well as by DNA single/double-strand breaks and DNA-strand crosslinks (see Chapter 22). Carcinogens such as *N*-methyl-*N′*-nitro-*N*-nitrosoguanidine and methyl methanesulfonate alkylate DNA to produce *N*-alkylated and *O*-alkylated purines and pyrimidines. Ionizing radiation and reactive oxygen species commonly oxidize guanine to produce 8-oxoguanine (Figure 24.12). Formation of DNA adducts may involve any of the bases, although the N7 position of guanine is one the most nucleophilic sites in DNA and, thus, one of the most frequently altered. As shown in Figure 24.12, (+)-benzo[*a*]pyrene-7,8-diol-9,10-epoxide-2 forms adducts mainly at guanine N2, whereas aflatoxin B1 epoxide, another well-studied rodent and human carcinogen, binds preferentially to the N7 position of guanine (Figure 24.12). For some carcinogens, there is a strong correlation between the formation of very specific DNA adducts and tumorigenicity. DNA damage can result in mutations, and the mechanisms through which this occurs are described in Chapter 22. Quantitation and identification of specific carcinogen–DNA adducts can be used as a biomarker of carcinogen exposure. In addition, the identification of specific carcinogen. DNA

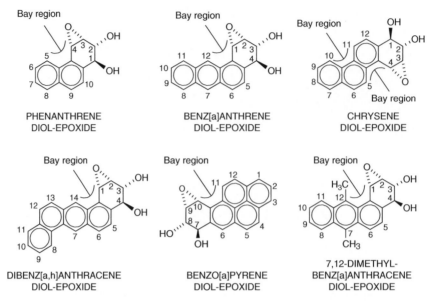

Figure 24.13. Examples of bay region diol epoxide ultimate carcinogenic species of various polycyclic aromatic hydrocarbons (PAHs). Polycyclic aromatic hydrocarbons are metabolized to highly electrophilic bay region diol epoxides that are considered to be the ultimate carcinogenic species. (Adapted from Conney, A. H. *Cancer Res.* **42**, 4875, 1982.)

adducts has allowed for the prediction of specific point mutations that would likely occur in the daughter cell, provided that there was no repair/misrepair of the DNA adduct in the parent cell. As will be discussed in a later section, some of these expected mutations have been identified in specific oncogenes and tumor suppressor genes in chemically induced rodent tumors, supporting the hypothesis that carcinogen binding resulted in the observed mutation. Specific point mutations found in the *p53* tumor suppressor gene in human tumors are consistent with those that would be expected from specific carcinogen exposure. For example, the mutation spectra in *p53* in human lung tumors of smokers, liver tumors of aflatoxin (dietary)-exposed individuals, and skin tumors from sun exposed areas of skin are consistent with the type of DNA damage inflicted by benzo[*a*]pyrene (component of cigarette smoke), aflatoxin, and UV solar radiation, respectively.

24.4.3   Complete Carcinogenesis Model

Exposure of rodents to either high or repeated doses of a carcinogen can produce tumors in the absence of any exogenous exposure to a tumor promoting agent. Such carcinogen-induced cancer is termed the complete carcinogenesis model. The 2-year rodent carcinogenesis bioassay is an example of a complete carcinogenesis model. It is thought that at high doses, genotoxic carcinogens have both initiator and promoter activity. This model is in contrast to the initiation-promotion carcinogenesis model that utilizes a single low dose of genotoxic carcinogen treatment and requires a tumor promoting agent to produce tumors (see Section 24.6.1).

TABLE 24.5. Evidence of Epigenetic Mechanisms of Carcinogenesis

1. Cancer is associated with altered differentiation.
2. The cancerous state of tumors is sometimes reversible.
3. Carcinogenesis is induced by nonmutagenic agents.
4. Not all carcinogens are mutagens.
5. Carcinogenesis is associated with changes in DNA methylation.
6. Cell transformation can occur at very high frequencies in vitro.

24.5 EPIGENETIC MECHANISM OF TUMORIGENESIS

Although the role for somatic mutation in carcinogenesis is indisputable, there is also substantial evidence for epigenetic mechanisms in tumor development. The major pieces of evidence that support an epigenetic mechanism of carcinogenesis are shown in Table 24.5. Epigenetic mechanisms are those that do not produce mutations, or alter the sequence of DNA; instead, they result in the altered expression of genes that influence cellular proliferation, differentiation, or apoptosis. These changes in gene expression provide the cell with a proliferative/survival advantage. Epigenetic mechanisms can involve perturbations in signal transduction pathways, alterations in chromatin remodeling including histone acetylation/methylation/ubiquitination, or alterations in the methylation of cytosine residues present in CpG islands within the promoter of specific genes (hypermethylation leads to gene silencing and hypomethylation leads to increased gene expression). As discussed in Section 24.3.2, agents such as hormones, tumor promoters, and immunosuppressive agents contribute to the cancer process through epigenetic mechanisms. The importance of epigenetic mechanisms in carcinogenesis is underscored by the fact that approximately 40% of all rodent carcinogens are not mutagenic in in vitro mutagenesis assays. Tumorigenesis in both human and experimental rodent models involves the cooperation of genetic and epigenetic mechanisms.

24.6 MULTISTAGE TUMORIGENESIS

Carcinogenesis in humans and laboratory animals is a complex multistep process. This process involves epigenetic events such as the inappropriate expression of certain cellular genes and genetic events that include the mutational activation of oncogenes and the inactivation of tumor suppressor genes. A number of in vitro and in vivo models have been important in the identification of epigenetic and genetic events associated with carcinogenesis. In experimental animal models, carcinogenesis can be divided into at least three stages termed *initiation, promotion*, and *progression* (Figure 24.14). Initiation is the first stage in multistage tumorigenesis and involves a genotoxic or mutagenic event. Thus, initiation results in a permanent heritable change in the cell's genome (mutated oncogene(s) and/or tumor suppressor gene(s)). This cell is often referred to as an "initiated cell." It is generally considered that the initiated cell is a stem cell within the affected organ.

Many factors impinge upon the formation of the initiated cell (Figure 24.14). For example, carcinogen uptake/pharmacokinetics/tissue disposition can clearly

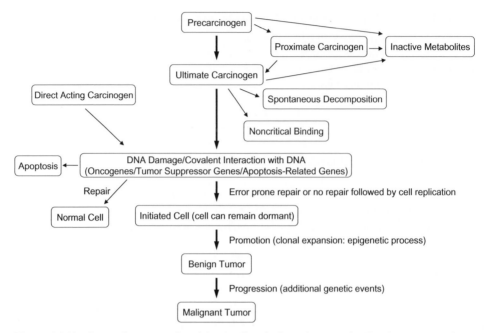

Figure 24.14. General aspects of multistage chemical carcinogenesis. Carcinogens can form reactive species that damage DNA. If damage occurs in a critical gene and either error-prone repair or no repair is followed by cell replication, a mutation can result. This mutated cell is now referred to as an "initiated cell." Tumor promoters allow for the clonal expansion of an initiated cell to produce a benign tumor. This benign tumor can progress to a malignant tumor, and this involves additional genetic changes in critical genes.

influence whether an initiated cell will be produced in vivo. The metabolism of the carcinogen to the ultimate carcinogen can be increased or decreased, depending on the expression level and/or polymorphisms in Phase 1 and 2 enzymes involved in its metabolism. Once formed, the ultimate carcinogen may spontaneously decompose, bind to noncritical sites within the cell, bind to cellular nucleophiles such glutathione, or, in the case of carcinogenesis, bind to critical sites in the DNA. The adducted/damaged DNA can be repaired (see Chapter 23) to produce an undamaged normal cell or the cell could undergo apoptosis. However, if there is an error in the repair of the damaged DNA or the DNA damage is not repaired before the cell replicates, an error in the newly synthesized DNA could occur and result in a mutation. When this mutation occurs in an oncogene/tumor suppressor gene, the cell is referred to as an initiated cell. This initiated cell may remain dormant (not undergo clonal expansion) and not form a tumor for the lifetime of the animal. However, under the influence of a repeated dose of a tumor promoter, the initiated cell can clonally expand to eventually produce a benign tumor. This is an epigenetic process termed tumor promotion. Tumor promoters favor the proliferation of initiated cells, leading to their selective clonal expansion. The development of a malignant tumor from a benign tumor encompasses a third stage of multistage carcinogenesis, termed progression. Tumor progression involves the acquisition of additional mutations.

24.6.1   Initiation–Promotion Model

Experimentally, the initiation–promotion process has been demonstrated in a variety of organs/tissues including skin, liver, lung, colon, mammary gland, prostate, and bladder as well as in cultured cells. Although tumor promoters have different mechanisms of action and many are organ-specific, all have common operational features (Figure 24.15). These features include the following: (1) Initiation at a subthreshold dose of initiating carcinogen will produce very few, if any, tumors (dormant initiated cells); (2) following a subthreshold of initiating carcinogen, chronic treatment with a tumor promoter will produce many tumors (clonal expansion of initiated cells); (3) chronic treatment with a tumor promoter in the absence of initiation will produce very few, if any, tumors (absence of initiated cells); (4) the order of treatment is critical; that is, you must first initiate and then promote; (5) initiation produces an irreversible change (mutation); and (6) promotion is reversible in the early stages; for example, if an equal number of promoting doses are administered, but the doses are spaced far apart in time, tumors would not develop or would be greatly diminished in number. Many tumor promoters are organ-specific and all require repeated doses to be active. 12-O-Tetradecanoylphorbol-13-acetate (TPA) belongs to a family of compounds known as phorbol esters. Phorbol esters are isolated from croton oil (derived form the seeds of the croton plant), and almost exclusively act as skin tumor promoters. Some hepatic tumor promoters include: Phenobarbital, DDT, chlordane, TCDD, and peroxisome proliferators Wy 24,643, clofibrate, and nafenopin. TCDD is also a promoter in lung and skin. Some bile acids are colonic tumor promoters, while various estrogens are tumor promoters in the mammary gland and liver.

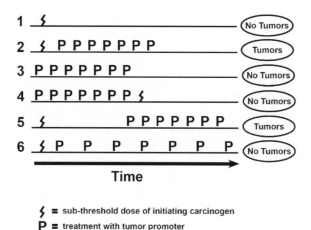

ξ = sub-threshold dose of initiating carcinogen

P = treatment with tumor promoter

Figure 24.15. Diagrammatic scheme of the initiation promotion model. Topical application of a subthreshold dose of an initiating carcinogen to mouse skin will result in no tumor formation; however, if this dose is followed by repetitive treatment with a tumor promoter, then tumors will develop. Initiation is an irreversible genetic event leading to the development of an "initiated cell" which can remain dormant until exposed to a tumor promoter. Tumor promoters produce the clonal expansion of the initiated cell to form a tumor.

TABLE 24.6. Some General Mechanisms of Tumor Promotion

Selective Proliferation of Initiated Cells
Increased responsiveness to and/or production of growth factors, hormones, and other active molecules
Decreased responsiveness to inhibitory growth signals
Perturbation of intracellular signaling pathways

Altered Differentiation
Inhibition of terminal differentiation of initiated cells
Acceleration of differentiation of uninitiated cells

Altered Apoptosis
Inhibition of apoptosis in initiated cells stimulation of the apoptosis of uninitiated cells

24.6.2 Mechanisms of Tumor Promotion

Tumor promoters alter pathways that cooperate with the genetic changes present in the initiated cell to produce the selective growth of that cell. Generic mechanisms of tumor promotion are shown in Table 24.6. Although the precise molecular/biochemical mechanisms of tumor promotion are unknown, we do know that certain receptors, growth factors, transcription factors, cytokines, and kinases are involved in tumor promotion induced by a particular tumor promoter.

TPA is a potent skin tumor promoter and mitogen that produces many of its effects by binding to the phorbol ester cellular receptor. This receptor has been identified as protein kinase C (PKC). PKC is a multigene family of serine/threonine kinases that is involved in the transduction of extracellular signals conveyed by growth factors, neurotransmitters, hormones, and other biological molecules. When TPA binds to the regulatory domain of certain isoforms of PKC, the enzyme becomes catalytically active and phosphorylates specific target proteins. It has been proposed that these phophorylated proteins regulate pathways that control cellular events important in proliferation and tumor promotion. Okadaic acid, another skin tumor promoter, functions through its ability to inhibit the serine/threonine protein phosphatases known as PP1 and PP2A. Phosphatase inhibition by okadaic acid provides further evidence for serine/threonine protein phosphorylation pathways in skin tumor promotion. While TPA treatment leads to the chronic activation of certain PKC family members, it also leads to the downregulation of some members. Both of these events are considered important in TPA-induced tumor promotion. Diacylglyerols (DAGs) are intracellular second messengers that are produced by receptor-mediated hydrolysis of phospholipids and are the endogenous activator of PKC. When applied to mouse skin, DAGs produce biochemical, mitogenic, and inflammatory alterations similar to TPA and are effective tumor promoters.

In humans, chronic inflammation of the liver is associated with hepatitis B virus (HBV) infections, and this inflammation is considered an important contributing event in HBV-induced liver cancer. In animal models, there is evidence that inflammation contributes to tumor promotion. Treatment of mouse skin with TPA produces an inflammatory response and increases the expression of proinflammatory mediators such as TNF-α, IL-1α GM-CSF, and cyclooxygenase-2 (COX-2). Genetically modified mice deficient in TNFα or COX-2 are resistant to TPA-induced tumor promotion. These results indicate that the inflammatory effects of TPA are important in tumor promotion.

TPA treatment also induces the expression of the mitogenic growth factor (TGFα), induces hyperplasia, alters differentiation, and modulates the trans-activation activity of the transcription factor AP-1. Benzoyl peroxide and chrysarobin are skin tumor promoters that are thought to function through the generation of oxygen free radicals. Oxidative stress has been shown to activate the JNK signaling pathway and increase the trans-activation activity of the c-Jun transcription factor.

In sun-exposed areas of human skin, UVR causes DNA damage and results in inactivating mutations in the tumor suppressor gene, p53, in keratinocytes. Thus, UVR functions as an initiator, and the majority of squamous cell carcinomas of the skin contain a mutated p53. Subsequent exposures of the skin to UVR cause additional DNA damage. In response to DNA damage, normal keratinocytes undergo apoptosis via a p53-dependent mechanism whereas the keratinocytes with mutant, inactive p53 cannot undergo apoptosis. To replace the normal cells that underwent UVR-induced apoptosis, the p53 mutant cells proliferate. Thus in skin, UVR also functions as a promoter by producing the selective proliferation of keratinocytes with a mutant inactive p53. UVR can also suppress the immune system, which may further enhance the development of skin cancers.

In the rodent liver initiation-promotion model, phenobarbital is a tumor promoter. Phenobarbital is a hepatic mitogen, and recent evidence indicates that it activates the transcription factor, constitutive androstane nuclear receptor (CAR). Additionally, CAR was found to be essential for liver tumor promotion in phenobarbital-treated mice. In phenobarbital promoted preneoplastic foci and hepatic tumors, phenobarbital can also suppress apoptosis and accelerate the growth of the foci or tumor. Withdrawal of phenobarbital treatment is accompanied by an increase in apoptosis. Similar results involving the inhibition of apoptosis have been reported for nafenopin, a peroxisome proliferator and hepatic tumor promoter.

Peroxisome proliferators are a diverse group of chemicals that include the fibrate class of hypolipidemic drugs (clofibrate, Wy 14,643, and nafenopin), plasticizers (di-2-ethylhexyl phthalate, DEHP), and herbicides (2,4-dichlorophenoxyacetic acid). Peroxisome proliferators cause an increase in the number and size of peroxisomes in the liver, kidney, and heart of rats and mice. Peroxisome proliferators are hepatic mitogens and tumor promoters. These agents function through their ability to bind and stimulate peroxisome proliferator-activated receptor-α (PPARα), a member of the steroid/thyroid receptor superfamily. PPARα forms a heterodimer with the retinoid RXR receptor, and this complex binds to the peroxisome proliferator response element (PPRE) in the promoters of specific genes, resulting in increased transcription. These PPARα responsive genes are involved in tumor promotion and appear to regulate cell proliferation, differentiation, and survival. PPARα knockout mice are refractory to peroxisome proliferation, peroxisome proliferator-induced changes in gene expression, and WY-14,643-induced hepatocarcinogenesis.

Halogenated hydrocarbons such as organochlorine pesticides, polybrominated biphenyls, and TCDD are representative of a large group of hepatic tumor promoters. TCDD is a potent promoter in rodent liver, where it binds to and activates the TCDD receptor, a transcription factor. The activated receptor regulates gene expression by binding to TCDD responsive elements in the promoter region of certain genes thought to be important in tumor promotion. Many of these gene products and their importance in tumor promotion remain to be elucidated.

Tumor promoters from different mechanistic classes inhibit gap junction intercellular communication (GJIC) or cell-to-cell communication. Cell communication can be mediated through gap junctions, which are formed when a connexon of one cell couples with a connexon in a contiguous cell to form a pore between the cells. This serves to synchronize functions of cells within a tissue. It is postulated that within any given tissue, normal cells hold initiated cells in check (prevent abnormal proliferation) through GJIC. When a tumor promoter blocks GJIC, the normal cells can no longer restrain the initiated cell and the initiated cell selectively proliferates and clonally expands.

24.7   TUMOR VIRUSES

Having described the importance of both genetic and epigenetic events in carcinogenesis, we will now discuss the genes involved. As described, chemical carcinogenesis involves the mutational activation of certain genes known as oncogenes, and the inactivation of tumor suppressor genes and these events are critical for tumor development. But how were these oncogenes first identified, and what is their exact function? Oncogenes were first identified as genes contained in acute transforming retroviruses that cause cancer primarily in birds and rodents. With the exception of a rare human T-cell leukemia, retroviruses are not considered to play a role in human cancer. However, because acute transforming retroviruses were instrumental in the identification of cellular genes (oncogenes) involved in cancer, it is instructive to understand how they contributed to our current knowledge. In contrast to retroviruses, oncogenic DNA viruses are associated with 20% of human cancer worldwide and function by inactivating tumor suppressor genes. These viruses will be briefly discussed in Section 24.7.3.

24.7.1   Acute Transforming Retroviruses

Retroviruses encode their genetic material in RNA. The viral genome is composed of a powerful retroviral promoter contained within the long terminal repeat (LTR) and three genes termed *gag* (group associated antigen), *pol* (polymerase), and *env* (envelope) (Figure 24.16). The life cycle of a retrovirus involves the following: (1) infection of the cell, (2) reverse transcription of the viral RNA genome into DNA, (3) integration of this DNA into the host genome, (4) transcription of this integrated viral-derived DNA driven by the powerful retroviral promoter to produce the viral mRNA and RNA genome, and (5) translation of viral RNA to produce viral proteins that package the RNA genome creating new infectious retroviruses that are then released from the cell.

Figure 24.16. Genome of a normal retrovirus and the Rous sarcoma retrovirus. Rous sarcoma virus (RSV) is an acute transforming retrovirus that contains an assimilated cellular gene called Src, which is responsible for the cancer-causing properties of the virus.

Acute transforming retroviruses are a unique subset of retroviruses that have acquired the ability to produce cancer in rodents and avian species in a relatively short time period (several weeks). Examples of acute transforming retroviruses include the Rous sarcoma virus (RSV), which produces sarcomas in chickens, and the Harvey murine sarcoma virus (Ha-MSV), which produces sarcomas in rats and mice. Studies utilizing RSV provided the first evidence that cellular genes, now termed oncogenes, exist in vertebrate cells.

RSV can produce sarcomas in chickens and also transform cells in culture. Cellular transformation in cultured cells involves the acquisition of characteristics of cancer cells (see Section 24.6.1 for more details). In the mid-1970s, a segment of the RSV genome that was not required for viral replication, but was essential for transformation, was identified. This segment of the RSV genome was hypothesized to contain a transforming gene, and this gene was termed *Src* (*Src* for sarcoma) (Figure 24.16). A working hypothesis was developed stating that upon RSV infection, the retrovirus incorporates into the host genome as described above. However, in the case of RSV, strong retroviral promoters drive transcription of Src to produce high levels of Src expression which results in cellular transformation. Accordingly, it was proposed that RSV-infected chickens would contain the *Src* gene in the chicken genome, while uninfected chickens would not. Using a probe that recognized the transforming segment of RSV (*Src*), Southern blot analysis was conducted. It revealed that the *Src* sequence was present as expected in infected chickens; surprisingly, it was also present in the genome of uninfected chickens! Subsequently, *Src* was detected in the genome of other uninfected vertebrate species as well as in *Drosophila*, indicating that the Src genetic sequence was conserved in evolution for millions of years! This conundrum was solved when it was determined that a cellular gene, a proto-oncogene termed c-*Src* (c for cellular), had been transduced (a rare recombination event between the retroviral genome and the *Src* gene) into the genome of the RSV retrovirus. Thus, the cellular Src gene had been assimilated from the host somewhere back in time, into the virus. It is the assimilated cellular Src gene that is responsible for the oncogenic activity of RSV. This discovery was tremendously important because it demonstrated that a single gene could transform a cell and that this gene was contained within the vertebrate genome. This led many investigators to examine other acute transforming retroviruses for the presence of transduced cellular genes. Approximately 30 distinct cellular oncogenes were discovered due to their transduction into acute transforming retroviruses. The acute transforming retroviruses were instrumental in the identification of numerous cellular oncogenes.

Cellular proto-oncogenes/oncogenes are labeled with the prefix c- (i.e., c-*Ha-Ras* and c-*Src*), where c-indicates a cellular gene within the cellular genome and *Ha-Ras* and *Src* are the gene names. This designation distinguishes cellular genes from their viral counterparts where a v-prefix is used—for example, v-*Src*, where v-indicates that the gene is present (tranduced cellular gene) in the viral genome. If the v- or c-designation is not indicated, then it is understood the reference is to the cellular gene. With the exception of RSV, all of the acute transforming retroviruses are replication-deficient because a portion of their genome has been replaced with the transduced cellular gene. Their replication is dependent upon co-infection with a helper retrovirus, which provides the proteins encoded by the deleted genes in the acute transforming retrovirus.

24.7.2   Weak Transforming Retroviruses

Another group of retroviruses, known as weak transforming retroviruses, also produce cancer predominately in avian and rodent species. However, the time required for cancer development is longer than that of the acute transforming retroviruses and occurs over a period of many months to a year. These weak oncogenic retroviruses do not contain cellular genes; instead they disrupt normal regulatory elements that control the expression of cellular proto-oncogenes. These viruses insert into the host DNA near a proto-oncogene where their strong retroviral promoters drive transcription of a cellular proto-oncogene. Avian leukosis virus (ALV) produces B-cell lymphomas in birds, and it was determined that the virus inserts itself adjacent to the c-*Myc* gene in the host genome and causes a 50- to 100-fold increase in the expression of c-Myc. This event is termed insertional mutagenesis, or retroviral insertion. Retroviral insertion and subsequent activation of a proto-oncogene has led to the identification of additional cellular oncogenes.

24.7.3   DNA Oncogenic Viruses

Several DNA viruses are associated with, and are important in, the etiology of certain human cancers. For example, chronic infection with hepatitis B or C virus (HBV, HCV) is associated with liver cancer, infection with human papillomavirus (HPV) type 16 and type 18 is associated with cervical cancer, and infection with Epstein–Barr virus (EBV) is associated with Burkitt's lymphoma. About one-fifth of human cancer worldwide is associated with DNA oncogenic viruses. In 2006, the FDA approved a vaccine against HPV 16 and 18, which are responsible for about 70% of cervical cancer cases.

Unlike retroviruses, the normal life cycle of DNA viruses, like HBV and HPV, does not involve integration into the host DNA to replicate their genome. Most DNA viruses do not possess all of the genes necessary for DNA synthesis. Upon infection, the virus expresses its "early" genes and these cause the infected quiescent cells to enter S-phase and begin DNA replication. The virus then uses the host cell's replication machinery to provide an adequate supply of precursors and enzymes needed for viral replication. Subsequently, additional viral genes, referred to as "late" genes, are expressed and encode both capsid proteins and proteins involved in the lytic response. After virus production and assembly within the host cell, cell lysis occurs, releasing infectious virus (lytic response).

In some rare cases, certain DNA viruses can cause cell transformation. Transformation is accompanied by integration of the viral genome into the cell's genome and by the loss of the ability of the virus to produce a lytic response. It is the virus's genes that drive the cancer process. Several of the viral "early genes" such as *SV40 large T-antigen, polyoma large and middle T-antigen, adenovirus E1A and E1B*, and *papillomas virus E6 and E7*, have transforming ability. As will be discussed in the tumor suppressor gene section, some "early gene" protein products of DNA viruses have the ability to bind to and inactivate important growth regulatory proteins encoded by the tumor suppressor genes, *p53* and *Rb*.

24.8 CELLULAR ONCOGENES

As mentioned above, specific genes found in normal cells, termed proto-oncogenes, can be mutational targets for chemical carcinogens. Proto-oncogenes are highly conserved in evolution, and their expression is tightly regulated. Their protein products generally function in the control of normal cellular proliferation, differentiation, or apoptosis. Gain-of-function mutations in these proto-oncogenes produce dominant transforming oncogenes that have an important role in cancer development.

24.8.1   Cell Transformation; Role of Carcinogens and Oncogenes

Treatment of cell lines in culture with carcinogenic and mutagenic agents can induce cell transformation. The characteristics of transformed cells in culture resemble those of cancer cells in culture. The general characteristics of transformed cells are: (1) anchorage independence (ability to grow in soft agar), (2) loss of contact inhibition (ability to pile up and form foci), (3) immortality, (4) a reduced requirement for growth factors and/or serum, and (5) the ability to produce tumors when injected into immunodeficient mice (mice that lack a thymus and therefore do not have T-lymphocytes). As described previously, chemical mutagens and carcinogens produce DNA damage in the form of point mutations, frameshift mutations, and chromosome aberrations in mammalian cells. While specific carcinogen-induced DNA adducts were isolated in the 1970s, it was not until the early to mid-1980s that specific genes mutated by chemical carcinogen treatment were identified. A critical technical advance, termed "transfection," occurred in the early 1970s and aided in the identification of cellular oncogenes. This technique allowed researchers the ability to transfer genes/DNA into cultured cells.

In the late 1970s, researchers treated a mouse cell line with the known mutagen and carcinogen, 3-methylcholanthrene (3MC), a component of coal tar (remember Sir Percival Pott!). Treatment resulted in cell transformation. DNA was extracted from the 3MC transformed cells and transfected into nontransformed NIH3T3 cells (mouse fibroblast cell line). DNA from chemically induced transformed cells, but not DNA from untreated cells, resulted in the transformation of the recipient NIH3T3 cells. This experiment and others indicated that normal cellular genes could be altered/mutated by carcinogen treatment, and these altered/mutated genes could produce cell transformation. Importantly, DNA extracted from a variety of human tumors could also transform NIH3T3 cells. Could the cellular genes that were identified in acute transforming retroviruses be responsible for the transformation of the NIH3T3 cells? The answer was a resounding yes! For example, the Ras oncogene was identified as the cellular gene responsible for the transformation of NIH3T3 cells following transfection with human DNA isolated from a human bladder carcinoma cell line. Soon after, it was discovered that the Ras oncogene present in the human bladder carcinoma line had a single point mutation that was responsible for Ras's transforming activity.

Subsequently, Ras was found to be mutated in chemical carcinogen-induced rodent tumors and in many types of human cancer. Ras is a member of a larger group of normal cellular genes, termed proto-oncogenes, that can be altered to become oncogenes. Of the 200 oncogenes that have been identified in the past three decades, approximately 30 of these cellular oncogenes are frequently linked to various human cancers. Some oncogenes, such as members of the Ras family, are frequently mutated in a variety of different types of cancers including pancreas, thyroid, colorectal, lung, liver, and acute myelogenous leukemia. Others, such as the c-abl/bcr fusion protein, are specific to one type of cancer—in this case, chronic myelogenous leukemia.

24.8.2   Activation of Proto-oncogenes to Oncogenes

Proto-oncogenes can be activated to oncogenes either qualitatively or quantitatively. Proto-oncogenes can be activated to oncogenes by mutation, chromosomal translocation, gene amplification, or promoter insertion (Figure 24.17). Frequent qualitative changes include: (i) point mutations in the coding region of a proto-oncogene, (ii) generation of a hybrid or chimeric gene that results from chromosomal translocation, and (iii) sequence deletion. The best example of an oncogene that is activated by point mutation is Ras. The specific location of these point mutations within the coding region of the *Ras* gene, as well as the chemical carcinogens that produce these mutations, will be described in a later section. An example of an oncogene that is activated by chromosomal translocation and results in the production of a hybrid protein is *ABL*. In 95% of the cases of human chronic myelogenous leukemia (CML), an abnormal chromosome, termed the Philadelphia chromosome, is present. The Philadelphia chromosome results from a translocation between chromosomes 9 and 22. The translocation juxtaposes a portion of the tyrosine kinase proto-oncogene *ABL*, located on chromosome 9 with a portion of chromosome 22

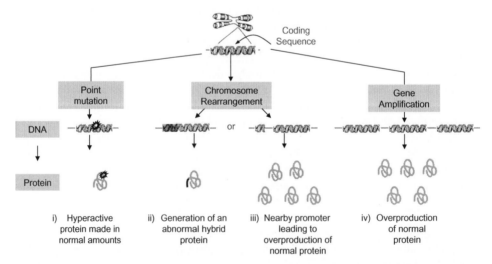

Figure 24.17. Activation of a proto-oncogene to an oncogene. Proto-oncogenes can be activated to oncogenes by; i) point mutation to produce an abnormal protein, ii) chromosomal rearrangements leading the generation of an abnormal hybrid protein, iii) chromosomal rearrangements leading to overproduction of a normal protein or iv) by gene amplification leading to the overproduction of a normal protein.

where the BCR gene is located. *BCR* encodes a protein that has serine/threonine kinase activity and also has a region homologous to guanine nucleotide exchange proteins. The translocation results in a fusion gene that produces a Bcr/Abl hybrid mRNA, resulting in a hybrid protein, where the amino terminus region is derived from the Bcr protein and the carboxy terminus region from the Abl protein. This fusion protein possesses both growth-promoting properties and elevated tyrosine kinase activity. Chromosomal translocations resulting in fusion proteins are especially prominent in leukemias and lymphomas.

Proto-oncogenes can also be activated by quantitative changes that led to the aberrant overexpression of an unaltered normal proto-oncogene. Common quantitative changes include gene amplification (increase in the number of copies of a given gene within the chromosome) and chromosomal translocation of a proto-oncogene. In some mammary carcinomas, *ERBB2/HER2* is amplified up to 30 times. Amplification of MYC occurs in some human carcinomas and leukemias and N-MYC is amplified in neuroblastomas. In Burkitt's lymphoma, the *MYC* gene is translocated and inserted near an immunoglobulin locus. This juxtaposition allows for the deregulation of *MYC* expression. Its expression is now under the control of the immunoglobulin promoter, allowing it to be constitutively expressed at high levels. Translocation of *MYC* from chromosome 8 to an immunoglobulin region in either chromosome 14, 22, or 2 is a consistent feature found in the great majority of Burkitt's lymphoma tumor biopsies (Figure 24.18). Other mechanisms that result

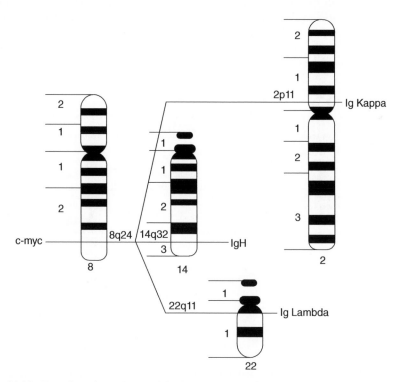

Figure 24.18. Translocation of c-myc in Burkitt's lymphoma. c-Myc is translocated to an immunoglobulin region on either chromosome 22, 14, or 2, and its expression is now under the regulation of the immunoglobulin promoter.

in increased proto-oncogene expression can involve alterations in chromatin such as histone acetylation and the hypomethylation of cytosine residues in CpG islands within the promoter region.

24.8.3   Oncogenes and Signal Transduction

Signal transduction pathways are used by cells to receive and process information to produce a cellular response. Signal transduction pathways represent the cellular circuitries that are involved in conveying specific information from the outside to the inside of the cell. This information must be conveyed across the cell membrane and then through the cytoplasm to the nucleus, where specific genes are expressed that carry out the cellular response (Figure 24.19). Generic components of the circuits include growth factors, growth factor receptors, GTPases, kinases, and transcription factors. Cell proliferation is regulated by extracellular signals known as growth factors, and these factors mediate their proliferative effect through binding to and activating membrane-bound receptors. These receptors convey this information to GTPases and then to kinases that phosphorylate and activate transcription factors. Transcription factors regulate the expression of specific genes that produce the evoked biological response—in this example, cell proliferation. Oncogenes encode proteins that are critical components of signaling circuits. If a component of the circuit is altered, then the entire cellular circuit is altered. It is not difficult to imagine how an alteration in a pathway that regulates an important cellular process such as proliferation could have profound effects on cellular homeostasis. This is in fact the molecular basis of how oncogenes contribute to the cancer process.

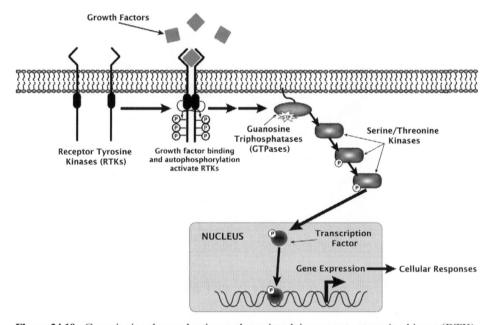

Figure 24.19. Generic signal transduction pathway involving receptor tyrosine kinase (RTK). An extracellular growth factor signal is conveyed via receptors, GTPases (Ras), kinases, and ultimately to transcription factors that alter gene expression and produce a cellular response.

TABLE 24.7. Human Oncogene Classification

Oncoprotein Families	Oncogenes
Growth factors	PDGF, HGF, TGF-α, VEGF, WNT-1, IGF-2
Receptor tyrosine kinases (RTKs)	ERBB1,[a] ERBB2,[b] KIT, RET, MET[c]
Nonreceptor tyrosine kinases	SRC, ABL, YES, LCK
Guanosine triphosphatases (GTPases)	H-RAS, K-RAS, N-RAS
Serine/threonine kinases	RAF-1, B-RAF, AKT, PIM-1, BCR
Transcription factors	MYC, FOS, JUN, ETS, REL, MYB, GLI, E2F1
Survival proteins	BCL-2, AKT, E2F1, MDM2

[a]ERBB1 is also known as the EGF receptor (EGF-R).
[b]ERBB2 is also known as Her2 or Neu receptor.
[c]MET is also known as the HGF receptor (HGF-R).

24.8.4 Oncogene Classification

Oncogenes can be classified with respect to their biological function. As shown in Table 24.7, many protein products of oncogenes are involved in signal transduction and are represented by growth factors, growth factor receptors, small-molecular-weight GTPases, nonreceptor tyrosine kinases, serine/threonine kinases, and transcription factors. In addition, a number of oncogene protein products function to regulate cell survival (also known as anti-apoptotic proteins). Examples of some oncogenes and their functions are described below.

24.8.4a   Growth Factors as Oncogenes. Transforming growth factor alpha (TGFα) binds to the epidermal growth factor receptor (EGFR) and stimulates epithelial cell proliferation and survival. Because TGFα stimulates cell proliferation, it is termed a mitogen. TGF-α is overexpressed in many carcinomas (lung, breast, prostate, etc.) that express EGFR. Thus, in these tumor cells, TGF-α functions in an autocrine manner to stimulate the proliferation of the tumor cells that produced it. TGF-α expression is also increased in some chemically induced rodent liver and skin tumors. Insulin-like growth factor 2 (IGF-2) is a ligand for the insulin like growth factor receptor-1R (IGF-1R). IGF-2 is overexpressed in some tumors, such as colorectal tumors, and functions as an autocrine factor to promote cell proliferation and survival. Tumor cells often overexpress multiple growth factors, and these cooperate to function in an autocrine manner to regulate tumor cell proliferation and survival. Some tumor promoters stimulate the expression of growth factors such as TGF-α.

24.8.4b   Receptor Tyrosine Kinases as Oncogenes. Many growth factors mediate their proliferative effects through receptors that possess intrinsic tyrosine kinase activity. These receptors are referred to as receptor tyrosine kinases (RTK). RTKs consist of three general domains: an extracellular N-terminal domain that binds to its specific ligand; a transmembrane domain that spans the plasma membrane; and an intracellular C-terminal domain that contains intrinsic tyrosine kinase activity. EGFR is the prototype of RTK signaling. Binding of epidermal growth factor (EGF) or TGFα to EGFR causes EGFR dimerization. This is accompanied by activation of its tyrosine kinase activity. The activated kinase domains then trans-

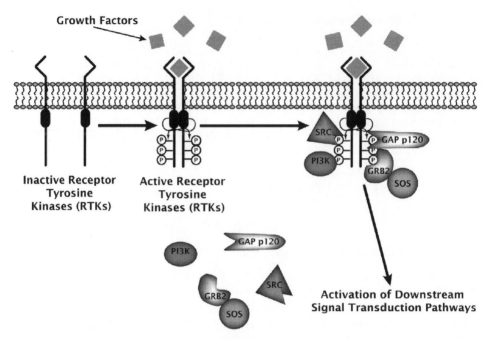

Figure 24.20. Receptor tyrosine kinase signaling. Activation of RTKs results in autophosphorylation, and these phosphorylated tyrosines serve as docking sites for proteins (PI3K, GRB2, GAPp120, SRC) containing SH2 domains. Once docked, these proteins become active, thus resulting in the activation of downstream signaling proteins.

phosphorylate tyrosine residues within the intracellular domain of each receptor (Figure 24.20). Cytosolic proteins containing specialized regions, termed src-2 homology domains (SH2 domains), recognize these phosphorylated tyrosine residues on the intracellular domain of the receptor and bind/dock. Some examples of proteins that contain SH2 domains and bind/dock to phosphotyrosines of the RTK include: enzymes such as phospholipase Cγ (PLCγ), phosphatidylinositol-3-kinase (PI3K), SRC, GTPase activating protein (GAP p120), the tyrosine phosphatase SHP2, and the adaptor proteins, SHC and GRB2. These adapter proteins do not have catalytic activity, but function by recruiting additional proteins to the activated receptor. GRB2 contains SH3 domains that recognize proline rich motifs in signaling proteins. GRB2 is bound via its SH3 domains to the protein Son of sevenless (SOS), which is a guanine nucleotide exchange factor (GEF). GRB2, when docked on the receptor, brings SOS into the proximity of Ras, which is associated with the inner leaflet of the plasma membrane. SOS participates in the activation of RAS, as will be discussed in Section 24.8.4d.

There are at least 52 members of the RTK family, and approximately 30 members have been shown to be activated to oncogenes by mutation, gene rearrangements/translocations, or amplification. For example, EGFR, also known as ERBB1, is overexpressed (gene amplification) in numerous epithelial cancers including non-small-cell lung carcinoma, breast cancer, and ovarian cancer. ERBB2/HER2, another member of the EGFR/ERBB family, is overexpressed in 30% of breast cancers. RET is an RTK that is required for the development of kidneys and the enteric system as well as neuronal differentiation and survival. In response to ligand binding, RET forms a heterodimer with a glycosylphosphatidylinositol (GPI)-linked cell surface receptor. Somatic rearrangements result in fusions between the N-terminus of various proteins and the tyrosine kinase domain of RET, leading to ligand-independent constitutive activation of the receptor. This rearrangenement is important in the development of thyroid carcinoma. Lastly, more than 30 different gain-of-function somatic mutations have been identified in KIT (RTK receptor), and these mutant forms of KIT are associated with the following human malignancies: gastrointestinal tumors, acute myeloid leukemias, myelodysplastic syndromes, and small-cell lung carcinoma.

24.8.4c Non-Receptor Tyrosine Kinases as Oncogenes.

The SRC family of non-receptor tyrosine kinases (SRC, FES, FGR, FPS, LCK, and YES) contain both SH2 and SH3 domains and are associated with the inner surface of the plasma membrane. Many of these non-receptor tyrosine kinases are rapidly activated by a variety of transmembrane signaling receptors, including RTKs. Some of these kinases are constituitively active in various tumor types. For example, SRC kinase activity is elevated in some human colon cancers. As described above, the ABL proto-oncogene is a cytoplasmic/nuclear tyrosine kinase. Following a chromosomal translocation, *ABL* forms a hybrid gene with the *BCR* gene to produce a fusion hybrid protein with increased tyrosine kinase activity and growth promoting properties. This hybrid protein is present in 95% of human chronic myelogenous leukemia cases.

24.8.4d Guanosine Triphosphatases as Oncogenes.

In humans, the Ras superfamily of small guanosine triphosphatases (GTPases) contain over 150 members. Based on sequence and functional information, these are divided into the following five major branches: RAS, RHO, RAB, RAN, and ARF. Within the RAS subfamily/branch, only *H-RAS, K-RAS*, and *N-RAS* are frequently mutated in human cancer and carcinogen-induced animal tumors. Approximately 25% of all human tumors contain mutated *RAS*. In humans, members of the Ras subfamily (*H-*, *K-*, and *N-RAS*) are mutated in particularly high frequency in tumors of the pancreas (90%), thyroid (55%), colorectal (45%), lung (adenocarcinoma 35%), liver (30%), and acute myelogenous leukemia (30%).

H-RAS, K-RAS, and *N-RAS* genes code for highly related proteins of 188–189 amino acid residues, generically known as p21 (21 kD). *Ras* genes are present in all eukaryotic organisms from yeast to humans. Mammalian *RAS* can rescue Ras deficient yeast cells, and yeast Ras can transform NIH3T3 cells. The fact that *Ras* genes are highly conserved in evolution suggests that the protein product they encode plays an essential role in normal cellular physiology.

RAS proteins function as membrane associated binary molecular switches operating downstream of a variety of membrane receptors including receptor tyrosine kinases (i.e., PDGF, EGFR) (Figure 24.21), cytokine receptors without intrinsic tyrosine kinase activity (IL-12), T-cell receptors, and some receptors that are coupled to heterotrimeric G-proteins. RAS proteins are membrane-bound proteins that bind guanosine triphosphate (GTP) and guanosine diphosphate (GDP) (Figure 24.21). RAS proteins are in the "off" position when bound to GDP and in the "on" position when bound to GTP. When stimulated by an upstream signal, GDP is exchanged

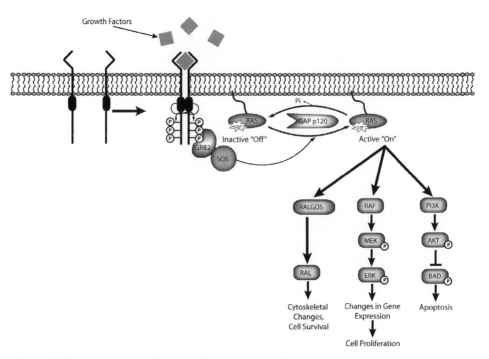

Figure 24.21. Activation of Ras signaling pathways. RAS proteins function as membrane-associated molecular switches operating downstream of a variety of membrane receptors. RAS proteins are in the "off" position when bound to GDP and in the "on" position when bound to GTP. When stimulated by an upstream signal, GDP is exchanged for GTP on the Ras protein. This exchange reaction is mediated by the guanine nucleotide exchange factor, SOS. RAS bound to GTP (on position) undergoes a conformational change allowing it to interact with its effector proteins (RAF, PI3K, RALGDS) in the cell. The effector proteins transmit the signal downstream to eventually produce a biological response. RAS is turned off through the hydrolysis of GTP to GDP. This is mediated through RAS's intrinsic GTPase activity with assistance from the GAPp120 protein (GTPase activating protein).

for GTP on the Ras protein. This exchange reaction is mediated by SOS, a member of a family of proteins referred to as guanine nucleotide exchange factors (GEFs). RAS bound to GTP (on position) undergoes a conformational change allowing it to interact with its effector proteins in the cell. The effector proteins transmit the signal downstream to produce the appropriate biological response. RAS is turned off through the hydrolysis of GTP to GDP. This is mediated through RAS's intrinsic GTPase activity with assistance from the GAPp120 protein (GTPase activating protein), which increases the rate of RAS-mediated GTP hydrolysis by 2–3 orders of magnitude.

Several downstream effectors of RAS have been identified. The three best-characterized ones are RAF, phosphatidylinositol-3-kinase (PI3K), and RalGDS (Figure 24.21). Ras-GTP binds to and stimulates RAF, a cytoplasmic serine/threonine kinase. RAF phosphorylates MEK1 (MAP kinase-extracellular signal regulated kinase), a dual-specificity kinase that phosphorylates threonine and tyrosine residues on ERK1 (p44) and ERK2 (p42) (extracellular regulated MAP kinase). ERK1/2 are serine/threonine kinases that have numerous cellular substrates, including cPLA2 and p90 ribosomal S6 kinase (RSK) and numerous transcription factors, such as ELK-1, SAP-1, and ETS. Importantly, RAS signaling leads to the increased transcription and expression of a variety of genes including Cyclin D1. The induction of Cyclin D1 by RAS is important in propelling cells from G1 into the S-phase of the cell cycle through the activation of CDK4/6 and phosphorylation of RB (see Section 24.9.1). Cyclin D1 is overexpressed in numerous tumors through RAS-dependent and independent pathways. When overexpressed, Cyclin D1 is considered an oncogene.

PI3K is another important RAS effector. PI3K is a lipid kinase and is composed of a regulatory/adaptor subunit (p85) and a catalytic subunit (p110). RAS-GTP binds to the catalytic subunit and stimulates PI3K activity. PI3K phosphorylates the number 3 position of the inositol ring of phosphatidylinositol (PIP2), which is located on the inner leaflet of the plasma membrane. This lipid product, phosphatidylinositol (3,4,5) triphosphate (PIP3), recruits cytoplasmic proteins that have a pleckstrin homology (PH) domain to the membrane. AKT (also known as PKB) is a serine/threonine kinase that contains a PH domain that targets it to the plasma membrane, where it is activated by phosphorylation by other PH containing kinases. Activated AKT then phosphorylates numerous key proteins involved in promoting cell survival, cell proliferation, and cell growth (protein synthesis) (Figure 24.22).

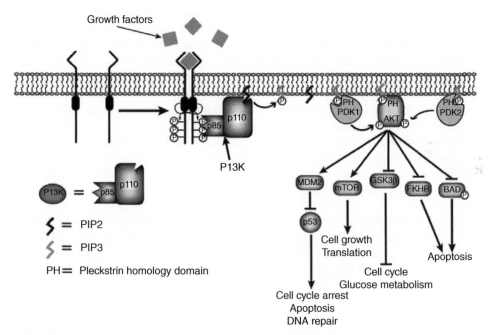

Figure 24.22. Phosphatidylinositol-3-kinase (PI3K) signaling pathway. PI3K can be activated by RTKs (or Ras) to produce phosphatidylinositol (3,4,5)-triphosphate (PIP3), which serves as a docking site for proteins containing a pleckstrin homology (PH) binding domain. PIP3-dependent kinase 1 (PDK1), PIP3-dependent kinase 2 (PDK2) and AKT (also known as PKB) all contain a PH domain. Subsequent to docking, AKT is activated by phosphorylation by PDK1/2 and, in turn, phosphorylates and regulates numerous proteins including glycogen synthase kinase 3 (GSK3), MDM2, forkhead (FKHR transciption factor), mTOR (regulates protein synthesis), and BAD (kinase).

RAL nucleotide guanine dissociation stimulator (RALGDS) is a guanine nucleotide exchange factor (GEF) that functions as an important mediator between Ras signaling and RAL activation. RAL is another small guanosine triphosphatase (GTPase), similar to RAS. RALGDS catalyzes the exchange of GDP for GTP on the RAL protein, activating its ability to regulate downstream target proteins. Activation of RAL is important in the regulation of cellular processes such as endocytosis, exocytosis, actin cytoskeletal rearrangements, and cell survival. Notably, RALGDS was recently discovered to be important in tumorigenesis involving mutant Ras, using the mouse skin model of tumorigenesis.

Oncogenic point mutations in RAS most often occur in the 12th, 13th, or 61st codon. These mutations impair RAS's intrinsic GTPase activity and also render GAPp120 ineffective at increasing GTP hydrolysis. The net result of these oncogenic mutations is that RAS remains in the GTP bound "on" state producing constitutive stimulation of its downstream pathways. Mutated *Ras* has been detected in animal tumors induced by diverse chemical carcinogens. As described in a previous section, many chemical carcinogens bind covalently to DNA; and if the damaged DNA is replicated/misrepaired, this can produce specific mutational alterations in the DNA. The study of the *Ras* proto-oncogene as a target for chemical carcinogens has revealed a correlation between specific carcinogen–DNA adducts and specific activating mutations of *Ras* in chemically induced tumors. For example, 7,12-dimethyl benz[*a*]anthracene, a polycyclic aromatic hydrocarbon carcinogen, is metabolically activated to a bay region diol epoxide that preferentially binds to adenine residues in DNA. Skin tumors isolated from mice treated with DMBA contain a mutated H-*Ras* oncogene with an A to T transversion of the middle base in the 61st codon of H-*Ras*. Therefore, the identified mutation in *Ras* is consistent with the expected mutation predicted from the DMBA-DNA adduct formation. Likewise, rat mammary carcinomas induced by nitrosomethylurea contain a G to A transition in the 12th codon of H-*Ras*, and this mutation is consistent with the modification of guanine residues by this carcinogen. In many animal tumor models, mutations in *Ras* have been detected soon after carcinogen treatment, indicating that mutational activation of *Ras* by specific chemical carcinogens in rodent tumor models is an early and causal initiating event in carcinogenesis.

24.8.4e Serine/Threonine Kinases as Oncogenes. BRAF, a member of the RAF family, is a cytosolic serine/threonine kinase that is frequently mutated (60%) in human malignant melanoma. The gain-of-function mutation results in a BRAF protein with elevated kinase activity which is thought to promote cell proliferation. AKT is a serine/threonine kinase (discussed in previous section), whose kinase activity is elevated in numerous tumors because of alterations in upstream regulatory proteins, including mutation of RAS, over-active RTKs, mutated PI3K or loss of PTEN. PTEN is a tumor suppressor gene that dephosphorylates phosphatidylinositol (3,4,5) triphosphate (PIP3) to inactivate its downstream signaling pathways. The AKT gene is also amplified in ovarian, breast and some gastric tumors. AKT regulates pathways involved in survival, proliferation and cell growth (protein synthesis).

24.8.4f Transcription Factors as Oncogenes. FOS is a phosphoprotein that heterodimerizes with JUN to form the transcription factor, AP-1. AP-1 binds to the promoters of genes containing an AP-1 consensus sequence [also referred to as the

TPA responsive element (TRE)] and increases their expression. FOS is over-expressed in the majority of human osteosarcomas, while JUN is overexpressed in some lung cancers. In Burkitt's lymphoma, the transcription factor MYC is translo-cated and inserted near an immunoglobulin locus, where it falls under the regulation of an immunoglobulin promoter (Figure 24.18). This results in constitutive expres-sion of high levels of MYC. In addition, various members of the MYC family are amplified in a variety of human and rodent tumors.

24.8.4g Oncogenic Proteins Involved in Cell Survival. BCL-2 has an anti-apoptotic function. In certain B-cell lymphomas, BCL-2 undergoes chromosomal translocation resulting in its overexpression and inhibition of apoptosis. MDM2, another oncogene involved in cell survival, is a protein that targets the tumor sup-pressor gene, p53, for proteosomal degradation. p53 regulates apoptosis in response to DNA damage and oncogene activation. MDM2 is amplified in some human cancers, resulting in the downregulation of p53 and survival of cells with DNA damage and/or activated oncogenes. As described, the serine/threonine kinase, AKT, is an oncogene that plays a critical role in cancer cell survival.

24.8.5 Oncogene Cooperation

Cells harvested directly from an animal and placed in culture (referred to as primary cells) are not immortal, and these cells undergo a limited number of cell divisions before becoming senescent (permanent growth arrest). These cells cannot be trans-formed by a single oncogene, however, certain oncogenes can cooperate with one another to transform primary rodent cells. For example, transfection of *Myc* or *Ras* alone into rat primary embryo fibroblasts does not produce transformation. However, when both *Myc* and *Ras* are simultaneously introduced, the primary cells become transformed and are tumorigenic. Experiments like these demonstrated that certain oncogenes can cooperate with others to transform primary rodent cells. For example, oncogenes such as *H-Ras, K-Ras, N-Ras, Src*, and *Polyoma Middle T*, can cooperate with any one of the oncogenes *Myc, N-Myc, L-Myc, Adeno E1A, Polyoma Large T, SV40 Large T*, and *Papillomavirus E7* to transform rat embryo fibroblasts. Oncogene cooperation has also been demonstrated in vivo through the use of transgenic mice containing two oncogenic transgenes. For example, mice doubly transgenic for *Myc* and *Ras* (under the regulation of the mouse mammary tumor virus promoter) display a synergistic increase in the incidence of mammary carcinomas compared to mice that contain only *Ras* or *Myc*.

While primary rodent cells can be transformed by two cooperating oncogenes, human cells require the co-transfection of at least three genes. Transfection of normal human epithelial cells with the catalytic subunit of telomerase in combina-tion with the *SV40 Large-T Antigen* and mutant *H-Ras* results in transformation and tumorigenic conversion. Telomerase is expressed in germ cells and some cancer cells. Telomerase inhibits senescence by preventing the shortening of chromosome telomeres. As described below, *SV40-large-T antigen* is a viral oncoprotein that binds to and inactivates the tumor suppressor genes RB and p53. The fact that it takes multiple oncogenes to transform primary human cells supports the early epidemio-logical studies that indicated that multiple hits or mutations are required for cancer development in vivo.

24.9 TUMOR SUPPRESSOR GENES

Gain-of-function mutations in proto-oncogenes result in their activation, while loss-of-function mutations in tumor suppressor genes result in their inactivation. Tumor suppressor genes are sometimes termed anti-oncogenes, recessive oncogenes, or growth suppressor genes. Tumor suppressor genes often encode proteins that function as negative regulators of cell proliferation or positive regulators of cell death. In addition, some tumor suppressor genes function in differentiation and cellular nutrition assessment. Tumor suppressor genes were first identified in rare familial cancer syndromes, and it is now known that some are frequently mutated in sporadic cancers through somatic mutation. Major tumor suppressor genes, their proposed function, and the cancer syndrome they are associated with are shown in Table 24.8. When tumor suppressor genes that negatively regulate cell proliferation are inactivated by allelic loss, point mutation, chromosome deletion or epigenetic silencing, the result is uncontrolled cell proliferation. Generally, if one allele of a tumor suppressor gene is inactivated, the cell is normal (this gene is referred to as haplosufficient). However, when both alleles are inactivated, the ability to control cell proliferation is lost. In some cases, a single "inactivated" allele of certain tumor suppressor genes such as p53 can produce an intermediate phenotype (this gene is referred to as haploinsufficient).

The concept that cancer is a recessive trait and involves loss of gene function arose from somatic cell fusion experiments conducted by H. Harris and colleagues in the late 1960s. When tumor cells were fused with normal cells, the resulting hybrids were nontumorigenic. These cell hybrid experiments suggested that within the hybrid cell, the normal cell contributed genes to the tumor cell that imposed growth restraints on the latter and also suggested that cancer is a recessive trait.

TABLE 24.8. Human Tumor Suppressor Genes

Gene Name	Familial Cancer Syndrome	Protein Function	Sites/Types of Commonly Associated Neoplasms
TP53	Li–Fraumeni syndrome	Transcription factor	Most human cancers
RB1	Hereditary retinoblastoma	Transcriptional modifier	Retinoblastoma, osteosarcoma
APC	Familial adenomatous polyposis	β-catenin degradation	Colon, stomach, intestine
CDKN2A (p16^{INK4A})	Familial malignant melanoma	Cyclin-dependent kinase inhibitor	Melanoma, pancreas
CDKN2A (p14^{ARF})		p53 stabilizer	Melanoma
PTCH	Gorlin syndrome	Transmembrane receptor	Basal cell carcinoma, ovary, heart, medulloblastoma, meningioma
PTEN	Cowden syndrome	PIP3 phosphatase	Hamartoma, glioma, uterus
TGFBR2		Transmembrane receptor	Colon, stomach, ovarian

These genes were called tumor suppressor genes because the transferred "genes" suppressed growth and/or tumorigenicity.

Evidence for tumor suppressor genes in human cancer came from the studies of Knudson. In the early 1970s, Knudson postulated that two mutational events or "hits" are necessary for the development of retinoblastoma, a rare tumor of the retina that occurs in 1/20,000 infants and children. Approximately 60% of retinoblastomas are sporadic (nonhereditary) and 40% are familial (hereditary). Knudson observed that the majority of familial cases developed bilaterally (both eyes were affected) and much earlier than the sporadic form which developed unilaterally (one eye affected). To explain these differences between the familial and sporadic forms, Knudson proposed a model in which two hits were necessary for retinoblastoma to develop. He proposed that in the familial cases, one hit was inherited in the germ line and present in all somatic cells and the other hit occurred as a somatic mutation in the retinal lineage. In contrast to the familial form, he proposed that the sporadic form requires two somatic mutations or hits within the same cell in the retinal lineage. Knudson's two-hit hypothesis for cancer causation was eventually substantiated by molecular and cytogenetic studies. This led to the identification of the first tumor suppressor gene, termed the retinoblastoma gene (RB), and the discovery that both copies of the gene (two mutational events) are inactivated and/or deleted in retinoblastoma tumors. Thus, a child that carries an inherited mutant RB allele in all somatic cells is normal except they have an increased risk of developing cancer. This child is heterozygous at the RB locus (one mutant and one normal allele). If the normal allele is inactivated by mutation, or is replaced with the mutant allele through mitotic recombination or gene conversion, then this cell is now homozygous for the mutant allele, resulting in a predisposition for cancer development. Mitotic recombination and gene conversion result in the replacement of the normal allele and neighboring chromosomal region with the mutant allele and its neighboring chromosomal area. This is referred to as loss of heterozygosity (LOH) because both chromosomes now share the same genetic sequence. Mitotic recombination occurs at a frequency of 10^{-4}–10^{-5} per generation per cell, whereas the frequency of mutation of a gene per generation is 10^{-6}. Therefore, mitotic recombination/gene conversion, rather than mutational inactivation of the remaining normal tumor suppressor allele, is more common in tumorigenesis.

24.9.1 Retinoblastoma Gene and the Cell Cycle

The retinoblastoma (*RB*) gene is a large gene of 190 kilobases. Individuals with an inherited *RB* mutant allele are susceptible to development of retinoblastoma, osteosarcomas, and soft tissue sarcomas. All these cancers require loss of the wild-type allele. In sporadic cancers including lung, bladder, esophageal, and breast, *RB* can be inactivated by deletion, nonsense point mutations, or promoter methylation. *RB* encodes a 105-kD nuclear protein (pRB) that controls cell cycle progression. pRB regulates the movement of cells through G_1 into the S phase of the cell cycle (see Figures 24.23 and 24.24). Regulation of $G_1 \rightarrow S$ phase by pRB represents a key control point in the cell cycle because once the cell has passed the restriction point in G_1 (a point in late G_1), the cell moves into S phase and is now irreversibly committed to replicate its DNA and progress through the entire cell cycle. The activity of pRB is regulated by phosphorylation on numerous sites within the protein; in G_0

and early G_1 pRb is hypophosphorylated and in late G_1, S, G_2 and M phases of the cell cycle pRb becomes increasingly phosphorylated or hyperphosphorylated. The hypophosphorylated form of pRb is the active form of pRb that suppresses cell proliferation. It is this form of pRb that binds to numerous cellular proteins (>150 proteins), many of which are transcription factors. Phosphophorylation of pRb inactivates its function.

The best-characterized pRb-binding proteins belong to the E2F family of transcription factors. With respect to the cell cycle, E2Fs are involved in the transcription of genes that are required for DNA synthesis (S-phase genes). The hypophosphorylated form of pRb suppresses cell cycle progression by binding to and inhibiting the transcription activity of E2F-1, -2, and -3. E2F1-3 bind to the promoters of their target genes. pRb binds to E2Fs and attracts histone deacetylases (HDACs) to the pRb/E2F/promoter complex. HDACs removes acetyl groups from histones, producing a chromatin configuration that is not compatible with RNA polymerase and transcription. Upon growth factor-induced mitogenic stimulus, the expression of Cyclin D1 is induced (often via a Ras-dependent pathway), and it forms a complex with cyclin-dependent kinases 4 and 6 (CDK4/CDK6) (Figure 24.23A,B). CDKs are serine/threonine protein kinases that are involved in controlling cell cycle progres-

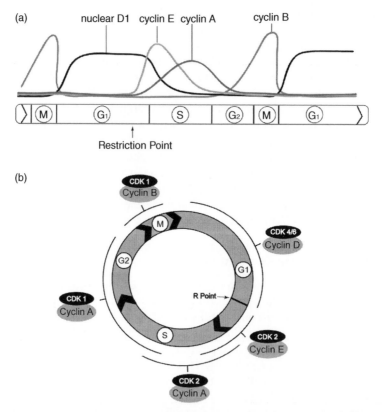

Figure 24.23. Cyclin expression is regulated according to the cell cycle. (A) The regulatory subunits of the cyclin-dependent kinases (CDKs) are referred to as "cyclins" because these proteins are synthesized and degraded during the cell cycle. (Adapted from Weinberg RA Biology of Cancer 2007.) (B) Specific cyclins bind to and activate specific CDKs, and this propels the cell through specific stages of the cell cycle.

sion and are constitutively expressed. The regulatory subunits of CDKs are called "cyclins" because these proteins are synthesized and degraded during the cell cycle (Figure 24.23A). CDKs are only active when bound to their specific cyclin. Cyclin-Cdk complex activity is greatly increased by phosphorylation of the "T-loop" of CDKs by CAK (CDK activating kinase). Thus, when Cyclin D binds to CDK4 or CDK6, these kinases are phosphorylated by CAK; this results in the further activation of CDK4 and CDK6, which then phosphorylate pRb in early to mid G_1 and thereby lead to the expression of cyclin E (Figure 24.24). Cyclin E then forms a complex with CDK2, leading to the activation of CDK2, and to the further phosphorylation of pRb. In addition the increased levels of the CDK4/6-Cyclin D complex also function as a sink for the CIP/KIP family of cyclin-dependent kinase inhibitors thereby relieving their inhibition on the existing CDK2-cyclin E complex. This results in hyperphosphorylation of pRB (Figure 24.24). Hyperphosphorylated pRB can no longer bind to E2F1-3. When E2Fs are bound to their target gene promoters in the absence of pRb, they then attract histone acetylase. Histone acetylation produces chromatin modifications compatible with E2F-mediated transcription of S-phase genes including dihydrofolate reductase, thymidine kinase, Cyclin E, and E2F1. The cell now moves from G_1 into S phase of the cell cycle. During the S and G_2/M phases, Cyclin A and Cyclin B are sequentially expressed, resulting in CDK2 and CDK1 activation, respectively (Figure 24.23B). These CDKs then propel the cell through S and G_2/M phases, eventually undergoing mitosis to produce a daughter cell. pRb is dephosphorylated at the termination of M-phase and is once again active and able to form complexes with E2Fs.

24.9.2 Cyclin-Dependent Kinase Inhibitors and the pRb Circuit

The phosphorylation of pRb by cyclin D/CDK4 and cyclin D/CDK6 can be blocked by the INK4 (<u>In</u>hibitor of CD<u>K4</u>) family of cyclin-dependent kinase inhibitors (Figure 24.24). The family consists of $p16^{Ink4a}$, $p15^{Ink4b}$, $p18^{Ink4c}$, and $p19^{Ink4d}$. These INK4 proteins block cell-cycle progression in G1 by inhibiting the Cyclin D-CDK4/6-mediated phosphorylation of pRB. $p16^{Ink4a}$ is a tumor suppressor gene that is inactivated in many different types of cancers. $p16^{Ink4a}$ expression is frequently silenced by promoter hypermethylation, and the loss of $p16^{Ink4a}$ expression results in deregulation of cell proliferation. The CIP/KIP family is another family of CDK inhibitors and is composed of $p21^{CIP1}$, $p27^{Kip1}$, and $p57^{KIP2}$. These proteins are potent inhibitors of Cyclin E/Cyclin A-CDK2 and Cyclin A/Cyclin B-CDK1 kinase activity (Figure 24.24). The CIP/KIP family can also inhibit Cyclin D/CDK4/6 kinase activity and are thus considered broad-spectrum inhibitors. At low intracellular protein levels, members the CIP/KIP family function as assembly factors for the cyclin D-CDK4/6 complexes. As will be described below, the p53 tumor suppressor gene can positively regulate pRb function through the induction of $p21^{CIP1}$, which prevents pRb phosphorylation, thereby blocking cell cycle progression.

24.9.3 pRB Is Altered in Tumorigenesis by Multiple Mechanisms

TGFβ, a potent growth inhibitory cytokine, inhibits cell proliferation by increasing the expression of $p15^{Ink4b}$, $p21^{CIP1}$, and $p57^{CIP2}$. As described above, these cyclin-dependent kinase inhibitors block pRB phosphorylation and cell cycle progression. The TGFβ circuit is impaired in some cancers, resulting in the inability of the cells to respond to the growth inhibitory effects of TGFβ. pRB can also be inactivated

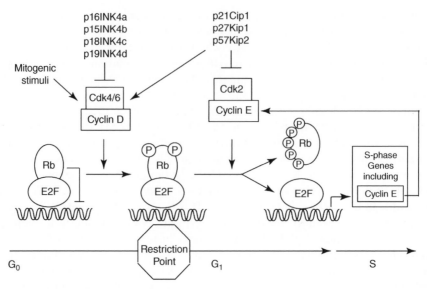

Figure 24.24. Retinoblastoma and cell cycle. Rb regulates the movement of cells through G_1 into the S phase of the cell cycle. Regulation of $G_1 \rightarrow$ S phase by Rb represents a key control point in the cell cycle because once the cell has passed the restriction point in G_1 (a point in late G_1), the cell moves into S phase and is now irreversibly committed to replicate its DNA and progress through the entire cell cycle. In early G1 phase, Rb binds to E2F and blocks E2F's ability to transcribe S-phase genes. Phosphorylation of Rb by Cdk4/6 and Cdk2 releases E2F from Rb and allows E2F to transcribe S-phase genes. The cell then enters S phase. (Adapted S. Ortega et al. *Biochimica et Biophysica Acta.* **1602**, 73–87, 2002.)

by proteins produced by DNA tumor viruses. As mentioned in Section 24.7.3, some DNA viruses have the ability to override the restriction point in G_1 and cause cellular proliferation. Transforming proteins encoded by DNA tumor viruses such as SV40 large T antigen, papilloma virus E7, and adenovirus type 5 E1A bind specifically to the hypophosphorylated form of pRb and block its ability to suppress cell proliferation. The binding of these proteins to hypophosphorylated pRb is a critical event in the transforming activity of SV40, adenovirus and papilloma virus.

The pRb circuit can be altered by numerous mechanisms including the mutation or deletion of *RB1*, hypermethylation of the RB1 promoter, inactivation of p16[Ink4a], overexpression of Cyclin D, loss of TGFβ function, or inactivation of pRB by DNA oncogenic viruses. It is noteworthy that most, if not all, cancers involve some alteration in the cyclin D/CDK4/6/Ink4a/RB/E2F circuit highlighting the importance of the "Rb circuit." Carcinogens and tumor promoters can also result in alterations in the Rb circuit. For example, carcinogen-induced mutations in Ras, or tumor promoter-induced increases in a growth factor such as TGFα, result in increased expression of Cyclin D.

24.9.4   *p53*

p53 (also termed Trp53 or TP53) is a 12.5-kb gene that encodes a 53-kD protein. *p53* is mutated in 50% of all human cancer and is considered to be the most frequently known mutated gene in human cancer. The majority (~80%) of *p53*

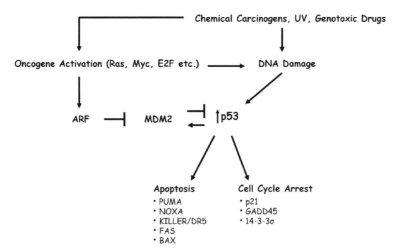

Figure 24.25. p53 regulates apoptosis and cell cycle progression. In response to DNA damage or oncogene activation, the p53 protein undergoes posttranslational modifications that increase its stability and activity. p53 accumulates in the cell and can regulate the expression of genes involved in apoptosis and cell cycle arrest. Oncogene activation is also believed to activate components of the DNA damage response pathway to further increase p53. MDM2 is a feedback inhibitor of p53 and targets p53 for proteasomal degradation.

mutations are missense mutations, and *p53* is mutated in approximately 80% of nonmelanoma skin cancers, 70% of small-cell lung cancer, and 50% of colon, bladder, ovary, and head and neck cancers.

p53 is a transcription factor that exists in the cell as a homotetramer. p53 regulates genes involved in producing cell cycle arrest, senescence, apoptosis, and angiogenesis (Figure 24.25). In normal cells, p53 protein levels are low; however, in response to DNA damage, hypoxia, or oncogenic stress, p53 protein levels rapidly accumulate. The accumulated p53 protein is active and, depending upon cellular circumstances, can lead to cell cycle arrest, apoptosis, senescence or inhibition of angiogenesis. With respect to DNA damage, p53 mediates a pause in cell cycle progression through the upregulation of p21^{CIP1}, GADD45, and 14-3-3 (checkpoint response) (see Chapter 23). This pause prevents the replication of damaged DNA and allows time for DNA repair. Once repaired, the cell resumes its progression through the cell cycle. If the DNA damage is too severe, p53 will induce apoptosis through the upregulation of genes that stimulate both the extrinsic and intrinsic apoptotic pathways (Figure 24.25). Oncogenic stress can lead to apoptosis (see p19^{Arf} section below) or to the permanent withdrawal from the cell cycle, termed senescence. Senescence likely represents a mechanism to prevent clonal expansion of potential tumor cells. Little is known regarding the genes that p53 regulates to induce senescence.

p53 has been called the "guardian of genome" because it has roles in the G1, S, and G2 checkpoints and can induce apoptosis and senescence. Inactivation of p53 disrupts these functions and leads to instability of the genome and survival of tumor cells. The p53 protein is composed of 393 amino acids. Surprisingly, p53 function can be inactivated by single missense point mutations, which can occur at over 100 different codons. However, 98% of *p53* mutations in tumors occur within codons 110–

307, which encode the p53 DNA-binding domain (amino acids 100–292). Mutational hot spots in p53 occur at codons 175, 245, 248, 249, 273, and 282. It has been proposed that the mutation spectrum of *p53* in certain human cancers may aid in the identification of the specific carcinogen that produced the genetic damage. In other words, some carcinogens produce signature mutations in p53. For example, certain mutations in p53 reflect oxidative damage, while other mutations in p53 found in hepatocellular carcinomas from individuals exposed to aflatoxin display a mutation spectrum characteristic of aflatoxin-induced DNA damage. In sun-exposed areas where skin tumors develop, p53 mutations are characteristic of UV-light-induced DNA damage; and finally, the mutation spectrum in p53 induced by (+)-benzo[*a*]pyrene-7,8-diol-9,10-epoxide-2 in cells in culture is similar to the mutational spectrum in p53 found in lung tumors from cigarette smokers. Thus, certain carcinogens produce a "mutational signature" in p53 that may have implications for tumor etiology. In addition to mutational inactivation, p53 function can also be inactivated by DNA tumor viruses. DNA oncogenic viral proteins, such as SV40 large T, adenovirus E1A and E1B, and human papillomavirus 16 E6, all bind to the wild-type p53 and inhibit the transcriptional function of p53.

Regulation of p53 protein levels and p53 transcription activity is complex, and approximately 10 regulatory feedback loops have been identified. p53 undergoes extensive posttranslational modification, involving phosphorylation and acetylation. DNA damage results in the activation of a number of kinases that phosphorylate p53, and this increases its short half-life of 15 min to 3–4 hr in addition to activating its transcription function. p53 protein levels are regulated by MDM2 and the tumor suppressor p14^{ARF} (p19^{Arf} in mice). MDM2 negatively regulates p53 protein by mediating its nuclear export and ubiquitination, resulting in proteosomal degradation of p53. MDM2 also decreases p53 transcriptional activity by directly binding to p53. MDM2 is part of a p53 negative feedback loop, because it is a p53 target gene. A critical function of the tumor suppressor, p14^{ARF}, is to neutralize Mdm2, thus resulting in accumulation of p53. p14^{ARF} is upregulated by oncogenic stress generated by activated oncogenes including RAS and MYC.

24.9.5   p14^{ARF}

p14^{ARF} is encoded by the same locus as p16^{INK4}; however, it has a different first exon, and the remaining shared exons are read in an <u>a</u>lternative <u>r</u>eading <u>f</u>rame (ARF). The INK4A/ARF locus is one of the most frequently mutated in human cancer. Why two distinct tumor suppressor genes evolved to share the same locus is an enigma! As described earlier, p53 can regulate pRb function through the induction of p21^{CIP1} which prevents pRb phosphorylation, thereby blocking cell cycle progression. While p53 can regulate pRb function, the opposite is also true. That is, the pRb pathway can regulate the p53 pathway, and this occurs via the induction of p14^{ARF}. The phosphorylation of pRb, combined with subsequent release of E2F, is an important event in the G1 to S phase transition and cell proliferation. Alterations in the pRb circuit, such as the overexpression of cyclin D or the inactivation p16^{Ink4a}, result in the hyperphosphorylation of pRb and abnormally high levels of E2F activity. Abnormally high levels of E2F result in increased expression of p14^{ARF}. This, in turn, inactivates Mdm2 protein, resulting in the accumulation of p53 and the activation of an apoptotic response or cell cycle arrest/senescence. Oncogenic RAS and MYC activity can also increase the expression of p14^{ARF}. Cells with a functional p53 circuit have a mechanism to eliminate cells that possess alterations in the cyclin D/CDK4/6/Ink4a/RB/E2F circuit. Therefore, the pathogenesis of tumor development in most tissues requires that both the p53 and pRb circuits be inactivated.

24.10 MUTATOR PHENOTYPE/DNA STABILITY GENES

Mutational inactivation of genes involved in genome maintenance, such as those involved in cell cycle checkpoints, DNA repair, mitotic recombination, and chromosomal segregations, results in an increased rate of mutation. This condition is often referred to as a mutator phenotype or genomic instability. When genes involved in the above processes are inactivated, the mutation rate increases for all genes. However, when the mutation occurs in an oncogene or tumor suppressor gene, a selective proliferative advantage is conferred. Genes involved in DNA repair, genomic maintenance, and chromosomal stability are sometimes referred to as stability genes because they keep genetic alterations in the genome to a minimum. There are a number of familial cancer syndromes that are associated with germ-line mutations in genes involved in genome maintenance, including genes involved in base excision repair (BER), nucleotide excision repair (NER), mismatch repair (MMR), and chromosomal stability (see Chapter 23). Individuals that inherit these defective genes are predisposed to cancer development. Examples of stability genes that are inactivated within the germ line are ATM, BRCA1, BRCA2, XPA, MSH2, and WRN (Table 24.9). It appears that for nonhereditary tumor types, somatic mutations in genes that control gross chromosomal stability are more frequently altered than those involved in BER, NER, and MMR. Mutational inactivation of genes controlling chromosomal stability results in chromosomal rearrangements, LOH, and aneuploidy. Candidate genes responsible for these gross chromosomal changes include those involved in cell cycle checkpoints and telomere maintenance. As tumor progression ensues, the cancer cell genome becomes increasingly unstable, and this results in further mutations in oncogenes and tumor suppressor genes.

TABLE 24.9. Human Genomic Stability Genes

Gene Name	Familial Cancer Syndrome	Protein Function	Sites/Types of Commonly Associated Neoplasms
ATM	Ataxia telangiectasia	Response to dsDNA breaks	Leukemias, lymphomas
BRCA1/BRCA2	Hereditary breast cancer	Repair of dsDNA breaks	Breast, ovarian
MSH2	Human non-polyposis colon cancer	DNA mismatch repair	Colon, uterus
MLH1	Human non-polyposis colon cancer	DNA mismatch repair	Colon, uterus
NBS1	Nijmegen breakage syndrome	Processing of dsDNA breaks	Lymphomas, brain

24.11 INTERACTION OF ONCOGENES AND TUMOR SUPPRESSOR GENES

As mentioned earlier, epidemiological studies have revealed that the incidence of cancer increases exponentially with age. From these studies, it has been suggested that three to seven mutations or "hits" are necessary for cancer development. Definitive proof that these hits occur in oncogenes and tumor suppressor genes and that these genes cooperate in tumorigenesis comes from molecular studies in human cancer and mouse cancer models.

24.11.1 Colonic Tumorigenesis in Humans

As shown in Figure 24.26, the development of human colon cancer involves the mutational activation of oncogenes and the specific deletion and inactivation of tumor suppressor genes. The development and progression of a benign adenoma to a malignant carcinoma involves the inactivation of APC (<u>a</u>denomatosis <u>p</u>olyposis <u>c</u>oli) and p53 tumor suppressor genes and the activation of the RAS oncogene. There is also a frequent loss of heterozygosity (LOH) on the long arm of Chromosome 18. However, the identity of a tumor suppressor in this region has not been conclusively identified. Within the colonic crypts, stem cells are located at the bottom of the crypts. These stem cells are stimulated to proliferate by the extracellular signal Wnt, which is produced by the adjacent stroma cells. WNT stimulates the accumulation of β-catenin, which, in turn, binds to and activates TCF/LEF transcription factors, which produces a mitogenic response. As the cells move upward from the base of the crypt, the APC gene is expressed and functions to negatively regulate levels of β-catenin, thereby turning off cell proliferation. When APC is inactivated, β-catenin levels continue to accumulate, resulting in abnormal, uncontrolled cell proliferation. β-Catenin itself has been found to be mutated in some tumors, and these mutations make β-catenin resistant to the effects of APC. Thus mutated β-catenin can function as an oncogene. Hypomethylation of DNA occurs early in tumor development, and these epigenetic changes presumably lead to alterations in gene expression and contribute to colonic tumorigenesis. The changes shown in Figure 24.26 represent the more common sequence of changes observed during colon tumorigenesis. However, not all tumors demonstrate all of these changes and not all demonstrate this sequence of events. What appears to be most critical in tumor development is the accumulation of critical mutagenic events.

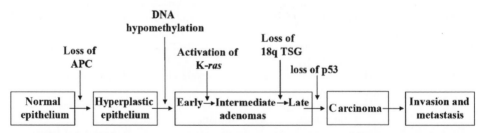

Figure 24.26. Genetic model for colorectal carcinogenesis. (Adapted from Fearon, E. R., and Vogelstein B. A genetic model for colorectal tumorigenesis. *Cell* **61**, 759–767, 1990.)

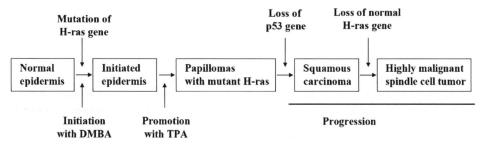

Figure 24.27. Mouse skin model of multistage carcinogenesis. Genetic changes associated with mouse skin tumorigenesis.

24.11.2   Skin Tumorigenesis in Mice

The mouse skin model of multistage carcinogenesis is one of the better-defined in vivo models of epithelial neoplasia where oncogenic Ras mutations precede p53 and INK4A/ARF mutations during tumor development and progression. A number of molecular events that are associated with the operational multistage terms initiation, promotion, premalignant progression, and malignant conversion have been described in this model (Figure 24.27). DMBA-induced oncogenic Ras mutation is the initial critical event responsible for the development of squamous papillomas, and it occurs at nearly 100% frequency in TPA-promoted squamous papillomas. Some papillomas undergo malignant progression to SCCs, which is accompanied by the inactivation of p53. Progression to a highly invasive spindle cell carcinoma is often accompanied by the inactivation of the Ink4a/Arf locus. The progression of papillomas to malignant tumors has also been characterized phenotypically by inappropriate expression of membrane receptor/adhesion molecules, keratins, growth factors, and cyclins/cyclin-dependent kinases.

24.12   GENETICALLY MODIFIED MOUSE MODELS

Transgenic, gene knockout, and gene knockin mice allow for the study of the function of genes in vivo in a complex living animal. Further refinements allow for a conditional knockout in which the gene of interest is deleted in a specific tissue or the gene of interest can be deleted at a specific time point, for example, in the adult mouse or during development. Genetically modified mice provide powerful tools for cancer research, and many such mice have provided and will continue to provide mechanistic information on the role of specific genes in the development of cancer.

24.13   CONCLUSIONS

Cancer susceptibility is determined by complex interactions between age, environment, and an individual's genetic make-up. The basis for understanding individual susceptibility to cancer will likely come from characterizing gene–environment interactions and the role of SNPs in both coding and promoter regions of certain genes within the human genome. Genomic technology, bioinformatics, molecular epidemiology, and genetically modified mouse models will undoubtedly play major roles in this important endeavor. Understanding how the protein products of genes are organized into pathways and how these pathways are deregulated in cancer will provide the framework for further understanding of the cancer process. This knowledge in terms of toxicology and environmental health will provide the basis for the mechanistic identification of human environmental carcinogens and for a rational approach for the estimation of individual risk of cancer development.

SUGGESTED READING

Cancer Facts and Figures 2007, American Cancer Society.

Eleventh Report on Carcinogens, U.S. Department of Health and Human Services, Public Health Service, National Toxicology Program 2005 (www.niehs.nih.gov).

Hananhan, D., and Weinberg, R. A. Hallmarks of cancer. *Cell* **100**, 57–70, 2000.

Overall evaluations of carcinogenicity to humans. International Agency for Research on Cancer (IARC) (www.iarc.fr).

Vogelstein, B., and Kinzler, K. W. Cancer Genes and the Pathways they Control. *Nature Med.* **10**, 789–799, 2004.

Weinberg, R. A. *The Biology of Cancer*, Garland Science, Taylor and Francis Group, New York, 2007.

CHAPTER 25

Genetic Toxicology

R. JULIAN PRESTON

National Health and Environmental Effects Research Laboratory, U.S. Environmental Protection Agency, Research Triangle Park, North Carolina 27711

25.1 INTRODUCTION AND HISTORICAL PERSPECTIVE

Genetic toxicology is a branch of the field of toxicology that assesses the effects of chemical and physical agents on the hereditary material (DNA and RNA) of living cells. The field can be considered to have its roots in 1927 with the publication of the seminal paper by H. J. Muller on the induction of phenotypically described mutations in *Drosophila* by X-rays. In his studies, he showed not only that radiation exposures could increase the overall frequencies of mutations but also that the types of mutations induced were exactly the same in effect, or phenotype, as those occurring in the absence of radiation exposure. This observation exemplifies the fact that an induced mutagenic response must be assessed in relation to background mutations. This requirement will be discussed further later in this chapter.

Karl Sax (1938) extended Muller's observations by showing that X-rays could induce chromosome alterations in plant pollen cells. Sax and his colleagues in an elegant series of studies showed that at least two critical lesions in a target were necessary for the production of exchanges within and among chromosomes. We know now that the "lesions" referred to by Sax are DNA double-strand breaks, base damage, or multiply damaged sites. The important conclusion that was derived from these studies is that the two lesions necessary for chromosome aberration formation are independently produced leading to nonlinear dose-response curves. The relevance of this conclusion will become more apparent in subsequent discussions in this chapter.

Charlotte Auerbach and colleagues in 1946 described studies (actually conducted in 1941) that demonstrated that mustard gas could induce mutations and that these were similar at the whole-organism level in *Drosophila* to those induced by X-rays. Thus, research began in the new area of chemical mutagenesis that ran in parallel with radiation mutagenesis studies.

The next event of particular significance to the field of genetic toxicology was the development, by William Russell in 1951, of a mouse mutagenesis assay that could

Molecular and Biochemical Toxicology, Fourth Edition, edited by Robert C. Smart and Ernest Hodgson
Copyright © 2008 John Wiley & Sons, Inc.

be used to establish if previously described data on radiation-induced mutations in *Drosophila* could be replicated in a mammalian system. The mouse tester strain developed for the specific locus assay contained recessive mutations at seven loci coding for visible mutations such as coat color, eye color, and ear shape. This strain could be mated with irradiated wild-type mice to identify induced heritable recessive mutations at wild-type loci in irradiated mice. Mutations were indeed detected following X-ray exposure of mouse germ cells at frequencies that were similar to those induced by X-rays in *Drosophila*. Subsequent studies showed that mutations could be induced at these seven loci by a range of chemical agents, including the highly effective ethylnitrosourea (the so-called "super mutagen"). Over the next 20 years, genetic toxicologists collected qualitative and quantitative information based on phenotype about induction of mutations, largely by radiation, in germ and somatic cells. Progress was greatly facilitated by the ability to culture mammalian cells in vitro and to assess chromosome alterations in mammalian cells, particularly in human lymphocytes following mitogen stimulation in culture. The potential utility of mutagenicity data for genetic and cancer risk assessments was enhanced by the development of an extrapolation approach for estimating human responses in vivo from rodent data obtained in vivo and in vitro and human data obtained in vitro. This utility was further enhanced by the knowledge that most tumors contained chromosome alterations, that the formation of tumors required mutations, and that cancer was a clonal process, with the addendum that tumors originated from single cells. In the 1970s, two significant advances allowed for a clearer understanding of the mechanisms of mutagenesis and the potential role of mutagens in carcinogenesis. James and Elizabeth Miller and their colleagues demonstrated that chemical carcinogens could react to form stable (covalent) derivatives with DNA, RNA, and proteins both in vitro and in vivo. They further demonstrated that for their formation these derivatives could require the metabolism of the chemical to primary and subsequent metabolites. Thus, the requirement for metabolism for some carcinogens was established (see Chapter 24). In vivo such metabolism is endogenous, but for most in vitro cell lines metabolic capability has been lost. To overcome this, Heinrich Malling and colleagues developed an exogenous metabolizing system based upon a rodent liver homogenate (S9). This system has proven to be of utility, but with some drawbacks related to species- and tissue-specificity.

The second, significant development occurring in the 1970s was that of Bruce Ames and colleagues, who designed an assay with the bacterium *Salmonella typhimurium* that can be used to detect chemically induced reverse mutations at the histidine locus. Not only was the assay able to play an important role in hazard identification, but also mutagens were predicted to be carcinogens on the basis that cancer required mutation induction. The next few years saw the development of about 200 short-term assays that were used to screen for potentially carcinogenic chemicals via the assessment of mutagenicity, DNA damage, or other genotoxic activity. Several international collaborative studies were conducted to establish the sensitivity and specificity of a set of short-term assays as well as to compare results obtained by different laboratories (International Programme on Chemical Safety, 1988). The outcome was that most assays, to a greater or lesser degree, were able to detect carcinogens or noncarcinogens at only about 70% efficiency. In many ways, this mutagenicity–carcinogenicity relationship became somewhat of a hindrance to the field of genetic toxicology because it provided a study framework that was too rigid. The comparison of results for different assays in a standard battery by Tennant et al. (1987) emphasized this point by showing that mutagenicity only predicted a portion (~60%) of known carcinogens. This finding, in part, initiated a downturn in enthusiasm for short-term tests used solely for detecting carcinogens. Subsequently, the lack of a tight correlation between carcinogenicity and mutagenicity (and the converse noncarcinogenicity and nonmutagenicity) was determined to be due to the presence of chemicals that were not directly mutagenic but instead induced genetic damage necessary for tumor development indirectly by, for example, clonally expanding preexisting mutant cells (i.e., tumor promotion). Such chemicals were categorized under the somewhat misleading term *nongenotoxic*. Although a chemical can be carcinogenic via a non-DNA-reactive, mutagenic process, mutations have to be produced to move cells along the multistep pathway to tumor formation. With the appreciation that non-DNA-reactive, mutagenic chemicals could be carcinogenic, there was a greater concentration on studies that were designed to identify cellular processes, other than DNA-reactivity, with the potential to be involved in carcinogenicity (for example, cytotoxicity with regenerative cell proliferation, mitogenicity, or receptor-mediated processes). Examples of mechanisms involved in the carcinogenicity of non-DNA-reactive, mutagenic chemicals are described in Chapter 24.

In the past 15 years, the field of genetic toxicology has moved into the molecular era. The potential for advances in our understanding of basic cellular processes and how they can be perturbed has become enormous. The ability to manipulate and characterize DNA, RNA, and proteins has been at the root of this advance in knowledge. However, the development of sophisticated molecular biology techniques does not in itself imply a comparative advance in the utility of genetic toxicology and its application to risk assessment. Knowing the types of studies to conduct and knowing how to interpret the data developed remains as fundamental as always. Finer and finer detail of the understanding of cellular processes and their alteration by mutation can perhaps lead to a reduction in the utility of some currently used genotoxicity assays. The following sections provide descriptions of standard genotoxicity assays and their use in hazard identification. In addition, a considerable amount of information on the mechanism of induction of mutations and chromosomal alterations by environmental chemicals can be generated from these same genotoxicity assays. This will be enhanced over the next several years.

25.2   GENETIC TOXICOLOGY AND RISK ASSESSMENT (GENERAL CONSIDERATIONS)

The process of estimating human cancer risk following exposure to environmental chemicals includes the following steps: hazard identification, dose–response assessment, exposure assessment, and risk characterization. Genotoxicity assays were originally developed for the purposes of hazard identification—namely, Is a particular chemical likely to be carcinogenic based on its genotoxicity profile? The ability to also obtain data on the mechanisms by which chemicals can induce mutations, for example, and possibly also how they might induce tumors, has enhanced the utility of these genotoxicity assays. This enhanced utility has considerable merit because, for example, the U.S. Environmental Protection Agency's (U.S. EPA) *Guidelines for Carcinogen Risk Assessment* (2005) encourages the incorporation of mechanistic data into the risk assessment process, particularly at the dose–response characterization step. Thus, the use of genetic toxicology assays can provide data beyond hazard identification for incorporation into the overall risk characterization component of the cancer risk assessment process (National Research Council, 1994). In much the same way, a hazard identification step is necessary for the conduct of a genetic risk assessment, with emphasis on the use of germ cell assays for this purpose. The understanding of underlying mechanisms of mutation, which can be a predictor of birth defects and other congenital diseases, can also enhance the quantitative assessment of genetic risk from exposure to environmental chemicals.

25.3   GENOTOXICITY ASSAYS

The aim of this section is not to provide detailed protocols of the range of prokaryotic and eukaryotic assays used to detect the presence or absence of a genotoxic response. Such descriptions can be found in the published literature—for example, in the series of reports of the US EPA's Gene-Tox Committees. Rather, the principles that need to be addressed by the assays and the current approaches for obtaining underlying mechanistic data are presented.

25.3.1   DNA Damage and Repair Assays

Any one of a number of DNA damage and repair assays can be used to provide preliminary information on the potential mutagenicity of a chemical. Briefly, DNA damage can be quantitated by the assessment of specific DNA adducts using HPLC-based methods when information on the chemical is known, and for DNA adducts in general using ^{32}P-postlabeling techniques. DNA strand breakage can be measured by a DNA elution assay or by the relatively recently developed single-cell gel electrophoresis (Comet) assay. The advantage of the latter is that observations can be made at the level of the single cell. The Comet assay is based on the fact that DNA containing strand breaks migrates more rapidly in a gel when electrophoresed. The length of the DNA "tail" drawn by electrophoresis from the cell is a measure of the DNA damage present. The conditions under which the electrophoresis is conducted determines the nature of the DNA damage detected: double-strand breaks at neutral pH and single-strand breaks at high pH. One of the current concerns with the Comet assay is that it can be overly sensitive and might lead to apparent DNA damage under non-damage-inducing exposure conditions.

DNA repair can be detected by the same assays (alkaline elution and Comet) that are used to detect DNA damage, but with various time intervals between the end of exposure and the time of sampling. An additional assay for assessing DNA repair has been utilized for many years, namely the unscheduled DNA synthesis (UDS) assay. The endpoint measured is the uptake of tritiated thymidine into DNA repair sites following exposure. The sensitivity of the assay is relatively low, especially in vivo, and this limits its predictive value for hazard identification. Thus, a negative response in a UDS assay has very limited value because of this low sensitivity.

The assessment of DNA repair per se provides rather little information on the quantitative assessment of the potential subsequent mutagenicity of a chemical because it is the frequency of misrepair that is of consequence to the cell with regard to an adverse outcome. Misrepair can include incorrect rejoining of DNA at the sites of strand breaks or the loss of DNA sequences or incorporation of an incorrect base during DNA repair.

25.3.2 Gene Mutation Assay: Salmonella Typhimurium

The *Salmonella* (Ames) assay utilizes a set of tester strains that have been constructed to select for different classes of reverse mutations at the histidine locus. The tester strains are unable to grow in the absence of histidine, but a specific reverse mutation at the histidine locus (depending on the particular tester strain) leads to histidine independence. Thus, after treatment with a test chemical, the bacteria are plated on histidine-deficient growth medium and the colonies that develop are scored as induced mutants. The strains commonly used together with the genetic alteration detected are shown in Table 25.1. An *R* factor plasmid that contains several mutant genes in DNA damage response pathways was incorporated to provide a greater sensitivity to mutation induction than with the original parental strains.

The assay can be conducted with an exogenous metabolizing system (S9), allowing for the detection of mutagens that require metabolic activation to a DNA-reactive intermediate (e.g., polycyclic aromatic hydrocarbons). The *Salmonella* assay has been conducted with literally hundreds of chemicals, such that various calculations of specificity and sensitivity for detecting carcinogens and noncarcinogens can be performed for various chemical classes and modes of action for inducing DNA damage.

More recently, it has been feasible to collect data on the molecular nature of the revertants induced and thereby better understand the mechanism of their induction and allow for interspecies extrapolation. The sequencing of the histidine revertants leads to the generation of mutation spectra for a chemical and its metabolites and for simple and complex mixtures. Of particular importance for assessment of effects at low exposure levels, it is possible for many of the cases to distinguish between background and induced mutations, thereby enhancing the sensitivity of the assay. Comparisons of mutation spectra for a specific chemical across species from *Salmonella* to humans suggest that a similarity of formation of mutations holds across all species. Thus, data generated for prokaryotes have some application in predicting effects in humans. A similar assay to that developed for *Salmonella* is available for

TABLE 25.1. Strains Commonly Used in the Salmonella/ Histidine Reversion (Ames) Assay for Detecting Mutations at the Histidine Locus

Strain	Mutations Detected
TA 1535	Base-pair substitution
TA 1537	Frameshift
TA 1538	Base-pair substitution
TA 98	TA 1538 with *R* factor
TA 100	TA 1535 with *R* factor
TA 102	Mutants from oxidative DNA damage and DNA crosslinks

Escherichia coli, employing reversion at the tryptophan locus. The WP2 strains used can only detect base-pair substitutions, given that the original trp mutation is a base-pair substitution.

There are some obvious limitations to the use of prokaryotic assays for predicting responses in mammals, given the difference in cellular and genomic complexity between prokaryotes and eukaryotes. However, for hazard identification purposes, these prokaryotic assays are a useful component.

25.3.3 Yeast Assays

Forward and reverse mutation assays for a range of selectable loci can be conducted with *Saccharomyces cerevisiae* as well as with other yeast strains and fungal species (e.g., *Neurospora, Aspergillus* and *Ustilago*) in much the same way as for bacteria. However, yeast have multiple chromosomes, haploid and diploid karyotypes, mitotic and meiotic cycles, and, of increasing significance, a considerable homology—as regards cell cycle control, DNA repair capacities, and other housekeeping processes—to those in mammalian species including humans. Thus, mutations in genes specific for these pathways can be assessed and can in some cases be considered as viable biomarkers of tumor responses. Mutational spectra can also be developed as an aid to interspecies extrapolation of response.

In general, *S. cerevisiae* exhibits a high degree of homologous, mitotic recombination. For example, transferred plasmids have a high probability of incorporation into targeted genes; much higher than for mammalian species. The advantage is that assay systems can be developed for assessing mitotic crossing-over and mitotic gene conversion as additional cellular responses to chemical exposure. As an example, Schiestl and colleagues have demonstrated that chemicals that are not DNA-reactive (e.g., polychlorinated biphenyls and 2,3,7,8-tetrachlorodibenzo-*p*-dioxin) can induce deletions via intrachromosomal recombination in an assay established to detect such events (the DEL assay).

25.3.4 Mammalian Cell Assays (In Vitro)

25.3.4a Gene Mutation Assays. As noted above, information on mutation induction can be obtained with prokaryotic and lower eukaryotic systems that can enhance the ability to predict effects in humans. However, because of differences in genome organization and genetic complexity, assays using mammalian cells have been developed. In vitro, mammalian cell assays most frequently utilize cells that are either immortalized spontaneously (e.g., Chinese hamster ovary cells, mouse lymphoma cells) or virally transformed, as in the case of human nontumor cell lines, such as lymphoblastoids. The benefit of using such immortalized cells is that they can be cloned as single cells, a necessary requirement for the quantitation of mutations.

The loci that can typically be used for mutation assessment are ones that can be selected for, although PCR methods currently available and under development will allow for analysis of any gene for which the DNA sequence is known. The selection procedure requires that a locus be heterozygous or hemizygous such that only a single mutation is required for the selectable phenotype to be assessed. Examples of such loci are: the *hprt* (hypoxanthine-guanine phosphoribosyl transferase) gene that is X-linked and is thus hemizygous in males, and effectively hemizygous in females because of X-chromosome inactivation; the thymidine kinase locus that is heterozygous in a mouse lymphoma cell line (L5178Y) and a human lymphoblastoid cell line (TK6); and the *aprt* (adenine phosphoribosyl transferase) gene that is heterozygous in a selected CHO cell line. Because the mutations induced by the great majority of chemicals involve errors of DNA replication on a damaged template, at least part of the treatment period should be for cells in the DNA synthesis (S) phase.

The CHO/*HPRT* assay is presented in somewhat more detail as an example. After treatment with a chemical being assayed for its mutagenic potential, cells are subcultured for an expression period of about 6–8 days during which time any residual HPRT protein is degraded. Mutant clones are identified as those that grow in the presence of the selective agent 6-thioguanine. Wild-type cells are killed by the toxic metabolite, 6-thioguanine monophosphate, which they can form via the normal *hprt* gene; *hprt* mutant cells cannot perform this metabolic step. The analysis of the mutation spectra for a range of chemical and physical agents has shown that the assay can detect point mutations and small to mid-sized deletions. However, large deletions covering many megabases of DNA are not recovered, largely because viability genes are located in the proximity of the *hprt* gene. Because the *hprt* gene is X-linked, compensation by an active, homologous chromosome does not take place. The identification of mutations at the DNA sequence level allows for a mutation frequency, as opposed to a mutant frequency, to be assessed. The distinction is important because a mutant is a phenotypic descriptor whereas a mutation is a genotypic descriptor; thus, a single mutation can give rise to many mutants by clonal expansion. It is also feasible to assess mutations at the *hprt* locus in primary cells, especially peripheral T-lymphocytes of rodents or humans. An advantage with human lymphocytes is that the true mutation frequency can be quantitated by eliminating mutant clones on the basis of the cells T-cell receptor genotype—mutants derived from a single induced mutation will all have the same T-cell receptor genotype.

Other assay systems employing autosomal loci are required if the detection of very large deletions is required. An example of such an assay is that utilizing the heterozygous thymidine kinase locus (*tk+/−*) in the L5178Y mouse lymphoma cell line. The selection for mutants relies on the phenotype of trifluorothymidine resistance. A further utility of this assay has been proposed, namely that fast-growing mutant colonies are produced by cells containing gene mutations, whereas slow-growing colonies arise from cells containing chromosome mutations, particularly deletions. The distinction is perhaps less clear-cut than this, and molecular characterization of mutants provides the definitive description of the induced mutations.

25.3.4b Cytogenetics Assays

(i) Structural and Numerical Chromosomal Alterations. Chromosome aberration in vitro assays can be conducted with any higher eukaryotic cell type that divides or can be induced to divide in tissue culture. As mentioned above, the majority of cell lines routinely used are immortalized cells (e.g., CHO cells, human lymphoblastoid lines), although human peripheral lymphocytes can be stimulated to divide in vitro by the addition of a mitogen such as phytohemagglutinin. A diagram of a

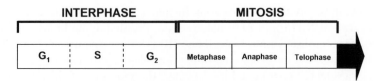

Figure 25.1. Mitotic cell cycle.

mitotic cell cycle is shown in Figure 25.1 for defining the various cell cycle stages referred to in this section and Section 25.3.5b on in vivo cytogenetic assays.

The endpoints utilized for assessing chromosomal alterations are chromosome structural alterations for clastogenicity (chromosome breaking ability) and numerical alterations (most effectively gains of chromosomes). The classical method is to assess both structural and numerical alterations in metaphase cells using a general stain such as Giemsa and light microscopy. As an alternative, chromosomal breakage and chromosome loss events can be assessed by the presence of micronuclei. Acentric chromosome fragments or whole chromosomes that fail to segregate at anaphase and are not incorporated into either daughter nucleus become bounded by a membrane and remain as a micronucleus in the cytoplasm through one or more cell cycles. The frequencies of micronuclei are most accurately assessed in cells that have undergone one cell division, since they are not produced in the absence of a cell division and are reduced in frequency through loss (or extrusion) at the second or subsequent division after formation. Cytochalasin B can be used to inhibit cytokinesis, and thus micronuclei can be conveniently assessed in binucleate cells, namely ones that have undergone a single division (as contrasted with mononucleate, no division; tri- and higher nuclear numbers, more than one cell division).

Another relatively recent modification of the assay is to stain micronuclei with a fluorescent antibody (CREST) to the centromeric region of chromosomes and view through a fluorescence microscope. A micronucleus that contains a whole chromosome will fluoresce, whereas an acentric fragment (i.e., with no centromere) will not fluoresce (Tucker et al., 1997). It should be noted that the assessment of micronuclei only identifies subsets of cytogenetic alterations, acentric fragments for clastogenicity, and chromosome loss for aneuploidy. Conventional metaphase analysis can be used for identifying all classes of chromosomal alteration and chromosome gains or losses.

A broad range of fluorescence-based molecular techniques have been developed in the past 10 years or so, and these have considerably enhanced our ability to assess cytogenetic alterations. It is quite feasible to identify specific gene regions, repetitive DNA sequences, and specific whole chromosomes using fluorescent DNA probes for the appropriate chromosomal units. Such fluorescent staining is called "painting." Thus, chromosomal structural alterations can be readily detected. For example, reciprocal translocations (exchanges of chromosomal material between two chromosomes) can be identified using fluorescence in situ hybridization (FISH) whole chromosome painting techniques. An exchange can be readily seen as a color switch along a chromosome. This is of particular importance, since this class of aberrations is generally transmissible from one cell generation to the next and many tumors are characterized by the presence of specific reciprocal translocations. With conventional metaphase analysis, the aberration types that are most readily observed are cell lethal and are not per se a risk factor. Reciprocal translocations are not readily observed by conventional staining techniques unless they are the result of very unequal exchanges of chromosomal material. An additional advantage of being able to assess transmissible cytogenetic alterations is that the accumulated effects of longer-term exposures can be measured, although this is of particular utility with in vivo exposures.

Additional, somewhat more complex assessment of genomes can be conducted by chromosome banding techniques, comparative genomic hybridization (CGH), and spectral karyotyping (SKY) (Preston, 2005a). CGH is of particular utility when attempting to identify chromosomal differences between two genomes, such as those of normal and tumor cells, or between mutant clones and wild-type cells. The technique is basically not informative for standard genotoxicity assays. The more recent addition of chromosome array-based analysis has considerably enhanced the sensitivity of CGH. SKY uses computerized representations of karyotypes and allows for the identification of each pair of human chromosomes. Again, the greater utility of SKY is for identifying karyotypic changes in tumors, but there could be an application for identifying genomic instability that can develop at some time after a particular exposure. It has been shown that genomic instability is a hallmark of tumor formation. Adding its assessment to standard assays for chromosomal alterations or mutations would appear to have significant benefit for the risk assessment process. However, the high cost of this assay would prohibit its routine use.

As mentioned above, many aberration types are cell lethal, with the loss of chromosomal material at cell division being a major cause of the lethality. Thus, a requirement for measuring the induced frequency of total aberrations is that the sampling of cells has to be at the first mitosis after the exposure. In addition, the exposure needs to be of a duration of less than that of a cell cycle for the particular cell type being used. The great majority of chemicals produce aberrations via the conversion of DNA damage (e.g., DNA adducts, crosslinks) into an aberration as an error of replication on a damaged DNA template. Thus, to maximize the sensitivity of a cytogenetic assay, cells should receive at least part of their exposure while they are in the DNA synthesis (S) phase of the cell cycle.

(ii)   Sister Chromatid Exchange (SCE) Assay. It is possible to identify apparently reciprocal exchanges between the two sister chromatids of a chromosome by differential staining methods. Use of these methods results in the two chromatids of a chromosome being distinguished by their different staining patterns. The general approach is shown in Figure 25.2. Visualization of chromatid exchanges can be achieved by (a) treatment with Hoechst 33258 stain, DNA elution with saline sodium citrate under black light, and staining with Giemsa or (b) use of an antibody to BrdU.

SCE appear to be produced largely as errors of DNA replication on a damaged template or by homologous recombination. The fact that they appear to be reciprocal in nature means that they probably do not represent a risk factor to cells. An increase in SCE can be via processes that do not involve directly induced DNA damage, such as alterations in S-phase kinetics, making the interpretation of observed increases somewhat difficult. In the review of a significant body of test data for a range of chemicals, the SCE assay appears to provide rather equivocal information. For this reason, it is not currently recommended for inclusion in a standard genotoxicity test battery.

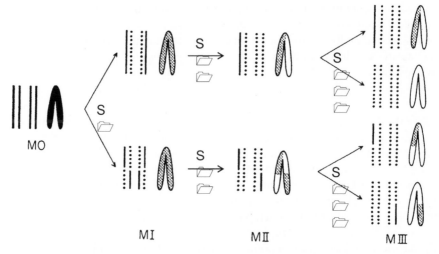

Figure 25.2. Sister chromatid exchange formation. For differential staining of the sister chromatids, bromodeoxyuridine (BrdU) is incorporated into replicating DNA in place of thymidine (TdR). The solid line represents the original TdR-containing DNA strand and the dotted line represents the DNA strand containing BrdU. An SCE is shown to occur at the first S phase (S☞) but will be observed at the second mitosis (M☞☞).

25.3.5 Mammalian Cell Assays (In Vivo)

25.3.5a Gene Mutation Assays. As with the in vitro assays discussed above, the conduct of an in vivo gene mutation assay currently requires a selectable marker. The majority of human studies have been conducted on peripheral lymphocytes using the *HPRT* gene. The same assay can be conducted with rodent lymphocytes, but these cells are rather difficult to clone in vitro. Additional information can be obtained in human studies by assessing the T-cell antigen receptor status of mutant clones; the presence of an identical sequence in two or more clones indicates that these were derived by clonal expansion from a single mutant cell. Thus, an induced mutation frequency can be obtained by correcting the observed mutant frequency for clonal expansions. An earlier version of the human *HPRT* assay relied on autoradiographic identification of mutant cells (i.e., those that could grow in 6-thioguanine) rather than cell cloning assays. The problem with such an assay is the inability to distinguish induced mutations from mutants derived from the clonal expansion of the original mutant cell, leading to concerns with interpretation of the data in terms of significant increases in mutations.

Other genetic markers have been utilized for the assessment of gene mutations in vivo in rodents. These include the *Dlb*-1 locus (binding site for Dolichosbiflorus agglutinin) and the *lacI* or *lacZ* transgene in mice or rats. There are a number of advantages to the use of the transgenic mouse assays, namely that mutagenicity can be assessed in any tissue, especially target tissues for tumor formation. There are drawbacks; for example, only point mutations and small deletions can be detected. Large deletions that remove the viral packaging sites are not recovered. Second, in vivo cell replication is required for the formation of the mutation in vivo; and third, the *lac* gene is not an endogenous gene and is not transcribed. As with in vitro assays, PCR-based mutation analysis is reaching the levels of sensitivity at which any gene is potentially available for the assessment of mutations. Mutational spectra can be assessed; and for comparisons between rodents and humans, currently the *hprt* gene assay represents the most viable system.

25.3.5b Cytogenetic Assays

Structural and Numerical Alterations. The basic principles underlying in vivo cytogenetic assays are the same as those for in vitro assays, discussed above (Section 25.3.4b). The cell types that are available for the analysis of chromosome alterations are those that are cycling cell populations, such as bone marrow and spermatogonial cells. For acute exposures, analysis of metaphase cells at their first metaphase after S-phase treatment and analysis of all aberration types is appropriate for hazard evaluation. For chronic exposures, it is inappropriate to analyze unstable aberration types; the informative approach is to analyze stable types (e.g., reciprocal translocations and inversions) using FISH.

For numerical chromosomal alterations, it is possible to analyze interphase cells using FISH with whole chromosome paints, or chromosomal specific probes. Thus, it is feasible to assess nondisjunctional events in any cell type, including target cells for tumorigenesis. Other fluorescence-based methods described in the section on in vitro methods (Section 25.3.4b) can be used in vivo, but they are of rather limited value for basic mutagenicity assays.

A micronucleus assay can be conducted with bone marrow cells, peripheral lymphocytes or germ cells (spermatids) of mice. Rats can be used for micronucleus assays with bone marrow or spermatogenic cells; but the use of peripheral lymphocytes requires surgical removal of the spleen, since this is the predominant site of removal of micronucleated erythrocytes in rats.

Bone marrow cells, following exposure to the test chemical, are collected and stained with Wright–Giemsa or acridine orange, for example. Micronuclei are assayed in polychromatic erythrocytes (PCE), the most recently matured erythrocytes. Cytotoxicity is assessed as the ratio of PCE to normochromatic erythrocytes (NCE) in treated versus control animals. NCE are the more mature erythrocytes. A lower PCE/NCE ratio in the treated animals than in the controls suggests cytotoxicity. There are specific criteria for establishing that a cytoplasmic inclusion is a micronucleus (Hayashi et al., 1994). As mentioned above, the micronucleus assay detects a subclass of cytogenetic alterations (deletions and chromosome loss) and is useful for hazard identification but not for identifying chromosome aberration spectra.

25.3.6 Germ Cell Assays

Several assays have been developed for assessing genetic alterations in germ cells. The information developed can be used for hazard identification in the cancer risk assessment process or, more appropriately, in the development of a genetic risk assessment. The decision of when to conduct a germ cell assay is an important one because the assays are generally expensive and time-consuming. An exception is the use of *Drosophila* in assays that can serve as a substitute for mammalian assays.

25.3.6a Drosophila *Sex-Linked Recessive Lethal Test (SLRL).* For the SLRL, the mutations being assessed are observed in the progeny of chemically treated males or females and thus are heritable effects. The longer the time interval between treatment and mating, the further back in the spermatogenic cycle were the cells at

the time of treatment, thereby allowing for the assessment of effects in all germ cell stages. The particular genotypes of the flies used allows for the detection of recessive lethal mutations on the X chromosome by an absence of wild-type males in the F_2 generation. Another feature of the assay is that *Drosophila* has a xenobiotic metabolizing system that is quite similar to that in mammals.

25.3.6b Dominant Lethal Test (Rodent). The dominant lethal assay is designed to detect induced mutations in germ cells that manifest themselves as embryonic lethality. The test system can be applied to mice and rats, and assessment can be made of effects upon male and female germ cells. Generally, male mice are utilized. Following exposure of male mice to the test compound, the mice are mated with untreated virgin females. Serial mating over several weeks with virgin females allows for assessment of effects on all the different germ cell stages.

For determining dominant lethal effects in exposed males, for example, the number of dead implants as a proportion of total implants is measured between days 13 and 19 of gestation. The test is not very sensitive for detecting mutagenic responses, and so it is not particularly valuable for detecting potential carcinogens. It has a greater utility for detecting chemicals that *are* germ cell mutagens (Ashby et al., 1996).

25.3.6c Heritable Translocation Test (Rodent). The heritable translocation test is based on the observation that carriers of reciprocal translocations have reduced litter sizes (semisterility) as a consequence of lethal segregants (duplications and deficiencies) of the translocations. Thus, the assay requires the assessment of responses in the F_2 of a treated animal. Reciprocal translocations are induced in parental germ cells, leading to an F_1 animal carrying a particular translocation in every cell, and resulting in the F_2 offspring showing semi-sterility. It is possible also to assess translocations by direct observation of diplotene/diakinesis spermatocytes. As with the dominant lethal test, the heritable translocation test is not useful as a screen for carcinogens because of its rather low sensitivity. It can be used for identifying chemicals that *are* germ cell mutagens.

25.3.6d Mouse Specific Locus Test. The mouse-specific locus test was introduced in the Section 25.1 because it represented the initial use of the mouse for assessing induced germ cell mutations. The test relies upon the use of a tester stock that is homozygous recessive for 7 loci or 5 loci, depending on the particular version of the test. The treated individuals are wild-type at the same loci. Thus, any mutation induced at any one of the wild-type specific locus genes will be observed as a phenotypic mutant in the offspring following mating to the tester stock. The mutants are saved for further testing. In fact, more recently, a number of the mutations have been characterized at the DNA level. The information generated can be used for comparison with data obtained in other assay systems, as well as for genetic risk assessment. Clearly, the assay is labor-intensive and very expensive. Its main use is for specific instances where germ cell mutagenicity has to be assessed for genetic risk assessment purposes.

25.3.7 Summary of Genotoxicity Assays

The preceding sections provide a brief overview of the conduct and utility of prokaryotic and eukaryotic genotoxicity assays. The intent is to introduce the basic principles for each assay, the use of the data, and how current advances, particularly in molecular biology, can enhance the predictive value of an assay. There is a long list of other assays that are available; the ones selected here are those that are the most frequently used and consequently have the larger available databases. The question that will be addressed in Section 25.3.9 is, How can a group or battery of these tests be optimized to assess mutagenicity and predict potential carcinogenicity?

25.3.8 Modifiers of Genetic Response

There are a number of cellular factors that can modify the mutagenic response to exposure to environmental chemicals, either increasing it or decreasing it. For the present discussion of genetic toxicology assays, it is important to pay regard to these potential modifiers in the interpretation of the data generated. The three modifiers, for which the most information is available and for which modifications can result in substantial effects, are DNA repair, gene expression changes, and cell cycle control characteristics. Details of these three cellular responses to exposure to chemicals can be found in other chapters in this volume: DNA repair (Chapters 22 and 23), gene expression (Chapters 2–6), and cell cycle control (Chapter 23 and 24). Some general considerations are discussed here.

As discussed above, the great majority of mutagenic chemicals produce gene mutations as the result of errors of replication on a damaged DNA template. Thus, the alterations are produced during the S phase of the cell cycle (S-phase dependent). This is in contrast to ionizing radiation and the small number of radiomimetic (radiation-like) chemicals that produce genetic alterations by errors of DNA repair in G_1 and G_2 and errors of replication in S; these agents are said to be S-phase independent. The important point is that the mutation frequency (for gene mutations and chromosome alterations) following exposure to mutagenic chemicals will be dependent upon the amount of DNA damage that is present in a particular genomic region at the time of replication of that region. The damage remaining will be greatly influenced by the effectiveness of DNA repair and the time between the induction of the damage and the start of replication. In a sense, mutation frequency is a result of the outcome of the race between DNA repair and replication. The outcome can be modified by increasing the rate of repair, thereby reducing mutation frequency or reducing repair efficiency and thereby increasing mutation frequency. Thus, the repair characteristics of cells used for genotoxicity assays need to be known in order to be able to interpret the outcome of the assay.

Another way of influencing the repair-replication race is to delay the start of DNA replication, thereby providing a longer time for repair to take place before replication begins. Cells have developed an extraordinary array of genes for controlling the cell cycle to ensure an orderly progression and to reduce the potential for genetic alterations to occur. For example, in response to DNA damage, cells can activate checkpoints at critical stages in the cell cycle (e.g., G_1/S, S, or G_2/M) that prevent cells from progressing through the cell cycle for a defined period of time, In this way, there is a longer period available for DNA repair prior to replication, leading to a commensurate reduction in mutation frequency. Thus, the cell cycle characteristics of cells used for genotoxicty assays as well as their responses to DNA damage need to be known, because this can influence the outcome of the assay and the interpretation of the data.

Finally, alterations in the expression of specific genes involved in DNA damage responses to chemical exposures can influence the outcome of an assay and the interpretation of the data generated. It is not normal practice to assess gene expression changes as part of a genotoxicity assay, but the development of gene microarray techniques and the relative ease of assessing whole genome responses to exposure might make this a more available approach. The aim might be to use a DNA damage response microarray and profile the changes in gene expression in response to exposure to a test chemical. An aim could be to determine that there was no unusual response for a specific chemical or cell line that might impact the response or interpretation of the assay data.

The modifiers of response become of greater interest as genotoxicity assays are used much more frequently for generating mechanistic data rather than for the yes/no type of response used for hazard identification.

25.3.9 Test Batteries for Detecting Carcinogens

The development of the *Salmonella* mutagenicity assay for detecting mutagens— and, by association, carcinogens—was viewed initially as a potential "magic bullet" for hazard identification. It quickly became apparent that this association was more applicable to positive results, but even in this case the predictivity for carcinogenicity in humans (or rodents) was far from 100%. Thus, the use of a battery of tests was proposed that would include mammalian assays, both in vitro and in vivo, in addition to prokaryote ones. Subsequently, as noted above, since a significant proportion of chemicals can be carcinogenic in the absence of direct mutagenicity, additional assays have been developed for assessing the potential for a chemical to be a "nongenotoxic" carcinogen. Some mechanisms of "nongenotoxic" carcinogens are discussed in Chapter 24.

A variety of test batteries have been described, each with merit and potential drawbacks associated with predictivity for carcinogenicity. The particular scheme presented in Figure 25.3 is of utility because it allows for the detection of both somatic and germ cell mutagens (as needed) and also provides a preliminary approach for identifying the mode of action for potential carcinogens. A more detailed description of the testing approach following this scheme can be found in Ashby et al. (1996). The basic components (or required tests) for initial hazard identification of this and the great majority of test batteries for assessing genotoxicity are a microbial assay such as the Ames test, a mammalian in vitro gene mutation or cytogenetic assay, and an in vivo mammalian assay for chromosomal effects. The aim of this group of assays is to screen products for mutagenicity in the decision-making process in product development. In order to receive approval for commercialization, regulatory agencies require additional tests to be conducted. These can vary according to the particular nature of the human exposure. For example, a negative outcome in an initial test could require an evaluation of mutagenicity in a second tissue, depending upon the level of potential human exposure. The interpretation of the test data from such a battery can be complicated by conflicting outcomes—for example, a positive result for an in vitro cyto-

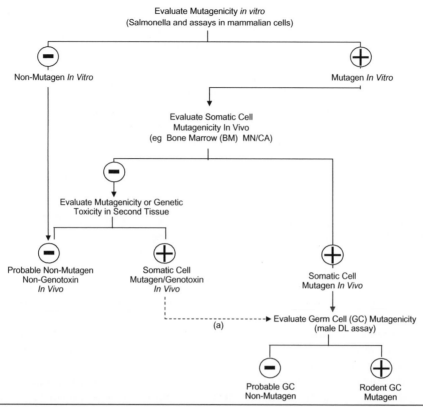

Figure 25.3. Scheme for detecting mutagens in vitro, somatic and germ cell mutagens in vivo, rodent genotxins, and potential rodent carcinogens. Such data contribute to the recognition of germ cell mutagens and nonmutagens, genotoxic and nongenotoxic carcinogens, and non-carcinogens. TG, transgenic assay; CA, chromosomal aberration assay; UDS, unscheduled DNA synthesis; MN, micronucleus assay; HT, heritable translocation assay; DL, dominant lethal assay; SL, specific locus assay; PB/HPRT, peripheral blood HGPRT/HPRT assays in rodents/humans. (a) observation of genotoxicity in vivo for BM nonmutagens does not necessarily trigger the need for GC studies; the genetic event monitored, the magnitude of the induced effect, and the dose level required influence this decision. [From Ashby et al. (1996) with permission.]

genetics assay and a negative result for an in vivo cytogenetics assay. A number of possible explanations are available. High concentrations of some chemicals in vitro can lead to chromosomal alterations by processes unrelated to direct interaction with DNA. Also, the frequency of false positives among the different genotoxicity assays is variable but generally between 10% and 20% (Kirkland and Dean, 1994), with in vitro cytogenetic assays being at the upper end of this range. Despite the collection of enormous quantities of data with genotoxicity assays, caution is still required in their interpretation, given that they are generally conducted at high exposure levels, frequently using transformed cells, and the assays have been developed to be sensitive for endpoint induction. However, for hazard identification it is prudent to use a conservative approach that is somewhat more likely to identify false positives than false negatives. On the other hand, the next steps of risk assessment after hazard identification require a quantitative analysis that could be unreasonably compromised by the use of data from overly sensitive assay systems.

25.4   USE OF MECHANISTIC DATA IN CANCER AND GENETIC RISK ASSESSMENTS (SPECIFIC CONSIDERATIONS)

The concept of how data from genotoxicity assays can be used in both cancer and genetic risk assessments was introduced in Section 25.2. This concept is expanded here to provide a broader rationale for the collection of genotoxicity data in support of these risk assessment processes.

The previous section (25.3) described the use of genotoxicity assays for the identification of mutagens and, thus, of one class of potential carcinogens ("genotoxic" carcinogens). In addition, it has been appreciated for some time that data from assays measuring mutagenicity can be used as a component of quantitative cancer risk assessment. This stems, in part, from the knowledge that cancer is a genetic disease, whereby a series of mutations is required for passage of a cell from normal to transformed (i.e., multistage carcinogenesis; see Chapter 24). These required mutations are most frequently in proto-oncogenes or tumor suppressor genes leading, for example, to cell cycle perturbations or genomic instability, both of which are hallmarks of tumor cells. A more recent concept of the changes required to transform a cell into a malignant tumor has been provided by Hanahan and Weinberg (2000) in their article entitled "The Hallmarks of Cancer." In this approach, there are six acquired characteristics that have to be obtained by a cell during the process of tumor formation. These characteristics are: evading apoptosis, self-sufficiency in growth signals, insensitivity to anti-growth signals, tissue invasion and metastasis, limitless replicative potential, and sustained angiogenesis. Each of these steps requires mutations as at least part of the acquisition of any one of the characteristics. Thus, it can be argued that data on mutagenicity can be used as an effective surrogate for direct tumor assessment, at least as far as shape of the dose–response curve is concerned, at exposure levels below those at which increases in tumors can be detected, but at which detection of increases in mutagenicity are practical. This forms the basis for the approach to cancer risk assessment presented in the U.S. EPA's *Guidelines for Carcinogen Risk Assessment* (2005).

In this approach it is proposed that mechanistic data be used to support the assessment of cancer risk at low (environmental or occupational) exposure levels, in the absence of human or laboratory animal tumor data in this exposure range. This represents a significant advance over previous practice whereby, for mutagenic chemicals, a default approach was used that required a linear extrapolation from the lowest observed tumor incidence for a particular chemical, without some regard to known mode or mechanism of action. In some cases, a linear response is the appropriate one even when regard is paid to mode or mechanism of action (e.g., for the production of point mutations involved in tumor production). On the other hand, nonlinear responses can be predicted and assessed for other mechanisms of action (e.g., when chromosomal alterations are involved in tumor production). For the majority of non-DNA reactive (indirectly mutagenic) chemicals, a nonlinear or threshold response may best describe the shape of the tumor dose–response curve over the low exposure range. This proposition is supported by the available data on mode or mechanisms of action of this broad family of chemicals.

The framework for the new EPA *Guidelines for Carcinogen Risk Assessment* is based on the use of the mode of action for a chemical or chemical class that describes the necessary, but not sufficient, steps whereby a tumor can be formed. Modes of action include DNA-reactivity, receptor-mediated processes, and cytotoxicity together with regenerative cell proliferation and mitogenicity. Each mode of action is described by a set of key events that are necessary to convert a normal cell into a transformed cell and ultimately a metastatic tumor. Each of the key events is a measurable endpoint and, thus, has the potential to be used in the quantitative assessment of tumor risk by being used essentially as a surrogate for the tumor itself. A comparison of key events that are observed in a rodent model and in humans for a specific chemical can be used for establishing the human relevance of the rodent tumor data to humans using a human relevance framework. The aim is to establish plausibility of the mode of action in humans, not necessarily to have it based on the measurement of all key events in humans. This approach is designed to provide a clear path for establishing the need (or not) of having to conduct a full quantitative risk assessment for a particular chemical being considered.

Recently, considerably increased attention has been paid to the so-called epigenetic processes that are involved in maintaining genome stability and the control of the sequence of cellular housekeeping processes (i.e., DNA replication, DNA repair, cell cycle control, and gene expression). Epigenetics broadly refers to the study of heritable changes in gene expression that occur in the absence of a change in DNA sequence. These changes in gene expression can be the result of DNA methylation and/or chromatin remodeling that involves histone modificiations. A significant feature is that a mutation in a key protein for an epigenetic process that might result from exposure to environmental chemicals can produce genome-wide responses.

There remains a clear need to enhance the database for both the mode and, particularly, the mechanism of action for tumorigenicity for the great majority of chemical (and physical) agents. However, the availability of new techniques for assessing effects at the cellular and molecular level should allow for the rapid expansion of this database. In parallel, there will be an improvement in the reliability of cancer risk assessments for human exposures. The area of whole-cell analysis of gene transcription using gene microarray methods, translation using mass spectrometry-based approaches and other so-called "omics" techniques, will certainly enhance our ability to develop informative bioindicators of tumor responses and the mutagenic precursor steps along the pathway. It might well be that these cell profiling techniques will be added to a battery of gentoxicity assays in the not-too-distant future, once a functional change can be associated with changes in gene expression or protein levels. Additional information on these new genomic assessment techniques can be found in Chapters 2–6.

25.5   NEW RESEARCH DIRECTIONS

The myriad of new research tools available to probe the consequences of the inter-action of chemicals with cellular macromolecule has allowed for a new enthusiasm in addressing the following question: How do chemicals induce tumors and other adverse health outcomes? Some of these recent advances are noted here for illustra-tive purposes.

In order to extrapolate from tumor data collected in animal studies to responses in humans, it is greatly advantageous to demonstrate that a similar etiology or underlying mechanism is involved. A number of new approaches facilitate such an endeavor, and three examples are presented here.

As noted above, sophisticated techniques are available for the analysis of muta-tions at the gene and chromosomal level. Such analysis can be used to define the mechanisms of formation of specific human tumors. In addition, it can be used for the development of informative bioindicators of a tumor response. These include ligation-based PCR technique and fluorescence in situ hybridization (FISH) that can be used in conjunction with clones that cover limited genomic regions to identify specific chromosomal breakpoints (Preston, 2005a).

To provide global comparisons between the gene expression patterns of normal and tumor cells, for example, cDNA microarrays have been developed. The approach is to hybridize expressed genes in the form of fluorescent cDNAs to an array of hundreds to thousands of fragments of known genes or expressed sequence tags (ESTs). The cDNAs can be derived from treated versus untreated cells or tumor versus nontumor cells. The resultant hybridization patterns can be compared for alterations in expression. The analysis of the very large quantities of data generated by these techniques has proven to date to be quite difficult to conduct. However, there is an increasing emphasis being placed on the bioinformatics component of whole genome analysis (Ness, 2006).

In parallel with these advances in mRNA detection and assessment, significant advances in the detection, characterization, and quantitation of proteins have been made. This is not only a result of the rapidly expanding database of protein sequences but also of the extraordinary advances in sensitivity of mass spectrometry tech-niques. It is proposed, with some real sense of expected success, that protein char-acterization at the whole genome level will be forthcoming (Domon and Aebersold, 2006).

There is an ever-growing appreciation of the importance of stem cell alterations in disease processes, and the recent observations on the role of cancer stem cells in the development of tumors have enhanced this view (Burkert et al., 2006). The ability to culture adult stem or progenitor cells in vitro makes it feasible to conduct genotoxicity assays on the "relevant" cell type for the outcome being predicted—cancer in the present context. The more proximate the endpoint being assayed is to the endpoint of concern, the more reliable the extrapolations that have to be made (Preston, 2005b). Some of the recent advances in whole genome assessment for characterizing tumors and the effects of chemical exposures should be conducted on cancer stem cells because it is these that appear to be the site of tumor initiation. For example, using tumor cell samples in general might well cloud the picture of the essential steps in tumor formation.

The examples noted above serve to highlight where advances in the understanding of basic cellular housekeeping processes and the consequences of their perturbation can have a very significant impact on determining the mechanism of adverse health outcomes and the subsequent assessment of risk. Thus, studies of mutagenicity play an important role in hazard identification, dose–response assessment, and the characterization of risk.

25.6 CONCLUSIONS

The field of genetic toxicology has had an overall life of about 70 years, during which time it has served to address both practical and basic scientific issues. Much has been learned about the relative potency of chemicals and radiation to induce gene mutations and chromosome aberrations. To a great extent, this information has been obtained from short-term assays using prokaryote and eukaryotic cells in vitro and insect and mammalian cells in vivo. The suggestion that carcinogens are mutagens led to the development of genotoxicity test batteries that were designed to detect potential carcinogens. The lack of a high level of predictivity was due to there being a large class of carcinogens that are not directly mutagenic. Assays for the detection of this class have been developed based upon the range of their different mechanisms of carcinogenicity. New molecular techniques have enhanced the utility of genotoxicity assays by allowing for a much clearer understanding of mechanisms of induction of gene mutations and chromosome alterations. Such data can be used to help reduce uncertainty in cancer and genetic risk assessments. The ability to sequence DNA, for example, allows for the detection of the specific nature of mutations. Fluorescence in situ hybridization techniques make the analysis of transmissible chromosome alterations feasible. The current view that cancer (and to some extent other diseases) is a genetic disease, requiring several mutations to convert a normal cell to a transformed one, has allowed for a consideration of the association between induced genetic alterations and tumors in laboratory animal models and humans. The way forward for the field of genetic toxicology would seem to be to provide data that can be used for the description of tumor dose–response curves (or other disease outcomes) below exposure levels at which tumors can be reliably observed. In addition, by understanding the mechanism of formation of tumors in different species, a greater reliability can be placed on extrapolating from tumor data collected in an animal model to predicted responses in humans. The expectation is that the uncertainty in cancer (and perhaps noncancer) risk assessments will be reduced.

ACKNOWLEDGMENTS

I wish to thank Drs. James Allen and Andrew Kligerman for their very helpful review of this chapter. This manuscript has been reviewed in accordance with the policy of the U.S. Environmental Protection Agency, although it does not necessarily reflect the regulatory policies of the Agency.

SUGGESTED READING

Ashby, J., Waters, M. D., Preston J., Adler, I. D., Douglas, G. R., Fielder, R., Shelby, M. D., anderson, D., Sofuni, T., Gopalan, H. N. B., Becking, G., and Sonich-Mullin, C. IPCS harmonization of methods for the prediction and quantification of human carcinogenic/mutagenic hazard, and for indicating the probable mechanism of action of carcinogens. *Mutat. Res.* **352**, 153–157, 1996.

Burkert, J., Wright, N. A., and Alison, M. R. Stem cells and cancer: An intimate relationship. *J. Pathol.* **209**, 287–297, 2006.

Domon, B., and Aebersold, R. Mass spectrometry and protein analysis. *Science* **312**, 212–217, 2006.

Hanahan, D., and Weinberg, R. A. The hallmarks of cancer. *Cell* **100**, 57–70, 2000.

Hayashi, M., Tice, R. R., MacGregor, J. T., Anderson, D., Blakey, D. H., Kirsh-Volders, M., Oleson, F. B. Jr., Pacchierotti, F., Romagna, F., Shimada, H., Soutou, S., and Vannier, B. In vivo rodent erythrocyte micronucleus assay. *Mutat. Res.* **312**, 293–304, 1994.

Ashby, J., DeSerres, F. J., Shelby, M. D., Margolin, B. H., Ishidate, M., and Becking, G. (Eds.). International Programme on Chemical Safety, *Evaluation of Short-Term Tests for Carcinogens, Vols. I and II*, University Press, Cambridge, MA, 1988.

Kirkland, D. J., and Dean, S. W. On the need for confirmation of negative genotoxicity results in vitro and on the usefulness of mammalian cell mutation tests in a core battery: Experiences of a contract research laboratory. *Mutagenesis* **9**, 491–501, 1994.

National Research Council, Committee On Risk Assessment of Hazardous Air Pollutants. *Science and Judgment in Risk Assessment*, National Academy Press, Washington, D.C., 1994.

Ness, S. A. Basic microarray analysis: Strategies for successful experiments. *Methods Mol. Biol.* **316**, 13–33, 2006.

Preston, R. J., Mechanistic data and cancer risk assessment: The need for quantitative endpoints. *Environ. Mol. Mutagen* **45**, 214–221, 2005a.

Preston, R. J. Extrapolations are the Achilles heel of risk assessment. *Mutat. Res.* **589**, 153–157, 2005b.

Tennant, R. W., Margolin, B. H., Shelby, M. D., Zeiger, E., Haseman, J. K., Spalding, J., Caspary, W., Stasiewicz, S., anderson, B., and Minor, R. Prediction of chemical carcinogenicity in rodents from in vitro genetic toxicity assays. *Science* **236**, 933–941, 1987.

Tucker, J. D., Eastmond, D. A., and Littlefield, L. G. Cytogenetic end-points as biological dosimeters and predictors of risk in epidemiological studies. *IARC, Scientific Publication* **142**, 185–200, 1997.

U.S. Environmental Protection Agency (U.S. EPA), Guidelines for Carcinogen Risk Assessment. Risk Assessment Forum. U.S. Environmental Protection Agency, Washington, D.C., 2005. www.epa.gov/cancerguidelines

CHAPTER 26

Molecular Epidemiology and Genetic Susceptibility

RUTH M. LUNN

National Institute of Environmental Health Sciences, Research Triangle Park, North Carolina 27709

MARIANA C. STERN

Department of Preventive Medicine, Keck School of Medicine, University of Southern California, Los Angeles, California 90089

26.1 INTRODUCTION

Epidemiology is a public-health-related discipline that evaluates the determinants and distribution of disease or other health-related events in specified populations. Molecular epidemiology uses knowledge on disease mechanisms and tools, such as biological markers, that develop from the basic experimental sciences and incorporates them into analytical epidemiological studies, in order to better identify and understand the exposure–disease relationship. Successful molecular epidemiological studies combine the use of state of the art tools from the basic sciences and rigorous epidemiological methods.

Although biological markers such as the measurement of toxins or infectious agents in bodily fluids has been used in traditional epidemiological studies for several decades, the discipline of molecular epidemiology, which was proposed in the 1980s, expanded on this, by attempting to use biomarkers not only to measure exposure but also to measure the steps that occur between exposure and disease and the factors that modify the disease pathway. Thus, the discipline of molecular epidemiology overlaps both public health and experimental sciences. Molecular epidemiology contributes to public heath by improving exposure assessment (primary prevention), by identifying early endpoints of disease, which can be used for screening purposes or clinical trials, and by identifying individuals who are at the greater risk for disease, and it can help improve risk assessment by providing better exposure–response relations and information on susceptibility. Moreover, by incorporating biomarkers that can identify exposures, alterations, and susceptibility at the cellular and molecular level, molecular epidemiology allows for a mechanistic understanding of the disease process. One successful example of molecular epide-

Molecular and Biochemical Toxicology, Fourth Edition, edited by Robert C. Smart and Ernest Hodgson
Copyright © 2008 John Wiley & Sons, Inc.

miological research is the study of exposure to aflatoxin, a toxin present in moldy foodstuffs, and its relationship to liver cancer. The use of biomarkers facilitated the identification of aflatoxin as a known human carcinogen by improving exposure assessment; it also contributed to the identification and measurement of a synergistic interaction with hepatitis B virus infection to cause liver cancer and has provided mechanistic understanding on aflatoxin-induced liver cancer in humans.

Environmental exposures contribute greatly to disease burden. For example, 60–90% of cancer is thought to be attributed to environmental exposures (including lifestyle factors and exposure to environmental toxicants). However, the impact of environmental toxicants on our health is heterogeneous, which can be partly explained by differences in genetic susceptibility. The contribution of the constitutive genetic variation present in human populations to the observed variability of health determinants can be studied using different approaches. Studies that aim to understand the contribution of genetic variability at the population level are typically done using an epidemiological approach. The discipline of molecular epidemiology provides an ideal framework for the investigation of the combined effect of environmental toxicants and genetic factors. This has been facilitated by many milestones reached in the past two decades, such as advances in our understanding of the human genome, the development of more efficient high-throughput technologies for processing and analyzing biological markers, and the development of sophisticated analytical strategies to handle the complexity of the generated data. The past two decades have seen an exponential growth in the publication of epidemiological studies focusing on the analysis of genetic susceptibility to various diseases, such as cancer, diabetes, obesity, and others.

In the present chapter we will first review basic concepts of epidemiological research and follow with a discussion of different types of biomarkers used in molecular epidemiology, and finally we will concentrate on the study of genetic susceptibility. In our discussion we will put special emphasis in the challenges that using such markers introduce when used at the population level.

26.2 BASIC CONCEPTS IN EPIDEMIOLOGY

The objective of epidemiological studies is to determine whether a causal relationship exists between a putative risk or protective factor and disease. Thus, most studies are designed to test a hypothesis: Does exposure (e.g., consumption of cooked meats) cause disease (e.g., colon cancer)? After the study is conducted, the data are analyzed and the magnitude of the putative association between exposure and disease is calculated. The study is then evaluated to see whether the association (if observed) could be a result of random or systematic errors or whether other exposures or factors are influencing the exposure of interest (i.e., effect modifiers).

The objective of this section is to provide a brief introduction to the general concepts of epidemiology, in order to understand the discipline of molecular epidemiology. This section discusses (1) epidemiological study designs, (2) measures of occurrence and association, (3) random error (statistical testing, confidence intervals and statistical power), (4) systematic biases, (5) confounding, and (6) effect modification. Throughout the chapter, when available, we will use as a theme examples of diet and colon cancer. In particular, when available, we will use examples of meat consumption and heterocyclic amines (HCAs, are carcinogens formed when cooking meats at high temperatures) to illustrate the concept; the same examples will be repeated throughout the section (i.e., examples used in designing studies will be repeated in the other section when applicable).

26.2.1 Designing Epidemiological Studies

Epidemiological studies can be categorized into three major types: (1) descriptive or studies based on routine data, (2) analytical observational studies in which the researcher observes the occurrence of disease in two groups already segregated by disease or exposure, and (3) experimental studies in which the researcher randomly assigns individuals to the exposure in question.

26.2.1a   Descriptive Studies. Descriptive epidemiology focuses on the distribution of disease within a population by person, place, or time, and the data are usually obtained from routine data-collection systems. The studies can be carried out at the individual level or at an aggregated level if the group rather than the individual is the unit of study. Examples of descriptive epidemiology are (a) case reports or case series (a report of disease occurring in an individual or group of individual) and (b) studies reporting disease mortality or incidence by personal factors (such as age, ethnicity, sex, social class, occupation) or other personal variables (such as country and time).

One main advantage of descriptive studies is that they can study a large population of people for low costs, and thus they are useful for generating hypothesis and for public health planning. The major disadvantage is that they are not informative for establishing causality because they are only looking at a restricted range of variables, which may be proxy measurements for other factors.

Ecological studies are descriptive studies that make observations at the group level—for example, cities, factories, or countries—rather than at the individual level. They usually use data collected from routine data collection systems or from surveys. Disease is usually measured by incidence or mortality, and exposure is also usually measured by overall index. The major problem with these types of studies is that information on other exposures is not available and that exposure assessment is done at the population, and not individual, level. Thus, these studies are subjected to what is known as the "ecological fallacy," which means that the observation of a relationship at a population level does not imply that the same observation will be observed at an individual level because we do not know for certain that the subjects who are exposed to a specific agent are the same subjects who developed the disease. Nonetheless, this type of study design is very instrumental to generate new hypotheses that can then be tested using other study designs.

Example 1. Kono (2004) conducted an ecological study in Japan that compared consumption of selected nutrients with annual colon cancer incidence and mortality across time. Annual *per capita* consumption data of selected nutrients and foods was obtained from nationwide surveys, which were conducted from 1950 to 2000 and contained data from randomly selected houses across the nation. The food group identified as meat contained poultry and fish. Consumption of red meat was not available for the early years. Annual age-adjusted rates of cancer incidence (from cancer registries) were available for the period 1975 to 1997, and age-adjusted mortality rates were available for the period 1950 and 2000. Means of height and body weight for different age groups were available from nutritional surveys. (Results of this study are discussed in Section 26.2.2b.) *Source*: Kono (2004).

26.2.1b Observational Analytical Studies. Analytical studies, which can be observational or experimental, evaluate the determinants or causes of disease. Because of the ethical constraints for conducting experimental studies (discussed below), most epidemiological studies are observational. The major drawback of observational studies is that the observed groups may differ in many other characteristics in addition to the one that is being studied; thus the role of a specific exposure may be more difficult to establish than in experimental studies. There are three major types of observational studies: (1) follow-up or cohort studies, (2) case–control studies, and (3) cross-sectional studies.

Follow-up or Cohort Studies. Follow-up studies enroll a defined population of subjects (referred to as a "cohort") without disease, obtain exposure information on the subjects at the time of recruitment, and follow the subjects overtime for the development of disease. The incidence of disease in the exposed subgroup is compared to the incidence of disease in the unexposed subgroup (see Section 26.2.2 for a discussion of measures of associations). The cohort can be chosen from the general population (e.g., residents of a community), from a narrow defined population (e.g., nurses or doctors), or based on high exposure to the exposure of interests (e.g., occupational cohorts). The comparison group, typically the unexposed group, can be either internal (members of the cohort who are not exposed to the agent of interest) or external (members who are not part of the original cohort, such as the general population). External comparisons are more common in some occupational studies because the cohort includes only exposed individuals. In these studies the incidence or mortality of disease in the cohort is compared to the incidence or mortality of disease in the general population (in the same geographical area) or to other workforces not exposed to the agent of interest. Most cohort studies are prospective, that is, they follow individuals into the future; however, some studies, "historical cohorts," are constructed retrospectively through existing records of previous exposure. The study subjects are then traced to the present time or can be followed further into the future.

The major advantage of cohort studies is that exposure is measured before disease occurs and thus provides strong evidence of causality, given that the exposure of interest will be unlikely to be affected by disease status. Other advantages of cohort studies are that they allow measurement of disease in the exposed and unexposed population, can measure multiple outcomes, can evaluate rare exposures, and are not subjected to some types of biases, such as recall bias (see Section 26.2.4 for a description of biases). Disadvantages include (a) the requirement of large number of subjects, (b) expense, (c) requirement of an extensive time to set up and follow up, (d) cannot be used to evaluate rare diseases, and (e) can be associated with some type of biases such as selection bias.

Example 2. The Cancer Prevention Cohort Study (CPS II) mortality cohort consists of 1.2 million adults residing in 50 states, the District of Columbia, and Puerto Rico. Individuals in this cohort completed self-administered questionnaires in 1982 with information on race, diet, exercise, medical history, and other lifestyle factors. Findings on the relationship between colorectal cancer and meat consumption on a subset of this cohort, the CPS II Nutritional Cohort, were published in 2005 by Chao et al. The CPS II Nutritional Cohort was limited to 148,601 individuals aged 50 to 74 that resided in 21 states with a cancer registry and who provided information on meat consumption at two time periods, 1982 and 1992/1993. The 2005 publication reports the results of 1667 incident cases of colon or rectal cancer that were identified from the time of enrollment in 1992/1993 through August 2001. (Results of this study are discussed in Section 26.2.2b.) *Sources*: Calle et al. (2002) and Chao et al. (2005).

Case–Control Studies. These are retrospective studies that enroll individuals with disease (cases) and without disease (controls) at the beginning of the study. The prevalence of the exposure of interest is then compared between the controls and the cases. Cases may be incident, which are new cases that are identified in a defined time period, or prevalent, which are all the cases that have occurred up to a certain point in time. Controls can be selected from the population (such as the same geographical region as cases) or from the same hospitals as the cases; hospital controls are individuals who have disease other than the disease of interest. When using hospital-based controls, care should be taken not to select individuals with diseases that may share the same risk factors as the disease under study. To make subjects as comparable as possible, with the exception of the disease of interest, controls are often "matched" (either individually or by frequency) to the cases for age, sex, ethnicity, and other relevant variables.

Advantages of case-control studies are that they are cheaper and less time-consuming than cohort studies, they allow for the evaluation of multiple exposures, and they are good for the study of rare diseases. Disadvantages are that they do not provide information on incidence of disease, they are not good for evaluating rare exposures, they assess exposure after disease has occurred, and they are subjected to many types of "biases," such as selection and recall (see Section 26.2.4 for a description of biases).

Example 3. Butler et al. (2003) conducted a population-based case–control study that evaluated levels of HCAs, meat intake according to doneness and cooking method, and the risk of colon cancer. The study population consisted of participants selected from 33 counties in North Carolina who were part of the North Carolina Colon Cancer Study. Cases included 274 blacks and 346 whites, between the ages of 40 and 84 with invasive adenocarcinoma of the colon diagnosed from 1996 to 2000. Controls, 426 blacks and 611 whites, were randomly selected from the North Carolina Division of Motor Vehicles (under 65) and the Center for Medicare and Medicaid services (over 65). Exposure was assessed using a food-frequency questionnaire. Meat intake frequency data, cooking method, and level of doneness was used to estimate exposure values for three specific HCAs. (Results of this study are discussed in Section 26.2.2b.). *Source*: Butler et al. (2003).

Cross-Sectional Surveys. These are observational studies in which subjects are contacted at a fixed point in time without regard to either exposure or outcome. The exposure(s) and outcome(s) are then measured in the subjects, and the subjects are assigned to categories based on that information. Cases are a prevalent series of individuals with the disease or outcome, and controls are the remaining population. Cross-sectional studies often deal with exposures that cannot change, such as blood type or when there is a good correlation between past and present exposure. The association between the putative risk factors is evaluated by either (1) comparing the prevalence ratio of the prevalence of the factor of interest in those exposed versus those unexposed or (2) comparing the odds of exposure in the cases versus the odds or exposure in the controls.

The major advantages of cross-sectional surveys are that they are relatively inexpensive and easy to design and can evaluate multiple exposures and outcomes. The major disadvantages are that they provide limited information for establishing causality because they are looking at prevalence cases and do not establish a time sequence of events. They also are not useful for studying rare diseases, diseases of short duration, and rare exposures.

*26.2.1c **Experimental Studies.*** In experimental or interventional studies, the researcher assembles a cohort of individuals (with or without disease, depending on the objective of the study), randomly assigns them to an intervention or control group, and then follows the cohort to see whether the intervention prevents or cures disease. The studies may enroll people without disease (usually called "field study") to evaluate potential preventative strategies or individuals with disease to evaluate the efficacy of new treatments (usually called "clinical trials"). The major limitations of these studies are the costs and the ethical constraints, given that these studies are limited to evaluating exposures that either prevent or treat disease. Therefore, to carry out these studies, there should be evidence from laboratory, animal, or observational studies that the exposure may be beneficial.

Example 4. The Women's Health Initiative is a long-term national health study sponsored by the National Institutes of Health that consists of observational studies and randomized clinical trials to evaluate strategies for preventing heart disease, breast and colorectal cancer, and osteoporotic fractures in postmenopausal women. For the Dietary Intervention Trial, 48,835 postmenopausal women, aged 50 to 79, were enrolled between 1993 and 1998; 20% of them were minorities. The women were randomly assigned to the dietary intervention group (40%, $N = 19,541$) or control group (60%, $N = 29,294$) and were followed to see whether the dietary intervention reduced the development of colorectal cancer, breast cancer, and cardiovascular disease. The dietary intervention involved a series of seminars, and the objective was to reduce fat to 20% of energy intake and increase consumptions of fruits, vegetables, and grains; the controls only received dietary guidelines. At the end of year six, few women were able to reach the target goal of 20% energy from fat; however, there was a significant difference in fat consumption between the intervention group and the controls. Beresford et al. (2006) reported that at the end of an average of 8.1 years of follow-up, the risk of colorectal cancer was similar in the dietary intervention group and the controls. *Source*: Beresford et al. (2006).

26.2.2   Measuring Occurrence and Associations

After a study has been conducted, the data are analyzed to see whether an association is observed. This section briefly defines the two major measures of occurrence (incidence and prevalence) and discusses measures of associations in detail, focusing on the most common measures used in the epidemiological study designs described above, such as correlation coefficient, relative risk, and odds ratio.

26.2.2a Measures of Occurrence. Measures of occurrence are usually reported in descriptive epidemiological studies. The incidence of disease is the number of new cases occurring during a defined time period, whereas prevalence is the number of individuals with disease at a specified time period. The formula for incidence and prevalence rates are as follows:

Incidence rate: $\dfrac{\text{Number of new cases of a disease in a specified time period}}{\text{Population at risk during that time period}}$

Prevalence rate: $\dfrac{\text{Number of existing cases of disease at a single time point}}{\text{Total population at that point in time}}$

26.2.2b Measures of Association. Ecological studies usually measure the strength of an association between the exposure and disease (both of which are measured quantitatively) by calculating a correlation coefficient, denoted by r or by fitting a regression line, which predicts incidence or mortality as a function of the exposure. The first step is to plot (called a "scattergram") the amount of exposure versus disease mortality or incidence. Regression analysis is then performed to find the best line that fits the data. The correlation coefficient (which varies between -1 and $+1$) measures the degree of scatter around the regression line. The closer r is to -1 (negative correlation) or $+1$ (positive correlation), the stronger the relationship is between the two variables (e.g., exposure and disease). Correlation coefficients are often used to measure the strength of an association between exposure and disease but are limited because the magnitude of r is dependent on the range of the exposure variable and because correlation coefficients are not meaningful to describe a relative effect. Relative effects can be estimated from the slope and intercept of the regression line.

Example 1 (continued). In the ecological study conducted by Kono et al., Pearson correlation coefficients were calculated for annual data on consumption of selected nutrients and food and colon cancer rate with a lag time of 20 years. A positive correlation was found between consumption of fat ($r = 0.97$ for men and 0.98 for women), meat ($r = 0.98$ for men and 0.97 for women) and alcohol ($r = 0.96$ for men and 0.98 for women) and cancer incidence; similar results were found for cancer mortality. Negative correlations were found for consumption of miso and cereal. The high correlation found for consumption of alcohol and fat may be somewhat misleading because there were only small variations in consumption whereas there was a much larger variation in average intake of meat and cancer incidence.

Cohort studies measure the strength of an association between an exposure and outcome of interest or disease by comparing the incidence of disease in the exposed population to the incidence of disease in the unexposed population. Absolute measures of association calculate the difference of disease between the exposed and unexposed groups. "Excess or attributable risk" is the number of extra cases of disease that the exposure is responsible for, assuming that the relationship between exposure and disease is causal.

Relative measures of exposure (referred to as "relative risks") are measures of the strength of an association. The relative risk (RR) compares the likelihood of developing disease in the exposed group to the likelihood of developing disease in

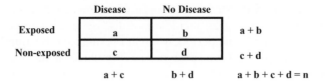

Cohort Studies: Risk Ratio (RR)

$$RR = \frac{\text{Risk in the exposed group (a/a+b)}}{\text{Risk in the unexposed group c/c+d}}$$

$$RR = ad\backslash bc$$

Case-control Studies: Odds Ratio (OR)

$$OR = \frac{\text{Odds of exposure among cases (a/c)}}{\text{Odds of exposure among controls (b/d)}}$$

$$OR = ad\backslash bc$$

Figure 26.1. Measuring the strength of an association relative risk (RR).

the unexposed group. Figure 26.1 depicts a 2×2 table, which is used in both cohort studies and case–control studies (see below) and contains the four "cells" that are needed to calculate the RR in (a) exposed people with disease, (b) exposed people without disease, (c) unexposed people with disease, and (d) unexposed people without disease. The RR is calculated by comparing the incidence rate of disease in the exposed group [a/(a + b)] to the incidence rate of disease in unexposed group [c/(c + d)]. An RR of 1 means that there is not an association between exposure and disease, an RR greater than 1 means that there is a positive association between exposure and disease (for example, an RR of 4 means that there is a fourfold increase in risk of disease among the exposed population), and an RR less than 1 means that there is an inverse association between exposure and disease; for example, an RR of 0.5 means that the exposure is protective of disease and that the exposed population is 50% less likely to get disease.

As mentioned in Section 26.2, case–control studies do not provide information of disease incidence in the exposed and unexposed population because subjects are selected based on disease status. For these studies, the RR is estimated by calculating the odds ratios of exposure (OR), which is the ratio of the odds of exposure in cases or the diseased group (a/c) compared to the odds of exposure in the controls or nondiseased group (b/d) (see Figure 26.1). For both cohort studies and case–control studies, the equations can be simplified as ad/bc. The OR is equal to the RR when the cases and controls are representative of the population and the disease is rare (thus the number of cases are a negligible part of the population). (See Section 26.2.3 for examples of RRs and ORs from the literature.)

Intervention studies usually measure "relative or absolute risk." Analysis can either be done on (1) intention to treat, which compares outcome on the subjects assigned to each groups regardless of whether they continued with the intervention strategy, or (2) randomized treatment, which measures the outcome observed in the subjects being treated. The preferred method is intention to treat, which avoids bias that can arise from different levels of participation due to "loss of follow-up" (the loss of contact with subjects); however, the true effects of the intervention may sometimes be diminished.

26.2.3 Evaluating Random Error: Confidence Intervals, Statistical Testing, and Statistical Power

Because case–control and cohort studies are the most commonly used studies in epidemiology, the remaining subsections will focus on these designs. Once an association between exposure and disease has been calculated, the next step is to evaluate the study to determine whether the calculated relative risk could result from error. The error could be random (due to chance), systematic (resulting from the study design) (discussed in Section 26.2.4), or due to confounding (discussed in Section 26.2.5).

The effect of chance is usually evaluated by statistical testing or by calculating confidence intervals. Statistical testing is used to determine whether the null hypothesis, which states that there is no association between exposure and outcome, is operating. The *P* value, calculated from statistical model, is the probability that the null hypothesis is true. The value of 0.05 is usually selected as the criterion to distinguish significant associations ($P \leq 0.05$) from nonsignificant association ($P > 0.05$). Type 1 or alpha errors occur when the null hypothesis is rejected but is actually true (which is 5% when $P = 0.05$), and type II or beta errors occur when the null hypothesis is accepted but is not true. *P* values depend on the magnitude of the association and the sample size of the population.

Confidence intervals (CIs) are usually the preferred means for evaluating chance in epidemiological studies because they are more informative than *P* values. CIs are the interval (usually 95%) around the risk estimate (which is a point estimate) that represent the upper and lower values of the true risk estimate. CIs provide information on both the precision of the point estimate and statistical significance. Wide confidence intervals indicate that there is a high degree of uncertainty on the accuracy of the point estimate and is usually a result of small study populations. CIs that exclude 1 means that there is a 95% probability that the null hypothesis is not operating.

When evaluating statistical significance, it is important to consider the statistical power of the study. The power of the study is the probability of finding a significant finding (e.g., $P < 0.05$ or CIs that exclude 1) when there is a true association. The power of the study depends on (a) the magnitude of the anticipated effects (e.g., RR) (the greater the effect, the greater the power to detect a significant finding) and (b) the width of the confidence interval, which depends on the study size (the bigger the study, the smaller the error) and the confidence interval chosen (e.g., 95% versus 99%). Statistical models are available for calculating study power. In general, larger studies are needed to see smaller effects. The use of RR, ORs and CI is illustrated below in the examples of the cohort and case-control studies identified in Section 26.2.1.

Example 2 (continued). In the cohort study conducted by Chao et al. (2005), RRs to measure the association between meat consumption and the risk of colorectal or rectal cancer were calculated using a regression model. Part of the study focused on long-term meat consumption, which was assessed by measuring meat consumption in 1982 and in 1992/1993. Individuals who were in the highest quintile of meat intake in both 1982 and 1992/1993 were compared to individuals in the lowest quintile for both those time periods. Long-term consumption of red meat (high intake) was associated with an RR of 1.43 (95% CI = 1.00 to 2.05) for rectal cancer, meaning that consumption of red meat increased the risk of developing rectal cancer 43% and that the association was close to statistical significance. Long-term consumption of processed meat was associated with a RR of 1.50 (95% CI = 1.04 to 2.17) for colon cancer, meaning that consumption of process meat was associated with a 50% increased risk of developing colon cancer and that the association appears to be statistically significant.

Example 3 (continued). The population case–control study conducted by Butler et al. evaluated the association between meat consumption and intake of heterocyclic amines with colon cancer. Exposure variables were categorized by quintiles based on the distribution of controls. ORs were calculated by logistic regression, comparing the fifth to the first quintile level of intake. Elevated ORs of colon cancer were observed for the highest quintile of consumption of red meat (OR = 2.5, 95% CI = 1.6 to 3.8), well or very well done red meat (OR = 2.1, 95% CI = 1.4 to 3.1), pan-fried red meat (OR = 2.5, 95% CI = 1.7 to 3.5), pan-fried chicken (OR = 1.5, 95% CI = 1.1 to 2.1), and intake on one specific HCA, 2-amino-3,4,8-trimethylimidzo[4,5-*f*]quinoxaline, DiMeIQx (OR = 1.8, 95% CI = 1.1 to 3.1). Associations were not observed for other categories of meat consumption such as baked red meat (OR = 1.1, 95% CI = 0.7 to 1.6) or broiled chicken (OR = 0.8, 95% CI = 0.5 to 1.2).

26.2.4 Evaluating Systematic Biases

In addition to chance, systematic biases can also affect the relationship between an exposure and disease. Biases lead to an incorrect estimate of the relationship between the exposure and disease that is an incorrect measure of the relative risk. Some biases will result in an effect being observed (i.e., statistically significant RR) when there is not a causal relationship, whereas other biases will result in obscuring a causal relationship between exposure and disease (refer to as biasing toward the null hypothesis). In an individual study, biases can be introduced during the selection of the subjects, follow-up of disease status, or exposure assessment. Biases can also occur in the evaluation of a causal relationship across studies.

26.2.4a Selection Bias. Selection bias occurs when the study subjects are not representative of the general population, such that the relationship between the exposure and disease is different for the participating subjects compared to the nonparticipants. Selection bias can be more problematic in case–control studies and are typically introduced by flawed control selection. Therefore, it is important that controls are selected from the same population as the cases using sampling methods that are representative of the cases. Some examples of selection bias are the use of volunteers, self-selection (subjects seek out the investigator), selection of controls from a different population than the cases, and low participation rates or difference in the participation rates between controls and cases. Some examples of selection bias that can occur in cohort studies are the "healthy worker effect" and loss or incomplete follow-up of the subjects. The healthy worker effect occurs when workers, who are usually healthier than the general population, are compared to the general population. Selection bias can bias the RR in either direction, toward the null hypothesis or toward finding an effect that is not real.

Example 3 (continued). Butler et al. also addressed the issue of selection bias in the case–control study on meat consumption. In this study, there was a 16% greater response rate among cases and they could not exclude the possibility of bias in their study.

26.2.4b *Measurement or Information Bias.* Measurement bias occurs when the classification of exposure or disease is not valid, which can occur because of errors by the observer, the respondent, or the instrument used to measure exposure or disease (e.g., questionnaire, laboratory assay). Misclassifications can be "nondifferential" or "differential." Nondifferential misclassification occurs when the exposure or disease is classified incorrectly in the same proportions in the two comparison groups—for example, in both cases and controls. This type of error will usually bias the RR toward the null hypothesis and make it harder to detect a true effect and can also hinder the ability to observe exposure response relationships. Nondifferential misclassification bias can occur in both cohort and case–control studies. The use of pathological or histological reports to classify disease can reduce disease misclassification, whereas the use of "biomarkers" to measure exposure or outcome can reduce exposure misclassification (see Section 26.3 for a discussion of biomarkers).

Example 3 (continued). The case–control study by Butler et al. was able to reduce misclassification error resulting from the level of doneness by using photographs in their exposure selection. They cited Keating et al. (2000), which is a study that compared HCA levels in home-cooked meat samples. Keatings found a significant different in HCA levels in three doneness levels when doneness was classified by photograph analysis but not when doneness was classified by a self-reported questionnaire. *Source*: Keating et al. (2000).

Differential misclassification occurs when the classification of disease is dependent on the exposure status or the classification of exposure is dependent on the disease status. Differential misclassification can bias the RR in either direction, and often the direction is unknown. Some examples of differential misclassification of exposure are "recall bias" and "observer bias." Recall bias, which is limited to case–control studies, occurs when the cases remember exposure differently than healthy controls; this type of bias usually results in finding a greater effect than what is real. Observer bias can occur if the observers, such as study interviewers, incorrectly assign exposure because they know the outcome status of an individual, or it can occur in the follow-up of disease if the observer knows the exposure status of the subject. Ideally, the observer should be blind to the outcome or exposure status of the study subjects.

The third major category of bias, "Publication bias," applies to the evaluation of a putative risk factors across studies that are reported in the published literature. Positive findings are probably published more often than negative studies, and thus a meta analysis or review of the literature would be biased toward finding a greater effect than the true effect.

26.2.5 Evaluating Confounding

Confounding occurs when an observed association between the suspected risk factor and outcome is due to a different risk factor, which is the cause of the disease.

For a variable to be a confounder, it must (1) be associated with the exposure under study in the population from which the cases are derived, (2) be associated with the outcome, either as a causal factor or as proxy for an unknown cause (e.g., social class), and (3) not be an intermediate step in the causal pathway between exposure and disease. It is important to evaluate known risk factors for a disease when designing a study, so that confounding can be evaluated. Some potential confounders that are specific to dietary studies and colorectal cancer risk include energy intake, energy-adjusted fat intake, and fiber intake. Confounding can be addressed in the study design or in the analysis, if it is measured. The study can be restricted to exclude individuals exposed to the confounder (e.g., nonsmokers), or controls and cases can be matched on potential confounders. In experimental studies, putative confounders can be controlled for by randomization. Confounders can also be controlled for in the analysis by either stratification or statistical modeling (e.g., regression modeling). Both methods will provide an adjusted risk estimate—that is, a risk estimate that controls for the influence of the confounder. Stratification is a technique that divides the cases and controls into categories (stratum) of the confounding variable and then calculates the RR or OR of the putative risk factor in each stratum. The results can be pooled together using the appropriate weighting to obtain an adjusted risk estimate. An example of a statistical test that calculates a weighted or adjusted risk ratio is the Mantel–Hanzel odds ratio.

Example on Confounding. The concept of confounding is probably best illustrated by using the well-established example of a known confounder, cigarette smoking, on the evaluation of alcohol consumption as a potential risk factor for lung cancer. Figure 26.2 depicts the results from a theoretical cohort study. If one looks at the total population, the RR for alcohol consumption is 1.5, suggesting that alcohol

Total Study Population

Alcohol Consumption	Lung cancer	No lung cancer	Total
Yes	72	1128	1200
No	48	1152	1200

$$RR = \frac{72/1200}{48/1200} = 1.5$$

Non-Smokers

Alcohol Consumption	Lung cancer	No lung cancer	Total
Yes	5	475	480
No	7	713	720

$$RR = \frac{5/480}{7/720} = 1.1$$

Smokers

Alcohol Consumption	Lung cancer	No lung cancer	Total
Yes	67	653	720
No	41	439	480

$$RR = \frac{67/720}{41/480} = 1.1$$

Figure 26.2. Example of confounding by cigarette smoking on evaluating the relationship between alcohol consumption and lung cancer.

consumption is a risk factor for lung cancer. However, if the population is divided (stratified) according to smoking status, then the RR is decreased to 1.1 in both strata, suggesting that alcohol consumption is not a risk factor and that the elevated RR was observed because of confounding by cigarette smoking. Following up on the studies on colorectal cancer and meat intake mentioned in previous examples, while there are many important potential confounders (such as energy intake) for studying the relationship between meat consumption and colorectal cancer, definite confounders have not been identified, and most of the potential confounders probably only affect (increase or decrease) the magnitude of the risk estimate.

26.2.6 Evaluating Interaction or Effect Modification

Interaction occurs when the effect of the putative risk factor is dependent upon the presence of another factor (that is an effect modifier). Because many diseases (especially chronic disease) are multifactorial, the evaluation of multiple risk factors and their interactions in etiological studies is very important and thus should be considered in the study design. The concept of interaction can apply to both the individual level, which is the biological interplay of two or more agents, and the population level, which is assessed using epidemiological studies. Interactions can be additive, synergistic, or antagonistic. Synergistic interactions between two risk factors occur when the risk of disease due to exposure to both risk factors is greater than the expected due to the sum or multiplication of the two individual agents. Some examples are (a) exposure to Aflatoxin and HBV infection vis-à-vis liver cancer risk and (b) exposure to asbestos and smoking vis-à-vis lung cancer risk. Genetic background has been shown to influence disease risk, and also modify environmental–disease associations, which is referred to as gene × environment interactions (see Section 26.4).

To evaluate whether a factor is an effect modifier, one can stratify by the suspected modifier and calculate the relative risk for the putative risk factor. If the RR risk varies substantially across stratum, then an effect modification should be suspected and evaluated using statistical models, which calculate departure from additive or multiplicative effects. When studying effect modifiers, the impact of sample size on study power is very important; therefore, large studies are needed to detect statistically significant interactions.

Figure 26.3 illustrates a hypothetical example of interaction between intake of well-done meats, intake of dairy products, and colon cancer risk. The best candidates for interaction with meat intake and colon cancer risk are probably genetic susceptibility factors, which are discussed in detail in Section 26.4. To our knowledge, nongenetic interactions with HCA intake and colon cancer risk have not been studied. Animal studies have shown that dairy products were protective of aberrant crypts (a preneoplastic colonic lesion) induced by 2-amino-1-methyl-6-phenylimidazo(4,5-b)pyridine (PhIP, a specific HCA). In this hypothetical example, a modest risk is observed for HCA intake and colon cancer in the total population (OR = 1.47). However, when the population is stratified according to dairy consumption, an increased risk for HCA intake is only observed in the strata of low dairy consumption (OR = 2.43), suggesting that the effect of high intake of HCAs from meat may only be relevant when intake of dairy products is low. *Source*: Tavan et al. (2002).

Total Study Population

HCA Intake	Cases	Controls
High	225	320
Low	325	680

$$OR = \frac{225 * 680}{320 * 325} = 1.47$$

High dairy intake

HCA Intake	Cases	Controls
High	90	175
Low	160	250

$$OR = \frac{90 * 250}{175 * 160} = 0.80$$

Low dairy intake

HCA intake	Cases	Controls
High	135	145
Low	165	430

$$OR = \frac{135 * 430}{145 * 165} = 2.43$$

Figure 26.3. Hypothetical example of effect modification between consumption of dairy products, HCA intake, and colon cancer risk.

26.3 BIOMARKERS USED IN MOLECULAR EPIDEMIOLOGY

Molecular epidemiology incorporates biological markers into epidemiological designs. Biomarkers, as defined by the World Health Organization, are "any parameter than can be used to measure an interaction between a biological system and an environmental agent, which may be chemical, physical or biological" (WHO International Programme on Chemical Safety, Biomarkers and Risk Assessment: Concepts and Principles, Environmental Health Criteria 155 World Health Organization, Geneva, Switzerland. 1993). As depicted in Figure 26.4, biomarkers can be used to measure exposure (including internal dose and biologically effective dose), effect (including early responses, alterations in structure or function, and disease), and effect modifiers (or susceptibility), and disease. They can also be used to measure confounders. However, the division between different classes of biomarkers is not always clear, and the same biomarker can be used in some studies to measure exposure and in other studies be used to measure effect. The use of biomarkers can increase the sensitivity and specificity of exposure assessment, provide earlier detection of disease, and evaluate gene environmental interaction.

Case–control studies are probably the most common study design used in molecular epidemiology. Biomarkers can also be employed in cohort studies; however, because these studies require large number of subjects, they are not always amenable to using laboratory assays that are costly or complex to complete. Nested case–control studies combine some of the advantages of both cohort and case–control studies and are often used in molecular epidemiology. In this study design, a cohort is assembled, and biological samples (for example blood, urine, DNA) are archived on each individual. Individuals are followed up for disease occurrence, and individuals who develop the disease become the cases. Controls are then selected at random from

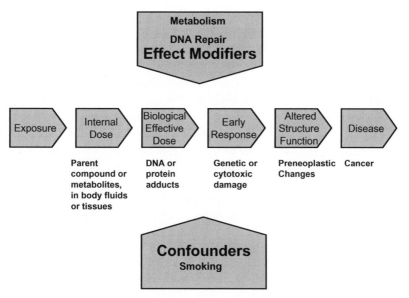

Figure 26.4. Molecular epidemiology diagram illustrating biomarkers in the disease pathway.

the cohort; the controls are usually matched on certain variables as the cases such as age, race, gender, and time of sample collection; often more than one control is selected for each case. The archived samples of the cases and controls are then assayed for the desired biomarker. This design offers the prospective nature, similar to cohort studies (exposure is assessed before disease); but because all samples do not have to be assayed, it reduces some of the cost associated with cohort studies.

This section describes the identification and evaluation of biomarkers (Section 26.3.1) and the major types of biomarkers—exposure, effect, and susceptibility (Sections 26.3.2 to 26.3.4); the subdiscipline of genetic susceptibility is discussed in greater detail in Section 26.4.

26.3.1 Identification and Validation of Biomarkers

To identify valid biomarkers for epidemiological studies, two main questions must be asked: (1) Does the biomarker measure what it is supposed to measure? and (2) Will the biomarker provide the information on exposure or disease that is required for that study? A positive answer to both questions is needed to determine whether a biomarker has been "validated" for epidemiological studies. Once a "validated" biomarker has been selected, the third issue to consider is biospecimen collection. Considerations must be given to what biological specimens are needed, how they should be stored and processed, and for how long they will be viable. This third issue is important for avoiding potential misclassification (e.g., due to sample degradation) and confounding (e.g., due to presence of other factors during biospecimen collection).

In identifying and validating a new biomarker several important questions need to be considered, including:

- Does the biomarker measure exposure accurately or does it cross-react with other potential risk factors?
- Will the biomarker capture exposure during a relevant period of time or is it more transient and will only capture exposure in the past few hours?
- What biospecimens need to be collected to assess this biomarker? Are they easy and affordable to collect, such as urine or blood?
- For how long does the biomarker stay viable in the biospecimen collected? Can it be frozen, and, if so, at what temperature, and for how long?
- When collecting biospecimens for biomarker analyses, what other data should be collected to address possible confounders to the measurements? In the example of HCA intake and colorectal cancer risk, some possible confounders to consider would be other sources of HCAs, such as cigarette smoking, and factors that may alter HCA metabolism, such as antioxidants and dietary fiber.

Lastly, as with all studies involving human subjects, any ethical implications of biomarker testing should be considered and properly addressed.

The following section discusses the three main steps of the biomarker validation process, or the "life cycle" of a biomarker: (1) its identification and development, (2) its validation using epidemiological methods, and (3) its application in etiological or clinical studies.

26.3.1a   Identification and Development of Biomarkers. The identification and development of assays for a biomarker occurs in an experimental setting, usually as part of basic research studies. A laboratory assay that can identify a biological process or measure a specific exposure (external or internal) must first be developed. Next, when appropriate, dose–response studies are done to determine the range and level of detection. Analyses are done to determine how many subjects may be misclassified as "negative" when being true positives (i.e., a false negative), a concept we refer as "sensitivity"; and analyses are done to determine how many subjects may be misclassified as "positive" when indeed they are negatives (i.e., false positive), a concept we refer as "specificity." The reliability of the biomarker analysis should be assessed to determine day-to-day variability and variability across different laboratories. Finally, the optimal conditions for collection, storage, and processing samples should be determined, such as the half-life and overall stability of the marker.

26.3.1b   Applied Transitional Studies. Once a biomarker has been identified and an assay for it has been developed in the laboratory, the relationship between the biomarker and the event that it tries to capture (i.e., exposure, disease, or susceptibility) needs to be studied. These studies are usually conducted among healthy subjects using a cross-sectional or a short-term longitudinal (e.g., prospective) study design. Similarly to laboratory-based studies, transitional studies will also try to assess the specificity, sensitivity, and variability of results obtained using the biomarker being validated. However, transitional studies use a population-based approach. This approach allows one to learn about interindividual variability, identify potential confounders and effect modifiers, and also test the feasibility of using this biomarker in the field. The endpoint of this type of study is the biomarker measurement.

26.3.1c *Validation in Applied Epidemiological Studies.* The final stage in biomarker validation is to apply it in the context of an etiological or clinical study. Common study designs used for these validations are case–control or nested case–control studies. These studies measure the prevalence of the exposure or effect in the study population, refine measurements of inter- and intra-variability, determine sensitivity and specificity at the population-based level, and identify more potential confounders and effect modifiers.

Example. Validation of a potential biomarker to measure HCA adducts in blood.

1. A laboratory-based assay needs to be developed in the lab to measure HCA adducts.
2. The assays' specificity and sensitivity is determined by using human samples spiked with the HCA-adduct and experimental models (i.e., rodents), fed known doses of the HCA of interest. Spiked human samples are used because HCAs are very potent mutagens and because some specific HCAs have been classified *as reasonably anticipated to be human carcinogens* by the National Toxicology Program. Therefore human studies involving voluntary exposure to purified HCAs are prohibited.
3. The biomarker is validated in a cross-sectional population-based study to determine if there are correlations between self-reported HCA exposure (for example, food frequency questionnaire using photographs similar to the case-control study reported by Butler et al.) and HCA measurements.
4. The biomarker is applied to a population etiological study focused on investigating the role of HCA exposure to cancer. Due to concerns of biases introduced by cancer status ("reverse causation"), a nested case–control study design is preferred.

26.3.1d *Sources of Biomarker Variability.* Inter-individual variability can be due to differences in ethnicity, gender, diet, age, and health status. Intra-individual variability can also be due to health status, diet, menstrual cycle, and season. Whereas inter-individual variability can sometimes be accounted for in the analysis phase, intra-individual variability is more difficult to adjust for. Therefore, careful consideration should be given to this issue. Other sources of variability result from measurement error, which can occur because of sample degradation, variability in sample handling, changes in protocols used for biospecimen processing, and biomarker detection (sometimes called "laboratory drift"). All of the above can be prevented or minimized by using biomarkers that have been properly validated, by using reliable laboratory controls and consistent methodologies, by using duplicates or triplicates during biomarker assay runs, and by collecting as much relevant information in the field as possible in order to identify potential sources of variability and confounding.

26.3.2 Biomarkers of Exposure

In traditional epidemiological studies, exposure is usually assessed by questionnaires, job titles, job exposure matrixes (conducted by occupational hygienists based on job descriptions and sometimes environmental measurements), environmental measurements (for example, concentrations in ambient air, water, or soil), or personal exposure monitoring. All of these measurements are subjected to error leading to misclassification, which decreases the power of the study to detect a significant association between exposure and disease.

Biomarkers of exposure measure the presence or magnitude of previous exposures to an environmental or endogenous agent on the individual level. The use of sensitive, specific, and validated markers can reduce exposure misclassification and thus increase the power to detect a causal association. Furthermore, the use of biomarkers of exposure may help identify sources of exposure that would have otherwise gone undetected, commonly referred to as "diffuse exposures." Misclassification of exposure in molecular epidemiology studies that use biomarkers is related to the validity, reproducibility, and stability of the marker. The stability of the marker is a critical issue because exposures need to be assessed that occurred at the relevant time related to disease development. Another important consideration is the tissue that is sampled. Molecular epidemiology studies are usually done on biological samples that are less invasive to obtain, such as urine, saliva, and blood; however, these may not always be the biologically relevant tissue to assess exposure, which depends on the route of administration, location of metabolism, and target organ. Finally, potential biases should also be considered, such as "reverse causation," particularly a problem in case–control studies, where the disease status may induce or alter the biomarker being measured, thus leading to erroneous results. This section discusses how biomarkers are used for studying the following types of exposure: (1) internal dose, (2) biologically effective dose, and (3) endogenous exposure. Candidate biomarkers of exposure to HCAs will also be discussed as examples.

26.3.2a Biomarkers of Internal Dose. Whereas exposure is the potential or opportunity of an agent to be delivered from the external environment to the internal body, dose is the amount of the agent that is actually deposited within the body. Biomarkers of internal dose usually measure the presence of the agent or its metabolites in the body, and thus they demonstrate that the agent has been absorbed and distributed. A variety of tissue or fluids from the body have been used to measure the dose, including blood, urine, feces, breast milk, amniotic fluid, hair, nails, saliva, adipose tissue, and exhaled air. Biomarkers of internal dose, which include measurement of a chemical or its metabolite, metal, viral DNA, viral protein, and many others, have been used in human and animals studies. Table 26.1 contains a list of some specific biomarkers that have been commonly used in molecular epidemiology. As mentioned above, a critical factor on the utility of a biomarker is the stability of the biomarker or toxin in the body; many toxins are cleared rapidly (i.e., they have a short half-life) and thus measurements are of recent exposure, whereas past exposures may be more relevant for disease with long latency such as cancer. The half-life of a biomarker is particularly a concern when trying to measure dietary exposures, which are more prone to vary from day to day. Therefore, a biomarker that only captures exposure within the past 12 hr may not be relevant for past dietary exposures of seldom-eaten foods. Archived samples (e.g., from nested case–control studies) can sometimes be used to measure biomarkers of past exposures.

TABLE 26.1. Selected Examples of Biomarkers of Exposure

Type of Biomarker	Example
	Biomarkers of Internal Dose
Chemicals	*Parent Compound*
	Organochlorine compounds
	TCDD in adipose tissue
	Volatile organic chemicals in exhaled air
	Metabolites
	Aflatoxin urinary metabolites
Chemical mixtures	Cotinine levels in urine (a compound present in tobacco)
Metals	Lead blood levels
Biological agent	HBV viral DNA
	HBsAg viral protein
	Anti-HBV core antigens
	Biomarkers of Biologically Effective Dose
Chemicals	4-Aminobiphenol protein adducts (usually serum proteins)
	Aflatoxin B_1-N^7-guanine adducts in urine
	Endogenous Exposure
Hormones	Estrogen, catechol estrogen-3,4-quinones

HBV, hepatitis B virus; HBsAg, hepatitis B virus surface antigen; TCDD, 2,3,7,8-tetrachlorodibenzo-*p*-dioxin.

26.3.2b Biologically Effective Dose. Biologically effective dose is the amount of the agent or its active metabolites that has interacted with cellular macromolecules at a target site or surrogate. Because these biomarkers provide evidence of potential toxicological damage, they are probably more informative of disease risk than markers of internal dose, because they capture the balance between absorbed agent ingested, distributed, and detoxified. Specific examples of biomarkers of exposure are described in Table 26.1. The most common markers that have been used in cancer studies are protein or DNA adducts formed from the interaction of chemical carcinogen or their metabolites. If DNA adducts are not repaired, they have the potential to cause mutations that can lead to disease such as cancer; however, adducts may also be repaired and thus the presence of adducts does not ensure that disease will occur. Similar to markers of internal dose, the persistence (i.e., stability) of the biologically effective marker in the body is important in measuring the time of critical exposure. In general, biologically effective dose markers are able to integrate a longer time period of exposure than markers of internal dose. Persistence depends on the stability of the agent (usually a chemical) and the rate of DNA repair (for DNA adducts) or protein turnover (for protein adducts). DNA adducts have been measured in blood, urine, tissue, and tumors. Protein adducts and DNA measured in nontarget tissue are usually considered to be "surrogate markers."

Examples of Biomarkers of Exposure. In the studies described in Section 26.1, biomarkers were not used to measure meat consumption or HCA intake. In the case–control study reported by Butler et al. (Example 1.3), HCA intake was assessed by using meat consumption data, cooking methods, and level of doneness. A food-frequency questionnaire was used to assessed meat intake and cooking methods, and photographs were to assess level of doneness. However, the potential for misclassification of exposure is probably high due to inconsistent reporting, difficulty in quantifying cooking doneness, day-to-day variation in the diet, and incomplete information on cooking practices (which can effect HCA levels), thus making in harder to detect an association between HCAs and cancer. The development of biomarkers of exposure to specific HCAs (internal or biologically effective) is being pursued. An intervention study in humans has demonstrated that PhIP could be measured in urine with 12 hr of consumption of broiled beef. Levels of PhIP declined to baseline within 48–72 hr, demonstrating the transient nature of the biomarker. The latter presents a problem when trying to use this biomarker to capture HCA intake, which may not be consumed on a daily basis. In addition to the studies that used urinary biomarkers, other studies in humans have evaluated potential HCA biomarkers such as HCA serum albumin and hemoglobin adducts, as well as HCA levels in hair, but these markers have not been properly studied or validated. The importance of using validated markers was discussed in Section 26.2.1.

Biomarkers of exposure have been very effective in the study of other chemicals, such as arylamines, potent carcinogens that are present in cigarette smoke and various occupational exposures such as the dye industry. Using a bladder cancer case–control study in Los Angeles County, Skipper and colleagues measured 3- and 4-aminobiphenyl (ABP) hemoglobin adducts as a biomarker of exposure to arylamines. They found an overall strong association between high levels of ABP adducts and bladder cancer status; but more interestingly, they still observed a strong association among nonsmokers, which suggests that other sources of arylamines might contribute to bladder cancer. This study is an example of how the use of biomarkers of exposure can help not only to measure but also to better understand which exposures are relevant for disease etiology. *Sources*: Alexander et al. (2002), Kobayashi et al. (2005), Strickland et al. (2002), and Skipper et al. (2003).

26.3.2c Endogenous Exposure. Biomarkers can also be used to measure endogenous exposures that may result in disease. For example, sex hormones such as estrogens and androgens are known risk factors for breast cancer. Some examples of candidate biomarkers of exposure to study the relationship between endogenous estrogen exposure and cancer are serum estrogen levels, estrogens metabolites (catechol estrogen quinones), and estrogen adducts. Other examples of endogenous exposure are related to age-related diseases. In this case, the marker of exposure (such as oxidative stress and nitric oxide) can be the same as markers of effect from exogenous exposure.

26.3.3 Biomarkers of Response or Outcome

Biomarkers of response, also referred to as biomarkers of effect or disease, measure an event resulting from a toxic interaction, which is predictive of an adverse health effect or is an adverse health effect. The development of biomarkers of effect and their use in studies have many public health applications, including providing information of the etiology of disease, early detection of disease (screening), and evaluating the efficiency of therapeutic agents. Biomarkers of effect also are useful in

TABLE 26.2. Selected Examples of Biomarkers of Effect Relevant for Carcinogenesis

Class of Biomarker	Type of Effect	Specific Examples
Genetic-related damage	Gene mutation	HPRT, GPA, p53, K-ras
	Gene expression and gene expression arrays	Tumor suppression genes (e.g., p53), oncogenes (e.g., K-ras), enzymes (e.g., aromatase, receptors
	Cytogenetic effects	Chromosomal aberrations, micronuclei formation
Oxidative stress	Oxidative DNA damage	8-OHdG
	Lipid peroxidation	MDA, HNE, acrolein, and isoprostanes
	Proteins	Glutathione and glutathione disulfide
Internal parameters	Hormones	Estrogen, androgens
	Growth factors	Insulin-like growth factors, TGF-β
Inflammation	Peptides	Cytokines
Cell proliferation	Histology	Aberrant crypt foci (colon polyps precursors)
	Protein expression	PCNA expression
Altered structure or function	Histological preneoplastic changes	Polyps (mucosal lesions) adenomas
Cell invasion	Protein expression	VEGF expression

MDA, malondialdehyde; HNE, 4-hydroxy-2-neonal; 8-OHdG, 8-hydroxy-20-deoxyguanosine; PCNA, proliferating cell nuclear antigen; TGF-β, transforming growth factor beta; VEGF, vascular endothelial growth factor.

mechanistic studies because these markers usually measure the sequence of events in the disease process. In these studies, association between different types of markers (such as exposure, early effect/late effect, clinical disease, and effect modifiers) can be measured. Often case–case study designs, comparing exposed and unexposed individuals with disease, are used to evaluate the association between biomarkers in the disease pathway. This type of study design has been especially useful in studying complex, chronic disease such as cancer.

Similar to exposure markers, these markers can be evaluated in a wide range of body fluids or tissues, and the biological relevance of the tissue as well as the ease of obtaining the tissue are important factors to consider. There are many examples of markers of effects, which have been developed for many different types of diseases or disease processes; thus it is beyond the scope of this section to discuss these types of markers in detail, but instead this section will briefly describe some general principles of markers of effect and give a few examples to illustrate these principles. Table 26.2 summarizes selected examples of biomarkers of effect relevant for carcinogenesis, which were selected because of the role these markers play along the carcinogenic pathway of tumor development. In a similarly fashion, one could think of relevant biomarkers of effect for other diseases, by taking into account the disease process. For example, nonspecific markers of inflammation such as protein levels of C-reactive (CRP) protein and exhaled nitric oxide may serve as biomarkers for heart disease and asthma, respectively.

26.3.3a Stage of Events. In chronic or multistage disease processes such as cancer, biomarkers can be developed to measure a wide spectrum of events form the earliest steps, such as genetic changes, to later events such as preneoplastic lesions (see Figure 26.4). Because mechanisms are not fully elucidated for many diseases, the division between biologically effective dose biomarkers (damage that may or may not be related to disease) and biomarkers of early events (damage predictive of disease) is not always clear. For example, single-strand breaks have sometimes been used as a marker for a biologically effective dose in some studies and a marker of early effect in others.

26.3.3b Types of Markers. Biomarkers of early events have been developed for many diseases. Many of the molecular epidemiology studies evaluating early markers of effects have been cancer studies and have focused on markers related to genetic damage, such as gene mutations, changes in gene expression, and cytogenetic effects. Other types of markers that are used to study cancer and other diseases include markers of oxidative stress, inflammation, and internal parameters, such as growth factors and hormone levels.

26.3.3c Specificity. Most biomarkers of effect are nonspecific for the toxin under study; however, there are a few examples where a chemical-specific effect has been observed, and thus the biomarker of effect can be considered a "fingerprint of exposure." Mutational spectrum studies have identified specific mutation changes that are exposure-specific. For example, human studies have reported a high prevalence of a specific mutation in the *p53* gene (G to T transversion at codon 249) in liver tumors from individuals exposed to aflatoxin, whereas the mutation is not frequently observed in liver tumors of unexposed individuals. This mutation has been hypothesized to be both a "fingerprint" of aflatoxin exposure and an important step in the carcinogenic pathway. Conversely, many makers of effects, such as glycophorin A mutations and chromosomal aberrations, are not specific to the environmental toxin.

26.3.3d Preclinical Biomarkers. Preclinical biomarkers are used in clinical trials for screening purposes or to evaluate therapies. Also, these markers are useful to shorten the time of disease development in nested case–control studies within a cohort. These markers are used to predict disease and are often called surrogate endpoints, intermediate endpoints, or early-outcome predictors. It is not always necessary that they be in the direct pathway of exposure to disease (i.e., may be a surrogate for a step in the pathway). Biomarkers of early detection differ from biomarkers of risk in that they require a higher sensitivity for predicting disease. Some examples of biomarkers that have been used for early detection of disease include (a) prostate-specific antigen (PSA) for prostate cancer screening and (b) the detection of polyps in colon cancer screening.

Candidate Biomarkers of Effects for HCA-Induced Colon Cancer. Although there are many potential candidate biomarkers of effect for colon cancer in general, such as mutations in *APC tumor suppressor gene*, MMR (mismatch repair genes), and K-*ras oncogene*, we are not aware of any molecular epidemiology studies evaluating the association of biomarkers of effects and exposure to specific HCAs, such as PhIP. However, studies in animals or human cell lines treated with PhIP have provided information on disease pathways and on possible candidate markers of effects that could be developed for human studies (see Figure 26.4). Studies in human cells have shown that PhIP causes gene mutations, micronuclei formation, chromosomal aberrations, sister chromatid exchange, DNA damage, and unscheduled DNA synthesis. *HPRT* mutations observed in human lymphoid cell lines were mainly G to T transversions. All of these endpoints (if they are present from exposures to humans from cooked meat and if they are validated) represent potential markers for human studies and probably would be useful in studies evaluating the mechanism of HCA toxicity in humans. Studies in animals have detected mutations in APC (all of which involved a guanine deletion at a 5′–GGGA–3′ site) or *β-catenin* genes in colon tumors induced by PhIP. It is not known whether these mutations would be observed in humans or animals fed well-done meats. *Source*: NTP (2004).

26.3.4 Biomarkers of Susceptibility

Biomarkers can also be used to identify factors that increase the likelihood that an individual will develop disease. This is an important area of research in molecular epidemiology as it becomes more evident that not all risk factors will contribute to disease equally across the human population. Therefore, in order to determine whether an environmental agent is related to disease, those factors that are also required for disease development need to be taken into account. Otherwise, many disease risk factors may go undetected. Examples of susceptibility factors that can be ascertained using biomarkers are some viral infections, which may predispose to specific diseases (for example, HIV infection and Kaposi sarcoma) or HBV infection and liver cancer. Biomarkers can also be used to measure dietary factors that can contribute to disease. The most common susceptibility factor studied using a molecular epidemiological approach are hereditary factors, which are discussed in the following section.

26.4 BIOMARKERS OF GENETIC SUSCEPTIBILITY

Our understanding of the contribution of genetic variation to disease susceptibility in the human population is still in its infancy. There are several examples of rare mutations that can explain various diseases and syndromes, such as *Xeroderma pigmentousm* and *Ataxia telangectasia*, which can be caused by single-point mutations in DNA repair genes. However, in contrast with these rare syndromes, most common diseases and the majority of cancers are likely to be caused by a combination of environmental exposures and their interplay with multiple genetic variants, referred to as "susceptibility genes." Unlike rare mutations, common genetic variants or polymorphisms are present in the human population at a frequency of at least 1% and are usually of low penetrance. Common genetic variants, mostly single nucleotide polymorphisms (SNPs), occur throughout the genome. A 1999 genome survey suggests that, on average, each gene contains approximately four variants in its coding region, with allelic frequencies of 1–5%.

The challenge for molecular epidemiological studies on genetic susceptibility consists of identifying which of these genetic variants can contribute to disease risk and also identifying which ones can modify the effect of environmental exposures. Until now, the approach that most studies have taken to answer these questions has been the *"candidate gene approach."* The underlying hypothesis of this approach is that genetic variants that alter gene function will contribute to disease development if they occur in genes that are relevant for disease causation. Given the multiple biological pathways that are relevant for disease causation, genetic susceptibility genes can have an effect at different points in the exposure–disease continuum. In the studies of meat intake and colorectal cancer mentioned in earlier examples, the suspected carcinogens are HCAs and polycyclic aromatic hydrocarbons (PAHs). Susceptibility genes could affect the early steps in disease development, with examples being variants in genes that metabolize HCAs and PAHs, such as P450 enzymes *CYP1A2* and *CYP1A1* and many others. Susceptibility genes might also be found among those genes that participate in the repair of damage induced by HCAs, such as genes that participate in the nucleotide excision (NER) or base excision repair (BER) pathways (e.g., XPD, XRCC1). Variants in genes that participate in cell cycle control and apoptosis might also constitute good susceptibility gene candidates, given the contribution of alterations in these processes to the proliferation of accumulated mutations and growth of premalignant lesions. Finally, variants in genes that convert premalignant lesions (e.g., polyps) to cancerous lesions (e.g., carcinomas), such as variants in genes that are relevant for angiogenesis, might also be good candidates as susceptibility genes. To date, most epidemiological studies on genetic susceptibility have been done on cancer, focusing on variants in carcinogen metabolism enzymes as putative susceptibility genes. These include enzymes that participate in the activation of carcinogens, such as P450 enzymes, *N*-acetyl-transferases, cyclooxygenases, and sulfo-transferases, as well as enzymes that are known to detoxify carcinogens, such as glutathione-*S*-transferases, glucuronidases, and catalases. With the increased availability of re-sequencing efforts to identify genetic variation, the list of genes studied in association studies has expanded to other biological processes, such as DNA repair, apoptosis, cell cycle, and hormonal pathways.

The main prediction of molecular epidemiological studies that use the candidate gene approach is that subjects who carry susceptibility genes will be more likely to develop the disease of interest than subjects who do not carry these variants. Epidemiological studies such as case–control or nested case–control studies are conducted to test this prediction by comparing the prevalence of a given genetic variant among subjects who developed the disease to that among comparable subjects who did not. In these studies, genetic variant is the putative risk factor or protective factor of disease and is treated similarly as exposure. Using this approach, epidemiologist first choose "candidate genes" that are selected based on their known role in the disease of interest. Next, genetic variants that affect gene function are identified within each candidate gene. Once identified, genetic data are collected from all participant subjects from an epidemiological study specifically designed to investigate the disease of interest. Once data are collected, the contribution of each genetic variant to disease risk is studied using standard epidemiological methods. These steps are described in detail below.

26.4.1 Types of Genetic Variants

With the completion of the human genome project, efforts shifted to the identification of genetic variation across the genome. Currently, there are several efforts addressing genetic variations and various databases and publications that report the identified variants are available (see Suggested Reading for more information). However, the existing databases and reports do not cover all genes, nor all possible ethnicities; therefore, in many instances, re-sequencing of the gene of interest to identify all potential genetic variants may still be necessary.

Whereas most genetic variation results from SNPs, polymorphisms can also be due to deletions, insertions, and microsatellites. Insertions and deletions usually lead to frameshift mutations that can lead to truncated proteins. Microsatellites can affect protein function by leading to deletions and expansions. The impact of SNPs on gene function will depend on the location of the base change and the nature of the change. SNPs that occur in coding regions (cSNPs), or 5' and 3' regulatory regions, are more likely to impact protein function than SNPs in introns. However, the latter can have an effect when they occur in splice sites. cSNPs that induce amino acid changes in the protein (nonsynonymous cSNPs) are more deleterious than those that do not (synonymous cSNPs), which are usually silent. The impact of missense nonsynonymous SNPs on gene function will depend on the nature of the amino acid change. Nonsynonymous SNPs that induce a change of one amino acid for another of similar polarity ("conservative" cSNPs) are likely to have less impact than those that induce changes of very different amino acids ("nonconservative" cSNPs). Nonsense nonsynonymous SNPs are highly deleterious because they lead to truncated proteins.

One of the challenges in molecular epidemiological studies of genetic susceptibility that use the "candidate gene approach" is the selection of genetic variants in the gene of interest. Two of the main criteria to consider are the impact of the variant on gene function and the allelic frequency of the variant allele. If variants are too rare in the population, it becomes more unlikely to detect any potential associations due to lack of statistical power. In vitro or in vivo (e.g., animal experiments) studies that directly assess the functional impact of a variant gene by testing the effect of each allele on either gene or protein function are very instrumental in understanding which genetic variants may constitute good susceptibility gene candidates. Such studies may include analyses of protein expression, biochemical analysis of protein activity, gene expression, protein–protein or protein–DNA binding or carcinogenesis experiments using knock-in mice. However, these types of studies are vary labor-intensive and/or expensive and not many genes have been evaluated to date. In the absence of experimental data *in silico* (i.e., computer-based), methods to predict functional impact can be used. These methods rely on various algorithms such as SIFT, PolyPhen, and CODDLE, which take into account (a) sequence conservation across species and (b) the location of the variant in the gene, among others, and provide predictions on how likely any given variant is to impact gene or protein function. (Resources to access these programs are listed at the end of the chapter.) The ENCODE project, launched by the National Human Genome Research Institute, proposes to identify all functional elements in the human genome using a combination of experimental and computational approaches and make these information publicly available. This resource will be instrumental in complementing the available *in silico* methods for SNP impact prediction.

An alternative approach to the analysis of individual candidate gene variants is to use as the unit of analysis gene "haplotypes" instead of single SNPs. This overrides the need of knowledge on the functional consequences of genetic variants. Several studies of linkage disequilibrium across the human genome have shown that most of the genetic variation can be characterized by a handful of common haplotypes. The latter can be identified by SNPs that uniquely "tag" them, commonly referred to as haplotype-tagging SNPs (htSNPs). Provided that the proper htSNPs are selected, most of the genetic arquitecture of a given gene can be captured by a few SNPs per gene. To aid these efforts, the International HapMap project was launched in 2002, with the goal of cataloguing human sequence variation and haplotype structure across various ethnicities, making all data available to the public.

26.4.2 Analytical Methods

Once candidate polymorphisms, or haplotypes, are selected for analyses, the main goal is to determine whether such genetic variant is associated with disease risk. The statistical methods used for these analyses are similar to those used for traditional epidemiological analyses that evaluate associations between a putative risk factor and disease risk (refer to Section 26.2.2). Similarly, the interpretation of the observed measures of effect (e.g., ORs) are no different from any other risk factors. Therefore, if an OR = 1.5 is observed for a given gene variant, this is interpreted to mean that this variant increases the likelihood of developing disease by 50% among carriers of the genetic variant when compared to subjects who carry the more common allele. As with analyses of nongenetic risk factors, careful examination of potential factors that might confound the association between the candidate gene or haplotype and disease must be considered. An example of a confounder factor of genetic studies is age, which is a known risk factor for several diseases such as cancer and neurological diseases. Many genotypes may have an impact on longevity due to their role in several diseases. Therefore, if carriers of a particular polymorphism died earlier than noncarriers, the distribution of genotypes for this gene among older subjects may not be representative of the general population. A gene–disease association study including several age groups, without adjusting for age, may incorrectly conclude that carriers of the variant gene are associated with disease. Restricting analyses to older subjects would indicate that such association does not exist. Another common factor that can confound gene–disease associations is ethnicity/race, which is discussed in Section 26.4.4.

26.4.3 Gene–Environment Interactions

Whereas there are several examples of diseases that are considered purely "genetic," most diseases arise due to the complex interplay between genes and the environment. A classic example is the Mendelian disease phenylketonuria (PKU), where carriers of a rare mutation in one gene are unable to metabolize phenylalanine, leading to an accumulation of this amino acid in the brain, with severe consequences. If dietary exposure to this amino acid is prevented, the presence of the gene has no effect. Similarly, intake of phenylalanine has no deleterious effects among subjects who lack this rare variant. In a similar fashion, most common variants in candidate genes contribute to disease development by increasing the susceptibility of carriers to specific environmental toxicants. For example, subjects who carry variants that increase the activation of HCAs may be more prone to develop colorectal cancer when the intake of well-done cooked meats is high, as compared with those subjects who carry variants that are less efficient for the activation of this carcinogen.

Effect of HCA intake among the total study population ignoring genotype

HCA Intake	Cases	Controls
High	850	400
Low	2,000	3,000

$$OR = \frac{850 * 3,000}{400 * 2,000} = 3.19$$

Effect of HCA intake among subjects with CYP1A1 increased activity (Val/Val carriers)

HCA Intake	Cases	Controls
High	625	125
Low	500	750

$$OR = \frac{625 * 750}{125 * 500} = 7.50$$

Effect of HCA intake among subjects with CYP1A1 normal activity (Ile/Ile or Ile/Val carriers)

HCA intake	Cases	Controls
High	225	275
Low	1,500	2,250

$$OR = \frac{225 * 2,250}{275 * 1,500} = 1.23$$

Figure 26.5. Hypothetical example of gene–environment interaction between HCA intake, *CYP1A1 Ile462Val* gene variant, and colon cancer risk.

In other words, the magnitude of the effect of the environmental exposure varies depending on the presence or absence of a given genetic variant. In epidemiological terms, this phenomena is known as a gene–environment interaction, which is one example of effect modification, as we describe in Section 26.2.6. We give a numeric example of a gene–environment interaction in Figure 26.5. In this example we see that when we ignore genotype, intake of HCA has a moderate association with colon cancer risk. However, when we take into account the genotype of a polymorphism in a gene that participates in the activation of HCA in the colon lumen, we see that the effect of HCA seems to be restricted to subjects who carry the variant allele for this gene, among whom the effect is very strong (~7-fold increased risk).

The HCAs example is one of several types of gene–environment interactions. The classic PKU example described above is an example where both gene and exposure are necessary for disease to occur. Several other types have been described in the epidemiological literature, which vary in the relationship of gene and exposure to disease and to each other. Overall, the molecular epidemiological studies in cancer published so far suggest that most common genetic variants tend to have either no effect or very modest effects on cancer risk when environmental factors are ignored. However, when relevant environmental exposures are taken into account, it can often be observed that the presence of genetic susceptibility genes may considerably increase the magnitude of the effect of the exposure, in many instances allowing the detection of exposure–disease associations that may otherwise go undetected. For example, an elevated risk ratio may not be observed between an exposure and disease when one looks at the total population; however, if the population is stratified according to genetic variability relevant to that exposure and disease, the association may become apparent in the strata of the individuals who carry the variants conferring susceptibility. It has become increasingly clear that to better understand the role of various environmental factors, it is necessary to take into account the underlying genetic variation of those genes that play key roles in the mechanism of action of the exposures of interest. Furthermore, it is necessary to study the combined effect of most genes along a pathway, instead of just focusing on one gene at a time. These "pathway-driven" analyses are becoming more common and are likely to become the standard in analysis of genetic susceptibility, further contributing to the more mechanistic ambitions of the molecular epidemiology field.

26.4.4 Special Considerations

Not unlike other nongenetic risk factors studied in epidemiological studies, analyses of gene–disease associations are also susceptible to various biases and confounders. Unfortunately, there are too many inconsistent results in the molecular epidemiological literature. We briefly describe below the main factors that contribute to such inconsistencies.

26.4.4a Population Admixture. One of the main concerns in epidemiological analyses of genetic susceptibility is the fact that genetic variation differs across ethnicities. Similarly, the incidence rate of many diseases also differs across ethnicities. When both these factors vary together, "confounding by ethnicity" or bias from "population stratification" may occur, leading to false positives. A classic example that illustrates this type of bias is the study reported by Knowler in 1988 on non-insulin-dependent diabetes mellitus (NIDDM) among Native Americans of the Pima and Pagago tribes in the Gila River Indian Community described in Example 5. When this bias occurs, it may be erroneously concluded that a given genetic variant is associated with disease, when in fact it is not.

Example 5. Among residents of the Gila River Indian Community, the prevalence of a haplotype in the immunoglobulin G (Gm haplotype) was compared between subjects diagnosed with NIDDM and healthy controls. An inverse association was observed between the Gm haplotype and disease, suggesting that carriers of this haplotype are protected against development of diabetes. Given that subjects in this population have varying levels of Caucasian heritage, the authors reanalyzed their data, taking into account genetic background. They observed that whereas among the Caucasian population the prevalence of the Gm haplotype was close to 66%, among full-heritage Native Americans it was only 1%. At the same time, whereas the prevalence of NIDDM among Caucasians was close to 15%, among full-heritage Native Americans it was 40%. When analysis of gene–disease association were done stratifying by an index of Native American heritage (i.e., how many Native American ancestors each subject had), no association was observed between the Gm haplotype and NIDDM. This example shows that ethnicity was a confounder in this gene–disease association. *Source*: Knowler et al. (1988).

Whereas the impact of population admixture, or population stratification, to false positives in case–control studies is still under debate, every effort to minimize it should be taken. These efforts include the use of homogeneous populations or the use of family relatives as controls (e.g., siblings in case–sib studies). When these approaches are not possible, results should always be controlled for ethnicity. In particular, genomic-based approaches to identify different subpopulations of diverse ethnic origin should be considered [a discussion on these genomic-based methods can be found in Thomas (2004); see Suggested Reading].

26.4.4b   Multiple Comparisons. The current availability of higher throughput and more affordable genotyping technologies has caused an exponential growth in the number of genes tested for association for any given disease. This has afforded researchers the capability of testing multiple genes in a study. Unfortunately, testing multiple genes with a low probability of contributing to disease may lead to false positives simply due to random chance and multiple comparisons. Statistical methods that take into account the likelihood of any given gene to be associated with disease based on biological plausibility and prior knowledge have been proposed to correct for multiple comparisons. Ultimately, as with any other epidemiological study, it is desirable to be able to replicate findings in different study populations. Although lack of replication may not always indicate a false positive, it may also reflect heterogeneity across studies that might be indicators of relevant differences between the two populations. Therefore, each individual study can be thought as one data point for a future meta-analysis that will draw conclusions on the role of any given gene and disease.

26.4.4c   Lack of Statistical Power. Molecular epidemiological studies that are based on small numbers may erroneously conclude that there is lack of an association between a gene and disease, simply due to lack of statistical power to detect an association. Conversely, small studies may also lead to false positives because the error in classification of a few subjects may have a large effect of the risk estimate. Therefore, careful power calculations should always be done prior to doing any study in order to understand the magnitude of effect that can be expected, given the study population available and the prevalence of the genetic and environmental factors under study.

26.5   SUMMARY AND CONCLUDING REMARKS

The past two decades have seen an exponential growth in the number of molecular epidemiological studies. With the increased incorporation of biological measurements in most epidemiological studies, the distinction between "classical" epidemiology and molecular epidemiology starts to disappear. Whereas traditional epidemiology was concerned with the identification of the causes and determinants of disease, molecular epidemiological approaches can increase the power to identify cause of disease and can also identify the steps leading from the causes to the outcome. This attempt to understand not only the causes but also the mechanisms of disease is perhaps the best goal and ambition of molecular epidemiology. The post-Human Genome era has seen exponential growth in the development of high-throughput techniques to obtain biological information on large numbers of subjects and large numbers of molecules, such as mRNA expression array, protein arrays, and high-throughput genotyping. As we learn how to take advantage of these powerful techniques, we face the challenge of developing newer statistical techniques to handle the complexity of the data, along with the challenge of interpreting our findings in the light of the available biological knowledge. The latter highlights the importance of fostering strong collaborations between basic researchers, biostatisticians, and epidemiologists in order to achieve the ambitions of this field.

SUGGESTED READINGS

Readings Related to Examples Cited in Text

Alexander, J., Reistad, R., et al. Biomarkers of exposure to heterocyclic amines: Approaches to improve the exposure assessment. *Food Chem. Toxicol.* **40**(8), 1131–1137, 2002.

Beresford, S. A., Johnson, K. C., et al. Low-fat dietary pattern and risk of colorectal cancer: The Women's Health Initiative Randomized Controlled Dietary Modification Trial. *JAMA* **295**(6), 643–654, 2006.

Butler, L. M., Sinha, R., et al. Heterocyclic amines, meat intake, and association with colon cancer in a population-based study. *Am. J. Epidemiol.* **157**(5), 434–445, 2003.

Calle, E. E., Rodriguez, C., et al. The American Cancer Society Cancer Prevention Study II Nutrition Cohort: Rationale, study design, and baseline characteristics. *Cancer* **94**(9), 2490–2501, 2002.

Chao, A., Thun, M. J., et al. Meat consumption and risk of colorectal cancer. *JAMA* **293**(2), 172–182, 2005.

Keating, G. A., Sinha, R., et al. Comparison of heterocyclic amine levels in home-cooked meats with exposure indicators (United States). *Cancer Causes Control* **11**(8), 731–739, 2000.

Knowler, W. C., Williams, R. C., Pettitt, D. J., and Steinberg, A. G. Gm3;5,13,14 and type 2 diabetes mellitus: an association in American Indians with genetic admixture. *Am. J. Hum. Genet.* **43**, 520–526, 1988.

Kobayashi, M., Hanaoka, T., et al. 2-Amino-1-methyl-6-phenylimidazo[4,5-b]pyridine (PhIP) level in human hair as biomarkers for dietary grilled/stir-fried meat and fish intake. *Mutat. Res.* **588**(2), 136–142, 2005.

Kono, S. Secular trend of colon cancer incidence and mortality in relation to fat and meat intake in Japan. *Eur. J. Cancer. Prev.* **13**(2), 127–132, 2004.

NTP 2004. *Selected Heterocyclic Amines*. Report on Carcinogens, 11 edition, US Department of Health and Human Services, Public Health Services, National Toxicology Program, 2004.

Skipper, P. L., Tannenbaum, S. R., Ross, R. K., and Yu, M. C. Nonsmoking-related arylamine exposure and bladder cancer risk. *Cancer. Epidemiol. Biomarkers Prev.* **12**, 503–507, 2003.

Strickland, P. T., Qian, Z., Friesen, M. D., Rothman, N., and Sinha, R. Metabolites of 2-amino-1-methyl-6-phenylimidazo(4,5,-b)pyridine (PhIP) in human urine after consumption of charbroiled or fried beef. *Mutat. Res.* 2002. 163–173, 506–507.

Tavan, E., Cayuela, C., et al. Effects of dairy products on heterocyclic aromatic amine-induced rat colon carcinogenesis. *Carcinogenesis* **23**(3), 477–483, 2002.

Basic Concepts in Epidemiology

Dos Santos Silva. *Cancer Epidemiology: Principles and Methods*, IARC Scientific Publications, Lyon, France, 1999.

Gordis, L. *Epidemiology*, W. B. Saunders, Philadelphia, 2000.

Rothman, K. J., and Greenland, S. *Modern Epidemiology*, Lippincot-Raven, Philadelphia, 1998.

Schlesselman, J. J. *Case–Control Studies, Design, Conduct, Analysis*, Oxford University Press, New York, 1982.

Molecular or Genetic Epidemiology

Buffler, P. A. *Mechanisms of Carcinogenesis: Contributions of Molecular Epidemiology*, IARC Scientific Publications #157, Lyon, France, 2004.

Khoury, M., Beaty, J., and Cohen, B. H. *Fundamentals of Genetic Epidemiology*, Oxford University Press, New York, 1993.

Perera, F. P. Molecular cancer epidemiology: A new tool in cancer prevention. *J. Natl. Cancer Inst.* **75**(5), 887–898, 1987.

Shields, P. G. *Cancer Risk Assessment*. Taylor & Francis, London, 2005.

Thomas, D. *Genetic Epidemiology and Statistical Genetics: Bridging the Gap*. Oxford University Press, 2004.

Toniolo, P., Boffetta, P., Shuker, D. E. G., Rothman, N., Hulka, B., and Pearce, N. *Application of Biomarkers in Cancer Epidemiology*, IARC Scientific Publications #142, Lyon, France, 1997.

Vineis, P., Malats, N., Lang, M., d'Errico, N., Caporaso, N., Cuzick, J., and Boffetta, P. *Metabolic Polymorphisms and Susceptibility to Cancer*, IARC Scientific Publications #148, Lyon, France, 1999.

Resources for SNP Identification

http://www.ncbi.nlm.nih.gov/SNP/
http://snp500cancer.nci.nih.gov/home_1.cfm
http://www.genome.utah.edu/geneshps
http://www.hapmap.org/

Resources for *in Silico* Analyses of SNPs

http://www.proweb.org/coddle/
http://blocks.fhcrc.org/sift/SIFT.html
http://genetics.bwh.harvard.edu/pph/
ENCODE Project: http://www.genome.gov/10005107

Journals that Publish Molecular Epidemiology Articles

Cancer Epidemiology, Biomarkers and Prevention (official journal of the Molecular Epidemiology working group of the American Association of Cancer Research)

Carcinogenesis (has a special section on Molecular Epidemiology)

American Journal of Epidemiology

Epidemiology

■■■■■ CHAPTER 27

Respiratory Toxicity

JAMES C. BONNER

Department of Environmental and Molecular Toxicology, North Carolina State University, Raleigh, North Carolina 27695

27.1 INTRODUCTION

Because of the essential function of the respiratory system in transporting atmospheric oxygen to the bloodstream, it is the primary interface for exposure to a vast array of inhaled natural and man-made toxicants. Unlike the protective keratinized epithelium of the skin, the epithelial lining of the respiratory tract is vulnerable to injury, by inhaled toxicants. To counterbalance this vulnerability, the mammalian lung has evolved exquisite cellular defense mechanisms to cope with the clearance of inhaled agents and finely tuned repair mechanisms to restore the delicate architecture of the lung after injury. Nevertheless, the response to injury by inhaled toxicants often does not resolve but instead leads to pathologic outcomes characterized by aberrant tissue remodeling. Genes and environment are intertwined in the response to injury, and genetic susceptibility is a central factor in the development of lung diseases caused by exposure to toxicants. In this chapter, the response of the respiratory system to toxic exposures will be discussed in the context of anatomical structure, respiratory physiology, and regulation of cellular function.

27.2 ANATOMY AND FUNCTION OF THE RESPIRATORY TRACT

The respiratory system has classically been divided in to the upper respiratory tract and the lower respiratory tract. The general features of upper and lower regions of the respiratory tract are illustrated in Figure 27.1. The upper respiratory tract consists of the mouth, nose, and pharyngeal region. Air enters the respiratory systems of mammals through the nose or mouth in humans, whereas some species such as rodents are obligate nasal breathers. The primary functions of the upper respiratory tract are olfaction, temperature equilibration and humidification of inspired air, and uptake of inhaled particles and irritant gases. The lower respiratory tract begins distal to the pharyngeal region and consists of the tracheobronchial region and the

Molecular and Biochemical Toxicology, Fourth Edition, edited by Robert C. Smart and Ernest Hodgson
Copyright © 2008 John Wiley & Sons, Inc.

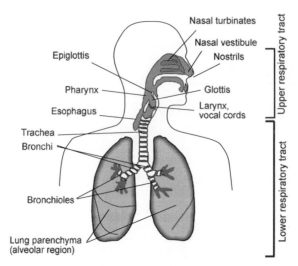

Figure 27.1. Anatomy of the human respiratory system. (Adapted from LifeART illustration series, Lippincott Williams & Wilkins, Hagerstown, MD, 1994. This figure was completely redrawn by the author from materials cited.)

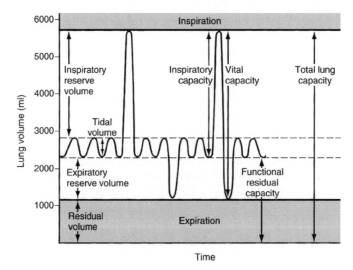

Figure 27.2. Diagram showing respiratory functional volumes during normal breathing and maximal inspiration or expiration. (From Guyton, A. C., and Hall, J. E. *Textbook of Medical Physiology*, 10th ed., W.B. Saunders, Philadelphia, 2000.)

pulmonary parenchyma or alveolar region. The major functions of the lower respiratory tract are gas exchange and defense against inhaled toxicants.

The mammalian lung evolved as an elegant and efficient means of harvesting oxygen from the atmosphere and the transfer of the metabolic gas CO_2 from the blood to the exhaled air. Because the goal of the lung is to provide oxygen to the tissues and remove CO_2, the inflow and outflow of air between the atmosphere and the alveoli, termed *ventilation*, is a critical function. A simple method for studying pulmonary ventilation is to record the volume movement of air into and out of the lungs, a process called *spirometry*. Four different pulmonary lung volumes are illustrated in Figure 27.2 and are defined as follows: (1) The *tidal volume* is the volume of air inspired or expired with each normal breath; (2) the *inspiratory reserve volume* is the maximum extra volume of air that can be inspired over and above the normal tidal volume; (3) the *expiratory reserve volume* is the maximum extra volume of air that can be expired by forceful expiration after the end of a normal tidal expiration; and (4) the *residual volume* is the volume of air remaining in the lungs after the most forceful expiration. Two or more of the volumes together are called *pulmonary capacities*, which are described as follows: (1) The *inspiratory capacity* equals the *tidal volume* plus the *inspiratory reserve volume*; (2) the *functional residual capacity* equals the *expiratory reserve volume* plus the *residual volume*; (3) the *vital capacity* equals the *inspiratory reserve volume* plus the *tidal volume* plus the *expiratory reserve volume*; and (4) the *total lung capacity* is the maximum volume to which the lungs can be expanded with the greatest possible effort and is equal to the *vital capacity* plus the *residual volume*.

Toxicity to the respiratory tract can interfere with this gas-exchange function by (1) altering the tone of the airways resulting in decreased airflow; (2) damaging the delicate architecture of the alveolar/capillary barrier of the deep lung, resulting in impaired gas exchange; or (3) causing tissue damage that leads to chronic structural changes in the lung, resulting in decreased lung volumes or lung mechanics (the degree of elasticity of the lung). The total lung capacity and vital capacity are reduced in fibrotic lung disease in which the lung becomes smaller and stiffer. This type of change in the lung is called *restrictive lung disease* and is marked by smaller lung volumes and little change in airflow. In an emphysematous lung, on the other hand, the total lung capacity may increase as a result of the breakdown of alveolar walls and loss of elastin fibers that allow the lung to deflate on exhalation, but vital capacity is often reduced due to airway collapse during exhalation. This type of change in the lung is called *obstructive lung disease* and is marked by reduced airflow. The breathing patterns of exposed mice are sometimes used in toxicology to evaluate the irritancy of an airborne chemical toward the respiratory tract. Breathing patterns are also used in toxicology studies to record the total amount of an airborne pollutant to which an animal is exposed by inhalation.

Measures of the lung to expand and fill with air during inspiration and to deflate during exhalation are termed lung mechanics. These properties depend on the elasticity of the lung and the caliber of the airways. Lung mechanics are commonly reported as compliance and resistance. *Compliance* is the volume change per unit pressure change. *Resistance* is the pressure difference per change in airflow. Compliance and resistance are generally measured during steady tidal breathing by recording transpulmonary pressure, tidal volume, and airflow rate. Because the direct measurement of transpulmonary pressure requires invasive procedures such as an intrapleural catheter, indirect measures approximating compliance and resistance are typically applied to humans using *plethysmography*.

The most common test of lung function in humans is the *forced exhalation test*, which evaluates both lung volumes and flow performance. After inhaling maximally, the individual exhales as rapidly and deeply as possible to reach residual volume. The volumes of air exhaled per unit time and the airflow during exhalation are recorded. The forced expiratory volume measured in one second (FEV_1), along with the FEV_1 as a percentage of the forced vital capacity, is typically reported. The shape of the flow/volume curve provides invaluable information on the type of lung obstruction in airways. Reductions in peak flow at high lung volumes are generally indicative of large airway obstruction. Reductions of flow at lower lung volumes are generally indicative of small airway obstruction. The equivalent of the forced exhalation tests as performed in humans can be conducted in animals using negative airway driving pressures.

27.2.1 Upper Respiratory Tract as a Site of Toxicity

The upper respiratory tract, particularly the nose, has a unique anatomy that performs normal physiologic functions as well as innate defense against inhaled toxicants. The nose extends from the nostrils to the pharynx. Inspired air enters the nose through the nostrils. The nasal cavity is divided longitudinally by a septum into two nasal compartments. In most mammalian species, each nasal cavity is divided into a dorsal, ventral, and middle (lateral) meatus by two turbinate bones, the nasoturbinate and maxilloturbinate. These turbinates project from the dorsolateral and ventrolateral wall of the cavity, respectively. In the posterior portion of the nose, the ethmoid recess contains the ethmoturbinate. The nasal cavity is lined by a vascular mucosa that consists of four distinct types of epithelia. In rodents, these epithelia are (1) the stratified squamous epithelium that lines the nasal vestibule and the floor of the ventral meatus in the anterior portion of the nose; (2) the nonciliated, pseudostratified, transitional epithelium that lies between the squamous epithelium and the respiratory epithelium and lines the lateral meatus; (3) the ciliated respiratory epithelium that lines the remainder of the nasal cavity anterior and ventral to the olfactory epithelium; and (4) the olfactory epithelium (neuroepithelium) that lines the dorsal meatus and ethmoturbinates in the caudal portion of the nose. The relative abundance and exact locations of these upper respiratory epithelium differ among mammalian species.

The olfactory epithelium is composed of basal, neuronal (olfactory), and sustentacular (support) cells (Figure 27.3). The portion of each olfactory cell that responds to the olfactory chemical stimuli is the cilia. The odorant substance first diffuses into the mucus that covers the cilia and then binds to specific receptor proteins in the membrane of each cilium. Next, receptor activation by the odorant activates a multiple molecules of the G-protein complex in the olfactory epithelial cell. This, in turn, activates adenylyl cyclase inside the olfactory cell membrane, which, in turn, causes formation of a greater multitude of cAMP molecules. Finally, the cAMP molecules trigger the opening of yet an even greater multitude of sodium ion channels. This amplification mechanism accounts for the exquisite sensitivity of the olfactory neurons to extremely small amounts of odorant. The olfactory epithelium is an important target of certain inhaled toxicants. Certain metals, solvents, proteins, and viruses are transported to the brain via transport from the olfactory epithelium to the olfactory tract and exert neurotoxicity.

As air passes through the nose, essential normal respiratory functions are performed by the nasal cavities. These functions are collectively referred to as the air conditioning function of the upper respiratory tract. First, the air is warmed and is almost completely humidified by the extensive mucosal surfaces of the nasal conchae and septum. Second, the nose plays a critical role in trapping the majority of inhaled

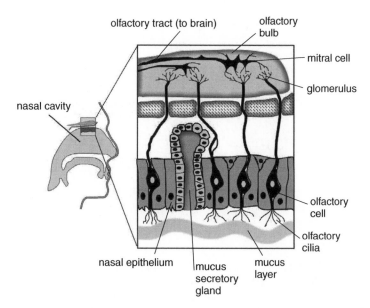

Figure 27.3. Anatomy of the olfactory apparatus in the nasal cavity of the upper respiratory tract. (Adapted from LifeART illustration series, Lippincott Williams & Wilkins, Hagerstown, MD, 1994. This figure was completely redrawn by the author from materials cited.)

particles before they reach the lower lung. Hairs at the entrance to the nostrils and nasal vestibule are important for filtering out large particles. Inspired air is further filtered by *impaction* of particles on the surfaces of the nasal turbinates (also referred to as nasal conchae), which are folds of osseous tissue lined with ciliated and mucus-producing epithelial cells. Impaction means that particles in the inspired air passing through the nasal passageways collide with many obstructions, including the nasal turbinates and septum, and stick in the mucous lining of the nasal epithelium. The term "turbinates" was derived from the fact that these structures cause turbulence of the inspired air. As air encounters the turbinates, it changes direction of movement. However, particles that are suspended in the air have greater mass than the air itself and continue forward, impacting the surface of the obstruction. The particles are then trapped in the mucus coating and transported by the beating cilia of the nasal epithelium in a unidirectional manner to the pharynx, where they are swallowed. The trapping and removal of particles through nasal turbulence and impaction is so effective that almost no particles larger than 5 µm in diameter enter the lungs through the nose.

The nasopharynx and the oropharynx are posterior to the nose and mouth, respectively, and join to form the pharynx. The pharyngeal cavities are lined by a mucociliary epithelium coated with mucus. Humans and rodents handle inspired air in the upper respiratory tract differently. In rodents, the nose is the only site of entry into the respiratory system for inhaled materials. For this reason, rodents are referred to as obligate nose breathers. Humans breathe in air either through the mouth or the nose. Therefore, the mouth is also an important site of entry for inhaled materials in humans. Large particles (>5 µm) inhaled orally tend to impact the wall of the pharynx.

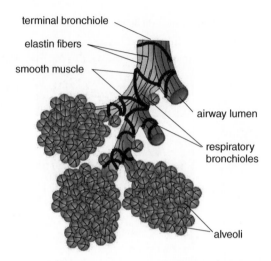

terminal bronchiole

elastin fibers

smooth muscle

airway lumen

respiratory
bronchioles

alveoli

Figure 27.4. Anatomy of the distal airways and functional units (alveoli) in the lower respiratory tract. (Adapted from LifeART illustration series, Lippincott Williams & Wilkins, Hagerstown, MD, 1994. This figure was completely redrawn by the author from materials cited.)

27.2.2 Lower Respiratory Tract as a Site of Toxicity

The lower respiratory is comprised of the conducting airways (trachea, bronchi, and bronchioles) and the lung parenchyma, which consists primarily of gas exchange units (alveoli) (Figure 27.1). The trachea, bronchi, and bronchioles conduct air to the pulmonary parenchyma. The trachea extends from the larynx distally, where it divides to form the two main bronchi, which enter the right and left lungs. The bronchi bifurcate to form bronchioles and continue to progressively bifurcate in a tree-like fashion to form bronchioles of decreasing diameter. The most distal conducting segment of the tracheobronchial tree is called the terminal bronchiole, which bifurcates to form respiratory bronchioles that contain some alveolar ducts and terminate in clusters of alveolar sacs. This distal region of the lung where airways transition to alveoli is illustrated in Figure 27.4.

27.2.3 Airways of the Lower Respiratory Tract

The trachea and bronchi contain bands of cartilage in the airway wall that prevent collapse. In contrast, the walls of the intrapulmonary airways do not contain cartilage but are supported by flexible elastic fibers and bands of smooth muscle. These smaller airways are susceptible to bronchoconstriction in diseases such as asthma, where allergens provoke a neurogenic response that results in airway smooth muscle contraction. The airways of the lung are lined with pseudostratified, columnar cells that are predominantly ciliated or mucus-producing serous cells (Figure 27.5). Together, these epithelial cell types contribute to the clearance of particles from the airways through a mechanism termed the "mucociliary escalator." Throughout the tracheobronchial region, mucociliary clearance is an important clearance and defense mechanism for moving inhaled particles up the airway tree where they are expelled from the trachea and swallowed. In humans, submucosal glands in the

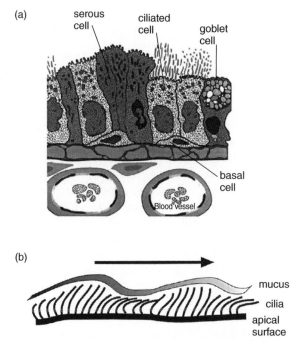

Figure 27.5. Structure and dynamics of the airway epithelium. (A) Illustration of tracheal and bronchial epithelial cell types. (B) The mucociliary escalator wherein epithelial cell cilia move in a low-viscosity periciliary layer to propel mucus with their tips.

bronchi also contribute to mucus production, especially during chronic irritation such as cigarette smoking or during chronic respiratory viral infections. In the terminal bronchioles, the population of mucus and serous cells gradually transition to nonciliated cuboidal (Clara) cells. Clara cells have relatively high levels of cytochrome P450 enzymes and may be selectively damaged by toxicants that require metabolic activation by P450s. The airway epithelium in large and small airways also contain triangular-shaped basal cells that are thought to give rise to ciliated and mucus cells following toxicant injury. All airway epithelial cells are attached basally to a basement membrane or lamina composed of extracellular matrix.

The airway epithelium forms a continuous lining for the conducting airways. The varied composition of the epithelium allows it to perform a variety of functions. First, the epithelium along with its apical mucus layer and its basal lamina comprise an important barrier against inhaled toxicants and xenobiotics. The apical surfaces of the airway epithelial cells are connected by tight junctions and effectively provide a barrier that isolates the airway lumen. Second, the various airway epithelial cells produce a mixture of secretions composed of (a) an aqueous "sol" phase containing proteins, lipids, and ions and (b) a gel phase containing mucus. Third, ciliated cells comprise the largest proportion of exposed cells in the normal airway; as discussed above, they propel the mucus within the airway lumen proximally, thereby mediating clearance of inhaled particles and debris (Figure 27.5). Fourth, the airway epithelium exhibits repair following injury, thereby establishing normal airway architecture. Fifth, the airway epithelium can produce a variety of soluble mediators (cytokines, growth factors, protease, and lipid mediators) that modulate the responses of other lung cells including airway smooth muscle cells, fibroblasts, immune cells, and phagocytes.

27.2.4 Parenchyma of the Lower Respiratory Tract

The primary function of the lung parenchymal region is gas exchange. The major structures are the respiratory bronchioles, alveolar ducts, alveolar sacs, and alveoli. The respiratory bronchioles are lined with cuboidal ciliated and Clara cells and have alveoli opening into their lumina. Therefore, the respiratory bronchioles function both as conducting passages and as a gas exchange region. A number of mammalian species, including humans, have respiratory bronchioles, whereas other species, including rats, have no respiratory bronchioles. In the latter case, the terminal bronchioles end in alveolar ducts. Alveolar ducts are tubular structures whose walls are covered by alveoli. The alveoli open polyhedral chambers lined with thin type I epithelial cells interspersed with cuboidal type II cells.

Type I cells comprise 8–11% of the structural cells found in the alveolar region and yet cover 90–95% of the alveolar surface. Their major function is to allow gases to equilibrate across the air–blood barrier and to prevent leakage of fluids across the alveolar wall into the lumen. The type I epithelium is particularly sensitive to damage from a variety of inhaled toxicants due to their large surface area. Moreover, their repair capacity is limited because they have few organelles associated with energy production and macromolecular synthesis.

Type II cells comprise 12–16% of the structural cells in the alveolar region, but cover only about 7% of the alveolar surface. They are cuboidal cells with a microvillus surface and unique organelles (called lamellar bodies) that store surfactant. The major function of type II cells is to secrete surfactant to lower the surface tension in the alveoli, thereby reducing the filling of the alveolar compartment with fluid and alveolar collapse. Type II cells also serve as a progenitor cell for type I cells, which cannot replicate. Therefore, type II cells are critical to alveolar epithelial repair after injury. An interesting third pneumocyte, called the brush cell, is sparsely distributed and appears at alveolar duct bifurcations. Even though this cell type was identified decades ago, essentially nothing is known about its function.

The wall of the alveolus is composed of the alveolar epithelium, a thin layer of collagenous and elastic connective tissue interspersed with fibroblasts (termed the pulmonary interstitium), and a network of capillaries lined by endothelial cells. This distance between the alveolar space and the capillary lumen is known as the air–blood barrier. The air–blood barrier is a multilayered structure (~0.4μm in thickness) that consists of an alveolar type I cell, alveolar basement membrane, interstitial space, endothelial basement membrane, and a capillary endothelial cell (Figure 27.6). CO_2 and O_2 are exchanged between air and blood by diffusion across the air–blood barrier. A number of factors determine how rapidly a gas will pass through the respiratory membrane. First, the rate of gas diffusion through the membrane is inversely proportional to the thickness of the membrane. In situations where the thickness of the membrane increases, gas exchange between the air and blood is decreased. For example, edema fluid in the alveolar space results in gases requiring passage not only through the cellular membrane but also through a fluid layer. Another example is thickening of the lung interstitial space between the

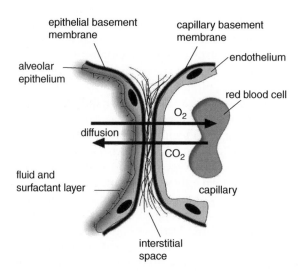

Figure 27.6. Ultrastructure of the alveolar respiratory membrane shown in cross section. (Adapted from Guyton, A. C., and Hall, J. E. *Textbook of Medical Physiology*, 10th edition, W.B. Saunders, Philadelphia, 2000. This figure was completely redrawn by the author from materials cited.)

alveolar membrane and the blood capillary membrane during pulmonary fibrogenesis. In general, any factor that increases the thickness of more than two to three times normal can significantly interfere with normal respiratory exchange of gases.

The pulmonary interstitium consists of extracellular matrix (collagen and elastin) and resident interstitial cells. Although the amount of collagen and elastin in the pulmonary parenchyma is small, these structural proteins are key to normal pulmonary mechanics. Increases or decreases in these proteins lead to impairment as in pulmonary fibrosis or emphysema. The major resident cell type is the fibroblasts, although interstitial macrophages, lymphocytes, plasma cells, and mast cells are also present in the interstitium. Resident interstitial cells comprise about 35% of the structural cells in the alveolar region. However, during an inflammatory response the relative abundance of these interstitial cells, as well as neutrophils, monocytes, and lymphocytes infiltrating from the blood, greatly increases.

Capillary endothelial cells comprise 30–42% of cells in the alveolar region and comprise the walls of the extensive network of blood capillaries in the lung parenchyma. The endothelium forms a continuous, attenuated cell layer that transports respiratory gases, water, and solutes. However, it also forms a barrier to the leakage of excess water and macromolecules into the pulmonary interstitial space. Pulmonary endothelial cells, like type I cells, are vulnerable to injury from inhaled substances and substances in the systemic circulation. Injury to the endothelium results in fluid and protein leakage into the pulmonary interstitium and alveolar spaces, resulting in pulmonary edema.

The lung parenchyma is constantly surveyed by mobile phagocytes, which provide an essential defense against inhaled foreign materials. The most common of these is the alveolar macrophage, which patrol the surface of the alveolar spaces. Their major defense roles are phagocytosis, killing, and clearing of microorganisms, such as bacteria, as well as phagocytosis and clearance of a wide variety of inhaled particulate matter. After engulfing a foreign particle, clearance may be accomplished via the *mucociliary escalator*. Particle-laden macrophages migrate from the alveolar spaces to the distal airways and are taken up the airway by the unidirectional beating of cilia on the airway epithelium, which moves the macrophages and their cargo up and out of the lungs where they are expelled into the pharynx and swallowed. In addition to the mucociliary escalator, macrophages may also clear engulfed foreign material by migrating into the pulmonary interstitium and into the lymphatics. Phagocytosis triggers the release of cytokines and chemokines, growth factors, proteolytic enzymes, and reactive oxygen species. These mediators recruit and activate other cells that participate in either the resolution of an inflammatory response or structural alterations in lung tissue that lead to a pathologic disease outcome. The macrophage population in the lung is replenished by recruitment of bone-marrow-derived monocytes via the bloodstream (Figure 27.7). Once in the lung, monocytes proliferate in response to specific growth factors and mature into alveolar macrophages under the direction of other specific cytokines and differentiation factors. Other subpopulations of macrophages that are less recognized are the pulmonary interstitial macrophage that resides beneath the epithelial lining and the intravascular macrophage that is present in humans and some other mammals, but not rodents.

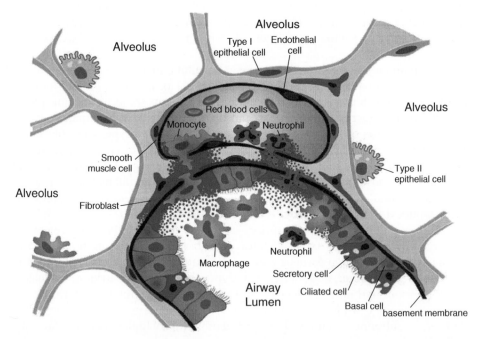

Figure 27.7. Macrophages migrate to a site of injury in the wall of a distal airway. Alveolar macrophages normally residing in the alveolar spaces migrate to chemoattractants released by injured epithelial cells. Other leukocytes such as monocytes and neutrophils also respond to chemotactic molecules and migrate from the blood across the pulmonary interstitium.

27.2.5   Circulatory, Lymphatic, and Nervous System of the Lung

The respiratory systems communicates with other organ system primarily through direct connections with the circulatory, lymphatic, and nervous systems. The *circulatory system* bridges the closely adjacent heart and lungs, and the entire cardiac output of the heart enters the lung to be replenished with oxygen. Oxygen-poor blood from the right ventricle travels through the pulmonary arteries to supply capillary beds of the respiratory bronchioles, alveolar ducts, and alveoli, where gas is exchanged. Oxygenated blood then returns to the heart via pulmonary venules in the lung parenchyma that merge into pulmonary veins that feed the left atrium. A second arterial system, the bronchial system, supplies oxygenated blood from the left ventricle of the heart through the aorta and the bronchial arteries to the large airways of the lung, the pleura, and large pulmonary vessels.

The *pulmonary lymphatic system* is a vascular network that serves to remove excess fluid from the connective tissue spaces of the lung parenchyma. The lymphatic system is also important in clearing particulate material from the lung to the lymph nodes. The lymphatic system in the lung is divided into superficial and deep portions, but these two portions are connected. The superficial portion is located in the connective tissue of the pleura lining the lung. The deep portion is in the connective tissue surrounding the bronchovascular tree. The two portions connect in the interlobular septa. The lymphatic vessels are structurally similar to thin-walled veins. The presence of valves in the lymphatic vessels and the movement of the lung during respiration promote the flow of lymph from the periphery and pleura toward the hilus. Afferent lymphatics from the lung drain into the tracheobronchial lymph nodes. Lymph from tracheobronchial and hilar nodes drain into the thoracic, right, and left lymphatic ducts and from these ducts drain into the systemic venous system.

The respiratory tract contains both sensory (afferent) and motor (efferent) innervation. Both the parasympathetic and sympathetic portions of the autonomic *nervous system* provide the motor innervation. Preganglionic parasympathetic fibers descend in the vagus nerves to ganglia located around airways and blood vessels. The postganglionic fibers innervate the smooth muscle of the airways and blood vessels, bronchial glands, and epithelial mucus cells. In general, the same structures are also innervated by postganglionic fibers from the sympathetic ganglia. Vagal stimulation causes airway constriction, dilation of the pulmonary circulation, and increased glandular secretion. Conversely, sympathetic nerve stimulation causes bronchial relaxation, constriction of pulmonary blood vessels, and inhibition of glandular secretion. Sensory receptors that respond to irritants or mechanical stress are located throughout the respiratory tract. Stimulation of these receptors leads to reflex responses such as stimulation of nasal receptors to cause sneezing. Three principal vagal sensory reflexes and their corresponding receptors are known as (1) bronchopulmonary stretch receptors, (2) irritant receptors, and (3) C-fiber receptors. Stretch receptors are associated with smooth muscle of the trachea and bronchi, are stimulated by lung inflation, and normally function to terminate inspiration. Rapidly adapting irritant receptors are located in the epithelium of extrapulmonary —and, to a lesser extent, intrapulmonary—bronchi. They respond to a variety of stimuli, including inhalation of irritant gases and mechanical stimulation of the airways, to cause bronchoconstriction, cough, and increased mucus secretion. C-fiber receptors are located both in the parenchyma and along conducting airways. Bronchial C-fibers along conducting airways respond to stimuli near the bronchial arterial system and, when stimulated, cause airway constriction. Pulmonary C-fibers may contribute to the sensation of dyspnea that accompanies pulmonary edema, pneumonia, and inhalation of noxious gases. Stimulation of both pulmonary and bronchial C-fibers causes a reflex increase in airway secretion.

27.3   TOXICANT-INDUCED LUNG INJURY, REMODELING, AND REPAIR

The respiratory tract is exposed to many environmental factors (particles, gases, infectious microbial agents) and has evolved sophisticated defense and repair systems. The response to injury by foreign agents that enter the respiratory tract involves *host recognition* of the toxic insult followed by *acute inflammation* and then *tissue remodeling* that can result in one of two general outcomes. First, an inflammatory response may resolve and lead to a tissue repair process where normal respiratory architecture and function are restored. Alternatively, an inflammatory response may not resolve but instead may progress to an abnormal tissue remodeling response leading to diseases such as fibrosis and emphysema. This general concept is illustrated in Figure 27.8.

A number of factors determine whether tissue in the lung parenchyma is successfully repaired after injury or whether an inflammatory response progresses to a pathologic outcome. As mentioned previously, the alveolar region is especially vulnerable to damage due to the delicate architecture of the type I epithelium and blood capillary endothelial membranes. An appropriate balance of catabolic and

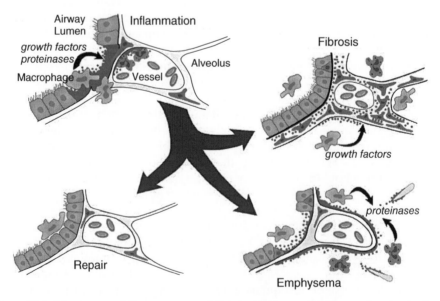

Figure 27.8. Tissue remodeling outcomes following injury and inflammation in the lower respiratory tract that result in repair or disease. (See text for details.)

anabolic activities involving cytokines, growth factors, lipid mediators, and proteases is required for tissue repair. Following injury, damaged cells are triggered to undergo apoptosis by specific cytokine signals (e.g., TNF-α) released from phagocytes, or the damaged cells themselves release apoptotic factors in an autocrine manner. At the same time, extracellular debris is degraded by specific proteases (e.g., elastases, collagenases) released by infiltrating neutrophils and mononuclear cells. Resident macrophages then mediate the clearance of the resulting cellular and extracellular debris. The rebuilding process involves a precise balance of growth and differentiation factors to stimulate repopulation of the epithelium and restore connective tissue, vascular tissue, and nerves. An overexuberant production of growth factors (e.g., TGF-β, CTGF, PDGF) may lead to a fibrotic response characterized by increased fibroblasts and collagen. Alternatively, an imbalance in protease/antiprotease systems may lead to a progressive degradation of structural proteins (e.g., elastin) in the alveolar wall and cause emphysema.

Injury to the airway epithelium also may lead to repair or disease outcomes. Airway epithelial injury is common in individuals with viral bronchitis and small patches of epithelium that are damaged by the ensuing airway inflammatory response that accompanies viral infection. Columnar epithelial cells of the airways rapidly fill in denuded areas, and this repair response is facilitated by the action of specific matrix metalloproteinases. However, pathologic phenotypes may arise instead of repair, and this depends on the insulting chemical or biological agent, the dose and duration of exposure, and genetic susceptibility. For example, a mucus cell hyperplastic response results during chronic obstructive pulmonary disease caused by cigarette smoke, or in asthma caused by an allergic response to a variety of allergens. In this case, the normal balance of mucus-producing serous cells and ciliated cells is perturbed; and over time, mucus-producing goblet cells predominate (Figure 27.9). Another pathologic airway epithelial phenotype is squamous metaplasia, where flattened squamous epithelial cells repopulate the airway lumen. This phenotypic change may occur after high-dose occupational exposure to gases such as chlorine.

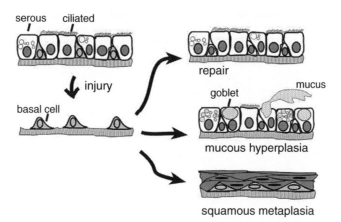

Figure 27.9. Different phenotypic outcomes of airway epithelium after injury. Repair usually follows intermittant or low dose exposures to toxicants. Mucous hyperplasia occurs after chronic injury, while squamous metaplasia occurs after high-dose occupational exposures.

27.3.1 Oxidative Stress and Lung Injury

There is considerable evidence that links oxidants to the development of a number of human lung diseases. Oxidants can be generated by endogenous mechanisms involving NADPH oxidase systems in phagocytic cells, as well as xanthine oxidase. Lung oxidants can also be increased from exogenous sources, such as inhaled air pollution (gases or particles) and cigarette smoke. The intake of oxygen through the lungs is required for aerobic life, and yet conversion of oxygen to reactive oxygen species (ROS) can have profound detrimental effects on tissues of the respiratory tract.

Oxygen is converted to the oxidizing agent superoxide anion (O_2^-) by cellular NADPH oxidase systems (prinicipally in phagocytes) and by xanthine oxidase (Figure 27.10). Hydrogen peroxide (H_2O_2) is formed by the further oxidation of O_2^- by superoxide dismutase (SOD):

$$2O_2^- + 2H^+ \rightarrow H_2O_2 + O_2$$

Three types of SODs occur: (1) extracellular (EC)-SOD, (2) manganese SOD found in the mitochondria, and (3) copper–zinc SOD found in the cytosol and nucleus. H_2O_2 can be converted to the highly toxic hydroxyl radical ($^{\cdot}OH$) via the iron-catalyzed Fenton reaction:

$$Fe^{2+} + H_2O_2 \rightarrow Fe^{3+} + {}^{\cdot}OH + OH^-$$

Moreover, H_2O_2 in the presence of O_2^- and a divalent metal can also produce $^{\cdot}OH$ via the iron-catalyzed Haber–Weiss reaction:

$$Fe^{2+} + H_2O_2 \rightarrow Fe^{3+} + {}^{\cdot}OH + OH^-$$
$$Fe^{3+} + O_2^- \rightarrow Fe^{2+} + O_2$$
$$O_2^- + H_2O_2 \rightarrow {}^{\cdot}OH + OH^-$$

At least three different sources of ROS contribute to oxidative stress in the respiratory system. First, the initial inflammatory response involving macrophage-mediated phagocytosis involves the release of O_2^-, which is primarily generated by the NADPH oxidase system in these cells. This "respiratory burst" likely evolved as

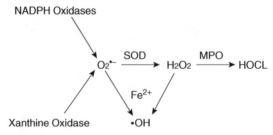

Figure 27.10. Conversion of superoxide anion (O_2^-) to hydroxyl radical ($^{\cdot}OH$) in the presence of iron, or to hydrogen peroxide (H_2O_2) by superoxide dismutase (SOD), which is converted to hypochlorous acid by myeloperoxidase (MPO).

a microbial killing mechanism. However, the phagocytosis of a variety of particles and fibers also activate macrophages to undergo a respiratory burst. Some of these agents (e.g., asbestos fibers) are not easily cleared by macrophages and therefore stimulate chronic activation of phagocytes. Other resident lung cells in the immediate vicinity of a respiratory burst (e.g., airway epithelial cells) may be damaged by macrophage-derived ROS, resulting in the release of cytokines and chemokines that signal the migration of other inflammatory cells (monocytes, lymphocytes, and neutrophils) from the blood across the pulmonary interstitium to the site of injury. Second, particles containing metal oxides generate an additional burden of reactive oxygen species via the Fenton and Haber–Weiss reactions shown above. Third, ozone and nitrogen dioxide gases in the environment are reactive species that further contribute to oxidative stress. Finally, cigarette smoke contains a multitude of oxidizing compounds, including nitrogen oxides, quinones, semiquinone radical, and hydroquinone moieties. The combined burden of ROS from these sources results in pulmonary cell damage by lipid peroxidation of the polyunsaturated fatty acids in cell membranes.

While the majority of toxic agents that generate ROS enter via inhalation, some chemicals cause injury via entry through the circulation. Ingestion of the herbicide paraquat causes pulmonary fibrosis by accumulating in the epithelium of the lung and causes oxidant formation as a result of redox cycling (Figure 27.11). The reduction of paraquat (PQ^{2+}) by NADPH is dependent on cytochrome P450 reductase present in the endoplasmic reticulum and mitochondria. The intravenous administration of the chemotherapeutic drug bleomycin also causes lung fibrosis via redox cycling, although it is strictly Fe-dependent. Because bleomycin binds avidly to DNA, much of the site-directed ˙OH formation leads to DNA damage. Finally, ionizing radiation directly produces ˙OH without the necessity for a transition metal.

The lung is also a source of reactive nitrogen species (RNS). Nitric oxide (NO˙) is a vaso- and bronchodialator that has beneficial functions in regulating airway tone. Nitric oxide synthases (NOS) generate NO˙ either constitutively or upon induction with inflammatory cytokines or bacterial endotoxin. Under conditions of oxidative stress, NO˙ can react with O_2^- to form peroxynitrite ($ONOO^-$), a highly toxic RNS. The $ONOO^-$ molecule rapidly diffuses into cells to cause nitration of tyrosine residues. Tyrosine is a key component of membrane kinase receptors and is phosphorylated in response to growth factors and cytokines to mediate downstream intracellular signaling that results in a variety of cellular functions. Tyrosine nitration competes with and interferes with tyrosine phosphorylation, thereby causing aberrant cell signaling patterns. Tyrosine nitration is associated with a variety of lung diseases, including asthma, fibrosis, COPD, and cancer.

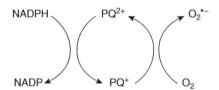

Figure 27.11. Formation of superoxide anion (O_2^-) by the herbicide paraquat (PQ^{2+}) via redox cycling.

27.3.2 Antioxidant Mechanisms in the Lung

There are a number of enzymatic and nonenzymatic antioxidant defense mechanisms that counterbalance the deleterious effects of oxidative stress in the lung. Several general concepts have emerged regarding the action of antioxidants. First, there is specificity in the scavenging ability of various antioxidants. Nearly all cells contain a number of enzymatic scavengers, including *superoxide dismutase* (SOD), *catalase*, and *glutathione redox systems* which degrade specific oxidants in specific ways Figure (27.12). For example, SOD converts superoxide anion to H_2O_2, whereas catalase specifically degrades H_2O_2. Second, there is usually a selective localization of antioxidants within cells. For example, manganese SOD is located adjacent to mitochondria, whereas copper–zinc SOD is located primarily in the cytoplasm. Third, antioxidant levels and activities are dynamic. Antioxidants can be inactivated by oxidants. For example, H_2O_2 can inactivate SOD and superoxide anion can inac-

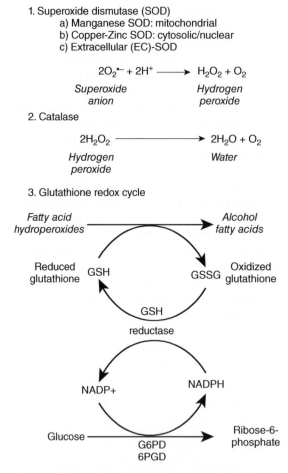

Figure 27.12. Lung intracellular antioxidant enzyme systems. The three major systems: Superoxide dismutase, catalase, and the glutathione redox cycle are shown. (See text for details.)

tivate catalase. Moreover, oxidative stress can induce the transcription of genes that encode oxidant-generating systems such as NADPH oxidase or alternatively induce antioxidant enzymes as a protective feedback mechanism. Catalase, SOD, and enzymes of the glutathione redox cycle are primary intracellular antioxidant defense mechanisms that eliminate oxygen radicals and hydroperoxides that may pose a threat to the cell by oxidizing cellular structures. Catalase is located primarily in peroxisomes which contain many of the enzymes that generate H_2O_2 in aerobic cells. Catalase reduces H_2O_2 to water and oxygen. In the lung, catalase is present mainly in type II cells, Clara cells, and macrophages.

The glutathione redox cycle is a central mechanism for reduction of intracellular fatty acid hydroperoxides and complements catalase as a reducing system for H_2O_2. Glutathione metabolism also degrades large molecule lipid peroxides formed by free radical action on polyunsaturated lipid membranes. The key enzyme in the glutathione redox cycle responsible for the reduction of hydroperoxides is glutathione peroxidase. Nonstressed cells contain a high intracellular ratio of reduced glutathione (GSH) to oxidized glutathione (GSSG). This high GSH/GSSG ration ensures availability of GSH for reduction of hydroperoxides via the glutathione redox cycle. Regeneration of GSH requires nicotinamide-adenine dinucleotide phosphate (NADP)-reducing equivalents that are supplied through the glucose-6-phosphate dehydrogenase (G6PD) activity in the hexose monophosphate shunt. NADPH is an important antioxidant molecule as it is the cofactor for the regeneration of GSH from its oxidized form, GSSG. The NADPH, in turn, is regenerated by the enzyme glucose-6-phosphate dehydrogenase as part of the hexose monophosphate shunt pathway of energy metabolism.

The lung also possesses nonenzymatic antioxidants such as vitamin E, beta-carotene, vitamin C, and uric acid. Vitamin E is lipid-soluble and partitions into lipid membranes, where it is positioned optimally for maximal antioxidant effectiveness. Vitamin E converts superoxide anion, hydroxyl radical, and lipid peroxyl radicals to less reactive oxygen metabolites. Beta-carotene also accumulates in cell membranes and is a metabolic precursor to vitamin A. Furthermore, it can scavenge superoxide anion and react directly with peroxyl-free radicals, thereby serving as an additional lipid-soluble antioxidant. Vitamin C is widely available in both extracellular and intracellular spaces where it can participate in redox reactions. Vitamin C can directly scavenge superoxide and hydroxyl radical. Uric acid formed by the catabolism of purines also has antioxidant properties and primarily scavenges hydroxyl radical and peroxyl radicals from lipid peroxidation.

27.3.3 Oxidants and Cell Signaling in the Lung

Maintenance of intracellular redox homeostasis is essential to cellular function and survival. A low level of intracellular ROS, particularly H_2O_2, is required for cell survival and proliferation. For example, polypeptide growth factors (EGF, PDGF) stimulate mitogenesis by binding to specific cell-surface receptors that possess intracellular tyrosine kinase activity. The activation of these receptor tyrosine kinases depends on the phosphorylation of specific tyrosine residues. The phosphorylation event is constitutively inhibited by the action of protein tyrosine phosphatases (PTPs), which serve to dephosphorylate tyrosine residues. This is where H_2O_2 plays a critical role. H_2O_2 inhibits the activity of PTPs, thereby favoring phosphorylation of tyrosine residues. Therefore, ROS are not only injurious byproducts of cellular metabolism but also essential participants in cell signaling and regulation. The paradox in the roles of ROS as essential biomolecules in the regulation of cellular functions and as toxic byproducts of metabolism are due to differences in the concentrations of ROS produced. Oxidative stess is generated when ROS production overwhelms the antioxidant defense. ROS can elicit a variety of cellular responses ranging from protective to injurious, depending on the level of oxidative stress. A hierarchical oxidative stress model is illustrated in Figure 27.13. The three tiers of oxidative stress shown in this figure are defined by the ratio of reduced to oxidized glutathione (GSH/GSSG). In tier 1, mild oxidative stress activates antioxidant and phase II enzymes via the antioxidant response element (ARE), which is activated by the transcription factor Nrf2. In this response level to oxidative stress, Nrf2 escapes from a cytosolic protein termed Keap1, which normally serves to mediate Nrf2 proteasomal degradation and prevent its nuclear accumulation. Once Nrf2 dissociates from Keap1, it is transported to the nucleus, where it binds the ARE in the promoter region of phase II enzymes, leading to their transcriptional activation. Phase II enzymes include heme oxygenase (HO-1), NAD(P)H-quinone oxidoreductase 1 (NQO1), glutathinone-S-transferase (GST), superoxide dismutase (SOD), and glutathione peroxidase (GPx). Tier 1 antioxidant defenses are critical for protecting against airway inflammatory responses in diseases such as asthma.

The failure of antioxidant mechanisms to correct redox disequilibrium could lead to the escalation of oxidative to tier 2. Tier 2 cellular responses are characterized by the activation of cellular signaling pathway such as stress-activated kinases (p38 MAP kinase and JNK) along with activation and nuclear translocation of transcription factors NF-κB and STAT-1. NF-κB-induced transcriptional activation leads to the production of a number of pro-inflammatory cytokines, including the neutrophil chemoattractant IL-8. STAT-1 activation stimulates the increased production of CXC-motif chemokines that function in lymphocyte recruitment and activation. Therefore, tier 2 oxidative responses result in an inflammatory response in the lung.

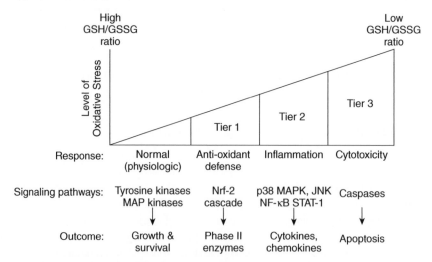

Figure 27.13. Biologic responses of cells to oxidative stress. (Adapted from Li, N. and Nel, A. E. *Antioxidant & Redox Signaling* **8**, 88–98, 2006.)

Tier 3 of oxidative stress involves a cytotoxic response and mitochondrial membrane damage that lead to the activation of caspases that mediate programmed cell death (apoptosis) or necrosis, depending on the severity of oxidative insult.

27.3.4 Respiratory Tract Injury from Inhaled Particles and Fibers

A variety of naturally occurring and man-made particles and fibers pose a threat to the respiratory tract, including air pollution particulates, allergens such as pollen, soot from the industrial burning of oil, diesel exhaust particles, metal oxides, particles in cigarette smoke, and asbestos fibers. Many of these, including air pollution particulates and cigarette smoke, are complex mixtures of organic and inorganic chemical substances. Several different types of particles and fibers will be discussed below to highlight the heterogeneous nature of particle exposure in environmental or occupational settings.

The inhalation of urban air particles has been associated with increased morbidity and mortality principally due to the physiologic impact on the pulmonary and cardiovascular systems. Adverse respiratory effects in exposed human populations include increased asthmatic episodes and an increase in the prevalence of chronic bronchitis. Moreover, chronic exposure to air pollution particles could contribute to the increasing prevalence of chronic obstructive pulmonary disease (COPD). A mixture of organic and inorganic agents contribute to the composition of air pollution particles, including transition metals released during the burning of petrochemicals, polycyclic aromatic hydrocarbons derived from diesel exhaust, and endotoxins from bacterial sources. Many of these constituents are known to stimulate a variety of intracellular signaling pathways that mediate cellular stress responses leading to the pathologic phenotypes that characterize airway remodeling.

Many of the pathophysiologic effects of inhaled particles are due to oxidative stress. The oxidative potential of air pollution particles is generally attributed to transition metals (e.g., zinc, copper, vanadium, iron), which can induce generation of ROS either via Fenton-like reactions or by stimulating an oxidative burst in leukocytes that engulf particles via phagocytosis. Particles from a number of sources have been shown to induce oxidant generation that is associated with metal content of the sample. A number of studies have shown that residual oil fly ash (ROFA), a PM source rich in metal content, induces oxidant generation in alveolar macrophages and epithelial cells. Transition metals are primary inorganic constituents of air pollution particles that generate ROS production. Organic constituents of some particles also generate ROS in some particles. For example, polycyclic aromatic hydrocarbons (PAH) in diesel exhaust particles or semiquinone radicals in cigarette smoke or diesel exhaust can participate in redox cycling leading to sustained production of ROS. ROS generated by particle exposure can, in turn, act as signaling intermediates to activate intracellular signaling targets, including receptor tyrosine kinases, mitogen-activated protein (MAP) kinases, and transcription factors that, in turn, lead to transcriptional activation and the expression of genes involved in airway inflammation and remodeling (Figure 27.14). Chronic exposures lead to disease outcomes such as asthma, bronchitis, and COPD.

Specific types of particle constituents have been identified as mediators of occupational lung diseases. For example, the occupational exposure to oil fly ash generated as a waste product during the industrial burning of fuel oil in power plants is

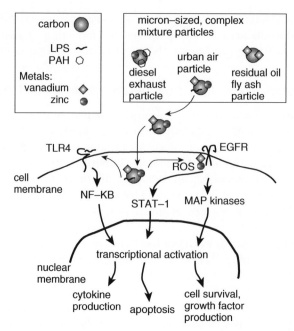

Figure 27.14. Activation of intracellular signaling pathways and cytokine gene expression by air pollution particles containing transition metals and endotoxin. (From Walters, D. M., and Bonner, J. C. Signal transduction and cytokine expression in particulate matter (PM)-induced airway remodeling. In: Foster, W. M., and Costa, D. L. (Eds.). *Air Pollutants and the Respiratory Tract*, 2nd edition, Taylor & Francis, New York, 2005, pp. 233–258.)

due primarily to vanadium pentoxide, a transition metal that is richly abundant in oil deposits. The occupational exposure to grain dust causes asthma and bronchitis, due principally to endotoxin (a pro-inflammatory bacterial wall component) that adheres to the grain dust particles. Occupational exposure to beryllium, a metal used widely in the manufacture of nuclear weapons, causes a severe hypersensitivity pneumonitis in a small percentage of susceptible individuals. Therefore, the composition of certain particles determines toxicity and the specific type of lung disease that occurs following exposure.

27.3.5   Particle and Fiber Deposition and Clearance

While chemical composition is important in determining the toxicity of particles and fibers, it is equally or more important to determine where a particle or fiber will deposit in the respiratory tract and how long it will stay there. The quantity and location of particle deposition in the respiratory tract depends on factors related to both the exposed individual and the inhaled particles. The mechanism of deposition is determined by the physical (size, shape, and density) and chemical (hygroscopicity and charge) characteristics of the inhaled particles. Particle deposition is also affected by biological factors inherent to the exposed individual such as breathing pattern (volume and rate), route of breathing (mouth versus nose), and the anatomy of the airways.

Particle deposition in the respiratory tract occurs primarily by three mechanisms: *impaction, sedimentation*, and *diffusion*. Impaction and sedimentation depend on the particle's *aerodynamic diameter*. The aerodynamic diameter of a particle is the size parameter of greatest importance for deposition considerations. It is equal to the diameter of a unit-density sphere having the same terminal settling velocity as the particle in question. *Impaction* is the collision of a moving particle with a static structure. It occurs when an inhaled particle has too much momentum to change course with the directional change in airflow and, as a result, impacts against the airway surface. Most inhaled particles greater than 5 μm in aerodynamic diameter impact on the surfaces of the pharyngeal tracheobronchial regions of the respiratory tract and do not reach the distal lung. *Sedimentation* occurs by the gravitational settling of particles on a respiratory tract surface. Diffusion takes place when a particle reaches an airway surface by random *Brownian movement*. This is an important mechanism for particles with diameters in the nanometer range that reach the distal lung (terminal bronchioles and alveoli), where there is almost no airflow. The combined processes of diffusive and sedimentary deposition are important for particles in the range 0.1–1 μm. Impaction and sedimentation predominate above, and diffusion predominates below this range (Figure 27.15). Air pollution particulate matter has been characterized into three categories: (1) ultrafine (<0.1 μm diameter), also referred to as nanosized particulates, (2) fine particles (0.1- to 2.5-μm diameter), and (3) course particles (>2.5-μm diameter). Epidemiological studies suggest that ultrafine and fine particles are better correlated with adverse health effects when compared to course particles.

Other mechanisms of deposition are *interception* and *electrostatic charge*. *Interception* is most important for fibers and occurs when an inhaled fiber contacts an

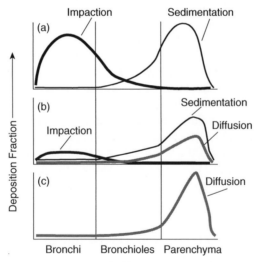

Figure 27.15. The deposition site of particles in the lung depends on particle size. Shown are predicted deposition patterns of (a) 5-μm particles, (b) 1-μm particles, and (c) 0.1-μm particles. (Adapted from Bennett, W. D., and Brown, J. S. Particulate dosimetry in the respiratory tract. In: Foster, W. M., and Costa, D. L. (Eds.). *Air Pollutants and the Respiratory Tract*, 2nd edition. Taylor & Francis, New York, 2005, pp. 21–73. This figure was completely redrawn by the author from materials cited.)

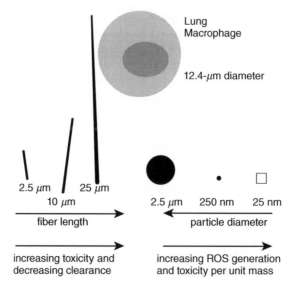

Figure 27.16. Size and shape characteristics of particles that determine deposition, clearance, and toxicity in the respiratory tract. In general, smaller particles are more toxic whereas longer fibers are more toxic. A lung macrophage is shown as a size reference. (See text for details.)

airway wall or when long, thin fibers intercept the airway bifurcations. The likelihood of interception is enhanced with increasing fiber length. Such is the case for chrysotile asbestos fibers, which primarily deposit at alveolar duct bifurcations in the distal lung. While spherical particles with a diameter greater than 5 μm do not reach this region of the lung due to impaction in the upper respiratory tract, asbestos fibers with a length exceeding 20 μm can reach the alveolar region of the lung and persist for months or even years due to the inability of macrophages to effectively clear such long, thin structures. Therefore, for particles, decreasing diameter correlates with increasing toxicity because (1) smaller particles in the submicron or nanometer range reach the alveolar region (distal lung) and (2) an equivalent mass of smaller particles have a greater surface area per unit mass and therefore a greater potential to generate ROS and oxidative stress (Figure 27.16). For fibers, increasing length corresponds to greater toxicity given that the fiber is respirable (i.e., one that can be inhaled and deposit in the alveolar region of the lung) because longer fibers are more difficult to clear from the lung and persist in the interstital tissues to cause damage.

27.3.6 Respiratory Tract Injury from Gases and Vapors

The respiratory toxicity of gases and vapors is determined for several different physical and chemical properties. These factors include (1) chemical *dose*, (2) *water solubility* (hydrophilicity versus lipophilicity), and (3) *chemical reactivity*.

The *dose* of the toxic substance to which an individual is exposed generally determines severity of injury. For example, the occupational exposure to gases (e.g., chlorine, ammonia, HCl) can be divided into three levels of dose exposure that directly correlate to the degree of airway injury and severity of symptoms. Low exposures to these gases cause sneezing, rhinitis, and sore throat. Repeated low exposures or mild exposure causes persistent cough and bronchitis. Accidental exposure to high concentrations of these gases cause laryngeal edema, acute respiratory distress syndrome, and possibly death. Moreover, high-dose exposure to toxic gases and vapors can cause substantial airway epithelial cell damage in the respiratory tract, resulting in a reactive airways dysfunction syndrome (RADS) that is characterized by denuded epithelium and airway luminal fibroproliferative lesions.

The *solubility* of an inhaled substance influences the deposition pattern in the respiratory tract and the site of injury. Gases such as formaldehyde, hydrochloric acid, and sulfur dioxide are taken up by the mucosal surfaces of the upper respiratory tract and will exert most of their toxicity in the nasal region. This is because a highly water-soluble chemical will tend to leave the air in the respiratory tract and enter the mucous lining. The relative amount of a chemical in two compartments (e.g., air and mucous lining) at equilibrium is called the *partition coefficient* for that chemical in those two compartments. Gases and vapors with low water solubility such as nitrogen dioxide and ozone tend to deposit farther down the respiratory tract and cause damage to terminal and respiratory bronchioles. Selected water-soluble and insoluble gases are listed in Table 27.1.

The *reactivity* of an inhaled chemical refers to a more unstable conformation (high-energy state) such as formaldehyde that can easily bond with other molecules. Also, chemical reactivity often means that the reactive substance has the ability to generate reactive oxygen or reactive nitrogen species as a consequence of its reaction with other molecules.

A number of gases also cause *irritant responses* in the respiratory tract. These can be classified as *sensory irritants, pulmonary irritants*, or *bronchoconstrictors. Sensory irritants* stimulate the trigeminal nerve endings in the upper respiratory tract, leading to a burning sensation in the nose. Examples of sensory irritants are ammonia, acrolein, formaldehyde, and sulfur dioxide. A *pulmonary irritant* is one that stimulates sensory receptors (C-fiber receptors) in the airways, thereby increasing respiratory rate and causing dyspnea (difficulty in breathing) or rapid, shallow breathing. Examples of pulmonary irritants are phosgene, nitrogen dioxide, ozone, and sulfuric acid mist. *Bronchoconstrictors* stimulate nerve endings in airways to cause contraction of airway smooth muscle, thereby narrowing the airway lumen and increasing resistance to airflow in conducting airways. Examples of bronchoconstrictors are

TABLE 27.1. Water Solubility, Irritant Classification, and Site of Injury for Selected Highly Reactive Gases

Gas/Vapor	Water Solubility	Irritant Class	Site of Injury
Chlorine	High	Sensory	Nasal, trachea
Formaldehyde	High	Sensory	Nasal, trachea
HCL	High	Sensory	Nasal, trachea
Ammonia	High	Sensory bronchoconstrictor	Nasal, trachea
Phosgene	Low	Pulmonary	Terminal bronchioles
Ozone	Low	Pulmonary	Terminal bronchioles
Nitrogen dioxide	Low	Pulmonary	Terminal bronchioles
Sulfur dioxide	High	Sensory, bronchoconstrictor	Nasal, trachea

sulfur dioxide and ammonia. In addition to gases and vapors, a number of particles and allergens stimulate bronchoconstriction in individuals with asthma.

27.4   OCCUPATIONAL AND ENVIRONMENTAL LUNG DISEASES

While the respiratory system is well-equipped to defend against exposure to a vast array of toxic substances, the intricate cellular and molecular mechanisms designed to repair injured lung tissues often fail, resulting in a number of chronic lung diseases, including cancer, fibrosis, asthma, hypersensitivity pnuemonitis, and chronic obstructive pulmonary disease (COPD), which is a combination of bronchitis and emphysema.

27.4.1   Pulmonary Fibrosis

Fibrotic interstitial lung diseases are restrictive disorders that involve the proliferation of myofibroblasts in the interstitium, including the alveolar walls and perivascular and peribronchial tissues. These include the chronic progressive disorders such as idiopathic pulmonary fibrosis (fibrotic disorders of unknown etiology) and a number of fibrotic diseases that occur as a result of environmental or occupational exposures, including asbestosis and silicosis. These chronic diseases are thought to arise from prolonged, low-grade injury that feature smoldering lesions within the lung that develop over a period of years. Fibrosis can also occur following acute lung injury, as in the adult respiratory distress syndrome, where destruction of the epithelium and endothelium destroys the permeability barrier and permits flooding of the airspaces with proteinaceous edema and the infiltration of myofibroblasts into airspaces. Myofibroblasts may also infiltrate into small airways of the lung, as occurs during the progression of obliterative bronchiolitis following tissue rejection after lung transplantation. Finally, it is becoming increasingly clear that more subtle fibrotic reactions are associated with airway remodeling in diseases such as asthma, chronic bronchitis, and chronic obstructive pulmonary disease.

The cellular responses in fibrotic lung diseases are orchestrated by a variety of cytokines, growth factors, proteases, and lipid mediators. In general, widely different types of fibrogenic agents (bleomycin, paraquat, radiation, asbestos fibers) cause lung injury via the generation of reactive oxygen species, although the precise mechanisms of ROS generation may vary (Figure 27.17). ROS contribute to injury by causing epithelial cell apoptosis directly or by stimulating factors such as TNF-α that are pro-apoptotic. ROS also activate macrophages and lymphocytes to produce a variety of cytokines, including pro-inflammatory cytokines such as IL-1β, IL-6, IL-8, IL-13, and TNF-α. Many of these cytokines are multifunctional and contribute to inflammatory cell recruitment and activation, and they also stimulate leukocytes, epithelial cells, smooth muscle cells, endothelial cells, and fibroblasts to synthesize other cytokines and growth factors that participate in the fibrotic response to injury. For example, IL-13 produced by lymphocytes stimulates epithelial cells to produce PDGF, a potent mitogen and chemoattractant for lung fibroblasts and myofibroblasts (Figure 27.18). Activated macrophages secrete the pro-inflammatory cytokine TNF-α that, in turn, stimulates the production of TGF-β to promote collagen deposition by myofibroblasts. Moreover, TGF-β acts on

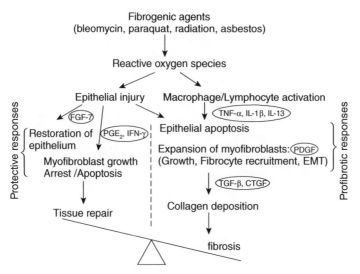

Figure 27.17. Mechanisms of pulmonary fibrosis. (See text for details.)

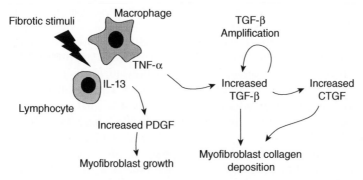

Figure 27.18. Example of cytokine-growth factor cascades that contribute to myofibroblast proliferation and collagen production during fibrogenesis.

fibroblasts to amplify its own production and well as stimulate CTGF, another factor that contributes to myofibroblast activation and collagen production. During fibrogenesis, protective mechanisms include (a) restoration of epithelium by factors such as FGF-7 and (b) myofibroblast growth arrest and apoptosis by PGE$_2$ and interferons. These protective mechanisms play a role in suppressing the severity of the fibrotic response and are important in the resolution of partial resolution of some fibrotic responses.

Chronic exposure to high concentrations of airborne inorganic dusts can result in a form of interstitial lung disease referred to as *pneumoconiosis*. These are fibrotic disorders confined to the lower respiratory tract, primarily the terminal bronchioles and alveolar walls. Inorganic dusts are normal components of the ambient air. However, pneumoconiosis are not diseases of the general population but are instead confined to individuals chronically exposed to high concentrations of these dusts in the occupational setting. Three major classes of inorganic dusts that can cause

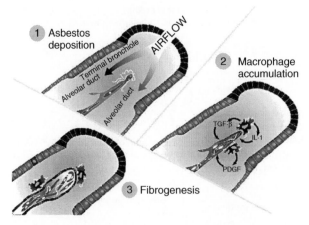

Figure 27.19. Illustration of early asbestos-induced fibroproliferative lesions in a rat inhalation model. (1) Inhaled fibers deposit at alveolar duct bifurcations. (2) Within 24 hr, macrophages accumulate at sites of fiber deposition and become activated by fibers to secrete growth factors. (3) Within 72 hr fibroblasts proliferate.

interstitial lung disease are asbestos (the cause of "asbestosis"), silica ("silicosis"), and coal workers' "pneumoconiosis."

Rodent models of fibrotic disease that have been developed to better understand the cellular and molecular mechanisms involved in the disease process. These include (1) intratracheal or intravenous exposure to bleomycin, (2) high-dose gamma radiation to cause lung injury, and (3) asbestos inhalation. The bleomycin model is by far the most widely used animal model of fibrosis. Asbestos inhalation has proven useful as a model to correlate site of fiber deposition in the lung with early lesion formation. Figure 27.19 illustrates the deposition of asbestos fibers at alveolar duct bifurcations in the distal lung followed by macrophage accumulation at sites of deposition that occurs within 24 hr post-exposure. Macrophages activated by asbestos fibers produce cytokines (e.g., IL-1β, TNF-α) and growth factors (e.g., PDGF, TGF-β1) that stimulate lung myofibroblasts and fibroblasts to proliferate and deposit collagen to form early fibroproliferative lesions within the interstitium of the bifurcation.

27.4.2 Asthma

Asthma is an *obstructive* allergic airways disease that is hallmarked by *acute bronchospasm* of airways called an "asthma attack," but also features *chronic airway inflammation and remodeling* that occurs over a period of years. Both the acute bronchospasm and the chronic airway remodeling process contribute to the obstructive nature of this disease (Figure 27.20). Asthma is common. Over the past few decades, asthma prevalence has increased and affects up to 10% of the population in developed countries. The reason for the increase in asthma is not entirely clear, but children are primarily affected. Some cases of asthma in children resolve, while some individuals with asthma develop irreversible changes in lung function due to the chronic airway remodeling process. A variety of allergens cause asthma and act as sensitizing agents (Table 27.2). High-molecular-weight allergens from plants,

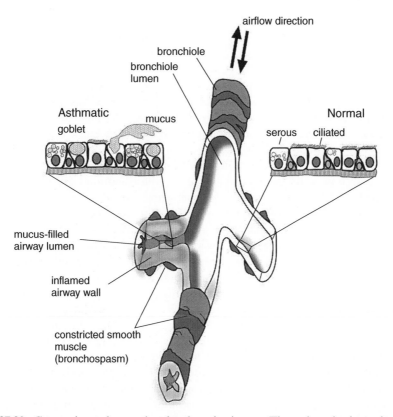

Figure 27.20. Comparison of normal and asthmatic airways. The asthmatic airway is characterized by thickened airway smooth muscle that constricts to cause airway narrowing and obstruction. Mucus secreted by epithelial (goblet) cells also contributes to the obstruction of the airway lumen. (Adapted from LifeART illustration series, Lippincott Williams & Wilkins, Hagerstown, MD, 1994. This figure was completely redrawn by the author from materials cited.)

TABLE 27.2. Agents that Cause or Exacerbate Asthma

Agent
Allergens Causing Asthma
Cockroach
House dust mite
Cat
Plant debris
Endotoxin
Molds
Exacerbation of Asthma
Environmental tobacco smoke
Sulfur dioxide
Nitrogen dioxide
Ozone
Particulates

animals, bacteria, or mold trigger and IgE-mediated immune response. Most cases of asthma are caused by indoor allergens, particularly components of chitin exoskeleton from house dust mite or cockroaches. Low-molecular-weight allergens such as metals, anhydrides, and penicillin generally act as "haptens" (incomplete allergen) and must combine with serum proteins to elicit an allergic response.

Asthma features a chronic airway remodeling response that is characterized by (1) eosinophilic inflammation, (2) airway smooth muscle thickening, (3) mucus cell hyperplasia and mucus hypersecretion, and (4) subepithelial fibrosis. Allergic diseases, including asthma, are thought to result from a dysregulated immune response to commonly encountered antigens in genetically predisposed individuals. Immunological research into the mechanisms of allergy has identified cytokine production by T-helper 2 (Th2) effector lymphocytes as being critical for orchestrating allergic inflammation rich in eosinophils. Upon recognition their cognate antigen, Th2 lymphocytes produce cytokines that regulate IgE synthesis, growth and activation of eosinophils and mast cells, and expression of endothelial cell adhesion molecules. The first step in the allergic immune response is the uptake and presentation of allergen by antigen-presenting cells called dendritic cells. Macrophages and B lymphocytes may also serve as antigen-presenting cells. Following recognition and uptake, dendritic cells migrate to the T-cell-rich area of draining lymph nodes, display an array of antigen-derived peptides on the surface of major histocompatibility complex (MHC) molecules, and acquire the cellular specialization to select and activate naive antigen-specific T cells. Allergen targeting to the dendritic cells occurs via membrane-bound IgE. Dendritic cells interact with many cell types, including mast cells, epithelial cells, and fibroblasts. Mediators released by these cells can activate the dendritic cells so that it is induced to mature and attract memory Th2 cells through release of Th2-selective chemokines. Mature effector Th2 cells play a central role in asthma pathogenesis by releasing cytokines (e.g., IL-13) that stimulate eosinophil recruitment, smooth muscle cell cytokine and chemokine production, and goblet cell hyperplasia (Figure 27.21).

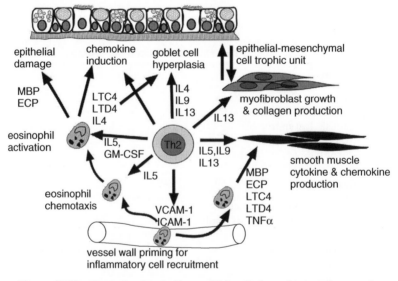

Figure 27.21. Central role of effector Th2 cells in asthma pathogenesis.

27.4.3 Hypersensitivity Pneumonitis

Hypersensitivity pneumonitis (HP), also known as *extrinsic allergic alveolitis*, is an exaggerated adaptive immune response that occurs in susceptible individuals. Unlike asthma, which affects the airways of the lung, HP is caused by the inhalation of allergens that elicit lymphocytic inflammation in the peripheral airways and surrounding interstitial tissue. Although individuals that develop HP produce exuberant levels of antibody against the offending antigen, the progression of HP proceeds. The immunopathogenesis of HP involves three major steps. First, there is an *acute phase* that features lymphocyte and macrophage accumulation and activation. $CD8^+$ cytotoxic lymphocytes and monocytes accumulate widely in the alveolar spaces. After inhalation soluble antigens bind IgG antibody and immune complex initiate the complement cascade, and the resulting C5 activates macrophages. Activated resident macrophages in the lung then secrete chemokines that attract circulating T lymphocytes and monocytes. In contrast to IgE-mediated allergic responses (i.e., asthma), the IgG-mediated response in HP does not feature eosinophilic inflammation. Second, a *subacute phase* occurs when the monocytes mature into foamy macrophages and develop into *granulomas* widely dispersed throughout the lung. Third, a *chronic phase* begins as the inflammatory cells produce growth factors such as TGF-β that, over time, stimulate fibroblasts and myofibroblasts to proliferate and deposit collagen to form scar tissue. The end result is interstitial fibrosis.

A variety of agents cause HP, and many of these agents are encountered occupationally (Table 27.3). Most of the causative allergens have been recognized in a wide variety of occupations. Consequently, once the problem has been identified,

TABLE 27.3. Causative Agents of Hypersensitivity Pneumonitis

Agent	Source	Disease
Microbes		
Thermophilic actinomycetes	Moldy plant materials	Farmer's lung
Saccharopolyspora rectivirgula	Moldy hay	Farmer's lung
Thermoactinomycetes vulgaris	Compost	Mushroom worker's lung
Aspergillus species	Tobacco mold	Tobacco worker's lung
Penicillium chrysogenum	Moldy wood dust	Woodworker's lung
Bacteria and fungi	Metal-working fluids	Machine operator's lung
Animals		
Avian proteins	Bird excreta, feathers	Bird fancier's lung
Animal fur protein	Animal fur	Furrier's lung
Plants		
Soybean	Soybean hulls	Soybean worker's lung
Coffee	Coffee bean dust	Coffee worker's lung
Chemicals		
Isocyanates	Paints, plastics	Paint-refinisher's lung
Anhydrides	Plastics	Chemical-worker's lung
Pyrethrum	Insecticides	Insecticide lung
Metals		
Cobalt	—	Hard metal lung disease
Beryllium	—	Chronic beryllium disease

Source: Patel et al., *J. Allergy Clin. Immunol.* **108**, 661–670, 2001.

the exposure can be controlled and the disease prevented. As a result of this reduction in occupational exposures, the disease is now less common than it was 20 years ago. It is also important to recognize that many existing exposures also occur at home. They are especially associated with pet birds, contaminated humidifiers, and heavy concentrations of indoor molds. Many of the occupational environments that cause HP also contain airborne gram-negative endotoxin in addition to allergens. As a consequence of this mixed exposure, workers can develop HP, which is a lymphocytic disease, along with COPD, which is a neutrophilic disease, or both.

27.4.4 COPD

Chronic obstructive pulmonary disease (COPD) is a progressive disease that encompasses both *chronic bronchitis* and *emphysema*. Chronic bronchitis is defined clinically as the presence of chronic productive cough for 3 months in each of 2 successive years in a person in whom other causes of cough have been excluded. Emphysema is defined anatomically by abnormal permanent enlargement of the airspaces distal to the terminal bronchioles, accompanied by destruction of alveolar walls without obvious fibrosis. Reduced airflow in COPD is caused by increased resistance to airflow due to inflammation, fibrosis, and goblet cell metaplasia, along with smooth muscle cell hypertrophy in small airways. A major factor in reducing airway function is loss of elastic recoil due to inflammation and the progressive loss of elastin-dependent attachment of alveoli to bronchioles.

The major cause of COPD is cigarette smoking, particularly the emphysema component of the disease. A variety of particulate and gases in cigarette smoke mediate cause the formation of ROS, leading to remodeling of the airways and lung parenchyma. Cigarette smoking causes increased mucous gland size (hypertrophy) and goblet cell number (hyperplasia) in large airways. Smaller airways of smokers show goblet cell metaplasia with mucous plugging, inflammatory infiltrates, airway fibrosis, and loss of airway elastin. All of these changes contribute to airway narrowing and increased resistance to airflow. Cigarette smoking also interferes with ciliary action, damages the airway epithelium, and impairs the ability of alveolar macrophages to clear bacteria, thereby predisposing the individual to infection and exacerbation of bronchitis. Individuals with emphysema almost always have bronchitis. However, bronchitis frequently occurs in the absence of emphysema, and a number of inhaled substances (e.g., particles, metals, irritant gases, microbial infections) cause chronic bronchitis without causing emphysema.

The progressive destruction of elastin in the distal lung is the major cause for loss of alveolar walls in emphysema. The loss of elastin is thought to be due to an imbalance between proteases released from inflammatory cells and antiproteases in the lung that normally serve to protect against excessive proteolytic degradation. This is referred to as the *protease–antiprotease hypothesis* for the development of emphysema. This hypothesis is supported by the following: (1) individuals with a genetic deficiency in alpha1-antitrypsin (the major inhibitor of neutrophil elastase) are predisposed to the emphysema whether they smoke or not, and (2) emphysema can be induced experimentally in animals by the intratracheal instillation of neutrophil elastase. Moreover, ROS released by cigarette smoke inactivates the inhibitor of elastase (alpha1-antitrypsin) by oxidizing amino acids within the active site of the enzyme. While elastase appears to be a major protease involved in emphy-

sema, other proteases may be contributory, including cathepsins, matrix metalloproteinases, and collagenases. In addition to alpha1-antitrypsin, other antiproteases
include secretory leukoprotease inhibitor (SLPI), tissue inhibitors of metalloproteases (TIMPs), and alpha2-macroglobulin.

27.4.5 Lung Cancer

Lung cancer is the leading cause of death from cancer in the United States. The major
risk factor for lung cancer is tobacco smoke. A higher incidence of lung cancer in
men versus women is directly correlated with a higher rate of smoking in men versus
women. However, an increase in smoking women has led to a doubling in lung cancer
incidence in women over the past 20 years, whereas a decline in smoking among men
has led to a slight decline in lung cancer incidence in men over the same period. In
addition to cigarette smoke, occupational exposures to a variety of agents, including
arsenic, asbestos, polycyclic aromatic hydrocarbons, chromium, and nickel are also
associated with increased incidence of lung cancer. The major histopathologic types
of human lung cancer are squamous cell carcinoma (29%), adenocarcinoma (32%),
small cell carcinoma (18%), and large cell carcinoma (9%). Adenocarcinomas and
large cell carcinomas tend to occur more in the peripheral lung, whereas squamous
cell carcinomas and small cell carcinomas are more likely to occur in the central lung
adjacent to large airways. Mesothelioma is an unusual type of lung cancer that develops along the pleural lining of the lung as a result of asbestos exposure. The major
histopathologic types of lung cancer found in rats and mice include adenomas, adenocarcinomas, and squamous cell carcinomas. The mouse has proven to be a suitable
model for the study of lung cancer progression, and most agents that cause cancer
in humans also cause cancer in mice (Table 27.4). The molecular mechanisms of
carcinogenesis are discussed in detail in Chapter 25.

**TABLE 27.4. Pulmonary Tumor Response of Laboratory
Rodents to Inhalation of Known Human Pulmonary
Carcinogens**

Agent	Human	Rat	Mouse
Chemicals			
Arsenic	+	ND	ND
Asbestos	+	+	±
Beryllium	+	+	±
Chromium	+	±	ND
Coal tar	+	+	+
Mustard gas	+	ND	±
Nickel	+	+	±
Soots	+	+	ND
Vinyl chloride	+	+	+
Environmental Agents			
Tobacco smoke	+	+	±
Radon	+	+	−

Key: +, positive, ND, no data, ±, limited data, −, negative.

Source: Hahn, F. F. Lung carcinogenesis. In: Kitchum, K. T. *Carcinogenicity*, (Ed.). Marcel Dekker, New York, 1998.

SUGGESTED READING

Crystal, R. G., and West, J. B. (Eds.). *The Lung: Scientific Foundations*, Raven Press, New York, 1991.

Foster, W. M., and Costa, D. L. (Eds.). *Air Pollutants and the Respiratory Tract*, 2nd ed., *Lung Biology in Health and Disease*, Vol. 204, Taylor & Francis, New York, 2005.

Gardner, D. E., Crapo, J. D., and McClellan, R. O. (Eds.). *Toxicology of the Lung*, 3rd ed., Raven Press, New York, 1999.

Guyton, A. C., and Hall, J. E. (Eds.). *Textbook of Medical Physiology*, 9th ed., W.B. Saunders, Philadelphia, 2000.

Lambrecht, B. N., Hoogsteden, H. C., and Diamant, Z. (Eds.). *The Immunologic Basis of Asthma, Lung Biology in Health and Disease*, Vol. 174, Taylor & Francis, New York, 2003.

■■■■■ CHAPTER 28

Hepatotoxicity

ANDREW D. WALLACE

Department of Environmental and Molecular Toxicology, North Carolina State University, Raleigh, North Carolina 27695

SHARON A. MEYER

College of Pharmacy, University of Louisiana at Monroe, Monroe, Louisiana 71209

28.1 INTRODUCTION

The liver is a frequent target tissue of toxicity from specific members of all classes of toxicants and natural toxins. Experimental hepatotoxicity of the rat from solvent CCl_4 is a prototype model system for study of toxicodynamics. In humans, drug-induced liver injury (DILI) remains one of the major reasons that new drugs fail to meet regulatory approval or are removed from the market. For example, the type II diabetes drug troglitazone (Rezulin) was removed from the market after causing acute hepatocellular injury in susceptible individuals as determined by serum enzyme measurements during treatment. In a number of cases, prolonged treatment led to complete liver failure. Liver damage from repeated exposure to toxic doses of ethanol is a major human health problem. The risk of chemically induced hepatotoxicity is increased by multiple, simultaneous exposures that can occur with chemical mixtures in the environment or from multiple, concurrent therapies. The rising prevalence of viral liver disease is another interacting factor that can be expected to aggravate the hepatotoxicity of chemicals. Understanding the molecular and cellular mechanisms mediating these toxic effects within the context of the function and structure of the liver is important knowledge necessary for predicting avoidable toxicity and adverse interactions.

Parodoxically, the structures and functions of the liver predispose it to chemical toxicity. Hepatocytes are exposed to orally administered xenobiotics without systemic modification or dilution because they flow directly to the liver by the portal venous blood that delivers absorbed nutrients from the gastrointestinal tract. Xenobiotics may be extensively metabolized by the liver so that little of the parent xenobiotic enters the systemic circulation, a phenomenon known as the first pass effect. The movement of xenobiotics into the liver coupled with an opposing move-

---

Molecular and Biochemical Toxicology, Fourth Edition, edited by Robert C. Smart and Ernest Hodgson
Copyright © 2008 John Wiley & Sons, Inc.

ment of bile from its site of synthesis at the hepatocyte canalicular membrane to the duodenum establishes a enterohepatic circulatory loop. Toxicants caught in the enterohepatic circulation have slower clearance rates, and hepatocytes are repeatedly exposed to the toxicants. Routing of portal venous blood to the afferent vasculature also has the adverse effect of decreasing pressure and oxygenation on the arterial side of the hepatic sinusoids compared to capillary beds of other tissues. The predominance of hepatocyte enzymes such as the cytochrome P450s (CYPs) and flavin-containing monoxygenase (FMOs) metabolize hydrophobic xenobiotics, a necessity for their renal and biliary excretion into aqueous compartments. These enzymes can generate locally reactive toxic metabolites if uncoupled from protective conjugating reactions. Also, the highly active, anabolic activities that occur within hepatocytes make this cell type especially susceptible to adverse effects of toxicants that act as antimetabolites or compromise mitochondrial energy production (see Chapter 17). The importance of these diverse liver functions for metabolite availability to critical organs such as brain causes severe effects of liver toxicants to be threatening to the viability of the organism.

28.2 LIVER ORGANIZATION AND CELLULAR COMPONENTS

The liver is a heterogeneous tissue composed primarily of parenchymal hepatocytes plus several additional nonparenchymal cell types. In addition, hepatocytes are functionally diverse depending upon their positional relationship to the afferent blood supply of the sinusoids. The liver parenchyma is comprised of a mesh of plates, each being the thickness of one hepatocyte (Figure 28.1). Liver plates are separated by sinusoids, an arrangement ensuring that two surfaces of each hepatocyte are in close proximity to the blood supply. In two-dimensional microscopic sections, this organization appears as rows of hepatocyte cords separated by nearly parallel sinusoids. A specialized region of the plasma membrane, the canalicular membrane, runs as a band around each hepatocyte midway between the sinusoidal

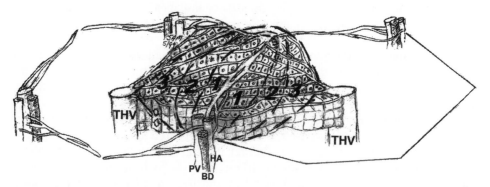

Figure 28.1. Comparison of structural liver lobule with functional acinar regions. The liver lobule is centered on the terminal hepatic venule (THV), also called the central vein, and assumes a roughly hexagonal shape with its vertices at the portal triads, which contain the portal vein (PV), hepatic artery (HA), and bile duct (BD). The liver acinus is centered upon the tract of blood vessels that branch from the hepatic artery and portal vein of the portal triads. Hepatocytes within the acinus are grouped within functional zones 1, 2, and 3 located at increasing distances from the vascular tracts that interconnect adjacent portal triads.

surfaces. Canalicular regions from neighboring hepatocytes align and are joined by tight intercellular junctions to form bile canaliculi.

Several organizational models of liver structure have been proposed. The most frequently used is the morphologically defined lobule. The hepatocyte lobule is visualized in two dimensions with light microscopy as a hexagon centered on the terminal branch of the hepatic vein. The first few hepatocytes of cords radiating from this terminal hepatic venule (THV) or central vein are classified as perivenous or centrilobular. At the vertices of the imagined hexagon are the portal triads, cross-sectioned connective tissue tracts in which are embedded a bile duct (BD) with its distinctive cuboidal epithelial cells and preterminal branches from the hepatic artery (HA) and portal vein (PV). Hepatocytes located around the portal triads and following the circumference of the hexagonal lobule are classified as periportal. Midzonal hepatocytes intervene between centrilobular and periportal hepatocytes. An earlier concept defined the hepatocyte heterogeneity with respect to the functional acinus. Acinar hepatocytes are grouped into three zones that are layered concentrically around the distributing vessels. The similarity of intrazonal hepatocytes is determined by their exposure to a nearly equivalent blood composition, while interzonal hepatocyte differences reflect gradients in blood constituents known to exist across the sinusoids. Although conceptually different, the nomenclature from the acinar model continues to be used. A rough equivalence exists between lobular periportal and acinar zone 1 hepatocytes, centrilobular and zone 3 hepatocytes, and midzonal and zone 2 hepatocytes. There is an increase in the size of hepatocytes from zone 1 to zone 3, and an increase in the DNA content with diploid (2N) cells in zone 1 and increasing polyploidy (4N) cells in zone 3. Hepatocytes found in zone 3 are thought to represent terminally differentiated hepatocytes unable to undergo cell division.

28.2.1 Regional Differences of Lobular Function

Gradients in hepatocyte function occur over the periportal to perivenous distance. Periportal (zone 1) hepatocytes are exposed to the highest concentrations of oxygen, hormones, nutrients, and xenobiotics coming from the intestine or the systemic circulation. Periportal hepatotoxicity suggests that the parent compound may itself be toxic as zone 1 is exposed to the highest concentrations. The partial pressure of oxygen (pO_2) is about 65 mm Hg (corresponding to about 9–13% O_2) in this zone. Cells are rich in mitochondria with more and longer cristae, and oxidative energy metabolism predominates. Glucose for systemic delivery to extrahepatic sites is primarily synthesized in periportal hepatocytes due to their enrichment in gluconeogenic enzymes such as phosphenolpyruvate carboxykinase (PEPCK). Co-localization of amino acid transaminases in periportal hepatocytes provides one source of gluconeogenic substrates. By contrast, perivenous (zone 3) hepatocytes are exposed to blood with reduced oxygen content and increased concentration of metabolic products. Perivenous pO_2 is about 30–40 mm Hg (4–5% O_2), and mitochondria are matrix-rich and cristae-poor. Enzymes of anaerobic glycolysis and lipogenesis are preferentially located in these cells.

The regional expression of xenobiotic metabolizing enzymes determines the zone-specific localization of toxicant damage. Many hepatotoxicants elicit perivenous damage because CYPs are preferentially localized in centrilobular hepatocytes. For example, acetaminophen causes hepatotoxicity because the reactive metabolite NAPQI is produced in this region by CYP2E1. High levels of CYPs coupled with a reversed distribution for conjugating glutathione (GSH) makes perivenous hepatocytes especially susceptible to reactive electrophiles. Differential distributions also exist for glutathione S-transferases, UDP-glucuronosyltransferases, epoxide hydrolase, and alcohol dehydrogenase, which have higher perivenous activities, while sulfotransferase is predominant in periportal hepatocytes. Ammonia detoxication to urea occurs periportally, but is of limited capacity. Residual ammonia reaching centrilobular hepatocytes is incorporated into glutamine by perivenous glutamine synthetase.

Mechanisms maintaining metabolic regional hepatocellular specificity differ depending upon the process. The periportal to perivenous gradient of carbohydrate metabolism can be reversed by changing the direction of blood flow and thus appears dependent upon substrate availability and posttranscriptional signaling events. Repression of periportal transcription by hormonal influences, especially thyroid and growth hormones, appears to be responsible for CYP distribution. However, plasticity exists for these systems as changes in zonal size occur to accommodate functional needs. In contrast, glutamine synthetase expression is a stable phenotype of perivenous cells.

28.2.2 Liver Stem Cells

Unlike most mature cells, differentiated hepatocytes retain the ability to proliferate. The basal rate of hepatocyte cell proliferation is very low, but under conditions necessitating an increase in liver capacity, hepatocytes will act as facultative stem cells. For example, hepatocytes will enter the cell cycle almost synchronously after surgical resection until liver functional capacity is restored. Also, chemicals like phenobarbital will cause an increase in liver mass, in part, through hyperplasia. During continuous exposure to these chemicals, hepatocytes transiently undergo cell proliferation, but thereafter return to a basal rate while augmented liver mass is maintained through chemically mediated inhibition of cell loss. In contrast, high doses of hepatotoxicants inhibit hepatocyte proliferation. If a growth stimulus is received during exposure to hepatotoxicants, another cell type will proliferate. These so-called hepatic progenitor cells (HGCs) or "oval" cells are named after their distinctive oval nuclei and are derived from the portal areas. They are bipotential in that they can differentiate into hepatocytes or bile duct epithelial cells; hence, they are believed to be a true liver stem cell. The fetal liver is also a source of progenitor cells with the ability to colonize livers damaged by hepatotoxic compounds in rat transplantation experiments. A well-defined model for HGC proliferation is surgical resection of liver of rats during treatment with the potent hepatotoxicant, 2-acetylaminofluorene (2-AAF).

28.2.3 Nonparenchymal Liver Cells

Approximately 30% of liver cells are nonparenchymal cells. Cell types located in the interlobular portal tracts include bile duct epithelial cells and connective tissue fibroblasts. Those located within the lobule interior are the rare fixed, natural killer lymphocytes (pit cells) and the more abundant endothelial cells lining the sinusoids, Kupffer cells, and stellate cells (also known as fat-storing or Ito cells). The recent availability of isolation and cell culture techniques for these cell types has allowed characterization of toxicant effects on nonparenchymal cells that are consistent with the tissue pathology. An example is the susceptibility of sinusoidal endothelial cells to toxicity by chemicals such as monocrotaline, which causes veno-occlusive disease due to fibrosis.

Endothelial cells of the sinusoids are highly fenestrated. This porosity plus the absence of highly structured connective tissue, as found in underlying basement membrane of true capillaries, determines the high permeability of hepatic sinusoids. Plasma proteins and small particles pass into the space of Disse, the interval between hepatocytes and sinusoids, but particles larger than 150 nm (e.g., chylomicrons, erythrocytes, viruses) are impeded. Sinusoidal endothelial cells participate in the liver inflammatory response through secretion of cytokines and expression of adhesion molecules for neutrophils. Both sinusoidal endothelial cells and Kupffer cells are active in endocytosis, including that for clearance of modified plasma proteins such as oxidized lipoproteins. Kupffer cells are the fixed macrophages of the liver located within the sinusoidal lumen, where they are active in phagocytosis of particulate matter. They are important for clearance of systemic endotoxin and initiation of the liver inflammatory response through secretion of tumor necrosis factor-*a* (TNF*a*). Gadolinium chloride (GdCl$_3$) and dextran sulfate, inhibitors of Kupffer cell activation, have become important experimental tools for determination of the role of inflammation in chemically induced hepatotoxicity.

Quiescent, perisinusoidal hepatic stellate cells (HSC) accumulate retinoids (vitamin A and its metabolites) in large cytoplasmic droplets and store about 80% of body's retinoids. However, these cells also express the muscle protein desmin and become activated to myofibroblasts upon exposure to hepatotoxicants. Activated HSC synthesize extracellular matrix proteins, especially when stimulated by cytokines and growth factors (see Section 28.4.2).

28.3 TYPES OF CHEMICALLY INDUCED LESIONS

Classification of chemically induced hepatotoxicity is primarily based upon pattern of incidence and histopathological morphology. *Intrinsic* hepatotoxic drugs demonstrate a broad incidence, dose–response relationship and will usually give similar results in humans and experimental animals. The incidence of liver damage from *idiosyncratic* hepatotoxicants is limited to susceptible individuals and results from hypersensitivity reactions or unusual metabolic conversions that can occur due to polymorphisms in drug metabolism genes (see Chapters 11 and 13).

Morphological liver damage differs qualitatively depending upon duration of exposure. The hallmarks of acute exposure are impaired parenchymal cell function and viability that are manifested histopathologically as *steatosis, fibrosis, cholestasis*, and *necrosis*. Many toxicants induce steatosis—that is, hepatocellular lipid accumulation to greater than 5% of weight. This response is generally considered to be reversible; however, when severe, it is associated with more progressive injury as, for example, with the anticonvulsant, valproic acid. Fibrosis is thought to occur as part of the wound healing response to hepatocellular damage and is reversible. Cholestasis occurs when impaired hepatocyte function results in inhibition of bile formation and secretion. A consequence of cholestasis is impaired clearance of bilirubin, a yellow degradation product of heme whose accumulation leads to jaundice. Damage to bile duct epithelium can also result in acute cholestasis. A notable example of human cholestasis resulted from accidental exposure to 4,4′-diamino-diphenylmethane, an epoxy resin hardener that was a bread contaminant causal to the skin rash known as Epping jaundice.

Loss of viability or necrosis is the most severe form of acute hepatoxicity. Necrosis can be focal or diffuse throughout the lobule, massive to include many lobules, or region-specific with potential bridging across equivalent regions of adjacent lobules depending upon severity. Regional hepatotoxicant damage is due to lobule compartmentalization of mediators of toxicity, such as higher oxygen concentration periportally, and the centrilobular localization of CYPs. Acute, drug-induced hepatitis follows necrosis as activated neutrophils infiltrate; and, if damage is sublethal, regenerative cell proliferation of residual hepatocytes ensues. A more limited form of lethality is apoptosis or programmed cell death. Apoptosis occurs in isolated hepatocytes and without an inflammatory reaction. Acute exposure to lower doses of necrogenic toxicants is associated with apoptosis, whereas drugs that increase liver size, such as phenobarbital, inhibit apoptosis.

The pathology following chronic exposure to hepatotoxicants includes a broad variety of lesions. Steatosis and cholestasis can result from chronic, as well as acute, exposures. For example, both acute and chronic exposures to the antipsychotic/antiemetic drug chlorpromazine (Thorazine) may induce cholestasis. Immune-mediated hepatotoxicity in susceptible individuals is also observed upon repeated exposure to some chemicals. The most well-studied example is hepatitis from reexposure to the general anesthetic halothane that presumably results from immunization against liver proteins adducted to halothane metabolite formed upon initial exposure. *Fibrosis* often accompanies repeated hepatocellular damage and can result in extracellular matrix deposition in the perisinusoidal space, around the terminal hepatic venule, and in portal tracts. *Cirrhosis* results when exaggerated deposition of collagen and associated matrix proteins occurs in interconnected bands of fibrous scar tissue. Cirrhosis impairs regeneration, is not reversible, and results in severe liver dysfunction. Fibrosis is also associated with damage to biliary epithelium and periductal cirrhosis can result from prolonged exposure.

Vascular Lesions. Chemical damage to the hepatic efferent vasculature results in various disorders causing impairment of sinusoidal perfusion. Obstruction of sinusoidal drainage through terminal hepatic venules related to endothelial cell damage and fibrosis occurs in veno-occlusive disease. Blockage of flow through larger hepatic veins from thrombosis occurs in Budd–Chiari syndrome. Both of these vascular disorders, as well as sinusoidal dilation, have been associated with chronic exposure to oral contraceptives.

Neoplasia. Approximately one-fourth of the chemicals tested in the National Toxicology Program rodent bioassay are liver carcinogens; however, the significance of these high-dose exposures to human risk has been questioned. The livers of the C3B6F1 tester strain of male mice are very susceptible to chemical induction of tumors. In humans, the mycotoxin aflatoxin B_1 is highly associated with hepatocellular carcinoma, especially in combination with hepatitis B viral infection. Alcoholic beverage consumption is listed as a known human carcinogen in the *11th Report on Carcinogens* based, in part, upon its epidemiological association with liver cancer, presumably due to chronic exposure to ethanol. Tumors of the vasculature endo-thelia (angiosarcomas) have occurred with industrial exposure to vinyl chloride and administration of the radioactive contrast agent, thorium dioxide. Further discus-sion of chemical carcinogenesis can be found in Chapter 24.

28.4 MECHANISMS OF CHEMICALLY INDUCED HEPATOTOXICITY

The most direct mechanism of hepatotoxicity is through specific interaction of a chemical with a key cellular component and consequential modulation of its func-tion. More common mechanisms, however, involve secondary effects of toxicant interaction. These include: depletion of cellular molecules, such as ATP and GSH; free radical and oxidant damage, in particular to membrane lipids; covalent binding of reactive metabolites to critical cellular molecules; and collapse of regulatory ion gradients. The following discussion will highlight how these cellular and molecular mechanisms contribute to specific types of chemically induced hepatic lesions.

28.4.1 Fatty Liver

Lipid accumulates in hepatocytes primarily as triglyceride and occurs when there is an imbalance between uptake of extrahepatic triglyceride and precursors compared to hepatic secretion of triglyceride-containing lipoprotein and fatty acid catabolism. The primary source of extrahepatic lipid to the liver is nonesterified fatty acids that have been mobilized from adipose tissue and bound to serum albumin. A secondary source is residual triglyceride of chylomicron remnants that originate from intestinal processing of dietary lipid. Within the hepatocyte, fatty acids are synthesized *de novo* and esterified when acetyl CoA and glycerol are available from glycolysis. Intrahepatic catabolism of fatty acids occurs primarily through mitochondrial β-oxidation with a smaller contribution from peroxisomal oxidation. Fatty acid β-oxi-dation is limited by translocation of fatty acyl-carnitine derivatives across the inner mitochondrial membrane. Hepatocyte triglycerides are combined with phospholip-ids, cholesterol, cholesterol esters, and glycosylated apolipoproteins, and are then secreted as very low density lipoproteins (VLDL). Further processing of VLDL occurs in the systemic circulation to yield other lipoprotein classes of lesser triglyc-eride content that become a source for apoprotein recycling by the liver.

This multicomponent process, summarized in Figure 28.2, provides for many sites of chemical interaction. Excessive free fatty acid (FFA) delivery to the liver seems to be one component in the mechanism in steatosis induced by CCl_4, ethionine, and phosphorus. These agents trigger lipolysis in the adipose tissue and cause a dramatic increase in circulating FFA. Under these conditions, triglycerides accumulate in the liver due to saturation of the rate-limiting step of synthesis of VLDL. Mobilization of depot fat from adipose tissue is under the control of the pituitary–adrenal axis, the stimulation of which triggers massive release of catecholamines leading to acti-vation of hormone-sensitive lipase in adipose tissue and mobilization of FFA. Pro-longed systemic stress by chemicals such as DDT, nicotine, and hydrazine also stimulate this system. Apparently, a similar mechanism causes fatty liver in starved animals treated with a large dose of ethanol.

Intrahepatic events contributing to increased fatty acid availability include enhanced *de novo* synthesis and impaired β-oxidation, whereas increased esterifica-tion would lead to triglyceride accumulation. Each of these processes is sensitive to

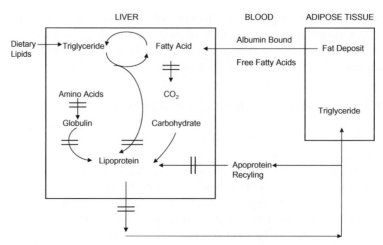

Figure 28.2. Triglyceride cycle in the pathogenesis of fatty liver. The scheme shows the transport of free fatty acids from fat deposits.

the cellular redox state and thus thought susceptible to ethanol through elevation of NADH. Increased diglyceride made available through ethanol induction of phosphatidic acid phosphatase may also enhance esterification. Valproic acid steatosis is related to inhibition of β-oxidation. The synthesis of apoprotein by hepatocytes is a highly endergonic process; and depletion of ATP, such as that which occurs with ethionine, an antimetabolite of methionine that competes for S-adenylation, is causal to fatty liver. Triglyceride accumulation would also result from impaired assembly and secretion of VLDL. Acetaldehyde produced from ethanol oxidation has been shown to adduct tubulin and inhibit microtubule polymerization. Microtubules are required for trafficking of VLDL-containing vesicles to the plasma membrane for exocytosis. Similarly, components of the hepatocyte secretory process are involved in fatty liver by orotic acid and CCl_4.

28.4.2 Fibrosis

Hepatic fibrosis is seen with chronic exposure to hepatoxicants that cause increasing damage to hepatocytes and is part of the wound healing response. Chronic fibrosis leads to severe disruption of the liver architecture by the deposition of extracellular matrix (ECM). Advanced fibrosis disrupts the proper blood flow and results in scarring of the liver that can lead to irreversible liver damage known as cirrhosis. Chronic exposure to the hepatoxins CCl_4, monocrotaline, and alcohol are examples of compounds that cause excessive fibrosis.

Recent evidence indicates that hepatic stellate cells (HSC) located in the space of Disse play a critical role in the progression to fibrosis. HSC undergo both changes in number and structure when activated. HSC undergo proliferation and change from star-shaped cells to fibroblast-like cells that synthesize and secrete large amounts of ECM including collagen, proteoglycans, and glycoproteins. Activation of HSC occurs in a process involving a number of HSC-activating factors including insulin-like growth factor (IGF), tumor necrosis factor alpha (TNFα), transforming growth factor beta (TGFβ), and fibroblast growth factors. These activating factors are produced from damaged hepatocytes, infiltrating immune cells, and HSC themselves. For example, ethanol exposure leads to Kupffer cell release of TGFβ, which activates HSC to myfibroblasts and secretion of ECM. Cessation of hepatic injury to hepatotoxins can cause significant reversal of fibrosis. Recovery from fibrosis is thought to involve HSC becoming quiescent or undergoing apoptosis, along with breakdown of the EMC and hepatocyte regeneration.

28.4.3 Cholestasis

Hepatocyte formation of bile depends upon energy-dependent transport of bile acids into the canalicular lumen, a multicomponent process involving vectoral movement across the cell (Figure 28.3). The osmotic pressure generated by canalicular bile acids is the driving force for the fluid component of bile that helps propel bile flow. Bile acids are taken up from the blood against a concentration gradient by transport proteins located on the hepatocyte sinusoidal membrane. The best characterized of these is the sodium taurocholate cotransporting polypeptide (NTCP, gene symbol SCL10A1), whose energy is derived from the downhill gradient of Na^+ that is maintained by the plasma membrane Na^+,K^+-ATPase. Intracellular binding proteins coordinate transcellular transport of bile acids, and cytoskeletal elements mediated trafficking within vesicles to the canalicular membrane. Bile acids excreted at canalicular membrane occurs via ATP-binding cassette (ABC) transporters such as the bile salt export pump (Bsep), which catalyzes the rate-limiting, concentrative movement of taurocholine (BA^-) into the canalicular space.

Another sinusoidal transporter catalyzes Na^+-independent uptake of organic anions and is instrumental for biliary clearance of glucuronidated and sulfated steroids, the diagnostic chemical bromosulfophthalein (BSP) and possibly bilirubin. Canalicular transport of glucuronidate and GSH conjugates is coupled to ATP

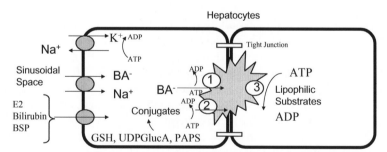

Figure 28.3. Transport of bile acids and other constituents across the hepatocyte. The Na^+ dependent bile salt (taurocholate) transporter (BA^-) is shown on the sinusoidal membrane that utilizes the Na^+ gradient maintained by the NAK pump, shown here on the lateral aspect of the plasmalemma. Bile salt transcellular transport involves microtubules, which then deliver substrate to the canalicular bile salt transporter (1). Bilary excretion of GSH, glucuronate (GluA), and sulfate conjugates of compounds such as 17β-estradiol (E2), bilirubin, and bromosulfothalein (BSP) is catalyzed by the multispecific organic anion transporter (MOAT; 2). Both 1 and 2 are members of the ABC family of ATP-dependent transporters that also includes P-glycoprotein (3), another canalicular transporter catalyzing excretion of lipophilic compounds such as the chemotherapeutic drug, daunorubicin.

hydrolysis catalyzed by the multispecific organic anion transporters of the multidrug resistance associated protein (MRP) family. For example, MRPs catalyze the excretion of BSP-GSH and bilirubin glucuronidate after intracellular conjugation. Another member of the multidrug resistance protein family, the P-glycoprotein product of the mdr1 gene, has also been localized to the canalicular membrane and catalyzes ATP-dependent efflux of a wide variety of hydrophobic polycyclic compounds.

Cholestatic hepatotoxicants that target these molecular and cellular events are well known and some have deleterious effects on multiple components. Compounds with global effects on membrane permeability, such as chlorpromazine through its cationic detergent action, can inhibit Na^+,K^+-ATPase and collapse the Na^+ gradient necessary for sinusoidal bile acid uptake. A detergent effect of lithocholate on canalicular membranes is also evident from ultrastructural morphology. Cholestasis from the 17β-glucuronide conjugate of estradiol has been related to its activity as a substrate of both MRPs and P-glycoprotein. The cholestatic effects of cholchicine, a microtubule poison, and phalloidin and cytochalasin B, inhibitors of microfilament function, emphasize the importance of cytoskeletal elements in bile formation.

Toxicity to bile duct epithelial cells also causes cholestasis. Necrosis of these cells can result in sloughing of an obstructive cast into the duct lumen. The experimental chemical α-naphthylisothiocyanate (ANIT) is known to target bile duct epithelial cells in addition to hepatocytes. Bile duct epithelial toxicity is GSH-dependent and is thought to depend upon concentrative accumulation within bile ducts subsequent to hepatocanalicular transport of the ANIT-GSH conjugate.

28.4.4 Necrosis and Apoptosis

The ultimate event in cell death is irreversible loss of plasma membrane integrity. Previous mechanistic studies have emphasized that the inability to maintain transmembrane calcium gradients is the early determinant, initiating an irreversible sequence of events leading to cell death. However, new microscopic techniques have enabled observation of temporal relationships of multiple events at the level of a single toxicant-exposed cell. These elegant studies have demonstrated that loss of mitochondrial function with an associated depletion of ATP can precede elevation of cytosolic Ca^{2+}. Thus, ATP depletion and disruption of Ca^{2+} homeostasis are both important determinants of lethality; and, when extreme, each may act independently. Alternately, sublethal responses of one may lead to and synergize with the other such that the temporal order is inconsequential. A variety of different types of primary toxic effects would then be expected to converge upon mitochondrial function and/or maintenance of low cytosolic Ca^{2+} levels by the ATP-dependent Ca^{2+} pumps of the plasma membrane and endoplasmic reticulum (ER).

Phosphorylation of ADP to ATP by mitochondria is driven by an electrochemical proton gradient established across the inner mitochondrial membrane as a consequence of vectoral transport of protons from NADH and succinate during oxidation by the respiratory chain (see Chapter 17). Hence, lipophilic weak acids or bases (such as 2,4-dinitrophenol) that can shuttle protons across membranes will dissipate the proton gradient and uncouple oxidation from ADP phosphorylation. Intramitochondrial ADP can be rate-limiting as demonstrated by inhibition of the mitochondrial adenosine nucleotide carrier by atractyloside. Inhibition of ATP synthesis also is a consequence of the mitochondrial permeability transition (MPT). Exposure to several different types of toxicant effectors opens a mitochondrial high conductivity pore that nonspecifically allows passage of molecules with molecular weights <1.5 kD to result in the MPT. The involvement of such a pore in necrosis is suggested by the protection against lethality from several toxicants by cyclosporin A, an inhibitor of the MPT.

Induction of the MPT requires Ca^{2+}; conversely, plasma membrane and ER Ca^{2+} pumps require ATP. One of the earliest effects of CCl_4 is metabolism-dependent inhibition of the ER Ca^{2+} pump through oxidation of a critical –SH residue. A similar oxidant-sensitive –SH residue is present in the plasma membrane Ca^{2+} pump. Thus, the reciprocal requirements of Ca^{2+} for the MPT and ATP for cellular Ca^{2+} pumps provide a mechanism for linkage of ATP depletion and disruption of Ca^{2+} homeostasis as mediators of cell death. Most proximal toxic effects that lead to hepatocellular mitochondrial and Ca^{2+} dysfunction can be generalized into two classes, those involving production of oxidative stress or formation of reactive metabolites.

28.4.5 Oxidative Stress

Cellular metabolism of dioxygen (O_2) is tightly regulated to yield primarily $4e^-$ reduction to $2O^{2-}$. However, some leakage of the $1e^-$ partial reaction products, $O_2^{\cdot-}$ superoxide radical and hydrogen peroxide (H_2O_2), occurs from processes such as mitochondrial respiration and cytochrome P450-catalyzed oxidation. Peroxisomal oxidative metabolism of fatty acids is also a significant source of endogenous cellular H_2O_2. Metabolism of superoxide to H_2O_2 by superoxide dismutase and reduction of H_2O_2 via GSH peroxidase and catalase complete the reduction to H_2O. Under normal conditions, these pathways complemented by additional hepatocellular antioxidants such as GSH, vitamin E, and ascorbate have sufficient capacity to prevent deleterious oxidation of other cellular constituents (see Chapter 20).

Exposure to certain hepatotoxicants can enhance the rate of production of reactive oxygen intermediates such that they exceed the capacity of these protective pathways. A structurally diverse group of compounds cause proliferation of peroxisomes within rodent hepatocytes and hence elevate endogenous levels of H_2O_2 leaked from this organelle. Peroxisome proliferators include the hypolipidemic agent clofibrate, plasticizer diethylhexyl phthalate (DEHP), and herbicides 2,4-dichloro- and 2,4,5-trichloro-phenoxyacetic acid (2,4-D and 2,4,5-T). The liver is particularly susceptible to injury from iron overload because it is a major site of iron storage. Iron is usually stored in hepatocytes associated with proteins ferritin and hemosiderin; however, catalytically active iron can be released from storage by superoxide. Ferritin iron is also reduced and mobilized by NADH, which may play a role in ethanol-induced oxidative stress in liver. Released iron can react with reducing agents and O_2 to form superoxide. Superoxide and H_2O_2 react with iron to form highly reactive hydroxyl radicals (˙OH). Copper ions can also catalyze the Fenton reaction to produce ˙OH. Superoxide is also produced by single electron transfer to O_2 by nonoxidant free radicals. Formation of the trichloromethane radical $CCl_3^{\cdot}$ from cytochrome P450-dependent metabolism of CCl_4 contributes to hepatocyte oxidative damage through this mechanism.

Certain phenolic derivatives are conducive to $1e^-$ reduction, especially as catalyzed by microsomal NADPH-cytochrome P450 reductase, with the formation of

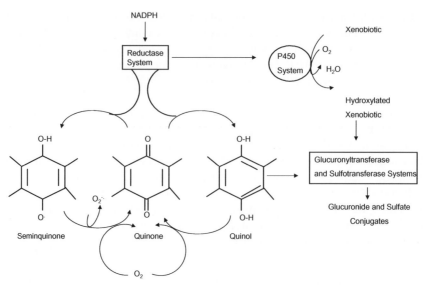

Figure 28.4. Mechanism of quinone drug redox cycling by liver microsomal NADPH : cytochrome P450 reductase. A one-electron pathway competes with a two-electron process (catalyzed by DT-diaphorase) and is coupled with the monoxygenation and glucuronidation and sulfation systems.

oxygen-centered free radical anion. The $1e^-$-reduction products of these can then transfer an electron to O_2 resulting in the formation of superoxide with reoxidation to the parent compound (Figure 28.4). This type of reaction is particularly destructive because the parent compound undergoes redox cycling through repeated rounds of $1e^-$ reactions. Superoxide, continuously produced during redox cycling, will be detoxified through H_2O_2 and GSH peroxidase until NADPH becomes limiting for GSSG reduction. GSSG is then effluxed from the hepatocyte and total glutathione levels fall. Compounds that undergo redox cycling include the *p*-quinones, *o*-quinones (catachols), quinoneimines, and quinone methides and include such hepatotoxicants as metabolites of estrogen, acetaminophen, eugenol, and bromobenzene. The antimicrobial agent nitrofurantoin and the bipyridyl herbicides diaquat and paraquat also cause hepatotoxicity through redox cycling. A protective mechanism against quinone redox cycling is afforded by NAD(P)H : quinone oxidoreductase (DT diaphorase or NQO1), which competes for the quinone substrate and catalyzes $2e^-$ reduction to the hydroquinone.

Extrahepatocellular factors also contribute to oxidative toxicity in liver. Xanthine oxidase is primarily located in sinusoidal endothelial cells and catalyzes $1e^-$ reduction of O_2. Kupffer cells produce reactive oxygen species as part of their phagocytic function. Both activated neutrophils and macrophages recruited to liver upon chemically induced damage produce a respiratory burst of superoxide through the activity of membrane-bound NADPH oxidase. Also, much of the damage associated with reperfusion of liver after ischemia is associated with production of reactive oxygen species.

Several adverse consequences result from exposure of the cell to oxidative stress. Cellular proteins with critical –SH residues and membrane polyunsaturated fatty acids appear to be especially susceptible to oxidative damage that can lead to necrosis. Reactive oxygen species can directly oxidize protein –SH groups to form disulfides, an effect that is exaggerated by impaired repair due to depletion of GSH consequential to its consumption by GSH peroxidase. Two critical –SH containing proteins whose functions are impaired by oxidation are the plasmalemma and ER Ca^{2+} pumps. Such damage to these activities would provide a mechanism for oxidant-induced derangement of cellular Ca^{2+} homeostasis. Mitochondrial ATP synthesis may also be directly affected because sulfhydryl oxidation is an inducer of the MPT.

Oxidant radicals, O_2^- and $^{\cdot}OH$, as well as other free radicals such as semiquinone and $CCl_3^{\cdot}$, are very active in initiating peroxidation of polyunsaturated lipids. Lipid peroxidation is a multiphasic process and involves initiation, propagation, and termination (Figure 28.5). The hydrogen atoms on methylene carbons of unsaturated fatty acids are highly susceptible to free radical attack. The abstraction of methylene hydrogen atom yields lipid free radical. Propagation continues as a lipid free radical abstracts methylene hydrogen from a neighboring lipid to generate a second free radical until termination reactions yield nonradical products. These lipid radicals are unstable and undergo a series of transformations, including rearrangement of double bonds, to give conjugated dienes. Lipid free radicals react rapidly with O_2 to form lipid peroxyl radicals, which decompose to aldehydes, with the most abundant being malondialdehyde and 4-hydroxy-2,3-nonenal.

Subcellular membranes are rich in unsaturated fatty acids esterified to phospholipid and are thus obvious targets of lipid peroxidation. Membrane lipid peroxidation will result in loss of both structural integrity and function of the affected organelles. Also, lipid modification in the annula surrounding membrane transport proteins is known to modulate their activity. Thus, lipid peroxidation can significantly impair membrane maintenance of ion gradients, such as the mitochondrial proton gradient necessary for ATP synthesis and plasma membrane and ER Ca^{2+} gradients.

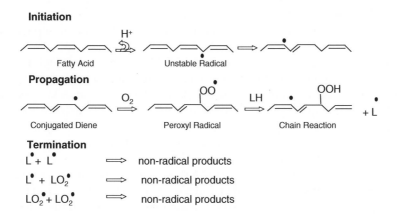

LH = polyunsaturated; L$^{\bullet}$ = lipid radical; LO$_2^{\bullet}$ = lipid peroxyl radical

Figure 28.5. Initiation, propagation, and termination reactions in lipid peroxidation. (Adapted from Bus, J. S. and Gibson, J. E. *Rev. Biochem. Toxicol.* **1**, 125, 1979.)

Deleterious effects of reactive electrophiles are generally attributed to two general mechanisms, through binding to nucleophilic sites of cellular macromolecules and depletion of GSH. Toxicity from macromolecule adduction is best understood for binding to nucleophilic sites of DNA and proteins. Although electrophilic adduction of nucleotides can contribute to cytolethality through formation of DNA single-strand breaks, it is of greatest toxicological interest with respect to mutagenesis and carcinogenesis as discussed in Chapters 22 and 24.

Metabolism-dependent binding of toxicants to cellular proteins is a common observation. However, determination of the functional consequences of binding and relating these to toxic effects is challenging. In those cases where general binding of reactive metabolites to protein creates protein adducts known as neoantigens, toxicity is mediated by sensitization and the subsequent inflammatory response. The anesthetic halothane is a prototypical hepatotoxicant with this type of mechanism. Another situation in which clear effects result from binding is that resulting from suicide inhibition of CYP isozymes by irreversibly bound reaction intermediates at the active site. Diallyl sulfide, a component of garlic, is a suicide inhibitor of CYP2E1; and 4-hydroxy-2,3-nonenal, a byproduct of lipid peroxidation, has recently been shown to inhibit a several CYP isozymes through adduction of the prosthetic heme group. The functional consequence of CYP inhibition through adduction will depend on whether activation or detoxication reactions are affected.

Acetaminophen provides the most thoroughly studied example of hepatotoxicity associated with protein adduction. Although acetaminophen primarily is conjugated with glucuronide and sulfate and excreted, metabolism to a reactive electrophile, N-acetyl-p-benzoquinoneimine (NAPQI) by CYP becomes significant when the conjugation pathways are saturated (see Chapters 12 and 18). Through the use of antibodies that recognize protein-bound NAPQI, several specific adducted proteins have been identified. More recently, application of two-dimensional gel electrophoresis and matrix-assisted laser desorption ionization (MALDI) mass spectrometry has expanded this list to include 23 adducted mouse liver proteins, including mitochondrial proteins involved in energy production and cellular ion control. However, even with knowledge of the biochemical function of these proteins, the molecular mechanism involving a critical target protein(s) causal to acetaminophen hepatotoxicity remains to be determined.

Numerous hepatotoxicants are converted to reactive toxic electrophiles by CYP. CYP metabolism of chloroform to phosgene and bromobenzene to its 3,4-epoxide are classic examples. In addition, reactive electrophiles are generated upon activation of hepatotoxicants by other metabolic enzymes. Alcohol dehydrogenase catalyzes formation of acetaldehyde from ethanol and acrolein from allyl alcohol. Thioacetamide hepatotoxicity can result from flavin monooxygenase-catalyzed activation through S-oxidation, and aflatoxin B_1 epoxidation can be catalyzed by lipoxygenase. Glucuronidation of the carboxylic acid of diclofenac, a nonsteroidal antiinflammatory drug, facilitates its adduction of protein through acylation.

Electrophiles not only complex with macromolecular nucleophilic sites, but also complex with those of small molecules. The most important of these is the –SH group of GSH. Conjugation of electrophiles with GSH is an important detoxication reaction catalyzed by GSH transferases (see Chapter 12). Although GSH concentration in hepatocytes is high relative to other tissues (4–8 mM), depletion of GSH can occur upon acute exposure to protoxicants whose electrophilic metabolites conjugate GSH, such as the acetaminophen metabolite NAPQI. GSH lost to conjugation is stoichiometric and thus occurs at high doses and with a threshold. When GSH levels are depleted to ~25% or less, significant risk of oxidative damage ensues as GSH becomes limiting for GSH peroxidase activity. Because GSH conjugates are actively effluxed from the hepatocyte, recovery of GSH requires *de novo* synthesis through the sequential activities of γ-glutamylcysteine synthetase and GSH synthetase. GSH synthesis also requires an adequate supply of cysteine. These properties enable the use of several experimental tools to indicate GSH depletion as an effector of chemically induced hepatotoxicity. Potentiation will be observed with diethyl maleate, which depletes GSH through direct conjugation, and buthionine sulfoximine, an inhibitor of γ-glutamylcysteine synthetase. Also, overnight food withdrawal, as is frequently used in acute oral exposure studies, will deplete GSH and enhance toxicity. Conversely, *N*-acetylcysteine serves as a source of cysteine and is currently used clinically to treat acetaminophen toxicity.

28.4.6 Enterohepatic Circulation

The process of bile acid formation in the liver, movement to the gallbladder, and then movement to the duodenum is a pathway that can concentrate toxicants in hepatocytes. From the gallbladder where they are stored, bile acids are released into the gut to aid in the digestion of lipophobic substances. Most of the released bile acids are reabsorbed in the intestines and are returned to the liver by the hepatic portal circulation.

Evidence suggests that the mushroom toxin phalloidin can be a substrate for bile salt transporters in the membrane of hepatocytes. Phalloidin is actively taken up by hepatocytes and acts to block the normal actin remodeling of the cytoskeleton, leading to hepatocellular necrosis. The α-amanitin toxins from poisonous mushrooms are a group of bicyclic octoapeptides that are extremely toxic to hepatocytes. After ingestion, amatoxins are rapidly absorbed and delivered to hepatocytes, where they are effectively taken up by hepatocellular tranposrters. Amatoxins are able to inhibit RNA polymerase activity by forming a complex that causes a decrease in protein synthesis and results in hepatocyte necrosis. Amatoxins that are released from hepatocytes in the bile undergo reuptake in the gut, and enterohepatic circulation of amatoxins effectively concentrates the toxins in hepatocytes. Similarly, methyl mercury and arsenic undergo enterohepatic circulation, which limits their clearance and contributes to toxicity.

28.5 INTERACTIONS

28.5.1 Metabolism-Dependent

Exposure to two or more chemical agents can result in altered expression of hepatotoxicity. However, qualitatively different interactions may be achieved, depending upon the relative timing of exposures. Simultaneous exposure to competing substrates of a specific CYP isozyme will often slow metabolism and can be protective against reactive metabolite-mediated hepatotoxicity. Alternately, pretreatment with one agent may induce metabolic enzymes that either protect against or potentiate the hepatotoxicity of a second chemical, depending upon whether induction results in increased detoxication or activation. Acetaminophen and ethanol are two hepatotoxicants that illustrate these relationships. If ethanol precedes an acute dose of acetaminophen, but is not present simultaneously, potentiation of acetaminophen hepatotoxicity results. Prior exposure to ethanol induces CYP2E1, an isoform that activates acetaminophen to NAPQI. However, when acetaminophen and ethanol are administered concurrently, ethanol protects against hepatotoxicity because both are CYP2E1 substrates. The CYP2E1 suicide inhibitor, diallyl sulfide, also protects against acetaminophen toxicity. Prior exposure to ethanol and other inducers of CYP2E1 such as isopropanol, acetone, and the antimicrobial agent isoniazid also potentiate the hepatotoxicity of CCl_4. Several organochlorine pesticides are also potent inducers of CYP isozymes and consequently modify hepatotoxicant potency. Examples are potentiation of hepatotoxicity of acetaminophen by mirex and CCl_4 by DDT. Hepatotoxicity of CCl_4 is also enhanced by chlordecone pretreatment, but in this situation the mechanism is related to suppression of tissue repair. For pharmaceuticals, interactions involving hepatic CYP3As are of concern because a large number of drugs are metabolized by this enzyme. Such interactions can alter the pharmacokinetics of a drug to effect a change in therapeutic efficacy. Well-studied examples are interactions with 3A substrates erythromycin and ketaconozole and inducer rifampin.

28.5.2 Protective Priming

Preexposure to a sublethal dose of CCl_4 has been shown to protect against lethality from a second, independently lethal dose, a phenomenon called "autoprotection." Comparable histopathological damage to liver is seen independent of the early priming dose, which discounts a lessening of reactive metabolite formation by a previously inhibited CYP as the protective mechanism. The kinetics of autoprotection correlate with regenerative cell proliferation induced by the priming dose. This observation has led to the hypothesis that newly born hepatocytes are more resistant to the necrogenic effects of CCl_4. Importantly, protective priming also occurs with heterologous pairs of chemicals as shown by protection against acetaminophen hepatotoxicity by pretreatment with a sublethal dose of thioacetamide. Although the molecular basis of protective priming is unknown, it represents another instance in which chemicals interact to modulate heptotoxicity.

28.6 DETECTION AND PREDICTION OF HEPATOTOXICITY

28.6.1 Clinical

Necrosis of hepatocytes leads to discharge of intracellular contents into the serum. Some of these provide noninvasive parameters of hepatotoxicity that can be used in the clinical setting. The most important and specific serum marker of hepatocellular damage is the enzyme alanine aminotransferase (ALT; also called serum glutamate pyruvate transaminase, SGPT). Another commonly used, but less specific, indicator is aspartate aminotransferase (AST; also called serum glutamate oxaloacetate transaminase, SGOT). In milder forms of hepatocellular injury, serum levels of these enzyme may be elevated to as much as 3 times normal values. Similar values are also characteristic of more serious cholestatic injury, but this can be distinguished by coincident elevation in the levels of alkaline phosphatase. With severe hepatocellular necrosis, serum aminotransferases may be elevated as much as 50 times normal. Since albumin and prothrombin are synthesized by hepatocytes, serum albumin concentration and prothrombin clotting time are useful parameters of hepatocellular toxicity. Elevated serum levels of bilirubin and bile acids are indicative of either hepatocellular or bile duct epithelial damage. More specific tests for liver function include rate of biliary clearance of injected BSP and determination of urinary metabolites of probes for hepatocyte metabolic enzymes, such as caffeine for cytochrome 450 1A2.

28.6.2 Experimental

The serum enzymes ALT, AST, and alkaline phosphatase are also important indicators of chemically induced heptotoxicity in experimental animal studies. An additional heptocyte enzyme, sorbitol dehydrogenase (SDH), is also used frequently as a more sensitive complement to the aminotransferases. In experimental studies, histopathology is the most definitive indicator of injury and is highly informative about mechanistic events.

Additional information about mechanisms of chemically induced toxicity to liver cells is obtained in experiment studies through use of ex vivo systems. Isolated perfused liver is especially useful for assessment of cellular metabolite fluxes and bile flow. Isolated hepatocytes are readily obtained by collagenase perfusion and can be used in suspension or in monolayer cultures (see Chapter 8). Although isolated hepatocytes lose activity and inducibility of the certain CYPs, Phase II conjugating enzymes, and transporters with time, they offer a cellular context in which to evaluate mechanism of toxic effects over the short term. The availability of the hepatocyte collagen-sandwhich-configuration (SC) model system offers a powerful in vitro model system. These SC hepatocyte models, with intact canalicular lumens, have provided a valuable system for mechanistic studies of bile formation, drug, and toxin tranport. Nonparenchymal cells can also be obtained by collagenase perfusion, and isolated cell types can be isolated with additional separation procedures. Co-cultures can then be constructed to determine cellular interactions in mediating chemical toxicity. This approach has been especially fruitful in understanding mechanisms involved in hepatotoxicity initiated through toxicant effects on inflammatory cells.

Isolated subcellular fractions, especially microsomes, are most frequently used to obtain metabolite profiles of drugs and identify metabolizing enzymes. Production of a metabolite with structures of known high reactivity may indicate potential toxicity through a previously known mechanism. Preparations made from cells engineered to express specific drug metabolizing enzymes are commercially available and demonstrate the potential for metabolism in vivo by that enzyme. Liver slices are also frequently used in metabolite profiling now that newer methods for preparation and maintenance have improved their viability, although loss of CYP activity limits their use to short-term studies. These precision-cut liver slices offer the simplicity of an in vitro system, but with retention of cellular context, normal tissue architecture, and cell–cell interactions. Mechanistic knowledge gained from these in vitro systems has been instrumental for prediction of parent drug toxicity across species and in susceptible individuals and of interactions with other drugs.

Recently the term toxicogenomics has been coined to describe the identification of genomic biomarkers of toxicity. Toxicogenomics encompasses information about a toxicant from genomic-scale mRNA profiling (by microarray analysis), cell-wide or tissue-wide profiling (proteomics), genetic susceptibility, and computational modeling (see Chapters 3–5). The advances in molecular biology have allowed for the determination of gene expression profiles of hepatocytes after treatment with hepatotoxic compounds such as CCl_4, acetaminophen, and a number of therapeutic agents. Use of microarrays may allow the identification of useful biomarkers that would be predictive of future liver responses to toxicant insults. Identification of the gene expression profiles associated with the future development of hepatic steatosis, fibrosis, cholestasis, and necrosis would be useful in the rapid evaluation of potentially hepatotoxicant agents.

28.7 COMPOUNDS CAUSING LIVER INJURY

28.7.1 Drugs

Several therapeutic agents used as tranquilizers, antidepressants, anticonvulsants, antibiotics, and anti-inflammatory drugs, as well as some prescribed for cardiovascular, endocrine, rheumatic and neoplastic diseases, have been found to cause hepatotoxicity. Recent episodes of liver toxicity have prevented development of fialuridine for treatment of hepatitis B and the nonsteroidal antiinflammatory drugs (NSAIDS), bromfenac and benoxaprofen. The NSAID diclofenac has also been shown to be metabolized to a number of reactive metabolites in hepatocytes that are thought to be responsible for its toxicity. Hepatotoxicity has led to roglitazone being withdrawn from the market for treatment of diabetes and has limited the usage of tacrine for Alzheimer's disease. One of the most extensively studied and widely available hepatotoxic drugs is acetaminophen. Although therapeutic doses are well below toxic levels, alcohol potentiation of acetaminophen hepatotoxicity is a concern. Protein adduction by reactive electrophilic metabolite NAPQI, depletion of GSH, and redox cycling are all contributors to acetaminophen hepatotoxic effects and have been discussed in detail above.

The antipsychotic chlorpromazine is a prototype heptotoxicant for production of cholestasis. Pleiotropic effects of chlorpromazine on membrane permeability and associated ion gradients and microfilament-mediated canalicular contraction have been attributed to detergent effects. Valproic acid, an anticonvulsant, is associated with microvesicular steatosis. Inhibition of mitochondial fatty acid β-oxidation is an important component of this toxic effect and is apparently related to carnitine availability as evidenced by the protection afforded by L-carnitine supplements. The hypolipidemic drugs clofibrate, fenofibrate, and gemfibrozil are peroxisome proliferators in rodent liver, but not in humans. Isoniazid, an antibiotic used to treat tuberculosis, exhibits an approximately 1% incidence of hepatotoxicity. Although toxicity is known to be metabolism-dependent and protein adduction has been well-correlated with toxicity, the identity of the reactive metabolite and target protein(s) is unknown.

28.7.2 Ethanol

Alcoholic liver disease is a progressive, multifaceted disorder. Thus, it is not surprising that a large number of molecular targets are affected by ethanol. Several points of ethanol interference with hepatic lipid metabolism have been identified above (Section 28.4.1). The role of Kupffer cells, through their secretion of TNFα, has been clearly demonstrated in ethanol-induced necrosis and inflammation and may be related to ethanol effects of their responsiveness to endotoxin. Kupffer cells isolated from ethanol-treated rats also have an increased capability to produce superoxide. Considerable evidence suggests that protein adduction and oxidative stress are causal to ethanol hepatotoxicity. Microtubule adduction by acetaldehyde, the product of ethanol oxidation by alcohol dehydrogenase, has a role in impaired VLDL secretion. Prooxidants are primarily from partial reactions of the ethanol-inducible CYP2E1 with secondary contributions from mobilization of active iron and free radical reactions originating with formation of 1-hydroxyethyl radical and xanthine oxidase metabolism of acetaldehyde. In addition, ethanol inhibits signaling by the receptor for the hepatomitogen, epidermal growth factor, and thus may have detrimental effects on tissue repair.

28.7.3 Halogenated Aliphatic Hydrocarbons

Exposures to hepatotoxic levels of halogenated aliphatic hydrocarbons have been reduced considerably with improved industrial hygiene and use restriction. Important exposure scenarios that remain involve the use of fluorinated general anesthetics, replacement of ozone-depleting refrigerants with hydrochlorofluorocarbons (HCFCs), and storage of high volumes of certain haloalkanes in EPA Superfund sites. Hepatotoxicity from fluorinated anesthetics is thought to result from hypersensitivity to metabolism-based adduction of liver proteins. The brominated, fluorinated anesthetic halothane has had the highest incidence of toxicity and has largely been replaced by safer, nonbrominated analogs. The reactive metabolite of halothane, trifluoroacetyl chloride, is formed by CYP-catalyzed oxidative debromination. A similar mechanism has been suggested for a recent occurrence of hepatotoxicity in industrial workers repeatedly exposed to HCFCs. HCFCs have also been shown to be peroxisome proliferators in rodent liver. Of the high-volume chemicals at Superfund sites, those with hepatotoxicity include CCl_4, chloroform, trichloroethylene, and tetrachloroethylene. The human hepatocarcinogen, vinyl chloride, is also a Superfund site contaminant.

28.7.4 Pesticides

Liver failure from acute exposures to pesticides is rare as lethality usually results from toxicity to other organ systems. Fatty liver, centrilbular necrosis, and cholestasis with destruction of bile duct epithelia have resulted from exposure to the herbicide paraquat and elevated serum enzymes have been detected in workers employed in the production of chlordecone. The fumigant ethylene dibromide causes experimental hepatotoxicity similar to CCl_4. The herbicides 2,4-D, 2,4,5-T, dicamba, lactofen, and tridiphane are peroxisome proliferators in rodent liver. Exposure to the pesticide endosulfan has been shown to result in hepatocellular necrosis, elevated serum levels, and induction of microsomal enzyme activity. The most consistent effect of pesticides in liver is induction of CYP isozymes, especially the organochlorine insecticides such as endosulfan. Thus, although hepatotoxicity as an independent toxic effect of pesticides is uncommon, the potential for toxic interactions is significant.

28.7.5 Toxins

The plant species yielding natural hepatotoxins include *Lantana camara, Sassafras albium*, and more than 200 species of *Crotalaria*. A large number of species of the genera *Senecio, Heliotropium, Cyanoglossum*, and *Symphytum* are sources of pyrrolizidine alkaloids, etiological agents of veno-occlusive disease noted with excessive consumption of some herbal teas. Medicinal plants containing the diterpenoid glycoside, atractyloside, have caused fatalities associated with centrilobular hepatic necrosis. Atractyloside is a specific inhibitor of the mitochondrial adenosine nucleotide carrier and an inducer of the MPT. Unripe akee fruit of the *Blighia surpida* tree of Jamaica leads to acute steatosis from hypoglycin A, an inhibitor of fatty acid oxidation. Hepatotoxic effects of tannins at high doses and of phalloidin and α-amanitin from poisonous mushrooms are also known. Though controversial, certain extracts of kava (*Piper methysticum*), used for its sedative properties, contain kava-lactones that may be hepatotoxic. Several metabolic products of bacteria and lower fungi cause liver injury. An important endogenous hepatotoxicant is endotoxin lipopolysaccharide (LPS) from enteric Gram-negative bacteria that is normally present in the systemic circulation at low levels, but whose concentration can increase when gastrointestinal integrity is compromised. The mycotoxin aflatoxin B1 from *Aspergillus* is acutely hepatotoxic as well as carcinogenic. Molecular epidemiology has associated a site-specific mutation of the p53 tumor suppressor gene characteristic of aflatoxin B1 exposure with incidence of human hepatocarcinogenesis in the People's Republic of China.

28.8 CONCLUSION

Much is known about the biochemical toxicology of hepatotoxicants, yet much remains to be learned. Hepatotoxicity resulting in either cell necrosis, fibrosis, or fatty infiltration is known to be a widespread phenomenon, potentially of importance to human health. It is caused by numerous drugs and environmental agents, and its incidence is expected to increase as confounding viral liver disease becomes more prevalent. Much is known about mechanisms based upon comprehensive studies with a few prototypical chemicals—namely, CCl_4, ethanol and acetaminophen—which support a convergence of varied primary effects on the ultimate failure of mitochondrial function and Ca^{2+} homeostasis. The extensive metabolic activity of the liver exposes its cells to a continuous flux of prooxidants. The importance of metabolic activation for the production of reactive metabolites is well-recognized. The high specific activity of these enzymes in the liver insures that the potential for metabolism-dependent toxic effects is extensive.

The recognition of the roles of the various liver cell types in mediation of toxicity is a model demonstrating the value of cellular toxicology in mechanistic research. The delineation of key metabolic pathways has been made possible by the availability of systems engineered to express individual molecular pathway components. This knowledge has been readily applied to drug discovery and development. Perhaps in no other tissue has the knowledge gained from fundamental research in biochemical toxicology had such obvious implications for protecting human health.

SUGGESTED READING

Alison, M. R., and Lovell, M. J. Liver cancer: The role of stem cells. *Cell Prolif.* **38**, 407, 2005.

Bataller, R., and Brenner, D. A. Liver fibrosis. *J. Clin. Invest.* **115**, 209, 2005.

Benyon, R. C., and Arthur, M. J. Mechanisms of hepatic fibrosis. *J. Pediatr.L Gastroenterol. Nutr.* **27**, 75, 1998.

Benyon, R. C., and Iredale, J. P. Is liver fibrosis reversible? *Gut* **46**, 443, 2000.

Cohen, S. D., Hoivik, D. J., and Khairallah, E. A. Acetaminophen-induced hepatotoxicity. In: Plaa, G. L., and Hewitt, W. R. (Eds.). *Toxicology of the Liver*, 2nd ed., Taylor & Francis, Washington, D.C., 1998, p. 159.

Day, C. P., and Yeaman, S. J. The biochemistry of alcohol-induced fatty liver. *Biochim. Biophys. Acta* **1215**, 33, 1994.

Fontana, R. J., and Watkins, P. B. Genetic predisposition to drug-induced liver disease. *Gastroenterol. Clin. North Am.* **24**, 811, 1995.

Hinson, J. A., Reid, A. B., McCullough, S. S., and James, L. P. Acetaminophen-induced hepatotoxicity: Role of metabolic activation, reactive oxygen/nitrogen species, and mitochondrial permeability transition. *Drug Metab. Rev.* **36**, 805, 2004.

Hodgson, E., and Meyer, S. A. Pesticides. In: Sipes, I. G., McQueen, C. A., and Gandolfi, A. J. (Eds.). *Comprehensive Toxicology*, Vol. **9**, *Hepatic and Gastrointestinal Toxicology*, Elsevier Science Ltd., New York, 1997, p. 369.

Ishak, K. G., and Zimmerman, H. J. Morphologic spectrum of drug-induced hepatic disease. *Gastroenterol. Clin. North Am.* **24**, 759, 1995.

Jaeschke, H., Smith, C. W., Clemens, M. G., Ganey, P. E., and Roth, R. A. Mechanisms of inflammatory liver injury: Adhesion molecules and cytotoxicity of neutrophils. *Toxicol. Appl. Pharmacol.* **139**, 213, 1996.

James, L. P., Mayeux, P. R., and Hinson, J. A. Acetaminophen-induced hepatotoxicity. *Drug Metab. Dispos.* **31**, 1499, 2003.

Kourie, J. I. Interaction of reactive oxygen species with ion transport mechanisms. *Am. J. Physiol.* **275**, C1, 1998.

Lemasters, J. J. V. Necrapoptosis and the mitochondrial permeability transition: Shared pathways to necrosis and apoptosis. *Am. J. Physiol.* **276**, G1, 1999.

Malarkey, D. E., Johnson, K., Ryan, L., Boorman, G., and Maronpot, R. R. New insights into functional aspects of liver morphology. *Toxicol. Pathol.* **33**, 27, 2005.

Mehendale, H. M., Roth, R. A., Gandolfi, A. J., Klaunig, J. E., Lemasters, J. J., and Curtis, L. R. Novel mechanisms in chemically induced hepatotoxicity. *FASEB J.* **8**, 1285, 1994.

Oinonen, T., and Lindros, K. O. Zonation of hepatic cytochrome P-450 expression and regulation. *Biochem. J.* **329**, 17, 1998.

Park, B. K., Kitteringham, N. R., Maggs, J. L., Pirmohamed, M., and Williams, D. P. The role of metabolic activation in drug-induced hepatotoxicity. *Annu. Rev. Pharmacol. Toxicol.* **45**, 177, 2005.

Qiu, Y., Benet, L. Z., and Burlingame, A. L. Identification of the hepatic protein targets of reactive metabolites of acetaminophen in vivo in mice using two-dimensional gel electrophoresis and mass spectrometry. *J. Biol. Chem.* **273**, 17940, 1998.

Susick, R., Moss, N., Kubota, H., Lecluyse, E., Hamilton, G., Luntz, T., Ludlow, J., Fair, J., Gerber, D., Bergstrand, K., White, J., Bruce, A., Drury, O., Gupta, S., and Reid, L. M. Hepatic progenitors and strategies for liver cell therapies. *Ann. NY Acad. Sci.* **944**, 398, 2001.

Waring, J. F., Cavet, G., Jolly, R. A., McDowell, J., Dai, H., Ciurlionis, R., Zhang, C., Stoughton, R., Lum, P., Ferguson, A., Roberts, C. J., and Ulrich, R. G. Development of a DNA microarray for toxicology based on hepatotoxin-regulated sequences. *EHP Toxicogenomics* **111**, 53, 2003.

Waring, J. F., Jolly, R. A., Ciurlionis, R., Lum, P. Y., Praestgaard, J. T., Morfitt, D. C., Buratto, B., Roberts, C., Schadt, E., and Ulrich, R. G. Clustering of hepatotoxins based on mechanism of toxicity using gene expression profiles. *Toxicol. Appl. Pharmacol.* **175**, 28, 2001.

Watkins, P. B. Role of cytochromes P450 in drug metabolism and hepatotoxicity. *Semin. Liver Dis.* **10**, 235, 1990.

Watkins, P. B. Idiosyncratic liver injury: Challenges and approaches. *Toxicol. Pathol.* **33**, 1, 2005.

Watkins, P. B., and Seeff, L. B. Drug-induced liver injury: Summary of a single topic clinical research conference. *Hepatology* **43**, 618, 2006.

Zimmerman, H. J., and Ishak, K. G. General aspects of drug-induced liver disease. *Gastroenterol. Clin. North Am.* **24**, 739, 1995.

CHAPTER 29

Biochemical Mechanisms of Renal Toxicity

JOAN B. TARLOFF

Department of Pharmaceutical Sciences, Philadelphia College of Pharmacy, University of the Sciences in Philadelphia, Philadelphia, Pennsylvania 10104

ANDREW D. WALLACE

Department of Environmental and Molecular Toxicology, North Carolina State University, Raleigh, North Carolina 27695

29.1 INTRODUCTION

Nephrotoxicity can be a potentially serious complication of drug therapy or chemical exposure. Although in most instances the mechanisms mediating nephrotoxicity are unclear, susceptibility of the kidney to toxic injury appears to be related, at least in part, to the complexities of renal anatomy and physiology.

The focus of this chapter is threefold: (1) to review components of renal physiology contributing to susceptibility to chemically induced nephrotoxicity, (2) to examine current methodologies for assessment of nephrotoxicity, and (3) to provide examples of a few specific nephrotoxicants, emphasizing mechanisms thought to contribute to the unique or selective susceptibility of specific nephron segments to these toxicants.

29.2 FUNDAMENTAL ASPECTS OF RENAL PHYSIOLOGY

29.2.1 Structural Organization of the Kidney

Upon gross examination, three major anatomical areas of the kidney are apparent: cortex, medulla, and papilla (Figure 29.1). The cortex is the outermost portion of the kidney and contains proximal and distal tubules, glomeruli, and peritubular capillaries. Cortical blood flow is high relative to cortical volume and oxygen consumption; the cortex receives about 90% of total renal blood flow. A blood-borne toxicant will be delivered preferentially to the renal cortex and therefore has a greater potential to influence cortical, rather than medullary or papillary, functions.

Molecular and Biochemical Toxicology, Fourth Edition, edited by Robert C. Smart and Ernest Hodgson
Copyright © 2008 John Wiley & Sons, Inc.

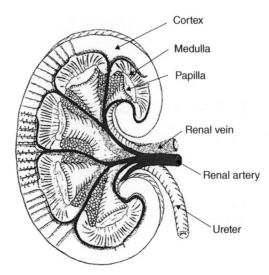

Figure 29.1. Sagittal section of a human kidney, showing major gross anatomical features and renal blood supply.

The renal medulla is the middle portion of the kidney and consists of the loops of Henle, vasa recta, and collecting ducts. Medullary blood flow (about 6% of total renal blood flow) is considerably lower than cortical flow. However, by virtue of its countercurrent arrangement between tubular and vascular components, the medulla may be exposed to high concentrations of toxicants within tubular and interstitial structures.

The papilla is the smallest anatomical portion of the kidney. Papillary tissue consists primarily of terminal portions of the collecting duct system and the vasa recta. Papillary blood flow is low relative to cortex and medulla; less than 1% of total renal blood flow reaches the papilla. However, tubular fluid is maximally concentrated and the volume of luminal fluid is maximally reduced within the papilla. Potential toxicants trapped in tubular lumens may attain extremely high concentrations within the papilla during the process of urinary concentration. High intraluminal concentrations of potential toxicants may result in diffusion of these chemicals into papillary tubular epithelial and/or interstitial cells, leading to cellular injury.

29.2.2 Nephron Structure and Function

The nephron is the functional unit of the kidney and consists of vascular and tubular elements (Figure 29.2). Both elements of the nephron have multiple specific functions, any one or more of which may be influenced by toxicants.

29.2.3 Renal Vasculature and Glomerular Filtration

The renal vasculature serves to (1) deliver waste and other materials to the tubule for excretion, (2) return reabsorbed and synthesized materials to the systemic circulation, and (3) deliver oxygen and metabolic substrates to the nephron. Vascular components of the nephron include afferent and efferent arterioles, glomerular capillaries, peritubular capillary network, and the vasa recta (Figure 29.2).

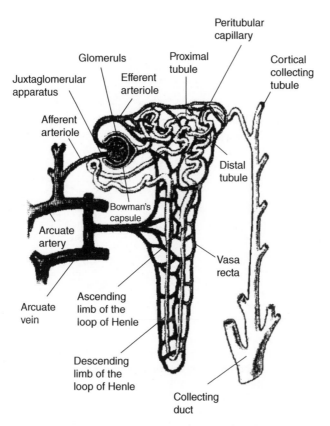

Figure 29.2. Schematic representation of nephron and vasculature. The glomerulus is positioned between afferent and efferent arterioles, and the juxtaglomerular apparatus is the point of contact between the vascular pole and distal tubule of the nephron. A capillary network surrounds tubular structures. (From Guyton, A. C., and Hall, J. E., *Textbook of Medical Physiology*, 11th ed. Elsevier Saunders Company, Philadelphia, 2006. Reproduced with permission.)

29.2.3a Arterioles. The glomerulus is unique in that it is the only capillary bed in the body positioned between two vasoactive arterioles (Figure 29.3). The afferent and efferent arterioles control blood flow and hydrostatic pressure through the glomerulus. Afferent and efferent arterioles are vascular smooth muscle cells, innervated by sympathetic nerve fibers, and contract in response to nerve stimulation, endothelin, angiotensin II, antidiuretic hormone (ADH, vasopressin), and other stimuli. Additionally, the afferent and efferent arterioles form the vascular component of the juxtaglomerular apparatus, an important mechanism in the regulation of glomerular filtration rate.

29.2.3b Morphologic Basis for Glomerular Filtration. Urine formation begins at the glomerulus where an ultrafiltrate of plasma is formed. The process of filtration is governed by physical processes that determine fluid movement across capillary beds: transcapillary hydrostatic (hydraulic) pressure and colloid oncotic (osmotic) pressure. Glomerular filtration rate (GFR) can be represented mathematically as follows:

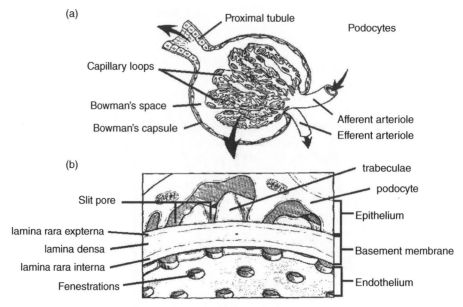

Figure 29.3. (A) Glomerular capillaries are positioned between two arterioles that regulate hydrostatic pressure and blood flow through the capillaries, thereby controlling glomerular filtration rate. (B) The glomerular basement membrane consists of three layers: the capillary endothelium, the capillary basement membrane, and podocytes from epithelial cells of Bowman's capsule. The capillary basement membrane also contains three layers: the lamina rara interna (blood side), lamina densa, and lamina rara externa (epithelial side). (From Guyton, A. C., and Hall, J. E. *Textbook of Medical Physiology*, 11th ed. Elsevier Saunders Company, Philadelphia, 2006. Reproduced with permission.)

$$\mathrm{GFR} = K_f (P_{\mathrm{GC}} + \pi_{\mathrm{BS}}) - (P_{\mathrm{BS}} + \pi_{\mathrm{GC}})$$

where P represents hydrostatic pressure and π indicates colloid oncotic pressure in the glomerular capillaries (GC) and Bowman's space (BS). GFR is determined primarily by the ultrafiltration coefficient (K_f) and by the difference between mean capillary pressure (ΔP_{GC}) and mean plasma oncotic pressure ($\Delta\pi_{\mathrm{GC}}$). K_f is a measure of capillary permeability and is determined by the total surface area available for filtration and the hydraulic permeability of the capillary wall. Filtration is favored when transcapillary hydrostatic pressure exceeds plasma oncotic pressure. Plasma oncotic pressure does not remain constant along the length of the glomerular capillary, but increases as water is filtered from plasma. When plasma oncotic pressure is equal to or exceeds transcapillary hydrostatic pressure, glomerular filtration ceases.

Glomerular filtrate passes through three barriers before entering Bowman's space: (1) capillary endothelium, (2) capillary basement membrane, and (3) visceral epithelium (podocytes) (Figure 29.3). These three filtration barriers contain anionic molecules, such as glycosaminoglycans and heparan sulfate, which are thought to restrict the filtration of anionic molecules. In general, for a given molecular size, filtration of anionic dextran sulfate is less than that of neutral dextran, suggesting that fixed negative charges of the glomerular capillary may retard or hinder the filtration of anions via electrostatic repulsion.

The glomerular capillary endothelium behaves as a porous membrane and forms an effective filtration barrier only to molecules larger than 50–70 Å. In comparison, hemoglobin and albumin in solution have molecular radii of approximately 32 and 36 Å, respectively. The glomerular basement membrane is a trilamellar structure consisting of the lamina rara interna (glomerular capillary side), lamina densa, and lamina rara externa (Bowman's space side), as shown in Figure 29.3. The glomerular basement membrane is an effective barrier for filtration: digestion, removal, or neutralization of fixed anionic sites of the glomerular basement membrane results in increased permeability to large, anionic molecules such as ferritin and albumin. The visceral epithelium is an unusual epithelium consisting of a cell body (podocyte) from which many trabeculae extend. These trabeculae, in turn, are in contact with the lamina rara externa of the glomerular basement membrane via many pedicles (foot processes). The distance between foot processes varies between 25 and 60 nm and forms the basis of the filtration slit (Figure 29.3). Foot processes also greatly increase the capillary surface area available for filtration.

29.2.3c Mesangial Cells. The exact function(s) of mesangial cells is (are) unknown. These cells are thought to provide mechanical support for the loops of the glomerular capillary tuft. As modified smooth muscle cells, mesangial cells contain contractile proteins and receptors for substances such as angiotensin II, vasopressin, and catecholamines. Contraction of mesangial cells results in reduction of glomerular capillary surface area, thereby reducing glomerular filtration rate. Additionally, mesangial cells have phagocytic properties. Experimental studies utilizing injection of tracer materials (such as ferritin and colloidal carbon) have demonstrated uptake of these tracers by mesangial cells. In some forms of immune-complex glomerulonephritis, mesangial cells appear to play an important role in the sequestration and elimination of immune complexes deposited within the glomerular capillary tuft.

29.2.3d Peritubular Capillaries. Postglomerular blood flows into peritubular capillaries or the vasa recta (Figure 29.2). The peritubular capillaries perfuse the renal cortex and contain blood in which plasma colloid oncotic pressure is elevated due to concentration of plasma proteins during glomerular filtration. This elevated plasma oncotic pressure provides a driving force for fluid and electrolyte reabsorption from renal interstitium into the peritubular capillary network. In addition, the peritubular capillaries deliver oxygen and nutrients to tubular epithelial cells.

29.2.3e Vasa Recta. The arrangement of postglomerular capillary loops into the vasa recta (Figure 29.2) is an efficient arrangement for delivery of blood-borne nutrients to medullary and papillary tubular structures. Additionally, relatively low blood flow in medulla and papilla ensures hypertonicity in these areas. Indeed, a common complication of vasodilator therapy is a reduction in medullary hypertonicity, resulting in impaired urinary concentrating ability.

29.2.3f Juxtaglomerular Apparatus and Renin Secretion. The juxtaglomerular apparatus is a specialized nephron portion consisting of both tubular and vascular elements. Within an individual nephron, cells of the afferent arteriole come into contact with the distal tubule at the juxtaglomerular apparatus. Cells of the distal tubule form the macula densa, the tubular component of the juxtaglomerular apparatus. Both afferent arteriolar and distal tubular cells contain renin. Renin is secreted by arteriolar cells in response to sympathetic nerve stimulation and/or decreased stretch, as in hypotension. Additionally, in response to signals at the macula densa (i.e., increased distal tubular flow rate, reduced Na^+/Cl^- flux at the distal tubule, and/or increased tubular fluid concentrations of Na^+/Cl^-), cells of the juxtaglomerular apparatus increase renin secretion, resulting in increased formation of angiotensin II. By constricting afferent arterioles, as well as mesangial cells, angiotensin II reduces single nephron blood flow and glomerular filtration rate.

29.2.4 Tubular Function and Formation of Urine

The renal tubule begins as a blind pouch surrounding the glomerulus. Renal tubules consist of multiple segments, shown schematically in Figure 29.4. These tubular elements selectively modify the composition of glomerular ultrafiltrate, enabling conservation of electrolytes and metabolic substrates while allowing elimination of waste products. For example, renal tubules reabsorb 98–99% of filtered electrolytes

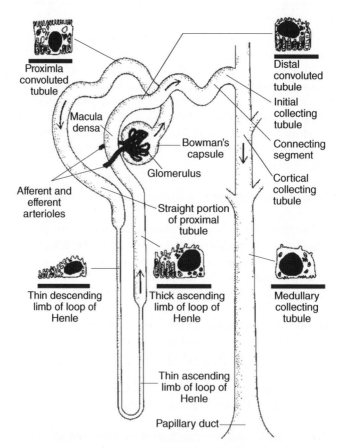

Figure 29.4. Schematic representation of a renal tubule. The epithelial cells have differing morphology and function along the length of the tubule. (From Vander, A. J. *Renal Physiology*, 3rd ed., McGraw-Hill, New York, 1985. Reproduced with permission.)

and water, and virtually 100% of filtered glucose and amino acids. Additionally, renal tubules participate in the reabsorption of bicarbonate and secretion of protons, thereby participating in acid–base balance.

29.2.4a   *Proximal Tubule.*

The initial tubular segment, the proximal tubule, reabsorbs about 70% of water filtered at the glomerulus. In other words, for each 100 ml of glomerular ultrafiltrate formed, only 30 ml will enter the loop of Henle. The proximal tubule consists of the proximal convoluted (pars convoluta) and proximal straight (pars recta) segments (Figure 29.4). Filtered bicarbonate, proteins, and glucose are reabsorbed primarily by the proximal convoluted tubule, whereas organic ion secretion occurs primarily in the proximal straight tubule.

Water reabsorption is an isosmotic, passive process driven primarily by reabsorption of Na^+ occurring via Na^+,K^+-ATPase. The proximal tubule contains numerous active transport systems capable of driving concentrative, uphill transport of many metabolic substrates, including amino acids, glucose, and citric acid cycle intermediates. Proximal tubular transport systems also scavenge virtually all of the filtered proteins by specific endocytotic protein reabsorption mechanisms. In addition, proximal tubular brush border enzymes metabolize proteins and/or peptides to constituent dipeptides and amino acids. Proximal tubular transport systems mediate the reabsorption of sulfate, phosphate, calcium, magnesium, and other ions following glomerular filtration. An important excretory function of the proximal tubule is secretion of weak organic anions and cations by specialized transport systems that drive concentrative, uphill movement of these substances from postglomerular blood into proximal tubular cells followed by excretion into tubular fluid.

29.2.4b   *Loop of Henle and Countercurrent System.*

In addition to conservation of electrolytes and metabolic intermediates, mammalian nephrons are adapted for maximal conservation of water. The thin descending and ascending limbs of the loop of Henle and the thick ascending limb of the loop of Henle are critical to the processes mediating urinary concentration (Figure 29.4). The descending and ascending limbs of the loop of Henle have differential permeabilities to water and electrolytes, establishing medullary hypertonicity necessary for urinary concentration. The loop system removes an additional 10–20% of fluid filtered at the glomerulus so that, of 100 ml of ultrafiltrate initially formed, only 10–20 ml are delivered to the distal tubule (Figure 29.5).

Four factors contribute to the ability to concentrate urine: (1) Active reabsorption of Na^+, K^+, and Cl^- without water reabsorption by the thick ascending limb of the loop of Henle results in interstitial hypertonicity and hypoosmotic tubular fluid. (2) Selective permeability to water, but not small electrolytes, in the descending thin limb of the loop of Henle allows passive reabsorption of water, facilitated by interstitial hypertonicity. (3) Relatively low medullary blood flow maintains medullary hypertonicity, allowing continued elaboration of concentrated urine. (4) In the presence of ADH, the distal tubule and collecting ducts are permeable to water so that water may diffuse out of the tubular lumen into the medullary, and papillary interstitium. Because of the ability of the thick ascending limb of the loop of Henle to move solutes but not water into the medullary interstitium, the medullary, and papillary interstitium are hyperosmotic and hypertonic compared to plasma and cortical interstitium (Figure 29.5).

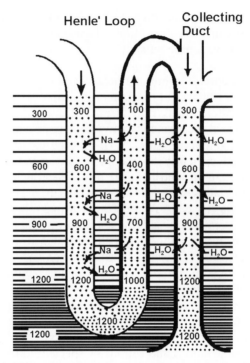

Figure 29.5. Schematic illustration of passive and active movement of electrolytes and water following glomerular filtration. Concentrations of tubular urine and peritubular fluid are given in milliosmoles per liter. Horizontal lines represent areas of increasing interstitial osmolarity. Tubule segments edged with thick black lines are either impermeable to water (loop of Henle) or display antidiuretic hormone-dependent water permeability (collecting duct).

29.2.4c Distal Tubule. Fluid entering the distal tubule is hypoosmotic compared to blood plasma. The distal tubule further modifies tubular fluid by reabsorbing most of the remaining intraluminal Na^+ in loose exchange for K^+ and H^+. Both H^+ and K^+ secretion are driven by Na^+ reabsorption. In turn, Na^+ reabsorption is determined both by plasma Na^+ concentration and distal tubular flow rate. As flow rate is increased, either by inhibition of reabsorptive processes in the proximal tubule and/or loop of Henle or by enhanced glomerular filtration, the amount of Na^+ reabsorbed by the distal tubule is increased. Since a loose exchange of Na^+ for K^+ and H^+ occurs, increases in Na^+ reabsorption concomitantly increase K^+ and H^+ secretion. Consequences of increased K^+ and H^+ secretion include hypokalemia and metabolic alkalosis, frequent complications of diuretic therapy in which distal tubular flow rate is markedly increased and Na^+ reabsorption is increased. During transit through the distal tubule, tubular fluid volume is reduced an additional 20–30%: Of the original 100 ml of glomerular filtrate, only about 5 ml enters the collecting system (Figure 29.5).

29.2.4d Collecting Duct. The collecting tubule and duct performs final regulation and fine-tuning of urinary volume and composition. Active transport systems in the collecting tubule reabsorb Na^+ and secrete K^+ and H^+. Additionally, the combination of medullary and papillary hypertonicity generated by countercurrent multiplication and the action of ADH combine to enhance water permeability of the collecting duct system. Water permeability of the collecting tubule and duct is dependent on the presence of antidiuretic hormone, and agents that interfere with the secretion or action of antidiuretic hormone may impair urinary concentrating ability. Additionally, maximal urinary concentration depends upon medullary and papillary hypertonicity. Thus, agents that increase medullary blood flow may impair urinary concentrating ability by dissipating the medullary osmotic gradient. In the presence of antidiuretic hormone, intraluminal volume may be reduced to leave a scant 0.5 ml of 100 ml of original glomerular filtrate (Figure 29.5).

29.3 FACTORS CONTRIBUTING TO NEPHROTOXICITY

Several factors contribute to the unique susceptibility of the kidney to toxicants (Table 29.1). First, renal blood flow is high relative to organ weight. For an organ constituting less than 1% of body weight, the kidneys receive about 25% of the resting cardiac output. Thus, the kidneys will receive higher concentrations of toxicants (per gram of tissue) than poorly perfused tissue such as skeletal muscle, skin, and fat. Renal blood flow is unequally distributed, with cortex receiving a disproportionately high flow compared to medulla and papilla. Therefore, a blood-borne toxicant will be delivered preferentially to the renal cortex and thereby have a greater potential to influence cortical, rather than medullary or papillary, functions.

Second, the processes involved in forming concentrated urine also will serve to concentrate potential toxicants present in the glomerular filtrate. Reabsorptive processes along the nephron may raise the intraluminal concentration of a toxicant from 10 mM to 50 mM by the end of the proximal tubule, 66 mM at the hairpin turn of the loop of Henle, 200 mM at the end of the distal tubule, and as high as 2000 mM in the collecting duct. Progressive concentration of toxicants may result in intraluminal precipitation of poorly soluble compounds, causing acute renal failure secondary to mechanical obstruction. Additionally, high intraluminal concentrations of a toxicant may enable passive diffusion of toxicant from tubular lumen into epithelium. The potentially tremendous concentration gradient for passive diffusion between lumen and cell may drive even a relatively nondiffusible toxicant into tubular cells. Furthermore, the proximal tubular epithelium is "leaky"; that is, it is fairly permeable to solutes and water. In contrast, the distal tubule is a relatively "tight" epithelia and less permeable to solutes and water than the proximal tubule. Thus, high intraluminal concentrations of toxicants may result in fairly high concentrations in proximal tubular epithelium and selectively injure the proximal tubule.

TABLE 29.1. Factors Influencing Susceptibility of the Kidney to Toxicants

High renal blood flow
Concentration of chemicals in intraluminal fluid
Reabsorption and/or secretion of chemicals through tubular cells
Biotransformtion of protoxicants to reactive intermediates

Third, active transport processes within the proximal tubule may further raise the intracellular concentration of an actively transported toxicant. During active secretion and/or reabsorption, substrates generally accumulate in proximal tubular cells in much higher concentrations than present in either luminal fluid or peritubular blood.

Fourth, certain segments of the nephron have a capacity for metabolic bioactivation. For example, the proximal and distal tubules contain isozymes of the cytochrome P450 monooxygenase system that may mediate intrarenal bioactivation of several protoxicants. Additionally, prostaglandin synthetase activity in medullary and papillary interstitial cells may be involved in cooxidation of protoxicants, resulting in selective papillary injury.

29.4   ASSESSMENT OF NEPHROTOXICITY

Evaluation of the effects of a toxicant on renal function can be accomplished by several methods. The method used depends on the complexity of the question to be answered. For example, unanesthetized or anesthetized animals may be used to determine if a chemical has an effect on kidney function. Investigations also may be conducted in vitro to examine specific biochemical or functional lesions. Finally, histopathologic techniques can provide a great deal of information about structural integrity.

29.4.1   In Vivo Methods

Initial evaluation of the effect of a chemical on renal function often is performed in intact, unanesthetized animals. An advantage of this type of study is that a great deal of information concerning overall renal function may be obtained relatively quickly using noninvasive methods. A major limitation of these noninvasive tests is lack of specificity: These tests provide no information concerning an intrarenal site of toxicity, but rather assess overall kidney function.

The standard battery of noninvasive tests includes measurement of urine volume and osmolality, urinary pH, and excretion of Na^+, K^+, glucose, and protein (Figure 29.6). An increase in urine volume and decrease in osmolality following toxicant administration may suggest an impaired ability to concentrate urine or interference with antidiuretic hormone secretion or action. Alternately, a toxicant may produce an abrupt decline in urine volume, possibly due to intraluminal obstruction by precipitation of the toxicant or cellular debris following tubular injury. Abnormalities in urine osmolality may indicate toxicant effects on renal medullary function. Alternations in the ability to acidify or alkalinize urine may suggest alterations in distal tubular function. Since both glucose and protein are reabsorbed almost completely by the proximal tubule, glucosuria and/or proteinuria may indicate proximal tubular dysfunction. However, glucosuria also may be secondary to toxicant-induced hyperglycemia so that measurement of serum glucose concentrations may be warranted. Proteinuria may reflect either proximal tubular or glomerular damage. Excretion of high-molecular-weight proteins, such as albumin, is suggestive of glomerular injury whereas excretion of low-molecular-weight proteins, such as β_2-microglobulin, is more suggestive of proximal tubular injury. Serial blood samples may be obtained

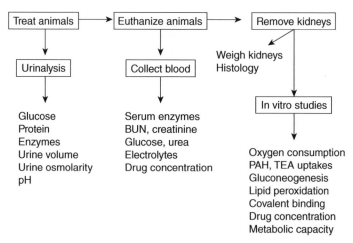

Figure 29.6. Representative design of experimental protocols to test renal function following drug or chemical administration.

and blood urea nitrogen and plasma creatinine concentrations determined. Elevations in blood urea nitrogen and/or plasma creatinine concentrations often suggest decreased glomerular filtration rate. However, increased blood urea nitrogen and creatinine concentrations may be secondary to dehydration, hypovolemia, muscle injury, or protein catabolic states and therefore may not necessarily reflect renal damage.

Attempts have been made to develop noninvasive tests that might provide more specific information about the site of injury. For example, excretion of enzymes in urine may reflect renal injury. Urinary excretion of enzymes of renal origin (enzymes that are specific to the kidney, such as maltase, γ-glutamyl transpeptidase, or trehalase) could indicate specific destruction of renal proximal tubules, whereas alkaline phosphatase in urine could arise from renal or prerenal (e.g., hepatic) damage (Figure 29.7).

The determination of renal function in anesthetized animals provides specific information on the effects of chemicals on glomerular filtration rate and renal blood flow. In addition, the ability of the kidney to reabsorb or secrete electrolytes may be determined by fractional clearance of Na^+, K^+, HCO_3^-, Cl^-, and so on. Fractional clearance involves comparison of electrolyte clearance to the clearance of a substance such as inulin, which is removed from plasma by glomerular filtration. Thus, fractional clearance takes glomerular filtration rate into account, allowing comparisons of electrolyte transport between treated and control animals even if renal hemodynamics have changed. Nephron function may be assessed by free water clearance, representing the ability of the kidney to remove almost all Na^+ from urine.

Recently the term toxicogenomics has been coined to describe the effort to identify genomic biomarkers of toxicity (see Chapters 3–5). Toxicogenomics encompasses information about a toxicant from genomic-scale mRNA profiling (by microarray analysis), cell-wide or tissue-wide protein profiling (proteomics), urine or blood metabolite profiles (metabonomics or metabolomics), genetic susceptibility, and computational modeling. Advances in molecular biology have allowed for the

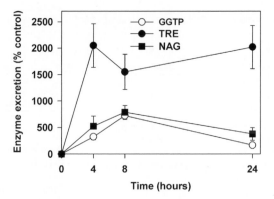

Figure 29.7. Urinary excretion of proximal tubular enzymes following *para*-aminophenol (PAP) administration to female Sprague–Dawley rats. Rats received PAP (300 mg/kg ip) and urine was collected for 24 hr following treatment. Urine was assayed for activities of γ-glutamyl transpeptidase (GGTP, ○), trehalase (TRE, ●), and *N*-acetyl-β-glucosaminidase (NAG, ■). Data are expressed as a percentage of enzyme excretion in rats treated with saline. Each data point represents the mean ± standard error of at least four determinations. GGTP and TRE are enzymes located primarily on the brush border of proximal tubule cells, and NAG is an intracellular enzyme.

determination of kidney gene-expression profiles after treatment with nephrotoxic agents such as gentamicin, puromycin, and cisplatin. The initial changes in mRNA levels expressed in toxicant exposed kidney cells may be the earliest events in the progression of cellular injury. The use of proteomics in the study of nephrotoxicity allows for the identification of protein isoforms and posttranslational modifications after toxicant exposure. Proteomic analysis by two-dimensional electrophoresis and mass spectrometry can be done using specific kidney cells or the proteins found in urine. Metabonomics has been used to characterize the changes in the metabolite profile in urine or blood after toxicant exposure using nuclear magnetic resonance (NMR), mass spectrometry (MS), gas chromatography/mass spectrometry (GC/MS), and high-performance liquid chromatography (HPLC). Presently, it is not evident if metabonomic profiling offers a more sensitive measure of nephrotoxicity than traditional methods, but future studies may provide profiles representing both the location and mechanism of toxicity.

Using the toxicogenomic techniques described, some significant data have been obtained; but to determine the functional significance of these data, additional studies need to be done. Use of these new technologies and their specificity and sensitivity to assess nephrotoxicity remains controversial, but in the future they may allow the identification of useful biomarkers that would be predictive of future responses to toxicant insults and the region of the kidney affected. Identification of specific biomarkers associated with the future development of loss of kidney function, fibrosis, or recovery would be useful in the rapid evaluation of potentially nephrotoxic agents.

29.4.2 In Vitro Methods

A variety of in vitro techniques may be employed to help elucidate mechanisms of chemical-induced nephrotoxicity. Toxic effects of chemicals may be evaluated in vitro in tissue obtained from naive animals by adding the toxicant in vitro and/or in tissue obtained from animals treated in vivo with the proposed toxicant. These approaches may be used to distinguish between an effect on the kidney due to direct chemical insult and secondary effects, such as those due to extrarenal metabolites and/or alterations in pharmacokinetics or hemodynamics.

Renal cortical slices have been used extensively to evaluate the influence of nephrotoxicants on the transport of organic anions such as para-aminohippurate (PAH) and organic cations such as N-methylnicotinamide (NMN) and tetraethyl-ammonium (TEA). Reabsorptive transport in slices may be evaluated using the nonmetabolized amino acid analog, α-aminoisobutyric acid, and the nonmetabo-lized sugar, α-methyl-D-glucose. The ability of renal cortical slices to produce ammonia and glucose in vitro can provide specific information about metabolic alterations produced by a toxicant. Renal cortical slices also have been used to investigate biochemical alterations following toxicant exposure, such as lipid per-oxidation (reflected by malondialdehyde (MDA) production), glutathione (GSH) depletion, ATP depletion, and oxygen consumption. Subtle biochemical effects of toxicants measured in vitro may provide important information regarding mecha-nisms of cytotoxicity that are difficult to assess in intact animals.

Several laboratories have developed methods to physically separate and char-acterize the metabolic capacity of glomeruli, proximal tubules, and distal tubules. These techniques can offer insight into biochemical changes associated with site-specific nephrotoxicity. Additionally, micropuncture and microperfusion experiments have been utilized to help identify specific loci of action of nephrotoxicants.

29.4.3 Histopathology

Finally, histopathologic examination of tissue can reveal structural changes that have occurred in response to nephrotoxicants, often allowing identification of affected areas of the nephron (Figure 29.8). For instance, light microscopy can iden-tify changes in renal morphology caused by chemicals, such as the presence of

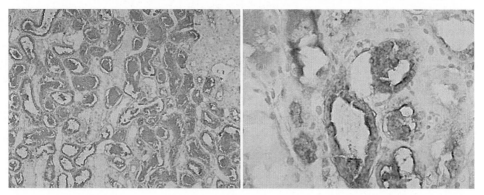

Figure 29.8. Kidney tissue from a rat treated with para-aminophenol (300 mg/kg ip). Tissue was stained with hematoxylin and eosin and was subjected to terminal deoxynucleotidyl transferase-diaminobenzidine (TUNEL) staining to reveal DNA strand breaks. Darker gray areas represent tubules that stain positive for DNA strand breaks. Final magnification was 100× (left panel) and 360× (right panel).

intraluminal protein casts, sloughed brush border, and crystals or stones in the kidney and urine. Many histochemical techniques are available to evaluate renal responses to toxicants. Electron microscopy provides information concerning subcellular localization of tubular injury. Electron microscopy has been employed extensively in efforts to understand the changes in glomerular structure that might account for changes in permeability following nephrotoxicant exposure.

29.4.4 Compensation for Renal Damage

Fortunately, the kidney has a remarkable ability to compensate for the loss of renal functional mass. Within a short time after unilateral nephrectomy, the remaining kidney hypertrophies such that overall renal function appears normal by standard clinical tests. Compensation becomes a problem when evaluating the effects of nephrotoxicants; specifically, changes in kidney function may not be detected until the ability of the kidney to compensate is exceeded. Then, within a short period of time, an animal might develop life-threatening renal failure.

29.5 SITE-SPECIFIC NEPHROTOXICITY

Many compounds have been implicated as nephrotoxicants (Table 29.2). Only rarely have specific receptors for specific nephrotoxicants been identified. Rather, in many cases it appears that toxicants exert multiple effects on intracellular systems. This is not to say, however, that there are not specific targets for certain nephrotoxicants in the kidney. For example, the proximal convoluted tubule seems to be more susceptible than other nephron segments to certain metals, such as chromium. The straight portion of the proximal tubule seems to be more susceptible to damage due to halogenated hydrocarbons (i.e., hexachlorobutadiene and dichlorovinyl-L-cysteine). Some agents, such as analgesic mixtures (usually aspirin, phenacetin, and caffeine) taken over long periods, can produce a unique toxicity characterized by renal medullary and papillary necrosis. Histological evaluation following intoxication with analgesic mixtures reveals damage to the ascending limbs of the loop of Henle. Likewise, fluoride ion and outdated tetracyclines produce damage in this area.

29.5.1 Glomerulus as a Site of Toxicity

Aminoglycoside antibiotics (Figure 29.9), such as gentamicin, amikacin, and netilmicin, are powerful drugs for the treatment of serious gram-negative infections. However, about 10% of patients treated with aminoglycosides will develop moderate but significant declines in glomerular filtration rate and elevations in serum creatinine concentration. The therapeutic utility of aminoglycosides is limited by nephrotoxicity, ototoxicity and neuromuscular junction blockade. Aminoglycoside nephrotoxicity is characterized by proximal tubular necrosis, proteinuria, and a profound decline in glomerular filtration rate.

Gentamicin-induced declines in glomerular filtration rate could be due to alterations in one or more factors determining glomerular filtration rate (i.e., changes in K_f, P_{GC}, π_{GC}). Micropuncture experiments indicate that, following low dosages of gentamicin in rats, both single-nephron glomerular filtration rate and whole kidney glomerular filtration rate are markedly reduced. However, glomerular plasma

TABLE 29.2. Segments of the Nephron Affected by Selected Toxicants

Glomerulus
Immune complexes
Aminoglycoside antibiotics
Puromycin aminonucleoside
Adriamycin
Penicillamine

Proximal Tubule
Antibiotics
  Cephalosporins
  Aminoglycosides
Antineoplastic agents
  Nitrosoureas
  Cisplatin and analogs
Radiographic contrast agents
Halogenated hydrocarbons
  Chlorotrifluoroethylene
  Hexafluropropene
  Hexachlorobutadinene
  Trichloroethylene
  Chloroform
  Carbon tetrachloride
Maleic acid
Citrinin
Metals
  Mercury
  Uranyl nitrate
  Cadmium
  Chromium

Distal Tubule/Collecting Duct
Lithium
Tetracyclines
Amphotericin
Fluoride
Methoxyflurane

Papilla
Aspirin
Phenacetin
Acetaminophen
Nonsteroidal anti-inflammatory agents
2-Bromoethylamine

flow rate and glomerular capillary hydrostatic and colloid oncotic pressures are unaltered by low-dose gentamicin treatment, suggesting that changes in renal hemodynamics do not contribute significantly to the gentamicin-induced decline in glomerular filtration rate. In contrast to the lack of effect on renal hemodynamics, gentamicin treatment produces a significant and marked reduction in capillary ultrafiltration coefficient (K_f) sufficient to account for the decline in single nephron

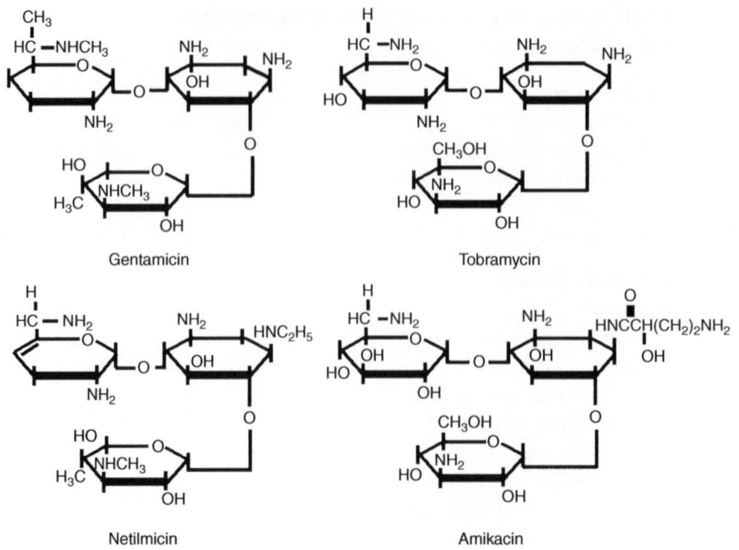

Figure 29.9. Chemical structures of several aminoglycoside antibiotics.

glomerular filtration rate. Thus, gentamicin may affect the structural integrity of glomerular capillaries, resulting in a reduction in the surface area available for filtration. Ultrastructural examination of glomeruli following gentamicin treatment indicates a reduction in the size and number of endothelial fenestra, an effect that would reduce capillary ultrafiltration coefficient and, hence, decrease glomerular filtration rate.

Aminoglycosides and polycationic compounds, along with the observed ultrastructural alterations of glomerular capillaries following gentamicin treatment, would be consistent with interactions between cationic aminoglycoside molecules and anionic sites on the glomerular capillaries. However, there is little correlation between net positive charge on various aminoglycoside molecules and capacity to reduce the size and number of endothelial fenestra. For example, netilmicin, gentamicin, and tobramycin all contain five ionizable groups. Based on electrostatic interactions alone, each compound would be expected to bind equally to glomerular anionic sites and have equal propensities to reduce capillary ultrafiltration coefficient and, hence, glomerular filtration rate. Determinations of whole kidney glomerular filtration rate and electron microscopic examination of glomerular endothelial fenestra indicate the following rank order of toxicity: gentamicin > tobramycin > netilmicin. Thus, factors other than or in addition to electrostatic interactions may contribute to aminoglycoside-induced changes in glomerular structure and function.

A prominent component of aminoglycoside nephrotoxicity is acute proximal tubular necrosis. Interestingly, the rank order of toxicity for both tubular damage and reduction of glomerular filtration rate is similar: gentamicin > tobramycin > netilmicin. This suggests the possibility that reduced glomerular filtration rate may be secondary to tubular damage. Several laboratories have suggested that at least a portion of glomerular alterations produced by nephrotoxic aminoglycosides may be related to activation of the renin–angiotensin system. With extensive proximal tubular necrosis, normal reabsorptive processes may be impaired. Indeed, nephrotoxic aminoglycosides produce a nonoliguric renal failure, suggesting that tubular reabsorption of water and electrolytes may be reduced. As previously discussed, increased flow rate at the macula densa is a stimulus for secretion of renin and, hence, formation of the vasoconstrictor, angiotensin II. In support of an involvement of angiotensin II in aminoglycoside-induced glomerular alterations, maneuvers designed to suppress angiotensin II formation (e.g., inhibition of renin secretion by sodium loading and converting enzyme inhibitor therapy) attenuate aminoglycoside-induced decrements in glomerular filtration rate. Thus, direct glomerular toxicity, as well as activation of angiotensin II, may contribute to aminoglycoside-induced alterations in glomerular filtration rate.

29.5.2 Proximal Tubule as a Site of Toxicity

The proximal tubule is the most frequently identified site of toxicant-induced renal dysfunction. The reasons for this enhanced susceptibility may relate, in part, to clearly identifiable proximal tubular functions (e.g., gluconeogenesis, ammoniagenesis, organic ion transport, glucose reabsorption) that can be assessed easily in vivo or in vitro following toxicant exposure. Thus, proximal tubular damage may be easier to detect than damage to the loop of Henle or distal tubule, where functions are more integrated and less easily identified and assessed. Additionally, enhanced susceptibility of the proximal tubule relative to other nephron segments may be related to one or a combination of the following factors: (1) Increased intraluminal concentration of potential toxicants, coupled with epithelial permeability to small, lipophilic organic compounds, may contribute to intracellular accumulation of potential toxicants due to passive diffusion. (2) Active transport functions, both secretory and reabsorptive, may result in high intracellular concentrations of toxicants. (3) Xenobiotic metabolism, mediated by cytochrome P450 monooxygenases in the proximal tubule, is capable of biotransforming protoxicants to reactive, toxic intermediates.

29.5.3 Aminoglycoside Nephrotoxicity: Role of Proximal Tubular Reabsorption

As discussed previously, aminoglycoside therapy is frequently limited by nephrotoxicity. Nonoliguric acute renal failure appears within 5–7 days after aminoglycoside therapy is initiated. Increased urine volume appears secondary to a concentrating defect and may precede increases in blood urea nitrogen and serum creatinine concentrations, delaying the recognition of acute renal failure. A small number of patients may have permanent deficits in renal function despite discontinuation of aminoglycoside therapy. Aminoglycoside-induced alterations in renal function include (1) reduced organic anion and cation secretion, (2) proteinuria, primarily involving low-molecular-weight proteins, (3) enzymuria, including excretion of alanylaminopeptidase and lysosomal enzymes (e.g., N-acetyl-β-glucosaminidase), and (4) decreased glomerular filtration rate. Morphologically, aminoglycoside nephrotoxicity is characterized by the presence of myeloid bodies, secondary lysosomes containing electron-dense lamellar membranous structures, and widespread proximal tubular necrosis (Figure 29.10).

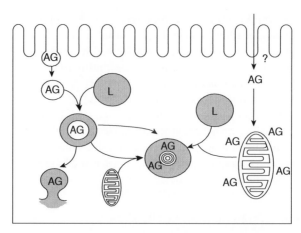

Figure 29.10. Postulated pathways of aminoglycoside-induced cellular injury. On the left, aminoglycoside (AG) enters the cell by pinocytosis and endocytosis, subsequently fusing with a primary lysosome (L). Aminoglycosides may interfere with normal lysosomal function, forming myeloid bodies (center). Additionally, aminoglycosides may destabilize lysosomes, leading to release of intralysosomal enzymes (lower left). Intracellular aminoglycosides may produce direct injury to intracellular organelles such as mitochondria. (Adapted from G.J. Kaloyanides and E. Pastoriza-Munoz, *Kidney Int.* **18**, 571–582, 1980.)

29.5.3a Renal Handling of Aminoglycosides. Aminoglycoside antibiotics are

organic polycations and carry net positive charges (Figure 29.9). These compounds have relatively low volumes of distribution, and the primary route of elimination is by renal excretion. Gentamicin, a typical nephrotoxic aminoglycoside, is freely filtered at the glomerulus and appears to be reabsorbed via active transport processes at the proximal tubular brush border.

Intracellular accumulation of gentamicin appears to occur following binding to plasma luminal membrane sites and incorporation of bound drug into apical vesicles (Figure 29.10). Major binding sites are acidic phospholipids of the renal brush border membranes, and initial binding is driven by anionic–cationic interactions. Megalin, a transmembrane protein, may mediate internalization and trafficking of gentamicin to endosomes. Small-molecular-weight proteins such as lysozyme appear to compete with gentamicin for intracellular uptake; small proteins both inhibit cortical accumulation of gentamicin and attenuate nephrotoxicity, suggesting a correlation between uptake and toxicity. However, cortical concentration and nephrotoxicity are not tightly correlated for all aminoglycosides. For example, using equivalent dosage regimens, gentamicin and netilmicin accumulate in renal cortical and medullary tissue to a similar extent, yet only gentamicin produces nephrotoxicity, suggesting that factors other than, or in addition to, cortical accumulation are important in gentamicin-induced nephrotoxicity.

Gentamicin nephrotoxicity is attenuated by treatment with other aminoglycosides or with organic polycations such as spermine, polyaspartate, and polylysine. Initially, competition for binding and intracellular uptake was the postulated mechanism of interaction between gentamicin and polycations. However, recent studies indicate that intracortical concentrations of gentamicin are not reduced in the presence of polycations, suggesting that mechanisms other than competitive uptake contribute to the amelioration in gentamicin nephrotoxicity.

29.5.3b Biochemical Mechanisms in Aminoglycoside Cytotoxicity. The sequence of biochemical events leading to gentamicin-induced proximal tubular dysfunction is unknown. Perhaps owing to its polycationic structure, gentamicin interferes with a number of intracellular proteins and macromolecules, producing a variety of biochemical effects (Table 29.3). Several mechanisms have been proposed to account for gentamicin cytotoxicity, including (1) lysosomal damage, (2) altered phospholipid metabolism, (3) inhibition of critical intracellular enzymes, (4) inhibition of mitochondrial respiration, (5) lipid peroxidation, and (6) misreading of mRNA.

Lysosomal alterations and the presence of myelin bodies and cytosegresomes are characteristic of aminoglycoside nephrotoxicity. Aminoglycosides accumulate in lysosomes and may inhibit lysosomal enzymes, contributing to aminoglycoside-induced phospholipidosis. It is not entirely clear how lysosomal accumulation of aminoglycosides may contribute to aminoglycoside-induced toxicity. It is possible that accumulation of aminoglycosides within lysosomes may, due to mechanical or chemical injury, promotes lysosomal labilization. With lysosomal rupture, lysosomal enzymes (i.e., acid hydrolases) may be released intracellularly and attack intracellular organelles leading to cell injury and necrosis. However, lysosomal alterations and myelin bodies are found following therapy with other cationic, amphophilic compounds that do not produce nephrotoxicity. Similarly, embryonic rat fibroblasts

TABLE 29.3. Membrane Alterations in Aminoglycoside Nephrotoxicity

Plasma Membrane Alterations
1. Phospholipid alterations
   Aminoglycoside binding to membrane acidic phospholipids
   Changes in content and metabolism of membrane phospholipids
   Displacement of calcium bound to phospholipids
   Formation of myeloid bodies
2. Transport defects
   Inhibition of Na^+,K^+-ATPase
   Alterations in organic acid and base transport
   Renal wasting of potassium, magnesium, and calcium
   Polyuria

Mitochondrial Membrane Alterations
1. In vivo gentamicin
   Decreased state 3 and DNP-uncoupled respiration
2. In vitro gentamicin
   Increased state 4, decreased state 3 respiration
   Decreased DNP-uncoupled respiration
   Alterations in magnesium-controlled, monovalent cation permeability pathway
   Inhibition of calcium transport
   Formation of hydrogen peroxide

Lysosomal Membrane Alterations
1. Lysosomal instability in vivo and in vitro
2. Diminished phospholipase and sphingomyelinase activities

incubated with gentamicin display lysosomal aberrations (including myelin bodies) with no loss of viability. Thus, lysosomal alterations and myelin bodies alone are not sufficient to produce cytotoxicity. However, lysosomes may act as intracellular storage depots for gentamicin. When intralysosomal concentrations of gentamicin reach critically high values, lysosomes may rupture (Figure 29.10), leading to very high intracellular gentamicin concentrations that may alter critical cellular functions.

Phospholipidosis is an early cellular alteration induced by gentamicin (Table 29.3) and is characterized by an increase in total phospholipid content with no change in relative amounts of individual phospholipids. Both in vivo and in vitro, gentamicin inhibits a variety of enzymes involved with phospholipid metabolism, including lysosomal phospholipases A and C and extra-lysosomal phosphatidylinositol-specific phospholipase C. However, cultured rat fibroblasts incubated with gentamicin display phospholipidosis with no loss in cell viability. The exact role of phospholipidosis in gentamicin-induced proximal tubular necrosis, therefore, is not clear.

Gentamicin inhibits several intracellular enzymes, including Na^+,K^+-ATPase, and alters intracellular functions including Na^+-dependent glucose transport and antidiuretic hormone stimulation of adenylate cyclase (Table 29.3). The significance of these gentamicin-induced alterations to cytotoxicity, however, is uncertain.

In vivo, gentamicin reduces mitochondrial state 3 but not state 4 respiration (Table 29.3) and reduces mitochondrial respiratory control ratio and kidney ATP concentrations. In vitro, gentamicin inhibits state 3 and stimulates state 4 respiration independent of ADP but does not completely release respiratory control or decrease ADP:O ratio (Table 29.3). Gentamicin in vitro displaces Mg^{2+} from inner mitochondrial membranes, alters Ca^{2+} fluxes across mitochondrial membranes by competing with Ca^{2+} or membrane binding and transport sites, and causes mitochondrial swelling (Table 29.3). Taken collectively, these data suggest that gentamicin-induced mitochondrial damage may contribute, in part, to the progression of gentamicin-induced proximal tubular injury.

Gentamicin nephrotoxicity is associated with lipid peroxidation and an increase in unsaturated fatty acids in renal cortex, indicative of oxidative stress. However, it is not clear whether lipid peroxidation is causally related to, or is secondary to, nephrotoxicity. Conflicting results have been reported concerning the effects of antioxidants on gentamicin-induced nephrotoxicity. For example, one laboratory has noted that although gentamicin-induced lipid peroxidation can be prevented by antioxidant therapy (diphenyl-phenylenediamine, vitamin E), the induction of nephrotoxicity (reflected by blood urea nitrogen concentration, serum creatinine, or histopathology) is not altered. In contrast, another laboratory has reported that antioxidants (e.g., dimethylthiourea, deferoxamine, dimethyl sulfoxide, sodium benzoate) confer marked protection against gentamicin-induced nephrotoxicity (increases in blood urea nitrogen and serum creatinine concentrations, proximal tubular necrosis). Thus, the role of lipid peroxidation in gentamicin nephrotoxicity remains controversial.

The antimicrobial mechanism of aminoglycoside antibiotics involves binding to ribosomes, followed by misreading of mRNA and cessation or alteration of protein synthesis. A similar mechanism may mediate nephrotoxicity. However, gentamicin binding to ribosomes has not been investigated following gentamicin therapy, nor has fidelity of mRNA transcription or protein synthesis been determined. Thus, the mechanisms contributing to gentamicin-induced nephrotoxicity are unresolved. Rather than any single event mediating gentamicin nephrotoxicity, it is likely that a number of different mechanisms probably contribute to gentamicin-induced toxicity.

29.5.4 Cephalosporin Nephrotoxicity: Role of Proximal Tubular Secretion

Cephalosporins are broad-spectrum antibiotics similar in structure to penicillin. For several cephalosporins, therapy is limited by the development of nephrotoxicity. Cephaloridine-induced nephrotoxicity has been examined extensively in laboratory animals and is characterized by an increase in blood urea nitrogen concentration within 24–48 hr, reductions in PAH and TEA transport, and inhibition of glucose production following treatment.

29.5.4a   Renal Handling of Cephalosporins. Cephaloridine is zwitterionic, and the principal route of elimination is by the kidneys. Cephaloridine clearance approximates inulin clearance, indicating absence of net secretion for cephaloridine. However, in rabbits, cortex/serum (C/S) ratios for cephaloridine are greater than unity, and higher than simultaneously determined C/S ratios for PAH and inulin, indicating that intracellular accumulation of cephaloridine occurs. Furthermore, inhibitors of organic anion transport such as PAH, penicillin, and probenecid inhibit both intracellular accumulation and nephrotoxicity of cephaloridine. Toxicity correlates with maturation of organic anion transport: Neonatal rabbits have low rates of PAH transport compared to adults, accumulate less cephaloridine than do adults, and are not susceptible to cephaloridine nephrotoxicity. Stimulation of neonatal renal organic anion transport by pretreatment with penicillin increases cortical accumulation of cephaloridine and susceptibility to nephrotoxicity. Additionally, renal cortical cephaloridine concentrations are increased by inhibitors of organic cation transport (mepiperphenidol, TEA, cyanine 863), and cephaloridine nephrotoxicity is exacerbated by these inhibitors. Taken together, these data suggest that, owing to its zwitterionic charge, cephaloridine is actively accumulated into proximal tubular cells via the organic anion transport system (inhibited by probenecid, PAH) and that a portion of cephaloridine efflux occurs via the organic cation transport system (inhibited by mepiperphenidol, cyanine). Once cephaloridine is transported into proximal tubular cells, it diffuses across the luminal membrane into tubular fluid only to a limited extent (Figure 29.11). Thus, active transport of cephaloridine into proximal tubular cells results in extremely high intracellular cephaloridine concentrations compared to other organs, which, in turn, contributes to selective nephrotoxicity.

However, cortical concentration does not appear to be the sole determinant of toxicity for other cephalosporin antibiotics, since several cephalosporins reach high cortical concentrations without producing nephrotoxicity. For example, both cephaloglycin and cephaloridine produce nephrotoxicity whereas cephalexin is not nephrotoxic. Cortical concentrations of cephaloridine, cephaloglycin, and cephalexin are approximately equal initially (0–2 hr) following treatment with these antibiotics. While cortical cephaloridine concentration does not decline, cortical concentrations of both cephaloglycin and cephalexin decline in a similar fashion over 3 hr. Thus,

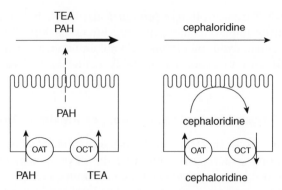

Figure 29.11. Schematic representation of proximal tubular transport and urinary excretion of para-aminophippurate (PAH), tetraethylammonium (TEA), and cephaloridine in rabbit kidney. (Left) PAH and TEA are excreted following both filtration and active secretion by the proximal tubule. PAH is transported across the basolateral membrane by organic anion transporter(s) (OAT) and TEA is secreted by organic cation transporter(s) (OCT). Intracellular concentrations of PAH and TEA may become great enough to drive passive diffusion from intracellular fluid to tubular fluid. Alternately, anion and cation exchangers may facilitate movement across the luminal membrane. (Right) Cephaloridine is excreted primarily following filtration. Active cortical uptake of cephaloridine, inhibited by probenecid and PAH, indicates a secretory component for cephaloridine transport. However, diffusion of cephaloridine from proximal tubular cell to lumen is restricted, leading to high intracellular concentrations of cephaloridine. Some efflux of cephaloridine from proximal tubular cells appears to be mediated by organic cation transporter(s) because inhibitors of this transport system potentiate cephaloridine nephrotoxicity.

neither cortical accumulation nor retention completely accounts for cephalosporin-induced nephrotoxicity, suggesting that some component of molecular structure also contributes to toxicity.

29.5.4b Biochemical Mechanism of Cephalosporin Cytotoxicity.

Although the role of renal tubular transport in cephaloridine nephrotoxicity has been well-defined, the exact molecular mechanisms mediating cephaloridine nephrotoxicity are less well understood. Several mechanisms have been postulated to mediate cephaloridine nephrotoxicity, including (1) production of a highly reactive acylating metabolite(s) by cytochrome P450-dependent monooxygenases, (2) production of mitochondrial respiratory toxicity, and (3) production of lipid peroxidation.

Cobaltous chloride and piperonyl butoxide, inhibitors of cytochrome P450-dependent monooxygenases, reduce cephaloridine nephrotoxicity in mice and rats, suggesting that a cytochrome P450-dependent reactive intermediate mediated cephaloridine nephrotoxicity. However, more recent studies fail to support the hypothesis of cytochrome P450-dependent metabolism for the following reasons: (1) Piperonyl butoxide protects against cephaloridine nephrotoxicity in rabbits, whereas cobaltous chloride does not protect. (2) In rabbits, piperonyl butoxide protection against cephaloridine nephrotoxicity may be related to reduction of cephaloridine uptake in renal cortex, rather than inhibitory effects on cytochrome P450. (3) Inducers of renal cytochrome P450 activity in rats (e.g., β-naphthoflavone, *trans*-stilbene oxide, polybrominated biphenyls) do not potentiate cephaloridine nephrotoxicity. (4) Phenobarbital, a renal cytochrome P450 inducer in rabbits, potentiates cephaloridine nephrotoxicity in rabbits but also increases renal cortical accumulation of cephaloridine. (5) A glutathione conjugate of cephaloridine has not been identified in kidneys of cephaloridine-treated animals.

Mitochondrial damage may at least partially mediate cephaloridine cytotoxicity. This hypothesis is supported by the following evidence: (1) Ultrastructural damage to renal mitochondria has been observed following cephaloridine administration. (2) Mitochondrial respiration is depressed following either in vivo or in vitro exposure of rabbit kidney tissue to cephaloridine and is characterized by marked inhibition of ADP-dependent respiration using succinate as substrate. (3) Functional impairment of renal mitochondria occurs as early as 1–2 hr following cephaloridine administration. (4) Nephrotoxic cephalosporins (cephaloridine, cephaloglycin) produce similar patterns of respiratory depression, whereas non-nephrotoxic cephalosporins (e.g., cephalexin) do not alter mitochondrial function. Although the relatively early depression in mitochondrial respiration is suggestive of a pathogenic mechanism in cephaloridine nephrotoxicity, other biochemical changes following cephaloridine treatment have also been detected 1–2 hr following drug administration, including lipid peroxidation (see below). Further studies are needed to document a cause–effect relationship between cephaloridine-induced mitochondrial dysfunction and proximal tubular necrosis.

Recently, lipid peroxidation has been proposed as a mediator of cephaloridine cytotoxicity (Figure 29.12). Several lines of evidence support the involvement of peroxidative injury in cephaloridine nephrotoxicity: (1) Renal cortical concentra-

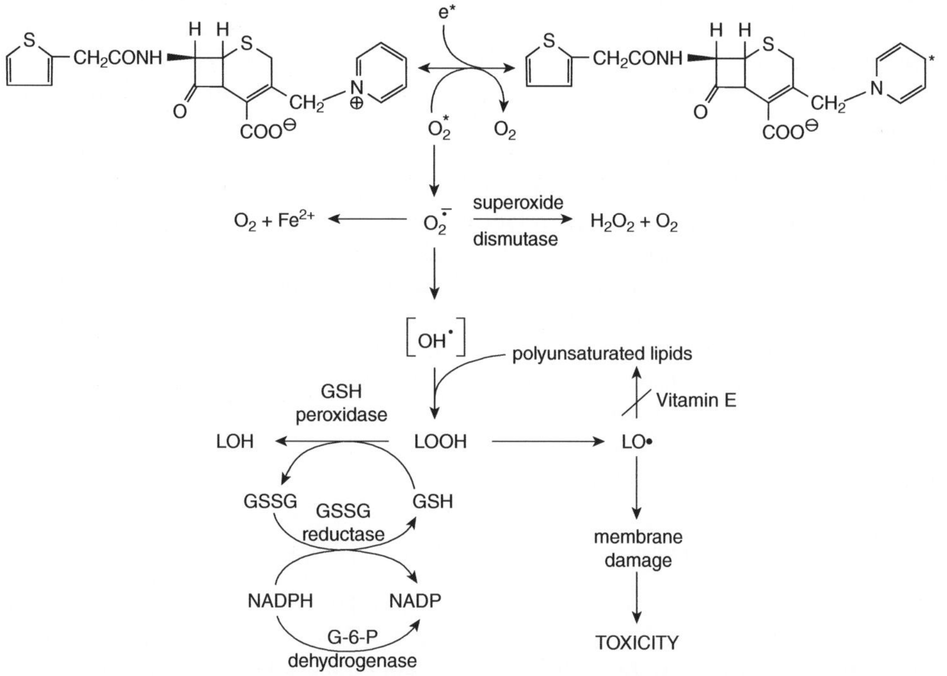

Figure 29.12. Proposed mechanism of cephalosporin-induced nephrotoxicity. LOOH, lipid hydroperoxide; LOH, lipid alcohol; LO•, lipid radical; G-6-P, glucose-6-phosphate.

tions of conjugated dienes, products of lipid peroxidation, are increased 1–2 hr following cephaloridine administration. (2) Rats fed vitamin E- and/or selenium-deficient diets are more susceptible to cephaloridine nephrotoxicity than rats fed vitamin E and/or selenium adequate diets. In addition, cephaloridine produces a dose-related depletion of renal cortical reduced glutathione, accompanied by an increase in renal cortical concentrations of oxidized glutathione. These observations are suggestive of oxidative stress and are consistent with the postulated role of lipid peroxidation in cephaloridine nephrotoxicity.

Lipid peroxidation may be initiated by one electron reduction and redox cycling of the pyridinium ring of cephaloridine, resulting in the reduction of molecular oxygen to superoxide anion (Figure 29.12). In support of this mechanism, cephaloridine undergoes anaerobic reduction by isolated renal cortical microsomes with production of superoxide anion radicals and hydrogen peroxide. However, the postulated involvement of the pyridinium ring is not supported by the observation that cephalothin, which is structurally identical to cephaloridine with the exception that it lacks a pyridinium ring, also induces lipid peroxidation (as reflected by malondialdehyde production).

In vitro exposure of renal cortical slices to cephaloridine results in time- and concentration-dependent increases in lipid peroxidation, as reflected by malondialdehyde production. Furthermore, the onset of cephaloridine-induced malondialdehyde production preceded cephaloridine-induced inhibition of organic ion accumulation, suggesting that cephaloridine-induced lipid peroxidation mediates the effects of cephaloridine on organic ion transport. Additionally, antioxidants (e.g., promethazine, N,N'-diphenyl-p-phenylenediamine) block the effects of cephaloridine on lipid peroxidation and on organic ion transport, suggesting a cause–effect relationship between cephaloridine-induced lipid peroxidation and inhibition of organic ion transport.

Glutathione is known to play a critical role in the detoxification of lipid hydroperoxides by acting as a cosubstrate for glutathione peroxidase, which catalyzes the conversion of lipid hydroperoxides to lipid alcohols (Figure 29.12). Incubation of renal cortical slices with cephaloridine depletes renal cortical glutathione content in a concentration- and time-dependent manner. Furthermore, depletion of renal cortical glutathione content by cephaloridine precedes the earliest detectable increase in cephaloridine-induced lipid peroxidation, a phenomenon that is consistent with the role of glutathione in the detoxification of lipid hydroperoxides. However, antioxidant treatment does not protect against cephaloridine-induced glutathione depletion. The absence of antioxidant protection against cephaloridine-induced glutathione depletion may suggest that these antioxidants act subsequent to the formation of lipid hydroperoxides or that cephaloridine-induced glutathione depletion is not related to lipid peroxidation.

29.5.5 Chloroform Nephrotoxicity: Role of Metabolic Bioactivation

Chloroform is a nephrotoxicant that most likely undergoes metabolic bioactivation within the kidney. Chloroform ($CHCl_3$), a common organic solvent widely used in the chemical industry, produces hepatic and renal injury in humans and experimental animals. Renal necrosis due to chloroform is sex- and species-specific: For example, male mice exhibit primarily renal necrosis whereas female mice develop primarily hepatic necrosis following chloroform administration. Nephrotoxicity may be related, in part, to high intraluminal concentrations of chloroform attained following glomerular filtration and fluid reabsorption in the proximal tubular, coupled with relatively high permeability of the proximal tubule compared to other nephron segments.

Tissue injury by chloroform is probably not due to chloroform per se, but is mediated by a chloroform metabolite. The initial step leading to chloroform-induced tissue injury is believed to be the biotransformation of chloroform to a reactive intermediate, phosgene ($COCl_2$), by cytochrome P450-dependent monooxygenases (Figure 29.13). Formation of phosgene may proceed through oxidative dechlorination involving oxidation of the C–H bond of chloroform, producing the trichloromethanol (CCl_3–OH) intermediate, a highly unstable species that would spontaneously dechlorinate to phosgene (Figure 29.13). Phosgene is a highly reactive intermediate and may react with intracellular macromolecules to induce cell damage.

Kidneys have relatively low xenobiotic-metabolizing enzyme activities, and chemically induced nephrotoxicity has been assumed to be produced by toxic intermediates generated in the liver and transported to the kidney. If a single hepatic metabolite of chloroform produced both kidney and liver injury, species, strain, and sex differences in susceptibility to chloroform nephro- and hepatotoxicity should be similar. However, species, strain and sex differences in susceptibility to chloroform nephrotoxicity are not consistent with those of chloroform hepatotoxicity. In addition, several modulators of tissue xenobiotic-metabolizing activities alter

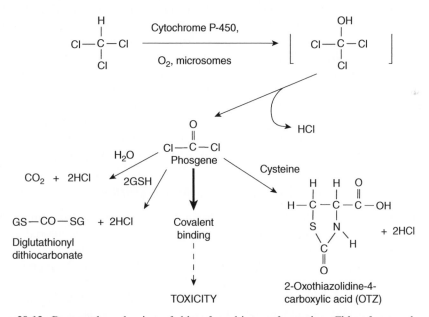

Figure 29.13. Proposed mechanism of chloroform biotransformation. Chloroform undergoes cytochrome P450-catalyzed conversion to trichloromethanol (CCl_3-OH), which spontaneously decomposes to form phosgene. Phosgene is highly reactive and may be detoxified by reacting with sulfhydryl-containing chemicals (cysteine, glutathione [GSH]). Alternately, phosgene can react with sulfhydryl groups on protein, leading to covalent binding and possibly to toxicity.

chloroform nephrotoxicity and hepatotoxicity differently. Since chloroform-induced kidney injury does not parallel liver damage, it is unlikely that hepatic metabolism of chloroform mediates renal toxicity. Furthermore, the highest concentration of cytochrome P-450 in the kidney is found in proximal tubular epithelial cells, consistent with preferential necrosis of the proximal tubule following chloroform treatment.

The concept that kidney injury is produced by a chloroform metabolite generated in the kidney has been demonstrated directly using in vitro techniques. In order to avoid hepatic metabolism of chloroform, renal cortical slices from naive animals were incubated with chloroform in vitro. Under these conditions, the only site of metabolism of chloroform is the kidney. In vitro exposure to chloroform produced toxicity in kidney slices from male, but not from female, mice. Furthermore, ^{14}C-labeled chloroform was metabolized to $^{14}CO_2$ and covalently bound radioactivity by male, but not female, renal cortical microsomes. In vitro metabolism of chloroform by male, but not female, renal slices is consistent with reduced susceptibility of female mice to in vivo chloroform nephrotoxicity. Metabolism required oxygen, an NADPH regenerating system, was dependent on incubation time, microsomal protein concentration, and substrate concentration, and was inhibited by carbon monoxide. The negligible degree of chloroform metabolism and toxicity in female mice is consistent with lower renal cytochrome P450 concentrations and activities in female versus male mice.

Chloroform containing deuterium ($CDCl_3$) is metabolized in the liver to phosgene at approximately half the rate of $CHCl_3$ metabolism to $COCl_2$. $CDCl_3$ is also less hepatotoxic than chloroform. Since the C–D bond is stronger than the C–H bond, these data suggest that cleavage of the C–H bond is the rate-limiting step in the activation of chloroform. $CDCl_3$ is also less toxic to the kidney than chloroform. This deuterium isotope effect on chloroform-induced nephrotoxicity suggests that the kidney metabolizes chloroform in the same manner as the liver (e.g., by oxidation to phosgene). Indeed, rabbit renal cortical microsomes incubated in media supplemented with L-cysteine metabolized ^{14}C-labeled chloroform to radioactive phosgene-cysteine 2-oxothiazolidine-4-carboxylic acid (Figure 29.13). These in vitro data collectively support the hypothesis that mouse and rabbit kidneys biotransform chloroform to a metabolite (phosgene) that mediates nephrotoxicity.

29.5.6 Mercury Nephrotoxicity: Role of Molecular Mimicry

Mercury is one of many heavy metals, including lead, chromium, and uranium, that are nephrotoxic. Mercury is found in the environment, and many industrial settings and exposure may occur from dietary sources such contaminated water or food items such as large predator fish. Mercury can exist as elemental (Hg), mercury salts ($HgCl_2$), or organic mercury (R–Hg). In the body, elemental mercury is a cation (Hg^{2+}) that binds to sulfhydryl-containing molecules including glutathione (GSH), cysteine, homocysteine, and metallothionein. Within the kidney, inorganic and organic mercury accumulates rapidly in the proximal tubule, which is the site of nephrotoxicity.

The uptake of mercury in the proximal tubule can occur by two mechanisms, with one being from the luminal membrane and another from the basolateral membrane. Luminal uptake involves the activity of γ-glutamyltransferase and cysteinylglycinase enzymes. By this mechanism, glutathione conjugated Hg^{2+} in the luminal side of the membrane is degraded by γ-glutamyltransferase, resulting in thiol conjugates of Hg^{2+} (Cys–S–Hg–S–Cys). Rapid uptake of Cys–S–Hg–S–Cys in the proximal tubule cells occurs because this molecule is structurally similar to the amino acid cystine. The Cys–S–Hg–S–Cys conjugate mimics the structure of cystine, and amino acid transporters mistakenly recognize this conjugate and effectively transport it into the cell. Uptake of mercury also occurs from postglomerular blood into proximal tubular cells on the basolateral side of the membrane. Basolateral uptake is thought to occur by one or more organic anion transporters (OATs), with several mercuric glutathione conjugates acting as suitable substrates.

The nephrotoxicity of mercury is characterized by increasing excretion levels of alkaline phosphatase and γ-glutamyltransferase enzymes and elevated water, amino acids, and albumin in the urine. Mercury-exposed kidney cells also exhibit increased expression of the metal-binding protein metallothionein and initially elevate levels of glutathione and other glutathione-dependent enzymes. Intracellular toxicity of mercury occurs due to its high affinity for thiol-containing proteins that can lead to oxidative stress involving (a) mitochondrial dysfunction, altering the normal flux of calcium, sodium, and potassium ions, and (b) disruption of ATP production. Several studies have also shown mercury exposure also leads to immune-cell-mediated glomerular nephritis. Thiol-containing metal chelating agents such as *meso*-2,3-dimercaptosuccinic acid (DMSA) or 2 2,3-dimercapto-1-propanesulfonic acid (DMPS) are often utilized as antidotes for mercury poisoning and allow for excretion of mercuric conjugates in the urine.

29.5.7 Loop of Henle as a Site of Toxicity

The loop of Henle has not been identified as a primary target for toxicant-induced injury. This lack of susceptibility may be related, in part, to difficulties in identifying and localizing abnormalities in structure and function of this nephron segment. Segments within the loop of Henle (i.e., thin descending and ascending limbs) are composed of thin, poorly characterized epithelial cells, and damage may be quite difficult to detect. Additionally, function of the loop structures is integrated closely with functions of other nephron segments, making it difficult to identify a specific effect in the loop of Henle. Functional abnormalities of the loop of Henle, such as those induced therapeutically by thiazide diuretics, manifest primarily as an impairment of urinary concentrating mechanisms and increased excretion of Na^+. Polyuria and hyposthenuria can occur following renal vasodilation, glomerular damage, distal and collecting duct damage, lack of antidiuretic hormone, or several other factors. Thus, identification of structural and functional abnormalities of the loop of Henle may hinder recognition of toxicant-induced damage in this nephron segment.

In contrast to ambiguities in vivo, specific damage to the thick ascending limb of the loop of Henle has been demonstrated in vitro using isolated perfused kidneys from rats. Within 15–30 min of in vitro perfusion, a marked concentrating defect, as well as progressive increase in Na^+ excretion, is observed in isolated perfused kidneys. Histologically, damage is localized to the thick ascending limb and is characterized by mitochondrial swelling, cytoplasmic disruption, and progressive nuclear pyknosis. Selective damage to the medullary thick ascending limb in the isolated perfused kidney relates to the delicate balance between low O_2 tension and increased

O_2 demand in this nephron segment. Medullary damage in the isolated perfused kidney can be attenuated by reducing tubular work and O_2 consumption, via diuretics (e.g., furosemide, ouabain) or by increasing O_2 supply by inclusion of an oxygen carrier (e.g., hemoglobin, erythrocytes) in the perfusate. Attenuation of damage by diuretics or ouabain relates to the primary work of the medullary thick ascending limb. In this "diluting segment," oxygen consumption correlates strongly with Na^+ and Cl^- reabsorption. Ouabain and diuretics (such as furosemide) inhibit reabsorption of electrolytes, thereby reducing oxygen demand by the thick ascending limb. Alternatively, inclusion of oxygen carrier in perfusate increases oxygen delivery to the thick ascending limb, allowing supply and demand to match more closely. Thus, damage to the thick ascending limb in the isolated perfused rat kidney is related to relative hypoxia in this nephron segment coupled with high oxygen demand mandated by high rates of electrolyte reabsorption.

Theoretically, reductions in renal medullary perfusion in intact animals might produce selective damage to the medullary thick ascending limb. However, in vivo, the primary site for anoxic and/or hypoxic damage is the proximal straight tubule, rather than ascending thick limb. Therefore, the relevance of the selective injury in the in vitro isolated perfused kidney to the situation observed in vivo is questionable. Nevertheless, elucidation of the mechanism of thick ascending limb damage in the isolated perfused kidney has been important in defining the relationship between oxygen supply and demand with regard to cell injury in medullary structures.

29.5.8   Distal Tubule as a Site of Injury

Distal tubular injury is difficult to detect compared to proximal tubular nephrotoxicants. Again, lack of highly specialized functions and poorly characterized tubular ultrastructure may contribute to this difficulty. Additionally, compared to the proximal tubule, the distal tubule is a "tight epithelium" with high electrical resistance. Therefore, the distal tubule may be intrinsically less permeable to intraluminal toxicants than the proximal tubule, despite significantly higher concentrations at the distal tubule compared to proximal. Thus, distal tubular cells may not be exposed to high intracellular concentrations of toxicants as are proximal tubular cells. Also, in contrast to the proximal tubule, the distal tubule has negligible mixed function oxidase activity so that conversion of protoxicants to reactive intermediates via this mechanism is unlikely in the distal tubule.

One compound that has been associated with distal tubular injury is amphotericin B, a polyene antifungal agent used in the treatment of systemic mycoses caused by opportunistic fungi. Clinical utility of amphotericin B is limited by its nephrotoxicity, characterized functionally by polyuria resistant to antidiuretic hormone administration, hyposthenuria, hypokalemia, and mild renal tubular acidosis.

Amphotericin B is highly lipophilic (Figure 29.14) and interacts with membrane lipid sterols, such as cholesterol, to disrupt membrane permeability. Because amphotericin is freely filtered, it achieves high concentrations in distal tubular fluid and easily forms complexes with cholesterol and other lipids present in distal tubular luminal membranes. Amphotericin effectively transforms the "tight" distal tubular epithelium into an epithelium leaky to water, H^+ and K^+. Functional abnormalities observed with amphotericin B are attenuated when the antifungal agent is administered as an emulsion formulation whereby amphotericin is incorporated into lipid

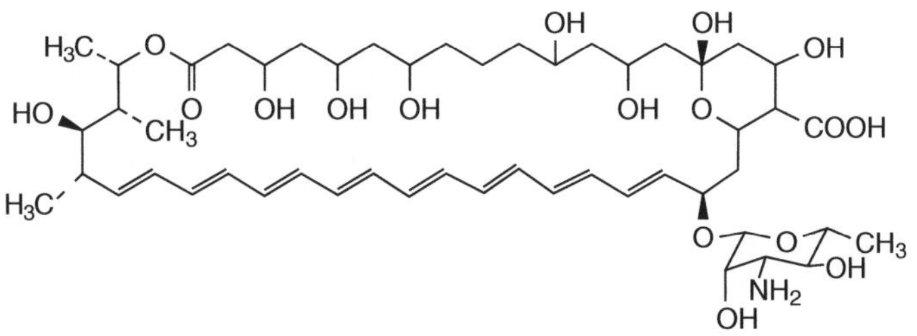

Figure 29.14. Structure of amphotericin B.

micelles. Antifungal activity of emulsion-formulated amphotericin B is equivalent to the standard non-emulsion formulation, whereas polyuria and hyposthenuria are significantly reduced by emulsion formulation.

29.5.9 Collecting Tubules as a Site of Injury

Chronic consumption of large dosages of combination analgesics, typically phenacetin- and/or caffeine-containing preparations, may be associated with renal papillary necrosis. This injury is described as an ischemic infarction of the inner medulla and papilla of the kidney. Renal function may be compromised modestly by a loss of concentrating ability; or, in severe cases, anuria, sepsis and rapid deterioration of renal function may occur. Morphologically, there is loss of renal papilla (containing terminal collecting ducts), medullary inflammation and interstitial fibrosis, and loss of renomedullary interstitial cells.

A variety of non-narcotic analgesics have been implicated in the etiology of renal papillary necrosis, including acetaminophen, aspirin, acetanilid, and nonsteroidal antiinflammatory agents such as ibuprofen, phenylbutazone, and indomethacin. Although these agents are dissimilar structurally and chemically, they share a common mechanism of action, acting as analgesics by inhibiting prostaglandin synthesis. Specifically, these analgesics inhibit cyclooxygenase activity but not prostaglandin hydroperoxidase activity of prostaglandin H synthase complex. In the kidney, prostaglandin H synthase activity is distributed asymmetrically, with highest activity in renal medulla and lowest activity in renal cortex. Additionally, the prostaglandin hydroperoxidase component of prostaglandin H synthetase complex is capable of metabolizing xenobiotics, including acetaminophen and phenacetin, to reactive intermediates capable of covalent binding to cellular macromolecules. Thus, renal papilla may be injured selectively by non-narcotic analgesic agents due to the combination of high concentrations of potential toxicants present in tubular fluid and specialized enzymes capable of biotransforming protoxicants to active intermediates.

In laboratory animals, papillary necrosis due to non-narcotic analgesic has been extremely difficult to produce. However, papillary necrosis has been demonstrated following administration of 2-bromoethylamine; 2-bromoethylamine has been used to demonstrate the role of urinary concentrating mechanisms in the etiology of 2-bromoethylamine-induced papillary necrosis. Maneuvers that produce large volumes of dilute urine, such as diuretic therapy, lack of antidiuretic hormone (Brattleboro rats), or volume expansion with 5% glucose, prevent papillary necrosis due to

2-bromoethylamine. In contrast, maneuvers that restore urinary concentrating ability, such as exogenous antidiuretic hormone replacement in Brattleboro rats, render these animals susceptible to 2-bromoethylamine-induced papillary necrosis. Thus, papillary necrosis due to 2-bromoethylamine is critically dependent on the ability to produce concentrated urine.

Biochemical mechanisms responsible for papillary necrosis are not entirely clear. Several mechanisms have been proposed, including: (1) direct cellular injury, (2) reduction or redistribution of blood flow, (3) prostaglandin inhibition, and (4) free radical formation. A decrease in papillary blood flow would lead to ischemia of deeper portions of the kidney, ultimately leading to cellular necrosis. Renal prostaglandins and thromboxanes are proposed to play a role in control of intrarenal blood flow. Salicylates and nonsteroidal anti-inflammatory agents inhibit renal cyclooxygenase activity, preventing prostaglandin synthesis. As vasodilators, prostaglandins in medulla and papilla may have an important role in maintaining normal blood flow. However, it is not entirely clear that papillary and medullary blood flow are altered early in the development of papillary necrosis. Additionally, acetaminophen has been implicated in the etiology of papillary necrosis, yet it is relatively ineffective as an inhibitor of prostaglandin synthesis, calling into question the relationship between inhibition of prostaglandin synthesis and papillary necrosis. Free radicals (generated from prostaglandin H synthetase-catalyzed cooxidation of xenobiotics) and/or covalent binding of activated metabolites to cellular macromolecules also may contribute to papillary necrosis. However, the relevance of these mechanisms to papillary necrosis is, as yet, unclear.

29.5.10 Crystalluria

Kidney stones are not uncommon and are thought to occur when urine becomes supersaturated with poorly soluble drugs or chemicals. Clinically, most kidney stones are comprised of calcium oxalate monohydrate (COM) crystals. However, crystalluria has been observed with drugs such as older sulfonamides and endogenous chemicals such as uric acid. The consequences of crystal deposition within the kidney can range from simple irritation to necrosis and fibrosis. The mechanism whereby crystals form and become impacted within tubules remains unclear. At least two theories have been put forward to explain intrarenal deposition and impaction of crystals such as COM. In the free-particle model, calcium oxalate is proposed to undergo *de novo* nucleation within the tubular lumen. As the nascent particle travels through the remaining portions of the tubule, aggregation and crystal growth contribute to formation of a particle large enough to become trapped and occlude fluid movement in the tubule. The rate of calcium oxalate crystal growth is estimated at 1–2 μm/min. Since transit time from glomerulus to renal pelvis is estimated at 3–4 min, it is questionable if a spontaneously formed and freely moving calcium oxalate crystal would have sufficient time to achieve a size large enough to permit trapping. To overcome this uncertainty, the fixed particle model has been proposed. In this scenario, small COM nuclei become attached to damaged epithelial cells. Therefore, some component of renal injury is necessary to support kidney stone formation. From a theoretical standpoint, spontaneous crystal formation would be likely only at points along the tubule where there is substantial narrowing (i.e., transition to thin loop of Henle) or concentration (terminal collecting duct).

Rats receiving ethylene glycol in their drinking water display hyperoxaluria and COM crystals after 2–4 weeks of treatment. Ethylene glycol is metabolized to several products, including oxalic acid. Rats receiving 0.75% ethylene glycol in drinking water display hyperoxaluria within 1 week and crystalluria within 2–4 weeks of treatment. Following filtration, oxalate is reabsorbed by proximal tubule cells. These cells may be exposed not only to oxalate but also to COM because intracellular calcium may allow oxalate crystallization to occur. Studies in cultured cells have shown that proximal tubular cells (LLC-PK$_1$) but not distal tubular cells (MDCK) display LDH leakage and necrosis when exposed to COM. Oxalate does not appear to contribute to the toxicity in these cultured cells because toxicity was not observed when cells were incubated with sodium oxalate and EDTA, preventing formation of COM crystals. Crystals are also observed in kidneys from hydroxyl-L-proline-treated rats, and these crystals are found in all parts of the kidney and in all tubule segments. Hydroxy-L-proline is a physiological precursor for oxalic acid and sodium oxalate. Rats treated with hydroxyl-L-proline also showed symptoms of renal injury including LDH, hydrogen peroxide, and 8-isoprostane excretion. A component of crystalluria may be related to reactive oxygen intermediates and oxidative stress, possibly a response to oxalate exposure of renal epithelial cells.

As a component of oxidative stress in renal epithelial cells, expression of crystal-binding proteins may be enhanced. For example, hyaluronan (HA) is a glycosaminoglycan found at the apical surface of MDCK cells. HA is also present in the inner medullary interistitum of the kidney, and its expression is upregulated during inflammation. In at least one study of COM crystalluria, crystals co-localized with HA in rat kidneys, suggesting that HA may act as a crystal-binding protein in the renal medulla. However, other crystal-binding proteins such as phosphatidylserine, collagen, and nucleolin-related protein may also contribute to kidney stone formation.

Osteopontin (OPN) is a controversial protein that may be involved in crystalluria as either an inhibitor of crystallization or a promoter of crystal formation and retention. In bone, OPN inhibits hydroxyapatite formation during mineralization. Some studies have suggested that OPN in the kidney keeps oxalate in solution, preventing formation of COM crystals. However, in mice, OPN seems to be required for crystallization. Mice with targeted disruption of the OPN gene developed calcium oxalate crystals in the distal tubule and collecting duct when treated with 1% ethylene glycol in drinking water. Wild-type mice had similar levels of urinary calcium and oxalate but failed to develop detectable intrarenal crystals. Urine from both knockout and wild-type mice contained calcium oxalate crystals, but those crystals either failed to develop or did not impact in kidneys of wild-type mice. It is interesting that OPN expression was markedly increased in wild-type mice after 4 weeks of hyperoxaluria, and it may be argued that crystals did not form or impact in these mice because of this upregulation.

29.6 SUMMARY

Susceptibility of the kidney to chemically-induced toxicity is related, at least in part, to several unique aspects of renal anatomy and physiology. By virtue of high renal blood flow, active transport processes for secretion and reabsorption, and progressive concentration of the glomerular filtrate following water removal during the formation of urine, renal tubular cells may be exposed to higher concentrations of potential toxicants than are cells in other organs. Additionally, intrarenal metabolism, via cytochrome P450 or prostaglandin H synthetase, may contribute to the generation of toxic metabolites within the kidney.

The precise biochemical mechanisms leading to irreversible cell injury and nephrotoxicity are not well-defined. Many diverse biochemical activities occur within the kidney, and interference with one or more of these functions may lead to irreversible cell injury. Rather than any one single mechanism mediating chemically induced nephrotoxicity, it is likely that a toxicant alters a number of critical intracellular functions, ultimately leading to cytotoxicity and cellular necrosis.

SUGGESTED READING

Bridges, C. C., and Zalups, R. K. Molecular and ionic mimicry and the transport of toxic metals. *Toxicol. Appl. Pharmacol.* **204**, 274–308, 2005.

Dekant, W. Chemical-induced nephrotoxicity mediated by glutathione *S*-conjugate formation. *Toxicol. Lett.* **124**, 21–36, 2001.

Eikmans, M., Baelde, H. J., de Heer, E., and Bruijn, J. A. RNA expression profiling as prognostic tool in renal patients: toward nephrogenomics. *Kidney Int.* **62**, 1125–1135, 2002.

Gibbs, A. Comparison of the specificity and sensitivity of traditional methods for assessment of nephrotoxicity in the rat with metabonomic and proteomic methodologies. *J. Appl. Toxicol.* **25**, 277–295, 2005.

Humes, H. D., and Weinberg, J. Toxic nephropathies. In: Brenner, B. M., and Rector, F. C., Jr. (Eds.). *The Kidney*, 3rd ed., W.B.Saunders, Philadelphia, 1986, p. 1491.

Inui, K. I., Masuda, S., and Saito, H. Cellular and molecular aspects of drug transport in the kidney. *Kidney Int.* **58**, 944–958, 2000.

Lee, W., and Kim, R. B. Transporters and renal drug elimination. *Annu. Rev. Pharmacol. Toxicol.* **44**, 137–166, 2004.

Schnellmann, R. G. Toxic responses of the kidney. In: Klaassen, C. D. (Ed.). *Casarett and Doull's Toxicology: The Basic Science of Poisons*, 6th ed., McGraw-Hill, New York, 2001, pp. 491–514.

Van Vleet, T. R., and Schnellmann, R. G. Toxic nephropathy: Environmental chemicals. *Semin. Nephrol.* **23**, 500–508, 2003.

Zalups, R. K. Molecular interactions with mercury in the kidney. *Pharmacol. Rev.* **52**, 113–143, 2000.

CHAPTER 30

Biochemical Toxicology of the Peripheral Nervous System

JEFFRY F. GOODRUM

(retired) Department of Pathology and Laboratory Medicine, University of North Carolina at Chapel Hill, Chapel Hill, North Carolina 27599

ARREL D. TOEWS

Department of Biochemistry and Biophysics, University of North Carolina at Chapel Hill, Chapel Hill, North Carolina 27599

THOMAS W. BOULDIN

Department of Pathology and Laboratory Medicine, University of North Carolina at Chapel Hill, Chapel Hill, North Carolina 27599

30.1 INTRODUCTION

All parts of the nervous system are susceptible to toxic injury. When toxic injury alters the function of the peripheral nervous system (PNS), the disease process is referred to as *toxic neuropathy*. Therapeutic drugs are the usual cause of toxic neuropathy, with industrial and environmental agents being a well-recognized but less common cause. Toxic neuropathy may lead to serious and sometimes permanent disability. Recognition of the mechanisms responsible for toxic injury of the PNS is critically important if we are to better predict which of the new chemicals constantly being introduced into our pharmacopoeia, workplace, and environment have the potential for causing neuropathy. A list of the more common therapeutic drugs and environmental agents causing neuropathy is provided in Table 30.1. Only a few of these toxicant-induced neuropathies have been studied in detail, and thus our understanding of the biochemical toxicology of these neuropathies is very limited. This chapter reviews the biochemical mechanisms implicated in axonal degeneration and segmental demyelination, which are the two common noncarcinogenic responses of the PNS to toxic injury.

The PNS is defined as that part of the nervous system external to the brain and spinal cord (Figure 30.1). As such, the PNS includes the cranial nerves, dorsal and ventral spinal roots, spinal nerves and their branches, and ganglia. The primary function of the PNS is to convey sensory and motor information, including informa-

Molecular and Biochemical Toxicology, Fourth Edition, edited by Robert C. Smart and Ernest Hodgson
Copyright © 2008 John Wiley & Sons, Inc.

TABLE 30.1. Common Therapeutic Drugs and Environmental Agents Causing Neuropathy

Drugs	Environmental Agents
Amiodarone	Acrylamide
Chloroquine	Allyl chloride
Cisplatin	Arsenic
Colchicine	Buckthorn toxin
Dapsone	Carbon disulfide
Disulfiram	Chlordecone
Glutethimide	Dimethylaminopropionitrile
Gold compounds	Diphtheria toxin
Hydralazine	Ethylene oxide
Isoniazid	n-Hexane
Misonidazole	Methyl n-butyl ketone
Nitrous oxide	Lead
Nitrofuratoin	Mercury
Nucleoside analogs	Methyl bromide
(antiretrovirals)	
Paclitaxel (taxanes)	Polychlorinated biphenyls
Phenytoin	Tellurium
Pyridoxine (vitamin B6)	Tetrachlorobiphenyl (TCB)
Sodium cyanate	Thallium salts
Suramin	Trichloroethylene
Thalidomide	Vacor
Vincristine	Zinc pyridinethione

tion from the autonomic nervous system, between the central nervous system (CNS: brain and spinal cord) and the rest of the body. Single nerve fibers convey the information between the periphery and the CNS. The nerve fibers are bundled together to form individual peripheral nerves (Figure 30.2).

Each *nerve fiber* is composed of an axon and its supporting Schwann cells. The *axon* is a long, highly specialized cytoplasmic process of the neuron and conducts the nerve impulse. Axons measure 0.2 to 22 μm in diameter and may be over 1 m in length (Figure 30.3). The neuronal cell bodies giving rise to these peripheral nerve axons may be located in the CNS (e.g., spinal-cord motor neurons innervating skeletal muscle) or PNS (e.g., sensory neurons innervating skin and other organs). The sensory neurons of the dorsal root ganglia have an axon which branches, with one axonal process traveling to the CNS via a spinal nerve root and another axonal process traveling to the periphery via a peripheral nerve.

Schwann cells, which are supporting cells analogous to the glial cells of the CNS, envelop all axons of the PNS. The Schwann cells covering the larger-diameter axons show further specialization by covering the axon with a *myelin sheath*, which is an elaboration of each enveloping Schwann cell's plasmalemma (Figure 30.4). The myelin sheath permits much more rapid transmission of a nerve impulse by the nerve fiber. The differentiation of the Schwann cell into a myelinating Schwann cell is dictated by the axon.

The biochemistry of neurons and Schwann cells is discussed below. It is, however, important to realize that there are certain other structural features of the PNS which

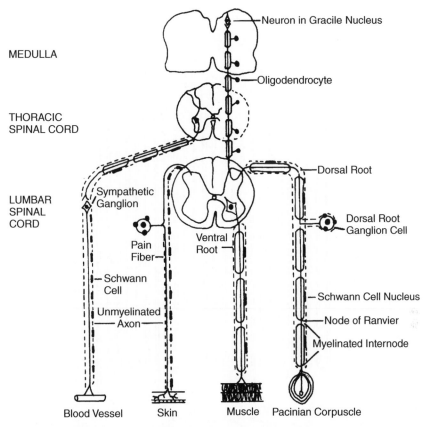

Figure 30.1. Peripheral nervous system. The major components of the PNS are illustrated in this diagram. (From H. H. Schaumburg et al., *Disorders of Peripheral Nerves*, F.A. Davis Co., Philadelphia, 1983.)

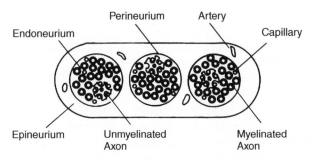

Figure 30.2. Peripheral nerve. This diagram shows a peripheral nerve in cross section. The nerve contains three bundles (fascicles), with each fascicle containing a mixture of myelinated and unmyelinated axons. (From H. H. Schaumburg et al. *Disorders of Peripheral Nerves*, F.A. Davis Co., Philadelphia, 1983.)

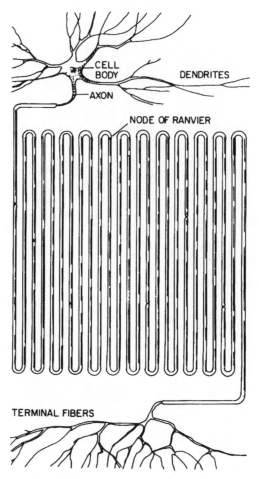

Figure 30.3. Neuron. This diagram is drawn to scale for a neuron with a 1-cm axon (the axon is folded for diagrammatic purposes). Many PNS axons may be more than a meter in length. (From C. F. Stevens. The neuron. *Sci. Am.* **241**, 54, 1979.)

greatly influence the effects of systemically administered toxicants. Of great importance is the fact that the PNS has a *blood–nerve barrier* (BNB), which is analogous to the blood–brain barrier (BBB) of the CNS. The BNB regulates the chemical composition of the extracellular (interstitial) fluid surrounding the individual nerve fibers within a peripheral nerve. The barrier is located at the walls of the nerve capillaries (endoneurial capillaries) and at the perineurium, which is a multilayered cellular sheath that surrounds each peripheral nerve. The endothelial cells lining the endoneurial capillaries form a "nonporous" barrier between the blood and the interstitial fluid within the nerve, while the perineurial cells form a nonporous barrier between the interstitial fluid outside and inside the nerve (Figure 30.2). These cellular barriers greatly restrict the movement of macromolecules, ions, and water-soluble nonelectrolytes, but not of lipid-soluble molecules, into peripheral nerve. In addition to these nonporous structural barriers, the endothelial cells of the BNB have carrier-mediated transport systems for selectively moving certain required molecules (e.g., glucose) across the barrier into the nerve. The combina-

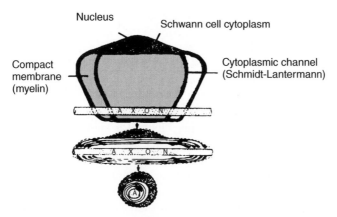

Figure 30.4. Myelinating Schwann cell of PNS. The same Schwann cell is shown unwrapped (top), in longitudinal section (middle), and in cross section (bottom). Note the channels of cytoplasm (Schmidt–Lantermann clefts) and the large expanses of compacted cell membranes (myelin). (These drawings are not to scale.) (From Raine, Morphology of myelin and myelination. In: P. Morell (Ed.). *Myelin*, 2nd ed., Plenum Press, New York, 1984.)

tion of these structural barriers and carrier-mediated transport systems permits the BNB to closely regulate the composition of the interstitial fluid surrounding the individual nerve fibers within each peripheral nerve.

The importance of the BNB in protecting the PNS from toxic injury is illustrated by the toxic neuropathy associated with diphtheria. Diphtheria is a human disease caused by an infection of a wound or the upper respiratory tract by the bacterium, *Corynebacterium diphtheriae.* The polypeptide toxin released by some strains of this bacterium is capable of entering the myelin-forming cells (Schwann cells of PNS; oligodendrocytes of CNS), inhibiting myelin synthesis, and producing breakdown of the myelin sheath (demyelination). It has long been recognized that the demyelination associated with diphtheria is remarkably localized to the ganglia of the PNS and completely spares the CNS. Experimental studies in animal models have revealed that the BBB, which effectively excludes the toxin from the CNS, is responsible for the absence of demyelination in the CNS. Likewise, the BNB excludes the toxin from much of the PNS. However, the ganglia of the PNS do not have a BNB and are exposed to the blood-borne diphtheria toxin. It is this absence of a BNB in the PNS ganglia that explains the localization of the demyelination to the PNS ganglia in diphtheria neuropathy.

30.2   SPECIALIZED ASPECTS OF NEURONAL METABOLISM

Discussion of vulnerability of the axon and its terminal to toxicants must be based on an appreciation of specialized aspects of the metabolism of the neuron. This cell type has been discussed in the chapter dealing with the CNS. The concepts of chemically mediated neurotransmission, second messenger systems, specialized ion channels, and conduction of waves of depolarization with consequent high demand for energy to restore Na^+ and K^+ ion gradients are applicable to the PNS as well as the CNS. Of particular relevance is the fact that the neuromuscular junction is cholinergic and there is a wide range of toxic agents that act at this site; for example, many insecticides such as organophosphorous compounds inhibit cholinesterase.

Another aspect of the PNS which renders its neurons particularly sensitive to certain types of toxic insult is the sheer size of many of the motor and sensory neurons. Neurons have a highly asymmetric distribution of cytoplasm and plasma membrane; in the larger PNS neurons the volume of the axon and its surface area exceeds that of the cell body by several orders of magnitude (Figure 30.3). The cell body, however, contains most of the machinery for synthesis and degradation of macromolecules. Thus, maintenance of the vast expanse of the axon and of the distant and metabolically active nerve terminals is absolutely dependent on the continued delivery (*axonal transport*) of materials to and from the cell body. The processes involved in axonal transport involve mechanisms similar to those utilized by all eukaryotic cells to transport materials from their site of synthesis to their site of utilization and, eventually, to their site of degradation. However, the unique structure of neurons makes the significance of these transport processes, relative to other metabolic processes, much greater than in most other cell types. It is considered likely that many toxicant-induced metabolic insults may directly or indirectly affect axonal transport. Thus, this process is a focus of many studies exploring the mechanisms by which certain toxicants bring about axonal degeneration.

30.2.1   Anterograde Axonal Transport

Anterograde axonal transport (movement from the cell body to the nerve ending) involves movement of macromolecules at one of several relatively discrete velocities. The extremes of this spectrum are referred to respectively as rapid and slow axonal transport. *Rapid transport* involves the movement of cytoplasmic vesicles at rates from 200 to 400 mm/day. Included are vesicles involved in neurotransmission (and containing, among other components, various enzymes of neurotransmitter metabolism), vesicles presumed to be carrying various surface membrane components, and agranular endoplasmic reticulum. It has long been suspected that microtubules form the framework on which such vesicles move, and it has now been clearly demonstrated that a microtubule-stimulated kinase, kinesin, is involved in conversion of the energy of ATP to anterograde movement of vesicles along the microtubules. Mitochondria, and possibly certain other subcellular organelles, are transported at slower net rates than are the rapidly transported vesicles. This is explained without presumption of another anterograde transport system. These slower rates of transport are explained by variation in the time of association with the transport system among different classes of particles. Thus, vesicles carrying certain enzymes involved in neurotransmitter synthesis remain strongly associated with the transport vector and move rapidly while mitochondria spend less time attached to the vector and therefore have a slower net transport rate.

Slow axonal transport involves primarily components of the cytoskeleton and its protein precursors. Slow transport can be resolved into two rates, one of less than 1 mm/day and one of several mm/day. Components traveling at the slower rate include tubulin and its associated proteins, along with a group of three other structural proteins known as the neurofilament triplet. Components traveling at the faster rate include some tubulin, as well as actin and various soluble enzymes. It has been noted that the rate of rapid transport is constant in an animal over a range of axon types and is relatively constant among various species if correction is made for temperature differences; in contrast, slow transport has some dependence on the length or diameter of the axon.

The nature of the slow transport vector and its mechanism has been the subject of intense debate for decades, but a consensus has recently been reached. Direct visualization of the movement of cytoskeletal elements within axons has demonstrated that these structures have an instantaneous transport rate that is equivalent to rapid transport and that is powered by the fast transport motors. Like mitochondria, the rapid movement of cytoskeletal elements is interrupted by long stationary periods, leading to a very slow net transport rate. Thus there is now a unified theory of all of anterograde axonal transport.

30.2.2 Retrograde Axonal Transport

Retrograde axonal transport (from nerve ending back to cell body) has been studied in a number of systems and consists of a single phase of transport of vesicles at a rate slightly more than half that of rapid anterograde transport. Another microtubule-stimulated kinase, dynein (distinct from kinesin), is specifically involved in using energy from ATP to bring about movement of vesicles from the nerve terminal back to the cell body. The retrogradely traveling vesicles are different from those moving anterogradely and appear related to those formed in connection with endocytic and degradative pathways. Both anterograde and retrograde transport are dependent on local oxidative metabolism within the axon for a continued supply of ATP and have certain requirements for calcium ions.

30.3 SPECIALIZED ASPECTS OF SCHWANN CELL METABOLISM

Schwann cells in contact with axons may undergo further specialization and make myelin. Myelin is a greatly extended and modified plasma membrane wrapped around the axon in a spiral fashion (Figure 30.4). Myelin accounts for almost half of the protein and an even larger proportion of the lipid of the sciatic nerve of mammals (because of its size and accessibility, this nerve is the object of most studies of PNS myelination). As is the case in the CNS, much of the myelin in the PNS is deposited in a restricted period of development; for example, axons in the sciatic nerve of the rat go from being unmyelinated at birth to being well-myelinated at two weeks of age. It should be noted that in many cases, rapid synthesis of myelin continues well beyond the time when, by morphological criteria, much of the deposition of myelin appears to be nearing completion. This is because body growth continues long after peripheral nerves are myelinated and functional. As limbs grow, so do the peripheral nerves, and the Schwann cells must produce new myelin to accommodate the elongating peripheral nerves.

Schwann cells share with neurons the characteristic of having a cell body that must support an enormous amount of peripherally located cell components: myelin in the case of the Schwann cell, and axon in the case of the neuron (Figures 30.3 and 30.4). Thus, as in the case of axonal transport in neurons, it is assumed that highly specialized metabolic processes are involved in supporting the topologically distant myelin. In addition, myelin itself has a unique composition and structure (see Section 30.3.1), suggesting additional possible sites of action of toxicants.

30.3.1 Myelin Composition and Metabolism

Myelin is modified plasma membrane. Myelin of the PNS resembles that of the CNS with respect to lipid composition. There is an enrichment in such specialized lipids as cerebroside and ethanolamine plasmalogen, and the high content of cholesterol plays an important role in control of membrane fluidity. The protein composition of PNS myelin is, however, distinct from that of CNS myelin. A single protein, P_0, accounts for half of all protein of PNS myelin. Of the other proteins present, most are expressed in the CNS as well as the PNS but in quantitatively different amounts. Prominent among these proteins are myelin basic proteins and myelin-associated glycoprotein.

Both the lipid and protein components of the myelin sheath are involved in metabolic turnover. Although this metabolism is less vigorous than that of the plasma membrane of other cell types, myelin is far from being metabolically inert. It is now known that there are certain components of myelin which turn over extremely rapidly; prominent in the PNS are the phosphate groups of P_0 protein and of the polyphosphoinositides. It should also be noted that the tight apposition of the cytoplasmic faces of the membranes (viewed in the electron microscope as the "dense" line of the repeat structure) is often split by a cytoplasmic channel (Schmidt–Lantermann cleft), which may be a metabolically active compartment. The possibility that this region of cytoplasmic contact with myelin is active metabolically is a relatively new concept, since for many years it was thought that myelin was, metabolically and structurally, a highly stable structure. Regions of metabolic activity in the membrane may also be suggestive of pores or channels, and this hypothesis is reinforced by data indicating the presence in compact myelin of carbonic anhydrase and other enzymes which might be related to transport. The observation, detailed below, that certain compounds may induce vacuolation (edema) in myelin raises the possibility that these compounds may act by interfering with some mechanism of water and/or ion transport involved in the structural stability of myelin membrane.

30.4 TOXIC NEUROPATHIES

30.4.1 Selective Vulnerability

The neuron and the Schwann cell are the principal cell types in the PNS. There are great morphological, biochemical, and functional differences between neurons and Schwann cells, and this is reflected in the considerable variation in their vulnerability to toxic injury. Some toxic neuropathies are characterized primarily by injury of the neuron, its axon, or its terminal, as evidenced by the presence of axonal degeneration in peripheral nerve, while other toxic neuropathies are characterized primarily by Schwann cell injury, as evidenced by the presence of demyelination. Those neuropathies characterized by axonal injury are often categorized as "axonal neuropathies," whereas those characterized by demyelination are categorized as "demyelinating neuropathies."

30.4.2 Characteristics of Axonal Neuropathies

The vast majority of neurotoxic agents that affect the PNS preferentially cause axonal injury rather than Schwann cell injury. Axonal injury is usually manifested as *axonal degeneration*, a pathologic process characterized by complete dissolution of the axon. If the degeneration involves a myelinated axon, then the myelin sheath enveloping the degenerating axon also breaks down. This myelin breakdown, which occurs in the context of axonal degeneration, is not considered demyelination, since "demyelination" refers to loss of the myelin sheath from an intact axon.

Detailed studies of the PNS in animal models of toxic neuropathy reveal that in most instances the axonal degeneration initially involves only the distal end of the axon, with the more proximal axon and neuronal cell body remaining intact (Figure 30.5B). In addition, these studies reveal that usually this distal axonal degeneration is initially limited to the longest and largest-diameter axons, which are those innervating the distal extremities. This distal axonal degeneration is often referred to as a "dying-back" type of axonal degeneration because continued exposure to the toxic agent results in progression of the degeneration to more proximal portions of the axon and may eventually lead to degeneration of the entire axon and the neuronal cell body. Fortunately, if exposure to the toxic agent is ended before the axonal degeneration extends centripetally to involve the proximal axon and neuronal cell body, the intact proximal axon may regenerate and extend distally to reestablish contact with the periphery and permit return of nerve function (Figure 30.5C). This potential for axonal regeneration and return of function is unique to the PNS; there is no significant regeneration of axons within the CNS.

Some of the postulated mechanisms for this increased vulnerability of the distal ends of the largest and longest axons to toxicant-induced axonal degeneration are discussed in the following sections. Of considerable interest is whether the dying-back axonal degeneration is due to direct toxic injury of the neuronal cell body or its axon. Because the neuronal cell body is responsible for maintaining both itself and its axon, some hypotheses suggest that the toxic agent initially compromises the metabolism of the cell body, such that the cell body can no longer adequately maintain its entire axon. It is further reasoned that the distal axon, being farthest removed from the neuronal cell body, would be most severely affected by the metabolic compromise and would be the first part of the neuron to degenerate.

Competing hypotheses suggest that the distal axonal degeneration is due to the toxic agent having a direct effect on the axon or its terminal. The vulnerable region might be at the distal axon, thus directly accounting for the localization of the degeneration to the distal axon. It is also possible that the deleterious effect of the toxicant (e.g., destruction of an axolemmal component) is uniformly expressed along the axon but that the limited capacity of axonal transport to replace the damaged component results in the more distal axon being less likely to be repaired.

A few toxic agents, such as mercury, initially produce degeneration of the neuronal cell body rather than the distal axon (Figure 30.5D). These toxic neuropathies are sometimes referred to as "neuronopathies" to emphasize that the cell body of the neuron is the initial site of degeneration. Degeneration of the axon also occurs in these toxic neuronopathies, since the axon is totally dependent on the neuronal cell body for survival. In contrast to dying back neuropathies, neuronopathies have no potential for axonal regeneration and recovery of function, since an intact cell body is a prerequisite for axonal regeneration.

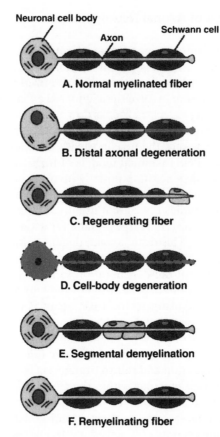

Figure 30.5. (A) Normal myelinated axon in PNS. (B) Distal axonal degeneration (dying-back neuropathy). The distal end of the axon has degenerated. The myelin covering the degenerated axon has also broken down as a consequence of the axonal degeneration. The proximal axon, its overlying myelin sheath, and the neuron's cell body remain intact. (C) Regenerating myelinated axon. Subsequent to distal axonal degeneration, there is potential for regeneration of the axon and the return of nerve function. Note that Schwann cells remyelinate the regenerated axon. (D) Cell-body degeneration (neuronopathy). The neuron's cell body and its entire axon have degenerated. There is no potential for regeneration when the cell body degenerates. Note that the myelin also breaks down as a consequence of the degeneration of the underlying axon. (E) Demyelination. Several myelin internodes (segments), each representing the myelin of one Schwann cell, have undergone degeneration (segmental demyelination). Note that the underlying axon remains intact despite breakdown of the myelin internodes. (F) Remyelinated axon. Subsequent to demyelination, there is remyelination of the segments of demyelinated axon. The remyelinated segments characteristically have shorter internodal distances than the original myelin internodes. Remyelination permits return of function to the affected axon.

A variety of morphologic abnormalities in the axon or neuronal cell body may precede axonal degeneration in toxic neuropathies. These associated morphologic abnormalities are often very characteristic of a particular toxic agent or group of agents and are of great interest because they may give clues as to the basic mechanisms by which a particular toxic agent produces axonal injury and degeneration.

Among the best studied of these morphologic abnormalities associated with toxic neuropathies are the large masses of neurofilaments that accumulate locally within axons during intoxication with 2,5-hexanedione (2,5-HD) and related γ-diketones, as well as with acrylamide, β,β-iminodipropionitrile (IDPN), and carbon disulfide.

30.4.3 Biochemical Mechanisms of Axonal Degeneration

The following sections discuss the major categories into which PNS toxicants have traditionally been grouped and the major hypothetical mechanisms of action. These categories are not mutually exclusive; a toxicant may manifest characteristics of more than one type and have more than one mechanism of action. The effects manifested by a toxicant may also be dose-dependent, further complicating easy categorization. The discussion in the following sections is meant more to give insight as to how relevant hypotheses are framed rather than to imply that the mechanisms of action of toxicants inducing neuropathies are well understood.

30.4.3a Inhibitors of Energy Production. Most of the published biochemical investigations of toxicant-induced axonopathies are interpreted in terms of the specialized aspects of neuronal metabolism. The preferential vulnerability of neurons with very long axons is of special relevance to this focus. These neurons have a great metabolic demand to maintain the ion gradients necessary for nerve conduction. The other main energy demand is for synthesis and axonal transport of the macromolecules needed for maintenance of the axon and nerve endings. It is well known that the process of rapid axonal transport is absolutely dependent on production of ATP, which in turn is dependent of glycolysis, oxidative phosphorylation, and oxygen.

Based on these current concepts of energy requirements in the axon, a hypothesis has been proposed to account for the preferential vulnerability of the distal regions of long axons to certain toxicants. It has been suggested that the supply of enzymes for glycolysis may be a rate-limiting step in energy production because the glycolytic enzymes must be supplied to the axon by slow axonal transport. If a particular toxicant has partially inactivated an enzyme, the cell body may be unable to respond with an adequate increase in production and axonal transport of the needed enzyme, causing the distal region of the axon to become energy-deficient. This hypothesis assumes that the demand for production of energy is relatively constant along the axon and that successively fewer enzyme molecules are delivered to progressively distal regions. Although postulated to be the mechanism of action of several neurotoxicants causing distal axonal degeneration, this hypothesis has not been adequately tested for any particular set of compounds. In this connection, a promising model utilizes the rodenticide Vacor (*N*-3-pyridylmethyl-*N'*-*p*-nitrophenylurea, a structural analog of nicotinamide). Ingestion by humans brings on the onset of neuropathy within hours. Studies with rats have shown that although axonal transport does not appear affected in the proximal region of the sciatic nerve, there is a considerable reduction of material reaching the distal nerve regions. It is assumed that the block in axonal transport is connected with the development of neuropathy and that it could well be causal. The defect in axonal transport is not known, but the possibility that this nicotinamide analog might perturb energy metabolism by inhibiting NAD-NADH utilizing enzymes is of interest. Relevant to this argument are the recent observations that the timing and extent of axonal degeneration following physical damage to axons in culture is critically dependent on NAD levels within the axon.

30.4.3b Inhibitors of Protein Synthesis.

As mentioned above, some neurotoxicants initially produce degeneration of the neuronal cell body rather than the distal axon. These neuropathies are referred to as "neuronopathies." In some cases the toxic agent appears to act by inhibition of protein synthesis. Examples include the antimitotic doxorubicin and the plant lectin ricin. Doxorubicin binds to DNA and blocks RNA transcription, while ricin binds to the 60s ribosomal subunit and blocks RNA translation. In both cases the resulting block in protein synthesis leads to neuronal degeneration. Doxorubicin cannot cross the BNB, so that the neuropathy is restricted to those regions of the PNS where there is no barrier (see Section 30.1). Ricin is unique in that the toxin gains access to the neuronal soma by being taken up by the axon and carried by retrograde axonal transport to the cell body. Ricin neuropathy is thus restricted to those neurons whose axons have taken up ricin.

Methylmercury and trimethyltin also produce neuronopathies. Methylmercury inhibits protein synthesis, but experimental evidence suggests that additional or alternative mechanisms must exist to account for the neuronal degeneration. The mechanism of action of trimethyltin is unknown.

30.4.3c Structural Components of the Axon as Sites of Vulnerability.

The structural components intrinsic to the axonal transport system are also a point of vulnerability. A prominent example of this is the sensorimotor neuropathy that may complicate treatment of leukemia or lymphoma with the antineoplastic agent vincristine. This alkaloid functions as a mitotic spindle inhibitor by binding to tubulin. This same action of binding to tubulin leads to inhibition of axonal transport in experimental systems and could account for the neuropathy in humans. Two other major structural components of the axon that are potential targets of neurotoxicants are discussed below.

Neurofilaments. Neurofilaments are presumed to be the target site for several neurotoxicants of environmental significance. Exposure to certain toxicants induces focal accumulations of neurofilaments within neurons and their processes. Accumulation may be within the cell body, or in proximal, middle, or distal regions of the axon. Within myelinated axons, these accumulations of neurofilaments generally occur at multiple, paranodal sites and produce axonal swellings. This common feature of multifocal accumulations of neurofilaments suggests that these toxicants share a common mechanism of action.

Based on the observation that many of the toxicants causing axonal accumulations of neurofilaments react covalently with proteins, including neurofilament proteins, a "unified" hypothesis has been advanced to explain the pathogenesis of this class of neuropathies. Briefly, it is suggested that certain toxicants cause a specific covalent modification of neurofilament proteins. These modifications are postulated to destabilize the cytoskeletal framework. As the neurofilaments become increasingly modified, they eventually become nontransportable and accumulate. The exact position along the nerve at which transport fails would depend on the reactivity, concentration, and time course of administration of the toxicant. The somewhat different ultrastructural pathology seen with each of these agents is presumed to be a result of the specific covalent modification of neurofilaments that each toxicant makes.

This hypothesis is best supported in the case of the hexacarbons. Hexacarbons can react both in vitro and in vivo with lysine ε-amino moieties of proteins to yield pyrrole adducts, which permit secondary auto-oxidative crosslinking. It has been postulated that pyrrole derivitization and secondary crosslinking of neurofilaments are the initiating events in hexacarbon neuropathy. Although this mechanism has been challenged, many studies support this mechanism for the hexacarbons. However, this hypothesis offers no explanation for the early effects on retrograde transport reported in hexacarbon neuropathy.

Axonal Smooth Membranes. Several neurotoxicants, including *p*-bromophenyl-acetylurea, zinc pyridinethione, acrylamide, and certain organophosphorous compounds, cause distal axonal swellings composed partly or entirely of accumulations of vesiculotubular membranous structures derived from the endoplasmic reticulum. These neuropathies manifest a distal "dying-back" degeneration of nerve fibers. Although a range of axonal transport abnormalities have been reported in these neuropathies, it is unknown whether these accumulations of membrane reflect a direct effect of the toxicant on axonal membranes, a direct effect on the transport mechanism, or merely a nonspecific reactive response of the axon. For acrylamide it has recently been hypothesized that it acts through forming protein adducts that disrupt transmitter release at nerve terminals.

30.4.3d   Schwann Cell Effects on Axons. Myelinating Schwann cells exert considerable control over the axons which they ensheath, affecting the number of neuro-filaments and their degree of phosphorylation, as well as mediating Na^+ channel localization at nodes of Ranvier. Alterations in Schwann-cell integrity due to various insults or diseases similarly produce alterations in axonal integrity and function, including changes in axonal diameters and neurofilament spacing and phosphorylation, as well as abnormal axonal transport. In addition, axons associated with abnormal Schwann cells are prone to degeneration, particularly at their distal ends.

30.4.4   Characteristics of Demyelinating Neuropathies

In contrast with the large number of toxic agents that produce axonal degeneration in the PNS, only a few agents selectively cause Schwann-cell injury. Toxic injury to Schwann cells is restricted, with rare exceptions, to Schwann cells ensheathing myelinated axons and is manifested morphologically as *demyelination*—that is, loss of myelin from an intact axon (Figure 30.5E). Less commonly, the toxic injury results in fluid accumulation within the myelin sheath (intramyelinic edema). The degenerating myelin is catabolized by Schwann cells and by macrophages that are attracted to the degenerating myelin. Although the demyelinated axons retain their connections with the periphery, the loss of continuity of the myelin sheath results in loss of the axon's ability to convey nerve impulses. Shortly after the onset of demyelination, Schwann cells begin proliferating and cover the demyelinated segment of axon. Within a few days, these Schwann cells begin producing a new myelin sheath to cover the demyelinated segments. If there is cessation of exposure to the toxic agent, the process of demyelination ends and the reparative process of remyelination is able to restore function to the demyelinated nerve fibers (Figure 30.5F).

30.4.5   Biochemical Mechanisms of Demyelination

Several aspects of Schwann cell metabolism emerge as potential points of vulner-ability to toxicants. The Schwann cell perikaryon (cell body) supports an enormous peripheral structure, the myelin sheath, which, if unwrapped, would dwarf the body of the Schwann cell (see Figure 30.4). Thus, as in the case of axonal transport in neurons, there may be specialized processes involved in supporting the topologically distant myelin. Furthermore, myelin has a specialized lipid and protein composition and a relatively rigid and ordered structure as compared to other membranes. Meta-bolic perturbations that potentially cause alterations in the composition of lipids and proteins assembled to form this membrane may cause destabilization and collapse of the myelin membrane. In this context, myelin might be much more vulnerable than plasma membranes of other cells.

The toxic agents that selectively damage myelin can usefully be separated into (a) those that bring about myelin alterations without apparent injury to the myelina-ting cell and (b) those that injure both the myelinating cell and its myelin. Triethyltin and hexachlorophene are examples of the former. These neurotoxicants character-istically produce a reversible vacuolation (edema) of the myelin sheath through splitting of the myelin lamellae at the intraperiod line. The intramyelinic edema is much more prominent in CNS myelin than in PNS myelin. The specificity of these myelinotoxic agents suggests that these toxicants are located primarily in myelin (this would also be consistent with the known lipid solubility of triethyltin and hexachlorophene). It has been suggested that triethyltin produces edema by a direct action on myelin membranes, analogous to its action in mitochondria where it acts as an anion exchanger and collapses the proton gradient across the mitochondrial membrane. The specific mechanisms, however, by which these myelinotoxic agents produce intramyelinic edema are not yet established.

Diphtheria toxin, inorganic lead, and tellurium are considered toxicants that cause demyelination through injury to the myelinating cell. The mechanism of action of diphtheria toxin was discussed in Section 30.1. The mechanism by which inorganic lead causes Schwann cell injury and demyelination is not well understood, but may be related to uncoupling or inhibition of oxidative phosphorylation secondary to interference of lead with some aspect of ion transport across the mitochondrial membrane.

The study of tellurium neuropathy has provided some insights as to how a toxic agent may cause demyelination. Inclusion of elemental tellurium in the diet of 20-day-old rats results in a rapid demyelination of peripheral nerves which is maximal at about five days after starting the diet (the CNS does not show demyelination). Available evidence suggests that the active agent is tellurite (Te^{4+}) and that its primary metabolic action is inhibition of the conversion of squalene to squalene epoxide, thereby causing marked accumulation of squalene and almost complete inhibition of cholesterol synthesis. The inhibition of the squalene epoxidase reaction occurs in all tissues including brain, raising the question as to why the demyelination is limited to the PNS. An explanation may lie in the fact that, in young animals, the rate of membrane deposition and of cholesterol synthesis is normally much higher in peripheral nerve than in brain and other tissues. Thus, the rate of accretion of membranes deficient in cholesterol is more rapid in PNS myelin than in brain myelin. Furthermore, plasma membrane systems of other cells are presumably not as vulnerable to a deficiency in cholesterol because they are not as highly ordered and cholesterol-rich as myelin.

30.4.6   Molecular Approaches to Assessing Neurotoxic Insults to the PNS

Behavioral and functional assessment of neurotoxic insults will continue to be of value, as will examination of morphological alterations. The same will continue to be true for metabolic and molecular biological studies examining specific metabolites and/or steady-state mRNA levels of target genes. However, the development and standardization (in recent years) of powerful tools of "global" cellular/molecular analyses, specifically genomics, proteomics, and metabolomics (sometimes collectively grouped as "trinomics"), allows collection of a wealth of additional information regarding the actual molecular bases of these insults. The challenge, of course, is to properly analyze and interpret the vast amount of data available from these approaches. The "global" nature of such collective data has the potential to reveal previously undetected interactions among various genes, their protein products, and/or levels of related metabolites. Ultimately, a neurosystems biology approach, which integrates genomic, transcriptomic, proteomic, and metabolomic information with cellular and organismal (anatomical, electrophysiological, and behavioral) analyses, should allow for overall modeling of normal and abnormal nervous system function. These approaches should also be useful in elucidating potential "biomarkers" for perturbations of the Schwann-cell–axon relationships critical for normal nervous system function, as well as suggesting potential therapeutic approaches to minimize damage and/or promote repair of injured nervous tissue. While global gene and protein expression profiles do have considerable potential, more detailed follow-up studies will always be necessary to confirm and fully document the suggested alterations.

Such a multifaceted approach to the full spectrum of PNS metabolism and function should strengthen our ability to explore the critical features of the cellular elements that comprise the PNS, their vulnerability to insult, and the consequences of any such perturbations on normal PNS function. In addition, because the PNS has a much greater potential for regeneration and functional recovery than the CNS, perhaps this better understanding of the degenerative and regenerative events in the PNS will be useful in developing strategies to promote regeneration and recovery of function in the damaged CNS.

30.5   CONCLUSION

The intent of this chapter is to provide an appreciation for the unique features of the PNS, which may make it vulnerable to toxicant-induced injury, and to discuss possible ways in which that injury may come about. The reader should remember that the categories and conceptualizations used in this chapter (e.g., axonal neuropathy versus demyelinating neuropathy; toxicants affecting energy metabolism versus structural components) are not mutually exclusive. Few toxicants have such specific effects that they fall exclusively in one category. Placing a toxicant in one category or another is necessary and useful for framing testable hypotheses, but this categorization should not prevent the consideration of alternative possibilities. We would also reemphasize that in very few cases have the specific mechanisms of action been well established for PNS toxicants. Furthermore, in those instances where a specific biochemical alteration has been demonstrated (e.g., crosslinking of proteins by hexacarbons, or inhibition of cholesterol synthesis by tellurium), a causal link to the neuropathy is yet to be firmly established. The recent development of methodologies for "global analyses" of gene and protein expression should be helpful, both in gaining understanding of mechanisms and in formulating approaches for screening agents that may be neurotoxic. The references cited below are intended to provide the reader with an introduction to the literature on neuronal (axonal) and Schwann cell (myelin) metabolism and on PNS toxicants.

ACKNOWLEDGMENTS

The authors gratefully acknowledge the important contribution of the late Dr. Pierre Morell, to this chapter. Dr. Morell co-authored this chapter in previous editions of this textbook.

SUGGESTED READING

Allt, G., and Lawrenson, J. G. The blood–nerve barrier: Enzymes, transporters and receptors— a comparison with the blood–brain barrier. *Brain Res Bull.* **52**, 1–12, 2000.

Brown, A. Axonal transport of membranous and nonmembranous cargoes: A unified perspective. *J. Cell Biol.* **160**, 817–821, 2003.

Coleman, M. Axonal degeneration mechanisms: Commonality and diversity. *Nat. Rev. Neurosci.* **6**, 889–898, 2005.

Dunckley, T., Coon, K. D., and Stepahan, D. A. Discovery and development of biomarkers of neurological disease. *Drug Discovery Today* **10**, 326–334, 2005.

Dyck, P. J., and Thomas, P. K. (Ed.). *Peripheral Neuropathy*, 4th ed. Elsevier Saunders, Philadelphia, 2005.

Graham, D. Neurotoxicants and the cytoskeleton. *Curr. Opin. Neurol.* **12**, 733–737, 1999.

LoPachin, R. M., and DeCaprio, A. P. Protein adduct formation as a molecular mechanism of neurotoxicity. *Toxocol. Sci.* **86**, 214–225, 2005.

Martini, R. The effect of myelinating Schwann cells on axons. *Muscle and Nerve* **24**, 456–466, 2001.

Morfini, G. A., Stenoien, D. L., and Brady, S. T. Axonal transport. In: Siegel, G. J., Albers, R. W., Brady, S. T., and Price, D. L. (Eds.). *Basic Neurochemistry—Molecular, Cellular and Medical Aspects*, 7th ed., Elsevier, Amsterdam, pp. 485–501, 2006.

Pleasure, D. Peripheral neuropathy. In: Siegel, G. J., Albers, R. W., Brady, S. T., and Price, D. L. (Eds.). *Basic Neurochemistry—Molecular, Cellular and Medical Aspects*, 7th ed., Elsevier, Amsterdam, pp. 617–628, 2006.

Quarles, R. H., Macklin, W. B., and Morell, P. Myelin formation, structure and biochemistry. In: Siegel, G. J., Albers, R. W., Brady, S. T., and Price, D. L. (Eds.). *Basic Neurochemistry— Molecular, Cellular and Medical Aspects*, 7th ed., Elsevier, Amsterdam, pp. 51–71, 2006.

Spencer, P. S., and Schaumburg, H. H. (Eds.). *Experimental and Clinical Neurotoxicology*, 2nd ed. Oxford University Press, New York, 1999.

Toews, A. D., Harry, G. J., and Morell, P. Schwann cell neurotoxicity. In: Aschner, M., and Costa, L. G. (Eds.). *The Role of Glia in Neurotoxicity*, 2nd eds. CRC Press, Boca Raton, FL, pp. 41–59, 2005.

Toews, A. D., and Morell, P. Molecular biological approaches in neurotoxicology. In: Tilson, H. A., and Harry, G. J. (Eds.). *Neurotoxicology*, 2nd eds. Taylor and Francis, Philadelphia, pp. 1–35, 1999.

Umapathi, T., and Chaudhry, V. Toxic neuropathy. *Curr. Opin. Neurol.* **18**, 574–580, 2005.

Waters, M. D., Olden, K., and Tennant, R. W. Toxicogenomic approach for assessing toxicant-related disease. *Mutat. Res.* **544**, 415–424, 2003.

■■■■ CHAPTER 31

Biochemical Toxicology of the Central Nervous System

BONITA L. BLAKE

Departments of Pharmacology and Psychiatry, University of North Carolina at Chapel Hill, Chapel Hill, North Carolina 27599

31.1 INTRODUCTION

The central nervous system (CNS) consists of the brain and the spinal cord. Throughout, the system works on a basic mechanism of excitation balanced with inhibition. These functions are carried out primarily by the amino acid neurotransmitters glutamate and GABA, respectively. Other neurotransmitters, such as acetylcholine and dopamine, are somewhat more localized in different regions; these allow for a diversification of CNS functions, such as movement control or emotional responses. In addition to the neurotransmitters, peptides (e.g., endorphin), small molecules (e.g., nitric oxide), and even immune mediators (e.g., cytokines) modulate neural activity. The mechanisms by which the nervous system processes chemical and electrical information have been covered in Chapter 16 of *A Textbook of Modern Toxicology* (see Suggested Reading), which the reader is encouraged to review for background on this topic.

Neural tissue contains many different types of cells. Because each is so specialized in its structure and function, the cellular responses to a single neurotoxicant can be very different. In addition, each type of cell communicates in network fashion with the other types, so that damage to one can affect the function of the others. This cellular interplay contributes to the principle that the effects of a single large dose of neurotoxicant can be very different from that caused by multiple small doses. These issues often make it difficult to identify and delimit the primary effects of toxicants in the nervous system. In fact, neurotoxic effects may be so subtle as to go unrecognized; yet, in combination with other factors such as age, genetics, or stress, they contribute to permanent neurodegeneration. This chapter attempts to summarize the current understanding of neurotoxicant action in the CNS, placing it in the context of multiple interacting factors that affect human health.

Molecular and Biochemical Toxicology, Fourth Edition, edited by Robert C. Smart and Ernest Hodgson
Copyright © 2008 John Wiley & Sons, Inc.

31.2   CNS SITES OF TOXIC ACTION

31.2.1   Neuronal Targets

Some neurons are more sensitive than others to the effects of a variety of toxicants; that is, they display a selective vulnerability to neurotoxicants. For example, mitochondrial respiratory complex inhibitors such as cyanide and 3-nitropropionic acid are toxic to all cell types; yet within the CNS, neurons in the basal ganglia (a group of regions that collectively control motor behavior) appear to be particularly sensitive to these agents. In most cases, selective vulnerability to neurotoxicants arises because of a unique combination of factors that predispose a cell type or region to particular insults. These factors may include the presence of certain ion channels, receptors or uptake sites, the activity level of xenobiotic metabolizing or antioxidant enzymes, the expression profile of neurotrophic factors or their receptors, and so on. Three CNS sites highly vulnerable to neurotoxicant effects are described separately below.

Hippocampus. Highly vulnerable CNS neurons are often large, with long myelinated axons that extend from one brain region to another. In general, neurons with these characteristics have a larger surface area and a high energy requirement, and they are critically dependent on axonal transport and cytoskeletal integrity. The pyramidal neurons of the hippocampus belong to this class. Pyramidal neurons also contain very little of the calcium-binding protein calbindin, a protective factor in many neurons. Neighboring granule neurons, on the other hand, express high levels of calbindin and are more resistant to damage cause by seizures, ischemia (loss of blood supply), and the effects of Alzheimer's disease. As will be described in a later section, calcium is a major factor in neuronal destruction, and the ability to buffer calcium often determines survival or death.

The hippocampus is selectively targeted by the neurotoxicant trimethyltin (TMT), although other regions that have connections with the hippocampus may be affected to a lesser extent. Within this region, inflammation and degeneration of pyramidal cells occurs at low doses of TMT, with effects on granule cells at higher doses. The membrane protein stannin, isolated from neurons damaged by TMT, is expressed preferentially in the TMT-sensitive neurons of the hippocampus, as well as in the kidney and immune system (also selectively targeted by TMT). Antisense oligonucleotides to stannin protect cultured hippocampal cells from TMT-mediated neurodegeneration, suggesting that stannin expression is necessary for the development of toxicity. Hippocampal neurotoxicity by TMT may also occur indirectly by the subsequent inflammatory response mounted by activated microglia.

Cerebellum. Two cerebellar neuron types, Purkinje cells and granule cells, have very different characteristics and are differentially, but highly, vulnerable to chemical insults. Purkinje neurons are very large cells that synthesize and release GABA and are the only neurons in the cerebellum that send their processes to other regions. Granule neurons are small neurons that release glutamate to activate the Purkinje cells. Both Purkinje and granule neurons express receptors for glutamate; however, the subunit composition of N-methyl D-aspartate (NMDA, named for a compound to which it binds) type of glutamate receptor is different. NMDA receptors are critical to glutamate neurotransmission. The subunits confer different properties to the sensitivity and activity of NMDA receptors; this is thought to be part of the reason for the differential susceptibilities of these neurons to toxicants.

Purkinje Neurons. Purkinje neurons develop prior to birth, whereas granule cells develop postnatally, and then migrate to a specific layer within the cerebellum. The late migration of granule cells, combined with low levels of calcium-binding proteins, antioxidants, and DNA repair enzymes, make these neurons among the most sensitive in the entire CNS to the effects of a variety of xenobiotics. These include metals and alkyl metals (e.g., methylmercury), halogenated hydrocarbons of various structural classes (e.g., 2-halopropionic acids, methylhalides, organochlorine pesticides, polychlorinated biphenyls), and organophosphates. This temporal difference in cerebellar neuron development also means that certain developmental toxicants such as methoxyazomethanol (MAM), when administered in utero, selectively affect Purkinje cells. Alternatively, perinatal or postnatal exposures are more likely to affect granule cells.

Basal Ganglia. The area known as the basal ganglia (Figure 31.1) consists of a circuit of interconnected brain regions involved in motor control. Among them, the substantia nigra is a region that sends a dense supply of dopamine to other areas of the brain. Dopamine neurons are particularly sensitive to inhibitors of mitochondrial respiration. The caudate putamen receives most of the dopamine that originates from the substantia nigra. The GABAergic (i.e., neurons producing and

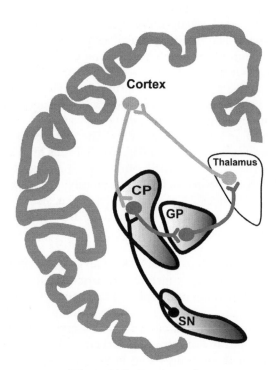

Figure 31.1. Basal ganglia.

releasing GABA) neurons of the globus pallidus and caudate putamen are highly vulnerable to insult, even more so than their neighboring cells, the acetylcholine-releasing interneurons. Interneurons are cells that exist entirely within a region to modulate the activity of neurons that send axons to other areas (projection neurons). The loss of oxygen or glucose and other metabolic insults are particularly deadly to GABAergic projection neurons. Part of this susceptibility lies in their intrinsic membrane properties that cause them to depolarize when ATP production is low. While interneurons express specific potassium channels that allow them to maintain their resting membrane potential when energy is depleted, GABAergic projection neurons lack these channels. Consequently, the loss of membrane potential, resulting in irreversible depolarization in the projection neurons, does not occur in interneurons. Like the cerebellar neurons, differential sensitivity to glutamate is also a factor in the enhanced vulnerability of GABAergic projection neurons. Because of their differing receptor subunit composition, projection neurons are more likely to depolarize at lower concentrations of glutamate released by afferents (incoming axons) from the cortex. As described later, glutamate receptor activity is pivotal in the neurodegenerative mechanism of many neurotoxicants and is intimately related to mitochondrial impairment.

31.2.2 Glial Targets

Increasingly, glial cells are emerging as active participants in neural function, with complex roles in both normal and pathological conditions. Glial cells are nine times more prevalent than neurons in the CNS, and this critical mass appears to be necessary for the function and survival of neurons. Neurons and glia support each other in a dynamic and interactive relationship; chemicals that disrupt this dialogue are likely to be neurotoxic. The interplay between neurons and glia begins early in development, when neurons are differentiating and migrating to their final destinations. Glial cells secrete trophic factors that protect and support the growth of neurons, and they express intercellular adhesion molecules that are used for recognition and guidance by migrating nerve cell progenitors. In adulthood, glial functions include myelination, immune protection, metabolic support, maintenance of ionic balance, and participation in neuronal signaling. These processes are also vulnerable to the action of neurotoxicants, with damage to glial cells invariably disrupting the function of neurons. In contrast, glial cells often participate indirectly in the outcome of neurotoxicity by exacerbating, delaying, or modifying the expression of neurotoxicity. The roles of the specific CNS glial subtypes (i.e., astrocytes, microglia, and oligodendrocytes) diverge in adulthood to assume these widely varied functions. Each cell type is discussed separately below.

Astrocytes. Astrocytes constitute about 85% of all glial cells in the CNS; thus their numbers surpass all other cell types, including neurons. Occupying this large fraction of total neural mass, it is perhaps not surprising that astrocytes maintain connections with both microvasculature and neurons. This positioning puts them in immediate contact with the incoming source of most neurotoxicants, the blood. Consequently, part of the protective function of astrocytes is their critical role in the formation of the blood–brain barrier (further discussed in Section 31.4.2.2). Astrocyte endfeet (the widened ends of astrocyte projections) wrap around capillaries and encapsulate synapses. They actively participate in buffering extracellular levels of neurotransmitter and ionic composition. Astrocytes express receptors for neurotransmitters (e.g., GABA, acetylcholine), enabling them to respond quickly to neuronal signaling events, and then transmit information widely through an astrocyte network. In this network, glia are connected to each other by gap junctions (areas of specialized cytoplasmic connections that allow direct biochemical and electrical communication). Astrocytes also participate in cellular communication by secreting vasoactive and neuromodulatory factors such as prostaglandins, arachadonic acid, and steroids.

An important protective function of astrocytes is the synthesis and release of a variety of neuroprotective factors. Despite this protection, under certain conditions, astrocytes may be harmful to neurons. For example, astrocytes and microglia secrete a calcium-binding protein known as S100B. At low concentrations, S100B promotes neuron survival and neurite outgrowth and enhances synaptic function. This protein has been shown to lessen the neurotoxic effects of trimethyltin. Astrocytes leak S100B when they are damaged or killed, causing it to accumulate rapidly in the extracellular space, even up to levels that can be detected in cerebrospinal fluid. Extracellular S100B is deadly: At tissue concentrations in the range of 100 nM and above, it stimulates the production of proinflammatory cytokines and nitric oxide, leading to neuronal dysfunction and death. Methylmercury (MeHg) may be demethylated in astrocytes to the less toxic inorganic mercury, but MeHg blocks glutamate transporters on these cells, increasing levels of extracellular glutamate and allowing partially for the toxicity that accompanies MeHg exposure. Some neurotoxicants depend directly on activation by astrocytes. Trichloroethylene and 1-methyl-1,2,3,6-tetrahydropyridine (MPTP), for example, are bioactivated within astrocytes by the metabolic enzymes β-lyase and monoamine oxidase B, respectively.

Following a mild chemical, inflammatory, or other insult, astrocytes proliferate in the affected area and undergo multiple changes in gene expression and morphology in a process known as reactive gliosis. Hypertrophy of astrocyte extensions is accompanied by upregulation of insoluble, filamentous structural proteins such as glial fibrillary acidic protein (GFAP). Since GFAP is reliably upregulated by insults and is easy to detect using immunohistochemical techniques, it has become a widely used marker for the detection of neurotoxic events. In most cases, reactive gliosis is transient and reversible. Persistent glial activation responses are seen with neurodegenerative disorders such as Alzheimer's and Parkinson's diseases, suggesting the presence of continuing or ongoing neural injury. On the other hand, the continuing damage of chronic alcohol consumption inhibits the activation responses of both astrocytes and microglia.

Reactive gliosis serves a protective role by creating a dense "glial scar" within the vacant space previously occupied by dying neurons. The scar surrounds the lesion site to keep inflammatory processes localized, and it helps to reseal the blood–brain barrier after it has been broken by injury. Recent studies using GFAP-deficient and other genetically engineered mice have shown that reactive astrocytes help to preserve motor function after mild or moderate spinal cord injury, although the mechanism underlying this effect is not understood. The downside of glial scarring in the CNS is the impenetrable physical and biochemical barrier it forms against neurons that might otherwise attempt to regenerate their processes.

Microglia. Microglial cells are the primary immune mediators in the nervous system. Under resting (i.e., noninflammatory) conditions, they constitute approximately 10% of all glia in the CNS and release growth factors that promote survival of neurons. Microglia are extremely sensitive to changes in their microenvironment; even the mildest neurotoxic insult is likely to be accompanied by some level of microglial response. Like astrocytes, microglia react to injury by altering their morphology and producing neuroactive factors such as cytokines and other inflammatory mediators. If not tightly regulated, however, this process can be a source of further injury. Several of the neuroactive factors released by activated glial cells can be found in Section 31.4.2.

Activation of microglia consists of two phases. First, the microglia proliferate and migrate to the site of injury, where they become hypertrophied and begin releasing inflammatory mediators. They also express certain phenotypical marker proteins that toxicologists use to identify microglial activation as a measure of neural insult. Markers for activation include the complement receptor CR3, also known as OX-42 or CD-11b; the chemokine receptor MCP-1, also known as CCL2; major histocompatibility complex (MHC) class II, also known as OX-6; the lysosome antigen ED-1; and plant lectin RCA-1. If the activating injury is severe enough to cause neuronal death, microglia advance to the second phase of activation, by transforming into phagocytic cells (macrophages). These cells continue to express a similar pattern of microglial markers. Microglia-derived phagocytes possess cytotoxic enzymes and reactive oxygen species to aid in efficiently clearing away dead and dying tissue that might prolong the inflammatory response.

Oligodendrocytes. Oligodendrocytes are the myelinating glial cells of the CNS, constituting approximately 5% of all glia. Unlike Schwann cells of the peripheral nervous system that form myelin sheaths around one axon, oligodendrocytes extend several processes to myelinate multiple axons. Myelin is inhibitory to neurite outgrowth and neuron regeneration in the CNS, and the underpinnings of this effect are only beginning to be clarified. Oligodendrocytes express transmembrane proteins that are recognized by receptors on neuronal membranes, such as Nogo, myelin associated glycoprotein (MAG), and oligodendrocyte–myelin glycoprotein (OMgp). These proteins are constitutively expressed and bind extracellularly to specific neuronal receptors. These receptors, in a complex with nerve growth factor receptor and other proteins, signal the activation of the small G-protein Rho. Rho plays a key role in myelin-dependent inhibition of axon regeneration through its regulation of actin and microtubule dynamics.

Mature oligodendrocytes are targeted by the copper-chelating agent cuprizone. When fed to rodents, this chemical induces widespread apoptosis of oligodendroglia followed by massive demyelination. At the same time, oligodendrocyte progenitor cells continue to proliferate and invade demyelinated areas, so that upon termination of cuprizone administration, nearly complete remyelination is accomplished in a few weeks. Cuprizone toxicity is used to study cellular and molecular regulation of demyelination/remyelination, as well as functional recovery from demyelination.

Oligodendrocytes are extremely sensitive to insults that promote oxidative stress, such as inflammation and excessive glutamate signaling. Oligodendrocyte precursors are particularly susceptible to developmental exposure to lead at low doses.

Lead interrupts differentiation of these progenitor cells and also affects the enzymes that synthesize and metabolize galactolipids, the primary lipid component of myelin. Finally, chronic ethanol exposure has injurious effects on mature oligodendrocytes, myelin, and developing oligodendrocyte precursors. Demyelination and loss of oligodendrocytes is typically found in the postmortem hippocampus of alcoholics. In addition, the corpus callosum, a large bundle of myelinated nerve fibers that connect the two hemispheres, loses density and shrinks in size. Chronic administration of ethanol to rats produces changes in cell surface expression of glycoproteins, suggesting that cell-to-cell recognition mechanisms are attenuated. This would affect both the migration of oligodendrocyte precursors and the interaction of mature oligodendrocytes with neuronal axons.

31.2.3 Vascular and Extracellular Targets

Perfusion of the CNS by blood flow and neuronal activity are tightly coupled. Functional neuroimaging techniques utilize this coupling to identify areas of the brain that respond hemodynamically to neural activity. Cerebrovascular tone is regulated by direct innervation from neurons, and indirectly through signaling to astrocytes. Neurons, astrocytes, microvessels, and the surrounding extracellular matrix can be thought of as an integrated neurovascular unit. The architecture and workings of the unit form the structural and functional basis of the blood–brain barrier (BBB). The BBB has an important role in controlling the access of potential neurotoxicants to the CNS and will be discussed in Section 31.4.2.2.

The dynamic interaction between elements of the neurovascular unit can be disrupted by neurotoxicants. For example, during development, astrocytes promote the differentiation and maturation of endothelial cells in cerebral microvessels. Particularly during the early postnatal period, when microvessels are actively growing and the BBB is maturing, susceptibility to lead exposure is enhanced. Acute exposure during this time causes hemorrhages in the cerebellum and cerebral edema at high doses and increased BBB permeability at lower doses. As development continues, vessel growth slows and the BBB matures, although previous lead exposure leaves it semipermeable. These maturation events correlate with decreasing susceptibility to acute lead dosing. Experiments with co-cultured cells confirm that lead inhibits astroglia-induced microvessel formation in vitro. This loss of cellular interaction may be related to the demonstrated ability of lead to alter the expression of extracellular recognition proteins and their posttranslational modifications.

It is becoming increasingly clear that physical interconnections between neurons, glia, and vascular cells play a crucial role in neural function. The requirement for dynamic cellular interactions within the nervous system is enormous, due in part to the constantly changing morphology of synapses. Adhesive contacts near synaptic junctions must be modified rapidly and efficiently, whereas more stable contacts, such as those between astrocytes and endothelium, must be maintained. The study of extracellular signaling molecules in the CNS is an emerging field, and neurotoxicant effects on matrix function have not been well-studied. Inflammatory conditions within the CNS are responsible for much of the damage to the extracellular environment. Thus, it follows that xenobiotics affecting microglia activation or BBB function would be deleterious to the extracellular matrix. This type of damage has been observed with trimethyltin, a neurotoxicant that incites a powerful inflammatory response leading to neurodegeneration in the hippocampus and other brain regions. Matrix breakdown and remodeling enzymes such as metalloproteases and tissue plasminogen activators are upregulated by acute doses of trimethyltin. It has been suggested that neuronal migration failures induced by developmental exposure to chlorpyrifos and methylmercury are caused by similar changes in the expression ECM proteins. Methylmercury also may have direct effects on extracellular matrix protein function, through its affinity for cysteine sulfhydryl residues. Matrix metalloproteases have a "cysteine switch" that coordinates zinc and maintains the enzyme in an inactive state. Hypothetically, methylmercury could disrupt this interaction and thus activate the enzyme. This and other potential effects of neurotoxicants on the ECM remain largely unexplored. Toxic disruption of the matrix could be an important, yet unrecognized, mechanism in the response of the CNS to chemical injury.

31.3   FACTORS AFFECTING NEUROTOXICANT SUSCEPTIBILITY

31.3.1   Endogenous Factors

An important concept to emerge from the previous section is that neurotoxicants may produce a considerable amount of neural injury indirectly, by stimulating endogenous CNS tissue responses that result in glial activation and oxidative stress. Paradoxically, these processes are often more damaging to tissue than the neurotoxicant itself. The degree of neural destruction in these cases depends on the robustness of the reactive tissue response. When the innate immune response is highly sensitive, greater microglial proinflammatory damage is more likely than when the immune response is less sensitive to stimuli. The character of neurotoxic effect is thus shaped by endogenous factors that are, in turn, controlled by factors such as genetics, hormonal status, age, and preexisting disease.

Endogenous susceptibility factors are highly interrelated. One example of the interrelatedness is demonstrated in the roles of estrogen, aging, and inflammation. The neuroprotective effects of estrogen have been established in a variety of paradigms, including toxicant-based models of Parkinson's disease. Neuroprotection by estrogen has been proposed as one of the factors leading to the lower incidence of this disease in women than in men. Estrogens are thought to act, in part, through modulating glia, thereby decreasing their proinflammatory responses and increasing their anti-inflammatory properties. The symptoms of PD worsen rapidly with aging and the arrival of menopause in women, whereas estrogen replacement has been shown to alleviate the severity of PD symptoms. These findings strongly suggest that loss of the neuroprotective effect of estrogen exacerbates the progression of PD. One might be tempted deduce that neurotoxicants are more harmful to men than to women, and for agents that damage the dopaminergic system (e.g., MPTP, rotenone, manganese, methamphetamine) there is epidemiological support for a small gender difference in sensitivity. On the other hand, neurotoxicants with (putatively) more severe effects in women include chronic ethanol intoxication, sarin nerve gas, and the syndrome of multiple chemical sensitivities.

31.3.1a   Genetic Determinants of Susceptibility. Genetic control over the susceptibility to neurotoxicant action occurs both before and after an agent enters the nervous system. While many toxicants have been characterized respect to genetic susceptibility outside the nervous system, only a few have been studied in the context of CNS toxicity. A notable exception is methylmercury toxicity, associated with polymorphisms in genes encoding glutathione S-transferase (GST) and other glutathione-related enzymes, heme pathway enzymes, and in the gene encoding brain-derived neurotrophic factor (BDNF). Polymorphic GST expression also affects ethylene oxide and methylene chloride susceptibility. Several polymorphic gene products have been linked to the CNS effects of ethanol, including aldehyde dehydrogenase, CYP2E1, GABA receptors, and dopamine and serotonin transporters.

Polymorphisms of human cytochrome P450 genes are among the most abundant of those important to neurotoxicology. The most easily recognized is *CYP2D6*. The CYP2D6 enzyme is responsible, at least in part, for the oxidation of over 25% of all drugs used clinically, including dextromethorphan, chlorpheniramine, amphetamine, and numerous antipsychotics and antidepressants. This enzyme is over- or underexpressed due to single-nucleotide polymorphisms in 7–10% of all Caucasians, with lower percentages for non-Caucasians. Although controversial, several studies have demonstrated an overrepresentation of poor metabolizers of debrisoquine (phenotypic marker of *CYP2D6* polymorphism) in patients with Parkinson's disease. Another P450, CYP3A4, may be the single most important pharmaceutical-metabolizing enzyme in the human body. Its substrates include calcium channel blockers, benzodiazepines, opioids, barbiturates, and antidepressants. Several polymorphisms have been identified in this gene, however, induction and inhibition of CYP3A4 by foods and coadministered drugs is thought to play the greatest role in interindividual variability.

Studies associating gene polymorphisms with both history of exposure to toxic chemicals and neurodegenerative diseases support the idea that many neurodegenerative disorders do not arise from a single factor, but from a combination of genetic and environmental factors. Several genes are clearly candidates for this type gene–environment interaction, because they are involved in xenobiotic metabolism (e.g., CYP2D6 and other cytochromes P450, GST, N-acetyltransferase, MAO-A & -B) or cellular response to injury (interleukins, tumor necrosis factor alpha) or are neuroprotective (BDNF, metallothioneins, aromatase). The enzyme paraoxonase-1 (PON1) is an inducible esterase that hydrolyzes lipid peroxides (indicators of oxidative stress), but also detoxifies the bioactive oxon metabolites of organophosphates. Acetylcholinesterase and PON1 are found at the same locus on chromosome 17. Several human PON promoter and coding region polymorphisms result in altered enzyme activity or level of expression, and these appear to relate to oxidative stress. Recently, genetic variations in paraoxonase were correlated with memory loss and physiological changes in the cortex and hippocampus of people exposed to chronic low levels of organophosphate pesticides. Furthermore, a meta-analysis of studies in patients with Parkinson's disease found an association with some PON1 polymorphisms, and it ruled out others as risk factors for the disease. Correlating genetic information with a history of environmental exposure in studies such as these helps to identify susceptible individuals, and it facilitates discovery of the important interactions between a toxicant and biological pathways. This knowledge then can be used to find ways to control toxicant exposure and its biological consequences.

31.3.1b Epigenetic Determinants. Although "epigenetics" is often used to describe essentially all nongenetic susceptibility factors, in this section we use the term specifically in reference to DNA modifications that result in alternate states of chromatin structure, thereby altering gene activity. Like the nervous system itself, these modifications form a type of "exposure memory" that can be long-lasting, yet still allows for further adaptations and plasticity. It is important to realize that compared to the DNA sequence, epigenetic codes are far more sensitive to environmental effects and, furthermore, that epigenetic changes are easier to reverse. The role of epigenetics may be underappreciated in neurobiology, as well as in the noncarcinogenic toxicity of chemicals.

Two well-known and critical forms of DNA modification are histone tagging and DNA methylation. Histone tagging refers to the acetylation, methylation, phosphorylation, and ubiquitylation of histone protein, collectively forming a "histone code" that directs gene expression by interacting with the transcriptional machinery. Among these modifications, the best characterized is acetylation, regulated by the activities of histone acetyltransferases (HATs) and histone deacetylases (HDACs). The widely used anticonvulsant drug valproic acid was recently found to be a potent inhibitor of HDAC1. Valproic acid and other HDAC inhibitors protect cells from excitotoxicity, presumably through the upregulation of genes encoding stress proteins and other neuroprotective factors.

DNA methylation of cytosine residues is critical to the flexibility of CNS gene expression, and thus it is normally tightly controlled. Diet, hormones, and xenobiotics influence DNA methylation. While some chemical mediators directly methylate DNA, others affect regulatory enzymes to produce hypo- or hypermethylation. Many of the neuroactive agents that interfere with methylation appear to alter the activity of methionine synthase, an enzyme central to the methyl donor pathway in the brain. These include ethanol, thimerosal, elemental mercury, and other metals such as aluminum and lead. Another route for disruption of DNA methylation is through the altered expression or activity of methyltransferases. In one of several ways, AH receptor agonists might interfere with reproductive function, a mixture of these agents (i.e., PCBs, dibenzodioxins and dibenzofurans) severely reduces expression of DNA methyltransferase in the hypothalamus of postnatal rats. On the other hand, methylhalides and tricholorofon promote DNA hypermethylation in the cerebellum by inhibiting the repair enzyme O^6-methylguanine DNA methyltransferase.

Epigenetic regulation is essential for normal neurodevelopment and function. Several heritable diseases that involve chromatin remodeling through epigenetic mechanisms strongly affect the development and performance of the CNS. A few of these are Fragile X, Rett, and Prader–Willi syndromes. In addition, abnormalities in DNA methylation have been observed in the brains of patients with neurodegenerative disorders such as Parkinson's and Alzheimer's diseases. Considering the strong link between epigenetic processes and the CNS abnormalities, there is probably much more to be learned about where neurotoxicants fit into this picture.

31.3.2 Susceptibility in Development and Aging

Some of the earliest processes in brain and spinal cord development involve the proliferation and migration of progenitor cells. Although the rate of development varies from region to region, the signals that guide these events are extremely sensitive to chemical or physical disruption. Neurotoxicant exposures at this stage of development are likely to cause anatomical malformations (dysmorphogenesis). At later stages, when newly differentiated cells undergo exuberant growth of processes and establishment of synaptic contacts, neural plasticity is at its peak. With tightly controlled apoptosis and pruning, redundant or unnecessary connections are eliminated. Exposure to neurotoxicants during this period, while less likely to cause overt structural dysmorphogenesis, can bring about ultrastructural, neurochemical, and behavioral effects. The behavioral effects may range from subtle (e.g., mild cognitive or mood disorders) to explicit (e.g., delayed development, sensorimotor defects). The potential for neurotoxicants to cause functional behavioral disability without structural malformation is known as behavioral teratogenesis. Developmental sensitivity to neurotoxicants continues well after birth because the period of maturation in the CNS is prolonged. Importantly, a fully functioning blood–brain barrier is not established until 6 months after birth, and myelin deposition is incomplete until adolescence. These lowered defenses allow metals, ions, and hydrophilic compounds to enter the CNS, whereas a fully competent adult nervous system would exclude toxicants.

Metals and hydrophilic agents are actually found in higher concentrations in the fetal brain than in the maternal brain, an accumulation that may explain the recent finding that young children exhibit behavioral deficits after exposure in utero to levels of lead that have been formerly considered safe. Lead is thought to perturb several essential cellular processes in development by substituting for other polyvalent cations, such as calcium and zinc. Very low levels of lead interfere with zinc-finger proteins, transcription factors that control the expression of many genes in differentiating cells. When exposed to lead in maternal milk, the brains of young rodents undergo an immediate but transient increase in transcription of the gene encoding beta-amyloid precursor protein (β-APP), the principal protein implicated in Alzheimer's disease. Surprisingly, a delayed upregulation of this protein occurs again nearly two years later, suggesting that low-level developmental lead exposure may have latent and less-than-obvious effects on the CNS in adulthood. Together, these and other recent findings have led many scientists to two conclusions about the effects of lead in development: first, that there is no threshold for the harm caused by lead exposure; and second, that there may be a fetal basis for neurodegenerative adult diseases like AD.

The idea that the effects of neurotoxicant exposure during development may be manifested later in life is supported by several lines of evidence. Separate studies have shown, for example, that neonatally administered dieldrin, manganese, and paraquat produce latent and persistent defects in dopamine transporter activity, long after levels of these agents can no longer be detected in the body. The resulting dopaminergic dysfunction in adulthood is frequently accompanied by behavioral effects that were not observed at earlier ages. It has been suggested that these long-term alterations in dopamine signaling may lead to a silent but enhanced vulnerability of the dopamine system to further insult. A challenge may be needed to expose dormant vulnerabilities. Postnatal exposure to methylmercury predisposes adult rats to self-injurious behavior, a trait only seen when the animals are dosed with amphetamine. Similarly, CNS disorders that primarily bear upon the elderly may represent, in effect, a convergence of circumstances. For one, aging itself alters pharmacokinetic factors, neuronal density, and neurotransmitter activity and reduces neuroprotective mechanisms. Second, genetics, physiological stressors (e.g., inflammation), and early life exposure to toxicants or other environmental factors are also implicated. One or more of these factors, superimposed upon a background of declining functional and protective reserves, may be all that is needed to produce the progressive and irreversible injury of neurodegenerative disease.

31.3.3   Interacting Endogenous and Environmental Factors

Other sections have described environmental interactions with endogenous mechanisms. A few more "lifestyle" effects bear mention. For example, every toxicologist is aware that the metabolism (and thus toxicological effects) of xenobiotics can be modulated by the presence of other agents. Other toxicants, dietary factors, pharmacological agents, and smoking affect an individual's response to neurotoxicants. These agents do not have to be present simultaneously; a history of exposure is often enough. Recently, the role of stress as a factor in toxicant susceptibility has come to the forefront. Within the CNS, psychological stress is a particularly powerful sensitizing agent for many adverse chemical effects. In rats subjected to repeated rounds of alcohol intoxication and withdrawal, the anxiety that occurs during withdrawal increases with each round. Repeated stresses and repeated alcohol withdrawals can substitute for one another in producing a state of extreme sensitivity to stressors, and sensitivity remains intensified even after a long period of abstinence. Studies such as these suggest that stress forms a type of "memory" in the brain for any adverse event that might come its way. This "stress memory" apparently can be preserved from the prenatal stage into adulthood, since maternal stress has been shown to affect responses to neurotoxicants in adult offspring. The biochemical basis for this memory is still unknown. It has been shown, however, that stress can induce changes in blood–brain barrier permeability, can enhance toxicant accumulation in the CNS, and can activate cytokines. Furthermore, stress may induce the expression of oxidative enzymes such as NOS and cyclooxygenase-2 (COX-2) that produce cell-damaging reactive oxygen species in the brain. These effects would enhance the neurotoxicity of exogenous chemicals.

Diet can also influence susceptibility to neurotoxicants. The nutritional status of young children affects the severity of lead toxicity, and thiamine and other dietary deficiencies are tightly interrelated with chronic alcoholism. Studies have shown that elderly patients consuming diets with relatively high saturated and trans fats along with higher copper showed more rapid deterioration of cognitive abilities (i.e., copper required the high fat to have effect). Unusual diets may also have neurotoxicological consequences; this is a possible cause of a syndrome similar to amyotrophic lateral sclerosis (ALS) and Parkinson's disease in the Chamorros natives of Guam. The true origin of the disease is unknown, although the lifestyle of the Chamorros seems to be a significant risk factor. Of all of the potential causes of the disease, it is the unusual diet of the Chamorros people that has captured the most attention. One hypothesis has centered on cycad palm seeds that the natives use to make flour and medicines. Cycad seeds contain at least three types of neurotoxins. The most heavily studied, although still controversial, of these is β-methylamino-alanine (BMAA), an amino acid that activates glutamate receptors and is thus potentially excitotoxic. Traces of BMAA have been identified in brain tissue from Chamorros natives. Although BMAA produces ALS- and Parkinson-like effects in monkeys, it was estimated that humans would have to ingest 125 kg (about 276 pounds) of flour per week in order to achieve a similar dose. Recently, attention has turned to another peculiarity of the Chamorros diet, the highly prized delicacy of the fruit bat. These cycad seed-eating bats are prepared by boiling them (head, wings, hair, and all) in coconut milk. Based on tissue analyses, it appears that the fruit bats (also known as flying foxes) may bioaccumulate BMAA from the cycad seeds in their flesh and, in so doing, provide the Chamorros with a large dose of the toxin in a single meal. Despite the hypotheses, however, the role of dietary BMAA in ALS-PDC continues to be debated and the origin of this mysterious disease is, shall we say, still in the air.

31.4 MECHANISMS OF NEUROTOXICITY AND NEUROPROTECTION

31.4.1 Neurotoxic Processes

Some neurotoxicants directly interfere with the function of the nervous system by interacting with neurotransmitter-mediated processes. For example, the synaptic concentration of neurotransmitter may be altered by toxic agents that disrupt neurotransmitter synthesis (e.g., atrazine and dithiocarbamates decrease norepinephrine synthesis), release (e.g., heptachlor and deltamethrin induce dopamine release), reuptake (e.g., methylmercury blocks astrocyte glutamate transporters), and metabolism (e.g., organophosphates block the breakdown of acetylcholine by acetylcholinesterase). In addition, neurotransmitter receptors, ion channels and intracellular signaling molecules are frequent targets of natural toxins, metals, and other agents. Several examples of toxicant action at these sites have already been mentioned in this chapter. Others are covered in Chapter 19 and in *A Textbook of Modern Toxicology*, Chapter 16 (see Suggested Reading). In this section, two of the most critical aspects of neural function, cytoskeletal integrity and cellular respiration, are discussed in relation to toxic action.

31.4.1a Disruption of Critical Structure and Function. The cytoskeleton bears a heavy burden in maintaining neuronal function, because it orchestrates anterograde and retrograde transport, provides structural support, and participates in intracellular signaling through protein–protein interactions. The three primary components of the neuron cytoskeleton exist as monomers and polymers of different proteins and serve different functions. The smallest are the microfilaments, composed of G-actin (monomer) and F-actin (polymer). Actin microfilaments form a dense network underneath the cell membrane and participate in cellular signaling and small shape changes that occur with synaptic plasticity. Neurofilaments are highly stable polymers of neurofilament protein that provide mechanical support and determine axon caliber and soma size. Neurofilament protein is regulated by phosphorylation, and its expression can be modified by neuronal signaling adaptations that promote larger-scale stabilization or pruning of axons. The largest of the cytoskeletal elements, the microtubules, are dynamic polymers of tubulin that guide anterograde and retrograde transport and participate in the growth and plasticity of dendrites. Microtubule-associated proteins (MAPs) regulate microtubule polymerization and stabilize directional growth that allows for dendritic branching.

Axons are particularly vulnerable to cytoskeletal insults because of their long length and larger diameter, requiring a high level of mitochondrial activity and

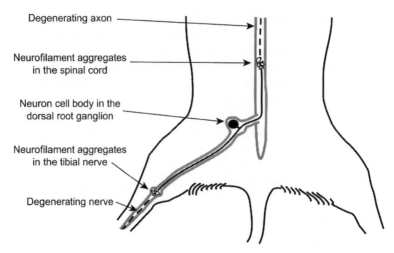

Figure 31.2. Central-peripheral distal axonopathy.

neurofilament content. A common axonal response to neurotoxic challenge is central-peripheral distal axonopathy, also known as "dying back" neuropathy (Figure 31.2). Swellings develop along the axon, along with neurofilament accumulation and loss of microtubules. In sensory neurons, the initial damage is typically observed distally in the peripheral nervous system. As toxicity continues, the damage advances proximally to affect fibers that extend into the CNS, where the cellular environment does not permit regeneration. Carbon disulfide, the first industrial solvent to be considered a neurotoxicant, displays this type of pathology. Neurofilaments are covalently crosslinked with one another by carbon disulfide, *n*-hexane, and methyl *n*-butyl ketone. Aluminum, ethylene oxide, and acrylamide also adduct neurofilaments but do not lead to crosslinking. Acetaldehyde, the first metabolite of ethanol, interferes with microtubule polymerization by forming adducts with free lysine groups. On the other hand, methylmercury has a high affinity for thiol groups and thus adducts microtubule cysteine residues. As described below, many additional cellular macromolecules containing critical thiol residues are also adducted by MeHg, including those involved in antioxidant processes.

The CNS requires a large amount of energy for normal function, and about 20% of all glucose and oxygen utilized in the body is used in the brain. Despite this need, neurons are unable to store energy substrates to a significant extent, and thus they rely on astrocytes for a steady glucose supply. Astrocytes in close contact with blood vessels transport glucose and release glucose and lactate to be taken up by neurons. Under conditions of ischemia, when the blood supply to the brain is blocked, this complex transfer process is inhibited. Because of the high energy requirements of the CNS, agents that interfere with the transport or utilization of blood-borne glucose and oxygen (hypoxia) can produce energy starvation within a few minutes. Nitrates that cause methemoglobinemia, carbon monoxide, and uncouplers of oxidative phosphorylation are examples of toxic agents that rapidly cause metabolic failure resulting in neuronal death. The complex cellular processes that are involved in this type of toxicity are further discussed in the next section.

31.4.1b   *Molecular and Biochemical CNS Responses to Cytotoxic Stress.*

A common consequence of exposure to many toxic compounds is excessive neuronal stimulation, an effect that is frequently manifested clinically as seizures. Some toxicants generate seizures directly, while others may do so indirectly through serious damage to the liver or kidneys, for example. Seizures are initiated when a sudden imbalance in the excitatory activity of glutamate and the inhibitory activity of GABA provokes a net increase in excitation. Consequently, glutamatergic neurons fire repetitively in unison, releasing large amounts of glutamate into synapses. Under normal signaling conditions, glutamate transporters found on the presynaptic membrane and on neighboring astrocytes clear away the released glutamate. These transporters are quickly overwhelmed, however, when glutamatergic neurons fire repetitively, *en masse*. The barrage of the neurotransmitter that descends on glutamate receptors incites depolarization of postsynaptic neurons. This excessive glutamate receptor stimulation, known as excitotoxicity, is lethal to postsynaptic neurons and to the cells surrounding them.

Cell death by excitotoxic mechanisms is implicated in the mechanism of several neurotoxicants, even many that do not necessarily cause seizures. This is because neural injury, whether induced by neurotoxicants, hypoxia/ischemia, or physical insult, can promote the accumulation of extracellular glutamate by multiple mechanisms, including enhanced release, blockade of astrocyte glutamate transporters, reversal of glutamate transporters, and extravasation of glutamate from blood. In the case of mitochondrial toxicants like those discussed above, deficient ATP production results in failure of the Na^+, K^+-ATPase pump and inability to maintain the sodium concentration gradient. Glutamate transporters are dependent upon external sodium, thus collapse of the gradient impairs the removal of extracellular glutamate. The sodium that then enters the neuron effectively depolarizes the membrane, potentially causing release of more glutamate by the postsynaptic neuron.

In very intense insults (i.e., high doses of neurotoxicant), the influx of sodium is accompanied by passive entry of chloride in an attempt to maintain ionic equilibrium. As the intracellular level of sodium chloride rises, the osmotic gradient created promotes diffusion of water into the cell. Cellular swelling and disruption of organelle function ensue, causing the neuron to burst. An inflammatory response is then set in motion, including microglial activation, release of cytokines, and infiltration of circulating immune cells. In severe cases, the neural immune response may drive further damage. Otherwise, this immediate form of cell death may be limited to a few cells, depending on the intensity of the stimulus, the cell type, or the brain region.

The activation of glutamate receptors is a necessary step in neurotoxicant-mediated excitotoxicity, and the ionotropic NMDA receptor plays a critical role. Selective inhibition of this receptor prevents cell death by excitotoxicity, attributed to the ability of NMDA antagonists to block calcium influx through the NMDA receptor. Although calcium is a necessary component of normal intracellular signaling, pathological elevation of intracellular calcium (iCa^{2+}) is the cornerstone of excitotoxicity. Thus, if neurons depolarized by glutamate survive the initial osmotic insult, death by calcium dysregulation will ensue. Following, or concomitant with, glutamate receptor activation, neuronal depolarization stimulates voltage-gated

Ca^{2+} channels to open as well, allowing more Ca^{2+} to enter the neuron. Then, locally increased levels of iCa^{2+} stimulate endoplasmic reticulum receptors to release more calcium from internal storage sites into the cytoplasm (Ca^{2+}-mediated Ca^{2+} release). Adding to this rapidly growing and self-perpetuating iCa^{2+} overload is the opening of the mitochondrial permeability transition pore (PTP), which results in Ca^{2+} release from mitochondria and loss of any remaining ATP production. Furthermore, a net increase in ROS occurs with calcium overload, partially due to generation by Ca^{2+}-activated enzymes such as phospholipase A2, xanthine dehydrogenase, and nitric oxide synthase. Together, excessive iCa^{2+} and ROS oxidize and/or dysregulate multiple intracellular signaling components such as proteases, kinases, DNA endonucleases, and lipid membranes. Neurons at this stage die by oncosis (necrosis). The loss of cellular membrane activates microglia, which adds to the oxidative stress and widens the area of damage.

Apoptotic cell death is also frequently observed in excitotoxicity, although usually this occurs at later time points, or at the fringes of a localized area of damage. In this scenario, some cells survive long enough to maintain mitochondrial ATP production and membrane integrity. Eventually, the mitochondria orchestrate programmed cell death through release of cytochrome c, procaspases, and apoptosis-inducing factor (AIF). The process of apoptosis in neurons is similar to other cells, the details of which can be found in Chapter 17. In neural tissue damaged by excitotoxicity, the presence of neuronal death occurring by both apoptosis and oncosis highlights a simple but important message: Under the same toxicological conditions, different neurons can die by different death mechanisms.

31.4.2   Neuroprotection

31.4.2a   Antioxidants, Innate Immunity, and Neurotrophism. In the CNS, modest antioxidant defenses provide protection from the effects of neurotoxicant-initiated oxidative stress. Glia have a particularly important role in neuroprotection from oxidant damage; however, as described below and throughout this chapter, they also may also be cause some of the damage. The most prominent neuroprotective elements in the CNS include superoxide dismutases (CuZnSOD and MnSOD), glutathione (GSH), and ascorbate, which are present in millimolar quantities. Neurons produce inadequate amounts of precursors for GSH synthesis, and thus they must rely on that supplied by astrocytes (Figure 31.3). In addition, glia are able

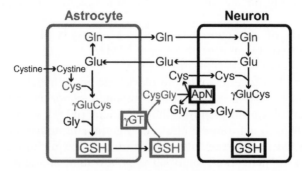

Figure 31.3. Interaction between astrocytes and neurons.

to accelerate GSH synthesis under stressful conditions. Accordingly, glia are generally less susceptible than neurons to reactive species such as H_2O_2 and peroxynitrite. Both neurons and glia have transporters that concentrate ascorbate, which is higher in cerebrospinal fluid than in plasma. In the presence of iron, ascorbate itself is capable of stimulating oxidative damage by reducing Fe^{3+} to Fe^{2+}, which is more oxidatively active. This may be a factor in the enhanced susceptibility of dopaminergic neurons to oxidative stress: These neurons are particularly rich in iron and ascorbate.

The extensive oxygen consumption by the CNS produces large quantities of superoxide as a byproduct of oxidative phosphorylation. Under physiological conditions, superoxide is quickly disproportioned by the ubiquitous SOD enzymes to H_2O_2 and oxygen, and the high levels of H_2O_2 thus generated are capable of producing hydroxyl radicals by the iron-catalyzed Fenton reaction. Catalase and glutathione peroxidase (GPx) are present in neurons and glia to dispose of hydrogen peroxide, and both enzymes are upregulated upon exposure to H_2O_2. Despite its presence in the CNS, however, catalase levels are relatively low compared to liver or kidney. In addition, catalase expression in the brain may be limited primarily to the catalase I isoform, a notion supported by the rapid inhibition of catalase activity upon administration of the selective catalase I inhibitor and broadleaf herbicide aminotriazole. On the other hand, a primary neuroprotective role for H_2O_2 clearance by GPx has been demonstrated repeatedly in transgenic mice overexpressing or lacking GPx1.

Enzymes that repair oxidative damage to DNA and proteins are present in the CNS and seem to be essential for normal function. However, their responses can be quickly overwhelmed in situations like excitotoxicity, described above. In this case, proteins are often oxidatively damaged at multiple sites and then "tagged" with ubiquitin for removal by the proteosome. Often, heavily oxidized proteins are resistant to proteolytic attack; they form insoluble aggregates that may serve a neuroprotective role by sequestering oxidized proteins. On the other hand, if the formation of insoluble complexes is very severe, the aggregates themselves may be toxic to neurons. The formation of protein aggregates has been observed in several neurodegenerative disorders such as Alzheimer's, Parkinson's, Huntington's, and prion diseases, suggesting that oxidative stress and protein processing may underlie their pathologies.

Metallothioneins help restrict the formation of oxidative stress by binding heavy metals. Physiologically, these proteins store intracellular zinc and some copper in mitochondria, in lysosomes, and in the nucleus, where they are thought to support the activity of zinc-dependent transcription proteins. Metallothioneins are also expressed extracellularly and within neurons; however, their levels are highest in astrocytes, endothelial cells, and the choroid plexus. In heavy metal toxicity, astrocyte mitochondria may accumulate manganese up to 200 times the concentration of lead or manganese found extracellularly. Upon exposure to stress, injury, or toxic metals, these proteins are upregulated as part of the activated microglial response. As such, they enhance the secretion of anti-inflammatory cytokines, growth factors, and their receptors. Metallothioneins reduce oxidative stress and improve vascular remodeling and neuronal repair following injury. Metallothionein-deficient mice are phenotypically similar to wild-type mice under physiological conditions, but their neural and clinical responses to neurotoxicants or physical injury are much more severe. It has thus been suggested that metallothioneins are essential for coping with brain damage.

As we have seen, the contribution of inflammatory processes to the overall outcome of neurotoxicant injury is considerable. The purpose of innate immunity in the CNS (i.e., immunity provided by microglia and to some extent astrocytes) is to defend the tissue against pathogens and clear up cellular debris. Additional cytokines from the circulation, and even activated T cells, may cross into CNS via the choroid plexus, a richly vascularized area of epithelial cells that secretes cerebrospinal fluid. Despite these defenses, typical pathogens like bacteria rarely gain access to the brain and spinal cord. Neurotoxicants however, are ubiquitous, yet the innate immune response seems to provide inadequate defense against chemical insults. This leaves the negative aspects of innate immunity (i.e., free radical production, bioactivation, etc.) often to outweigh the positive protective effects that might be gained from immune responses. The double-edged sword of glial activation suggests that much of the pathological changes observed in neurotoxicant-treated CNS are related less to the toxicant than to the *response* to the toxicant. An example that clearly demonstrates this idea is demonstrated by experiments in which dopaminergic neurons were co-cultured with or without microglia. Low concentrations of rotenone in the co-cultured cells induced reactive oxygen species through the action of microglial NADPH oxidase, whereas the neurons grown without glia showed a marked decrease in toxic indicators.

Innate immunity in the CNS is by no means entirely detrimental. Cytokines released by microglia serve to recruit immune system cells to the site of injury and clear up cellular debris. Some of these inflammatory mediators have dual roles as adaptive and maladaptive factors. For example, the cytokine interleukin-1 (IL-1) induces the activation of ROS-producing factors such as phospholipase A2, cyclooxygenase-2, nitric oxide, and prostaglandins. It also enhances its own release and the release of other cytokines such as IL-6 and tumor necrosis factor α (TNFα). These proinflammatory mediators contribute to neuronal death, sometimes causing collateral damage to otherwise unharmed tissue. On the other hand, IL-1 also stimulates the release of trophic factors such as nerve growth factor (NGF), transforming growth factor-β1 (TGF-β1), and other antiinflammatory cytokines.

In addition to inflammatory factors, glial cells produce many peptides that fulfill roles as neuroprotectants, neurotrophic factors, and neurotransmitters. As protective agents, these peptides trigger or strengthen endogenous defense mechanisms such as antioxidants and anti-apoptotic factors, while inhibiting growth-inhibitory factors. Many neuroprotective peptides (known as neuropeptides) are similar to neurotransmitters in that they exert their effects (usually) by binding to G-protein-coupled receptors. Because much of their effect is to modulate the signaling output of neurotransmitters, they are sometimes called neuromodulators. There are literally hundreds of neuropeptides; the most commonly known include oxytocin and vasopressin. Nearly all neuropeptides are expressed in a region-specific manner; thus they have differing neuroactive and behavioral roles. Pituitary adenylate cyclase-activating peptide (PACAP) and vasoactive intestinal peptide (VIP) are immunomodulators that have been studied widely for their ability to protect neurons from injury. VIP, for example, has been shown to promote the synthesis and release of GDNF while inhibiting proinflammatory mediators released by activated microglia.

Neurotrophic factors that are important in the CNS include NGF, brain-derived neurotrophic factor (BDNF), neurotrophin-3, glial-cell-derived neurotrophic factor (GDNF), and fibroblast growth factor (FGF). Neurotrophins bind to tyrosine kinase receptors and can be transported retrogradely and anterogradely. The protein kinase A (PKA) pathway is downstream of both types of receptor, and evidence has shown that increasing the cyclic AMP/PKA can enhance the expression of neurotrophin genes and increase cell surface levels of receptors and promote outgrowth of neuronal processes. The neurodevelopmental damage caused by methylmercury may be mediated in part its ability to deprive cerebellar granule neurons of trophic input. MeHg interferes with the synthesis of BDNF and with NGF expression. In addition, MeHg inhibits the survival-promoting activity of insulin-like growth factor (IGF-1) on immature granule cells. In the adult CNS, loss of trophic factor support can lead to oxidative stress and apoptosis.

31.4.2b   The Blood–Brain Barrier.   In most tissues of the body, the extracellular space is composed of fluid filtered from capillary blood plasma. Components of plasma enter the interstitial space by diffusion between, or transport through, endothelial cells that form the walls of the capillary. In the CNS, however, this relative freedom of communication between the blood and parenchyma is profoundly limited by the blood–brain barrier. Structural and functional specializations in and around the microvasculature of nervous tissue comprise the blood–brain barrier. For example, whereas the endothelial cells that line capillaries in most of the periphery contain permeable membrane domains called fenestrations, CNS capillaries do not. Furthermore, between CNS capillary endothelial cells are tight junctions, areas that resist diffusion of hydrophilic materials around the cells. This forces most molecules that do go through endothelial cells to be limited to gases such as oxygen and carbon dioxide, and small lipophilic compounds that move within membranes. Small hydrophilic molecules require more specific transit mechanisms such as endocytosis; this mechanism is also limited in brain compared to the periphery. Large molecules such as peptides are excluded from crossing the blood–brain barrier unless they are recognized by specialized carrier-mediated transport mechanisms. Brain capillaries are surrounded by perivascular cells such as astrocytes, pericytes, and microglia (Figure 31.4). Importantly, astrocytic endfeet are closely apposed to capillary walls, where specialized proteins are expressed to maintain communication between the capillary and the astrocytes. In this way, the activity of the neural tissue and the dynamics of capillary blood flow are coordinated.

The blood–brain barrier is a biochemical as well as a physical barrier. Brain endothelial cells create an enzymatic barrier composed of secreted proteases and nucleotidases, as well as intracellular metabolizing enzymes such as cytochrome P-450. Furthermore, γ-glutamyl transpeptidase, alkaline phosphatase, and aromatic acid decarboxylase are more prevalent in cerebral microvessels than in non-neuronal capillaries. The efflux transporter P-glycoprotein and other extrusion pumps are present on the membrane surface of endothelial cells, juxtaposed toward the interior of the capillary. Furthermore, CNS endothelial cells display a net negative charge at the interior of the capillaries and at the basement membrane. This provides an additional selective mechanism by impeding passage of anionic molecules across the membrane.

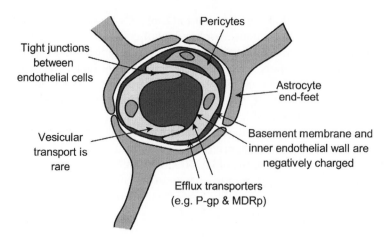

Figure 31.4. Blood–brain barrier.

Although most neural tissue is protected from exposure to blood–borne neuro-toxicants by the blood-brain barrier, there are a few areas of the brain that are not separated from the circulation. The circumventricular organs (CVOs) are neuro-endocrine-like regions (including parts of the hypothalamus) found at discrete locations adjacent to the ventricles of the brain. These areas serve as information relay stations between the brain and the general circulation. These areas contain rich capillary networks that monitor systemic homeostasis by sensing changes in circulating osmolarity, hormone status, and immune responses and respond by releasing neurotransmitters and neuropeptides. CVO capillaries are an exception to the rule that capillaries in the brain are unfenestrated (see above). They also lack tight junctions and are permeable to blood-borne factors that would be excluded elsewhere in the brain. Because this represents a vulnerable area, protective mecha-nisms separate these sites from the rest of the brain. A specialized border of glial cells isolates many of these areas; and, in addition, CVOs are rich in drug metaboliz-ing enzymes. The location of CVOs next to the ventricles of the brain means that they are bathed in cerebrospinal fluid (CSF). To avoid contaminating the CSF with blood-borne pathogens, CVOs are surrounded by an epithelial cell layer containing tight junctions and xenobiotic-metabolizing enzymes. This so-called blood–CSF barrier plays an important role in neuroendocrine signaling, neuroimmune and neuroinflammatory responses, drug metabolism, and protection from chemical-induced neurotoxicity. Neurotoxicants can and do circumvent the blood–brain barrier. Some agents have favorable physicochemical characteristics for crossing the barrier directly. Others utilize axonal transport mechanisms. For example, ultrafine particles of magnesium can enter the brain directly via the olfactory nerve. Impor-tant considerations in this type of transport include particle size, solubility, and other physical factors, as well as endogenous factors such as the viscosity of mucus. Inhala-tion is a frequent mode of manganese exposure and can enter the bloodstream in the respiratory system. From there, the properties of manganese may allow it to enter the brain via transport through cerebral capillaries or from the CSF across the choroid plexus. It is not clear to what extent to each of these three pathways contributes to the neurotoxicity of manganese to humans.

Another method of neurotoxicant entry into the CNS is by exploitation of a compromised blood–brain barrier. Aluminum itself compromises the blood–brain barrier; and reactive species, both directly and indirectly, can open the barrier. Blood-borne inflammation, possibly through the release of reactive oxygen species, is widely known to create a "leaky" blood–brain barrier. This allows inflammatory cells, endotoxins, and neurotoxicants to enter the CNS. Inflammation within the brain itself also affects blood–brain barrier permeability, an aspect that potentially links psychological stress to increased susceptibility to neurotoxicants. It has been shown that inflammatory responses are induced in animals under stressful conditions, such as electrical shock, restraint, or adverse social interactions.

31.5 THE DYNAMIC NERVOUS SYSTEM: ADAPTABILITY, PLASTICITY, AND REPAIR

One of the most remarkable aspects of neurons is their ability to continually adapt to changes in response to a constant barrage of incoming stimuli. Under normal functional conditions, the sensitivity of neurons to perturbations in their microenvironment is exquisite, so much so that even slight alterations in the type or intensity of signals may evoke structural or functional adaptations. To illustrate this point, suppose the activation of a neuron in the hippocampus results in its release of a nonexcitotoxic amount of glutamate onto a dendrite on a receiving neuron. Postsynaptic glutamate receptors detect the glutamate and respond accordingly by opening their ion channel or activating their associated G protein. The activation of these receptors on this single dendrite may not be strong enough to create an action potential that affects the entire neuron; that would need concurrent stimulation of many dendrites. The message, however, does not go unheeded. Instead, this local release of glutamate initiates the activation of multiple cellular signaling pathways, culminating with the insertion of even more glutamate receptors into the postsynaptic membrane. Upon subsequent episodes of glutamate release from the presynaptic neuron, the increased number of receptors will elicit a stronger intracellular response than the first one. In this manner, the responsiveness of the postsynaptic neuron is augmented by a single stimulatory event. Other signal-enhancing processes occur as well (see below), and often these are accompanied by structural changes, such as the formation of new areas of postsynaptic contact. These activity-dependent synaptic modifications occur constantly within the nervous system as we experience our surroundings. They are thought to form, in part, the cellular basis of learning and memory. Neural adaptations of this nature illustrate dramatically the concept that, even under normal conditions, one's nervous system is never the same from one moment to the next.

Part of the plastic response to receptor stimulation involves changes in gene expression. Out of the estimated 30,000 or so human genes, over 5000 are expressed in the brain, more than any other organ. The expression of many of these genes is tightly regulated by the physiological state of the neuron, and thus the proteome is being modified continuously to fit demands. The complexity is further amplified by signal-sensitive changes in posttranslational modifications such as alternative proteolytic processing and phosphorylation; often these are precursors to gene expression changes. All of these events are geared ultimately to maintain homeostasis; an adaptive attempt to return the system to its normal state by counterbalancing changes. In order to accomplish this mission, terminally differentiated neurons have co-opted the cell-cycle machinery to control synaptic plasticity. Many of the same genes that are altered in cells as they progress through the stages of division are used in neurons to provide the flexibility needed to adapt to changing conditions. This system usually works well; however, it also brings neurons closer to the same processes that orchestrate cell death. With advancing age and loss of the already-limited mechanisms of repair, the risk that these signals will cross over to maladaptive, rather than adaptive, effects grows more likely. Furthermore, while neurotoxicants promote acute and chronic damage, they may also cause subtle, latent effects that synergize with aging through the stress they place on the adaptive system. With the link between environmental exposure, genetics, and aging becoming ever clearer, toxicologists are challenged to delineate the contributions of each of these factors to the diseases of the adult central nervous system.

SUGGESTED READING

Arundine, M., and Tymianski, M. Molecular mechanisms of neurodegeneration in excitotoxicity. *Cell Calcium* **34**, 325–337, 2003.

Blake, B. L. Toxicology of the nervous system. In: Hodgson E. (Ed.). *A Textbook of Modern Toxicology*, 3rd ed. John Wiley & Sons, Hoboken, NJ, 2004, pp. 279–297.

Campbell, A. Inflammation, neurodegenerative diseases and environmental exposures. *Ann. NY Acad. Sci.* **1035**, 117–132, 2004.

Fonnum, F., and Lock, E. A. The contributions of excitotoxicity, glutathione depletion and DNA repair in chemically induced injury to neurons: Exemplified with toxic effects on cerebellar granule cells. *J. Neurochem.* **88**, 513–531, 2004.

Halliwell, B. Oxidative stress and neurodegeneration: Where are we now? *J. Neurochem.* **97**, 1634–1658, 2006.

Krantic, S., Mechawar, N., Reix, S., and Quirion, R. Molecular basis of programmed cell death involved in neurodegeneration. *Trends Neurosci.* **28**, 670–676, 2005.

Massaro, E. J. (Ed.). *Handbook of neurotoxicology*, Humana Press, Totowa, NJ, 2002.

Nedergaard, M., Ransom, B., and Goldman, S. A. New roles for astrocytes: Redefining the functional architecture of the brain. *Trends Neurosci.* **26**, 523–530, 2003.

Tilson, H. A., and Harry, G. J. (Ed.). *Neurotoxicology*, 2nd ed., Taylor and Francis, Philadelphia, 1999.

Uversky, V. N. Neurotoxicant-induced animal models of Parkinson's disease: Understanding the role of rotenone, maneb and paraquat in neurodegeneration. *Cell Tissue Res.* **318**, 225–241, 2004.

Wallace, D. R. Overview of molecular, cellular, and genetic neurotoxicology. *Neurol. Clin.* **23**, 307–320, 2005.

Yuan, J., Lipinski, M., and Degterev, A. Diversity in the mechanisms of neuronal cell death. *Neuron* **40**, 401–413, 2003.

CHAPTER 32

Immunotoxicity[1]

MARYJANE K. SELGRADE

Immunotoxicology Branch, U. S. Protection Agency, Research Triangle Park, North Carolina 27711

DORI R. GERMOLEC

National Institute of Environmental Health Sciences, Research Triangle Park, North Carolina 27709

ROBERT W. LUEBKE

Immunotoxicology Branch, U.S. Environmental Protection Agency, Research Triangle Park, North Carolina 27711

RALPH J. SMIALOWICZ

(retired) Immunotoxicology Branch, U. S. Environmental Protection Agency, Research Triangle Park, North Carolina 27711

MARSHA D. WARD

Immunotoxicology Branch, U.S. Environmental Protection Agency, Research Triangle Park, North Carolina 27711

CHRISTAL C. BOWMAN

Immunotoxicology Branch, U.S. Environmental Protection Agency, Research Triangle Park, North Carolina 27711

32.1   INTRODUCTION

The immune system defends the body against infectious agents (bacteria, viruses, fungi, parasites), which are ubiquitous in our environment. It also has a role in defending the body against certain tumor cells that may arise spontaneously or as the result of environmental insults (viral, radiation, and chemical). A crucial part of this function is the ability to distinguish endogenous components ("self") from potentially harmful exogenous components ("non-self"). The immune system is composed of several tissues and cell types. Immune cells include a variety of leu-

---

[1]**Disclaimer**: This chapter has been reviewed by the National Health and Environmental Effects Research Laboratory, U.S. Environmental Protection Agency, and approved for publication. Approval does not signify that the contents necessarily reflect the views and policies of the Agency, nor does the mention of trade names or commercial products constitute endorsement or recommendation for use.

Molecular and Biochemical Toxicology, Fourth Edition, edited by Robert C. Smart and Ernest Hodgson
Copyright © 2008 John Wiley & Sons, Inc.

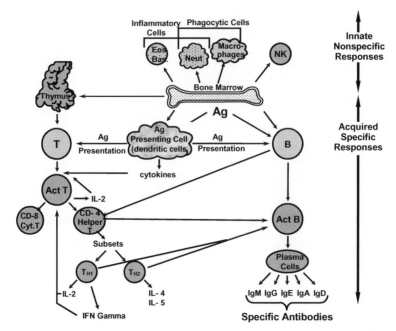

Figure 32.1. Schematic of the immune system. The immune system is composed of primary lymphoid organs (bone marrow and thymus), secondary lymphoid organs (not shown), and several cell types. In addition, a number of mediators including cytokines, antibodies, and complement regulate and/or are produced by the immune system. Abbreviations: Eos, eosinophil; Bas, basophil; Neut, neutrophil; NK, natural killer cell; Ag, antigen; T, T lymphocyte; B, B lymphocyte; Cyt T, cytotoxic T lymphocyte; Act B, activated B lymphocyte.

kocytes, which are derived from bone marrow, circulate in blood and lymph, and are distributed throughout the body in lymphoid and other tissues (Figure 32.1).

A properly functioning immune system is essential to good health. In some individuals the immune system is compromised by primary immune deficiencies caused by genetic defects or secondary immune deficiencies resulting from diseases (e.g., AIDS, leukemia) or drug therapies. These individuals are more susceptible to infectious diseases and certain types of cancer, the consequences of which can be life-threatening. On the other hand, the immune system mediates certain types of disease. It may react to foreign substances that would otherwise be relatively innocuous, such as certain chemicals, pollens, or house dust. The resulting allergic reactions can produce an array of pathologies ranging from skin rashes and rhinitis to more life-threatening asthmatic and anaphylactic reactions. In addition, although the immune system can generally distinguish between "self" and "non-self," in some circumstances it may mistakenly react against "self" components, resulting in another type of immune-mediated health effect, autoimmune disease.

Table 32.1 describes the major consequences that may result from interactions between toxic agents and the immune system. There is ample evidence that xenobiotic compounds and physical stressors, such as ultraviolet and ionizing radiation, can modulate immune function by acting as immunosuppressants, allergens, or potentiators of allergic disease. There is also growing evidence that xenobiotic exposures are associated with autoimmune disease. A number of resources, cited in "Suggested Reading" at the end of this chapter, provide information on the

TABLE 32.1. Potential Consequences of Interactions Between the Immune System and Toxic Chemicals

Nature of Interaction	Disease Enhanced
Suppression	Infectious
	Neoplastic
Stimulation	Allergic
	Autoimmune

physiology, biochemistry, and toxicology of the immune system. In this chapter we will provide a brief overview of the immune system and immunotoxicology and then describe a variety of biochemical mechanisms underlying the toxicity of representative immunotoxicants.

32.2   ORGANIZATION OF THE IMMUNE SYSTEM

32.2.1   Cells, Tissues, and Mediators

Cells of the immune system include several different types of leukocytes (white blood cells): neutrophils, eosinophils, and basophils/mast cells[2] (also collectively known as granulocytes or polymorphonuclear leukocytes), monocytes/macrophages,[2] natural killer (NK) cells (which are large granular lymphocytes), T and B lymphocytes, plasma cells (B cells that produce antibodies), and dendritic cells (Figure 32.1). Primary lymphoid tissues include the bone marrow, from which immune cells are derived, and the thymus, which has a major role in the differentiation of T lymphocytes. Secondary lymphoid tissues include the spleen, lymph nodes (scattered throughout the body), tonsils, and adenoids. There are also lymphoid aggregates in the lung, gut, and skin, the three major portals of entry for environmental agents. These localized aggregates are referred to, respectively, as bronchus-, gut-, and skin-associated lymphoid tissue (BALT, GALT, SALT). From the preceding description it is clear that the immune system provides a diffuse target for toxic insult.

The activation, maturation, differentiation, and mobilization of immune cells are controlled by cytokines (e.g., interleukins, interferons, and chemokines), which are soluble mediators produced by immune cells and/or by cells outside the immune system (e.g., epithelial cells and cells of the nervous system). Other soluble (humoral) mediators produced by immune cells include antibodies (immunoglobulins) and complement proteins (plasma proteins produced by monocytes and macrophages as well as hepatocytes). Mediators are important in the implementation and regulation of immune responses.

32.2.2   Innate Responses

Immune responses are divided into innate responses directed nonspecifically against foreign substances and acquired responses directed against specific antigens (see Table 32.2 for important definitions). Innate immunity is generally viewed as the

[2]Different designations for similar cells found in blood (to left of slash) versus tissue (to right of slash).

TABLE 32.2. Important Definitions

Term	Definition
Antigen	Molecules (proteins or carbohydrates) that evoke specific immune responses; usually foreign to the host.
Antibody	Soluble proteins known as immunoglobulins; these molecules circulate freely and react specifically with invoking antigen; subclasses: IgM, IgG, IgE, IgD, and IgA.
CD + a number (cluster of differentiation)	Designates a particular cell surface molecule.
Complement	Series of nonimmunoglobulin plasma proteins sequentially activated by antigen–antibody complexes. They damage target cell membranes and are active in host defense.
Cytokines	Soluble substances (lymphokines and interleukins) that are secreted by cells and have a variety of effects on other cells.
Hapten	Molecule that that can react specifically with an antibody, but is too small to elicit an antibody response unless coupled to a protein.
Hypersensitivity	Excessive humoral or cellular response to an antigen leading to tissue damage.
Immunotoxicity	Undesired effects resulting from interactions of xenobiotics with the immune system.
Major histocompatibility complex (MHC) class I and class II molecules	Cell-surface molecules critical to antigen presentation.
Toll-like receptors (TLR)	Pattern recognition receptors; recognize features common to many pathogens; have important role in activation of the innate and adaptive immune responses.

provider of rapid, although usually incomplete, antimicrobial host defense, whereas acquired immunity is a slower but more definitive response. There is considerable interaction between these two types of immunity [see Hoebe et al., (2004)].

Macrophages and neutrophils are phagocytic cells that, as part of the innate response, engulf and in many cases destroy infectious agents and other foreign particles. This process can be nonspecifically enhanced by complement proteins or enhanced by antibodies bound to molecules on the surface of microbes or other targets (a process called opsonization). Macrophages are also activated by cytokines and, once activated, can kill certain tumor and virus-infected cells. Eosinophils contain lysozyme and other mediators and appear to be particularly important as a defense against helminthic parasites, which are not easily phagocytosed due to size. Eosinophils are also typically present in certain types of allergic responses. In general, leukocytes have a significant role in inflammation, a major component of the body's defense mechanism. In addition to an influx of these cells, inflammation is characterized by activation of clotting mechanisms, increased blood flow, and increased capillary permeability. These responses facilitate mobilization of immune cells to the site of injury and result in the swelling and reddening associated with inflammation.

TABLE 32.3. Toll-like Receptors and Known Ligands

TLR	Ligand
TLR-1	Triacyl lipoproteins; synergizes with TLR-2 (bacteria)
TLR-2	Peptidoglycan (gram-positive bacteria), viral glycoproteins, fungi, and others
TLR-3	Double-stranded RNA (viruses)
TLR-4	Lipopolysaccharide (gram-negative bacteria); viral glycol proteins
TLR-5	Flagellin (bacteria)
TLR-6	Diacyl lipoproteins (mycoplasma); synergizes with TLR-2
TLR-7	Single-stranded RNA (viruses)
TLR-8	Single-stranded RNA (viruses)
TLR-9	Unmethylated CpG DNA (bacteria, viruses)
TLR-11	Uropathogenic bacteria, protozoan proteins

Other components of the innate response include natural killer (NK) cells and a number of cytokines. NK cells lyse certain types of tumor cells and virally infected cells and are a rich source of "immune interferon" (interferon-γ), which stimulates macrophages and T cells; hence they are thought to play an important role in host resistance to both neoplastic and viral disease. Type I interferons (interferon α and interferon β) are produced by a number of different cell types and appear very rapidly after viral infection. Type I interferons inhibit viral replication, inhibit cell proliferation, and increase the lytic potential of NK cells and therefore play a role in controlling viral and neoplastic disease. Several cytokines are important in the initiation of inflammatory responses. Those that have received the most attention include tumor necrosis factor alpha (TNFα), interleukin (IL)-1, and IL-6. There are also a number of chemotactic cytokines (including IL-8), called chemokines, which help to mobilize immune cells to the site of injury.

Although the innate immune system does not recognize specific antigens, proteins known as pattern recognition receptors occur on macrophages, neutrophils, and dendritic cells and recognize features common to many pathogens. When the receptors bind to these highly conserved microbial constituents, a cascade of events is triggered that culminates in phagocytosis, chemotaxis, and production of molecules that influence the initiation and nature of subsequent adaptive immune responses. Toll-like receptors (TLR) are the most well studied of the pattern recognition receptors. Each of the 11 TLR in humans recognizes a particular molecular structure that is present in many pathogens (Table 32.3). Subsequently, pathogen recognition is converted into an effective host defense against initial infection. The complement system is one of the major mechanisms responsible for this early response, primarily as mentioned above by facilitating phagocytosis and pathogen membrane lysis.

32.2.3 Acquired Immune Responses

In addition to their phagocytic function, macrophages, along with dendritic cells (including Langerhans cells in the skin) and Kupffer cells in the liver, have an important role in the development of specific immune responses to pathogens in that they process and present antigens to T lymphocytes. The T-cell antigen receptor (TCR) recognizes and binds proteolytically processed short peptide fragments (antigens) bound to self major histocompatibility complex (MHC) molecules on the surface of an antigen-presenting cell. There are two major classes of MHC molecules that present different types of antigens to different types of T cells (described in more detail below): Class I MHC generally presents peptides derived from proteins produced within the cell (e.g., viral or tumor antigens) and is recognized by CD8 T cells. Class II MHC generally present peptides derived from proteins taken up by the cell and is recognized by CD4 T cells. In addition to the TCR, other molecules contribute to activation of T cells either by functioning as co-receptors that initiate nonspecific, co-stimulatory signal transduction events or by increasing the avidity of the interaction with the antigen-presenting cell. Protein–tyrosine phosphorylation is important in the initiation of cellular responses that follow TCR recognition of the MHC–antigen complex. The subsequent signal transduction events that lead to T-cell activation and the role of numerous co-receptors on the surface of both T cells and antigen-presenting cells have been the subject of intense study. The known signaling pathways rely on inositol 1,4,5-trisphosphate (IP_3) and diacylglycerol (DAG) second messengers. IP_3 triggers calcium mobilization, which leads to the activation of the transcription factor called nuclear factor of activated T cells (NF-AT). DAG activates protein kinase C (PKC), leading to activation of the transcription factor NFκB. Ras guanine nucleotide-releasing protein activates the Ras-mitogen-activated protein kinase (MAPK) pathway and ultimately the transcription factor AP-1. Signal transduction events are described in more detail in Section 32.4.1 in conjunction with the mechanisms associated with cyclosporin A immune suppression. [Also see Singer et al. (2003).] The net result of T-cell activation is clonal expansion (proliferation) and activation of T cells specific for a particular antigen.

There are two major divisions of T cells that are distinguished by expression of different cell surface markers, CD4 and CD8. Pre-T cells migrate from the bone marrow to the thymus, which plays a key role in T-cell differentiation. As relatively immature cells, T cells express CD3 associated with the TCR and both CD4 and CD8 molecules. As maturation progresses, these cells undergo both positive and negative selection. During positive selection, T cells are "screened" by the MHC molecules on the cortical epithelial cells of the thymus. Only cells that bind to MHC with a certain affinity survive. Cells that bind with higher or lower affinity undergo apoptosis (programmed cell death). As a result of this process, T cells become MHC-restricted; that is, they will only respond to antigen presented in association with MHC. Cells that survive positive selection are potentially able to respond to self proteins. However, as the cells move from the thymic cortex to the medulla, they undergo negative selection, during which self-reactive cells are removed or functionally inactivated. During the course of positive and negative selection, CD4+ and CD8+ cells downregulate the expression of one of these co-receptor molecules such that mature T cells express only CD4 or CD8. All continue to express CD3. Mature T cells leave the thymus and populate secondary lymphoid organs.

CD4 and CD8 T cells have different functions when activated. Cytotoxic T lymphocytes (CD8) lyse cells expressing specific viral or tumor antigens. CD8 cells have also been shown to downregulate (suppress) other immune responses under some circumstances, probably by the production of certain cytokines. Hence, they are sometimes referred to as suppressor T cells. CD4 helper T cells consist of two major subpopulations (Figure 32.2). These subpopulations appear to regulate different sets

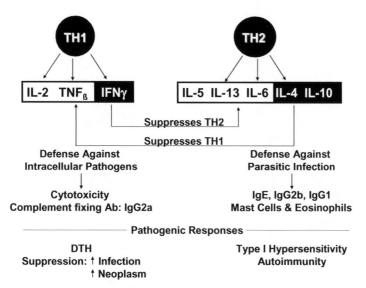

Figure 32.2. Subpopulations of T helper (Th) lymphocytes characterized by production of different cytokines. The two populations are mutually antagonistic and have different roles in immune defense and pathogenesis. Abbreviations: IL, interleukin; INF, interferon; TGF, transforming growth factor; DTH, delayed-type hypersensitivity; Ab, antibody.

of immune responses. T helper (Th)1 cells produce interleukin 2 (IL-2) and interferon gamma (IFNγ). These cells are involved in delayed-type hypersensitivity (DTH) responses, activate macrophages to destroy intracellular organisms more efficiently, and activate B cells to produce complement fixing and opsonizing antibody subtypes. Th2 cells produce a different array of cytokines, including IL-4 and IL-5, facilitate B-cell production of IgG1, IgA, and IgE antibodies associated with immediate hypersensitivity responses, and enhance eosinophil differentiation. Hence, Th2 cells may be particularly important in responding to certain parasitic infections, and they also play an important role in reactions to common allergens such as pollen and dust mite. Th1 and Th2 cells are mutually antagonistic in that cytokines produced by each subtype tend to downregulate the other. Recently, it has been suggested that there are also similar subpopulations of CD8 T cells distinguished by different cytokine profiles. There are also two broad categories of regulatory T cells, a thymus-derived population (CD4+ CD25+), and inducible antigen-specific populations characterized by the production of inhibitory cytokines such as IL-10 and TGFβ. These cells play a special role in controlling immune responses, including induction and maintenance of tolerance to self antigens and to antigens encountered via oral exposure. Tolerance is a process by which the immune system "actively ignores" an antigen to prevent inappropriate and detrimental responses, including allergy and autoimmune disease. Whereas these are clearly undesirable outcomes, too much tolerance may be hazardous because it does not support effective responses against tumor cells or invading pathogens. In some cases, tumors or pathogens persist by exploiting tolerance mechanisms.

B-cell receptor (membrane-bound immunoglobulin) recognizes native or denatured forms of proteins or carbohydrates in soluble, particulate, or cell-bound form. B cells are particularly important in defending the host against extracellular patho-

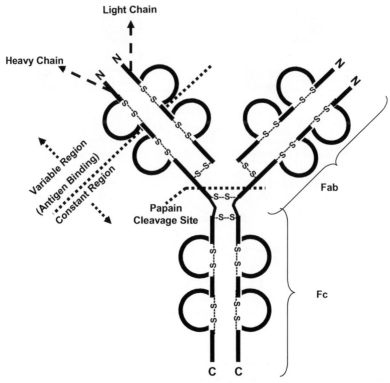

Figure 32.3. The structure of the antibody molecule. All immunoglobulins consist of a basic subunit made up of four polypeptide chains (two light and two heavy) bound together by disulfide bonds. The variable region of the molecule contains the antigen binding site. Papain digestion yields a constant fragment (Fc) and variable fragments (Fab).

gens and their toxins. Although B and T cells recognize distinct forms of antigen using very different receptors, the signal transduction events that result from the interaction of their antigen receptors with antigen are quite similar. B cells specific for a given antigen also expand clonally in response to that antigen, differentiate into plasma cells, and produce antibodies which specifically bind to the eliciting antigen. There are two types of antigens that stimulate B cells. T-independent antigens have a high density of repeating antigenic determinants (e.g., bacterial polysaccharides) capable of crosslinking immunoglobulin (Ig) receptors on the B-cell surface and thus activating the cell without T-cell help. Antigens that do not have repeating determinants are usually T-dependent; that is, T-cell help (via the secretion of important cytokines such as IL-2) is required in order to activate B cells. Most antigens belong to this latter category.

The immunoglobulin molecule structure is shown in Figure 32.3. All immunoglobulins consist of a basic unit of four polypeptide chains, two identical light chains and two identical heavy chains, held together by a number of disulfide bonds. The antigen binding region is located at the N-terminus. Papain digestion of the molecule results in two fragments at the antigen binding end (referred to as Fab) and a fragment (called Fc) at the C-terminus. The biologic functions of the antibody molecule derive from the properties of a constant (Fc) region, which is identical for antibodies of all specificities within a particular class (isotype defined below). The constant region occupies about three-quarters of the molecule. The specificity of an antibody for a particular antigen is determined by the amino acid sequence in the variable region, which differs from antibody to antibody. Phagocytic cells such as macrophages have Fc receptors, thus facilitating phagocytosis of antigens bound to antibody. In addition to facilitating phagocytosis, antibodies may, in the presence of complement, specifically lyse bacteria or cells bearing tumor or viral antigens on the surface, or neutralize viruses (preventing entry into host cells). Antibodies may also facilitate macrophage and NK-cell-mediated cytotoxicity.

There are several classes (called isotypes) of immunoglobulin molecules based on the structure of their heavy chains: IgM, IgG, IgA, and IgE. IgM is the predominant antibody in the primary immune response (following the initial exposure to an antigen). IgG appears later following a primary infection but is the predominant antibody in the secondary response (following subsequent exposure to the same antigen). Human IgG has been further characterized into four subclasses: IgG_1, IgG_2, IgG_3, and IgG_4. IgE acts as a mediator of allergy and parasitic immunity. IgA is found in secretions such as mucus, tears, saliva, and milk, as well as serum, and acts locally to block entrance of pathogens through mucous membranes. A fifth class, IgD, is mainly membrane-bound on B cells. Its function is unknown. Although a given B cell forms antibody to just one single antigen during its lifetime, it can switch to make a different class of antibody. For example, in allergic reactions, class switching within a given B-cell clone involves progression from synthesis of IgM to synthesis of IgG1, to synthesis of IgE.

Memory is built into the immune system such that specific immune T- and B-cell responses to most antigens are activated more rapidly in cases where the immune system has encountered the antigen previously. Memory is mediated by long-lived T and B cells. It is this rapid recall that is responsible for the success of vaccination in preventing subsequent infection. T-independent antigens are an exception in that they give rise to predominantly IgM responses; and relatively poor, if any, memory is generated by repeated exposure.

32.3 IMMUNOTOXICOLOGY

32.3.1 Identifying Immune Suppressants

There is ample evidence that a number of chemicals, certain microbial products (e.g., mycotoxins), and ionizing and ultraviolet radiation (UVR) can suppress various components of the immune system and enhance susceptibility to both infectious and neoplastic disease in laboratory animals (Table 32.4). In addition, a limited number of clinical and epidemiologic studies have reported suppression of immune function and/or increased frequency of infectious and/or neoplastic disease following exposure of humans to some of these agents.

Because of the complexity of the immune system, tiered approaches to testing chemicals for immune suppression have frequently been employed. In some cases the first level of the tier relies solely on structural endpoints including changes in the weight of thymus and other lymphoid organs, histopathology of these organs, or differential blood cell counts. However, although these nonfunctional endpoints

TABLE 32.4. Selected Examples of Chemicals and Radiation that Are Immunosuppressive

Chemical Class	Example[a]
Polyhalogenated aromatic hydrocarbons	TCDD, PCB, PBB
	Hexachlorobenzene
Aromatic hydrocarbons	Organic solvents
Polycyclic aromatic hydrocarbons	Benzo[a]pyrene[b]
Aromatic amines	Benzidines
Heavy metals	Lead, cadmium, methylmercury
Oxidant gases	NO_2, O_3, SO_2, Phosgene[b]
Organotins	DOTC[b], DBTC[b]
Radiation	Ionizing, ultraviolet
Mycotoxins	Aflatoxin[c]
	Ochratoxin A[c]
	Trichothecenes T-2 toxin[c]
Other	Asbestos
	Diethylstilbestrol (DES)
	Dimethylnitrosamine
	Tobacco and environmental smoke

[a]TCDD, 2,3,7,8-tetrachlorodibenzo-p-dioxin; PCB, polychlorinated biphenyls; PBB, polybrominated biphenyls; DMBA, 7,12-di-methylbenz[a]anthracene; DOTC, di-n-octyltin dichloride; DBTC, di-n-butyltin dichloride.
[b]Effects in humans are unknown; for all other compounds without superscripts, changes have been demonstrated in both rodents and humans.
[c]Effects in humans unknown; veterinary clinicians have noted immunosuppression in livestock ingesting mycotoxins at levels below those that cause overt toxicity.

may be effective in identifying immunotoxic effects following exposure to high doses of chemicals, these endpoints are not very accurate in predicting changes in immune function or altered susceptibility to infectious agents or tumor cells at lower chemical doses [see Luster et al. (1992, 1993) for more information]. Hence, the first tier of tests often includes a limited number of functional assays designed to assess (1) antibody-mediated responses, (2) T-cell-mediated responses, and (3) NK-cell activity. Occasionally, a test for macrophage phagocytosis is included. The most commonly used immune function assay in laboratory animals assesses the IgM antibody response in a mouse or rat following challenge with sheep red blood cells (SRBC). Because the sheep red blood cell is a T-dependent antigen, T and B cells, as well as antigen-presenting cells, must be functional to have a successful immunization. In addition, T-cell function may be determined by assessing the proliferation response (via incorporation of ^3H thymidine) of spleen or lymph node cells. These cells may be obtained from a previously immunized and treated animal and challenged with specific antigen in vitro, or they may be taken from nonimmunized mice and stimulated with nonspecific mitogens in vitro. NK activity is usually assessed by the cytotoxic action of lymphocytes (usually spleen cells) against target cells that are particularly sensitive to NK activity in vitro.

A second tier of more sophisticated tests can be conducted when a more in-depth evaluation of an immunotoxicant is desired (usually in a research setting). This tier may include an evaluation of animal resistance to challenge with infectious agents or transplantable tumor cells. More sophisticated immune system tests might involve flow cytometric analysis of cell surface markers to assess the proportion of T cells (total CD3+, CD4+, and CD8+ subsets) and B cells or to assess other cell surface markers associated with lymphocyte activation. Tier II can also include assessment of production or activity of an assortment of cytokines. As described below, sophisticated studies of some immunotoxicants have been done to assess the impact of immunotoxicants on the cell signaling that leads to lymphocyte activation.

Similar to animal studies, information obtained from routine hematology (differential cell counts) and clinical chemistry (serum immunoglobulin levels) may provide general information on the status of the immune system in humans. However, as with the animal studies, these may not be as sensitive or as informative as assays that target specific components of the immune system and/or assess function. Increased availability of monoclonal antibodies against specific leukocyte surface molecules and improved flow cytometry technology have made enumeration of lymphocyte subsets in peripheral blood possible. These assays are relatively accurate and reproducible, although, as with most immune tests, considerable variability exists within the human population. The assessment of certain lymphocyte surface antigens has been successfully used in the clinic to detect and monitor the progression or regression of leukemias, lymphomas, and HIV infections, all diseases associated with severe immunosuppression. However, the clinical significance of slight to moderate quantitative changes in the numbers of immune cell populations has not been established. There is consensus within the immunotoxicology community that tests that measure the response to an actual antigen challenge are likely to be more reliable predictors of immunotoxicity than flow cytometric assays for cell surface antigens because the latter generally only assesses the state of the immune system at rest. For ethical reasons it is difficult to conduct controlled studies of human immune responses to antigen challenge following exposure to toxic chemicals. One potential approach is to assess responses to vaccines in chemically exposed populations. This approach has been successfully used to assess the effects of stress on immune responses (Glaser, 2005).

32.3.2 Identifying Proteins and Chemicals that Cause or Exacerbate Allergic Disease

To date, toxicologists have primarily been concerned with two types of allergic effects: reactions in the respiratory tract that can lead to rhinitis and/or allergen-triggered asthma and reactions in the skin, referred to as contact hypersensitivity (CHS) or allergic contact dermatitis (ACD)—for example, poison ivy reactions. In each case the initial exposure primes the immune response (induction or sensitization). Reactions are manifested following subsequent exposures (challenge or elicitation). Chemicals may be involved in allergic reactions in the lung either directly as the allergen or indirectly by enhancing allergic sensitization to common allergens (e.g., dust mite, cockroach, pollen) or by exacerbating the symptoms associated with subsequent exposure. Agents that enhance sensitization are sometimes referred to as adjuvants.

Certain proteins and low-molecular-weight (<3000) compounds have the potential to cause allergic sensitization. Responses to subsequent inhalation exposure include pulmonary inflammation, increased mucus secretion, specific bronchial hyperreactivity to the offending allergen, and nonspecific bronchial hyperresponsiveness to challenge with an agonist such as methacholine. These symptoms are generally considered to be the hallmarks of allergic asthma and represent a significant health hazard. In humans, these reactions to allergen challenge can lead to chronic inflammation characteristic of chronic asthma. In both animal models and in humans, genetic predisposition is an important determinant in the development of allergy and asthma. Also, although allergens are important triggers for asthma, the majority of individuals who have allergies do not develop asthma.

The ability of proteins and low-molecular-weight chemicals to induce allergy varies. In general, the potential to induce this type of response and the relative potency of different proteins or chemical agents has been assessed in guinea pigs or mice sensitized by the respiratory route and monitored for the development of cytophilic antibody (IgG1 in guinea pigs; IgE in mice and humans) as well as increased respiratory rate following pulmonary challenge. These approaches are expensive and not amenable to routine testing. As more is learned about the mechanisms associated with allergic asthma (described in more detail in Section 32.5.3), it is hoped that better test methods will be developed. In humans, similar endpoints are used to assess individuals for allergic responses. The skin prick test, in which different proteins are injected under the skin, tests for the presence of cytophilic antibodies and helps to identify which proteins are causing a response in an individual. Under very controlled situations, patients may be exposed via the respiratory route to potential allergens (broncho-provocation test) and respiratory function monitored to pinpoint the offending allergen.

Because low-molecular-weight chemicals are haptens and must react with a host protein in order to be allergenic, efforts have also been made to identify chemicals with potential for allergenicity based on the probability that a given chemical structure will react with protein. Two classes of compounds, diisocyanates and acid anhydrides, fit this description; and some, but not all, chemicals in these two groups have been identified as respiratory allergens following occupational exposure of humans and/or experimental exposure of laboratory animals.

In laboratory rodent studies, air pollutants such as diesel exhaust and residual oil fly ash have been shown to behave as adjuvants (i.e., they enhance allergic sensitization to common allergens). These studies have been undertaken in a research setting using a variety of strategies. There is no standard approach to assessing chemicals for such effects. Likewise, epidemiology has shown an association between episodes of high air pollution and exacerbation of asthmatic symptoms requiring hospital emergency room visits. Again, however, there is no standard approach to testing chemicals for the capacity to exacerbate allergic symptoms in sensitized individuals.

In contrast to respiratory hypersensitivity, methods to assess low-molecular-weight chemicals (including drugs, pesticides, dyes, cosmetics, and household products) for the potential to induce contact sensitivity (dermatitis) are well established. These protocols assess the actual disease endpoint, skin irritation, following sensitization and challenge with the test agent. Two tests that have been commonly used for decades are the guinea pig maximization test and the Buehler occluded patch test. The endpoint that is assessed, erythema and edema (redness and swelling), is somewhat subjective (see Klecak for detailed descriptions of the guinea pig tests).

Recently, a more economical, less subjective test for contact hypersensitivity has been developed using mice. This test, the local lymph node assay (LLNA), assesses the proliferative response of lymphocytes in the draining lymph node following application of the agent to the ear and is based on our understanding of the immunologic mechanisms responsible for sensitization (see Section 32.5.5). The LLNA has been approved as a stand-alone alternative to the guinea pig tests by the Interagency Coordinating Committee on the Validation of Alternative Methods (ICCVAM).

32.4 MECHANISMS OF IMMUNE SUPPRESSION

From the preceding brief description of the immune system, it is clear that toxic substances may potentially disrupt this system at many levels. This section provides a more detailed description of the different mechanisms that have been shown to account for immunotoxicity of representative compounds.

32.4.1 Cyclosporin A and Glucocorticoids

Cyclosporin A (CsA) is an immunosuppressive drug derived from fungi that inhibits the early phase of T-cell activation following binding to an intracellular receptor. It is used in humans to control rejection of organ transplants and to treat certain autoimmune diseases. CsA has also been used extensively by immunologists as a tool to evaluate the mechanisms underlying T-cell functions and is sometimes used by immunotoxicologists as a positive control. As a result, the mechanisms of immunosuppression have been more clearly elucidated for CsA than for most other immunotoxicants.

Glucocorticoids (GC) produced endogenously in response to stress alter both innate and antigen-driven immunity. Cortisol (hydrocortisone) is the predominant GC in humans and most other species, whereas corticosterone is the most plentiful in mice and rats. Both natural and synthetic GC are used to reduce inflammation, manage autoimmune and hypersensitivity diseases, and treat certain neoplastic diseases. Certain inflammatory cytokines, including IL-1, IL-6 and TNFα stimulate (directly and indirectly) GC synthesis and release, a feedback mechanism to control inflammation.

Both CsA and GC enter cells by diffusion and mediate immunosuppressive effects by binding to receptors: cyclophilin in the case of CsA, and glucocorticoid receptor (GR) for GC. CsA prevents activation of lymphocytes. In contrast, GC prevent transcription of proinflammatory cytokines.

One of the best-characterized effects of CsA is inhibition of T-cell-receptor-mediated signal transduction pathways, resulting in decreased cytokine (IL-2, IL-3, IL-4, TNFα, and GM-CSF) production. Effects on IL-2 production have received the most research attention (Figure 32.4). In normal T cells, antigen binding to the T-cell receptor (TCR) leads to the phosphorylation and activation of phospholipase Cγ (PLCγ) by the enzyme ZAP 70. PLCγ then catalyzes the metabolism of phosphatidylinositol bisphosphate to diacylglycerol (DAG) and inositol triphosphate (IP$_3$). IP$_3$ stimulates the release of Ca^{2+} from intracellular stores, thus activating a serine/threonine phosphatase enzyme, calcineurin. Calcineurin then dephosphorylates a constitutively expressed, T-cell-specific cytosolic protein, nuclear factor of activated T cells (NF-AT$_c$). Dephosphorylation enables NF-AT$_c$ to enter the nucleus, where it combines with nuclear NF-AT$_n$ to form the active dimer NF-AT, a potent transcription factor that binds to the promoter region of the IL-2 gene. Also, DAG

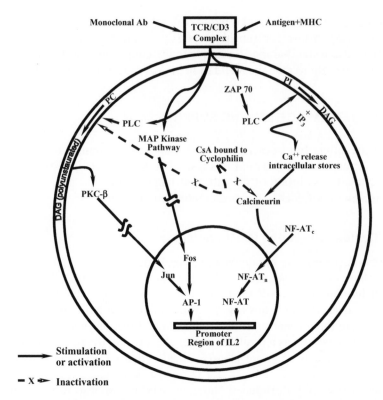

Figure 32.4. Cyclosporin A (CsA) disruption of signal transduction pathways leading to IL-2 production. CsA binds to cyclophilin in the cytoplasm. The complex disrupts at least two signaling pathways, decreasing activation of transcription factors AP-1 and NF-AT that lead to activation of genes involved in cytokine production. See text for detailed explanation. Abbreviations: TCR, T-cell receptor; PLC, phospholipase C; IP$_3$, inositol triphosphate; PKC, protein kinase C; DAG, diacylglycerol; NF-AT, nuclear factor of activation; PI, phophatidylinositol; PC, phosphatidylcholine.

stimulates PKC-β, which activates the transcription factor NF6B and the production of other proteins, including Jun. In the nucleus, Jun combines with Fos to produce the transcription factor AP-1. The third main signaling pathway involves activation of small GTP-binding proteins leading to production of several transcription factors via a MAP kinase cascade, including Fos. CsA has been shown to antagonize IL-2 gene transcription by interrupting at least two pathways: (1) inhibition of NF-AT$_c$ dephosphorylation by calcineurin, thus preventing translocation into the nucleus, and (2) blockade of PKC-β activity. Although CsA-mediated suppression of IL-2 synthesis requires binding of CsA to cyclophilin, it is the binding of the CsA/cyclophilin complex to calcineurin, and the ensuing inactivation of calcineurin, that prevents dephosphorylation and entry of NF-AT$_c$ into the nucleus. Likewise blockade of the PKC-β pathway of transcription factor production is not a direct effect of CsA on the enzymatic activity of PKC. Instead, activation of T-cell fatty acid metabolism is blocked by CsA/cyclophilin, preventing the normal switch from predominantly saturated fatty acids in the resting T-cell membrane to higher levels of polyunsaturated fatty acids in activated cells. For diacylglycerols to be effective stimulators of PKC activity, they must contain polyunsaturated fatty acids. The low content of unsaturated fatty acids in DAG of CsA-exposed T cells is thought to provide a poor stimulus for PKC-β activity. The ultimate effect is decreased availability of the protein Jun and hence decreased AP-1. CsA has also been shown to antagonize the binding of other transcription factors to enhancer elements in the IL-2 promoter region, including Oct/OAP and NFκB, probably via the same cyclophilin/calcineurin mechanism that inhibits NF-AT activity. In summary, CsA binds to a cytoplasmic receptor, cyclophilin. This drug–receptor complex subsequently disrupts at least two signal transduction pathways that begin with the T-cell receptor on the cell surface and lead to activation of genes and the production of cytokines that are critical in the early activation of T cells.

GR are found in the cytoplasm of many cells as a complex with molecular chaperones, including heat shock proteins and immunophilins. Following GC binding, the activated ligand–receptor complex dissociates from the chaperone proteins and is transported to the nucleus, where two GR molecules form a homodimer that binds to glucocorticoid responsive elements (GREs), leading to both increased production of proteins that inhibit both NF-kB and AP-1, and mitogen-activated protein kinase phosphatase-1 and production of proinflammatory products (e.g., cytokines, chemokines and their receptors, adhesion molecules). However, these effects on transcription typically occur at high concentrations of GC, and some have questioned the biological relevance of these modes of action at the relatively low therapeutic doses of GC required to reduce inflammation.

Certain proinflammatory genes lack GRE, but their transcription may still be inhibited by GC independent of GR–GRE interactions—for example, by direct inhibitory protein–protein interaction with transcription factors for proinflammatory genes (AP-1, and NF-κB). Furthermore, activated GR directly interfere with histone/DNA unwinding, leading to reduced transcription of proinflammatory genes; this mechanism is responsible for a significant portion of total GC effects on transcription. GR also directly decrease the stability of mRNA for certain proinflammatory enzymes associated with the MAP kinase signaling pathway and upregulate expression and prolong stability of an inhibitor of p38 MAP kinase. Hence, there are several mechanisms by which GC inhibit NF-κB, AP-1, and p38 MAP kinase and inhibit the production of proinflammatory cytokines.

32.4.2 Halogenated Aromatic Hydrocarbons

2,3,7,8-Tetrachlorodibenzo-*p*-dioxin (TCDD) is the most toxic of the group of structurally related compounds known as halogenated aromatic hydrocarbons (HAHs). These HAHs include the polychlorinated dibenzodioxins (PCDDs, of which TCDD is a member), polychlorinated dibenzofurans (PCDFs), and polychlorinated biphenols (PCBs). Based on the structure of these HAHs (see Figure 32.5), chlorine substitutions can exist in any or all of 8 or 10 positions in the compound. As a result, there are over 400 chlorinated HAH congeners possible. PCDDs and PCDFs are primarily formed as byproducts during the production of chemicals derived from chlorinated phenols. Because of their physical and chemical properties, PCBs have been used in a number of commercial applications. Approximately 30 of the chlorinated HAHs (7 PCDDs, 10 PCDFs, and 13 PCBs) display toxic properties similar to that of TCDD, albeit having lower potencies. Of the myriad toxicities associated with TCDD exposure of experimental animals,

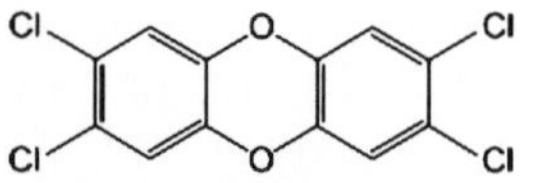

Figure 32.5. Molecular structures of chlorinated aromatic hydrocarbons. Chlorine substitutions can exist in any or all of 8 or 10 positions, resulting in over 400 possible congeners.

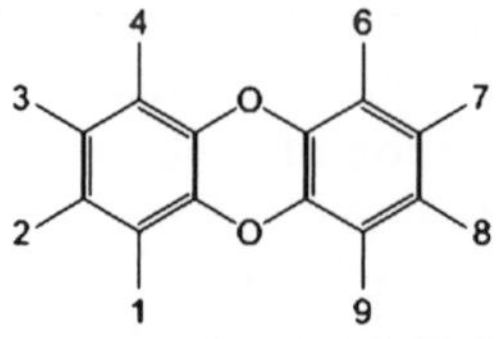

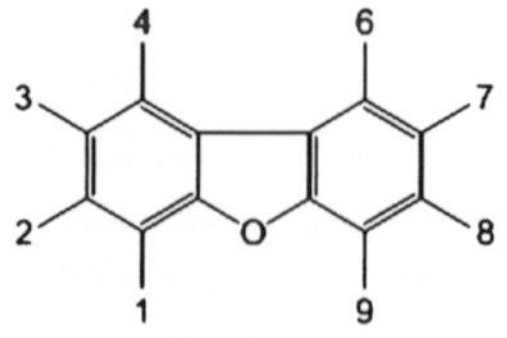

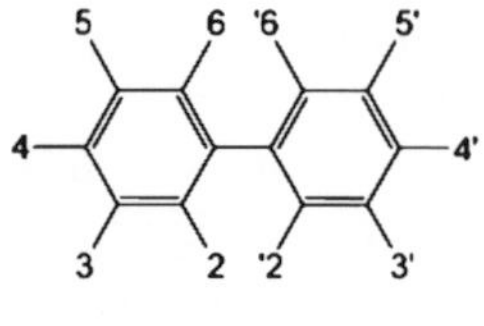

thymic and lymphoid tissue atrophy and immune responses are among the most sensitive endpoints.

In experimental mouse studies, TCDD exposure results in thymic atrophy and alterations in an array of adaptive immune responses including delayed-type hypersensitivity (DTH), cytotoxic T lymphocyte (CTL) activity, and T-cell-dependent antibody responses. In contrast, TCDD enhances neutrophil recruitment to the site of antigen challenge. Because both cell-mediated and humoral immunity are suppressed by TCDD and related HAHs, it is not surprising that administration of these compounds to mice results in increased susceptibility to challenge with viral, bacterial, or parasitic diseases, as well as syngeneic tumors.

Suppression of T-cell-mediated immunity (i.e., DTH response) occurs at lower doses of TCDD when rodents are exposed perinatally compared to exposure as adults. TCDD exposure during immune system development also results in a persistent suppression of immune function. For example, maternal exposure to submicrogram/kilogram doses on gestation day 14 results in suppressed DTH responses that persist in offspring for at least 19 months. Such exposure to TCDD results in alterations in fetal and neonatal CD4 and CD8 thymocyte precursor populations. This suggests that the mechanism underlying the immunotoxicity of perinatal exposure to TCDD may be interference with maturation of T cells in the thymus, including effects on positive or negative selection.

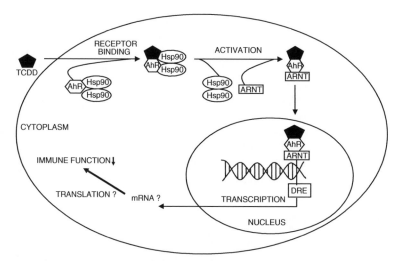

Figure 32.6. Model of AhR-mediated mechanism of action of TCDD and related HAHs. TCDD binds to the AhR in the cytoplasm. This complex interacts with another cytosolic protein (ARNT) and is translocated into the nucleus where it binds to specific DNA enhancer sequences upstream of TCDD responsive genes. Exactly which genes are important in the resulting immune suppression is unclear. Abbreviations: AhR, aryl hydrocarbon receptor; ARNT, AhR nuclear transporter; DRE, dioxin-responsive elements.

In adult mice the primary antibody response to sheep red blood cells (SRBC) is one of the most sensitive and reproducible immune endpoints that is affected by TCDD and related HAHs. For example, the TCDD dose that suppresses this response by 50% in mice is approximately 0.7 µg TCDD/kg. The antibody response to SRBCs is dependent on the collaborative interaction of antigen-presenting cells (dendritic cells) and T and B lymphocytes; all of these cells appear to be targets for TCDD.

Research on the mechanism(s) by which TCDD and related HAHs induce bio-chemical and toxic effects has primarily focused on the aryl hydrocarbon receptor (AhR) (Figure 32.6). The AhR functions as a ligand-activated transcription factor, which works in a fashion similar to that of steroid-receptor-mediated responses. Stereospecific binding of the chemical and AhR occurs within the cytoplasm of the cell, followed by interaction of the AhR–ligand complex with another cytosolic protein called the AhR nuclear transporter (ARNT). This ligand–receptor complex is then translocated into the nucleus, where it binds with high affinity to specific DNA enhancer sequences, called dioxin-responsive elements (DRE), upstream of TCDD responsive genes, initiating transcription of structural genes such as CYP1A1 in hepatocytes. These genes encode mRNA for production of the enzyme cytochrome P450-1A1 as well as other gene products that regulate differentiation and prolifera-tion of cells. However, the induction of P450 enzymes has not been found to play a role in the immunotoxicity of TCDD; immune suppression was observed in CYP1A1 deficient mice treated with TCDD. The molecular events that follow AhR activation and result in immune suppression are unclear, although modulation of immunophilin function and cross-talk between AhR and NFκB are potential candidates.

Two areas of investigation provide evidence of a central role for the AhR in TCDD-induced immunosuppression. First, differences in susceptibility to TCDD-induced immunosuppression have been reported in different inbred strains of mice associated with allelic variation at the Ah locus. Second, structure–activity relationships (SAR) for HAH-induced immunosuppression have been reported to be associated with the AhR binding affinity of TCDD relative to that of TCDD-like HAHs. Similarly, in vitro work with B-cell lines, which differ in their expression of AhR, has further demonstrated an association between AhR expression and sensitivity to TCDD. Also, inhibition of antibody secretion and Ig heavy-chain transcription, in a B-cell line that displays marked expression of AhR protein, was found to follow a rank order potency of HAHs related to the congener's binding affinity. This rank order potency of HAHs for immunosuppression correlated with the induction of CYP1A1 expression in these cells.

32.4.3 Ultraviolet Radiation

Ultraviolet radiation (UVR), a form of non-ionizing radiation, falls within the wavelengths of 100–400 nm in the electromagnetic spectrum. The wavelengths of concern with respect to immune suppression are in the mid-range, 280–315 nm (UVB). The sun is the principal source of UVR exposure for people, although sun lamps used in medical therapy and cosmetic tanning also represent exposure sources of concern. In addition to the immune system, the skin and eye are major targets for UVR.

Mice exposed to UVR have suppressed resistance to UVR-induced skin cancers, decreased responses to contact sensitizers, suppressed DTH responses to antigens injected into the footpad, and decreased resistance to a variety of intracellular infectious agents, including the causative agents of Lyme disease, leprosy, tuberculosis, herpesvirus infections, and leishmaniasis. IgM and IgG antibody responses are not affected. Contact hypersensitivity (CHS) responses (described in more detail in Section 32.5.5) are suppressed when a sensitizer is applied to UV-irradiated skin (local suppression) or when a sensitizer is applied at a site distant from that of UV radiation (systemic suppression). UVR-induced immune suppression appears to involve the induction of antigen-specific "suppressor" T cells. When lymphocytes from UVR-treated and immunized mice are adoptively transferred into untreated syngeneic mice that are subsequently challenged with the antigen, immune suppression similar to that observed in the donor is also observed in the recipient. When exposed to UVR at the time of initial immunization, mice become tolerant (immunologically unresponsive) to the immunogen; it is subsequently difficult to immunize them with the same antigen even in the absence of UVR. Again, this suggests the presence of suppressor cells. Two types of regulatory T cells have been associated with UVR-induced immune suppression, CD4+ CD25+ CTLA-4+ cells that secrete IL-10 and natural killer T cells (CD4+ T cells that also express markers normally found on NK cells). Suppression of contact sensitivity responses and the development of tolerance have also been demonstrated in humans following local exposure, and UVR-induced immune suppression is thought to play a role in the development of non-melanoma skin cancers in humans.

Because UVR does not penetrate below the skin, systemic effects must occur indirectly via mediators produced in the skin. Target cells in the skin include Langerhans cells (dendritic cells in the skin) and keratinocytes (Figure 32.7). As indicated in Section 32.5.5, exposure to contact sensitizers causes increased expression of MHC II molecules on Langerhans cells and increased expression of intracellular adhesion

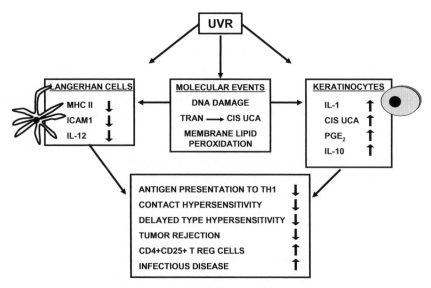

Figure 32.7. Mechanism underlying UVR-induced immune suppression. Uroconic acid, membrane lipids, and DNA are molecular targets. Langerhans cells and keratinocytes are the cellular targets. Mediators produced by skin cells have both local and systemic effects. Abbreviations: MHC, major histocompatibility complex; ICAM, intracellular adhesion molecule; UCA, urocanic acid; IL, interleukin; PGE, prostaglandin E.

molecules (ICAM), both of which facilitate antigen presentation. In UVR-exposed mice, however, the expression of MHC II and ICAM are suppressed. Antigen presentation (both locally via Langerhans cells and in the spleen) is altered in such a way that activation of Th1, but not Th2, cells is suppressed. In response to UVR, keratinocytes produce a number of mediators, which have immunosuppressive potential, including TNFα, IL-10, *cis*-urocanic acid, and prostaglandin E$_2$ (PGE$_2$). It is postulated that these affect immune sensitization locally and also spill into the circulation to cause systemic suppression. Several studies support this theory. When supernatants from keratinocyte cultures exposed to UVR were injected into mice, suppression of the DTH response occurred in a manner similar to that observed in UVR exposed mice. When IL-10 activity in the supernatant was blocked with antibody, immune suppression did not occur, indicating that keratinocyte-derived IL-10 could be responsible for the systemic suppression observed following UVR exposure. Likewise, suppression of DTH in UVR-irradiated mice was reversed by injecting mice with antibodies to IL-10. Studies also showed that injection of IL-12, which drives Th1 responses and counteracts IL-10, overcame UVR-induced immune suppression.

TNFα is thought to be an important mediator in the suppression of CHS. This hypothesis is supported by the finding that injection of antibodies to TNFα reverses the suppression of CHS that usually occurs following UVR exposure. Conversely, intracutaneous injection of recombinant TNFα at the site of sensitization suppresses the development of a contact sensitivity response. Also, the genetic difference between UVR-resistant and UVR-susceptible strains of mice appears to involve a polymorphism in a regulatory region of the TNFα gene.

Three epidermal photoreceptors have been identified that convert UVR energy into biologic signals that mediate immune suppression: DNA, urocanic acid, and membrane lipids. UVR exposure is known to induce DNA damage in the form of pyrimidine dimers. In studies that introduced a bacteriophage-derived pyrimidine dimer repair enzyme into keratinocytes (via liposomes), UVR-induced suppression of both DTH and CHS did not occur. Also, nonspecific enzyme-induced double-stranded DNA breaks caused suppression of CHS both locally and systemically and induced the production of IL-10. These data strongly suggest that DNA damage plays an important role in UVR-induced immune suppression. UVR exposure causes isomerization of urocanic acid in the skin from *trans*-urocanic acid to *cis*-urocanic acid. When *cis*-urocanic acid was painted on the skin or injected subcutaneously or intravenously, the resulting immune suppression mimicked that observed with UVR. DTH responses were completely restored in mice treated with antibody to *cis*-urocanic acid. Finally UVR perturbs cellular redox, leading to free radical formation and membrane lipid peroxidation. Treatment with antioxidants blocks UVR suppression of CHS and induction of tolerance. It is not clear whether DNA damage, urocanic acid isomerization, and lipid peroxidation work together or separately and what, if anything, determines the predominant pathway.

UVR is frequently referred to as a complete carcinogen because it has both initiation and promotion properties. Immune suppression may be responsible for promotion. It should be noted that many chemical carcinogens are also immunosuppressive. The observation that DNA damage induces immune suppression has ramifications that extend beyond UVR-induced immune suppression. Furthermore, similar mechanisms account for immune suppression observed after dermal exposure to jet fuels, particularly JP-8, suggesting that mechanisms described here are not unique to UVR.

32.4.4 Cyclophosphamide

The drug cyclophosphamide (Cy) is a cytostatic phosphamide derivative of bis(β-chloroethyl)amine (nitrogen mustard). Cy has been used extensively in humans, alone and in combination with other drugs, to treat a variety of neoplastic, lymphoproliferative, and autoimmune diseases and as an immunosuppressant to prevent rejection of foreign tissue grafts. Although used therapeutically, Cy is also a human carcinogen. Cy causes profound immunosuppression in rodents. Doses in the range of 30–100 mg/kg/day have often been used as a positive control for immunosuppression in immunotoxicology studies. Antibody production, DTH, cytotoxic T-cell activity, mixed lymphocyte responses, NK cell activity, and resistance to viral, parasitic, bacterial, or tumor cell challenge are all suppressed by Cy. Antibody-mediated immunity is more sensitive to suppression by Cy than cellular responses. However, if immunity is established prior to Cy exposure, the effect of daily Cy administration on antibody production is slight and is quickly reversed once the drug is withdrawn.

Cy immunosuppression is the result of cytotoxicity. Metabolites of Cy are alkylating agents which covalently crosslink DNA, leading to inhibition of DNA synthesis, cell cycle arrest, and, if DNA repair is unsuccessful, apoptosis. The ultimate result of these cytotoxic events is blockade of cell proliferation and clonal expansion, central components of the immune response. Cy itself has no alkylating or cytotoxic activity; rather, hepatic oxidation via microsomal P450 enzymes (specifically, the CYP2B isoform), followed by other enzymatic (primarily aldehyde dehydrogenase) oxidative steps, is required to produce reactive metabolites (Figure 32.8). The initial

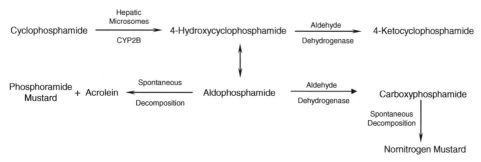

Figure 32.8. Metabolic activation pathway for cyclophosphamide. Acrolein, phosphoramide mustard, and nornitrogen mustard are the immunotoxic metabolites.

P450 oxidation product is 4-hydroxycyclophosphamide (4-OHCy), which has little immunosuppressive activity; 4-OHCy and its tautomeric form, aldophosphamide, are present in equilibrium. The 4-OHCy metabolite enters cells easily by diffusion and may represent a transport form of Cy. Acrolein, a toxic byproduct, and phosphoramide mustard, the metabolite believed to have the greatest biological activity, are formed by the spontaneous decomposition of aldophosphamide. Both of these metabolites are immunosuppressive. 4-Ketocyclophosphamide and carboxyphosphamide are formed by oxidation of 4-OHCy and aldophosphamide, respectively. These metabolites are the main decomposition products, have little immunosuppressive activity, and represent the major metabolic products excreted in the urine. Carboxyphosphamide, the major metabolic product of Cy, is further metabolized to the immunotoxic metabolite nornitrogen mustard, also present in the urine. Rates of metabolite formation and elimination are quite variable in humans and have been associated with aldehyde dehydrogenase gene variants that determine levels of ADH activity. Mouse strain differences in sensitivity to both the toxic and immunosuppressive effects of Cy exist as well.

In summary, highly reactive metabolites of Cy disrupt DNA synthesis, thus inhibiting cell proliferation. As a result, production of lymphocytes and accessory cells is suppressed, and host immunocompetence is compromised. As noted below and summarized in Table 32.5, for a number of immunosuppressive compounds a metabolic product rather than the parent compound is responsible for the toxicity.

32.4.5 Polycyclic Aromatic Hydrocarbons

PAHs such as benzo[*a*]pyrene (B[*a*]P) are metabolized to reactive electrophilic intermediates primarily in the liver, and reactive metabolites are transported by serum proteins to other tissues. These reactive metabolites, primarily diol epoxides, alter normal cellular function by binding covalently to RNA, DNA, and proteins. A number of other mechanisms have also been proposed to explain PAH-induced immunosuppression, including membrane perturbation resulting in altered signal transduction, calcium mobilization, gene expression, and/or cytokine production. In laboratory animal studies, there is a significant correlation between suppression of antibody forming cell responses and the carcinogenic activity of PAHs, suggesting that there may be a common mechanism.

TABLE 32.5. Examples of Immunotoxic Compounds Requiring Metabolic Activation

Class	Compounds
Mycotoxins	Aflatoxin
	Ochratoxin A
	Wortmannin
Solvents	Benzene
	Carbon tetrachloride
	Ethanol
	N-Hexane
	2-Methoxyethanol
PAHs	Benzo[a]pyrene
	Dimethylbenzanthracene
	3-Methylcholanthrene
Pesticides	Chlordane
	Malathion
	Parathion
Other	Cyclophosphamide
	Dimethylnitrosamine

PAHs have been shown to increase intracellular calcium levels in rodent and human lymphocytes. In T lymphocytes, PAH-induced activation of protein tyrosine kinases Fyn and Lck and increased tyrosine phosphorylation of PLCγ result in depletion of intracellular calcium stores. Activation of the T cell following antigen binding to TCR requires calcium mobilization (see Section 32.4.1, Figure 32.5). PAH apparently interferes with this process by depleting intracellular calcium stores, which causes premature signaling and leads to tolerance. In B lymphocytes, PAH also elevates intracellular calcium levels. P450 metabolism appears to play an important role in this altered cell signaling, as PAH metabolites are significantly more effective in elevating intracellular calcium levels than their parent compounds, and treatment with P450 inhibitors has been shown to prevent calcium increases. Prolonged elevation of intracellular calcium precedes apoptosis, and at higher concentrations, PAHs induce apoptosis in lymphoid precursors in the bone marrow, circulating B and T lymphocytes, and lymphocytes in the thymus, spleen and lymph nodes.

In vivo exposure to B[a]P has been shown to target both primary and secondary immune tissues, and to significantly alter lymphoid cell numbers and cell surface antigen expression in the spleen, thymus and bone marrow. In rodents, in vivo exposure to B[a]P inhibits both antibody- and cell-mediated immunity, as well as some aspects of innate immunity including macrophage phagocytosis and interferon production. Although DNA adduct formation is similar in immune and nonimmune tissues after in vivo exposure, in vitro cultures of murine splenocytes have little ability to generate B[a]P/DNA adducts, suggesting that hepatic bioactivation is an important mediator of B[a]P-induced immunotoxicity. In vitro studies using a murine splenocyte-rat hepatocyte coculture system demonstrated that biotransformation of B[a]P by Phase I enzymes is required in the suppression of T-dependent antibody responses against sheep red blood cells.

32.4.6   Organic Solvents

Organic solvents, which induce CYP2E1, are comprised of a few broad chemical classes, including hydrocarbons such as benzene and toluene, halogenated aliphatic compounds such as carbon tetrachloride and dichloroethane, aliphatic alcohols such as ethanol, and hydroxyethers such as 2-methoxyethanol. Industrial solvents are frequently mixtures of several compounds. The most frequent solvent-associated toxicity occurs from occupational exposure. A number of organic solvents have been examined for their effects on the immune system, and the requirement for their bioactivation to produce immunotoxicity has been well established.

Of the organic solvents that induce CYP2E1, benzene and its metabolites have been most extensively studied with respect to immunotoxicity because they have long been associated with hematologic and immunologic disorders, including leukemia in humans. A number of experimental studies suggest that benzene metabolites, including hydroquinone, catechol, and phenol, are responsible for its hematotoxicity. Benzene is metabolized by hepatic CYP2E1 primarily to phenol and in turn to hydroquinone and/or catechol. The phenolic metabolites preferentially accumulate in the bone marrow and lymphoid tissues of rodents. In the bone marrow, enzymatic conversion of both phenol and hydroquinone to more reactive binding species such as the semiquinone radical may be P450-independent and involve myeloperoxidases and prostaglandin synthetases. The semiquinone radical binds covalently to cellular proteins and forms DNA adducts, disrupting normal cellular functions such as cell division and mitochondrial RNA synthesis. Rapidly proliferating cells, such as lymphoid and myeloid progenitor cells in the bone marrow or clonally expanding lymphocyte subpopulations, are highly sensitive targets for such effects.

The earliest manifestation of benzene toxicity in exposed workers is a decrease in lymphocyte counts. A variety of blood disorders, including leukopenia, thrombocytopenia, granulocytopenia, and aplastic anemia, have been associated with benzene exposure. Studies in laboratory animals have demonstrated that treatment with benzene or its metabolites induces myelo- and immunosuppression. Benzene and its metabolites appear to be particularly cytotoxic to progenitor cells within the bone marrow, targeting the lymphocyte, monocyte, granulocyte, and erythrocyte lineages. There is also considerable evidence that benzene metabolites alter the stromal cell population in the bone marrow, which supports the differentiation and maturation of progenitor cells. In vivo exposure to benzene and/or its metabolites inhibits T-dependent antibody responses, B- and T-cell lymphoproliferative responses, and cytotoxic T-lymphocyte-mediated tumor cell killing, and it increases susceptibility to challenge with infectious agents.

Structure–activity studies suggest that the polyhydroxy metabolites of benzene have the most immunosuppressive activity and that benzene and phenol are significantly less toxic to both lymphoid and myeloid cells. Co-administration of thiol-reactive agents blocks hydroquinone-induced suppression of mitogen-stimulated lymphoproliferation and agglutination in rat splenocytes, indicating that oxidation to thiol-reactive quinones may be a critical step in bioactivation. Induction of CYP2E1 in rats by treatment with ethanol enhances the myelotoxicity of benzene and inhibition of CYP2E1 activity by administration of propylene glycol partially prevents benzene-induced immuno- and myelotoxicity.

TABLE 32.6. Some Mechanisms Associated with Immune Suppression

Mechanism	Example
Disruption of T-cell maturation in the thymus	In utero/perinatal TCDD
Inactivation of calcineurin disrupting signals needed for IL-2 gene transcription	Cyclosporin A
Covalent binding to DNA disrupting replication, cell proliferation, and function leading to cytotoxicity	Cyclophosphamide Polyaromatic hydrocarbons Organic solvents
Same as above targeting bone marrow progenitor cells	Benzene
Interaction with intracellular receptor which acts as a ligand-activated transcription factor the product of which affect immune response by unknown mechanisms	TCDD, HAHs
Blockade of fatty acid metabolism in the membrane, thus impairing diacyglycerocapacity to stimulate protein kinase C activity	Cyclosporin A
Cytokine-mediated alterations in antigen presentation, resulting in suppression of Th1 activation and skewing toward a Th2 response	Ultraviolet radiation
Depletion of intracellular calcium stores, premature signaling leading to tolerance	Polyaromatic hydrocarbons

32.4.7   Summary of Immunosuppressive Mechanisms

The various mechanisms associated with immune suppression described above are summarized in Table 32.6. Several generalizations can be made. At the molecular level, damage to DNA of immune cells can have profound effects on cellular proliferation needed for clonal expansion. DNA damage of nonimmune cells has also been shown to modulate cytokine responses in a manner that suppresses the Th1 response and activates T regulatory cells associated with immune tolerance. Alternately, toxic chemicals may interact with intracellular receptors and, in so doing, impact signal transduction pathways. In some cases, induction of transcription factors occurs; in other cases the signals needed to activate transcription factors are interrupted. The net result is modulation of cytokine, cell surface molecule, or receptor expression, leading to alterations in the course of the immune response. Depletion of intracellular calcium stores may also interfere with signal transduction needed for lymphocyte activation. There are a number of cellular targets. Alteration of T-cell responses (frequently as a result of changes in cytokine responses), changes in the function of antigen presenting cells, toxicity to stem cells in the bone marrow, and change in the production of mediators by nonimmune cells all provide mechanisms that can lead to immune suppression. At the developmental level, depletion of bone marrow stem cells or interference with maturation and selection of cells in the thymus can lead to permanent immune suppression. The modulation of immune responses may result in suppression of some components of the immune system without affecting or even enhancing other components of the immune response. This will become more evident as some of the mechanisms associated with immune stimulation are described below.

TABLE 32.7. Gell and Coombs Classification of Hypersensitivity Reactions

Type I	Cytophilic antibody (IgE) binds to mast cells; antigen binds to antibody and crosslinks receptors, causing mediator (histamine) release; immediate hypersensitivity
Type II	IgM- and IgG-mediated cytolysis of cells
Type III	IgM, IgG complexed with antigen (immune complexes) leading to inflammation
Type IV	T-cell-mediated; delayed-type hypersensitivity

32.5 MECHANISMS ASSOCIATED WITH HYPERSENSITIVITY

32.5.1 Classification of Immune-Mediated Injury Based on Mechanisms

Under certain circumstances, immune responses, rather than providing protection (and sometimes in addition to providing protection), produce damage in response to antigens that might otherwise be innocuous. These deleterious reactions are collectively known as hypersensitivity or allergy. Classically, hypersensitivity reactions have been divided into four types, originally proposed by Gell and Coombs (Table 32.7), based on differences in the underlying immune mechanisms, although in some cases the pathology may appear to be similar. The four types are not mutually exclusive. Often hypersensitivity responses are mixed.

Type I hypersensitivity (sometimes referred to as atopy) is mediated by antigen-specific IgE, which binds via the Fc receptor to mast cells and basophils. On subsequent exposure, the allergen binds to these cell-bound antibodies and crosslinks IgE molecules, causing the release of mediators such as histamine and slow-reacting substance of anaphylaxis (SRS-A). These mediators cause vasodilation and leakage of fluid into the tissues, plus sensory nerve stimulation (leading to itching, sneezing and coughing). Type I is frequently referred to as immediate-type hypersensitivity because reactions occur within minutes after exposure of a previously sensitized individual to the offending antigen. Examples of Type I reactions include allergic asthma, allergic rhinitis (hay fever), atopic dermatitis (eczema), and acute urticaria (hives). The most severe form is systemic anaphylaxis (sometimes seen in response to a bee sting or food allergen), which can result in severe airway obstruction and cardiovascular collapse leading to anaphylactic shock and potentially death.

Type II hypersensitivity is the result of antibody-mediated cytotoxicity that occurs when antibodies respond to cell surface antigens. Frequently, blood cells are the targets, as in the case of an incompatible blood transfusion or Rh blood incompatibility between mother and child. Antibodies or antigen–antibody complexes bind to the cell and activate the complement system, which leads to enhanced phagocytosis or lysis of the target cell. Autoimmune diseases, such as immune hemolytic anemia and thrombocytopenia, can result from drug treatments with penicillin, quinidine, quinine, or acetaminophen. Presumably these drugs interact with the cell membrane such that the immune system detects "foreign" antigens on the cell surface. Hemolytic disease may also have unknown etiologies, although the mechanism is a Type II response.

Type III reactions are the result of antigen–antibody (IgG) complexes that accumulate in tissues or the circulation, activate macrophages and the complement system, and trigger the influx of neutrophils, eosinophils, and lymphocytes (inflammation). This is sometimes referred to as the Arthrus reaction and can lead to conditions such as glomerular nephritis. Farmer's lung, a pneumonitis caused by molds, has been attributed to both Type III and Type IV reactions.

Unlike the preceding three types, Type IV or delayed-type hypersensitivity (DTH) involves T cells and macrophages, not antibodies. Activated T cells release cytokines that cause accumulation and activation of macrophages, which, in turn, cause local damage. This type of reaction is very important in defense against intracellular infections such as tuberculosis, but is also responsible for contact hypersensitivity responses (allergic contact dermatitis) typified by the poison ivy response. Inhalation of beryllium can result in a range of pathologies including acute pneumonitis, tracheobronchitis, and chronic beryllium disease, all of which are associated with Type IV reactions. As the name indicates, the expression of Type IV responses following challenge is delayed, occurring 24–48 hr after exposure.

Type IV responses can be subdivided into three groups based on knowledge of T-cell biology obtained since Gel and Coombs devised their classification. CD4Th1 cells respond to soluble antigens presented by MHC II and activate macrophages to produce the classic tuberculin type reaction. CD4Th2 cells also respond to soluble antigens presented by MHC II but activate eosinophils leading to inflammation typically seen in allergy and asthma. CD8 cells respond to cell-associated antigens presented by MHC I and are directly cytotoxic.

32.5.2 Features Common to Respiratory and Dermal Allergies

The remainder of this section will focus in more detail on Type I (immediate) and Type IV (delayed) hypersensitivity with emphasis on the lung and skin as target organs. In both cases the adverse effects of hypersensitivity develop in two stages: (1) Induction (sensitization) requires a sufficient or cumulative exposure dose of the sensitizing agent to induce immune responses that cause no obvious symptoms. (2) Elicitation (challenge) occurs in sensitized individuals upon subsequent exposure to the allergen (immunogen) and results in adverse antigen specific responses that include inflammation. Because inflammation also occurs as an innate response, a distinction must be made between an irritant and a sensitizer. An irritant is an agent that causes local inflammatory effects but induces no immunological memory. Therefore, on subsequent exposures, local inflammation will again result, but there is no enhancement of the magnitude of the response and no change in the dose required to induce the response. In allergic inflammation, there may be no response to a sensitizer during the induction stage, but responses to subsequent exposures are exacerbated. The dose required for elicitation is usually less than that required to achieve sensitization. It should be noted that some sensitizers also produce irritation, particularly at higher doses.

The immunogens involved in hypersensitivity may be proteins or haptens. Although most proteins can stimulate an immune response, only some proteins cause allergic responses. However, no amino acid sequence motifs specific for allergenicity have been identified. Protein sensitizers are generally hydrophilic, are heat- and digestion-stable, and may have enzymatic activity. Examples include house dust mite, cat, and cockroach allergens. A hapten is a small molecule (≤3000 molecular weight) that is not itself immunogenic. After conjugation to a carrier protein, it becomes immunogenic and induces hapten-specific antibody, which can then bind the hapten in the absence of the carrier. Therefore, the hapten or its metabolite must have a reactive site for conjugation with a protein. This conjugate complex is processed and presented to lymphocytes. When conjugated with proteins residing in the respiratory system or in the skin, haptens become sensitizing agents. Haptens associated with allergy include chemicals that are generally man-made substances (e.g., trimellitic anhydride and toluene diisocyanate) and metals.

The events that initiate hypersensitivity responses in the respiratory tract and skin have homologous features. Like any adaptive immune response, the antigen must be presented to the immune system and ultimately to T lymphocytes. Dendritic cells (DC) have been identified as highly efficient antigen-presenting cells (APC) in both the epidermis (where they are called Langerhans cells) and the respiratory tract. These cells function to process the antigen and migrate to the draining lymph node, where the antigen is presented in the context of the MHC to lymphocytes. In the respiratory tract both APC activity and T-cell proliferation in response to stimulatory signals are usually inhibited by signals from resident macrophages and other adjacent (primarily epithelial) cells. As the DC mature and migrate to the draining lymph nodes, they become more effective at antigen presentation and less efficient at antigen uptake and processing. The environment in which initial interactions with T cells occur appears to play a role in immune outcome, specifically whether a Th1 (delayed-type) or Th2 (immediate-type) response ensues. As described in Section 32.5.8, cytokines produced in the local milieu can influence antigen presentation to favor either Th1 or Th2 responses.

32.5.3 Respiratory Allergy/Allergic Asthma

A major determinant in occupationally and environmentally induced respiratory diseases is immediate (Type I) bronchial hypersensitivity to specific allergens, often referred to as atopy. Atopy, which occurs in genetically predisposed individuals, is the presence of cytophilic, antigen-specific immunoglobulin IgE (or IgG1 in guinea pigs) that can lead to allergic rhinitis. A significant number of asthmatic individuals are atopic and have respiratory allergies that may act as stimuli for asthma attacks. In these individuals, antigen challenge causes an immediate (Type I) response. Between 2 and 8 hr after this event, a more severe and prolonged (late-phase) reaction occurs, which is characterized by mucous secretion, bronchoconstriction, airway hyperresponsiveness to a variety of nonspecific stimuli (e.g., histamine, methacholine, cold air), and airway inflammation characterized by eosinophils. Late-phase responses may last up to 12 hr and do not appear to be mediated by IgE. Th2 cells and associated cytokines (particularly IL-5 and IL-13) and eosinophils are thought to play a significant role.

Protein allergens have been shown to produce sensitization and respiratory allergic responses (both early- and late-phase) in humans, guinea pigs, and mice. Additionally, they have been strongly associated with asthma morbidity. Of the low-molecular-weight chemicals that have been associated with allergic asthma, certain diisocyanates and acid anhydrides have received the most attention. Whereas protein allergens characteristically result in antigen-specific IgE, the presence of allergen-specific IgE in low-molecular-weight chemical respiratory allergy has not been universally demonstrated. In addition, the early-phase response does not always occur in individuals with occupational asthma triggered by these low-molecular-weight compounds. The mechanisms underlying low-molecular-weight chemical respiratory allergy are still under investigation.

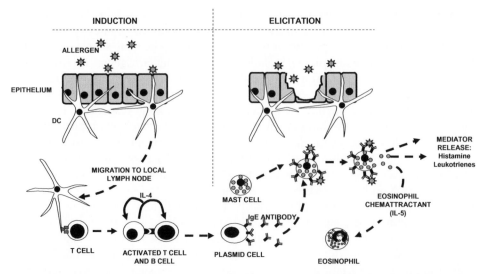

Figure 32.9. Schematic representation of Type I hypersensitivity. Induction: Resident respiratory tract dendritic cells (DC) take and process antigen, mature, migrate to the draining lymph nodes, and present antigen to T lymphocytes. Activated T-lymphocytes, in turn, activate B-cell differentiation into antibody-producing plasma cells. IL-4 promotes Ig isotype class switching from IgM to IgE and promotes mast cell development. IgE is associated with mast cells. Elicitation: Allergen crosslinks the mast-cell-bound IgE, thereby causing the release of preformed mediators and cytokines. (See Table 32.7.) Inflammation and bronchoconstriction occur.

Mechanisms underlying induction of IgE have been studied extensively. There are two primary requirements for B-cell Ig isotype class switching from IgM to IgE: (1) the presence of the Th2-associated cytokine IL-4 (or IL-13) and (2) direct cell-to-cell interaction via CD40 expressed on B-cells and CD40 ligand (CD40L), expressed on T-cells, basophils, and mast cells. IL-4 exhibits autocrine activity in Th2 cell differentiation and promotes mast cell development. Bone-marrow-derived immature mast cell precursors localize under the epithelium of mucosal areas (respiratory tract and gut) and the skin, where tissue-specific maturation and expansion occur. These mast cells contain preformed mediators and are capable of producing other effecter molecules including IL-4 and IL-5. The IgE-armed mast cells in the respiratory tract are then set for elicitation of antigen-specific allergic events (Figure 32.9). Upon subsequent exposure, the specific allergen crosslinks the mast-cell-bound IgE, resulting in the release of the preformed mediators and newly synthesized substances (summarized in Table 32.8). Chief among the preformed mediators is histamine, which acts through receptor ligation to cause increased vascular permeability, smooth muscle contraction, vascular constriction, and mucus production. In addition to histamine, there are a variety of preformed cellular chemotactic factors and enzymes. Products from the activation of two metabolic pathways of membrane-derived arachidonic acid, the lipoxygenase and cyclooxygenase pathways, provide other important mediators of allergic inflammation. The Th2-associated cytokine IL-5, secreted by mast cells, has been shown to be essential in proliferation and maturation of eosinophil precursors as well as in viability and eosinophil granule protein release and chemotaxis. Eosinophils that are recruited into the lung release their

TABLE 32.8. Mediators of Allergic Inflammation

Source	Mediator	Action
Mast cells	Histamine	Smooth muscle contraction
		Increased vascular permeability
		Vascular constriction
		Mucus production
	Enzymes: Tryptase, lysosomal hydrolases, proteoglycans	Tissue damage
	Arachidonic acid metabolites	
	Lipoxygenase pathway:	Smooth muscle contraction
	5-Hydroxyeicosatetraenoic acid	Increased vascular permeability
	(5HETE) and leukotriene B_4	Mucus production
	(LTB$_4$)	Chemotactic activity
	Cyclooxygenase pathway:	
	Prostaglandins (PGE$_1$, PGE$_2$, PGF$_{2a}$,	Increased vascular permeability
	PGD$_2$)	Smooth muscle contraction
	Prostacyclins	Vasodilation
	Thromboxanes (Thromboxane A_2)	Bronchoconstriction
Eosinophils	Major basic protein (MBP)	Microbial killing, tissue damage
	Eosinophil peroxidase (EPO)	
	Eosinophil cationic protein (ECP)	
	EPO/H$_2$O$_2$/halide	Mast cell degranulation

own set of mediators, which are thought to be important in late phase responses. Furthermore, cytokines (IL-5, IL-3, and GM-CSF) secreted by activated eosinophils found at the site of allergic inflammation may result in a positive-feedback loop.

32.5.4 Immediate-Type Hypersensitivity Skin Reactions

The respiratory tract is not the only site of atopic responsiveness. Atopic dermatitis/immunologic contact urticaria is an IgE-mediated inflammatory skin disease and is mechanistically generated in much the same manner as IgE-mediated respiratory allergy described in Section 32.5.3. The allergens involved are usually of high molecular weight and are not absorbed through the skin. Therefore a prerequisite would be skin damage such as irritant dermatitis. The symptoms (redness, pruritis, and skin lesions) occur within minutes to an hour or so following contact by a sensitized individual. Although clinically atopic dermatitis/immunologic contact urticaria is phenotypically similar to contact hypersensitivity (described in Section 32.5.5), these conditions are kinetically and mechanistically quite different.

32.5.5 Allergic Contact Dermatitis (Contact Hypersensitivity)

Allergic contact dermatitis (ACD) occurs as the result of repeated dermal exposure to low-molecular-weight chemicals (haptens), generating a Type IV response. Unlike other types of chemical hypersensitivity, where sensitization may occur following several routes of exposure, induction of ACD is limited to skin exposure. During the induction (sensitization) phase of this response, the chemical (hapten) couples to carrier proteins on dermal and epidermal cells to become fully immunogenic. The Langerhans cells take up and process the antigen and migrate to the regional draining lymph node, where they present the antigen to lymphocytes. Activation and rapid proliferation of lymph node cells ensues, resulting in the production of effector and memory T cells which travel back to the skin. During the elicitation phase, effector Th1 and CD8 cells are responsible for the erythema, edema, and pruritus that characteristically appear 24–72 hr post-exposure.

The application of hapten to the skin initiates a cascade of events (Figure 32.10). Epidermal keratinocytes secrete inflammatory cytokines (including IL-1β, IL-6, IL-12, TNF-α, and granulocyte-macrophage stimulating factor (GM-CSF)) that facilitate the maturation and mobilization of Langerhans cells, which also produce cytokines in an autocrine fashion. During Langerhans cell maturation, expression of cell-surface molecules (including MHC class I and II) and adhesion and co-stimulatory molecules is enhanced, thus facilitating antigen presentation and sub-sequent T-cell activation and clonal expansion. T cells activated in this manner express a skin homing receptor, cutaneous lymphocyte-associated antigen (CLA). Although T cells are thought to be the key effector cells in the development of CH, both T and B lymphocytes proliferate in response to contact sensitizers. As indicated in Section 32.3.2, enhanced proliferation in the draining lymph node is the endpoint used in the LLNA to assess chemicals for the potential to cause ACD.

Re-exposure to the relevant hapten triggers the same cytokine responses that occur following induction and elicits a response characterized by rapid recruitment and activation of specific T cells at the site of hapten challenge (Figure 32.10).

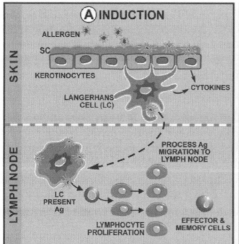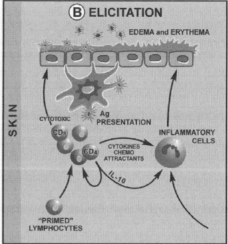

Figure 32.10. Schematic representation of allergic contact dermatitis. Induction: Low-molecular-weight allergen (hapten) binds to host protein. Immature Langerhans cells (LC) take up and process haptenated protein. Simultaneously, keratinocytes and LC release cyto-kines that promote LC migration to the draining lymph node and maturation into effective antigen-presenting cells. LC–T-cell interactions result in lymphocyte proliferation of "primed" effector lymphocytes with allergen-specific memory and skin homing receptor. Elicitation: Initial events are the same as for induction. "Primed" lymphocytes home to the area in a very specific response. CD8 cells are directly cytotoxic to antigen-bearing keratinocytes. Th1 cells are cytotoxic in the presence of IFNγ and release chemokines that promote inflammation and activation of mast cells. CD4-regulatory T cells secrete IL-10 to limit the reaction.

Hapten-specific CD8+ T lymphocytes, likely the major effector population, are directly cytotoxic to chemical exposed keratinocytes and also release cytokines that boost the inflammatory response. In addition, Th1 cells release a number of cytokines and chemokines that promote inflammation and activate mast cells and, in the presence of IFNγ, are also capable of killing keratinocytes. Although the hapten may persist in skin for some time, this reaction is self-limited. CD4+ regulatory T cells that secrete IL-10 (similar to those described in relation to UV-induced immune suppression in Section 32.4.3) appear to play an important role in this regulation.

Clinical manifestation, histopathology, and immunohistology of ACD are virtually indistinguishable from irritant responses (nonallergic), and in fact most haptens act as irritants at higher doses. One difference is that the erythema, edema, and pruritus that are characteristic of the ACD elicitation phase occur 24–72 hr after epicutaneous application of the allergen, whereas irritant responses occur much sooner. Irritant responses exhibit the same dose response regardless of the number of previous exposures, whereas the dose required to elicit a response to challenge in ACD declines with subsequent exposures. Among the myriad of mediators induced by both reactions, IL-1β appears to be specific for ACH and is not observed with nonspecific irritation. Finally, irritants do not induce hapten-specific immunologic responses.

32.5.6 Delayed-Type Hypersensitivity in the Lung

Type IV responses are not limited to the skin. Hypersensitivity pneumonitis (HP) (extrinsic allergic alveolitis), an allergic lung syndrome considered to be a mix of Type III (elevated levels of IgG to the antigen) and Type IV (CD8+ lymphocytes) hypersensitivity responses, is caused by inhalation exposure to a wide variety of organic dusts. These dusts contain antigenic substances, including fungal/bacterial components, serum proteins, and some chemicals. HP is relatively rare and has not been studied as extensively as immediate-type respiratory disease, but it is a chronic disease that can progress to disabling or even fatal lung disease. HP is characterized by the predominance of mononuclear inflammation of the lung interstitium, terminal bronchioles, and alveoli. The most significant clinical feature of HP is granuloma formation, possibly progressing to lung fibrosis (hardening of the lungs). Historically, HP has predominantly resulted from occupational exposures and therefore has a variety of names based on occupation or antigen association (e.g., farmer's lung). As with ACD, there is a nonallergic respiratory illness caused by inhalation of organic dust called organic dust toxic syndrome (ODTS), which is superficially similar to HP. ODTS is usually an acute self-limited illness resulting from heavy airborne organic dust exposure. ODTS differs from HP in several ways: Lung lavage fluid has an increased number of neutrophils, characteristic of nonspecific inflammation, rather than a mononuclear cell influx. Repeated exposures do not lead to chronic lung disease. Disease results from a single (rather than repeated) high dose exposure and is self-limited.

Occupational exposure to beryllium, a hapten, by inhalation of fumes/dust and/or by skin contact may result in one of two conditions that primarily affect the lungs. In acute berylliosis, which may occur following a high concentration exposure, the metal acts as a direct chemical irritant, causing a nonspecific inflammatory reaction (acute chemical pneumonitis). However, a small percentage of those exposed develop beryllium-specific T-cell-mediated hypersensitivity (Type IV) with proliferation and accumulation of CD4+ T cells in the lung. Although the factors that determine disease progression are not clear, it is thought that in some individuals, antigen-nonspecific inflammatory events may induce changes in lung permeability and production of proinflammatory cytokines and growth factors that lead to granulomatous disease called chronic beryllium disease (CBD) or berylliosis. In addition, there is a genetic component to the disease such that a single amino acid residue of the MHC class II may dictate beryllium presentation and, potentially, disease susceptibility.

32.5.7   Food Allergy

Recently, food allergy has become of interest to toxicologists with the advent of genetically engineered crops because of concern that an allergenic protein might be inadvertently introduced into the food supply [See Bernstein et al. (2003)]. While the normal response to orally introduced antigens is induction of tolerance, some proteins elicit undesirable immune responses in susceptible individuals. The involvement of the immune system distinguishes food allergy from food intolerance, which may result in similar symptoms but is not immune-mediated. The most common foods provoking allergic reactions are eggs, milk, soy, wheat, peanuts, tree nuts, fish, and shellfish. Food allergy is more prevalent in children, the majority of whom lose their sensitivity to milk, eggs, soy, and wheat by age 5. Allergies to peanuts, tree nuts, fish, and shellfish frequently last into adulthood. There is a distinct genetic component to food allergy, and individuals with food allergy are frequently asthmatic as well. The characteristics of a protein that render it allergenic are as yet unclear, although most food allergens tend to be 10–70 kD in size, resistant to enzymatic and acid degradation (indigestible), heat stable, and glycosylated.

Generally, immune responses to food can be divided into IgE- and non-IgE-mediated reactions, although a given adverse reaction may involve multiple mechanisms (any of those listed in Table 32.7). IgE-mediated (immediate-type hypersensitivity) reactions to food occur in a variety of target organs, manifesting as hives, swelling or redness in the skin, gastrointestinal symptoms such as vomiting, pain, or diarrhea, respiratory difficulties such as wheezing or allergic rhinitis, and hypotension or cardiac arrhythmia in the cardiovascular system. A severe, systemic IgE-mediated response may result in anaphylaxis, a potentially fatal reaction. Other IgE-mediated reactions include oral allergy syndrome, atopic dermatitis, and eosinophilic gastrointestinal disorders. Cell-mediated or non-IgE antibody-mediated reactions to food are frequently delayed or chronic and include food-protein-induced enterocolitis, proctitis, and enteropathy, as well as celiac disease.

The mechanisms underlying oral tolerance, which protects most individuals from developing food allergies, are not well understood but undoubtedly involve some of the regulatory T cells mentioned earlier in this chapter. An understanding of the mechanisms underlying oral tolerance is important not only for understanding the pathogenesis of food allergy, but also because there is a desire to exploit this phenomenon in the development of therapeutics to treat autoimmune disease and asthma.

32.5.8   Adjuvants

The word adjuvant is derived from the Latin *adjuvare*, which means "to help." Adjuvants enhance (or help) the immune response to antigens and are often included in vaccine preparations in order to generate a more robust antibody response and longer immunologic memory. During infection, an effective antigen-specific memory response is initiated by substances associated with the presence of infectious agents, such as microbial products. Therefore, many adjuvants consist of bacteria or bacterial components, which are effective activators of macrophages and dendritic cells. Activated macrophages and dendritic cells are highly stimulatory to T cells, so a macrophage that has encountered an adjuvant while acquiring antigen can instigate a stronger response in the T cell specific for that antigen. Most often, microbial components are toll-like receptor ligands, such as bacterial CpG DNA, double-stranded RNA, or LPS (see Section 32.2.2).

Importantly, adjuvants may be major contributors to the development of inappropriate immune responses. Being associated with infection or other damage, they provide a "danger signal," which may prompt misdirected or more aggressive responses against bystander antigens. Many autoimmune diseases have strong associations with viral infections, which may have an adjuvant effect. In other cases, antigens may be encountered in the environment along with substances that have adjuvant effects. Food allergy (see Section 32.5.8) and allergic airway disease are good examples. It is unclear to what degree the food matrix influences the allergenicity of ingested proteins. Inhaled antigens are encountered within complex bioaerosols. Indoor air includes material from both bacterial (LPS) and fungal sources (glucans) that cause nonspecific inflammation and may act as adjuvants. A number of animal studies have found that the fungal cell wall component $(1\rightarrow3)\beta$-D-glucan induces nonspecific inflammatory responses, including neutrophilic lung infiltration and an increase in TNFα. Others have shown that $(1\rightarrow3)\beta$-D-glucan can enhance allergic responses to known antigens, including antigen-specific IgE and IgG1 as well as lung eosinophils. These studies suggest that $(1\rightarrow3)\beta$-D-glucan may act as an adjuvant during the initiation phase of allergic responses. Frequently, allergens purified from sources such as mold, dust mite, or cockroach are less potent than crude extracts from these same organisms.

Toxicologists are particularly interested in the possibility of air pollutants (e.g., diesel exhaust, ozone, nitrogen dioxide, sulfur dioxide, particulate matter) enhancing both the induction and expression of allergic lung disease. A number of epidemiological studies have found that the incidence and severity of asthma attacks increase as levels of air pollution increase. Many of these effects are thought to be a result of direct irritation of an already diseased tissue (i.e., enhanced expression of disease in individuals who are already asthmatic). However, some human and primate data and a number of animal studies have demonstrated that air pollutants can also act as adjuvants to enhance allergic sensitization. These studies have demonstrated not only pollutant-enhanced IgE production and bronchial reactivity, but also enhanced production of Th2 cytokines. Most of these air pollutants cause nonspecific inflammation, which may establish a local environment in the lung that favors Th2 sensitization. In responses to air pollution exposure, epithelial cells lining the lung and alveolar macrophages produce several mediators that suppress Th1 (but not Th2) responses including PGE$_2$, IL-6, and NO. Hence, exposure to toxicants could possibly exacerbate allergic responses and allergic asthma in two ways, by acting as an adjuvant to enhance sensitization or by aggravating inflammatory responses that occur at the time of elicitation.

32.5.9 Summary of Hypersensitivity Mechanisms

In summary, there are analogous events in the various hypersensitivity profiles such as antigen processing and presentation to the T-cells of the draining regional lymph nodes by antigen-presenting cells. Likewise, inflammation is a common component in the expression of hypersensitivity responses. In most cases, hypersensitivity (an adaptive immune response) can be distinguished from nonspecific inflammation (irritation), based on the need for multiple exposures and the tendency for the effective dose to drop with subsequent exposures. The kinetics of events and symptoms differ with the type of hypersensitivity induced. Additionally, a complex network of cytokines, chemokines, and adhesion and co-stimulatory molecules, the nature of which is not entirely understood, is involved in the various hypersensitivity responses. In all cases, T cells appear to have an important role, although different types of hypersensitivity are characterized by different types of T-cell responses. The net result is that the immune system can be stimulated to induce tissue damage, and toxic chemicals may act directly as antigens in this process or may exacerbate induction or expression of allergic responses induced by other antigens.

32.6 MECHANISMS ASSOCIATED WITH AUTOIMMUNE DISEASE

32.6.1 Autoimmune Diseases

Autoimmune diseases result from a breakdown of immunological tolerance leading to immune responses against self-molecules. The mechanisms for this self-reactivity are the same as those associated with responses to foreign antigens, including activation of the innate immune systems and an excessive production of inflammatory mediators and/or an activation of the adaptive immune response with T and B lymphocytes responding to self-antigens. In many instances the events that initiate the immune response to self-antigens are unknown; however, epidemiologic studies suggest associations with specific genetic loci and environmental factors, including exposures to certain drugs, chemicals, and infectious agents. Autoimmune disorders can affect virtually any site in the body and present as a spectrum of diseases ranging from organ-specific, in which antibodies and T cells react to self-antigens localized in a specific tissue, to systemic, which are characterized by reactivity against a specific antigen or antigens spread throughout various tissues in the body. In addition to disease traditionally labeled as autoimmune, there are suggestions that a number of common health problems, such as atherosclerosis, schizophrenia, inflammatory bowel disease, and aspects of male and female infertility may have an autoimmune component. Overall, women have a significantly higher risk of developing an autoimmune disease than do men. However, for some autoimmune diseases, such as anklyosing spondylitis and adult-onset diabetes, there appears to be a higher risk among men. The immunomodulating influences of estrogens, androgens, and gonadotrophins are likely to be important in the disproportionate risk.

32.6.2 Mechanisms of Autoimmunity

Lymphocytes that recognize self-antigens with high affinity undergo negative selection in the bone marrow (B cells) and thymus (T cells) and are eliminated via apoptosis (central tolerance). However, autoreactive B and T cells constitute a normal part of the immune cell pool and natural autoantibodies are observed in sera from "normal, healthy" individuals. The presence of these cells presents a background level of risk for autoimmune disease. Normally, these lymphocytes are controlled by a variety of cell types and soluble mediators that maintain peripheral tolerance. One such regulatory check is that two signals from antigen-presenting cells are required to initiate lymphocyte proliferation, one consisting of antigen–MHC molecules via the TCR and the other consisting of a nonspecific (co-stimulatory) signal. In the absence of a co-stimulatory signal, self-reactive cells are rendered unresponsive (anergic). However, in some instances, these autoreactive cells can recognize self-antigens against a background that overcomes anergy and makes the antigens appear immunogenic, such as in the presence of polyclonal activation following infection or chemical-induced inflammation. In these cases, tolerance can be broken and an autoimmune pathology may result.

Following activation, regulation of the immune response is mediated by a number of inhibitory pathways. Defects in these regulatory pathways prevent restoration of normal immune homeostasis and contribute to the pathogenesis of autoimmune disease. Changes in apoptotic pathways, leading to inappropriate cell death or survival or disturbances in the clearance of apoptotic cells, have been suggested as the underlying mechanisms for several autoimmune diseases including rheumatoid arthritis (RA), systemic lupus erythematosis (SLE), and Hashimoto's thyroiditis. Several populations of T cells that regulate such responses have been identified (regulatory T cells): Some of these arise during maturation in the thymus, and others may be induced by exposure to antigen in the periphery. Regulatory CD4/CD25 T cells that express the transcription factor Foxp3 and the cell-surface receptor GITR suppress effector T-cell activity and may also modulate antigen-presenting cell function. Dysfunction of regulatory T-cell activity through deletion or mutation of Foxp3, IL-2, or IL-2 receptors leads to severe antigen-triggered lymphadenopathy and multi-organ autoimmune disease. A decrease in regulatory NK T cells and suppression of IL-4 production by these cells has been associated with disease onset in a mouse model for diabetes.

The effector mechanisms responsible for tissue destruction and pathology may differ substantially from the mechanisms that trigger the disease. It is generally thought that organ-specific autoimmune diseases are characterized by Th1-biased cell-mediated responses, and systemic autoimmune diseases tend to skew more toward antibody-mediated Th-2-type responses. However, some autoimmune syndromes, such as multiple sclerosis, are not easily classified, because they demonstrate both organ-specific and systemic components. Autoantibodies appear to be a common feature of all autoimmune diseases; and although they may not be the primary factor in tissue damage, they are important in the clinical diagnosis of autoimmune disease. Autoantibodies can react with cell-surface antigens, cell-surface receptors and intracellular constituents. These antibodies can (a) activate the complement cascade and induce cell lysis (hemolytic anemia), a Type II response, (b) form immune complexes that deposit in tissues (lupus nephritis), a Type III response, or (c) modulate receptor function (Grave's disease and autoimmune thyroiditis). Cell-mediated responses can be effected directly by autoreactive CD8 cytotoxic T cells or indirectly via release of proinflammatory cytokines and other soluble mediators by activated T cells and macrophages.

32.6.3 Genetic Predisposition

Familial studies suggest a clear association between genetics and essentially all autoimmune diseases. Concordance rates between identical twins can range from approximately 9% to 40%, depending on the disease. This lack of absolute concordance suggests that a significant component of the etiology of autoimmune disease may be due to environmental effects. In the majority of autoimmune diseases, a multigenic process with multiple susceptibility loci working in concert has been suggested. Specific loci that suppress autoimmune phenotypes have also been reported.

In theory, all genes coding for products that are involved in the induction and maintenance of self-tolerance and in regulating immune effector functions as well as organ-specific functions may be involved in defining individual susceptibility. The most clearly established genetic association is with specific alleles within the MHC gene complex. However, with rare exceptions, a specific MHC haplotype is not sufficient for development of autoimmune disease.

There are also a number of non-MHC genes that may contribute to autoimmunity. Candidates include genes for receptors that provide stimulatory signals for the immune response, including the TCR, Fc receptor, and cytotoxic T-lymphocyte antigen-4 (CTLA-4) prominently expressed on T regulatory cells. Genes that regulate programmed cell death and genes that affect the function or the level of expression of regulatory or effector molecules of inflammation, fibrosis, or other pathologic processes involved in autoimmune disease development have also been observed. TNFα polymorphisms have been implicated as independent susceptibility factors for RA and SLE. These polymorphisms may be directly involved in disease pathogenesis, because TNFα is known to be a strong inflammatory factor and has been a successful target for therapeutic intervention and long lasting immune response modification. Finally, polymorphisms in genes associated with nonimmune parameters, such as drug-metabolizing enzymes, may result in differential susceptibilities to drug or chemical-induced autoimmunity.

32.6.4 The Role of Environmental Factors in Autoimmune Disease

Environmental factors have been shown to be important triggers for expression of autoimmunity and have been suggested to both induce onset and modulate disease severity. Lifestyle factors such as diet, smoking, therapeutic and recreational drug use, infection with certain bacteria and viruses, and exposure to UV radiation and environmental chemicals have all been implicated in the pathogenesis of autoimmune diseases.

For some infectious agents, the causal link between infection and disease is fairly well established. Many peptide fragments of microbial agents are homologous with host proteins, and the induction of an immune response to these antigens results in cross-reactivity with self-antigens and the induction of autoimmunity. One example of this "molecular mimicry" is a membrane protein on the β-hemolytic streptococcus bacterium, which has a high degree of homology with cardiac myosin. Antibodies that target the bacterium also cross-react with cardiac muscle and induce rheumatic fever. Infection or direct tissue trauma may expose self-antigens (cryptic or hidden antigens) that generally have relatively low circulating levels or are anatomically sequestered in specific tissues (e.g., myelin basic protein or thyroglobulin). This "bystander effect" may be important in tissues where vascular and/or cellular basement membranes constitute an effective barrier that usually prevents access by autoreactive cells. Infections also can produce an adjuvant effect. For example, *Mycoplasma arthriditis* superantigen has been shown to stimulate cytokine production, upregulate MHC Class II expression, and stimulate TH cells, resulting in polyclonal B-cell activation and/or differentiation of antigen-specific B cells.

For other environmental factors, the associations with autoimmunity have been suggested from epidemiologic studies and/or animal models or in vitro studies. Dietary factors such as iodine, nutritional supplements such as omega fatty acids, and food contaminants such as those found in the cooking oil associated with toxic oil syndrome have been linked with autoimmune disorders. Table 32.9 lists a number

TABLE 32.9. Chemicals and Drugs Associated with Autoimmunity[a]

Autoimmune Syndrome	Compound
SLE	Trichloroethylene
	Silicone
	Aromatic amines
	Formaldehyde
	Hydralazine
	Procanimide
	Chlorpromazine
	Silica
	Isoniazid
	Penicillamine
Hemolytic anemia	Methyldopa
	Penicillin
	Sulfa drugs
Thyroiditis	Iodine
	Lithium
	PCBs, PBBs
	Penicillamine
Thrombocytopenia	Interferon-α
	Gold Salts
	Rifampin
	Acetazoldomide
	Quinine
Scleroderma	Vinyl chloride
	Trichloroethylene
	Spanish toxic oil
	Tryptophan
	Silicone
	Interleukin-2
	Diphenylhydantoin
Hepatitis	Halothane
	Interferon-α
	Ethanol
	Phenobarbital

[a]This table is not meant to be a comprehensive review, but rather is a sampling of the extensive literature associating drug and chemical exposures with autoimmunity.

of chemical and therapeutic agents that have been linked to autoimmunity. A number of these agents act via the chemical-induced alteration of self-peptides. Metals such as mercury induce autoimmune disease via the creation of new high-affinity binding sites for MHC molecules on chemical-bound self-peptides, allowing activation of previously anergized T cells. In addition, many drugs induce autoimmunity via formation of hapten-induced autoantibodies (Type II responses). Compounds such as penicillin and halothane induce reactions in which hapten-specific T cells provide help to antibody-producing B cells that recognize the modified hapten but not the native form of the self-protein.

32.6.5   Summary of Autoimmune Mechanisms

In summary, the mechanisms associated with xenobiotic induction or exacerbation of autoimmune disease are in many ways similar to those associated with other types of immune-mediated injury, with the chief difference being the endogenous, rather than exogenous, nature of the antigen. Although there are now a variety of animal models designed to mimic different types of autoimmune disease, much of the information we have to date comes from associations based on human epidemiology. Unlike immune suppression where there is very little epidemiology, the need to extrapolate from animal data to human effects is not great. However, the nature of epidemiologic studies makes it very difficult to make definitive cause–effect statements about etiology, much less about mechanisms underlying the disease. The importance of gene–environment interactions in the development of this complex group of diseases makes dissecting the role of xenobiotics in the development and exacerbation of autoimmune disease a challenging problem for toxicologists.

SUGGESTED READING

Bernstein, J. A., Leonard Berstein, I. L., Bucchini, L., et al. Clinical and laboratory investigation of allergy to genetically modified foods. *Environ. Health Perspect.* **111**, 1114, 2003.

Germolec, D. 2005. Autoimmune diseases, animal models. In: Vohr, H.-W. (Ed.). *Encyclopedic Reference of Immunotoxicology*, Springer-Verlag, Berlin, pp. 75–79.

Glaser, R. Stress-associated immune dysregulation and its importance for human health: a personal history of psychoneuroimmunology. *Brain Behav. Immun.* **19**, 3, 2005.

Hoebe, K., Janssen, E., and Beutler, B. The interface between innate and adaptive immunity. *Nat. Immunol.* **5**, 971, 2004.

Janeway, C. A., Jr, Travers, P., Walport, M., and Shlomchik, M. J. Immunobiology: *The Immune System in Health and Disease*, 6th ed., Current Biology/Garland Publishing, London, 2005.

Luebke, R. W., House, R. V., and Kimber, I. *Immunotoxicology and Immunopharmacology*, 3rd ed., Taylor and Francis/CRC Press, Boca Raton, FL, 2006.

Luster, M. I., et al. Risk assessment in immunotoxicology I. Sensitivity and predictability of immune tests. *Fundam. Appl. Toxicol.* **18**, 200, 1992; and II. Relationships between immune and host resistance tests. *Fundam. Appl. Toxicol.* **21**, 71, 1993.

National Autoimmune Diseases Coordinating Committee, Autoimmune Diseases Research Plan, NIH/NIAMS, 2002, 6. http://www.niaid.nih.gov/dait/pdf/ADCC_Report.pdf.

Rose, N. R., and MacKay, I. R. *The Autoimmune Diseases*, Elsevier, St. Louis, 2006.

Selgrade, M. J. K. In: Hodgson, E. (Ed.). *Immune System in A Textbook of Modern Toxicology*, 3rd ed. John Wiley & Sons, New York, 2004, pp. 327–342.

Singer, A. L. et al. Signaling via the T cell receptor. *Sci. STKE 2003*, (2003), cm7. http://stke. sciencemag.org/

CHAPTER 33

Reproductive Toxicology

JOHN F. COUSE

Taconic Inc., Albany Operations, Rensselaer, New York 12144

33.1 INTRODUCTION

The evidence of declining birth rates in developed and developing countries alike is indisputable. This trend over the past 30 years is assumed to be the result of changing socioeconomic factors and behaviors, such as an increased proportion of women in the work force and expanded access to contraception. However, a rising demand for assisted fertility treatment among couples and reports of comparable declines in birth rates among wild-life populations have prompted some to postulate that adverse biological factors, such as environmental toxicants, may be contributing to a general deterioration in reproductive health. There are data to support a link between toxicant exposure and reduced fecundity in certain human cohorts and wild-life populations following inadvertent exposure, but these are relatively isolated incidents and the evidence is by no means definitive. Therein lies the charge to current and future reproductive toxicologists.

The survival of all species depends on the integrity of its reproductive system. *Reproductive toxicology* may be defined as the study of the effects of physical and chemical agents on the reproductive and neuroendocrine systems of adult males and females, as well as those of the embryo, fetus, neonate, and prepubertal animal. This chapter focuses primarily on the potential sites of toxic insult in the reproductive systems of adult mammals, the biochemical mechanisms of such toxicants, and the manifestations that may result. The latter part of the above definition is a subspecialty of developmental toxicology (Chapter 34) and is discussed only in brief.

33.2 GENERAL PRINCIPLES OF REPRODUCTIVE TOXICOLOGY

The ability to predict the biochemical mechanisms and manifestations of a toxicant requires a thorough understanding of the physiological sites at which an organ, tissue, or biological system is most vulnerable. This is especially difficult in the study of reproductive toxicants because fertility depends on the integrated functions of three heterogeneous organ systems: the central nervous system (e.g.,

Molecular and Biochemical Toxicology, Fourth Edition, edited by Robert C. Smart and Ernest Hodgson
Copyright © 2008 John Wiley & Sons, Inc.

hypothalamus), the endocrine system (e.g., pituitary, gonads), and the specialized organs of the male and female reproductive tracts. Among these tissues is a remarkable breadth of specialized cell types that must (i) provide for the generation and delivery of haploid germ cells, (ii) provide for the homeostasis and function of the internal and external genitalia, (iii) induce and mediate the sexual behaviors necessary for mating, (iv) allow for fertilization and implantation of the embryo in the womb, (v) maintain pregnancy and induce parturition, and (vi) provide nourishment to offspring. Toxic insult to one particular cell type can potentially undermine the whole system and reduce the reproductive capabilities of an individual. The dependence of the reproductive system on endocrine signaling introduces additional complexity, because dysfunction in an endocrine organ is more likely to manifest in a distant target tissue. Finally, the ultimate effects of a reproductive toxicant may not emerge until a prolonged period, even years, after exposure, making it quite difficult to discover or establish a link between exposure and effect.

33.2.1 General Mechanisms of Reproductive Toxicants

The general principles of pharmacology that concern routes of exposure, metabolic activation, and distribution of toxicants to target tissues are all applicable in reproductive toxicology. The general principles of pharmacokinetics that dictate how toxicants may cause cellular dysfunction and/or death (as discussed in Chapters 16–22) also apply in the case of reproductive toxicants. The relatively high rate of mitotic and meiotic activity in germ cells, continuous proliferation and differentiation of certain somatic cell types in the gonads, and dramatic tissue remodeling that occurs in the uterus and vagina make these tissues especially susceptible to toxicants that disrupt the cell cycle or cause genetic damage. The dependency of reproductive functions on the central nervous and endocrine systems also make it especially vulnerable to agents that inhibit the synthesis or action of neurotransmitters and hormones.

33.2.2 Cell Signaling and Endocrine Disruption

The mechanisms of intercellular signaling are generally described as *endocrine, paracrine*, or *autocrine* (Figure 33.1). All three occur within and among the organs of the reproductive system and are paramount to their integrated function. The gonadal-derived steroid hormones (or sex steroid hormones) are critical signaling molecules within the gonads (i.e., auto- and paracrine signaling) as well as modulate the functions of neuroendocrine and peripheral reproductive tissues via endocrine signaling pathways.

Several xenobiotics, known as endocrine disrupting chemicals (EDCs), can disrupt steroid signaling and thereby impair reproductive function. The steroid hormones and their respective intracellular receptors act as ligand-induced transcription factors to regulate gene expression and ultimately the phenotype of target cells (Chapter 19). The toxic effects of EDCs are due to their capacity to magnify, inhibit, or alter the timing of an endogenous steroid hormone signal. Specific postulated mechanisms by which EDCs exert such effects include (i) binding and activation of the intracellular receptor (i.e., agonist activity), (ii) competitive inhibition with the endogenous hormone for binding to the intracellular receptor (i.e., antago-

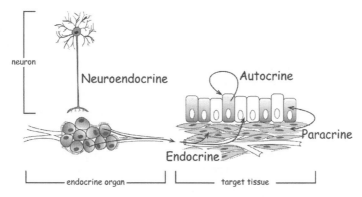

Figure 33.1. Primary pathways for intercellular communication.

nist activity), (iii) altering the synthesis of endogenous steroids, (iv) altering the metabolism or clearance of endogenous steroids, or (v) modification of intracellular receptor levels within target cells. The reproductive toxicities and manifestations of selected illustrative EDCs are described in Table 33.1.

33.3   SEXUAL DIFFERENTIATION

The reproductive capabilities of an individual depend on proper differentiation and development of reproductive and neuroendocine tissues during embryonic, fetal, and neonatal life. Sexual differentiation of the reproductive system is a highly organized process that requires precise spatial and temporal signaling via numerous endocrine, paracrine, and autocrine factors. Hence, it is not surprising that the reproductive system is most sensitive to toxic insult during development, most especially to endocrine-disrupting chemicals.

The gonads are the first visible reproductive structures to differentiate in a developing fetus. In both sexes, the gonads arise from common bi-potential anlages that are first colonized by primordial germ cells migrating from the alantois. *Genotypic* sex is determined by the complement of sex chromosomes present at fertilization. The *Y* chromosome carries the testis-determining factor (TDF) or *SRY* gene, and its presence (i.e., *XY* genotype) dictates the primordial gonads to differentiate as testes. In the absence of a *Y* chromosome (i.e., *XX* genotype), the primordial gonads differentiate as ovaries. Hormones produced by the differentiating gonad then determine the phenotype of the internal and external genitalia. Sertoli cells of the fetal testes secrete the peptide hormone Müllerian-inhibiting substance (MIH), which causes atrophy of the fetal female reproductive tissues (Müllerian ducts) that would otherwise differentiate to form the oviducts, uterus, and upper vagina. During this same period, Leydig cells of the fetal testes secrete testosterone to induce differentiation of the fetal male reproductive anlagen (wolffian ducts) into the male reproductive tract. Differentiation of the external male genitalia particularly requires local conversion of the testosterone to the more potent androgen, dihydrotestosterone (DHT), by the enzyme 5α-reductase. In the absence of testosterone and DHT, the wolffian ducts atrophy. Therefore, differentiation of the male reproductive tract requires the fetal testes to create the appropriate hormonal milieu of MIH and

TABLE 33.1. Examples of Known and Potential Endocrine Disrupting Chemicals (EDCs) and Their Postulated Mechanisms

I. Steroid Receptor Agonists
Estrogen receptor
  Diethylstilbestrol
  Methoxychlor
  Bisphenolic compounds
  Genistein
  Zearalenone
Progesterone receptor
  Dimethistrone
  Norethindrone

II. Steroid Receptor Antagonists
Estrogen receptor
  Tamoxifen
  ICI-164,384 (Faslodex)
Androgen receptor
  Vinclozolin metabolites
  Flutamide
  Cyproterone acetate
Progesterone receptor
  RU-486 (mifepristone)

III. Alter Endogenous Steroid Synthesis
Aminoglutethiamide (CYP11A1 inhibitor)
  Danazole (HSD3B inhibitor)
  Ketoconazole (CYP17A1 inhibitor)
  Anastrazole (Arimidex®) (CYP19A1 inhibitor)
  Letrozole (Femara®) (CYP19A1 inhibitor)
  Finasteride (SRD5A inhibitor)
  TCDD (inhibits testosterone synthesis)
  Ethane dimethantesulfonate (Leydig cell toxicant)
  Pthalates (Granulosa cell and Sertoli cell toxicants)

IV. Alter Metabolism of Endogenous Steroids
TCDD (increases metabolism of estradiol)

V. Modification of Intracellular Receptor Levels
TCDD (downregulates estrogen receptor)
ICI-164,384 (increases estrogen receptor turnover)

androgens, whereas differentiation of the female reproductive tract appears to be the "default" phenotype and proceeds normally in the absence of gonadal hormones.

Sexual differentiation of the central nervous system also occurs during fetal development, and it is equally fundamental to reproductive function during adulthood. Adult males exhibit a relatively continuous pattern of gonadotropin secretion from the hypothalamic–pituitary axis and are therefore fertile throughout their reproductive lifetime. In contrast, adult females exhibit a cyclical pattern of gonadotropin secretion and are fertile only during the transient period that follows ovulation during each reproductive cycle. This fundamental difference is due to irreversible sexually dimorphic "imprinting" of the central nervous and neuroendocrine tissues during fetal and neonatal development. Masculinization of the developing male brain is generally attributed to the actions of estradiol generated via local aromatization of circulating testosterone from the fetal testes. This *organizational* effect of estradiol on the male brain can be manipulated in laboratory animals by removing the gonads during fetal development, resulting in XY males that exhibit feminine patterns of gonadotropin secretion and sexual behavior during adulthood. In contrast, exposure of female (XX) fetuses to exogenous testosterone or estradiol leads to male-like patterns of gonadotropin secretion and behavior during adulthood.

The dominance of steroid hormone signaling during differentiation and development of the reproductive systems make this process especially vulnerable to endocrine disruption. An imbalance in androgen/estrogen ratio during fetal development may have permanent consequences on reproductive function during adulthood. This was tragically illustrated by the clinical use of diethylstilbestrol (DES), an orally active synthetic estrogen, in pregnant women thought to be at risk of miscarriage from the 1950s to 1971. DES had no lasting effects in the mothers, but is linked to several reproductive abnormalities, including a rare form of vaginal cancer, in a percentage of offspring that were exposed during fetal development (Chapter 34). Among the many lessons gleaned from the unfortunate history of DES is the need for reproductive toxicologists to consider all periods of potential exposure, from fetal development to adulthood, when evaluating a toxicant. The tissues and integrative signaling pathways of the reproductive system are unique in that they continue to organize and mature after birth. Relative to adulthood, exposure to a reproductive toxicant during prenatal, neonatal, or pubertal development is more likely to cause irreversible effects on fertility.

33.4 NEUROENDOCRINE REGULATION OF REPRODUCTION

The reproductive capabilities of both sexes depend on proper function of the hypothalamic–pituitary–gonadal (HPG) axis. The hypothalamus forms the major portion of the ventral region of the diencephalon at the base of the brain and serves as the interface between the central nervous system (CNS) and endocrine organs. The pituitary is an endocrine organ that lies in a small, bony cavity located just below the hypothalamus and is attached to the latter organ via either (a) neurons, in the case of the neurohypophysis (posterior pituitary), or (b) a network of capillaries (portal vessels), in the case of the adenohypophysis (anterior pituitary). By integrating external stimuli to the appropriate secretory patterns of endogenous hormones, the HPG axis is fundamental to proper gonadal function, sexual behavior, parturition, and lactation.

As depicted in Figure 33.2, a series of signaling pathways and endocrine feedback loops among the component organs of the HPG axis are central to maintaining homeostasis and function of the gonads and reproductive tract tissues. Specialized neurons of the hypothalamus secrete a small peptide known as gonadotropin-releasing hormone (GnRH) into the portal blood vessels that bathe the anterior pituitary. The gonadotrope cells of the anterior pituitary respond to GnRH via the G-protein-coupled GnRH receptor and are stimulated to synthesize and secrete the gonadotropins follicle-stimulating hormone (FSH) and luteinizing hormone (LH).

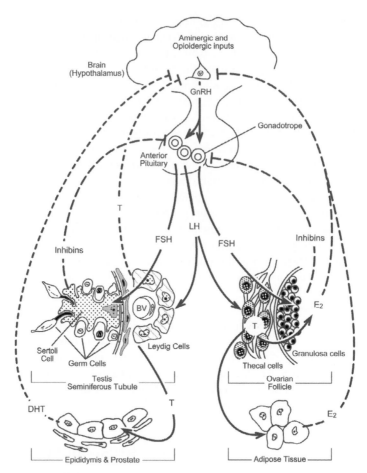

Figure 33.2. Endocrine feedback loops of the mammalian hypothalamic–pituitary–gonadal (HPG) axis. (Adapted from La Barbera A. R. Differentiation and function of the female reproductive system. In: Boekelheide, K., Chapin, R. E., Hoyer, P. B., and Harris, C. (Eds.). *Comprehensive Toxicology, Vol. 10, Reproductive and Endocrine Toxicology*, Elsevier, New York, 1997, pp. 255–272; and Creasy, D. M., and Foster, P. M. D. Male reproductive system. In: Haschek, W. M., Rousseaux, C. G. and Wallig, M. A. (Eds.). *Handbook of Toxicologic Pathology*, 2nd ed., Academic Press, San Diego, 2002, pp. 785–846. E_2, estradiol; T, testosterone, DHT, dihydrotestosterone, FSH, follicle stimulating hormone; LH, luteinizing hormone.

FSH and LH enter the circulation and stimulate the gonads to perform two principal functions, gametogenesis and hormone synthesis. After reaching a threshold level in the circulation, the gonadal hormones then feedback upon the hypothalamus and pituitary to decrease further GnRH and gonadotropin secretion. These classic *negative-feedback* loops are obligatory to maintaining the appropriate levels of FSH and LH in the circulation. The gonadal peptides known as INHIBINS specifically regulate FSH secretion, whereas gonadal-derived steroid hormones, primarily androgens (i.e., testosterone) in males and estrogens (i.e., estradiol) in females, regulate LH secretion. In some species, however, estrogens and androgens are equally effective in controlling LH secretion in males. Most evidence indicates the hypothalamus as the dominant site of action for the gonadal steroids, although some species exhibit an important role for steroid hormone actions directly at the level of the anterior pituitary.

GnRH secretion from the hypothalamus occurs in a pulsatile pattern that is sex-specific and reflected in the patterns of gonadotropin and gonadal hormone secretion. The underlying mechanism or "pulse generator" in the CNS that determines the pattern of GnRH secretion is poorly understood, but is obligatory to maintaining pituitary function. Variations in the amplitude and frequency of hypothalamic GnRH secretion are hypothesized to allow the single releasing hormone to differentially regulate FSH and LH secretion. Any disturbance to the "pulse generator" that disrupts the pattern of GnRH secretion will at least transiently shut down the reproductive system and cause infertility. For example, a toxic insult that decreases the frequency or magnitude of GnRH pulses will cause insufficient stimulation of the anterior pituitary, leading to reduced levels of circulating FSH and LH and subsequently diminished gonadal function. In contrast, toxicants that severely increase the frequency or magnitude of GnRH pulses to the extent that a nadir no longer occurs lead to constant perfusion of the pituitary with stimulating hormone. If allowed to continue, the gonadotropes of the pituitary downregulate the GnRH receptor and becomes refractory to further stimulation, resulting in severely reduced gonadotropin secretion and gonadal quiescence.

33.4.1 Neuroendocrine Control of the Ovarian Cycle in Females

Sexual differentiation of the HPG axis in females provides for a cyclical rather than continuous pattern of gonadotropin secretion. Cycle duration varies among species, but the principal endocrine and physiological changes are comparable (Figure 33.3). The human menstrual cycle is 28–30 days and marked by shedding of the endometrial uterine lining (i.e., menses). In contrast, rodent species exhibit a cycle of 4–5 days with no menstruation; hence it is termed an estrous cycle. The initial period or *follicular* phase of the ovarian cycle is marked by maturation of ova and concomitant synthesis of estradiol in the ovaries, which stimulates proliferation of the uterine endometrium. As the ova approach maturation, circulating estradiol levels climax and the hypothalamic–pituitary axis enters a transient period during which it responds to estradiol in a positive rather than negative manner. This *positive-feedback* of estradiol leads to the release of a large bolus of gonadotopins (primarily LH) from the pituitary and is the hallmark of the female cycle. The gonadotropin "surge" stimulates ovulation of the mature ova from the ovary and marks the end of the follicular phase and beginning of the *luteal* phase of the ovarian cycle. Although poorly understood, data clearly indicate that a sustained period of elevated estradiol, and progesterone in some species, is obligatory to generation of the gonadotropin surge. GnRH-secreting neurons in the hypothalamus are thought to lack estrogen receptors, prompting speculation that estradiol stimulates GnRH secretion indirectly via actions on surrounding afferent neurons. A direct effect of estradiol on the pituitary that transiently enhances its response to GnRH may also be involved.

The ovarian cycle is closely tied to circadian rhythms. Elegant studies in rodents have demonstrated that the gonadotropin surge is under 24-h photoperiodic control and that inducing sleep by administration of an anesthetic (e.g., pentobarbital) on the afternoon preceding ovulation causes a delay in gonadotropin surge and ovula-

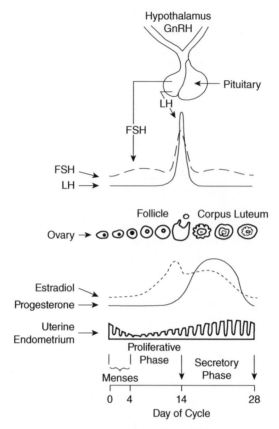

Figure 33.3. The female reproductive cycle. FSH, follicle stimulating hormone; GnRH, gonadotropin-releasing hormone; LH, luteinizing hormone. (Reproduced with permission from Thomas, M. J., and Thomas, J. A. Toxic responses of the reproductive system. In: Klaassen, C. D. (Ed.). *Casarett and Doull's Toxicology*, 6th ed. McGraw-Hill, New York, 2001, pp. 673–709.

tion by exactly 24 hr. In turn, female rats kept in an environment that lacks a defined photoperiod will become acyclic, underscoring the role of environmental cues in the ovarian cycle and the importance of the hypothalamus as an interface between the CNS and reproductive endocrine systems.

33.4.2   HPG Axis as a Target for Toxicants

The complex nature of the HPG axis and its dependence on a variety of hormones and endocrine pathways makes it especially susceptible to toxicants. Physical agents or xenobiotics that alter the synthesis or action of a particular hormone are likely to have profound downstream effects that may ultimately compromise fertility.

33.4.2a   GnRH Analogs. Constant perfusion of the pituitary with GnRH eventually leads to a refractory state and cessation of further gonadotropin secretion, which then leads to reduced gonadal steroid synthesis and decreased activity among the secondary sex organs. This physiological phenomenon is often exploited for the clinical treatment of women exhibiting irregular menstrual cycles or ovarian dysfunction, for which treatments with a GnRH agonist (e.g., leuprolide acetate) are prescribed to transiently shut down the HPG axis. More recently, GnRH analogs that act as antagonists (e.g., Antagon) have been developed and are advantageous over the earlier agonists because they do not elicit a transient period of heightened gonadotropin secretion and hypergonadism. These compounds are also employed in the treatment of certain steroid-dependent breast and prostate cancers.

33.4.2b   Disruption of Neurotransmitters. Several different types of neurotransmitters influence the GnRH-secreting neurons in the hypothalamus. Catacholamine neurotransmitters such as epinephrine and norepinephrine primarily exert a direct stimulatory action on GnRH secretion. Consequently, toxicants that disrupt (i) norepinephrine synthesis, such as dopamine β-hydroxylase inhibitors (e.g., the pesticide Thiram), (ii) norepinephrine postsynaptic action, such as α-adrenergic antagonists (e.g., the pesticide chlordimeform), or (iii) or presynaptic storage, such as certain alkaloids (e.g., Reserpine), can potentially inhibit GnRH secretion and, hence, ovulation. The role of other monoamine neurotransmitters, such as dopamine and serotonin, are less clear because each have been shown to have stimulatory and inhibitory effects on the LH surge. Opioid receptor agonists, such as morphine, are also known to inhibit LH secretion in both sexes and inhibit the preovulatory gonadotropin surge in females. Chronic exposure to tetrahydrocannabinol, the psychoactive component of marijuana, also decreases gonadotropin secretion via actions in the hypothalamus.

33.4.2c   Steroids and Endocrine-Disrupting Chemicals. Given the significant role of steroid hormones in modulating hypothalamic and pituitary function (Figure 33.2), xenobiotics that possess progestin, androgenic, or estrogenic activities, whether agonistic or antagonistic, or affect the synthesis of sex steroids could be expected to have profound effects on the HPG axis. In fact, the underlying principle of estrogen-based (e.g., ethynylestradiol) oral contraceptives in females is to maintain suprabasal levels of estrogenic activity on the hypothalamic–pituitary axis, thereby ensuring constant negative-feedback and inhibition of the preovulatory gonadotropin surge. A similar phenomenon is thought to underlie the reduced conception rates of sheep and cattle that occur while feeding on forage that is rich in estrogenic chemicals known as phytoestrogens. In contrast, xenobiotics that act as antagonists for endogenous estradiol (e.g., clomiphene citrate) or testosterone (e.g., the pesticide vinclozolin) can inhibit negative feedback in the hypothalamus of females and males, respectively, leading to increased gonadotropin secretion and hyperstimulation of the gonads (i.e., hypergonadotropic hypergonadism).

33.5   MALE REPRODUCTIVE SYSTEM

The male reproductive tract consists of the (i) testes, which serve as the site of gamete and steroid hormone production, (ii) excurrant ducts and epididymes for the transport, maturation, and storage of sperm, (iii) accessory sex glands for the production of seminal fluid, and (iv) penis, for copulation and delivery of mature sperm to the female reproductive tract.

33.5.1 Testes

The testes and the penis make up the external genitalia of the male reproductive system. Among mammalian species, the anatomical location of the testes varies from that of a pendulous external scrotum (e.g., humans), or near the ventral abdominal wall (e.g., hedgehogs), to close to the kidneys (e.g., elephants). Each testis contains two principal functional compartments, the seminiferous tubules and the interstitial spaces, encased within a fibrous capsule (Figure 33.4). The seminiferous tubules are luminal structures ensheathed by myoid cells and arranged as coils in the testis, but, when extended, range from 20–50 m in length depending on the species. The luminal side of the tubule is lined with the *seminiferous epithelium*, which is composed of

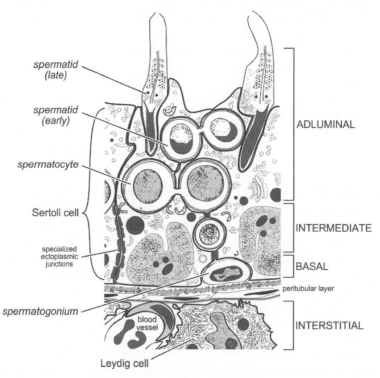

Figure 33.4. Diagramatic representation of a cross section of seminiferous tubule and interstitial tissue in the testis. The seminiferous tubules are luminal structures ensheathed by a peritubular layer formed from myoid cells. The *seminiferous epithelium* lines the luminal side of the tubule and is further divided into the basal, intermediate, and adluminal zones. The seminiferous epithelium is composed of Sertoli cells in close association with germ cells as they advance through the stages of maturation—that is, spermatogonia, spermatocytes and spermatids, respectively. A single Sertoli cell nurses several maturing germ cells simultaneously and mediates the migration of germ cells from the basal to adluminal zone as they mature. Specialized ectoplasmic junctions between Sertoli cells form the blood–testis barrier. Leydig cells outside the tubules in the vascularized interstitial zone composed primarily of Leydig cells. (Adapted from Foster, P. M. D. Testicular organization and biochemical function. In: Foster, P. M. D., and Lamb, J. C. (Eds.). *Physiology and Toxicology of Male Reproduction*, Academic Press, San Diego, 1988.

germ cells at advancing stages of maturation and the somatic Sertoli cells. The ends of each tubule are open and terminate in the Rete testis, which connect to a series of anasomizing tubules (ductus efferentes) that merge and serve to transport sperm from the testis to the epididymis. Between the seminiferous tubules are the interstitial spaces of the testis, which contain blood and lymph vessels, macrophages, and Leydig cells.

Testes perform two primary functions that are vital to male fertility: the generation and maturation of viable germ cells (i.e., spermatogenesis) and synthesis of sex steroids and peptide hormones to properly regulate the HPG axis as well as maintain the androgen-dependent accessory sex organs. Testes possess an array of xenobiotic metabolizing enzymes, including cytochrome P450, epoxide hydrolase, esterase, and alcohol and aldehyde dehydrogenases. Therefore, both extra- and intratesticular metabolism and activation of xenobiotics must be considered when evaluating a potential gonadal toxicant in males.

33.5.1a Spermatogenesis. Spermatogenesis is the process of germ cell division and differentiation to produce spermatozoa, the mature male germ cell. All spermatogenesis occurs within the seminiferous tubules of the testes, where germ cells compose almost 90% of the seminiferous epithelium (Figure 33.4). Male germ cells progress through three respective stages toward maturation: *spermatocytogenesis, meiosis*, and *spermiogenesis*. During the process of maturation, the germs cells migrate from the basal, through the intermediate and into the adluminal zones, allowing for their eventual release (i.e., *spermiation*) into the tubule lumen.

During *spermatocytogenesis*, the first stage of spermatogenesis, stem cell *spermatogonia* located along the basal lamina divide by mitosis to continue a lineage of stem cells (type A) as well as produce spermatogonial germ cells (type B) that are committed to maturation. Type B spermatogonia remain attached to each other via intercellular bridges that are thought to facilitate synchronous development. Following multiple rounds of mitosis, type B spermatogonia become *primary spermatocytes* and proceed to *meiosis*, during which the chromosomes are duplicated and then twice divided (reductive and equatorial, respectively) to yield haploid *spermatids*. During the final stage, *spermiogenesis*, the round-shaped spermatids differentiate to form *spermatozoa*, which are easily recognized by (a) their streamlined head that contains a densely packed nucleus and enzyme-filled acrosome and (b) a whiplike tail for motility. Condensation of the nuclear material in *spermatozoa* requires the histones to be replaced by specialized structures known as protamines. Once fully differentiated, the spermatozoa are released into the tubule lumen in a process termed *spermiation*. The length of spermatogenesis varies among species: 74 days in humans, 61 days in dogs, and 60 days in rats.

Differences in metabolic function and location within the seminiferous epithelium determine the susceptibility of male germ cells to toxicants. Spermatogonia possess significant DNA repair capabilities and are therefore relatively resistant to toxic insult, despite being on located on the "blood" side of the blood–tubule barrier and likely exposed to a greater array of toxicants. However, actively dividing spermatogonia are especially susceptible to damage by ionizing radiation, ischemia, and certain chemotherapeutic agents (e.g., adriamycin and busulfan). Late-stage spermatocytes, which are moved by Sertoli cells to the "privileged" side of the blood–tubule barrier, are less vulnerable to irradiation but more sensitive to hypothermia and a greater array of chemotherapeutics (e.g., cyclophosphamide, platinol, 5-fluoruracil). The condensed packing of the nuclear material in spermatids and spermatozoa precludes all DNA repair capabilities, making these stages particularly vulnerable to DNA damaging agents as well as compounds capable of alkylating protamines (e.g., ethylene oxide, methanesulfonate, acrylamide). Glutathione (GSH) is a primary detoxifying mechanism for these and other reactive agents in testes; and depletion of GSH stores, and hence limiting endogenous defense mechanisms, is thought to be a common mechanism of several germ cell-specific toxicants in males.

Once released into the tubule lumen, spermatozoa flow with the tubule fluid to the Rete testis, where they exit the testis and enter the caput portion of the epididymis. In lower vertebrate species (e.g., fish and amphibians), sperm at this stage are fully motile and competent to fertilize. In contrast, in higher vertebrate species, sperm exiting the testis must acquire the ability to swim and fertilize during transit through the epididymes. Mature spermatozoa that have traveled through the epididymis are reportedly less sensitive to alkylating agents because disulfide bond formation among nuclear protamines is complete. However, potent alkylating agents (e.g., triethanolamine) may still cause chromosomal abnormalities even in caudal epididymal sperm.

Chemicals that specifically effect sperm motility could be expected to be most harmful to sperm during transit either in the epididymis or in the female reproductive tract following ejaculation. Sperm motility is highly dependent on the hydrolysis of ATP as an immediate energy source. Therefore agents that disrupt glycolysis will likely cause severe detriment to sperm motility and their ability to fertilize. The industrial chemical α-chlorohydrin is thought to impair sperm motility and velocity in rodents via inhibition of gyceraldehyde-3-phosphate dehydrogenase (GAPD) in the glycolytic pathway.

Due to the differential sensitivity of male germ cells at progressive stages of maturation, the detrimental effects of a single toxic insult on sperm count or fertility may be either immediate or markedly delayed. For example, transient exposure to a toxicant that specifically targets primary spermatocytes will not immediately effect fertility because there is no insult to the mature spermatozoa in the epididymis that will compose the next ejaculate. However, a drop in fertility is likely to occur in the future when the exposed crop of spermatocytes cells would have eventually matured. In contrast, a toxicant that specifically targets epididymal sperm would cause immediate infertility in an exposed male, but fertility would eventually return as the insensitive stages of germ cells mature and replace the stores. This principle is the basis for *dominant lethal* studies in which male rodents are first transiently exposed to a toxicant and then serially mated or sampled for sperm production over a prolonged period of time to determine which stage of germ cell is most susceptible (Figure 33.5). It should be noted that although germ cells at progressive stages of spermatogenesis exhibit differential susceptibility to toxic insult, increased doses of an agent will undoubtedly increase the number of stages affected.

33.5.1b   Sertoli Cells. Sertoli cells are the somatic cellular component of the seminiferous epithelium. These are highly specialized cells that employ tight intercellular cytoskeletal adhesions, known as ectoplasmic specializations, to form a syncytium around the lumen of the tubule. This "blood–lumen" barrier divides the

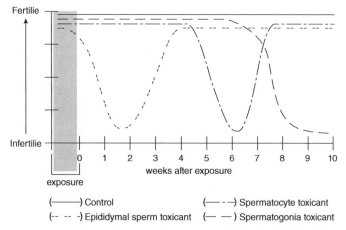

Fertilie

Infertilie

0 1 2 3 4 5 6 7 8 9 10

weeks after exposure

exposure

(———) Control

(-- -- -) Epididymal sperm toxicant

(——- -—) Spermatocyte toxicant

(——— —) Spermatogonia toxicant

Figure 33.5. Transient exposure to a germ cell-specific toxicants produces predicted periods of reduced fertility in males. Males are exposed to a toxicant over a period of one week and then their fertility is monitored by breeding over a period of several weeks. A toxicant that specifically targets epididymal sperm results in almost immediate infertility that recovers over time. A toxicant that specifically targets spermatocytes results in a delay to infertility. A toxicant that specifically targets spermatogonia results in an even greater delay to infertility and may result in permanent infertility.

seminiferous epithelium into a basal and adluminal compartment, the latter being immunologically privileged and isolated from serum-borne components (Figure 33.4). *Spermatocytogenesis* occurs on the basal side of this barrier, while *meiosis* and *spermiogenesis* occur on the adluminal side. A toxicant's capacity to affect sperm cells of different stages may depend on its ability to cross the blood–tubule barrier.

Along with creation of the blood–lumen barrier, Sertoli cells function in (i) the structural and nutritional support of maturing germ cells, (ii) the movement of maturing germ cells from the basal to adluminal compartments, (iii) the *spermiation* of mature spermatids, (iv) the secretion of luminal fluids, (v) the removal of waste and residual bodies left behind following spermiation, and (vi) the synthesis of INHIBINS. A single Sertoli cell may simultaneously "nurse" as many as 20–30 germ cells of varied maturational stages, and it communicates with each via intimate contact and specialized gap junctions. Hence, the rate of sperm production is directly correlated with the size of the Sertoli cell population within the testes (Figure 33.6). Any toxicant that disrupts Sertoli cell development or function will therefore undoubtedly compromise spermatogenesis. This is especially critical because the Sertoli cell population is established during testis differentiation and puberty and is generally nonrenewable thereafter. If Sertoli cells are prohibited from developing and proliferating during testis differentiation in the fetus, the gonad may fail to develop as a testis but rather as an ovary, leading to XY "sex reversal." Most evidence indicates that FSH is an important mitogen for Sertoli cells. In contrast, thyroid hormone appears to inhibit Sertoli cell proliferation and hypothyroid conditions during prepubertal periods, due either to a physiological defect or to exposure to goitrogens (e.g., 6-propyl-2-thiouracil), can lead to an increased Sertoli cell population and above average rates of sperm production.

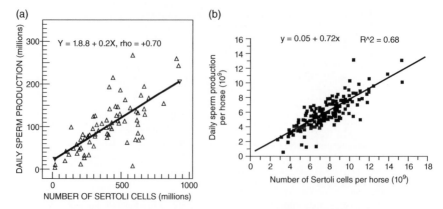

Figure 33.6. The relationship of Sertoli cell number with sperm production in testes of (A) human or (B) horse. (Part A is adapted from Johnson, L., Zane, R. S., Petty, C. S., and Neaves, W. B. Quantification of the human Sertoli cell population: Its distribution, relation to germ cell numbers, and age-related decline. *Biol. Reprod.* **31**, 785–795, 1984. Part B is adapted from Johnson, L., Carter, G. K., Varner, D. D., Tayler, T. S., Blanchard, T. L., and Rembert, M. S. The relationship of daily sperm production with number of Sertoli cells and testicular size in adult horses: Role of primitive spermatogonia. *J. Reprod. Fertil.* **100**, 315–321, 1994.)

Several toxicants are known to disrupt Sertoli cell functions and lead to transient episodes of subfertility/infertility. Cadmium and the natural contraceptive, gossypol, are thought to disrupt the ectoplasmic specializations that provide for the blood–luminal barrier. Other toxicants disrupt Sertoli cell–germ cell junctions and thereby cause germ cell "sloughing," in which immature spermatocytes and spermatids become detached from the seminiferous epithelium. This is often indicated by the presence of immature germ cells possessing fragments of Sertoli cells in the tubule lumen. Toxicants that specifically target vimentin filaments (e.g., colchicine and 2,5-hexanedione) are thought to cause germ cell sloughing. Other toxicants (e.g., phthalates) may lead to a similar phenomenon known as germ cell "shedding," which occurs when almost fully differentiated spermatids are prematurely released into the lumen and can be differentiated from "sloughing" by the lack of attached Sertoli cell fragments. In contrast, exposure to boric acid can prevent spermiation and disrupt spermatogenesis by causing prolonged retention of mature sperm by Sertoli cells.

Sertoli cell functions are dependent on hormonal signaling. The paracrine actions of testosterone from neighboring Leydig cells are obligatory to Sertoli cell function and spermatogenesis. FSH also acts directly on Sertoli cells and facilitates spermatogenesis. Therefore, any toxicant that inhibits either testosterone synthesis in Leydig cells, testosterone action in Sertoli cells, or gonadotropin synthesis or secretion from the hypothalamic–pituitary axis is likely to impair spermatogenesis.

33.5.1c Leydig Cells and Steroidogenesis. The other primary function of the testes is the production of steroid hormones (i.e., steroidogenesis), predominantly testosterone. As the major circulating androgen in males, testosterone is necessary to (i) sustain spermatogenesis, via paracrine actions on Sertoli cells, (ii) maintain the functions of the internal and external genitalia, (iii) stimulate development of male sexual characteristics (e.g., pubic hair, increased muscle mass, lowered voice), and (iv) maintain appropriate gonadotropin levels via negative-feedback on the hypothalamic–pituitary axis. Therefore, any toxicant that compromises the steroidogenic functions of the testes will have far-reaching detrimental effects on the male reproductive tissues and fertility.

Leydig cells are the primary steroidogenic cell type in the testis and are located in the vascularized interstitial compartments. Leydig cells constitutively express the LH receptor and require LH to induce expression of the enzymes necessary for steroid biosynthesis. Cholesterol is the parent substrate for all steroid biosynthesis in mammalian tissues and is primarily obtained by Leydig cells via *de novo* synthesis, although uptake from the circulation or hydrolysis of intracellular lipids can serve as additional sources. Steroidogenesis begins with the transport of cholesterol to the inner mitochondrial membrane, where CYP11A1 is located. This process is carried out by the steroid acute regulatory protein (StAR) and is the rate-limiting step in the synthesis of all active steroid hormones. The cholesterol is then converted to testosterone via a cascade of enzymatic reactions within the mitochondria and smooth endoplasmic reticulum of Leydig cells (Figure 33.7). In addition, certain tissues of the internal and external genitalia express the enzyme 5α-reductase to provide for local conversion of testosterone to the more potent androgen agonist, DHT. Also, Sertoli cells in some species possess markedly high levels of CYP19A1 and therefore have the capacity to convert androgens to estrogens.

Toxicity to Leydig cells will primarily manifest as decreased testosterone production, leading to secondary effects such as reduced weights among secondary sex organs and declining sperm production, all in a context of normal or even elevated circulating gonadotropins. Certain toxicants inhibit steroidogenesis by targeting specific steroidogenic enzymes, often causing accumulation of the precursor substrate in the biosynthetic pathway. For example, ketoconazole (antifungal) inhibits CYP17A1 activity and androstenedione synthesis, while finasteride (drug) inhibits 5α-reductase and DHT synthesis. In contrast, other toxicants are cytotoxic to Leydig cells (e.g., ethane dimethane sulfonate) and cause an acute death shortly after exposure, which leads to more prolonged complications to fertility. Certain species and strains of rodents are also especially susceptible to the development of Leydig cell tumors following chronically increased LH secretion from the pituitary.

33.5.2 Efferent Ducts and Epididymes as Targets for Toxicants

The efferent ducts primarily serve to transport spermatozoa from the testis to the epididymis, as well as reabsorb luminal fluids and in effect concentrate the ejaculate. The epididymes also function as transport to the final storage location in the distal regions as well as serve as a site for final maturation. Sperm are transported through the epididymis via rhymic contractions of the smooth muscle wall. The time for spermatozoa to pass through the epididymis may vary but is precise among species, and it is required for the spermatozoa to acquire motility and the capacity to fertilize. Any agent that accelerates this transit time is likely to cause reduced sperm numbers and compromised maturation. Androgens, most especially DHT, are essential to ensuring the optimal rate of sperm transport. Therefore, androgen receptor antagonists (e.g., cyproterone acetate, flutamide) or 5α-reductase inhibitors (e.g.,

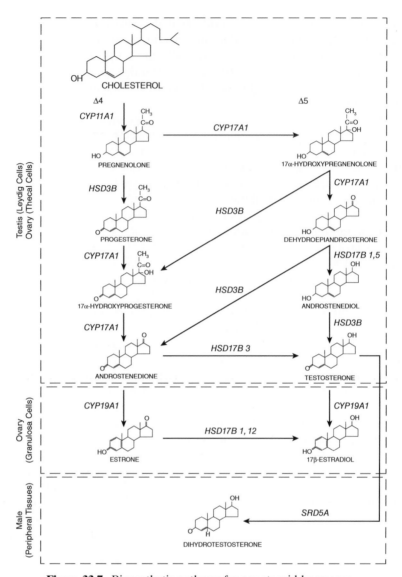

Figure 33.7. Biosynthetic pathway for sex steroid hormones.

finasteride) can disrupt the time of epididymal sperm transport and thereby their maturation. Estrogens, too, can accelerate the rate of epididymal sperm transport in mice. Several xenobiotics are also known to alter epididymal function, including certain fungicides (e.g., benzimidazole and ornidazole).

33.5.3   Male Accessory Sex Glands

The accessory sex glands of the male reproductive tract include the prostate, seminal vesicles, and bulbourethral glands. These are highly androgen-dependent tissues that function primarily to produce the seminal fluids that compose semen. As in the epididymes, these tissues primarily rely on the conversion of circulating testosterone to DHT by 5α-reductase and are therefore sensitive to a similar array of xenobiotics. Little is known about toxicants that directly target any one of these tissues; however, the function and physiology of these tissues have become very useful indices in the experimental study of toxicants that effect the HPG axis or steroidogenesis. Because these tissues rely on testicular androgen synthesis for maintenance and function, changes in organ weights and function serve as effective biological markers of insufficient steroidogenesis. Furthermore, the androgen response of these tissues appears to be fixed or "imprinted" during early development and can be altered by developmental exposure to estrogens, making such changes in adult animals a potential indicator of early exposure to estrogenic chemicals.

33.6 FEMALE REPRODUCTIVE SYSTEM

The female reproductive system consists of a pair of gonads (ovaries), each attached to the abdominal wall via a specialized ligament (mesovarium) and joined to the reproductive ducts via an oviduct or a fallopian tube, depending on the species. The oviducts serve as the site of fertilization and transport for the developing embryo to the uterus. The uterus is a hollow, muscular organ in the female pelvic region that functions as the site of embryo implantation and pregnancy. The caudal end of the uterus forms a canal (i.e., cervix) that connects to the external genitalia or vagina. The cervix and upper vagina serve as the birth canal during parturition, while the lower vagina serves as the copulatory organ. The mammary glands (or breasts) are also considered part of the female reproductive system and function to provide nutrition to offspring during the neonatal period. Whereas the male reproductive system is responsible for providing and delivering a haploid germ cell, the female reproductive system is responsible for these functions as well as providing the site of fertilization and pregnancy, delivery of offspring, and nourishment during neonatal development. Therefore, a toxic insult to fecund females will likely have drastic consequences on the reproductive capabilities of a population or species.

33.6.1 Ovaries

The ovaries perform two major functions: (i) the storage, maturation, and expulsion of healthy haploid germ cells (i.e., oocytes) for fertilization, and (ii) the synthesis and secretion of hormones to prepare the reproductive tissues for the establishment and maintenance of pregnancy, to properly regulate gonadotropin secretion from the hypothalamic–pituitary axis, to induce appropriate sexual behaviors, and to provide lactation. The two primary functional units in the ovaries are the maturing follicles and the corpora lutea.

33.6.1a Embryonic Development of Female Germ Cells. *Oogenesis* is the process by which haploid female germ cells are generated and become competent for fertilization. This process begins in the developing female embryo, during which embryonic germ cells called *oogenia* migrate from the yolk sac to the bi-potential gonad in the genital ridge. The oogenia then undergo numerous mitotic divisions to populate the developing ovarian tissue but, for reasons that remain unknown, cease proliferating and enter meiosis to become primordial *oocytes*. The primordial oocytes remain arrested in prophase I of meiosis throughout the remainder of ovarian development and will not resume meiosis until shortly before ovulation several years later. Inexplicably, oogenia that fail to enter meiosis and large numbers of primary oocytes in the fetal ovary die before birth in a process termed *atresia*. In human ovaries, atresia is responsible for a decline from 7 million to approximately 2 million oogenia between 20 weeks gestation and birth. Additional episodes of atresia between birth and puberty further reduce the pool of oocytes to approximately 400,000, representing a loss of over 90% of the original pool. This process is conserved among mammalian species and appears to involve activation of the natural intracellular pathways for apoptosis.

Therefore, unlike males, females are born with a finite stock of germ cells that cannot be replenished. A toxicant that alters the size of this pool, perhaps by inhibiting mitosis or increasing the rate of atresia in the embryo, will have irreversible effects on reproductive capabilities during adulthood and likely lead to premature ovarian failure. Polycyclic aromatic hydrocarbons (PAH) in cigarette smoke are well-known ovotoxicants that specifically target primordial ooctyes in rodents and humans. Years of epidemiological study indicate a strong correlation between habitual cigarette smoking and premature menopause, which is brought about by oocyte depletion in the ovaries. Laboratory studies in rodents have demonstrated that PAHs specifically induce apoptosis in primordial and primary oocytes by inducing the expression of *BAX*, a pro-apoptotic factor, via activation of the aryl-hydrocarbon receptor complex. Another toxicant known to specifically induce apoptosis in primordial and primary oocytes is 4-vinylcyclohexane (VCH) and its epoxide derivatives, which are byproducts of rubber, plastic, and pesticide manufacturing. Studies indicate a dose-dependent correlation between VCH exposure and premature ovarian failure in rodents. Ionizing radiation and chemotherapeutic agents (e.g., cyclophosphamide) also specifically target primordial ooctyes in the ovary.

33.6.1b Folliculogenesis and Steroidogenesis. Menarche or vaginal opening marks the commencement of reproductive capabilities in humans and rodents, respectively. From this point forward, primordial oocytes are continuously recruited to mature with each menstrual/estrous cycle. The process of follicle recruitment and growth toward ovulation is referred to as *folliculogenesis* (Figure 33.8). Only one or a small fraction of oocytes from each recruited cohort will grow and survive to ovulation, while the remainder will die by atresia. Hence, human females successfully ovulate only 400–500 oocytes during a lifetime. Cyclical recruitment, ovulation, and atresia combine to further reduce the pool of oocytes to approximately 25,000 by 35 years of age and less than 1000 by age 50. Shortly thereafter, the menstrual cycle becomes irregular and ceases permanently due to the depletion of oocytes (i.e., menopause).

The frequency of the menstrual/estrous cycle varies among species. For example, monoestrous species (e.g., dogs, cats) exhibit 1–2 cycles per year, whereas polyestrous species (e.g., rodents, primates) exhibit more frequent cycles of shorter duration. Primordial follicles, each consisting of an oocyte surrounded by a single layer of flattened somatic cells known as granulosa cells, are recruited to become primary follicles, a transition marked by increased oocyte size, formation of a glycoprotein matrix (zona pellucida) around the oocyte, and transformation to cuboidal-shaped

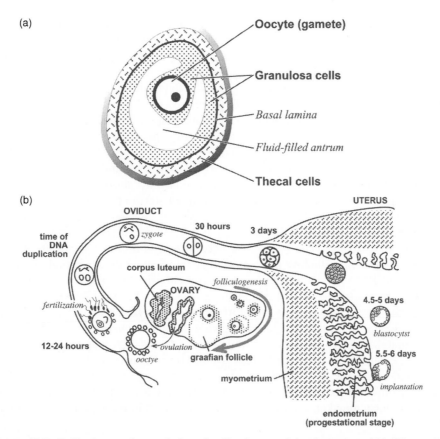

Figure 33.8. Folliculogenesis, ovulation, fertilization, and implantation. (A) Diagramatic drawing of a preovulatory (i. e., Graafian) follicle. The follicle is composed of a single maturing oocyte closely surrounded by layers of granulosa cells. The *cumulus* granulosa cells are those closest to the oocyte and forming the cellular stalk that is attached to the *mural* granulosa cell layers closer to the follicular wall (*basal lamina*). Preovulatory follicles are also characterized by a large fluid-filled cavity or antrum. Thecal cells lie outside the follicular wall and are the vascularized portion of the follicle. (B) Diagramatic drawing of folliculogenesis and ovulation in the ovary, oocyte fertilization, and maturation of the zygote as it migrates through the oviduct and into the uterine lumen, where it will be implanted into the uterine wall. (Adapted from Williams & Wilkins, *Medical Embryology*, 1975, p. 30.)

granulosa cells. The process involved in primordial follicle recruitment is unclear but not dependent on pituitary gonadotopins. Once recruited, primary follicles embark on an irreversible path toward maturation, during which the oocyte grows to a maximum size of 70–100 μm in diameter, the granulosa cells proliferate to form several concentric layers that surround the oocyte, and a second layer of specialized somatic cells known as theca form around the whole follicle (Figure 33.8). The thecal layer is the only vascularized portion of the resulting pre-antral follicle, and the granulosa cells and oocyte are encased in a semipermeable basal lamina. Survival and continued growth of the follicle from this point forward is gonadotropin-dependent. FSH stimulates massive proliferation and fluid production among the granulosa cells, leading to a dramatic increase in follicle size and formation of a fluid-filled antral cavity. The resulting preovulatory or "Graafian" follicle is marked by a centrally located oocyte surrounded by two to three layers of specialized granulosa cells (i.e., cumulus cells) that are connected to granulosa cells along the follicle wall (i.e., mural cells) by a cellular stalk (Figure 33.8).

Folliculogenesis and steroidogenesis are concomitant, interdependent processes in the ovary. Preovulatory follicles are the primary endocrine structures during the follicular phase of the ovarian cycle (Figure 33.3) and secrete substantial amounts of estradiol, which is necessary to induce the ovulatory gonadotropin surge from the pituitary, activate sexual behaviors, and prepare the uterine endometrium for embryo implantation. The biosynthetic pathway for steroid production in preovulatory follicles is generally the same as that employed in testes (Figure 33.7) except that much higher levels of CYP19A1 in the ovaries provide for aromatization of androgens to estrogens. The *two-cell/two-gonadotropin* model is generally used to describes estradiol production in preovulatory follicles, on the basis that granulosa cells lack CYP17A1 and therefore depend on thecal cells as the sole source of androgen precursors for aromatization (Figure 33.9). In brief, LH stimulates thecal cells to produce androgens that then diffuse across the basal lamina of the follicle and into the granulosa cell compartment, where they are converted to estrogens by the enzymatic activities of CYP19A1 and hydroxysteroid-17β-dehydrogenases, both under the positive regulation of FSH. Estradiol then diffuses back across the basal lamina and enters the circulation.

The enormous degree of cellular proliferation and differentiation, tissue remodeling, and biochemical and molecular changes involved in the transition of primor-

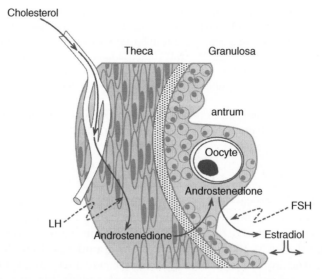

Figure 33.9. The two-cell–two-gonadotropin model of estradiol synthesis in ovarian follicles. (Adapted from and reproduced with permission from Carr, B. R., MacDonald, P. C., and Simpson, E. R. The role of lipoproteins in the regulation of progesterone secretion by the human corpus luteum. *Fertil. Steril.* **38**, 303–311, 1982.)

dial to preovulatory follicles undoubtedly makes this process susceptible to toxic insult, yet few toxicants that specifically target growing follicles are known. This may be because toxicants that target growing follicles but cause no detriment to primordial oocytes are likely to have only a transient effect on fertility that is reversed once exposure to the toxicant is ceased, and may therefore be overlooked. Nonetheless, toxicants that disrupt meiosis (in oocytes) or mitosis (in granulosa cells), alter the development of the vasculature in the thecal layer, or block the enzymes and components necessary for steroidogenesis are all potentially detrimental to folliculogenesis.

33.6.1c   Ovulation and Luteinization. *Ovulation*, the term used to describe expulsion of the mature oocyte from the ovary, marks the end of the *follicular* phase and the beginning of the *luteal* phase of the ovarian cycle (Figures 33.3 and 33.4). Ovulation is induced by a large surge of gonadotropin (primarily LH) from the hypothalamic–pituitary axis. In mammals that spontaneously ovulate (e.g., rodents, primates), the gonadotropin surge and subsequent ovulation follows a defined duration of steroid (estradiol and progesterone) actions on the hypothalamic–pituitary axis. In contrast, reflex ovulators (e.g., cats, rabbits) require mating to generate the gonadotropin surge that induces ovulation.

Ovulation is a highly complex process that involves dynamic changes in cellular biochemistry and gene expression. Most follicles do not survive folliculogenesis but instead die by atresia before reaching the mature, preovulatory stage. Only those follicles that have maintained the capacity to respond to FSH and synthesize estradiol will acquire expression of LH receptors and thereby be capable of responding to the ovulatory LH surge.

The gonadotropin surge induces an inflammatory-like reaction among the granulosa and thecal cell layers of preovulatory follicles, leading to proteolytic degradation of the follicle wall and connective tissues that are proximal to the ovarian surface, eventually allowing the follicle to rupture and the oocyte to be expelled. Prostaglandins, which are classically associated with inflammation, are synthesized at enormous rates in the granulosa cells during this process and are required for follicle rupture. Hence, nonsteroidal anti-inflammatory drugs (NSAIDs), such as indomethacin and naproxen, that block prostaglandin synthesis can inhibit ovulation. Progesterone synthesis is equally increased in ovulating follicles and also required for follicle rupture. Therefore, agents that reduce progesterone synthesis via inhibition of CYP11A1 (e.g., aminoglutethiamide, cyanoketone) or hydroxysteroid-3β-dehydrogenase (e.g., danazol) or via progesterone receptor action (e.g., RU486) will also inhibit ovulation.

Following ovulation, the remaining granulosa cells in the follicle undergo terminal differentiation to form the *corpus luteum*. Because a corpus luteum is formed at each ovulation site on the ovarian surface, these prominent structures are often used in laboratory animals to count the number of ovulations that have occurred in a particular cycle. The corpus luteum is primarily an endocrine structure that secretes enormous amounts of progesterone, estradiol, and INHIBINS that function to (i) reduce the secretion of gonadotropins from the hypothalamic-pituitary axis via negative feedback, and thereby halt further maturation of follicles in the ovary, and (ii) increase the secretory activity of the uterine endometrium in preparation for implantation of the embryo. In the absence of fertilization, the corpus luteum regresses, causing circulating progesterone levels to rapidly decline and causing the uterine lining to shed (i.e., menstruation), allowing the ovarian cycle to resume.

33.6.2 Fertilization

Along with inducing ovulation, the gonadotropin surge is also required to stimulate resumption of meiosis in the oocyte. Following the gonadotropin surge but prior to ovulation, the oocyte resumes meiosis to the first meiotic division and expulsion of the first polar body and then arrests again at meiotic metaphase II. Fusion with a sperm (*fertilization*) induces the oocyte to proceed with the final stages of meiosis and produce (a) the female haploid pronucleus for fusion with the sperm nucleus and (b) the second polar body. Throughout this period the oocyte is especially susceptible to toxicants that disrupt microtubule function (e.g., nocodazole), potentially leading to aneuploidy in the resulting zygote. Furthermore, the ovulated oocyte has a finite window during which fertilization must occur, and thus toxicants that hinder transport through the oviduct or delay fertilization lessen the possibility for establishing pregnancy (Figure 33.8).

33.6.3 Uterine Cycle, Implantation, and Pregnancy

The uterus or womb serves as the site of pregnancy. It is a luminal organ composed of two principal layers, an inner epithelial lining (endometrium) and an outer musculature (myometrium). Both undergo dramatic biochemical and morphological changes that are dictated primarily by the balance of steroid hormones from the ovary. During the follicular phase of the ovarian cycle, heightened estradiol levels from the ovary stimulate proliferation of the uterine endometrium and increase activity in the myometrium. Hence, this is referred to as the *proliferative* phase of the uterus (Figure 33.3). Following ovulation and the ovarian shift to progesterone secretion by the corpus luteum, secretory activity among the uterine endometrial glands increases and the myometrium becomes quiescent to provide a more conducive environment to embryo implantation, known as the *secretory* phase (Figure 33.3). This sequential exposure to estradiol and progesterone is obligatory to preparing the uterus for pregnancy. Once again, in the absence of fertilization, the corpus luteum regresses, circulating progesterone levels decline, and menstruation ensues.

Implantation of an embryo into the uterine wall and the establishment of pregnancy is a complex process that depends on temporally and spatially regulated proliferation, differentiation, migration and remodeling of numerous heterogeneous cell types in both the embryo and uterus. Homeostasis and maintenance of pregnancy continues this dependency on hormones produced by the mother, placenta and fetus. Several pituitary-like hormones are produced by the syncytiotrophoblast of the fetus, namely choriogonadotropin, which feeds back upon the ovary and signals for the preservation of the corpus luteum and therefore continued progesterone synthesis. Depending on the species, progesterone synthesis is then shifted to the fetal placenta later in the pregnancy.

Given this dependence on steroid hormones, endocrine-disrupting xenobiotics that affect the synthesis or actions of estradiol or progesterone are likely to disrupt embryo implantation or pregnancy. Indeed, the abortifacient action of the "morning after pill" that is used for emergency contraception is specifically due to its properties as a progesterone receptor antagonist. Compounds that inhibit progesterone synthesis, such as aminoglutethiamide and cyanoketone, may have similar effects. Some emergency contraception regimens call for the ingestion of large doses of estradiol to impair implantation, perhaps by accelerating the rate of embryo transport and thereby hindering implantation (Figure 33.8). For example, Methoxychlor, a pesticide with estrogenic activity, has been shown to accelerate the rate of embryo transport through the uterus in laboratory animals.

33.6.4 Parturition

Pre- and postmature childbirths are each estimated to occur in almost 10% of pregnancies in the United States. The physiological processes of *parturition* (birth) must be precisely timed and effective to avoid harm to the mother and fetus. Like pregnancy, parturition relies on the sending and receiving of numerous hormonal signals between mother and fetus and is therefore especially susceptible to toxic insult.

The signal to initiate parturition is thought to originate in the fetus via secretion of a hypothalamic hormone. This signal then stimulates a sequence of events in the mother that ultimately allow the uterus to transform from the relatively quiescent state of pregnancy to a highly active muscle capable of producing oscillating contractions with force sufficient to expel the fetus. In a classic example of reproductive toxicity, parturition fails to occur in pregnant sheep that have ingested skunk cabbage (*Veratrum californicum*) during pregnancy because of the severe head and hypothalamic deformations that the toxicant causes in the fetus. Several environmental chemicals are also linked to altered gestation length in rodents and humans. For example, some women that experience premature parturition are reported to possess above-average blood concentrations of the organochlorine, dichlorodiphenyltrichloroethane (DDT). Laboratory studies indicate that DDT may increase the frequency of uterine contractions, possibly via enhancing depolarization of the myometrial cell membranes. Prostaglandin synthesis in both uterine and fetal tissues is also critical to parturition. Agents that block prostaglandin synthesis, such as NSAIDs, can delay parturition, whereas labor can be artificially induced with intravaginally applied prostaglandin agonists.

33.6.5 Lactation

The development of commercial substitutes for breastmilk has lessened the frequency of breastfeeding in humans, but this does not apply to domesticated and wild animals. Much attention is given to the issue of toxicant transfer from mother to offspring via breast milk (Chapter 34), yet reproductive toxicologists must also consider the impact that impaired lactation or lactational behavior may have when evaluating changes in the fecundity of a population.

The breast and mammary glands develop under the influence of numerous hormones, but unlike most tissues, this process occurs largely after birth. During puberty, estradiol from the ovaries stimulates growth of the mammary gland ducts and surrounding stromal tissues. During pregnancy, ovarian-derived hormones, mainly estradiol and progesterone, once again stimulate further growth and vascularization as well as differentiation of the alveoli structures that produce the milk. However, milk production (*lactogenesis*) is inhibited until after parturition by the elevated levels of estradiol and progesterone that remain during pregnancy. Following birth and the precipitous decline in sex steroid levels, prolactin from the anterior pituitary stimulates the gland to begin milk production. Mechanical stimulation of the nipple or breast by the offspring during nursing then provides neural signals to release prolactin and oxytocin from the pituitary, which provide for lactogenesis and milk expulsion, respectively (Figure 33.10).

Toxicants that block the synthesis or actions of estradiol or progesterone are likely to compromise the development and differentiation of the mammary glands during puberty or pregnancy. In addition, continued exposure to exogenous estrogens or estrogenic xenobiotics during the postnatal period may inhibit lactation. This approach is often used clinically in women that opt not to breastfeed. Because prolactin secretion from the pituitary is under the negative control of dopamine, toxicants that increase dopamine secretion or action (e.g., bromocriptine) may inhibit lactation, while toxicants that block dopamine synthesis or action (e.g., certain neuroleptics) can cause hyperprolactinemia, leading to gynecomastia and galactorrhoea.

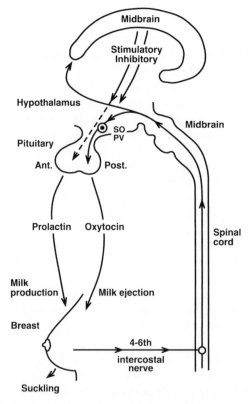

Figure 33.10. Suckling reflex. Ant., anterior pituitary; Post., posterior pituitary. (Reproduced with permission from *Maternal–Fetal Medicine: Practices and Principles*.) (CREASY Robert K., RESNIK Robert [edited by], Maternal-Fetal Medicine: Principles and Practice, Philadelphia: Saunders, c1989. p. 160.)

33.7 GENERAL CATEGORIES OF REPRODUCTIVE TOXICANTS

This interdependency among the tissues of the HPG axis and reproductive tract provides a means by which a reproductive toxicant can directly disrupt the function of a single primary target organ but cause more obvious manifestations in another reproductive organ or tissue—that is, having a secondary, or even tertiary, effect. Yuan and Foley devised a scheme that divides reproductive toxicants among three categories based on their effect in the gonads and internal/external genitalia. Type I reproductive toxicants are those that cause inactivity or quiescence in both the gonads and reproductive organs. The primary effect of a Type I toxicant is likely to involve inhibition of gonadotropin secretion via actions on the hypothalamus or pituitary, or impairment of gonadal steroidogenesis. Because steroidogenesis and folliculogenesis are concomitant processes in the ovary, a Type I toxicant may also prevent steroid production in the ovary by inhibiting follicle growth as well *versus* direct inhibitory actions on the steroidogenic pathway. Type II reproductive toxicants are described as those causing inactivity in the gonads but preserved health and perhaps even hyperactivity in the genitalia. These are most often endocrine disrupting chemicals that act as agonists in the sex steroid signaling pathways. For example, an estrogenic toxicant will activate estrogen-mediated negative feedback in the hypothalamic–pituitary axis, thereby leading to reduced gonadotropin secretion and hence reduce ovarian function, but will directly stimulate the estrogen sensitive organs of the reproductive tract (i.e., uterus, vagina) and cause these tissues to appear relatively normal or even have enhanced functions. Endocrine disrupting chemicals able to activate the androgen signaling pathways will have similar effects in males, causing reduced testicular function but providing maintenance of the sex organs. In contrast, a Type III reproductive toxicant causes hyperactivity among both the gonads and reproductive tract tissues. These are usually peptide compounds able to stimulate excess secretion of GnRH from the hypothalamus (i.e., GnRH agonists) or preparations of exogenous gonadotropins or gonadotropin-like peptides and are often used in the clinical treatment of infertility. Examples of Types I, II, and III reproductive toxicants are listed in Table 33.2.

TABLE 33.2. Examples of Type I, II and III Reproductive Toxicants[a]

Toxicity	Manifestation	Example
I	Quiescent gonads and reproductive tissues	Narcotics, marijuana, lead
II	Quiescent gonads but hyperactive reproductive tissues	Estrogen agonists (e.g., diethylstilbestrol, bisphenol A) Androgen agonists (e.g., testosterone) Progesterone agonists (e.g., dimethistrone)
III	Hypergonadism and hyperactive reproductive tissues	Transient exposure to GnRH agonist (e.g., Lupron), FSH agonist (e.g., pregnant mares serum gonadotropins), LH agonist (e.g., human choriogonadotropin)

[a]Adapted and modified from Yuan, Y. D., and Foley, G. L. Female reproductive system. In: Haschek, W. M., Rousseaux, C. G., and Wallig, M. A. (Eds.). *Handbook of Toxicologic Pathology*, Vol. 2, Academic Press, San Diego, 2002, pp. 847–894.

33.8 SUMMARY

Fertility and reproduction call upon more individual physiologies than any other life function. Hence, the target sites at which a toxicant can potentially disrupt reproductive functions are numerous. This level of complexity poses a great challenge to any investigative toxicologist trying to determine if a link exists between reduced fertility and toxicant exposure. Although not discussed in this chapter, a sizeable battery of assays and biological tests have been designed and used for the study of reproductive toxicants.

SUGGESTED READING

Andersen, M. E., Thomas, R. S., Gaido, K. W., and Conolly, R. B. Dose–response modeling In reproductive toxicology in the systems biology era. *Reprod. Toxicol.* **19**, 327–337, 2005.

Ballatori, N., and Villalobos, A. R. Defining the molecular and cellular basis of toxicity using comparative models. *Toxicol. Appl. Pharmacol.* **183**, 207–220, 2002.

Boekelheide, K., Chapin, R. E., Hoyer, P. B., and Harris, C. *Comprehensive Toxicology, Vol. 10: Reproductive and Endocrine Toxicology*, Elsevier Science, New York, 1997.

Collins, T. F. History and evolution of reproductive and developmental toxicology guidelines. *Curr. Pharm. Des.* **12**, 1449–1465, 2006.

Cooke, B. A. In vitro models for the investigation of reproductive toxicology in the testis. *Adv. Exp. Med. Biol.* **444**, 95–102, 1998.

Haschek, W. M., Rousseaux, C. G., and Wallig, M. A. *Handbook of Toxicologic Pathology*, 2nd ed., Academic Press, San Diego, 2002.

Hoyer, P. B. Reproductive toxicology: Current and future directions. *Biochem. Pharmacol.* **62**, 1557–1564, 2001.

Kimmel, C. A., and Makris, S. L. Recent developments in regulatory requirements for developmental toxicology. *Toxicol. Lett.* **120**, 73–82, 2001.

Neubert, D. Reproductive toxicology: The science today. *Teratog. Carcinog. Mutagen.* **22**, 159–74, 2002.

Safe, S. H., Pallaroni, L., Yoon, K., Gaido, K., Ross, S., Saville, B., and McDonnell, D. Toxicology of environmental estrogens. *Reprod. Fertil. Dev.* **13**, 307–315, 2001.

CHAPTER 34

Developmental Toxicology

JOHN F. COUSE

Taconic Inc., Albany Operations, Rensselaer, New York 12144

34.1 INTRODUCTION

Less than half of all human pregnancies result in the birth of a normal, healthy infant. An estimated one-third of all pregnancies are spontaneously aborted, and among those that produce offspring, 6–7% and 14% exhibit major and minor birth defects, respectively, within the first year. Abnormal neurological functions, which are not always apparent within the first few years of life, are estimated to occur in 16–17% of offspring. These adverse outcomes to pregnancy have an enormous impact on individual and public health.

The historical record of birth defects among humans and animals extends as far back as 6500 B.C. Prior to the modern age, birth defects were attributed to a diverse array of causes, including the manifestation or foretelling of astrophysical events, insalubrious maternal impressions or emotions, interbreeding between humans and animals, retributions from God, and physical trauma to the mother or a narrowness of the womb. The Scientific Revolution of the sixteenth and seventeenth centuries, along with the later recognition of Mendel's Laws of Inheritance, first published in 1865, led to genetics as the widely accepted cause of birth defects. Today, genetic causes are estimated to account for 15–25% of birth defects among human offspring.

Evidence of birth defects due to adverse environmental conditions (e.g., trauma, hyperthermia) began to be reported in the scientific literature in the 1800s. Maternal conditions such as Rubella infection or dietary deficiencies in vitamin A were soon recognized to cause severe birth defects in humans and animals. The notion of chemical or physical agents as a cause of birth defects remained largely marginalized until 1961, when thalidomide, an over-the counter sedative, was linked to a distinct syndrome of severe malformations among children in Germany. Today, adverse maternal factors or exposure to infectious, physical, or chemical agents are esti- mated to cause 5–10% of all birth defects. More than one-third of the approximately 3000 chemicals that have been tested are teratogenic in animals. The potential of human exposure to an ever-increasing number of chemicals and agents summons the need for continued expansion of developmental toxicology.

Molecular and Biochemical Toxicology, Fourth Edition, edited by Robert C. Smart and Ernest Hodgson
Copyright © 2008 John Wiley & Sons, Inc.

34.2   OVERVIEW OF DEVELOPMENT

A fundamental understanding of development is vital to gaining insight into the processes by which toxicants cause birth defects. Development is a continuous process beginning with fertilization and proceeding through puberty and the acquisition of reproductive function, if not arguably through to senescence. The following discussion focuses primarily on gestational development, which for the sake of study is divided into defined stages based on descriptive changes in the conceptus. However, birth should not be necessarily recognized as a sharp developmental boundary because elements of the central nervous, immune, metabolic, and reproductive systems continue to develop for years after birth.

34.2.1   Embryonic Period

The *embryonic* period is the first major stage of gestational development and includes *fertilization, blastocyst* formation, *implantation*, and *organogenesis*. This period is approximately two-thirds the total gestational time in rodents, whereas in monkeys and humans, it comprises a much smaller fraction (approximately one-quarter) of total gestation.

The embryonic period begins with *fertilization*, defined as penetration of a haploid female germ cell (ovum) by a haploid male germ cell (sperm), followed by fusion of the genetic material from each germ cell to form a diploid *zygote*. Fertilization takes place in the oviduct of the female reproductive tract and must occur generally within 24 hr of ovulation because the fertility of the ovum rapidly degrades thereafter. Fertilization requires a series of biochemical changes in both types of germ cells. The fertilizing sperm cell must undergo capacitation and acrosomal reaction to penetrate the glycoprotein shell (*zona pellucida*) and plasma membrane of the ovum. This is then followed by the cortical reaction in the ovum to harden the zona pellucida and prevent entry of additional sperm. At the time of ovulation, ooctyes are arrested in metaphase II of meiosis and must resume meiosis to form the haploid female pronucleus and second polar body upon fertilization. Fusion of the male and female pronuclei produces the zygote, which immediately undergoes a series of rapid cell divisions (*cleavage*) to form a *blastocyst*, all while musculature contractions in the oviduct direct its passage toward the uterus.

The blastocyst is a hollow, fluid-filled ball of approximately 1000 cells. The cells that form the outer layer are referred to as *trophoblasts* and will ultimately develop as extraembryonic tissues (e.g., placenta), while the cells of the *inner cell mass* are omnipotent (i.e., stem cells) and form the embryo. Depending on the species, the blastocyst arrives at the uterus within 5–10 days of fertilization, whereupon it "hatches" from the zona pellucida and implants into the uterine wall, which has been preconditioned by ovarian-derived steroid hormones (see Chapter 33). Shortly after implantation, the inner cell mass undergoes *gastrulation* to form a trilaminar embryo composed of three primary germ layers, the *ectoderm, mesoderm*, and *endoderm*.

Organogenesis, the final embryonic stage, is the period during which the primary germ layers of the gastrulated embryo undergo massive differentiation and morphogenesis to form the rudiments of bodily structures and organs. This is undoubtedly the most complex and dynamic stage of development. *Morphogenesis* of each structure and organ is the collective result of dramatic rates of cell proliferation, cell migration, cell–cell interactions, and programmed cell death (*apoptosis*). Fundamental to organogenesis is the process of *induction*, during which a cohort of cells (the inducers) influences the behavior, position, and phenotype of another (the induced). Induction generally relies on intercellular signaling via (i) direct cell contact between the inducer and induced cells, (ii) diffusible substances that are secreted by the inducer cells and act in a *paracrine* fashion on the induced cells (see Figure 33.1), or (iii) interactions of induced cells with an extracellular matrix that is fabricated by the inducer cells. These processes provide a "cascade" effect that ultimately allows for ordered development of an organ or limb. The intercellular signaling pathways and morphogens that govern cell fate and tissue patterning are remarkably conserved among divergent metazoan species, including *Drosophila* (fruit fly), roundworm (*Caenorhabditis*), zebra fish, frogs, chickens, rodents, and humans. Furthermore, within a species, this assortment of intercellular signaling pathways is used repeatedly in different spatial and temporal patterns to dictate the development of the different bodily structures. Intercellular signaling via Hedgehog, Notch-Delta signaling, Wingless-Int, nuclear hormone receptors (e.g., retinoids), nuclear factor κB, transforming growth factor beta, epidermal growth factor, and fibroblast growth factor are all critical to embryonic development. The extraordinary degree of conservation in the developmental processes of varied species validates the continued use of experimental animals in the study of development and development toxicology to gain insight into the vulnerabilities of the human conceptus.

34.2.1a Sonic Hedgehog Signaling in Limb Morphogenesis. Sonic hedgehog (SHH) is one of three known mammalian homologs of the Hedgehog signaling molecule that was first discovered as a key determinant of the segmented body plan in fruit flies. The *Sonic hedgehog* gene encodes a secreted protein (SHH-N) that acts as a morphogen during patterning of the neural tube and limb buds of vertebrates. The general mechanism of SHH signaling is shown in Figure 34.1. Inducer cells synthesize SHH as a precursor protein that requires autocatalytic cleavage to produce the functional ~20-kD N-terminal portion (SHH-N), which then requires covalent binding of a cholesterol molecule to the C-terminus for activation. Secretion of SHH-N requires the synthesizing cell to express a membrane-bound protein named *Dispatched-1* (DISP1). Therefore, *paracrine* actions of SHH-N on neighboring cells are dependent on co-expression of DISP1, whereas *autocrine* actions of SHH-N within the synthesizing cell do not. Secreted SHH-N binds to the *Patched-1* (PTCH1) receptor on the membrane surface of target cells. In the absence of SSH-N, PTCH1 constitutively inhibits the activity of *Smoothened* (SMO), a seven-pass transmembrane protein. However, when SSH-N is available, it interacts with PTCH1 and releases its repression of SMO, allowing for the activation of intracellular transcription factors, such as GLI1 and GLI2, which then translocate to the nucleus and induce expression of hedgehog target genes.

During limb formation, SHH-N is synthesized and secreted by a small cohort of inducer cells in the posterior margin of the limb bud, a region referred to as the zone of polarizing activity (ZPA) (Figure 34.1). Studies in mice indicate that the type of digit formed at the distal end of the limb is determined by both the concentration and duration of SHH-N exposure to the cells at the distal end of the

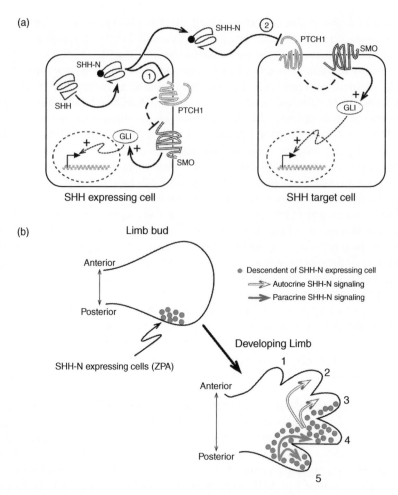

Figure 34.1. Sonic hedgehog signaling and its role in limb bud differentiation. (A) Schematic diagram of *autocrine* (1) and *paracrine* (2) signaling mechanisms for the activated form of Sonic hedgehog (SHH-N). SHH, inactive form of Sonic hedgehog prior to autocatalytic cleavage and binding of cholesterol molecule; PTCH1, *Patched 1* receptor; SMO, *Smoothened*. (B) Schematic of the postulated mechanism by which SHH-N determines digit identity in the developing limb bud. A cohort of cells (zone of polarizing activity) in the posterior portion of the early limb bud express and secrete SHH-N. The prolonged exposure of these cells to high concentrations of SHH-N (i.e., *autocrine* signaling) dictates their development into digits 5 and 4. Digit 3 is composed of some cells that were originally part of the ZPA, but not all, and therefore relies on both *autocrine* and *paracrine* SHH-N signaling. Digit 2 is formed solely via *paracrine* SHH-N signaling, whereas differentiation of digit 1 is independent of all SHH-N signaling. (Adapted from data in Harfe, B. D., Scherz, P. J., Nissim, S., Tian, H., McMahon, A. P., and Tabin, C. J. Evidence for an expansion-based temporal Shh gradient in specifying vertebrate digit identities. *Cell* **118**, 517–528, 2004.)

developing bud. As shown in Figure 34.1, digits 5 and 4 are generated from cells that were once part of the ZPA and are therefore exposed to autocrine SHH-N activity of the highest concentration and longest duration. In contrast, digit 3 is composed of cells that are exposed to SHH-N at much lower levels and of shorter duration. Furthermore, not all of the cells in digit 3 are derived from the ZPA but instead were neighboring cells. Therefore, digit 3 requires both autocrine and paracrine SHH-N signaling or morphogenesis. No ZPA-derived cells are incorporated into digit 2, making this digit totally dependent on paracrine SHH-N signaling of much lower concentration and time; and development of digit 1, the farthest from the ZPA, is apparently independent of SHH signaling.

34.2.1b Homeobox (HOX) Genes and Body Segmentation. The homeobox (HOX) family of proteins is a highly conserved set of transcription factors, each encoded by an individual gene, which dictate body patterning during embryogenesis. Each HOX protein possesses a homeobox domain that provides for specific binding to consensus enhancer sequences located within the regulatory regions of target genes. The homeobox (*hox*) genes were first discovered as a single cluster in *Drosophila* and later as a set of four clusters in mammals. Humans and mice possess four clusters of *HOX/Hox* genes of 9–13 genes per cluster, and each cluster is located on a separate chromosome. *HOX* genes of the same number among the four clusters are considered paralogs and expressed in similar and overlapping patterns. As shown in Figure 34.2, the *HOX* genes are expressed in a specific pattern along the anterior to posterior axis of the developing embryo, and this pattern parallels their

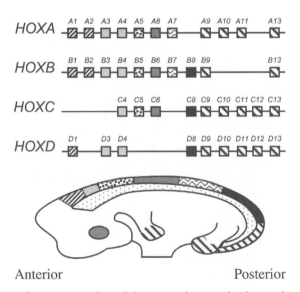

Figure 34.2. Schematic representation of the genomic organization and embryonic expression pattern of the mammalian *HOX* gene clusters. HOX genes in the 3′ regions of the cluster are generally expressed in the anterior portions of the embryo, whereas the HOX genes in the 5′ regions of the clusters are expressed in the posterior portions and limb buds of the embryo. (Adapted and reproduced with permission from Daftary, G. S., and Taylor, H. S. Endocrine regulation of HOX genes. *Endocr. Rev.* **27**, 331–355, 2006.)

3' to 5' chromosomal alignment. The anterior expression boundary of each HOX gene is conspicuous and then forms a tail-like expression gradient toward the posterior. The anterior limit of the adjacent 5' HOX gene overlaps with the fading posterior expression of the upstream gene, thereby generating a region in which multiple HOX proteins are present at distinct levels. The combination of *HOX* genes expressed within a defined area is referred to as the HOX code and determines the identity of that particular body segment. In the event of a mutation or repression of a particular *HOX* gene, the balance of overlapping expression (or HOX code) is disrupted. This allows the adjacent 3' HOX gene to have a greater than normal influence, causing the body segment to develop a phenotype that is more similar to the adjacent anterior segment. This phenomenon is referred to as a *homeotic transformation* and often results in the duplication of body segment or part.

Several members of the nuclear receptor superfamily of ligand-activated transcription factors have been shown to regulate the expression *HOX/Hox* genes. For example, retinoids act via the retinoic acid (RAR) and retinoid (RXR) family of receptors to regulate the expression of HOX gene paralogs in the 3' regions of each cluster, most especially *Hoxa1, Hoxb1*, and *Hoxd4*. These genes are expressed early during embryogenesis and are especially critical to the development of the head and central nervous system. Likewise, estrogens acting via the estrogen receptor (ER) have been shown to disrupt the pattern of *Hoxa* gene expression during development of the murine female reproductive tract. Not surprisingly, developmental exposure to exogenous retinoids or estrogens causes severe malformations in humans and laboratory animals (Section 34.4).

34.2.2 Fetal Period

The conclusion of organogenesis marks the beginning of the *fetal* period. This occurs at approximately 56 days post-fertilization in humans, 45 days in monkeys, and 15–17 days in rodents. This second and final gestational stage is characterized by differentiation and specialization of cells and tissues (i.e., *histogenesis*), functional maturation of organ systems, and overall growth of the fetus, and it ends with birth. The rate of overall growth during the fetal period is dramatic. For example, the human fetus exhibits a greater than sevenfold increase in body length (from 5 to 36 cm) and 425-fold increase in body weight (from 8 to 3400 g) during the second and third trimesters of pregnancy. Continued cell proliferation is paramount to overall growth. Equally dramatic rates of cytodifferentiation, marked by increased golgi apparatus, endoplasmic reticulum, and the appearance of biosynthesis enzymes and specialized intracellular signaling pathways, provide for specialization of cellular functions. Programmed cell death (*apoptosis*) of transient cell types is also critical to morphogenesis and functional maturation by allowing for the shaping and remodeling of digits and the palate, the formation of hollow structures in the neural tube and blood vessels, and remodeling of particular brain regions. Ultimately, it is during the fetal period that limbs are elongated and the long bones become ossified, the fetal kidneys begin functioning to produce fluids that are needed to contribute to the amniotic volume, the eyes and ears become functional, erythropoeisis commences in the fetal liver and spleen but is eventually shifted to the bone marrow, head and body hair become visible, epithelial cells in the lung begin secretory activity, the internal and external genitalia differentiate, substantial deposits of white and brown adipose tissue are formed, and the central nervous system acquires the ability to induce rhythmic breathing-like contractions of the lungs as well as regulate body temperature.

34.3 WILSON'S PRINCIPLES OF TERATOLOGY

Teratology is the study of the mechanisms and manifestations of abnormal development and congenital malformations due to all possible causes, including genetic, maternal, toxic, or other factors. Hence, teratology is a branch of both developmental biology and pathology. Conversely, *developmental toxicology* is more limited to the pharmacokinetics, mechanisms, pathogenesis, and manifestations that follow exposure of the conceptus to *teratogenic agents*, whether physical, chemical, or infectious.

In the early 1970s, James G. Wilson formulated six principles of teratology based on accumulated findings and experimental data concerning malformations caused by gestational exposure to chemical, physical, or biological agents in humans and animals (Table 34.1). Although scientific progress over the past 30 years has called for some modification, these tenets continue to serve as valuable guides in developmental toxicology.

34.3.1 Role of Gene–Gene and Gene–Environment Interactions

This principle is based on numerous observations that species, strains, and even littermates (or siblings) exhibit differential susceptibility to the same developmental toxicant. These differences are attributed to variations in biochemical and morphological attributes that are genetically determined. Disparities in which a toxicant is absorbed, metabolized, or eliminated by both the mother and conceptus, as well as its ability to interact with certain cell types or components, are thought to underlie the variations in susceptibility.

Perhaps the best-known example of an *interspecies* difference in the susceptibility to a developmental toxicant is that of thalidomide. Human, nonhuman primate, and

TABLE 34.1. Wilson's General Principles of Teratology

1. Susceptibility to teratogenesis depends on the genotype of the conceptus and the manner to which this interacts with environmental factors.
2. Susceptibility to teratogenic agents varies with the developmental stage at the time of exposure.
3. Teratogenic agents act through specific mechanisms on developing cells and tissues to initiate abnormal embryogenesis (pathogenesis).
4. The final manifestations of abnormal development are death, malformation, growth retardation, and functional disorder.
5. The access of adverse environmental influences to developing tissues depends on the nature of the influences (agent).
6. Manifestations of deviant development increase in degree as dosage increases from the no-effect to the totally lethal level.

Source: Wilson, J. G. (1977) Curent status of teratology. In: Wilson, J. G., and Fraser, F. C. (Eds.). *Handbood of Teratology*, Vol. 1: *General Principles and Etiology*, Plenum Press, New York, 1977, pp. 47–74.

rabbit embryos are extremely susceptible to the teratogenic effects of thalidomide exposure (described in Section 34.4.5), whereas no effects are observed in certain strains of rats, hamsters, and mice, even when exposed to doses as high as 8000-fold the lowest teratogenic dose in humans. Another example of interspecies variation is the unusually high susceptibility of mouse embryos to cleft palate induction by glucocorticoids, while other mammalian species are relatively resistant. It must be emphasized, however, that differential sensitivities to a particular teratogen, as well as the labeling of a species or strain as either "resistant" or "sensitive," apply only to that particular agent.

Some chemicals cause comparable malformations in all species or strains tested and are therefore referred to as *universal teratogens*. These compounds are generally believed to disrupt certain fundamental physiological processes, such as DNA or cellular replication. Retinoids are recognized as universal teratogens based on their ability to regulate and disrupt the *HOX* code pattern of gene expression that is critical to determining body pattern of vertebrate and invertebrate species.

Intraspecies differences in the susceptibility to a particular teratogen are also observed and likely due to genetic variations. For example, the A/Jax strain mice are markedly more susceptible to glucocorticoid-induced cleft palate than the C57BL/6J and CBA strains. This divergence is postulated to be due to a varied number of glucocorticoid receptors in the developing maxillary processes and rates of glucocorticoid metabolism among the different strains. Variations in birth defects induced by benzo[*a*]pyrene among offspring in the same litter of mice are associated with genotype. Mutations in developmental genes may make an individual more susceptible to a teratogenic agent, thereby increasing the risk of a certain birth defect that already occurs spontaneously at a low frequency. For example, mothers who smoke and possess a variant allele of the transforming growth factor-alpha gene are reported to exhibit an increased risk of having offspring with cleft palate.

34.3.2 Susceptibility Varies with Gestational Age

Wilson's hypothetical curve of susceptibility over the course of gestational development is shown in Figure 34.3. This curve focuses primarily on structural malformations and therefore highlights the period of organogenesis as being most susceptible

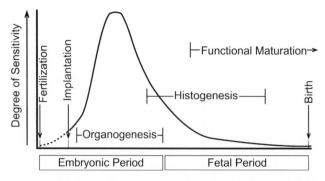

Figure 34.3. Sensitivity of the developing mammalian fetus to structural defects by xenobiotics over the course of gestation. (From Wilson, J. G. *Environment and Birth Defects*, Academic Press, New York, 1973.)

to perturbation by teratogenic agents. Although this is true, the hypothetical curve should not be interpreted to indicate that the other periods of gestational development—fertilization, implantation, and functional maturation—are refractory to the effects of developmental toxicants.

Embryonic lethality is generally considered to be the primary manifestation of developmental toxicity during preimplantation. At this stage the embryo is largely composed of totipotent cells and therefore believed to be capable of overcoming some degree of perturbation, and those that survive may manifest only a delay in growth and development. However, certain toxicants are known to cause malformations when exposure occurs prior to implantation exposure, including ethylene oxide, retinoic acid, and cyproterone acetate.

As mentioned, the embryo is most likely to exhibit structural malformations if exposure to the developmental toxicant occurs during organogenesis. The extreme rates of cell proliferation, migration, differentiation, and cell interactions that provide for morphogenesis of the organs and bodily structures make this period highly susceptible to perturbation. There are peak periods of susceptibility for each forming structure over the course of organogenesis. For example, hamster embryos exposed to retinoic acid around gestational day 7 are likely to exhibit spina bifida but normal limbs, whereas exposure on gestational day 10 leads to shorter limbs but no obvious neural tube defects.

The manifestations of toxicant exposure during fetal development are most often characterized as retarded growth or a lack of functional maturation. Structural malformations that do result from fetal exposure to a toxicant are more often attributed to toxicant-induced deformation of a once-normal structure. The central nervous system is a prime target during this period, and fetal exposure to developmental toxicants can lead to mental retardation as well as deficits in behavioral, mental, and motor skills.

34.3.3 Developmental Toxicants Act via Specific Mechanisms

Teratogens act via specific *mechanisms* within developing cells to initiate abnormal development at the tissue and organ level (*pathogenesis*). The proposed mechanisms of developmental toxicants are largely similar to those of most general toxicants (Table 34.2). However, the enhanced rate at which cellular proliferation, differentiation, and functions occur during embryonic and fetal development makes these processes substantially more prone to toxic insult. The average cell-cycle length during certain embryonic periods is as short as 3–5 hr, considerably faster than any adult tissue. Disruption of cell proliferation, or even cell death, can lead to the loss or malfunction of inductive sequences and the failure of a developmental cascade. The teratogenic effects of thalidomide, which include severe malformations of the limbs but no obvious defects in the brain, illustrate the specificity of teratogenic agents. The manifestations are clearly due to thalidomide's capacity to disrupt specific biochemical pathways.

34.3.4 Final Manifestations of Developmental Toxicity

The four general outcomes of abnormal development are *death, malformation, growth retardation*, and *functional deficiency*. As discussed earlier, the exact

TABLE 34.2. General Mechanisms and Pathogenesis of Developmental Toxicants

Mechanisms (Early Events that Initiate Altered Development)
Genetic mutations and altered nucleic acid integrity
Chromosomal breaks, nondisjuction, mitotic interference
Diminished supplies of precursors and substrates
Altered energy sources
Altered osmobalance
Enzyme inhibition

Pathogenesis (Manifestations that Follow the Above Mechanisms)
Excessive or reduced cell death
Inhibition of cell proliferation/differentiation
Failed cell–cell interactions
Reduced biosynthesis
Impeded morphogenesis
Mechanical disruption of tissues

Final Manifestations
Death
Malformation
Growth retardation
Functional disorder

manifestation likely depends on the gestational period during which exposure occurs. For example, a single compound may cause death if exposure occurs during early embryogenesis but lead to structural malformations if exposure is delayed until organogenesis. The above endpoints do not always represent a continuum of increasing toxicity; that is, death is not always simply the effect of increasing the dose of a compound that at lower doses causes only malformations or growth retardation. A reagent may induce death in a substantial portion of exposed embryos but otherwise cause no apparent malformations in those that survive, and it would therefore be referred to as *embryolethal* rather than teratogenic.

The manifestations of toxicant exposure during gestation are not always immediately apparent at birth, or even shortly thereafter. For example, the rare form of vaginal adenocarcinoma associated with fetal exposure to diethylstilbestrol (DES) does not usually manifest until 18–24 years of age (described in Section 34.4.3). Additionally, fetal exposure of rats to ethanol is reported to shorten their life span by 7–20 weeks. This latency period between gestational exposure and manifestation is especially true for birth defects in neurological function, estimated to occur in 16–17% of human offspring. The functional maturation of the central nervous system that occurs during the fetal period is a prime target of developmental toxicants. The absence of any obvious structural abnormalities conceals the functional defect for possibly months to even years. Furthermore, birth defects of the human central nervous system are difficult to model in experimental animals, limiting their use in the evaluation and study of potential human teratogens that affect the central nervous system.

34.3.5   Nature of the Toxicant Determines Access to the Conceptus

As with any toxicant, access to the target tissue is essential to the action of a teratogen. The two major routes by which a reagent gains access to the conceptus are direct penetration of maternal tissues or via indirect transmission.

Ionizing radiation, microwaves, ultrasound, or hypothermia are the major physical agents that can affect the fetus via direct transmission through maternal tissues. In general, the dose required for a physical agent to cause detriment to the fetus surpasses that required to induce maternal toxicity. Mechanical impact or changes in temperature, unless extreme, are likely minimized by the hydrostatic pressure of the womb and maternal homeostatic capabilities.

Chemical reagents generally reach the embryo or fetus via the maternal circulation. As with any toxicant, the possibility and degree of insult depends on whether a sufficient concentration of toxicant is able to gain access to the developing target tissue. This is determined by the physical properties of the agent and its rate of uptake, distribution, metabolism, and excretion by the maternal, placental, and embryonic compartments. Pregnancy leads to measurable changes in maternal physiology, such as altered gastrointestinal, cardiovascular, and hepatic functions, that may influence the level of active toxicant that reaches the conceptus. The metabolic capacity of the fetal liver is generally considered to be quite limited relative to the mother. Therefore, absent or inefficient biochemical pathways for deactivation toxicants in the fetal liver may allow for a tetragon to accumulate to harmful levels. Likewise, a reagent that requires metabolic activation may have to first pass through the maternal liver to become toxic.

In general, the placenta is a poor barrier to xenobiotics and allows for bidirectional transfer of most substances. The majority of molecules cross the placenta via diffusion along a concentration gradient, the rate of which is a product of the agent's size, charge, lipid solubility, and affinity for other biomolecules.

34.3.6   Teratogenesis Is a Threshold Concept

The maternal capacities that provide a homeostatic environment and metabolic deactivation of potential toxicants, along with the repair and regenerative capabilities of the embryo/fetus, are believed to impart a threshold phenomenon to developmental toxicity. The supposition of a threshold implies that a maternal dose exists at which a toxicant will elicit no adverse effect on the conceptus. This is in contrast to the threshold principle of carcinogenesis, which assumes that exposure to any amount of carcinogen, even a single molecule, can potentially lead to cancer.

The typical dose–response curve for a developmental toxicant is steep and covers less than one to three orders of magnitude below the dose that kills or malforms half the embryos. However, the threshold and shape of a dose–response curve for a particular teratogen may differ, depending on the gestational time of exposure and the type of embryotoxicity that is measured. Furthermore, the dose at which an agent causes malformations may be above, below, or equivalent to the *embryolethal threshold*. Toxicants that cause little or no maternal toxicity, even at elevated doses, but adversely affect the conceptus are especially dangerous (e.g., thalidomide, diethylstilbestrol).

34.4   SELECTED EXAMPLES OF DEVELOPMENTAL TOXICANTS

Approximately 1200 chemicals have been shown to be teratogenic in experimental animals. However, less than 40 physical, chemical, or infectious agents are known to produce birth defects in humans (Table 34.3). Five xenobiotics known to be

TABLE 34.3. Known Developmental Toxicants in Humans

Physical Agents
Hyperthermia
Radiation (x-rays)

Infectious Agents
Viral
   Cytomegalovirus
   Herpes simplex virus I and II
   Rubella virus
   Varicella virus
   Venezuelan equine encephalitis virus
Bacterial
   Treponema pallidum (syphilis)
Protozoan
   Toxoplasma gondii (toxoplasmosis)

Maternal Metabolic Diseases
Alcoholism
Diabetes
Folic acid deficiency
Phenylketonuria
Sjögren's syndrome
Virulizing tumors

Xenobiotics
Androgenic agents
Antibiotics (e.g., tetracycline)
Anticancer drugs (e.g., cyclophosphamide, busulfan)
Anticonvulsants (e.g., valproic acid)
Antithyroid drugs
Chlorobiphenyls
Cigarette smoke
Cocaine
Ethanol
Ethylene oxide
Diethylstilbestrol
Iodides
Lithium
Metals (e.g., mercury, lead)
Retinoids
Thalidomide

developmental toxicants in humans were selected for further description based on their historical impact and diversity of mechanisms and pathogenesis.

34.4.1 Anticonvulsants

The availability of anticonvulsant drugs over the past 30 years has been of enormous therapeutic benefit to persons suffering from epilepsy. However, a number of these drugs have come under scrutiny as being teratogenic in humans, including carbamazepine, phenytoin, and valproic acid (VPA) (Figure 34.4). This remains a contro-

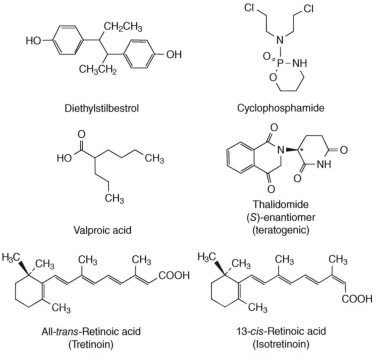

Figure 34.4. Chemical structures of selected known developmental toxicants in humans.

versial issue because persons suffering from epilepsy may be genetically prone to having children with deformities unrelated to their use of anticonvulsants during pregnancy. Epidemiological and laboratory evidence to date indicates carbamazepine and phenytoin are weak developmental toxicants in humans, whereas the evidence of teratogenic effects due to VPA exposure is more convincing.

The teratogenic syndrome of in utero VPA exposure in humans includes neural, craniofacial, cardiovascular, and skeletal defects. A similar teratology is exhibited in rodents, rabbits, and nonhuman primates. The most dramatic of malformations associated with gestational VPA exposure is spina bifida, a neural tube defect that is estimated to occur in 1–2% of VPA exposed infants. Mice are the only known animal model to exhibit a comparable effect of VPA on neural tube development.

The mechanisms by which VPA disrupts neural tube development remain largely unknown. Neural tube closure during organogenesis requires continuous cell proliferation of the neuroepithelia, and in vitro studies indicate that VPA causes cell cycle arrest in neuronal cells. These findings suggest that VPA either directly disturbs the proliferative ability of neuronal cells or perhaps alters their sensitivity to neurotrophic growth factors.

34.4.2 Cyclophosphamide

Cyclophosphamide is an alkylating agent employed as an anticancer agent and immunosuppressant to prevent the rejection of organ transplants (Figure 34.4). In humans, exposure to cyclophosphamide during the first trimester is associated with ectrodactyly (absence of fingers or toes), syndactyly (webbing of fingers or toes), cardiac abnormalities, and growth retardation. Cyclophosphamide is also extremely teratogenic in several animal species, where it reportedly causes fetal death, gigantism, dwarfism, cleft palate, and neural tube defects, depending on the time of exposure.

Cyclophosphamide requires metabolic activation to form the teratogenic metabolites, phosphoramide mustard, and acrolein (Figure 34.5). The former metabolite produces single-stranded DNA breaks, DNA–DNA and DNA–protein crosslinking. In contrast, acrolein preferentially binds to proteins, and it concentrates in the

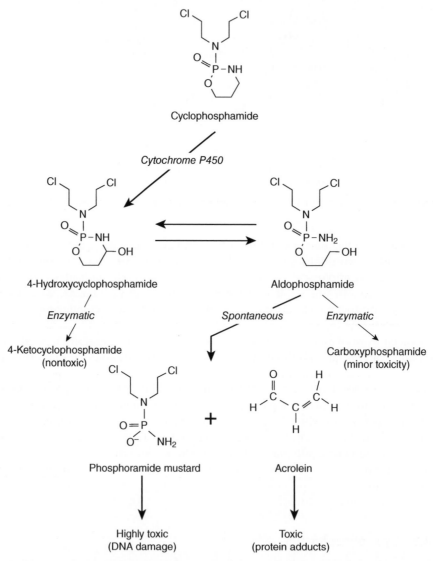

Figure 34.5. Metabolic pathway for the bioactivation of cyclophosphamide. (Adapted from Mirkes, P. E. Cyclophosphamide teratogenesis: A review. *Teratog. Carcinog. Mutagen.* **5**, 75–88, 1985.)

embryonic yolk sac. Studies in laboratory animals indicate that a quite divergent teratogenic syndrome results when each metabolite is present alone.

34.4.3 Diethylstilbestrol

Diethylstilbestrol (DES) is a potent, orally available synthetic estrogen used clinically for numerous indications, including the treatment of various steroid-dependent cancers (e.g., prostate and breast) and the suppression of lactation and postmenopausal osteoporosis (Figure 34.4). The teratogenic properties of DES are associated with its use as a prophylactic agent for premature labor and spontaneous abortion in pregnant women, which began in the 1940s in the United States and other developed nations throughout the world. In the early 1970s, physicians at Massachusetts General Hospital diagnosed a small cohort of young women with clear cell adenocarcinoma of the vagina, an extremely rare form of cancer, and established a putative link to DES exposure during fetal development. The risk of vaginal cancer among women exposed prior to the 18th week of gestation is estimated to be less than 1%. Nonetheless, DES remains the only known transplacental carcinogen in humans. Several noncancerous abnormalities have since been described in DES "daughters," including a malformed "T-shaped" uterus, uterine hypoplasia, and cervical abnormalities. These reproductive tract deformities are thought to contribute to an increased incidence of ectopic pregnancy, spontaneous abortion, and premature delivery that is reported among women exposed to DES during fetal development. Males exposed to DES during fetal development are reported to be at increased risk of epididymal cysts and reduced fertility, although these findings remain controversial. Several animal models of perinatal DES exposure, most especially rodents, have been developed over the past 30 years and shown to effectively mimic the carcinogenic and teratogenic effects.

DES is a nonsteroidal estrogen, and its teratogenic and carcinogenic properties may occur via distinct mechanisms. The majority of laboratory evidence indicates that the structural abnormalities of the female reproductive tract induced by fetal DES exposure are due to its estrogenic activity—that is, via its properties as an endocrine disrupting chemical (as described in Chapter 33). Recent evidence indicates that the HOX code for differentiation of the female reproductive ducts during organogenesis is highly conserved between humans and mice. Perinatal exposure to DES in mice leads to a posterior shift in the HOX code, such that the *Hoxa9* expression that is normally localized to the developing oviduct now occurs in the uterus, *Hoxa10* expression is likewise shifted to a more caudal position in the developing tract, and the usually high caudal *Hoxa11* expression is altogether diminished (Figure 34.6). DES induction of this posterior shift of the *Hox* code in the developing mouse uterus is postulated to cause portions of the cervical and vaginal epithelium to improperly develop a uterine-like phenotype, which may render these tissues more susceptible to cancer. Genetically modified mice that lack estrogen receptor alpha (ERα), a member of the family of nuclear receptors that act as ligand-induced transcription factors, are resistant to the teratogenic effects of DES, including the posterior shift in *Hoxa* gene expression, supporting the role of hormonal DES actions as a mechanism for teratogenesis in the female reproductive tract. However, DES and its phase I metabolites (e.g., 4-hydroxyestradiol) have also been shown to be directly mutagenic, a property that is putatively

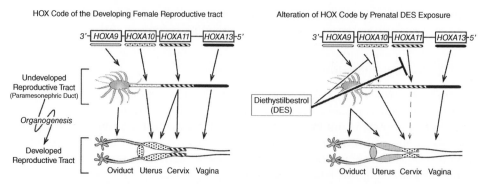

Figure 34.6. Schematic representation of the HOX code during differentiation of the mammalian female reproductive ducts. (Left) HOXA genes are expressed in an anterior to posterior pattern that parallels their 3′–5′ arrangement in the cluster. The level of *HOXA* gene expression in each segment of the developing reproductive tissues influences the type of structure that is formed. (Right) Studies indicate that developmental exposure to the estrogen diethylstilbestrol (DES) causes decreased expression of *HOXA10* and *HOXA11*, leading to a posterior shift in the HOX code. As a result, portions of the fully developed cervix and upper vagina exhibit characteristics of the uterus, an event known as a *homeotic transformation*. (Adapted from Daftery, G. S., and Taylor, H. S. Endocrine regulation of HOX genes. *Endocr. Rev.* **27**, 331–355, 2006.)

unrelated to its hormonal activity and hypothesized to provide for its carcinogenic actions.

34.4.4   Retinoids

Vitamin A (retinol) and other retinoid compounds are essential for normal embryonic development (Figure 34.4). These compounds act as critical morphogens during mammalian embryogenesis via members of the family of ligand-activated transcription factors known as the retinoic acid (RARα, RARβ, RARγ) and retinoid X (RXRα, RXRβ, RXRγ) receptors. Vitamin A cannot be synthesized *de novo* and must therefore be consumed in food or supplements. A diet that is insufficient in vitamin A during pregnancy is associated with embryonic defects of the cardiovascular, skeletal, and central nervous systems. Similarly, genetically modified mice that lack one or more of the RAR or RXR subtypes exhibit distinct birth defects that are comparable to those attributed to vitamin A (i.e., ligand) deficiency.

Retinoids were found to be teratogenic in humans in the early 1980s. Isotretinoin or *cis*-retinoic acid (Figure 34.4), marketed as Accutane, is a systemic medication for the treatment of recalcitrant cystic acne. Despite the enforcement of considerable regulations on the prescribed indications for Accutane by the United States Food and Drug Administration, an estimated 1000 infants were exposed during gestation. The resulting teratogenic syndrome included craniofacial abnormalities (e.g., misshapen head, small lower jaw, microtia, clefting) and defects of the thymic, cardiovascular, and central nervous systems. The association of gestational exposure to isotretinoin with defects in limb development and cognitive abilities remains controversial. Studies indicate the lowest teratogenic dose of isotretinoin in humans is half that prescribed for the treatment of acne (40 mg/day). Few women who take

Accutane during pregnancy are expected to give birth to children free of birth defects. There are also a limited number of reports describing birth defects attributed to the ingestion of massive amounts of vitamin A (retinol) or the use of topical retinoid preparations, primarily tretinoin (all-*trans*-retinoic acid; Figure 34.4) during pregnancy.

Retinoids are potent teratogens in every animal species tested, including mice, rats, guinea pigs, hamsters, rabbits, dogs, chicks, pigs, and monkeys. The time of gestational exposure profoundly influences the form of birth defect, from embryo lethality to the numerous malformations described above. The peak period of sensitivity for a given tissue appears to be during development of the primordial structures.

The teratogenic mechanisms of retinoids remain unclear, but mounting evidence indicates that exogenous retinoids during pregnancy disrupt the level or temporal pattern of morphogenic endogenous retinoid actions in the developing embryo. As discussed in Section 34.2.1b, retinoids are potent regulators of *HOX* genes, most especially those in the 3′ end of the *HOX* gene clusters, and are therefore involved in anterior patterning—that is, the head, brain, and central nervous system. Genetically modified mice that lack one or more of the RAR or RXR subtypes are providing insight into the precise role of retinoid signaling during development. For example, mice lacking functional RARγ are resistant to the teratogenic effects of retinoid exposure on the skeleton but still exhibit craniofacial anomalies, implicating a roll for this receptor subtype in the former pathogenesis. In contrast, RARβ-null mice exhibit full susceptibility to the teratogenic effects of retinoids, suggesting that this receptor has little involvement.

34.4.5 Thalidomide

Thalidomide is perhaps the most infamous of pharmaceutical teratogens (Figure 34.4). In the late 1950s, the once over-the-counter sedative and antiemetic was advocated for the treatment of "morning sickness" in pregnant women in Germany and other European countries. The FDA did not approve the used of thalidomide in the United States despite it having few reported adverse effects in the mothers. However, in the early 1960s, German physicians soon recognized that thalidomide used during pregnancy was linked to a tragic syndrome of severe birth defects, predominantly partial (phocomelia) to complete (amelia) reduction of the limbs. Other malformations were also noted to involve the eyes and ears (i.e., microtia), kidneys, heart and gastrointestinal tract. Although thalidomide was immediately removed from the market, it is estimated that approximately 8000 children suffered malformations due to gestational exposure.

The teratogenic actions of thalidomide require exposure to occur during discrete gestational "windows." Limb defects result when the embryo is exposed between the 22nd and 36th days, post-conception. More precisely, exposure during days 27–30 predominantly leads to defects of the arms, whereas exposure during days 30–33 leads to defects in both the arms and legs. A single dose, generally 50–100 mg, is believed to be sufficient to induce the teratogenic syndrome if taken at the appropriate stage of pregnancy. The numerous experimental investigations that followed the human tragedy indicate the rabbit as the most suitable animal model and that a comparable syndrome of birth defects is induced with a dose approximately

100-fold greater than that required in humans. However, hamsters, mice, and some species of rat are largely resistant.

A diverse array of mechanisms are proposed for the teratogenicity of thalidomide, ranging from (a) biochemical alterations involving the synthesis of vitamin B, nucleic acids, and glutamic acid to (b) oxidative phosphorylation, disruptions of cellular processes including cell–cell communication, or inhibition of nerve and vascular growth during initiation of limb development. A recent hypothesis that is gaining considerable attention proposes that thalidomide inhibits the expression of specific developmental genes that require constitutive binding of Sp1, a ubiquitous transcription factor, to their promoter. Thalidomide is believed to intercalate with GC-rich regions of DNA, such as the GC box elements (GGGCGG) that provide for Sp1 binding. Hence, thalidomide is postulated to inhibit Sp1 binding and thereby repress the expression of insulin-like growth factor 1 (*Igf1*) and fibroblast growth factor 2 (*Fgf2*), the products of which are known to promote vascularization in the developing limb bud by inducing the expression of alpha-5 and beta-3 integrins (Stephens and Fillmore, 2000). An alternative hypothesis proposes that thalidomide increases the level of oxidative stress within the developing tissues of sensitive species, thereby leading to misregulation of the redox-sensitive transcription factor, NF-κB, which subsequently causes reduced expression of factors that are critical to limb development, such as fibroblast growth factor-8 (*Fgf8*) and sonic hedgehog (*Shh*) (Hansen and Harris, 2004). The apparent lack of any effect of thalidomide on *Hox* gene expression is proposed to provide for the relatively normal patterning of fingers or toes that occurs at the end of the shortened, malformed limb.

Recently, thalidomide has gained approval for use in the United States as an immunomodulatory agent in the treatment of HIV-associated diseases, arthritis, myeloma, and diabetic retinopathy. The FDA has instituted the System of Thalidomide Education and Prescribing Safety (STEPS) to prevent accidental exposures during pregnancy.

34.5   SUMMARY

The frequency and range of adverse outcomes to pregnancy have an incalculable impact on individual humans and public health overall. That infectious, physical, and chemical agents cause developmental defects in animals, including humans, is now universally accepted. However, the biological mechanisms that underlie the teratogenic and toxic properties of such agents during development remain largely unknown. This fact, along with an ever-increasing number of chemicals and agents to which humans of all ages are potentially exposed, commands the continuation and expansion of current research efforts in developmental toxicology.

SUGGESTED READING

Boekelheide, K., Chapin, R. E., Hoyer, P. B., and Harris, C. *Comprehensive Toxicology, Vol. 10: Reproductive and Endocrine Toxicology*, Elsevier Science, New York, 1997.

Collins, T. F. History and evolution of reproductive and developmental toxicology guidelines. *Curr. Pharm. Des.* **12**, 1449–1465, 2006.

Cuomo, V., De Salvia, M. A., Petruzzi, S., and Alleva, E. Appropriate end points for the characterization of behavioral changes in developmental toxicology. *Environ. Health Perspect.* **104** (Suppl. 2), 307–315, 1996.

Daftery, G. S., and Taylor, H. S. Endocrine regulation of HOX genes. *Endocr. Rev.* **27**, 331–355, 2006.

Finnell, R. H., Gelineau-van Waes, J., Eudy, J. D., and Rosenquist, T. H. Molecular basis of environmentally induced birth defects. *Annu. Rev. Pharmacol. Toxicol.* **42**, 181–208, 2002.

Flick, B., and Klug, S. Whole embryo culture: An important tool in developmental toxicology today. *Curr. Pharm. Des.* **12**, 1467–1488, 2006.

Hansen, J., and Harris, C. A novel hypothesis for thalidomide-induced limb teratogenesis: Redox mis-regulation of the NF-κB pathway. *Antioxid. Redox Signal.* **6**, 1–14, 2004.

Kimmel, C. A., and Makris, S. L. Recent developments in regulatory requirements for developmental toxicology. *Toxicol. Lett.* **120**, 73–82, 2001.

Rutledge, J. C. Developmental toxicity induced during early stages of mammalian embryogenesis. *Mutat. Res.* **396**, 113–127, 1997.

Stephens, T. D., and Fillmore, B. J. Hypothesis: Thalidomide embryopathy-proposed mechanism of action. *Teratology* **61**, 189–195, 2000.

Villavicencio, E. H., Walterhouse, D. O., and Iannaccone, P. M. The sonic hedgehog–patched-gli pathway in human development and disease. *Am. J. Hum. Genet.* **67**, 1047–1054, 2000.

CHAPTER 35

Dermatotoxicology

NANCY A. MONTEIRO-RIVIERE

Center for Chemical Toxicology Research and Pharmacokinetics, North Carolina State University, Raleigh, North Carolina 27606

35.1 INTRODUCTION

One of the principal portals of entry by which environmental toxicants can enter the body is through the skin. Skin has emerged as an organ of interest because it is a highly visible organ that serves as an interface with the environment. This makes skin vulnerable to the damaging effects of toxicants. The paradigm of structure–function relationship is central to modern biology and infers, based on our current knowledge, that the structural and functional relationships of the skin are likely to be complex. Anatomical factors may affect the barrier function, causing an altered rate of absorption. Therefore, the anatomical complexity of this organ must be fully understood before percutaneous penetration, metabolism, and cutaneous responses to specific toxicants be investigated.

35.2 FUNCTIONS OF SKIN

Skin, the largest organ of the body, is considered to be the barrier between the well-regulated "milieu interieur" and the outside environment. Skin is a very heterogeneous, yet integrated, dynamic organ that has a myriad of biological functions that go far beyond its role as a barrier to the external environment. One of its primary roles is to act as a communicator between the interior and exterior environments. It acts as an environmental barrier by protecting major internal organs, as a diffusion barrier that helps to minimize insensible water loss that could result in dehydration, and as a metabolic barrier that can metabolize a compound to more easily excretable products after absorption has occurred. The skin or integument participates in thermoregulation where blood vessels constrict to retain heat and dilate to dissipate heat. In addition, hair in humans and the fur of lower mammals act as insulation devices, while sweating facilitates heat loss by evaporation. Skin can serve as an immunological affector axis by having Langerhans cells process antigens and

Molecular and Biochemical Toxicology, Fourth Edition, edited by Robert C. Smart and Ernest Hodgson
Copyright © 2008 John Wiley & Sons, Inc.

can act as an effector axis by setting up an inflammatory response to a foreign insult. Skin has a well-developed stroma, which supports all other organs. In addition, the skin has receptors that sense the modalities of touch, pain, and heat. The skin can function as an endocrine organ. It participates in the synthesis of vitamin D and is (a) a target for androgens, which regulate sebum production, and (b) a target for insulin, which regulates carbohydrate and lipid metabolism. The skin has numerous sebaceous glands that can secrete sebum, a complex mixture of lipids that function as antibacterial agents or as a water-repellent shield in some animals. In addition, the skin contains both apocrine and eccrine sweat glands that produce a secretion that contains scent that functions in territorial demarcation. The integument also plays a role in the biosynthesis of keratin, collagen, melanin, lipids, and carbohydrates, as well as in respiration and biotransformation of xenobiotics.

In order to have a basic understanding and appreciation of how chemicals may interact with skin, its anatomy, physiology, and chemical composition must be fully grasped. All of the aforementioned biological functions and structural adaptations have a substantial impact on the skin's barrier properties and the rate and extent of percutaneous absorption. When a compound or toxicant is applied topically, it must penetrate through several cell layers of the skin in order to be absorbed by the capillaries for systemic distribution. Alternatively, it may have a direct effect on the keratinocytes themselves. Skin can be anatomically divided into two principal components, the outermost epidermis and the underlying dermis.

35.3   EPIDERMIS

The epidermis consists of keratinized stratified squamous epithelium derived from ectoderm and forms the outermost layer of the skin. Two primary cell types based on origin, the keratinocytes and nonkeratinocytes, comprise this layer. The classification of epidermal layers from the outer or external surface is as follows: stratum corneum (horny layer), stratum lucidum (clear layer), stratum granulosum (granular layer), stratum spinosum (spinous or prickle layer), and stratum basale (basal layer). In addition to the keratinocytes, the nonkeratinocytes include the melanocytes, Merkel cells, and Langerhans cells that reside within the epidermis but do not participate in the process of keratinization. The epidermis is avascular and undergoes an orderly pattern of proliferation, differentiation, and keratinization that as yet is not completely understood. Various skin appendages, such as hair, sweat and sebaceous glands, digital organs (hoof, claw, nail), feathers, horn, and glandular structures are all specializations of the epidermis. Beneath the epidermis is the dermis, or corium, which is of mesodermal origin and consists of dense irregular connective tissue. A thin basement membrane separates the epidermis from the dermis. Beneath the dermis is a layer of loose connective tissue, the hypodermis (subcutis), which is superficial fascia that contains elastic fibers and aids in binding the skin to the underlying fascia and skeletal muscle.

In general, the basic architecture of the integument is similar in all mammals. However, differences exist in the thickness of the epidermis and dermis, the number of cell layers, and the blood flow patterns between species. Additionally, differences exist within the same species at different body sites. Skin is thickest over the dorsal and lateral surfaces and thinnest on the ventral and medial surfaces of the body. The stratum corneum is thickest in glabrous skin regions such as the palmar and plantar surfaces where considerable abrasive action occurs. The epidermis is thin in areas where there is a heavy protective coat of hair or fur. Understanding these variations in the skin are important in studies involving biopharmaceutics, dermatological formulations, cutaneous pharmacology, and dermatotoxicology.

35.3.1   Epidermal Keratinocytes

35.3.1a   Stratum Corneum. The stratum corneum is the outermost layer of the epidermis and consists of several layers of completely keratinized flattened cells without nuclei or cytoplasmic organelles (Figure 35.1). The most superficial layers of the stratum corneum that undergo constant desquamation are called the stratum disjunctum. The stratum corneum cell layers vary in thickness between body sites and between species. These individual cells are highly organized and stacked upon one another to form vertical interlocking columns and having a flattened 14-sided polygonal (tetrakaidecahedron) shape. This structure provides a minimum surface-to-volume ratio, which allows for space to be filled by packing without interstices that yield a tight, water-impermeable barrier. Between the stratum corneum cells, the intercellular substance derived from the lamellar granules forms the intercellular lipid component of a complex stratum corneum barrier, which prevents both the penetration of substances from the external environment and the loss of body fluids. A plasma membrane and a thick submembranous layer that contains involucrin encompass these cells. This protein is synthesized in the lower stratum spinosum

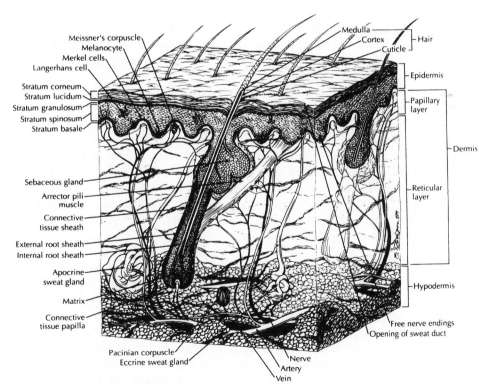

Figure 35.1. Schematic diagram of the integument depicting skin in various regions of the body. [Reprinted from Monteiro-Riviere (1991).]

layers and crosslinked in the stratum granulosum by an enzyme that makes it highly stable. Involucrin provides structural support to the cell, thereby allowing the cell to resist invasion by microorganisms and destruction by environmental agents, but does not appear to regulate permeability.

35.3.1b Stratum Lucidum. The stratum lucidum consists of a thin translucent, homogeneous line between the stratum granulosum and stratum corneum layers (Figure 35.1). It is only found in specific areas of the body where the skin is exceptionally thick and lacks hair (e.g., plantar and palmar surfaces). This stratum consists of several layers of fully keratinized, closely compacted, dense cells devoid of nuclei and cytoplasmic organelles. Their cytoplasm contains protein-bound phospholipids and eleidin, a protein that is similar to keratin but has a different histologic staining affinity.

35.3.1c Stratum Granulosum. The next layer down is the stratum granulosum, which consists of several layers of flattened cells lying parallel to the epidermal–dermal junction (Figure 35.1) containing irregularly shaped, nonmembrane-bounded, electron-dense keratohyalin granules. These granules contain profilaggrin, a structural protein and a precursor of filaggrin, and are thought to play a role in keratinization and barrier function. An archetypal feature of this layer is the presence of membrane-bound lamellar granules also known as Odland bodies, lamellated bodies, or membrane-coating granules. These granules increase in number and size as they move toward the cell membrane, where they fuse and release their lipid contents by exocytosis into the intercellular space between the stratum granulosum and stratum corneum. These lipids are responsible for coating the cell membrane of the stratum corneum cells and are the primary component of the barrier to chemical absorption across the skin. These lipids include the ceramides, cholesterol, fatty acids, and small amounts of cholesterol esters, as well as hydrolytic enzymes such as acid phosphatases, proteases, lipases, and glycosidases. The content and mixture of lipids can vary between species.

35.3.1d Stratum Spinosum. The next deeper layer of the epidermis is the stratum spinosum, or "prickle cell layer" (Figure 35.1) that consists of several layers of irregular polyhedral-shaped cells in which the uppermost layers contain small lamellar granules. Desmosomes connect these cells to the adjacent stratum spinosum cells and to the stratum basale cells below. The most notable characteristic feature of this layer is the numerous tonofilaments, which differentiates this layer morphologically from the other cell layers.

35.3.1e Stratum Basale. The next layer, the stratum basale, consists of a single layer of cuboidal or columnar-shaped cells that are (a) attached to the underlying irregular basement membrane by hemidesmosomes and (b) attached laterally to each other and to the overlying stratum spinosum cells by desmosomes (Figure 35.1). The basal cells continuously undergo mitosis, which causes the daughter cells to be distally displaced, keeping the epidermis replenished as the stratum corneum cells are sloughed from the surface epidermis. Depending on the region of the body, age, disease states, and other modulating factors, cell turnover and self-replacement in normal human skin is thought to take approximately one month. The mitotic rate increases after mechanical-induced (tape stripping, incisions) or chemical-induced injuries. In addition, cell turnover can vary greatly between animal species, with rodents having a much more accelerated turnover. The epidermal kinetics of cell proliferation in pigs and humans are very similar, having a 30-day epidermal cell turnover time. Chemicals related to genotoxicity causes damage to these cells. Chemical carcinogens that transform basal keratinocytes may result in squamous cell carcinomas.

35.3.2   Epidermal Nonkeratinocytes

35.3.2a   Melanocytes. Melanocytes are located within the basal layer of the epidermis, the external root sheath and hair matrix of hair follicles, the sweat gland ducts, and the sebaceous glands (Figure 35.1). Melanocytes possess dendritic processes that extend between adjacent keratinocytes and clear cytoplasm except for pigment-containing ovoid granules commonly referred to as melanosomes. Following melanogenesis, the melanosomes emigrate to the tips of the dendritic processes; the tips pinch off and are phagocytized by the adjacent keratinocytes. They are randomly distributed within the cytoplasm of the keratinocytes and sometimes localized over the nucleus, forming a cap-like structure that protects the nucleus from ultraviolet radiation. Skin color is determined by several factors, such as the number, size, distribution, and degree of melanization.

35.3.2b   Merkel Cells. These cells are located in the basal region of the epidermis in both hairless and hairy skin, with their long axis running parallel to the skin surface (Figure 35.1). Ultrastructurally, Merkel cells possess a lobulated and irregular nucleus with a clear cytoplasm, lack tonofilaments, and are connected to adjacent keratinocytes by desmosomes. Their cytoplasm is vacuolated on their dermal side and contain spherical electron-dense granules. Merkel cells are associated with axonal endings, and as the axon approaches the epidermis, it loses its myelin sheath and terminates as a flat meniscus on the basal aspect of the stratum basale cell. Merkel cells act as slow-adapting mechanoreceptors for touch.

35.3.2c   Langerhans Cells. Langerhans cells are found in the upper stratum spinosum layers and have long dendritic processes that traverse the intercellular space to the granular cell layer (Figure 35.1). They have a clear cytoplasm containing organelles and an indented nucleus, but lack tonofilaments and desmosomes. They are very apparent in toluidine blue-stained sections embedded in epoxy and appear as dendritic clear cells in the suprabasal layers of the epidermis. A unique characteristic of this cell is the presence of distinctive rod- or racket-shaped granules within the cytoplasm, called Langerhans (Birbeck) cell granules, which may function in antigen processing. Langerhans cells are derived from bone marrow and are functionally and immunologically related to the monocyte–macrophage series. They play a major role in the skin immune response because they are capable of presenting antigen to lymphocytes and transporting them to the lymph node for activation. They are considered to be the initial receptors for cutaneous immune responses (such as delayed-type hypersensitivity) and to contact allergens, and they can play an initiating role in some forms of immune-mediated dermatologic reactions.

35.3.3 Keratinization

The process in which the epidermal cells differentiate and migrate upward to the surface epithelium is referred to as keratinization. This is designed to provide a constantly renewed protective surface. As the cells proceed through the terminal differentiation stage, many cellular degradation processes occur. The spinosum and granular layers have lost their proliferative potential and thus undergo a process of intracellular remodeling. The cytoplasmic volume increases; and tonofilaments, keratohyalin granules, and lamellar granules also become abundant. Keratin is the structural protein abundantly synthesized by the keratinocytes that consists of many different molecular types. A loose network of keratins K5 and K14 are located within the basal cells. The active stratum spinosum cells secrete K1 and K10 and contain coarser filaments than those in the stratum basale. As the cells become flatter, their cellular contents increase, the nuclei disintegrate, and the lamellar granules discharge their contents into the intercellular space coating the cells. The nucleus and other organelles disintegrate, and the flattened cells become filled by filaments and keratohyalin. This envelope consists of the precursor protein involucrin and the putative precursor protein cornifin-α/SPRR1. The final product of this epidermal differentiation and keratinization process can be thought of as a stratum corneum envelope consisting of interlinked protein-rich cells containing a network of keratin filaments surrounded by a thicker plasma membrane coated by multi-lamellar lipid sheets. This forms the typical "brick and mortar" structure in which the lipid matrix acts as the mortar between the cellular bricks. This intercellular lipid mortar constitutes the primary barrier and paradoxically the pathway for penetration of topical drugs through skin.

35.3.4 Basement Membrane

The basement membrane zone or epidermal–dermal junction is a thin extracellular matrix that separates the epidermis from the dermis. Ultrastructurally, it can be divided into four component layers: (1) the cell membrane of the basal epithelial cell, which includes the hemidesmosomes; (2) the lamina lucida (lamina rara); (3) the lamina densa (basal lamina); and (4) the sub-basal lamina (reticular lamina), which contains a variety of fibrous structures (anchoring fibrils, dermal microfibril bundles, microthread-like filaments). In addition, it has a complex molecular architecture comprised of many large macromolecules that play a key role in adhesion of the epidermis to the dermis. The molecular components of the epidermal basement membrane, common to all basement membranes, include type IV collagen, laminin, entactin/nidogen, and heparan sulfate proteoglycans. Constituents limited to the skin include bullous pemphigoid antigen (BPA), epidermolysis bullosa acquisita (EBA), fibronectin, GB3, L3d, and 19DEJ-1. The basement membrane has several functions: It helps to maintain epidermal–dermal adhesion, as a selective barrier between the epidermis and dermis by restricting some molecules and permitting the passage of others, influences cell behavior and wound healing, and serves as a target for both immunologic and nonimmunologic injury. Pertinent to toxicology, the basement membrane is the target for specific vesicating agents such as bis(2-chloroethyl)sulfide and dichloro(2-chlorovinyl)arsine, which causes blisters on the skin after topical exposure.

35.3.5 Dermis

The dermis lies directly under the basement membrane and consists of dense irregular connective tissue with a feltwork of collagen, elastic, and reticular fibers embedded in an amorphous ground substance of mucopolysaccharides. The predominant cells of the dermis are fibroblasts, mast cells, macrophages, plasma cells, chromatophores, fat cells, and extravasated leukocytes often found in association with blood vessels, nerves, and lymphatics. Sweat glands, sebaceous glands, hair follicles, and arrector pili muscles are present within the dermis. Arbitrarily, the dermis can be divided into a superficial papillary layer that blends into a deep reticular layer. This layer is thin and consists of loose connective tissue, which is in contact with the epidermis and conforms to the contour of the basal epithelial ridges and grooves. When it protrudes into the epidermis, it gives rise to the dermal papilla. When the epidermis invaginates into the dermis, epidermal pegs are formed. The thicker reticular layer is made up of irregular dense connective tissue with fewer cells and more fibers.

One of the primary components of the dermis is the extensive network of capillaries that help to regulate body temperature. Blood flow through skin can vary 100-fold, depending on environmental conditions, making it one of the most highly perfused organs in the body. The dermis participates in several functions: mechanical support, exchange of metabolites between blood and tissues, fat storage, protection against infection, and tissue repair.

Deep to the reticular layer of the dermis is the hypodermis (subcutis) consisting of very loose connective tissue with adipose cells. This layer helps to anchor the dermis to the underlying muscle or bone. This thermal barrier and mechanical cushion is sometimes considered to be a site that acts as a depot or reservoir for certain toxic compounds.

35.3.6 Appendageal Structures

Appendageal structures commonly found within the skin are the hairs, hair follicles, associated sebaceous glands, apocrine and eccrine sweat glands, and arrector pili muscles. Hairs are formed by epidermal invaginations. These keratinized structures traverse the dermis and may extend into the hypodermis. The free part of the hair above the surface of the skin is the hair shaft, and the part deep within the dermis is the hair root, which forms an expanded knob-like structure called the hair bulb. This is composed of a matrix of epithelial cells in different stages of differentiation. Hair is composed of three concentric epithelial cell layers: the outermost thin cuticle, a densely packed keratinized cortex, and a central medulla of cuboidal cells. The hair follicle consists of four major components: (1) internal root sheath (internal root sheath cuticle, granular layer, pale epithelial layer); (2) external root sheath (several layers similar to the epidermis); (3) dermal papilla (connective tissue); and (4) hair matrix (comparable to the stratum basale of the epidermis).

The process of keratinization is continuous in the surface epidermis, while in the hair follicle the matrix cells undergo periods of quiescence during which no mitotic activity occurs. This cyclic activity of the hair bulb allows for the seasonal change in the hair coat of domestic animals. The length of time to regrow new hairs depends on the growth stage of the hair follicle. The period of the hair cycle where cells are mitotically active and grow is called anagen. After this growth phase, catagen occurs where metabolic activity slows down and the hair bulb atrophies. The final stage called telogen occurs when the hair follicle enters a resting or quiescent phase and growth stops. In this stage, the base of the bulb is located at the level of the sebaceous canal. As the new hair grows beneath the telogen follicle, it gradually pushes the old follicle upward toward the surface, where it is eventually shed. This intermittent mitotic activity and keratinization of the hair matrix cells constitute the hair cycle. It is controlled by several factors, including length of daily periods of light, ambient temperature, nutrition, and hormones, particularly estrogen, testosterone, adrenal steroids, and thyroid hormone. If a chemical's mechanism of action requires interaction with an active metabolic process, then toxicity may be exerted only during the anagen growth phase. Exposure at other times may not elicit any response. The regulation of hair growth by endocrine factors may provide another mechanism for chemical interactions.

Bundles of smooth muscle fibers commonly seen attached to the connective tissue sheath of the hair follicle toward the papillary layer of the dermis are referred to as arrector pili muscles (Figure 35.1). The hair sits at an obtuse angle to the skin surface, and the arrector pili is situated at the lower portion of the follicle. Contraction of this muscle in cold weather will elevate the hairs, forming "goose pimples" on the skin. These muscles in humans are well-developed and are supplied by postganglionic adrenergic sympathetic nerve fibers. Also, contraction of this muscle plays a role in emptying the sebaceous glands.

Many cytotoxic chemicals (e.g., cancer chemotherapeutic drugs and immunosuppressants like cyclophosphamide) whose mechanism of action is to kill dividing cells will produce hair loss (alopecia) as an unwanted side effect of nonselective activity (e.g., thallium). Damage may occur with the use of alkylating agents such as anti-metabolites or colchicine, which affect the matrix cells. Alkali and oxidizing agents such as peroxides cause keratolytic damage.

Inorganic constituents of human hair are receiving attention because of the potential in diagnostic medicine. Trace elements present in hair may come from the water supply, which provides calcium and magnesium. Transition metals like iron and manganese can deposit on the hair from the general water supply, and copper can come from swimming pools. Metals can also originate from sweat deposits, diet, air pollution, and metabolic problems. In addition, metal contamination in hair can come from hair products like dandruff shampoos that can add zinc or selenium and lead from lead acetate hair dyes. Accumulations of heavy metals in hair are usually at low levels; however, if the concentrations are well above the norm, then it can be utilized as a diagnostic tool. There has been a good correlation of cadmium levels in hair with other target organs. Learning disorders such as dyslexia in children can be detected early by cadmium analysis of hair due to the high cadmium levels that exist in dyslexic children. In cases of arsenic poisoning, hair has served as a good source for the localization of arsenic. Also, mercury can be detected in hair when subjects have been exposed to a diet containing high levels of mercury. Iron deficiencies can be detected by analyzing the iron content of hair. Hair analysis is a useful tool for toxicology for diagnosing other illnesses or conditions of exposure.

35.3.7 Sebaceous Glands

These glands are usually found all over the body and are associated with hair follicles (Figure 35.1). They are evaginations of the epithelial lining and histologically
are simple, branched, or compound alveolar glands containing a mass of epidermal
cells enclosed by a connective tissue sheath. They produce an oily secretion called
sebum, which is derived from disintegrating glandular cells and contain a mixture
of lipids that can vary between species. These lipids act as antimicrobial agents
and, in hairy mammals, as a waterproofing agent. In lower mammals, these glands
can become specialized and are often associated with a pheromone-secreting role,
making their exact composition very species-specific. In humans, the principal lipids
are squalene, cholesterol, cholesterol esters, wax esters, and triglycerides. In addition,
the density of these glands can vary between anatomical site and between individuals. These cells move inward through mitotic activity and accumulate lipid droplets
to release their secretory product, sebum, by the holocrine mode of secretion.
During early adolescence, hormonal activity increases human sebum production,
which results in acne vulgaris. Sebum production is thus involved in the screening
of anti-acne drug candidates. Some toxicants that interact with sebaceous gland
function can induce comedons, an acne-like response, to cause a condition termed
chloracne. Several chloracnegens have been considered a hazard in the occupational
setting. They include chloronaphthalenes, polychlorinated biphenyls, tetrachloroaz-
oxybenzene, tetrachloroazobenzene, polychlorinated dibenzodioxins, polychlorinated dibenzofurans, and polychlorinated biphenyls. Many of these chloracnegens
induce cytochrome P-450-mediated microsomal monooxygenase activity.

35.3.8 Apocrine Sweat Glands

The apocrine sweat glands are much larger than the eccrine sweat glands and are
located in specific body areas such as the axillary, pubic, areolae, and perianal
regions (Figure 35.1). Microscopically, they consist of simple sacular or tubular
structures with a coiled secretory portion located in the deep dermis and a straight
duct that runs parallel to the hair follicle and penetrates the follicular epidermis to
open alongside the follicle at the surface. Apocrine sweat glands release a milky oily
fluid that contains a mixture of lipids, proteins, lipoprotein, and saccharides. When
surface bacteria metabolize this secretion, they produce a body scent in humans
and other mammals that may be related to communications between species. These
glands may also act as a sex attractant and in domestic species, such as the dog and
cat, can serve as a territorial marker (pheromone-secreting role).

35.3.9 Eccrine Sweat Glands

Eccrine (merocrine) sweat glands are simple tubular glands that open directly onto
the skin surface (Figure 35.1). In humans, they are found over the entire body
surface except for the lips, external ear canal, clitoris, and labia minora. Myoepithe-
lial cells located in the secretory portion of these glands are specialized smooth
muscle cells, which, upon contraction, aid in moving the secretions toward the duct.
The eccrine sweat gland duct is comprised of two layers of cuboidal epithelium

resting on the basal lamina that opens in a straight path onto the epidermal surface. Some investigators postulate that the duct of these glands provide an alternate pathway for polar molecules, normally excluded by the stratum corneum, to be absorbed through skin.

The exocrine gland, whose principal function is thermoregulation, is one of the major cutaneous appendages and is functionally very active in humans. Sweating in humans refers to a distinct physiological function of excreting body fluids to the surface of the skin. This is necessary for fluid and electrolyte homeostasis. Physiologically stressed individuals can excrete 2 liters/hr to support evaporative heat loss. Only the higher primates and horses have a built-in mechanism that can accommodate this large volume loss without circulatory collapse. The secretory portion secretes isotonic fluid that is low in protein and similar in ionic composition and osmolarity to plasma. On passage down the duct portion, it becomes hypotonic and reabsorption of sodium chloride, bicarbonate, lactate, and small amounts of water occur. Abnormality in this fluid and electrolyte transport system leads to cystic fibrosis. In fact, analysis of this secretion is a prime diagnostic tool for this disease.

35.4 ANATOMICAL FACTORS TO CONSIDER IN MODEL SELECTION

One of the most important questions in dermatotoxicology is, What model system should be used? What is the most appropriate model? Of course, humans would be ideal and the most logical choice; however, the study of potent or toxic compounds would be unethical. Because of the ethical considerations and inability to conduct such research and testing in humans, animal models have been developed and extensively utilized. The global animal rights movement has had a major impact on the type of testing that it done by scientists. In response, new in vitro testing methods have been developed to study irritation, absorption, corrosion, and toxicity. Toxicologists thus often utilize animal models and extrapolate the data to humans. What key factors are important in selecting animal models?

35.4.1 Species Differences

In general, the basic architecture of the integument is similar in all mammals, although differences exist in the number of stratum corneum cell layers and in the thickness of the epidermis (Table 35.1). The number of epidermal cell layers varies among species and body sites. The mouse, rat, and rabbit have 1–2 viable cell layers, whereas the pig has 4–5 viable epidermal cell layers. The content and mixture of lipids can vary between species and are considered to function as the primary component of the permeability barrier. In humans, 50% of the total lipid mass consists of ceramides, 20–27% cholesterol, 10% cholesterol esters, and 35% fatty acid content. Alternatively, pig epidermis contains only 1–2% cholesterol esters. Lipid extraction studies were conducted on the abdominal, inguinal, and back regions of pigs to determine the concentrations of lipids at different body sites. It was shown that the relative proportions of individual lipids (ceramides 1–6, fatty acids, cholesterol, triglycerides, and cholesterol esters) were similar across all body regions.

TABLE 35.1. Comparative Stratum Corneum Thickness, Epidermal Thickness, and the Number of Epidermal Cell Layers in the Ventral Abdomen (VAB) and at the Thoracolumbar Junction (TLJ) of Several Species (H&E, Paraffin Sections)

Species	Area	Cell Layers Mean ± SE	Epidermis (µm) Mean ± SE	Corneum (µm) Mean ± SE
Monkey	TLJ	2.67 ± 0.24	26.87 ± 3.14	12.05 ± 2.30
	VAB	2.08 ± 0.08	17.14 ± 2.22	5.33 ± 0.40
Pig	TLJ	3.94 ± 0.13	51.89 ± 1.49	12.28 ± 0.72
	VAB	4.47 ± 0.37	46.76 ± 2.01	14.90 ± 1.91
Rabbit	TLJ	1.22 ± 0.11	10.85 ± 1.00	6.56 ± 0.37
	VAB	1.50 ± 0.11	15.14 ± 1.42	4.86 ± 0.79
Rat	TLJ	1.83 ± 0.17	21.66 ± 2.23	5.00 ± 0.85
	VAB	1.44 ± 0.19	11.58 ± 1.02	4.56 ± 0.61
Mouse	TLJ	1.75 ± 0.08	13.32 ± 1.19	2.90 ± 0.12
	VAB	1.75 ± 0.25	9.73 ± 2.28	3.01 ± 0.30

Source: Modified from Monteiro-Riviere N. A., et al. *J. Invest. Dermatol.* **95**, 582–586, 1990.

35.4.2 Body Sites

There are variations in the thickness of the epidermis and dermis within species in different regions of the body (Table 35.1). Skin is the thickest over the dorsal and lateral surfaces of limbs, and thinner on the ventral and medial surfaces of limbs. The back (thoracolumbar lumbar junction) is usually thicker than the abdomen. In areas possessing high hair density, the epidermis is thin; whereas in glabrous areas such as mucocutaneous junctions, the epidermis is thicker. The palmar and plantar surfaces consist of extremely thick stratum corneum because it is an area where abrasive action occurs.

35.4.3 Hair Follicles

Along with body site differences, there are differences in the density and arrangement of hair follicles (hf) between species. Hair density in pigs and humans is sparse (11 hf/cm^2), whereas the Fisher-344 rat has 289 hf/cm^2, mouse has 658 hf/cm^2, and the nude mouse has 75 hf/cm^2. Not only is the density different in different species, but also the arrangement of hair follicles can vary. In young pigs, hf occurs in groups of three, while the horse and cow have individual follicles that are evenly distributed. Dogs have compound follicles consisting of a single primary hair and a group of smaller secondary hairs. The cat possesses a single, large primary guard hair surrounded by clusters of 2–5 compound hair follicles. The anatomical arrangement and density of hair follicles are often ignored but extremely important when conducting in vitro diffusion cell experiments utilizing dermatomed skin or epidermal membranes. Also, when animal skins are obtained from an abattoir, the de-hairing procedure (scalding) often significantly damages the skin. Skin should be used immediately after necropsy and before extensive processing begins. In these preparations, holes may appear where the hair shafts once were.

TABLE 35.2. Comparative Blood Flow Measurements at the Ventral Abdomen and at the Thoracolumbar Junction in Five Species

Species	TLJ	VAB
Monkey	2.40 ± 0.82	3.58 ± 0.41
Pig	2.97 ± 0.56	10.68 ± 2.14
Rabbit	5.46 ± 0.94	17.34 ± 6.31
Rat	9.56 ± 2.17	11.35 ± 5.53
Mouse	20.56 ± 4.69	36.85 ± 8.14

Source: Modified from Monteiro-Riviere, N. A., et al., *J. Invest. Dermatol.* **95**, 582–586, 1990.

35.4.4 Blood Flow

It is important to be aware of the anatomical and physiological differences in blood flow between species and within different body sites of the same species. Cutaneous blood flow varies at different body sites. The ventral abdomen has a higher blood flow than the back (thoracolumbar junction) (Table 35.2). Laser Doppler velocimetry (LDV) or laser Doppler perfusion imaging (LDPI) are noninvasive techniques used to assess cutaneous blood flow. These techniques can evaluate the vascular response during acute inflammation, ultraviolet light, or drug therapy such as nitroglycerin and corticosteroid. In addition, the LDPI provides a two-dimensional mapping of peripheral blood perfusion which allows visualization of spatial tissue blood flow distribution in wound healing studies, in irritant or allergenic patch testing, and in vasoactive drug evaluation studies.

A comprehensive study comparing the epidermal histologic thickness and the cutaneous blood flow as assessed by LDV was conducted in nine species (mouse, rat, rabbit, cat, dog, pig, cow, horse, and monkey) at five cutaneous sites (buttocks, abdomen, skin over the humeroscapular joint, skin over the thoracolumbar junction and ear). Blood flow did not correlate to skin thickness across species and body sites but rather were independent variables, suggesting that they must be evaluated separately in pharmacology, dermatology, and toxicology studies.

35.4.5 Age

The cutaneous barrier function may also be affected with age. The ability to differentiate actinically damaged skin from chronologically aged or environmentally influenced such as chronic sun exposure, wind, cold, low humidity, chemical exposure, or physical trauma is difficult. Skin of aged individuals have been characterized as dry and wrinkly. Aged skin is microscopically different and is associated with vascular thickening and a decrease in lipid content. All of these changes may affect the clearance and absorption of topically dosed chemicals. Age-related differences have been observed in the absorption of 14 different pesticides, where the absorption of some compounds increased while in others it decreased. This indicates that the physiological properties of the compound are extremely important in assessing the percutaneous absorption of a compound.

35.4.6   Disease State

Skin diseases including essential fatty acid deficiency and ichthyosis may also affect the transdermal delivery of a compound. Studies have shown that the epidermal barrier function is altered by abnormal lipid composition in noneczematous atopic dry skin. Numerous other dermatologic conditions affect the anatomical structure and function of skin, which may impact on the nature of the toxic responses seen.

35.4.7   Metabolism

Skin plays a role in drug and chemical metabolism and therefore is an active site for xenobiotic metabolizing enzymes. The metabolites generated by a compound in skin can diffuse through the skin to become absorbed by the cutaneous vasculature and exert its influence systemically. This results in a first-pass metabolic effect for some topically applied chemicals. There are multiple isozymes of cytochrome P450 present in skin. The precise location of these drug-metabolizing enzymes is controversial. Studies have demonstrated that the epidermis contains the highest specific activity of monooxygenases. P-450 dependent monooxygenase activity in the skin has been shown with a variety of substrates including as benzo[a]pyrene, ethoxy coumarin, ethoxyresorufin, aldrin, and parathion. Human skin contains 5α steroid reductase activity that metabolizes testosterone. Other studies have demonstrated cortisone conversion to cortisol.

The skin has a well-documented role in vitamin D metabolism. 7-Dehydrocholesterol (provitamin D3) is activated by exposure to ultraviolet radiation in the skin to previtamin D3, which isomerizes to vitamin D3. Recently, further metabolism of 24,25-dihydroxyvitamin D to biologically active 1,25-dihydroxyvitamin D has also been demonstrated in skin, a conversion previously assumed to occur only in the kidney.

Subcellular localization studies have identified P-450-dependent monooxygenase activity in adult hairless mice sebaceous glands. Phase II conjugation pathways have also been identified in skin. Extracellular enzymes including esterases are present in skin, which has been utilized to formulate lipid-soluble ester prodrugs which penetrate the stratum corneum and then are cleaved to release active drug into the systemic circulation. Finally, co-administration of enzyme inducers and inhibitors modulate cutaneous biotransformation and thus alter the systemic toxicity profile. These metabolic interactions that occur in skin have attracted a great deal of research attention and clearly illustrate that skin is more than a passive barrier to toxin absorption.

35.4.8   Summary

Once information is gathered concerning the physical and physiological differences between species, one then has a basis to select the correct species for study. In order to comprehend how a toxicant or drug can permeate the skin, the vehicle and its components, disease state, hair follicle density, age, thickness, body site, blood flow, and metabolism must be considered because they can modulate chemical absorption and affect the skin barrier. Toxicants, "inert" components of chemical mixtures, and vehicles as well as products generated by the skin such as sweat, sebum, dead cells, and metabolic byproducts can alter the physicochemical properties of skin which can influence the rate of a toxicant's penetration or toxicological activity. Many studies do not consider these factors. For example, hair follicle growth cycle and body site were significant factors in hydrocortisone absorption. The challenge for investigators is to determine which key factors have the greatest impact on a specific chemical's transdermal passage or toxicity before it can exert a toxic effect.

35.5 PERCUTANEOUS ABSORPTION AND PENETRATION

The tendency for a toxicant to traverse the skin is a primary determinant of its dermatotoxic potential. That is, a chemical must penetrate the stratum corneum in order to exert toxicity in lower cell layers. The quantitative prediction of the rate and extent of percutaneous penetration (into skin) and absorption (through skin) of topically applied toxicants is complicated by the biological complexity discussed above.

The skin is generally considered to be an efficient barrier preventing absorption (and thus systemic exposure) of most topically administered toxicants. It is relatively impermeable to most ions and aqueous solutions. It is, however, permeable to a large number of lipophilic solids, liquids, and gases, suggesting that the term barrier is inappropriate for these substances. A number of agricultural workers have experienced acute dermal poisoning from direct exposure to pesticides such as parathion during application, or from secondary exposure by contact with vegetation previously treated with such insecticides. In fact, percutaneous absorption of nicotine in tobacco workers led to the concept of the modern transdermal nicotine "patch" delivery systems.

As discussed above, compared to most routes of drug absorption, the skin is by far the most diverse across species (e.g., sheep versus pig) and body sites (e.g., human forearm compared to scalp). The stratum corneum affords the greatest deterrent to absorption. Although highly water retarding, the dead, keratinized cells are also highly water absorbent (hydrophilic), a property that keeps the skin supple and soft as these cells absorb water on its way to being evaporated from the surface. This is the mechanism of action of most cosmetic moisturizers. Additionally, sebum appears to augment the water-holding capacity of the epidermis, but has no appreciable role in retarding the penetration of xenobiotics.

A number of studies have demonstrated that disruption of the stratum corneum removes all but a superficial deterrent to penetration. This is supported by "tape-stripping" experiments, in which an adhesive (cellophane tape) is placed on the skin repeatedly to remove progressive layers of the corneum. The skin ultimately loses its ability to retard penetration and compound flux increases greatly. This may be noninvasively assessed by measuring the skin's ability to prevent insensible evaporative water loss from the body to the environment by utilizing water as a marker of molecular transport across the epidermal barrier. This is performed by measuring transepidermal water loss (TEWL) with an instrument called an evaporimeter. TEWL increases greatly when the stratum corneum is either stripped, removed by extracting the intercellular barrier lipids with solvents such as acetone, or damaged in response to cutaneous toxicants.

The stratum corneum has been estimated to contribute 1000 times the diffusional resistance to chemical penetration as the layers beneath it, except for extremely lipid-soluble compounds with tissue/water partition coefficients greater than 400. As in most other epithelial tissues, the two other layers of the skin (dermis and subcutaneous tissue) offer little resistance to penetration. Once a substance has penetrated the outer epithelium, these tissues are rapidly traversed. This may not be true for highly lipid-soluble compounds, because the dermis may function as an additional aqueous barrier preventing a chemical that has penetrated the epidermis from being absorbed into the blood.

35.5.1 Dermatopharmacokinetics

The rate of diffusion of a topically applied toxicant across the rate-limiting stratum corneum is directly proportional to the concentration gradient across the membrane, the lipid/water partition coefficient of the drug, and the diffusion coefficient for the compound being studied. This can be summarized by Fick's law of diffusion in the equation

$$\text{Rate of diffusion} = \frac{DP}{h}(\Delta C)$$

where D is the diffusivity or diffusion coefficient for the specific penetrant in the membrane being studied, P is the partition coefficient for the penetrant between the membrane and the external medium, h is the thickness or actual length of the path by which the drug diffuses through the membrane, and ΔC is the concentration gradient across the membrane. The diffusivity is a function of the molecular size, molecular conformation, and solubility in the membrane milieu, as well as the degree of ionization. It should be noted that if the compound is dosed in an organic vehicle, the vehicle itself may penetrate into the intercellular lipids of the stratum corneum and change the estimated diffusional coefficient. The partition coefficient reflects the ability of the penetrant to gain access to the lipid membrane. Depending on the membrane, there is a functional molecular size/weight cutoff that prevents very large molecules from being passively absorbed across any membrane. The total flux of drug across a membrane is dependent upon the area of membrane exposed and thus is usually expressed in terms of cm^2. This relationship, which works well in an in vitro experiment, is only an approximation in vivo because penetration may be slow and a long period of time is required to achieve steady state.

If the lipid/water partition coefficient is too great, the toxicant may be retained in the membrane rather than traverse it and thus some fraction of compound will actually not be available for diffusion through the system. However, passage through the skin generally correlates with experimentally determined lipid/water partition coefficients in octanol/water and olive oil/water. In some cases where the specific lipid composition of the membrane is known (e.g., the stratum corneum from a specific species), the slurry of the actual lipids may be employed. This is becoming more sophisticated with the advent of advanced organ culture techniques where, for example in skin, lipid membranes very similar to in vivo with regard to composition, structure, and function can be prepared in culture and used to study drug transport.

In dermatotoxicology, the amount of toxicant per area of skin (e.g., mg/cm^2), rather than the amount of toxicant per unit of body weight (e.g., mg/kg), used in oral and parenteral studies is the primary determinant of dose. This explains why infants, with a relatively small ratio of skin surface area to body mass, are particularly prone to systemic toxicity from topical poisons when large areas of skin are exposed. This is further potentiated in neonates, who do not have a fully developed cutaneous barrier.

Occlusion of the skin, seen with application of water-impermeable drug vehicles or patches, alters the rate and extent of toxicant absorption. As the skin hydrates, a threshold is reached where transdermal flux dramatically increases (approximately 80% relative humidity). When the skin becomes fully hydrated under occlusive conditions, flux can be dramatically increased. This occlusive effect must be accounted for when extrapolating toxicology studies conducted under occlusive conditions to field scenarios where the ambient environmental conditions are present. Hydration may also markedly affect the pH of the skin, which varies between 4.2 and 7.3. Therefore, dose alone is often not a sufficient metric to describe topical doses when the method of application and surface area become controlling factors. Dose must be expressed as mg/cm^2 of exposed skin.

The dermis is a highly vascular area, providing direct access for systemic absorption once the epithelial barrier has been passed. The blood supply in the dermis is under complex, interacting neural and local humoral control whose temperature-regulating function can have an effect on toxicant distribution and absorption by altering blood supply to this area. The absorption of a chemical possessing vaso-active properties would be affected through its action on the dermal vasculature; vasoconstriction would retard absorption and increase the size of a dermal depot, while vasodilation may enhance absorption and minimize any local dermal depot formation. For a systemic toxicant, vasodilation would potentiate activity while a vasoconstriction might blunt the response. However, if the chemical is directly toxic to the skin, the reverse occurs with vasoconstriction preventing removal of drug to the systemic circulation away from cutaneous toxic sites, thereby potentiating local effects.

35.5.2 Routes of Absorption and Penetration

The appendages of the skin that extend to the outer surface may play a role in the penetration of certain compounds and also be selective targets after topical exposure. Anatomically, percutaneous absorption might occur through different routes. The majority of nonionized, lipid-soluble toxicants appear to move through the intercellular lipid pathway between the cells of the stratum corneum (Figure 35.2). Movement across keratinocytes does not generally occur. Some absorption may occur through the appendages such as hair follicles or sweat ducts. Very small and/or polar molecules appear to have more favorable penetration through appendages or other diffusion shunts, but only a small fraction of drugs are represented by these molecules. However, the epidermal surface area is 100–1000 times the surface area of the skin appendages, depending on species. The only exception to this rule is particulate exposures (microspheres, liposomes, nanoparticles), which may lodge in the opening to hair follicles and provide a unique access to the dermal circulation.

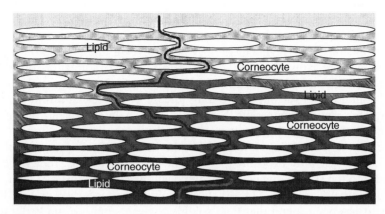

Figure 35.2. Schematic representation of the barrier property of skin composed of protein-aceous keratinocytes (corneocytes) embedded in an extracellular nonhomogenous matrix of lipid. Arrow depicts intercellular lipid pathway.

In addition to movement through shunts, polar substances may diffuse through the outer surface of the protein filaments of the hydrated stratum corneum, while nonpolar molecules dissolve in and diffuse through the nonaqueous lipid matrix between the protein filaments. The rate of percutaneous absorption through this intercellular lipid pathway is correlated to the partition coefficient of the penetrant, as presented above in Fick's law.

Penetration of drugs through different body regions varies due to the anatomical factors discussed previously. In humans, it is generally accepted that for most non-ionized toxicants, the rate of penetration is in the following order: scrotal > forehead > axilla = scalp > back = abdomen > palm and plantar. The palmar and plantar regions are highly cornified and their much greater thickness (100–400 times that of other regions) introduces an overall lag time in diffusion. In addition to thickness, differences in stratum corneum cell size and hair follicle density may also affect absorption of some molecules. The scalp should thus be considered in a different light than the rest of the body. Differences in cutaneous blood flow in different body regions may be an additional variable to consider in predicting the rate of percutaneous absorption.

35.5.3 Factors and Chemicals that Affect Percutaneous Absorption

The most damaging substances routinely applied to skin include soaps and detergents. Whereas organic solvents must be applied in high concentrations to damage skin, 1% aqueous solutions of detergents and surfactants (e.g., sodium lauryl sulfate) increase the penetration of toxicants through human epidermis.

Organic solvents can be divided into damaging and nondamaging categories relative to their effects on the barrier properties of skin. Methanol, acetone, ether, hexane, and mixed solvents such as chloroform:methanol or ether:ethanol are in the damaging category because they are able to extract epidermal lipids and proteolipids which alter permeability. Another mechanism for this solvent effect is that the solvents themselves partition into the intercellular lipid pathway, changing its lipophilicity and barrier property that result in an increased diffusion coefficient. Use of more polar or amphoteric solvents may enhance the penetration of polar molecules, in some cases, by forming "ion pairs" that have a greater ability to penetrate the lipid domain. In contrast, solvents such as higher alcohols, esters, olive oil, and so on, do not appear to damage skin appreciably. However, the penetration rate of solutes dissolved in them is often reduced. This is best explained by partitioning and retention of the penetrant into the nonabsorbed solvent, preventing release of the toxicant into the stratum corneum. Thus, one can appreciate that for a specific chemical, the rate of penetration can be drastically modified by the solvent system used.

These phenomenon brings into question the wisdom of using organic solvents to decontaminate skin after exposure to lipophilic toxicants, because they will be easily absorbed into the skin and may enhance toxicant absorption through their interaction with intercellular lipids. This practice should be strongly discouraged. Not surprisingly, it has been found that lipid-soluble toxicants may be markedly resistant to washing within a short time after application due to subsequent depot formation. For example, 15 min after application, a substantial portion of parathion cannot be removed from exposed skin by soap and water.

In environmental exposures, the chemical may come into contact with the skin as a mixture or in contaminated soil. In mixtures, other components may function as solvents and modulate the rate of absorption. This may be a determining factor in the toxicity of a complex chemical mixture. Our laboratory has extended this concept to classify chemical mixtures based on how components (mixtures of defined solvents, surfactants, reducing agents, vasoactive compounds) may modulate the absorption or direct cutaneous toxicity of suspected toxicants using a classification paradigm termed Mechanistically Defined Chemical Mixtures (MDCM). This approach would allow complex environmental mixtures to be assessed for the presence of such modulating compounds to triage them according to potential toxic potential. In soil, a large fraction of the toxicant may remain bound to soil constituents thereby reducing the fraction absorbed.

35.5.4 Experimental Techniques Used to Assess Absorption

It is difficult to say that there is a single animal species that exactly matches the morphology and physiology of human skin. Even the genetically closest primate model is very different due to the presence of hair. Rodents are often used so that data can be compared to other routes of administration because the predominant species used in routine toxicology testing are rats and mice. Rats and mice also behave very different in carcinogenesis studies involving initiation and promotion. Rabbits and guinea pigs have been used in dermatotoxicology testing often because of immunological considerations and to compare to an archival database. If a single species had to be selected, being the closest match to human skin based purely on a morphological basis, the domestic pig would be the species of choice. The skin surface area to body weight ratio in the pig compares favorably to humans, an attribute not present in smaller laboratory animal species. Although generalizations are difficult, human skin appears less permeable than the skin of the cat, dog, rat, mouse, or guinea pig. The skin of pigs and some primates serve as useful approximations to human skin, but only after a comparison has been made for each specific substance. The major determinants of species differences are keratinocyte thickness, cell size, hair density, lipid composition, and cutaneous blood flow.

Whole-animal studies assess the percent of the applied dose absorbed into the body using classic techniques of bioavailability, where absorbed chemical is measured in the blood, urine, feces, and tissues with mass balance techniques. Recently, methods have been developed to assess absorption by measuring the amount of chemical in the stratum corneum because it is the driving force for diffusion. Cellophane tape strips are collected 30 minutes after chemical exposure and the amount of drug assayed in these tape strips correlates to the amount systemically absorbed. If the focus of the research is to determine the amount of chemical that has penetrated into skin, core biopsies may be collected and serially sectioned, and a profile of the chemical as a function of skin depth may be obtained.

Regulatory agencies are constantly seeking new in vitro methods to assess toxicity and improve the accuracy of predictions relating to the safety of new drugs or chemicals. In vitro approaches are often used to assess topical penetration. Most employ diffusion cell systems that sandwich skin sections between a donor and a receiver reservoir. The chemical is placed in the donor compartment (epidermis), and compound flux into the receiver (dermal) solution is monitored over time. This system can use a variety of "skin" sources, ranging from full-thickness specimens (epidermis and dermis) to epidermis alone to various "artificial" membranes such as lipid layers. In most skin studies, the donor compartment is open to the ambient air. A diffusion cell in which the receiver solution is a fixed volume is termed a static cell. A cell in which the receiver solution flows through the dermal reservoir is termed a "flow-through" system. This mimics the in vivo setting where blood continuously removes absorbed compound. Various cell and organ culture approaches have also been developed which assess absorption across cultured epidermal and/or dermal membranes. The rate of steady state flux can directly be used to calculate a permeability constant for the chemical under study using Fick's law above.

In vitro studies should be conducted at 30°C to 37°C. Debate exists as to the choice of receptor fluids to use. In pharmaceutical studies involving relatively hydrophilic drugs, saline is often the receptor of choice. In contrast, toxicological investigations generally involve the assessment of lipophilic compounds, which require a receptor fluid in which the penetrant is soluble (e.g., albumin based buffers, solvent/saline mixtures). These systems have also been used to assess cutaneous metabolism, although maintenance of viability using oxygenated perfusate and glucose-containing receptor fluid is then required.

The next level of in vitro systems employed is the use of isolated perfused skin flap preparations that are surgically prepared vascularized skin flaps harvested from pigs and then transferred to an isolated organ perfusion chamber. This model allows absorption to be assessed in skin that is viable and anatomically intact and that has a functional microcirculation. Studies conducted to assess the percutaneous absorption of drugs and pesticides in this model compared to humans show a high correlation. Validation of these in vitro methods is a prerequisite for regulatory acceptance.

35.6 DERMATOTOXICITY

If a cytotoxic chemical is capable of being absorbed across the stratum corneum barrier, it has the potential to cause toxicity to the skin. Chemical-induced damage to skin can be assessed by determining which key anatomical structures or physiological processes are perturbed after exposure to topical compounds.

A large number of cutaneous irritants specifically damage the barrier properties of skin, which results in an irritation response. These include the organic solvents discussed above that extract the intercellular lipid and perturb the skin's barrier. Some chemicals digest or destroy the stratum corneum and underlying epidermis. Those that caused chemical burns are called corrosives and include strong acids, alkalis, and phenolics. They essentially attack the epidermal barrier and chemically destroy the underlying viable cell layers. The best treatment in these cases is to dilute and remove the offending agents by flushing with water or aqueous solvents. The exceptions are (a) CaO (quicklime), which violently reacts with water and generates heat, thereby causing further thermal damage, and (b) metallic (e.g., tin, titanium) tetrachloride compounds, which hydrolize to hydrochloric acid, thereby causing further damage. These types of reactions are easy to assess using in vitro models such as the Corrositex® system, which detects macromolecular damage to a collagen matrix, resulting in a chemical color change in an associated detector system.

In other cases, epidermal cells are affected, which may then initiate other sequelae. If an absorbed chemical is capable of interacting with the immune system, the resulting manifestations will be dependent upon the type of immunologic response elicited (e.g., cellular versus humoral, acute hypersensitivity, etc.). It should be stressed that immune cells (e.g., Langerhans cells, lymphocytes, mast cells) may modulate the reaction or the keratinocytes themselves may initiate the response. This is in contrast to the previous role of epidermal keratinocytes in the skin, which was believed to be related only to maintaining the biophysical integrity of the skin by producing keratins and lipids for the formation of an intact stratum corneum barrier. Keratinocytes may thus act as the key immunocyte in the pathophysiology of allergic contact and irritant contact dermatitis. Figure 35.3 illustrates the pathways that may trigger the production of proinflammatory cytokines after topical exposure to chemicals. Direct chemical-induced irritation of keratinocytes may also initiate this cytokine cascade without involvement of the immune system, blurring the distinction between direct and indirect cutaneous irritants.

35.6.1 Mechanisms of Keratinocyte Mediation of Skin Irritation/Inflammation

Evidence over the past decade has shown that keratinocytes are dynamic contributors that can produce growth factors, chemotactic factors, and adhesion molecules and thus act as initiators of inflammation. Environmental cues such as contact allergens (urushiol, nickel sulfate), toxic chemicals (croton oil, TPA, SLS), or physical stimuli (ultraviolet radiation) cause perturbations of the skin barrier that can induce a complex sequence of events known as the acute inflammatory response (Figure 35.3). This biochemical cascade is responsible for localizing tissue damage to the site of injury and coordinating the wound healing mechanism of the body in an attempt to repair damage. The classic signs of this phenomenon include erythema, edema, heat, and pain. These symptoms are indicative of a process whereby resident cells release mediators that increase blood flow and capillary permeability to the injured area. Langerhans cells were once thought to be the only active instigator of this response. Keratinocytes, which make up 95% of the epidermal cell population,

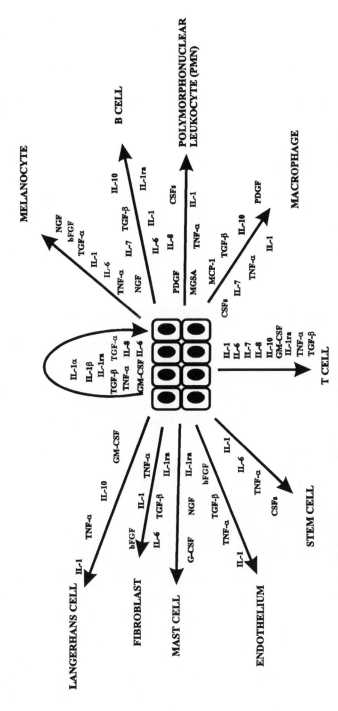

Figure 35.3. Schematic illustrating the role keratinocytes as modulators of cutaneous irritation.

were found to act as signal transducers capable of the initiation and amplification of the acute inflammatory response by producing cytokines that may affect the surrounding tissue and incite migrating immune cells to differentiate. Tumor necrosis factor alpha (TNFα) and interleukin 1 (IL-1) are considered to be the two primary proinflammatory modulators of this response.

During the initiation phase of cutaneous inflammation, external stimuli as mentioned above can trigger a cutaneous inflammatory response by directly inducing epidermal keratinocytes to produce specific proinflammatory cytokines and adhesion molecules (Figure 35.3). It is known that keratinocytes in conjunction with Langerhans cells are responsible for (a) the release of many acute inflammatory cytokines such as TNFα and IL-1 and (b) expression of the surface protein intercellular adhesion molecule 1 (ICAM-1). IL-1 and TNFα secretion leads to activation of the dermal endothelial cells' production of the surface leukocyte adhesion ligands ICAM-1, endothelial cell adhesion molecule 1 (ELAM-1), and vascular cell adhesion molecule 1 (VCAM-1) and consequent recruitment and sequestration of mononuclear cells from the circulation. Concomitant release of IL-8 and IL-1 by the keratinocytes promotes T-lymphocyte migration along a chemotactic concentration gradient to the epidermis.

Further interaction of recruited T lymphocytes with keratinocytes leads to amplification of cytokine production by epidermal cells as well as promoting T-cell proliferation. Other immune-mediated processes can occur, resulting in the production of antigen-specific memory T cells. If memory T cells recognize this same antigen in the skin, then this reaction amplifies to a fulminating immune-mediated inflammatory response. If immune recognition does not occur, the response is limited. Although there are many cytokines that play a significant role in the evolution and perpetuation of an immuno-inflammatory response, TNFα is considered to be the primary instigator.

The main sources of TNFα in the skin are mast cells, resident keratinocytes and Langerhans cells. Mature mast cells contain significant quantities of preformed TNFα in their granules that are released in an IgE-dependent manner. Investigators studying the inflammatory response in mouse skin revealed by Western blot analysis that exposure to acetone and tape stripping increased TNFα expression by 72% at 2.5 hr after treatment. The mRNA levels were highest at 1 hr after acetone, and then they decreased to control levels by 8 hr. These studies indicate an immediate response of TNFα to irritants that perturb the epidermal barrier.

It must be stressed that the primary mechanism of many topical irritants (e.g., organic solvents, corrosives) is the impairment to the stratum corneum barrier properties discussed above, reflected by an increase in transepidermal loss (TEWL). If the stratum corneum barrier is perturbed, the feedback response mediated by cytokines (especially TNFα) may be initiated whereby regeneration of the barrier occurs. However, additional responses to these inflammatory mediators may in themselves launch an irritation response mediated by the keratinocytes or lead to an immune reaction if the antigen is recognized. Regardless of the initiating mechanism, the sequelae to many irritants is the same, namely, epidermal cell death.

35.6.2 Cell Death: Apoptosis or Necrosis

There are two primary mechanisms of cell death that occur in response to an irritant response: apoptosis or necrosis (Figure 35.4). Apoptosis is a form of programmed

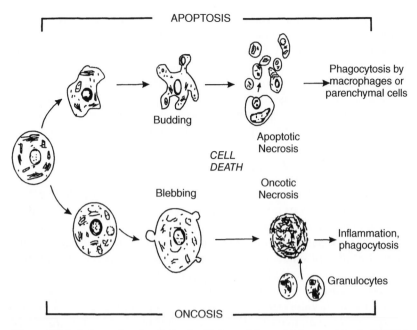

Figure 35.4. Schematic of cell death. [Adapted from Majno and Joris (1995).]

cell death that occurs when the offending toxin interferes with the cell's metabolism without immediately rupturing cell membranes or mitochondria. Similarly, a toxicant that damages DNA or results in slower damage to a cell's metabolic function may also lead to apoptotic cell death. Finally, some cells undergo a genetically programmed cell death, such as that involved in the normal turnover of basal keratinocytes, via apoptosis. Apoptotic cell death minimizes any further inflammatory response because cellular contents are not released. This is also a protective response to some toxicants; for example, in the case of carcinogens, cells with damaged DNA will undergo a programmed death rather than proliferate and transmit the damaged genome. In contrast, if the offending toxic agent compromises the cell's membrane or mitochondrial integrity and causes ischemia, the cell may lyse, which leads to necrosis, now termed oncosis. This response releases cellular contents into the surrounding tissue, which may illicit an inflammatory response as described earlier.

The primary, and most widely accepted, method to differentiate apoptosis from necrosis is histology or supplementary transmission electron microscopy. There are no generally accepted specific histochemical markers for apoptosis. Apoptotic cells are characterized by condensation of chromatin, nuclear fragmentation or karyorrhexis, nuclear membrane dissolution, cell membrane budding, and cell shrinkage (Figure 35.4). Necrosis is characterized by cell swelling, mitochondrial damage, cell membrane blebbing, karyolysis, or pyknosis. Apoptosis tends to occur in single cells, whereas necrosis tends to involve multiple cells. Finally, apoptotic cells tend to be phagocytized by macrophages or parenchymal cells (e.g., liver). Necrotic cells induce inflammation and are phagocytized by granulocytes.

As can be appreciated from the above discussion, cutaneous irritants may induce a wide range of responses in the skin, ranging from mild irritation to complete destruction with widespread necrosis. The degree of damage seen is dependent upon the chemical type of irritant involved (e.g., induce release of cytokines from kerati-nocytes, damage a cell's metabolic processes, corrosion and immediate cell lysis, and necrosis), the dose, the ability to penetrate the stratum corneum barrier, the dura-tion of exposure, and its tendency to induce an immune response. Chemicals that are carcinogens can also transform basal keratinocytes into squamous cell carcino-mas, a response fully discussed in another chapter of this text. Because of this widespread range of responses possible, toxicologic tests designed to detect cutane-ous toxicity are varied and not always appropriate for all types of toxins.

35.6.3 Irritancy Testing Protocols

Direct irritation may thus be defined as an adverse effect of chemicals directly applied to the skin that does not involve prior sensitization and thus initiation by an immune mechanism. Irritation is usually assessed by a local inflammatory response characterized by erythema (redness) and/or edema (swelling). Other responses may be present that do not elicit inflammation such as an increase in skin thickness. Irritant reactions may be classified as acute, cumulative, traumatic, or pustular. How-ever, two classifications are generally used by toxicologists. Acute irritation is a local response of the skin usually caused by a single agent that induces a reversible inflam-matory response. Cumulative irritation occurs after repeated exposures to the same compound and is the most common type of irritant dermatitis.

There are several types of irritancy testing protocols that are used to comply with federal and international safety regulations. The classic Draize test was developed in 1944 to measure acute primary irritation. The test compound is applied in an occluded fashion to a clipped area of abraded and intact skin of at least six albino rabbits and evaluated 24 hr and 72 hr after patch removal. The degree of erythema and edema, ranging from one to four, is recorded to reflect severity of the irritation. Because these tests are occluded, irritancy is potentiated due to hydration, which reduces the skin barrier. The Draize test may be modified to assess sensitization by preexposing animals to a sensitizing dose of the study chemical and then rechal-lenging the animals at a later date to illicit the immune-mediated response.

Studies have correlated the Draize scores with biophysical estimates of cutaneous barrier function and erythema. Erythema may be assessed using a variety of differ-ent color measuring instruments. These systems are based on reflectance principles and operate by irradiating skin with specific light wavelengths and measuring the color of the reflected light. These systems can also detect toxin reactions that alter the melanization process and produces altered skin pigmentation. Erythema, which results from increased blood flow, may be directly assessed using noninvasive laser Doppler velocimetry discussed earlier.

A focus of recent research has been to develop humane alternatives to the Draize test. The majority of these approaches use a skin organ culture that attempts to provide the stratum corneum barrier and viable epidermal cells that can react to penetrated compounds. Because these are essentially organ culture systems, cell viability may be assessed by sampling the bathing culture medium. Glucose utiliza-tion is easily assayed. Neutral red (3-amino-7-dimethyl amino-2-methylphenazine hydrochloride) can be used to probe lysosomal integrity, and thiazol blue (MTT) is a mitochondrial enzyme substrate that can assess mitochondrial function. Leakage enzymes such as lactate dehydrogenase (LDH) can be assayed to detect cell membrane damage. Reactive products (cytokines, prostaglandins) produced by keratinocytes from an irritant response may be detected in the culture medium. Alternatively, histochemistry can be performed on examined tissue samples and specific enzymes that more closely reflect the mechanism of the offending chemical. The isolated perfused skin model described above has also been a useful model for the assessment of direct chemical toxicity caused by numerous chemicals.

In addition to the above testing protocols, the local lymph node assay (LLNA) has been validated and accepted to assess the skin sensitization potential of chemicals in animals. This does not replace the guinea pig maximization test but is considered to be an equivalent. This in vivo method helps to reduce the number of animals used for contact sensitization activity. This test is based on the principle that sensitizer can induce a primary proliferation of lymphocytes in the lymph node draining the site of chemical application. This provides a quantitative measurement in which the proliferation is proportional to the dose applied.

35.6.4 Phototoxicity

There are additional environmental factors that lead to cutaneous irritation other than chemical toxicants. The most prevalent is ultraviolet (UV) radiation from sunlight. UV radiation encompasses short-wave UVC (200–290 nm), mid-wave UVB (290–320 nm), and long-wave UVA (320–400 nm). This topic is becoming increasingly important with the recent identified increase in UVB exposure in certain geographical regions apparently secondary to atmospheric ozone depletion. Although the ozone layer surrounding the earth filters out all UVC radiation, enough UVB and UVA reach the earth's surface to pose chronic and acute health hazards. Most known biological effects, including sunburn and skin cancer, may be attributed to UVB exposure. The most frequent and familiar phototoxic reaction is sunburn, the excessive cutaneous exposure to UVB radiation. Phototoxicity is a nonimmunological UV-induced response. Sunburn is characterized by erythema, edema, vesication, increased skin temperature, and pruritis followed by hyperpigmentation.

The most characteristic feature of sunburn is the presence of "sunburn cells" in the stratum basale layer of the epidermis (Figure 35.5). These cells are dyskeratotic with a bright eosinophilic cytoplasm and a pyknotic nucleus in hematoxylin and eosin (H&E)-stained tissue sections. This is an excellent example of cell death by apoptosis (Figure 35.5), in this case induced by UV exposure rather than by chemical toxicants. Ultrastructurally, these cells possess cytoplasmic vacuoles and condensed filament masses mixed with remnants of other cytoplasmic organelles (Figure 35.6). In vitro models used to study phototoxicity (UVA and UVB) include *Candida albicans*, photohemolysis of red blood cells, tissue culture, and isolated human fibroblasts. Such simple models lack the complexity of living skin systems and are often unreliable. In vivo models include guinea pigs, rabbits, mice, and opossum. However, the skin of most animal models is not anatomically comparable to human skin. Studies have shown that UVB-induced erythema and sunburn cell expression in pigs are comparable to humans, making the pig an accepted animal model for studying UV-induced phototoxicity.

The optimal in vitro cutaneous model should possess viable cells and structures similar to intact skin as well as a functional vasculature. This would allow all

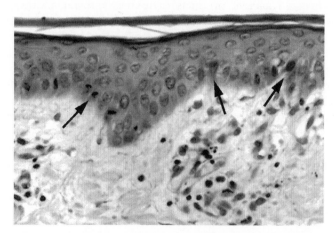

Figure 35.5. Light micrograph of an isolated perfused porcine skin flap irradiated with 1260 mJ/cm^2 depicting sunburn cells in the stratum basale layer (*arrows*) (560×). (Source: Modified from Monteiro-Riviere, N. A., et al., *Photodermatol. Photoimmunol. Photomed.* **10**, 235–243, 1994.)

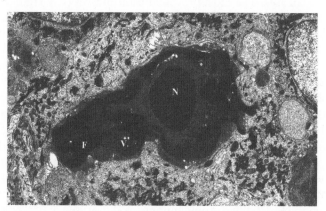

Figure 35.6. Electron micrograph of a sunburn cell located in the stratum basale layer of pig skin exposed 8 hr after being irradiated with 1260 mJ/cm^2. Note pyknotic nucleus (N), cytoplasmic vacuoles (V), and condensed filaments (F) (7,760×). (Source: Modified from Monteiro-Riviere, N. A., et al., *Photodermatol. Photoimmunol. Photomed.* **10**, 235–243, 1994.)

manifestations of the pathogenesis of UVB exposure to be investigated. Our laboratory has utilized the previously described isolated perfused skin model to study UV phototoxicity. In addition to morphologically assessing the dose-dependent formation of SBCs, decreased glucose utilization was observed accompanied by increased vascular resistance and decreased cell proliferation by assaying proliferating cell nuclear antigen (PCNA). Levels of prostaglandin E_2 in the perfusate from UVB exposed skin flaps also increased in a dose-dependent manner reflecting the cutaneous irritation response.

UV radiation generates oxidative stress in the skin creating photodamage. A photon of radiation interacts with *trans*-urocanic acid in skin to generate a singlet oxygen. Singlet oxygen can initiate an oxygen radical cascade with resulting oxidation of nucleic acids, proteins, and lipids, which can then cause cancer and photoaging. Vitamin C, a predominant antioxidant in skin, can protect against this damage. Most animals can make vitamin C, but humans have lost their ability due to genetic changes. Therefore, humans have to introduce vitamin C and E into their diets or by taking vitamin supplements. These vitamins help to neutralize oxidative stress. The absorption, distribution, and metabolism by the body limits the amount that can be delivered to the skin. We have shown that topical application of vitamin C and E stabilized by ferulic acid in pigs can achieve greater protection against oxidative stress than by ingestion and should protect against photoaging and skin cancer.

In addition to the direct effects of UV light on skin, some chemicals may be photoactivated to toxic intermediates, which then cause skin toxicity. Exogenous phototoxic chemicals can be found in therapeutic, cosmetic, industrial, or agricultural formulations. These include, among others, tetracyclines, furosemide, chloroquin, organic dyes, furocoumarins, polycyclic aromatic hydrocarbons, and some nonsteroidal anti-inflammatory agents. Phototoxicity may also be aggravated by natural plant products and endogenous substances resulting from disorders of intermediates of heme biosynthesis such as porphyrins produced by inherited or acquired enzymatic defects.

Several types of phototoxic events may occur, those that are oxygen-dependent and non-oxygen-dependent. When a reaction involves oxygen, the molecules can absorb the photons and transfer the energy to oxygen molecules, thereby generating singlet oxygen, superoxide anions, and hydroxyl radicals that cause damage to the skin. This is what happens to protoprophyrins when they are irradiated. A photochemical reaction involves absorbance of photons by the chemical resulting in an excited state that will react with the target molecules to form photoproducts. A good example of this type of reaction is 8-methoxypsoralen (xanthotoxin), which reacts with specific sites on the DNA by forming covalent bonds between the pyrimidine base and the furocoumarin. This phototoxic reaction may be restricted to areas of skin exposed to UV light has been used as a therapeutic technique to treat accessible skin tumors and psoriasis. With other compounds like chlorpromazine and protriptyline, the molecules absorb photons to form stable photoproducts that then induce cutaneous toxicity. The sequelae to these photoxic reactions is similar to that described earlier for cutaneous irritation.

35.6.5 Vesication

The final type of chemical toxicity that will be presented are the vesicants, chemicals that cause blisters on the skin. There are two classes of blisters that implicate different mechanisms of vesication. Intraepidermal blisters are usually formed due to the loss of intercellular attachment caused by cytotoxicity or cell death. The second class occurs within the epidermal–dermal junction (EDJ) due to chemical-induced defects in the basement membrane components. The classic chemical associated with EDJ blisters is the chemical warfare agent sulfur mustard (bis-2-chloroethyl sulfide; HD). HD is a bifunctional alkylating agent that is highly reactive with many biological macromolecules, especially those containing nucleophilic groups such as DNA and proteins.

The HD-induced dermal lesion is characterized by vesication and slow wound healing. Our laboratory has shown that the epidermal–dermal separation associated with vesication occurs in the upper lamina lucida of the epidermal basement membrane after HD-induced dermal injury in the perfused skin model. Previously, it was believed that alkylation of DNA with subsequent DNA crosslinks or breaks was the primary and initial event responsible for HD cutaneous toxicity. Thus, a hypothesis regarding DNA alkylation, metabolic disruption, and proteolytic activity has been proposed. In this scenario, DNA repair processes are induced, including the activation of poly(ADP-ribose) polymerase, which consumes NAD^+ as a substrate. As repair continues, NAD^+ is depleted, which decreases epidermal glycolysis and stimulates activation of the $NADP^+$-dependent hexose monophosphate shunt, resulting in protease release. Extracellular proteases digest dermal tissue, causing cell death, inflammation, and blister formation. Recently, results from several laboratories failed to support this hypothesis. Inhibitors of NAD^+ synthesis did not cause vesication. Additionally, when nicotinamide was used to increase NAD^+ levels, there were normal levels of glycolysis, although cell death and microblister formation still occurred.

Our laboratory has shown that gross blisters and microvesicles were present 5 hr after HD exposure in the isolated perfused skin model. This suggests that a basement membrane component is the target of HD-alkylation. Other DNA alkylating agents do not cause vesication and the other pathological changes characterized by HD. Therefore, cell death may not directly cause HD-induced vesication, but may only be associated with blister formation due to direct toxicity to epidermal cells. In fact, HD may have unique protein targets in the basement membrane zone, which leads to diminished stability of the epidermal–dermal junction leading to vesication. Alkylation of laminin would delay wound healing requiring therapeutic strategies to remove this damaged laminin scaffolding before normal skin regeneration can occur.

35.7 DERMAL TOXICITY OF NANOPARTICLES

Nanomaterials are structures with characteristic dimensions between 1 and 100 nm; when engineered appropriately, these materials exhibit a variety of unique and tunable chemical and physical properties. These characteristics have made engineered nanoparticles central components in an array of emerging technologies. Currently, they have widespread potential applications in material science, engineering, and medicine, but the toxicology of these materials has not been thoroughly evaluated under likely environmental, occupational, and medicinal exposure scenarios. The toxicology of materials such as titanium dioxide, fullerenes (C_{60}) or (buckyballs), single-walled carbon nanotubes (SWCNT) and multi-walled carbon nanotubes (MWCNT), dendrimers, and quantum dots (QD) has been of great interest over the last few years. The dermal toxicity of MWCNT has been addressed in primary human epidermal keratinocytes (HEK), which showed MWCNT within the cytoplasmic vacuoles of (HEK). Transmission electron microscopy depicted numerous vacuoles within the cytoplasm containing MWCNT of various sizes, up to 3.6 μ in length. Twenty-four hours after exposure, 59% of the keratinocytes contained MWCNT, compared to 84% by 48 hr at the high dose. Viability decreased with an increase in MWCNT concentration, and IL-8, an early biomarker for irritation, increased with time and concentration. This data showed that MWCNT, not derivatized nor optimized for biological applications, were capable of both localizing and initiating an irritation response in skin cells. These data are suggestive of a significant dermal hazard after topical exposure to select nanoparticles should they penetrate the stratum corneum barrier. Proteomic analysis has also been conducted in HEK exposed to MWCNT, which showed both an increase and decrease in expression of many proteins relative to controls. These protein alterations suggested dysregulation of intermediate filament expression, cell-cycle inhibition, altered vesicular trafficking/exocytosis, and membrane scaffold protein downregulation. Other studies have shown significant cellular toxicity when unrefined SWCNT were exposed to immortalized nontumorigenic human epidermal (HaCaT) cells, suggesting that carbon nanotubes may be toxic to epidermal keratinocyte cultures.

There are conflicting reports as to the potential toxicity of fullerenes such as C_{60}. Although C_{60} itself has no solubility in water, it has been shown to agglomerate with either organic solvent inclusion or after partial hydrolysis to create water-soluble species n-C_{60}. These agglomerations have been shown to have exceptionally low mobility in aqueous solutions but have been proposed to have high cellular toxicity. By functionalizing a fullerene with carboxyl or hydroxyl groups, their cytotoxic response can decrease by several orders of magnitude. Larger particles of zinc and titanium oxide used in topical skin-care products can penetrate the stratum corneum barrier of rabbit skin, with highest absorption occurring from water and oily vehicles. This may also be suggestive to manufactured nanoparticles.

QD or semiconductor crystals have great potential for use as diagnostic and imaging agents in biomedicine and as semiconductors in the electronics industry. QD of different sizes, shapes, and surface coatings can penetrate intact skin at an occupationally relevant dose.

This finding is of importance to risk assessment for exposures to nanoscale materials, because it indicates that the dermal route should not be overlooked.

35.8 CONCLUSION

Skin is a complex organ that is a focus of toxicology research. Because it is a primary portal for entry of numerous toxicants, its barrier properties have been extensively studied. Similarly, the ease by which chemicals can be exposed to skin increases the chances that direct dermatotoxicity may occur. Its role as an immune organ further increases its susceptibility to damage when allergens are encountered. In order to fully understand the toxicologic effects in skin, one must take into account the numerous anatomical and physiological factors that differentiate species and body regions and are important in proper experimental design.

SUGGESTED READING

Barry, B.W. *Dermatological Formulations: Percutaneous Absorption*, Marcel Dekker, New York, 1993.

Barker, J. N. W. N., Mittra, R. S., Griffiths, C. E. M., Dixit, V. M., and Nickoloff, B. J. Keratinocytes as initiators of inflammation. *The Lancet* **337**, 211–214, 1991.

Bos, J. D. *Skin Immune System (SIS)*, 2nd ed., CRC Press, Boca Raton, FL, 1997.

Drill, V. A., and Lazar, P. *Cutaneous Toxicity*, Raven Press, New York, 1984.

Elias, P. M. Epidermal lipids, barrier function, and desquamation. *J. Invest. Dermatol.* **80**, 44–49, 1983.

Feldman, R. J., and Maibach, H. I. Regional variations in percutaneous penetration of ^{14}C cortisol in man. *J. Invest. Dermatol.* **48**, 181–183, 1967.

Levin, S., Bucci, T. J., Cohen, S. M., Fix, A. S., Hardisty, J. F., LeGrand, E. K., Maronpot, R. P., and Trump, B. F. The Nomenclature of Cell Death: Recommendations of and ad hoc committee of the society of toxicologic pathologists. *Toxicol. Pathol* **27**, 484–490, 1999.

Majno, G., and Joris, I. Apoptosis, oncosis, and necrosis; An overview of cell death. *Am. J. Pathol.* **146**, 3–14, 1995.

Marzulli, F. N., and Maibach, H. I. *Dermatotoxicology*, 5th ed. Taylor & Francis, Washington, D.C., 1996.

Monteiro-Riviere, N. A., Bristol, D. G., Manning, T. O., and Riviere, J. E. Interspecies and interregional analysis of the comparative histological thickness and laser Doppler blood flow measurements at five cutaneous sites in nine species. *J. Invest. Dermatol.* **95**, 582–586, 1990.

Monteiro-Riviere, N. A. Comparative anatomy, physiology, and biochemistry of mammalian skin. In: Hobson, D. W. (Eds.). *Dermal and Ocular Toxicology: Fundamentals and Methods*, CRC Press, New York, 1991, Chapter 1, pp. 3–71.

Monteiro-Riviere, N. A., Inman, A. O., and Riviere, J. E. Development and characterization of a novel skin model for cutaneous phototoxicology. **10**, 235–243, 1994.

Monteiro-Riviere, N. A. Anatomical factors affecting barrier function. In: Marzulli, F. N., and Maibach, H. I. (Eds.). *Dermatotoxicology*, Taylor & Francis, Washington, D.C., 1996, Chapter 1, pp. 3–17.

Monteiro-Riviere, N. A. Integument. In: Dellmann, H. D., and Eurell, J. (Eds.). *Textbook of Veterinary Histology*, 5th eds., Williams & Wilkins, Baltimore, 1998, Chapter 16, pp. 303–332.

Monteiro-Riviere, N. A., Inman, A. O., Mak, V., Wertz, P., and Riviere, J. E. Effect of selective lipid extraction from different body regions on epidermal barrier function. *Pharm. Res.* **18**, 992–998, 2001.

Monteiro-Riviere, N. A., Nemanich, R. J., Inman, A. O, Wang, Y. Y., and Riviere, J. E. Multi-walled carbon nanotube interactions with human epidermal keratinocytes. *Toxicol. Lett.* **155**, 377–384, 2005.

Monteiro-Riviere, N. A., and Ryman-Rasmussen, J. P. Toxicology of nanomaterials. In: Riviere J. E. (Ed.). *Biological Concepts and Techniques in Toxicology: An Integrated Approach*, Taylor and Francis, New York, 2006, Chapter 12, pp. 217–233.

Mukhtar, H. *Pharmacology of the Skin*, CRC Press, Boca Raton, FL, 1992.

Nickoloff, B. J. The cytokine network in psoriasis. *Arch. Dermatol.* **127**, 871–883, 1991.

Rietschel, R. L., and Spencer, T. S. *Methods for Cutaneous Investigation*, Marcel Dekker, New York, 1990.

Webster, R. C., and Maibach H. I. Percutaneous absorption in diseased skin. In: Surber, C., and Maibach, H. I. (Eds.). *Topical Corticosteroids*, Karger Publishing, Basel, Switzerland, 1992, pp. 128–141.

Printed and bound by CPI Group (UK) Ltd, Croydon, CR0 4YY